SILICATE POWDER PATTERNS

The Geological Society of America, Inc.
Memoir 122

Calculated X-ray Powder Patterns for Silicate Minerals

I. Y. BORG
Lawrence Radiation Laboratory,
University of California
Livermore, California

D. K. SMITH
Lawrence Radiation Laboratory,
University of California
Livermore, California

Printed by The Geological Society of America
in cooperation with the
Mineralogical Society of America

1969

Printed in The United States of America

*The printing of this volume
has been made possible through the bequest of
Richard Alexander Fullerton Penrose, Jr.,
and is partially supported by a grant from
The National Science Foundation.
The Lawrence Radiation Laboratory assisted
very materially in the publication of this book
by supplying camera-ready copy
for photo-offset reproduction.*

ACKNOWLEDGMENTS

The efforts which culminated in this Memoir have covered a long span of time. Suggestions and ideas from many individuals have been incorporated into this study and the computer program used to produce the patterns. To identify all the individuals would result in a very long list. Special thanks should go to several, however, whose contributions played a significant role in this study. These include Tibor Zoltai, University of Minnesota; William Parrish, Philips Research Laboratories and NASA Howard McMurdie, Howard Swanson, Eloise Evans and Stanley Block, ASTM Associateship and the National Bureau of Standards; David Stewart, U. S. Geological Survey; J. V. Smith, University of Chicago; G. V. Gibbs, Virginia Polytechnic Institute; and H. D. Megaw, Cavendish Laboratory, Cambridge University. In addition, many individuals released crystal structure data to us prior to its formal publication in order that this set of calculated patterns could be completely up to date at press time. We are especially grateful for this consideration and would like to thank Joan R. Clark, J. J. Papike, and D. E. Appleman, U. S. Geological Survey; Y. Takéuchi, Mineralogical Institute at Tokyo; C. T. Prewitt, E. I. duPont de Nemours; C. W. Burnham, Harvard University; A. A. Colville, California State College; G. V. Gibbs, P. Ribbe and G. V. Novak, Virginia Polytechnic Institute; J. H. Fang, Southern Illinois University; W. Leimer, Continental Oil Co.; L. Finger, Geophysical Laboratory, Carnegie Institute; J. McCauley, Army Materials & Mechanics Research Center, Massachusetts; W. Dollase, University of California at Los Angeles; R. Young, Georgia Institute of Technology.

This work was performed under the auspices of the U. S. Atomic Energy Commission at the Lawrence Radiation Laboratory (LRL), University of California, Livermore, California. We are particularly indebted to Russell Peterson and Harold Blum of the L. R. L. Computer group for expediting the computing phase of the work. We are similarly grateful to Yvonne Colombo and Alfred Horn who helped assemble the voluminous data.

The manuscript was prepared for photographic reproduction with the assistance of the L. R. L. Technical Information Department. The efforts of Vivian Mendenhall, in particular, has helped to produce this Memoir in the shortest possible time.

We are indebted to the John Simon Guggenheim Memorial Foundation for expediting the last phases of manuscript preparation. We are also grateful to the Pennsylvania State University, College of Earth & Mineral Suences, where one of us (D. K. S.) is currently employed, for aid in preparing the final version of Table 43.

CONTENTS

X

PART I. CALCULATION OF POWDER PATTERNS

INTRODUCTION

The x-ray diffraction method is a very powerful technique in modern research. Besides providing the best means for the identification of a crystalline compound, it is invaluable in all materials research, relating physical and chemical properties to the crystalline state.

To effectively utilize the powder method, one often requires a standard reference pattern for direct comparison. Such standard patterns are obtained by using the best material available, carefully preparing it into a special sample holder, and recording the diffraction effects by means of film or counting devices. The most common recording techniques currently in use are the Debye-Scherrer camera and the counter diffractometer. Because the Guinier camera is increasing in usage, it should be considered among the standard methods. Even with modern precision equipment the experimental pattern is usually less than perfect, suffering from sample problems, geometric aberrations of the recording apparatus, and limitations of the detection system.

The basic requirement for a standard pattern is that it be truly representative of both the material and the recording method. The pattern must be reproducible for different mounts of the same sample both by different experimentalists and on different units of similar apparatus. In practice, these conditions are not easy to achieve. Samples are rarely ideal and suffer many defects, including nonrandom orientation of crystallites, inhomogeneity of chemical or structural composition, small or imperfect crystallites, primary and secondary extinction, strain, absorption (or the lack of it), microabsorption, and admixtures with other phases. Apparatus may not be precisely aligned, or the detection system may respond imperfectly. Even when all the conditions are optimum, the very weak portions of the pattern are difficult to observe because of the background. In compounds of extreme complexity, the pattern contains many regions with broad bands of diffracted intensity which cannot be resolved into the individual components. Thus no aspect of obtaining a good reference pattern is without problems.

Many efforts have been made to establish catalogs of standard reference patterns. The best known and most generally useful such catalog is the X-ray Powder Data File produced by the Joint Committee on Powder Diffraction Standards of the American Society for Testing and Materials. This group has supported the preparation of reference patterns in many laboratories throughout the world, as well as the abstraction of patterns which have appeared in the open literature. The File now includes around 15,500 patterns, and is indispensible in the x-ray laboratory.

The patterns recorded in the literature and in the File were obtained under a variety of conditions by individuals with a range of competence, resulting in standards with a wide variation in quality. The quality is not often known, although the File does identify patterns which were recorded under known established conditions. Because of the prevalence of preferred orientation of sample crystallites, biased intensity values are undoubtedly the most common defect present in the standard patterns. Many of the more complex patterns are incorrectly indexed, an error which does not interfere with identification but hinders other applications of the data such as the accurate determination of the lattice constants.

With the advent of high-speed computers, it has become practical to calculate a theoretical x-ray powder pattern based upon the known atomic arrangements. Such a pattern is free of the many sample defects. Instrumental aberrations can be approximated in the calculation, and the resulting pattern should be very close to the ideal experimental pattern. The calculational approach was begun independently by Zoltai and Jahangabloo (1963), Jietschko and Parthé (1965) and Smith (1963). Initial calculations included only the "integrated intensity" of each diffraction maxima in a pattern assuming complete resolution. The integrated intensity is independent of the instrumental aberrations and is related to the area under the diffraction profile. Various attempts were made to compensate for the effects of overlap of adjacent maxima, the most successful of which was by Smith (1967). At present, a pattern may be calculated which is very close in most details to an experimental diffractometer trace.

Calculated x-ray patterns offer many advantages over experimental patterns, and when used in conjunction with experimental patterns, they are most valuable. Calculated patterns based on good structure data can be used as identification standards and as guides for indexing the experimental patterns. The d-spacings are based on refined cell data, and the intensity comparisons resolve questionable matches. The intensities may readily be placed on their absolute scale by applying an "absolute scale factor" to the listed relative intensities (100 maximum), such scaled values are potential standards for quantitative analysis. Calculated patterns may be prepared for intermediate members of isomorphous series when the structures of the end members are known. They may also be prepared for ordered and disordered structures, oriented and stressed samples, substitutions in known structures, and even for hypothetical structures. Comparisons between calculated and experimental patterns often help verify or disprove the structure model and thus aid in the planning of more advanced crystallographic studies. Presentation of tabulated patterns, as in this memoir, is only the initial step in the potential development of calculated patterns.

In order to establish the validity of using theoretical patterns as reference standards, many comparisons were made between calculated and experimental patterns for compounds whose crystal structures were well established. Some materials whose structures were less well known were also included. Initially, calculated patterns were compared with experimental patterns prepared by the ASTM Associateship at the National Bureau of Standards. Some of these comparisons may be seen in the work of Smith (1968). For compounds whose structures were well established, the comparisons supported both the validity of the calculational approach and the experimental methods established at the National Bureau of Standards. Where the crystal structure was questionable, the degree of fit between the two patterns was definitely poor. The complexity of the compound had little influence on the quality of the fit as very excellent matches were obtained with triclinic crystals having moderately large unit cells. Based on the success of these tests, the Joint Committee on Powder Diffraction Standards (ASTM) began a program to augment the File with calculated patterns. Calculated patterns will be included in the File for the first time in 1969.

The ASTM is concentrating their calculational efforts on compounds of high symmetry which for the most part are not minerals. To extend the calculations to low-symmetry compounds and to study the details of patterns of a complex group of minerals, an independent program of calculated patterns for the feldspars was undertaken by Borg and Smith (1968, 1969). In this study calculated patterns were compared with experimental patterns obtained by several individuals, especially D. B. Stewart of the U. S. Geological Survey and William Parrish of Philips Research Laboratories and the National Aeronautics and Space Administration. The calculated patterns proved to be close to the experimental ones (Borg and Smith, 1968; 1969) and were invaluable aids in identifying details in the latter.

The successful nature of the feldspar study and the availability of the computing facilities at the Lawrence Radiation Laboratory prompted the·continuation of the program to other important groups of mineral silicates. The project grew rapidly. The literature was finally surveyed for all structure analyses of mineral silicates. A written request was made directly to individuals known to be studying mineral structures in order to include the most recent and often yet to be published crystal structure analyses in our file. As the data accumulated, it was soon realized that the calculated patterns were too numerous to present in short individual articles. As a consequence, the decision was made to compile all the meaningful patterns into one unit. It is hoped that its usefulness will match that of its predecessor, the Peacock Atlas of ore minerals by Berry and Thompson (1962).

3

PRINCIPLES OF THE CALCULATION OF PATTERNS

The patterns included in this Memoir are based on the Debye-Scherrer and diffractometer methods. They may also be applied to Guinier patterns. A few aspects of basic diffraction theory are necessary to explain the methods used in generating the calculated patterns, and they will be briefly treated here. For further detail standard references may be consulted, especially the International Tables Volumes II and III (1959, 1962).

Any diffraction pattern may be considered to consist of two parts: (1) the positions of the diffraction maxima, and (2) the intensities of these maxima. The positions of the maxima are related to the size and shape of the crystal lattice on which the crystal structure is built. This dependence is most commonly expressed by the Bragg relation

$$\sin \theta_{hkl} = \frac{\lambda}{2d_{hkl}} \tag{1}$$

where θ_{hkl} is the Bragg angle, λ is the x-ray wavelength, and d_{hkl} is the interplanar spacing for the atomic planes represented by the Miller indices h, k, and l (nh, nk, nl). The relationship between d_{hkl} and the parameters which describe the direct lattice is very cumbersome for low-symmetry systems (see Azaroff and Buerger, 1958, p. 52). It is more simply stated in terms of the reciprocal cell parameters, where

$$d^2_{hkl} = h^2 a^{*2} + k^2 b^{*2} + l^2 c^{*2} + 2hka^*b^* \cos \gamma^*$$
$$+ 2hla^*c^* \cos \beta^* + 2klb^*c^* \cos \alpha^*. \tag{2}$$

The quantities $a^*, b^*, c^*, \alpha^*, \beta^*$, and γ^* are the reciprocal lattice constants, and are found from the real cell parameters through the equations

$$a^* = \frac{bc \sin \alpha}{V} \qquad \cos \alpha^* = \frac{\cos \beta \cos \gamma - \cos \alpha}{\sin \beta \sin \gamma}$$

$$b^* = \frac{ac \sin \beta}{V} \qquad \cos \beta^* = \frac{\cos \alpha \cos \gamma - \cos \beta}{\sin \alpha \sin \gamma} \tag{3}$$

$$c^* = \frac{ab \sin \gamma}{V} \qquad \cos \gamma^* = \frac{\cos \alpha \cos \beta - \cos \gamma}{\sin \alpha \sin \beta}$$

$$V = abc(1 - \cos^2 \alpha - \cos^2 \beta - \cos^2 \gamma + 2 \cos \alpha \cos \beta \cos \gamma)^{\frac{1}{2}}$$

where a, b, c, α, β and γ are the real cell parameters and V is the volume of the unit cell.

Using these relations or others which may be derived from them, it is possible to relate the positions of all the powder pattern maxima observed to the crystal lattice with no knowledge of the intensities of the maxima. Miller indices are often assigned to the maxima in a powder pattern by: (1) determining the cell constants either through single crystal studies or by guessing a set of cell constants which "fit" the ex-

4

perimental powder pattern; (2) calculating the $d_{hk\ell}$ set from the chosen
cell constants; and (3) matching them with the measured d-values. In
regions of the patterns where the maxima are clearly resolved no problems
are usually encountered; but where many maxima partially or fully overlap,
it is not easy to decide which Miller indices should be assigned. At this
stage, some knowledge of the intensities of the individual maxima is
valuable.

The intensities of the maxima are functions of the kind and arrangement of atoms in the unit cell. The relation is generally expressed as

$$I_{hk\ell} = \kappa pLPAT \left| F_{hk\ell} \right|^2 \qquad (4)$$

where $I_{hk\ell}$ is the intensity of the diffraction maxima, κ is a scale factor
usually used to place the data on a relative scale, P is the multiplicity
factor, L is the Lorentz factor, p is the polarization factor, and $F_{hk\ell}$ is
the structure factor. The structure factor is the term which expresses
the effect of the atomic arrangement, and is

$$F_{hk\ell} = \sum_{j=1}^{N} f_j \exp[2\pi i (hx_j + ky_j + \ell z_j)] \qquad (5)$$

where N is the number of atoms in the unit cell, f_j is the atomic scattering factor, and x, y, and z, are the position coordinates for the jth atom.
For relative intensities it is only necessary to evaluate the set of expres-
sions represented by Eq. (4). The various correction factors are applied,
and a κ is chosen to scale the set to a maximum value, usually 100.

The relative intensity values obtained from Eq. (4) are satisfactory
for the interpretation of patterns of single-phase samples. When two or
more phases are present, the diffracted energy produced by each phase is
in proportion to the scattering power and amount of that phase seeing the
x-ray beam modified by a complicated function of the absorption coefficients of the mixture (Klug and Alexander, 1956, p. 410). To include the
effect of the variations in scattering power of different materials, the
x-ray intensities must be placed on an absolute scale. To obtain such
absolute intensities, all the factors contributing to the diffraction energy
and the method of recording must be taken into account. Using diffractometer geometry, Eq. (4) may be rewritten (Cullity, 1956, p. 389) as

$$I_{hk\ell} = \left[I_0 A \left(\frac{e^2}{mc^2} \right)^2 \frac{\lambda^3}{32\pi r} \right] \left[\frac{1}{2\mu V^2} \right] \left[p \frac{1 + \cos^2 2\theta}{\sin^2 \theta \cos \theta} \left| F_{hk\ell} \right|^2 e^{-2B \frac{\sin^2 \theta}{\lambda^2}} A(\theta) \right] \qquad (6)$$

where
      I_0 = the intensity of the incident beam/unit area,
      A = the cross-section area of the incident beam subtending the
         sample,

e = the electronic charge,

m = the electronic mass,

r = the distance to the detector slit,

μ = the linear absorption coefficient,

B = the temperature parameter, and

A(θ) = the angular-dependent portion of the absorption factor.
For diffractometer samples with μ greater than around 25 cm^{-1} A(θ) is essentially unity, and for the diffractometer patterns included in this Memoir no corrections for A(θ) have been required.

Equation (6) has been bracketed into three parts. The first term includes only physical constants and constants based on the recording equipment. The second term is independent of the angular position of the diffraction maxima, but is dependent on the sample. The third term contains the angularly dependent relations of the sample and the sample geometry. For relative intensity values, only the final term need be evaluated before the maxima are scaled to the desired values. However, to place the data on an absolute scale, the first two terms also need to be evaluated. The first term, in general, will always be the same for any given study because the same recording instrument will usually be used for all patterns which must be compared. Consequently, this term may be ignored under such conditions. Where different instruments may, for some reason, be used, this factor must be considered. The second term is a function of the sample, and can be evaluated readily. It is the important term when placing the intensities on an absolute scale when combined with the unnormalized intensities of the third part.

In the calculation of the patterns for this Memoir, the third part of Eq. (6) is evaluated for each hkℓ in the pattern. The set of values is normalized to a maximum of 100. The inverse of this "normalizing factor" combined with the quantity $1/(2\mu V^2)$ may then be used to place the normalized intensities on an absolute scale. Because the equations for combining absolute intensities of mixtures involve the effective linear absorption coefficients of the mixture (Klug and Alexander, 1956, p. 410), an "absolute scale factor," designated ASF, is calculated for each pattern which is defined as $1/(V^2N)$ where N is the "normalizing factor." The omission of the $1/(2\mu)$ factor facilitates the use of the equations for mixtures.

Thus to place an individual pattern on its absolute scale, the tabulated normalized intensities must be multiplied by (ASF)/2μ.

Calculational standards for quantitative analysis may be prepared in several ways, depending on the method of analysis (Zoltai and Jahanbagloo, 1963). Once the absolute intensities have been determined, they may be adjusted proportionally to the amount present in a mixture and appropriately modified by the effective absorption coefficient. A series of standards for different ratios of phases may be readily calculated.

Standards may be calculated for a fixed quantity of an internal standard in the mixture or for a fixed total sample mass subtending the x-ray beam. The effort involved in calculating analytical standards is negligible compared to preparing experimental standards, and their inherent accuracy is equal or greater than the experimental standards providing the structures of the individual phases are accurate.

To obtain relative integrated intensity values for the Debye-Scherrer method, Eq. (6) requires an absorption factor because the $A(\theta)$ term is significant. Relative intensity values, only, were calculated for this method. These values are obtained from the integrated values by using the absorption tables for a cylindrical sample tabulated in the International Tables, Volume II (1956, p. 295).

The integrated intensity is related to the area included under the diffraction profile. If the profile is fully resolved, its area may be measured by counting squares on the chart paper, planimetering the diffraction trace, or point counting. Even where the profile is resolved, it is not easy to decide what background to subtract because of the long tails which contain a significant portion of the area. Where the profiles are closely spaced, the tails of one maximum affect the adjacent maxima, and it is usually difficult to determine exactly how much the effect is. By far, the majority of compounds produce patterns with some degree of overlap. For these reasons, an attempt was made to generate the actual diffractometer trace by assuming a diffraction profile for each maximum and determining the intensity distribution when all the profiles were combined.

The analytical function used to approximate the diffraction profile is known as the Cauchy distribution. It is also known as the Lorentzian distribution or the Witch of Agnesi. For this work it is of the form

$$I(\theta) = \frac{2I_{int}}{\pi\omega}\left[\frac{1}{1 + 4\left(\frac{\theta - \theta_0}{\omega}\right)^2}\right] \tag{7}$$

where $I_{(int)}$ is the integrated intensity, ω is the full width at half intensity, and θ_0 is the Bragg angle associated with $I_{(int)}$. Equation (7) is normalized so that $\int I(\theta)\,d\theta = I_{(int)}$. The so-called half width ω, width at half intensity, varies with diffraction angle both because of the Bragg condition and the aberrations.

The Cauchy approximation to the diffraction profile has come from studies of the spectral distribution of the primary x-ray beam. The tail regions of experimental profiles are matched fairly closely by the function, but the peak portion tends to be more nearly Gaussian. The aberration functions (Klug and Alexander, 1956, p. 246) are exponential or Gaussian, and the convolution of these functions with the Cauchy spectral distribution yields a Cauchy-shaped curve for the observed profile. In actual practice, individual curve fitting with such a corrected curve has produced fits with less than 1% mismatch, so the approximation is satisfactory to the first order.

To determine the ω dependence on θ, several experimental traces were prepared from different materials using diffractometers of several manufacturers. For these data two empirical $\omega(\theta)$ relations were determined, one for well-crystallized silicon and the other for 0.03 μ corundum (Linde A). These curves are shown in Fig. 1. These two curves are used to generate a family of parallel curves by proportionately interpolating or extrapolating from the two given curves. Thus, to determine the $\omega(\theta)$ relation for a pattern a point is designated (ω, θ), and the specific $\omega(\theta)$ curve passing through this point is generated by the computer. It has been our practice to designate ω at $40°2\theta$. This angle was chosen because it is around the center of the region of interest in many patterns. Because of inadequate dispersion of α_1 and α_2 in this region, the actual ω is difficult to measure and is usually done by determining $\omega/2$ between the peak and the low-angle side of the profile.

The pattern generation proceeds by calculating a $I(\theta)$ function for both α_1 and α_2 of each diffraction maximum in the integrated intensity set. Each profile is calculated over a range that is 50ω to include over 99% of the area under the profile, then the contribution of each profile is summed at 2θ intervals of $0.02°$. The inclusion of the long tails is necessary because they approach zero very slowly. After the complete $I(\theta)$ trace has been calculated, the intensity and positions of the peak associated with each reflection are determined. The computer accomplishes this step by starting at the 2θ for each reflection and following up the $I(\theta)$ curve to the first maximum encountered.

To directly simulate the $I(\theta)$ trace for either the Debye-Scherrer or Guinier methods requires different $\omega(\theta)$ functions. Such a trace would represent the graphical trace obtained by scanning the film pattern with a microdensitometer and recording the optical density as a function of position. Usually, studies with films are made by visual comparisons because of both the greater speed and the general unavailability of a suitable microdensitometer. Also, if the profiles are reasonably normal, the eye tends to integrate the intensities. For these reasons, direct simulation of peak intensities for these patterns have not been considered in the present study.

CALCULATION OF THE TABLES

The intensity data listed in the tables were obtained by the methods outlined in the previous section. The column labeled I(INT) contains the relative integrated intensities normalized to a maximum value of 100.

8

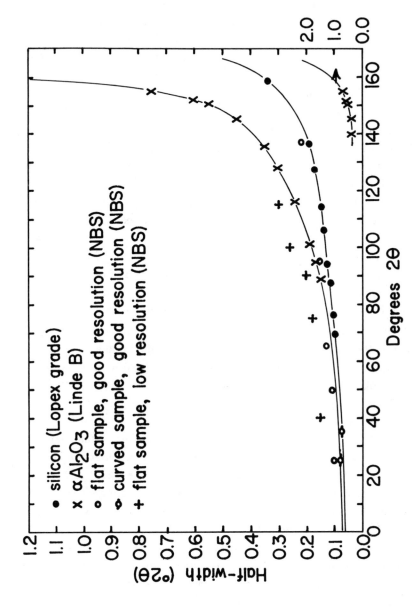

FIGURE 1. Width of diffraction profile at half intensity as a function of 2θ. The identified points are experimental. The two solid curves are the functions used by the computer.

9

In those instances where two or more intensities superimpose exactly and whose sum becomes the strongest intensity, the values have been scaled so that this sum has a value of 100. Because of the lack of an angularly dependent absorption correction for the diffractometer, the I(INT) column applies directly to diffractometry.

The column labeled I(DS) consists of the integrated intensity values corrected for absorption assuming a cylindrical sample with a 0.3-mm diameter. This set has also been normalized. The strongest reflection is considered to be identical to the one in the I(INT) column. Occasionally, due to the absorption effect, a different reflection may be stronger. Renormalization was not made in these instances.

The columns labeled I(PK) and PEAK refer to the peak intensity from the simulated diffractometer trace and the angular position of the peak respectively. PEAK values were determined to the nearest $0.02°2\theta$. This approach has led to some problems. Usually, only the α_1 intensity will be recorded in the listings. Occasionally, the α_1 of one reflection is superimposed on the α_2 of an adjacent strong reflection. Under these conditions the α_2 contributions are included in the peak intensity. Thus, a very weak α_1 could have a strong peak value in the tables. Where two adjacent maxima overlap such that no minimum exists in the $I(\theta)$ curve between the respective 2θ positions they will be considered as the same peak. Maxima of nearly equal intensities will resolve with small 2θ separations, whereas maxima with markedly different intensities may not resolve until the 2θ separation is quite large. If, during the peak search, the separation of calculated and plotted position is found to be greater than ω, then that peak is considered to be unresolvable, and its I(PK) value has been set equal to zero. These problems are inherent in experimental traces, and no further treatment of the data is justified.

To aid in the reading of the tables, PEAK and I(PK) values are given only for the first reflection in a group when the maxima are completely overlapped. Consequently, a blank space in the PEAK and I(PK) columns implies overlap with the value listed immediately above. Thus, the I(PK) column should correspond to a list of experimental peak intensities for a comparable sample.

If all the calculated intensities were tabulated for each pattern, many of the tables would be unwieldy. Patterns consisting of up to 1000 lines have been calculated for many of the low-symmetry compounds which include a few that terminated around $65°2\theta$ (the limiting number of maxima calculated by the program was set at 1000). For this reason, many of the patterns have been annotated. Maxima whose d is above some limit d_1, for example 1.0, have been omitted if I(INT) is less than some cutoff I_1, for example 1.0. Maxima whose d is less than d_1 have been omitted if I(INT) is less than I_2, for example

0.5. The abridging factors used for each pattern are included in the descriptive mineral tables preceding the tabulated I and θ values. The example given above would be listed $I_1:d_1:I_2 = 1.0:1.0:0.5$.

Attempts have been made to produce annotated lists of between 200 and 300 maxima. The abridging factors have been chosen to allow some of the stronger back reflection maxima to be accepted even if their intensity is somewhat less than the weaker front reflection maxima that were arbitrarily excluded to keep the number of reflections within the desired range. The inclusion of the reflections in the back reflection portion of a pattern serves as an aid to indexing the maxima that could prove most useful in least-squares fitting of cell parameters. Very weak front-reflection maxima are generally of minor value except where space group assignment or recognition of superlattices or domain structures depend on the presence of certain reflections, e.g. anorthite. For these compounds, the weak reflections have been retained.

It will be noted that the tables contain zero peak intensities. Because the tables were prepared directly from the computer printouts, a standard format was used for all the lists. Intensities were always rounded to the nearest integer, and zeros mean intensities less than 0.5.

In preparing the simulated $I(\theta)$ trace, criteria for deciding the appropriate half width were established based on our experience with experimental traces. An effort has been made to calculate the pattern with the half width that most closely approximates the best experimental pattern which could be expected for the particular material. With a better pattern as a guide, the user should be capable of mentally modifing the pattern to compare it with a lower quality experimental trace. For materials such as quartz, beryl, topaz and other gem-like minerals, $(\omega, 2\theta) = (0.08°, 40.00°)$ has been used. For the kaolins and chlorites, $(0.15°, 40.00°)$ has been used. Other traces have been calculated using $(0.11°, 40.00°)$. The actual (ω, θ) used for each pattern is given in the figure caption.

Most patterns were calculated for copper radiation. Where the $FeO+Fe_2O_3+MnO$ was greater than 15% by weight, the pattern was calculated for Fe radiation. The chosen radiation is given in the caption and applies to both the figure and the table. This change of radiation makes direct comparison of related traces somewhat more difficult (for example within the pyroxene group), but the user should be able to mentally make the necessary shifts. Successful comparisons between Cu-calculated and Cr-experimental patterns have been made (Parrish, 1968).

To allow the placing of the integrated intensity values on an absolute scale, the absolute scale factor, ASF, (as defined in the previous section) and the linear absorption coefficient μ are included in Tables 1-42.

Some general comments about the behavior of the $\omega(\theta)$ functions for the Debye-Scherrer and Guinier methods will assist the user in applying

11

the data presented in this Memoir. The widths of Debye-Scherrer profiles are affected by many factors — most notably the sample diameter, the
absorption, beam divergence, and the normal beam spreading at high
Bragg angles. These effects result in profiles in the low-angle region
which are about twice as broad as those in the 75° to 90°2θ range. In the
back reflection region, the beam spreading results in profiles several
times as wide as the 75°-90° region. A detailed discussion of the effects
is presented in Peiser, Rooksby, and Wilson (1956, p. 91). Overlap is
still a very significant problem and comparisons of calculated intensities
with measured patterns are not straightforward. Some success has been
obtained by using the absorption corrected intensities, I(DS), together
with the diffractometer trace, which is used as a guide to the degree of
overlap. This approach is satisfactory in the front reflection region, but
is less successful in the back reflection region where resolution is
generally better and absorption effects emphasize the intensities in the
Debye-Scherrer method.

For the Guinier method, the variability of the camera geometry
results in somewhat different $\omega(\theta)$ functions for different cameras. In
general, the $\omega(\theta)$ function is dependent on the position of film focusing
circle with respect to the zero point, the thickness of the specimen, the
divergence of non-equitorial x-rays, and the $\alpha_1 \alpha_2$ dispersion (deWolff,
1948). A well-aligned camera will result in very sharp diffraction maxima
narrowest around 30-45°2θ and increasing slightly toward both lower and
higher angles. Because of the extreme narrowness of the maxima overlap
is minimized, and comparisons may usually be made directly with calculated, integrated intensity values, mentally taking into account the minor
broadening effects. Because of the psychological visual integration on the
part of the observer, comparisons should present little difficulty. For
heavily absorbing samples, the region of the film diametrically opposite
the sample on the focusing circle will show the lowest absorption effect.

The choice of all structure parameters was usually based on the
values quoted in the paper cited. The few exceptions include the variation
in f-curves and the unit cell parameters when more recent data warranted
the adjustment (for example, quartz).

The proper spelling of some mineral names has been a problem.
Where the name has been used by the author in English, we have followed
the author, e.g. analcime. Where the name must be translated, especially
from the Russian, we have tried to find the English equivalent, but in some
cases, e.g. kainosite, we have had to arbitrarily select the spelling among
different translations.

CRITERIA FOR SELECTION OF STRUCTURAL ANALYSES

In general, the philosophy followed in considering structures for inclusion in this Memoir is that a good structure should be included regardless of its rarity, and that a questionable structure could raise more problems than it might solve. Date of the source data is not a suitable criterion as evidenced by the close similarity of patterns calculated from Warren and Bragg (1928) and Clark, Appleman, and Papike (1968) for diopside or the Taylor (1930) and Knowles, Rinaldi, and Smith (1965) analcime. The importance to the mineralogist, petrologist, and geochemist was taken into account for structures of somewhat intermediate accuracy (i.e. structures done with 2-dimensional methods). Doubtful structures were omitted. Incomplete structures were, in general, omitted. For example, among the zeolites many structure analyses do not include location of all the zeolitic water and alkalis. If only a few atoms were missing, the compound was included if the analysis was otherwise acceptable. Lack of quoted temperature factors was not considered an exclusion criteria because reasonable estimates can be made, and incorrect estimates will have little effect on the low-angle region which is most important for identification purposes.

The criteria for quality of the structure determination are somewhat arbitrary. They take the method (2-dimensional, 3-dimensional, film, counter), the R-factor, and the completeness of the structure into account. The R-factor used was the one quoted by the author. If several values were quoted, e.g. for weighting effects, the R-value for all reflections was used. An R-factor less than 0.10 was considered good regardless of the method. R-factors between 0.15 and 0.10 were considered intermediate and acceptable if the nature of the structure (its complexity) and the method suggested the analysis was carried as far as possible with an apparently adequate set of input data. Structures with the R-factor over 0.15 were considered weak structures and, in general, omitted. The rating for each structure has been indicated in the caption; one implies a good structure, two an intermediate one, and three a weak one. The rating is given in parenthesis following the quoted R-factor, i.e. R = 0.05(1).

Because the symmetry positions for each space group are maintained on a file tape, only those orientations used in the International Tables Volume I (1952) were available for most space groups. This condition sometimes required redefinition of the crystallographic axes from those given in the paper cited. Where this change has been made, both original and new orientations are listed in the pattern caption. The Miller indices in the table then refer to the new axes as does any other crystallographic data quoted such as cleavage. Certain alternate settings are commonly used in preference to the standard settings in the International Tables

(1952) e.g. $P2_1/a$ and Pbnm, and they are added to the symmetry file. Redefinition of the crystal axes for these orientations was not required.

INTERPRETATION

Undoubtedly the most common problem in powder samples is the tendency of crystallites of many minerals to have a nonspherical shape which causes the powder to assume a preferred orientation in the sample holder whether it is flat or cylindrical. The nature and degree of the preferred orientation is clearly related to the crystallite habit and/or cleavage. The effect is to enhance intensities of certain families of reflections and diminish others relative to intensities appropriate to randomly oriented powders. There are numerous methods described in the literature that are designed to minimize preferred orientation (McCreery, 1949; Flörke and Saalfeld, 1955; Bloss, Frenzel and Robinson, 1967) which help considerably. However, it is questionable that it can be completely eliminated in all materials, even with very careful sample preparation. As a consequence, "ideal" experimental patterns are probably only rarely obtained from powders of fibrous silicates such as crocidolite or pectolite or from platy varieties among the mica and clay minerals.

As an aid in interpreting experimental patterns compared with their calculated counterparts, a description of the cleavage and habit is given for each mineral. In many cases, the information was supplied by the authors who performed the structure determination or refinement. In other cases it was taken from Dana's System of Mineralogy (6th Ed. 1892) after comparing axial ratios and verifying that the choice of crystal axes was the same as that used in the calculations. If the orientation of the cell differed from that deduced from axial ratios given in Dana, appropriate transformations were made on the quoted cleavage, etc. to relate it to the new choice of cell. Uncertainties in the correct transformation in a few triclinic minerals precluded designation of either cleavage or habit. In other instances lack of data, e.g. for synthetic materials, similarly precluded designation.

With any degree of preferred orientation, intensities of all orders of diffraction parallel to a single cleavage will be enhanced in powder patterns; conversely, intensities of maxima normal to that cleavage will be diminished relative to that observed in random mounts. In the case of a sample with two perfect cleavages, not only is there enhancement of maxima parallel to the cleavages but also of those cozonal with both cleavages, e.g. $0k\ell$ in the case of perfect 00ℓ and $0k0$ cleavages or $0k0$ and $h00$ in the case of two $\{hk0\}$ cleavages. Intensities of other maxima normal to both cleavages are reduced with maximum diminition associated with those normal to the zone axis common to the two, e.g. hk0, h0ℓ and h00 (max); and 00ℓ in the two examples above.

The intensities of other h k ℓ maxima are also altered relative to the ideal, but the amount is sensitive to their specific intercepts, the angle between the cleavages, and the relative perfection of the cleavages if there is more than one.

The criteria for the acceptance of a structure for the calculation of a powder pattern have been discussed. They are not infallible, and some of the structure determinations used may be inaccurate. A short discussion of the possible effects of errors in the source data on the powder pattern may be in order. It must be emphasized that any errors in a structure analysis are magnified in the corresponding powder pattern because intensities in a powder pattern are based on the <u>square</u> of the structure factor. Furthermore, the variation in multiplicity among the different classes of reflections results in the errors being weighted differently throughout the pattern.

One input error may be incorrect cell parameters. At least in the recent literature it is probably the least common because cell dimensions can be measured with no knowledge of the cell contents and there are many reliable experimental techniques in use. The direct effect of inaccurate lattice constants is to cause the positions of the calculated diffraction maxima to be shifted from their true position. Where adjacent maxima overlap, the degree of resolution and consequently the peak intensity will be affected by such shifts.

If the atom position parameters or their effective scattering powers are incorrect, the intensities of the diffraction maxima may be markedly affected. It is well known in crystal structure analysis that structure factors are less sensitive to the positions of light atoms than to the heavier ones. Atoms such as Li or Be could be omitted or misplaced with little effect in a silicate, but positional errors as low as 0.01A for Ca or Fe would produce significant intensity changes. The effect of partial substitution of atoms on the same site depends on the difference in scattering factors and the state of ordering. In those instances where atoms of similar scattering power substitute, e.g. Al and Si or Fe^2, Fe^3, and Mn, ordered distributions would be virtually indistinguishable from the disordered state. On the other hand, substitution of atoms of widely differing scattering powers would be sensitive to any state of order. Even simple disordered substitution of such atoms as Mg/Fe in the olivines has such a strong effect that the patterns of the end members can hardly be said to resemble each other.

Early structure analyses often assumed an ideal composition. More modern studies consider the chemistry very carefully and attempt to include the minor constituents. The technique is to weight the atomic scattering factor for a site in proportion to the estimated site occupancy. If the minor constituents were ignored, the difference between the assumed and the true scattering power of a site will cause errors in the intensities.

The latter point should be expanded somewhat. Where a structure analysis is attempted with incorrect assignment of scattering power, the least-squares fitting procedures will "correct" the error by adjusting the effective temperature factor. The resulting structure factors and corresponding intensities would not necessarily be in error.

Thus, where an author assumed a proportional site occupancy that differed in some way from that suggested by the chemistry, we adhered to the author's choice in our calculations.

Another variable, in addition to the correct weighted scattering factor, is the state of ionization. Where possible, the author's choice was followed. If the state of ionization used by the author was not represented among the form factors in the program file, the closest alternative was used. The more ionized state was used in general. Errors caused by an incorrect choice only affect the low angle portion of the pattern. For CuKα this range would be below $2\theta = 35°$, and it often contains the stronger maxima used for initial identification. Thus, some of the key intensity values may be questionable. The strongest intensities, however, should be unaffected because these maxima are usually produced by most of the atoms in the structure scattering in phase. Consequently, it makes little difference to which atom the bonding electrons are assigned; the full contribution is still included in the peak's intensity. Above 35° 2θ the effect of the ionic state is negligible as can be readily seen by examining the various form factor curves in the International Tables, Volume III (1962, p. 201).

The role of the temperature factor is quite the opposite. An incorrect value will result in larger intensity deviations at the higher angles than at the lower angles. However, as used today, the temperature parameter is obtained from fitting procedures on the data, and any systematic errors in the data are compensated by modifications of the temperature parameters. Thus, uncompensated absorption, wrong choices in atomic form factors, wrong designation of site occupancy, imperfect or disordered crystals, and many other aspects may be "corrected" by the free use of temperature factors. In using the temperature factors quoted by the authors such compensating effects are also included in the calculated patterns. Although the physical interpretation of the data is imperfect, the resulting pattern should be closer to the experimental data. Where temperature factors have been estimated, this fact is noted in the pattern caption, and the pattern should be interpreted accordingly.

SILICATE CLASSIFICATION AND ORGANIZATION

In order to present the patterns included in this Memoir in some systematic manner, the decision was made to follow the Zoltai (1960) classification of tetrahedral structures. This choice was made because of the increasing acceptance of the role of all tetrahedrally coordinated groups in controlling the properties of a crystal containing mixed poly-hedral groups. The classification tends to bring closely related com-pounds into subgroups. This proximity allows ready comparisons between potentially related powder patterns which may serve as guides in the interpretation of a new but similar pattern.

This classification does separate some minerals which are often considered inseparable. The aluminosilicates mullite and sillimanite are considered as chain structures, whereas the closely related andalusite and kyanite retain their better known position among the orthosilicates. On the other hand, beryl and cordierite are associated with the zeolites which is reasonable in view of the common exchangable properties. Thus, in spite of some undesirable groupings, the advantages of this classifica-tion are many, and it has been followed without exception.

Not all subgroups in the Zoltai classification are represented among the patterns in this Memoir. However, most of the important groups and subgroups are present, and the following table gives the sequence of appearance.

1. Isolated groups of tetrahedra
    a. Single tetrahedra (TO_4)
    b. Double tetrahedra (T_2O_7)
    c. Large and mixed groups of tetrahedra
2. One-dimensional non-terminated groups of tetrahedra
    a. Single chains
        pyroxenes (TO_3)
        others
    b. Double chains
        amphiboles (T_8O_{22})
        others
    c. Single rings
3. Two-dimensional non-terminated groups of tetrahedra
    a. Single sheets (T_4O_{10})
        Micas
        Kaolin clays
        others
    b. Double sheets (T_8O_{16})
4. Networks with shared corners (TO_2)
    Feldspars

Zeolites

Silica minerals

others

5. Mixed types

Following the patterns and tables is an index based on the five strongest diffraction maxima. The choice of a five-line index comes from the new trend in the ASTM File to list more than three lines organized according to the Hanawalt Index (see ASTM File, 1968, p. vii-viii for description). Not only does the large number of patterns included in the File require more than three lines to narrow the identification to only a few compounds, but the complexity of many patterns, especially among the silicates, precludes a definitive three-line index. The index tabulated here is based on peak intensity values, the values most easily interpreted in identification procedures and most compatible with the present experimental patterns in the File. However, the listing of the strongest peaks is in no way intended to supplant the existing ASTM file based on experimental patterns. It is meant rather as auxiliary information to aid in identification.

REFERENCES—PART I

American Society for Testing and Materials, 1968, Index (Inorganic) to the powder diffraction file: ASTM Publication PDIS-181, Philadelphia, Pa.

AZAROFF, L. V. and BUERGER, M. J., 1958, The powder method in x-ray crystallography: McGraw-Hill, New York, p. 342

BERRY, L. E. and THOMPSON, R. M., 1962, X-ray powder data for the ore minerals: The Peacock Atlas: Geol. Soc. America Memoir 85, Colorado Building, Boulder, Colorado, p. 281

BLOSS, F. D., FRENZEL, G., and ROBINSON, P. D., 1967, Reducing preferred orientation in diffractometer samples: Am. Mineral., v. 52, p. 1243-1247

BORG, I. Y. and SMITH, D. K., 1968, Calculated powder patterns: Part I – five plagioclases: Am. Mineral., v. 53, p. 1709-1723

BORG, I. Y. and SMITH, D. K., 1969, Calculated powder patterns. Part II. Six potassium feldspars and barium feldspar: Am. Mineral., v. 54, p. 163-181

CLARK, J. R., APPLEMAN, D. E., and PAPIKE, J. J., 1968, personal communication

CULLITY, B. D., 1956, Elements of x-ray diffraction: Addison-Wesley, Reading, Mass., p. 514

DANA, E. S., 1892, Dana's system of mineralogy, 6th Ed. Appendix I 1899; II, 1909, III. 1915: John Wiley, New York

FLÖRKE, O. W. and SAALFELD, H., 1955, Ein Verfahren zur Herstellung texturfreier Röntgen-Pulverpräparate: Z. Krist., v. 106, p. 460-466

International tables for x-ray crystallography, Volume I: 1952, The
    Kynoch Press, Birmingham, England, p. 588

International tables for x-ray crystallography, Volume II: 1956, The
    Kynoch Press, Birmingham, England, p. 444

International tables for x-ray crystallography, Volume III: 1962, The
    Kynoch Press, Birmingham, England, p. 362

JIETSCHKO, W. and PARTHÉ, E., 1965, A FORTRAN IV program for
    the intensity calculation of powder patterns: University of
    Pennsylvania, The School of Mechanical Engineering, Philadelphia,
    Pa.

KLUG, H. P. and ALEXANDER, L. E., 1956, X-ray diffraction pro-
    cedures for polycrystalline and amorphous materials: John Wiley,
    New York, p. 716

KNOWLES, C. R., RINALDI, F. F., and SMITH, J. V., 1965, Refinement
    of the crystal structure of analcime: The Indian Mineral., v. 6,
    p. 127-140

McCREERY, G. L., 1949, Improved mount for powdered specimens used
    on the Geiger-counter x-ray spectrometer: J. Am. Ceram. Soc.,
    v. 32, p. 141-146

PARRISH, W., 1968, X-ray analysis of lunar material: Philips Labora-
    tories Technical Report No. 227, Philips Laboratories, Briarcliff
    Manor, New York

PEISER, H. S., ROOKSBY, H. P., and WILSON, A. J. C., 1955, X-ray
    diffraction by polycrystalline materials, Reinhold, New York,
    p. 725

SMITH, D. K., 1963, A FORTRAN program for calculating x-ray powder
    diffraction patterns: UCRL-7196, Lawrence Radiation Laboratory,
    Livermore, Calif.

SMITH, D. K., 1967, A revised program for calculating x-ray powder
    diffraction patterns: UCRL-50264, Lawrence Radiation Laboratory,
    Livermore, Calif.

SMITH, D. K., 1968, Computer simulation of x-ray diffractometer traces:
    Norelco Reporter, v. XV, p. 57

TAYLOR, W. H., 1930, The structure of analcite ($NaAlSi_2O_6 \cdot H_2O$): Z.
    Krist., v. 74, p. 1-19

WARREN, B. E. and BRAGG, W. L., 1928, The structure of diopside
    $CaMg(SiO_3)_2$: Z. Krist., v. 69, p. 168-193

deWOLFF, P., 1948, Multiple Guinier cameras: Acta Cryst., v. 1,
    p. 207-211

ZOLTAI, T., 1960, Classification of silicates and other minerals with
    tetrahedral structures: Am. Mineral., v. 45, p. 960-973

ZOLTAI, T. and JAHANGABLOO, I. C., 1963, Powder Diffraction: New
    Dimensions. Encyclopedia of X-rays and Gamma Rays, Editor,
    G. L. Clark, Reinhold, pp. 814-816.

PART II.   RESULTS OF CALCULATIONS

MINERALS INCLUDED IN TABLES 1-42

     The minerals have been put into groups of three to five. In the interest of conservation of space. certain subgroups with few representatives are grouped with representatives of other subgroups. Calculations for each group of five are preceded by a descriptive table (Tables 1-42). The minerals described in the tables are as follows:

GROUP 1. Isolated groups of tetrahedra

     a. Single Tetrahedra (TO_4)

         Table 1 - Simple Silicates with Single Tetrahedra

             Zircon

             Sphene

             Topaz

             $\gamma\text{-}Ca_2SiO_4$

             $\beta\text{-}Ca_2SiO_4$

         Table 2 - Olivine Structures

             Forsterite

             Hyalosiderite

             Hortonolite

             Fayalite

             Monticellite

         Table 3 - Garnets

             Grossular

             Pyrope (synthetic)

             Pyrope

             Andradite

             Uvarovite

         Table 4 - Chondrodite - Humite Group

             Humite

             Norbergite

             Chondrodite

             Clinohumite

         Table 5 - Miscellaneous Silicates with Single Tetrahedra

             Kyanite

             Andalusite

             Yoderite

             Eulytite

         Table 6 - Miscellaneous Silicates with Single Tetrahedra

             Staurolite

             Afwillite

             Bultfonteinite

             Dumortierite

             Spurrite

b. Double Tetrahedra (T_2O_7)

c. Large and Mixed Groups of Tetrahedra

GROUP 2. One dimensional non-terminated structures of tetrahedra

a. Single Chains

27

ORGANIZATION OF THE TABULATED POWDER PATTERN DATA

In the ensuing discussion, it is convenient to distinguish between (1) the descriptive tables (Tables 1-42) containing information pertinent to the source data and (2) the results of the intensity and 2θ calculations, henceforth called tabular and graphical powder pattern data.

Each descriptive table (Tables 1-42) is followed by tabulated and graphical data pertinent to the group of minerals described in the table. A simulated diffractometer trace precedes the tabulated intensity and spacing data for each mineral. The figure caption contains the wave length and half-width $(\omega, 2\theta)$ used in the calculations together with I(PK) maximum. The latter is the I(PK) of the strongest peak in the plot shown. In most cases, I(PK) maximum = 100 and thus, all I(PK) values can be read directly from the accompanying tabulated I and d-spacing data. For illustrative purposes in some instances, it was advantageous to exaggerate the vertical scale and bring out weak peaks. In such cases, I(PK) maximum > 100. However, the accompanying tabulated data in these cases remain scaled to I(PK) = 100.

Because of space limitation, almost all figures contain one or two truncated peaks. As all peaks are listed in the tabulated data, it is a simple matter to determine their normalized I(PK) value. The peaks in the figures are not indexed. Again, reference to the tabular data will supply the h k l of reflections contributing to that peak.

CONCISE EXPLANATION OF TERMS AND ABBREVIATIONS USED IN TABLES 1-42 AND ASSOCIATED CALCULATED POWDER PATTERNS

2THETA	Twice the Bragg angle associated with $hk\ell$ reflection referred to the wave length of the α_1 component.
PEAK	Position (2θ) of a peak in the complete simulated pattern which is associated with the theoretical position of $I(INT)_{hk\ell}$. It includes contributions of all adjacent α_1 and α_2 maxima.
I(INT)	Integrated intensity.
I(PK)	Peak intensity for flat plate samples (diffractometry) resulting from dispersion of the integrated intensity over Cauchy profiles of specified half-width, ω.
I(DS)	Integrated intensity corrected for absorption appropriate to cylindrical samples (Debye-Scherrer method). No dispersion of the intensity over a profile is assumed.
I(PK) MAX.	Intensity of the strongest peak in the graphically displayed, calculated powder pattern (usually truncated), e.g., 700 indicates a 7-fold expansion of the vertical scale relative to I(PK) = 100 of the tabulated powder pattern.
M1, M2 A1, A2 T1, T2	Cation site designations — in most instances arbitrarily assigned to differentiate several sites in a structure.
Z	Number of formula units per unit cell.
R	Reliability factor — a measure of the difference between calculated and observed structure factors in the crystal structure analysis used to generate the powder pattern.
B	Isotropic temperature factor.
μ	Linear absorption coefficient.
ASF	Absolute scale factor — a constant used as a multiplier to restore the tabulated and normalized I(INT) to an absolute scale.
Abridging factors	Numerical criteria used to abridge the complete listing of $I(INT)_{hk\ell}$. $I_1:d_1:I_2$ where I_1 is the minimum intensity a reflection with a d-spacing greater than d_1 must have to be listed. I_2 is the minimum intensity a reflection with a d-spacing between d_1 and 0 must have to be listed.
ω	Width of the peak at half maximum intensity, a function of 2θ and has the stated value at 40°2θ.

TABLE 1.   SIMPLE SILICATES WITH SINGLE TETRAHEDRA

Variety	Zircon	Sphene	Topaz	γ-Ca_2SiO_4	Larnite
Composition	$ZrSiO_4$	$CaTiSiO_5$	$Al_2SiO_4(OH)_2$	γ-Ca_2SiO_4	β-Ca_2SiO_4
Source	Ilmen Mts, Ural	Lindvikskollen, Norway	South America	Synthetic	Synthetic
Reference	Krstanović, 1958	Zachariasen, 1930	Ladell, 1965	Smith, Majumdar & Ordway, 1965	Midgley, 1952
Cell Dimensions					
$\underline{a}$ Å	6.6164	6.56 (6.55 kx)	4.6499	5.091	5.514
$\underline{b}$ Å	6.6164	8.71 (8.7 kx)	8.7968	11.371	6.757
$\underline{c}$	6.0150	7.45 (7.43 kx)	8.3909	6.782	9.315
α	90	90	90	90	90
β deg	90	119.72	90	90	94.55
γ	90	90	90	90	90
Space Group	$I4_1/amd$	$C2/c$	$Pbnm$	$Pbnm$	$P2_1/n$
Z	4	4	4	4	4

33

TABLE 1. (cont.)

Variety	Zircon	Sphene	Topaz		Larnite
Method	2-D, film	2-D, film	3-D, counter	3-D, film	2-D, film
R & Rating	0.07 (2)	--- (3)	0.04 (1)	0.10 (2)	0.15 (av.) (3)
Cleavage and habit	{100} imperfect Elongated to c axis	{110}distinct;{100}, {1̄12}imperfect	{001} perfect Prismatic		{100} prominent (?) Granular
Comment	B estimated	B estimated		Olivine structure	B estimated
μ	394.4	343.5	109.6	250.8	268.6
ASF	1.559	0.1716	0.7161×10^{-1}	0.1169	0.6483×10^{-1}
Abridging factors	0.1:0:0	0.5:0:0	0.5:0:0	1:1.2:0.5	1:1.2:0.5

34

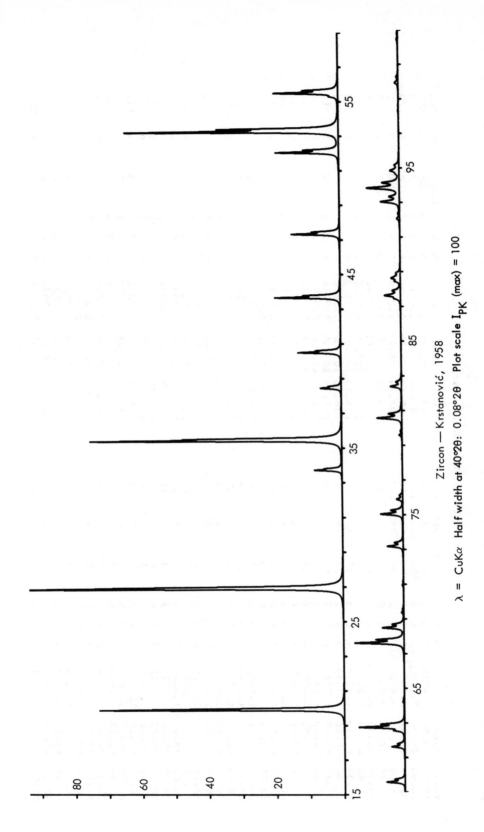

Zircon — Krstanović, 1958

$\lambda = CuK\alpha$ Half width at $40°2\theta$: $0.08°2\theta$ Plot scale I_{PK} (max) = 100

ZIRCON - KRSTANOVIĆ, 1958

2THETA	PEAK	D	H	K	L	I(INT)	I(PK)	I(DS)
19.93	19.94	4.451	1	0	1	64	73	46
26.93	26.92	3.308	2	0	0	100	100	100
33.73	33.72	2.655	2	1	1	9	8	12
35.45	35.46	2.530	1	1	2	80	75	112
38.45	38.46	2.339	2	2	0	7	6	11
40.50	40.50	2.225	2	0	2	14	13	23
43.68	43.68	2.071	3	0	1	22	19	41
47.33	47.34	1.9188	1	0	3	17	14	35
52.06	52.06	1.7552	3	2	1	23	19	54
53.29	53.30	1.7175	3	1	2	82	64	200
55.30	55.30	1.6598	4	0	0	2	3	6
55.51	55.50	1.6541	2	0	4	24	19	62
59.57	59.58	1.5505	4	1	1	7	6	21
61.62	61.62	1.5037	4	2	0	5	4	16
62.56	62.56	1.4836	3	3	2	3	3	10
62.75	62.74	1.4795	4	2	2	17	14	53
64.21	64.20	1.4494	4	3	1	1	1	2
67.61	67.60	1.3844	2	2	4	22	15	74
68.48	68.48	1.3690	3	2	3	9	6	32
69.36	69.36	1.3537	3	3	3	1	1	3
70.93	70.94	1.3275	4	2	2	1	0	2
73.17	73.16	1.2924	5	0	1	7	5	26
75.02	75.02	1.2649	4	2	4	10	7	41
75.87	75.88	1.2529	2	2	4	3	2	11
78.22	78.22	1.2211	4	1	3	0	0	1
79.56	79.56	1.2038	3	2	4	1	1	6
80.56	80.56	1.1914	5	1	2	12	7	54
81.20	81.20	1.1836	5	2	1	0	0	1
82.38	82.38	1.1696	1	0	5	5	3	25
87.45	87.46	1.1144	4	4	0	3	2	14
87.62	87.62	1.1127	4	3	2	8	5	39
88.44	88.44	1.1044	4	0	4	1	2	5
88.44		1.1044	4	3	3	2	3	8
88.61	88.62	1.1027	5	0	3	5	1	23
92.04	92.04	1.0704	6	1	1	2	1	9
93.02	93.02	1.0617	5	3	2	11	6	60
93.83	93.84	1.0546	4	2	4	19	10	106
94.66	94.68	1.0476	5	2	1	1	1	8
94.83	94.84	1.0461	6	2	0	5	3	29
98.29	98.28	1.0184	5	4	1	0	0	2
99.92	99.92	1.0061	3	3	4	2	1	14
102.44	102.44	0.9881	6	2	2	1	0	4
103.58	103.58	0.9802	6	1	3	7	3	43
104.63	104.62	0.9733	1	1	6	2	1	13
106.30	106.30	0.9626	5	3	3	2	1	10
107.34	107.34	0.9561	6	3	1	1	0	10
111.16	111.16	0.9337	7	0	1	1	0	5
113.09	113.08	0.9232	4	4	4	6	2	44
114.17	114.16	0.9185	5	4	3	1	2	11
116.85	116.86	0.9041	6	4	0	6	0	40
117.99	118.00	0.8986	3	1	6	7	3	55
119.11	119.10	0.8935	7	2	1	2	1	12
119.11		0.8935	5	1	5	9	5	70
119.84	119.82	0.8901	4	3	5	4	1	31
119.84		0.8901	5	0	5	2	1	13
120.03	120.02	0.8893	6	0	4	0	0	3
120.99	120.98	0.8850	6	3	3	3	1	23
127.51	127.52	0.8588	6	2	4	3	3	12
128.56	128.56	0.8550	7	0	3	10	3	80
129.35	129.16	0.8521	1	0	7	3	0	10
131.96	131.96	0.8433	3	3	6	0	1	2
133.33	133.34	0.8389	6	3	0	3	0	24
134.69	134.70	0.8346	7	0	1	1	0	5
137.03	137.28	0.8278	7	1	2	15	4	133
137.28		0.8270	8	0	0	2	1	15
137.95	137.98	0.8252	2	1	7	4	0	33
142.62	142.62	0.8131	8	1	1	1	1	22
142.62		0.8131	7	4	0	1	1	14
147.47	147.48	0.8024	8	2	2	7	1	8
148.32	148.46	0.8007	3	0	7	7	1	64
149.99	149.98	0.7974	8	0	2	0	0	1
152.30	152.30	0.7933	5	1	6	25	4	257

ZIRCON - KRSTANOVIĆ, 1958

2THETA	PEAK	D	H	K	L	I(INT)	I(PK)	I(DS)
158.62	159.10	0.7838	5	4	5	3	2	32
159.10		0.7832	6	4	4	13		140
161.54	161.60	0.7803	6	5	3	9	1	92
162.09	161.98	0.7798	6	6	0	9	1	93
163.61	163.56	0.7782	3	2	7	6	1	60

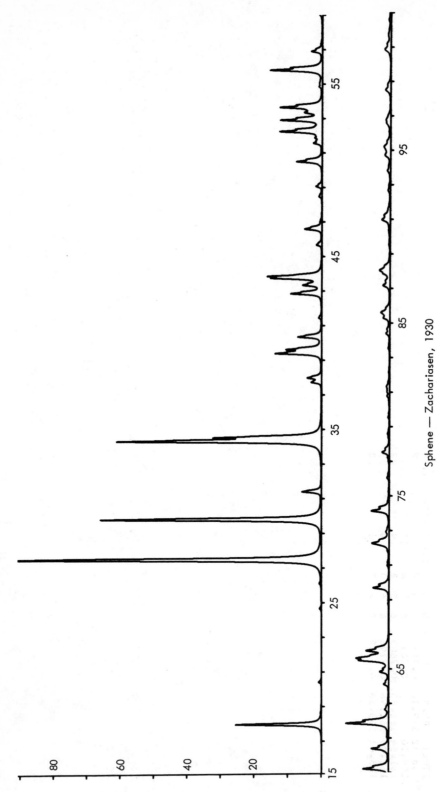

Sphene — Zachariasen, 1930

$\lambda = CuK\alpha$ Half width at $40°2\theta$: $0.11°2\theta$ Plot scale $I_{PK} (max) = 100$

38

SPHENE - ZACHARIASEN, 1930

2THETA	PEAK	D	H	K	L	I(INT)	I(PK)	I(DS)
17.94	17.94	4.940	-1	1	1	22	26	15
20.37	20.38	4.355	0	2	0	1	1	1
24.62	24.62	3.613	0	2	1	1	1	0
27.43	27.44	3.248	1	1	1	100	100	100
29.77	29.78	2.999	-2	0	2	70	66	77
31.38	31.38	2.849	-2	0	0	6	6	7
34.24	34.28	2.617	-2	1	1	1	61	2
34.51	34.50	2.613	-1	3	0	66	32	85
37.70	37.70	2.597	0	2	2	29	3	38
37.96	37.96	2.384	2	2	0	3	3	5
39.36	39.36	2.368	-1	3	1	5	4	7
39.60	39.60	2.287	-1	1	3	16	14	24
40.32	40.32	2.274	1	3	2	11	11	16
41.43	41.44	2.235	0	4	1	8	7	13
42.82	42.82	2.177	-3	1	0	1	1	2
43.31	43.32	2.110	-2	2	2	11	9	19
43.72	43.72	2.087	-2	2	1	6	6	11
43.83	43.82	2.069	-3	1	1	16	15	28
45.65	45.66	2.064	0	4	1	10	16	17
46.57	46.58	1.9855	2	2	1	2	2	4
46.97	46.98	1.9486	-3	1	3	7	5	12
48.45	48.44	1.9327	0	2	3	1	1	1
49.04	49.04	1.8773	-1	3	3	1	1	2
50.48	50.48	1.8561	-2	0	4	2	2	5
51.54	51.54	1.8064	0	4	2	10	8	21
51.84	51.84	1.7716	-1	1	4	3	3	7
52.21	52.22	1.7620	-1	4	2	2	2	4
52.88	52.88	1.7504	2	0	0	17	13	37
53.30		1.7299	2	4	2	17	12	37
53.35	53.34	1.7172	-2	3	1	3	5	7
53.63	53.62	1.7158	-1	1	3	17	13	38
54.83	54.84	1.7075	-2	2	4	1	1	3
55.08	55.08	1.6729	-1	5	1	1	1	1
55.78	55.78	1.6659	0	3	0	22	16	53
56.87	56.88	1.6466	-3	3	4	5	3	11
59.19	59.20	1.6175	0	0	1	11	8	29
		1.5596	1	5	1			
59.25	60.36	1.5583	2	4	1	1		2
60.35		1.5323	0	4	3	1	5	2
60.37	61.06	1.5320	-4	2	2	7		18
61.06	61.84	1.5163	0	2	4	1	1	2
61.82		1.4994	-4	0	4	5	13	14
61.85	62.64	1.4988	-1	3	3	15		42
62.63	64.10	1.4819	-3	0	4	2	1	4
64.09	64.82	1.4517	0	1	0	2	2	7
64.82	65.48	1.4371	-3	5	0	4	3	12
65.48	65.62	1.4243	4	0	0	12	9	35
65.60	66.04	1.4218	-1	3	1	9	10	28
66.04		1.4135	3	3	1	8	7	25
66.09		1.4126	-3	2	3	2		6
69.67	69.68	1.3484	-2	5	1	8	5	26
69.84	69.86	1.3455	-3	2	1	1	3	2
70.96	70.96	1.3270	-2	5	3	1	1	3
71.81	71.82	1.3134	-3	4	1	1	1	2
72.13	72.24	1.3083	-4	4	3	2	5	8
72.24		1.3066	-2	6	2	8		27
72.90	72.90	1.2965	-5	1	3	1	1	4
74.12	74.12	1.2781	-1	3	5	10	5	34
76.03	76.02	1.2507	-1	3	4	4	1	5
77.51	77.50	1.2305	-2	4	6	4	2	15
78.55	78.56	1.2167	2	0	1	1	1	2
79.52	79.52	1.2043	0	6	3	1	0	3
80.75	80.76	1.1890	-4	0	6	1	1	5
81.15	81.16	1.1842	-2	2	6	1	1	3
81.31	81.30	1.1823	-5	6	2	1	1	5
82.14	82.14	1.1724	1	7	1	1	0	5
83.51	83.52	1.1566	-4	2	3	3	1	4
84.37	84.38	1.1470	2	2	6	2	2	10
85.28	85.28	1.1371	-5	1	4	4	3	16
85.44	85.60	1.1354	-3	5	5	4		5
85.61		1.1335	-2	6	1	4		17
87.13	87.16	1.1177	-1	3	5	1	2	6
87.15		1.1174	0	5	5	3		15
87.84	88.02	1.1104	-1	7	3	3	3	15

SPHENE — ZACHARIASEN, 1930

2THETA	PEAK	D	H	K	L	I(INT)	I(PK)	I(DS)
88.02	89.50	1.1086	-5	3	5	6	0	26
89.50	90.34	1.0941	-4	4	1	1		3
90.21	90.96	1.0873	-6	0	4	1	2	2
90.93		1.0806	0	2	2	1		6
90.95	91.20	1.0804	-4	6	4	5	2	21
91.15		1.0785	-0	6	3	1		3
91.17		1.0783	-0	0	6	1	2	4
92.66	92.66	1.0648	-6	0	0	2	1	8
93.79	93.80	1.0550	-6	2	2	2	1	11
94.66	94.70	1.0475	-5	5	4	3	2	14
94.76		1.0467	0	2	3	2		9
95.14	95.18	1.0436	-4	6	6	2	2	7
95.21		1.0430	-4	4	4	3		17
96.05	96.04	1.0361	2	6	4	1	0	3
96.57	96.60	1.0319	0	8	2	3	1	13
96.75	96.74	1.0304	5	1	1	1	1	11
98.47	98.48	1.0170	2	8	0	3	2	15
98.51		1.0167	4	6	0	2		9
100.81	100.84	0.9996	-6	0	6	2	2	11
100.86		0.9992	-3	3	7	4		20
101.62	101.72	0.9938	-5	1	7	1	1	7
101.77		0.9928	4	4	2	2		8
102.57	102.58	0.9871	-1	1	7	1	1	9
103.85	103.90	0.9784	-2	8	4	2	0	3
104.71	104.72	0.9728	-6	4	4	1	0	5
105.21	105.20	0.9695	3	5	3	2	1	13
105.64	105.64	0.9663	0	4	6	1	1	6
107.21	107.46	0.9569	-1	5	5	1		8
107.45		0.9554	-3	9	1	3		17
108.98	109.00	0.9462	-2	7	7	2	1	10
110.21	110.24	0.9391	-2	8	5	3	1	15
110.29		0.9387	-4	6	4	2		10
112.19	112.20	0.9281	6	0	6	2	1	11
112.25		0.9277	-7	1	5	1		4
113.48	113.48	0.9211	-4	8	2	2	0	9
116.36	116.36	0.9065	-4	4	8	2	0	13
117.17	117.18	0.9025	-5	7	3	2	1	14

2THETA	PEAK	D	H	K	L	I(INT)	I(PK)	I(DS)
119.53	119.54	0.8915	-2	5	1	2	1	13
120.81	121.14	0.8858	-7	2	8	1	1	8
121.13		0.8844	-7	3	3	4		25
123.31	123.34	0.8752	-4	0	4	1	1	6
123.35		0.8750	-3	9	6	2		12
124.32	124.40	0.8711	-4	4	3	2	1	14
124.50		0.8704	-6	5	0	2		13
125.22	125.02	0.8675	-5	7	5	1		5
126.23	126.36	0.8636	-5	5	7	1	1	11
126.38		0.8630	-1	7	1	4	1	9
127.56	127.58	0.8586	-6	6	2	2		29
127.73		0.8580	3	1	7	1		7
129.63	129.66	0.8512	-7	7	3	2	1	16
129.73		0.8508	0	4	7	1	0	5
130.87	131.10	0.8469	1	9	3	2	1	16
131.09		0.8462	-6	3	5	5	0	36
131.35		0.8453	-0	2	8	2		12
132.65	132.62	0.8410	-2	4	2	1	1	10
134.42	134.44	0.8355	-2	10	8	2	0	12
135.26	135.24	0.8329	3	4	1	4	0	11
136.06	136.04	0.8306	-7	9	0	1	1	29
137.48	137.46	0.8265	-6	1	1	2	0	4
138.64	139.22	0.8233	-7	6	6	5	1	18
139.21		0.8218	-8	3	7	2		37
140.52	140.58	0.8183	-7	0	5	5	1	18
140.70		0.8179	4	1	5	1		8
141.83	142.10	0.8151	5	6	3	1	2	5
142.06		0.8145	-5	3	3	5		41
142.12		0.8143	6	0	9	3		5
142.16		0.8143	0	8	2	4		40
144.34	144.44	0.8091	-8	2	8	3	1	31
144.50		0.8088	-7	8	6	1		28
145.90	145.26	0.8057	-4	8	8	1	1	7
146.48	147.20	0.8044	6	4	6	6	1	5
147.18		0.8030	-6	8	8	2		48
148.04	148.30	0.8012	-1	9	5	2		15
148.48		0.8003						21

40

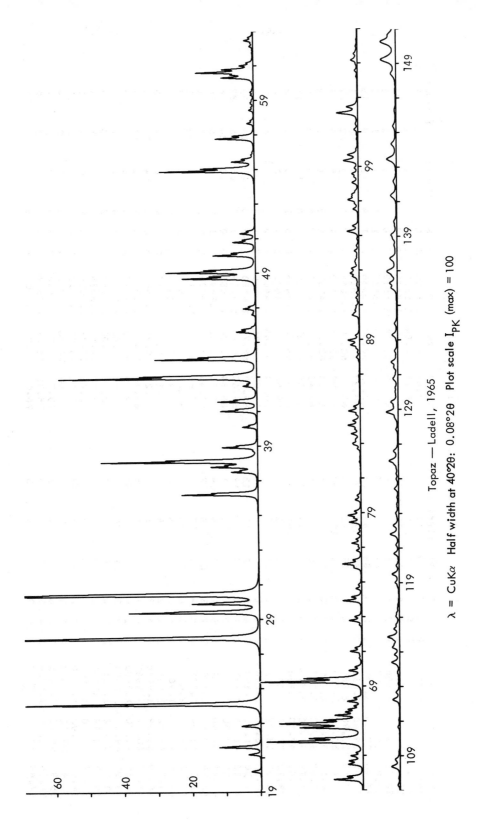

Topaz — Ladell, 1965

λ = CuKα Half width at 40°2θ: 0.08°2θ Plot scale I$_{PK}$ (max) = 100

41

TOPAZ - LADELL, 1965

2THETA	PEAK	D	H	K	L	I(INT)	I(PK)	I(DS)
20.17	20.18	4.398	0	2	0	3	3	2
21.16	21.16	4.195	0	2	1	3	4	3
21.60	21.60	4.111	1	1	0	11	12	10
22.81	22.82	3.896	0	2	1	5	6	5
24.09	24.08	3.692	1	1	1	67	70	65
27.90	27.90	3.195	1	2	0	84	87	83
29.40	29.40	3.036	0	2	2	38	38	37
29.90	29.90	2.986	1	2	1	19	20	19
30.42	30.42	2.936	1	1	1	100	100	100
36.18	36.18	2.480	1	3	0	24	22	25
37.49	37.50	2.397	1	0	3	8	8	8
37.79	37.80	2.379	1	3	1	14	14	15
38.09	38.10	2.360	0	2	3	50	46	52
38.70	38.70	2.325	2	1	0	1	1	1
38.91	38.92	2.312	1	1	3	11	10	12
40.08	40.08	2.248	2	1	0	5	4	5
41.00	41.00	2.199	0	4	1	12	11	13
41.56	41.56	2.171	2	1	1	13	12	14
42.29	42.30	2.135	1	3	2	1	1	1
42.45	42.45	2.127	2	2	1	4	4	4
42.94	42.94	2.105	1	2	3	66	58	72
43.08	43.04	2.098	0	0	4	1	1	2
44.02	44.02	2.055	2	2	0	34	30	38
45.59	45.60	1.9881	1	4	0	7	6	8
45.75	45.72	1.9813	2	3	0	2	4	2
46.59	46.58	1.9478	0	4	2	1	1	2
46.93	46.92	1.9345	1	1	4	4	4	5
48.69	48.70	1.8685	1	3	3	27	22	30
49.05	49.04	1.8557	2	3	2	32	26	36
50.02	50.02	1.8218	1	5	0	15	13	17
50.77	50.78	1.7966	2	1	4	8	7	9
51.27	51.28	1.7803	1	4	3	6	5	6
54.90	54.90	1.6710	2	2	4	36	28	43
55.43	55.43	1.6563	1	5	3	8	7	10
55.82	55.82	1.6455	1	1	0	1	1	1
56.76	56.76	1.6204	4	0	3	15	11	18
57.65	57.64	1.5977	2	4	0	3	2	3
58.41	58.42	1.5785	1	0	5	2	1	1
58.78	58.84	1.5695	0	4	1	1	1	1
58.85		1.5679	2	2	1	1		1
59.28	59.28	1.5575	0	0	4	2	1	1
59.44	59.44	1.5537	1	1	5	1	1	1
60.30	60.30	1.5336	2	1	4	12	9	15
60.37		1.5319	1	1	5	22	17	27
60.61	60.60	1.5265	2	0	2	5	4	6
60.99	60.98	1.5179	0	2	4	1	1	2
62.11	62.12	1.4931	2	1	3	1	1	5
62.45	62.46	1.4857	1	4	5	4	3	5
63.59	63.60	1.4619	3	3	0	12	9	15
64.46	64.46	1.4443	3	0	1	5	4	6
64.52		1.4430	1	4	2	2	2	2
64.67	64.66	1.4402	1	2	5	2	2	3
64.95	64.96	1.4345	3	1	1	4	4	6
65.79	65.78	1.4183	2	5	3	41	28	53
66.60	66.60	1.4029	3	1	0	25	18	32
66.84	66.84	1.3985	0	4	6	28	24	36
67.31	67.30	1.3899	3	3	2	10	8	14
67.45	67.50	1.3873	2	4	1	1	6	1
67.63	67.64	1.3841	0	6	2	1	5	1
67.65		1.3837	2	1	5	4	4	5
67.83	67.84	1.3805	3	3	2	1		1
68.11	68.10	1.3755	1	0	4	1	3	8
69.24	69.24	1.3557	3	3	3	6	4	8
69.44	69.44	1.3524	3	1	1	44	30	58
69.89	69.88	1.3447	0	2	6	1	17	2
70.61	70.62	1.3327	1	6	3	5	4	7
70.99	71.00	1.3265	3	2	0	1	1	1
72.50	72.50	1.3026	2	6	2	5	3	7
72.76	72.76	1.2985	3	4	3	1	1	1
73.91	73.92	1.2812	2	2	6	5	4	7
74.60	74.60	1.2710	2	5	1	9	5	12
75.79	75.80	1.2540	1	4	4	3	2	4
75.88	75.88	1.2527	3	4	2	3	2	4
76.03	76.02	1.2507	2	6	3	7	6	10

TOPAZ – LADELL, 1965

2THETA	PEAK	D	H	K	L	I(INT)	I(PK)	I(DS)
76.79	76.80	1.2401	2	6	0	2	1	3
77.22	77.22	1.2343	2	3	5	1		1
77.23		1.2343	3	3	4	1	3	
77.78	77.78	1.2268	2	6	1	2	1	2
78.44	78.44	1.2182	1	3	6	5	3	8
78.83	78.84	1.2132	1	7	0	5	3	7
78.85		1.2128	3	0	4	1	4	2
79.73	79.72	1.2017	0	6	4	3		4
81.93	81.92	1.1749	1	5	5	1	2	1
82.68	82.68	1.1662	2	5	4	1	0	2
82.91	82.98	1.1635	1	5	4	2	1	2
83.00		1.1625	4	0	0	3	2	4
83.15	83.16	1.1608	1	0	7	1		1
83.46	83.54	1.1571	2	4	5	1	2	6
83.54		1.1562	3	5	6	4	3	1
83.92	84.02	1.1520	3	1	1	1		9
84.03		1.1508	1	3	4	6	4	1
84.35	84.30	1.1472	3	4	6	1		2
84.66	84.66	1.1438	4	4	1	6	2	8
84.85	84.86	1.1417	4	1	6	2	3	3
87.95	87.96	1.1093	2	3	5	1	1	2
88.47	88.48	1.1041	0	6	5	4	1	7
88.66	88.66	1.1023	3	2	0	4	3	6
88.93	88.94	1.0996	0	8	0	4	3	6
90.39	90.40	1.0856	4	2	2	2	2	3
90.51		1.0845	3	4	4	2	1	2
90.92	90.92	1.0807	2	3	0	2	1	3
91.39	91.38	1.0764	1	5	5	1	1	1
92.08	92.08	1.0701	2	8	0	1	1	1
92.19	92.18	1.0690	2	7	2	1	1	1
92.58	92.62	1.0655	4	1	3	2	2	3
92.63		1.0651	3	6	0	4		3
92.79	92.80	1.0637	0	8	2	1	3	6
93.04	93.06	1.0615	1	8	5	1	2	1
93.05		1.0614	3	3	5	2		2
93.48	93.48	1.0577	2	1	7	2	1	3
93.60	93.58	1.0566	3	6	1	1	1	2

2THETA	PEAK	D	H	K	L	I(INT)	I(PK)	I(DS)
94.08	94.10	1.0525	0	4	7	1	1	1
94.10		1.0523	4	4	6	2		2
94.79	94.80	1.0465	4	3	3	3	1	1
95.23	95.22	1.0428	4	2	3	7	3	11
96.51	96.50	1.0324	3	6	2	3	2	5
97.04	97.04	1.0281	4	2	3	5	3	8
97.09		1.0277	4	4	0	1		2
97.24	97.24	1.0265	1	4	7	2	2	4
98.04	98.06	1.0203	0	2	8	2	2	4
98.06		1.0201	0	4	2	1		2
98.56	98.56	1.0163	1	1	8	3	2	6
99.32	99.32	1.0105	3	2	6	8	4	14
99.38		1.0101	4	1	4	1		2
99.66	99.66	1.0080	4	3	7	2	3	3
100.56	100.56	1.0014	2	3	7	1	1	1
101.12	101.12	0.9974	2	6	5	2	0	3
102.07	102.10	0.9907	4	2	4	1	1	22
102.10		0.9904	2	5	6	13	6	
104.20	104.20	0.9762	3	7	0	1	0	1
104.54	104.54	0.9739	0	8	4	1	1	2
105.15	105.16	0.9699	4	5	0	4	2	7
105.31		0.9689	1	5	7	1		1
105.97	105.96	0.9647	4	4	3	1	1	2
107.27	107.28	0.9565	1	9	0	1	1	2
108.34	108.36	0.9500	4	1	5	3	2	5
108.40		0.9497	3	6	4	3		5
110.64	110.64	0.9366	2	8	3	3	1	5
112.20	112.22	0.9280	0	6	7	2	2	2
112.22		0.9279	2	6	6	2		3
112.26		0.9277	1	4	8	2		3
113.14	113.14	0.9229	4	4	5	1	0	2
113.75	113.82	0.9197	0	8	1	1	1	1
113.84		0.9193	5	1	6	1		
114.39	114.38	0.9164	1	7	9	6	2	10
114.83	114.84	0.9141	1	0	9	1	2	1
115.24	115.26	0.9121	0	2	9	4	2	8
115.38		0.9113	2	5	7	2		3

TOPAZ — LADELL, 1965

2THETA	PEAK	D	H	K	L	I(INT)	I(PK)	I(DS)
115.86	115.86	0.9090	2	3	8	7	3	13
116.55	116.54	0.9056	4	6	1	2	1	3
117.05	117.04	0.9031	5	1	2	1	1	2
117.23	117.22	0.9023	1	8	5	1	1	3
119.00	119.00	0.8940	4	0	6	3	2	5
119.48	119.48	0.8918	3	8	1	4	1	8
120.66	120.66	0.8865	5	5	0	1	1	7
121.79	121.80	0.8816	5	3	0	7	2	13
122.23	122.30	0.8797	0	10	0	1	2	2
123.01	123.00	0.8764	4	4	5	2	1	4
123.38	123.40	0.8749	0	10	1	1	1	1
123.92	123.92	0.8727	1	3	9	1	0	1
124.41	124.40	0.8707	3	4	7	3	1	6
125.27	125.28	0.8673	5	2	6	1	1	2
125.28		0.8673	2	7	8	1		1
126.00	126.00	0.8645	3	1	6	5	3	3
126.02	126.02	0.8644	1	8	0	2		10
126.03		0.8643	0	10	0	10		3
127.82	127.82	0.8576	2	9	3	2	1	3
128.82	128.82	0.8540	3	8	1	10	3	20
129.20	129.42	0.8527	1	7	7	2	2	1
129.36		0.8521	5	4	1	1		3
130.01	130.00	0.8498	1	2	6	1	1	3
130.24	130.24	0.8491	2	2	9	2	1	7
130.74	130.74	0.8473	3	6	6	3	1	3
131.43	131.40	0.8450	5	3	3	5	2	11
131.71	131.70	0.8441	1	0	9	1		9
133.26	133.26	0.8391	0	6	10	4		10
133.27		0.8390	1	6	8	5	2	14
135.99	136.00	0.8308	4	1	7	7	2	10
136.27	136.26	0.8300	2	3	9	5	3	6
136.90	136.94	0.8282	4	4	6	3		8
136.99		0.8279	2	9	4	4		1
137.72	137.70	0.8258	1	10	3	1	2	2
138.30	138.46	0.8242	0	2	10	1	1	4
138.83	139.06	0.8228	3	9	1	2		
138.84		0.8228	2	10	0			
139.05		0.8222	5	5	0	1		2
139.07		0.8221	1	1	10	4		9
141.23	141.34	0.8165	5	1	4	2	2	4
141.36		0.8162	4	7	3	6		13
143.42	143.44	0.8112	0	10	4	3	3	2
143.45		0.8112	0	1	9	11		23
144.87	144.88	0.8079	3	4	8	1	0	2
145.36	145.36	0.8068	5	5	2	4	1	8
147.34	147.34	0.8026	4	3	7	1	0	2
148.36	148.38	0.8006	4	6	5	4	2	8
148.43		0.8004	3	7	6	2		5
149.08	149.20	0.7992	1	10	4	2	4	3
149.20		0.7989	3	0	9	18		39
149.25		0.7988	4	8	0	1		2
150.24	150.26	0.7970	4	5	6	11	4	24
150.66	150.62	0.7962	3	6	7	8	2	17
150.97		0.7957	3	1	9	1		1
153.71	153.72	0.7910	3	8	5	14	3	29
154.03		0.7905	1	7	4	3	6	7
154.64	155.06	0.7895	2	9	6	4		10
154.76		0.7893	2	10	3	3		7
154.80		0.7893	4	0	10	1		3
155.09		0.7888	2	5	3	29		62
155.54		0.7881	5	1	0	1		3
156.51	156.46	0.7867	1	11	9	4	3	8
156.95	156.92	0.7861	0	6	0	14	3	31
156.98		0.7861	2	1	10	3		6
158.53	158.58	0.7840	3	2	0	5	4	11
158.64		0.7838	0	4	10	23		51
161.66	161.66	0.7802	5	3	5	3	1	6
162.69	162.74	0.7791	4	4	7	4	1	9
163.07		0.7787	0	0	8	2		4

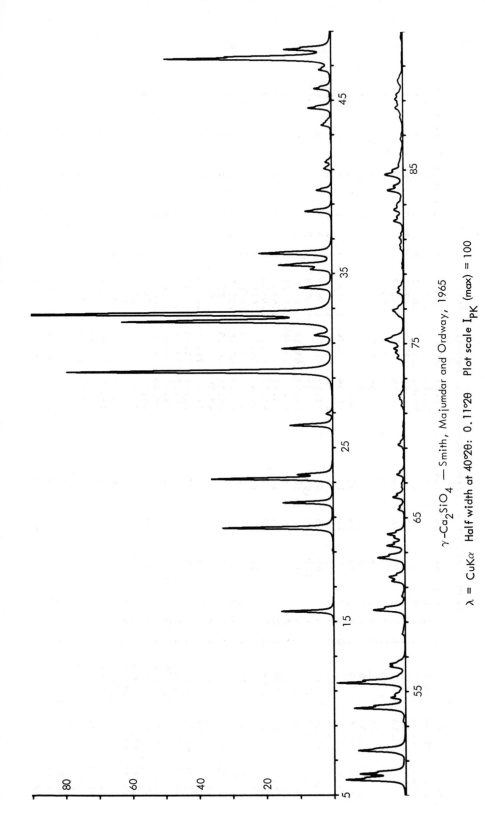

γ–Ca$_2$SiO$_4$ — Smith, Majumdar and Ordway, 1965

λ = CuKα Half width at 40°2θ: 0.11°2θ   Plot scale I$_{PK}$ (max) = 100

GAMMA CA2SIO4 - SMITH, MAJUMDAR AND ORDWAY, 1965

2THETA	PEAK	D	H	K	L	I(INT)	I(PK)	I(DS)
15.57	15.58	5.685	0	2	0	12	16	7
20.36	20.38	4.357	0	2	1	27	33	19
21.81	21.82	4.072	1	1	1	13	15	9
23.18	23.18	3.833	1	1	0	32	36	24
23.44	23.44	3.793	1	2	0	8	11	6
26.26	26.26	3.391	0	0	2	12	13	10
26.91	26.92	3.310	1	2	1	2	2	1
29.35	29.36	3.040	1	3	0	76	79	69
30.67	30.68	2.912	0	2	2	14	15	14
31.44	31.44	2.843	0	4	0	5	5	4
32.24	32.24	2.774	1	3	1	61	63	60
32.66	32.66	2.739	2	0	2	100	100	100
34.17	34.18	2.622	0	4	0	10	10	10
35.23	35.24	2.545	2	0	0	6	6	6
35.48	35.48	2.528	1	2	2	16	16	17
36.13	36.16	2.484	2	1	0	6	22	7
36.16		2.482	2	4	0	18		19
38.57	38.56	2.332	1	1	2	9	8	10
39.79	39.78	2.264	2	3	1	5	5	6
41.03	41.04	2.198	2	1	2	3	2	3
41.41	41.42	2.179	0	4	2	2	2	3
43.55	43.56	2.076	1	5	0	3	3	4
43.78	43.76	2.066	1	0	3	1	2	2
44.53	44.54	2.033	1	1	3	8	7	11
45.65	45.66	1.9855	1	5	1	6	5	8
46.74	46.74	1.9419	1	2	3	4	4	5
47.39	47.40	1.9166	2	2	2	60	50	86
47.93	47.94	1.8964	2	4	1	16	14	23
49.89		1.8263	2	0	3	10	18	16
49.92	49.92	1.8252	0	6	1	13		20
50.25	50.24	1.8141	1	3	3	16	14	24
51.57	51.60	1.7708	1	5	2	10	14	15
51.61		1.7694	3	3	0	10		16
54.04	54.04	1.6955	0	0	4	19	16	32
54.63	54.64	1.6784	3	1	0	5	5	9
54.86	54.78	1.6719	2	1	3	1	3	2
55.47	55.50	1.6551	2	4	2	4	21	6
55.50	55.50	1.6543	0	6	1	24		41
56.43	56.44	1.6293	3	1	1	5	5	10
56.55	56.58	1.6261	3	2	0	2	5	3
56.60		1.6248	0	2	4	1		2
58.30	58.30	1.5813	3	2	1	4	1	7
59.64	59.70	1.5489	1	7	0	10	10	20
59.70		1.5476	3	3	0	4		7
60.49	60.48	1.5293	1	5	3	10	1	9
61.34	61.34	1.5100	3	1	1	5	4	10
61.60	61.60	1.5042	3	2	2	5	5	5
62.57	62.68	1.4833	1	6	2	10	8	5
62.69		1.4808	3	1	3	10		20
63.38	63.38	1.4662	3	3	2	7	5	15
63.82	63.86	1.4571	3	4	0	2	3	4
63.87		1.4562	0	4	4	2		4
64.06	64.04	1.4523	3	0	4	5	5	9
65.46	65.46	1.4246	3	4	1	3	2	6
66.16	66.16	1.4111	2	0	4	5	4	11
67.46	67.46	1.3871	1	6	2	3	2	7
69.16	69.18	1.3572	3	0	3	1		3
69.19		1.3566	2	5	3	1	1	2
71.44	71.44	1.3194	0	5	4	2	1	4
71.82	71.82	1.3133	1	5	3	2	1	6
74.15	74.16	1.2777	3	3	0	3	2	7
74.48	74.48	1.2727	4	2	0	2	3	4
74.69	74.70	1.2697	2	7	2	1		4
74.71		1.2695	1	8	1	4	6	3
75.09	75.22	1.2640	2	4	4	4		10
75.21		1.2623	3	5	2	6		15
75.27		1.2615	2	6	3	2		4
76.90	76.90	1.2387	1	3	5	4	3	10
77.94	77.94	1.2248	3	4	3	3	2	7
78.24	78.20	1.2207	2	8	1	1	2	3
80.44	80.44	1.1928	3	1	4	4	1	4
80.63	80.66	1.1905	2	1	5	1	1	2
81.08	81.08	1.1851	4	1	2	2	1	5
82.04	82.04	1.1736	3	2	4	2	3	5

GAMMA CA2SIO4 — SMITH, MAJUMDAR AND ORDWAY, 1965

2THETA	PEAK	D	H	K	L	I(INT)	I(PK)	I(DS)
82.05		1.1735	3	7	0	3		9
82.67	82.68	1.1662	4	2	2	4	3	11
82.74		1.1654	3	5	3	1		2
82.74		1.1654	2	8	2	1		2
83.07		1.1616	4	4	0	2		5
83.82	82.94	1.1532	1	9	2	9	2	26
84.56	83.82	1.1450	4	4	2	1	5	
84.69	84.72	1.1435	3	3	4	4	5	11
84.73		1.1430	1	7	4	7		20
85.28		1.1371	0	10	0	1		3
85.31	85.30	1.1367	4	3	2	1	1	2
85.42		1.1356	1	5	5	1		2
85.91	85.90	1.1303	0	0	6	1	1	3
86.76	86.76	1.1214	0	10	1	1	1	4
87.82	88.02	1.1106	0	2	6	1		2
88.02		1.1086	0	2	6	1		4
88.37	88.58	1.1051	3	4	4	1	2	2
88.56		1.1032	2	4	5	2		6
88.59		1.1030	0	6	5	2		6
89.06	89.06	1.0983	1	1	6	4	2	13
89.30	89.32	1.0960	4	5	1	2	2	4
89.96	90.14	1.0897	3	8	0	1	2	5
90.00		1.0893	0	8	4	1		2
90.15		1.0879	2	7	3	2		7
92.61	92.66	1.0653	1	1	6	1	0	2
93.79	93.80	1.0550	3	6	1	1	1	5
94.33	94.32	1.0503	0	4	6	1	0	2
95.09	95.10	1.0440	4	6	0	2	0	3
95.79	95.80	1.0382	3	10	1	1	1	7
95.89		1.0374	4	8	2	1		2
96.40	96.38	1.0332	3	0	3	1	1	3
98.02	98.06	1.0204	4	0	5	2	1	4
98.35	98.54	1.0179	3	2	4	4	2	8
98.54		1.0164	2	2	6	1		13
98.62		1.0158	0	10	3	1		2
98.88	98.88	1.0138	4	1	4	1	2	2
98.99		1.0131	1	11	0	1		3
99.12		1.0121	2	6	5	1		2
99.56	99.54	1.0088	4	6	1	3	1	9
100.34	100.54	1.0030	5	1	1	1	1	4
100.56		1.0014	2	8	0	2		6
101.19	101.18	0.9968	4	5	4	1		3
101.76	101.76	0.9928	3	5	3	1	1	2
101.77		0.9928	1	9	4	1		4
103.13	103.14	0.9833	5	3	0	1	1	4
103.17		0.9830	1	11	2	1		2
104.89	105.00	0.9716	5	5	0	4	1	13
104.98		0.9710	3	9	2	2		7
105.02		0.9708	0	6	6	4		2
105.93	105.94	0.9649	3	7	4	3	1	12
106.98	107.00	0.9583	4	4	4	3	1	10
107.16		0.9572	4	6	3	1		3
109.29	109.30	0.9444	5	3	2	2	1	5
109.30		0.9444	0	10	4	2		7
109.45		0.9435	2	10	3	1		3
110.22	110.30	0.9390	4	8	1	1	1	4
110.32		0.9385	0	12	1	1		3
110.48		0.9375	0	1	6	1		4
110.83	110.76	0.9356	3	10	1	1		9
112.18	112.22	0.9281	3	5	5	2	1	9
112.70	112.74	0.9253	5	1	3	1	0	5
112.79		0.9248	3	6	4	2	1	8
113.11	113.12	0.9231	1	3	6	1	1	3
113.56	113.50	0.9207	5	5	1	1	1	3
114.26	114.32	0.9171	0	4	7	1		4
114.34		0.9167	3	8	4	1		4
115.02	115.06	0.9132	4	8	2	2	1	5
115.13		0.9126	0	12	2	1		2
116.24	116.24	0.9071	2	6	6	3	1	11
118.50	118.50	0.8963	5	5	2	4	1	17
119.18	118.96	0.8931	3	4	2	1	1	3
120.90	120.92	0.8854	2	10	4	3	1	13
121.07		0.8847	2	8	6	1		2
121.50	121.46	0.8828	3	11	0	1	1	5

GAMMA CA2SIO4 - SMITH, MAJUMDAR AND ORDWAY, 1965

2THETA	PEAK	D	H	K	L	I(INT)	I(PK)	I(DS)
122.03	122.00	0.8805	2	12	1	1	1	5
123.62	123.64	0.8739	0	12	3	1	1	6
124.19	124.24	0.8716	3	10	3	1	1	3
124.24		0.8714	0	10	5	1		6
124.50	124.72	0.8703	5	1	4	1	1	3
124.68		0.8696	1	11	4	1		7
124.76		0.8693	3	5	6	2		10
126.44	126.48	0.8628	2	4	7	2	1	3
126.46		0.8627	5	7	0	1		6
126.47		0.8627	0	6	7	1		6
127.36	127.36	0.8593	2	5	5	2	1	7
127.43		0.8591	1	13	2	1		11
128.48	128.50	0.8552	5	3	1	2	1	7
129.78	129.80	0.8506	5	3	4	1	0	3
130.56	130.58	0.8480	4	10	0	1	1	10
130.62		0.8477	0	0	8	2		7
132.53	132.56	0.8414	4	10	1	1	1	3
133.26	133.24	0.8391	3	1	7	1	1	17
134.26	134.32	0.8360	4	2	6	3	1	9
134.42		0.8355	1	13	5	2		4
135.06	135.08	0.8335	4	6	1	2		9
135.29		0.8329	6	2	9	6	1	30
135.87	135.88	0.8311	1	9	6	1	1	4
137.10	137.10	0.8276	4	8	4	1	1	6
137.46		0.8266	2	12	3	1		3
138.06	137.96	0.8249	4	3	6	1	0	7
139.35	139.38	0.8214	3	3	7	1	1	3
139.40		0.8213	3	12	1	1		5
141.04	141.24	0.8170	2	6	7	1	1	15
141.21		0.8166	1	3	8	3		27
142.00	142.02	0.8146	6	2	2	5	1	18
142.65	142.76	0.8131	6	4	0	3	1	6
142.93		0.8124	0	4	8	1		5
143.00		0.8122	5	1	5	3		15
145.98	146.52	0.8055	1	13	3	3	1	20
146.53		0.8043	2	0	8	4		4
146.62		0.8041	5	8	2	1		

2THETA	PEAK	D	H	K	L	I(INT)	I(PK)	I(DS)
150.31	150.38	0.7969	4	11	1	2	0	12
150.49		0.7965	1	14	1	1		3
151.93	151.82	0.7940	4	10	3	2		9
152.61	152.74	0.7928	5	9	0	1	0	5
152.95		0.7922	4	5	6	1		5
157.86	159.06	0.7849	1	8	6	1	1	8
158.13		0.7845	3	11	4	3		17
159.26		0.7830	3	11	5	5		30
159.44		0.7828	5	4	5	1		5
159.80		0.7824	2	11	5	2		3
164.34	164.72	0.7775	6	3	3	3	0	9
164.74		0.7771	4	8	5	3		18
164.95		0.7769	3	12	3	4		22

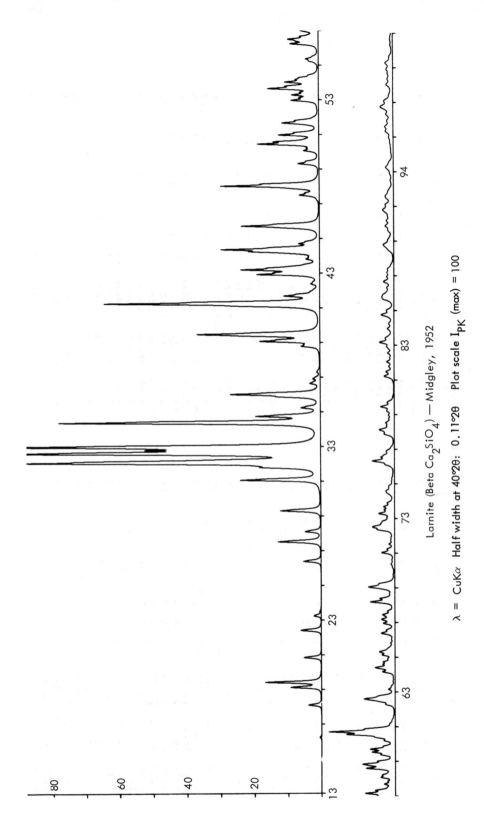

Larnite (Beta Ca_2SiO_4) — Midgley, 1952

λ = CuKα Half width at 40°2θ: 0.11°2θ Plot scale I_{PK} (max) = 100

49

LARNITE (BETA CA2SIO4) - MIDGLEY, 1952

2THETA	PEAK	D	H	K	L	I(INT)	I(PK)	I(DS)
18.07	18.08	4.904	-1	0	1	3	4	2
19.10	19.10	4.643	0	1	0	8	9	5
19.39	19.40	4.574	1	1	0	14	17	9
20.81	20.82	4.264	1	0	1	5	6	3
22.38	22.38	3.969	-1	1	1	6	6	4
23.22	23.24	3.827	0	0	2	2	2	2
26.36	26.36	3.378	1	1	2	5	5	4
27.49	27.50	3.241	-1	1	2	13	13	11
28.08	28.08	3.175	0	2	1	5	5	4
29.27	29.28	3.048	1	2	1	12	12	11
31.04	31.04	2.878	1	0	3	25	24	24
31.77	31.78	2.814	0	1	3	13	18	13
32.01	32.02	2.793	-1	1	3	88	91	87
32.55	32.56	2.748	2	0	0	100	100	100
32.75	32.76	2.732	0	2	2	35	52	35
32.93	32.94	2.717	2	1	0	97	99	98
34.33	34.34	2.610	1	1	3	85	78	90
34.72	34.72	2.582	-1	2	3	19	20	20
35.22	35.22	2.546	0	1	0	5	6	6
35.85	35.86	2.503	-2	1	1	11	14	12
35.99	36.00	2.493	-2	1	2	27	27	29
36.62	36.62	2.452	-1	2	2	3	3	3
36.88	36.88	2.435	-2	2	3	2	2	2
38.76	38.76	2.321	1	0	4	5	6	6
39.05	39.06	2.305	-2	0	2	20	18	24
39.37	39.37	2.287	-2	2	2	17	37	21
39.45	39.46	2.282	0	2	3	31	31	38
41.21	41.20	2.189	0	3	1	66	64	84
41.66	41.66	2.166	2	1	2	11	11	14
41.93	41.92	2.153	-1	1	3	2	2	2
42.36	42.36	2.132	-2	2	0	2	3	3
42.90	42.90	2.106	-2	2	1	21	19	28
43.17	43.18	2.094	-1	1	4	26	24	35
43.38	43.28	2.084	1	3	2	3	16	7
43.79	43.80	2.065	1	2	3	3	4	5
44.13	44.13	2.051	2	2	1	5	21	6
44.21	44.21	2.047	-1	1	3	18		25

2THETA	PEAK	D	H	K	L	I(INT)	I(PK)	I(DS)
44.34	44.34	2.041	-2	1	1	24	30	34
44.68	44.68	2.026	0	3	2	6		8
45.59	45.68	1.9881	1	1	4	7	23	10
45.68	45.68	1.9843	-2	2	2	25		35
47.48	47.48	1.9133	0	2	4		6	10
48.00	48.00	1.8938	2	2	2	35	30	53
49.29	49.30	1.8471	-2	0	4	8	6	13
50.04	50.04	1.8212	0	3	3	5	5	8
50.41	50.42	1.8085	-2	2	2	23	19	37
50.57	50.54	1.8033	0	1	5	4	14	6
50.95	50.96	1.7907	-1	0	5	15	12	25
51.56	51.64	1.7711	3	1	0	2	11	3
51.64	51.64	1.7684	3	0	1	13		22
51.84	51.78	1.7621	-3	1	1	2	7	3
52.94	52.94	1.7280	-2	1	5	11	8	18
53.26	53.26	1.7186	3	1	1	9	8	16
53.63	53.62	1.7079	0	0	4	6	15	10
53.99	54.00	1.7074	2	3	3	14		25
54.01		1.6968	-2	3	1	7	10	12
54.25	54.24	1.6963	-3	1	2	5		9
55.22	55.24	1.6892	0	4	0	6	6	10
55.34	55.34	1.6620	4	4	4	3	3	5
56.23	56.24	1.6587	-2	3	2	3	4	6
56.50	56.50	1.6345	0	0	3	11	8	20
56.75	56.64	1.6275	-4	2	5	11	9	20
56.98	56.98	1.6207	2	4	0	1	6	2
57.14	57.14	1.6147	-1	4	4	5	5	9
57.32	57.30	1.6106	3	2	0	8	9	16
57.67	57.66	1.6059	-3	1	1	2	6	4
58.05	58.06	1.5971	-1	2	4	4	3	7
58.17	58.18	1.5874	-1	4	1	5	4	9
58.57	58.58	1.5846	1	4	4	2	4	4
58.82	58.82	1.5746	-1	3	1	11	9	22
59.35	59.36	1.5686	3	2	1	11	10	21
59.52	59.52	1.5558	-2	2	3	4	5	8
59.70	59.70	1.5518	1	3	3	6	7	12
59.70		1.5476	0	0	6	7	8	13

LARNITE (BETA CA2SIO4) — MIDGLEY, 1952

2THETA	PEAK	D	H	K	L	I(INT)	I(PK)	I(DS)
60.18	60.20	1.5363	-1	4	2	3	3	6
60.54	60.54	1.5282	1	3	4	15	15	31
60.69	60.70	1.5246	3	0	3	15	20	31
60.71		1.5242	2	2	2	4		8
61.15	61.14	1.5142	1	4	1	1	2	2
62.26	62.28	1.4898	3	1	2	3	3	2
62.52	62.58	1.4843	3	1	6	5	9	7
62.59		1.4828	-1	4	3	9		11
63.13	63.14	1.4714	-3	2	3	3	2	19
64.15	64.16	1.4505	2	1	5	5	4	5
64.40	64.40	1.4455	-1	4	3	5	6	11
64.72	64.72	1.4391	2	4	0	7	4	14
65.12	65.12	1.4312	-2	4	1	4	4	9
65.27		1.4283	-2	3	4	5	5	11
65.34	65.30	1.4269	-1	3	6	3	5	6
66.38	66.38	1.4077	-1	3	5	2	4	5
66.94	66.94	1.4070	0	2	6	4		9
67.18	67.16	1.3967	-2	0	6	5	4	11
67.45	67.44	1.3922	3	1	0	4	4	9
68.18	68.18	1.3873	-1	2	6	4	4	9
68.94	69.04	1.3742	-4	0	6	12	7	27
69.02		1.3609	2	3	1	2	8	6
69.03		1.3595	-3	0	5	2		5
69.07		1.3593	2	4	2	8		18
70.34	70.34	1.3587	2	5	1	2		4
71.01		1.3373	0	4	3	1	1	2
71.02	71.02	1.3262	-2	5	1	6	4	16
71.51		1.3182	4	4	6	2	2	6
71.52	71.52	1.3050	2	0	7	7	5	17
72.35	72.36	1.3028	-1	5	1	7	7	17
72.49	72.50	1.3017	0	1	7	2		4
72.56	72.80	1.2976	0	5	2	4	4	9
72.83	73.28	1.2908	-2	2	6	7	5	18
73.27	73.68	1.2847	2	4	3	5	3	11
73.67	74.56	1.2738	-4	2	1	1	3	4
74.42		1.2729	-4	4	1	1		3
74.57		1.2715	-4	2	3	3		7

2THETA	PEAK	D	H	K	L	I(INT)	I(PK)	I(DS)
74.87	74.80	1.2672	4	1	2	1	2	3
76.11	76.12	1.2496	0	5	4	3	3	8
76.31	76.32	1.2468	-3	4	3	8	7	22
76.33		1.2466	-2	4	4	1		4
76.66	76.54	1.2419	3	4	1	3	4	8
76.82	76.88	1.2398	3	4	0	1	2	4
76.91		1.2385	0	5	3	1		4
77.85	77.86	1.2259	-4	0	4	4	3	11
77.98	78.10	1.2242	-1	2	7	2	5	4
78.12		1.2224	3	4	2	5		14
78.51	78.50	1.2173	2	2	6	3	3	9
78.96	78.96	1.2115	-1	3	6	2	2	6
79.16	79.20	1.2089	-4	3	3	1	2	4
79.23		1.2079	-2	5	1	6		3
79.45	79.44	1.2052	-2	4	2	1	4	17
81.17	81.20	1.1840	4	5	2	4	3	3
81.21		1.1835	-2	5	2	1		13
81.48	81.46	1.1803	3	3	4	2	2	4
81.77	81.76	1.1768	3	2	5	3	2	5
82.03	82.02	1.1737	-4	3	1	3	2	8
82.52	82.52	1.1679	0	5	4	2	2	10
82.91	82.90	1.1634	2	5	2	6	4	5
83.15	83.14	1.1607	0	0	8	3	2	18
83.94	83.94	1.1518	-1	0	3	1		10
84.29	84.22	1.1479	4	3	7	5	4	4
84.73	84.74	1.1430	0	5	1	2	2	16
84.75		1.1428	-2	5	5	4		5
85.64	85.76	1.1333	1	5	3	1	4	12
85.68		1.1328	-4	1	1	4		4
85.78		1.1318	3	4	3	2		11
85.89	85.88	1.1305	-1	4	6	2	3	6
86.31	86.30	1.1262	0	6	0	3	2	9
86.63	86.66	1.1228	3	3	6	1	3	4
86.67		1.1224	-4	4	3	4		13
87.15	87.16	1.1174	-3	0	7	3	2	9
87.35	87.36	1.1153	2	4	5	2	2	7
88.60	88.76	1.1029	1	1	8	2	3	5

51

LARNITE (BETA CA2SIO4) - MIDGLEY, 1952

2THETA	PEAK	D	H	K	L	I(INT)	I(PK)	I(DS)
88.69		1.1020	-5	0	1	2		5
88.79		1.1010	-2	0	8	1		13
89.86	89.86	1.0907	-2	5	4	4	3	12
90.18	90.16	1.0876	3	5	0	4	2	4
90.45	90.46	1.0851	5	1	0	1	2	4
90.66	90.78	1.0831	4	3	4	1	3	8
90.78		1.0819	-2	5	7	3		13
90.84		1.0814	-1	5	5	4		4
91.10	91.08	1.0790	-3	2	5	1	3	6
91.49	91.58	1.0754	-5	1	2	2	3	9
91.61		1.0744	3	5	1	6		12
92.48	92.48	1.0665	-4	4	1	3	1	4
93.07		1.0613	-1	2	1	1	2	4
93.24	93.24	1.0598	2	5	8	1		14
94.14	94.14	1.0520	4	1	4	4	1	4
94.65	94.68	1.0476	-5	2	1	1	1	5
94.92	94.96	1.0454	5	2	5	1	1	5
95.32	95.32	1.0421	5	6	0	1	2	16
95.56	95.58	1.0401	-2	0	8	4	3	12
95.95	95.92	1.0369	-1	4	7	3	2	5
96.30	96.30	1.0340	-1	6	3	1	2	11
96.72	96.72	1.0307	2	3	3	3	4	21
96.78		1.0302	5	2	1	6		11
97.11	97.06	1.0276	-4	4	1	3	3	6
98.00	98.02	1.0206	-4	2	3	2	2	16
98.64	98.66	1.0157	-5	2	6	4	2	8
98.88	98.90	1.0139	5	2	5	2	2	12
99.27	99.26	1.0109	5	0	3	3	2	13
100.32	100.32	1.0032	-3	5	4	4	1	5

TABLE 2. OLIVINE STRUCTURES

Variety	Forsterite	Hyalosiderite	Hortonolite	Fayalite	Monticellite
Composition	$(Mg_{0.9}Fe_{0.1})_2SiO_4$	$(Mg_{0.535}Fe_{0.456} Mn_{0.006}Ca_{0.02})_2 SiO_4$	$(Mg_{0.49}Fe_{0.49} Mn_{0.01}Ca_{0.01})_2 SiO_4$	$(Fe_{0.92}Mg_{0.04}Mn_{0.04})_2 SiO_4$	$MgCaSiO_4$
Source	Minas Gerais, Brazil	Skaergaard intrusion, Greenland	Camas Mòr, Scotland	Obsidian Cliff, Yellowstone, Wyo.	Crestmore, Calif.
Reference	Birle, Gibbs Moore & Smith, 1968	Ibid.	Ibid.	Ibid.	Onken, 1964
Cell Dimensions					
a̲ Å	4.762	4.787	4.785	4.816	4.822
b̲ Å	10.225	10.341	10.325	10.469	11.108
c̲ Å	5.994	6.044	6.038	6.099	6.382
α	90	90	90	90	90
β deg	90	90	90	90	90
γ	90	90	90	90	90
Space Group	Pbnm	Pbnm	Pbnm	Pbnm	Pbnm
Z	4	4	4	4	4

TABLE 2. (cont.)

Variety	Forsterite	Hyalosiderite	Hortonolite	Fayalite	Monticellite
Site Occupancy	——— M_1 and M_2 sites considered to be completely disordered ———				
Method	3-D, counter	3-D, counter	3-D, counter	3-D, counter	3-D, counter
R & Rating	0.08 (1)	0.08 (1)	0.07 (1)	0.07 (1)	0.069 (1)
Cleavage and habit	{010}distinct; {001} less so. Equidimensional	——— {010} distinct;{100} less so ——— Equidimensional or flattened parallel to {010} or {100} ———			
μ	172.4	226.7	249.2	250.1	194.3
ASF	0.8350×10^{-1}	0.7864×10^{-1}	0.8202×10^{-1}	0.1100	0.9781×10^{-1}
Abridging factors	0.4:0:0	0.4:0:0	0.4:0:0	0.4:0:0	0.5:0:0

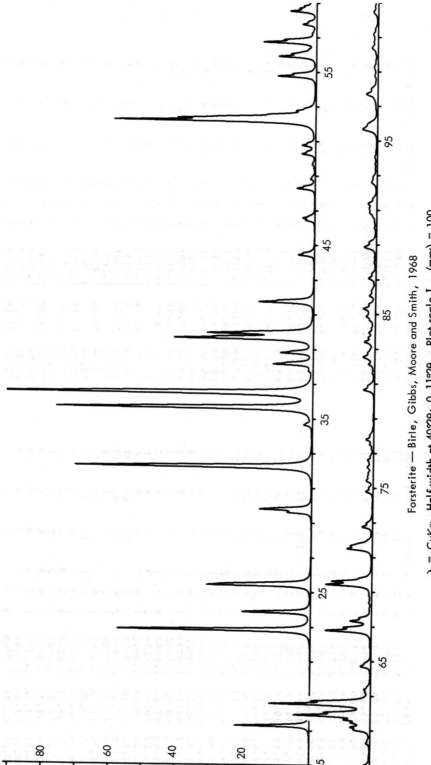

Forsterite — Birle, Gibbs, Moore and Smith, 1968

λ = CuKα Half width at 40°2θ: 0.11°2θ Plot scale I$_{PK}$ (max) = 100

55

FORSTERITE — BIRLE, GIBBS, MOORE AND SMITH, 1968

2THETA	PEAK	D	H	K	L	I(INT)	I(PK)	I(DS)
17.28	17.28	5.127	0	2	1	17	22	13
22.80	22.80	3.896	0	2	0	48	58	39
23.84	23.84	3.729	1	0	1	17	20	14
25.40	25.40	3.504	1	1	1	24	31	21
29.60	29.60	3.016	1	2	0	6	7	5
29.79	29.78	2.997	0	2	1	14	16	12
32.21	32.22	2.777	0	3	0	69	71	64
34.64	34.64	2.587	1	2	1	2	3	2
35.60	35.60	2.520	0	3	1	76	77	75
36.46	36.46	2.462	1	1	2	100	100	100
37.75	37.76	2.381	2	0	0	1	2	1
38.15	38.14	2.357	0	4	0	12	12	12
38.79	38.80	2.319	2	1	0	10	10	10
39.61	39.62	2.274	1	3	1	43	41	45
39.90	39.90	2.257	2	1	1	30	32	32
41.72	41.72	2.163	2	1	1	18	17	19
44.44	44.44	2.037	1	3	2	5	5	6
44.56	44.54	2.032	2	2	1	0	3	0
46.44	46.44	1.9538	2	3	0	3	3	3
46.58	46.58	1.9482	0	4	2	3	4	3
48.27	48.28	1.8837	1	5	0	6	5	8
48.81	48.82	1.8643	2	0	2	1	1	2
49.66	49.66	1.8342	2	1	2	1	1	1
50.27	50.28	1.8134	1	1	3	5	4	6
50.76	50.76	1.7971	1	5	1	4	4	6
52.16	52.16	1.7521	2	2	2	71	60	91
52.40	52.30	1.7447	2	4	0	21	41	27
52.75	52.74	1.7339	2	4	1	4	6	6
54.75	54.76	1.6751	2	3	2	14	11	19
55.89	55.90	1.6437	1	6	0	14	11	19
56.15	56.04	1.6367	2	3	2	3	8	4
56.71	56.72	1.6218	1	3	3	20	16	28
57.21	57.20	1.6087	1	1	4	1	1	1
57.76	57.76	1.5948	1	6	2	5	4	7
58.52	58.52	1.5759	0	4	4	10	8	14
58.82	58.80	1.5687	3	1	0	4	4	5
60.26	60.26	1.5344	3	0	1	1	1	2

2THETA	PEAK	D	H	K	L	I(INT)	I(PK)	I(DS)
61.00	61.04	1.5175	3	1	1	4	5	5
61.06		1.5163	3	2	0	3		4
61.17	61.18	1.5138	2	1	3	4	6	5
61.44	61.44	1.5078	2	1	2	5	6	7
61.60	61.60	1.5042	2	5	1	1	8	8
61.86	61.86	1.4985	0	0	4	29	23	43
61.97		1.4961	1	4	3	1	1	1
62.50	62.50	1.4847	0	3	2	0	30	2
63.20	63.20	1.4700	3	2	1	6	30	61
63.36	63.36	1.4666	3	2	2	1	2	2
64.69	64.70	1.4397	2	2	3	2	2	2
64.76		1.4383	3	3	0	4	3	6
66.75	66.76	1.4002	0	1	4	1	14	2
66.77		1.3999	1	7	1	0		19
67.31	67.32	1.3898	3	3	2	12	7	13
67.88	67.88	1.3795	3	1	3	8	1	14
68.39	68.38	1.3706	2	5	2	9	1	1
68.79	68.80	1.3635	1	7	1	0	1	2
69.40	69.40	1.3530	3	2	2	18	14	1
69.42		1.3527	2	6	1	1		31
69.60	69.60	1.3496	3	1	3	6	12	2
71.48	71.48	1.3187	4	3	1	10	8	11
71.61	71.64	1.3166	1	4	4	5	7	17
71.76		1.3142	3	4	1	3		8
72.75	72.74	1.2988	0	4	3	5	3	3
73.08	72.96	1.2937	0	7	3	0	2	8
74.77	74.80	1.2686	1	2	4	1	2	1
74.79		1.2682	1	6	2	3		2
75.38	75.40	1.2598	2	1	4	3	2	5
75.46		1.2586	2	6	1	2		4
75.70	75.68	1.2553	3	5	0	0	2	3
75.83	75.86	1.2535	0	8	1	1	2	1
75.86		1.2530	1	6	3	0		2
76.19	76.20	1.2485	2	4	4	2	2	2
76.24		1.2477	2	7	0	1		4
76.60	76.60	1.2428	3	0	3	2	2	4
77.26	77.26	1.2338	3	1	3	1	1	1

FORSTERITE - BIRLE, GIBBS, MOORE AND SMITH, 1968

2THETA	PEAK	D	H	K	L	I(INT)	I(PK)	I(DS)
77.50	77.50	1.2306	3	4	2	1	1	2
77.80	77.80	1.2266	2	2	3	3	2	6
78.18	78.02	1.2215	2	7	1	1	1	1
80.63	80.64	1.1905	4	0	0	5	3	10
80.75		1.1890	2	3	4	1		2
81.62	81.62	1.1786	0	8	2	3	1	2
82.12	82.12	1.1727	1	5	4	4	2	5
82.51	82.54	1.1680	3	3	3	4	3	8
82.58		1.1673	0	2	5	2		4
82.99	82.98	1.1625	1	0	5	1		3
83.40	83.40	1.1578	3	5	2	5	3	10
83.93	83.92	1.1519	2	7	2	3	2	5
84.40	84.42	1.1467	1	7	3	1	1	1
84.64	84.66	1.1441	1	8	1	3	2	4
84.84	84.86	1.1418	3	6	1	3	3	6
84.99		1.1402	2	6	3	1		2
85.31	85.30	1.1368	2	4	4	6	4	12
86.07	86.06	1.1287	2	8	0	3	2	7
86.49	86.32	1.1243	4	3	0	1		1
87.06	87.06	1.1184	3	4	3	4	2	9
87.96	87.96	1.1092	2	8	1	1	1	2
88.83	88.84	1.1006	1	3	5	5	4	11
88.89		1.1000	2	1	5	3		6
89.96	89.96	1.0897	4	1	2	1	1	2
90.35	90.36	1.0859	0	4	5	0	1	1
90.83	90.84	1.0815	4	2	2	3	2	7
91.02	91.10	1.0798	4	4	0	2	2	4
91.11		1.0789	0	8	3	1		3
91.36	91.36	1.0766	3	7	0	4	3	8
92.55	92.60	1.0658	3	2	4	2	2	5
92.65		1.0650	2	1	5	2		4
92.88	92.86	1.0629	3	5	3	1	1	1
93.41	93.38	1.0583	2	7	3	1	1	1
93.66	93.66	1.0563	2	8	2	1	1	2
94.06	94.06	1.0526	4	3	2	1	1	3
95.64	95.66	1.0394	1	9	2	10	5	22
95.79		1.0382	3	3	4	3		6
96.85	96.86	1.0296	4	5	0	0	1	2
97.68	97.68	1.0231	1	7	4	8	4	19
98.76	98.76	1.0148	4	5	1	1	1	4
99.21	99.26	1.0114	1	5	5	1	1	1
99.28		1.0108	0	10	1	1		4
100.35	100.36	1.0030	4	2	3	2	2	2
100.36		1.0028	3	4	4	4		8
100.89	100.70	0.9990	0	8	4	1	1	2
101.13	101.14	0.9973	3	0	6	3	1	7
102.44	102.44	0.9880	2	8	0	4	1	9
103.22	103.42	0.9827	2	8	3	2	1	5
103.41		0.9815	0	6	5	1		6
103.54		0.9806	0	2	6	1		1
104.63	104.64	0.9733	2	1	6	5	2	13
104.79		0.9722	1	9	2	1		2
105.28	104.98	0.9691	1	0	3	1		2
106.05	106.06	0.9642	4	6	1	2	1	4
106.35	106.40	0.9623	3	5	4	4	1	1
106.89	106.88	0.9589	2	7	4	1	1	2
107.93	107.96	0.9525	3	1	5	5	1	3
108.22	108.48	0.9507	1	10	2	2	1	1
108.36		0.9499	4	4	3	3		2
108.48		0.9492	2	5	5	2		3
108.63		0.9483	5	1	0	5		2
108.97	108.92	0.9463	3	8	2	1	1	3
109.73	109.74	0.9419	2	10	0	3	1	9
109.98	110.08	0.9404	1	3	5	1	1	2
110.64	110.64	0.9367	0	8	6	3	1	7
110.69		0.9364	5	2	1	0		1
111.45	111.46	0.9321	4	0	4	4	2	12
111.68	112.04	0.9308	0	4	6	1	3	2
111.75		0.9304	4	6	2	2		3
112.06		0.9288	4	1	0	6		16
112.14		0.9283	4	4	1	4		1
112.72	112.46	0.9252	5	2	1	1	1	1
113.46	113.46	0.9212	5	3	5	0	1	6
114.19	114.18	0.9175	5	3	0	2	1	5

FORSTERITE — BIRLE, GIBBS, MOORE AND SMITH, 1968

2THETA	PEAK	D	H	K	L	I(INT)	I(PK)	I(DS)
114.62	114.64	0.9152	4	5	3	1	1	3
114.68		0.9149	1	11	0	1		2
115.18	115.10	0.9123	0	10	3	1	1	2
116.18	116.32	0.9074	2	6	5	0	2	1
116.32		0.9067	2	2	6	6		16
117.38	117.38	0.9016	2	8	4	4	1	10
117.85	117.84	0.8993	4	3	4	1	1	3
118.01		0.8985	2	2	0	0		1
118.50	118.44	0.8963	3	4	5	1	0	3
119.99	120.00	0.8895	2	3	6	1	1	2
121.13	121.14	0.8844	3	9	2	5		14
121.45	121.58	0.8830	5	4	1	0	1	1
121.56		0.8826	1	5	6	1		2
123.10	123.48	0.8760	4	4	4	3	2	8
123.34		0.8750	1	11	2	1		3
123.52		0.8743	3	7	4	6		16
125.36	125.42	0.8669	3	2	4	1	0	2
126.17	126.52	0.8638	5	5	0	1	2	2
126.32		0.8632	4	8	1	1		3
126.52		0.8625	0	6	6	7		21
127.41	127.04	0.8591	2	11	1	1	1	4
128.07	128.08	0.8567	5	1	3	5	1	14
128.55	128.64	0.8550	5	5	1	2	2	6
128.66		0.8546	0	12	0	1		4
129.22	129.24	0.8526	2	10	3	4	2	13
129.41		0.8519	2	10	3	3		8
130.58	131.10	0.8479	5	2	0	1	1	3
130.67		0.8476	2	9	4	1		3
131.05		0.8463	6	0	4	2		7
131.13		0.8460	0	12	2	1		10
132.16	131.70	0.8426	3	1	6	6	1	7
133.74	133.82	0.8376	4	8	2	2	1	4
134.65	134.78	0.8348	3	6	5	1	1	14
134.83		0.8342	3	2	6	5	2	15
134.96		0.8338	2	11	2	2		1
135.70	136.22	0.8316	1	2	7	7		3
136.18		0.8302	3	8	4	6	3	18

2THETA	PEAK	D	H	K	L	I(INT)	I(PK)	I(DS)
136.25	136.88	0.8300	5	5	2	8		26
137.00		0.8279	3	10	1	1	2	4
138.69	139.22	0.8232	4	0	9	1	0	2
139.19		0.8218	0	12	2	1		4
140.52	140.56	0.8183	4	6	0	1	1	3
140.55		0.8183	1	3	9	5		16
141.64	141.36	0.8155	4	9	1	1	1	3
142.57	143.14	0.8132	1	7	6	1	1	2
143.02		0.8122	0	4	7	0		11
143.56	143.52	0.8109	2	6	6	3	1	11
144.02		0.8099	1	12	0	3		2
146.74	147.14	0.8039	3	11	0	0	1	7
147.02		0.8033	2	1	7	2		4
147.18		0.8030	3	4	6	1		8
147.98	147.98	0.8013	5	1	4	2	1	8
149.43	150.06	0.7985	5	7	0	1	2	4
150.00		0.7974	2	10	4	9		31
150.12		0.7972	2	12	1	2		6
150.67		0.7962	2	11	3	2		6
150.78		0.7960	2	2	7	7		5
152.55	153.58	0.7929	5	5	3	3	1	4
153.38		0.7915	5	7	1	1		8
153.51		0.7913	6	1	0	2		6
153.70		0.7910	3	10	3	5		17
155.96	157.36	0.7875	3	9	4	3	1	3
156.71		0.7864	4	7	4	3		9
157.22		0.7857	0	12	3	7		26
158.13		0.7845	6	1	3	2		8
159.74	161.00	0.7825	6	5	4	6	1	22
160.38		0.7817	3	5	6	6		21
160.90		0.7811	4	5	5	5		19
161.08		0.7809	1	11	4	3		10
162.01		0.7798	2	7	6	4		13
162.30		0.7795	7	1	1	7		2
162.54		0.7793	0	10	5	7		25
164.13		0.7777	6	2	1	10		34
164.45		0.7774	1	8	6	1		4

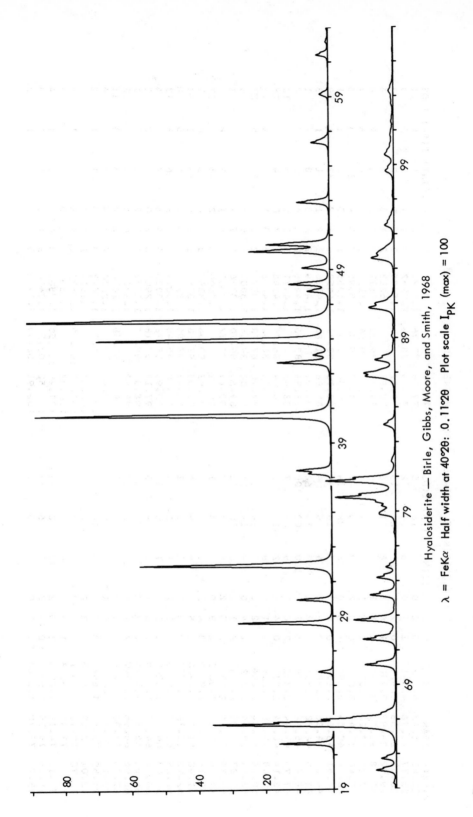

Hyalosiderite — Birle, Gibbs, Moore, and Smith, 1968

λ = FeKα Half width at 40°2θ: 0.11°2θ Plot scale I_{PK} (max) = 100

HYALOSIDERITE — BIRLE, GIBBS, MOORE AND SMITH, 1968

2THETA	PEAK	D	H	K	L	I(INT)	I(PK)	I(DS)
21.61	21.62	5.162	0	2	0	11	16	6
25.77	25.78	4.341	1	1	0	3	4	2
28.56	28.58	3.924	0	2	1	23	28	15
29.92	29.92	3.750	1	0	1	8	11	6
31.88	31.88	3.525	1	1	1	47	57	34
37.21	37.22	3.034	1	2	1	5	7	4
37.40	37.40	3.019	1	3	0	9	10	7
40.54	40.54	2.794	0	3	1	83	88	74
43.61	43.62	2.606	1	2	0	15	16	14
44.05	44.06	2.581	2	0	1	4	5	4
44.88	44.88	2.536	1	3	1	69	70	68
45.98	45.98	2.479	2	1	2	100	100	100
47.73	47.74	2.392	0	4	0	6	7	7
48.14	48.14	2.373	1	1	2	12	12	13
49.08	49.08	2.331	2	1	0	8	8	9
50.04	50.04	2.289	1	2	2	25	24	27
50.44	50.44	2.272	2	4	0	19	19	21
52.87	52.88	2.174	1	1	1	10	10	12
56.34	56.34	2.051	3	2	0	6	5	7
59.04	59.12	1.9645	1	3	2	1	3	1
59.13		1.9619	2	4	0	2		3
61.40	61.40	1.8960	1	5	0	4	4	6
62.16	62.16	1.8751	2	0	2	1	1	1
63.29	63.30	1.8449	2	1	2	1	1	1
64.03	64.02	1.8260	1	1	3	7	6	10
64.70	64.70	1.8089	1	5	1	5	4	8
66.63	66.64	1.7624	2	2	2	70	54	104
66.96	66.96	1.7547	2	4	0	21	22	32
67.34	67.32	1.7459	1	2	3	3	5	5
70.13	70.12	1.6850	2	4	1	12	9	20
71.59	71.60	1.6549	0	6	1	13	10	22
72.01	72.00	1.6466	2	3	3	1	2	2
72.70	72.70	1.6331	1	3	3	17	13	29
73.42	73.42	1.6193	1	6	0	0	1	1
74.15	74.16	1.6056	1	5	2	11	8	18
75.16	75.16	1.5872	0	4	3	8	6	14
75.77	75.78	1.5763	3	1	0	4	3	7

2THETA	PEAK	D	H	K	L	I(INT)	I(PK)	I(DS)
77.76	77.76	1.5421	3	0	1	1	1	1
78.79	78.84	1.5252	3	1	1	1	1	9
78.87		1.5239	3	2	0	5	5	3
78.91		1.5233	2	1	3	2		3
79.30	79.30	1.5171	1	4	2	4	4	4
79.53	79.54	1.5133	2	5	1	4	6	8
79.77	79.78	1.5095	0	4	4	25	18	46
79.96	79.96	1.5065	1	6	1	0	11	1
80.70	80.70	1.4950	0	6	2	32	21	59
81.86	81.86	1.4776	3	2	1	1	1	3
81.97		1.4759	3	3	0	0		1
83.84		1.4488	2	4	2	1		2
83.96	83.96	1.4471	0	3	3	5	3	10
86.74	86.76	1.4096	3	1	2	14	9	28
86.92	86.92	1.4073	3	3	1	8	9	17
87.70	87.70	1.3973	3	2	2	10	6	21
89.08	89.08	1.3801	1	5	3	1	1	3
89.69	89.70	1.3726	1	7	1	0	0	1
90.67		1.3610	2	6	0	2	8	5
90.72	90.72	1.3604	3	2	2	13		27
91.02	90.94	1.3569	3	4	0	3	6	7
93.58	93.58	1.3281	1	3	4	14	7	31
93.97	93.84	1.3238	3	1	3	4	5	10
95.48	95.48	1.3079	0	6	3	6	3	13
95.95	95.72	1.3030	0	4	4	2	2	4
97.18	97.18	1.2906	8	0	1	1	0	2
98.56	98.62	1.2772	1	7	2	5	3	12
98.62		1.2766	0	4	2	5		11
99.55	99.58	1.2678	2	6	2	2	3	4
99.64		1.2670	1	5	4	1		2
100.14	100.14	1.2623	3	5	1	5		2
100.21		1.2617	1	6	3	3		2
100.69	100.70	1.2573	1	4	4	2	1	4
100.88		1.2556	2	7	0	1		2
101.49	101.50	1.2501	3	0	3	1	1	4
102.52	102.54	1.2410	3	1	3	1	1	3
102.91	102.90	1.2376	3	4	2	1	1	2

HYALOSIDERITE - BIRLE, GIBBS, MOORE AND SMITH, 1968

2THETA	PEAK	D	H	K	L	I(INT)	I(PK)	I(DS)
103.15	103.26	1.2356	3	5	1	3	1	1
103.27		1.2346	2	5	3	3		8
107.94	108.04	1.1969	2	3	4	1	3	2
108.03		1.1962	4	0	0	7		18
109.31	109.32	1.1867	0	8	2	1	1	8
110.11	110.12	1.1809	1	5	5	3	1	8
110.82	110.94	1.1759	0	2	3	2	3	5
110.94		1.1750	3	3	3	6		16
112.44	112.46	1.1646	3	5	2	11	4	29
112.61		1.1634	1	1	7	1		3
113.23	113.20	1.1593	2	7	2	3	2	9
113.94	113.98	1.1546	1	7	3	1	1	2
114.36	114.38	1.1518	1	8	2	3	1	7
114.89	114.98	1.1484	3	6	1	3	2	8
115.01		1.1476	2	6	3	2		7
115.54	115.54	1.1443	2	4	4	12	4	33
116.90	116.90	1.1359	2	8	0	5	2	14
117.89	117.88	1.1299	3	0	0	0	0	1
118.72	118.72	1.1251	4	3	3	6	2	17
120.25	120.26	1.1163	2	8	1	3	1	8
121.68	121.70	1.1085	1	9	5	9	3	27
122.19	122.14	1.1057	4	1	2	5	3	15
123.86	123.88	1.0970	1	9	1	1	1	3
124.49	124.52	1.0938	0	4	5	1	1	3
125.84	125.86	1.0872	4	2	2	10	3	32
125.99		1.0864	0	8	3	1		2
126.21	126.26	1.0854	4	4	0	5	4	15
126.73	126.70	1.0829	3	7	0	10	4	32
129.00	129.04	1.0724	2	4	4	4	2	13
129.05		1.0722	2	5	5	3		8
129.70	129.50	1.0694	3	5	3	1	1	5
129.96		1.0682	4	8	1	1		3
131.15	131.14	1.0631	2	8	2	3	1	10
132.33	132.32	1.0582	4	3	2	1	1	5
133.06	132.96	1.0553	2	5	1	1	1	3
135.34	135.38	1.0464	1	9	2	31	6	105
135.83	135.82	1.0446	3	3	4	11	6	36

2THETA	PEAK	D	H	K	L	I(INT)	I(PK)	I(DS)
139.96	139.98	1.0302	1	7	4	28	5	98
143.17	143.30	1.0202	4	5	1	6	1	20
143.74	143.92	1.0185	1	5	5	3	2	10
144.03		1.0177	0	10	1	6		20
147.17	147.24	1.0091	3	4	4	7	1	24
147.40		1.0085	4	2	0	2		8
148.26	148.00	1.0063	0	0	6	2		9
149.50	149.50	1.0033	3	8	0	9	1	33
153.01	153.42	0.9955	1	10	1	1	2	2
153.34		0.9948	2	4	5	16		58
155.93	156.68	0.9897	3	8	1	1	2	2
156.21		0.9892	2	8	3	12		47
156.62		0.9885	0	6	5	13		48
157.04		0.9877	0	2	6	7		27
158.49		0.9853	4	3	3	1		5
160.46	161.90	0.9822	4	6	0	4	2	14
161.35		0.9810	0	8	4	4		17
161.79		0.9803	1	1	6	38		145
163.11		0.9786	2	9	2	5		18
164.47		0.9769	0	10	2	1		3

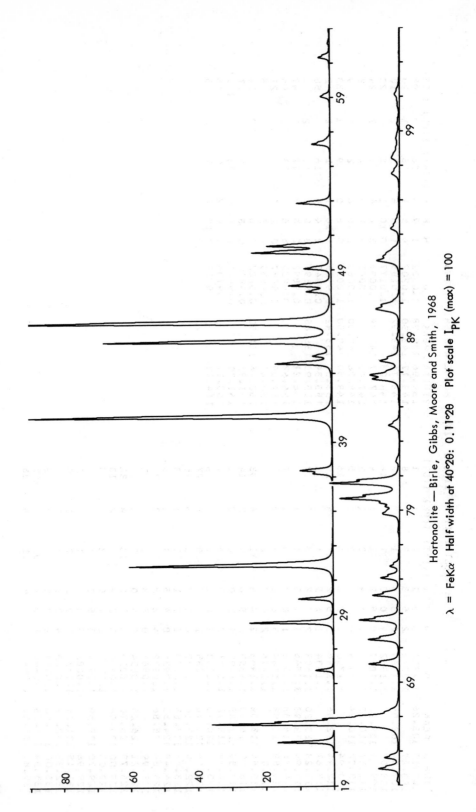

Hortonolite — Birle, Gibbs, Moore and Smith, 1968

λ = FeKα · Half width at 40°2θ: 0.11°2θ · Plot scale I$_{PK}$ (max) = 100

62

HORTONOLITE - BIRLE, GIBBS, MOORE AND SMITH, 1968

2THETA	PEAK	D	H	K	L	I(INT)	I(PK)	I(DS)
21.58	21.58	5.170	0	2	0	12	17	6
25.75	25.76	4.344	1	1	0	4	5	2
28.53	28.54	3.929	0	2	1	20	25	13
29.90	29.90	3.753	1	0	1	8	10	5
31.85	31.86	3.527	1	1	1	49	60	34
37.17	37.18	3.037	1	2	1	4	6	4
37.36	37.36	3.022	0	0	2	8	10	7
40.49	40.50	2.797	1	3	0	85	90	75
43.56	43.56	2.609	0	2	2	16	17	15
43.98	43.98	2.585	0	4	0	5	6	4
44.84	44.84	2.539	1	3	1	67	67	65
45.93	45.94	2.481	1	1	2	100	100	100
47.71	47.72	2.393	2	0	0	7	7	7
48.06	48.06	2.377	0	4	1	12	12	13
49.05	49.06	2.332	2	1	0	8	8	9
49.99	50.00	2.291	1	2	2	24	23	26
50.37	50.38	2.275	1	4	0	19	19	21
52.84	52.84	2.176	2	1	1	11	10	13
56.27	56.28	2.053	1	3	2	6	5	8
58.99	59.04	1.9660	2	3	0	1	3	1
59.04	59.04	1.9645	0	4	2	2		3
61.31	61.32	1.8986	1	5	0	4	3	6
62.12	62.12	1.8763	2	0	2	1	1	1
63.25	63.24	1.8461	2	1	2	1	1	1
63.96	63.96	1.8277	1	1	3	8	6	11
64.61	64.62	1.8113	1	5	1	5	4	8
66.57	66.58	1.7637	2	2	2	70	55	107
66.89	66.88	1.7563	2	4	0	21	23	32
67.27	67.26	1.7476	1	2	3	3	5	5
70.05	70.06	1.6866	2	4	1	12	9	20
71.47	71.48	1.6574	0	6	0	13	9	21
71.94	71.94	1.6480	2	3	2	1	1	2
72.61	72.62	1.6348	1	3	3	17	12	28
73.30	73.30	1.6216	1	6	0	0	1	1
74.04	74.04	1.6076	1	5	2	11	8	20
75.05	75.06	1.5891	0	4	3	8	6	14
75.73	75.74	1.5770	3	1	0	4	3	7
77.72	77.72	1.5428	3	0	1	1	1	1
78.75	78.80	1.5259	3	1	1	5	5	9
78.82		1.5247	3	2	1	2		3
78.83		1.5245	2	1	3	2		4
79.21	79.20	1.5185	2	4	2	4	4	8
79.43	79.44	1.5150	2	5	1	4	6	7
79.68	79.68	1.5110	0	0	0	25	18	47
80.57	80.56	1.4971	0	6	2	32	21	61
81.80	81.82	1.4784	1	2	1	1	1	3
81.89		1.4771	2	3	0	0		1
83.74	83.90	1.4503	0	2	4	1		2
83.90	83.90	1.4480	3	3	0	5	3	10
86.59	86.60	1.4116	1	7	0	14	8	29
86.85	86.82	1.4082	3	0	1	8	9	18
87.63	87.64	1.3981	3	1	2	10	6	23
88.95	88.94	1.3817	1	5	3	1	1	3
89.53	89.54	1.3746	1	7	1	0	0	1
90.53	90.66	1.3626	2	6	0	3	7	6
90.65		1.3613	3	2	2	12		27
90.94	90.88	1.3578	3	4	0	3	6	8
93.46	93.46	1.3294	1	4	3	14	7	33
93.88	93.70	1.3248	3	1	1	4	5	9
95.31	95.32	1.3097	3	3	6	6	5	13
95.81	95.76	1.3045	0	4	3	2	1	4
96.98	96.98	1.2926	0	8	0	1	0	1
98.38	98.52	1.2789	1	7	2	1	2	2
98.51		1.2777	2	0	4	5		13
99.39	99.42	1.2693	2	6	2	4	2	11
99.52		1.2681	1	2	4	2		4
100.03	100.02	1.2634	2	1	5	1		2
100.04		1.2632	3	0	3	1	1	2
100.54	100.54	1.2586	1	4	4	2		5
100.71		1.2571	2	7	0	1	1	2
101.40	101.40	1.2509	3	0	3	1	1	4
102.43	102.44	1.2418	3	3	3	1	1	3
103.12	103.12	1.2359	2	5	3	3	1	8
107.80	107.98	1.1980	2	3	4	1	3	2

HORTONOLITE - BIRLE, GIBBS, MOORE AND SMITH, 1968

2THETA	PEAK	D	H	K	L	I(INT)	I(PK)	I(DS)
107.97		1.1967	4	0	0	7		19
109.07	109.08	1.1885	0	8	2	2	1	5
109.92	109.94	1.1823	1	5	4	3	1	8
110.65	110.82	1.1771	0	2	5	2	3	5
110.82		1.1758	3	3	3	6		18
112.29	112.30	1.1656	3	5	2	11	4	31
112.45		1.1646	1	1	5	1		3
113.71	113.70	1.1561	1	7	3	1	1	2
114.11	114.12	1.1535	1	8	1	2	1	7
114.72	114.78	1.1495	3	6	1	3	2	9
114.82		1.1489	2	6	3	2		7
115.36	115.36	1.1454	2	4	4	11	4	34
116.66	116.66	1.1374	2	8	0	5	2	16
117.79	117.82	1.1306	4	3	0	1	0	2
118.56	118.58	1.1260	3	4	3	5	2	17
120.00	120.02	1.1177	1	1	1	3	1	9
121.47	121.48	1.1096	1	3	5	9	2	27
122.09	121.96	1.1063	4	2	2	5	2	16
123.54	123.56	1.0987	1	9	1	1	0	3
124.26	124.30	1.0950	0	4	3	1	1	3
125.68	125.74	1.0879	0	8	3	1	4	2
125.72		1.0878	4	2	2	10		33
126.08	126.16	1.0860	4	4	0	5	4	16
126.49	126.46	1.0840	3	7	0	10	4	34
128.82	128.84	1.0733	2	2	4	4	2	13
128.84		1.0703	2	1	5	3		9
129.48	129.30	1.0703	3	5	3	1	1	4
129.80		1.0689	2	8	2	1		3
130.83	130.82	1.0645	4	3	2	3	1	10
132.17	132.20	1.0589	4	3	2	1	1	5
132.82	132.76	1.0562	2	5	2	1	1	3
134.95	134.98	1.0479	1	9	4	30	6	108
135.60	135.52	1.0455	3	3	4	11	5	38
139.58	139.58	1.0315	1	7	4	28	5	102
142.93	143.46	1.0210	4	5	1	5	2	19
143.36		1.0197	1	5	5	3		10
143.49		1.0193	0	10	1	6		21

2THETA	PEAK	D	H	K	L	I(INT)	I(PK)	I(DS)
146.85	146.92	1.0100	3	4	4	7	1	26
147.17		1.0091	4	2	3	2		8
147.87	147.64	1.0073	0	0	6	2	1	9
149.04	149.02	1.0044	3	8	0	8	1	33
152.31	152.94	0.9969	1	10	1	1	2	3
152.87		0.9958	2	4	5	15		60
155.35	156.00	0.9908	3	8	1	1	2	2
155.56		0.9904	3	8	1	12		47
155.98		0.9897	0	6	5	12		48
156.48		0.9887	0	2	6	6		26
158.11		0.9859	4	3	3	1		6
159.94	161.24	0.9830	0	6	4	3	2	14
160.45		0.9822	1	8	4	3		14
161.10		0.9813	1	1	6	36		148
162.14		0.9799	2	9	2	4		19
163.26		0.9784	0	10	1	1		2
164.36		0.9771	1	9	3	3		13

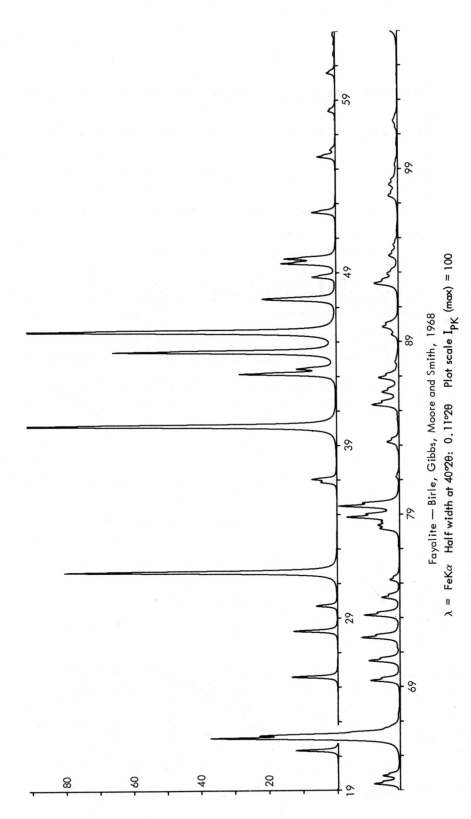

Fayalite — Birle, Gibbs, Moore and Smith, 1968

λ = FeKα Half width at 40°2θ: 0.11°2θ Plot scale I_{PK} (max) = 100

FAYALITE — BIRLE, GIBBS, MOORE AND SMITH, 1968

2THETA	PEAK	D	H	K	L	I(INT)	I(PK)	I(DS)
21.31	21.32	5.234	0	2	0	9	12	5
25.56	25.58	4.375	1	0	1	10	14	6
28.21	28.22	3.972	0	2	1	10	13	7
29.68	29.68	3.780	1	1	1	5	6	3
31.60	31.60	3.555	1	2	0	65	80	46
31.70		3.544	1	1	1	6		4
36.83	36.84	3.064	1	2	1	4	5	3
37.01	37.02	3.049	0	0	2	6	8	5
40.06	40.06	2.826	1	3	0	94	100	83
43.11	43.12	2.635	0	2	2	27	29	26
43.41	43.42	2.617	0	0	2	10	12	9
44.36	44.36	2.564	1	2	1	64	66	63
45.53	45.52	2.502	1	2	0	100	99	100
47.40		2.408	2	0	1	12	22	13
47.46	47.46	2.405	2	3	0	13		13
48.72	48.72	2.347	0	4	1	7	7	8
49.51	49.52	2.312	2	1	2	16	16	18
49.79	49.78	2.300	1	2	2	15	15	16
52.46	52.46	2.190	1	4	1	8		9
55.68	55.68	2.073	2	1	2	6	5	7
56.08	56.08	2.059	2	3	1	2	2	2
58.34	58.34	1.9861	0	4	0	2	2	3
58.47	58.46	1.9819	2	1	2	0		1
60.54	60.54	1.9202	1	5	0	3	2	4
61.62	61.62	1.8898	2	0	2	1	1	1
62.73	62.74	1.8598	2	1	1	1	1	1
63.34	63.34	1.8437	1	2	3	9	8	14
63.81	63.82	1.8315	1	5	1	6	5	9
65.99	66.00	1.7775	2	1	2	69	56	106
66.22	66.14	1.7721	2	4	2	20	41	31
69.34	69.34	1.7017	0	6	1	11	9	19
70.48	70.48	1.6775	2	3	1	12		21
71.25	71.26	1.6618	2	3	3	1	1	1
71.83	71.82	1.6503	1	5	3	15	11	26
73.13	73.14	1.6249	1	5	3	15	11	26
74.16	74.16	1.6055	0	4	3	7	5	13
75.18	75.18	1.5868	3	1	0	4	3	7
78.09	78.16	1.5366	2	1	3	1	6	3
78.15		1.5357	1	1	1	5		10
78.20		1.5348	3	1	0	4		3
78.36	78.36	1.5322	2	4	2	3	6	7
78.52	78.54	1.5295	2	5	0	22	5	5
78.82	78.82	1.5247	0	6	1	1	16	41
79.46	79.46	1.5145	3	2	0	1	18	50
81.14	81.14	1.4884	0	6	2	6	1	2
82.79	82.80	1.4639	3	1	1	1	1	1
83.17	83.18	1.4584	1	7	0	14	4	12
84.49	84.50	1.4398	3	1	0	8	1	1
85.33	85.34	1.4283	1	5	0	10	8	28
86.07	86.08	1.4184	3	1	1	3	5	17
86.89	86.90	1.4076	1	6	2	3	6	22
87.80	87.80	1.3960	2	6	3	2	1	4
89.38	89.38	1.3765	3	2	1	8	2	7
89.83	89.84	1.3709	3	4	2	2	5	19
90.04	90.04	1.3684	3	4	0	14	4	4
92.34	92.34	1.3419	1	3	3	3	7	34
92.88	92.90	1.3358	2	4	1	6	3	6
92.93		1.3352	3	4	1	3		8
93.95	93.96	1.3240	0	6	1	6	3	13
94.57	94.56	1.3175	0	4	3	3	1	6
95.41	95.42	1.3086	0	8	0	1	0	2
97.43	97.44	1.2882	2	0	4	6	3	16
98.06	98.06	1.2820	2	6	2	6	3	14
98.42	98.34	1.2786	2	1	4	1	2	3
98.90	98.88	1.2740	2	5	0	5	1	2
99.23	99.22	1.2708	1	7	0	7	1	3
99.27		1.2705	1	0	4	3		1
100.40	100.40	1.2599	2	1	3	0	0	2
101.40	101.42	1.2509	3	1	3	1	1	4
101.77	101.76	1.2476	2	5	0	2	1	6
101.83		1.2471	3	5	2	3		6
107.02	107.04	1.2040	4	0	0	1	2	17
107.21		1.2026	0	8	2	6		4
108.32	108.32	1.1941	1	5	4	2	1	6

FAYALITE - BIRLE, GIBBS, MOORE AND SMITH, 1968

2THETA	PEAK	D	H	K	L	I(INT)	I(PK)	I(DS)
109.54	109.54	1.1850	3	3	3	6	2	16
109.89	109.84	1.1825	1	0	5	1	2	2
110.86	110.88	1.1755	3	5	2	11	4	33
110.94		1.1750	3	1	1	3		4
111.26	111.20	1.1728	2	7	2	3	4	7
112.12	112.12	1.1667	1	8	2	2	1	5
113.09	113.14	1.1602	2	6	3	3	2	9
113.15		1.1598	3	6	1	2		6
113.76	113.76	1.1558	2	4	4	11	4	32
114.68	114.68	1.1498	2	8	0	5	2	14
117.01	117.02	1.1352	3	4	3	4	1	13
117.90	117.90	1.1299	2	8	1	3	2	10
119.61	119.62	1.1199	1	3	5	8	2	24
120.75	120.76	1.1135	4	1	2	4	1	13
122.21	122.22	1.1056	0	4	4	1	1	3
123.39	123.44	1.0994	3	1	5	3	1	11
124.24	124.38	1.0951	4	2	0	11		38
124.40		1.0943	3	7	0	10	6	34
124.49		1.0938	4	4	0	5		17
126.85	126.98	1.0823	2	1	5	2	1	6
126.99		1.0817	3	2	1	3		10
127.45	127.42	1.0795	3	5	3	1	1	5
128.08	128.20	1.0766	4	4	1	2	1	5
128.24		1.0759	2	8	2	3		12
131.86	131.88	1.0602	1	9	2	29	6	102
133.40	133.40	1.0539	3	3	4	11	3	39
135.09	135.28	1.0474	2	9	0	1	1	2
135.22		1.0469	0	10	0	4		13
136.07	136.46	1.0437	4	5	0	1		5
136.44		1.0424	1	7	4	25	5	92
139.48	139.58	1.0318	0	10	1	5	1	18
139.75	140.24	1.0309	4	1	3	1	2	2
140.15		1.0296	1	5	5	3		10
140.41		1.0288	4	5	1	4		16
143.79	143.88	1.0184	3	4	4	3	1	12
144.45	144.56	1.0165	0	0	6	2	1	6
144.54		1.0162	4	2	3	1		5

2THETA	PEAK	D	H	K	L	I(INT)	I(PK)	I(DS)
145.23	145.22	1.0143	3	8	0	5	1	20
148.90	148.94	1.0048	2	4	5	11	2	45
150.56	150.98	1.0008	2	8	3	10	2	40
151.05		0.9997	0	6	5	9		36
151.89		0.9979	0	2	6	7		30
155.27	155.72	0.9910	4	6	0	2	2	9
155.46		0.9906	2	9	2	2		8
155.72		0.9901	1	1	6	26		104
156.81		0.9882	1	9	3	1		3
162.90	163.68	0.9789	1	6	5	2	1	7
163.47		0.9781	4	6	1	13		54
163.88		0.9776	3	5	4	4		17

67

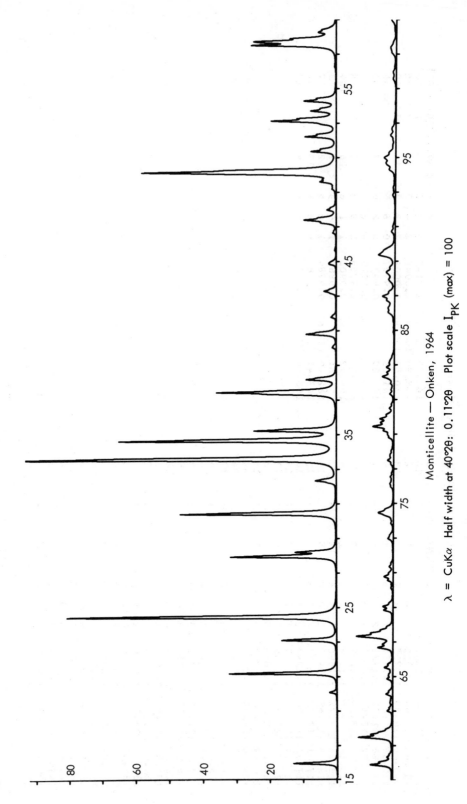

Monticellite — Onken, 1964

λ = CuKα Half width at 40°2θ: 0.11°2θ Plot scale I_{PK} (max) = 100

68

MONTICELLITE - ONKEN, 1964

2THETA	PEAK	D	H	K	L	I(INT)	I(PK)	I(DS)
15.94	15.96	5.554	0	2	0	10	13	7
20.06	20.06	4.423	1	1	1	2	2	1
21.19	21.20	4.190	0	2	1	27	32	21
23.10	23.10	3.847	1	0	1	14	17	11
24.43	24.46	3.641	1	2	0	18	81	15
24.46		3.635	1	1	1	56		47
27.94	27.94	3.191	0	1	2	29	32	26
28.19	28.20	3.163	1	2	1	10	12	9
30.41	30.42	2.937	1	3	0	45	47	42
32.33	32.32	2.767	0	2	2	5	6	5
33.56	33.56	2.668	1	3	1	100	100	100
34.63	34.64	2.588	0	4	1	65	65	67
35.21	35.22	2.546	1	2	2	24	25	25
37.34	37.44	2.406	1	4	0	17	36	18
37.44		2.400	2	2	0	27		29
38.16	38.16	2.356	2	1	0	9	9	10
40.01	40.00	2.252	1	4	1	1	1	1
40.76	40.78	2.212	2	1	0	3	9	4
40.79		2.210	2	1	1	7		8
41.76	41.76	2.161	1	3	2	2	2	2
43.15	43.26	2.095	0	4	2	2	4	2
43.26		2.090	2	4	2	3		4
44.82	44.88	2.020	2	3	0	3	2	4
44.88		2.018	1	0	0	2		2
46.62	46.62	1.9463	1	1	3	1	1	1
47.20	47.22	1.9239	1	5	0	2	5	3
47.21		1.9237	2	5	1	2		2
47.38	47.38	1.9171	1	1	3	10	10	12
47.95	47.96	1.8954	2	1	2	3	3	4
49.17	49.18	1.8513	0	6	0	1		1
49.59	49.60	1.8368	1	6	3	5	5	6
49.60	50.14	1.8206	2	4	2	24	58	33
50.14		1.8177	2	2	0	56		77
51.34	51.34	1.7780	0	6	1	9	8	13
52.20	52.20	1.7507	2	6	1	11	9	16
53.11	53.12	1.7228	1	4	3	23	19	33
53.64	53.70	1.7070	2	3	2	1	8	1

2THETA	PEAK	D	H	K	L	I(INT)	I(PK)	I(DS)
53.70	54.28	1.7054	1	5	2	8	10	12
54.27		1.6888	0	6	3	12		18
57.50	57.50	1.6013	0	0	4	31	25	49
57.73	57.72	1.5955	0	4	3	22	25	35
57.80		1.5938	1	1	0	1		2
57.92	57.88	1.5908	3	1	1	3	14	4
58.24	58.26	1.5827	2	5	1	4	5	7
58.30		1.5813	2	4	2	2		3
58.39	58.40	1.5790	2	1	3	1	4	2
59.85	59.86	1.5440	3	1	1	2	7	2
59.87		1.5435	3	4	0	7		12
60.30	60.30	1.5335	0	2	4	1	2	1
60.32		1.5332	2	2	3	14		1
61.46	61.46	1.5073	1	7	0	2	10	23
62.99	63.00	1.4744	3	3	0	1	2	4
63.44	63.46	1.4650	2	5	2	3	1	1
63.96	63.96	1.4542	2	3	3	2	2	5
64.85	64.84	1.4366	3	1	0	1	2	4
65.13	65.04	1.4310	2	6	1	3	1	2
65.51	65.50	1.4237	3	3	2	6	2	6
66.65	66.66	1.4020	1	3	4	1	5	11
66.95	66.84	1.3965	0	6	0	4	3	2
67.24	67.30	1.3911	3	2	2	13	11	7
67.31		1.3898	3	4	2	2		24
67.67	67.66	1.3834	0	4	4	2	4	3
67.68		1.3832	2	4	3	2		4
68.82	68.82	1.3629	1	7	1	1	3	7
69.04	69.02	1.3592	3	2	3	3	3	4
70.09	70.10	1.3414	1	6	3	4	1	2
70.54	70.54	1.3339	2	6	2	2	2	4
70.75	70.76	1.3305	2	0	4	2	2	5
70.79		1.3298	1	4	4	2		1
71.33	71.32	1.3211	2	2	3	2	3	3
72.95	72.96	1.2957	2	5	3	2	1	4
73.07		1.2939	2	2	4	1	2	1
73.83	73.84	1.2824	3	0	3	1		3
74.32	74.44	1.2752	3	4	2	2	5	4

MONTICELLITE - ONKEN, 1964

2THETA	PEAK	D	H	K	L	I(INT)	I(PK)	I(DS)
74.40		1.2740	3	1	3	2		4
74.45		1.2732	0	8	2	4		9
75.92	75.96	1.2522	2	3	4	1	1	1
75.97		1.2515	1	5	4	1		1
76.11	76.12	1.2496	3	2	3	1	1	2
76.51	76.52	1.2440	0	2	5	1	1	3
77.25	77.26	1.2339	1	0	5	2	1	2
77.47	77.48	1.2310	1	8	2	2	2	5
77.82	77.80	1.2264	1	1	5	1	1	2
77.99	78.00	1.2241	2	7	2	1	1	1
78.78	78.80	1.2137	3	6	0	2	2	4
78.93	78.96	1.2118	3	1	3	1	2	3
79.18	79.20	1.2086	0	6	4	1	3	2
79.19		1.2085	2	6	3	2		5
79.41	79.42	1.2057	3	5	0	5	6	12
79.43		1.2055	4	0	0	3		6
79.61	79.64	1.2032	2	8	0	2	4	2
79.87	79.86	1.1999	2	4	4	6	4	12
80.48	80.48	1.1923	3	6	1	2	1	4
81.30	81.30	1.1824	2	8	1	2	1	5
81.90	81.90	1.1752	1	1	6	2	1	3
82.29	82.30	1.1706	1	3	5	6	3	13
82.84	82.84	1.1643	3	4	3	4	2	8
83.23	83.10	1.1598	0	4	5	1	2	3
84.87	84.88	1.1415	2	5	4	1	0	1
85.91	86.02	1.1303	1	8	3	1	2	7
86.01		1.1292	3	7	0	3		2
86.17	86.26	1.1276	1	4	5	1	2	2
86.27		1.1265	2	1	4	1		2
86.42		1.1250	1	9	2	2		2
86.68	86.72	1.1223	3	3	5	2		6
86.71		1.1219	4	1	2	5		12
86.93	86.94	1.1197	1	9	0	2	4	4
87.82	87.92	1.1107	3	2	3	2	2	5
87.93		1.1095	3	4	4	2		3
88.30	88.34	1.1058	4	4	0	1	3	3
88.33		1.1055	2	2	5	1		2

2THETA	PEAK	D	H	K	L	I(INT)	I(PK)	I(DS)
88.37		1.1052	4	2	2	3		7
89.33	89.34	1.0957	1	7	4	9	5	22
89.47		1.0943	3	10	1	4		9
90.69	90.68	1.0828	0	3	4	2	1	5
92.80	92.80	1.0637	4	0	6	1	1	4
93.26	93.24	1.0596	3	5	0	2	1	3
94.27	94.30	1.0509	3	6	5	2	2	4
94.29		1.0507	3	8	0	4		5
94.55	94.66	1.0485	3	4	3	2	3	5
94.69		1.0473	2	8	4	4		9
94.94	94.96	1.0453	4	5	1	1		4
94.95		1.0451	2	4	5	2		6
94.99		1.0448	4	2	2	1		2
95.00		1.0447	0	1	6	1		2
95.07		1.0442	4	1	3	1		3
95.29	95.26	1.0423	2	9	2	1	2	3
95.72	95.68	1.0388	1	1	9	1	1	9
96.28	96.28	1.0342	2	6	6	3	2	9
96.73	96.64	1.0306	1	2	3	2	1	4
99.54	99.54	1.0089	4	10	0	2	1	6
99.95	99.92	1.0058	2	5	5	2	1	5
101.24	101.36	0.9965	2	10	1	2	2	6
101.37		0.9955	3	1	0	1		3
102.39	102.40	0.9884	1	11	5	1	1	3
102.94	102.96	0.9847	1	0	0	3	1	10
103.07		0.9838	3	2	3	2		3
104.11	104.12	0.9767	1	11	1	1		3
105.22	105.24	0.9695	5	1	6	2	0	2
106.42	106.60	0.9618	2	8	0	6	0	6
106.58		0.9608	2	2	4	1	2	4
106.60	106.94	0.9607	4	1	4	4		13
106.94		0.9586	2	6	4	1		3
106.99		0.9582	5	1	1	1	2	5
108.33	108.36	0.9501	5	4	3	2	1	5
108.61	108.62	0.9484	4	5	5	3	1	4
109.89	110.06	0.9409	1	5	6	1	1	3
110.08		0.9398	5	2	1	1		3

MONTICELLITE - ONKEN, 1964

2THETA	PEAK	D	H	K	L	I(INT)	I(PK)	I(DS)
110.78	110.78	0.9359	3	9	2	3	1	10
111.24	111.20	0.9333	5	3	0	1	1	4
113.26	113.34	0.9223	0	6	6	3	2	11
113.37		0.9217	3	7	1	3		9
113.38		0.9217	2	11	1	1		3
115.14	115.40	0.9125	4	6	3	1	1	2
115.34		0.9116	2	10	0	1		4
115.45		0.9110	5	4	0	1		2
115.59	115.86	0.9103	4	8	0	1		2
115.88		0.9089	4	4	4	3	1	8
116.69	116.70	0.9049	2	9	4	1		2
116.75		0.9046	3	10	1	1	1	2
117.46	117.42	0.9012	4	1	1	1	1	5
118.48	118.54	0.8964	1	11	3	1	0	2
119.22	119.18	0.8929	1	1	7	1	0	3
121.18	121.18	0.8842	3	1	6	1	0	2
121.54	121.58	0.8827	4	5	4	1	1	4
122.74	123.18	0.8775	3	8	4	2	2	8
123.05		0.8763	5	5	1	2		6
123.13		0.8759	3	2	6	2		6
123.20	123.20	0.8755	5	1	3	3		9
123.22		0.8755	2	8	5	1		2
123.26		0.8754	4	8	2	1		3
124.41	124.42	0.8707	1	3	7	4	1	12
124.82	124.90	0.8691	1	7	6	1	1	5
125.20		0.8676	5	2	3	1		5
126.54	126.80	0.8624	4	9	0	1		2
126.81		0.8614	2	6	6	2		8
129.04	129.20	0.8532	4	3	3	1	2	5
129.15		0.8528	1	13	1	1		2
129.19		0.8527	4	3	5	2		9
129.25		0.8525	2	10	4	5		18
130.31	129.76	0.8488	5	5	2	1	1	5
131.45	131.50	0.8450	0	12	3	2	1	6
131.62		0.8444	3	4	6	1		4
132.55	132.92	0.8414	0	8	6	1	1	3
132.90		0.8402	1	13	0	2		8

2THETA	PEAK	D	H	K	L	I(INT)	I(PK)	I(DS)
133.63	133.62	0.8379	0	10	5	3	1	12
133.96		0.8369	4	8	3	1		2
134.86	134.82	0.8341	1	13	1	1	1	2
134.87		0.8341	2	12	2	1		4
138.46	138.52	0.8238	3	5	6	2		14
138.72		0.8231	5	1	4	2		6
141.74	142.46	0.8153	4	5	5	1	1	7
141.77		0.8152	2	4	7	1		4
142.44		0.8136	1	13	2	8		31
143.42	143.28	0.8112	3	8	1	1	1	5
143.83		0.8103	4	10	0	1		4
145.94	145.96	0.8056	5	3	3	3	0	10
150.70	151.06	0.7961	2	5	7	1	2	6
150.83		0.7959	3	12	1	5		20
151.03		0.7955	1	2	6	2		7
151.12		0.7954	6	2	0	1		3
151.49	151.44	0.7947	1	9	6	6	2	25
152.23	152.34	0.7934	0	14	0	4	2	17
152.47		0.7930	3	0	7	1		3
152.51		0.7930	3	10	4	1		3
153.00	153.30	0.7921	4	6	5	5	2	20
153.39		0.7915	2	10	2	4		16
153.47		0.7914	4	10	1	1		3
153.61		0.7911	5	4	4	2		8
153.69		0.7910	3	1	7	2		8
153.91		0.7907	4	8	4	1		6
154.65		0.7895	4	2	6	5		21
154.79		0.7893	6	2	1	3		12
159.37	159.88	0.7829	1	14	0	1	0	4
159.79		0.7824	1	13	3	3		12
160.55		0.7815	1	11	5	5		9
162.49	162.60	0.7793	6	0	2	2	0	7
162.55		0.7793	1	2	8	1		5
163.86		0.7779	3	12	2	7		30
164.43		0.7774	6	1	2	1		3

TABLE 3. GARNETS

Variety	Grossular	Pyrope	Pyrope	Andradite	Uvarovite
Composition	$Ca_3Al_2(SiO_4)_3$	$Mg_3Al_2(SiO_4)_3$	$(Mg_{1.6}Fe_{1.2}Ca_{0.2})Al_2(SiO_4)_3$	$(Ca_{2.5}Mg_{0.02}Fe^2_{0.04})(Fe^3_{1.98}Al_{0.02})(SiO_4)_3$	$(Ca_{2.9}Mg_{0.1})(Cr_{1.8}Al_{0.2}Fe^3)(SiO_4)_3$
Source	Xalostoc, Mexico	Synthetic	Blue Mt. Lake, New York	Val Malenco, Italy	Outokumpu, Finland Deer, Howe & Zussman 1, Table 19, 1962
Reference	Prandl, 1966	Gibbs & Smith, 1965	Euler & Bruce, 1965	Quareni & dePieri, 1966	Novak & Gibbs, 1968
Cell Dimensions					
a Å	11.855	11.459	11.556	12.061	11.934
b Å	11.855	11.459	11.556	12.061	11.934
c Å	11.855	11.459	11.556	12.061	11.934
α	90	90	90	90	90
β deg	90	90	90	90	90
γ	90	90	90	90	90
Space Group	Ia3d	Ia3d	Ia3d	Ia3d	Ia3d
Z	8	8	8	8	8

TABLE 3. (cont.)

Variety	Grossular	Pyrope	Pyrope	Andradite	Uvarovite
Method	3-D, Neutron dif-frac. & x-ray	3-D, counter	3-D, counter	3-D, film	3-D, counter
R & Rating	0.047 (1)	0.11 (1)	0.067 (1)	0.113 (2)	0.024 (2)
Cleavage and habit	{110} rare ——————— Granular ———————				
Comment					New chemical analysis indicates site occupancy used in refinement may be slightly revised
μ	1681.0	234.0	342.2	464.4	387.9
ASF	0.2415	0.1328	0.1956	0.2230	0.2290
Abridging factors	0.3:0:0	0.2:0:0	0.2:0:0	0.1:0:0	0.1:0:0

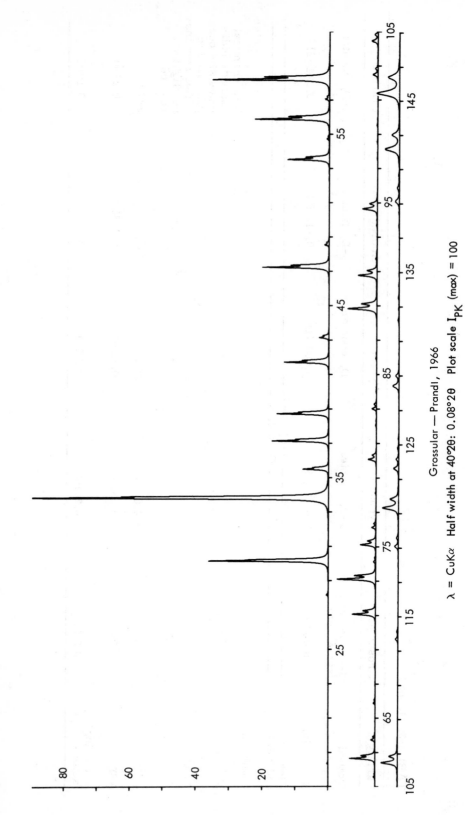

Grossular — Prandl, 1966

λ = CuKα   Half width at 40°2θ: 0.08°2θ   Plot scale I$_{PK}$ (max) = 100

74

GROSSULAR – PRANDL, 1966

2THETA	PEAK	D	H	K	L	I(INT)	I(PK)	I(DS)
28.14	28.14	3.168	3	2	1	1	1	0
30.13	30.12	2.964	4	0	0	36	36	29
33.78	33.78	2.651	4	2	0	100	100	100
35.49	35.48	2.527	3	3	2	8	8	9
37.12	37.12	2.420	4	2	2	17	17	21
38.69	38.70	2.325	4	3	1	16	16	21
41.69	41.70	2.164	5	2	1	15	14	21
43.13	43.12	2.096	4	4	0	3	3	5
47.22	47.22	1.9231	6	1	1	12	20	22
47.22	47.22	1.9231	5	3	2	11		19
48.53	48.52	1.8744	6	2	0	2	1	35
53.51	53.50	1.7111	4	4	4	15	13	2
54.70	54.70	1.6766	5	4	3	1	1	68
55.88	55.88	1.6440	6	4	0	27	22	3
57.04	57.04	1.6133	7	2	1	0	1	1
57.04	57.04	1.6133	6	3	3	44	35	117
58.18	58.18	1.5842	6	4	2	1	1	2
61.54	61.54	1.5056	6	5	1	10	8	31
62.64	62.64	1.4819	8	0	0	2	1	5
63.72	63.72	1.4592	7	4	1	1	1	3
65.86	65.86	1.4169	6	5	3	0	0	1
67.96	67.96	1.3781	7	5	0	0	0	1
70.03	70.04	1.3423	8	4	2	10	7	39
71.06	71.06	1.3254	8	4	1	17	12	70
73.09	73.10	1.2935	6	6	4	1	1	3
74.10	74.10	1.2784	8	6	5	7	5	30
75.11	75.10	1.2637	7	8	1	1	1	4
76.11	76.10	1.2496	9	5	4	1		4
78.09	78.10	1.2227	8	7	2	2	0	2
80.06	80.06	1.1975	10	2	0	0	2	9
82.02	82.02	1.1738	8	6	2	2	0	9
83.00	83.00	1.1625	7	5	5	1	1	2
83.00	83.00	1.1625	8	6	3			9
85.91	85.92	1.1303	7	2	1	1	0	3
88.82	88.82	1.1007	10	6	4	6	9	33

2THETA	PEAK	D	H	K	L	I(INT)	I(PK)	I(DS)
88.82	90.76	1.1007	10	6	4	10	6	56
90.75	91.72	1.0822	10	4	2	11	0	62
91.72	93.66	1.0733	8	7	3	1	0	4
93.66	94.62	1.0561	9	6	3	9	4	2
94.63	98.52	1.0478	8	6	0	0	0	55
98.52	99.50	1.0166	10	6	1	0	0	3
99.50	102.46	1.0092	11	4	1	3	1	
102.46	104.46	0.9879	8	8	4	4	2	21
104.45	106.46	0.9745	12	2	0	4	5	31
106.46		0.9616	10	6	2	5		42
106.46		0.9616	12	2	2	6		45
113.67	113.68	0.9201	11	6	3	1	1	9
113.67		0.9201	9	7	6	1		7
119.07	119.08	0.8936	12	4	4	3	1	27
121.31	121.32	0.8836	12	4	0	10	5	91
121.31		0.8836	11	6	0	4		36
123.61	123.60	0.8740	12	4	2	4	1	40
124.78	124.78	0.8693	11	7	0	1	0	8
128.39	128.40	0.8556	8	8	8	6	2	60
132.20	132.20	0.8425	13	5	2	1	0	6
132.20		0.8425	9	9	0	1		6
133.52	133.52	0.8383	10	10	0	10	0	10
134.87	134.88	0.8341	12	6	3	1	0	
137.67	137.66	0.8260	14	3	1	0	0	4
139.12	139.12	0.8220	12	8	0	0	1	53
140.62	140.62	0.8181	11	8	5	5	0	5
142.18	142.18	0.8142	12	8	2	0	4	140
142.18		0.8142	14	4	0	12		66
145.45	145.46	0.8066	12	6	6	6		63
145.45		0.8066	10	10	4	18	7	222
150.96	150.96	0.7957	13	7	2	6	0	76
153.02	153.02	0.7921	12	8	4	1	0	12
160.37	160.36	0.7817	13	6	1	1		17
160.37		0.7817	14	5	1	0		15
163.49	163.48	0.7783	14	6	0	1	0	6
			15	2	1	1		15
			14	4	0	1		17

76

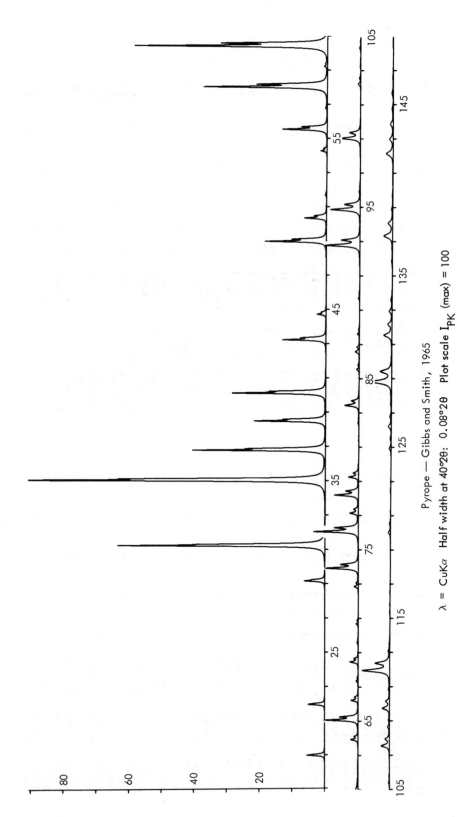

Pyrope — Gibbs and Smith, 1965

$\lambda = CuK\alpha$ Half width at 40°2θ: 0.08°2θ   Plot scale I_{PK} (max) = 100

PYROPE - GIBBS AND SMITH, 1965

2THETA	PEAK	D	H	K	L	I(INT)	I(PK)	I(DS)
18.95	18.96	4.678	2	1	1	4	5	3
21.92	21.92	4.051	2	2	0	4	6	5
29.13	29.14	3.063	3	2	1	6	6	5
31.19	31.20	2.865	4	0	0	59	63	53
34.99	34.98	2.562	4	2	0	100	100	100
36.76	36.76	2.443	3	3	2	39	41	41
38.45	38.46	2.339	4	2	2	22	22	24
40.09	40.08	2.247	4	3	1	30	29	34
43.21	43.20	2.092	5	2	1	14	13	17
44.70	44.70	2.026	4	4	0	3	3	4
48.96	48.96	1.8589	6	1	1	13	19	18
48.96		1.8589	5	3	2	8	7	10
50.32	50.32	1.8118	6	2	0	0	0	11
51.65	51.66	1.7682	5	4	1	2	2	1
54.24	54.24	1.6895	6	3	1	16	14	3
55.51	55.52	1.6540	4	4	4	1	1	26
56.76	56.76	1.6205	5	4	3	46	38	1
57.99	57.98	1.5891	6	4	0	1	1	77
59.20	59.20	1.5594	7	2	1	1	59	1
59.20		1.5594	5	5	2	72	2	125
60.40	60.40	1.5313	6	4	2	3	11	5
63.91	63.92	1.4553	7	3	2	14	2	26
65.06	65.06	1.4324	8	0	0	3	1	3
66.20	66.20	1.4105	7	4	1	1	3	5
67.32	67.32	1.3896	8	2	0	4	1	2
68.44	68.44	1.3696	6	5	3	1	0	7
69.55	69.56	1.3505	8	2	2	1	1	1
70.66	70.66	1.3321	7	4	3	2	1	2
72.83	72.84	1.2975	7	5	2	15	10	4
73.91	73.92	1.2812	8	4	0	20	14	32
76.06	76.06	1.2503	8	4	2	4	3	45
77.12	77.12	1.2357	7	6	1	11	7	8
78.18	78.18	1.2215	6	6	4	2	3	26
79.24	79.24	1.2079	7	5	4	2	0	5
79.24		1.2079	8	5	1	0	4	6
82.39	82.38	1.1695	8	4	4	4		1
83.43	83.44	1.1575	8	3	5			10

2THETA	PEAK	D	H	K	L	I(INT)	I(PK)	I(DS)
83.43	83.43	1.1575	9	4	1	3	3	8
85.51	85.52	1.1346	10	1	1	1	1	3
86.55	86.54	1.1236	10	2	0	2	1	5
87.59	87.58	1.1130	9	4	3	1	0	2
89.66	89.66	1.0926	7	6	5	0		1
89.66		1.0926	10	4	0	0		1
92.76	92.76	1.0639	8	6	4	9	11	25
92.76		1.0639	10	4	2	10		30
94.84	94.84	1.0461	10	7	3	17	9	49
95.88	95.88	1.0374	8	8	0	12	0	2
99.01	99.02	1.0128	8	7	2	1	6	38
102.17	102.18	0.9899	9	7	0	0	1	2
102.17		0.9899	7	7	6	1	0	2
103.24	103.24	0.9826	10	6	0	1	0	1
104.30	104.30	0.9755	11	4	1	5	0	2
105.37	105.38	0.9685	10	6	2	1	3	18
107.53	107.54	0.9549	8	8	4	5		2
107.53		0.9549	12	0	0	11		3
108.62	108.62	0.9484	9	8	1	10		18
109.72	109.72	0.9419	12	2	0	1	1	2
110.82	110.82	0.9356	11	5	2	0	2	38
111.93	111.94	0.9294	12	2	2	0	0	34
113.06	113.06	0.9294	10	6	4	1	9	3
115.33	115.32	0.9234	12	3	1	0		1
116.48	116.48	0.9116	11	6	1	1	0	1
120.00	120.00	0.9059	12	4	0	16	0	4
120.00		0.8894	11	6	3	8	1	2
122.43	122.42	0.8894	9	9	2	0	0	5
124.91	124.92	0.8789	9	8	5		1	2
124.91		0.8687	13	2	1		0	11
126.19	126.18	0.8687	10	7	5	3	1	3
127.48	127.48	0.8638	12	4	4	1	0	17
128.80	128.80	0.8589	12	5	3	4	7	67
128.80		0.8541	12	6	0	16		32
131.51	131.50	0.8541	10	8	4	8	2	2
132.91	132.90	0.8402	11	8	1	0	1	2

77

PYROPE - GIBBS AND SMITH, 1965

2THETA	PEAK	D	H	K	L	I(INT)	I(PK)	I(DS)
132.91		0.8402	11	7	4	2		9
135.80	135.80	0.8313	10	9	3	1	0	4
137.31	137.30	0.8270	8	8	8	9	3	41
138.86	138.86	0.8227	11	8	3	0	0	2
138.86		0.8227	13	4	3	0		2
142.11	142.12	0.8144	13	5	2	4	2	19
142.11		0.8144	14	1	1	0		1
142.11		0.8144	9	9	6	3		15
142.11		0.8144	10	7	7	0		1
143.83	143.84	0.8103	10	8	6	0	1	2
143.83		0.8103	10	10	0	3		12
145.63	145.62	0.8063	12	7	3	2	0	8
149.49	149.48	0.7984	13	6	1	1	1	7
149.49		0.7984	10	9	5	1		5
149.49		0.7984	11	7	6	0		2
151.59	151.60	0.7945	12	8	0	10	2	49
153.85		0.7907	13	5	4	1		3
153.85	153.86	0.7907	11	8	5	2	1	11
156.31	156.32	0.7870	14	4	0	9	5	47
156.31		0.7870	12	8	2	23		117
159.04	159.02	0.7833	13	6	3	2	1	11
159.04		0.7833	14	3	3	0		1
162.15	162.16	0.7797	10	10	4	22	9	112
162.15		0.7797	14	4	2	53		271
162.15		0.7797	12	6	6	15		76

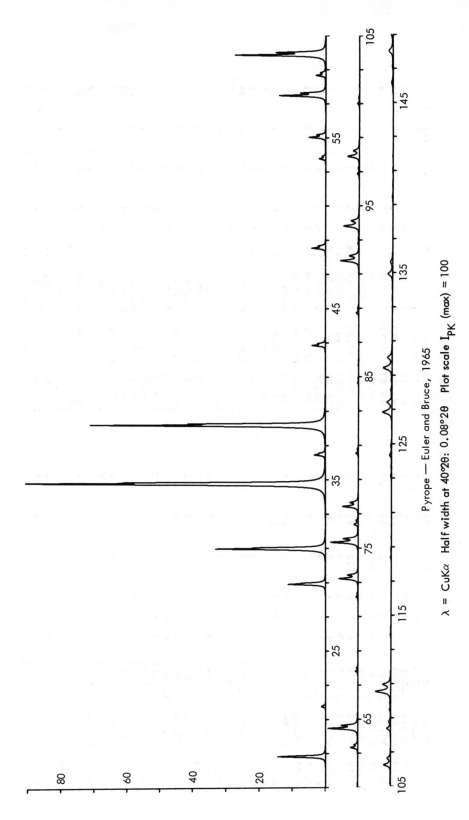

Pyrope — Euler and Bruce, 1965

λ = CuKα Half width at 40°2θ: 0.08°2θ Plot scale I_{PK} (max) = 100

79

PYROPE - EULER AND BRUCE, 1965

2THETA	PEAK	D	H	K	L	I(INT)	I(PK)	I(DS)
18.79	18.80	4.718	2	1	1	12	15	6
21.73	21.74	4.086	2	2	0	11	11	1
28.88	28.88	3.088	3	2	1	11	11	9
30.93	30.92	2.889	4	0	0	32	33	28
34.69	34.68	2.584	4	2	0	100	100	100
36.44	36.44	2.464	3	3	2	3	3	3
38.12	38.12	2.359	4	2	2	72	71	80
42.82	42.82	2.110	5	2	1	4	4	6
48.52	48.52	1.8746	6	1	1	2	4	2
48.52		1.8746	5	3	2	3	2	5
53.75	53.76	1.7038	6	3	1	2	2	4
55.01	55.00	1.6680	4	4	4	6	5	11
57.46	57.46	1.6025	6	4	0	18	14	34
58.66	58.66	1.5726	7	2	1	2	3	4
58.66		1.5726	5	5	3	1		2
59.84	59.84	1.5442	6	4	2	34	27	69
59.84		1.5442	6	5	1	1	2	3
63.31	63.32	1.4676	6	3	2	1		3
63.31		1.4676	7	0	0	1		3
64.45	64.44	1.4445	8	0	0	12	9	28
67.79	67.78	1.3812	6	5	3	1	1	3
72.13	72.12	1.3085	7	5	2	1	1	2
73.19	73.20	1.2920	8	4	0	9	6	23
75.31	75.30	1.2609	8	4	2	13	8	35
76.36	76.36	1.2461	7	6	1	1	1	2
76.36		1.2461	9	2	1	1		2
76.36		1.2461	6	5	5	0		1
77.40	77.40	1.2319	6	6	4	7	5	21
80.52	80.52	1.1919	9	3	2	1	1	2
80.52		1.1919	7	6	3	0		2
80.52		1.1919	10	1	1	1		2
84.62	84.62	1.1442	7	7	2	0	0	1
84.62		1.1442	7	7	2	0		2
88.70	88.70	1.1018	9	5	1	0	1	2
88.70		1.1018	10	3	5	0		2
88.70		1.1018	7	6	6	0		2
91.76	91.76	1.0729	10	4	4	3	6	13
91.76		1.0729	8	6	6	7		25

2THETA	PEAK	D	H	K	L	I(INT)	I(PK)	I(DS)
92.78	92.78	1.0638	10	3	3	0	0	1
92.78		1.0638	9	6	1	9		2
93.80	93.80	1.0549	10	4	2	0	5	33
96.87	96.86	1.0295	11	2	1	0	1	1
96.87		1.0295	9	6	3	0		1
97.89	97.90	1.0214	10	5	0	7	4	28
100.99	101.00	0.9983	11	3	2	0	1	1
100.99		0.9983	9	7	2	0		1
100.99		0.9983	10	5	3	0		1
105.17	105.18	0.9698	9	6	0	1	0	4
106.23	106.24	0.9630	12	0	0	4	2	17
106.23		0.9630	8	8	4	3		12
108.36	108.36	0.9499	12	2	2	0	1	1
109.44	109.44	0.9435	11	5	1	0	0	1
109.44		0.9435	10	7	0	7	5	34
110.52	110.52	0.9373	10	6	4	4		17
110.52		0.9373	12	2	2	0		2
113.82	113.82	0.9193	10	7	1	0	0	2
113.82		0.9193	11	6	1	0		2
118.36	118.36	0.8969	9	7	3	0	0	2
118.36		0.8969	13	2	1	0		2
123.10	123.10	0.8761	10	7	5	0	0	2
123.10		0.8761	12	4	4	2		9
123.10		0.8761	12	6	0	3		15
124.32	124.32	0.8711	10	8	1	5	3	30
126.83	126.82	0.8613	12	6	2	0		2
126.83		0.8613	10	9	1	0		2
128.11	128.10	0.8566	11	6	5	0	0	2
128.11		0.8566	13	3	2	0		2
128.11		0.8566	12	6	2	0		2
129.41	129.42	0.8519	12	6	3	8	3	46
133.49	133.50	0.8384	10	9	0	0	0	9
134.91	134.92	0.8340	8	8	8	5	1	29
139.40	139.40	0.8212	13	5	2	0	0	2
140.99	141.00	0.8171	10	9	6	0	0	2
146.14	146.14	0.8051	11	9	2	0	0	3

PYROPE - EULER AND BRUCE, 1965

2THETA	PEAK	D	H	K	L	I(INT)	I(PK)	I(DS)
146.14		0.8051	10	9	5	0		3
146.14		0.8051	14	3	1	0		3
146.14		0.8051	11	7	6	0		3
146.14		0.8051	13	6	1	0		3
148.01	148.02	0.8013	12	8	0	6	1	38
152.09	152.10	0.7937	12	8	2	9	2	61
152.09		0.7937	14	4	0	4		30
154.35	154.34	0.7900	14	3	3	0	0	2
154.35		0.7900	13	6	3	1		4
156.82	156.82	0.7863	14	4	2	16	4	109
156.82		0.7863	12	6	6	8		55
156.82		0.7863	10	10	4	8		55

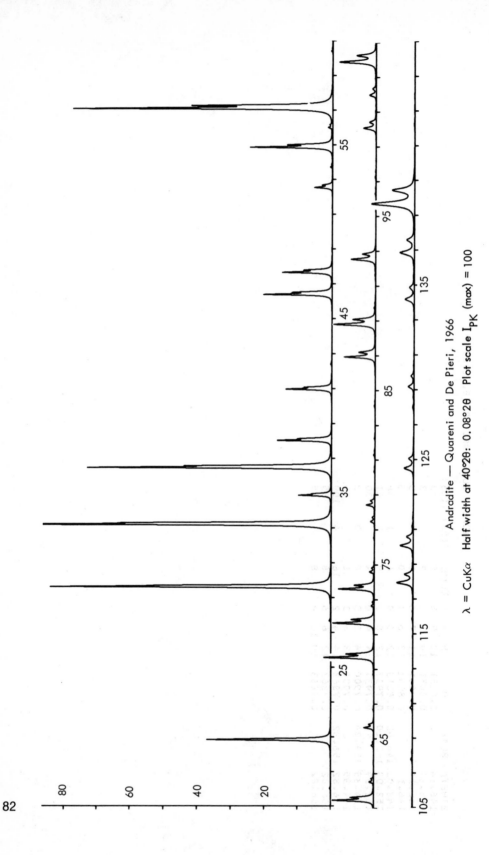

Andradite — Quareni and De Pieri, 1966

$\lambda = CuK\alpha$ Half width at 40°2θ: 0.08°2θ   Plot scale I_{PK} (max) = 100

ANDRADITE - QUARENI AND DE PIERI, 1966

2THETA	PEAK	D	H	K	L	I(INT)	I(PK)	I(DS)
18.00	18.00	4.924	2	1	1	0	0	0
20.81	20.82	4.264	2	2	0	31	37	17
27.65	27.66	3.223	3	2	1	0	0	0
29.60	29.60	3.015	4	0	0	78	84	67
33.19	33.20	2.697	4	2	0	100	100	100
34.86	34.86	2.571	3	3	2	10	10	10
36.46	36.46	2.462	4	2	2	73	73	83
38.01	38.00	2.365	4	3	1	17	16	21
40.95	40.96	2.202	5	2	1	15	14	20
42.36	42.36	2.132	4	4	0	0	0	1
46.37	46.36	1.9566	6	1	1	12	21	19
46.37		1.9566	5	3	2	11		18
47.64	47.64	1.9070	6	2	0	17	15	28
48.90	48.90	1.8611	5	4	1	6	6	12
52.52	52.52	1.7409	4	4	4	1	1	1
53.69	53.70	1.7057	5	4	3	1	1	1
54.84	54.84	1.6726	6	4	0	29	25	60
55.98	55.98	1.6413	5	5	2	0	1	2
55.98		1.6413	6	3	3	3		
57.10	57.10	1.6117	6	4	2	94	78	204
60.38	60.38	1.5317	6	5	1	0	0	1
60.38		1.5317	7	3	0	0		0
61.45	61.44	1.5076	8	0	0	16	12	39
62.51	62.50	1.4846	7	4	1	2	1	4
63.56	63.56	1.4626	8	2	0	0	0	0
64.59	64.60	1.4416	6	5	3	1	1	3
65.63	65.62	1.4214	8	2	2	3	3	9
65.63		1.4214	6	6	0	1		2
66.65	66.64	1.4021	7	5	2	0	0	1
68.67	68.66	1.3656	8	4	0	21	15	63
69.67	69.67	1.3485	8	4	2	18	12	55
71.65	71.66	1.3160	7	6	1	1	1	3
72.63	72.64	1.3006	6	6	4	0	0	0
72.63		1.3006	8	5	3			
73.61	73.60	1.2857	6	6	4	16	11	50
74.58	74.58	1.2713	8	5	1	1	1	3
74.58		1.2713	7	5	4	1	1	3

2THETA	PEAK	D	H	K	L	I(INT)	I(PK)	I(DS)
76.51	76.52	1.2440	8	3	3	0	0	1
77.47	77.48	1.2310	9	4	1	2	1	6
78.43	78.42	1.2183	8	5	1	2	2	7
78.43		1.2183	9	7	1	2		1
80.33	80.32	1.1942	10	1	1	0	0	2
81.28	81.28	1.1827	8	6	2	1	0	2
82.22	82.22	1.1715	9	4	3	0	0	2
84.10	84.10	1.1500	9	6	1	0	0	
86.92	86.92	1.1198	10	4	0	6	9	24
86.92		1.1198	8	6	4	10		40
88.79	88.78	1.1010	10	4	2	22	13	93
89.72	89.72	1.0920	8	7	3	1	0	2
91.59	91.60	1.0745	9	6	3	0	0	1
92.53	92.52	1.0661	8	8	0	14	7	61
95.34	95.34	1.0419	10	5	3	0	0	1
95.34		1.0419	7	6	6	0		0
96.28	96.28	1.0342	8	6	6	1	1	2
97.22	97.22	1.0267	11	4	1	0	0	1
97.22		1.0267	8	7	5	1		5
100.06	100.06	1.0051	12	0	0	7	4	34
100.06		1.0051	8	8	4	1		1
101.01	101.00	0.9982	9	7	2	0	0	1
101.01		0.9914	9	8	1	4		21
101.96	101.96	0.9848	12	2	2	0	2	1
102.92	102.92	0.9848	11	5	2	2	0	1
102.92		0.9783	10	7	1	0		
103.88	103.88	0.9783	12	2	4	11	11	58
103.88		0.9783	10	6	4	12		66
104.84	104.82	0.9719	12	3	1	1	0	1
107.76	107.76	0.9535	12	4	0	0	0	6
110.74	110.74	0.9361	11	6	3	1	1	5
110.74		0.9361	9	7	6	1		4
111.74	111.74	0.9305	10	8	2	1	1	8
112.75	112.74	0.9250	9	8	5	0	0	1
114.79	114.80	0.9143	11	7	4	2	0	1
115.82	115.82	0.9091	12	4	4	1	0	4

ANDRADITE - QUARENI AND DE PIERI, 1966

2THETA	PEAK	D	H	K	L	I(INT)	I(PK)	I(DS)
116.87	116.88	0.9040	12	5	3	0		1
117.92	117.92	0.8990	12	6	0	0	5	21
117.92		0.8990	10	8	4	3		58
118.98	118.96	0.8940	13	3	2	9		1
120.06	120.06	0.8891	12	6	2	0	4	66
121.14	121.14	0.8844	11	8	1	10	0	1
121.14		0.8844	11	7	4	0		5
124.48	124.48	0.8704	8	8	8	1	3	54
125.62	125.62	0.8659	11	8	3	8	0	1
127.96	127.96	0.8571	13	5	2	0	0	4
127.96		0.8571	9	9	6	1		4
129.15	129.16	0.8528	14	2	0	1	2	5
129.15		0.8528	10	6	6	2		13
129.15		0.8528	10	10	0	6		18
130.37	130.36	0.8486	12	7	3	0		3
132.87	132.88	0.8403	14	3	1	0	0	2
134.16	134.16	0.8363	12	8	0	9	3	67
135.48	135.48	0.8323	11	8	5	1	0	4
135.48		0.8323	13	5	4	0		2
136.83	136.82	0.8284	12	8	2	10	4	78
136.83		0.8284	14	4	0	5		35
139.63	139.64	0.8206	10	10	4	11	13	84
139.63		0.8206	12	6	6	10		77
139.63		0.8206	14	4	2	27		210
144.18	144.18	0.8095	14	5	1	0	0	2
144.18		0.8095	13	7	2	1		5
145.81	145.80	0.8059	12	8	4	2	0	3
151.17	151.18	0.7953	15	2	1	4	0	4
151.17		0.7953	14	5	3	1		2
151.17		0.7953	13	6	5	0		5
151.17		0.7953	11	10	3	3		1
153.18	153.16	0.7918	14	6	0	0	0	2
155.33	155.32	0.7885	13	8	1	0	0	2
155.33		0.7885	11	8	7	0		1
160.28	160.28	0.7818	15	3	2	2	0	15
160.28		0.7818	11	9	6	1		8

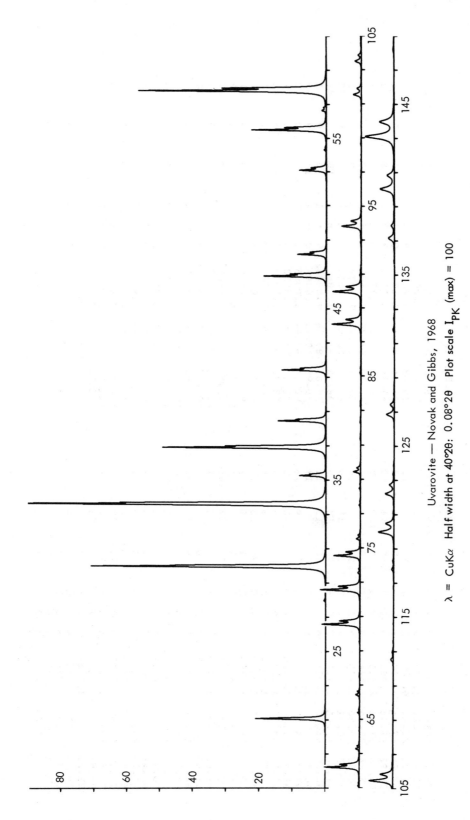

Uvarovite — Novak and Gibbs, 1968

$\lambda = CuK\alpha$ Half width at $40°2\theta$: $0.08°2\theta$ Plot scale I_{PK} (max) = 100

UVAROVITE - NOVAK AND GIBBS, 1968

2THETA	PEAK	D	H	K	L	I(INT)	I(PK)	I(DS)
18-19	18.20	4.872	2	1	1	0	0	0
21-04	21.04	4.219	2	2	0	18	21	10
27-95	27.96	3.189	3	2	1	0	1	0
29-92	29.92	2.983	4	0	0	68	71	60
33-55	33.56	2.669	4	2	0	100	100	100
35-24	35.24	2.544	3	3	2	100	100	100
36-87	36.86	2.436	4	2	2	51	49	58
38-43	38.42	2.340	4	3	1	16	15	19
41-40	41.40	2.179	5	2	1	15	13	19
46-89	46.90	1.9360	6	1	1	12	19	18
46-89		1.9360	5	3	2	11		17
48-18	48.18	1.8869	6	2	2	10	9	16
49-45	49.46	1.8415	5	4	1	0	0	0
53-12	53.12	1.7225	4	4	4	10	8	18
54-31	54.30	1.6877	5	5	0	1	1	1
55-48	55.48	1.6549	6	4	0	28	23	55
56-63	56.62	1.6240	5	5	2	1	1	2
56-63		1.6240	6	3	3	1		1
57-76	57.76	1.5947	6	4	2	71	57	148
61-09	61.08	1.5156	6	5	1	0	0	0
61-09		1.5156	7	3	0	0		0
62-17	62.18	1.4917	8	0	0	14	11	33
63-25	63.24	1.4690	7	4	1	2	1	4
65-37	65.36	1.4264	6	5	3	1	1	2
66-41	66.42	1.4064	6	6	0	0	1	1
66-41		1.4064	8	2	2	0		3
67-45	67.46	1.3873	7	5	4	0	0	1
69-50	69.50	1.3513	7	6	3	0	0	1
70-52	70.52	1.3343	8	4	2	17	12	46
72-53	72.54	1.3021	8	4	4	18	12	52
73-53	73.54	1.2869	9	2	1	0	1	1
73-53		1.2869	7	6	1	1		2
74-52	74.52	1.2722	6	6	4	12	8	36
75-51	75.52	1.2580	8	5	1	1	1	3
75-51		1.2580	7	5	4	1		3
77-48	77.48	1.2309	9	3	2	1	0	2
78-45	78.46	1.2180	8	4	4	1	0	2

2THETA	PEAK	D	H	K	L	I(INT)	I(PK)	I(DS)
79.43	79.42	1.2055	9	4	1	2	2	5
79.43		1.2055	8	5	3	2	0	6
81.36	81.36	1.1816	10	1	1	0	0	1
81.36		1.1816	7	7	2	0	0	1
82.33	82.32	1.1702	10	2	0	0	0	1
83.29	83.28	1.1591	9	4	3	0	0	1
85.21	85.20	1.1379	7	6	5	6	9	2
88.08	88.08	1.1080	10	4	0	0	0	23
88.08		1.1080	8	6	4	6	0	37
89.03	89.04	1.0986	9	4	1	16	9	1
89.03	89.98	1.0894	10	2	2	1	0	63
90.94	90.94	1.0805	8	7	3	0	0	2
92.85	92.86	1.0632	9	6	3	0	0	1
93.81	93.80	1.0548	8	8	0	12	6	48
96.69	96.68	1.0309	10	5	3	0	0	1
97.65	97.64	1.0233	8	6	6	0	0	1
98.61	98.62	1.0159	11	4	6	1	0	1
101.52	101.52	0.9945	12	0	1	0	2	2
101.52		0.9945	8	8	4	5	0	22
102.50	102.50	0.9877	9	8	1	0	0	1
103.48	103.48	0.9810	12	2	1	4	2	19
104.46	104.46	0.9744	10	7	1	0	0	1
105.45	105.44	0.9680	10	6	4	8	7	41
105.45		0.9680	12	2	2	2	0	39
106.44	106.44	0.9617	12	3	3	8	0	1
106.44		0.9617	9	8	1	0	0	1
109.45	109.46	0.9435	12	4	2	0	0	1
112.52	112.52	0.9263	9	9	2	0	0	1
112.52		0.9263	11	6	3	1	1	5
112.52		0.9263	9	7	6	1	0	5
113.56	113.56	0.9207	10	8	2	0	0	1
114.60	114.60	0.9153	9	8	5	0	0	1
116.72	116.72	0.9047	11	7	2	2	1	1
117.80	117.80	0.8996	12	4	4	4	5	9
119.98	119.98	0.8895	10	8	4	9	0	54
119.98		0.8895	12	6	2	3		20
121.09	121.08	0.8846	13	3	3	0	0	2

UVAROVITE - NOVAK AND GIBBS, 1968

2THETA	PEAK	D	H	K	L	I(INT)	I(PK)	I(DS)
122.21	122.20	0.8798	12	6	2	7	3	43
123.34	123.34	0.8750	11	7	4	1	0	4
123.34		0.8750	11	8	1	0		1
126.84	126.84	0.8613	8	8	8	7	2	43
128.05	128.04	0.8568	11	8	3	0	0	2
130.51	130.52	0.8481	13	5	2	0	0	1
130.51		0.8481	9	9	6	0		2
131.78	131.78	0.8439	10	10	0	1	1	8
131.78		0.8439	10	8	6	0		3
133.08	133.08	0.8397	12	7	3	0	0	1
135.75	135.76	0.8315	11	7	6	0	0	1
135.75		0.8315	14	3	1	0		2
137.13	137.14	0.8275	12	8	0	7	2	46
138.56	138.56	0.8235	11	8	5	0	0	2
138.56		0.8235	13	5	4	0		1
140.02	140.02	0.8196	14	4	0	5	4	37
140.02		0.8196	12	8	2	11		77
143.09	143.10	0.8120	14	4	2	21	9	155
143.09		0.8120	10	10	4	7		53
143.09		0.8120	12	6	6	7		51
148.17	148.16	0.8010	13	7	2	1	0	9
148.17		0.8010	11	10	1	0		1
152.00	152.00	0.7938	12	9	0	0	0	1
156.38	156.38	0.7869	14	5	3	0	0	2
156.38		0.7869	15	2	1	1		7
156.38		0.7869	11	10	3	0		1
156.38		0.7869	13	6	5	1		8
158.89	158.82	0.7835	14	6	0	0	0	1
161.73	161.72	0.7801	11	8	7	0	0	1
161.73		0.7801	13	8	1	0		3

TABLE 4. CHONDRODITE - HUMITE GROUP

Variety	Norbergite	Humite	Chondrodite	Clinohumite
Composition	$Mg_2SiO_4 \cdot MgF_2$	$3Mg_2SiO_4 \cdot Mg(OH,F)_2$	$2Mg_2SiO_4 \cdot Mg(OH,F)_2$	$4Mg_2SiO_4 \cdot Mg(OH,F)_2$
Source	Nicol quarry, Franklin, N.J.	not given	Tilly Foster Mine, New York	Vesuvius, Italy
Reference	Gibbs & Ribbe, 1969	Taylor & West, 1928	Taylor & West, 1928	Taylor & West, 1928
Cell Dimensions				
$\underline{a}$ $\mathring{A}$	4.7104	4.748 (4.738 kx)	10.29 (10.27 kx)	10.29 (10.27 kx)
$\underline{b}$ $\mathring{A}$	10.272	10.25 (10.23 kx)	4.742 (4.733 kx)	4.755 (4.745 kx)
$\underline{c}$	8.7476	20.90 (20.86 kx)	7.89 (7.87 kx)	13.71 (13.68 kx)
α	90	90	90	90
β deg	90	90	109.03	100.825
γ	90	90	90	90
Space Group	Pbnm	Pbnm	$P2_1/a$	$P2_1/a$
Z	4	4	2	2

TABLE 4. (cont.)

Variety	Norbergite	Humite	Chondrodite	Clinohumite
Method	3-D, counter	2-D, film	2-D, film	2-D, film
R & Rating	0.052 (1)	--- (3)	--- (2)	--- (2)
Cleavage and habit	{001} poor Varied habit	{001} distinct Elongated to $\underline{a}$ or $\underline{c}$ axis	{001} poor Varied habit	{001} poor Varied habit
Comment		B estimated	B estimated	B estimated
μ	92.6	95.2	92.8	96.3
ASF	0.7389×10^{-1}	0.5554×10^{-1}	0.3134×10^{-1}	0.5601×10^{-1}
Abridging factors	1:0:0	1:1:0.5	1:1:0.5	1:1:0.5

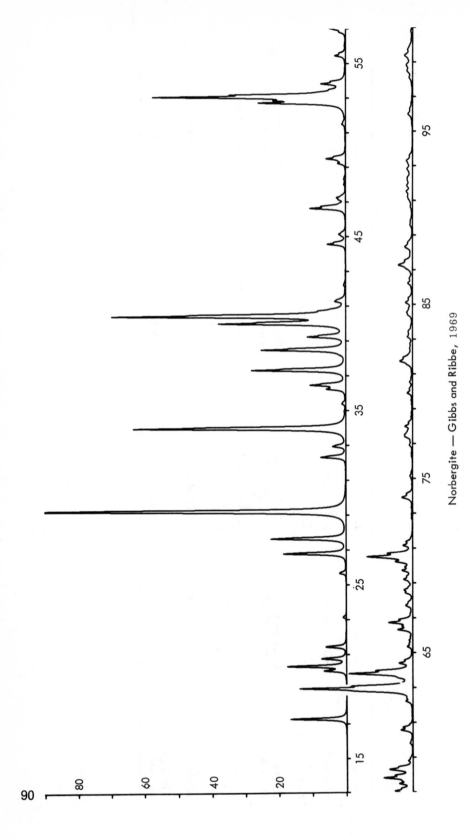

Norbergite — Gibbs and Ribbe, 1969

λ = CuKα Half width at 40°2θ: 0.11°2θ Plot scale I$_{PK}$ (max) = 100

NORBERGITE — GIBBS AND RIBBE, 1969

2THETA	PEAK	D	H	K	L	I(INT)	I(PK)	I(DS)
17.25	17.26	5.136	0	0	1	14	17	13
20.03	20.04	4.429	0	2	0	5	7	5
20.29	20.30	4.374	0	1	1	15	18	14
20.74	20.74	4.282	1	0	0	6	7	6
21.41	21.42	4.147	1	1	0	5	6	5
25.64	25.64	3.471	1	0	2	2	2	2
26.75	26.76	3.330	0	2	2	18	19	18
27.62	27.62	3.227	1	1	2	22	22	22
29.16	29.16	3.060	1	2	0	100	100	100
32.29	32.30	2.770	0	0	2	8	8	8
32.91	32.92	2.719	0	2	2	4	4	4
33.92	33.92	2.640	2	1	1	68	63	69
35.37	35.38	2.536	0	2	2	1	1	1
36.20	36.20	2.479	1	0	3	6	6	6
36.43	36.44	2.464	0	4	1	11	11	11
37.28	37.28	2.410	1	3	1	31	28	32
38.44	38.44	2.340	1	3	3	28	25	29
39.21	39.22	2.296	2	1	0	12	11	13
39.95	39.96	2.255	1	4	0	42	38	44
40.36	40.36	2.233	1	2	3	79	70	83
40.71	40.70	2.215	2	4	0	4	8	4
41.25	41.26	2.187	0	0	4	3	3	3
44.54	44.54	2.033	2	1	2	7	5	7
45.02	45.02	2.012	0	4	4	1	2	1
45.11	45.12	2.008	2	3	3	2	2	2
46.59	46.60	1.9476	1	1	4	13	11	14
46.77	46.72	1.9405	2	0	3	2	7	2
47.12	47.12	1.9272	0	3	4	2	2	2
47.23	47.22	1.9229	2	2	3	2	3	2
49.20	49.20	1.8503	1	2	4	3	2	3
49.46	49.46	1.8409	1	5	1	7	6	8
51.48	51.48	1.7737	2	3	2	1	1	1
52.69	52.70	1.7357	2	4	0	32	26	36
52.89	52.84	1.7296	1	5	3	3	2	3
53.02	53.02	1.7257	2	3	0	72	57	81
53.48	53.46	1.7120	0	6	1	1	5	1
53.80	53.80	1.7025	2	4	1	9	7	10
55.43	55.44	1.6561	0	2	5	4	2	5
56.80	56.80	1.6195	1	1	3	2	2	2
56.95	57.04	1.6154	2	3	3	2	5	2
57.03		1.6133	2	4	0	6		7
57.45	57.46	1.6026	0	6	1	5	4	5
57.78	57.78	1.5942	1	1	5	11	9	13
58.21	58.26	1.5834	2	6	0	3	7	3
58.25		1.5825	1	6	1	1		2
58.28		1.5819	1	5	4	6		7
58.77	58.76	1.5698	1	4	4	2	1	2
59.79	59.80	1.5454	3	0	1	1	1	1
60.53	60.54	1.5282	1	1	5	5	4	6
60.70	60.70	1.5245	2	5	1	1	3	1
62.19	62.20	1.4915	2	4	4	2	2	3
62.76	62.90	1.4791	1	3	3	16	34	19
62.90		1.4763	0	6	2	40		48
63.71	63.78	1.4594	2	5	0	2		3
63.78		1.4579	0	4	4	25	19	31
64.38	64.38	1.4459	0	4	5	5	4	6
65.32	65.32	1.4272	3	3	2	2	2	2
65.69	65.68	1.4202	3	2	1	1	1	1
66.30	66.30	1.4086	3	3	3	6	4	8
66.70	66.70	1.4010	1	7	0	11	7	13
67.67	67.70	1.3834	1	7	1	2	2	2
67.72		1.3824	3	0	3	1		2
68.41	68.42	1.3701	3	1	3	3	2	4
68.57	68.58	1.3674	2	5	3	2	3	2
69.18	69.18	1.3568	3	2	2	3	3	3
69.69	69.70	1.3481	0	6	4	4	3	4
70.20	70.20	1.3396	3	4	0	7	5	8
70.48	70.48	1.3349	2	4	5	20	14	25
71.14	71.14	1.3241	3	4	1	4	3	5
73.88	73.88	1.2817	1	8	0	5	3	6
76.89	76.90	1.2388	1	8	2	1	1	1
77.38	77.40	1.2322	2	4	3	1	2	1
77.39		1.2320	0	8	1	2	2	2
77.49	77.48	1.2307	2	1	6	6	2	3

NORBERGITE – GIBBS AND RIBBE, 1969

2THETA	PEAK	D	H	K	L	I(INT)	I(PK)	I(DS)
77.97	77.98	1.2243	1	4	6	4	2	5
79.89	79.90	1.1997	3	5	2	2	2	3
80.52	80.52	1.1919	1	8	2	1	1	2
81.52	81.70	1.1797	1	7	4	2	4	3
81.70		1.1776	4	0	0	6		8
81.85		1.1758	1	1	7	1		2
83.12	83.12	1.1611	3	0	5	3	2	3
83.46	83.40	1.1572	1	6	0	1	1	2
84.38	84.52	1.1469	3	5	3	2	2	3
84.52		1.1454	2	7	3	2		3
85.06	85.08	1.1394	3	2	5	2	2	2
85.10		1.1391	1	3	7	2		3
86.19	86.20	1.1273	2	8	0	3	1	4
87.08	87.26	1.1181	2	0	1	1	4	3
87.25		1.1164	2	4	6	8		11
88.29	88.30	1.1059	3	3	5	4	2	6
89.55	89.56	1.0936	1	7	5	1	1	2
90.60	90.62	1.0836	3	5	4	1	2	1
91.08	91.22	1.0792	4	3	2	2	2	2
91.22	91.50	1.0779	1	8	4	2	2	3
91.51		1.0752	1	9	2	2	2	3
91.85	91.84	1.0721	3	7	0	2	2	3
92.30	92.30	1.0680	4	2	3	2	2	3
92.80	92.80	1.0636	3	4	5	3	2	4
93.28	93.26	1.0594	1	1	8	4	2	5
95.97	95.96	1.0368	1	9	3	3	1	5
98.46	98.46	1.0171	2	3	8	2	2	4
98.84	98.80	1.0142	2	4	7	2	2	2
99.37	99.36	1.0102	1	7	6	6	2	9
99.92	99.70	1.0060	0	8	8	1	2	2
100.75	100.76	1.0000	0	10	2	3	1	5
101.60	101.60	0.9940	3	4	0	1	1	2
101.83	101.90	0.9923	4	3	4	2	1	2
102.27	102.26	0.9893	1	9	4	4	2	6
102.68	102.66	0.9864	3	4	6	6	1	6
106.04	106.04	0.9642	4	5	3	4	1	2
108.71	108.76	0.9478	1	1	9	1	2	2

2THETA	PEAK	D	H	K	L	I(INT)	I(PK)	I(DS)
108.74	108.82	0.9476	4	8	5	2		4
108.82		0.9472	4	6	2	2		2
110.73	110.74	0.9361	2	10	1	3	1	5
111.33	111.16	0.9328	5	1	1	2	1	3
112.25	112.26	0.9278	5	6	7	1	1	3
112.72	112.70	0.9252	2	4	8	2	1	2
113.41	113.58	0.9215	0	6	8	1		3
113.58		0.9206	4	6	3	2	3	3
113.61		0.9205	2	10	5	7		11
114.22	114.44	0.9173	5	0	2	1		2
114.45		0.9161	4	11	0	2	2	3
114.47		0.9160	1	1	6	5		8
116.38	116.38	0.9064	3	6	6	1	1	2
118.56	118.56	0.8960	2	10	3	1	1	3
119.01	119.06	0.8939	3	1	8	2	1	2
119.46	119.46	0.8918	2	8	6	1	1	4
120.57	121.00	0.8869	4	6	4	2		2
120.81		0.8858	0	6	5	1		3
120.99		0.8850	2	2	9	2		8
122.12	122.12	0.8801	3	9	3	5	1	8
125.10	125.18	0.8680	3	3	8	2	1	4
125.29		0.8672	5	5	3	1		2
125.91	126.22	0.8648	2	10	4	1		2
126.20		0.8637	3	7	6	3		4
126.43		0.8628	4	4	6	1		2
126.61		0.8622	5	1	4	1		2
128.18	128.24	0.8563	5	5	0	1	1	2
128.27		0.8560	0	12	0	2		3
129.41	129.46	0.8519	0	12	1	2	1	4
129.60		0.8513	4	8	1	1		2
130.40	130.52	0.8485	2	6	5	1	1	2
130.53		0.8480	4	4	9	1		2
130.82	131.38	0.8471	3	2	8	1	1	2
131.33		0.8453	4	2	7	1		2
131.37		0.8452	0	0	6	6		11
131.90	131.92	0.8435	3	10	9	6	2	6
132.85	132.82	0.8404	5	5	2	4	1	6

NORBERGITE - GIBBS AND RIBBE, 1969

2THETA	PEAK	D	H	K	L	I(INT)	I(PK)	I(DS)
136.56	136.58	0.8291	2	10	5	5	1	9
137.38	137.34	0.8268	5	1	5	4	1	8
138.20	138.14	0.8245	3	10	1	1	1	2
139.01	139.36	0.8223	3	5	8	1	2	2
139.26		0.8216	5	5	3	5		9
139.40		0.8213	3	8	6	7		12
139.87	139.90	0.8200	2	0	10	2	2	4
141.45	141.46	0.8160	4	9	1	1	1	2
141.47		0.8159	3	2	9	6		11
147.35	147.42	0.8026	3	11	0	2	0	3
148.65	148.62	0.8000	3	10	4	3	1	6
150.02	149.96	0.7974	5	5	4	2	1	5
151.47	151.38	0.7948	3	6	8	3	1	6
153.72	153.94	0.7910	4	5	7	2	0	4
154.62	154.72	0.7895	5	7	1	3	1	5
157.11	157.20	0.7859	0	2	11	3	0	5
157.83		0.7849	2	8	8	2		3
159.47	159.96	0.7828	6	1	0	2	0	3
159.70		0.7825	1	10	7	1		2
161.67	162.86	0.7802	4	3	8	2	1	4
161.81		0.7801	5	7	2	2		4
162.55		0.7793	1	13	0	1		3
162.84		0.7790	0	6	10	3		5
163.10		0.7787	1	9	8	6		11
163.91		0.7779	3	8	7	1		2
164.22		0.7776	2	11	5	3		5
164.37		0.7775	4	8	5	1		3

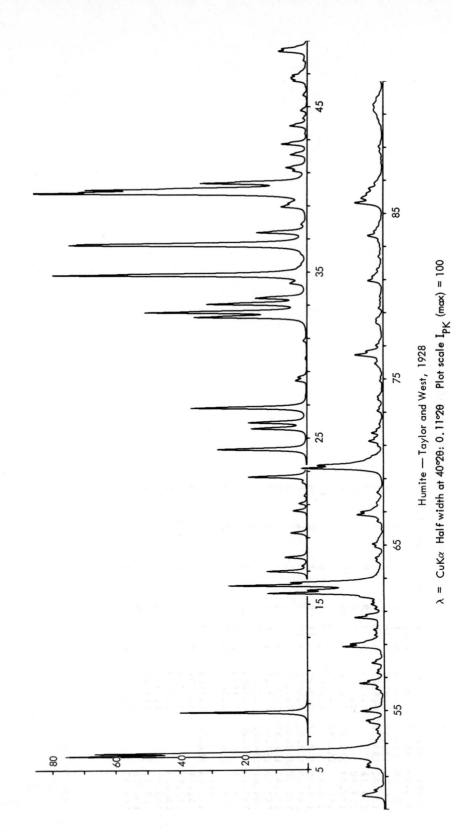

Humite — Taylor and West, 1928

λ = CuKα Half width at 40°2θ: 0.11°2θ Plot scale I$_{PK}$ (max) = 100

94

HUMITE – TAYLOR AND WEST, 1928

2THETA	PEAK	D	H	K	L	I(INT)	I(PK)	I(DS)
8.45	8.46	10.450	0	0	2	22	40	19
16.95	16.96	5.225	0	0	4	8	13	7
17.80	17.82	4.978	1	1	1	5	7	4
19.27	19.28	4.601	0	2	2	3	5	3
20.60	20.60	4.308	1	1	3	3	4	3
21.04	21.04	4.220	1	0	1	3	3	1
22.64	22.64	3.923	1	1	0	13	18	11
24.27	24.30	3.664	1	1	3	8	28	7
24.31		3.659	0	0	6	14	17	12
25.55	25.56	3.483	1	2	1	12	18	11
25.91	25.92	3.436	1	1	4	13	18	12
26.80	26.80	3.324	1	2	5	27	36	24
28.42	28.42	3.137	1	1	5	3	4	2
28.63	28.64	3.115	1	3	0	2	3	2
32.25	32.26	2.773	1	1	6	27	35	24
32.54	32.54	2.749	1	0	1	40	51	36
33.04	33.04	2.709	1	1	0	25	31	22
33.40	33.40	2.680	1	3	6	11	16	10
33.46		2.676	1	2	2	2		1
34.29	34.30	2.612	0	0	7	3	5	3
34.74	34.80	2.580	0	2	8	3	80	3
34.79		2.577	1	2	7	66		60
36.59	36.60	2.454	1	1	3	54	75	49
36.65		2.450	1	3	4	25		23
37.36	37.36	2.405	0	4	3	13	16	12
37.86	37.86	2.374	2	0	0	1	2	1
38.91	38.92	2.313	2	1	0	4	8	4
38.92		2.311	1	3	5	2		2
39.12	39.14	2.301	0	4	4	2	6	2
39.15		2.299	2	1	1	2		2
39.73	39.74	2.267	1	2	7	83	95	77
39.94	39.94	2.255	1	1	8	50	70	47
40.34	40.34	2.234	2	0	3	27	33	26
41.09	41.10	2.195	2	1	3	3	4	3
41.29	41.28	2.185	1	4	5	5	7	5
42.08	42.08	2.145	1	4	3	4	5	4
42.72	42.72	2.115	2	1	4	7	8	7
43.82	43.82	2.064	0	4	6	5	6	5
44.27	44.28	2.044	1	1	9	1	1	1
44.74	44.74	2.024	2	1	5	2	2	2
45.67	45.68	1.9847	1	4	5	1	1	1
46.54	46.54	1.9496	2	3	0	4	4	4
46.67	46.68	1.9445	1	0	7	1	5	1
46.76	46.76	1.9412	2	3	1	3	5	3
46.91	46.90	1.9353	0	2	10	2	5	2
47.13	47.12	1.9268	2	1	6	1	2	1
48.32	48.36	1.8821	1	1	0	4	9	4
48.36		1.8804	1	1	10	5		5
48.52	48.50	1.8745	1	5	1	4	8	4
49.80	49.82	1.8294	0	4	8	6	7	6
49.88		1.8266	2	3	4	2		2
50.17	50.16	1.8169	1	5	3	3	4	3
51.57	51.58	1.7707	1	5	4	5	6	5
51.78	51.72	1.7640	1	0	11	4	6	4
52.33	52.32	1.7469	2	0	11	5		5
52.50	52.48	1.7415	2	2	7	2		2
53.18	53.18	1.7208	2	4	0	100	100	100
53.84	53.84	1.7012	2	3	9	42	91	42
54.25	54.26	1.6895	0	6	6	1	3	1
54.37	54.38	1.6860	2	3	3	2	3	2
54.97	54.96	1.6691	0	6	2	4	5	4
56.31	56.32	1.6324	2	3	10	3	6	3
56.64	56.64	1.6237	0	6	7	8	8	8
57.26	57.26	1.6076	2	6	4	2	3	2
57.87	57.86	1.5921	0	5	5	8	8	8
58.87	58.86	1.5674	2	1	7	2	2	2
59.00	59.02	1.5641	1	3	11	4	4	4
59.07		1.5625	3	1	0	14	13	14
59.57	59.56	1.5506	2	1	8	2	11	2
59.88	59.88	1.5433	2	2	3	2		2
60.18	60.18	1.5364	3	0	3	3	3	3
60.58	60.62	1.5271	1	6	4	1	3	1
60.62		1.5262	0	4	8	2	2	2
60.62		1.5261	3	1	3	4	9	4

95

HUMITE – TAYLOR AND WEST, 1928

2THETA	PEAK	D	H	K	L	I(INT)	I(PK)	I(DS)
61.14	61.14	1.5145	2	5	3	2	3	2
61.42	61.42	1.5083	3	2	1	3	4	3
61.87	61.88	1.4984	3	1	4	5	8	5
62.12	62.12	1.4929	0	0	14	39	37	42
62.38	62.30	1.4874	2	5	4	3	24	3
62.59	62.60	1.4828	0	4	7	54	49	57
64.03	64.02	1.4530	1	4	11	1	2	2
64.22	64.22	1.4491	2	4	8	2	3	2
64.87	64.88	1.4361	3	3	0	3	3	3
65.04	65.04	1.4327	3	3	1	2	4	2
66.41	66.42	1.4065	3	3	3	2	2	2
66.80	66.82	1.3993	1	7	0	7	8	8
66.84		1.3986	1	5	10	2		
66.97	66.98	1.3961	1	7	1	3	7	3
67.55	67.54	1.3855	3	1	7	1	2	1
68.16	68.16	1.3746	2	6	2	2	2	2
69.63	69.64	1.3490	3	2	7	30	26	33
69.78	69.80	1.3465	3	1	0	10	21	11
70.97	70.98	1.3269	0	6	10	1	1	1
71.23	71.24	1.3227	3	4	3	3	2	3
71.74	71.74	1.3145	3	3	14	2	4	5
72.27	72.26	1.3062	0	0	16	4	1	1
73.08	73.08	1.2937	0	2	15	1	2	2
73.74	73.74	1.2838	2	4	11	2	2	2
75.77	75.92	1.2543	2	1	14	1		
75.92		1.2523	3	1	10	2	3	1
76.38	76.44	1.2458	2	5	10	1	9	6
76.44		1.2450	1	3	15	1		6
76.45		1.2448	1	4	14	5		6
77.02	77.02	1.2370	1	8	0	2	2	3
78.33	78.42	1.2196	0	6	12	2	2	2
78.44		1.2182	3	5	4	2		2
78.60	78.58	1.2160	3	0	11	1	2	2
79.26	79.26	1.2076	3	1	11	2	1	1
79.72	79.72	1.2018	2	5	11	1	1	1
80.92	80.92	1.1870	4	0	0	6	5	7

2THETA	PEAK	D	H	K	L	I(INT)	I(PK)	I(DS)
81.06	81.58	1.1853	2	3	14	1		2
81.60	82.38	1.1788	3	5	6	2	2	2
82.38	82.38	1.1696	1	5	14	2	2	2
83.61	83.64	1.1555	2	6	10	1	5	2
83.64	84.08	1.1552	3	5	7	5		6
84.09	84.08	1.1501	2	7	7	1		1
85.62	85.62	1.1334	2	4	14	13	9	16
85.98	85.90	1.1296	3	5	8	3	7	3
86.03		1.1291	4	2	4	3		3
86.18	86.18	1.1275	2	8	0	3	5	6
86.52	86.48	1.1239	1	8	17	3	4	4
86.92	86.78	1.1198	1	5	15	1	3	1
89.03	89.02	1.0986	3	4	11	1	1	1
90.47	90.48	1.0848	2	6	12	1	1	2
90.77	90.78	1.0821	3	6	7	1	2	1
91.17	91.20	1.0783	4	2	0	3	3	4
91.31		1.0771	4	4	0	2		3
91.55	91.56	1.0748	3	7	1	2		2
91.71		1.0734	3	0	1	1	3	1
91.97	91.90	1.0710	1	3	18	2	2	2
93.74	93.74	1.0554	1	9	6	1	1	1
95.77	95.78	1.0383	1	9	7	2	4	9
96.19	96.10	1.0350	3	0	14	2	3	2
97.59	97.60	1.0238	0	10	3	2	2	3
97.96	97.96	1.0209	1	9	8	4	3	6
98.12		1.0196	1	9	8	3		4
99.15	99.14	1.0118	4	3	10	1	1	2
100.75	100.78	1.0000	3	4	15	1	4	1
100.77		0.9999	3	4	14	6		8
100.81		0.9996	1	9	9	1		1
101.33	101.32	0.9958	3	8	0	5	3	6
102.55	102.54	0.9873	1	7	15	1	1	2
103.83	103.86	0.9786	1	9	10	1	1	1
104.56	104.58	0.9737	4	6	1	1	1	1
105.00	105.20	0.9709	3	0	17	1	2	1
105.18		0.9697	1	1	21	2		3
105.22		0.9695	0	10	7	1		2

HUMITE — TAYLOR AND WEST, 1928

2THETA	PEAK	D	H	K	L	I(INT)	I(PK)	I(DS)
105.67	105.58	0.9666	3	1	17	1	2	1
106.17	106.14	0.9634	4	2	12	1	1	1
106.66	106.64	0.9603	0	6	18	1	1	2
107.12	107.18	0.9575	3	8	6	2	1	1
107.20		0.9569	1	2	21	1		2
107.24		0.9567	2	7	14	1		1
107.93	107.94	0.9525	1	8	14	2	1	2
108.37	108.36	0.9499	1	10	7	4	2	6
109.10	109.24	0.9456	5	1	0	1	1	1
109.26		0.9446	5	1	1	1		2
110.04	110.06	0.9401	2	10	1	2	1	2
110.59	110.56	0.9370	5	1	3	2	2	3
110.84	110.98	0.9355	3	7	11	1	1	1
111.74	112.04	0.9305	3	8	8	1	4	1
111.75		0.9304	5	1	4	1		2
111.83		0.9300	2	4	19	1		1
111.90		0.9296	2	5	18	1		1
112.00		0.9291	4	0	14	6		8
112.03		0.9289	2	7	15	1		1
112.25	112.44	0.9277	1	6	22	2	5	3
112.45		0.9267	4	6	7	10		14
114.68	114.78	0.9149	5	3	0	1	1	1
114.78		0.9144	1	11	0	1		1
116.07	116.12	0.9079	3	4	17	1	1	1
116.98	117.00	0.9035	2	2	21	9	3	13
117.00		0.9034	2	9	11	1		1
117.76	117.74	0.8997	2	8	14	5	3	8
118.23	118.22	0.8975	2	10	7	3	2	5
119.09	119.12	0.8935	1	7	18	1	1	2
119.44	119.68	0.8919	3	6	15	1	2	1
119.61		0.8911	5	2	7	4		5
119.70		0.8908	0	4	22	1		1
119.77		0.8904	5	4	0	2		3
119.82		0.8902	2	6	18	1		2
121.44	121.46	0.8831	3	9	7	2	1	4
123.41	123.80	0.8748	5	3	7	1	2	2
123.53		0.8743	1	11	7	1		1

2THETA	PEAK	D	H	K	L	I(INT)	I(PK)	I(DS)
123.73	124.00	0.8735	4	4	14	3		4
124.02		0.8723	3	7	14	2	2	4
125.92	126.10	0.8648	1	7	19	1	2	1
126.05		0.8643	0	10	13	1		2
126.12		0.8640	4	8	3	2		4
126.78	127.20	0.8615	5	1	10	2	3	2
127.20		0.8599	0	6	21	10		16
128.78	128.82	0.8542	0	12	0	2	1	3
128.82		0.8540	2	3	22	1		1
129.26	129.34	0.8525	4	8	5	2	1	3
129.42		0.8519	3	4	19	1		1
129.51		0.8516	3	5	18	1		1

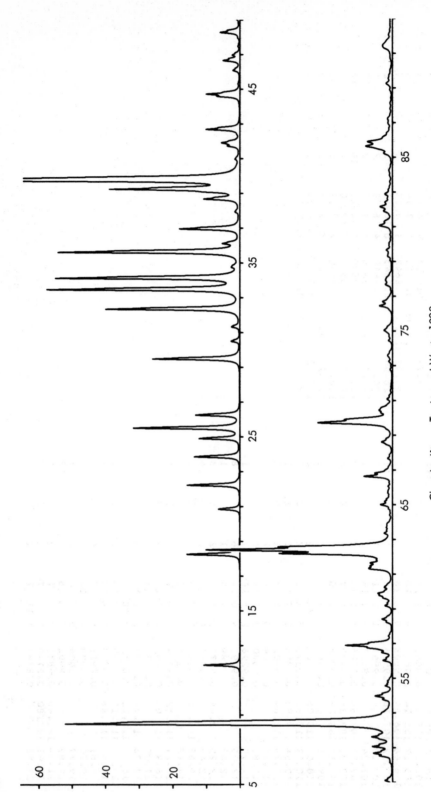

Chondrodite — Taylor and West, 1928

λ = CuKα Half width at 40°2θ: 0.11°2θ Plot scale I$_{PK}$ (max) = 100

98

CHONDRODITE – TAYLOR AND WEST, 1928

2THETA	PEAK	D	H	K	L	I(INT)	I(PK)	I(DS)
11.85	11.86	7.459	0	0	1	13	11	12
18.22	18.24	4.864	-2	0	1	22	16	21
20.82	20.82	4.263	1	1	0	9	6	9
22.19	22.20	4.002	0	1	1	23	16	22
23.84	23.84	3.729	0	0	2	20	14	19
24.88	24.88	3.576	-2	0	2	6	12	6
24.88		3.575	2	0	1	12		12
25.48	25.48	3.492	1	1	1	49	32	48
26.22	26.22	3.395	-2	1	1	14	13	6
26.22		3.395	1	1	0	42		14
29.46	29.46	3.029	-1	1	2	4	26	41
30.47	30.48	2.931	0	1	2	4	3	4
31.30	31.30	2.855	-2	1	2	68	40	68
32.32	32.32	2.767	-3	1	1	100	58	100
33.45	33.46	2.677	3	1	0	95	55	95
34.10	34.10	2.627	-1	1	2	1	3	1
34.73	34.74	2.581	-2	0	3	98	54	99
34.73		2.581	2	0	2	7		8
35.60	35.60	2.519	-3	1	2	7	5	7
36.09	36.10	2.486	0	0	3	26	18	26
36.93	36.94	2.432	-4	0	2	2		2
36.93		2.432	4	0	0	18	1	19
37.91	37.92	2.371	0	2	1	65	11	67
38.64	38.64	2.328	3	1	3	74	39	76
39.22	39.22	2.295	-1	1	3	74	91	76
39.73	39.74	2.267	-2	1	3	93		96
39.73		2.267	2	1	2	2	100	2
39.83	39.82	2.261	-4	1	1	5	2	5
40.95	40.96	2.202	0	1	3	1	4	1
41.70	41.70	2.164	-4	1	2	9	6	9
41.70		2.164	4	1	0	19	10	20
41.89	41.90	2.155	1	2	1	12	7	12
42.66	42.66	2.118	-4	0	3	13	10	14
44.57	44.58	2.031	-1	2	2	3	2	3
44.70	44.70	2.025	1	1	3	1		1
46.06	46.06	1.9689	-2	0	4			
46.06	46.06	1.9689	2	0	3			

2THETA	PEAK	D	H	K	L	I(INT)	I(PK)	I(DS)
46.63	46.62	1.9462	-3	2	1	10	5	11
46.95	46.96	1.9335	4	1	1	4	2	4
48.26	48.26	1.8841	-5	1	2	12	6	13
49.07	49.08	1.8548	-5	1	2	1	2	4
50.12	50.12	1.8184	2	1	4	1	1	1
50.50	50.50	1.8057	-1	1	0	7	4	7
50.67	50.66	1.7999	5	1	2	8	6	9
51.04	51.04	1.7877	4	0	1	11		12
51.48	51.50	1.7735	3	1	4	6	4	5
51.60	51.60	1.7696	-3	1	2	8	6	9
52.35	52.36	1.7461	-2	2	3	92	97	101
52.35		1.7461	2	2	3	92		101
52.44		1.7434	-4	1	1	70		77
53.00	52.98	1.7261	-5	0	3	2	3	3
53.75	53.76	1.7038	-6	0	2	4	3	4
53.76		1.7038	-6	0	2	1		1
53.96	53.96	1.6977	-4	2	2	3		3
54.54	54.54	1.6810	-3	2	3	6	5	7
54.83	54.82	1.6728	-4	1	4	4	2	5
56.01	56.00	1.6405	5	2	3	6	2	3
56.47	56.48	1.6282	-1	0	3	2	3	7
56.73	56.74	1.6213	-6	0	3	6	3	6
56.73		1.6213	6	1	0	5		6
56.88	56.88	1.6174	3	2	1	7	8	8
57.42	57.42	1.6034	-6	2	3	24	14	27
58.21	58.22	1.5835	3	0	4	1	1	2
58.38	58.38	1.5794	-4	3	1	3	2	4
58.45		1.5777	2	3	0	3	3	5
59.12	59.16	1.5602	-5	2	1	4		1
59.62	59.62	1.5495	0	3	1	1		5
59.75	59.76	1.5463	-1	3	1	5	2	5
59.88	59.88	1.5434	-5	2	3	4	3	5
60.21	60.22	1.5356	-6	1	3	6	4	7
60.28		1.5341	1	3	1	3	4	4
61.28	61.30	1.5114	1	1	3	11	2	13
61.35		1.5098	-4	0	5	5	7	6

99

2THETA	PEAK	D	H	K	L	I(INT)	I(PK)	I(DS)
61.35		1.5098	4	0	3	2		2
61.46	61.46	1.5074	-1	2	4	3	7	3
61.61	61.62	1.5040	5	2	0	4	7	5
61.65		1.5033	-2	3	1	1		2
61.65		1.5033	2	0	0	4		5
62.17	62.18	1.4918	0	0	5	70	34	81
62.40	62.40	1.4869	-6	0	4	51	55	60
62.40		1.4868	6	0	1	51		60
62.43		1.4862	-3	2	4	3		3
62.54	62.56	1.4839	-3	1	5	15	34	18
63.35	63.34	1.4669	-1	3	2	4	2	5
63.68	63.68	1.4602	-5	2	3	3	2	3
63.78		1.4580	5	1	2	1	2	1
63.91	63.86	1.4553	0	2	2	2		2
64.39	64.38	1.4457	-2	3	2	1	1	2
64.74	64.76	1.4386	4	1	1	1	1	3
64.98	64.98	1.4339	-3	3	3	8	4	9
65.31	65.30	1.4274	4	2	0	3	2	3
65.66		1.4208	3	3	2	2	1	3
66.05	66.04	1.4133	-7	1	2	21	8	25
66.62	66.62	1.4025	-6	2	1	3		3
67.66	67.66	1.3836	-5	1	5	3	3	4
68.46		1.3693	3	1	4	6		8
68.58	68.58	1.3671	-2	3	3	24	22	29
69.70	69.70	1.3480	-2	3	3	24		29
69.70		1.3480	-4	2	1	24		29
69.77		1.3468	-6	2	3	19		23
70.27	70.28	1.3383	-6	0	5	3	4	4
70.27		1.3383	6	2	0	2		3
70.41	70.42	1.3361	-6	1	0	3	4	4
70.41		1.3361	6	2	1	2		4
70.56		1.3336	7	0	2	3		3
71.07	71.06	1.3253	-4	3	0	3	2	3
71.07		1.3253	4	3	3	1		1
71.56	71.56	1.3174	-3	3	3	1	1	2
72.75	72.76	1.2987	-1	2	5	2	1	2
73.23	73.22	1.2915	1	3	3	2	1	3
73.53	73.56	1.2869	5	2	2	2	1	2
73.63		1.2854	5	1	3	1		1
74.43	74.44	1.2735	-4	1	5	2	1	3
74.43		1.2735	4	2	3	1		1
74.90		1.2667	4	3	1	1		3
75.02	75.02	1.2650	-3	1	6	5	2	3
76.42	76.42	1.2452	-4	1	6	5		6
76.43		1.2452	-1	4	1	5		6
76.70	76.66	1.2414	-8	1	3	5	3	4
77.65	77.66	1.2286	-1	3	4	3	3	6
77.78	77.92	1.2268	5	3	0	2	2	2
77.94		1.2247	-5	2	5	2		6
78.06		1.2231	3	2	4	2		3
79.06	79.06	1.2102	-1	2	5	1	1	2
79.29	79.30	1.2073	-5	1	6	2		3
79.41		1.2058	0	4	0	2		4
79.65	79.64	1.2026	-5	3	3	3	2	4
80.05	80.04	1.1977	-7	1	5	2	1	3
80.52	80.52	1.1918	6	0	3	3	1	3
81.04	81.04	1.1855	0	4	0	11	4	14
81.15		1.1842	-4	3	4	2		2
81.88	81.88	1.1754	3	1	6	6	3	8
82.13	82.12	1.1725	-1	3	5	7	4	9
82.42	82.38	1.1691	-7	2	4	2	2	2
82.61	82.62	1.1669	1	3	5	2	1	2
82.86	82.86	1.1640	3	3	4	1	2	3
82.86		1.1640	-6	2	5	2		2
82.86		1.1640	6	2	2	2		2
82.90		1.1635	5	3	2	2		2
83.32	83.30	1.1588	-6	3	1	1	1	2
83.94	83.94	1.1518	-2	4	4	1	1	2
84.09	84.24	1.1501	7	1	2	1	1	2
84.25		1.1484	-4	3	5	2		2
85.20	85.22	1.1379	-5	3	4	6	3	9
85.46	85.62	1.1352	5	1	7	2	8	3
85.50		1.1348	7	2	1	2		2
85.62		1.1335	-4	2	6	11		15

CHONDRODITE - TAYLOR AND WEST, 1928

2THETA	PEAK	D	H	K	L	I(INT)	I(PK)	I(DS)
85.62		1.1335	4	2	4	11		15
85.77		1.1318	-6	3	3	2		9
85.89	85.88	1.1306	-8	2	2	10	7	14
86.39	86.38	1.1253	-2	4	2	2	2	2
86.39		1.1253	-1	4	1	2		3
86.97	86.96	1.1193	4	0	5	2		2
87.11	87.22	1.1178	-8	2	1	1	1	1
87.23		1.1166	-2	3	5	1	1	2
88.13	88.10	1.1075	1	3	5	2		1
88.87	88.88	1.1002	5	3	5	1	1	2
89.24	89.24	1.0966	-3	1	7	6	2	3
89.74	89.52	1.0918	4	3	3	1	1	8
90.78	90.78	1.0820	-8	2	4	1	1	2
91.28	91.44	1.0773	-2	4	4	2	3	3
91.28		1.0773	-2	4	2	2		3
91.35		1.0767	-4	4	1	4		3
91.45		1.0757	-7	3	2	4		5
91.48		1.0755	-9	1	4	2	1	5
92.05	92.06	1.0703	-1	1	7	1		2
92.06		1.0702	-7	3	1	1		2
93.15	93.26	1.0605	-5	3	5	2	1	3
93.27		1.0595	3	3	4	2		2
93.29		1.0593	-7	3	3	1		2
93.93	93.92	1.0538	9	1	0	8	2	11
94.24	94.22	1.0511	1	3	5	1	2	1
94.77	94.76	1.0467	7	1	3	7	2	10
96.99	97.00	1.0285	-9	1	5	7	2	11
97.20	97.24	1.0268	-3	0	6	1	2	2
97.24		1.0266	-10	1	3	1		5
97.24		1.0266	-10	0	2	3		2
98.03	98.00	1.0204	-6	3	3	2	1	2
98.65	98.66	1.0156	-2	4	4	2	1	2
98.65		1.0156	-2	4	2	1		2
99.42		1.0098	-3	3	3	2		3
99.49	99.46	1.0092	-2	2	6	1	1	2
99.71	99.72	1.0077	-10	0	5	1		2
100.80	101.04	0.9996	-4	3	6	6	5	9

2THETA	PEAK	D	H	K	L	I(INT)	I(PK)	I(DS)
100.80		0.9996	4	3	4	6		9
100.94		0.9987	-7	1	7	6		9
101.03		0.9980	-1	3	6	1		2
101.08		0.9977	-8	3	2	8		11
101.19		0.9969	-1	2	7	1		2
101.33		0.9959	-5	2	1	1		1
102.97	102.98	0.9844	-4	0	8	1	1	2
103.11		0.9835	9	2	0	1		1
103.97	104.06	0.9777	7	2	3	1	1	1
104.05		0.9771	0	3	6	1		2
104.66	104.68	0.9731	-6	4	2	1	1	1
104.66		0.9731	-6	4	1	1		2
104.71		0.9728	-10	0	5	1		1
104.71		0.9728	10	0	0	1		1
105.05	105.04	0.9705	-6	2	4	1	1	1
105.09		0.9703	-9	1	6	1		1
105.64	105.44	0.9668	-8	2	6	1	0	1
106.11	106.14	0.9638	8	3	0	1	1	1
106.26		0.9628	-9	2	5	1		1
106.32		0.9625	3	3	5	1	1	1
106.48	106.46	0.9614	3	2	6	1	1	2
107.30	107.34	0.9564	-2	1	8	1		2
107.30		0.9564	2	1	7	1		2
107.65	107.84	0.9542	-8	1	7	1	2	2
107.65		0.9542	8	3	3	1		2
107.73		0.9537	-6	0	8	1		1
107.73		0.9537	6	0	5	1		5
107.86		0.9529	-10	1	5	1		5
107.86		0.9529	10	1	0	3		2
108.07		0.9516	-6	3	6	3		3
108.62	108.64	0.9484	1	3	6	2	1	3
108.72		0.9477	2	4	4	1		1
108.90	108.92	0.9467	7	2	2	1	1	2
108.93		0.9465	5	2	5	2		3
109.37	109.36	0.9439	1	1	5	1	1	3
109.70	109.74	0.9421	-10	2	3	2	1	1
109.70		0.9421	-10	2	2	3		4

CHONDRODITE - TAYLOR AND WEST, 1928

2THETA	PEAK	D	H	K	L	I(INT)	I(PK)	I(DS)
110.02	110.12	0.9402	-1	5	1	3	2	2
110.18		0.9393	-1	1	8	3		5
110.37		0.9382	-7	2	7	1		1
111.33	111.34	0.9328	-1	5	1	4	1	6
112.18	112.40	0.9281	0	4	5	11	4	16
112.31		0.9274	-10	2	4	1		2
112.40		0.9269	-6	4	4	8		13
112.40		0.9269	6	4	1	8		13
112.45		0.9266	10	0	1	1		3
112.59		0.9259	-8	3	5	1		2
114.14	114.18	0.9177	-3	3	7	1	1	1
114.16		0.9176	-7	3	6	1		2
114.21		0.9173	-11	1	1	1		2
114.75	114.92	0.9145	-9	2	6	2	1	1
114.92		0.9137	-3	5	1	3		3
114.96		0.9135	-4	3	7	1		2
115.66	115.70	0.9100	-3	2	8	1	1	1
115.81		0.9092	-4	2	8	1		1
117.00	117.14	0.9034	5	1	6	1	3	1
117.10		0.9029	-2	2	8	9		14
117.10		0.9029	2	2	7	9		14
117.25		0.9022	-1	3	7	1		1
117.48	117.52	0.9011	-8	2	7	6	4	10
117.48		0.9010	8	2	3	6		10
117.58		0.9006	-5	2	8	1		1
117.71		0.9000	2	2	5	2		4
117.71		0.9000	-10	2	0	2		4
117.73		0.8999	7	2	4	1		1
119.82	119.86	0.8902	-2	5	3	3	2	5
119.82		0.8902	2	5	3	3		5
119.90		0.8899	-4	5	1	3		5
120.27	120.32	0.8882	9	2	2	1		2
120.35		0.8879	7	3	3	3		4
121.04	121.02	0.8848	-6	2	8	1	1	2
121.04		0.8848	1	1	1	1		1
121.04		0.8848	6	2	5	1		1
121.11		0.8845	-11	1	5	1		2

2THETA	PEAK	D	H	K	L	I(INT)	I(PK)	I(DS)
121.57	121.56	0.8825	6	3	4	1	1	2
121.87		0.8812	-3	5	4	1		2
122.30	122.28	0.8794	-2	4	6	1	1	1
122.95	122.96	0.8767	-9	3	5	3	1	5
122.98		0.8765	4	0	7	1		2
123.54	123.78	0.8742	-10	0	2	3		5
123.54		0.8742	10	0	1	1		2
123.76		0.8733	-4	5	5	3		4
123.83		0.8730	-4	4	6	1		1
123.84		0.8730	4	4	4	3		4
125.38	125.64	0.8669	-11	2	2	1	1	2
125.55		0.8662	3	5	5	1		2
125.65		0.8658	-8	3	1	2		3
126.03	126.32	0.8644	-8	4	5	4		6
126.37		0.8631	5	3	3	1		1
126.37		0.8631	3	2	2	1		1
126.90	127.10	0.8611	-10	1	4	1		2
126.92		0.8610	-11	2	4	1		1
127.09		0.8603	-5	5	5	1		1
127.10		0.8603	-6	0	0	10	3	17
127.25		0.8598	6	1	1	10		17
127.72	127.68	0.8580	-3	3	9	3		2
127.86		0.8575	-7	3	7	3	3	5
127.87		0.8574	-12	0	1	5		8
129.07	129.16	0.8531	-5	1	5	4		1
129.25		0.8525	-1	5	5	1		6
129.42		0.8519	5	5	0	1		1
129.42		0.8519	-12	0	4	1		1
130.06	130.10	0.8497	-12	3	2	2	1	3
130.06		0.8497	-10	3	4	1		2
130.30		0.8488	-8	4	4	1		2
131.66	131.68	0.8443	-5	5	3	4		7
131.80		0.8438	-12	1	3	3	1	5
132.37	132.38	0.8419	1	2	8	1		1
132.45		0.8417	-11	2	5	1	1	2
132.81	132.76	0.8405	-6	4	6	1		2

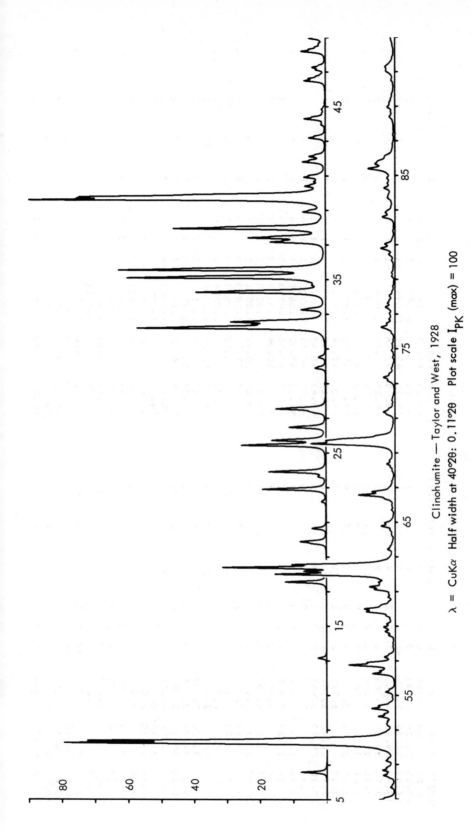

Clinohumite — Taylor and West, 1928

$\lambda = CuK\alpha$ Half width at $40°2\theta$: $0.11°2\theta$ Plot scale I_{PK} (max) = 100

CLINOHUMITE – TAYLOR AND WEST, 1928

2THETA	PEAK	D	H	K	L	I(INT)	I(PK)	I(DS)
6.56	6.56	13.466	0	0	1	4	8	4
13.14	13.14	6.733	0	0	2	2	3	1
17.53	17.54	5.054	-2	0	1	7	12	6
19.87	19.88	4.464	-2	0	2	4	8	3
20.63	20.64	4.303	-1	1	0	3	4	1
22.88	22.88	3.884	0	1	2	14	19	12
23.61	23.62	3.765	-1	1	2	1	3	1
23.89	23.90	3.721	-2	1	3	5	17	4
23.90		3.720	2	0	2	8	25	7
25.42	25.42	3.501	-1	1	2	19	16	16
25.70	25.70	3.463	-2	1	1	4	11	4
26.45	26.46	3.367	2	0	0	7	5	6
27.30	27.38	3.264	0	1	4	8	15	7
27.38		3.254	0	0	3	2	3	2
27.52	27.52	3.238	2	0	3	1	2	1
29.87	29.88	2.989	-1	1	3	11	57	10
30.48	30.48	2.930	-1	1	1	3	22	2
32.19	32.20	2.778	-2	1	4	2	28	1
32.40	32.40	2.761	-3	1	0	46	57	41
32.54	32.54	2.749	-1	1	5	11	22	10
33.24	33.24	2.693	3	0	1	19	28	17
34.25	34.26	2.616	0	1	4	5	7	5
35.10	35.10	2.554	3	1	3	34	39	30
35.56	35.56	2.523	-3	1	2	52	60	47
37.14	37.14	2.419	3	1	3	53	63	48
37.39	37.40	2.403	3	0	0	13	17	12
37.39		2.403	-4	0	3	7	23	6
37.94	37.94	2.370	4	0	5	12	46	11
38.88	38.88	2.314	-1	1	0	40	7	37
39.63	39.64	2.272	-1	2	5	5	90	5
39.63		2.272	-2	1	4	40	74	37
39.81	39.80	2.263	-2	1	1	39	6	37
40.37	40.38	2.232	-4	0	4	48		45
40.64	40.64	2.218	-4	2	0	4	4	3
40.89	40.90	2.205	-1	1	2	2	7	2
41.78	41.78	2.160	1	2	2	5	7	5

2THETA	PEAK	D	H	K	L	I(INT)	I(PK)	I(DS)
42.09	42.10	2.145	-4	1	3	3	5	2
42.10		2.145	4	1	1	2		1
43.16	43.16	2.094	-1	2	3	5	5	4
43.95	43.96	2.059	-1	2	6	1		1
44.27	44.28	2.044	-4	0	5	5	6	5
46.46	46.46	1.9527	-3	2	1	6	6	6
46.62	46.58	1.9466	-1	2	4	1	5	1
47.10	47.10	1.9277	1	2	0	2	3	2
47.25	47.24	1.9219	-2	1	6	1	4	1
48.17	48.18	1.8875	-5	1	1	7	7	7
48.43	48.42	1.8777	4	0	3	1	3	1
48.65	48.64	1.8700	0	2	4	4	2	4
48.92	48.92	1.8601	4	0	1	4	5	1
49.66	49.66	1.8343	-5	1	3	1	1	1
50.23		1.8148	3	1	4	1		4
50.34	50.34	1.8110	-1	2	3	3	4	1
50.64	50.64	1.8011	5	1	2	3		3
51.18	51.20	1.7833	0	1	7	2	3	3
52.21	52.22	1.7505	-2	2	1	50	100	50
52.21	52.34	1.7504	-2	2	5	50		50
52.35		1.7460	-4	2	4	39	93	39
53.33	53.34	1.7163	3	1	3	1		1
53.76	53.76	1.7035	-3	2	7	5	5	5
54.19	54.22	1.6910	4	0	5	2	7	2
54.22	54.36	1.6902	0	2	3	2		2
54.23		1.6901	-4	2	1	3	5	4
54.42	54.42	1.6846	-6	0	3	1		1
55.18	55.18	1.6632	-3	1	5	3	3	3
56.24	56.24	1.6343	-6	0	1	3	7	3
56.24		1.6341	6	0	3	4		4
56.69	56.74	1.6222	5	1	6	3	14	3
56.74		1.6211	3	1	4	3		3
57.21	57.20	1.6088	1	2	6	3	3	2
58.45	58.46	1.5776	-5	1	0	3	2	3
58.68	58.68	1.5719	1	1	0	2	3	2
58.93	58.92	1.5659	-1	3	3	3	4	3
59.15	59.10	1.5606	-1	3	1	1	3	1

CLINOHUMITE - TAYLOR AND WEST, 1928

2THETA	PEAK	D	H	K	L	I(INT)	I(PK)	I(DS)
59.59	59.60	1.5500	-4	2	5	1	3	1
59.82	59.82	1.5448	-3	1	8	3	9	7
59.90	59.98	1.5428	0	3	2	6	9	7
59.99		1.5407	-4	0	8	3		3
60.00		1.5405	4	0	6	2		2
60.67	60.66	1.5252	-5	2	3	2	2	2
61.10	61.10	1.5155	1	3	2	5	6	6
61.23	61.26	1.5124	-2	3	1	1	8	1
61.24		1.5124	2	3	0	2		2
61.27		1.5117	-1	2	7	2		2
61.53		1.5059	5	2	1	3	5	3
61.97	61.52	1.4962	0	0	9	41	36	43
62.16	61.98	1.4920	-1	3	3	2	28	2
62.34	62.14	1.4881	-6	0	6	30	52	32
62.36	62.36	1.4878	6	0	3	29		31
62.58	62.52	1.4830	-5	2	4	1	31	1
64.76	64.76	1.4383	-3	3	1	4	4	2
66.53	66.54	1.4043	1	3	4	1	11	5
66.55		1.4039	-7	1	2	1		1
66.81	66.74	1.3990	-3	3	3	11		12
68.33	68.34	1.3717	-5	1	8	1	7	1
68.36		1.3711	-1	3	5	1	2	1
69.49	69.50	1.3515	-2	3	5	14	25	16
69.49		1.3514	2	3	4	15		16
69.61	69.62	1.3494	-4	3	1	11	21	12
69.77	69.77	1.3468	-6	2	4	2	21	2
69.77		1.3467	6	2	1	2		2
69.78		1.3466	3	3	3	4		5
71.20	71.34	1.3231	-4	3	1	1	4	1
71.32		1.3212	-6	0	8	2		2
71.34		1.3210	6	0	5	2		2
73.13	73.14	1.2930	-4	2	8	1		1
73.33	73.34	1.2899	-3	1	10	3	2	4
74.88	74.88	1.2670	-1	3	8	1	3	1
76.08		1.2500	-7	2	2	1	1	1
76.21	76.22	1.2481	-4	1	10	2	4	3
76.23		1.2479	4	1	8	2		3
76.68	76.68	1.2416	-8	1	1	2	2	2
77.02	77.02	1.2370	-3	1	9	1	1	1
77.39	77.38	1.2321	-1	3	7	2	2	3
77.77	77.76	1.2270	-5	2	8	1	3	2
78.05	78.04	1.2233	4	0	7	1	2	1
79.15	79.16	1.2090	3	2	8	1	1	1
80.42	80.42	1.1931	-5	1	10	1	1	2
80.77	80.78	1.1887	0	4	0	6	4	7
82.56	82.58	1.1675	-3	2	10	1		1
82.57		1.1674	3	4	3	4		5
82.94	82.82	1.1631	-7	1	5	1	3	1
83.68	83.70	1.1547	6	2	5	1	2	1
83.84		1.1529	5	1	8	1		1
84.26	84.26	1.1482	-5	3	6	4	3	5
84.81	84.82	1.1421	3	1	10	2	2	3
85.36	85.38	1.1362	-4	2	10	6	8	8
85.37		1.1361	4	2	8	6		8
85.72		1.1324	-4	4	3	1		1
85.72		1.1323	2	4	2	1		2
85.82	85.64	1.1313	-8	2	2	5	6	6
88.93	88.94	1.0996	-3	1	12	3		4
90.97	91.06	1.0802	-2	4	5	2	2	2
90.98		1.0801	2	4	4	1		2
91.09	91.09	1.0791	-4	4	1	2		3
91.25	91.26	1.0776	-7	3	2	2	3	3
94.26	94.50	1.0510	3	3	8	1		3
94.50		1.0489	9	1	1	5		6
95.90	95.92	1.0373	7	1	7	4	2	5
96.18	96.18	1.0350	-9	1	7	1	4	6
97.02	97.04	1.0282	-10	0	2	1	2	2
97.64	97.64	1.0234	-3	1	10	3	1	2
99.28	99.28	1.0108	-7	1	3	4	3	4
100.46	100.46	1.0021	-4	3	10	4	4	5
100.47		1.0020	4	3	8	1		5
100.92	100.86	0.9988	-8	1	3	3	3	6
102.27	102.30	0.9892	-8	1	13	1	1	3
104.17	104.20	0.9764	-6	4	1	1	1	1

CLINOHUMITE - TAYLOR AND WEST, 1928

2THETA	PEAK	D	H	K	L	I(INT)	I(PK)	I(DS)
104.67	104.64	0.9730	10	0	2	1		1
105.08	105.04	0.9703	7	2	7	1		1
105.31	105.36	0.9689	0	3	11	1		1
105.61	105.58	0.9670	6	0	10	1		1
106.85	106.88	0.9592	-2	1	14	1		1
106.85		0.9591	2	1	13	1		1
107.43	107.50	0.9556	-8	1	11	1	2	1
107.45		0.9554	8	1	7	1		1
107.50		0.9551	-1	1	14	2		3
107.79	107.80	0.9534	-10	1	7	2	2	2
107.80		0.9533	10	1	2	2		2
108.16	108.16	0.9511	1	3	11	1	1	2
108.57	108.60	0.9487	5	2	10	1	1	1
108.88	108.90	0.9468	1	5	0	1	1	2
109.08		0.9457	-1	5	1	1		1
109.40	109.38	0.9438	-10	2	2	1	1	2
110.07	110.06	0.9399	-1	5	2	1	1	2

TABLE 5. MISCELLANEOUS SILICATES WITH SINGLE TETRAHEDRA

	Kyanite	Andalusite	Yoderite	Eulytite
Composition	Al_2SiO_5	Al_2SiO_5	$Mg_2Al_{5.3}Fe_{0.5}Ca_{0.2}Si_4O_{20}H_2$	$Bi_4Si_3O_{12}$
Source	Celo Mines, Burnsville, N. C.	Minas Gerais, Brazil Harvard No. 88421	Tanganika, Africa	Synthetic
Reference	Burnham, 1963b	Burnham & Buerger, 1961	Fleet & Megaw, 1962	Segal, Santoro & Newnham, 1966
Cell Dimensions				
$\underline{a}$	7.1192	7.7942	8.035	10.300
$\underline{b}$ Å	7.8473	7.8985	5.805	10.300
$\underline{c}$	5.5724	5.559	7.346	10.300
α	89.98	90	90	90
β deg	101.21	90	105.63	90
γ	106.01	90	90	90
Space Group	$P\bar{1}$	Pnnm	$P2_1/m$	$I\bar{4}3d$
Z	4	4	1	4

TABLE 5. (cont.)

Variety	Kyanite	Andalusite	Yoderite	Eulytite
Site Occupancy			M_1 {Al 0.66 / M_2 {Mg 0.25 / M_3 {Fe 0.06 / Ca 0.03	
Method	3-D, counter	3-D, counter	3-D, film	Neutron diffraction
R & Rating	0.086 (1)	0.057 (1)	0.11 (2)	0.04 (1)
Cleavage and habit	{100} perfect;{010} Tabular parallel to {100}	{010} perfect Columnar & fibrous parallel to c axis	None, {001}&{110} partings. Anhedral, somewhat elongated to c axis	{$\bar{1}$11} very poor Equidimensional
Comment			Similar, but not identical to kyanite	B estimated
μ	118.7	101.7	131.4	1203.4
ASF	0.5212×10^{-1}	0.7480×10^{-1}	0.6076×10^{-1}	2.465
Abridging factors	1:1:0.5	0.5:0:0:0	1:1:0.5	0.1:0:0

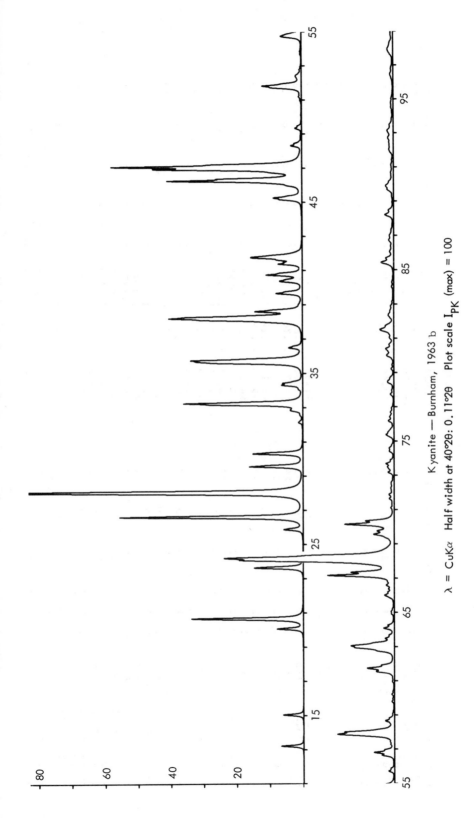

Kyanite — Burnham, 1963 b

$\lambda = CuK\alpha$ Half width at $40°2\theta$: $0.11°2\theta$ Plot scale I_{PK} (max) = 100

KYANITE – BURNHAM, 1963b

2THETA	PEAK	D	H	K	L	I(INT)	I(PK)	I(DS)
13.19	13.20	6.704	1	-1	0	7	7	6
15.01	15.02	5.898	-1	1	0	6	6	6
20.04	20.04	4.428	1	-1	0	9	8	8
20.60	20.64	4.307	0	1	1	20	34	19
20.64		4.301	1	1	1	20		19
23.60	23.62	3.766	-1	1	1	8	15	8
23.61		3.765	-1	2	0	8		8
23.67		3.756	0	2	0	1		1
25.86	25.86	3.443	-1	-1	1	7	6	7
26.57	26.58	3.352	-2	0	1	70	56	69
27.94		3.190	2	0	1	42		42
27.99		3.185	-1	1	1	24		24
28.00	28.00	3.184	0	-2	1	30	100	30
28.01		3.183	-2	1	1	45		45
29.54	29.54	3.022	0	2	1	12	16	12
29.54		3.021	-2	2	1	9		9
30.28	30.28	2.949	1	-2	2	11	15	11
30.29		2.948	-2	2	0	9		10
32.11	32.10	2.786	1	1	0	2	1	2
32.78		2.730	0	0	2	2	4	2
32.82	32.82	2.726	-1	0	1	31		32
33.14		2.701	-2	1	2	29	36	30
33.23		2.694	-2	-1	1	1		1
34.05	34.06	2.631	2	0	1	3	6	3
34.28		2.614	0	1	2	5		5
34.29	34.30	2.613	-1	1	2	28		30
34.42	34.40	2.603	-1	3	0	19	6	20
35.57		2.521	0	1	1	29	32	31
35.64	35.66	2.517	-1	-1	1	2		2
35.74	35.72	2.510	1	0	1	4	34	4
36.46	36.48	2.462	-1	2	1	6	4	7
36.47		2.462	2	-2	1	24		26
38.08	38.18	2.361	-2	0	2	29	40	31
38.09		2.356	1	-1	2	24		26
38.17		2.351	-2	3	0	24		26

2THETA	PEAK	D	H	K	L	I(INT)	I(PK)	I(DS)
38.57	38.58	2.332	1	-3	1	10	14	10
38.59		2.331	0	-3	1	10		11
39.63	39.64	2.272	-1	-3	2	7	8	8
39.65		2.271	0	-2	2	5		5
40.32	40.36	2.235	3	0	0	3	7	3
40.34		2.234	0	3	0	4		4
40.37		2.232	-2	3	1	4		4
40.72	40.72	2.214	-3	2	1	8	11	9
40.72		2.214	2	2	0	8		8
41.35	41.36	2.182	-3	2	1	4	7	4
41.36		2.181	-2	-2	1	6		7
41.60	41.62	2.169	1	1	2	7	8	8
41.73	41.76	2.163	-2	-1	1	12	16	13
41.78		2.160	2	1	1	12		13
45.13	45.14	2.007	0	3	1	8	7	10
45.23	45.24	2.003	-3	-1	1	2	9	2
46.13	46.24	1.9659	-3	3	0	63	41	73
46.24		1.9618	-1	3	2	62		72
46.91	46.92	1.9351	1	0	2	62	45	72
47.07	47.06	1.9291	-3	0	0	2	58	2
48.29	48.30	1.8829	-2	4	0	2	3	3
48.30		1.8828	0	4	0	2		3
49.31	49.32	1.8465	-1	4	1	12	2	15
51.70	51.80	1.7666	3	1	1	13	12	16
51.80		1.7635	-4	1	1	1		2
52.33	52.36	1.7467	0	2	3	1	2	1
52.42		1.7440	-2	2	3	3		4
54.67	54.74	1.6773	-1	1	3	5	6	5
54.72		1.6761	1	1	2	3		3
54.73		1.6756	0	-2	3	5		7
54.80		1.6738	-2	1	3	3		4
55.68	55.70	1.6494	-2	2	3	1	2	2
55.70		1.6489	-1	-2	2	2		2
56.61	56.62	1.6243	3	-1	2	6	4	7
56.78	56.78	1.6199	-4	1	2	4	6	5
56.79		1.6198	-4	3	0	4		5
57.24	57.24	1.6079	-2	3	0	1	1	1

KYANITE — BURNHAM, 1963

2THETA	PEAK	D	H	K	L	I(INT)	I(PK)	I(DS)
57.74	57.84	1.5952	2	2	2	2	17	3
57.75		1.5950	3	-2	2	2		4
57.82		1.5933	-1	4	2	3		7
57.83		1.5931	1	1	3	7		9
57.84		1.5928	1	-4	2	5		6
57.87		1.5921	-2	0	2	5		6
57.90		1.5913	0	-4	2	5		6
57.91		1.5909	-4	2	2	5		6
57.99	57.98	1.5891	-3	-3	2	3	15	4
58.64	58.64	1.5728	-3	1	3	8	3	10
58.65		1.5726	3	4	1	2		3
58.65		1.5458	-2	-4	1	2		2
59.77	59.78	1.5063	0	5	0	1	1	2
61.51	61.54	1.5056	-1	3	3	6	6	8
61.54		1.5048	-2	3	1	2		3
61.58		1.5015	1	5	1	2		3
61.73	61.72	1.5012	-1	5	1	6	8	7
61.74		1.4778	-2	-5	1	5		7
62.83	62.94	1.4762	-1	-3	3	4	12	5
62.91		1.4759	-1	-1	3	3		4
62.92		1.4757	3	3	0	5		6
62.93		1.4739	-4	-1	2	5		7
63.01	63.02	1.4734	2	-5	1	6	13	8
63.04		1.4733	-4	4	1	1		2
63.04		1.4731	0	-5	1	6		8
63.05		1.4645	-2	1	3	3		3
63.46	63.46	1.4524	1	-2	3	3	3	4
64.06	64.06	1.4357	-3	5	0	6	3	8
64.89	64.90	1.4167	0	3	3	2	1	2
65.87	65.98	1.4145	-5	1	1	2	3	3
65.98		1.4075	-4	-2	1	3		4
65.99		1.4034	-4	2	1	2		3
66.36	66.36	1.3931	-1	1	0	2	20	56
66.58	66.58	1.3886	1	5	0	1	13	2
67.13	67.14	1.3786	-2	-3	3	1	47	1
67.32	67.32							
67.38								
67.93	68.00							

2THETA	PEAK	D	H	K	L	L(INT)	I(PK)	I(DS)
67.98		1.3778	1	4	2	34		47
67.99		1.3776	3	-5	2	1		2
67.99		1.3775	3	-4	2	37		52
68.09	68.12	1.3758	-4	4	2	35	51	49
68.12		1.3753	-2	-4	2	37		52
69.45	69.54	1.3522	-1	5	2	1	6	2
69.51		1.3512	0	-5	3	3		4
69.53		1.3509	-5	3	1	3		5
69.55		1.3505	-5	1	2	2		2
69.56		1.3504	3	2	2	2		2
69.56		1.3503	-1	-2	2	1		2
69.74	69.74	1.3472	4	-2	0	2	5	2
69.76		1.3470	-4	-2	2	2		2
70.12	70.12	1.3408	5	0	0	31	15	45
71.61	71.62	1.3165	3	-1	3	2	2	3
71.86	71.84	1.3126	-4	-1	3	2	2	3
73.14	73.14	1.2928	-1	0	4	2	2	4
73.47		1.2878	1	0	4	3		4
73.66	73.66	1.2850	-3	0	4	3	3	4
74.37	74.38	1.2744	-5	4	0	2	2	2
74.38		1.2743	-2	4	1	1		2
74.54	74.56	1.2719	-2	6	1	1		1
75.25	75.34	1.2617	5	-3	1	2	2	3
75.36		1.2601	-4	-3	1	2	2	3
76.15	76.16	1.2490	2	4	3	4		5
76.65	76.66	1.2421	4	3	3	1	2	2
77.73	77.70	1.2275	-2	-4	3	3	1	5
78.40	78.42	1.2188	-1	3	4	4	1	5
78.49		1.2175	-1	1	4	4	2	4
79.18	79.18	1.2086	0	6	0	2		2
79.94	79.98	1.1990	-1	-3	4	1	1	5
80.03		1.1980	-3	3	4	3	2	4
80.34	80.36	1.1940	-3	-5	3	2		3
80.38		1.1936	0	-5	5	2	3	3
81.44	81.48	1.1807	-1	5	5	4	4	6
81.50		1.1799	-1	3	3	4		6
81.54		1.1795	-5	5	0	2		4

KYANITE - BURNHAM, 1963

2THETA	PEAK	D	H	K	L	I(INT)	I(PK)	I(DS)
82.19	82.32	1.1719	0	3	4	1	2	2
82.29		1.1707	-4	6	1	1		2
82.38		1.1696	-2	-3	1	2		3
82.86	82.86	1.1640	5	2	0	1		2
83.99	84.00	1.1513	3	2	3	1	1	2
85.37	85.40	1.1361	-2	4	4	4	4	7
85.41		1.1356	0	-4	4	6		10
86.05	86.04	1.1289	4	-6	1	1	1	2
86.65	86.70	1.1226	5	-5	1	1	2	2
86.75		1.1216	-3	-5	1	1		2
88.18	88.18	1.1070	-6	4	0	4	3	7
88.19		1.1069	6	4	0	4		6
89.65	89.68	1.0926	3	4	2	2	2	4
89.66		1.0925	5	-4	2	2		4
89.84	89.88	1.0908	-6	4	2	2	2	4
89.87		1.0905	-4	-4	2	2		4
91.10	91.10	1.0790	-6	2	3	1	1	2
91.86	92.00	1.0721	0	1	5	2	1	3
91.98		1.0710	-1	-2	5	1		3
92.03		1.0705	-3	1	2	2		3
92.45	92.44	1.0668	5	1	0	2	1	3
92.66	92.68	1.0649	5	3	1	2	2	3
93.07	93.12	1.0612	3	5	1	3	2	3
93.16		1.0604	-6	-3	4	1		4
93.52	93.44	1.0574	3	-3	1	1	1	2
94.49	94.50	1.0490	1	-1	5	1	1	2
94.72	94.74	1.0471	-3	-1	5	1	1	2
97.81	97.92	1.0221	-1	-1	4	2	2	3
97.90		1.0213	-1	-7	2	1		3
98.00		1.0206	0	-7	2	2		3
98.80	98.80	1.0145	2	3	4	2	1	2
99.17	99.20	1.0117	-4	-3	4	1	1	3
99.96	100.10	1.0058	2	-5	4	3	2	5
100.07		1.0050	-4	4	4	2		3
100.22	100.20	1.0039	3	-7	2	2	2	3
100.26		1.0035	-4	7	0	2		4
101.09	101.10	0.9975	-7	2	0	1	1	2
101.10	101.86	0.9975	6	2	0	2	1	2
101.59		0.9940	-6	-1	2	1		1
101.80		0.9925	-7	3	0	1		2
101.87		0.9920	-7	1	2	1		2
102.08	102.10	0.9905	2	-1	5	1	1	2
102.43	102.42	0.9881	-4	-1	5	1		2
103.19	103.48	0.9829	3	5	2	1	3	1
103.25		0.9825	1	5	2	1		2
103.27		0.9824	0	7	2	1		2
103.49		0.9809	-2	-8	0	9		17
103.85	103.82	0.9785	-2	-6	3	1	2	1
104.48	104.48	0.9743	-6	5	3	1	1	3
105.35	105.52	0.9686	-6	5	4	1	4	2
105.41		0.9682	6	0	2	1		18
105.51		0.9676	4	0	4	10		1
105.52		0.9675	5	-6	2	1		1
105.67		0.9665	-6	6	2	1		2
105.71	105.94	0.9663	-7	0	1	1		3
105.83		0.9655	-2	-8	1	1		3
105.85		0.9654	2	-8	0	1		2
105.98		0.9646	-6	0	4	4		20
107.07	107.12	0.9577	7	0	0	2	1	4
107.13		0.9574	2	1	5	2		4
107.54	107.52	0.9549	-5	6	5	2		5
108.31	108.50	0.9502	1	6	3	2	1	3
108.47		0.9493	7	-2	1	1		2
108.63		0.9483	-7	-2	1	1		2
109.80	109.82	0.9415	-4	5	0	2	1	3
109.81		0.9414	0	8	0	2		3
109.83		0.9413	4	-4	4	1		2
110.22	110.20	0.9390	4	-4	4	1	1	1
110.47		0.9376	7	-1	1	1		2

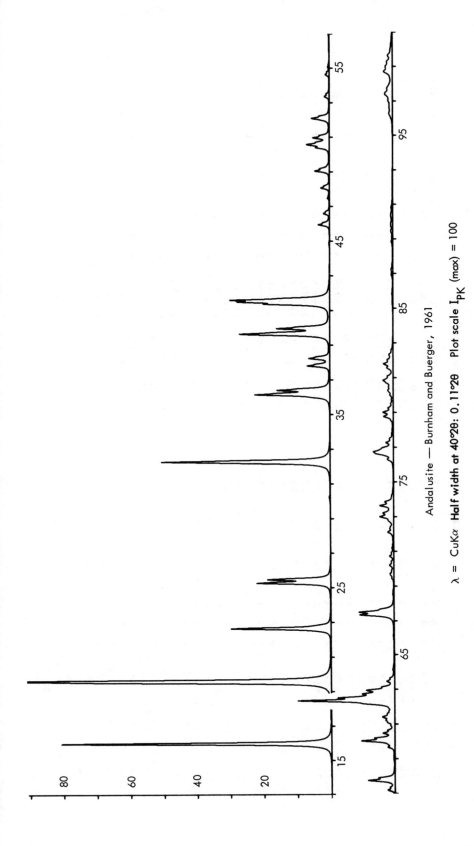

Andalusite — Burnham and Buerger, 1961

λ = CuKα **Half width at 40°2θ: 0.11°2θ**   Plot scale I_{PK} (max) = 100

ANDALUSITE – BURNHAM AND BUERGER, 1961

2THETA	PEAK	D	H	K	L	I(INT)	I(PK)	I(DS)
15.96	15.96	5.548	1	1	0	100	81	99
19.51	19.60	4.546	0	1	1	72	100	72
19.60	19.60	4.526	1	0	1	100	100	100
22.62	22.62	3.927	1	1	1	41	30	41
25.26	25.26	3.523	1	2	0	30	22	30
25.46	25.46	3.495	2	1	0	25	19	25
32.18	32.24	2.779	0	0	2	21	50	22
32.24	32.24	2.774	2	2	0	64		68
36.11	36.12	2.485	2	1	2	32	22	35
36.16		2.482	1	2	2	4		4
36.37	36.38	2.468	3	1	0	24	16	25
37.77	37.78	2.379	3	1	1	11	7	12
38.20	38.20	2.354	3	0	1	10	6	11
39.57	39.62	2.276	1	3	1	16	27	18
39.62		2.273	0	2	2	34		38
39.93	39.94	2.256	3	1	2	25	16	28
41.34	41.34	2.182	1	1	2	28	20	31
41.35		2.182	2	2	2	1		1
41.47	41.48	2.175	2	1	2	26	28	29
41.57	41.58	2.171	3	2	1	32	30	36
45.92	45.92	1.9746	4	0	0	6	3	7
46.57	46.58	1.9486	1	4	0	3	2	4
47.46	47.46	1.9142	4	1	0	1	1	1
48.05	48.06	1.8918	1	3	0	5	3	6
49.03	49.02	1.8564	3	3	0	9	4	10
49.23	49.16	1.8493	3	0	1	1	3	1
50.38	50.38	1.8099	1	0	3	7	4	8
50.55	50.56	1.8040	0	1	3	8	7	9
50.59		1.8028	0	4	1	4		4
50.94	50.94	1.7910	4	0	1	9	5	11
51.98	52.08	1.7576	1	1	3	2	5	2
52.07		1.7547	1	3	1	10		11
53.34	53.34	1.7161	3	3	1	3	2	4
54.62	54.62	1.6791	2	3	1	3	1	3
55.04	55.04	1.6670	4	2	1	1	0	1
57.17	57.18	1.6098	0	4	2	4	2	5
57.73	57.74	1.5955	4	0	2	17	8	21

2THETA	PEAK	D	H	K	L	I(INT)	I(PK)	I(DS)
59.67	59.68	1.5482	1	5	0	4	2	5
59.99	60.04	1.5408	2	3	3	1	10	5
60.04		1.5396	3	1	2	21		26
60.48	60.48	1.5293	4	0	0	6	4	8
60.91	60.92	1.5195	5	1	1	1	1	2
61.10	61.12	1.5153	2	4	3	3	2	3
61.22	61.22	1.5128	4	3	1	3	3	4
61.40	61.42	1.5086	2	1	2	3	4	3
61.45		1.5076	3	1	3	4		5
61.75	61.74	1.5009	3	0	1	2	2	2
62.36	62.36	1.4878	1	5	1	50	29	64
62.37		1.4875	4	3	2	7		8
62.64	62.52	1.4818	4	1	3	6	17	7
62.75	62.76	1.4794	5	1	2	14	10	18
62.98	62.96	1.4746	2	5	1	11	9	15
63.49	63.48	1.4640	5	1	1	4	2	6
64.18	64.18	1.4500	4	5	0	1	1	1
67.32	67.32	1.3897	4	4	0	22	10	30
67.47	67.48	1.3870	3	4	2	13	11	17
68.51	68.52	1.3684	3	5	0	1	0	1
69.43	69.44	1.3526	3	3	2	1	1	2
69.59	69.68	1.3498	3	5	0	1	1	3
69.69		1.3481	1	1	4	3		3
70.18	70.18	1.3399	5	1	0	4	1	4
70.70	70.70	1.3314	1	4	2	3	1	2
71.16	71.16	1.3238	5	3	0	1	1	5
72.09	72.10	1.3090	3	3	2	4	2	2
72.41	72.30	1.3039	3	3	1	1	1	12
72.97	72.98	1.2953	2	5	2	8	4	3
73.14	73.18	1.2928	2	1	4	2	3	3
73.23		1.2914	2	4	2	4		14
73.62	73.62	1.2856	5	2	2	10	4	2
73.87	73.84	1.2818	6	1	0	1	3	1
74.22	74.20	1.2767	2	4	3	4	1	5
76.28	76.28	1.2472	2	6	2	10	2	14
76.62	76.72	1.2425	2	2	4	11	6	16
76.73		1.2410	4	2	2			

ANDALUSITE - BURNHAM AND BUERGER, 1961

2THETA	PEAK	D	H	K	L	I(INT)	I(PK)	I(DS)
77.25	77.24	1.2340	6	2	0	5	2	7
77.76	77.76	1.2271	4	5	0	1	1	1
78.75	78.76	1.2142	3	1	2	5	3	7
78.76		1.2140	1	3	4	2		3
79.00	79.00	1.2110	3	4	4	4	3	6
79.23	79.22	1.2080	5	1	2	1	2	1
79.49	79.48	1.2047	6	3	1	1	1	1
79.96	80.00	1.1988	3	4	3	1	1	1
80.00		1.1983	4	5	1	1		2
80.17	80.26	1.1962	4	3	1	3	2	2
80.27		1.1949	5	4	0	2		3
80.44	80.44	1.1929	5	0	3	1	1	1
80.69	80.82	1.1897	0	6	2	3	3	5
80.83		1.1881	1	5	3	7		10
81.54	81.54	1.1795	5	1	3	4	2	6
81.76	81.78	1.1768	6	0	2	5	3	8
81.98	82.00	1.1743	3	6	0	1	2	2
82.31	82.30	1.1704	3	2	4	3	1	4
85.81	85.80	1.1315	4	0	4	1	0	1
86.90	86.92	1.1200	4	1	4	1	1	1
86.92		1.1198	5	4	2	1		1
87.22	87.20	1.1167	1	7	0	1	1	2
88.82	88.82	1.1007	1	0	5	1	0	2
89.42	89.42	1.0948	7	1	1	2	1	3
89.91	89.88	1.0901	1	2	5	1	1	1
90.29	90.30	1.0866	5	3	3	1	1	1
90.43	90.40	1.0853	6	4	0	1	1	2
90.83	90.80	1.0815	7	1	2	1	0	2
96.03	96.26	1.0362	1	7	2	2	1	2
96.19		1.0350	3	7	0	4		3
96.28		1.0342	1	5	4	1		1
96.55	96.58	1.0320	2	2	5	2	1	1
96.74	96.74	1.0305	5	1	2	2	1	5
96.99	97.02	1.0285	5	5	4	3	1	3
97.17	97.40	1.0271	6	2	3	1	2	1
97.37		1.0255	7	3	0	5		8
97.45		1.0249	7	1	2	3		5

2THETA	PEAK	D	H	K	L	I(INT)	I(PK)	I(DS)
97.53	97.68	1.0242	0	3	5	1	1	1
97.68	97.68	1.0231	4	5	3	1	2	1
97.94		1.0210	5	4	3	1		2
98.66	98.68	1.0155	1	4	1	1	3	2
98.67		1.0154	4	6	0	8		13
98.90	98.96	1.0137	3	1	5	3	2	5
99.27	99.24	1.0109	6	4	2	4	2	6
99.59	99.62	1.0085	7	3	1	1	1	2
99.67		1.0079	2	5	4	2		4
100.29	100.28	1.0034	6	5	0	1	0	1
100.30		1.0033	5	2	0	1		1
102.55	102.56	0.9873	0	8	0	5	1	9
103.37	103.38	0.9817	4	4	4	9	2	16
103.70	103.70	0.9795	1	8	0	1	1	1
104.48	104.48	0.9743	8	0	1	3	1	4
105.97	106.10	0.9646	1	8	1	1	0	1
106.11		0.9637	0	7	3	1		1
106.43	106.44	0.9617	4	7	1	1	1	2
106.95	106.94	0.9585	1	5	5	2	1	3
107.28	107.32	0.9565	7	3	3	1	1	2
107.44		0.9555	7	1	4	1		1
107.87	107.86	0.9529	3	8	1	1	1	2
107.91		0.9526	8	1	1	1		1
108.51	108.56	0.9490	6	0	4	1	1	2
108.58		0.9486	6	1	4	1		2
108.76	109.04	0.9475	7	3	3	1	1	1
109.06		0.9457	5	6	2	3		5
109.50	109.48	0.9432	2	8	1	1	1	2
109.67		0.9422	6	0	4	1		2
112.16	112.18	0.9282	2	6	4	4	1	7
112.48	112.60	0.9265	0	0	6	1	1	1
112.82	112.86	0.9246	6	6	0	2	1	3
113.18	113.18	0.9227	4	5	4	5	1	9
113.73	113.60	0.9198	4	4	4	1	1	1
114.85	114.90	0.9140	1	1	6	1	2	2
114.89		0.9138	7	2	0	2		3
115.63	115.62	0.9101	7	1	2	2	2	3

115

ANDALUSITE - BURNHAM AND BUERGER, 1961

2THETA	PEAK	D	H	K	L	I(INT)	I(PK)	I(DS)
116.11	116.10	0.9077	3	4	5	1	1	2
116.34		0.9066	4	4	5	1		1
116.96	117.30	0.9036	3	7	3	1	1	1
117.06		0.9031	1	5	5	3		5
117.28		0.9020	0	2	6	2		4
117.30		0.9019	5	7	1	2		3
117.36		0.9016	8	3	1	1		2
117.86	117.84	0.8993	5	1	5	4	1	7
118.10		0.8981	7	3	1	3		5
118.28	118.34	0.8973	7	5	3	1	1	1
118.35		0.8970	3	6	1	1		2
118.65	118.64	0.8956	2	1	6	1	1	2
118.67		0.8955	8	2	2	4		8
121.99	122.00	0.8807	4	8	0	1	0	3
123.67	123.70	0.8737	8	4	0	1	0	1
124.46	124.50	0.8705	1	7	4	1	0	3
124.96	125.02	0.8685	1	3	6	1	1	1
125.02		0.8683	5	7	2	2		1
125.88	125.54	0.8649	7	5	1	1	0	5
127.37	127.42	0.8593	8	4	3	2	0	3
127.93	128.34	0.8572	2	1	3	1	1	2
128.22		0.8562	9	9	0	1		3
128.35		0.8557	0	4	4	2		3
128.45		0.8554	6	4	4	2		3
129.86	129.86	0.8503	2	8	3	1	0	2
131.16	131.16	0.8459	9	2	0	2	0	5
132.20	132.20	0.8425	8	2	3	1	0	1
132.86	133.12	0.8404	7	6	1	1	1	1
133.11		0.8396	4	8	2	1		6
133.36		0.8388	0	4	6	3		2
134.01	133.94	0.8367	4	9	0	3	1	5
135.76	136.32	0.8315	3	7	0	1	1	1
136.23		0.8301	3	7	4	3		6
136.28	136.88	0.8299	2	6	5	1	1	1
136.83		0.8284	3	3	6	5		10
137.18		0.8274	6	6	3	1	1	2
137.61	137.96	0.8261	3	8	3	1	1	2
137.95		0.8252	7	3	4	10		20
138.41	138.96	0.8239	4	5	5	1	2	1
138.81		0.8228	5	4	5	2		3
138.88		0.8227	9	3	0	1		2
139.00		0.8223	9	1	2	7		14
139.01		0.8223	3	9	1	5		11
139.88	139.88	0.8200	2	4	6	14	3	28
139.98		0.8197	5	7	3	2		4
140.07		0.8195	8	3	3	1		2
140.43	140.54	0.8186	4	2	6	5	2	10
140.56		0.8182	2	9	2	6		11
141.09		0.8169	7	5	3	4		9
142.34	142.28	0.8138	9	3	1	4	1	8
142.47		0.8135	6	5	4	2		3
144.27	144.26	0.8093	9	5	2	3	0	5
146.27	146.28	0.8049	0	8	4	14	1	29
148.34	148.54	0.8006	1	8	0	2	0	3
148.55		0.8002	4	9	0	1		3
149.22	149.80	0.7989	5	8	2	8	1	17
149.82		0.7978	8	0	4	1		2
150.46		0.7966	3	9	2	1		1
150.80	150.80	0.7950	1	5	6	2	1	4
151.32		0.7946	8	5	2	1		
151.54		0.7925	7	7	0	1		3
152.75	152.66	0.7924	5	1	6	2	0	3
152.83		0.7902	0	1	7	2		5
154.22	154.24	0.7898	1	0	7	1	0	2
154.42		0.7879	7	10	1	4		8
155.70	155.70	0.7851	9	4	1	1	0	2
157.64	159.42	0.7846	7	7	1	4		9
158.04		0.7846	9	0	3	1		2
158.08		0.7829	2	6	6	8	1	16
159.38		0.7829	7	1	5	3		5
159.39		0.7807	5	2	6	10		21
161.21								

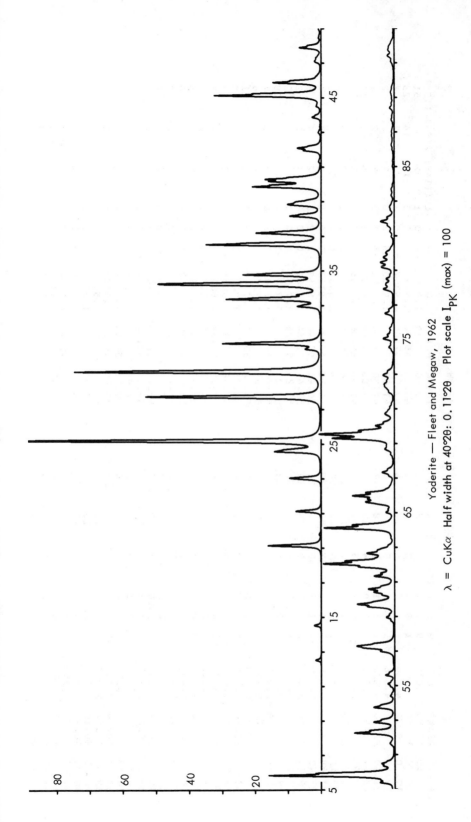

Yoderite — Fleet and Megaw, 1962

$\lambda = CuK\alpha$ Half width at $40°2\theta$: $0.11°2\theta$ Plot scale I_{PK} (max) = 100

117

YODERITE - FLEET AND MEGAW, 1962

2THETA	PEAK	D	H	K	L	I(INT)	I(PK)	I(DS)
12.50	12.50	7.074	0	0	1	2	2	1
14.50	14.50	6.104	-1	0	1	2	2	2
19.10	19.12	4.644	-1	1	0	6	16	6
19.13		4.636	-1	0	2	9		9
21.10	21.10	4.206	1	1	0	7	8	7
22.97	23.02	3.869	-2	0	1	4	10	4
23.02		3.860	0	0	2	6		6
24.55	24.56	3.623	-1	1	1	12	14	12
25.15	25.16	3.537	0	1	1	100	100	100
27.68	27.72	3.219	-2	1	0	32	53	32
27.73		3.215	-2	0	2	30		31
29.11	29.14	3.065	-1	1	2	9	74	10
29.13		3.063	-1	0	3	69		71
29.24		3.052	-2	0	2	13		13
30.46	30.46	2.932	1	0	2	4	5	4
30.78	30.78	2.902	0	2	0	32	30	34
32.93	32.94	2.718	0	2	1	7	7	8
33.34	33.34	2.685	2	0	1	32	29	34
33.59	33.60	2.665	-3	0	2	5	7	6
34.18	34.22	2.621	-1	1	2	39	49	42
34.23		2.617	3	1	1	27		30
34.75	34.76	2.579	1	2	0	27	24	29
36.49	36.50	2.460	-3	0	1	41	35	46
37.08	37.16	2.422	3	1	0	5	20	5
37.17		2.417	-2	1	2	20		23
38.15	38.14	2.357	-3	2	0	11	9	13
38.75	38.82	2.322	3	0	1	3	10	3
38.78		2.320	-2	2	1	3		3
38.82		2.318	2	0	2	7		8
39.84	39.84	2.261	-1	1	3	24	21	28
39.95	39.94	2.255	-1	2	1	2	15	2
40.15	40.16	2.244	0	2	2	16	16	18
40.26	40.26	2.237	2	1	1	11	17	12
41.93	41.94	2.153	-1	2	2	6	5	7
42.08	42.08	2.145	-2	1	3	7	7	8
43.85	43.86	2.063	1	1	4	3	3	4
44.49	44.50	2.035	-3	0	3	1	1	2

2THETA	PEAK	D	H	K	L	I(INT)	I(PK)	I(DS)
45.10	45.10	2.009	-4	0	1	42	32	51
45.85	45.86	1.9774	-1	1	3	19	14	23
47.04	47.08	1.9302	-4	0	2	1	2	2
47.09		1.9280	3	2	0	1		2
47.30	47.30	1.9201	-3	1	3	1	2	2
47.88	47.88	1.8983	-4	1	1	9	6	11
49.36	49.36	1.8445	-1	3	1	5	4	6
49.74	49.78	1.8316	-1	1	2	1	38	2
49.78		1.8302	-4	2	3	2		65
50.33	50.34	1.8113	0	2	3	51		4
50.46	50.46	1.8069	2	0	3	3		3
51.64	51.64	1.7686	-2	0	4	2		2
52.23	52.24	1.7499	0	0	1	15		20
52.24		1.7495	0	0	4	5		7
52.86	52.86	1.7306	4	1	0	5	6	6
52.88		1.7299	-1	1	4	8		6
53.72	53.72	1.7049	-2	3	2	4	6	11
53.78		1.7031	-1	2	3	2		3
54.38	54.38	1.6858	-1	0	4	2	1	2
54.94	54.94	1.6699	-3	1	3	2	1	3
55.07	55.08	1.6661	-4	2	1	1	2	3
55.59	55.60	1.6517	-1	3	2	2	3	5
56.97	56.98	1.6150	-3	2	1	4	4	6
57.18	57.28	1.6097	-4	2	1	5	5	12
57.27		1.6073	1	3	2	9		16
58.74	58.76	1.5705	4	2	2	1	1	2
58.93	58.94	1.5659	-4	1	4	2	3	3
58.94		1.5656	1	3	2	2		2
59.56	59.56	1.5509	-3	2	2	12		14
59.69	59.70	1.5478	-3	1	4	10	8	11
59.70		1.5476	-1	3	0	7	11	3
59.73		1.5468	3	0	1	2		1
59.79		1.5454	3	3	3	1		3
60.24	60.24	1.5350	3	0	3	3	4	5
60.40	60.40	1.5313	-2	3	4	5	6	7
60.57	60.58	1.5273	1	2	5	8	8	11
60.63		1.5260	-5	1	2	1		2

118

YODERITE — FLEET AND MEGAW, 1962

2THETA	PEAK	D	H	K	L	I(INT)	I(PK)	I(DS)
61.32	61.32	1.5106	-3	3	2	3	6	4
61.33		1.5103	0	3	4			8
61.86	61.88	1.4986	4	2	1	6	11	18
62.01	62.04	1.4953	5	1	0	13	22	8
62.04		1.4946	-4	3	3	6		29
62.20	62.20	1.4912	-5	0	3	20	15	10
62.58	62.64	1.4831	-2	3	3	7	8	4
62.65		1.4816	4	1	2	3		13
62.92	62.82	1.4759	-4	4	4	9	5	5
64.11	64.12	1.4512	0	4	0	3	21	51
65.01	65.02	1.4334	5	0	1	35	3	5
65.49	65.60	1.4240	1	3	3	3	9	6
65.59		1.4221	1	2	4	4		11
65.62		1.4214	2	1	4	8		6
65.76	65.78	1.4189	-1	1	5	4	8	4
65.82		1.4176	-2	0	5	2		3
65.97	65.96	1.4149	0	0	5	2	13	26
66.16	66.14	1.4111	-3	3	5	17	9	5
67.11	67.12	1.3936	-4	1	1	4	2	3
67.31	67.32	1.3898	-5	2	2	2	4	7
67.32		1.3712	-3	1	5	5	1	2
68.35	68.36	1.3666	-4	3	1	1	3	2
68.62		1.3656	5	2	3	3		4
68.67	68.66	1.3551	4	2	0	29	19	46
69.28	69.28	1.3507	-4	2	2	28	23	43
69.54	69.52	1.3426	0	4	4	8	6	12
70.02	70.02	1.3314	-1	3	2	3	3	5
70.69	70.70	1.3247	-4	0	4	2	1	3
71.11	71.10	1.3116	-2	4	5	1	2	2
71.92	71.98	1.3106	-2	4	1	1		2
71.99		1.3066	-1	2	5	2	3	3
72.24	72.30	1.3057	-2	2	5	2		3
72.30		1.3039	3	0	4	3	3	5
72.41	72.42	1.2966	4	1	3	1		2
72.89	72.94	1.2963	5	0	5	3	3	2
72.91		1.2954	-3	4	2	1		4
72.97		1.2746	5	1	0	3		2
74.36	74.38		-3	4	1	1	1	

2THETA	PEAK	D	H	K	L	I(INT)	I(PK)	I(DS)
74.52	74.54	1.2722	3	1	4	1	2	2
75.03	75.04	1.2648	3	4	0	3	1	10
75.44	75.44	1.2590	6	2	1	1	1	2
75.63	75.64	1.2563	-6	1	3	2	1	3
76.50	76.50	1.2442	-3	1	2	5	3	9
77.90	77.90	1.2253	-5	0	5	3	3	10
78.24	78.22	1.2208	3	4	1	3	2	5
78.49	78.48	1.2175	4	2	3	3	2	5
79.13	79.18	1.2092	5	3	0	4	4	7
79.18		1.2086	-4	2	5	5		8
79.46	79.44	1.2051	4	3	2	4	4	6
79.76	79.74	1.2013	-4	2	4	2	3	3
80.01	80.00	1.1982	-2	1	6	1	1	2
80.09		1.1972	6	1	4	1	1	2
80.93	80.94	1.1868	0	0	4	3	2	4
81.24	81.24	1.1831	-6	4	6	7	4	12
81.58	81.58	1.1791	0	3	1	2	1	3
81.81	81.80	1.1764	-4	0	4	3	1	3
82.47	82.48	1.1685	-2	5	2	5	1	6
82.65	82.64	1.1664	-4	3	5	6	1	2
83.70	83.68	1.1545	-1	5	2	3	1	4
88.37	88.38	1.1051	4	2	4	2		6
89.29	89.28	1.0962	1	5	2	2		2
91.05	91.06	1.0795	4	3	4	2	1	4
91.38	91.38	1.0764	6	3	0	3		6
91.74	91.74	1.0731	-4	2	6	1	2	2
91.78		1.0727	-5	4	0	1		2
91.94	91.92	1.0714	5	4	3	2	2	4
93.37	93.40	1.0586	3	3	2	1	1	2
93.45		1.0579	-1	2	6	1		2
94.18	94.22	1.0516	-2	0	7	1	1	2
94.45	94.58	1.0494	-7	2	3	3	2	3
94.59		1.0482	1	0	6	1		7
94.83	94.84	1.0461	-3	0	7	4	2	5
95.44	95.56	1.0411	-5	4	3	1		2
95.57		1.0400		5	2			7
95.80	95.86	1.0381	2	5	2	2	2	2

119

YODERITE - FLEET AND MEGAW, 1962

2THETA	PEAK	D	H	K	L	I(INT)	I(PK)	I(DS)
96.43	96.42	1.0330	7	2	0	3	2	6
97.47	97.46	1.0247	-3	1	7	2	1	2
98.01	98.08	1.0205	-5	2	6	1	2	3
98.10		1.0198	5	4	1	2		4
98.54	98.54	1.0164	1	5	3	3	2	5
98.98	98.98	1.0131	0	0	5	10	4	20
99.16		1.0117	-3	4	5	3		6
99.31	99.30	1.0106	0	0	7	1	4	3
99.45		1.0096	-7	2	4	1		6
100.15	100.14	1.0044	-8	0	2	4	2	8
101.59	101.60	0.9940	1	5	2	3	1	1
102.21	102.22	0.9897	-8	3	2	1	2	9
103.39	103.48	0.9816	-6	4	2	4	2	2
103.46		0.9811	-1	1	5	1		7
103.53		0.9806	4	5	4	3		2
103.86	103.88	0.9784	-4	4	4	1	2	3
103.93		0.9780	-1	2	7	1		3
104.01		0.9774	3	0	6	2		4
105.24	105.24	0.9699	3	4	4	2		5
105.26		0.9692	3	3	5	1	1	2
105.27		0.9691	1	0	7	1		1
105.52	105.62	0.9675	0	6	0	1		2
105.56		0.9672	8	0	0	1	2	3
105.69		0.9664	5	4	0	2		5
105.90		0.9651	-8	0	4	1		3
106.48	106.48	0.9615	-5	1	7	2		4
106.70		0.9601	-6	2	6	1	1	2
107.21	107.42	0.9569	-7	2	5	2		5
107.43		0.9556	-1	6	1	2		4
107.61	107.58	0.9544	0	2	7	2		3
107.67		0.9541	-8	1	0	1		2
108.39	108.38	0.9497	-5	3	6	1	1	3
108.83	108.82	0.9471	-5	1	1	1	1	5
111.03	111.08	0.9344	-1	6	2	2	1	3
111.07		0.9342	-5	4	5	3		6
111.73	111.74	0.9306	7	2	2	5	2	12
111.91		0.9296	6	0	4	1		1
112.07	112.10	0.9287	5	5	0	1	2	2
112.08		0.9287	-6	4	1	1		1
112.67	112.66	0.9254	8	0	4	1	1	3
112.68		0.9254	4	5	2	2		4
112.94	113.10	0.9240	-4	4	5	1	1	1
113.19		0.9227	-8	0	5	1		2
114.31	114.62	0.9168	-3	3	7	2	2	4
114.54		0.9156	5	1	5	2		4
114.55		0.9156	-3	0	8	1		1
114.64		0.9151	0	4	6	3		7
114.82		0.9142	-6	1	7	1		2
114.88	115.64	0.9139	8	1	1	1	1	3
115.62		0.9102	2	5	4	1		2
115.75		0.9095	-1	4	5	1		3
116.42	116.52	0.9062	-2	5	8	1	1	2
116.53		0.9056	4	4	4	1		1
116.58		0.9054	5	4	3	1		2
116.98	116.96	0.9035	-4	4	6	1	1	3
117.41	117.38	0.9014	4	6	5	1	1	1
118.35	118.76	0.8970	-7	2	6	2	1	4
118.75		0.8951	0	6	3	3		8
119.55	119.54	0.8914	-8	3	2	4	1	9
120.64	120.74	0.8866	3	4	5	1		2
120.79		0.8859	6	4	2	1		2
121.22	121.36	0.8840	-8	3	4	1		1
121.40		0.8832	5	2	3	2		4
121.76	121.74	0.8817	8	2	5	1	1	2
122.32	122.32	0.8793	-8	2	2	1	1	5
122.58		0.8782	-9	1	5	1		2
123.99	124.38	0.8724	0	2	7	1		1
124.09		0.8720	-7	0	7	1	1	2
124.17		0.8717	-4	6	1	1		1
124.34		0.8710	-9	0	6	2		5
124.42		0.8707	-5	3	7	3		3
125.47	125.84	0.8665	-1	1	7	1	1	7
125.51		0.8664	-5	8	0	1		2
125.78		0.8653	4	0	4	1		4

YODERITE - FLEET AND MEGAW, 1962

2THETA	PEAK	D	H	K	L	I(INT)	I(PK)	I(DS)
125.88		0.8649	-4	6	2	2		5
126.42	126.42	0.8629	6	1	5	1	1	2
126.61		0.8621	-8	1	0	1		1
126.65		0.8620	-6	5	6	1		3
126.82		0.8614	-9	3	4	1		3
127.24	127.20	0.8598	9	0	0	1		4
128.32	128.32	0.8558	-1	6	4	1	1	6
129.86	130.08	0.8504	-2	4	7	2	1	5
130.10		0.8495	-9	2	3	4		10
130.31		0.8488	0	6	4	2		6
130.93	131.14	0.8467	4	6	1	2	2	10
131.14		0.8460	-4	6	3	4		5
131.16		0.8459	-3	4	7	2		6
131.17		0.8459	0	2	8	4		9
131.66	131.68	0.8443	-9	2	1	1	2	1
132.00		0.8432	2	5	5	2		2
132.08		0.8429	-6	0	8	3		7
134.19	134.50	0.8362	5	3	0	3	1	3
134.38		0.8356	5	3	6	1		3
134.55		0.8351	-6	3	7	1		2
134.62		0.8349	-8	3	1	1		3
134.86	134.94	0.8341	-6	1	8	1	1	2
134.98		0.8338	4	4	5	3		3
135.52	135.56	0.8322	-1	6	4	1	1	7
135.66		0.8317	-4	4	6	1		2
135.92	136.52	0.8310	-7	4	5	1	2	3
136.48		0.8293	0	4	7	6		15
136.57		0.8291	3	2	7	3		8
136.61		0.8290	-2	3	8	1		4
137.23	137.68	0.8272	9	0	1	8	2	3
137.70		0.8259	-8	4	2	2		21
137.83		0.8255	-5	6	1	1		4
140.29	140.88	0.8189	9	1	2	11	3	30
140.62		0.8181	4	6	0	4		10
140.89		0.8174	-8	0	7	11		30
141.00		0.8171	-4	6	3	1		2
141.59		0.8157	2	0	8			

2THETA	PEAK	D	H	K	L	I(INT)	I(PK)	I(DS)
142.87	142.92	0.8125	8	1	3	2	1	4
142.88		0.8125	6	2	5	2		4
143.58	143.70	0.8109	2	7	0	1	2	3
143.61		0.8108	-2	7	1	1		2
143.66		0.8107	3	4	6	5		14
144.20	144.60	0.8094	-6	2	8	3	2	2
144.21		0.8094	-8	1	7	3		9
144.75		0.8082	-1	7	2	2		5
144.79		0.8081	-4	0	9	1		4
144.92		0.8078	7	2	7	1		2
145.26		0.8070	-1	4	6	1		2
145.37	145.90	0.8068	-2	6	5	1	1	4
145.76		0.8060	-1	4	7	2		4
145.90		0.8057	-9	0	0	4		10
146.27		0.8049	-8	4	4	3		9
146.86	146.86	0.8036	-8	4	4	3	1	8
147.26		0.8028	4	1	7	1		3
149.36	149.76	0.7986	0	6	6	1	1	3
149.60		0.7982	-5	3	8	2		6
149.68		0.7980	-9	1	6	2		3
149.70		0.7980	1	7	2	1		3
150.11	150.58	0.7972	8	3	2	1	1	3
150.50		0.7965	-1	1	9	4		12
151.42	151.52	0.7949	-8	1	6	1	1	2
151.51		0.7947	-5	5	6	3		10
151.64		0.7945	-10	1	3	1		3
151.76		0.7942	-9	3	4	2		5
154.46	154.68	0.7898	3	3	7	1	1	4
154.65		0.7895	3	3	0	5		13
159.47	161.52	0.7828	6	4	4	3	2	9
159.76		0.7824	-4	6	3	3		7
160.71		0.7813	9	1	2	5		14
161.00		0.7810	2	7	2	2		8
161.12		0.7808	2	1	7	2		6
161.21		0.7807	4	2	4	14		41
161.43		0.7805	-2	7	3	3		8
161.44		0.7805	-6	4	7	2		7

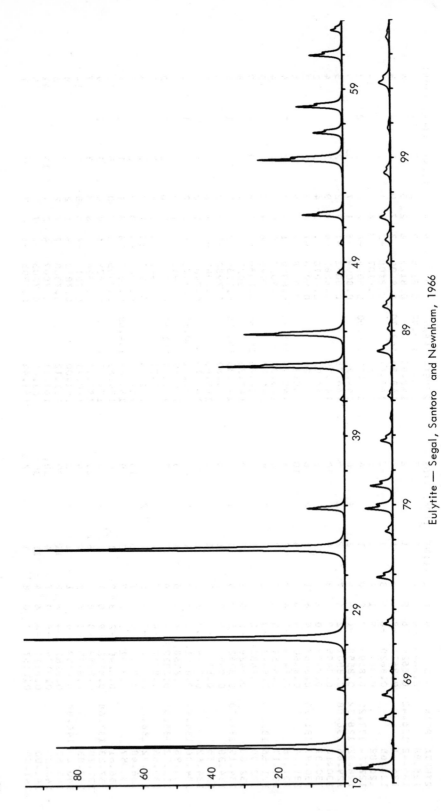

Eulytite — Segal, Santoro and Newnham, 1966

λ = CuKα   Half width at 40°2θ: 0.11°2θ   Plot scale I_{PK} (max) = 100

EULYTITE – SEGAL, SANTORO AND NEWNHAM, 1966

2THETA	PEAK	D	H	K	L	I(INT)	I(PK)	I(DS)
21.11	21.12	4.205	2	1	1	78	86	51
24.42	24.42	3.642	2	2	0	100	100	100
27.36	27.36	3.257	3	1	0	100	93	133
32.50	32.50	2.753	3	2	1	13	11	19
34.81	34.82	2.575	4	0	0			
39.08	39.08	2.303	4	2	0	1	0	3
41.07	41.08	2.196	3	3	2	46	37	100
42.98	42.98	2.102	4	2	2	16	30	37
44.83	44.84	2.020	5	1	0	22		52
48.36	48.36	1.8805	5	2	1	2	2	6
50.05	50.06	1.8208	4	4	0	2	1	4
51.70	51.70	1.7664	5	3	0	17	12	50
54.90	54.90	1.6709	6	1	1	13	25	44
54.90		1.6709	5	3	2	23		75
56.45	56.46	1.6286	6	2	1	13	9	44
57.98	57.98	1.5893	5	4	1	20	14	73
60.95	60.96	1.5187	6	3	1	15	10	59
62.41	62.42	1.4867	4	4	4	5	3	21
63.85	63.84	1.4566	7	1	0	6	12	26
63.85		1.4566	5	4	3	13		55
65.27	65.26	1.4284	6	4	0	1	1	4
66.67	66.68	1.4017	7	2	1	0	4	1
66.67		1.4017	6	3	3	6		29
66.67		1.4017	5	5	2	0		1
68.06	68.06	1.3764	6	4	2	6	3	27
69.43	69.44	1.3525	7	3	0	1	1	7
72.15	72.14	1.3081	6	5	1	4	3	23
72.15		1.3081	7	3	2	0		
73.49	73.50	1.2875	8	0	0	0	0	1
74.82	74.82	1.2678	7	4	1	9	5	50
77.46	77.46	1.2311	6	5	3	4	2	25
78.77	78.78	1.2139	8	2	2	9	8	53
78.77		1.2139	6	6	0	7		44
80.08	80.08	1.1974	8	3	1	5	7	29
80.08		1.1974	7	4	3	8		51
82.67	82.68	1.1662	7	5	2	7	3	48
83.96	83.96	1.1516	8	4	0		1	10
85.25	85.24	1.1374	9	1	0	3	1	19
86.53	86.54	1.1238	8	4	2	1	0	5
87.81	87.82	1.1107	7	6	1	1	4	6
87.81		1.1107	9	2	1	7		50
89.10	89.10	1.0980	6	6	4	3	1	22
90.38	90.38	1.0857	9	3	0	1	3	23
90.38		1.0857	8	5	1	6		4
90.38		1.0857	7	5	4	1		44
92.94	92.94	1.0624	9	3	2	0	1	5
92.94		1.0624	7	6	3	4		2
94.23	94.24	1.0512	8	4	4	5	2	29
95.51	95.52	1.0405	9	4	1	5	3	41
95.51		1.0405	7	7	0	4		41
98.10	98.10	1.0199	10	1	1	1	2	37
98.10		1.0199	7	7	2	3		23
99.39	99.40	1.0100	10	2	0	4	0	32
99.39		1.0100	8	6	2	0		4
100.69	100.70	1.0004	9	5	0	1	1	7
100.69		1.0004	9	4	3	2		22
103.31	103.32	0.9821	10	3	1	0	3	2
103.31		0.9821	7	6	5	3		33
105.96	105.96	0.9647	8	7	1	4	1	37
107.30	107.30	0.9563	10	4	0	4	0	42
107.30		0.9563	8	6	4	0		38
108.65	108.66	0.9482	9	6	1	0	1	3
110.01	110.02	0.9403	10	4	2	2	2	4
111.38	111.38	0.9325	11	1	0	0	2	22
111.38		0.9325	8	7	3	8		3
114.16	114.16	0.9176	11	2	1	4	1	82
114.16		0.9176	10	5	1	3		45
115.57	115.58	0.9104	8	8	0	2	0	29
117.00	117.00	0.9034	11	3	0	2	1	3
117.00		0.9034	9	7	0	2		24

123

EULYTITE - SEGAL, SANTORO AND NEWNHAM, 1966

2THETA	PEAK	D	H	K	L	I(INT)	I(PK)	I(DS)
119.92	119.92	0.8898	10	5	3	3	3	67
119.92		0.8898	11	3	2	5		65
119.92		0.8898	9	7	2	3		33
119.92		0.8898	7	7	6	2		21
121.41	121.40	0.8832	10	6	0	5	2	61
121.41		0.8832	8	6	6	2		26
122.92	122.88	0.8768	8	7	5	0		5
126.03	126.04	0.8644	9	6	5	5	0	66
127.63	127.64	0.8583	8	8	4	6	1	81
127.63		0.8583	12	0	0	3	2	35
129.27	129.28	0.8524	9	7	1	6	3	78
129.27		0.8524	9	7	4	3		40
129.27		0.8524	11	4	3	6		76
129.27		0.8524	11	5	0	1		17
132.66	132.68	0.8410	10	7	1	0	0	4
132.66		0.8410	10	5	5	1		14
132.66		0.8410	11	5	2	0		6
134.43	134.44	0.8354	10	6	4	4	1	64
134.43		0.8354	12	2	2	2		23
136.25	136.26	0.8300	12	3	1	7	1	98
136.25		0.8300	9	8	3	0		5
140.10	140.12	0.8194	10	7	3	6	1	100
140.10		0.8194	11	6	1	5		71
142.14	142.14	0.8143	12	4	0	5	1	81
144.28	144.28	0.8092	11	5	4	4	1	59
148.94	148.96	0.7994	9	9	2	1	2	15
148.94		0.7994	11	6	3	11		188
148.94		0.7994	9	7	6	10		163
151.52	151.54	0.7947	10	8	2	16	2	273
154.34	154.34	0.7900	12	5	1	7	1	66
154.34		0.7900	9	8	5	7		117
154.34		0.7900	11	7	0	5		87
161.11	161.22	0.7808	10	7	5	8	1	146
161.11		0.7808	13	2	1	11		195
161.11		0.7808	11	7	2	0		5

TABLE 6. MISCELLANEOUS SILICATES WITH SINGLE TETRAHEDRA

Variety	Staurolite	Afwillite	Bultfonteinite	Dumortierite	Spurrite
Composition	(Fe,Al,Ti)$_2$(Al,Mg Fe)$_9$(Si,Al)$_4$O$_{24}$H	Ca$_3$(SiO$_3$OH)$_2$·2H$_2$O	Ca$_4$[SiO$_2$(OH$_{1/2}$)$_2$]$_2$ 2H$_2$O	(Al,Fe)$_7$O$_3$BO$_3$ (SiO$_4$)$_3$	Ca$_5$(SiO$_4$)$_2$CO$_3$
Source	St. Gotthard, Switzerland	Scawt Hill, N. Ireland	Bultfontein Mine, Kimberley, S. Af.	not given	Scawt Hill, N. Ireland
Reference	Smith, 1968	Megaw, 1952	McIver, 1963	Golovastikov, 1965	Smith, Karle, Hauptman & Karle, 1960
Cell Dimensions					
a Å	7.8713	16.27	10.992	11.79	10.49
b	16.6204	5.632	8.185	20.209	6.705
c	5.6560	13.23	5.671	4.701	14.16
α deg	90	90	93.95	90	90
β	~90	134.8	91.32	90	101.32
γ	90	90	89.85	90	90
Space Group	C2/m	Cc	P$\bar{1}$	Pmcn	P2$_1$/a
Z	2	4	2	4	4

TABLE 6. (cont.)

Variety	Staurolite	Afwillite	Bultfonteinite	Dumortierite	Spurrite
Site Occupancy	{Fe 0.588, Al 0.292, Ti 0.038}; {Si 0.927, Al 0.064}; {Al(1A) 0.926, Mg 0.048}; {Al(1B) 0.927, Mg 0.051}; {Al(2) 0.934, Mg 0.049}; {Al(3A) 0.277, Fe 0.14}	{Al(3B) 0.185, Fe 0.095}; {Fe(U1) 0.056, Mn 0.024}; {Fe(U2) 0.030, Mn 0.008}			
Method	3-D, counter	3-D, film	2-D, film	3-D (?), film	3-D, film
R & Rating	0.08 (1)	0.245 (3)	0.099 (av.) (2)	0.14 (2)	0.19 (3)
Cleavage and habit	{010} distinct; {110} poor. Prismatic & flattened parallel to b axis	{101} perfect; {100} imperfect. Elongated to b axis	{100} perfect. Fibers parallel to b axis	{100} (?) distinct Needles parallel to c axis	{001} perfect; Granular
Comment					B estimated
μ	226.3	194.2	211.1	101.6	210.7
ASF	0.7363×10^{-1}	0.7588×10^{-1}	0.6413×10^{-1}	0.3471×10^{-1}	0.3390×10^{-1}
Abridging factors	1:1.3:0.5	1:1.3:0.5	1:0:0	1:1.3:0.5	1:1:0.5

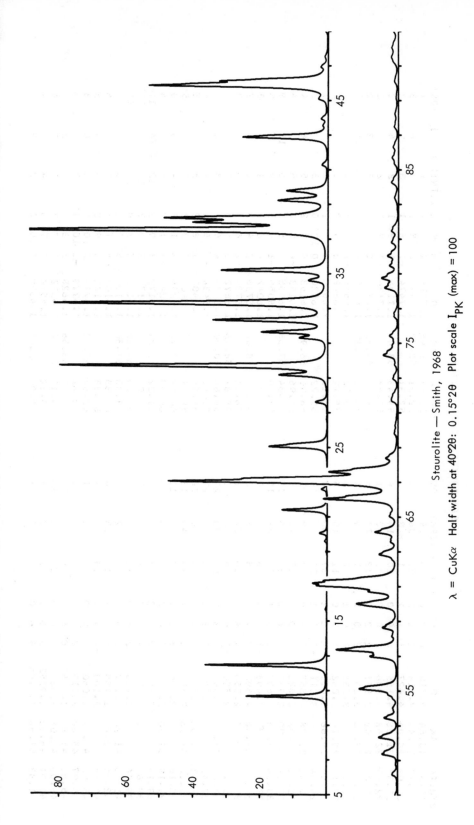

Staurolite — Smith, 1968

λ = CuKα Half width at 40°2θ: 0.15°2θ Plot scale I$_{PK}$ (max) = 100

STAUROLITE - SMITH, 1968

2THETA	PEAK	D	H	K	L	I(INT)	I(PK)	I(DS)
10.64	10.64	8.310	0	2	0	16	24	5
12.43	12.44	7.114	1	1	0	25	36	8
20.04	20.04	4.427	1	1	1	10	2	0
21.37	21.38	4.155	0	4	0	1	14	4
22.57	22.58	3.936	2	0	0	13	2	1
25.01	25.02	3.557	2	0	1	2	17	5
27.59	27.60	3.231	2	2	0	11	4	1
29.14	29.14	3.062	-2	0	1	37	15	5
29.64	29.66	3.011	1	2	1	34	80	17
29.64		3.011	-2	2	1	7		16
31.28	31.28	2.857	-2	4	0	17	8	3
31.61	31.62	2.828	0	0	2	31	20	8
32.29	32.30	2.770	0	6	0	39	34	15
33.24	33.24	2.693	-1	5	1	38	82	19
34.58	34.58	2.592	3	1	0	5	6	2
35.16	35.16	2.550	-2	4	1	16	32	8
35.16		2.550	-2	4	1	15		8
37.46	37.46	2.399	-1	3	2	50	100	27
37.46		2.399	3	3	2	50		27
37.91	37.92	2.371	3	3	0	33	40	18
38.16	38.16	2.356	-3	1	1	24	49	13
38.16		2.356	-3	1	2	20		11
39.19	39.20	2.297	-2	0	2	7	15	4
39.19		2.297	2	0	2	7		4
39.61	39.76	2.273	-1	7	0	2	12	1
39.76		2.265	2	6	0	11		6
41.25	41.26	2.187	3	3	1	14	2	1
42.84	42.84	2.109	-1	7	1	14	25	8
43.92	43.94	2.060	-3	5	0	2	2	9
45.07	45.08	2.010	-2	2	2	58	3	1
45.81	45.82	1.9789	0	6	2	27	53	38
46.09	46.08	1.9678	4	0	0	1	32	18
46.91	46.92	1.9352	3	5	1	5	2	2
51.33	51.34	1.7785	4	4	0	5	5	4

2THETA	PEAK	D	H	K	L	I(INT)	I(PK)	I(DS)
51.54	51.46	1.7718	1	7	2	1	4	1
52.31	52.32	1.7474	-2	8	1	3	6	3
52.31		1.7474	2	8	1	3		2
53.43	53.44	1.7134	-1	9	1	2	4	2
53.43		1.7134	1	9	1	2		2
54.78	54.80	1.6743	0	8	2	2	3	4
55.08	55.10	1.6658	2	2	3	5	11	4
55.12		1.6648	-2	2	2	2		2
55.12		1.6648	-3	5	2	2		1
55.22		1.6620	0	10	0	3		2
56.96	56.96	1.6153	4	0	2	5	8	4
56.96		1.6055	-4	0	2	5		4
57.34	57.36	1.6055	-1	5	3	10	18	9
57.34		1.6042	1	5	3	9		8
57.39		1.5736	4	4	0	4		3
58.61	58.62	1.5736	2	4	3	2	5	2
58.61		1.5672	-2	4	3	3		3
58.87	58.76	1.5434	5	1	0	1	3	1
59.88	60.00	1.5434	-4	6	1	3	12	3
59.99		1.5407	-2	8	2	5		4
59.99		1.5407	2	8	2	6		5
60.69	60.70	1.5246	3	1	3	4	9	4
61.02	61.04	1.5172	-1	9	2	5	24	5
61.02		1.5172	1	9	2	12		3
61.15	61.16	1.5143	5	3	0	13	25	11
61.33	61.30	1.5103	-5	1	1	13	22	11
61.33		1.5103	3	5	1	5		11
61.33		1.5102	-2	10	1	6		4
62.82	62.82	1.4779	-2	10	1	5	6	5
62.82		1.4779	5	1	1	3		5
63.55	63.56	1.4628	1	7	3	3	2	3
64.12	64.12	1.4512	-1	7	3	1	7	1
64.12		1.4512	1	7	3	5		4
65.03	65.04	1.4329	0	10	2	4	3	2

2THETA	PEAK	D	H	K	L	I(INT)	I(PK)	I(DS)
66.01	66.02	1.4140	0	0	4	31	22	30
67.01	67.02	1.3954	-4	6	2	50	68	50
67.01		1.3954	-4	6	2	50		50
67.58	67.58	1.3850	0	12	0	24	21	25
68.37	68.38	1.3708	-5	1	2	2	4	2
68.37		1.3708	-5	1	2	2		2
71.66	71.76	1.3158	-2	8	3	1		1
72.07	72.00	1.3094	3	11	0	1	2	1
72.94	72.96	1.2958	6	2	0	2	2	2
73.74	73.74	1.2837	6	2	4	2	2	2
73.74		1.2756	-1	5	4	1	1	1
74.29	74.30	1.2756	-3	11	1	3	5	4
74.29		1.2697	-3	11	0	3		3
74.69	74.48	1.2697	-4	10	0	1	3	1
74.86	74.84	1.2673	2	4	4	1	2	1
74.86		1.2673	-2	4	4	1		1
75.41	75.40	1.2594	0	6	4	3	3	4
76.52	76.54	1.2439	0	12	3	2	1	2
78.16	78.18	1.2218	-4	6	2	2	4	2
78.16		1.2218	-4	6	3	2		2
78.19		1.2215	-6	4	1	2		2
78.19		1.2215	-6	4	4	2		2
78.73	78.72	1.2145	-3	3	4	4	5	5
78.73		1.2145	-3	3	4	4		5
79.45	79.46	1.2052	-5	1	3	2	3	3
79.45		1.2052	-5	1	3	2		2
80.02	80.02	1.1980	5	9	0	5	3	6
80.79	80.80	1.1885	2	10	3	1	2	2
80.79		1.1885	-2	10	3	1		2
81.45	81.46	1.1806	5	3	3	2	2	2
81.45		1.1806	-5	3	3	1		2
83.36	83.36	1.1583	-4	10	2	1	1	1
83.36		1.1583	-4	10	2	1		1
84.25	84.26	1.1483	4	0	4	2	2	3
84.25		1.1483	-4	0	4	2		3
85.33	85.38	1.1366	2	14	0	1	1	1
85.70	85.68	1.1326	4	12	0	3	2	4

2THETA	PEAK	D	H	K	L	I(INT)	I(PK)	I(DS)
86.30	86.30	1.1263	-3	13	1	1	2	2
86.30		1.1263	3	13	1	2		2
87.46	87.46	1.1143	-2	14	1	1	1	2
87.46		1.1143	-2	14	1	1		2
87.96	87.70	1.1092	6	8	0	1		1
88.20	88.20	1.1068	4	4	4	1	1	1
88.20		1.1068	-4	4	4	1	1	1
89.15	89.14	1.0975	-1	3	5	1		1
89.15		1.0975	-1	3	5	1	1	1
90.08	90.10	1.0885	-6	8	1	1		2
90.08		1.0885	-6	8	1	1	1	2
91.21	91.50	1.0780	-2	2	5	1	2	2
91.21		1.0780	-2	2	5	1		1
91.32		1.0770	0	10	4	1		2
91.49		1.0754	3	11	3	1		2
91.49		1.0754	-3	11	1	1		3
92.04	92.04	1.0704	-5	11	1	2	2	2
92.04		1.0704	5	11	1	2		2
93.09	93.12	1.0611	-1	5	5	1		2
93.13		1.0608	-4	6	4	1		2
93.13		1.0608	-4	6	4	1		2
94.16	94.22	1.0518	-2	4	5	1	3	2
94.16		1.0518	-2	4	5	1		2
94.21		1.0514	4	12	2	2		3
94.21		1.0514	-4	12	2	2		3
95.28	95.26	1.0424	-6	4	3	1	2	2
95.28		1.0424	-6	4	3	1		2
96.37	96.40	1.0335	-5	3	4	3	4	4
96.37		1.0322	3	9	4	3		4
96.53		1.0322	-3	9	4	2		2
96.53		1.0268	-7	3	2	1		2
97.21	97.20	1.0268	-7	3	2	1	2	2
97.21		1.0229	1	15	2	2		3
97.70	97.70	1.0229	-1	15	2	2	3	3
97.70		1.0207	3	15	0	2		3
97.98	97.96							

129

STAUROLITE - SMITH, 1968

2THETA	PEAK	D	H	K	L	I(INT)	I(PK)	I(DS)
98.45	98.42	1.0171	-5	11	2	1	2	2
98.45		1.0171	-5	11	2	1		2
98.53		1.0165	4	14	0	1		2
99.03	98.96	1.0127	-1	7	5	1	2	2
99.03		1.0127	-1	7	5	1		2
100.72	100.70	1.0002	-7	7	1	1	1	1
100.72		1.0002	-7	7	1	1		1
101.20	101.18	0.9968	-7	5	2	1	1	2
101.20		0.9968	7	5	2	1		2
101.81	102.26	0.9924	0	12	4	1	5	2
102.24		0.9894	4	14	1	14		23
102.32		0.9889	-2	16	1	1		1
102.32		0.9889	-2	16	1	1		1
103.04	103.02	0.9839	-8	0	0	6	3	10
103.42		0.9813	3	13	3	1		1
103.42		0.9813	-3	13	3	1		2
104.36	104.64	0.9751	0	16	2	1	1	1
104.62		0.9734	-2	14	3	1		1
104.62		0.9734	-1	2	3	1		1
105.10	105.30	0.9702	1	17	0	1	2	1
105.24		0.9694	-8	0	1	1		2
105.24		0.9694	-8	0	1	1		2
105.51	105.44	0.9676	-6	10	2	1	2	2
105.51		0.9676	-6	10	2	1		2
107.12	107.34	0.9575	-1	9	5	1	3	2
107.12		0.9575	-1	9	5	1		2
107.32		0.9562	-1	17	1	1		2
107.32		0.9562	-1	17	1	1		2
107.35		0.9560	-6	8	3	1		2
107.46		0.9560	-6	8	3	1		2
107.46		0.9553	-6	2	4	1		1
109.41	109.48	0.9437	5	11	3	1	1	1
109.41		0.9437	-5	11	3	1		2
109.59		0.9427	0	0	6	1		1
112.85	113.12	0.9245	-4	6	5	1	2	1
113.06		0.9234	0	18	0	1		2

2THETA	PEAK	D	H	K	L	I(INT)	I(PK)	I(DS)
113.15	113.15	0.9229	-1	3	6	2		4
113.15		0.9229	-1	3	6	2		4
114.23	114.82	0.9172	-5	1	5	1	2	2
114.23		0.9172	-5	1	5	1		2
114.44		0.9161	3	17	0	1		3
114.85		0.9141	5	9	4	2		1
114.85		0.9141	5	9	4	2		5
115.69	115.74	0.9098	-5	10	5	1	2	4
115.69		0.9098	-2	10	5	1		2
115.77		0.9094	-7	9	2	1		1
115.77		0.9094	-7	9	2	1		3
115.81		0.9092	0	14	4	1		3
118.84	118.92	0.8947	5	15	1	1	1	2
118.84		0.8947	-5	15	1	1		2
118.86		0.8946	-7	7	3	1		2
118.86		0.8946	-7	7	3	1		1
119.34	119.28	0.8924	-0	6	6	2	1	4
120.67	121.28	0.8864	-2	16	3	1	1	2
120.80		0.8859	-2	14	4	1		1
120.80		0.8859	-2	14	4	1		3
121.23		0.8840	4	12	4	1		3
121.23		0.8840	-4	12	4	1		3
121.92	121.96	0.8810	-8	6	2	1	2	4
121.92		0.8810	-8	6	2	1		4
122.10		0.8803	-6	14	0	1		4
122.58		0.8782	5	13	3	1		1
122.69		0.8778	0	18	2	2		2
123.67	124.04	0.8737	-4	16	2	1	1	1
123.67		0.8737	-4	16	2	1		1
124.02		0.8723	-8	0	3	1		2
124.02		0.8723	-8	0	3	1		2
126.47	126.56	0.8627	-1	17	3	2	1	4
126.47		0.8627	-1	17	3	2		4
127.34	127.30	0.8594	-7	11	2	1	1	1
127.34		0.8594	-7	11	2	1		1
127.37		0.8593	-1	19	1	1		1
127.37		0.8593	1	19	1	1		1

STAUROLITE - SMITH, 1968

2THETA	PEAK	D	H	K	L	I(INT)	I(PK)	I(DS)
128.27	128.26	0.8560	3	11	5	2	1	4
128.27		0.8560	-3	11	5	2		3
128.83		0.8540	9	3	1	1		1
128.83		0.8540	-9	3	1	1		1
131.20	131.24	0.8458	9	5	0	1	1	3
134.09	134.60	0.8365	9	5	1	1	2	2
134.09		0.8365	-9	5	1	1		2
134.28		0.8359	4	18	0	1		2
134.62		0.8349	1	9	6	3		7
134.62		0.8349	-1	9	6	3		7
137.08	137.64	0.8276	3	15	4	2	2	5
137.08		0.8276	-3	15	4	2		5
137.59		0.8262	9	3	2	4		8
137.59		0.8262	-9	3	2	4		8
137.88		0.8254	-4	14	4	1		2
137.88		0.8254	4	14	4	1		2
142.78	142.90	0.8127	4	6	6	17	5	39
142.78		0.8127	-4	6	6	17		39
142.96		0.8123	5	13	4	1		1
142.96		0.8123	-5	13	4	1		1
143.02		0.8122	9	7	1	1		1
143.02		0.8122	-9	7	1	1		1
143.81		0.8103	9	5	2	1		2
143.81		0.8103	-9	5	2	1		2
144.92	145.06	0.8078	5	1	6	1	3	2
144.92		0.8078	-5	1	6	1		2
145.00		0.8076	8	0	4	7		17
145.00		0.8076	-8	0	4	7		17
145.65		0.8062	3	13	5	1		3
145.65		0.8062	-3	13	5	1		3
145.71		0.8061	6	16	1	1		3
145.71		0.8061	-6	16	1	1		3
146.20		0.8048	2	20	1	1		1
147.29	147.84	0.8027	2	10	6	14	5	34
147.59		0.8021	8	12	0	1		3
147.76		0.8018	2	14	5	1		3
147.76		0.8018	-2	14	5	1		3

2THETA	PEAK	D	H	K	L	I(INT)	I(PK)	I(DS)
147.84		0.8016	-4	18	2	2		34
147.84		0.8016	4	18	2	2		34
148.65		0.8000	-1	17	4	1		2
148.65		0.8000	1	17	4	1		2
151.80	151.52	0.7942	-8	12	1	3	2	8
151.80		0.7942	8	12	1	3		8
152.61		0.7928	8	4	4	1		2
152.61		0.7928	-8	4	4	1		2
153.08		0.7920	-6	8	5	2		4
153.08		0.7920	6	8	5	1		3
154.05		0.7904	9	9	0	1		2
156.44	159.70	0.7868	4	8	6	1	1	2
156.44		0.7868	-4	8	6	1		2
157.48		0.7854	-9	3	3	1		1
157.79		0.7849	5	11	5	3		8
157.79		0.7849	-5	11	5	4		10
159.44		0.7828	-9	9	1	4		9
159.44		0.7828	9	9	1	4		9
159.63		0.7826	-6	16	2	2		2
160.59		0.7814	-6	16	2	1		11
160.59		0.7814	-2	20	2	1		10
160.74		0.7813	2	20	2	1		11
160.74		0.7813	1	5	7	1		10
161.91		0.7800	-1	5	7	2		5
161.91		0.7800	-1	21	1	2		6
162.52		0.7793	-1	21	1	3		9
164.33		0.7775	0	12	6	1		3
164.33		0.7775	-2	4	7	2		4

131

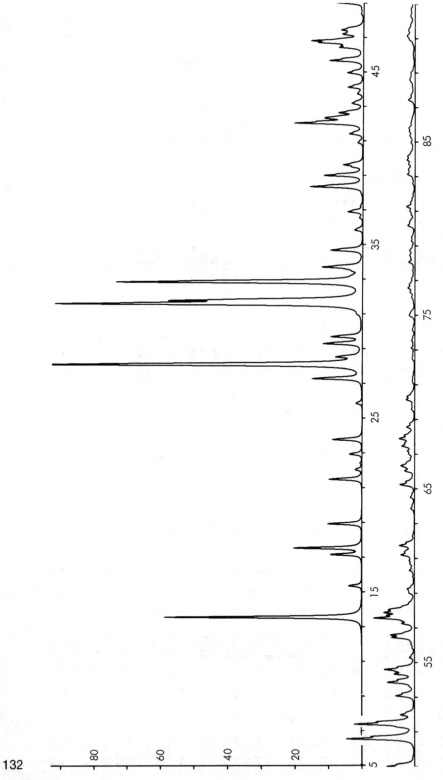

Afwillite — Megaw, 1952

λ = CuKα Half width at 40°2θ: 0.11°2θ Plot scale I$_{PK}$ (max) = 100

132

2THETA	PEAK	D	H	K	L	I(INT)	I(PK)	I(DS)
13.53	13.54	6.539	-2	0	2	46	59	36
15.34	15.34	5.772	2	0	0	3	4	6
17.13	17.14	5.173	-1	1	1	8	9	14
17.51	17.52	5.062	1	1	0	17	20	7
18.89	18.90	4.694	0	0	2	9	10	8
21.46	21.46	4.138	-1	0	2	9	10	2
22.01	22.02	4.035	-4	0	1	2	2	3
22.90	22.90	3.880	-3	1	1	3	2	8
23.75	23.76	3.743	-3	1	2	8	4	2
25.82	25.82	3.447	-3	1	1	2	9	14
27.25	27.26	3.269	-3	1	3	14	2	100
28.06	28.06	3.177	-4	0	4	100	15	6
28.50	28.50	3.129	3	1	0	6	100	12
29.26	29.26	3.049	-1	1	3	12	8	10
29.66	29.66	3.009	-2	1	4	9	12	100
31.55	31.56	2.833	-1	0	2	94	9	53
31.75	31.74	2.816	-3	1	4	49	92	1
31.76		2.815	0	2	0	1	58	86
32.81	32.82	2.727	-5	1	2	79	1	2
33.45	33.46	2.677	-5	1	4	2	73	13
33.68	33.68	2.659	-6	0	4	11	4	9
34.64	34.64	2.587	-2	2	1	8	12	3
34.65		2.586	3	1	2	3	10	3
35.83	35.82	2.504	-2	1	2	3	2	1
36.51	36.52	2.459	-6	1	2	1	1	6
36.87	36.88	2.436	-5	1	1	18	5	22
38.32	38.32	2.347	-1	1	4	13	16	16
38.97	38.98	2.309	0	0	4	3	12	4
39.48	39.48	2.281	-4	2	2	5	4	7
39.61	39.60	2.273	-4	2	3	2	6	2
40.87	40.88	2.206	-4	2	1	5	2	6
41.39	41.40	2.180	-6	0	6	24	4	31
42.00	42.00	2.149	-4	0	6	8	20	10
42.27	42.30	2.136	5	1	0	8	12	9
42.32		2.134	-4	2	4	7	7	10
42.61	42.60	2.120	-3	1	2	1	3	2
43.13		2.095	-7	1	5			

2THETA	PEAK	D	H	K	L	I(INT)	I(PK)	I(DS)
43.18		2.093	0	2	3	3		3
43.72	43.72	2.069	-2	2	2	2	2	3
44.09	44.10	2.052	-5	1	6	6	5	8
44.93	44.94	2.016	4	2	0	6	5	8
45.64	45.64	1.9861	2	2	0	13	10	18
46.40	46.40	1.9554	-7	1	6	8	7	11
46.49	46.52	1.9517	-7	1	2	1	7	2
46.65		1.9452	-6	2	3	3		4
46.68	46.68	1.9443	-8	0	6	11	14	16
46.78	46.78	1.9401	-6	2	6	11	16	15
47.14		1.9264	4	0	2	2		7
47.20	47.20	1.9241	6	1	0	5	6	10
47.43	47.42	1.9153	-3	3	1	7	7	2
47.51		1.9122	-1	3	1	1	8	2
48.98	49.00	1.8583	-6	2	5	1	21	14
48.99		1.8576	0	2	4	10	18	42
50.58	50.58	1.8029	-8	0	2	27	12	34
51.45	51.44	1.7746	-7	3	1	22	6	2
51.62	51.58	1.7692	-3	3	3	2	2	8
51.97	51.98	1.7579	2	2	3	5	8	8
53.06	53.06	1.7245	-3	3	3	5	6	2
53.07		1.7242	-6	2	6	1	9	2
53.09		1.7236	-4	2	6	1	2	2
53.59	53.60	1.7085	-9	3	0	11	7	18
53.85	53.84	1.7011	3	3	3	7	7	12
54.32	54.32	1.6873	-1	3	3	11	4	18
54.58	54.58	1.6799	-3	3	3	2	12	2
55.27	55.28	1.6606	-5	3	6	8	9	14
56.43	56.44	1.6291	-10	0	6	1	8	2
56.56	56.58	1.6257	-5	3	3	4	2	6
56.58		1.6252	-8	2	6	4	7	7
57.24	57.24	1.6082	5	3	2	16	4	28
57.56	57.56	1.6000	-4	2	2	10	12	17
57.87	57.88	1.5921	-7	1	8	2	9	4
57.95		1.5900	3	3	1	4		7
58.08	58.02	1.5867				1	8	4
58.44	58.40	1.5779					2	2

133

AFWILLITE – MEGAW, 1952

2THETA	PEAK	D	H	K	L	I(INT)	I(PK)	I(DS)
58.99	59.00	1.5645	-2	2	6	1	2	2
59.24	59.24	1.5585	-10	0	4	2	2	4
60.30	60.30	1.5336	-5	1	8	2	2	2
60.69	60.68	1.5247	-4	0	8	1	2	9
61.19	61.20	1.5133	3	1	4	5	4	3
61.44	61.36	1.5077	-3	3	5	2	3	11
61.73	61.74	1.5014	-8	2	2	6	5	3
63.86	63.86	1.4563	6	0	5	1	1	
64.50	64.52	1.4435	-7	3	6	5	1	2
65.23	65.24	1.4291	-11	1	8	2	4	10
66.02	66.02	1.4290	0	3	0	4		4
66.30	66.32	1.4138	-8	2	7	4	3	8
66.33		1.4086	-1	4	0	2	4	3
67.17	67.18	1.4080	0	4	1	3		6
67.58	67.46	1.3924	0	4	1	2	2	3
67.89	67.88	1.3871	-2	1	8	5	4	10
68.55	68.56	1.3850	-11	1	6	1		2
70.13	70.12	1.3794	1	1	6	6	5	12
70.28	70.30	1.3677	0	2	6	3	2	6
72.87	72.86	1.3408	-4	2	8	4	3	8
73.10	73.10	1.3383	-12	0	8	2	2	3
73.52	73.52	1.2969	2	0	6	1	1	1
74.11	74.12	1.2934	-9	3	6	1	1	2
74.19		1.2870	-2	4	10	2	1	3
74.93	75.00	1.2783	-2	2	9	1	1	1
75.04		1.2771	-6	2	10	2	2	4
76.03	76.02	1.2663	-7	1	7	1		3
76.34	76.34	1.2647	-10	2	9	1	1	2
76.55	76.56	1.2507	9	1	0	1	1	3
76.74	76.76	1.2464	-11	0	10	2	1	5
77.11	77.10	1.2435	5	3	2	2	2	3
77.56	77.56	1.2409	-7	1	8	1	1	3
78.04	78.04	1.2358	-12	0	10	1	1	2
78.41	78.44	1.2298	-1	1	8	1	1	7
78.47		1.2234	-6	4	5	3	2	6
		1.2186	-13	2	8	1		2
		1.2178	-2	2	8	2		4

2THETA	PEAK	D	H	K	L	I(INT)	I(PK)	I(DS)
78.68	78.70	1.2150	-5	3	8	1	2	2
78.78		1.2138	-12	2	6	1		3
79.28	79.28	1.2074	0	4	4	2	1	4
79.48	79.48	1.2049	-11	3	4	1	1	2
80.16	80.16	1.1963	-8	1	2	1	2	8
80.30		1.1946	-5	2	10	1		2
80.50	80.38	1.1921	-6	4	10	1	1	1
80.79	80.80	1.1886	-12	2	5	1	1	3
80.94		1.1868	-10	2	10	1	2	2
80.99	81.00	1.1861	-10	2	5	1	1	3
81.25	81.24	1.1830	2	4	2	3	3	8
81.74	81.70	1.1771	8	1	1	1	1	1
82.56	82.58	1.1675	8	0	2	1	2	2
83.11	83.12	1.1612	3	1	6	2		3
83.12		1.1610	-11	3	6	2		6
83.26		1.1595	-13	0	10	1		3
83.43	83.38	1.1575	10	2	7	1		3
83.70	83.72	1.1545	-6	3	0	1		3
83.73		1.1541	-7	3	0	1		2
83.94	83.96	1.1518	-1	3	10	3	2	7
84.11	84.16	1.1499	-9	3	7	1		4
84.18		1.1491	-9	3	9	1		4
84.94	84.96	1.1407	5	3	3	1		3
84.98		1.1404	-8	1	8	1	1	1
85.29	85.36	1.1370	-11	3	6	1		4
85.36		1.1362	-8	4	3	2		4
85.58	85.60	1.1339	1	3	6	2	2	7
85.79	85.78	1.1317	-12	0	10	3	2	4
86.19	86.06	1.1274	-14	1	6	1	1	9
87.43	87.44	1.1146	-13	4	4	3	2	3
87.45		1.1144	-5	5	9	1		1
87.94	87.94	1.1094	-8	4	4	1	1	2
88.21	88.22	1.1068	-4	7	1	1	1	2
88.45	88.60	1.1043	1	5	5	2		6
88.58		1.1030	-8	4	1	1		2
88.72		1.1017	-10	0	12	2		2
89.00	88.98	1.0990	-3	5	1	2	1	4

134

AFWILLITE - MEGAW, 1952

2THETA	PEAK	D	H	K	L	I(INT)	I(PK)	I(DS)
90.06	90.08	1.0887	-3	1	10	1	1	2
90.96	91.14	1.0802	-11	1	12	1	2	3
91.09		1.0791	-1	5	2	2		5
91.15		1.0785	8	2	3	2		5
91.49	91.48	1.0754	-11	3	3	1	2	2
91.57		1.0747	-8	0	12	2		4
92.29	92.28	1.0682	10	2	0	2	1	6
92.42		1.0670	-14	2	7	1		1
92.71	92.68	1.0644	4	4	3	1	1	2
92.72		1.0643	-5	5	3	1		2
93.29	93.30	1.0593	-5	5	2	1	1	2
93.50	93.50	1.0575	2	0	7	1	1	3
94.21	94.24	1.0514	-2	1	10	1	1	3
94.24		1.0512	-13	1	12	1		3
94.77	94.78	1.0466	-14	2	6	2	1	5
94.99	95.06	1.0448	-10	4	4	1	1	2
95.40	95.36	1.0414	-5	5	11	1	1	3
95.62		1.0394	-13	3	8	1		2
96.69	96.68	1.0309	-7	1	12	1	1	4
96.92	96.94	1.0291	-3	5	5	1		2
96.96		1.0288	7	1	4	1	1	3
97.37	97.32	1.0255	-11	3	2	1	1	3
98.46	98.48	1.0171	-4	2	6	1	1	4
98.46		1.0171	-10	4	3	1		2
99.68	99.71	1.0077	-7	5	3	1	1	2
100.20	100.22	1.0040	-8	2	12	1	1	2
100.30	100.33	1.0031	3	3	6	1		2
102.88	102.89	0.9850	-2	2	10	1	0	2
104.74	104.74	0.9726	-13	3	4	1	1	3
105.85	105.96	0.9654	-9	5	12	1		2
107.37	107.38	0.9559	-9	5	5	1	0	3
107.74	107.78	0.9536	-12	4	9	1	1	3
107.85		0.9530	-16	2	10	1		3
108.38	108.39	0.9497	-16	2	8	1	1	2
111.14	111.16	0.9338	6	0	6	1	1	4
111.40	111.52	0.9324	-2	6	1	1	1	3
111.89	111.90	0.9297	-13	1	14	1		2

2THETA	PEAK	D	H	K	L	I(INT)	I(PK)	I(DS)
111.94		0.9294	-17	3	8	1		3
113.61	113.66	0.9204	3	1	7	1	0	2
113.68		0.9201	-17	1	12	1		3
116.06	116.06	0.9079	-15	0	14	1	0	3
117.51	117.48	0.9009	-16	5	1	1	0	3
118.96	119.02	0.8941	-11	4	5	1	1	5
119.42	119.76	0.8920	-14	4	7	1	2	2
119.53		0.8915	7	5	7	1		4
119.73		0.8906	-1	5	5	2		8
119.81		0.8903	-9	9	9	2		6
120.47	120.58	0.8873	-16	1	4	1	1	4
120.53		0.8871	11	1	2	1		2
120.66		0.8865	2	4	7	1		3
120.68		0.8864	6	2	6	1		5
120.69		0.8863	5	5	3	1		2
121.97	122.00	0.8808	9	1	4	1	1	3
122.00		0.8807	-6	4	11	1		2
122.75	122.76	0.8775	-6	4	5	2	1	7
122.82		0.8772	13	1	0	1		5
123.16	123.18	0.8758	1	3	10	1		5
124.91	125.14	0.8687	-5	3	12	1		2
125.18		0.8677	-10	4	12	1		2
126.59	126.62	0.8622	2	6	3	3	1	10
126.62		0.8621	-14	4	11	1		2
128.64	128.68	0.8547	-11	5	3	2	1	6
128.75		0.8543	-8	4	12	1		2
129.81	129.76	0.8505	-18	2	12	2	1	6
131.04	131.04	0.8463	-10	4	1	2	1	6
131.05		0.8463	-16	6	4	1		4
131.86	131.74	0.8436	-8	6	3	1	1	3
133.73	133.68	0.8376	-4	4	11	1	0	3
134.49	134.56	0.8352	-17	3	12	1	1	3
134.66		0.8348	-18	0	6	1		4
135.41	135.36	0.8325	-8	6	7	1	1	3
135.80		0.8313	-4	6	0	1		4
138.15	138.46	0.8246	14	0	0	1	0	3
138.51		0.8237	-12	4	13	1		3

AFWILLITE – MEGAW, 1952

2THETA	PEAK	D	H	K	L	I(INT)	I(PK)	I(DS)
139.33	139.28	0.8215	-10	4	13	1	1	4
140.03	140.10	0.8196	-12	0	16	1	1	4
140.15		0.8193	-19	1	8	1		6
140.67	140.92	0.8180	-13	5	5	1	1	3
141.15		0.8167	0	2	11	1		3
141.22		0.8166	-13	1	16	1		3
141.71	141.70	0.8153	-18	2	14	1	1	3
141.82		0.8151	-19	1	14	1		3
144.17	144.14	0.8095	-14	4	13	1	0	3
144.75	144.80	0.8082	-14	4	3	1	1	4
145.69	145.90	0.8061	-2	6	7	1	1	4
145.96		0.8055	9	3	4	1		6
147.05	147.28	0.8032	-8	4	13	1	1	4
147.17		0.8030	2	2	10	1		3
147.27		0.8028	13	3	0	2		7
147.77	147.76	0.8018	-11	1	16	2	1	7
147.81		0.8017	1	3	10	1		5
148.48		0.8003	-18	2	6	1		3
148.53		0.8002	-13	5	4	1		3
149.54	149.56	0.7983	-13	3	15	1	1	2
149.81		0.7978	-10	6	8	1		3
150.55	150.66	0.7964	1	7	1	1	0	3
150.75		0.7960	10	2	4	1		4
151.68	151.70	0.7944	-3	7	1	1		3
153.54	153.66	0.7912	-10	6	3	1	0	6
153.57		0.7912	-3	7	3	1		3
153.62		0.7911	-1	5	9	1	1	2
153.66		0.7911	-8	6	9	1		6
154.14		0.7903	-7	3	14	1		3
156.35	157.78	0.7869	-12	2	16	1	1	7
156.46		0.7868	-1	7	3	5		22
156.55		0.7867	-4	0	14	1		3
157.34		0.7856	6	0	8	1		5
157.50		0.7853	-2	4	11	1		3
157.61		0.7852	-6	6	9	3		16
157.77		0.7850	-16	2	16	1		6
157.95		0.7847	3	1	10	3		13
158.03		0.7846	-12	4	14	1		5
159.18		0.7831	-17	3	5	1		3
159.77		0.7824	-5	5	11	1		5
159.87		0.7823	-10	6	9	2		10
159.92		0.7822	-4	4	12	1		5

136

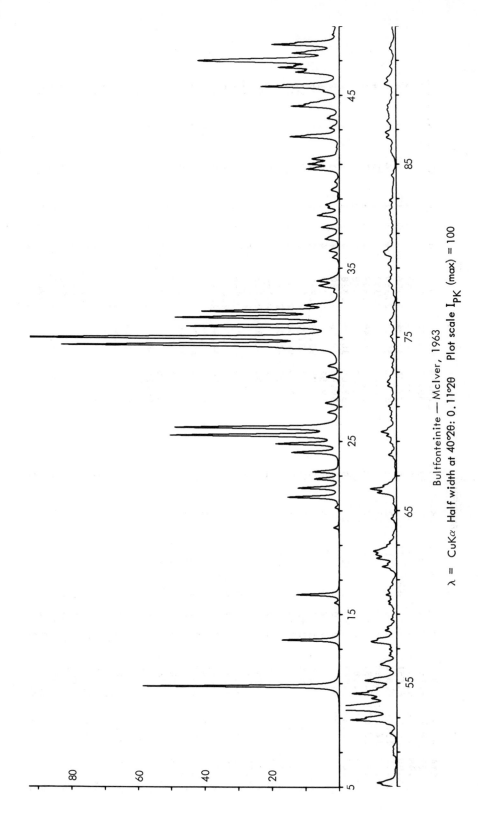

Bultfonteinite — McIver, 1963

λ = CuKα Half width at 40°2θ: 0.11°2θ Plot scale I$_{PK}$ (max) = 100

BULTFONTEINITE - MCIVER, 1963

2THETA	PEAK	D	H	K	L	I(INT)	I(PK)	I(DS)
10.83	10.84	8.166	0	1	0	44	59	29
13.49	13.50	6.557	1	1	0	6	17	4
13.50		6.551	-1	-1	0	7		5
15.65	15.66	5.656	0	0	1	1	2	1
16.12	16.12	5.495	2	0	0	10	13	7
19.99	20.00	4.437	-1	0	1	1	2	1
21.75	21.76	4.083	0	2	0	13	15	11
22.28	22.28	3.987	-2	0	1	11	12	9
22.80	22.80	3.897	2	0	1	6	8	5
23.21	23.22	3.828	1	2	0	4	8	3
23.23		3.826	-1	-2	0			
24.34	24.34	3.654	-2	-1	1	13	14	11
24.83	24.84	3.582	2	-1	1	17	19	15
25.32	25.32	3.515	-2	1	1	45	50	40
25.77	25.78	3.455	2	1	1	46	49	41
26.64	26.66	3.343	3	1	0	1	3	1
26.66		3.341	-3	-1	0	2		2
27.17	27.20	3.279	-2	2	0	2	4	1
27.20		3.275	-3	-2	0	2		2
28.71	28.70	3.107	3	0	1	4	4	3
29.33	29.32	3.043	-2	-2	1	3	3	3
30.53	30.54	2.925	-2	2	1	80	83	80
30.95	30.96	2.887	2	2	1	79	100	79
30.96		2.886	3	1	1	20		20
31.61	31.62	2.828	0	2	1	44	46	44
32.11	32.12	2.785	-2	0	2	48	49	49
32.47	32.48	2.755	2	0	2	40	41	41
32.48		2.754	1	2	1	1		1
32.80	32.84	2.728	-3	2	0	4	10	4
32.84		2.725	-3	-2	2	4		4
32.85		2.724	-1	1	2	2		2
33.61	33.62	2.664	-1	-3	0	2	2	1
33.89	33.98	2.643	1	3	0	2	6	1
33.97		2.641	-1	3	0	1		1
33.91		2.637	1	-1	2	4	7	4
34.23	34.24	2.617	-2	1	2	7		7
35.59	35.60	2.520	-3	-2	1	2	2	2

2THETA	PEAK	D	H	K	L	I(INT)	I(PK)	I(DS)
35.98	36.00	2.494	-4	0	1	3	3	3
36.66	36.66	2.449	-4	0	1	4	4	5
37.34	37.34	2.406	-4	-1	1	6	5	6
38.02	38.02	2.365	-4	1	1	4	7	4
38.02		2.364	-4	1	1	3		4
38.45	38.46	2.339	-1	-3	2	1	3	1
38.65	38.64	2.338	4	1	0	3		2
39.48	39.50	2.328	4	1	0	1	4	2
39.52		2.280	-4	2	0	1	2	1
39.98	39.98	2.278	0	2	2	1	1	2
40.70	40.70	2.253	-1	1	2	9	10	12
40.99	41.00	2.215	1	2	2	10	9	12
41.27	41.30	2.200	3	2	2	4	8	5
41.31		2.186	-3	3	0	5		6
42.55	42.56	2.184	5	3	1	8	15	12
42.57		2.123	-2	2	2	3		11
43.08	43.08	2.122	0	4	2	9	3	4
43.65	43.64	2.098	0	4	2	8	4	5
44.34	44.34	2.072	-1	0	2	3	14	22
44.55	44.54	2.041	0	3	0	16	9	9
45.19	45.20	2.032	1	0	2	7	5	5
45.46	45.46	2.005	-4	1	1	3	23	33
45.49		1.9935	4	1	1	24		3
45.56		1.9922	-5	1	1	3		3
46.24	46.30	1.9893	5	1	2	2	13	4
46.31		1.9618	-4	4	0	3		17
46.57	46.58	1.9590	-1	4	2	3	18	26
46.88	46.96	1.9484	5	2	1	19	42	8
46.89		1.9362	4	3	0	6		5
46.93		1.9361	-5	2	0	5		33
46.93		1.9346	4	3	0	4		4
46.98		1.9344	-1	4	1	3		32
47.04		1.9326	1	4	1	2		6
47.39	47.42	1.9302	-2	3	2	2	14	3
47.42		1.9168	4	1	2	11		15
47.43		1.9155	-4	2	2	2		3
		1.9150						

BULTFONTEINITE - MCIVER, 1963

2THETA	PEAK	D	H	K	L	I(INT)	I(PK)	I(DS)
47.91	47.92	1.8969	0	3	2	23	20	32
47.95		1.8955	2	-3	2	2		3
48.50	48.50	1.8755	-1	-1	1	2	2	2
49.08	49.10	1.8544	-1	4	1	4	5	6
49.20	49.20	1.8503	1	4	1	4	6	6
52.09	52.08	1.7543	-6	0	1	1	2	2
52.83	52.86	1.7313	-5	-1	2	5	14	8
52.85		1.7309	6	0	1	1		2
52.86		1.7304	1	-2	3	1		
52.87		1.7303	-2	1	3	1		2
53.11	53.00	1.7230	-6	-1	2	11	10	17
53.47	53.48	1.7121	0	-4	2	2	24	3
53.52		1.7108	5	3	0	23		36
53.57	53.60	1.7091	-5	3	1	4	26	7
53.61		1.7080	-2	1	0	11		17
53.63		1.7073	-6	1	1	3		4
54.10	54.10	1.6937	5	-1	2	6	8	10
54.15		1.6922	-3	0	3	2		
54.36	54.38	1.6864	6	-1	1	3	14	5
54.39		1.6854	-2	-2	3	13		21
54.65	54.68	1.6779	-3	-1	4	2	6	4
54.70		1.6766	-3	4	1	4		6
55.02	55.14	1.6766	3	4	1	2	10	3
55.07		1.6660	5	1	2	3		
55.14		1.6642	2	-0	3	9		14
55.26		1.6609	3	0	3	2		3
55.76	55.76	1.6471	3	-1	1	2	2	3
56.05	56.08	1.6394	4	4	0	4	5	6
56.11		1.6377	-4	4	0	4		6
56.59	56.60	1.6250	-6	-2	1	2	2	4
56.96	56.96	1.6151	-1	5	0	1	2	2
57.34	57.40	1.6054	-2	2	1	3	8	6
57.35		1.6052	6	-2	2	2		3
57.41		1.6036	0	4	1	6		
57.60	57.56	1.5989	-6	2	1	2	6	11
58.04	58.04	1.5878	-2	2	3	2	4	4
58.05		1.5874	-5	2	2	1		2

2THETA	PEAK	D	H	K	L	I(INT)	I(PK)	I(DS)
58.26	58.22	1.5822	6	2	1	2	2	4
58.69	58.68	1.5718	-4	3	1	2	1	3
59.19	59.18	1.5596	5	2	2	1	1	2
59.58	59.58	1.5504	4	-3	1	1	2	4
59.74	59.74	1.5466	-2	-3	1	3		3
60.10	60.10	1.5381	2	-5	1	1	3	5
60.37	60.36	1.5319	-5	0	1	2	3	4
60.83	60.84	1.5215	-7	-1	1	5		2
61.74	61.74	1.5012	-1	3	1	1	4	9
62.17		1.4919	-7	1	1	1	6	2
62.22	62.22	1.4907	-1	-1	3	5		10
62.45	62.44	1.4858	-2	5	1	5	6	9
62.56		1.4835	7	-1	1	3	7	5
62.64	62.64	1.4817	2	-5	1	1		5
62.75		1.4795	5	-3	2	2		3
62.99	62.98	1.4744	-7	1	1	3	3	5
63.55	63.56	1.4627	-4	-4	2	1	1	2
64.50	64.52	1.4435	4	-4	2	1	2	2
66.01	66.02	1.4140	0	0	4	7	6	14
66.24	66.24	1.4097	0	-1	4	8	8	16
68.21	68.22	1.3736	8	0	0	2	2	4
68.94	68.92	1.3609	-3	-5	1	1	2	2
69.29	69.30	1.3548	3	1	1	2	3	4
69.32		1.3544	-8	1	0	2		4
69.53	69.54	1.3508	8	-1	0	3	5	5
69.55		1.3504	-8	1	0	2		5
70.21	70.20	1.3394	-1	6	0	1	2	3
70.41	70.42	1.3361	-7	-3	1	1	2	3
70.80	70.82	1.3296	-2	5	2	1	2	3
71.02	71.00	1.3260	2	5	1	1		2
71.65	71.64	1.3160	7	-3	1	1	1	2
72.17	72.18	1.3078	-6	4	1	2	3	4
72.20		1.3073	-2	-6	1	1		3
72.42	72.42	1.3039	6	4	1	2	3	3
73.28	73.28	1.2907	2	-6	1	2	1	3
73.85	73.86	1.2821	-3	5	2	1	2	3
75.04	75.02	1.2647	-3	4	3	1	2	3

BULTFONTEINITE - MCIVER, 1963

2THETA	PEAK	D	H	K	L	I(INT)	I(PK)	I(DS)
75.73	75.76	1.2548	6	-2	3	1	2	3
75.94	75.96	1.2520	3	4	3	1	2	3
78.23	78.22	1.2209	0	3	4	2	2	4
78.69	78.78	1.2149	-7	-1	3	1	2	3
79.16	79.20	1.2089	-6	-5	1	2	2	4
79.77	79.92	1.2012	3	-5	3	1	4	3
79.87		1.1999	6	1	1	1		4
79.95		1.1989	-7	-1	3	2		3
80.85	80.84	1.1878	7	1	3	1	2	4
81.31	81.32	1.1822	-8	2	3	2	1	3
82.06	82.06	1.1733	7	1	3	1	2	3
82.31	82.30	1.1705	-2	5	3	1	2	5
82.65	82.62	1.1665	0	7	0	2	2	3
82.88	82.88	1.1638	2	5	3	1	1	6
84.35	84.34	1.1473	-8	-3	2	2	2	5
85.68	85.72	1.1328	-7	-3	3	2	1	3
86.03	86.02	1.1291	8	1	2	1	2	4
86.50	86.52	1.1241	-9	-1	2	1	2	3
86.54		1.1238	8	1	4	1		7
86.82	86.80	1.1208	-4	-6	3	3	3	4
86.86		1.1204	-2	3	2	2		4
87.30	87.48	1.1159	-8	-5	1	2	3	5
87.46		1.1143	7	-6	3	2		5
87.96	87.98	1.1092	2	-4	4	2	2	6
88.11	88.12	1.1078	-4	3	4	2	3	4
88.39	88.38	1.1049	4	3	2	1	3	6
89.61	89.62	1.0930	8	-4	4	3	4	7
89.62		1.0929	4	-2	5	1		3
89.63		1.0928	-2	2	3	2		4
89.98	89.92	1.0895	-6	4	3	2	3	5
90.62	90.64	1.0835	2	1	5	2	2	5
90.64		1.0832	2	-4	3	1		5
91.71	91.70	1.0734	6	-2	2	1	2	3
93.83	93.96	1.0547	-9	-3	2	1	2	3
93.97		1.0535	-5	6	2	1		4
94.88	94.86	1.0457	5	6	2	1	1	3
95.72	95.72	1.0388	9	-3	2	1	1	3

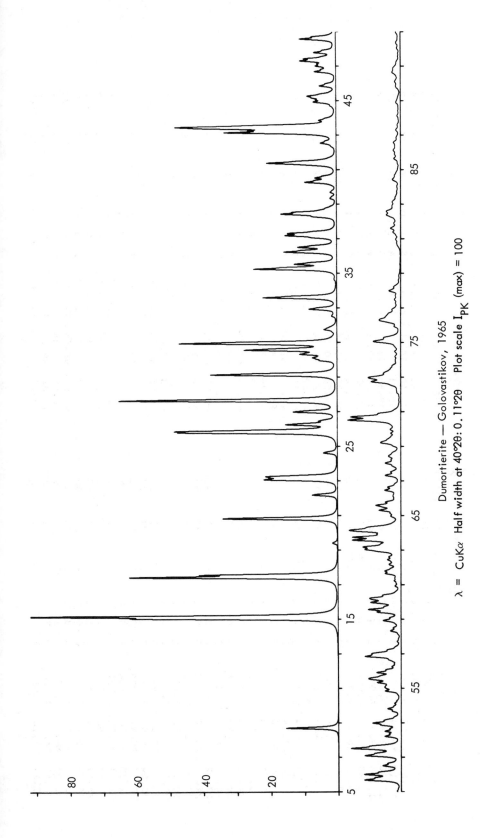

Dumortierite — Golovastikov, 1965

$\lambda = CuK\alpha$ Half width at $40°2\theta$: $0.11°2\theta$ Plot scale I_{PK} (max) = 100

DUMORTIERITE - GOLOVASTIKOV, 1965

2THETA	PEAK	D	H	K	L	I(INT)	I(PK)	I(DS)
8.68	8.68	10.184	1	1	0	15	16	15
8.74		10.104	0	2	0	1		1
15.02	15.04	5.895	2	0	0	53	62	53
15.13	15.14	5.849	1	3	0	100	100	100
17.40	17.42	5.092	2	2	0	68	62	68
17.54	17.54	5.052	0	4	0	37	41	37
19.37	19.38	4.579	0	1	1	2	2	2
20.79	20.80	4.268	1	1	1	32	34	32
20.82		4.262	0	2	1	12		12
22.16	22.16	4.008	2	0	1	9	8	9
23.03	23.06	3.858	2	2	1	17	22	17
23.05		3.855	1	3	0	9		9
23.17	23.16	3.836	2	4	0	16	21	17
23.24	23.24	3.823	1	5	0	15	22	16
24.60	24.60	3.616	2	2	1	5	4	5
25.77	25.78	3.454	1	4	1	53	48	56
25.84	25.84	3.442	2	4	1	34	49	35
26.23	26.24	3.395	0	6	0	18	16	19
26.44	26.44	3.368	0	4	1	5	6	5
26.96	26.96	3.304	2	3	1	17	13	18
27.62	27.62	3.226	1	5	1	88	65	93
29.11	29.12	3.065	0	5	1	52	38	55
29.94	29.94	2.982	3	4	1	5	5	6
30.04	30.10	2.972	2	4	1	3	7	3
30.10		2.966	1	5	1	5		6
30.30	30.30	2.947	4	0	0	12	11	13
30.54	30.54	2.924	2	6	1	37	28	39
30.92	30.92	2.889	3	2	1	66	47	70
31.73	31.74	2.818	3	3	0	5	4	5
32.51		2.752	3	5	1	2	2	2
32.91	32.92	2.719	2	5	1	12	9	13
33.57	33.58	2.667	1	6	1	32	22	35
35.22	35.22	2.546	4	8	1	36	25	40
35.51	35.51	2.526	0	17	0	17	13	19
36.21	36.22	2.478	4	1	1	23	16	25
36.49	36.50	2.460	0	7	1	16	12	18
37.17	37.16	2.417	3	5	1	18	16	20

2THETA	PEAK	D	H	K	L	I(INT)	I(PK)	I(DS)
37.31	37.28	2.408	1	7	0	14	15	15
38.26	38.28	2.350	0	0	2	8	8	9
38.40	38.40	2.342	5	1	0	3	17	3
38.41		2.342	4	3	1	19		22
38.53	38.50	2.335	0	1	2	6	13	6
39.30	39.32	2.290	1	1	2	1	2	2
39.32		2.289	0	2	2	3		3
39.66	39.66	2.270	2	7	1	14	2	16
40.25	40.26	2.239	4	4	1	9	10	10
40.50	40.50	2.225	0	8	1	6	7	7
41.25	41.36	2.187	1	1	1	8	21	9
41.32		2.183	2	0	2	23		
41.36		2.181	1	3	2	6		26
42.52	42.52	2.124	4	5	1	50	5	7
43.11	43.12	2.096	5	1	1	44	33	57
43.35	43.44	2.085	3	7	1	49	48	50
43.43		2.082	2	8	1	6		56
43.84	43.84	2.063	5	2	2	1	5	6
44.69	44.70	2.026	3	0	2	10	2	2
44.89	44.90	2.017	5	0	2	2	7	11
45.03	45.02	2.012	3	5	1	7	6	2
45.13	45.14	2.007	3	1	2	3	8	8
45.20	45.24	2.004	2	4	2	6	9	8
45.25		2.002	1	5	1	8		3
45.83	45.84	1.9782	3	2	2	10	5	8
46.65	46.66	1.9452	5	4	1	2	6	9
46.88	46.78	1.9364	3	8	2	7	5	12
46.98	46.98	1.9324	3	3	1	1	6	2
47.07	47.10	1.9289	6	2	1	16	6	9
47.11		1.9275	0	6	2	6		2
47.28	47.28	1.9210	2	5	2	10	11	18
47.40	47.40	1.9161	1	9	1	19	10	7
47.77	47.78	1.9023	5	9	1	1	7	12
48.55	48.56	1.8734	3	6	2	1	11	22
48.68	48.68	1.8689	5	4	1	18	7	1
49.02	49.02	1.8566	0	10	1	1	1	1
49.67	49.68	1.8340	1	10	1	18	11	21

DUMORTIERITE - GOLOVASTIKOV, 1965

2THETA	PEAK	D	H	K	L	I(INT)	I(PK)	I(DS)
49.72	50.00	1.8321	0	6	2	18	11	2
49.99	50.50	1.8228	6	7	1	8	6	21
50.50	50.50	1.8057	6	1	1	17	11	10
51.07	51.08	1.7867	5	6	1	3		21
51.14		1.7845	4	2	1	2	15	4
51.41	51.50	1.7759	4	8	2	24		3
51.50		1.7729	2	3	2			29
51.57		1.7709	2	10	1	2		3
52.20	52.20	1.7507	6	3	1	8	5	10
52.50	52.50	1.7414	2	7	2	7	5	9
52.86	52.86	1.7306	3	2	2	8	6	10
52.97	52.98	1.7270	4	6	2	10	9	12
53.18	53.12	1.7208	0	8	1	4	5	5
53.66	53.68	1.7064	6	4	1	3	3	4
53.80	53.80	1.7023	5	7	1	4	4	5
54.11	54.10	1.6934	1	11	1	3	2	3
54.83	54.84	1.6729	4	5	2	8	5	10
55.05	55.04	1.6668	4	10	2	5	5	6
55.12		1.6647	5	0	2	2		3
55.14		1.6643	3	11	0	1		1
55.33	55.32	1.6591	5	1	2	10	7	12
55.53	55.54	1.6536	3	7	2	9	10	11
55.59		1.6519	2	8	2	9		11
55.90	55.90	1.6433	2	11	1	10	7	12
55.93		1.6426	7	2	1	3		4
56.25	56.24	1.6340	3	3	0	3	3	4
56.81	56.84	1.6193	2	12	1	11	11	1
56.84		1.6184	5	8	1	15		19
56.93		1.6161	5	3	2	5		7
57.04	57.00	1.6132	4	6	1	5	9	6
57.22	57.20	1.6085	1	9	2	3	4	4
57.70	57.70	1.5964	6	4	2	4	1	2
58.31	58.32	1.5811	5	1	1	4	3	5
58.32		1.5807	3	1	1	1		1
58.81	58.90	1.5689	11	1	1	9	5	1
58.91		1.5664	7	2	0	1		11
59.40	59.40	1.5547	7	5	0	12	7	15
59.55	59.56	1.5510	6	6	0	11	10	14
59.88	59.98	1.5434	7	3	1	2	9	3
59.97		1.5412	1	13	0	15		19
60.16	60.20	1.5368	5	9	1	3	9	4
60.22		1.5353	6	7	1	12		15
60.41	60.38	1.5310	2	12	2	2	6	2
60.91	60.92	1.5196	1	10	2	3	2	3
61.21	61.22	1.5128	7	4	1	3	5	3
61.64	61.64	1.5034	6	1	2	8	5	10
61.77	61.78	1.5006	3	9	2	4		6
61.90		1.4977	2	2	3	1		2
62.15	62.16	1.4924	5	6	2	4	4	5
62.21		1.4911	6	2	2	2		3
62.44	62.44	1.4861	4	8	1	5		7
62.58	62.60	1.4831	2	10	1	8	4	4
62.92	63.02	1.4759	0	13	0	3		11
63.14		1.4737	8	0	1	14		19
63.19	63.18	1.4712	6	3	2	5	11	7
63.57	63.58	1.4703	3	12	1	7		10
63.75	63.74	1.4622	4	12	0	26	15	34
64.09	64.14	1.4587	5	10	1	12	15	16
64.14		1.4518	3	1	3	12	16	16
64.18		1.4506	2	4	3	13		17
64.95	64.96	1.4499	1	5	3	12		15
65.09	65.12	1.4346	7	6	1	4	3	5
65.30	65.30	1.4317	2	13	1	2	3	3
65.59	65.58	1.4277	3	10	2	11	7	15
65.97	65.98	1.4221	4	9	2	13	8	17
66.09	66.10	1.4148	8	4	0	1	2	2
66.45	66.46	1.4125	6	5	2	2	2	3
66.60	66.62	1.4057	2	11	1	9	5	12
67.30	67.30	1.4029	8	1	1	6	4	3
67.31		1.3900	7	7	2	4		8
67.86	67.86	1.3898	0	14	1	3	3	5
68.01	68.02	1.3799	7	0	3	5	5	3
68.05		1.3772	0	7	3	5		7
		1.3766	8	3	1	2		3

DUMORTIERITE - GOLOVASTIKOV, 1965

2THETA	PEAK	D	H	K	L	I(INT)	I(PK)	I(DS)
68.37	68.44	1.3708	4	2	3	3	4	4
68.45		1.3695	3	5	3	4		6
69.19	69.20	1.3567	7	2	2	8	6	11
69.27		1.3553	4	3	3	5		7
69.73	69.74	1.3474	7	1	0	3	3	5
69.96	69.96	1.3436	2	4	1	2	3	3
70.00		1.3430	1	4	3	1		2
70.41	70.50	1.3361	3	6	0	1	16	2
70.51		1.3345	4	4	3	32		44
70.68	70.70	1.3316	0	8	3	15	15	20
70.89	70.86	1.3282	8	5	1	3	7	4
72.09	72.10	1.3090	4	5	1	3	1	2
72.20	72.22	1.3073	9	1	3	1	1	1
72.69	72.72	1.2997	3	7	3	2	9	2
72.72		1.2993	4	11	2	12		17
72.74		1.2989	2	1	0	2		3
72.81		1.2979	5	13	2	2		3
72.88	72.96	1.2967	7	5	0	2	10	3
72.90		1.2964	4	14	1	2		2
72.98		1.2952	0	9	0	1		2
72.99		1.2951	6	8	1	8		12
73.02		1.2946	6	2	2	2		2
73.04		1.2943	5	2	3	1		6
73.29	73.20	1.2904	6	11	0	4	5	2
73.65	73.62	1.2850	0	15	3	1	3	1
73.66		1.2849	5	10	0	1		3
73.91	73.90	1.2813	5	3	2	1	2	2
74.00	74.00	1.2798	4	6	3	2	2	3
74.16	74.16	1.2775	1	9	3	2	2	2
74.47	74.46	1.2730	8	8	0	3	2	4
74.93	75.06	1.2663	2	13	2	9	8	1
75.02		1.2650	8	2	1	10		13
75.07		1.2643	2	15	1	2		14
75.15		1.2631	0	16	0	3		3
75.92	75.94	1.2522	9	2	1	2	2	4
76.18	76.28	1.2486	0	0	2	2	6	3
76.26		1.2475	7	10	1	10		14

2THETA	PEAK	D	H	K	L	I(INT)	I(PK)	I(DS)
76.35		1.2462	8	1	2	7		10
76.86	76.88	1.2392	8	2	3	1	3	1
76.92		1.2383	0	10	1	1		1
77.25	77.26	1.2339	6	12	1	1	2	1
77.27		1.2336	5	11	2	1		2
77.44	77.46	1.2313	3	13	1	2	2	3
77.54		1.2301	3	15	1	1		1
77.72	77.68	1.2277	8	3	1	7	2	2
77.97	77.98	1.2243	9	4	3	1	3	10
78.08		1.2229	6	1	3	1		1
78.59	78.58	1.2162	6	2	3	1	1	2
80.62		1.1906	6	4	2	1	1	2
80.74	80.74	1.1892	5	7	3	1		2
80.93	80.96	1.1869	4	13	3	1	1	1
80.99		1.1862	1	11	2	4		6
81.36	81.36	1.1817	9	6	0	4	2	6
81.58	81.60	1.1790	10	0	3	1	3	1
81.67		1.1779	4	0	4	1		3
81.90	81.86	1.1752	0	0	0	3	2	3
82.26		1.1711	10	2	0	1	4	2
82.36	82.36	1.1698	5	15	0	7		12
82.44		1.1689	0	11	3	1		2
82.47		1.1686	2	2	4	1		2
82.56	82.56	1.1675	1	14	0	3	4	6
82.57		1.1674	0	2	4	2		2
82.92	82.90	1.1633	6	14	0	1	3	2
83.06	83.10	1.1617	1	2	4	1	2	1
83.13		1.1610	4	16	0	3		3
83.41	83.34	1.1578	0	3	4	2	2	2
83.72	83.70	1.1543	8	10	0	1	1	4
84.04	84.02	1.1507	2	1	2	1	1	1
84.37	84.42	1.1470	1	17	4	4	2	7
84.46		1.1465	2	15	1	1		1
84.54		1.1460	8	7	2	1		1
84.58		1.1451	2	2	4	1		1
		1.1447	0	0	4	4		1
85.62	85.62	1.1335	7	10	2	4	2	6

DUMORTIERITE - GOLOVASTIKOV, 1965

2THETA	PEAK	D	H	K	L	I(INT)	I(PK)	I(DS)
85.85	85.88	1.1310	7	3	3	1		2
86.01	86.10	1.1293	6	14	1	1	2	2
86.10		1.1284	5	9	3	2	2	3
86.12		1.1281	9	3	2	1		1
86.22		1.1271	4	16	1	1		1
86.32	86.34	1.1261	2	12	3	1	2	1
86.49	86.56	1.1242	3	1	4	1	2	2
86.54		1.1237	2	4	4	1		2
86.57		1.1234	1	5	4	2		4
86.96	86.84	1.1193	8	8	2	1	1	1
87.22	87.28	1.1167	8	11	1	1	1	1
87.29		1.1160	9	4	2	3		4
88.11	88.10	1.1077	1	16	2	2	1	3
88.60	88.62	1.1029	10	18	0	1	1	2
88.85	88.88	1.1004	10	5	1	1	1	2
88.88		1.1001	9	9	1	1		2
88.99		1.0990	3	4	4	1		2
89.72	89.84	1.0920	0	18	1	1	1	1
89.75		1.0917	4	0	4	1		1
89.80		1.0913	8	9	2	1		1
89.86		1.0906	5	14	2	1		2
89.87		1.0905	2	6	4	2		1
90.08	90.08	1.0885	0	7	4	2	2	3
90.20		1.0874	1	18	1	4		7
90.62	90.62	1.0835	9	6	2	2	3	3
90.62		1.0834	5	16	1	1		2
90.75		1.0822	6	13	2	2		4
90.84	90.86	1.0814	4	15	1	2	2	2
91.70	91.70	1.0734	4	17	1	1	1	2
92.04	92.04	1.0704	2	1	4	2	1	3
92.32	92.34	1.0679	3	6	4	2	1	3
92.46		1.0666	9	11	0	1		1
92.93	92.96	1.0625	6	16	0	2	1	2
92.95		1.0623	7	12	2	1		2
93.04		1.0615	5	17	0	1		2
93.92	93.92	1.0539	10	0	2	1		1
94.16	94.16	1.0518	5	0	4	2		3
94.59	94.62	1.0482	10	2	2	3	2	5
94.70		1.0473	5	15	1	2		3
95.08	95.24	1.0440	2	17	2	3	3	4
95.22		1.0429	8	13	1	2		3
95.25		1.0426	6	14	2	2		4
95.42	95.44	1.0412	10	3	2	2		3
95.46		1.0409	4	16	1	3	3	5
95.64		1.0395	11	2	1	2		4
96.47	96.46	1.0327	8	11	2	1	1	2
97.87	97.90	1.0216	7	9	3	1	1	1
98.11	98.14	1.0198	10	5	3	2	1	3
99.90	100.10	1.0062	8	7	3	1	2	1
100.09		1.0048	11	7	0	3		6
100.51	100.48	1.0017	9	13	0	1	1	1
101.02	100.88	0.9981	11	6	1	3	1	4
101.32	101.34	0.9959	2	20	0	1	1	1
101.58	101.60	0.9941	9	3	3	2	1	4
102.03	102.02	0.9909	5	18	1	1	1	1
102.34	102.42	0.9887	8	8	3	3	1	4
102.44		0.9880	7	16	1	1		1
102.46		0.9879	3	10	4	1		2
103.09	103.36	0.9836	0	16	3	1	1	2
103.12		0.9834	4	9	4	4		7
103.36		0.9818	4	19	1	3		3
103.83	103.78	0.9786	2	11	4	4	1	6
104.17	104.32	0.9763	9	5	3	3	2	5
104.32		0.9754	11	0	0	1		2
104.34		0.9752	6	18	0	1		2
104.40		0.9748	7	11	3	3		7
104.66	104.66	0.9731	8	15	1	4	2	3
104.70		0.9729	9	11	2	2		1
104.94	105.04	0.9713	7	17	0	1	2	2
104.95		0.9712	5	8	4	1		2
104.98		0.9710	2	16	3	1		5
105.14		0.9700	0	19	2	3		2
105.28		0.9690						

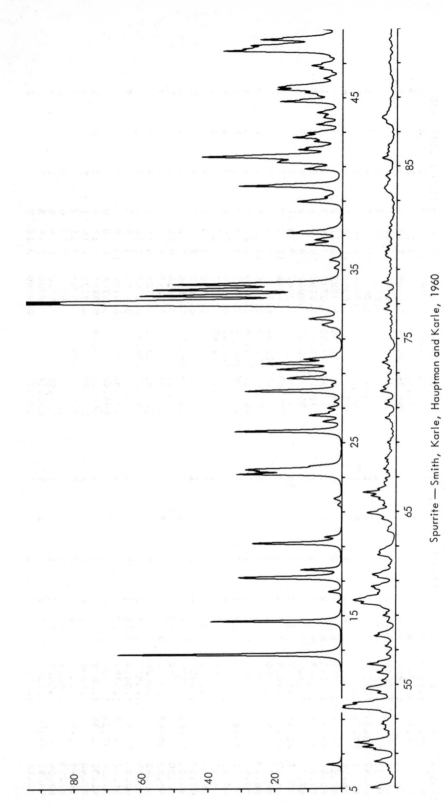

Spurrite — Smith, Karle, Hauptman and Karle, 1960

λ = CuKα Half width at 40°2θ: 0.11°2θ Plot scale I$_{PK}$ (max) = 100

SPURRITE - SMITH, KARLE, HAUPTMAN AND KARLE, 1960

2THETA	PEAK	D	H	K	L	I(INT)	I(PK)	I(DS)
6.36	6.36	13.885	0	0	1	5	5	3
12.74	12.74	6.942	0	0	2	82	67	54
14.66	14.66	6.038	0	1	1	49	39	33
15.76	15.78	5.617	1	1	0	2	1	1
16.36	16.38	5.413	-1	1	1	5	4	3
17.16	17.20	5.164	-2	1	1	29	31	21
17.23		5.143	2	0	0	24		17
17.64	17.64	5.023	1	1	1	15	12	11
19.16	19.18	4.628	0	0	3	30	27	22
19.21		4.615	-1	1	2	10		8
19.53	19.54	4.541	2	1	0	6	5	4
21.37	21.38	4.154	-2	1	2	2	1	1
21.70		4.091	1	1	2	2		1
21.76	21.76	4.081	-2	1	2	2	2	1
23.17	23.18	3.835	-2	0	3	41	31	33
23.33	23.34	3.809	0	0	4	25	26	20
23.44	23.44	3.792	2	0	2	24	29	19
23.48		3.785	-2	1	2	13		10
23.64	23.64	3.760	2	1	1	9	10	7
25.64	25.64	3.471	-1	1	4	48	32	40
26.21	26.22	3.398	1	1	3	8	6	7
26.57	26.58	3.352	0	2	0	14	10	12
26.76	26.76	3.329	2	1	2	4	4	4
26.99	27.00	3.301	-2	1	3	6	4	5
27.34	27.34	3.259	2	0	3	3	2	3
27.97	27.98	3.187	1	2	0	41	29	37
28.02		3.181	-1	2	1	6		5
28.32	28.32	3.149	-1	1	4	2	3	2
28.72	28.72	3.105	-1	2	2	24	16	22
28.80		3.097	-3	1	1	3		3
28.94	28.94	3.083	0	1	4	4	6	4
29.23	29.24	3.053	3	0	2	28	19	26
29.56	29.56	3.019	0	2	2	37	24	34
29.82	29.82	2.994	-3	1	2	16	12	14
30.10	30.10	2.966	-1	1	5	1	1	1
31.09	31.09	2.874	-1	2	4	2	2	2
31.69	31.70	2.821	1	1	4	4	4	4
31.80	31.82	2.812	-2	2	1	4	6	3
31.84		2.808	-2	2	0	4		4
32.15	32.16	2.781	-3	1	3	15	10	14
32.96	32.98	2.715	0	2	3	91	72	90
33.07	33.06	2.706	-2	2	2	98	100	97
33.11		2.703	-1	1	5	4		4
33.19	33.18	2.697	-2	0	5	100	99	100
33.50	33.50	2.673	1	2	3	90	60	90
33.84	33.84	2.646	2	0	4	86	56	88
34.06	34.16	2.630	3	1	2	19	50	19
34.16		2.622	-4	0	1	69		70
35.11	35.12	2.554	1	2	0	3	3	3
35.18	35.58	2.521	-3	1	4	6	4	6
36.46	36.46	2.462	-1	3	0	13	11	14
36.47		2.462	2	1	1	6		6
36.74	36.74	2.444	-3	0	1	11	8	12
36.77		2.442	-1	1	1	2		2
37.14	37.14	2.418	-3	2	1	28	17	31
37.28	37.24	2.410	-4	1	2	4	13	5
38.88	38.98	2.314	-3	0	6	16	14	4
38.96		2.310	3	2	1	16		18
39.00		2.308	-2	2	4	6		6
39.23	39.24	2.294	-2	2	3	5	5	6
39.26		2.293	-4	0	4	2		2
39.49	39.50	2.280	-1	1	5	2		2
39.85	39.88	2.260	-3	1	5	2	31	3
39.87		2.259	-3	2	3	51		59
40.86	40.86	2.207	0	3	1	19	11	22
41.23	41.24	2.187	1	3	0	27	20	32
41.30	41.32	2.184	-1	3	2	10	19	12
41.47	41.56	2.176	3	2	1	15	42	18
41.55		2.171	-1	2	1	46		55
41.59		2.169	4	1	1	25		30
41.97	41.98	2.151	4	0	4	20	13	24
42.11	42.08	2.144	-3	1	1	6	11	7
42.45	42.46	2.127	0	2	2	13	9	16
42.69	42.70	2.116	3	1	4	25	15	30

147

SPURRITE — SMITH, KARLE, HAUPTMAN AND KARLE, 1960

2THETA	PEAK	D	H	K	L	I(INT)	I(PK)	I(DS)
42.85	42.80	2.109	-1	3	2	3	11	4
42.95	42.96	2.104	-4	0	5	15	11	18
43.45	43.44	2.081	4	2	3	14	8	17
43.79	43.84	2.066	-4	1	1	1	6	
43.84		2.063	-1	3	6	8		10
43.94	43.94	2.059	1	3	2	2	5	3
44.11	44.14	2.051	-2	1	1	5	8	6
44.14		2.050	3	3	2	2		10
44.78	44.78	2.022	-3	1	0	8	19	44
45.09	45.14	2.009	-2	3	6	34	8	3
45.12		2.008	2	3	2	3		3
45.13		2.007	-1	3	5	3		4
45.18		2.005	-4	1	5	3		3
45.30	45.30	2.000	2	1	1	13	11	16
45.47	45.48	1.9928	-5	1	2	29	20	37
45.60	45.60	1.9875	4	1	3	10	19	13
45.64		1.9862	-2	0	7	3		
45.67		1.9846	-4	2	3	6		7
45.70		1.9835	0	0	7	12		15
46.02	46.12	1.9705	2	1	6	1	6	2
46.11		1.9667	5	0	6	8		10
46.49	46.50	1.9518	-3	2	1	4	3	5
46.68	46.68	1.9441	1	3	5	10	7	14
46.87	46.86	1.9366	-1	1	3	14	9	19
47.06	47.00	1.9292	-1	2	7	3	6	
47.38	47.38	1.9174	-4	0	6	4	4	5
47.71	47.72	1.9045	0	2	6	59	36	80
47.78		1.9020	0	1	7	14		18
47.87	47.84	1.8984	5	1	1	5	30	6
47.94	47.94	1.8959	4	0	4	3	27	
47.94		1.8958	3	1	5	20		27
48.03	48.04	1.8925	-4	2	4	30	27	41
48.15		1.8880	-2	2	2	39		5
48.37	48.38	1.8800	4	2	2	39	25	53
48.40		1.8792	0	3	4	3		3
48.47		1.8766	2	2	5	6		9
48.58	48.48	1.8723	3	3	0	3	18	4

2THETA	PEAK	D	H	K	L	I(INT)	I(PK)	I(DS)
48.97	49.02	1.8585	-3	3	2	3	7	5
49.01		1.8569	3	2	4	11		15
49.39	49.38	1.8435	-4	1	6	2	1	2
49.79	49.82	1.8288	-2	3	3	2	2	2
49.82	49.82	1.8222	-2	1	4	3	2	3
50.01	50.02	1.8144	-3	1	3	3	2	4
50.24	50.24	1.8055	5	1	7	7	6	10
50.51	50.52	1.8042	-3	1	3	7		7
50.54		1.7820	-4	2	5	10	6	15
51.22	51.22	1.7769	-5	2	1	13	10	19
51.38	51.38	1.7718	-5	1	2	17	12	2
51.53	51.62	1.7692	-5	1	1	5		25
51.65		1.7681	4	2	3	2		8
52.16	52.16	1.7519	-2	0	8	1	2	4
52.35	52.38	1.7460	-6	0	0	1	3	2
52.39		1.7449	-4	0	7	4		6
52.69	52.70	1.7356	0	0	8	6	4	8
52.81	52.82	1.7320	-1	2	7	3	4	5
53.58	53.64	1.7088	-2	2	7	14	16	21
53.63		1.7075	2	3	4	4		6
53.64		1.7071	0	1	7	13		19
53.67		1.7062	3	1	6	2		3
53.73	53.74	1.7045	5	2	1	9	16	13
53.79		1.7026	5	1	5	5		7
53.92	53.92	1.6989	1	1	3	5		5
53.92		1.6988	2	2	6	13	13	20
54.47	54.48	1.6829	-1	3	5	6	5	9
54.49		1.6824	5	2	4	2		4
54.71	54.72	1.6762	-5	0	4	11	9	17
54.78		1.6744	3	3	3	6		10
54.83		1.6730	-6	1	4	5		7
55.13	55.20	1.6644	-4	2	6	3	4	4
55.21		1.6622	6	0	1	5		9
55.48	55.48	1.6549	-4	3	2	3	2	3
55.65	55.68	1.6503	4	2	4	3	5	5
55.68		1.6493	4	3	1	3		4
55.70		1.6489	-1	4	1	5		8

SPURRITE – SMITH, KARLE, HAUPTMAN AND KARLE, 1960

2THETA	PEAK	D	H	K	L	I(INT)	I(PK)	I(DS)
55.92	55.84	1.6429	-3	2	7	3	4	5
56.15	56.16	1.6367	-1	4	1	7	8	11
56.15		1.6367	-3	1	8	2		3
56.16		1.6363	5	3	2	3		5
56.18		1.6358	-3	2	5	6		9
56.42	56.30	1.6294	0	4	2	1	5	2
56.69	56.74	1.6224	-1	3	6	2	3	4
56.74		1.6209	-1	4	2	3		5
57.20	57.20	1.6091	-5	2	5	5	3	8
57.78	57.80	1.5944	-2	4	1	7	6	12
57.80		1.5937	2	4	0	5		8
57.92	57.94	1.5908	3	3	5	2	5	4
58.03		1.5881	5	1	4	2		4
58.58	58.62	1.5744	-2	4	2	1	3	2
58.61		1.5737	-1	4	3	3		4
58.66		1.5725	2	4	1	1		2
58.98	58.98	1.5648	-1	2	8	2	4	3
59.66	59.68	1.5486	-2	0	9	6	10	10
59.69		1.5478	-6	2	2	9		15
59.73		1.5468	-4	2	7	5		9
59.83	59.86	1.5445	-6	1	1	6	13	10
59.85		1.5441	3	1	7	2		4
59.90		1.5427	2	0	2	8		13
59.91		1.5425	0	4	7	5		9
59.97		1.5413	1	2	9	2		4
60.27	60.26	1.5342	0	2	3	7	5	12
60.37	60.38	1.5319	-4	4	8	4	5	6
60.40		1.5313	-4	3	5	2		4
60.51	60.52	1.5287	-6	2	1	1	6	2
60.52		1.5285	-5	0	3	2		3
60.61	60.64	1.5263	-6	2	6	5	7	8
60.64		1.5258	6	2	0	1		3
60.65		1.5254	-5	3	0	3		5
60.76		1.5231	4	3	3	6		11
61.17	61.18	1.5137	6	2	1	8	5	2
61.29	61.34	1.5112	-3	3	4	1	7	14

2THETA	PEAK	D	H	K	L	I(INT)	I(PK)	I(DS)
61.37		1.5095	0	4	4	7		12
61.85	61.86	1.4987	-3	4	2	4	2	6
62.41	62.48	1.4868	-1	2	8	2	6	4
62.47		1.4853	-3	1	9	7		13
62.51		1.4846	-2	3	7	2		4
62.56		1.4835	0	3	6	1		2
62.64	62.64	1.4818	5	3	1	2	5	4
62.75	62.74	1.4793	5	1	5	5	5	9
62.81		1.4781	2	3	6	2		3
63.85	63.86	1.4566	-7	1	1	4	3	8
63.93		1.4549	-4	0	9	2		4
64.08	64.04	1.4520	-7	1	0	1	2	3
64.43	64.56	1.4449	-6	1	8	7	4	2
64.55		1.4425	-5	2	8	2		13
64.82	64.92	1.4371	-4	3	2	5	9	4
64.86		1.4363	5	3	0	3		9
64.91		1.4354	7	1	5	11		5
64.92		1.4351	0	4	4	3		21
65.33	65.36	1.4272	-3	4	4	3	4	6
65.36		1.4265	-7	1	2	4		7
65.47	65.52	1.4243	-4	2	6	2	4	4
65.70	65.70	1.4201	-2	4	5	6	5	15
65.81	65.90	1.4178	-5	3	5	8	8	4
65.90		1.4161	2	4	4	2		19
66.10	66.10	1.4124	-4	2	1	10	10	27
66.12		1.4120	-2	0	10	14		4
66.39	66.28	1.4068	3	1	8	2	5	6
66.55	66.58	1.4038	2	0	9	3	4	5
66.65	66.66	1.4020	0	4	5	1	4	2
66.68		1.4015	1	2	9	3		6
66.92	66.90	1.3970	0	2	3	3	3	9
67.26	67.26	1.3907	-6	4	6	4	4	10
67.45	67.46	1.3872	-1	2	8	2	4	7
67.48		1.3868	-3	3	9	1		2
67.59	67.60	1.3848	-3	3	6	3	3	5
67.88	67.88	1.3796	6	2	3	4	3	8
68.05	68.06	1.3765	5	1	6	4	3	8

SPURRITE — SMITH, KARLE, HAUPTMAN AND KARLE, 1960

2THETA	PEAK	D	H	K	L	I(INT)	I(PK)	I(DS)
68.19		1.3740	2	1	9	1		2
68.62	68.62	1.3664	-1	4	6	3	2	5
68.78	68.78	1.3638	-6	2	3	2	2	5
68.97	68.98	1.3604	-6	0	8	1	3	3
68.98		1.3603	6	3	0	2		5
69.49	69.48	1.3515	-2	4	6	2	1	5
69.76	69.76	1.3468	-3	4	8	2	2	4
70.40	70.40	1.3363	-4	0	10	6	3	12
71.20	71.20	1.3232	4	0	8	7	4	15
71.89	71.96	1.3122	-1	5	2	1	5	3
71.95		1.3112	-8	0	8	8		16
71.96		1.3110	-4	4	5	3		6
72.22	72.18	1.3069	-5	4	2	2	4	5
72.32		1.3054	4	4	3	2		5
72.53	72.52	1.3021	7	1	3	2	3	4
72.70	72.82	1.2995	5	4	0	2	3	3
72.82		1.2976	2	5	0	1		3
72.85		1.2973	6	3	2	1		2
72.86		1.2970	-5	3	7	4		3
73.25	73.26	1.2911	-8	0	4	1	3	8
73.28		1.2907	-1	4	7	2		3
73.60	73.60	1.2858	-7	1	7	5	4	5
73.61		1.2857	8	0	0	1		10
73.80	73.82	1.2828	0	3	10	2	3	2
73.87		1.2817	-2	3	9	2		3
74.05	74.04	1.2792	5	4	1	2	3	3
74.25	74.22	1.2762	-2	3	8	1	2	3
74.96	75.08	1.2658	-2	5	3	2	2	4
75.07		1.2643	2	5	2	2		4
75.21	75.22	1.2620	0	0	11	1	2	3
75.23		1.2620	-4	4	6	2		3
75.52	75.48	1.2578	8	0	1	1	2	3
75.66	75.68	1.2558	4	4	4	1	2	3
75.99	76.12	1.2512	1	4	7	1	2	3
76.11		1.2496	5	4	2	3		6
76.40	76.40	1.2455	1	2	10	1	2	3
76.45		1.2448	7	1	4	1		2

2THETA	PEAK	D	H	K	L	I(INT)	I(PK)	I(DS)
76.71	76.70	1.2413	-4	2	10	8	4	18
77.00	76.96	1.2374	-5	4	5	4	3	8
77.37	77.48	1.2323	-5	3	8	8	3	6
77.49		1.2308	4	2	8	6		14
78.13	78.22	1.2223	-7	3	4	1	4	3
78.22		1.2211	-8	2	7	11		25
78.56	78.44	1.2166	-5	2	7	5	3	4
78.66	78.66	1.2153	-8	2	1	1	2	3
78.67		1.2152	3	4	1	2		5
78.88	78.88	1.2125	-4	4	11	2		5
79.14	79.16	1.2092	-6	4	1	1		3
79.20		1.2084	-6	4	1	3		6
79.40	79.40	1.2058	7	3	1	4	2	8
79.74	79.82	1.2016	-4	1	9	3	3	7
79.83		1.2004	2	2	11	1		5
79.84		1.2003	-6	0	10	2		3
80.00	80.00	1.1983	-5	4	6	4	3	4
80.03		1.1980	-7	3	5	2		9
80.28	80.28	1.1948	-1	2	10	10		2
80.31		1.1945	-1	3	10	1		3
80.83	80.94	1.1880	3	1	7	3	2	6
80.93		1.1869	6	0	11	1		2
81.00		1.1860	-3	0	11	5		11
81.39	81.38	1.1813	0	2	3	1	2	5
81.70	81.68	1.1776	8	2	1	5	2	5
81.80		1.1764	7	3	3	1		5
82.28	82.30	1.1707	-3	5	5	2	1	3
82.63	82.54	1.1667	-7	3	8	2	1	6
83.03	83.02	1.1621	-6	5	3	1	1	6
83.52	83.70	1.1565	-2	5	6	3	2	6
83.56		1.1561	-2	0	11	1	4	3
83.59		1.1557	-8	2	6	2		6
83.68		1.1547	3	4	5	3		7
83.75		1.1539	2	5	5	2		6
83.86		1.1527	-6	3	3	3		5
84.04	84.02	1.1506	-3	1	12	3	3	6
84.08		1.1502	1	3	10	1		2

SPURRITE - SMITH, KARLE, HAUPTMAN AND KARLE, 1960

2THETA	PEAK	D	H	K	L	I(INT)	I(PK)	I(DS)
84.27	84.42	1.1481	8	2	2	2	3	5
84.42		1.1465	-1	4	9	2		5
84.43		1.1463	-8	0	8	6		15
84.89	84.90	1.1413	7	3	3	1	1	3
85.02	85.02	1.1399	2	4	8	2	1	4
85.40	85.46	1.1358	3	2	10	2	4	6
85.45		1.1353	8	0	0	4		11
85.46		1.1351	0	4	9	7		18
85.92	85.94	1.1302	-7	3	7	4	3	10
85.99		1.1295	5	1	2	2		5
86.11	86.18	1.1282	-5	5	5	2	3	5
86.20		1.1273	3	3	7	2		4
86.68	86.68	1.1222	-8	2	2	2	1	4
86.94	86.96	1.1196	-5	5	3	2	1	4
87.43	87.52	1.1146	-7	4	1	1	3	4
87.50		1.1139	0	6	1	5		13
87.54		1.1135	8	2	3	2		4
87.64	87.80	1.1125	-7	4	3	3	5	5
87.79		1.1110	-1	6	0	2		5
87.79		1.1109	1	4	9	2		5
87.86		1.1109	0	5	7	1		3
87.89		1.1102	5	5	1	3		7
88.11	88.06	1.1078	-7	1	10	2	3	4
88.30	88.30	1.1058	2	2	11	1	3	3
88.30		1.1058	9	1	1	1		6
88.39		1.1050	-5	3	10	2		3
88.48		1.1040	7	4	0	1		4
88.68	88.60	1.1021	-5	5	4	2	2	5
88.85	88.86	1.1004	-7	3	4	2	2	3
88.87		1.1002	-9	2	2	1		5
89.00		1.0989	-9	1	6	2		5
89.42	89.48	1.0948	-4	5	4	3	2	7
89.54		1.0937	0	2	12	2		6
90.12	90.34	1.0881	8	0	5	2	2	4
90.32		1.0863	0	6	3	4		12
90.54	90.62	1.0842	2	0	12	2	2	4

2THETA	PEAK	D	H	K	L	I(INT)	I(PK)	I(DS)
90.72	91.00	1.0825	-5	2	5	1	3	3
90.85		1.0813	-4	2	12	1		5
91.00		1.0799	-2	4	10	4		11
91.17	91.36	1.0784	-8	3	6	1		3
91.40		1.0762	2	4	9	4	2	10
91.78	91.78	1.0728	4	2	10	2	2	6
91.79		1.0726	5	2	0	1		4
92.30	92.34	1.0680	0	0	13	3		8
92.34		1.0677	-3	5	4	2	2	4
92.35		1.0676	3	5	10	1		3
92.66	92.64	1.0649	-2	5	8	1	2	4
92.88	92.86	1.0630	-6	5	1	1	2	4
92.98		1.0621	7	3	7	2		5
93.16	93.16	1.0605	-7	0	12	3	2	8
93.17		1.0604	-6	0	4	5		15
93.87	93.88	1.0543	-2	6	2	4		13
94.01		1.0531	2	6	6	1		4
94.39	94.48	1.0498	-5	2	12	2	2	7
94.50		1.0489	6	0	2	2		8
94.78	94.80	1.0466	-5	3	11	3		4
94.98		1.0449	-4	4	10	2		3
94.99	95.10	1.0448	-8	2	9	1		7
95.14		1.0436	-10	0	4	2		6
95.45	95.44	1.0409	-10	0	1	2	2	6
95.74	95.74	1.0386	4	4	8	3	1	8
96.08	96.04	1.0359	-2	3	13	1	1	4
96.42	96.50	1.0330	-9	3	3	2	2	5
96.51		1.0324	-9	3	3	3		8
96.58		1.0318	-1	2	13	1		3
96.61		1.0316	2	2	12	1		4

TABLE 7. SILICATES WITH PAIRS OF TETRAHEDRA

Variety	Fresnoite	Ilvaite (Lievrite)	Thortveitite	Kentrolite	Lawsonite
Composition	$Ba_2TiO\,Si_2O_7$	$Ca\,Fe_2^2Fe^3Si_2O_8$ (OH)	$Sc_2Si_2O_7$	$Pb_2(Mn_{1.2}Fe_{0.8})$ Si_2O_9	$Ca\,Al_2Si_2O_7$ $(OH)_2 \cdot H_2O$
Source	Fresno Co., California	———	Iveland, Saetersdalen Nedenäs, Norway	Långban, Sweden	not given
Reference	Moore & Louis-nathan, 1967	Belov & Mokejeva, 1954	Cruickshank, Lynton & Barclay, 1962	Gabrielson, 1965	Rumanova and Skipetrova, 1959
Cell Dimensions					
a Å	8.52	8.80	6.542	6.995	5.846 (8.788)
b Å	8.52	13.07	8.519	11.054	8.788 (5.846)
c Å	5.21	5.86	4.669	10.001	13.129 (13.129)
α deg	90	90	90	90	90
β deg	90	90	102.55	90	90
γ deg	90	90	90	90	90
Space Group	P4bm	Pbnm	C2/m	$C222_1$	Cncm (Ccmm)
Z	1	4	2	4	4

TABLE 7. (cont.)

Variety	Fresnoite	Ilvaite (Lievrite)	Thortveitite	Kentrolite	Lawsonite
Method	3-D, counter	2-D, film	3-D, film	2-D, film	2-D, film
R & Rating	0.111 (2)	--- (3)	0.138 (2)	--- (3)	0.167 (3)
Cleavage and habit	{001} perfect	{010}, {001} distinct; {100} less so. Prismatic	Prismatic cleavage. Needles elongated to $\underline{c}$ axis		{001}&{010}perfect {110} imperfect. Tabular parallel to {001}; Also prismatic.
Comment		B estimated Related to lawsonite	$8.89\%\ Y_2O_3$ and 2.83% Fe_2O_3 apparently ignored in refinement	B estimated	B estimated Cell parameters from ASTM card 13—533
μ	749.3	322.0	293.8	1565.6	142.8
ASF	0.9212	0.7662×10^{-1}	0.1853	0.3754	0.7188
Abridging factors	0.5:0:0	1:1:0.5	1:1:0.5	1:1:0.5	1:1:0.5

154

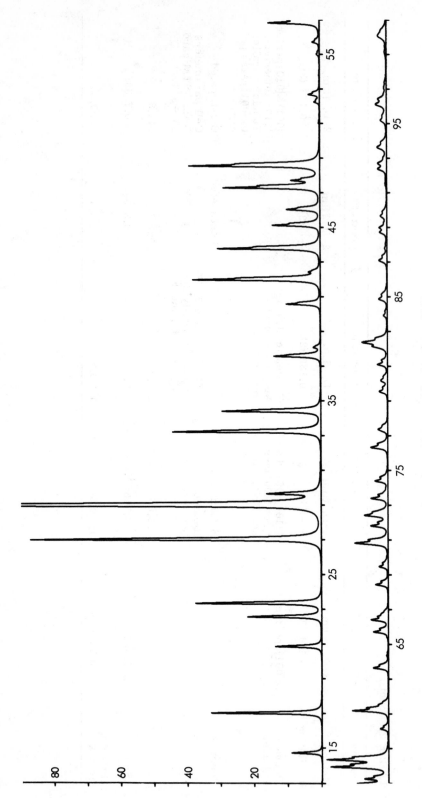

Fresnoite — Moore and Louisnathan, 1967

$\lambda = $ CuKα Half width at 40°2θ: 0.11°2θ Plot scale I_{PK} (max) = 200

FRESNOITE - MOORE AND LOUISNATHAN, 1967

2THETA	PEAK	D	H	K	L	I(INT)	I(PK)	I(DS)
10.37	10.38	8.520	1	0	0	3	0	0
14.69	14.70	6.025	1	1	0	4		1
17.00	17.02	5.210	0	0	1	14	16	6
20.83	20.84	4.260	2	0	0	6	7	4
22.54	22.54	3.941	1	1	1	10	11	7
23.33	23.34	3.810	2	1	0	17	19	12
27.01	27.02	3.298	2	1	1	41	44	37
29.01	29.02	3.076	2	2	0	100	100	100
29.63	29.64	3.012	3	1	0	7	8	7
33.22	33.22	2.694	3	1	1	23	22	29
34.36	34.40	2.608	2	2	1	14	15	2
34.40		2.605	2	0	2			18
37.55	37.56	2.393	3	1	2	7	7	10
37.58		2.391	3	2	1	1		2
38.05	38.04	2.363	3	2	0	1	1	2
40.56	40.56	2.222	2	2	2	6	5	10
41.94	41.98	2.152	3	1	1	2	19	3
41.98		2.150	3	2	2	21		37
42.40	42.40	2.130	4	0	0	18	2	3
43.77	43.78	2.066	4	1	0	9	15	35
45.11	45.12	2.008	3	3	0	3	7	17
45.99	46.02	1.9716	4	2	1	4	5	6
46.02		1.9704	4	1	2	18		7
47.28	47.28	1.9208	2	2	2	0	14	38
47.31		1.9197	3	0	1	5		0
47.69	47.70	1.9051	4	2	0	14	4	10
48.54	48.56	1.8738	3	3	0	12	20	31
48.57		1.8728	3	1	2			27
51.00	51.00	1.7893	4	2	1	0	0	0
52.22	52.22	1.7502	3	2	2	1	1	3
52.66	52.66	1.7367	0	0	3	2	2	6
53.75	53.74	1.7040	5	0	0	0	0	0
53.83		1.7017	4	3	0	0		0
54.90	54.98	1.6709	1	1	3	0	0	0
54.98		1.6687	5	1	0	1		1
55.69	55.70	1.6490	4	0	2	1	1	4

2THETA	PEAK	D	H	K	L	I(INT)	I(PK)	I(DS)
56.80	56.82	1.6196	5	0	1	0	8	0
56.80		1.6196	5	3	1	0		1
56.82		1.6189	4	1	2	10		29
57.24	57.24	1.6082	2	0	3	4	4	13
57.91	57.92	1.5911	5	1	1	5	9	15
57.93		1.5905	3	3	0	7		21
58.27	58.34	1.5821	2	1	3	1	9	3
58.34		1.5803	4	2	1	12		36
60.12	60.12	1.5378	5	3	0	2	1	6
61.17	61.16	1.5139	5	2	1	2	5	26
61.52	61.50	1.5061	4	4	0	8	2	5
61.59		1.5045	2	2	3	1		1
63.63	63.64	1.4612	5	3	0	3	2	11
63.70		1.4597	3	1	3	1		2
64.33	64.32	1.4469	6	4	1	0	0	1
65.70	65.70	1.4200	5	0	1	3	2	13
66.39	66.40	1.4069	5	3	1	3	3	13
66.41		1.4064	6	1	0	1		2
66.72	66.72	1.4007	3	2	3	1	1	4
66.79		1.3994	6	0	1	0		1
68.42	68.42	1.3700	6	2	1	3	2	12
69.42	69.44	1.3526	6	1	1	0	1	0
69.44		1.3523	6	2	0	2		9
69.75	69.64	1.3471	5	4	0	2	1	1
70.74	70.80	1.3306	4	1	3	0	5	6
70.81		1.3295	3	3	1	7		30
71.80	71.80	1.3136	3	3	3	4	3	18
72.40	72.40	1.3042	6	2	2	5	4	20
72.42		1.3039	4	4	2	1		3
72.51		1.3025	5	0	4	4		6
73.38	73.38	1.2892	1	0	1	4	3	19
73.49		1.2875	5	3	0	0		0
74.37	74.38	1.2744	5	3	2	3	2	15
74.46		1.2731	1	1	4	0		1
74.67	74.58	1.2701	6	3	0	0	1	1
76.31	76.32	1.2468	6	0	2	4	3	20
76.40		1.2456	2	0	4	1		4

FRESNOITE – MOORE AND LOUISNATHAN, 1967

2THETA	PEAK	D	H	K	L	I(INT)	I(PK)	I(DS)
77.25	77.36	1.2340	6	3	1	0	2	2
77.27		1.2337	6	1	2	1		3
77.36		1.2325	2	1	4	2		10
78.52	78.58	1.2171	7	0	0	0	0	0
78.58		1.2163	5	0	3	0		0
78.58		1.2163	4	3	3	0		0
79.47	79.54	1.2049	7	1	0	0	1	2
79.47		1.2049	5	1	3	2		0
79.54		1.2041	5	2	0	1		10
80.14	80.16	1.1966	6	1	3	1	1	7
80.22		1.1955	5	2	2	0		4
81.06	81.08	1.1852	2	2	4	1	2	0
81.09		1.1850	7	0	1	0		14
81.17		1.1839	5	4	2	3		0
81.37	81.32	1.1815	3	0	4	0	1	0
82.01	82.12	1.1739	6	4	1	0	2	0
82.12		1.1727	5	1	0	1		3
82.32	82.36	1.1703	3	5	1	4	4	18
82.38		1.1696	7	2	2	3		13
83.90	83.90	1.1523	5	2	0	3	1	19
84.84	84.84	1.1419	6	2	1	1	1	6
84.86		1.1416	7	2	1	2		13
84.95		1.1407	3	4	3	0		1
85.21	85.10	1.1378	4	4	3	0	1	3
87.02	87.08	1.1187	7	3	0	1	1	2
87.09		1.1181	5	3	3	2		5
87.76	87.76	1.1112	4	0	4	1	0	12
88.61	88.70	1.1027	7	1	0	0	1	4
88.70		1.1019	4	0	4	2		0
88.96	88.96	1.0993	6	1	3	1	1	14
89.53	89.64	1.0938	7	3	2	0	1	7
89.55		1.0936	5	1	1	0		0
89.55		1.0936	5	5	2	0		0
89.64		1.0928	3	5	0	2		2
89.83	89.88	1.0909	6	1	3	0	1	12
89.90		1.0903	6	1	3	0		0

2THETA	PEAK	D	H	K	L	I(INT)	I(PK)	I(DS)
91.42	91.48	1.0760	6	4	2	0	0	2
91.51		1.0752	4	2	4	0		2
92.34	92.36	1.0677	6	5	1	3	1	0
92.36		1.0675	7	2	0	0		20
92.65	92.68	1.0650	8	0	2	2	1	1
92.71		1.0644	6	2	3	1		10
93.58	93.62	1.0568	8	1	0	2	1	4
93.58		1.0568	7	4	0	0		10
93.65		1.0562	5	4	3	2		10
95.16	95.16	1.0434	8	0	1	2	1	15
95.33		1.0420	0	0	5	0		2
96.06	96.10	1.0357	7	4	1	3	2	20
96.10		1.0357	7	1	4	1		9
96.20		1.0348	8	3	3	0		0
96.20	96.40	1.0348	4	0	4	4		0
96.27		1.0343	5	3	5	0		0
97.06	97.06	1.0332	1	0	2	0	1	6
97.15		1.0279	8	2	0	1	0	6
97.21	97.38	1.0273	7	3	4	0		0
97.41		1.0268	5	1	5	0		1
98.93	99.10	1.0252	6	3	3	1	0	1
99.10		1.0135	8	2	1	1	0	1
99.90	100.06	1.0122	2	0	5	4		4
99.90		1.0062	6	5	2	3	1	3
100.05		1.0056	5	2	5	1		5
100.19		1.0051	2	1	5	2		17
102.10	102.18	1.0041	6	6	0	1	0	9
102.17		0.9904	7	5	0	1		0
102.17		0.9900	1	3	3	0		1
102.75	102.78	0.9859	5	5	6	1	1	4
102.77		0.9858	6	6	1	1		7
102.86		0.9852	8	0	2	0		3
102.92		0.9847	4	4	4	2		4
103.71	103.74	0.9794	2	2	5	1	1	2
103.73		0.9793	8	1	1	2		0
103.73		0.9793	7	4	4	3		22

FRESNOITE - MOORE AND LOUISNATHAN, 1967

2THETA	PEAK	D	H	K	L	I(INT)	I(PK)	I(DS)
103.88		0.9782	3	0	5	0		0
104.09	104.08	0.9769	6	4	3	1	1	7
104.67	104.80	0.9730	7	5	1	1	1	4
104.78		0.9723	5	3	4	1		11
104.85		0.9719	3	1	5	1		5
105.06	105.06	0.9705	7	2	3	1	1	10
106.64	106.72	0.9604	8	2	2	1	1	8
106.73		0.9599	6	0	4	2		18
107.71	107.72	0.9538	6	1	4	0	0	4
107.78		0.9534	3	2	2	0		1
107.92		0.9526	8	4	5	0		0
109.90	109.94	0.9409	9	1	0	0	0	4
109.97		0.9405	7	3	3	0		1
110.57	110.62	0.9370	8	4	1	0	1	3
110.60		0.9369	6	6	2	2		16
110.69		0.9364	6	2	4	1		7
110.76		0.9360	4	0	5	0		1
111.58	111.74	0.9314	9	0	1	0	1	0
111.60		0.9313	8	3	2	0		0
111.69		0.9308	5	4	4	2		15
111.76		0.9304	4	1	5	2		19
112.59	112.78	0.9259	9	5	1	0	1	3
112.61		0.9258	7	5	2	0		0
112.77		0.9249	3	3	5	1		13
112.92		0.9241	9	2	0	0		4
112.92		0.9241	7	6	0	0		0
112.99		0.9238	6	5	3	0		0
114.82	114.86	0.9142	4	2	5	0	0	1
115.78	116.08	0.9093	6	3	4	0	1	0
116.08		0.9079	8	0	3	3		23
117.05	117.12	0.9031	8	1	0	0	1	4
117.12		0.9028	8	1	3	1		11
117.12		0.9028	7	4	4	3		24
118.85	118.56	0.8946	8	4	2	0	0	2
119.90	119.94	0.8898	8	5	1	1	0	8
119.93		0.8897	9	0	2	0		0
120.03		0.8893	7	0	4	0		0

2THETA	PEAK	D	H	K	L	I(INT)	I(PK)	I(DS)
120.10	120.10	0.8890	5	3	5	0		0
120.10		0.8890	4	3	5	0		0
120.99	121.00	0.8850	9	3	1	5	1	45
121.01		0.8849	9	1	2	1		8
121.11		0.8845	5	5	4	1		2
121.11		0.8845	7	1	4	0		1
121.19		0.8842	5	1	5	1		6
124.35	124.50	0.8709	9	2	2	0		7
124.35		0.8709	7	6	2	0		1
124.46		0.8705	7	2	4	2		32
124.53		0.8702	5	5	5	3		19
124.77		0.8693	6	0	6	1	1	12
125.01	124.94	0.8683	7	0	7	0	1	8
127.01	127.10	0.8606	7	7	0	0	0	0
127.09		0.8604	7	5	3	1		10
127.33		0.8595	1	1	6	0		2
128.02	127.62	0.8569	4	4	5	0		1
129.00	129.08	0.8534	9	4	1	0	0	3
129.02		0.8533	8	5	2	1		8
130.21	130.42	0.8491	7	7	1	0	1	1
130.24		0.8490	9	3	2	2		22
130.35		0.8487	7	3	4	4		10
130.44		0.8484	5	5	5	2		18
130.61		0.8478	10	1	0	2		19
130.95		0.8466	2	1	6	1		13
132.71	132.78	0.8408	10	0	0	0		2
132.71		0.8408	8	6	0	3	1	38
132.95		0.8401	6	0	5	2		20
134.00	134.00	0.8368	10	1	1	2		26
134.15		0.8363	6	6	4	1		7
134.24		0.8360	6	1	5	0		1
134.43		0.8355	10	2	0	0		3
134.52		0.8352	8	4	3	0	1	6
137.11	137.38	0.8275	9	5	0	2	1	18
137.21		0.8273	9	1	3	1		9
137.49		0.8265	3	1	6	2		18
138.05	138.22	0.8249	10	2	1	0		1

FRESNOITE - MOORE AND LOUISNATHAN, 1967

2THETA	PEAK	D	H	K	L	I(INT)	I(PK)	I(DS)
138.21		0.8245	8	0	4	0		4
138.30		0.8242	6	2	5	1		15
139.50	139.66	0.8210	9	2	2	0	1	5
139.63		0.8206	8	1	4	1		16
139.63		0.8204	7	4	4	4		41
139.73		0.8204	5	5	5	2		26
140.93	140.40	0.8173	9	5	1	1	1	10
140.97		0.8172	7	7	1	1		1
141.52	141.46	0.8158	9	2	2	0	0	1
141.52		0.8158	7	6	3	0		8
141.83		0.8150	3	3	6	0		3
144.04	144.14	0.8098	10	0	2	0	0	1
144.04		0.8098	8	6	2	2		21
144.19		0.8095	8	2	4	1		15
145.63	145.68	0.8062	10	3	1	0	1	4
145.67		0.8062	10	1	2	5		64
145.94		0.8056	6	3	5	0		1
146.64	146.58	0.8041	4	0	6	1	1	9
147.90	148.36	0.8015	8	7	0	1	1	15
148.02		0.8013	8	5	3	2		20
148.38		0.8005	4	1	6	4		54
149.83	149.88	0.7977	9	3	3	8	1	102
150.22		0.7970	3	3	6	3		45
151.03	150.94	0.7955	10	2	2	0	1	4
151.20		0.7952	6	6	4	3		40
154.24	155.20	0.7901	4	2	6	1	0	8
155.17		0.7887	9	5	2	4		59
155.37		0.7884	7	1	5	5		4
155.53		0.7882	7	1	5	0		4
155.53		0.7882	5	5	5	1		10
155.85		0.7877	9	6	0	0		0
162.98	163.50	0.7788	9	6	1	1	0	7
163.05		0.7787	10	3	2	0		1
163.57		0.7782	7	2	5	5		71

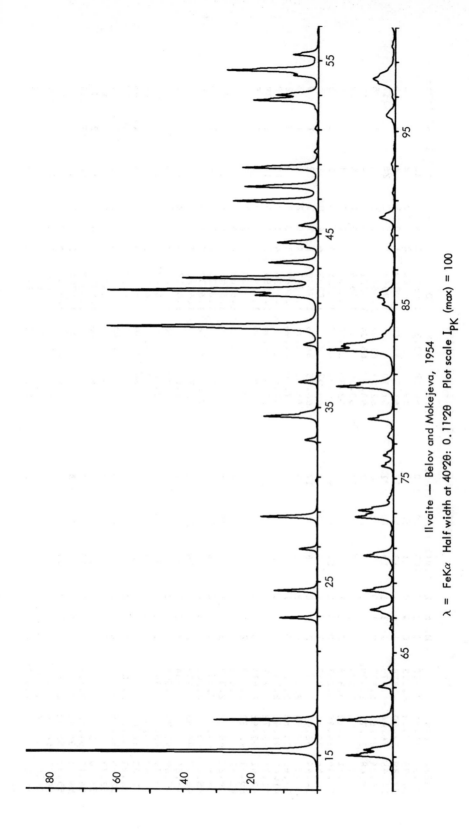

Ilvaite — Belov and Mokejeva, 1954

λ = FeKα Half width at 40°2θ: 0.11°2θ Plot scale I$_{PK}$ (max) = 100

ILVAITE - BELOV AND MOKEJEVA, 1954

2THETA	PEAK	D	H	K	L	I(INT)	I(PK)	I(DS)
15.24	15.24	7.300	0	1	0	100	100	100
17.04	17.04	6.535	1	0	1	31	31	33
22.89	22.90	4.878	1	2	0	12	11	17
24.46	24.46	4.570	1	1	1	15	13	21
26.84	26.86	4.170	2	0	0	7	6	10
28.68	28.72	3.909	1	2	0	5	17	8
28.71		3.904	1	3	0	17		28
33.11	33.12	3.398	2	0	1	5	4	10
34.46	34.48	3.267	0	4	1	21	16	44
34.66	34.66	3.249	1	3	1	4	5	9
36.44	36.58	3.096	2	3	0	8	6	17
38.58	38.58	2.930	0	4	2	6	4	13
39.65	39.66	2.854	0	3	1	78	63	187
41.42	41.42	2.737	2	1	1	23	19	59
41.71	41.72	2.719	1	2	1	90	63	232
42.41	42.42	2.676	3	2	0	58	40	152
43.31	43.32	2.623	2	4	0	20	15	54
44.22	44.22	2.623	3	0	1	2		4
44.47	44.48	2.572	1	1	2	5	4	13
45.45	45.46	2.558	2	2	2	18	12	50
46.77		2.506	1	5	0	9	6	25
46.86	46.86	2.439	2	0	2	16	25	48
47.69	47.70	2.434	3	2	1	27	22	81
48.79	48.80	2.394	2	4	1	34	23	103
49.69	49.68	2.344	1	3	2	36	23	112
51.03	51.04	2.304	3	3	1	2	1	6
52.21	52.22	2.247	3	3	0	2	1	6
52.69	52.68	2.200	2	4	2	2	1	3
52.77		2.181	0	4	0	31	20	109
53.00	53.00	2.178	4	1	0	1		4
54.11	54.12	2.169	2	2	1	19	13	65
54.42	54.42	2.128	1	4	2	10	8	36
55.32	55.32	2.117	4	2	0	46	27	165
58.24	58.24	2.085	1	6	1	13	8	47
59.05	59.06	1.9890	4	2	1	1	1	5
59.06		1.9644	4	3	0	9		34
		1.9638	4	3	0	16	14	65
59.38	59.38	1.9544	2	4	2	12	9	50
59.45		1.9522	2	6	0	2		8
61.01	61.10	1.9069	1	0	2	4	17	16
61.10		1.9043	1	5	2	28		117
62.64	62.66	1.8620	4	0	1	1	1	5
63.02	63.04	1.8521	2	6	1	4	4	18
63.04		1.8516	3	1	1	4		18
63.85	63.86	1.8306	3	5	3	1	1	5
64.01	64.02	1.8265	1	7	0	1	1	5
64.07		1.8249	4	0	0	1		5
66.35	66.36	1.7689	2	1	2	2	1	9
66.76	66.76	1.7593	1	3	3	3	1	5
67.30	67.44	1.7469	4	0	2	10	7	12
67.45		1.7436	1	3	2	18		48
68.55	68.56	1.7188	4	1	0	1	9	88
69.44	69.46	1.6994	5	0	0	1	1	5
69.47		1.6988	4	2	2	8		6
70.53	70.56	1.6766	0	6	3	11	9	42
70.57		1.6758	3	6	1	2		55
71.74	71.74	1.6520	4	3	0	1	1	8
72.67	72.78	1.6337	0	8	0	14	11	6
72.75		1.6322	5	2	1	11		72
72.80		1.6313	4	3	2	3		57
73.14	73.16	1.6246	2	6	2	17	11	16
73.16		1.6242	3	1	3	3		92
73.73	73.74	1.6134	3	2	3	8	2	16
75.69	75.70	1.5777	3	2	3	7	4	43
76.32	76.32	1.5667	2	4	3	2	3	37
77.34	77.40	1.5492	1	8	1	2	2	11
77.42		1.5479	4	6	0	17		13
78.40	78.40	1.5316	2	8	0	40	8	102
80.27	80.28	1.5017	3	6	2	47	17	243
82.37	82.36	1.4701	5	2	0	2	20	299
82.60	82.58	1.4667	6	0	0	26	16	13
82.71	82.70	1.4650	0	0	4	5	15	168
83.23	83.22	1.4575	6	1	0	7	4	31
85.00	85.00	1.4328	1	9	0		4	47

ILVAITE – BELOV AND MOKEJEVA, 1954

2THETA	PEAK	D	H	K	L	I(INT)	I(PK)	I(DS)
85.43	85.44	1.4269	0	8	2	10	5	64
85.53		1.4257	5	3	0	1		8
85.68	85.66	1.4236	4	7	0	4	4	27
88.13		1.3918	1	9	1	2		10
88.28	88.28	1.3900	6	3	0	4	2	27
89.04	89.04	1.3806	3	5	3	2	1	16
89.93		1.3698	5	4	0	3		19
89.99	90.00	1.3690	5	6	0	11	4	75
91.40	91.40	1.3525	6	3	1	2	1	16
92.79	92.80	1.3368	0	4	4	1	1	16
93.12	93.08	1.3331	5	6	1	2	1	12
93.94	93.94	1.3242	2	3	4	1	0	8
95.81	95.88	1.3045	6	1	1	1	3	8
95.85		1.3041	3	4	4	5		36
95.96		1.3029	3	6	3	4		32
96.96	96.98	1.2928	1	10	0	2	1	18
97.53		1.2872	1	9	2	2		13
97.75	97.78	1.2850	3	8	4	13	5	100
97.94	98.02	1.2832	3	2	3	6	7	51
98.05		1.2821	5	2	2	9		71
98.22		1.2804	4	7	2	2		14
99.26	99.24	1.2705	3	9	1	3	1	12
99.88	99.88	1.2647	1	5	4	7	1	27
101.18	101.18	1.2529	2	10	0	2	2	61
101.54	101.46	1.2497	6	5	0	3	2	21
102.56	102.56	1.2407	1	8	3	1	1	12
103.28	103.28	1.2345	7	2	0	1	0	12
105.43		1.2166	6	6	0	2		18
105.74	105.74	1.2141	6	1	4	4	1	37
107.71	107.78	1.1987	4	2	4	3	1	27
107.88		1.1974	5	9	0	8		15
108.95	108.96	1.1894	3	8	2	2	2	77
111.04	111.10	1.1743	4	3	4	6	2	59
111.18		1.1733	7	4	0	3		25
111.33	111.36	1.1723	6	5	2	4	2	10
111.68		1.1698	3	10	1	1		33
113.82	113.88	1.1553	1	9	3	3	0	10

2THETA	PEAK	D	H	K	L	I(INT)	I(PK)	I(DS)
115.10	115.10	1.1471	2	11	0	2	0	16
117.46	117.46	1.1325	6	1	3	3	1	32
119.41	119.42	1.1211	5	6	1	4	1	37
120.50	120.50	1.1150	2	7	4	12	2	127
122.22		1.1056	4	10	1	5		57
122.60		1.1036	0	4	5	2		18
122.67	122.66	1.1032	2	3	5	9	3	95
124.04	124.04	1.0961	1	11	2	3	1	34
124.74	124.74	1.0926	7	0	4	2	1	20
125.12		1.0907	0	8	4	1		12
125.50	125.46	1.0888	7	6	0	2		13
126.69	126.76	1.0831	3	9	3	1		17
127.15	127.16	1.0809	1	12	0	6		13
128.75		1.0736	3	2	5	1		44
128.83	128.84	1.0732	6	7	2	5		256
129.37	129.32	1.0708	0	12	1	3		36
129.54		1.0701	6	5	3	1		14
129.54		1.0701	2	4	5	6		63
132.23	132.26	1.0587	2	8	4	5		240
132.71	132.70	1.0559	7	2	4	1		60
132.91			5	3	5	3		15
134.59		1.0493	5	10	0	2		31
134.59		1.0493	8	3	1	1		37
134.63	134.62	1.0492	4	10	2	3		22
135.61	135.54	1.0455	4	11	0	5		38
136.41	136.42	1.0425	8	4	0	5		58
136.98	136.98	1.0405	2	12	0	3		66
138.05		1.0365	6	1	5	9		42
138.10	138.16	1.0333	6	10	1	6		108
139.05		1.0329	5	10	4	1		69
139.16	139.16	1.0327	6	6	3	20		17
139.22		1.0298	8	0	2	4		249
140.09	140.06	1.0266	8	2	1	12		56
141.04	141.12	1.0264	8	4	1	13		155
141.16		1.0244	1	9	4	7		169
141.81	141.82	1.0237	6	2	4			15
142.02			6	8	2			85
142.33	142.90	1.0227						

161

ILVAITE - BELOV AND MOKEJEVA, 1954

2THETA	PEAK	D	H	K	L	I(INT)	I(PK)	I(DS)
142.69		1.0217	4	2	5	4		46
142.89		1.0211	3	12	0	2		31
142.93		1.0209	4	7	4	8		107
143.03		1.0206	7	6	2	6		81
143.70	143.56	1.0187	3	10	3	9	3	119
147.43	147.52	1.0084	1	11	3	1	1	17
147.46		1.0084	6	3	4	9		122
148.44	148.40	1.0059	3	12	1	2	1	25
148.87	148.78	1.0048	2	6	5	4	1	59
148.91		1.0047	3	5	5	1		19
149.97	150.84	1.0022	8	3	2	4	3	54
150.82		1.0002	5	6	4	30		405
158.81	160.42	0.9848	5	11	0	1	1	13
158.87		0.9847	1	13	1	1		12
158.89		0.9846	4	11	2	4		57
160.46		0.9822	7	8	1	7		93
160.49		0.9822	8	4	2	6		89
160.68		0.9819	8	6	0	2		22
162.02		0.9800	7	5	3	2		22
164.71		0.9767	0	0	6	3		36

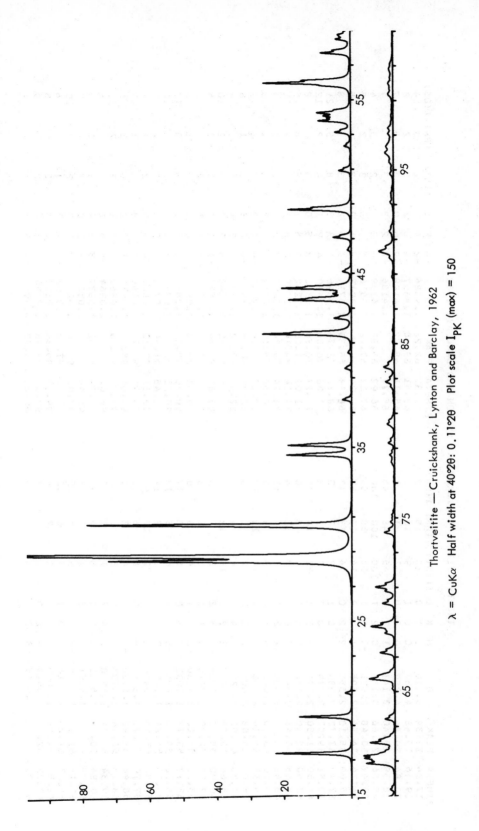

Thortveitite — Cruickshank, Lynton and Barclay, 1962

$\lambda = CuK\alpha$ Half width at 40°2θ: 0.11°2θ   Plot scale I_{PK} (max) = 150

THORTVEITITE – CRUICKSHANK, LYNTON AND BARCLAY, 1962

2THETA	PEAK	D	H	K	L	I(INT)	I(PK)	I(DS)
17.34	17.34	5.110	1	1	0	13	15	9
19.46	19.46	4.557	0	0	1	9	10	7
20.84	20.84	4.259	0	2	0	1	1	7
28.39	28.40	3.141	1	1	1	45	48	45
28.66	28.66	3.112	0	2	1	100	100	100
30.47	30.48	2.931	-2	0	1	56	53	59
34.54	34.54	2.595	-1	3	0	14	13	17
35.09	35.10	2.555	2	2	0	14	13	17
37.72	37.72	2.383	2	0	1	2	2	3
39.51	39.52	2.279	0	0	2	2	2	2
41.50	41.50	2.174	1	3	1	22	18	32
42.40	42.40	2.130	0	4	0	4	3	6
43.45	43.46	2.081	-2	0	2	9	12	14
43.48		2.080	2	2	1	7		11
43.80	43.80	2.065	3	1	0	3	4	5
44.13	44.14	2.050	-3	1	1	17	14	27
45.08	45.08	2.009	0	2	2	2	2	3
47.06	47.06	1.9295	0	4	1	4	4	8
48.66	48.66	1.8697	-2	2	2	16	12	29
50.90	50.90	1.7924	-1	3	2	3	2	3
52.30	52.30	1.7476	3	1	1	2	1	6
53.11	53.12	1.7230	-2	4	1	4	3	8
53.77	53.78	1.7032	3	3	0	8	6	17
54.06	54.06	1.6950	-3	3	1	6	5	12
54.25	54.24	1.6893	2	0	2	7	7	14
55.96	55.96	1.6418	1	3	2	22	17	46
57.70	57.70	1.5964	4	0	0	8	6	18
58.48	58.48	1.5768	-1	5	1	4	3	9
58.75	58.74	1.5703	2	2	2	3	3	6
59.34	59.34	1.5560	0	4	2	1	1	3
60.84	60.84	1.5213	0	5	2	8	6	18
60.93		1.5191	1	0	3	1		3
61.13	61.14	1.5146	-4	2	1	6	6	15
61.27	61.28	1.5116	3	3	1	5	6	12
61.59	61.58	1.5045	-2	0	3	2	2	4
62.03	62.04	1.4949	-4	2	2	1	5	3
62.05		1.4945	-3	3	2	6		15

2THETA	PEAK	D	H	K	L	I(INT)	I(PK)	I(DS)
62.33	62.22	1.4884	-2	4	2	2	3	4
63.41	63.40	1.4657	-4	0	0	2	2	8
65.71	65.72	1.4198	-0	6	0	7	5	17
65.77	65.88	1.4186	-2	0	3	3	3	7
66.02		1.4139	4	0	1	3		4
67.22	67.22	1.3916	1	5	2	2	3	8
67.24		1.3912	-3	1	3	3		5
68.34	68.34	1.3714	-1	3	3	2	2	8
68.73	68.74	1.3646	-1	5	2	8	5	21
70.06	70.06	1.3419	4	2	0	4	2	12
70.77	70.78	1.3302	-3	5	1	3	2	8
71.01	71.00	1.3262	-2	6	0	5	4	15
74.14	74.14	1.2778	2	0	3	4	2	11
74.74	74.74	1.2690	2	2	0	1	1	4
78.60	78.60	1.2162	-4	2	3	2	1	5
79.28	79.28	1.2074	-4	4	2	1	1	8
80.36	80.36	1.1939	-1	7	1	3	1	8
82.51	82.52	1.1681	-1	5	0	1	1	4
82.80	82.80	1.1648	1	3	4	3	2	10
83.61	83.64	1.1555	-5	1	1	1	1	5
83.66		1.1549	-5	5	2	2		9
89.91	89.92	1.0902	-6	0	3	2	2	6
90.30	90.30	1.0865	-6	5	1	3	3	12
90.31		1.0863	3	3	3	3		11
91.45	91.44	1.0758	-3	5	0	2	2	6
93.11	93.10	1.0609	4	6	2	3	1	7
94.74	94.74	1.0469	3	3	1	2	1	7
95.41	95.46	1.0414	-1	7	1	1	1	5
95.94	95.96	1.0369	0	8	3	3	1	11
96.01		1.0364	-5	1	4	1		6
96.34	96.32	1.0337	-3	5	0	1	1	5
96.49		1.0325	-6	2	1	1		5
97.06	97.06	1.0279	-5	3	2	2	1	10
97.40	97.40	1.0252	-6	2	1	1	1	6
99.37	99.34	1.0102	-2	8	1	2	1	8
99.98	99.96	1.0056	-4	7	2	2	1	8
100.81	100.80	0.9996	-5	5	2	1	1	7

THORTVEITITE - CRUICKSHANK, LYNTON AND BARCLAY, 1962

2THETA	PEAK	D	H	K	L	I(INT)	I(PK)	I(DS)
104.05	104.06	0.9771	-6	0	3	1	0	3
104.79	104.80	0.9722	2	8	2	1	1	5
106.28	106.34	0.9627	5	3	2	3	0	13
106.32		0.9624	-1	5	4	2	2	9
106.38		0.9621	5	5	1	3		15
108.69	108.70	0.9480	-2	8	1	2	1	8
115.35	115.70	0.9115	0	0	5	1	0	4
115.65		0.9100	-2	2	5	1		3
116.15	116.18	0.9075	-3	1	5	1	1	4
116.24		0.9071	7	1	0	1		5
117.53	117.52	0.9008	2	8	2	1	0	4
119.26	119.28	0.8928	6	0	2	1	0	6
119.88	119.88	0.8899	-4	8	1	2	1	11
120.40	120.38	0.8876	-7	3	1	2	1	12
120.80	120.82	0.8859	4	8	0	1	1	3
121.91	121.92	0.8810	5	7	0	2	1	12
122.39	122.42	0.8790	3	9	0	1	1	7
122.64		0.8779	-4	0	5	1		4
124.09	124.12	0.8720	-0	8	3	1	0	4
124.79	124.84	0.8692	-2	8	3	1	0	6
124.96		0.8685	-7	3	0	1		4
125.94	125.90	0.8647	-6	6	3	1	0	6
127.00	126.98	0.8607	-6	2	4	1	0	3
128.13	128.12	0.8565	-1	9	2	1	0	5
129.42	129.80	0.8519	0	10	0	2	0	7
129.79		0.8506	4	8	1	1		11
133.87	134.04	0.8372	-1	3	5	1	0	5
134.10		0.8365	-5	5	4	1		8
136.76	136.78	0.8285	4	2	4	2	0	11
138.71	138.92	0.8231	2	10	0	1	0	6
138.95		0.8225	7	3	1	2		11
139.53	139.80	0.8209	2	6	4	1	1	4
139.86		0.8201	-3	7	4	1		8
140.15	140.66	0.8193	-7	5	1	1	1	4
140.63		0.8181	-2	10	1	3		18
141.56	141.48	0.8157	2	8	3	1	1	8
141.92		0.8148	2	2	5	1		9

2THETA	PEAK	D	H	K	L	I(INT)	I(PK)	I(DS)
142.87	142.80	0.8125	3	5	4	1	0	7
143.84	144.54	0.8103	-8	0	2	1	1	8
144.44		0.8089	-4	8	3	2		17
144.53		0.8087	-1	9	3	1		5
144.59		0.8085	-7	5	2	1		8
144.60		0.8085	1	7	4	1		9
146.24	146.52	0.8049	-6	6	3	1	1	6
146.41		0.8046	-3	5	5	1		15
146.57		0.8042	7	5	0	3		23
147.47	147.46	0.8024	-8	2	1	2	1	17
147.56		0.8022	2	10	1	1		10
149.30	149.60	0.7987	-5	3	1	1	0	5
149.58		0.7982	8	0	0	2		12
149.71		0.7980	0	10	2	2		14
151.91	151.70	0.7940	4	8	2	1	0	7
156.37	157.08	0.7869	-7	5	3	2	0	18
157.04		0.7860	1	9	2	1		8
157.08		0.7859	3	9	0	1		6
158.08		0.7846	8	2	0	1		5
159.12		0.7832	5	7	2	1		8
159.24	161.22	0.7831	-2	8	3	1	0	4
161.10		0.7809	-6	2	4	6		42
162.03		0.7798	-8	0	3	1		5
162.68		0.7791	1	5	5	1		10
163.82		0.7780	-2	0	6	2		12
163.83		0.7780	0	8	4	2		18

166

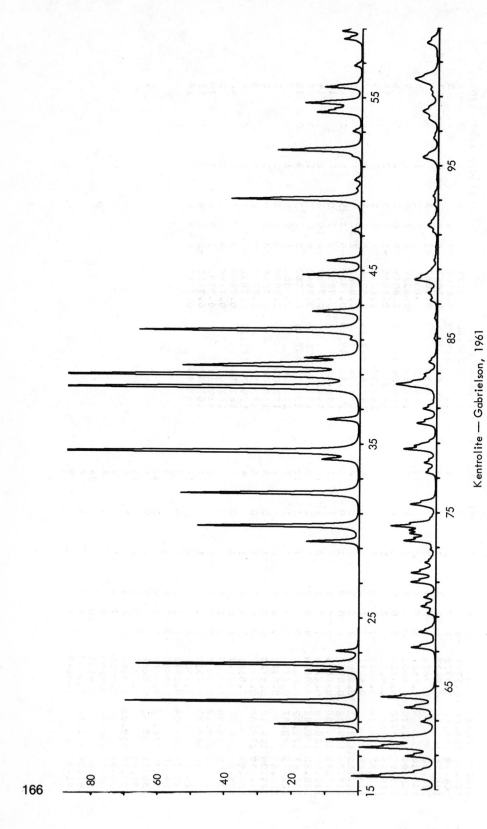

Kentrolite — Gabrielson, 1961

λ = FeKα   Half width at 40°2θ: 0.11°2θ   Plot scale I_{PK} (max) = 100

KENTROLITE - GABRIELSON, 1961

2THETA	PEAK	D	H	K	L	I(INT)	I(PK)	I(DS)
18.85	18.86	5.911	1	1	0	19	25	5
20.17	20.18	5.527	0	2	0	54	70	17
21.93	21.94	5.089	1	1	1	12	16	4
22.32	22.34	5.000	1	0	1	55	67	21
23.09	23.10	4.837	0	2	1	5	7	2
29.38	29.38	3.818	1	1	2	14	16	9
30.26	30.28	3.708	0	2	2	45	48	29
32.13	32.14	3.498	2	0	0	50	53	36
34.10	34.10	3.301	1	3	1	9	11	8
34.55	34.56	3.260	1	1	0	100	100	81
36.40	36.40	3.100	2	2	1	10	10	8
38.24	38.24	2.955	1	2	3	98	96	95
38.95	38.96	2.904	2	1	1	100	96	100
39.48	39.48	2.866	2	0	2	51	53	53
39.94	39.94	2.834	2	2	1	15	17	16
41.01	41.02	2.764	2	2	0	2	3	2
41.52	41.52	2.731	0	4	0	71	66	79
42.62	42.62	2.664	2	3	1	15	14	18
44.72	44.72	2.544	0	4	0	20	17	25
45.56	45.56	2.500	1	2	2	11	10	15
47.30	47.30	2.413	1	0	3	2	3	4
49.08	49.08	2.331	1	3	1	45	39	69
49.72	49.72	2.303	3	1	0	3	3	2
50.21	50.22	2.281	2	1	0	2	2	3
51.59	51.60	2.224	2	2	2	2	2	4
51.91	51.92	2.212	2	4	0	30	25	51
53.03	53.04	2.168	0	4	3	3	3	5
54.13	54.14	2.128	2	5	0	16	14	29
54.36	54.36	2.119	3	2	3	7	8	12
54.67	54.68	2.108	2	0	4	21	17	38
55.60	55.60	2.076	3	3	1	14	11	26
56.84	56.84	2.034	2	2	4	3	3	6
58.41	58.40	1.9840	1	5	0	7	6	14
58.85	58.86	1.9703	3	3	0	8	6	16
59.78	59.78	1.9425	1	5	2	33	25	71
60.10	60.08	1.9332	3	3	1	5	6	11
60.94	60.94	1.9088	2	2	4	11	9	25
61.45	61.46	1.8947	1	3	0	30	23	67
61.88	61.88	1.8828	3	1	0	37	33	84
61.95		1.8808	0	0	4	13		31
62.95	62.94	1.8540	0	6	1	5	4	13
63.39	63.40	1.8423	3	3	0	1	2	3
63.75	63.76	1.8331	2	4	2	13	9	30
64.36	64.36	1.8177	0	6	2	23	17	56
67.22	67.22	1.7488	4	0	0	11	7	28
68.10	68.10	1.7287	1	6	1	7	5	20
68.38	68.26	1.7226	4	0	1	1	3	3
69.19	69.20	1.7049	3	4	2	3	5	10
69.60	69.60	1.6962	4	2	0	6	5	18
70.11	70.12	1.6853	0	7	1	5	5	15
70.98	70.98	1.6673	2	0	5	9	4	26
71.00		1.6668	2	6	1	3	8	8
71.52	71.52	1.6565	0	4	5	11	8	32
71.80	71.68	1.6507	4	4	0	2	5	5
72.11	72.12	1.6446	4	3	1	4	3	13
72.44	72.42	1.6381	2	5	1	1	1	3
72.86	72.86	1.6300	2	7	1	2	2	6
73.37	73.38	1.6203	0	6	2	15	10	46
73.78	73.78	1.6125	0	6	5	10	8	30
73.98	73.98	1.6088	2	6	1	5	8	15
74.22	74.22	1.6043	3	5	0	19	14	59
74.68	74.66	1.5958	0	6	6	1	2	4
75.34	75.48	1.5840	4	2	5	2	8	6
75.47		1.5817	4	4	2	11		37
77.31	77.32	1.5498	2	5	2	5	3	16
77.87	77.86	1.5404	1	2	0	6	4	19
78.64	78.64	1.5276	3	7	1	16	10	57
78.96	78.82	1.5224	1	7	1	4	7	15
80.08	80.12	1.5047	3	1	6	1	6	5
80.12		1.5040	2	0	5	9		32
80.95	80.96	1.4912	1	2	3	7	3	26
82.23	82.36	1.4721	4	7	2	9	4	33
82.36		1.4702	1	4	5	17	12	66
82.76	82.74	1.4643	2	6	3	4	4	15

KENTROLITE - GABRIELSON, 1961

2THETA	PEAK	D	H	K	L	I(INT)	I(PK)	I(DS)
82.93	82.94	1.4619	4	4	1	4	4	4
83.63	83.64	1.4518	2	2	6	4	3	14
83.69		1.4510	1	5	5	1		4
84.98	84.98	1.4330	4	0	4	2	1	7
86.17	86.16	1.4171	4	4	2	2	1	7
87.20	87.20	1.4037	3	3	5	4	2	17
87.62	87.62	1.3983	1	7	1	5	3	21
88.38	88.40	1.3887	1	1	7	13	7	55
88.51		1.3871	4	2	4	2		9
89.52	89.52	1.3748	5	1	1	2	1	8
91.18	91.18	1.3551	1	6	5	3	2	13
91.54	91.54	1.3510	0	4	5	5	3	22
91.60		1.3502	4	4	3	2		6
91.98	91.76	1.3459	3	5	6	1		6
92.74	92.74	1.3374	5	1	2	3	2	12
93.24	93.24	1.3318	1	0	8	3	1	15
95.42		1.3086	0	8	2	4		21
95.48	95.48	1.3079	5	3	3	5	5	24
95.52		1.3075	3	7	0	1		7
95.52		1.3075	1	5	7	2		8
96.56	96.58	1.2969	5	3	1	1		7
96.60		1.2965	3	7	1	2	2	10
97.62		1.2863	2	2	7	3		15
97.74	97.74	1.2851	2	8	0	4	3	23
98.13	98.14	1.2813	5	1	3	7	5	36
98.19		1.2807	4	2	5	4		21
99.05	99.08	1.2725	2	3	6	1		8
99.09		1.2721	3	3	4	2	2	13
99.41	99.40	1.2691	0	4	7	3		13
99.81		1.2653	5	3	2	4		21
99.85		1.2650	3	7	2	3		16
100.00	100.00	1.2636	2	6	6	15	7	77
102.10	102.10	1.2447	2	8	2	10	4	56
103.88	103.88	1.2294	4	6	2	2	1	14
104.96	105.02	1.2204	1	1	8	10	5	58
105.10		1.2193	0	2	8	4		24
105.32	105.30	1.2175	5	3	3	6	5	32

2THETA	PEAK	D	H	K	L	I(INT)	I(PK)	I(DS)
105.36	105.36	1.2172	3	7	3	2	4	13
106.15	106.14	1.2109	7	1	1	9	2	54
107.42	107.68	1.2010	1	9	1	2		12
107.66		1.1991	2	8	4	9		13
108.46	108.48	1.1930	4	4	7	14	3	55
109.06	109.06	1.1885	4	6	5	7	5	86
109.48	109.42	1.1854	5	5	0	3	5	43
109.93	109.80	1.1822	5	0	6	6	3	16
110.63	110.64	1.1772	2	5	8	4	3	37
111.08	111.02	1.1740	5	5	1	4	2	26
112.05	112.26	1.1672	1	3	0	3	2	26
112.26		1.1658	6	0	8	1		21
113.28	113.64	1.1589	6	0	1	1	2	9
113.42		1.1580	3	0	4	4		9
113.63		1.1566	3	3	7	1		24
114.43	114.58	1.1514	2	2	8	8	3	9
114.57		1.1505	5	2	5	3		55
116.11	116.18	1.1407	6	1	2	8		21
116.18		1.1403	4	5	0	7		53
116.69	116.68	1.1372	5	3	6	3	4	49
116.74		1.1369	6	0	5	3		21
116.98	117.00	1.1354	2	1	1	4	3	20
117.31		1.1334	2	7	4	2		26
117.69	117.66	1.1311	6	1	4	4	2	14
118.05	118.02	1.1290	4	6	6	2	2	30
120.63	120.66	1.1142	5	3	5	4	1	16
121.00	121.00	1.1122	3	5	2	2	1	7
122.19	122.22	1.1058	6	2	6	1	2	26
122.25		1.1054	4	4	0	4		25
124.37	124.38	1.0963	0	1	10	3	3	8
124.37		1.0944	3	7	5	1		89
126.31	126.44	1.0849	4	2	7	4	2	31
126.46		1.0842	4	8	0	5		41
127.10	127.54	1.0812	2	8	5	1	4	8
127.28		1.0803	1	9	1	2		18
127.49		1.0793	0	10	2	8		62
127.50		1.0793	5	2	3	3		21

168

KENTROLITE – GABRIELSON, 1961

2THETA	PEAK	D	H	K	L	I(INT)	I(PK)	I(DS)
127.81	127.88	1.0778	4	8	1	5	4	36
128.38	128.38	1.0753	1	5	8	14	5	112
128.56		1.0744	2	6	7	5		36
130.34	130.30	1.0666	5	1	6	3	1	26
132.01	132.02	1.0595	4	8	2	10	2	86
132.99	133.00	1.0556	3	3	8	11	3	92
133.38		1.0540	2	10	0	5		40
134.88	135.04	1.0482	2	10	1	2	2	13
135.06		1.0475	1	7	7	3		28
135.15		1.0472	5	7	0	2		19
138.50	138.54	1.0351	1	9	5	21	4	182
138.70		1.0345	0	6	8	2		16
139.08	139.10	1.0332	3	9	3	9	4	78
139.62	139.60	1.0313	2	10	3	10	4	91
139.73		1.0310	4	8	3	2		15
139.73		1.0310	0	4	9	2		17
140.36	140.24	1.0290	5	3	6	2	3	22
140.42		1.0288	3	7	6	2		18
140.92	140.86	1.0271	4	4	7	6	2	56
141.62	141.58	1.0249	5	7	2	5	2	48
142.45	142.40	1.0224	6	4	3	10	2	88
144.03	144.20	1.0177	5	5	5	1	1	9
144.28		1.0170	4	0	8	3		29
147.08	147.12	1.0094	4	6	6	3	1	29
148.81	148.88	1.0050	2	10	3	4	1	35
150.84	151.04	1.0002	4	0	8	7	2	63
150.88		1.0001	0	0	10	9		84
151.35		0.9990	5	7	3	5		46
151.36		0.9990	3	1	9	1		13
152.47	153.16	0.9966	3	9	4	2	3	21
152.99		0.9955	5	1	0	15		146
153.38		0.9947	1	11	0	6		63
153.39		0.9947	4	8	4	3		30
155.30	155.16	0.9909	6	2	5	11	2	110
155.61		0.9903	7	1	1	1		15
155.89		0.9898	1	11	3	3		30
156.38		0.9889	2	4	9	2		17

2THETA	PEAK	D	H	K	L	I(INT)	I(PK)	I(DS)
157.51	157.92	0.9869	6	4	4	11	2	113
158.01		0.9861	3	5	8	10		104
158.01		0.9861	1	1	10	4		45
158.57		0.9852	6	6	0	1		15
159.22		0.9841	0	2	10	12		121
159.94		0.9830	1	5	9	2		20

170

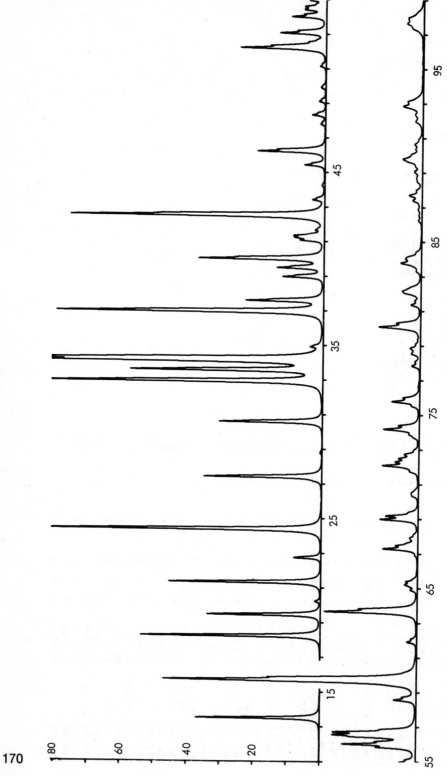

Lawsonite — Rumanova and Skipetrova, 1959

$\lambda = CuK\alpha$ Half width at $40°2\theta$: $0.11°2\theta$ Plot scale I_{PK} (max) = 100

2THETA	PEAK	D	H	K	L	I(INT)	I(PK)	I(DS)
13.48	13.48	6.564	0	2	0	31	37	19
18.21	18.22	4.867	1	1	0	48	54	30
19.43	19.44	4.564	1	1	1	31	34	20
20.19	20.20	4.394	1	0	1	1	2	1
21.31	21.32	4.167	0	2	1	44	46	28
22.72	22.72	3.910	1	1	2	8	8	5
24.36	24.36	3.651	0	0	2	83	84	55
27.38	27.38	3.254	1	2	2	37	35	26
30.56	30.56	2.923	2	0	0	34	31	24
32.88	32.88	2.721	1	1	4	96	84	70
33.53	33.54	2.670	1	3	2	64	58	47
34.07	34.08	2.630	2	0	4	89	87	66
34.21	34.20	2.619	0	1	0	86	100	64
34.90	34.90	2.568	1	3	1	3	4	2
36.90	36.92	2.434	1	0	2	47	80	36
36.92	36.92	2.432	2	1	2	51	80	39
37.55	37.56	2.393	1	2	1	27	23	21
38.94	38.94	2.311	2	2	1	15	12	11
39.45	39.46	2.282	2	1	2	16	14	13
39.96	39.96	2.254	0	2	5	46	37	37
40.09	40.06	2.247	1	3	3	8	26	2
41.05	41.06	2.197	0	4	0	8	7	6
41.22	41.22	2.188	0	0	6	8	9	6
41.32	41.32	2.183	2	0	4	6	9	4
41.64	41.64	2.167	2	4	1	98	3	2
42.46	42.46	2.127	2	4	3	4	76	80
43.39	43.40	2.083	0	2	4	1	4	3
44.20	44.20	2.047	1	3	2	8	1	1
45.40	45.40	1.9958	1	4	4	27	6	7
46.19	46.20	1.9635	0	6	3	2	20	23
46.31	46.30	1.9587	4	2	6	5	14	2
47.77	47.76	1.9025	3	1	0	3	2	5
48.30	48.30	1.8828	3	1	1	2	14	2
49.09	49.08	1.8543	1	3	5	2	4	2
49.91	49.90	1.8257	0	4	4	5	2	1
51.13	51.14	1.7849	2	2	5	3	2	2
52.17	52.18	1.7517	2	0	6	33	25	30
52.22	52.22	1.7501	1	1	7	3		3
52.40	52.30	1.7447	3	1	3	3	16	3
52.53	52.52	1.7407	2	4	1	3	5	3
53.04	53.04	1.7250	0	2	7	19	14	18
54.00	54.00	1.6966	1	5	4	14	10	13
54.40	54.42	1.6850	2	2	5	7	7	6
54.47		1.6832	3	3	0	3		3
54.61	54.60	1.6792	1	1	7	4	6	4
54.95	54.96	1.6695	1	5	4	7	5	6
55.80	55.82	1.6460	1	4	8	14	12	14
55.98	55.98	1.6411	0	8	0	25	22	24
56.40	56.42	1.6299	3	2	3	20	20	20
56.51	56.52	1.6272	2	2	6	23	25	22
56.68	56.68	1.6225	3	2	2	22	25	21
58.55	58.56	1.5751	1	0	2	10	7	10
59.58	59.66	1.5504	4	6	4	30	76	30
59.66		1.5485	0	4	6	100		100
61.90	61.90	1.4977	1	4	0	5	3	5
63.61	63.62	1.4615	4	0	5	42	28	45
63.69		1.4598	6	6	0	6		7
64.89	64.90	1.4357	3	1	6	3	2	3
65.13	65.20	1.4310	1	0	8	1	4	1
65.21		1.4295	0	2	7	5		5
65.36	65.36	1.4265	2	7	3	2		2
67.27	67.26	1.3906	1	8	1	18	4	19
67.90	67.90	1.3791	3	0	3	3	11	3
68.96	68.96	1.3607	1	8	2	19	3	22
69.18	69.16	1.3569	2	8	2	7	12	7
70.32	70.34	1.3375	3	6	2	3	10	3
70.44	70.46	1.3356	1	6	2	2	2	2
71.72	71.74	1.3148	0	8	4	4	3	4
72.06	72.06	1.3095	2	4	8	18	11	21
72.34	72.28	1.3051	3	6	1	2	7	2
72.46	72.46	1.3033	6	2	7	7	6	8
72.47		1.3030	4	6	4	4		4
72.75	72.70	1.2987	3	5	5	3	4	4
73.86	73.88	1.2820	2	7	2	2	3	4

LAWSONITE — RUMANOVA AND SKIPETROVA, 1959

2THETA	PEAK	D	H	K	L	I(INT)	I(PK)	I(DS)
74.16	74.16	1.2775	4	2	4	19	11	22
75.51	75.52	1.2579	0	2	10	3	3	4
75.76	75.76	1.2545	2	6	3	14	8	16
75.89		1.2527	1	2	7	1		2
77.82	77.82	1.2263	4	0	5	5	3	6
78.52	78.54	1.2172	0	6	6	4	3	5
78.59		1.2162	2	6	4	1		2
78.86	78.84	1.2128	3	4	4	4	3	5
79.94	80.06	1.1991	2	4	8	2	3	3
80.05		1.1976	2	0	10	23	13	29
80.15		1.1965	4	4	2	2		3
81.35	81.34	1.1818	1	7	3	6		7
82.03	82.14	1.1737	3	1	10	7	3	9
82.14		1.1724	4	4	3	1	6	8
83.28		1.1592	1	1	11	1		2
83.30	83.30	1.1590	5	1	0	1	2	1
83.61	83.76	1.1555	2	2	10	1		2
83.71		1.1544	0	6	7	1	6	2
83.76		1.1538	3	3	8	11		14
83.94	84.00	1.1518	0	2	11	2	5	3
84.13		1.1497	1	7	4	5		7
86.23	86.24	1.1270	4	4	10	2	1	3
86.86	86.86	1.1204	1	1	3	2	1	2
87.38	87.38	1.1151	2	7	5	6	3	8
87.69	87.68	1.1120	1	7	5	6	4	8
88.48	88.50	1.1041	4	4	5	3	2	4
88.65	88.66	1.1024	1	5	9	2	2	3
89.04	89.04	1.0985	0	8	0	5	3	7
89.63	89.76	1.0929	5	1	2	6	6	9
89.78		1.0914	4	0	8	11		15
90.36	90.36	1.0859	5	3	0	4	2	5
90.62	90.62	1.0834	0	8	2	3	2	5
91.94	91.96	1.0713	5	3	7	2	2	4
91.94		1.0713	1	1	6	1		2
92.03		1.0705	3	7	6	1		2
92.37	92.36	1.0674	1	1	12	3	2	4
92.82	92.82	1.0635	4	4	6	14	6	20

2THETA	PEAK	D	H	K	L	I(INT)	I(PK)	I(DS)
93.02	93.08	1.0617	0	5	11	4	5	6
93.18		1.0603	5	1	5	1		2
97.42	97.62	1.0251	2	8	1	4	5	5
97.62		1.0236	6	2	8	8		11
97.83	97.92	1.0219	4	6	2	4	5	5
97.88		1.0215	3	5	8	2		2
97.97		1.0208	4	3	7	1		2
98.00		1.0206	3	3	10	4		6
100.05	100.10	1.0051	4	2	9	2	2	3
100.11		1.0047	3	7	4	3	1	4
102.57	102.64	0.9872	2	4	11	1	1	1
102.64		0.9867	4	6	4	2		2
103.73	104.00	0.9792	3	7	5	5	4	8
103.97		0.9776	3	0	6	5		8
104.00		0.9775	4	4	10	3		5
104.12		0.9767	4	2	0	1		2
104.45	104.34	0.9745	2	6	10	1	2	2
104.47		0.9743	1	0	9	1		2
106.11	106.12	0.9638	3	6	0	3	1	4
106.21		0.9631	1	8	2	1		1
107.12	107.10	0.9575	8	0	5	1	0	1
107.78	107.80	0.9534	4	2	10	3	2	5
108.14	108.18	0.9512	6	2	0	2	2	2
108.25		0.9506	3	7	6	3		3
108.61	108.62	0.9484	1	8	2	2	1	3
108.70		0.9479	0	1	12	1		2
108.90		0.9467	5	8	7	1		2
109.81	109.90	0.9414	6	1	8	1	1	2
109.95		0.9406	1	9	3	4		7
110.88	110.90	0.9353	4	6	6	6	1	7
111.24	111.30	0.9333	5	5	4	5	2	6
111.33		0.9328	2	2	13	2		3
111.72	112.08	0.9306	2	8	6	2	4	2
111.92		0.9295	8	2	3	1		6
112.08		0.9286	2	4	12	4		2
112.71	112.46	0.9252	0	6	11	14	3	23
113.54	113.54	0.9209	1	1	14	2	1	4

LAWSONITE — RUMANOVA AND SKIPETROVA, 1959

2THETA	PEAK	D	H	K	L	I(INT)	I(PK)	I(DS)
114.15	114.16	0.9176	0	4	13	3	1	4
114.21		0.9173	1	5	12	1		2
115.08	115.08	0.9129	0	8	8	2		3
116.48	116.58	0.9059	4	6	7	1	1	2
116.54		0.9056	5	3	8	5		9
116.74		0.9046	4	2	11	2		3
116.83		0.9042	1	9	5	2		4
117.36	117.30	0.9017	2	9	7	4	2	7
119.32	119.36	0.8925	4	4	10	3	1	6
119.42		0.8920	3	1	13	1		2
119.85	119.84	0.8901	6	0	6	4	2	7
121.55	121.60	0.8826	6	4	2	2	1	3
122.44	122.68	0.8788	6	2	8	1	3	2
122.60		0.8781	4	8	0	7		12
122.75		0.8775	0	8	9	3		5
122.98	123.24	0.8765	0	6	12	1		2
123.07		0.8762	4	8	1	2	4	4
123.24		0.8755	2	4	13	5		9
123.34		0.8751	2	2	14	6		11
123.90	123.94	0.8728	6	4	3	4	4	6
124.00		0.8724	6	2	6	5		8
124.25	124.28	0.8714	2	8	8	8	4	14
124.33		0.8710	0	10	2	1		1
124.50		0.8704	4	8	2	5		10
126.93	127.32	0.8610	4	8	3	1	5	2
127.29		0.8596	4	4	4	21		38
127.46		0.8590	2	2	12	7		14
128.24	127.82	0.8561	3	9	3	1	3	2
130.28	130.30	0.8489	0	10	4	5	1	9
130.42		0.8484	5	7	2	1		1
130.46		0.8483	4	8	4	1		1
132.62	132.88	0.8411	3	1	14	3	2	6
132.83		0.8405	2	8	9	7		13
133.02		0.8399	2	10	1	1		1
133.06		0.8397	5	7	3	3		6
133.10		0.8396	2	6	12	1		2
133.46	133.40	0.8385	3	5	12	2	2	3

2THETA	PEAK	D	H	K	L	I(INT)	I(PK)	I(DS)
133.84	133.88	0.8373	5	5	8	1	2	1
134.00		0.8368	5	3	10	6		11
134.66	134.72	0.8348	2	10	2	2	2	5
135.12	135.12	0.8334	0	10	5	6	2	13
135.75	135.78	0.8315	5	1	11	2	2	3
135.76		0.8314	0	6	13	2		4
136.04		0.8306	1	9	11	1		3
136.35	136.90	0.8297	7	1	1	1	2	3
136.81		0.8284	3	9	5	3		5
136.97		0.8279	5	7	4	5		10
137.22		0.8272	2	2	14	1		1
138.52	138.76	0.8236	2	4	15	2	2	4
138.76		0.8230	6	2	8	8		16
139.66	139.62	0.8206	0	0	16	10	2	20
140.18		0.8192	1	5	14	1		2
141.15	141.18	0.8167	1	7	12	3	1	7
141.29		0.8164	4	2	13	1		3
142.43	142.88	0.8136	4	4	12	1	3	1
142.46		0.8135	5	7	5	6		11
142.63		0.8131	0	4	15	1		29
142.85		0.8126	4	6	10	14		21
143.42	143.48	0.8112	6	6	5	11	3	5
144.07	144.06	0.8097	6	5	9	2	3	7
144.09		0.8097	6	1	10	3		19
144.15		0.8095	2	8	10	9		5
146.41	147.10	0.8046	6	4	7	2	2	3
146.81		0.8037	1	9	9	1		15
147.09		0.8031	7	3	0	7		16
147.93	147.98	0.8014	2	10	5	8	2	19
148.80	148.84	0.7997	2	6	13	9	2	20
149.88	149.96	0.7976	6	6	3	10		7
150.12		0.7972	7	3	2	4		4
150.30		0.7969	5	5	7	2		2
150.34		0.7968	6	2	9	1		2
150.90	151.02	0.7958	0	10	7	1	2	13
151.00		0.7956	5	1	12	6		11
151.18		0.7953	4	8	7	5		

LAWSONITE — RUMANOVA AND SKIPETROVA, 1959

2THETA	PEAK	D	H	K	L	I(INT)	I(PK)	I(DS)
151.25		0.7951	3	1	15	2		3
153.93	154.26	0.7906	3	7	11	5	2	11
154.25		0.7901	1	11	1	12		25
154.31		0.7900	2	0	16	2		4
154.47		0.7898	0	6	14	1		2
155.95		0.7875	6	6	4	2		4
159.00	159.72	0.7834	2	4	15	3	3	6
159.27		0.7830	1	3	16	11		23
159.44		0.7828	6	4	8	2		5
159.77		0.7824	6	0	10	31		66
160.13		0.7820	5	5	10	1		2
160.32		0.7818	4	6	11	5		10
161.98		0.7799	1	7	13	16		35
162.89		0.7789	1	11	3	1		2
163.37		0.7784	5	7	7	4		9

TABLE 8. SILICATES WITH PAIRS OF TETRAHEDRA

Variety	Rinkite	Ca-Seidozerite	Wohlerite	Låvenite	Perrierite
Composition	Na(Na,Ca)$_2$(Ca,Ce)$_4$ (Ti,Nb)(Si$_2$O$_7$)$_2$O$_2$F$_2$	Na$_2$(Na,Ca)$_2$(Ca,Mn) TiZr$_2$O$_2$[Si$_2$O$_7$]$_2$ (F,OH)$_2$	Ca$_2$Na(Zr,Nb) Si$_2$O$_7$(O,F$_2$)	(Na,Ca)(Fe,Mn,Ca) (Zr,Ti)O(F,OH)Si$_2$O$_7$	Ce$_8$Fe$_2$Me$_4$Ti$_4$ [O$_4$(Si,Al)$_2$O$_7$]$_4$
Source	Greenland	N. Baikal Area, USSR	Langesund Fiord (Brevik)? Norway	Lovozersk Massif, Russia	Nettuno, Rome, Italy
Reference	Tê-yŭ, Simonov & Belov, 1965	Skszat & Simonov, 1966	Shibayeva and Belov, 1962	Simonov & Belov, 1960	Gottardi, 1960; Galli, 1965
Cell Dimensions					
a $\quad$ Å	18.28	5.54	10.80	10.54	13.61
b $\quad$ Å	5.59	7.10	10.26	9.90	5.62
c	7.38	18.36	7.26	7.14	11.67
α	90	90	90	90	90
β $\quad$ deg	~90	102.66	108.95	108.20	113.50
γ	90	90	90	90	90
Space Group	P2$_1$	P2/c	P2$_1$	P2$_1$/a	C2/m
Z	2	2	4	4	1

TABLE 8. (cont.)

Variety	Rinkite	Ca-Seidozerite	Wöhlerite	Låvenite	Perrierite
Site Occupancy	M_1 {Ca 0.75, Ce 0.25} M_2 {Ti 0.66, Nb 0.34} M_3 {Ca 0.83, Na 0.17}	M_1 {Zr 0.70, Mn 0.20, Fe 0.10} M_2 {Ti 0.78, Fe 0.22} M_3 {Ca 0.84, Mn 0.16} Na_1 {Na 0.78, Ca 0.22} Na_2 Na 1.0 Na_3 Na 1.0		M_1 Na 0.97 M_2 {Ca 0.70, Na 0.26} M_3 {Fe 0.31, Nb 0.09, Mn 0.33, Ca 0.24} M_4 {Zr 0.73, Ti 0.26}	M_1 {Ce 0.58, La 0.12, Y 0.04} $\&M_2$ {Ti 0.05, Ca 0.13, Na 0.08} M_3 {Fe^2 0.66, Ca 0.34} Me {Ti 0.74, Fe^3 0.10, Fe^2 0.03, Mg 0.12}
Method	2-D, film	2-D, film	2-D, film	2-D, film	3-D, film
R & Rating	0.163 (av.) (3)	0.12 (av.) (2)	0.182 (av.) (3)	0.20 (av.) (3)	0.119 (2)
Cleavage and habit	{100} perfect Tabular parallel to {100}	{001} (?)	{010} distinct Tabular parallel to {100}; also prismatic	{001} perfect Tabular parallel to {001};elongated to $\underline{a}$ axis	
Comment		Rinkite group	Related to cuspidine. Minor Nb ignored.	Related to cuspidine.	
μ	1361.0	265.7	224.4	301.7	907.7
ASF	0.1809	0.1940	0.1162	0.1508	0.3065
Abridging factors	1:1:0.5	1:1:0.5	1:1:0.5	1:1:0.5	1:1:0.5

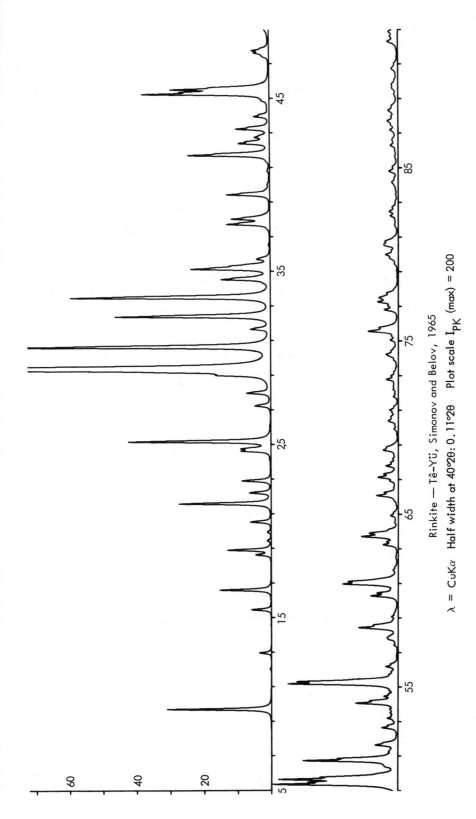

Rinkite — Tê-Yü, Simonov and Belov, 1965

λ = CuKα Half width at 40°2θ: 0.11°2θ Plot scale I_{PK} (max) = 200

177

RINKITE - TÊ-YÜ, SIMONOV AND BELOV, 1965

2THETA	PEAK	D	H	K	L	I(INT)	I(PK)	I(DS)
9.67	9.68	9.140	-2	0	0	14	16	2
12.93	12.94	6.843	-1	0	1	1	2	0
15.42	15.42	5.742	-2	0	1	2	3	1
16.57	16.58	5.346	1	1	0	7	8	3
18.59	18.60	4.769	2	1	0	2	2	1
18.87	18.88	4.699	3	0	1	4	7	2
18.87	18.88	4.699	-3	1	1	3		1
20.50	20.50	4.329	-1	1	1	2	3	1
20.50		4.329	1	1	1	1		1
21.55	21.56	4.119	3	1	0	14	14	8
22.17	22.18	4.005	2	1	1	2	3	1
22.17		4.005	-2	1	1	2		2
22.87	22.88	3.885	-4	0	1	2	4	1
22.87		3.885	-4	0	0	2		1
24.59	24.60	3.617	-1	1	2	2	5	1
24.59		3.617	-1	0	2	3		2
24.73	24.74	3.597	3	1	1	2	5	1
24.73		3.597	-3	1	1	2		2
25.15	25.16	3.538	-4	1	0	24	21	18
27.20	27.20	3.276	-5	0	1	1	2	1
27.20		3.276	4	1	1	2		2
27.94	27.94	3.190	-4	1	1	2	4	2
27.94		3.190	0	2	0	2		2
28.97	28.98	3.080	6	1	0	6	8	6
29.29	29.38	3.047	-1	1	2	36	100	35
29.29		3.047	1	2	1	49		49
29.39	29.39	3.037	-1	2	1	51		51
30.61	30.62	2.918	-2	2	1	23	39	25
30.61		2.918	2	1	1	24		26
31.63	31.64	2.826	-5	1	1	1	3	2
32.38	32.38	2.763	1	2	0	28	23	33
33.47	33.48	2.675	6	1	0	31	30	39
33.50	33.50	2.673	2	0	0	8		10
34.50	34.50	2.597	5	2	2	5	7	6
34.50		2.597	-5	0	2	4		6
35.11	35.12	2.554	4	1	2	7	12	10

2THETA	PEAK	D	H	K	L	I(INT)	I(PK)	I(DS)
35.11	35.11	2.554	-4	1	2	8		10
35.30	35.28	2.540	3	2	0	5	6	7
37.69	37.70	2.384	4	2	0	8	6	13
38.00	38.00	2.366	7	1	0	7	6	11
39.40	39.40	2.285	8	0	2	9	7	14
41.69	41.70	2.165	-2	2	2	8	12	14
41.69		2.165	2	2	2	8		14
42.37	42.36	2.132	-7	0	0	3	5	6
42.37		2.132	-7	1	0	3		6
42.71	42.72	2.115	8	2	1	3	2	5
43.20	43.20	2.093	3	2	0	4	5	7
43.20		2.093	-3	2	2	3		7
43.92	43.92	2.060	6	2	1	3	2	6
45.24	45.24	2.003	-6	2	2	14	19	29
45.24		2.003	-4	2	2	13		27
45.50	45.50	1.9917	-7	1	1	9	15	20
45.50		1.9917	7	1	1	9		20
47.61	47.60	1.9082	7	2	2	2	2	4
47.76	47.76	1.9026	5	2	1	1	3	3
47.76		1.9026	-5	2	1	1		3
49.35	49.36	1.8450	0	0	4	27	19	66
49.64	49.64	1.8350	8	1	2	12	18	30
49.64		1.8350	-8	1	2	12		29
49.84	49.76	1.8280	10	0	0	4	12	10
50.72	50.72	1.7984	6	2	2	10	14	26
50.72		1.7984	-6	2	2	11		28
51.22	51.22	1.7819	8	3	0	4	1	3
51.62	51.62	1.7691	3	2	0	4	3	12
52.63	52.64	1.7375	10	1	1	4	2	10
54.04	54.06	1.6955	9	2	2	3	6	7
54.04		1.6955	-9	0	2	2		7
54.06		1.6949	-7	2	2	3		7
54.44	54.44	1.6838	-3	1	2	1	2	4
54.44		1.6838	0	3	0	4		3
55.17	55.18	1.6633	-3	3	3	24	16	71
55.29	55.30	1.6601	5	3	0	10	15	31

RINKITE – TÊ-YÜ, SIMONOV AND BELOV, 1965

2THETA	PEAK	D	H	K	L	I(INT)	I(PK)	I(DS)
56.18	56.18	1.6359	-4	1	4	1		4
56.18		1.6359	-4	1	4	1		4
57.83	57.84	1.5929	11	1	0	1	2	4
58.43	58.42	1.5782	-6	0	4	4	6	14
58.43		1.5782	-1	2	4	4		14
60.27	60.26	1.5343	-1	2	4	3	4	12
60.27		1.5343	10	2	0	1		11
60.46	60.44	1.5299	10	2	0	1	3	16
60.95	60.96	1.5188	6	1	4	4	8	16
60.95		1.5188	-6	1	4	5		6
60.97		1.5184	-2	2	4	2		6
60.97		1.5184	-2	2	4	2		12
61.04		1.5168	7	3	0	3		6
61.11	61.10	1.5152	-11	0	2	2	7	6
61.11		1.5152	-11	0	2	2		6
63.21	63.22	1.4697	12	0	2	4	2	14
63.69	63.70	1.4599	6	3	2	4	6	14
63.69		1.4599	-6	3	2	4		14
63.72		1.4592	4	2	2	4		5
63.72		1.4592	-4	2	4	1		5
63.93	63.88	1.4549	7	1	4	2	4	6
63.93		1.4549	-7	1	2	2		6
63.93		1.4549	-7	1	2	3		6
66.05	66.06	1.4132	-10	2	2	3	3	11
66.05		1.4132	10	2	0	2		10
66.89	66.90	1.3975	0	4	0	5	1	7
67.11	67.12	1.3934	1	4	0	3	3	20
67.78	67.78	1.3814	2	1	2	2	2	13
68.68	68.68	1.3654	-12	1	2	2	2	7
68.68		1.3654	12	1	0	1		7
68.78		1.3637	13	1	0	2		5
70.32	70.38	1.3376	12	2	0	1	2	6
70.39		1.3364	4	0	0	2		8
72.76	72.76	1.2986	10	2	4	2	2	7
72.76		1.2986	-10	2	4	1		8
74.20	74.20	1.2769	8	2	4	1	2	7
74.20		1.2769	-8	2	4	1		7
75.03	75.02	1.2649	-10	1	4	1	1	5

2THETA	PEAK	D	H	K	L	I(INT)	I(PK)	I(DS)
75.54	75.54	1.2575	12	2	2	4	5	23
75.54		1.2575	-12	2	2	4		22
75.64		1.2561	13	2	0	1		6
76.79	76.78	1.2402	11	3	0	4	2	21
77.24	77.24	1.2341	5	3	4	3	3	15
77.24		1.2341	-5	3	4	3		15
77.38	77.48	1.2322	7	4	0	1		7
79.97	79.98	1.1987	-1	1	6	1	3	7
79.97		1.1987	1	1	6	1		15
80.49	80.56	1.1922	8	4	0	3	2	7
80.59		1.1910	2	1	6	1		7
80.59		1.1910	-2	1	6	1		7
82.20	82.20	1.1717	7	3	4	1		7
82.20		1.1717	-7	3	4	1		21
84.78	84.78	1.1425	-16	0	0	3	1	8
86.27	86.28	1.1265	-14	2	2	1	1	7
86.27		1.1265	14	2	2	1		7
86.67	86.56	1.1224	13	3	0	1		8
87.16	87.16	1.1174	-2	2	6	1	2	8
87.16		1.1174	-1	2	6	1		10
87.16		1.1174	1	2	6	1		11
87.69	87.68	1.1119	-1	4	4	2	1	11
87.69		1.1119	-2	4	4	2		15
88.30	88.30	1.1058	1	5	2	2	2	16
88.30		1.1058	-1	5	2	2		8
92.29	92.30	1.0681	-8	1	6	2	2	8
92.29		1.0681	-8	1	6	2		9
92.84	92.86	1.0633	-6	2	6	1	1	9
92.84		1.0633	6	2	6	1		10
92.84		1.0633		2	6	1		10
93.67	93.68	1.0560	-6	6	0	2	1	9
93.67		1.0560	6	2	6	1		9
94.42	94.42	1.0496	6	5	0	1	1	10

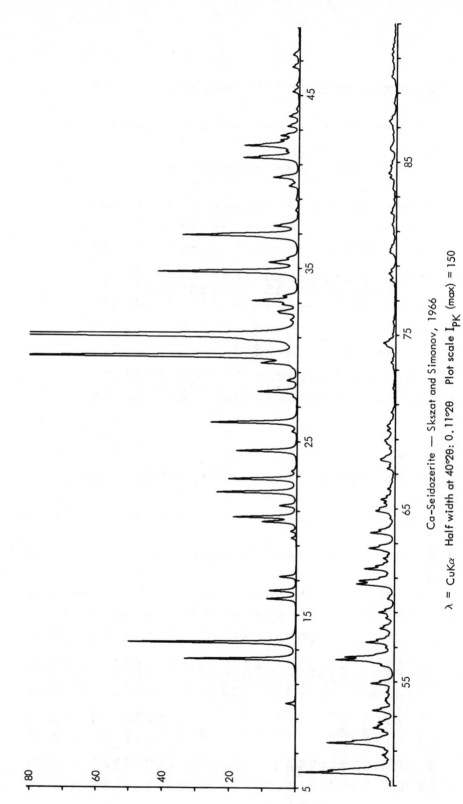

180

Ca–Seidozerite — Skszat and Simonov, 1966

λ = CuKα Half width at 40°2θ: 0.11°2θ Plot scale I$_{PK}$ (max) = 150

CA-SEIDOZERITE – SKSZAT AND SIMONOV, 1966

2THETA	PEAK	D	H	K	L	I(INT)	I(PK)	I(DS)
9.87	9.88	8.957	0	0	2	1	2	1
12.46	12.46	7.100	0	1	0	17	22	10
13.40	13.42	6.600	0	1	1	26	33	15
15.91	15.92	5.564	0	1	2	5	6	3
16.38	16.40	5.405	-1	0	0	4	5	3
17.19	17.20	5.155	-1	1	0	3	3	2
20.36	20.36	4.358	-1	1	1	6	7	4
20.63	20.64	4.301	1	1	0	10	13	8
21.28	21.28	4.171	-1	1	2	3	3	2
22.06	22.06	4.026	1	1	1	14	16	11
22.82	22.82	3.893	-1	0	4	12	14	9
24.45	24.46	3.637	1	1	4	11	12	9
26.08	26.08	3.414	-1	1	5	16	17	14
27.87	27.88	3.199	0	1	4	8	8	7
28.51	28.52	3.128	-1	1	4	2	2	2
29.51	29.52	3.024	-1	1	5	6	7	6
29.90	29.90	2.986	-1	2	1	9	9	9
29.90		2.986	0	2	0	50	60	48
31.11	31.12	2.872	1	2	1	100	100	100
31.22	31.22	2.863	-1	1	6	3	4	3
32.41	32.42	2.760	-2	0	2	2	4	3
32.50	32.50	2.752	0	1	6	2	3	2
32.89	32.90	2.721	1	2	0	2	3	3
33.12	33.12	2.703	-1	2	4	9	9	9
33.39	33.38	2.681	1	0	6	2	3	3
34.78	34.78	2.577	-1	1	4	29	28	32
35.26	35.34	2.543	-2	1	5	4	6	5
35.33		2.538	1	2	3	4	6	5
35.57	35.58	2.522	1	0	2	2	2	2
36.90	36.90	2.434	-1	2	5	24	23	28
37.43	37.44	2.400	1	1	6	5	5	6
39.74	39.75	2.266	-2	0	6	2	2	3
40.24	40.24	2.239	0	0	8	6	5	7
41.37	41.38	2.181	-2	2	1	12	11	17
41.76	41.76	2.161	1	3	1	2	3	2
41.97	42.08	2.151	-1	3	2	1	11	2
42.07		2.146	-2	2	3	10		13
42.08	42.40	2.145	-1	1	8	2		2
42.40	42.40	2.130	1	3	1	2	4	3
42.66	42.66	2.118	2	0	4	4	4	5
43.20	43.20	2.092	1	1	2	2	2	3
43.78	43.78	2.066	-1	3	5	2	2	2
45.20	45.20	2.004	-2	0	8	2	1	3
46.62	46.62	1.9466	-2	0	0	1	1	2
47.32	47.32	1.9193	2	2	2	24	19	39
49.73	49.72	1.8320	-1	2	7	6	13	10
49.96	49.86	1.8240	-1	0	10	1	1	2
50.62	50.62	1.8018	3	0	0	1	2	2
50.75	50.74	1.7975	-2	3	1	1	2	2
51.07	51.08	1.7869	-3	1	2	2	2	4
51.34	51.44	1.7780	-2	3	3	3	13	25
51.44		1.7750	0	4	0	15		4
52.12	52.12	1.7534	-1	2	9	2	2	3
52.32	52.32	1.7471	-2	3	1	2	4	2
52.32		1.7469	-2	1	9	1		2
52.34		1.7464	3	0	7	2	3	3
52.63	52.64	1.7375	0	2	9	1		2
52.67		1.7361	-2	2	5	5	4	8
53.33	53.32	1.7164	3	1	1	1	1	2
53.76	53.78	1.7036	3	1	2	1	2	2
53.88	53.90	1.7000	2	2	6	1	2	2
54.39	54.38	1.6852	-1	3	6	6	5	11
54.89	54.88	1.6713	-2	0	10	3	3	5
55.93	55.94	1.6426	-2	3	6	1	2	2
56.15	56.26	1.6367	-2	3	1	13	12	25
56.26		1.6337	-3	2	4	6		11
56.44	56.42	1.6288	-3	1	3	1	10	2
56.97	56.96	1.6151	-1	1	4	4	6	14
57.28	57.28	1.6071	-2	2	9	7	2	2
57.59	57.58	1.5990	1	0	10	1	3	3
58.06	58.08	1.5873	0	11	1	1	2	3
58.69	58.68	1.5718	-2	3	7	1	3	8
59.01	59.00	1.5641	2	0	8	4	3	8
59.79	59.80	1.5455	-3	1	8	2	1	3

CA-SEIDOZERITE - SKSZAT AND SIMONOV, 1966

2THETA	PEAK	D	H	K	L	I(INT)	I(PK)	I(DS)
60.57	60.64	1.5274	-1	0	12	2	8	4
60.64		1.5257	0	2	6	9		18
60.86	60.82	1.5209	2	2	7	5	7	10
61.52	61.52	1.5061	-1	1	11	8	6	17
61.84	61.82	1.4990	-3	2	7	2	2	4
62.13	62.12	1.4928	0	0	12	2	2	4
62.70	62.70	1.4806	3	2	3	4	5	8
62.71		1.4802	0	2	11	2		5
63.59	63.60	1.4618	-2	4	4	7	5	14
64.79	64.88	1.4376	-3	0	10	1	4	3
64.87		1.4361	2	4	2	4		10
65.33	65.32	1.4272	-1	4	6	2	2	5
65.53	65.52	1.4232	-2	2	11	3	3	6
66.13	66.12	1.4118	-3	0	9	2	1	4
67.30	67.30	1.3900	-3	2	2	2	2	5
67.77	67.84	1.3816	-4	0	2	3	3	6
67.85		1.3801	1	0	4	3		7
68.75	68.76	1.3643	1	0	12	1	2	3
68.88	68.90	1.3620	2	0	10	2	2	4
69.50	69.50	1.3513	4	0	0	2	2	4
70.53	70.54	1.3340	-1	5	4	1	1	3
73.41	73.40	1.2886	-4	0	8	1	1	3
74.40	74.40	1.2740	3	0	8	1	2	4
74.53	74.54	1.2721	-1	4	10	3	2	7
78.16	78.16	1.2218	4	0	4	2	1	5
78.54	78.54	1.2168	-2	4	10	2	1	6
79.76	79.78	1.2013	-2	2	11	1	1	4
82.05	82.04	1.1735	2	1	8	2	1	5
84.09	84.10	1.1501	1	6	1	1	1	6
84.33	84.34	1.1474	-1	0	16	2	2	4
84.78	84.78	1.1425	0	4	12	1	1	4
85.67	85.82	1.1330	-4	0	12	1	1	4
85.84		1.1311	-2	0	15	1		6
87.34	87.34	1.1155	-1	6	5	2	2	6
89.90	89.92	1.0903	-4	4	2	1	1	5
90.81	90.82	1.0817	-4	4	12	1	1	4
91.21	91.20	1.0780	2	2	13	1	1	4

2THETA	PEAK	D	H	K	L	I(INT)	I(PK)	I(DS)
91.51	91.50	1.0752	4	4	0	1	1	4
93.88	93.88	1.0542	-5	2	5	1	1	5
98.20	98.18	1.0191	-2	2	17	1	1	4
99.37	99.38	1.0102	0	6	3	1	1	3
101.41	101.42	0.9953	-3	6	9	1	0	3
102.28	102.22	0.9892	-2	6	2	1	1	2
105.16	105.14	0.9699	-5	6	11	1	1	4
106.04	106.12	0.9642	-1	6	16	1	1	2
106.13		0.9636	-1	4	3	1		5
107.13	107.14	0.9574	3	6	3	1	1	3
107.14		0.9573	0	2	17	1		3
107.52	107.52	0.9550	-4	6	12	1	1	3
107.99	108.00	0.9522	-2	7	1	1	0	2
108.47	108.42	0.9492	-2	7	3	1	1	2
109.64	109.62	0.9424	5	2	5	1	1	3

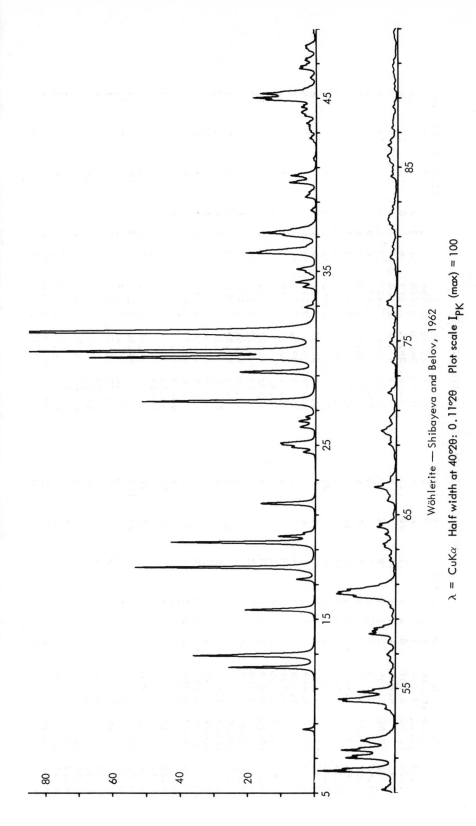

Wöhlerite — Shibayeva and Belov, 1962

λ = CuKα Half width at 40°2θ: 0.11°2θ Plot scale I$_{PK}$ (max) = 100

WOHLERITE - SHIBAYEVA AND BELOV, 1962

2THETA	PEAK	D	H	K	L	I(INT)	I(PK)	I(DS)
8.65	8.66	10.215	1	0	0	3	3	2
12.22	12.22	7.239	1	0	1	23	26	15
12.88	12.90	6.867	0	1	0	31	36	21
12.98		6.815	-1	0	1	13		9
15.51	15.52	5.706	0	1	1	19	21	13
15.60		5.677	-1	1	0	4		3
17.27	17.28	5.130	1	1	0	5	6	4
17.95	17.96	4.937	-2	0	1	52	54	39
19.35	19.40	4.584	1	1	1	12	43	9
19.40		4.572	2	0	0	34		26
19.75	19.76	4.492	-2	1	1	10	11	8
19.94	19.94	4.449	-1	1	1	2	4	2
21.60	21.62	4.110	0	2	1	14	16	12
21.66		4.099	-1	2	1	5		4
24.57	24.58	3.619	2	2	0	4	7	3
24.85	24.86	3.579	1	2	0	5		4
24.86		3.557	-2	2	1	2		2
25.01	25.02	3.542	-3	0	1	9	10	8
25.12	25.10	3.422	-1	1	2	8	11	7
26.02	26.02	3.379	-3	1	1	5		5
26.35	26.36	3.349	-1	2	1	5		5
26.36		3.243	-2	2	1	3		2
26.60	26.60	3.234	0	2	2	57	52	53
27.48	27.48	3.061	-1	3	1	1		1
27.56		3.057	0	0	3	15	23	14
29.15	29.18	2.976	-1	3	1	14	67	14
29.19		2.942	-1	0	2	79	87	78
30.00	30.00	2.935	-3	0	2	100	100	100
30.36	30.36	2.842	2	2	1	89	100	91
30.43		2.838	-2	3	0	5		5
31.45	31.48	2.837	3	2	0	42		43
31.49		2.719	3	0	1	4	3	4
31.51		2.628	3	1	1	5	4	5
32.91	32.92	2.608	-4	1	1	7	6	8
34.08	34.08	2.574	-1	2	2	2	2	2
34.36	34.36	2.565	0	4	0	3	3	3
34.83	34.84							
34.95	34.96							
35.11	35.12	2.554	-4	1	1	7	6	7
36.05	36.06	2.489	-1	3	1	25	21	28
36.22	36.14	2.478	-4	1	0	12	17	14
36.49	36.48	2.460	-3	3	1	5	3	6
37.00	37.08	2.427	-2	1	2	5	10	8
37.07		2.423	0	3	2	7		8
37.22	37.22	2.413	-2	3	0	12	17	14
37.23		2.414	3	3	1	7		8
37.43	37.42	2.401	-1	4	1	2	8	3
37.44		2.400	-2	4	1	3		4
38.49	38.50	2.337	1	1	2	4	2	2
39.27	39.28	2.292	-4	2	2	9		5
39.56	39.56	2.276	-2	3	1	8		2
40.11	40.12	2.246	2	0	3	1	8	11
40.13		2.245	-4	1	2	2		2
40.52	40.52	2.224	-3	2	1	2	8	11
40.64	40.62	2.218	-4	3	2	1	5	3
42.66	42.66	2.118	3	0	3	2	2	2
43.02	43.04	2.100	0	0	4	1	2	2
43.18	43.18	2.093	-2	4	1	2	2	2
43.36	43.36	2.085	-5	1	2	3	3	4
43.54	43.54	2.077	3	1	3	4	3	5
43.96	43.98	2.058	-1	4	2	1	4	2
44.12	44.16	2.051	4	3	0	2		3
44.17		2.049	-5	1	1	2		2
44.23		2.046	2	4	2	5	4	7
44.47	44.48	2.036	-5	2	2	18	15	25
44.89	44.90	2.017	2	3	3	15	19	21
45.02	45.02	2.012	-1	5	1	5	17	7
45.22	45.26	2.004	5	1	0	2		2
45.26		2.002	-4	2	1	16	5	22
45.26		2.002	3	4	2	5		7
46.69	46.70	1.9438	-4	3	2	2		3
46.71		1.9428	-1	2	3	2		3
46.84	46.82	1.9380	-5	2	2	2	5	3
46.96	46.94	1.9334	-3	4	2	2	3	2
47.17	47.18	1.9251	-5	2	2	4	3	6

WOHLERITE - SHIBAYEVA AND BELOV, 1962

2THETA	PEAK	D	H	K	L	I(INT)	I(PK)	I(DS)
47.29	47.28	1.9206	-4	2	3	2	3	2
47.72	47.78	1.9041	-2	3	0	2	2	2
47.77		1.9022	0	5	3	1		2
47.88	47.88	1.8982	-1	5	1	1	3	2
47.97		1.8949	-2	5	1	1		2
48.04	48.02	1.8922	-3	0	3	2	3	4
49.16	49.16	1.8519	-5	1	3	5		8
49.28	49.28	1.8434	2	1	0	1	4	2
49.40		1.8347	-4	3	3	1	3	2
49.65	49.64	1.8149	-2	0	1	2	2	2
50.23	50.22	1.8097	4	4	4	35	23	55
50.38	50.36	1.7899	-2	4	0	2	16	3
50.98	50.98	1.7863	3	5	2	18	15	29
51.09	51.10	1.7853	-1	1	2	4	13	7
51.12		1.7786	-4	3	2	2		3
51.32	51.42	1.7756	-3	5	1	1	16	2
51.42		1.7752	-5	3	2	4		6
51.43		1.7716	-4	5	2	17		28
51.54	51.56	1.7614	0	2	4	5	11	2
51.86	51.98	1.7601	2	5	2	1	10	8
51.90		1.7578	-2	5	3	7		2
51.98		1.7550	-1	4	3	1		12
52.07		1.7538	5	3	1	4	9	2
52.10	52.10		5	0	4	2	3	7
53.81	53.82	1.7022	-3	2	1	2	16	4
54.12	54.24	1.6931	0	1	4	1		2
54.24		1.6897	4	2	0	19		32
54.35	54.36	1.6865	-1	6	0	14	17	23
54.43		1.6843	-3	2	4	1		
54.78	54.78	1.6743	-6	2	2	16	11	27
55.88	55.88	1.6440	3	1	3	1	2	2
55.90		1.6434	1	3	3	1		2
56.58	56.60	1.6251	3	4	3	1	2	2
56.94	56.94	1.6158	6	2	0	3	3	5
57.57	57.62	1.5996	4	5	0	2	2	3
57.63		1.5980	5	4	4	2		3
58.10	58.12	1.5863	-1	3	2	5	8	9
58.12		1.5857	2	5	4	6		11

2THETA	PEAK	D	H	K	L	I(INT)	I(PK)	I(DS)
58.13	58.28	1.5856	4	3	2	1		3
58.19		1.5842	-3	2	1	1		2
58.31	58.44	1.5812	-4	5	4	3	7	6
58.43		1.5780	5	0	2	7	7	14
59.66	59.66	1.5485	-1	2	4	2	2	4
60.20	60.28	1.5358	-7	0	2	7	13	13
60.27		1.5343	0	3	4	2		4
60.27		1.5342	5	1	2	11		20
60.40	60.44	1.5312	0	6	2	4	18	7
60.43		1.5307	-2	6	2	11		20
60.53	60.54	1.5283	3	6	0	10	17	20
60.54		1.5281	-4	3	4	5		10
60.68	60.68	1.5249	-6	3	4	8	14	15
60.72		1.5241	-5	3	0	2		4
60.81		1.5219	-2	2	4	5		9
62.36	62.36	1.4878	-2	0	4	2	2	4
62.65	62.62	1.4815	-2	4	4	1	2	2
63.08	63.14	1.4724	-6	1	4	2	3	5
63.15		1.4710	3	0	3	1		3
63.30	63.30	1.4678	1	5	4	8	4	16
63.42		1.4653	-6	1	0	2		4
63.86	63.86	1.4564	-7	3	5	4	3	3
63.87		1.4561	5	6	1	2		8
64.17	64.28	1.4501	7	1	3	2	6	3
64.28		1.4478	-5	1	0	4		16
64.44	64.44	1.4447	-6	5	0	2	5	4
64.45		1.4445	-4	3	4	2		3
65.78	65.78	1.4184	2	0	0	4	3	8
65.88		1.4165	-4	2	5	2		3
66.16	66.18	1.4111	-7	6	2	2	2	4
66.29	66.30	1.4088	-2	2	0	2	2	3
66.45	66.56	1.4057	-2	6	5	2	3	4
66.57		1.4036	3	1	2	10	7	21
67.04	67.04	1.3949	-2	2	5	1	2	3
67.16	67.14	1.3926	-2	6	3	1	2	3
69.05	69.04	1.3591	-1	7	2	3	2	7
69.40	69.38	1.3531	3	1	4	1	2	3

WÖHLERITE - SHIBAYEVA AND BELOV, 1962

2THETA	PEAK	D	H	K	L	I(INT)	I(PK)	I(DS)
69.71	69.80	1.3478	6	1	2	2	5	4
69.80		1.3463	3	7	0	2		4
69.82		1.3460	-3	5	4	3		6
70.36	70.36	1.3368	-8	1	2	1	2	3
70.50	70.52	1.3346	-2	3	5	2	2	5
70.62		1.3326	-3	3	5	2		4
72.01	72.00	1.3102	-6	5	0	3	3	8
73.73	73.76	1.2838	-6	5	3	1	1	2
73.76		1.2677	3	3	4	1	2	3
74.83	74.88	1.2671	8	1	0	3		6
74.88		1.2547	-7	3	4	1	1	4
75.74	75.74	1.2439	-2	8	0	2	1	4
76.52	76.50	1.2367	-1	6	4	3	3	8
77.05	77.06	1.2342	-3	6	4	3	3	8
77.23	77.24	1.2013	6	4	2	1	2	3
79.76	79.76	1.1935	-8	4	2	1	2	3
80.39	80.40	1.1805	-1	0	6	2	2	6
81.46	81.46	1.1778	-1	8	0	2	2	7
81.69	81.70	1.1757	-3	8	2	3	3	7
81.86	81.98	1.1742	-5	0	6	3		5
81.99		1.1658	-5	6	4	2	2	3
82.70	82.70	1.1610	-8	3	4	1	1	12
83.13	83.12	1.1461	4	8	0	4	2	5
84.45	84.46	1.1406	-3	3	6	2	2	8
84.95	84.94	1.1365	3	5	4	3	2	9
85.33	85.34	1.1334	6	5	4	3	3	8
85.62	85.62	1.1271	-7	5	5	3		8
86.22	86.24	1.1268	-8	4	5	3	3	8
86.24		1.1231	4	4	2	1	2	9
86.60	86.52	1.1165	5	2	4	3	2	8
87.24	87.24	1.1041	-9	2	4	3	1	3
88.47	88.46	1.0946	3	8	2	1	1	3
89.45	89.48	1.0912	-5	8	2	1	1	3
89.80	89.78	1.0876	-1	9	0	1	1	4
90.18	90.20	1.0853	0	3	6	1	1	4
90.43	90.44	1.0670	4	4	4	1	1	2
92.42	92.44	1.0634	8	2	2	2	1	5
92.83	92.80							

2THETA	PEAK	D	H	K	L	I(INT)	I(PK)	I(DS)
93.48	93.62	1.0577	-8	5	4	1	1	3
93.62		1.0564	-10	2	2	2		4
94.86	94.98	1.0459	1	3	6	1	1	5
94.99		1.0448	6	1	4	2		3
95.37	95.32	1.0417	7	5	2	1	1	4
95.83	95.86	1.0379	9	4	0	1	2	4
95.92		1.0371	-7	3	6	2		5
96.08	96.06	1.0359	-9	5	2	1	2	3

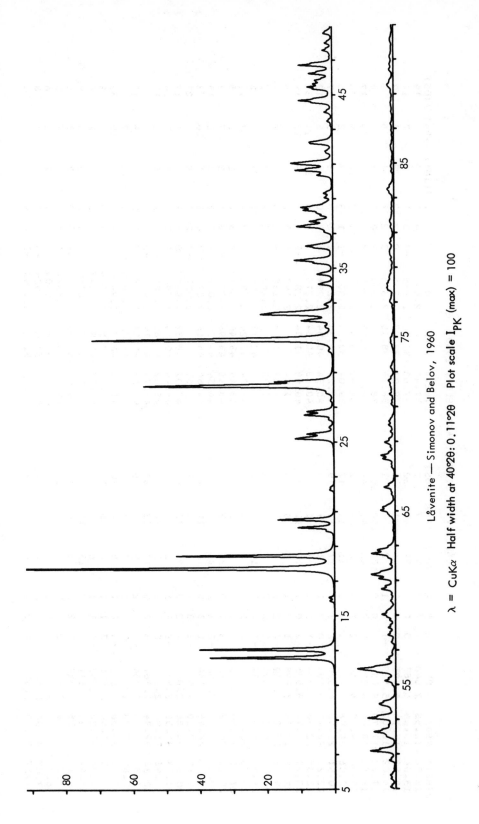

Låvenite — Simonov and Belov, 1960

λ = CuKα   Half width at 40°2θ: 0.11°2θ   Plot scale I_{PK} (max) = 100

187

LÅVENITE – SIMONOV AND BELOV, 1960

2THETA	PEAK	D	H	K	L	I(INT)	I(PK)	I(DS)
12.56	12.58	7.040	1	1	0	34	37	29
13.04	13.04	6.783	0	0	1	37	40	32
15.82	15.84	5.595	0	1	1	1	1	1
16.02	16.02	5.529	-1	1	1	2	2	1
17.70	17.70	5.006	-2	0	0	100	100	100
18.43	18.44	4.809	-2	0	1	46	47	48
19.86	19.88	4.468	2	1	1	3	3	2
20.06	20.06	4.423	-1	1	1	11	11	12
20.51	20.52	4.326	-2	1	1	17	17	19
22.21	22.22	3.998	-0	1	1	1	1	2
22.35	22.36	3.974	-1	2	1	1	1	1
25.17	25.18	3.535	2	0	1	12	11	15
25.44	25.44	3.498	-1	1	1	8	8	10
25.81	25.80	3.449	-2	2	1	1	1	2
26.25	26.26	3.391	-0	2	1	1	2	2
26.53	26.54	3.357	-1	0	2	10	9	13
26.73	26.74	3.332	-2	1	2	8	8	11
27.78	27.80	3.208	-0	1	2	1	2	2
28.19	28.22	3.163	3	1	0	39	57	57
28.23	28.46	3.158	-2	1	0	34	17	49
28.45		3.134	-1	3	0	16	2	23
30.19	30.20	2.957	-1	1	1	1		2
30.86	30.86	2.895	-1	3	1	89	72	142
31.57	31.58	2.832	-3	2	2	2	2	3
31.96	31.96	2.798	-0	2	2	11	10	18
32.32		2.767	3	2	0	18	22	30
32.36	32.34	2.764	-2	2	1	9		15
32.89	32.88	2.721	-4	0	1	3	3	5
34.06	34.06	2.630	3	1	1	5	4	10
34.63	34.64	2.588	-4	2	1	6	5	11
35.28	35.28	2.542	-1	3	1	6	6	11
35.44	35.44	2.531	-3	2	2	13	11	25
36.16	36.18	2.482	-0	4	0	6	6	11
36.26	36.26	2.475	2	0	2	6	8	11
36.31		2.472	-1	3	2	2		4
37.07	37.08	2.423	-1	1	2	2	3	4
37.37	37.40	2.405	-4	0	2	4	11	8

2THETA	PEAK	D	H	K	L	I(INT)	I(PK)	I(DS)
37.40		2.403	-1	4	0	9		19
37.68	37.68	2.385	-3	3	1	8	7	16
38.01	38.02	2.365	3	3	2	3	4	7
38.14	38.18	2.358	-3	2	1	1	5	3
38.19		2.355	-2	0	3	3		6
38.32	38.32	2.347	3	3	1	8	9	17
38.49	38.48	2.337	-1	1	2	10	9	20
39.04	39.04	2.305	-1	1	2	3	2	6
39.29	39.30	2.291	-1	1	2	3	3	7
40.34	40.34	2.234	4	2	0	6		12
40.63	40.64	2.219	-1	4	1	15	5	32
40.98	41.04	2.201	-2	4	1	9	9	19
41.04		2.197	-1	4	2	11		24
41.63	41.72	2.168	-3	1	3	2	2	4
41.72		2.163	-4	0	1	1		2
42.17	42.18	2.141	2	4	0	7	7	16
42.24	42.24	2.138	-1	2	3	5	7	11
43.45	43.46	2.081	3	3	1	2	2	5
43.98	43.98	2.057	-4	3	1	3	2	7
44.54	44.56	2.032	3	1	2	6	6	15
44.66		2.027	2	4	1	1		3
44.67	44.66	2.027	-3	2	2	9	10	21
45.32	45.32	1.9992	0	4	3	7	6	17
45.44	45.44	1.9943	4	3	0	6	7	14
45.62	45.60	1.9869	-2	4	2	5	6	13
45.74	45.74	1.9821	-5	1	2	1	5	3
45.82	45.82	1.9785	2	3	2	5	5	12
46.21	46.22	1.9628	5	1	0	10	7	24
46.70	46.70	1.9434	-4	2	2	14	10	36
47.56	47.56	1.9104	1	2	3	4	3	9
47.97	47.98	1.8947	-3	4	2	3	2	8
48.54	48.54	1.8739	2	0	3	3	3	9
49.08	49.08	1.8547	1	5	1	2	2	6
49.55	49.56	1.8381	-2	5	1	2	2	5
49.76	49.76	1.8309	-4	4	1	3	2	6
50.60	50.60	1.8025	-2	4	1	2	2	8
51.15	51.16	1.7843	-2	0	4	12	8	35

LÄVENITE – SIMONOV AND BELOV, 1960

2THETA	PEAK	D	H	K	L	I(INT)	I(PK)	I(DS)
51.91	52.00	1.7600	4	4	0	1	3	3
51.98		1.7576	3	3	2	2		7
52.26	52.26	1.7491	2	4	2	3	7	20
52.29		1.7480	5	1	1	2		8
52.55	52.54	1.7399	4	2	2	2	2	4
52.65	52.66	1.7368	2	1	3	3	2	5
53.05	53.06	1.7246	-4	4	2	2	8	35
53.09		1.7234	-4	3	0	3	12	3
53.79	53.86	1.7029	3	5	2	0	5	10
53.87		1.7004	-6	2	1	2	3	16
54.04	54.02	1.6956	-5	2	2	3	2	5
55.20	55.20	1.6626	-1	2	0	4	5	15
55.82	55.90	1.6456	6	1	4	4	7	22
55.91		1.6431	-3	5	0	2	13	40
56.42	56.42	1.6295	-3	4	2	2	4	12
56.81	56.80	1.6193	4	4	1	1	1	3
57.43	57.42	1.6032	0	6	1	1	1	4
57.84	57.84	1.5928	3	5	1	3	2	11
58.25	58.26	1.5824	-6	1	4	2	2	8
58.69	58.68	1.5718	1	1	0	2	2	7
58.88	58.90	1.5671	2	6	1	2	3	6
59.02	59.02	1.5638	5	3	3	1	4	12
59.02		1.5637	3	3	4	4	1	5
59.26	59.18	1.5580	4	3	2	2	3	6
59.81	59.82	1.5449	-3	3	4	4	2	6
60.16	60.16	1.5368	-5	1	4	2	1	5
60.48	60.48	1.5295	-6	5	2	4	5	19
60.52		1.5285	-4	1	2	2	2	6
61.09	61.10	1.5156	1	2	4	5	10	26
61.33	61.34	1.5101	5	1	1	2	7	35
61.42		1.5082	0	2	4	4	2	8
61.90	61.90	1.4977	-1	6	0	2	2	7
62.29	62.32	1.4892	-6	3	2	2	2	7
62.53	62.54	1.4841	-5	2	4	9	7	34
62.55		1.4837	0	6	2	2	3	12
62.77	62.72	1.4790	-7	1	1	3	5	6
62.79		1.4787	-2	6	2	2		
64.70	64.84	1.4394	-1	6	2	2	2	8
64.82		1.4371	-1	4	2	2		8
65.03	65.02	1.4330	3	5	4	4	4	16
65.17	65.18	1.4303	-6	3	1	1	1	6
65.29		1.4302	-6	4	1	1	1	4
66.33		1.4280	-3	4	0	4	4	6
66.34	66.34	1.4080	5	3	0	0	3	17
66.37		1.4072	-5	3	4	2	1	5
67.49	67.50	1.3866	5	5	2	0	1	6
67.99	68.00	1.3776	4	6	0	0	4	26
68.18	68.18	1.3742	7	2	0	5	4	17
68.66	68.66	1.3659	-3	2	2	3	3	7
68.93	68.94	1.3610	1	7	0	5	1	20
69.54	69.54	1.3506	-1	2	1	0	2	8
69.90	69.92	1.3446	2	1	5	5	2	7
69.93		1.3440	0	6	2	4	2	7
70.14	70.12	1.3406	6	3	1	5	1	6
70.26		1.3385	3	3	2	3	1	5
71.11	71.12	1.3246	-3	6	3	3	2	9
71.69	71.70	1.3154	-6	1	1	4	2	7
72.20	72.20	1.3073	-7	1	1	1	2	7
72.49	72.46	1.3028	0	8	0	0	3	6
76.99	77.00	1.2375	1	8	0	4	3	8
77.68	77.70	1.2282	4	8	0	4	3	13
77.81	77.80	1.2265	-3	5	4	4	2	14
78.05	78.02	1.2233	-5	5	3	4	2	5
78.17	78.16	1.2217	-8	3	2	2	2	8
78.37	78.38	1.2191	-4	7	1	4	2	5
79.44	79.44	1.2054	-1	6	2	4	1	9
79.76	79.86	1.2013	-2	8	0	4	1	6
79.87		1.2000	-3	1	2	4	1	7
80.39	80.38	1.1935	-8	1	4	4	1	11
81.26	81.30	1.1829	7	4	1	2	1	5
81.60	81.62	1.1788	6	5	6	0	1	5
83.26	83.28	1.1595	5	3	0	2	2	10
83.43	83.48	1.1575	4	3	4	4	2	10
83.51		1.1566	-3	2	6	6	2	9

LÅVENITE - SIMONOV AND BELOV, 1960

2THETA	PEAK	D	H	K	L	I(INT)	I(PK)	I(DS)
84.49	84.50	1.1457	-1	6	4	1	1	7
85.76	85.78	1.1319	-5	6	4	1	1	8
88.46	88.46	1.1043	4	7	2	1	1	8
89.03	89.06	1.0986	2	6	4	1	1	7
89.41	89.50	1.0950	-7	5	4	1	2	7
89.51		1.0940	-6	7	2	3		20
90.34	90.36	1.0861	0	7	4	2	1	11
90.41		1.0854	9	2	0	1		9
90.72	90.68	1.0825	-6	6	4	1	1	6
91.18	91.14	1.0782	-4	7	4	1	1	9
92.15	92.14	1.0695	-0	3	0	1	1	6
93.88	93.88	1.0542	9	3	0	1	1	7
94.62	94.60	1.0479	-10	1	2	1	1	7
98.08	98.06	1.0200	2	7	4	1	1	8
98.94	98.96	1.0134	-6	4	6	1	1	8
98.96		1.0133	-1	8	4	1	1	8
101.15	101.26	0.9972	-8	6	0	1	1	6
101.52	101.52	0.9945	-5	5	6	1	1	7
102.10	102.10	0.9904	6	3	4	1	1	6
102.46	102.46	0.9879	2	3	6	1	1	8
102.79	102.78	0.9857	-9	3	5	1	1	4
103.24	103.24	0.9826	8	4	2	1	1	5
104.10	104.10	0.9768	3	1	6	1	0	5
104.82	104.90	0.9720	3	9	0	1	0	5
106.05	106.08	0.9641	5	9	2	3	1	19
106.22		0.9631	-8	7	2	1		7
107.12	107.08	0.9574	-6	4	4	1	1	7
107.53	107.54	0.9549	-9	6	2	1	1	5
108.05	108.04	0.9518	-9	1	6	1	1	8
108.29	108.32	0.9503	0	10	2	1	1	8
108.51		0.9490	-2	10	2	1		7
109.95	110.08	0.9406	-4	4	7	1	1	6
110.10		0.9398	-11	2	2	1		11
110.61	110.52	0.9368	-2	4	7	1	1	8

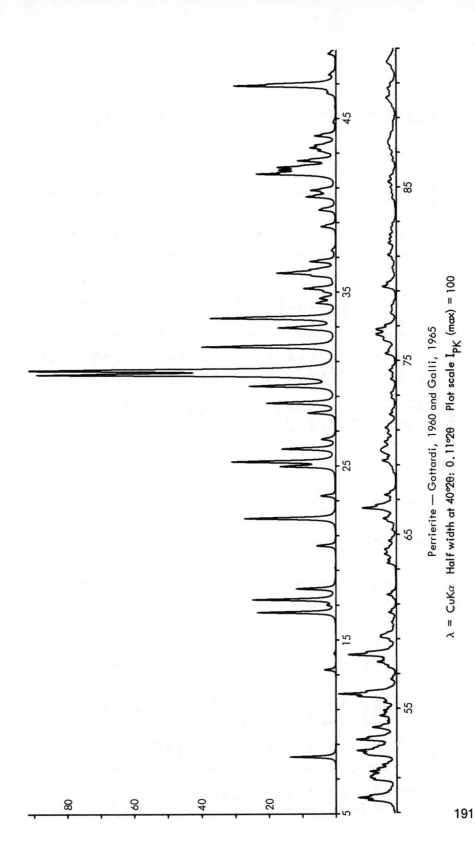

Perrierite — Gottardi, 1960 and Galli, 1965

λ = CuKα Half width at 40°2θ: 0.11°2θ Plot scale I_{PK} (max) = 100

191

PERRIERITE — GOTTARDI, 1960 AND GALLI, 1965

2THETA	PEAK	D	H	K	L	I(INT)	I(PK)	I(DS)
8.25	8.26	10.702	0	0	1	10	14	2
13.26	13.26	6.672	-2	0	0	3	3	1
16.55	16.56	5.351	0	0	2	19	24	8
16.98	16.98	5.219	-2	0	1	2	3	1
17.29	17.30	5.124	-1	1	0	21	25	9
17.91	17.92	4.948	-1	1	1	10	12	4
20.39	20.40	4.353	-1	1	1	5	6	3
21.94	21.96	4.047	-1	1	2	25	27	15
23.25	23.26	3.823	-2	0	3	4	5	3
24.94	24.94	3.567	0	0	3	15	17	11
25.21	25.22	3.530	-1	1	2	30	31	22
25.95	25.96	3.431	-3	1	1	16	16	12
26.50	26.50	3.361	-3	1	2	4	4	3
28.01	28.02	3.182	-1	1	3	8	8	7
28.58	28.58	3.120	-4	0	0	21	21	19
29.55	29.56	3.020	-3	1	1	26	26	24
30.19	30.20	2.958	-3	1	3	90	89	89
30.43	30.44	2.935	3	1	0	100	100	100
31.82	31.82	2.810	0	2	1	42	40	45
32.83	32.92	2.726	-1	1	4	5	17	6
32.92		2.718	-4	0	4	10		11
33.46	33.46	2.676	0	2	0	5		6
34.34	34.34	2.609	0	0	4	41	38	48
34.60	34.60	2.590	-4	0	4	6	6	7
35.00	35.00	2.562	-2	2	0	5	5	6
35.18	35.18	2.549	2	2	0	3	4	4
35.91	35.92	2.499	-1	1	4	10	10	13
36.07	36.08	2.488	3	0	2	6	8	8
36.28	36.28	2.474	0	2	2	18	18	24
36.73	36.74	2.444	-2	2	2	6	8	8
37.37	37.38	2.404	-5	1	2	9	8	12
38.74	38.74	2.322	2	2	1	1	1	2
39.70	39.70	2.268	-6	0	2	6	5	8
39.78	39.78	2.264	-2	2	3	1	5	9
40.46	40.46	2.227	-4	0	5	10	9	16
40.72	40.72	2.214	-6	0	1	3	5	5

2THETA	PEAK	D	H	K	L	I(INT)	I(PK)	I(DS)
40.84	40.84	2.207	0	2	3	7	8	11
41.77	41.78	2.161	-4	2	1	27	24	46
42.00	42.00	2.149	-4	2	2	16	17	26
42.17	42.16	2.141	-3	1	5	14	18	24
42.53	42.54	2.124	3	1	3	13	12	22
43.01	43.12	2.101	-1	1	5	2	5	3
43.11		2.097	-6	0	0	4		7
43.29	43.30	2.088	4	2	0	8	8	14
43.47	43.42	2.080	6	0	3	4	6	7
43.98	43.98	2.057	-4	2	4	8	6	14
44.75	44.76	2.023	-2	2	1	1	1	2
46.44	46.44	1.9538	4	0	4	2	3	4
46.85	46.86	1.9377	0	2	3	31	31	63
46.88		1.9363	-2	2	6	10		21
47.03	46.96	1.9304	-2	0	5	1	21	2
47.51	47.52	1.9121	-4	2	2	2	2	5
48.72	48.72	1.8674	5	1	1	3	2	6
49.38	49.40	1.8439	-1	3	1	1	2	2
49.69	49.70	1.8333	-7	1	2	9	9	20
49.85	49.84	1.8275	-7	1	3	10	11	22
49.98	49.98	1.8234	3	1	4	3	8	8
50.23	50.22	1.8147	-2	0	5	2	3	4
50.87	50.88	1.7935	-2	2	5	7	7	17
50.97		1.7902	-7	1	1	3	8	7
51.07	51.06	1.7870	-1	0	6	5		11
51.17	51.16	1.7837	-1	3	2	1	7	3
51.17		1.7836	0	1	6	3		7
51.36	51.36	1.7773	-1	1	6	8	8	19
51.55	51.52	1.7713	-7	1	2	3	6	8
51.75	51.74	1.7650	-6	2	2	4	4	9
52.37	52.42	1.7455	-4	2	5	5	10	11
52.43		1.7438	-6	0	3	3		8
52.56	52.56	1.7395	-6	2	1	4	12	9
52.58		1.7391	-6	1	6	8		19
52.83	52.82	1.7315	-3	3	1	5		11
53.18	53.20	1.7209	6	0	2	7	6	18
53.20		1.7202	4	0	4	7	12	17

PERRIERITE — GOTTARDI, 1960 AND GALLI, 1965

2THETA	PEAK	D	H	K	L	I(INT)	I(PK)	I(DS)
53.23		1.7195	1	3	2	3		7
53.36	53.32	1.7155	2	2	4	2	9	4
53.79	53.90	1.7027	0	2	5	2	7	6
53.89		1.6997	-8	0	3	8		19
54.24	54.24	1.6897	-8	0	2	2	3	6
54.56	54.56	1.6804	-6	2	0	6	5	15
54.86	54.86	1.6719	-6	0	7	4	4	11
55.33	55.34	1.6590	-4	0	3	4	4	12
55.68	55.68	1.6494	-3	3	1	13	12	36
55.83	55.82	1.6454	3	3	3	16	17	43
56.03	55.98	1.6400	-8	2	1	2	2	7
56.66	56.66	1.6231	4	1	6	2	9	6
57.36	57.36	1.6050	1	1	5	2	3	7
57.53	57.54	1.6007	-8	0	7	6	4	5
57.66	57.66	1.5974	-3	1	1	8	6	18
58.05	58.08	1.5876	7	1	1	2	6	23
58.06		1.5873	-6	2	5	12	14	6
58.09		1.5866	3	1	5	2		33
58.34	58.22	1.5804	-4	2	4	1	9	4
58.88	58.88	1.5671	-1	2	2	2	2	6
59.08	59.10	1.5622	-7	1	6	5	5	6
59.18	59.18	1.5598	-5	1	7	3	5	14
59.67	59.66	1.5483	6	0	3	1	1	9
59.94	59.94	1.5418	-5	3	2	2	2	3
60.50	60.50	1.5289	0	0	7	3	1	6
61.34	61.36	1.5100	-8	0	6	1	3	10
61.53	61.52	1.5059	0	2	6	3	3	4
63.34	63.34	1.4671	0	2	4	2		11
63.35		1.4669	7	4	2	1		7
63.55	63.52	1.4628	8	0	8	1	3	4
63.78	63.86	1.4580	-4	0	5	4	4	4
63.86		1.4564	-3	3	3	4		13
64.12	64.12	1.4509	3	1	1	3	2	13
64.62	64.62	1.4412	-7	1	7	2	1	10
65.25	65.26	1.4286	-4	2	2	2	3	7
65.74	65.76	1.4191	-2	2	4	3		11
65.89	65.90	1.4164	-8	2	1	4	4	15

2THETA	PEAK	D	H	K	L	I(INT)	I(PK)	I(DS)
66.49	66.50	1.4050	0	4	0	13	10	48
66.57		1.4035	1	1	1	2		5
67.26	67.26	1.3909	-8	2	5	5	4	18
68.57	68.58	1.3674	-6	2	7	1	1	5
69.22	69.22	1.3561	6	2	3	5	4	20
69.68	69.80	1.3482	7	1	7	1	5	4
69.73		1.3474	-7	3	2	3		12
69.79		1.3464	9	1	0	2		7
69.87		1.3451	-7	0	3	3		10
70.21	70.30	1.3394	-10	3	3	1	4	6
70.31		1.3378	0	0	8	4		6
70.77	70.78	1.3301	-8	2	6	3	3	17
70.86		1.3287	-7	3	1	2		11
71.11	71.12	1.3247	-1	1	3	3	3	7
71.16		1.3238	-7	1	8	2		12
71.26	71.26	1.3222	-7	3	0	1	2	6
71.88	71.88	1.3123	-5	3	4	3	2	5
72.83	72.84	1.2975	8	2	6	2	2	13
73.12	73.10	1.2931	-10	0	1	2	2	7
73.92	73.92	1.2811	4	4	0	2	2	7
74.41	74.40	1.2739	-4	4	3	3	2	10
75.37	75.38	1.2599	-6	2	8	4	3	14
75.44		1.2589	-8	2	7	1		19
75.76	75.60	1.2545	-3	1	9	2	2	5
76.13	76.16	1.2493	6	2	4	4	4	7
76.21		1.2481	4	4	1	1		18
76.26		1.2474	-3	3	7	3		7
76.44	76.50	1.2450	-3	3	4	7	6	12
76.52		1.2439	0	4	4	4		31
76.78	76.78	1.2403	7	3	5	4	6	16
76.82		1.2398	3	3	7	4		19
77.02	77.02	1.2371	-4	4	4	1	3	5
77.69	77.72	1.2281	-7	3	6	2	2	9
77.77		1.2269	-5	1	7	2		5
78.13	78.12	1.2222	-10	1	4	2	2	10
79.24	79.24	1.2079	-11	1	8	3	2	16
79.24	79.24	1.2079	0	2	8	3	4	17

PERRIERITE — GOTTARDI, 1960 AND GALLI, 1965

2THETA	PEAK	D	H	K	L	I(INT)	I(PK)	I(DS)
80.13	80.14	1.1967	-11	1	5	1	1	6
80.31	80.32	1.1944	-6	4	2	2	1	8
80.81	80.82	1.1883	-4	4	5	2	1	9
81.07	81.06	1.1851	-11	1	2	3	3	18
81.74	81.74	1.1772	-4	2	9	3	2	17
81.94	81.96	1.1747	-10	2	6	3	2	8
82.22	82.22	1.1715	-11	1	6	2	2	13
82.59	82.60	1.1672	-6	4	4	3	2	5
82.64		1.1666	-7	3	7	1		8
84.32	84.36	1.1476	8	2	3	1	1	6
84.67	84.66	1.1437	3	3	6	1	2	8
84.95	84.94	1.1407	10	2	0	2	2	12
85.28	85.28	1.1371	7	1	5	4	3	23
85.62	85.56	1.1334	-10	0	2	2	2	10
86.51	86.52	1.1240	-8	2	7	2	1	11
86.84	86.82	1.1207	10	0	8	2	2	11
87.18	87.22	1.1171	-7	1	2	2	2	10
87.25		1.1164	-1	3	10	1		7
87.43		1.1146	-9	3	8	1		7
87.52	87.52	1.1137	-8	0	0	2	3	11
87.66	87.66	1.1122	11	1	0	2	2	12
88.10	88.08	1.1078	-8	2	9	2	2	13
89.61	89.64	1.0930	-6	4	6	1	1	7
89.93	89.93	1.0899	-11	1	8	3	3	12
90.10	90.12	1.0883	6	4	2	2		15
90.12		1.0882	4	4	4	3		16
90.59	90.66	1.0837	-1	0	10	3	3	9
90.66		1.0830	-2	4	8	1		9
90.68		1.0829	-8	4	7	2		16
91.85	91.86	1.0722	-4	4	3	3		13
92.13	92.18	1.0696	-3	5	1	2	1	21
92.26		1.0685	3	5	9	4	3	23
93.13	93.08	1.0608	-3	3	1	1	1	7
93.65	93.66	1.0562	-4	0	12	1	1	9
95.46	95.52	1.0409	10	2	2	1	2	9
95.57		1.0400	-11	1	9	1		8
96.14	96.22	1.0353	-8	2	10	1	1	7
96.24		1.0345	0	4	7	2		11
96.54	96.54	1.0321	-11	3	4	2	2	16
97.23	97.20	1.0266	-3	1	11	1	1	8
97.85	97.84	1.0218	3	1	9	1	1	9
98.15	98.36	1.0194	-13	1	3	1	1	7
98.36		1.0178	-11	3	2	1		16
99.00	99.02	1.0129	11	1	5	2	2	14
99.24	99.30	1.0112	-3	5	5	2	2	9
99.51	99.48	1.0091	-11	3	6	1	1	11
100.73	100.74	1.0001	0	2	10	1	1	10
101.51	101.50	0.9946	-8	4	7	1	1	8
102.14	102.20	0.9901	-5	3	10	1	1	4
102.17		0.9899	10	0	4	1		4
102.19		0.9898	6	4	4	1		4
102.36	102.56	0.9886	-4	2	11	1	2	11
102.47		0.9879	-11	1	10	3		20
102.61		0.9869	7	3	5	3		19
103.28	103.28	0.9823	6	2	7	1	1	10
103.87	103.84	0.9783	3	3	8	2	2	4
104.98	105.28	0.9710	3	1	4	1	2	12
105.06		0.9705	11	3	0	2		9
105.22		0.9694	-10	4	2	1		24
105.32		0.9688	0	4	8	3		5
105.52		0.9675	-10	2	11	1		4
105.54		0.9674	-8	2	5	1		8
106.10	106.12	0.9638	-1	5	6	1	1	6
106.17		0.9634	-2	2	11	1		8
106.78	106.80	0.9596	-14	0	3	1	1	7
106.86		0.9591	-5	5	6	1		12
107.42	107.42	0.9556	-11	3	8	2	1	10
107.84	107.90	0.9531	-14	0	7	1	1	7
107.88		0.9528	-4	4	9	1		11
107.91		0.9526	-7	1	12	1		10
108.10		0.9515	-10	3	10	1		10
108.11		0.9514	-1	6	0	2		14
110.64	110.68	0.9367	-1	0	6	1	1	9
110.77		0.9359	3	1	10	1		

PERRIERITE - GOTTARDI, 1960 AND GALLI, 1965

2THETA	PEAK	D	H	K	L	I(INT)	I(PK)	I(DS)
111.17	111.20	0.9337	10	2	4	1	1	6
111.24		0.9333	7	5	6	1		4
111.86	111.90	0.9298	7	5	1	2	2	14
111.90		0.9296	3	5	5	3		22
112.08		0.9287	9	3	4	1		6
112.72	112.84	0.9252	-10	2	11	1	1	6
112.82		0.9246	-7	5	6	1		10
112.91		0.9242	-5	5	7	1		6
113.11	113.22	0.9231	-5	3	11	1	1	5
113.20		0.9226	0	6	2	1		9
113.43		0.9214	-11	3	9	1		9
113.82	113.86	0.9194	0	2	11	2	1	13
113.83		0.9193	-14	0	1	1		12
113.97		0.9186	-14	4	5	1		10
114.69	114.72	0.9149	-8	4	9	1	1	6
114.80		0.9143	-7	3	11	1		6
115.24	115.28	0.9121	-3	3	11	1	1	7
115.30		0.9118	13	1	1	1		8
115.31		0.9117	6	1	8	1		9
115.89	116.04	0.9088	11	1	4	1	2	5
115.92		0.9087	3	3	9	1		11
116.03		0.9081	-14	2	3	3		25
116.07		0.9079	-14	0	9	1		9
116.25		0.9070	-13	3	3	1		10
116.71	116.46	0.9048	-8	2	12	1	2	9
117.15	117.18	0.9026	-4	6	1	3	2	12
117.17		0.9026	-14	2	7	2		28
117.20		0.9024	11	6	2	1		16
117.33		0.9018	-4	5	7	1		8
118.23	118.28	0.8975	-7	6	2	1	1	8
118.32		0.8971	4	5	0	1		6
118.62	118.76	0.8957	-15	1	5	1	1	10
118.85		0.8946	-4	6	3	1		8
119.46	119.46	0.8918	0	0	12	1	1	8
119.73		0.8906	6	0	9	1		7
120.22	120.26	0.8884	6	4	6	1	1	5
120.29		0.8881	-4	0	13	1		8
121.06	121.18	0.8847	-11	1	12	1	2	12
121.10		0.8846	-11	3	10	2		15
121.21		0.8841	0	6	4	3		30
121.22		0.8840	-5	1	13	1		8
121.24		0.8839	2	6	3	1		7
122.83	123.10	0.8772	8	2	7	1	1	10
123.09		0.8761	10	4	2	3		27
123.59	123.94	0.8740	-1	5	8	1	1	9
123.67		0.8737	-14	2	1	1		6
123.79		0.8732	9	4	0	3		5
123.91		0.8727	-8	4	10	1		27
124.04		0.8722	-10	0	6	2		16
124.36	124.36	0.8709	-10	4	9	1	1	11
124.83		0.8691	-2	6	5	1		7
125.36	125.44	0.8669	-7	5	8	1	1	8
125.67		0.8658	-6	6	2	1		6
126.14	126.22	0.8639	-14	2	9	1	1	11
126.17		0.8638	2	2	11	2		18
126.27		0.8634	-4	6	5	1		9
126.48		0.8626	-6	6	1	1		11
127.43	127.72	0.8591	11	1	5	1	1	6
127.45		0.8590	-7	3	12	2		17
127.78		0.8578	4	4	8	3		26
128.49	128.40	0.8552	-6	6	4	1	1	11
128.80		0.8541	6	6	0	1		9
129.12	129.10	0.8530	8	4	5	1	1	11
129.36		0.8521	-15	1	9	1		10

TABLE 9.   COMPLEX SILICATES

Variety	Zoisite	Clinozoisite	Epidote	Ardennite	Kornerupine
Composition	$Ca_2(Al,Fe^3)_3Si_3O_{12}$ OH	$Ca_2(Al,Fe^3)Al_2Si_3$ $O_{12}OH$	$Ca_2(Al,Fe^3)Al_2$ $Si_3O_{12}OH$	$Mn(Mn,Ca)_2(AlOH)_4$ $[(Mg,Al,Fe^3)OH]_2(As,$ $V)O_4Si_3O_{10}(SiO_4)_2$	$Mg_3Al_6(Si,Al)_5$ $O_{21}(OH)$
Source	Conway Form., Shelburne Falls, Mass.	Willsboro, N.Y.	Kammegg, Haslital, Switzerland	Salm-Chateau, Ardennes, Belgium	Mautia Hill, Tanganyika
Reference	Dollase, 1968	Dollase, 1968	Cervan & Fang, 1968	Donnay & Allmann, 1968	Moore & Bennett, 1968
Cell Dimensions					
$\underline{a}$	16.212	8.879	8.879	5.8108 (8.7126)	16.100
$\underline{b}$ Å	5.559	5.583	5.603	18.5214 (5.8108)	13.767
$\underline{c}$	10.036	10.155	10.151	8.7126 (18.5214)	6.735
α	90	90	90	90	90
β deg	90	115.50	115.466	90	90
γ	90	90	90	90	90
Space Group	Pnma	$P2_1/m$	$P2_1/m$	Pmmn (Pnmm)	Cmcm
Z	4	2	2	2	4

TABLE 9. (cont.)

Variety	Zoisite	Clinozoisite	Epidote	Ardennite	Kornerupine
Site Occupancy	$\begin{matrix}Al(2) \\ Al(3)\end{matrix}\Biggl\{\begin{matrix}Al & 0.97 \\ Fe & 0.03\end{matrix}$	$Al(3)\begin{cases}Al & 0.96 \\ Fe & 0.04\end{cases}$	$Al(3)\begin{cases}Al & 0.70 \\ Fe^3 & 0.30\end{cases}$	$M_1\begin{cases}Mn & 0.88 \\ Ca & 0.12\end{cases}$ $M_2\ Mn\ 1.00$ $M_3\ Al_1\ 1.00$ $M_4\ Al_2\ 1.00$ $M_5\begin{cases}Mg & 0.50 \\ Al & 0.375 \\ Fe & 0.125\end{cases}$ $M_6\begin{cases}As & 0.90 \\ V & 0.10\end{cases}$	
Method	3-D, counter	3-D, counter	3-D, counter	3-D, counter	3-D, counter
R & Rating	0.044 (1)	0.031 (1)	0.066 (1)	0.070 (1)	0.113 (1)
Cleavage and habit	{100} good	{001} good	{001} good	{001} perfect	Prismatic cleavage. Fibrous
Comment	Mixed Si_2O_7-SiO_4 groups	Mixed Si_2O_7-SiO_4 groups	Mixed Si_2O_7-SiO_4 groups	Related to epidote by Ito-type twinning. Mixed SiO_4-Si_3O_{10} groups	Mixed Si_2O_7 and [Al,Si]$_2SiO_{10}$ groups. Fe^3 and B present were ignored.
μ	190.1	177.7	203.5	246.3	102.6
ASF	0.7117×10^{-1}	0.8002×10^{-1}	0.8668×10^{-1}	0.5249×10^{-1}	0.8104×10^{-1}
Abridging factors	1:1:0.5	1:1:0.5	1:1:0.5	1:1:0.5	1:1:0.5

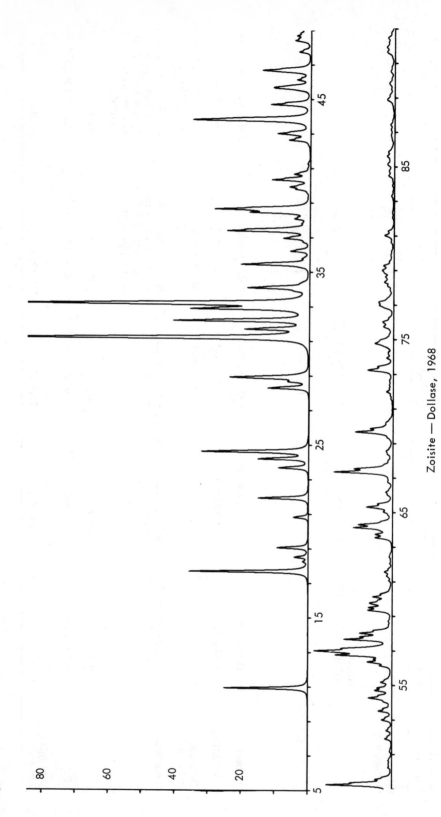

Zoisite — Dollase, 1968

λ = CuKα Half width at 40°2θ: 0.11°2θ Plot scale I_{PK} (max) = 100

198

ZOISITE - DOLLASE, 1968

2THETA	PEAK	D	H	K	L	I(INT)	I(PK)	I(DS)
10.91	10.92	8.106	2	0	0	19	25	13
17.66	17.66	5.018	0	2	1	29	36	23
18.23	18.24	4.863	0	1	1	1	2	1
18.49	18.50	4.794	1	1	1	3	4	3
19.04	19.04	4.658	1	0	2	8	9	6
20.80	20.80	4.267	2	0	2	4	5	3
21.91	21.92	4.053	4	0	0	13	15	11
23.65	23.66	3.758	3	0	1	8	9	7
24.18	24.18	3.677	3	0	2	13	15	12
24.61	24.62	3.615	3	1	1	29	32	26
28.28	28.28	3.153	4	0	2	12	12	11
28.65	28.66	3.113	2	1	2	5	6	4
28.85	28.92	3.092	4	1	0	10	24	9
28.91		3.085	5	0	1	16		15
31.18	31.18	2.866	0	1	1	100	100	100
31.67	31.68	2.823	1	1	3	17	19	17
32.18	32.18	2.779	5	0	0	40	41	41
32.86	32.86	2.723	2	0	2	33	36	34
33.12		2.702	6	0	3	10	89	10
33.13	33.18	2.702	5	1	0	23		24
33.18		2.698	2	2	1	66		68
34.07	34.08	2.629	5	0	1	19	19	20
34.34	34.32	2.609	6	2	1	21	3	23
35.42	35.42	2.532	3	0	3	21	21	23
35.76	35.76	2.509	0	1	4	1	2	2
36.20	36.20	2.479	1	0	4	6	6	7
36.72	36.72	2.446	5	1	2	2	2	2
36.94	36.94	2.431	0	2	2	8	8	9
37.37	37.38	2.405	1	0	2	26	25	30
37.49	37.46	2.397	2	0	4	4	19	5
37.78	37.78	2.379	6	1	2	4	5	5
38.07	38.08	2.362	6	0	1	5	18	20
38.43	38.44	2.340	4	1	3	17	28	32
38.63	38.62	2.329	2	2	2	28	4	5
39.77	39.78	2.264	2	2	4	4	6	6
39.91	39.90	2.257	7	0	1	5	11	15
40.32	40.32	2.235	4	2	1	13		

2THETA	PEAK	D	H	K	L	I(INT)	I(PK)	I(DS)
40.65	40.66	2.217	3	2	1	5	5	6
42.62	42.62	2.120	1	2	3	7	7	9
42.91	43.00	2.106	3	0	3	2	10	3
42.99		2.102	6	0	0	2		12
43.75	43.78	2.067	2	2	1	9	35	31
43.80		2.065	5	2	0	25		28
44.68	44.68	2.026	8	0	0	22	12	18
45.59	45.66	1.9880	3	0	3	14	11	3
45.63		1.9864	8	0	1	2		8
45.68		1.9843	5	0	4	6		9
46.11	46.12	1.9668	7	1	2	7	4	5
46.57	46.66	1.9484	2	0	5	4	14	8
46.65		1.9452	5	2	2	6		18
47.73	47.72	1.9039	8	0	2	14	4	4
48.40	48.40	1.8791	8	0	1	3	4	6
48.68	48.68	1.8688	5	1	3	5	5	5
48.86	48.82	1.8624	0	1	5	3	4	2
49.20	49.20	1.8503	1	2	4	1	3	34
49.53	49.52	1.8387	2	2	5	25	20	6
50.22	50.22	1.8151	2	1	4	5	5	2
51.88	51.88	1.7608	2	2	4	1	1	4
52.37	52.36	1.7456	6	1	4	3	2	3
52.99	52.98	1.7267	3	3	1	2	2	6
53.50	53.50	1.7114	4	1	5	4	3	8
54.04	54.06	1.6954	9	0	2	5	4	3
54.15	54.26	1.6923	4	1	1	2	3	2
54.26		1.6892	9	2	2	1	7	12
54.68	54.68	1.6770	7	1	6	8	5	8
54.84	54.84	1.6727	0	2	0	5	5	6
55.15	55.16	1.6638	6	0	6	2	3	3
56.12	56.12	1.6375	8	2	2	4	4	5
56.34	56.34	1.6315	5	1	5	3	8	13
56.72	56.74	1.6217	9	1	2	9	17	13
56.74		1.6209	0	3	3	12		19
56.97	56.96	1.6150	5	2	5	25	23	39
57.05		1.6129	1	3	3	2		5
57.11	57.12	1.6113	6	0	5	5	16	4

199

ZOISITE - DOLLASE, 1968

2THETA	PEAK	D	H	K	L	I(INT)	I(PK)	I(DS)
57.64	57.64	1.5979	3	0	6	18	15	30
57.97	58.00	1.5895	2	3	3	2	10	3
58.01		1.5885	5	3	1	11		17
59.31	59.32	1.5567	8	2	2	2	8	4
59.33		1.5564	10	1	0	7		11
59.48	59.48	1.5526	3	3	3	4	7	6
59.70	59.74	1.5476	6	1	5	3	7	6
59.76		1.5462	4	0	6	6		10
60.11	60.10	1.5380	10	1	1	7	6	12
61.04	61.04	1.5167	8	1	4	1	1	2
61.31	61.32	1.5108	6	3	3	1	2	2
61.34		1.5101	5	2	5	1		2
61.56	61.56	1.5050	4	3	3	3	2	5
63.52	63.52	1.4633	7	1	5	2	5	4
63.74	63.72	1.4589	9	0	4	5	6	9
64.11	64.10	1.4514	10	2	2	16	12	29
64.30	64.28	1.4474	7	2	4	6	10	11
65.02	65.02	1.4332	9	2	2	3	3	5
65.30	65.30	1.4276	0	2	6	11	8	19
66.01	66.02	1.4141	11	0	2	8	2	3
66.16	66.16	1.4113	2	2	6	2	2	2
66.74	66.76	1.4004	10	0	0	1	2	3
67.32	67.32	1.3897	0	4	0	26	18	49
67.40		1.3883	1	1	7	2		2
67.56	67.50	1.3853	3	2	6	1	12	4
68.35	68.40	1.3713	8	1	4	2	2	2
68.39	68.54	1.3704	11	1	2	1		2
68.51		1.3684	12	1	1	1	2	2
69.32	69.34	1.3543	5	3	4	1	2	3
69.65	69.64	1.3489	10	2	2	17	11	33
70.14	70.14	1.3406	9	1	5	1	2	2
71.97	71.96	1.3108	5	2	6	2	2	5
73.21	73.22	1.2913	10	2	1	9	8	18
73.24		1.2906	11	3	5	3		6
73.28		1.2906	4	3	1	2		3
73.92	73.92	1.2810	9	3	1	3	2	6

2THETA	PEAK	D	H	K	L	I(INT)	I(PK)	I(DS)
74.25	74.24	1.2762	5	1	7	2	2	3
74.62	74.76	1.2708	11	0	4	2	6	3
74.65		1.2703	1	2	7	1		2
74.75		1.2688	8	2	5	6		13
74.87		1.2671	5	4	5	1		2
75.28	75.26	1.2612	10	0	5	1	1	5
75.70	75.74	1.2553	5	3	8	3	3	5
75.76		1.2545	0	0	8	2		7
76.02	76.00	1.2508	9	3	0	3	3	5
76.82	76.96	1.2397	2	0	7	3	4	7
76.96		1.2379	5	4	0	3		5
77.11	77.14	1.2359	6	1	6	3	5	10
77.18		1.2348	6	0	7	2		6
78.09	78.10	1.2228	10	2	6	2	2	4
78.29	78.30	1.2201	10	3	1	3	3	4
78.40		1.2187	7	0	5	1		6
78.62	78.62	1.2159	6	3	1	2	3	2
78.98	78.98	1.2112	10	3	1	4	4	4
79.27	79.24	1.2075	9	2	7	3	3	8
80.61	80.62	1.1907	7	1	7	1	1	7
81.01	81.00	1.1859	5	1	3	5	3	3
83.28	83.28	1.1593	6	4	7	1	1	11
83.88	83.88	1.1525	6	2	3	1	1	3
84.45	84.46	1.1461	8	4	6	2	2	2
85.06	85.14	1.1394	10	1	1	1	2	6
85.13		1.1387	8	4	1	1		3
85.16		1.1383	5	4	4	1		3
85.60	85.60	1.1337	14	1	0	3	2	8
85.81	85.84	1.1314	8	4	5	1	2	3
87.16	87.14	1.1174	8	1	2	2	1	3
87.41	87.42	1.1147	6	2	8	3	1	4
88.84	88.88	1.1005	4	1	7	2	2	8
88.93		1.0996	9	0	3	1		6
89.48	89.48	1.0943	14	1	9	3	2	7
89.84	89.82	1.0909	1	1	6	1	1	3
90.50	90.50	1.0846	11	1	1	2	2	7
91.58	91.62	1.0746	15	0	1	2	2	6

ZOISITE – DOLLASE, 1968

2THETA	PEAK	D	H	K	L	I(INT)	I(PK)	I(DS)
91.67	91.91	1.0738	10	2	6	2	2	6
91.91	91.90	1.0716	3	1	9	2	2	6
92.46	92.46	1.0667	8	0	8	2	2	6
92.46		1.0666	1	4	6	1		3
93.78	93.80	1.0551	0	5	1	4	2	10
94.02	94.06	1.0530	6	2	8	3	3	7
94.54	94.54	1.0486	3	5	6	8	4	21
94.85	94.84	1.0460	5	5	1	2	4	6
96.03	96.06	1.0362	13	2	4	3	2	7
96.12		1.0355	3	5	3	1		4
96.36	96.36	1.0336	4	4	6	3	2	7
97.12	97.12	1.0275	11	2	6	1	1	3
97.94	97.96	1.0210	3	5	3	1	1	1
98.28	98.30	1.0184	7	3	7	2	2	2
98.31		1.0182	14	2	3	2		3
99.70	99.72	1.0077	9	4	4	3	2	5
99.90	99.90	1.0063	8	4	3	2	1	5
101.33	101.34	0.9958	10	2	8	1	1	2
101.99	102.00	0.9912	11	4	2	1	1	3
102.23	102.20	0.9895	8	3	7	1	1	3
102.63	102.78	0.9867	3	0	10	1	1	5
102.78		0.9857	13	3	6	2		5
103.03	103.08	0.9840	13	1	0	2	1	6
103.32	103.32	0.9820	14	3	4	1	1	2
104.04	104.04	0.9772	15	1	4	1	1	3
104.50	104.46	0.9742	4	0	10	1	1	3
105.17	105.20	0.9697	16	0	3	1	1	2
105.19		0.9696	6	3	8	1	1	4
105.33		0.9687	12	4	0	1	1	2
105.94	105.88	0.9649	9	4	5	1	1	5
106.77	106.78	0.9596	9	3	7	2	1	3
107.72	107.74	0.9538	1	3	9	1	1	6
108.02	108.08	0.9520	16	2	0	1	1	4
108.42	108.42	0.9496	11	3	6	2	1	5
109.07	109.14	0.9457	4	5	5	1	1	2
109.72	109.90	0.9419	9	5	1	1	1	3
109.90		0.9409	3	3	9	2	2	6

2THETA	PEAK	D	H	K	L	I(INT)	I(PK)	I(DS)
109.91		0.9408	6	0	10	2		6
110.33	110.36	0.9384	14	1	4	1	2	4
110.43		0.9378	11	1	4	1		4
110.47		0.9376	2	2	10	1		4
110.88	110.86	0.9353	16	2	5	1	1	4
111.12		0.9340	10	4	5	1		3
111.55	111.60	0.9316	5	5	4	2	1	6
111.61		0.9312	0	4	8	1		4
111.88	111.92	0.9297	9	5	2	1	1	3
112.25	112.28	0.9277	13	1	7	1	1	3
112.48	112.74	0.9265	0	6	0	2	1	4
112.73		0.9251	2	4	8	2	1	8
112.86		0.9244	8	3	8	1		3
113.60	113.84	0.9205	2	6	0	1	1	2
113.82		0.9193	4	2	10	1		4
113.83		0.9193	12	0	8	1		3
113.84		0.9193	0	4	6	1		3
114.11	114.26	0.9178	15	3	2	1	1	2
114.29		0.9169	10	5	0	1		4
114.64	115.00	0.9151	6	1	5	1	1	4
115.04	115.70	0.9131	10	6	2	1	1	4
115.72	115.66	0.9097	1	1	9	1	1	2
116.37	116.66	0.9065	10	0	2	1	1	2
116.58		0.9054	2	6	2	1		3
116.69		0.9049	17	1	6	1		3
116.76		0.9045	16	0	3	1		3
118.04	118.02	0.8984	17	2	1	1	1	2
118.40	118.46	0.8967	7	5	5	1	1	2
118.76	118.74	0.8950	5	4	8	1	1	5
119.55	119.58	0.8914	17	0	4	1	1	2
119.59		0.8912	13	2	7	1		4
120.03	120.08	0.8893	11	0	9	1	1	5
120.36	120.42	0.8878	17	2	2	1	1	2
120.46		0.8875	5	6	6	1		4
121.88	122.20	0.8812	1	3	10	1	1	2
122.06		0.8804	6	4	8	1		3

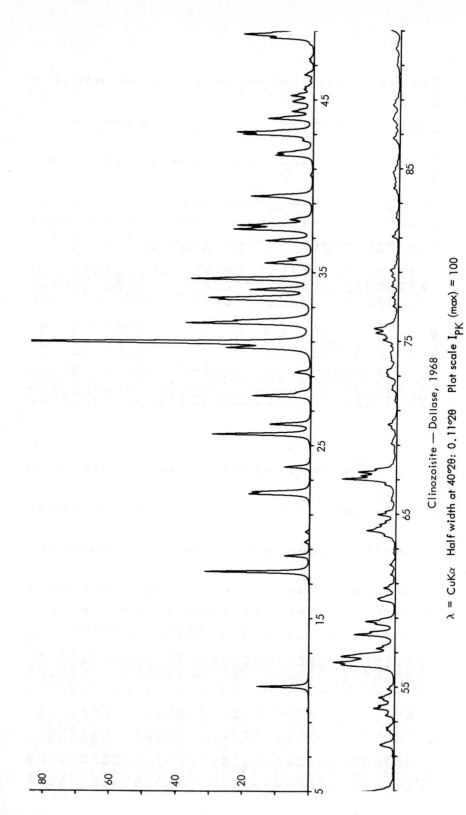

Clinozoisite — Dollase, 1968

λ = CuKα Half width at 40°2θ: 0.11°2θ Plot scale I$_{PK}$ (max) = 100

CLINOZOISITE - DOLLASE, 1968

2THETA	PEAK	D	H	K	L	I(INT)	I(PK)	I(DS)
11.03	11.04	8.014	-1	0	0	11	16	9
17.67	17.68	5.016	-1	0	2	26	31	21
18.59	18.60	4.768	0	1	1	6	7	5
19.40		4.572	-1	1	1	1	2	1
19.98	19.98	4.439	-2	0	1	1	1	1
22.17	22.18	4.007	-2	0	0	15	18	13
22.30	22.30	3.984	2	1	1	12	17	11
23.73	23.74	3.746	-1	1	2	7	8	6
25.61	25.62	3.475	-2	1	1	27	29	25
26.21	26.22	3.398	1	2	1	11	12	11
27.85	27.86	3.200	-2	1	2	17	17	16
28.06	28.06	3.177	3	0	0	3	5	3
29.20	29.20	3.055	-3	0	1	5	5	5
30.53	30.62	2.925	-3	1	1	2	5	2
30.61		2.918	-1	2	2	21	25	21
30.90	30.90	2.891	0	2	2	100	100	100
32.03	32.04	2.791	-1	0	3	37	38	37
32.21	32.20	2.777	0	1	3	17	23	18
33.40	33.42	2.680	2	1	3	26	28	27
33.52	33.50	2.671	3	1	1	17	31	18
33.98	33.98	2.636	1	2	3	17	19	18
34.00		2.634	-1	1	1	2		2
34.59	34.60	2.591	-3	2	1	38	36	41
34.65		2.586	-3	0	2	15		16
35.51	35.52	2.526	2	0	4	15	14	14
35.79	35.78	2.507	-1	2	2	7	7	7
36.81	36.82	2.439	-1	0	3	15	14	16
37.47	37.46	2.398	-3	2	2	25	23	28
37.70	37.70	2.384	0	2	1	22	22	25
38.05	38.04	2.363	-2	1	4	6	7	7
39.35	39.38	2.288	1	1	4	1		2
39.37		2.287	-2	1	2	2	18	2
39.38		2.286	-1	1	2	16		18
41.76	41.76	2.161	-4	0	3	12	11	14
41.91	41.90	2.154	-1	2	1	4	11	4
41.94		2.152	-1	0	4	4		5
42.61	42.62	2.120	0	1	2	3	3	3

2THETA	PEAK	D	H	K	L	I(INT)	I(PK)	I(DS)
42.96	42.96	2.104	-2	2	1	22	21	27
43.10	43.08	2.097	-2	2	3	16	23	19
43.85	43.90	2.063	-4	1	3	3	13	3
43.89		2.061	0	2	3	14		17
44.31	44.32	2.042	-2	0	3	7	7	9
44.60	44.60	2.030	-2	0	1	1	2	1
44.84	44.88	2.019	-3	2	1	4	6	5
44.89		2.017	4	0	2	8		10
45.22	45.22	2.004	-1	0	4	2	7	3
45.58	45.58	1.9885	-1	0	5	2	3	2
45.74	45.72	1.9821	3	0	5	1	3	1
46.07	46.08	1.9685	-1	0	2	3	1	4
46.43	46.44	1.9541	-3	2	5	1		2
47.19	47.20	1.9242	2	1	2	6	3	1
47.35	47.34	1.9180	4	1	3	8	1	1
48.22	48.22	1.8858	1	1	3	15	2	8
48.56	48.58	1.8732	2	2	4	5	13	10
48.57		1.8728	-1	1	4	3		5
48.71	48.76	1.8679	-2	2	6	4	21	21
48.77		1.8657	-1	2	2	1		7
48.78		1.8651	0	0	6	2		4
51.56	51.64	1.7711	-5	3	1	2	4	5
51.63		1.7687	-1	1	3	3		2
52.35	52.36	1.7462	0	1	6	7	1	3
52.49	52.50	1.7417	-4	2	1	2	2	3
52.63	52.64	1.7374	-2	3	3	3	2	4
53.33	53.34	1.7163	-4	1	2	7	3	11
53.74	53.74	1.7041	2	1	4	2	6	3
53.92	53.90	1.6988	0	0	6	3	5	5
54.23	54.24	1.6900	-2	1	2	4	4	6
54.36	54.36	1.6861	-5	0	6	3	5	6
54.86	54.86	1.6721	-3	1	6	9	16	14
56.18	56.26	1.6357	-1	3	1	13	18	19
56.26		1.6338	-5	0	3	12		18
56.39	56.40	1.6302	-1	3	3	4	16	6
56.49		1.6277	-4	2	4			
56.72	56.74	1.6215	-4	2	4	14	16	21

CLINOZOISITE - DOLLASE, 1968

2THETA	PEAK	D	H	K	L	I(INT)	I(PK)	I(DS)
56.80	57.22	1.6196	1	2	4	9		13
57.21		1.6088	2	3	1	3	6	4
57.21		1.6087	3	3	2	3		5
57.98	58.00	1.5894	0	3	3	4	12	7
58.01		1.5885	-4	0	6	13		20
58.61	58.62	1.5736	-1	1	5	6	6	10
58.75	58.76	1.5702	-3	3	1	7	9	12
60.00	60.02	1.5406	5	1	0	5	5	8
60.09	60.10	1.5383	4	1	1	5	5	8
60.71	60.72	1.5241	-3	3	3	4	4	7
61.50	61.50	1.5065	-4	2	5	2	2	3
62.06	62.07	1.4941	-5	2	2	2	2	3
63.72	63.74	1.4592	-6	0	4	4	4	7
64.02	64.02	1.4531	-5	2	4	10	9	17
64.11		1.4512	-2	2	4	3		6
64.39	64.38	1.4457	-2	2	6	5	5	9
64.95	64.96	1.4345	-3	2	6	8	6	14
65.77	65.78	1.4186	-4	0	3	2	2	3
66.68	66.70	1.4014	2	1	5	3	4	5
66.99	67.00	1.3957	0	4	0	23	16	40
67.00		1.3955	-2	1	7	2		3
67.30	67.40	1.3900	5	2	0	2	12	3
67.40		1.3883	4	2	2	15		27
69.07	69.06	1.3588	1	3	4	2	2	4
69.97	69.96	1.3434	5	0	1	2	2	3
70.17	70.16	1.3401	0	2	6	3	3	6
72.19	72.20	1.3074	-6	2	3	2	2	4
72.86	72.96	1.2970	-5	2	6	1	3	2
72.96		1.2955	-6	2	2	4		7
73.12	73.14	1.2932	-6	2	4	2	3	4
73.33	73.32	1.2899	-4	3	5	2	3	5
73.85	73.86	1.2821	-5	3	2	2	2	3
74.03	74.04	1.2794	2	4	1	2	2	3
74.22	74.24	1.2767	4	1	5	1	2	3
74.34	74.34	1.2748	0	0	7	2	2	4
74.72	74.72	1.2694	-3	1	8	2	2	5
75.04	75.04	1.2647	4	2	3	6	5	11

2THETA	PEAK	D	H	K	L	I(INT)	I(PK)	I(DS)
75.17		1.2628	4	0	4	1		3
75.41	75.46	1.2595	-6	2	1	2	7	5
75.43		1.2592	-3	4	2	5		11
75.47		1.2586	-5	4	1	1		7
75.48		1.2584	2	0	6	5		20
75.71	75.70	1.2551	-6	2	5	6	7	10
75.84		1.2534	-2	4	0	2		12
77.02	77.02	1.2371	3	4	0	3		3
77.51	77.52	1.2304	1	3	5	3	2	5
78.18	78.18	1.2216	2	4	3	4	2	8
78.34	78.38	1.2195	-1	4	4	2	2	7
78.73	78.76	1.2145	5	3	0	2	2	4
78.81		1.2133	4	3	2	3		7
79.11	79.10	1.2095	-6	1	7	3	3	6
81.05	81.04	1.1855	0	2	7	4	2	8
82.14	82.14	1.1725	-4	4	1	2	1	3
83.83	83.86	1.1530	-7	2	3	1	2	4
84.34	84.34	1.1473	-6	0	8	1	1	3
84.53	84.60	1.1452	4	4	0	3	2	5
84.76	84.76	1.1427	2	3	5	2	2	5
86.68	86.74	1.1223	0	1	8	2	2	3
86.75		1.1215	-7	1	0	3		7
87.36	87.36	1.1153	-7	1	7	2	1	5
87.81	87.88	1.1108	-5	2	8	2	2	5
87.89		1.1099	1	1	6	1		3
87.90		1.1098	-3	1	8	1		3
88.61	88.66	1.1028	-8	0	3	2	2	6
88.68		1.1020	-4	1	9	3		4
91.93	91.94	1.0714	0	0	8	1	2	7
91.99		1.0709	-1	0	7	4		3
93.01	93.20	1.0617	0	3	3	1		3
93.08		1.0612	-1	4	6	4	4	3
93.19		1.0602	-6	2	8	1		10
93.22		1.0599	-1	5	3	4		5
93.30		1.0592	0	2	0	1		9
94.44	94.54	1.0495	7	2	6	2	3	3
94.52		1.0488	0	5	3	1		5

CLINOZOISITE - DOLLASE, 1968

2THETA	PEAK	D	H	K	L	I(INT)	I(PK)	I(DS)
94.55		1.0485	-4	4	6	5		13
95.18	95.18	1.0432	-3	5	1	1	1	4
96.64	96.66	1.0313	-8	2	4	2	2	5
96.64		1.0313	-6	3	7	2		5
96.89	96.90	1.0293	-3	5	3	2		5
99.58	99.62	1.0086	-6	4	4	3	2	7
101.24	101.26	0.9965	6	2	3	1	1	3
101.46	101.50	0.9949	4	6	4	1	1	3
102.01	102.00	0.9910	-2	0	3	2	1	4
102.81	102.80	0.9855	-9	0	10	1	1	2
104.27	104.34	0.9756	0	3	4	1	1	2
104.35		0.9751	7	0	8	2		6
104.71	104.70	0.9728	-9	3	0	1	1	1
104.98	104.96	0.9710	-7	1	6	2	1	5
105.46	105.46	0.9679	5	4	7	1		4
105.53		0.9675	3	3	2	1		2
105.63		0.9668	-4	3	6	2		6
106.35	106.36	0.9622	-1	0	5	1	1	2
106.96	106.96	0.9584	-9	1	9	1	1	4
106.97		0.9583	7	2	0	1		3
107.99	108.28	0.9522	-2	3	2	1		1
108.07		0.9517	-3	2	9	2		6
108.28		0.9504	-4	5	10	1		3
108.80	108.76	0.9473	8	5	5	1	1	2
109.55	109.70	0.9429	4	4	0	1	1	3
109.70		0.9420	3	2	5	1		4
110.13	110.18	0.9396	-7	0	10	2		6
110.21		0.9391	-3	4	8	1		3
110.68	110.62	0.9364	4	4	4	2	1	5
110.99	111.02	0.9347	-5	5	4	1	2	3
111.00		0.9346	-2	4	1	1		4
111.37	111.38	0.9326	-8	4	8	1	2	3
111.46		0.9321	0	3	6	1		3
111.74	111.70	0.9305	0	6	0	1	1	3
112.89	113.18	0.9243	1	5	5	1	1	3
113.13		0.9230	3	5	0	1		3
113.28		0.9222	-6	2	10	1		2

2THETA	PEAK	D	H	K	L	I(INT)	I(PK)	I(DS)
114.43	114.52	0.9162	5	5	0	1	1	3
114.52		0.9157	4	5	2	1		3
114.68		0.9149	-1	6	2	1		2
115.19	115.22	0.9123	5	1	6	1		2
115.27		0.9119	0	6	2	1		3
116.24	116.36	0.9071	3	1	8	1	1	2
116.44		0.9061	-2	6	2	1		2
117.06	117.06	0.9031	-9	2	2	1		2
117.14		0.9027	-6	0	11	1	1	2
118.47	118.48	0.8964	7	2	3	1		2
119.10	119.10	0.8935	-2	6	1	1	0	3
121.00	121.18	0.8850	2	1	11	1		2
121.18		0.8842	2	5	5	1	0	2
121.38		0.8833	-7	0	4	1	1	3
121.53		0.8827	-2	5	7	1		2
123.80	123.88	0.8732	3	2	8	1		4
123.81		0.8731	2	6	2	2		3
123.87		0.8729	-10	0	3	1	1	2
123.99		0.8724	-2	6	4	1		4
124.74	124.44	0.8694	-2	4	7	1	1	3
125.22	125.26	0.8675	-2	1	9	1	1	2
125.55	125.80	0.8662	-7	1	11	2	1	2
125.79		0.8653	-8	4	3	2		8
126.62	126.52	0.8621	-9	3	6	1		5
127.86	127.86	0.8575	8	4	2	2	2	5
128.72	128.72	0.8544	-8	4	2	1	0	5
128.74		0.8543	-7	4	4	1	1	3
129.37	129.36	0.8521	-8	2	10	1		4
129.38		0.8521	-1	3	10	1	1	2
129.48		0.8517	-9	0	1	1		3

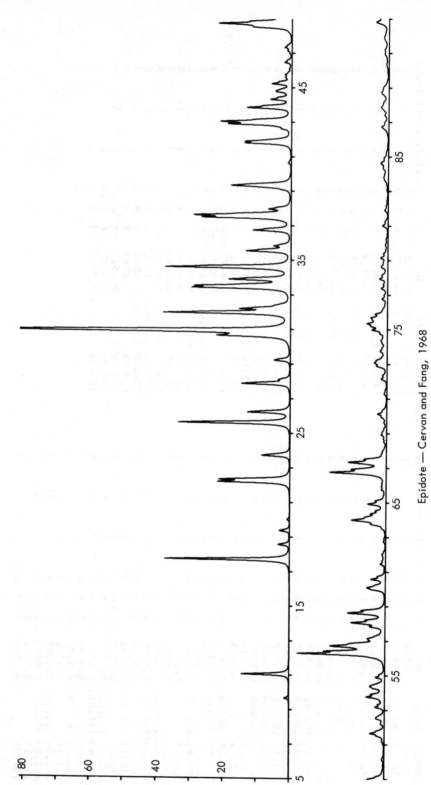

Epidote — Cervan and Fang, 1968

$\lambda = CuK\alpha$ Half width at 40°2θ: 0.11°2θ   Plot scale I_{PK} (max) = 100

206

EPIDOTE - CERVAN AND FANG, 1968

2THETA	PEAK	D	H	K	L	I(INT)	I(PK)	I(DS)
9.64	9.64	9.165	0	0	1	1	1	3
11.03	11.04	8.016	-1	0	1	10	14	7
17.67	17.68	5.014	-1	0	2	30	37	23
18.54	18.56	4.780	0	1	1	3	3	2
19.35	19.36	4.583	-1	1	0	3	3	2
22.16	22.16	4.008	-2	0	2	18	22	15
22.30	22.30	3.983	-2	0	1	16	21	13
23.69	23.70	3.752	0	1	2	7	8	6
23.79		3.736	-1	1	1	1		1
25.58	25.58	3.480	1	1	1	31	33	28
26.20	26.20	3.398	-2	0	2	12	13	11
27.84	27.84	3.202	-2	1	1	14	15	13
28.08	28.08	3.176	-2	0	3	3	4	2
29.21	29.22	3.055	0	0	3	5	5	5
30.53		2.925	-3	1	1	1		1
30.61	30.62	2.918	-1	2	0	17	22	17
30.88	30.88	2.893	1	1	2	100	100	100
31.92	31.92	2.801	-1	2	1	38	38	39
32.18	32.18	2.780	0	2	1	13	15	13
33.38	33.38	2.682	2	0	1	28	30	29
33.51	33.48	2.672	3	0	0	17	29	18
33.87	33.86	2.645	1	1	2	18	19	19
34.56	34.56	2.593	-1	1	3	39	38	42
34.63		2.588	-3	0	1	1		1
35.50	35.50	2.526	-2	0	2	14	13	16
35.80	35.80	2.506	4	0	4	4	5	5
36.72	36.72	2.446	-1	1	2	12	11	14
37.45	37.46	2.399	-1	2	3	26	27	30
37.60	37.58	2.390	-3	1	2	23	29	26
37.94	37.94	2.369	0	2	2	6	7	7
39.20		2.296	-2	2	1	6		6
39.29	39.30	2.291	-2	2	3	14	18	17
39.34		2.288	-1	1	4	2		3
39.35		2.288	-1	0	1	1		3
41.76	41.76	2.161	-4	2	3	14	14	18
41.85	41.86	2.157	-1	1	0	6	14	7
41.92		2.153	-4	0	3	2		3
42.59	42.60	2.121	0	1	4	2	3	3
42.86	42.86	2.108	-2	2	1	20	19	26
43.02	43.00	2.101	-2	2	3	17	21	22
43.81	43.82	2.065	-2	2	2	14	13	18
43.83		2.064	-4	1	3	2		2
44.30	44.30	2.043	-2	0	3	7	6	9
44.75	44.78	2.023	-3	2	1	3	5	5
44.81		2.021	-3	2	0	7		3
45.21	45.20	2.004	4	0	4	1	6	9
45.58	45.58	1.9886	-1	0	5	2	2	2
45.75	45.74	1.9815	-1	3	5	3	2	2
46.45	46.46	1.9532	-1	0	3	1	2	4
47.12	47.12	1.9271	-3	2	2	2	1	2
47.32	47.32	1.9192	2	2	4	9	2	3
48.48	48.54	1.8761	2	1	5	5	12	13
48.54		1.8740	-2	3	4	14		8
48.70	48.70	1.8682	-1	1	2	3	22	20
48.71		1.8681	-1	2	4	6		5
48.99	48.98	1.8678	3	0	2	3	5	8
51.50	51.50	1.8577	0	1	1	2	3	5
51.63	51.62	1.7736	-5	0	5	4	4	4
52.17	52.18	1.7688	-1	3	2	2	1	6
52.48	52.50	1.7518	-1	2	1	1	3	2
52.56	52.54	1.7421	0	3	5	2	3	3
53.16	53.16	1.7398	-4	2	2	4	3	3
53.75	53.74	1.7215	-4	3	1	6	6	5
53.92	53.88	1.7041	-2	2	5	1	4	10
54.25	54.36	1.6991	-2	0	4	3	4	2
54.34		1.6894	-5	1	2	3		5
54.42		1.6868	-5	0	6	1		5
54.88	54.88	1.6844	-3	0	3	10	3	2
56.20	56.22	1.6714	-1	0	6	13	26	17
56.23		1.6353	-5	1	3	12		21
56.23		1.6346	-1	3	3	3		19
56.40	56.38	1.6299	-4	2	0	14	18	6
56.66	56.68	1.6231	-4	2	4	14	16	23

207

EPIDOTE - CERVAN AND FANG, 1968

2THETA	PEAK	D	H	K	L	I(INT)	I(PK)	I(DS)
56.72	57.12	1.6216	1	2	4	8		13
57.04		1.6132	2	3	1	2	5	3
57.13		1.6110	3	2	2	4		6
57.81	57.82	1.5935	0	0	3	5	5	8
58.04	58.04	1.5878	-4	0	6	12	10	21
58.59	58.60	1.5742	-3	3	1	8	11	14
58.59		1.5741	1	1	1	7		11
59.96	60.06	1.5414	5	1	0	4	6	7
59.96		1.5392	4	1	5	6		10
60.56	60.56	1.5276	-3	3	2	5	4	8
61.45	61.44	1.5077	-4	2	3	2	2	4
61.99	61.98	1.4957	-5	2	5	1	1	2
62.73	62.74	1.4798	-6	2	2	1	1	2
63.73	63.74	1.4590	-6	0	6	4	5	8
63.83		1.4571	3	1	4	4		3
63.96	63.96	1.4543	-5	2	4	1	10	21
64.04		1.4528	2	2	4	11		6
64.34	64.34	1.4467	-2	2	6	3	5	9
64.50	64.50	1.4434	3	2	3	5	3	2
64.91	64.90	1.4354	-3	2	6	1	6	15
65.75	65.76	1.4190	-3	0	5	8	1	3
66.66	66.72	1.4019	2	1	5	2	17	7
66.72		1.4007	1	0	4	4		44
67.01	66.90	1.3953	-2	1	0	22	10	3
67.22	67.30	1.3915	1	1	7	2	11	4
67.31		1.3899	5	2	2	2		30
67.78	67.78	1.3813	4	2	0	15		
68.49	68.50	1.3688	-4	2	6	1	2	2
68.91	68.92	1.3614	-6	1	1	1	1	2
69.94	69.96	1.3439	1	3	4	2	2	4
70.11	70.12	1.3411	0	2	2	2	3	7
72.12	72.14	1.3085	-6	2	6	3		4
72.82	72.90	1.2976	-5	2	6	2	4	8
72.89		1.2966	-6	2	4	1	3	6
73.06	73.08	1.2940	-6	1	4	4		6
73.20	73.18	1.2919	-4	3	5	3		5
73.70	73.74	1.2843	-5	3	2	2		4
73.77	73.77	1.2833	2	4	4	1	2	3
74.20	74.22	1.2769	3	0	5	2	2	4
74.34	74.34	1.2749	0	1	7	2	2	6
74.75	74.76	1.2689	-3	0	8	3	5	11
74.96	74.96	1.2659	4	2	0	5	5	3
75.15	75.18	1.2631	0	0	4	1		5
75.17		1.2628	-3	4	2	2		3
75.31	75.32	1.2608	-5	3	1	6	6	13
75.33		1.2606	-6	2	1	1		3
75.47		1.2585	-2	0	6	1		13
75.67	75.66	1.2558	-6	2	5	6	5	3
75.86	75.86	1.2530	-2	4	8	3	4	13
76.76	76.76	1.2406	3	1	0	3	2	3
77.37	77.36	1.2324	1	3	2	3	2	6
77.92	77.92	1.2250	-4	2	3	4	2	6
78.57		1.2165	3	4	0	2	2	9
78.62	78.62	1.2154	5	3	2	2		4
78.65		1.2093	-6	1	6	3	2	8
79.13	79.14	1.1997	-6	2	2	1	1	2
79.89	79.88	1.1861	-6	4	7	3	2	7
80.99	81.00	1.1755	0	2	1	2	1	4
81.88	81.88	1.1537	2	4	3	3		3
83.77	83.78	1.1481	-7	4	0	4		3
84.27	84.30	1.1468	4	0	8	8	2	3
84.39	84.38	1.1443	-6	3	5	5		4
84.61	84.62	1.1224	2	1	8	8		4
86.67	86.70	1.1220	3	0	0	1	2	3
86.71		1.1151	0	7	7	7	1	8
87.38	87.38	1.1110	-7	1	7	2	1	5
87.79	87.84	1.1102	-5	2	8	8	2	5
87.86		1.1029	-1	0	9	9		4
88.60	88.70	1.1018	-8	3	3	3	2	9
88.71		1.0807	-4	1	2	2		3
90.92	90.94	1.0736	-8	4	6	6	1	3
91.69	91.90	1.0721	-3	0	7	7	2	3
91.86		1.0714	0	3	8	8		4
91.93		1.0638	1	0	6	6		
92.78	92.80	1.0638	-1	4	6	5	4	13

EPIDOTE - CERVAN AND FANG, 1968

2THETA	PEAK	D	H	K	L	I(INT)	I(PK)	I(DS)
92.80		1.0636	-1	5	3	4		11
93.06	93.10	1.0613	-6	2	8	1	3	3
93.17		1.0604	0	2	8	1		4
93.52	93.48	1.0573	-8	1	6	1	2	3
94.13	94.34	1.0521	0	5	3	1	4	4
94.33		1.0504	-4	4	6	5		15
94.36		1.0501	3	2	6	2		5
94.79	94.64	1.0465	-3	5	1	2	2	5
96.50	96.54	1.0324	-3	5	3	2	3	6
96.53		1.0322	-6	3	7	2		7
96.58		1.0318	-8	2	4	2		7
97.28	97.28	1.0262	-8	2	3	2	1	5
99.33	99.36	1.0104	-6	4	4	3	2	9
101.13	101.16	0.9973	6	4	3	1	1	3
101.19		0.9969	-4	4	3	1		3
101.89	102.04	0.9919	-6	4	10	1	1	2
102.05		0.9908	-2	0	4	1		5
102.81	102.78	0.9855	-9	3	0	1	1	2
104.14	104.18	0.9765	0	3	8	2	1	6
104.18		0.9763	-7	3	0	2		6
104.87	104.86	0.9717	-7	3	7	2	1	6
105.18	105.20	0.9697	5	4	2	2	1	5
105.18		0.9697	6	1	4	1		2
105.36		0.9685	3	3	6	1		2
105.46		0.9679	-4	3	5	2		8
106.25	106.26	0.9629	-9	1	9	2	1	5
106.98	106.98	0.9583	-1	0	6	1	1	3
107.00		0.9582	1	1	10	1		5
107.87	107.92	0.9523	-2	2	10	2	1	2
107.96		0.9505	-3	3	9	1		7
108.27	108.28	0.9505	-3	2	10	2		3
108.41		0.9496	-4	5	5	1	1	5
109.42	109.40	0.9437	8	4	5	1		5
109.45		0.9435	-2	2	0	1		2
110.39	110.56	0.9381	-5	4	4	1	2	4
110.56		0.9371	-5	5	1	2		7
110.58		0.9370	2	2	8	1		2

2THETA	PEAK	D	H	K	L	I(INT)	I(PK)	I(DS)
110.72	110.72	0.9362	2	4	6	1		4
110.87	110.87	0.9353	7	2	3	1		2
111.13	111.12	0.9339	-2	4	8	1	2	5
111.14		0.9338	0	6	0	1		4
111.33		0.9328	-8	3	6	1		3
112.28	112.24	0.9276	1	6	0	1		3
112.71	112.70	0.9253	1	5	5	1		4
113.28	113.22	0.9222	-6	2	10	1		3
113.99	114.04	0.9185	5	5	0	1		3
114.07		0.9180	-1	6	2	1		2
114.08		0.9180	0	6	2	1	1	4
114.66	114.52	0.9150	0	6	2	1	1	2
115.13	115.10	0.9126	5	1	6	1	1	2
115.56	115.80	0.9104	-4	1	11	1		2
115.82		0.9092	-2	6	2	1		3
116.20	116.20	0.9073	3	2	8	1	1	2
116.96	116.98	0.9036	-9	2	3	1		3
118.35	118.46	0.8970	7	2	1	1	0	2
118.45		0.8965	2	3	6	1	1	3
118.53		0.8961	4	4	8	2		6
120.46	120.48	0.8874	-6	4	5	1	1	4
120.72	121.06	0.8862	2	5	11	1	1	3
121.03		0.8849	-2	1	11	1		2
121.10		0.8845	7	5	7	1		4
121.31		0.8836	-7	0	5	1		4
123.13	123.32	0.8759	2	6	2	1	1	5
123.33		0.8751	-2	6	4	1		6
123.70	123.74	0.8736	3	2	8	2		3
123.85		0.8730	-10	0	3	1		2
124.35	124.32	0.8709	6	3	4	1	1	3
124.42		0.8707	2	4	7	1		11
124.95	124.96	0.8686	-2	4	9	1	1	6
125.47	125.50	0.8665	-8	4	3	3	1	5
126.47	126.42	0.8627	-9	3	6	2	1	2
127.71	127.76	0.8580	8	2	9	1	0	5
128.27	128.36	0.8560	2	2	9	1	1	2
128.37		0.8556	-8	4	2	2		7

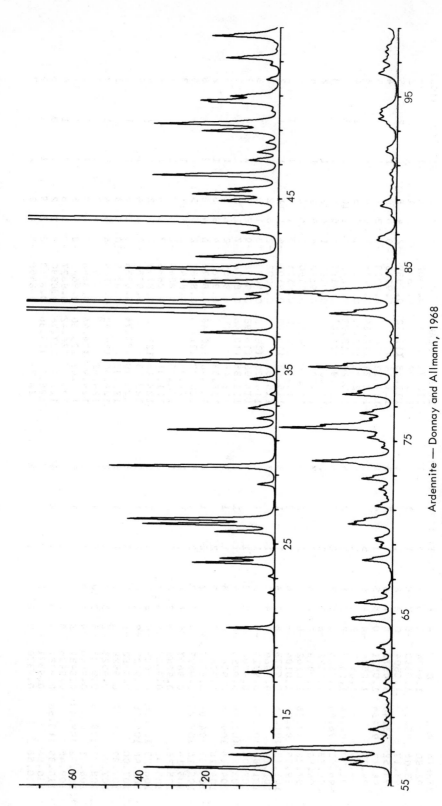

210

Ardennite — Donnay and Allmann, 1968

λ = FeKα Half width at 40°2θ: 0.11°2θ Plot scale I$_{PK}$ (max) = 150

ARDENNITE - DONNAY AND ALLMANN, 1968

2THETA	PEAK	D	H	K	L	I(INT)	I(PK)	I(DS)
12.00	12.00	9.261	0	2	0	18	26	9
12.76	12.76	8.713	0	1	1	6	9	3
17.55	17.56	6.346	0	2	1	4	6	2
20.11	20.12	5.544	1	1	0	7	10	4
22.16	22.16	5.037	0	3	1	1	1	1
23.89	23.90	4.678	1	1	1	13	16	9
24.13	24.14	4.630	0	4	0	8	11	6
25.68	25.68	4.356	0	2	2	9	12	7
26.11	26.12	4.285	1	2	1	21	26	15
26.39	26.40	4.241	1	1	2	21	29	16
26.45		4.231	0	3	2	3		4
28.43	28.44	3.942	0	1	3	3	4	2
29.47	29.48	3.806	1	3	1	30	33	23
31.56	31.56	3.559	0	3	3	20	22	16
32.25	32.26	3.486	1	0	2	4	4	3
32.83	32.84	3.425	1	1	3	5	6	4
32.99	32.98	3.409	0	5	1	2	3	2
33.65	33.66	3.344	1	4	1	2	1	1
34.52	34.52	3.262	1	2	2	5	5	4
35.53	35.54	3.173	0	4	2	34	35	31
37.19	37.20	3.035	1	3	2	15	15	14
38.44	38.44	2.940	1	5	0	100	100	100
38.92	38.94	2.905	2	0	0	38	58	38
38.94		2.904	0	0	3	20		20
39.43	39.44	2.869	0	1	3	5	6	5
40.12	40.12	2.822	1	5	2	19	19	20
40.68	40.70	2.785	2	2	0	7	10	7
40.87	40.88	2.772	2	0	3	14	30	15
40.89		2.771	0	2	3	16		17
41.12	41.12	2.756	2	0	1	5	8	5
41.60	41.60	2.726	2	1	1	16	16	18
42.99	43.00	2.642	2	2	3	7	7	8
43.23	43.24	2.628	0	3	3	2	4	2
43.68	43.76	2.602	1	6	1	51	90	58
43.75		2.598	1	0	3	60		68
44.83	44.84	2.538	1	5	3	12	12	14
45.20	45.22	2.519	0	6	2	11	17	13
45.24	45.54	2.517	2	3	1	8	10	9
45.54	45.54	2.501	1	2	3	10	25	11
46.34	46.32	2.461	2	4	0	26		32
46.34		2.460	2	0	3	1		2
47.21	47.22	2.417	1	4	3	6	6	8
47.69	47.70	2.394	1	2	4	3	4	5
48.25	48.26	2.368	2	2	1	3	3	3
48.90	48.90	2.339	1	7	1	17	15	21
49.30	49.30	2.321	0	5	1	27	25	35
50.12	50.12	2.286	1	5	3	1	2	1
50.59	50.60	2.266	0	7	0	13	14	18
50.69	50.68	2.261	0	2	4	9	16	12
50.94	50.94	2.251	2	3	1	10	10	13
51.92	51.92	2.211	2	5	2	3	3	4
52.77	52.78	2.178	0	1	4	2	3	3
53.16	53.16	2.163	2	6	0	13	11	18
54.46	54.46	2.116	2	1	4	16	14	23
54.69	54.56	2.107	0	6	2	3	9	3
56.18	56.24	2.056	2	7	1	2	8	4
56.23		2.054	0	6	3	5		7
		2.054	2	0	4	11		16
56.52	56.52	2.044	2	8	2	2	10	3
56.61		2.041	1	1	4	1		2
56.67	56.64	2.040	1	0	4	3	9	5
57.04	57.14	2.027	1	5	4	37	31	56
57.13		2.024	2	2	5	2		2
57.73	57.72	2.005	1	2	5	2	2	3
58.15	58.28	1.9918	1	6	1	4	5	7
58.29		1.9876	2	4	3	3		2
58.83	58.84	1.9710	0	8	4	3	1	5
60.26	60.26	1.9285	2	7	2	2	2	2
60.95	60.94	1.9087	0	9	1	2	2	3
62.07	62.08	1.8776	2	5	4	8	7	13
62.69	62.70	1.8608	0	9	2	2	2	3
63.02	63.02	1.8526	3	0	1	1	3	2
		1.8521	0	10	0	3		5
64.64	64.64	1.8106	2	8	0	7	8	12

ARDENNITE - DONNAY AND ALLMANN, 1968

2THETA	PEAK	D	H	K	L	I(INT)	I(PK)	I(DS)
64.65	64.76	1.8103	0	8	3	2	8	4
64.75		1.8079	3	3	1	5		9
65.61	65.62	1.7867	1	5	4	10	7	18
66.19	66.20	1.7727	2	8	1	3	2	5
68.07	68.08	1.7295	1	10	0	8	6	15
68.83	68.84	1.7127	2	2	1	3	3	6
69.21	69.22	1.7045	0	10	4	4	3	7
69.35	69.36	1.7013	3	5	2	2	4	5
70.17	70.18	1.6841	3	1	1	12	9	23
70.29	70.30	1.6816	0	7	4	3	8	4
70.50	70.48	1.6772	2	3	4	4	6	7
70.51		1.6770	0	3	5	2		3
71.22	71.22	1.6624	1	1	5	5	4	10
71.89	71.88	1.6490	2	9	1	2	2	3
72.21	72.22	1.6426	1	2	5	4	3	7
72.81	72.80	1.6311	2	4	1	9	6	18
73.79	73.82	1.6124	3	6	1	6	16	12
73.84		1.6114	3	0	3	16		32
74.62	74.62	1.5970	3	5	2	3	2	6
75.00	75.02	1.5901	1	11	1	1	3	5
75.14	75.20	1.5876	3	2	3	2	5	3
75.20		1.5864	0	8	4	5		10
75.73	75.74	1.5770	2	5	4	32	23	68
75.74		1.5768	0	5	5	4		8
76.30	76.28	1.5669	2	9	2	6	5	5
76.60	76.60	1.5618	0	10	0	6	7	13
76.61		1.5616	0	12	0	3		7
77.68	77.68	1.5434	0	0	3	16	10	35
77.99	77.86	1.5384	3	7	1	5	7	10
78.17	78.16	1.5354	3	6	2	1	1	2
78.99	79.02	1.5219	2	4	3	4	4	8
79.26	79.26	1.5176	0	6	4	25	17	56
79.27		1.5174	0	6	5	1		3
79.84	79.84	1.5081	1	10	3	1	1	3
81.06	81.06	1.4895	2	1	5	1	1	3
82.36	82.36	1.4702	2	10	2	22	13	52
83.38	83.40	1.4554	2	7	4	20	15	47

2THETA	PEAK	D	H	K	L	I(INT)	I(PK)	I(DS)
83.57	83.58	1.4527	4	0	0	28	23	69
84.70	84.70	1.4369	2	11	1	3	2	7
86.44	86.64	1.4135	0	3	6	1	4	3
86.63		1.4110	1	0	4	5		13
86.80	86.82	1.4088	1	1	6	3	3	7
88.06	88.08	1.3928	2	5	6	1	1	3
88.61	88.64	1.3858	0	5	5	2	3	4
88.63		1.3856	0	4	6	4		4
88.95	88.84	1.3816	2	11	2	2	3	11
89.45	89.50	1.3755	3	10	1	1	1	4
89.82	89.78	1.3711	1	10	4	1	1	3
90.20	90.20	1.3666	1	13	1	1	1	3
90.51	90.50	1.3629	2	0	2	2	1	6
91.28	91.28	1.3539	1	8	5	1	1	3
91.78	91.80	1.3482	3	5	4	4	2	11
92.06	92.02	1.3450	4	1	2	1	2	3
93.69	93.68	1.3269	3	12	1	7	4	18
94.04	94.02	1.3231	3	10	3	3	4	9
94.06		1.3230	0	14	0	1		4
94.25	94.26	1.3208	4	4	0	4	3	11
96.33	96.36	1.2992	2	0	2	3	4	9
96.36		1.2989	2	0	3	3		14
97.02	97.06	1.2923	3	1	5	5	3	9
97.08		1.2916	4	5	2	3		10
97.59	97.60	1.2866	4	2	2	3	3	8
97.62		1.2863	2	2	6	2		7
97.96	97.92	1.2830	3	2	5	3	2	8
99.20	99.22	1.2711	2	3	6	11	6	32
99.33		1.2698	3	7	4	3		10
99.40		1.2692	2	10	4	2		5
100.57	100.60	1.2584	4	6	2	4	2	12
100.67		1.2574	3	11	1	2		5
101.78	101.78	1.2475	3	4	5	2	1	5
102.65	102.66	1.2399	1	8	0	7	3	23
103.75	103.74	1.2305	2	1	4	1	1	5
106.77	107.04	1.2060	2	0	4	2	3	7
107.02		1.2040	2	14	0	5		16

ARDENNITE - DONNAY AND ALLMANN, 1968

2THETA	PEAK	D	H	K	L	I(INT)	I(PK)	I(DS)
107.90	108.00	1.1972	2	6	6	1	2	4
108.02		1.1964	1	15	1	3		10
109.14	109.16	1.1880	0	11	5	2		16
109.49	109.50	1.1854	1	11	2	5	2	6
109.66		1.1842	4	8	2	3		11
110.26	110.26	1.1798	0	5	7	2	1	7
110.39		1.1789	1	14	1	1		4
112.54	112.70	1.1639	1	15	3	2	2	6
112.67		1.1630	2	10	5	2		7
112.85		1.1619	3	0	6	2		6
113.69	113.72	1.1562	1	5	5	1	1	4
114.81	114.82	1.1490	4	5	0	1	1	14
115.74	115.76	1.1430	4	10	4	4	1	4
117.37	117.46	1.1330	4	8	3	1	2	5
117.41		1.1328	2	8	6	1		4
117.51		1.1322	1	6	7	3		10
117.79	117.76	1.1305	3	8	5	2	2	6
118.51	118.52	1.1263	0	7	7	2	1	7
118.74	118.72	1.1249	2	3	7	1	1	5
119.38	119.38	1.1213	1	10	6	4	1	15
120.55	120.56	1.1146	3	12	3	8	3	30
122.20	122.24	1.1057	1	7	7	2	2	8
122.21		1.1056	4	10	2	3		10
122.70	122.68	1.1030	2	13	4	2	2	8
123.28	123.36	1.1000	5	5	1	8	4	29
123.35		1.0996	0	11	6	1		5
123.36		1.0996	2	15	2	3		6
123.41		1.0993	4	7	5	3		12
123.59		1.0984	2	9	6	2		6
123.67		1.0980	4	3	5	5		5
124.63	124.64	1.0931	2	5	7	9	3	37
125.83	125.82	1.0872	0	1	8	1	1	4
127.00	127.04	1.0816	0	2	8	8	3	33
127.51	127.52	1.0793	5	6	3	7	4	27
127.57		1.0790	5	10	1	5		19
128.35	128.32	1.0754	2	16	0	6		25
128.52		1.0746	5	5	2	1		5

2THETA	PEAK	D	H	K	L	I(INT)	I(PK)	I(DS)
129.24	129.28	1.0714	4	0	4	4	2	18
129.46		1.0704	1	8	8	3		13
129.92	129.88	1.0684	4	5	5	4	3	15
130.37	130.38	1.0664	4	11	2	3	2	12
131.03	131.54	1.0636	4	10	3	4	5	16
131.10		1.0634	1	2	8	4		15
131.21		1.0629	5	1	1	1		6
131.22		1.0628	5	17	2	6		25
131.53		1.0616	3	10	5	6		41
132.43	132.48	1.0578	4	12	0	10	6	73
132.55		1.0574	1	16	3	18		28
132.82		1.0563	1	15	4	7		5
132.83		1.0562	5	7	1	1		22
134.19	134.38	1.0508	5	2	3	5	3	8
134.38		1.0501	2	7	7	2		40
134.58		1.0494	4	6	5	9		7
136.96	137.04	1.0405	1	12	2	2	1	17
137.14		1.0399	2	17	1	4		10
138.88	138.96	1.0338	3	15	1	5	1	22
139.11		1.0331	0	10	7	1		5
139.32		1.0324	3	3	7	2		7
140.51	140.62	1.0284	2	11	6	4	2	5
140.54		1.0284	1	7	8	1		15
140.61		1.0281	4	5	8	4		6
140.66		1.0280	4	12	2	1		7
141.05	141.08	1.0267	3	11	5	2	2	14
142.41	142.40	1.0225	3	14	3	3	1	6
142.51		1.0222	0	16	4	2		8
143.74	143.78	1.0185	5	6	6	1	1	5
143.85		1.0182	2	1	8	1		6
145.68	146.02	1.0131	4	3	6	4	3	18
145.83		1.0127	3	15	2	3		12
146.03		1.0121	4	10	4	12		56
146.71		1.0103	4	7	5	1		6
147.75	147.80	1.0076	3	8	7	2	2	10
147.79		1.0075	2	15	4	7		34
147.81		1.0075	0	15	5	2		11

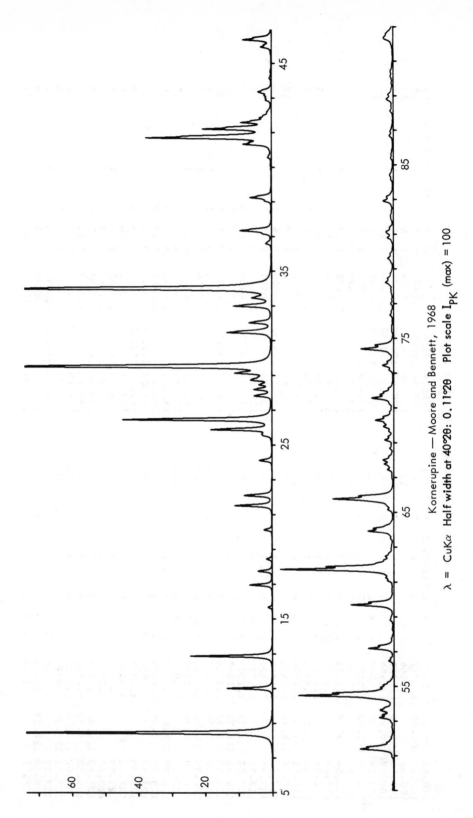

Kornerupine — Moore and Bennett, 1968

λ = CuKα Half width at 40°2θ: 0.11°2θ Plot scale I_{PK} (max) = 100

KORNERUPINE – MOORE AND BENNETT, 1968

2THETA	PEAK	D	H	K	L	I(INT)	I(PK)	I(DS)
8.44	8.44	10.463	1	1	0	75	100	68
10.98	10.98	8.050	2	0	0	10	14	9
12.85	12.86	6.883	1	2	0	19	24	17
15.63	15.64	5.663	1	1	1	1	1	1
16.93	16.94	5.232	2	2	0	5	7	5
17.72	17.74	5.000	3	1	0	4	4	3
18.41	18.42	4.814	0	1	1	1	2	1
20.10	20.10	4.413	1	2	1	2	2	2
21.49	21.50	4.132	2	3	0	10	11	9
22.07	22.08	4.025	2	0	1	7	8	6
22.12		4.015	3	1	1	1		1
24.09	24.10	3.691	1	3	1	3	4	2
25.52	25.60	3.488	3	3	0	2	3	2
25.62		3.475	4	2	0	1		1
25.86	25.86	3.442	0	0	2	17	18	16
26.44	26.44	3.367	1	4	0	43	44	42
27.81	27.82	3.206	2	0	2	5	5	5
28.17	28.18	3.165	5	1	0	5	5	5
28.44	28.44	3.135	3	3	1	3	4	3
28.71	28.72	3.107	1	1	2	5	6	5
28.80	28.78	3.097	5	0	0	6	6	5
28.89	28.88	3.088	3	3	1	2	4	2
29.11	29.12	3.065	0	2	2	9	11	9
29.50	29.50	3.025	0	2	2	92	90	90
31.45	31.44	2.842	5	1	2	13	13	13
32.02	32.02	2.793	3	1	2	7		7
32.98	32.98	2.714	1	5	0	12	11	12
33.44	33.44	2.677	1	3	2	5	6	5
33.98	33.98	2.636	5	3	1	100	93	100
36.58	36.58	2.455	5	2	0	2	2	3
37.15	37.16	2.418	0	4	1	2	3	2
37.33	37.32	2.407	2	4	2	10	9	10
39.02	39.04	2.306	5	2	2	6	2	2
39.23	39.22	2.295	0	6	1	1	6	6
41.54	41.54	2.172	0	2	3	9	1	1
42.31	42.32	2.134	0	6	2	9	9	9
42.69	42.70	2.116	6	4	0	43	38	45
42.75		2.113	1	5	2	4		4
43.19	43.20	2.093	5	5	0	23	21	24
43.57	43.56	2.076	2	3	2	10	9	10
43.84	43.84	2.063	6	2	1	3	4	3
44.86	44.86	2.019	6	2	1	2	2	2
45.13	45.12	2.007	5	3	2	1	2	2
45.28	45.34	2.001	0	3	1	1	4	1
45.34		1.9984	0	6	1	4		4
47.93	47.94	1.8962	9	2	0	4	3	4
48.36	48.36	1.8803	0	4	3	11	8	12
51.36	51.36	1.7774	9	5	1	12	10	13
51.47	51.48	1.7740	7	1	0	3	2	4
51.74	51.74	1.7652	0	8	0	3	4	2
53.18	53.18	1.7209	5	6	0	5	4	6
53.36	53.32	1.7154	7	3	2	1	4	2
53.57	53.62	1.7091	0	5	1	3		3
53.63		1.7075	5	1	4			
54.45	54.44	1.6837	7	0	0	36	28	41
54.63	54.58	1.6784	5	5	0	4	18	4
54.92	54.92	1.6704	6	2	3	3	4	4
55.05	55.06	1.6667	9	3	0	1	3	1
56.45	56.46	1.6286	5	7	1	2	2	2
57.16	57.16	1.6100	7	0	0	11	8	12
58.78	58.78	1.5695	10	0	0	1	1	1
59.66	59.66	1.5486	9	6	2	18	13	21
61.06	61.06	1.5163	6	6	2	1	2	
61.70	61.70	1.5022	7	3	3	49	34	59
62.47	62.46	1.4853	5	7	1	2	3	2
63.91	63.90	1.4554	9	1	0	11	8	13
64.05	64.06	1.4525	11	1	2	3	6	3
65.14	65.14	1.4308	10	0	4	2	3	3
65.63	65.76	1.4212	5	5	2	4	2	5
65.75		1.4190	7	3	3	26	18	32
67.43	67.44	1.3876	5	9	0	4	2	2
67.76	67.76	1.3817	6	6	3	4	3	5
68.01	68.00	1.3772	6	6	3	3	3	4
68.66	68.66	1.3658	0	8	3	1	1	2

215

KORNERUPINE - MOORE AND BENNETT, 1968

2THETA	PEAK	D	H	K	L	I(INT)	I(PK)	I(DS)
69.14	69.14	1.3575	0	6	4	4	4	5
69.65	69.66	1.3488	0	10	1	2	2	2
69.92	69.92	1.3443	5	7	3	4	3	5
70.07	70.12	1.3417	12	0	0	1	3	1
70.28	70.28	1.3382	10	4	2	8	6	10
70.76	70.76	1.3303	2	10	1	1	1	1
71.53	71.54	1.3179	10	6	0	2	7	3
71.55			0	4	4	8		11
71.91	71.90	1.3176	6	5	4	8	3	5
73.16	73.18	1.3118	12	2	1	2	1	2
73.43	73.44	1.2924	11	1	2	5	3	6
74.38	74.38	1.2883	0	10	2	16	10	21
76.34	76.34	1.2743	12	0	2	1	1	2
76.98	76.98	1.2464	5	1	5	1	1	2
78.20	78.20	1.2376	11	1	3	2	2	3
78.21			9	1	4	1		2
79.58	79.58	1.2212	0	8	4	2	2	3
80.78	80.78	1.2212	5	7	4	3	2	4
81.12	81.12	1.2035	11	3	3	3	2	4
82.89	82.90	1.1887	10	0	4	6	3	8
84.15	84.16	1.1845	5	11	1	1	1	2
86.66	86.66	1.1636	0	2	6	2	1	4
88.10	88.10	1.1494	11	1	4	4	2	3
88.78	88.78	1.1225	14	0	2	1	1	6
90.11	90.12	1.1079	5	9	2	4	2	2
92.30	92.34	1.1011	0	9	6	3	3	6
92.40		1.0883	5	1	6	3		5
93.58	93.58	1.0681	5	1	6	4	2	5
95.00	95.00	1.0672	11	9	1	1	1	6
95.51	95.48	1.0568	11	7	3	2	1	2
95.84	95.84	1.0447	10	6	4	1	2	4
96.16	96.16	1.0405	5	3	2	2		2
96.46	96.46	1.0378	11	3	6	1	1	2
97.41	97.42	1.0351	11	5	4	1	1	2
99.53	99.54	1.0328	11	9	2	3	2	5
100.86	100.86	1.0252	10	10	2	5	2	7
101.46	101.20	1.0089	5	13	1	1	1	2
		0.9992	10	10	2			
		0.9950	5	13	1			
103.13	103.12	0.9834	0	14	0	0	1	1
103.13	103.12	0.9833	14	6	0	1		1
104.67	104.68	0.9730	0	14	1	1	0	1
105.79	106.10	0.9658	16	4	0	3	3	5
106.08		0.9639	10	8	0	1		2
106.09		0.9639	11	7	4	4		7
106.21		0.9631	11	7	0	1		1
106.26		0.9628	0	10	5	1		1
106.89	106.86	0.9589	12	8	3	1	1	2
107.17	107.26	0.9571	11	9	3	1	1	1
107.24		0.9567	0	12	4	1		2
108.67	108.70	0.9481	12	0	6	1	1	1
109.38	109.38	0.9439	5	6	6	1	1	2
109.77	109.78	0.9416	5	2	5	1	1	2
111.28	111.28	0.9331	11	7	6	8	3	14
111.64	111.66	0.9310	9	3	6	1	2	2
112.29	112.30	0.9275	17	3	0	1	1	1
112.45		0.9266	0	4	7	1	1	1
113.55	113.58	0.9208	10	0	6	1	1	1
113.89	113.96	0.9190	4	14	2	1	1	1
114.58	114.60	0.9154	14	4	4	5	2	8
114.59		0.9154	11	11	2	1		2
115.05	115.04	0.9130	16	6	1	1	2	2
115.12		0.9127	10	2	6	2		2
115.62	115.54	0.9102	16	7	3	1		2
116.19	116.18	0.9073	15	7	2	2	1	3
116.41		0.9062	6	10	5	1	1	1
117.64	117.60	0.9003	10	12	2	1	0	1
118.94	118.94	0.8942	17	3	2	3	0	5
119.98	120.12	0.8895	10	6	6	4	2	8
120.13		0.8888	16	6	4	1		8
120.32		0.8826	15	3	4	1		2
121.55	121.62	0.8826	14	10	0	1	1	1
121.73		0.8818	5	11	5	1		1
123.50	123.52	0.8744	11	3	6	3	1	5
124.60	124.60	0.8700	0	10	6	4	1	7
125.02		0.8683	2	12	5	1		1

KORNERUPINE - MOORE AND BENNETT, 1968

2THETA	PEAK	D	H	K	L	I(INT)	I(PK)	I(DS)
126.93	126.94	0.8609	12	0	6	1	0	1
127.73	127.98	0.8580	11	13	0	1	1	1
127.99		0.8570	16	2	4	1		2
128.88	128.94	0.8538	5	15	2	3	1	2
129.25	129.54	0.8525	16	6	3	2	1	6
129.48		0.8517	7	11	5	1		4
129.65		0.8511	11	13	1	3		1
130.21	130.16	0.8491	0	14	4	2		5
132.22	132.40	0.8424	6	6	7	1	1	3
132.31		0.8421	7	9	6	1		2
132.39		0.8419	0	8	8	3		1
132.62		0.8411	16	0	2	1		6
133.04	133.22	0.8398	0	8	7	1	1	2
133.22		0.8392	10	14	0	2		1
133.23		0.8392	19	1	1	1		3
133.47	133.66	0.8384	18	4	2	1	1	1
133.68		0.8378	16	4	4	5		9
134.67	134.38	0.8347	5	7	7	2	1	3
135.02	135.64	0.8337	0	16	2	1	1	2
135.64		0.8318	11	9	5	5		10
136.08		0.8305	6	12	5	1		2
137.10	137.24	0.8276	6	10	6	1	1	1
137.25		0.8271	17	7	2	3		6
137.52		0.8264	7	15	3	1		1
139.35	139.18	0.8214	14	10	3	1	0	2
140.13	140.16	0.8193	6	16	0	1	0	2
141.06	141.06	0.8170	10	12	4	1	0	2
141.95	142.62	0.8147	15	11	0	1	1	1
142.53		0.8133	6	16	1	4		9
142.92		0.8124	17	3	4	2		4
144.14	144.38	0.8096	6	14	4	1	1	1
144.44		0.8089	19	3	2	2		3
145.50	145.72	0.8065	16	10	1	1	1	2
145.74		0.8060	5	13	5	3		5
146.10		0.8052	17	9	0	1		2
147.03	147.66	0.8033	14	0	6	1	2	2
147.11		0.8031	10	4	7	1		2

2THETA	PEAK	D	H	K	L	I(INT)	I(PK)	I(DS)
147.35		0.8026	11	1	7	1		3
147.67		0.8020	5	3	8	1		21
147.93		0.8014	11	13	3	1		2
149.77	149.80	0.7979	6	7	3	1	1	2
149.90		0.7976	17	7	3	2		3
150.71	150.78	0.7961	6	16	2	1		2
151.77	151.80	0.7942	0	14	5	2		4
151.92		0.7940	20	0	1	2		1
153.13	154.20	0.7919	11	3	7	3	1	5
153.41		0.7915	19	1	3	1		2
154.10		0.7904	0	6	8	7		13
154.60		0.7896	5	9	7	1		1
154.61		0.7886	14	12	2	1		1
155.21		0.7886	0	10	7	1		1
157.20	157.62	0.7858	8	16	1	1	1	14
157.48		0.7854	5	17	0	7		3
157.82		0.7849	16	4	5	1		3
157.84		0.7849	2	10	7	1		2
159.18	159.22	0.7831	17	9	2	4		8
159.91		0.7822	6	4	8	3		6
160.36		0.7817	14	10	4	3		6
160.73		0.7813	18	6	3	1		2
160.93		0.7810	5	5	8	1		1
161.79		0.7801	5	17	1	4		7
162.15		0.7797	14	8	5	1		2

TABLE 10. COMPLEX SILICATES

Variety	Neptunite	Zunyite	Pumpellyite	Vesuvianite
Composition	$LiKNa_2Fe_2Ti_2(Si_4O_{12})_2$	$Al_{13}Si_5O_{20}ClF_2(OH)_{16}$	$Ca_8Al_8(Mg,Fe,Al)_4(SiO_4)_4(Si_2O_7)_4(OH)_8(H_2O,OH)_4$	$Ca_{10}Al_4(Mg,Fe)_2Si_9O_{34}(OH)_4$
Source	San Benito, California	Zuni Mine, Silverton, Colorado	Hicks Ranch, Sonoma Co., California	Franklin, N.J. & Sanford, Maine
Reference	Cannillo, Mazzi & Rossi, 1966	Kamb, 1960	Gottardi, 1965	Warren & Modell, 1931
Cell Dimensions				
a Å	16.46	13.87	19.14 (8.81)	15.66 (15.63 kx)
b Å	12.50	13.87	5.94 (5.94)	15.66 (15.63 kx)
c Å	10.01	13.87	8.81 (19.14)	11.85 (11.83 kx)
α	90	90	90	90
β deg	115.433	90	97.60	90
γ	90	90	90	90
Space Group	C2/c	F$\bar{4}$3m	C2/m (A2/m)	P4/nnc
Z	4	1 (?)	1	4

TABLE 10. (cont.)

Variety	Neptunite	Zunyite	Pumpellyite	Vesuvianite
Site Occupancy				$M_1 \begin{cases} Mg & 0.5 \\ Fe & 0.5 \end{cases}$
Method	3-D, film	3-D, film	2-D, film	2-D, film
R & Rating	0.155 (2)	0.123 (2)	0.18 (3)	--- (3)
Cleavage and habit	{110} distinct Prismatic	{111} prob. Equidimensional	{001}&{100} mod. Fibrous & tabular	{100} poor Prismatic or granular
Comment	Chains of tetrahedra parallel to [110] and [$\bar{1}$10]	Isolated large groups of tetrahedra	Mixed SiO_4 and Si_2O_7 groups B estimated	B estimated
μ	325.5	91.8	166.1	259.5
ASF	0.2079	0.1127	0.5252×10^{-1}	0.1046
Abridging factors	1:1:0.5	0.2:0:0	1:1:0.5	1:1:0.5

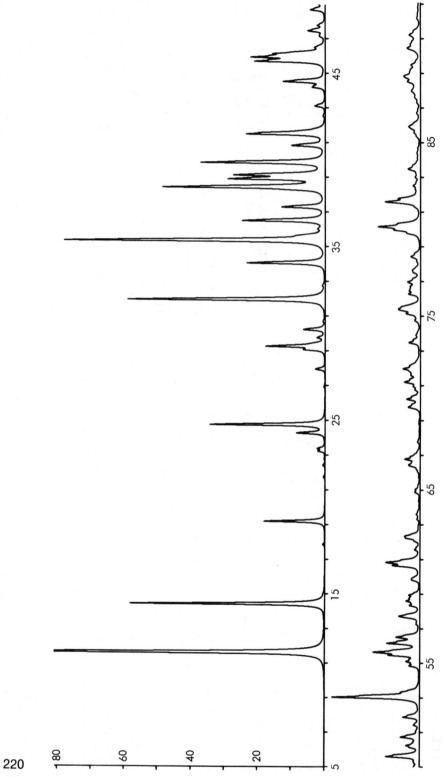

Neptunite — Cannillo, Mazzi and Rossi, 1966

λ = FeKα Half width at 40°2θ: 0.11°2θ Plot scale I$_{PK}$ (max) = 150

220

NEPTUNITE — CANNILLO, MAZZI AND ROSSI, 1966

2THETA	PEAK	D	H	K	L	I(INT)	I(PK)	I(DS)
11.61	11.62	9.567	-1	1	0	100	100	100
14.40	14.42	7.722	-1	1	1	41	39	46
19.16	19.16	5.817	1	1	1	13	12	18
23.20	23.22	4.814	-2	2	1	2	1	2
23.35	23.36	4.784	2	2	0	2	2	3
24.25	24.26	4.608	-1	1	2	6	6	9
24.73	24.74	4.520	0	0	2	28	23	46
27.92	27.92	4.012	1	3	0	2	2	5
29.04	29.06	3.861	-2	2	2	5	4	9
29.24	29.24	3.836	2	2	1	15	12	29
29.73	29.74	3.774	-1	3	1	2	2	4
30.20	30.20	3.716	-4	0	2	6	4	12
31.93	31.94	3.519	-1	3	1	53	39	116
34.03	34.04	3.308	-3	3	1	22	16	50
35.33	35.34	3.190	-1	3	2	41	52	100
35.34		3.189	-3	1	1	32	39	78
35.62	35.60	3.165	-1	1	3	4	6	10
35.99	36.00	3.133	-5	1	1	1	1	3
36.48	36.48	3.092	-1	1	3	24	16	60
37.26	37.26	3.030	-3	3	2	12	9	32
38.41	38.42	2.943	-2	2	3	46	32	124
38.88	38.88	2.908	2	2	2	26	19	72
39.11	39.10	2.892	5	1	1	24	18	66
39.83	39.84	2.841	1	3	2	37	25	103
40.81	40.82	2.776	3	3	1	10	7	28
41.47	41.48	2.734	-5	3	3	23	16	67
41.58		2.727	3	1	2	5	2	14
43.08	43.08	2.636	-2	4	2	3	3	10
44.18	44.18	2.574	-3	3	3	13	8	11
44.51	44.52	2.556	-1	3	3	21	14	42
45.68	45.68	2.494	-6	2	2	1	1	69
45.78	45.76	2.489	-4	4	1	20	15	4
45.92	45.92	2.481	-6	4	1	12	10	68
46.12	46.12	2.471	5	1	1	2	2	42
46.49	46.48	2.453	-3	1	4	1	1	6
47.10	47.12	2.423	-1	5	1	6	3	4
47.43	47.44	2.407	-4	4	2	6	3	19
47.75	47.74	2.392	4	4	0	2	2	8
48.66	48.66	2.350	-1	1	1	5	3	17
48.95	48.96	2.337	-1	5	1	1	1	5
49.21	49.22	2.325	-5	3	3	1	1	4
49.56	49.58	2.309	-5	1	4	2	7	7
49.58		2.309	-1	1	4	9		34
49.68		2.304	-2	3	4	1		5
50.11	50.12	2.286	1	2	3	2	2	9
50.22	50.22	2.281	-2	2	3	1	3	6
50.23		2.280	-2	4	1	1		4
50.44	50.44	2.272	-3	5	1	1	1	4
50.70	50.70	2.261	-7	1	2	6	4	22
51.40	51.40	2.232	-1	1	3	2	3	6
51.86	51.86	2.214	-7	1	3	6	3	22
53.01	53.02	2.169	-1	4	3	31	17	125
53.31	53.30	2.158	3	1	3	4	4	14
53.30		2.102	-1	5	2	3	2	14
54.83	54.84	2.094	4	2	0	4	3	17
55.08	55.08	2.081	-6	1	4	10	6	41
55.44	55.44	2.075	-1	3	4	9	9	37
55.62	55.62	2.075	-1	1	2	3		14
55.63		2.074	-5	3	4	2		8
56.14	56.14	2.057	-8	0	2	11	7	50
56.45		2.047	-5	3	4	1	5	6
56.50	56.50	2.045	6	2	1	7		29
57.25	57.26	2.020	-7	6	4	1		4
57.63		2.008	-2	6	1	3	1	14
57.70	57.70	2.006	-5	5	1	6	4	25
58.33	58.32	1.9865	-1	1	3	3		16
58.59	58.58	1.9784	-3	5	1	4	2	18
58.80	58.72	1.9719	-2	5	2	2	3	9
58.92	58.92	1.9681	-2	5	4	3	2	12
59.78	59.82	1.9425	5	0	4	1	2	7
60.63	60.64	1.9178	-2	6	2	10	6	46
60.83	60.82	1.9121	-2	6	1	10	7	46
61.54	61.54	1.8920	-7	3	0	3	1	8
62.22	62.24	1.8735	-1	1	5	3	3	16

221

NEPTUNITE - CANNILLO, MAZZI AND ROSSI, 1966

2THETA	PEAK	D	H	K	L	I(INT)	I(PK)	I(DS)
62.32	62.32	1.8709	-2	2	5	3	3	17
63.30	63.30	1.8448	1	5	3	3	1	6
63.58	63.58	1.8375	-7	1	4	1	1	7
64.85	64.86	1.8052	-9	1	2	2	1	11
66.36	66.42	1.7688	3	1	4	1	2	7
66.43		1.7670	-2	6	3	3		16
66.75	66.76	1.7595	2	0	6	6	3	33
67.74	67.74	1.7368	0	2	2	1	1	6
69.73	69.76	1.6933	-8	4	5	2	2	9
69.74		1.6932	1	1	1	2		11
69.79		1.6921	-7	1	5	2		13
70.20	70.22	1.6834	-4	4	1	4	3	25
70.23		1.6828	-8	4	5	1		9
70.75	70.76	1.6721	-7	5	3	2	1	11
70.96	70.96	1.6677	-4	0	6	2	2	10
71.19	71.20	1.6631	-7	5	5	3	3	15
71.20		1.6629	4	4	3	3		20
71.57	71.60	1.6554	-3	7	2	2	2	14
71.64		1.6539	-6	6	2	1		7
71.83	71.96	1.6502	-6	6	1	2	4	11
71.96		1.6476	3	5	3	6		34
72.24	72.12	1.6422	3	3	4	2	2	13
73.47	73.46	1.6185	7	5	0	5	2	30
73.91	73.90	1.6101	3	1	1	1	1	7
74.74	74.74	1.5948	-2	6	4	2	1	12
75.11	75.12	1.5880	-10	2	3	5	3	30
75.16		1.5871	2	6	3	1		7
75.34	75.38	1.5840	-7	5	4	3	4	21
75.38		1.5832	-6	2	6	2		10
75.42		1.5826	-10	2	2	4		26
75.52	75.50	1.5807	-7	1	6	2	4	16
76.27	76.28	1.5676	-3	7	3	5	2	32
76.42	76.44	1.5649	0	8	0	2	2	7
76.56	76.56	1.5625	0	0	8	2	2	15
76.69	76.70	1.5603	-3	5	5	2	2	16
76.95	76.96	1.5559	-1	1	6	4	3	26
77.01		1.5547	6	2	3	1		8

2THETA	PEAK	D	H	K	L	I(INT)	I(PK)	I(DS)
77.52	77.54	1.5462	-10	2	4	2	2	16
77.63	77.62	1.5444	-5	5	5	2	2	14
78.42	78.42	1.5313	-10	5	1	3	2	24
79.76	79.78	1.5097	7	1	1	4	3	31
79.95	79.96	1.5067	0	0	6	7	5	49
80.15	80.14	1.5036	-8	4	5	14	9	101
81.57	81.58	1.4818	8	8	2	18	7	128
81.91	81.76	1.4768	0	1	5	1	5	9
82.50	82.50	1.4681	-7	1	4	2	2	13
83.00	83.00	1.4609	-11	1	3	2	1	14
83.42	83.46	1.4548	7	1	2	2	3	14
83.46		1.4542	4	4	4	3		23
83.50		1.4537	3	5	4	1		10
85.55	85.56	1.4253	-9	5	4	2	1	15
85.90	85.90	1.4207	-2	6	4	2	2	18
85.95		1.4200	-11	1	1	2		17
86.69	86.68	1.4102	-7	7	1	2	1	13
87.29	87.30	1.4025	-11	1	5	2	1	13
87.60	87.60	1.3985	-8	6	4	1	2	11
87.95	87.98	1.3941	-7	1	7	3	2	21
88.05		1.3929	-4	8	3	2		16
88.50	88.52	1.3872	-6	2	7	3	3	23
88.52		1.3870	4	8	1	2		14
88.74	88.82	1.3842	-5	5	6	2		15
88.85		1.3829	-1	9	0	6	3	50
89.20	89.06	1.3785	-3	5	6	2	2	13
89.24		1.3781	9	5	0	2		18
89.48	89.46	1.3752	-1	9	1	3		26
90.36	90.44	1.3647	9	1	2	2	2	13
90.44		1.3637	6	2	4	6	3	52
90.85	90.66	1.3589	1	9	1	2	2	18
91.21	91.20	1.3547	-9	5	5	6	2	50
93.45	93.46	1.3295	-7	3	7	1	1	10
93.90	93.88	1.3246	-9	1	1	1	1	13
96.75	96.76	1.2949	9	5	3	3	3	25
98.67	98.70	1.2761	-9	7	2	2	1	16
98.94	99.04	1.2735	-5	9	2	1	2	10

NEPTUNITE - CANNILLO, MAZZI AND ROSSI, 1966

2THETA	PEAK	D	H	K	L	I(INT)	I(PK)	I(DS)
99.07		1.2723	-9	7	3	3		24
99.14		1.2717	2	6	5	1		10
100.94	100.90	1.2550	-9	7	1	1	1	12
102.14	102.16	1.2443	-9	7	4	1	2	12
102.15		1.2443	-8	8	2	5		45
102.27		1.2432	-5	1	8	2		20
102.56	102.48	1.2407	-12	4	2	2	2	23
103.14	103.22	1.2357	10	2	2	2	2	15
103.22		1.2350	9	1	3	2		19
103.43	103.44	1.2332	-2	10	1	1	1	11
104.40	104.42	1.2250	4	0	6	2	1	24
104.68	104.68	1.2227	4	8	3	2	1	23
105.55	105.54	1.2157	-4	2	8	1	1	12
108.90	108.96	1.1897	1	3	7	2	1	19
113.92	113.96	1.1547	-10	2	8	2	1	24
117.22	117.46	1.1340	-5	9	5	1	1	13
117.46		1.1325	7	7	3	2		27
117.50		1.1322	10	2	2	1		13
119.56	119.66	1.1202	-1	9	5	2	2	23
119.58		1.1201	-1	7	6	2		26
119.70		1.1194	-6	10	3	2		26
120.12	120.04	1.1171	-2	4	8	2	1	20
120.31		1.1160	6	10	0	1		15
121.98	121.96	1.1069	-14	2	6	1	1	13
122.31	122.70	1.1051	-9	9	2	1	1	16
122.68		1.1031	-7	7	7	3		33
124.12	124.44	1.0957	-8	8	6	2		25
124.44		1.0941	-14	2	1	3	1	33
125.59	125.72	1.0884	9	7	2	1		17
125.74		1.0877	-6	10	4	1	1	13
126.38	126.40	1.0846	0	8	6	1		18
126.42		1.0844	8	8	2	1		15
126.83	126.84	1.0824	6	10	1	1	1	18
127.82	127.76	1.0778	1	9	5	2	0	21
129.43	129.06	1.0705	-5	11	2	1	0	16

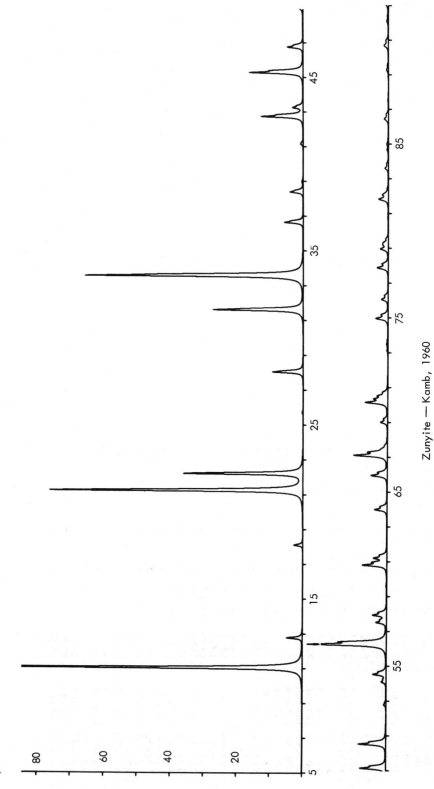

Zunyite — Kamb, 1960

λ = CuKα Half width at 40°2θ: 0.11°2θ Plot scale I_{PK} (max) = 100

224

ZUNYITE - KAMB, 1960

2THETA	PEAK	D	H	K	L	I(INT)	I(PK)	I(DS)
11.04	11.04	8.008	1	1	1	100	100	100
12.75	12.76	6.935	2	0	0	5	5	5
18.07	18.08	4.904	2	2	0	3	3	3
21.23	21.24	4.182	3	1	1	89	76	91
22.18	22.18	4.004	2	2	2	43	36	44
25.67	25.68	3.468	4	0	0	0	0	0
28.02	28.02	3.182	3	3	1	12	9	13
31.57	31.58	2.831	4	2	2	37	27	40
33.54	33.54	2.669	5	1	1	40	66	43
36.62	36.62	2.452	4	4	0	54	6	58
38.36	38.36	2.344	5	3	1	8	4	9
38.93	38.92	2.312	6	0	0	6	0	7
41.12	41.12	2.193	6	2	0	0	1	0
42.71	42.72	2.115	5	3	3	1	13	1
43.23	43.24	2.091	6	2	2	20	5	23
45.26	45.26	2.002	4	4	4	5	16	5
46.73	46.74	1.9422	7	1	1	26	5	30
46.73		1.9422	5	5	1	1		2
49.11	49.12	1.8535	6	4	0	6	8	7
50.50	50.50	1.8057	7	3	1	13	8	16
50.50		1.8057	5	5	3	1		2
52.75	52.76	1.7338	8	0	0	13	1	15
54.07	54.08	1.6945	7	3	3	1	2	2
54.51	54.52	1.6820	8	2	0	3	4	3
54.51		1.6820	6	4	4	6		8
56.23	56.22	1.6346	8	2	2	1	24	1
56.23		1.6346	6	6	0	41		49
57.49	57.50	1.6016	7	5	1	3	3	4
57.49		1.6016	5	5	5	5		6
57.91	57.92	1.5910	6	6	2	1	4	1
59.56	59.56	1.5507	8	4	0	8	0	10
60.79	60.78	1.5224	9	1	1	0	7	1
60.79		1.5224	7	5	3	4		5
61.19	61.20	1.5133	8	4	2	10	4	13
62.79	62.80	1.4785	6	6	4	7	0	9

2THETA	PEAK	D	H	K	L	I(INT)	I(PK)	I(DS)
63.98	63.98	1.4540	9	3	1	8	4	10
65.93	65.94	1.4156	8	4	4	10	5	13
67.08	67.08	1.3940	7	7	1	10	10	13
67.08		1.3940	9	3	3	11		14
68.99	69.00	1.3601	10	2	0	0	2	0
68.99		1.3601	8	6	2	5		6
70.12	70.12	1.3409	9	5	1	14	7	19
70.50	70.50	1.3346	10	2	2	0	3	4
70.50		1.3346	6	6	6	3		4
73.10	73.10	1.2934	9	5	3	3	0	1
73.47	73.46	1.2878	8	6	4	1	1	0
73.47		1.2878	10	4	0	0		1
74.94	74.94	1.2662	10	4	2	1	4	12
76.03	76.04	1.2506	11	1	1	9	2	4
76.03		1.2506	7	7	5	3		2
77.85	77.84	1.2259	8	8	0	8	3	11
78.93	78.94	1.2118	9	5	5	2	2	3
78.93		1.2118	11	3	1	0		1
78.93		1.2118	9	7	1	3		4
79.29	79.28	1.2072	8	8	2	2	2	3
79.29		1.2072	10	4	4	1		1
80.73	80.72	1.1893	10	6	0	0	0	0
80.73		1.1893	8	6	6	7		0
81.80	81.80	1.1764	9	7	3	1	3	1
81.80		1.1764	11	3	3	1		10
82.16	82.04	1.1722	10	6	2	2	2	1
83.58	83.58	1.1558	12	0	0	1	1	1
83.58		1.1558	8	8	4	1		3
84.65	84.64	1.1440	11	5	1	1	0	1
84.65		1.1440	7	7	7	0		1
85.00	84.98	1.1401	12	2	0	1	1	1
86.42	86.42	1.1250	12	2	2	3	1	4
86.42		1.1250	10	6	4	1		2
87.48	87.48	1.1141	11	5	3	0	0	0
87.48		1.1141	9	7	5	0		1

ZUNYITE - KAMB, 1960

2THETA	PEAK	D	H	K	L	I(INT)	I(PK)	I(DS)
89.25	89.26	1.0965	12	4	0	2	1	3
90.66	90.66	1.0831	12	4	2	2	1	4
90.66		1.0831	10	8	0	0		1
90.66		1.0831	8	8	6	1		2
92.08	92.08	1.0701	10	8	2	2	1	2
93.14	93.14	1.0607	13	1	1	0	2	1
93.14		1.0607	9	9	3	4		5
93.14		1.0607	11	5	5	1		2
93.14		1.0607	11	7	1	1		2
93.49	93.42	1.0576	10	6	6	0		0
94.91	94.94	1.0455	12	4	4	0	1	0
95.97	95.98	1.0367	13	3	1	1	0	0
95.97		1.0367	11	7	3	2	1	3
95.97		1.0367	9	7	7	1		2
96.33	96.32	1.0338	12	6	0	2		4
96.33		1.0338	10	8	4	3		5
97.75	97.76	1.0225	12	6	2	1	0	2
98.82	98.82	1.0143	13	3	3	1	1	2
98.82		1.0143	9	9	5	3		4
100.62	100.62	1.0010	8	8	8	1	0	2
101.70	101.70	0.9933	13	5	1	1	1	2
101.70		0.9933	11	7	5	1		1
102.06	102.06	0.9907	14	0	0	0	1	0
102.06		0.9907	12	6	4	1		2
103.52	103.51	0.9808	14	2	0	0	0	0
103.51		0.9808	10	10	0	0		0
103.51		0.9808	10	8	6	0		5
104.96	104.97	0.9711	14	2	2	3	1	3
104.97		0.9711	10	10	2	2		1
107.58	107.54	0.9549	11	9	3	1	0	0
107.54		0.9549	9	9	7	0		0
107.92	107.91	0.9526	12	8	0	2	0	1
107.91		0.9526	14	4	4	1		1
109.42	109.42	0.9437	10	10	4	5	2	8
109.41		0.9437	14	4	2	1		2
109.41		0.9437	12	6	6	4		6
110.54	110.54	0.9372	13	5	5	1	1	2

2THETA	PEAK	D	H	K	L	I(INT)	I(PK)	I(DS)
110.54	110.54	0.9372	13	7	1	2		4
110.54		0.9372	11	11	1	2		3
112.43	112.44	0.9267	12	8	4	1	0	2
113.59	113.60	0.9206	15	1	1	1	1	1
113.59		0.9206	11	9	5	2		3
113.59		0.9206	13	7	3	0		0
113.97	113.98	0.9186	14	4	4	1	1	2
113.97		0.9186	10	8	8	0		0
115.53	115.52	0.9106	14	6	0	0		0
116.71	116.72	0.9048	15	3	1	1	0	2
117.11	117.12	0.9029	10	10	6	0	0	1
117.11		0.9029	14	6	2	0		0
119.92	119.94	0.8898	11	11	3	1	1	2
119.92		0.8898	13	7	5	1		1
119.92		0.8898	15	3	3	1		1
119.92		0.8898	9	9	9	0		0
120.33	120.36	0.8879	12	10	0	1	1	1
120.33		0.8879	12	8	6	1		2
121.98	122.02	0.8807	14	6	4	0	0	0
121.98		0.8807	12	10	2	1		1
123.24	123.24	0.8755	15	5	1	1	1	3
123.24		0.8755	11	11	5	2		2
123.24		0.8755	13	9	1	1		1
125.38	125.40	0.8669	16	0	0	0	0	1
126.69	127.14	0.8618	15	5	3	0	1	1
126.69		0.8618	13	9	3	0		1
127.13		0.8602	14	8	0	2		3
127.13		0.8602	12	10	4	5		8
128.93	128.94	0.8536	16	2	0	4	1	0
128.93		0.8536	14	8	2	2		8
130.30	130.30	0.8488	10	10	8	1		4
130.30		0.8488	16	2	2	1		1
130.77	130.88	0.8472	13	11	7	8	2	14
130.77		0.8472	14	6	6	0	1	5
134.12	134.14	0.8364	13	9	5	1	2	2

ZUNYITE - KAMB, 1960

2THETA	PEAK	D	H	K	L	I(INT)	I(PK)	I(DS)
134.12		0.8364	15	5	5	3		5
134.12		0.8364	15	7	1	9		17
134.62	134.66	0.8349	16	4	2	3	2	5
134.62		0.8349	14	8	4	2		4
136.64	136.62	0.8289	12	10	6	1	0	2
138.20	138.22	0.8245	15	7	3	2	0	4
138.20		0.8245	11	9	9	0		0
140.93	140.94	0.8173	16	4	4	5	1	9
140.93		0.8173	12	12	0	3		5
142.64	143.24	0.8131	17	1	1	0	0	0
142.64		0.8131	13	11	1	1		1
142.64		0.8131	11	11	7	1		2
143.23		0.8117	16	6	0	2		4
143.23		0.8117	12	12	2	1		2
145.66	145.68	0.8062	14	8	6	11	1	20
145.66		0.8062	14	10	0	0		0
145.66		0.8062	16	6	2	0		0
147.59	147.60	0.8021	15	7	5	2	1	3
147.59		0.8021	17	3	1	1		2
147.59		0.8021	13	9	7	3		6
147.59		0.8021	13	11	3	0		0
148.25		0.8008	14	10	2	2		4
148.25		0.8008	10	10	10	2		4
151.05	151.08	0.7955	12	12	4	2	0	4
153.32	153.36	0.7916	17	3	3	4	0	8
153.32		0.7916	15	9	1	2		3
154.12		0.7903	16	6	4	0		0
154.12		0.7903	12	10	8	0		1
160.54	160.60	0.7815	13	11	5	1	1	1
160.54		0.7815	17	5	1	18		36
160.54		0.7815	15	9	3	10		20

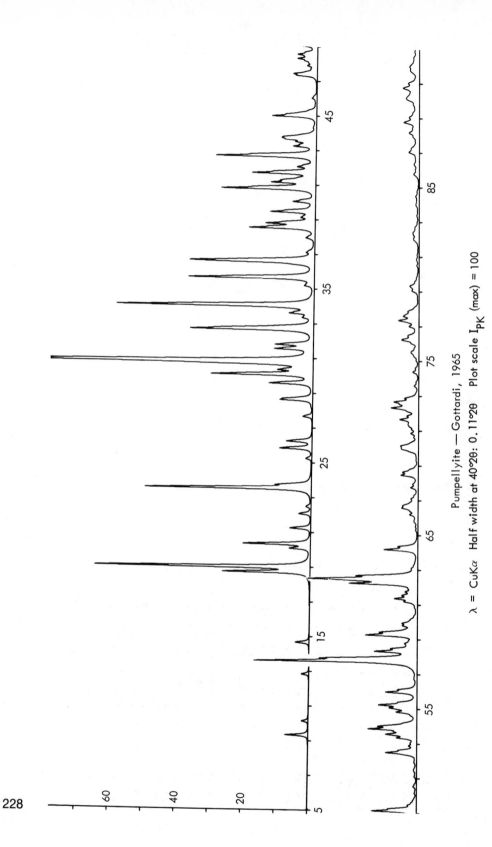

Pumpellyite — Gottardi, 1965

$\lambda = CuK\alpha$ Half width at $40°2\theta$: $0.11°2\theta$ Plot scale I_{PK} (max) = 100

228

PUMPELLYITE — GOTTARDI, 1965

2THETA	PEAK	D	H	K	L	I(INT)	I(PK)	I(DS)
9.31	9.32	9.486	2	0	0	5	7	4
10.12	10.12	8.733	0	0	1	2	2	1
12.83	12.84	6.895	-2	0	1	2	2	1
14.66	14.66	6.039	2	0	1	4	4	3
18.69	18.70	4.743	4	0	0	22	26	19
18.98	18.98	4.671	1	1	1	59	65	50
20.07	20.08	4.420	-4	0	1	5	6	4
20.32	20.32	4.366	0	0	2	18	20	16
21.23	21.24	4.182	-2	1	1	6	6	5
22.05	22.06	4.027	-3	1	1	3	4	3
22.46	22.48	3.954	4	0	1	2	2	1
23.51	23.52	3.781	2	0	2	49	50	44
23.73	23.72	3.746	3	1	1	7	11	7
25.23	25.24	3.527	-1	1	2	2	2	2
25.82	25.82	3.448	-4	0	2	10	10	9
26.23	26.24	3.395	1	1	2	8	8	7
27.65	27.66	3.224	-3	1	2	3	3	3
28.61	28.62	3.117	-5	1	1	9	10	9
28.70	28.68	3.108	-6	0	2	3	9	3
29.56	29.56	3.020	4	0	2	13	13	13
30.06	30.06	2.970	3	1	2	32	30	32
30.33	30.34	2.944	5	1	1	8	11	8
30.80	30.80	2.900	-2	0	3	100	100	100
30.89		2.892	-2	2	0	13		13
31.54	31.54	2.834	0	2	1	11	12	11
31.80	31.80	2.812	2	2	1	11	11	11
32.67	32.68	2.739	-6	0	2	40	37	41
33.34	33.34	2.685	2	0	3	3	4	3
33.60	33.60	2.665	2	1	3	7	7	7
34.03	34.04	2.632	-1	1	3	67	59	71
35.18	35.18	2.549	1	1	3	2	3	3
35.64	35.64	2.517	4	2	0	44	37	47
36.56	36.62	2.456	0	2	2	18	37	20
36.60		2.453	5	1	2	4		
36.63	36.62	2.451	-7	1	1	27	3	30
37.10	37.10	2.421	-2	2	2	2	3	2
37.94	37.94	2.369	-8	0	2	2	3	2

2THETA	PEAK	D	H	K	L	I(INT)	I(PK)	I(DS)
38.51	38.52	2.336	2	2	2	23	20	26
38.77	38.78	2.320	3	1	3	16	15	18
39.10	39.10	2.302	7	1	1			4
39.44	39.44	2.283	-5	1	3	16	13	18
40.03	40.04	2.250	-4	2	2	8	7	9
40.79	40.80	2.210	-8	0	2	34	28	40
41.15	41.16	2.192	-2	0	4	14	13	17
41.32	41.26	2.183	-2	2	3	6	11	7
41.69	41.68	2.165	0	2	3	23	19	27
42.05	42.04	2.147	6	2	1	6	6	7
42.66	42.66	2.117	-6	2	2	38	30	46
43.23	43.24	2.091	-4	2	4	8	8	9
43.49	43.50	2.079	0	2	3	5	5	7
43.65	43.66	2.072	-2	2	3	6	6	8
43.72	43.74	2.069	-2	2	4	6	10	7
43.81		2.065	-1	2	4	5		7
44.98	44.98	2.013	-6	2	2	17	13	21
45.20	45.10	2.004	-7	1	3	3	9	4
46.02	46.02	1.9706	-8	0	3	2	2	2
47.33	47.34	1.9191	-6	0	4	7	6	9
47.42	47.42	1.9154	-1	3	1	5	7	7
47.71	47.70	1.9046	-5	3	1	1	2	2
47.91	47.90	1.8972	0	0	4	3	3	4
48.09	48.20	1.8905	10	0	0	1	4	1
48.18		1.8869	9	1	1	3		4
48.32	48.32	1.8820	3	1	4	5	6	7
48.56	48.56	1.8732	6	2	2	7	6	9
49.12	49.12	1.8532	-8	2	0	17	14	22
49.15		1.8521	-8	2	1	2		3
49.46	49.46	1.8412	4	0	3	3	3	3
50.52	50.52	1.8051	10	0	1	1	1	2
52.38	52.50	1.7453	-9	1	3	3	9	4
52.49		1.7419	-7	1	4	10		14
52.50		1.7417	-5	3	3	1		2
52.88	52.88	1.7299	8	0	3	2	3	3
53.34	53.34	1.7161	5	1	4	7	6	10
53.55	53.54	1.7097	-4	2	4	9	9	13

PUMPELLYITE - GOTTARDI, 1965

2THETA	PEAK	D	H	K	L	I(INT)	I(PK)	I(DS)
53.55	53.86	1.7097	3	3	2	1	15	2
53.85	54.00	1.7010	5	1	1	19	12	27
53.97		1.6974	2	3	4	2		2
54.05		1.6953	-10	0	3	2		3
54.06	54.82	1.6948	6	0	5	4		6
54.06	55.06	1.6731	-3	1	1	7	6	11
55.06	55.22	1.6665	-11	1	1	9	8	12
55.24	55.96	1.6616	10	0	2	12	12	17
55.97	55.76	1.6416	-1	3	3	14	10	20
57.74	57.76	1.5953	5	3	1	8	49	11
57.76		1.5949	-7	2	4	63		94
57.76		1.5948	4	0	0	18	13	27
58.31	58.32	1.5810	12	0	5	3	3	2
58.64	58.64	1.5729	4	0	4	3	4	4
58.76	58.76	1.5700	-9	1	2	21	16	31
59.24	59.24	1.5585	-10	2	3	4		6
59.30		1.5571	3	3	3	3	12	5
59.44	59.40	1.5538	-12	0	2	4	5	6
59.78	59.80	1.5456	-5	3	1	1		2
59.81	59.92	1.5449	7	1	5	3	5	5
59.91	61.18	1.5425	10	2	2	3	4	5
61.17		1.5139	-2	2	6	2		3
61.17		1.5137	-7	1	0	8	7	12
61.35	61.34	1.5098	8	0	4	1	5	2
61.54	61.52	1.5055	0	2	5	1		2
61.59		1.5045	-8	0	5	1		4
61.74	61.74	1.5012	10	0	3	30	4	4
62.21	62.22	1.4909	-8	2	4	42	21	47
62.49	62.48	1.4850	0	4	0	5	33	67
62.51		1.4846	-4	2	5	5		9
63.09	63.08	1.4724	-1	3	4	1	2	2
64.17	64.18	1.4501	10	0	2	16	10	25
66.18	66.18	1.4109	0	2	6	3	3	5
66.44		1.4059	-12	0	2	1	5	9
66.55	66.56	1.4039	9	3	1	5		2
66.57		1.4036	3	3	1	1		2
66.67	66.68	1.4016	3	3	4	2	5	3

2THETA	PEAK	D	H	K	L	I(INT)	I(PK)	I(DS)
66.70	67.08	1.4011	-9	1	5	1	2	2
67.08	67.74	1.3940	-6	0	6	3	2	4
67.73	68.46	1.3822	2	4	4	8	5	6
68.45	68.62	1.3694	-12	0	3	3	5	14
68.60	69.98	1.3669	-11	1	3	5	5	6
69.94	70.02	1.3440	-13	1	6	1		8
69.97	70.12	1.3433	3	1	5	1		3
70.05	70.62	1.3421	-8	2	4	4	5	3
70.13	70.84	1.3407	-7	3	2	2	2	6
70.62	71.34	1.3326	4	4	4	3	2	3
70.85		1.3289	5	3	3	2	3	5
71.22		1.3228	0	4	3	3		5
71.33	71.64	1.3210	-2	2	6	2		4
71.40	72.32	1.3199	-12	2	6	1	6	2
71.65		1.3160	-2	2	2	8	8	15
72.22	72.66	1.3069	0	4	1	2		3
72.32	73.58	1.3055	-6	3	5	3		15
72.32	74.38	1.3054	-11	2	6	10	8	6
72.66	75.50	1.3001	-4	2	6	3	1	19
73.58		1.2861	-11	1	7	1	2	3
74.37	76.04	1.2744	2	0	4	1	2	6
75.35	76.24	1.2603	6	0	5	1		3
75.47	76.44	1.2586	-2	3	2	3		3
75.55	77.00	1.2574	-9	0	1	5	2	2
76.02	77.34	1.2508	-10	1	4	4	5	5
76.20	77.58	1.2483	14	0	5	1		9
76.26		1.2474	-15	2	7	1		7
76.54	77.86	1.2436	-12	2	4	2	3	3
76.99		1.2374	9	1	5	1	3	8
77.33	79.04	1.2329	14	2	0	2	6	8
77.35		1.2326	-8	4	2			11
77.59		1.2294	-2	1	7	6		2
77.65		1.2286	-1	1	4	1	5	2
77.70		1.2279	0	4	4	2		4
77.81		1.2264	-10	2	4	1		4
77.87		1.2257	-14	2	2	2		2
79.02		1.2107	-4	4	4	2	2	4

PUMPELLYITE – GOTTARDI, 1965

2THETA	PEAK	D	H	K	L	I(INT)	I(PK)	I(DS)
79.06		1.2101	8	2	5	2		3
79.32	79.30	1.2069	12	0	4	1	3	2
79.32		1.2068	-8	1	6	1		2
80.50	80.50	1.1922	7	1	6	1	1	2
80.86	80.86	1.1877	-14	2	3	1	1	4
81.97	82.02	1.1745	-6	4	4	2	2	2
82.38	82.38	1.1696	-8	0	7	1	1	2
83.94	84.04	1.1518	15	1	2	1	1	4
84.50	84.50	1.1456	11	3	6	2	1	5
85.75	85.76	1.1320	-13	3	3	1	2	3
86.25	86.34	1.1268	8	4	3	2	2	2
86.33		1.1259	-6	2	7	1		3
86.37		1.1255	2	2	5	1		3
87.01	87.04	1.1188	10	2	1	2	3	9
87.03		1.1187	5	5	4	1		3
87.20	87.20	1.1169	6	1	4	2	3	3
87.89	87.94	1.1100	9	4	6	1	2	2
87.92		1.1096	-16	2	1	1		2
88.16	88.16	1.1072	10	4	2	4	3	10
88.76	88.76	1.1013	-1	5	3	3	5	7
88.77		1.1012	16	2	0	6		14
89.76	89.80	1.0916	0	0	8	1	2	3
89.80		1.0912	8	2	6	2		3
90.04	90.14	1.0889	15	1	3	2	4	5
90.11		1.0883	-8	2	7	2		5
90.16		1.0878	4	2	7	2		5
90.26		1.0868	-7	5	1	2		5
90.73	90.74	1.0824	12	4	0	9	5	20
90.78		1.0819	-3	1	8	3		6
91.58	91.78	1.0746	3	5	3	1	3	3
91.69		1.0735	-12	4	2	1		3
91.71		1.0734	-17	1	3	2		5
91.81		1.0725	-15	3	1	2		5
91.96	91.98	1.0712	17	3	1	1	3	4
91.99		1.0708	-5	5	3	1		2
92.12		1.0697	-14	2	5	1		2
92.52	92.50	1.0661	9	3	5	3	2	6

2THETA	PEAK	D	H	K	L	I(INT)	I(PK)	I(DS)
93.36	93.36	1.0587	8	4	4	4	2	4
93.70	93.66	1.0557	10	4	3	2	2	4
95.39	95.40	1.0414	-10	2	7	2	1	3
96.09	96.10	1.0357	-16	2	4	1	2	3
97.49	97.74	1.0246	0	2	8	2		6
97.71		1.0229	2	4	6	1		3
97.93		1.0211	10	0	6	2		5
98.55	98.54	1.0163	-6	4	6	6	1	14
99.84	99.84	1.0067	-12	4	4	2	2	5
100.42	100.42	1.0024	-13	1	7	2	1	3
100.47		1.0021	-14	2	6	1		3
101.43	101.44	0.9951	-7	5	4	1		3
101.45		0.9950	18	0	2	1	1	3
101.73	101.76	0.9930	-19	1	3	1	1	1
102.09	102.32	0.9905	16	5	3	1		3
102.12		0.9903	5	5	0	1		2
102.16		0.9900	0	6	4	1		2
102.27		0.9892	-12	2	7	1		1
102.35		0.9887	8	4	5	3	2	9
102.71	102.72	0.9862	-8	2	8	1		1
102.77		0.9858	12	4	3	1		3
102.88		0.9850	-14	0	7	1		4
103.36	103.38	0.9818	-3	5	3	1	1	3
103.42		0.9813	9	3	6	1		3
103.55		0.9805	-11	5	1	1		4
104.11	104.10	0.9768	4	2	8	1	1	2
104.16		0.9764	-14	4	9	2		4
105.09	105.26	0.9703	0	4	3	2	1	4
105.27		0.9691	4	6	0	1		6
105.65	105.66	0.9667	15	5	3	2	1	5
105.79		0.9658	-13	3	6	1		2
106.17	106.18	0.9634	-2	6	3	1		3
106.42	106.68	0.9618	-3	3	8	2	1	6
106.57		0.9609	6	4	6	1		3
106.69		0.9601	-2	4	7	1		4
106.77		0.9596	-9	5	4	1		1
107.28	107.42	0.9565	7	1	8	1	3	2

PUMPELLYITE – GOTTARDI, 1965

2THETA	PEAK	D	H	K	L	I(INT)	I(PK)	I(DS)
107.38		0.9559	13	1	6	1		2
107.39		0.9558	-17	3	3	2		5
107.43		0.9556	-14	4	2	4		12
107.65		0.9542	17	3	1	2		
107.92	107.76	0.9526	2	0	9	1	2	4
108.58	108.58	0.9486	20	0	0	1	1	3
108.63		0.9483	7	3	7	1		3
108.96	109.04	0.9463	-7	5	5	1	1	2
109.15		0.9452	8	0	8	1		2
109.87	109.90	0.9410	6	2	8	1	1	3
109.93		0.9407	4	6	2	2		3
110.41	110.40	0.9380	16	2	4	3	1	4
110.65	110.66	0.9366	12	4	4	1	1	8
110.69		0.9364	-17	1	6	1		3
110.90		0.9352	-16	2	6	1		2
111.43	111.42	0.9322	14	0	2	1	1	4
111.64		0.9310	-6	6	9	1		3
112.30	112.44	0.9275	-4	2	3	1	1	1
112.46		0.9266	17	3	2	1		2
112.93	112.90	0.9241	-13	1	8	1	1	2
113.59	113.96	0.9206	8	4	6	1	1	3
113.92		0.9189	-8	4	7	1		4
113.97		0.9186	4	4	7	1		3
114.94	115.04	0.9136	8	6	0	1	2	4
115.04		0.9131	-12	2	8	2		5
115.07		0.9129	-18	0	6	4		10
115.64	115.48	0.9100	-19	1	1	1	2	2
115.65		0.9100	9	2	5	1		3
116.24	116.22	0.9071	2	2	8	2	1	3
116.24		0.9071	-11	5	3	1		3
116.73	116.72	0.9047	-13	3	7	1	1	5
117.17	117.18	0.9025	20	0	0	1	1	2
117.65	117.64	0.9003	-13	5	2	1	1	5
117.69		0.9001	3	5	6	1		2
118.15	118.14	0.8979	19	1	3	1	1	2
118.19		0.8977	-19	3	1	1		3
118.66	118.70	0.8955	-20	2	3	2	1	5

2THETA	PEAK	D	H	K	L	I(INT)	I(PK)	I(DS)
118.82	118.82	0.8948	-4	6	4	1		3
119.84	119.98	0.8901	-10	4	7	1	1	4
120.00		0.8894	14	2	6	1		4
120.32	120.40	0.8880	16	4	2	1	1	2
120.40		0.8876	15	1	6	1		3
121.12	121.18	0.8845	-2	0	8	1	1	4
121.21		0.8841	-20	5	5	1		3
121.63	121.76	0.8823	-11	1	7	1	2	2
121.71		0.8819	13	4	4	3		2
121.74		0.8818	-4	0	8	1		2
122.04	122.26	0.8805	-12	2	9	1	2	4
122.06		0.8804	-12	2	7	1		8
122.27		0.8795	0	4	8	2		3
122.71	122.86	0.8777	10	6	0	1	3	2
122.79		0.8773	10	4	6	1		34
122.87		0.8770	4	6	4	11		3
123.41	123.38	0.8748	21	1	1	3	3	10
123.78		0.8733	-20	2	10	2		5
124.23	124.42	0.8715	-3	0	10	2	2	2
124.38		0.8708	-10	6	3	4		14
124.48		0.8704	7	3	8	1		2
124.59		0.8700	13	3	6	1		3
124.72	124.94	0.8695	-15	5	1	2	2	6
124.92		0.8687	-20	0	3	2		6
125.06		0.8681	-22	0	2	1		2
125.09		0.8680	10	6	0	1		2
125.09		0.8680	7	1	9	1		2
125.26		0.8673	-21	1	1	1		2
125.26		0.8673	-5	5	4	1		2
126.38	126.40	0.8630	-1	5	2	1	1	2
126.43		0.8628	-2	6	6	1		3
126.68	126.94	0.8619	-16	6	5	1	1	5
126.89		0.8611	-18	4	8	1		3
127.38	127.62	0.8593	2	0	10	1	2	24
127.59		0.8585	-8	6	4	8		2
127.72		0.8580	1	1	10	1		

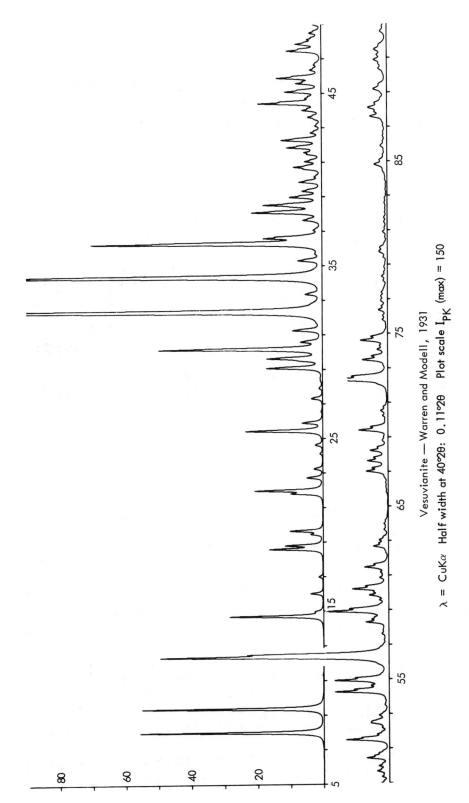

Vesuvianite — Warren and Modell, 1931

λ = CuKα Half width at 40°2θ: 0.11°2θ Plot scale I$_{PK}$ (max) = 150

VESUVIANITE – WARREN AND MODELL, 1931

2THETA	PEAK	D	H	K	L	I(INT)	I(PK)	I(DS)
7.98	7.98	11.073	1	1	0	25	37	13
9.35	9.36	9.450	1	1	1	25	37	13
14.68	14.68	6.029	2	1	0	14	19	9
15.99	16.00	5.537	2	2	0	2	3	1
18.56	18.56	4.777	3	0	0	8	11	6
18.76	18.78	4.725	3	1	1	6	8	4
19.41	19.42	4.569	3	2	0	2	3	1
19.61	19.62	4.523	2	2	2	5	7	4
21.77	21.78	4.078	4	0	0	11	14	8
21.95	21.96	4.045	1	4	1	3	3	3
22.69	22.70	3.915	4	1	2	1	3	2
23.20	23.20	3.830	3	0	3	1	2	1
24.59	24.60	3.617	3	1	1	13	16	11
25.40	25.42	3.503	2	2	2	4	4	3
25.87	25.88	3.440	4	1	3	2	2	2
27.28	27.28	3.266	5	0	0	10	11	10
29.05	29.06	3.071	4	1	0	10	11	9
29.61	29.60	3.015	3	0	2	31	33	29
30.14	30.14	2.963	4	2	1	4	4	3
30.57	30.56	2.922	1	1	4	6	6	5
31.23	31.24	2.862	4	1	3	76	100	76
32.30	32.30	2.769	4	3	0	24		24
32.31		2.768	5	2	2			
33.33	33.34	2.686	3	3	0	36	78	38
34.30	34.32	2.612	5	2	2	43		46
34.32		2.611	3	1	4	4		5
35.27	35.28	2.542	6	1	0	49	47	54
36.25	36.26	2.476	5	2	3	9	12	10
36.58	36.58	2.454	5	1	2	2	10	2
36.69	36.66	2.447	3	2	3	1	2	1
37.05	37.04	2.425	6	1	2	3	4	4
37.63	37.62	2.389	4	6	0	9	4	10
38.06	38.06	2.362	6	2	1	6	14	7
38.08		2.361	5	1	2			
38.41	38.50	2.342	4	5	3	10	12	12
38.51		2.336	4	1	4			
38.95	38.96	2.310	3	3	4	7	7	8

H	K	L	I(INT)	I(PK)	I(DS)	D	PEAK	2THETA
6	3	3	4	4	5	2.290	39.30	39.30
4	2	4	2	5	3	2.262	39.84	39.82
5	4	1	1		3	2.261		39.84
2	3	5	1	2	1	2.245	40.14	40.13
5	1	3	3	4	4	2.221	40.60	40.59
7	0	1	3	6	4	2.215	40.70	40.71
7	1	1	1	3	3	2.198	41.02	41.02
6	0	0	3	3	4	2.177	41.46	41.44
3	4	5	2	4	2	2.172	41.54	41.55
6	0	1	7	7	9	2.158	41.82	41.82
7	5	5	5	8	6	2.138	42.26	42.24
7	1	1	1		3	2.136		42.27
6	4	3	5	2	3	2.116	42.68	42.68
5	2	4	2	1	2	2.098	43.08	43.08
6	1	1	1	3	3	2.074	43.58	43.59
6	3	3	2	4	5	2.056	44.00	44.00
4	0	6	14	13	20	2.039	44.40	44.39
6	2	3	2	3	3	2.026	44.68	44.69
5	1	4	3	7	4	2.021	45.06	45.05
6	3	5	4		7	2.010		45.07
7	4	4	5	5	8	1.9897	45.56	45.55
5	3	1	10	9	14	1.9770	45.86	45.86
8	5	1	1	2	2	1.9584	46.34	46.32
8	0	3	7	7	11	1.9466	46.62	46.62
7	0	6	3	5	5	1.9150	47.44	47.43
2	2	4	1	3	2	1.8999	47.84	47.84
6	5	5	1		3	1.8899		48.10
5	0	5	2	2	3	1.8899	48.10	48.10
4	3	5	1	1	2	1.8763	48.48	48.47
5	1	5	2	2	4	1.8587	48.96	48.96
8	0	2	3	1	3	1.8457	49.34	49.33
8	2	1	1	1	4	1.8345	49.66	49.65
8	5	1	3	3	8	1.8239	49.96	49.96
7	1	6	2	2	3	1.8084	50.42	50.42
8	0	3	5	5	17	1.7993	50.68	50.69
7	1	4	2	9	8	1.7738	51.48	51.47
4	0	6	10	2	3	1.7633	51.80	51.80

VESUVIANITE - WARREN AND MODELL, 1931

2THETA	PEAK	D	H	K	L	I(INT)	I(PK)	I(DS)
52.43	52.46	1.7437	6	1	5	1	4	2
52.45		1.7430	8	1	3	2		4
52.53	52.52	1.7406	7	2	4	2		3
52.90	52.90	1.7294	9	1	0	2	4	3
54.26	54.26	1.6892	7	3	4	14	2	24
54.61	54.62	1.6791	8	0	2	3	11	5
54.91	54.92	1.6706	4	4	6	14	4	24
56.26	56.28	1.6338	5	3	6	23	11	42
56.28		1.6332	8	2	0	13	46	24
56.29		1.6328	9	2	4	25		46
56.29		1.6328	7	6	2	2		3
57.60	57.60	1.5988	8	2	4	1	1	2
58.28	58.28	1.5819	7	7	0	6	5	10
58.56	58.54	1.5749	6	0	6	2	3	4
58.88	58.90	1.5670	6	1	6	5	13	10
58.91		1.5664	6	6	4	11		21
59.16	59.06	1.5604	7	3	3	1	8	2
59.55	59.56	1.5510	7	5	4	2	2	3
59.85	59.84	1.5440	5	2	6	4	4	8
60.17	60.20	1.5365	6	4	6	1	7	2
60.21		1.5356	10	2	0	8		17
60.85	60.84	1.5210	9	5	0	2	2	4
61.46	61.46	1.5073	8	4	4	5	5	9
62.66	62.66	1.4813	7	0	8	3	3	8
67.01	67.00	1.3954	8	7	4	6	4	14
67.62	67.62	1.3842	8	3	0	4	4	9
68.22	68.22	1.3735	11	3	4	5	4	11
69.40	69.40	1.3531	9	5	8	9	6	21
70.52	70.52	1.3342	5	1	4	6	1	3
72.28	72.34	1.3060	4	4	8	6	13	14
72.33		1.3053	10	4	2	2		5
72.34		1.3051	11	4	0	12		28
72.35		1.3050	12	0	2	2		4
73.47	73.48	1.2878	9	1	6	7	5	18
74.07	74.06	1.2789	11	2	4	3	2	7
74.59	74.60	1.2712	6	2	8	8	6	20
75.79	75.80	1.2540	8	8	4	1	1	3

2THETA	PEAK	D	H	K	L	I(INT)	I(PK)	I(DS)
76.36	76.36	1.2461	11	3	4	2	2	5
76.85	76.92	1.2393	10	4	5	1	1	3
76.94		1.2381	12	3	3	1		3
77.36	77.36	1.2324	4	2	9	1	1	3
78.30	78.30	1.2200	7	2	8	1	1	3
79.71	79.76	1.2019	8	7	0	1	2	4
79.78		1.2011	13	1	0	1		3
80.89	80.88	1.1874	11	5	4	1	1	4
84.80	84.80	1.1423	12	3	4	1	3	4
84.81		1.1422	10	0	2	2		6
85.86	85.88	1.1308	8	4	8	1	1	4
87.58	87.58	1.1131	13	1	4	4	3	12
87.58		1.1131	11	7	4	3		9
88.05	88.10	1.1083	4	4	10	2	4	7
88.11		1.1077	13	3	6	6		17
88.64	88.62	1.1025	9	3	8	5	1	8
89.16	89.16	1.0974	5	2	10	5	3	15
89.25		1.0965	14	1	2	1		3
90.33	90.36	1.0862	9	8	6	1	2	3
90.37		1.0858	12	8	0	1		3
90.86	90.86	1.0812	7	7	8	4	3	15
91.47	91.46	1.0756	13	6	2	3	3	11
92.52	92.52	1.0661	10	2	8	2	1	7
94.79	94.80	1.0465	14	0	4	2	1	8
95.36	95.36	1.0417	15	1	0	1	1	4
95.91	95.90	1.0372	14	2	4	1	2	4
95.91		1.0372	10	10	4	2		7

TABLE 11. CLINOPYROXENES

Variety	Diopside	Jadeite	Acmite	Fassaite (Al-Augite)	Spodumene
Composition	$CaMg(SiO_3)_2$	$NaAl(SiO_3)_2$	$NaAl(SiO_3)_2$	$Ca(Mg,Fe^3,Al)(Si,Al)_2O_6$	$LiAl(SiO_3)_2$
Source	Gouverneur, N. Y.	Santa Rita Peak, New Idria, Calif.	Green River Form., Wyoming	Hessereau Hill, Oka, Quebec	Newry, Maine
Reference	Clark, Appleman & Papike, 1969; 1968	Prewitt & Burnham, 1966	Clark, Appleman & Papike, 1969; 1968	Peacor, 1967	Clark, Appleman & Papike, 1969; 1968
Cell Dimensions					
a Å	9.748	9.418	9.651	0.067	9.449
b Å	8.924	8.562	8.796	9.794	8.386
c	5.251	5.219	5.288	5.319	5.215
α	90	90	90	90	90
β deg	105.79	107.58	107.33	105.90	110.10
γ	90	90	90	90	90
Space Group	C2/c	C2/c	C2/c	C2/c	C2/c (C2)
Z	4	4	4	4	4

TABLE 11. (cont.)

Variety	Diopside	Jadeite	Acmite	Fassaite (Al-Augite)	Spodumene
Site Occupancy					
M2	Ca 1.0	Ca 0.02 Na 0.98	Na 1.0	Ca 0.975 Na 0.007 K 0.001 Mn 0.007	Li 1.0
M1	Mg 1.0	Mg 0.01 Al 0.99	Fe^3 1.0	Mg 0.570 Fe^2 0.063 Fe^3 0.159 Al 0.171 Ti 0.065	Al 1.0
T				Si 0.753 Al 0.247	
Method	3-D, counter	3-D, counter	3-D, counter	3-D, counter	3-D, counter
R & Rating	0.04 – 0.06 (1)	0.037 (1)	0.04 – 0.06 (1)	0.067 (1)	0.04 – 0.6 (1)
Cleavage and habit			{110} perfect Variable		
μ	179.6	105.0	208.8	241.4	98.9
ASF	0.1120	0.9844×10^{-1}	0.1033	0.1262	0.9839×10^{-1}
Abridging factors	0.5:0:0	0.5:0:0	0.3:0:0	0.5:0:0	0.5:0:0

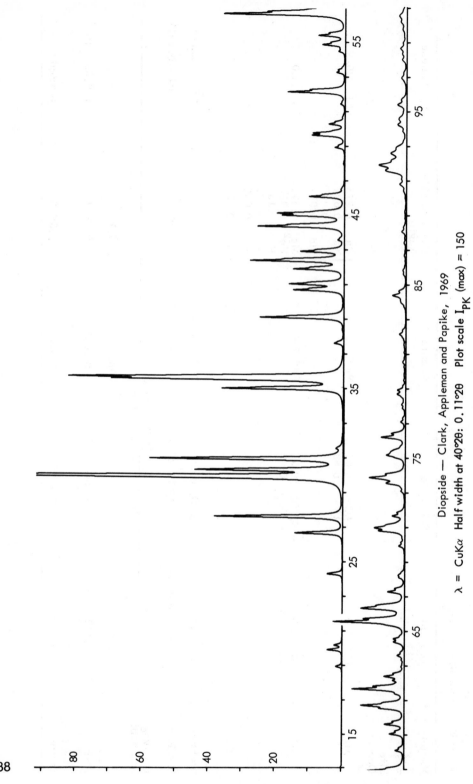

Diopside — Clark, Appleman and Papike, 1969

λ = CuKα Half width at 40°2θ: 0.11°2θ Plot scale I$_{PK}$ (max) = 150

DIOPSIDE — CLARK, APPLEMAN AND PAPIKE, 1969

2THETA	PEAK	D	H	K	L	I(INT)	I(PK)	I(DS)
18.90	18.92	4.690	2	0	0	1	1	1
19.88	19.88	4.462	0	2	0	1	1	2
20.15	20.16	4.402	-1	1	1	1	2	1
24.28	24.28	3.662	1	1	1	3	3	3
26.63	26.64	3.345	0	2	1	9	9	8
27.57	27.58	3.233	2	2	0	25	26	24
29.86	29.86	2.990	-2	2	1	100	100	100
30.26	30.26	2.951	-3	1	1	27	29	27
30.89	30.90	2.892	-3	1	0	39	38	39
31.52	31.52	2.836	-1	3	1	1	1	1
34.96	34.96	2.564	-1	3	1	25	24	27
35.45	35.50	2.530	-2	0	2	2	2	3
35.50		2.526	0	0	2	42	46	46
35.64	35.62	2.517	2	2	1	41	55	46
37.60	37.60	2.390	2	1	2	19	17	22
39.07	39.08	2.304	3	3	1	11	10	13
40.67	40.68	2.216	1	1	2	11	11	13
40.97	41.02	2.201	-2	0	2	21	19	14
41.02		2.198	0	2	2	10	9	26
41.88	41.88	2.155	3	1	0	21	17	12
42.36	42.36	2.132	-3	3	1	13	12	2
42.89	42.90	2.107	-4	2	1	8	13	26
43.56	43.56	2.076	0	4	0	9	7	17
44.35	44.34	2.041	2	2	2	2	1	11
44.95	44.98	2.015	-4	0	2	7	2	1
44.98		2.014	-2	0	2	3	7	1
45.12	45.10	2.008	-1	3	2	4	6	3
46.07	46.08	1.9684	-2	3	1	15	3	10
46.49	46.48	1.9515	3	1	1	2	1	5
48.91	48.92	1.8604	5	1	0	2	2	6
49.61	49.62	1.8359	-4	2	2	7	7	1
49.63		1.8353	2	2	2	1		5
49.75	49.74	1.8310	2	2	2	3	6	6
50.26	50.26	1.8137	4	1	1	4	3	1
51.42	51.42	1.7756	1	2	2	1	1	3
52.12	52.12	1.7533	4	5	0	15	11	22
53.25	53.26	1.7187	-5	1	2	2	2	3
53.41	53.40	1.7140	3	1	2	1	1	1
54.47	54.48	1.6832	-1	5	1	2	1	2
54.81	54.84	1.6734	-2	4	2	1	5	1
54.85		1.6723	0	4	2	5		8
55.38	55.38	1.6575	-5	1	3	7	5	11
56.60	56.62	1.6246	-2	3	1	16	24	25
56.64		1.6237	5	1	2	19		30
58.08	58.08	1.5868	6	0	3	2	2	4
59.03	59.04	1.5634	3	5	0	4	3	7
59.59	59.60	1.5500	-6	0	0	5	4	9
60.52	60.54	1.5285	-6	2	0	6	5	9
60.65	60.70	1.5256	4	0	2	3	4	5
60.70		1.5243	-5	3	1	7		12
61.37	61.38	1.5093	-1	3	2	2	2	5
61.64	61.64	1.5034	2	3	3	15	10	25
62.14	62.14	1.4924	0	4	3	2	2	4
62.38	62.38	1.4873	6	2	2	5	4	9
62.94	62.94	1.4754	-3	3	0	1	1	1
63.33	63.34	1.4673	-4	4	0	1	1	3
63.59	63.58	1.4620	-6	2	3	2	2	5
64.37	64.38	1.4460	5	3	1	3	3	3
64.55	64.54	1.4425	2	2	2	1	1	3
65.53	65.54	1.4232	-3	5	2	20	14	36
65.82	65.72	1.4178	1	6	1	1	8	2
66.30	66.30	1.4086	-7	5	0	11	9	19
66.37		1.4073	-2	5	2	5		9
67.27	67.26	1.3906	0	2	1	5	3	10
68.09	68.10	1.3759	-7	1	3	1	1	3
68.21	68.22	1.3736	-2	4	3	1	1	2
69.92	69.92	1.3442	0	4	3	2	2	4
70.78	70.80	1.3299	-7	1	2	7	5	14
70.93	70.96	1.3275	-5	2	1	5	6	9
70.99		1.3266	-4	1	2	1		2
71.08		1.3252	-3	3	0	1		3
71.65	71.64	1.3160	-5	4	3	4	3	8
73.08	73.08	1.2937	-4	4	3	1	1	2
73.52	73.54	1.2870	-3	1	4	4	4	8

DIOPSIDE — CLARK, APPLEMAN AND PAPIKE, 1969

2THETA	PEAK	D	H	K	L	I(INT)	I(PK)	I(DS)		2THETA	PEAK	D	H	K	L	I(INT)	I(PK)	I(DS)
73.59		1.2860	-1	1	4	2		5		92.61		1.0653	-5	7	1	5		12
73.84	73.86	1.2822	-2	6	2	4	7	8		92.75		1.0641	-9	1	2	5		1
73.88		1.2817	0	6	2	7		14		93.00	92.88	1.0619	7	1	1	1	2	3
73.97		1.2803	6	0	4	1		2		93.85	93.86	1.0544	-3	7	0	1	1	3
75.01	75.16	1.2651	-4	0	0	2	4	4		94.24	94.24	1.0511	8	2	4	4	2	9
75.14		1.2632	0	0	4	4		8		94.48	94.56	1.0491	-2	8	1	1	2	2
75.18		1.2628	-4	6	1	1		2		94.66		1.0476	3	1	5	2		4
75.29		1.2612	-7	1	1	2		3		95.38	95.38	1.0416	-3	3	2	1		3
76.20	76.20	1.2483	3	5	2	8	5	17		95.39		1.0415	5	3	3	3	2	9
77.08	77.08	1.2362	-1	7	1	2	1	3		96.71	96.96	1.0307	-5	7	2	1		2
77.88	77.88	1.2255	-7	3	1	1	1	2		96.95		1.0288	-1	7	3	2	1	6
78.08	78.08	1.2229	5	3	2	1	1	2		97.99	98.00	1.0207	-2	8	2	1		2
78.52		1.2171	-4	2	4	1	2	2		98.00		1.0206	2	2	2	1		3
78.68	78.68	1.2151	1	7	1	2		5		98.33	98.34	1.0180	2	4	4	3	2	2
78.92	78.92	1.2119	-7	1	3	1		3		98.64	98.64	1.0156	-9	3	1	1	1	8
79.49	79.48	1.2047	7	1	1	1	2	1		99.36	99.38	1.0103	3	3	4	3	1	2
80.54	80.54	1.1917	-3	3	4	1	0	1		99.75	99.76	1.0073	4	8	0	1	1	3
81.09	81.10	1.1849	1	5	3	1	1	1		100.22	100.22	1.0039	5	0	4	2	2	6
81.55	81.56	1.1793	-6	6	1	1	0	1		100.48	100.56	1.0020	5	7	1	2	1	4
82.13	82.14	1.1725	8	0	0	2	0	5		100.61		1.0011	1	1	4	3		5
83.94	83.98	1.1518	6	2	2	1	1	3		102.04	102.04	0.9908	6	2	5	1	1	7
84.08	84.08	1.1502	-6	2	3	2	1	3		103.00	103.04	0.9842	-2	6	4	1	1	2
84.38	84.38	1.1469	-6	4	4	4	3	9		103.09		0.9836	9	3	2	5	2	2
84.39		1.1468	-1	0	2	1		2		103.97	103.98	0.9777	-1	1	3	1		14
85.61	85.62	1.1335	-5	7	2	1	1	1		105.19	105.20	0.9697	-10	3	5	1	2	4
85.82	85.82	1.1313	-7	5	3	1	0	2		105.48	105.54	0.9677	8	0	2	1	1	4
86.69	86.68	1.1221	-5	5	3	1	0	1		106.07	106.10	0.9640	-9	3	3	4	2	11
87.40	87.40	1.1149	0	3	4	1		2		106.13		0.9637	-4	6	4	2		3
88.06	88.06	1.1082	2	6	3	2	1	5		106.33	106.32	0.9624	-5	7	3	3		6
90.43	90.42	1.0852	2	2	0	1	1	2		109.06	109.12	0.9458	8	2	2	2	2	6
90.80	90.80	1.0818	-8	8	3	3	1	6		109.51	109.50	0.9431	-4	8	2	2	0	2
91.53		1.0750	-2	2	1	5	4	14		110.12	110.12	0.9396	2	8	3	1	1	3
91.63		1.0741	8	5	0	5		13		110.14	110.44	0.9380	-2	0	1	2	1	7
91.91	91.90	1.0716	-7	5	3	8	6	19		111.14	111.12	0.9338	-10	0	5	1		3
92.06		1.0702	4	4	4	1		3		113.29	113.30	0.9221	-7	5	4	1	0	16
92.49	92.60	1.0664	3	1	4	1	3	3		113.30		0.9221	9	3	1	5	2	4
92.54		1.0660	-6	6	2	2		4		113.64	113.68	0.9203	3	5	4	1		4

DIOPSIDE — CLARK, APPLEMAN AND PAPIKE, 1969

2THETA	PEAK	D	H	K	L	I(INT)	I(PK)	I(DS)
114.20	114.28	0.9174	7	5	2	2	2	7
114.31		0.9168	2	2	1	2		11
114.57		0.9155	4	2	4	4		5
115.64	115.62	0.9101	-1	7	4	1	0	2
116.01	116.02	0.9082	-6	6	4	1	0	2
117.66	117.66	0.9002	-4	8	2	1	0	2
118.66	118.66	0.8955	-9	3	4	2	0	2
119.34	119.34	0.8924	0	10	0	4	1	7
119.81	119.82	0.8902	-10	4	1	1	1	3
120.50	120.50	0.8872	2	8	3	3	1	10
121.49	121.48	0.8829	-5	9	1	1	0	3
122.57	122.66	0.8782	9	1	2	1	1	2
122.65		0.8779	-8	2	5	3		11
122.91		0.8769	-11	1	1	1		3
123.17	123.10	0.8757	-10	0	4	1	1	2
124.19	124.32	0.8716	-4	0	6	2	1	6
124.27		0.8713	-2	10	1	1		4
124.34		0.8710	-3	1	6	1		3
124.56		0.8701	6	8	0	3		9
125.94	125.98	0.8647	10	4	0	2	1	6
126.62	126.60	0.8621	-10	4	3	1	1	6
126.71		0.8618	-1	9	3	1		3
128.05	128.06	0.8568	-5	1	6	1	0	2
129.95	129.98	0.8500	-2	8	4	1	0	3
130.29	130.30	0.8489	11	1	0	1	0	2
132.31	132.56	0.8421	0	6	6	1		11
132.52		0.8414	0	10	2	3		5
132.80		0.8405	1	9	3	1	1	6
133.62	133.52	0.8380	5	7	3	2	1	2
134.02	134.32	0.8367	-4	8	4	1		3
134.29		0.8359	0	6	5	4		5
134.89	134.94	0.8340	4	10	0	3	1	10
135.00		0.8337	-8	6	4	2		6
135.54	135.54	0.8321	4	6	4	3	1	10
137.27	137.36	0.8271	-11	3	3	1	1	4
137.64		0.8260	-1	6	1	1		4
138.27	138.28	0.8243	-9	7	1	1		4

2THETA	PEAK	D	H	K	L	I(INT)	I(PK)	I(DS)
138.95	138.94	0.8225	-8	8	1	1	1	5
140.33	140.88	0.8188	10	4	1	1	1	5
140.81		0.8176	-6	6	5	1		4
140.82		0.8176	-9	3	5	5		19
140.95		0.8173	-7	1	6	1		3
141.50	141.60	0.8159	-4	10	2	3	1	12
141.68		0.8154	-8	8	2	1		3
141.77		0.8152	-10	4	4	2		6
142.04		0.8145	6	8	2	1		3
143.80	143.80	0.8103	10	0	2	4	1	14
144.96	144.92	0.8077	-7	9	1	5	1	21
145.33		0.8069	9	7	0	1		3
146.86	147.50	0.8036	-1	1	5	2	1	6
147.46		0.8024	-11	1	5	2		10
152.19	152.62	0.7935	-12	2	1	8	2	32
152.25		0.7934	-10	6	2	2		7
152.68		0.7927	-5	7	5	7		30
152.69		0.7927	4	5	4	4		16
152.93		0.7923	-11	1	5	3		13
153.30		0.7916	9	1	2	1		2
153.44		0.7914	-10	3	3	1		5
153.61		0.7911	-7	6	6	1		4
155.02	155.22	0.7889	-6	4	3	1	1	3
155.22		0.7886	-8	8	4	3		10
155.28		0.7885	-3	9	4	1		5
155.43		0.7883	-1	9	4	1		3
155.50		0.7882	-11	3	4	1		4
155.71		0.7879	0	4	6	2	1	10
157.18	157.22	0.7858	-3	5	5	16	1	65
157.55		0.7853	3	11	0	1		4
158.33		0.7842	-12	2	3	7		29

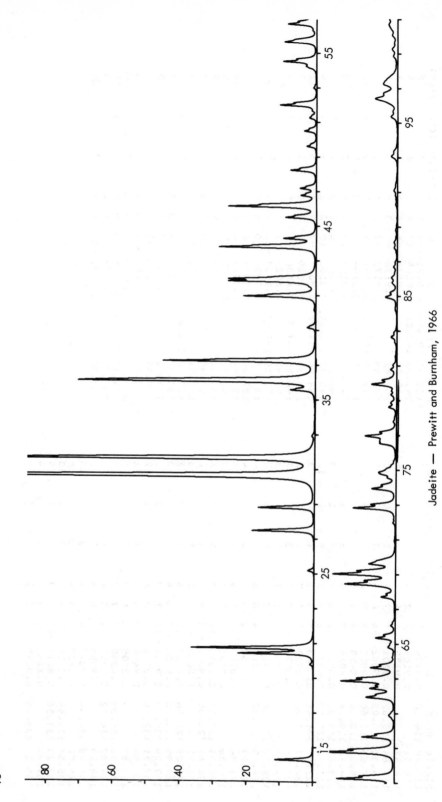

Jadeite — Prewitt and Burnham, 1966

$\lambda = CuK\alpha$ Half width at 40°2θ: 0.11°2θ Plot scale I_{PK} (max) = 150

242

JADEITE - PREWITT AND BURNHAM, 1966

2THETA	PEAK	D	H	K	L	I(INT)	I(PK)	I(DS)
14.28	14.30	6.196	1	1	0	6	8	5
19.76	19.76	4.489	2	0	0	1	1	0
20.41	20.42	4.347	-1	1	1	12	15	11
20.73	20.74	4.281	0	2	0	20	24	19
25.16	25.16	3.536	-1	1	1	1	1	1
27.46	27.46	3.245	0	1	1	12	12	12
28.79	28.80	3.098	2	2	0	11	11	11
30.62	30.62	2.917	-2	2	1	100	100	100
31.62		2.827	-3	1	1	30		30
31.64	31.64	2.825	3	1	0	44	72	44
35.55	35.56	2.523	-2	0	2	5	5	5
36.07	36.08	2.488	0	0	2	40	47	41
36.13		2.484	-1	3	1	17		18
37.22	37.22	2.414	2	2	1	32	30	34
39.14	39.14	2.300	1	3	1	2	2	2
40.95	40.96	2.202	3	1	1	16	14	17
41.82	41.82	2.158	1	3	1	18	17	19
41.97	41.94	2.151	0	2	2	12	17	12
43.78	43.78	2.066	-3	1	2	15	19	17
43.79		2.044	3	3	0	7		8
44.28	44.28	1.9924	-4	0	1	7	7	8
45.48	45.48	1.9879	-4	2	0	7	6	8
45.59	45.60	1.9662	4	0	1	1	4	1
46.13	46.12	1.9414	0	4	0	21	18	23
46.75	46.76	1.9255	2	0	2	3	3	4
47.16	47.16	1.8857	-1	3	2	4	3	4
48.22	48.22	1.8363	-2	4	1	6	5	7
49.60	49.60	1.8064	-5	1	1	2	2	3
50.48	50.48	1.7821	-4	2	2	3	3	3
51.22	51.24	1.7806	-3	3	1	1	1	1
51.26		1.7681	3	3	1	1		1
51.65	51.66	1.7574	2	1	2	1	7	1
51.99	52.00	1.7573	5	1	0	7	1	8
51.99		1.7269	1	3	2	2		2
52.98	52.98	1.6926	2	4	1	1	1	1
54.14	54.24	1.6900	-1	1	3	1	2	1
54.23			-5	1	2	2		2
54.51	54.50	1.6821	1	5	0	0	7	10
55.64	55.66	1.6503	-3	3	1	7	6	8
55.72		1.6484	3	1	2	2		3
56.32	56.32	1.6322	-2	4	1	1	6	1
56.68	56.68	1.6225	0	4	2	6	11	8
56.72		1.6216	-1	5	1	1		1
57.23	57.24	1.6082	-2	2	3	15	15	18
58.76	58.76	1.5700	-5	3	1	21	7	25
59.64	59.64	1.5490	4	4	0	9	1	11
60.81	60.88	1.5218	5	3	1	1	6	1
60.90		1.5198	5	1	2	1		1
61.91	61.92	1.4975	-6	0	0	6	5	7
61.96		1.4964	6	0	0	3		4
62.43	62.44	1.4863	3	5	0	6	4	8
62.56	62.60	1.4835	-4	5	1	1	11	1
62.85	62.86	1.4773	-1	3	3	15	8	18
62.93		1.4756	-5	3	2	1		1
63.02	63.02	1.4737	-6	2	1	7	5	2
63.62	63.62	1.4613	4	0	3	1	1	8
64.77	64.82	1.4380	2	4	2	1	4	2
64.84		1.4366	-5	1	3	1		1
65.34	65.34	1.4270	0	6	0	6	1	7
66.04	66.04	1.4135	-6	2	2	2	3	1
67.69	67.70	1.3830	4	2	2	4	10	5
68.32	68.44	1.3717	0	6	1	1	12	2
68.44		1.3696	-3	5	2	14		18
68.48		1.3689	3	5	0	1		1
69.00	69.02	1.3599	2	6	1	1	5	1
69.01		1.3597	5	3	2	18		23
69.10		1.3582	5	5	0	1		2
69.59	69.60	1.3497	1	5	3	7	1	9
69.70		1.3480	-2	4	3	3		4
71.97	71.96	1.3110	0	4	3	1	9	1
72.81	72.82	1.2979	-5	3	2	4	5	6
72.82		1.2977	-7	1	2	10		13
73.92	73.92	1.2810	-3	1	3	7		10
74.01		1.2797	3	1	3	1		1

JADEITE - PREWITT AND BURNHAM, 1966

2THETA	PEAK	D	H	K	L	I(INT)	I(PK)	I(DS)
74.56	74.56	1.2717	-1	1	4	2	2	2
74.72	74.86	1.2694	5	1	1	1	1	3
74.78		1.2684	7	1	0	2	4	2
74.86		1.2672	6	2	1	4		5
76.52	76.54	1.2438	0	0	4	4	3	6
76.65	76.66	1.2421	-2	0	2	3	3	4
76.96	76.96	1.2378	0	6	2	9	6	13
77.77	77.78	1.2270	-6	4	2	1	1	1
78.61	78.62	1.2159	-7	3	0	2	2	3
78.92	78.86	1.2119	1	7	1	1	1	1
79.32	79.32	1.2069	4	4	2	2	1	2
79.93	79.94	1.1992	3	5	2	9	5	12
80.78	80.78	1.1887	-1	7	1	2	1	3
82.64	82.64	1.1666	-1	7	1	5	1	3
84.94	84.94	1.1408	-6	0	4	1	2	7
85.78	85.78	1.1317	-8	2	1	1	1	2
86.07	86.06	1.1286	-6	4	3	1	1	2
86.68	86.68	1.1223	8	0	0	3	1	2
87.79	87.80	1.1109	-7	3	3	1	1	1
88.14	88.12	1.1074	-1	1	2	1	1	1
88.80	88.80	1.1009	6	2	2	1	0	1
90.27	90.28	1.0867	-4	4	4	1	0	1
90.81	90.82	1.0817	0	6	3	4	1	2
91.09	91.10	1.0791	2	2	1	3	1	4
93.05	93.06	1.0615	-8	2	3	3	1	5
95.34	95.34	1.0419	-7	5	2	1	1	2
96.21	96.40	1.0348	3	1	4	4	5	6
96.35		1.0336	-2	8	1	4		6
96.41		1.0331	-3	6	4	3	3	5
96.42		1.0300	-6	4	2	2	3	1
96.80	96.70	1.0300	4	4	3	6		9
97.24	97.26	1.0266	7	5	0	3		5
97.40		1.0253	-5	1	1	2	1	3
98.85	98.86	1.0141	7	1	2	1	1	1
99.27	99.26	1.0109	5	7	0	1	1	2
99.30		1.0107	-2	2	5	2	0	1
100.07	100.08	1.0050	2	8	1	2		4

2THETA	PEAK	D	H	K	L	I(INT)	I(PK)	I(DS)
100.45	100.70	1.0022	-5	1	5	1	2	1
100.55		1.0014	8	2	1	1		2
100.71		1.0003	5	3	3	3		5
101.02	101.02	0.9981	-1	7	4	2	1	3
102.31	102.32	0.9890	2	3	1	1		1
102.34		0.9888	3	7	1	1	1	1
103.98	103.98	0.9776	-9	3	2	2	1	3
104.32	104.38	0.9753	-8	4	0	1	1	1
104.51		0.9741	1	4	5	1	1	2
105.03	105.00	0.9707	-6	5	3	1	1	1
106.17	106.18	0.9634	-1	3	4	4	1	7
106.70	106.58	0.9600	-1	7	3	3		1
106.82		0.9593	5	7	1	2	1	3
108.12	108.36	0.9514	-5	3	5	1	1	2
108.36		0.9499	-6	2	3	2		3
109.18	109.20	0.9451	-4	6	4	1	1	2
109.65	109.66	0.9423	-9	3	3	3	1	5
110.05	110.04	0.9400	-10	0	0	1	1	3
110.60	110.56	0.9369	-5	7	2	2		4
114.05	114.06	0.9181	-10	2	1	1	0	2
115.46	115.48	0.9109	-2	8	3	3	1	3
116.33	116.34	0.9066	3	5	0	3	1	1
117.09	116.80	0.9030	-7	4	2	4	1	6
118.17	118.20	0.8978	10	0	1	2	0	2
118.78	118.84	0.8949	-7	2	7	2		1
118.85		0.8946	3	2	5	5	1	8
119.64	119.62	0.8910	-2	6	6	4		3
119.78		0.8904	3	6	4	3	1	1
120.89	121.20	0.8855	-6	4	2	2	1	1
120.96		0.8852	4	4	7	0		3
121.21		0.8841	3	7	0	4		6
122.84	122.80	0.8771	7	5	4	5	1	2
124.29	124.30	0.8712	-8	2	5	2	1	7
125.18	124.82	0.8677	2	0	6	4	1	1
125.84	125.82	0.8651	-4	1	6	1	0	3
127.00	126.98	0.8607	-10	4	2	1	0	2

JADEITE - PREWITT AND BURNHAM, 1966

2THETA	PEAK	D	H	K	L	I(INT)	I(PK)	I(DS)
128.21	128.48	0.8562	0	10	3	2	1	1
128.50		0.8552	-2	8	1	4		8
130.30	130.30	0.8488	-5	9	1	2		4
131.80	131.82	0.8438	0	10	1	1	0	1
132.59	132.64	0.8412	2	4	5	1	1	3
132.68		0.8409	-6	0	6	1		2
133.56	133.34	0.8381	9	1	2	1	0	1
133.89	133.86	0.8371	-2	10	3	1	0	3
135.16	135.20	0.8332	-1	9	3	2	1	4
135.24		0.8330	6	8	1	2		4
136.53	136.98	0.8292	0	0	6	1		1
136.97		0.8279	10	4	0	4	1	7
139.26	139.26	0.8216	-8	4	5	2	0	4
141.10	141.36	0.8169	-8	6	4	1	0	3
141.41		0.8161	-4	8	4	1		2
142.11	142.16	0.8144	-7	1	6	1	1	2
142.17		0.8142	-1	3	6	1		3
142.88	142.98	0.8125	11	1	0	1	1	3
142.97		0.8123	-11	3	1	1		5
143.83	144.52	0.8103	5	9	1	3		4
144.13		0.8096	0	10	2	2		7
144.52		0.8087	-9	3	3	4		3
145.43	145.46	0.8067	8	2	3	2	1	6
145.97	146.24	0.8055	-11	3	3	3	1	1
146.06		0.8053	-6	6	5	1		3
146.31		0.8048	-7	7	4	2		4
146.43		0.8045	5	7	3	2		10
147.35	147.28	0.8026	4	6	4	5	1	3
148.53	148.68	0.8002	1	1	6	1	1	5
148.66		0.8000	4	10	0	2		9
148.89		0.7995	-10	4	4	5		3
152.75	153.80	0.7926	-9	7	1	2	1	3
153.56		0.7912	4	8	3	1		2
153.84		0.7908	-8	0	6	3		7
153.84		0.7908	-8	8	2	1		1
155.93	156.76	0.7876	-8	8	1	1	1	1
156.56		0.7866	-4	10	2	5		10
157.82	157.78	0.7849	-1	0	5	2	1	5
157.88		0.7848	-12	7	2	1		1
157.95		0.7847	11	3	1	1		3
159.54		0.7827	-6	4	1	1		3
159.95		0.7822	10	4	4	2		4
162.51	163.26	0.7793	-7	1	7	2	0	3
163.41		0.7784	-5	5	5	10		21
164.22		0.7776	-8	2	6	1		2

245

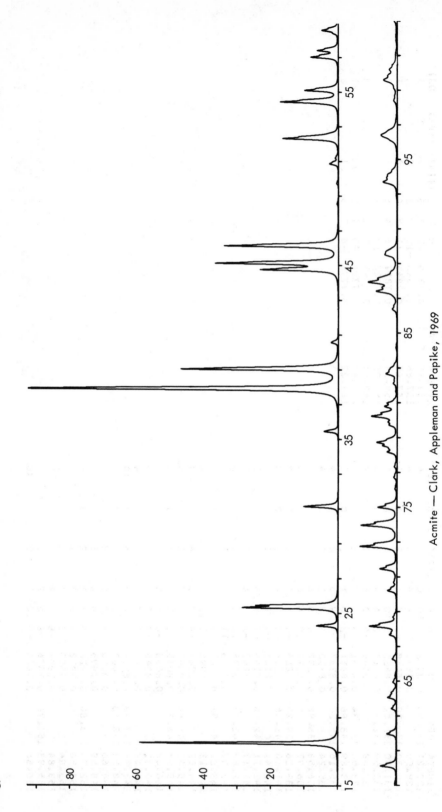

Acmite — Clark, Appleman and Papike, 1969

λ = FeKα Half width at 40°2θ: 0.11°2θ Plot scale I_{PK} (max) = 100

246

ACMITE — CLARK, APPLEMAN AND PAPIKE, 1969

2THETA	PEAK	D	H	K	L	I(INT)	I(PK)	I(DS)
17.50	17.52	6.362	1	1	0	43	59	28
24.26	24.26	4.606	-2	0	0	5	6	4
25.29	25.30	4.422	-1	1	1	21	29	16
25.43	25.44	4.398	0	2	0	17	25	13
31.11	31.12	3.609	1	1	1	9	10	8
35.43	35.44	3.181	-2	2	0	9	4	4
37.88	37.88	2.982	-2	2	1	100	100	100
39.01	39.02	2.899	3	1	0	47	47	48
40.54	40.54	2.794	-1	3	0	2	1	2
44.73	44.74	2.544	-1	1	2	24	23	28
45.10	45.10	2.524	0	0	2	38	37	45
46.11	46.12	2.472	2	2	1	37	34	44
48.52	48.52	2.356	1	3	1	1	0	1
49.70	49.70	2.303	4	0	0	1	1	1
50.84	50.84	2.255	-3	1	1	3	3	4
51.18	51.18	2.241	-3	3	1	1	1	1
52.23		2.199	0	4	1	1		
52.30	52.30	2.196	1	1	2	19	17	25
52.49	52.40	2.189	0	2	2	3	12	4
54.32		2.121	3	3	0	2		2
54.42	54.42	2.117	-3	1	2	20	17	28
55.09	55.08	2.093	-4	0	1	12	10	17
57.00	57.00	2.029	0	4	2	10	8	14
57.40	57.40	2.016	-2	2	2	7	6	11
58.57	58.56	1.9790	2	4	0	6	5	8
59.08	59.08	1.9632	-1	3	2	6	1	1
60.09	60.10	1.9333	-2	4	1	6	2	9
61.95	61.94	1.8809	-5	1	1	4	2	6
63.40	63.40	1.8421	-4	2	2	2	1	6
64.04	64.04	1.8256	3	3	2	3	3	3
64.32	64.32	1.8183	-3	3	2	2	0	6
64.87		1.8047	2	2	3	0		1
64.92	64.92	1.8034	5	1	0	1	1	2
65.31	65.30	1.7940	1	5	1	1	2	1
66.26	66.26	1.7711	1	3	3	2	3	3
67.65	67.66	1.7390	2	2	1	3	5	5
68.14	68.14	1.7280	4	5	0	12	8	20

2THETA	PEAK	D	H	K	L	I(INT)	I(PK)	I(DS)
68.66	68.66	1.7163	-1	1	2	2	2	2
70.19	70.20	1.6836	-3	1	2	1	3	7
70.66	70.66	1.6740	-3	1	3	4	1	2
71.44	71.44	1.6580	0	2	3	2	5	14
72.73	72.72	1.6326	-2	2	3	7	11	31
73.95	73.94	1.6094	-5	1	1	16	11	32
74.98	74.98	1.5905	4	4	0	9	6	17
76.04	76.04	1.5716	0	4	3	2	1	2
76.69	76.70	1.5603	5	1	1	3	1	3
77.30	77.30	1.5498	1	1	3	2	3	2
78.16	78.16	1.5355	-6	0	0	3	4	9
78.52	78.52	1.5296	-3	3	2	2	6	11
78.71	78.70	1.5265	3	5	0	6	2	13
78.80		1.5250	-3	5	1	1		6
79.70	79.72	1.5106	-6	2	2	3	3	4
79.86	79.88	1.5081	-4	2	3	2	8	12
80.21	80.20	1.5027	-1	3	0	6	4	3
80.76	80.76	1.4941	-4	2	2	1	3	1
80.96	80.96	1.4910	3	5	1	4	0	4
81.66	81.66	1.4806	-3	1	3	2	1	2
82.10	82.10	1.4740	0	3	3	4	3	4
82.64	82.66	1.4660	-5	1	2	2	2	2
82.75		1.4645	-6	0	3	4		6
82.99	82.84	1.4610	4	4	1	1	6	27
84.14	84.14	1.4447	-3	2	3	6	9	39
86.35	86.36	1.4147	5	2	2	3	6	7
87.38	87.38	1.4013	1	5	2	9	2	2
87.91	87.92	1.3946	1	3	4	7	4	15
88.27	88.12	1.3901	2	4	3	3	4	8
88.51	88.46	1.3871	-2	4	3	2	4	24
89.49	89.50	1.3751	-2	6	1	6	3	13
89.64	89.64	1.3733	-7	1	2	3	4	12
93.68	93.68	1.3270	-5	3	0	10	4	24
94.11	94.10	1.3223	-6	2	3	5	3	13
96.27	96.38	1.2997	-3	1	4	7	5	12
96.40		1.2984	-4	4	4	2		18
98.39	98.40	1.2789	-4	0	4	4	1	6

247

ACMITE — CLARK, APPLEMAN AND PAPIKE, 1969

2THETA	PEAK	D	H	K	L	I(INT)	I(PK)	I(DS)
99.12	99.16	1.2719	-2	6	2	2	4	4
99.56	99.56	1.2677	0	6	0	2	4	27
100.18	100.18	1.2620	0	0	4	10	2	14
100.86	100.86	1.2557	0	4	2	5	1	4
101.85	101.88	1.2469	-6	3	2	1	0	1
103.02	103.12	1.2367	-7	3	0	1	1	3
103.12		1.2358	4	6	1	2		7
104.03	104.04	1.2281	4	4	2	10	4	29
104.48	104.30	1.2244	-3	5	2	3	2	2
104.98	104.98	1.2203	-1	5	1	1	1	10
105.88	105.86	1.2130	0	7	1	3	0	1
106.37	106.36	1.2091	2	2	4	1	0	2
107.85	107.84	1.1977	1	7	1	4	1	12
108.85	108.86	1.1901	-8	0	2	1	0	2
113.04	113.30	1.1605	-8	2	1	3	2	9
113.29		1.1589	-6	6	0	6		18
113.58		1.1570	6	4	4	2		2
114.40	114.40	1.1516	8	0	0	7	1	7
117.07	117.40	1.1349	-1	7	2	3	1	3
117.20		1.1341	-2	0	4	1		1
117.34		1.1332	-5	5	3	0		4
117.42		1.1328	-2	4	2	1		2
118.29	118.28	1.1276	6	2	4	2	0	1
120.85	120.86	1.1130	1	3	1	0	0	8
122.08	122.26	1.1063	3	7	4	2	1	2
122.23		1.1055	-4	4	4	0		3
122.39		1.1047	-3	7	4	1		21
123.64	123.64	1.0982	2	2	4	6	0	8
124.35	124.32	1.0945	0	2	4	2	0	32
126.56	126.58	1.0837	-8	2	3	9	1	35
127.63	127.62	1.0787	-7	1	2	3	1	1
130.20	130.22	1.0672	-7	5	1	10	2	23
131.61	131.64	1.0612	-2	8	0	0	2	54
131.82		1.0603	-6	6	1	0		19
132.29	132.22	1.0584	-6	6	5	6	3	
133.42	133.84	1.0538	7	5	0	15	6	
133.45		1.0538	3	1	4	5		

2THETA	PEAK	D	H	K	L	I(INT)	I(PK)	I(DS)
133.69		1.0528	4	4	3	4		2
133.81		1.0523	-5	7	1	9		33
133.86		1.0521	-3	5	4	8		30
137.42	137.50	1.0389	7	5	2	6	2	23
137.63		1.0381	5	7	0	3		10
139.47	139.52	1.0319	2	8	1	7	1	26
139.87	140.00	1.0305	6	4	1	1	1	4
139.89		1.0305	-1	1	5	1		2
140.01		1.0301	2	6	3	1		4
140.85	140.86	1.0274	8	2	1	6	1	19
141.53	141.64	1.0252	-6	4	4	0	2	1
141.60		1.0250	-2	2	5	6		22
142.37	142.38	1.0226	5	3	3	13	3	49
142.46		1.0224	-5	7	3	3		4
143.01	142.98	1.0207	-1	7	3	7	3	27
143.55		1.0191	-7	3	4	1		3
145.46	145.30	1.0137	3	7	2	7	1	9
145.72		1.0130	-9	1	3	2		1
149.75	149.78	1.0027	-9	3	1	0	1	40
151.76	152.46	0.9982	-4	8	1	10	1	7
152.18		0.9972	6	6	2	2		2
152.41		0.9967	-8	4	3	0		11
153.71	154.28	0.9940	-6	6	3	3	1	1
154.20		0.9931	1	5	4	0		33
154.61		0.9922	4	8	0	8		2
156.07	155.92	0.9895	5	7	4	0	1	30
158.81	160.26	0.9848	0	2	1	7	1	48
159.30		0.9840	-1	7	5	1		5
160.53		0.9821	-7	3	3	9		36
160.89		0.9816	-2	6	4	0		2
163.46	163.36	0.9782	-1	3	5	30	1	128

248

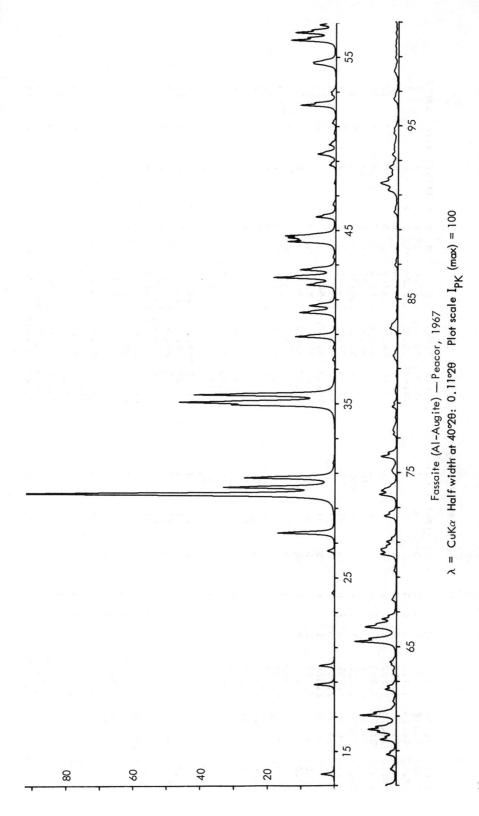

Fassaite (Al-Augite) — Peacor, 1967

λ = CuKα Half width at 40°2θ: 0.11°2θ Plot scale I_{PK} (max) = 100

249

FASSAITE (AL-AUGITE) - PEACOR, 1967

2THETA	PEAK	D	H	K	L	I(INT)	I(PK)	I(DS)
13.67	13.68	6.471	1	1	0	3	4	4
18.83	18.84	4.710	2	0	0	5	6	4
19.92	19.92	4.453	0	1	1	4	5	3
24.10	24.10	3.689	0	1	1	1	1	1
26.52	26.52	3.359	-2	2	0	2	2	2
27.54	27.54	3.236	-2	2	0	17	17	16
29.74	29.74	3.002	3	1	1	100	100	100
30.15	30.16	2.961	3	1	1	31	33	32
30.70	30.70	2.909	-3	1	1	27	27	28
34.90	34.90	2.568	-1	1	1	28	31	32
35.01	35.04	2.561	-2	0	2	1	46	1
35.05		2.558	0	0	2	42	42	48
35.48	35.48	2.528	2	2	1	44	42	51
37.52	37.52	2.395	-1	3	1	1	1	1
38.18	38.18	2.355	4	0	0	1	1	1
38.84	38.84	2.316	3	1	1	14	12	17
40.24	40.24	2.239	-1	1	2	12	11	16
40.60	40.60	2.220	-2	2	2	1	8	1
40.64	40.64	2.218	0	0	2	8		11
41.84	41.84	2.157	3	3	0	10	9	13
42.26	42.26	2.137	-3	3	1	21	19	29
42.71	42.72	2.115	-4	2	1	12	11	17
43.43	43.44	2.082	4	2	0	1	1	1
44.33	44.34	2.042	0	4	0	16	14	23
44.55	44.56	2.032	-4	0	2	13	14	19
44.67	44.66	2.027	2	0	2	9	15	14
45.77	45.76	1.9809	-1	3	2	7	6	11
46.47	46.46	1.9526	-2	4	1	1	1	2
47.72	47.72	1.9044	-5	1	1	1	1	1
48.76	48.76	1.8659	3	3	1	3	2	4
49.36	49.36	1.8447	2	2	2	2	5	4
49.41	49.40	1.8431	5	1	1	5		8
49.94	49.94	1.8247	-1	3	2	3	2	4
50.59	50.60	1.8026	2	4	1	1	1	1
51.18	51.18	1.7833	4	2	1	1		2
52.22	52.22	1.7502	-1	5	0	14	11	24
52.83	52.82	1.7315	-5	1	2	1	1	2
52.96	52.96	1.7274	3	1	2	1	2	2
54.49	54.62	1.6824	-1	5	1	2	7	3
54.60		1.6794	-0	2	4	6		11
54.69	54.68	1.6769	-3	1	3	5	7	10
55.98	55.98	1.6413	-2	1	3	18	14	34
56.42	56.42	1.6295	-5	3	1	16	12	29
56.86	56.86	1.6178	4	2	1	6	5	11
57.93	57.92	1.5906	5	1	0	2		4
58.77	58.76	1.5699	6	0	0	4	3	8
59.63	59.62	1.5493	3	1	2	7	5	13
60.06	60.06	1.5391	-6	0	2	6	6	12
60.22	60.22	1.5354	-4	0	2	8	8	15
60.38	60.38	1.5318	-5	3	1	3	6	7
61.02	61.02	1.5173	-5	1	2	1		2
61.03	61.84	1.5170	-1	3	3	15	11	30
61.85	62.52	1.4988	2	4	2	2	3	2
62.52	62.70	1.4843	0	6	0	5	3	10
62.72	63.44	1.4801	-3	1	3	1	3	1
63.43	63.58	1.4653	4	4	1	1	1	2
63.58	63.96	1.4622	-1	3	3	1	1	4
63.94	64.10	1.4546	-6	2	2	2	1	4
64.10	65.28	1.4516	-4	2	2	2	2	4
65.28	65.44	1.4281	5	3	1	19	13	41
65.41	66.10	1.4255	0	6	2	2	8	9
66.15		1.4124	-3	5	2	11	9	24
66.57	66.56	1.4113	1	5	2	5		10
67.67	67.68	1.4035	2	2	3	6	4	13
67.74		1.3834	-2	4	3	2	1	4
69.29	69.36	1.3821	-7	1	1	1		1
70.29	70.30	1.3538	0	4	3	3	1	3
70.42		1.3380	-7	3	2	7	5	17
70.46	70.58	1.3359	1	3	3	1		1
70.59		1.3352	3	1	3	1	1	1
70.75	70.76	1.3332	6	2	1	5	5	2
71.02	71.00	1.3305	7	1	0	1		12
51.18	71.18	1.3262	-5	3	3	4	3	3
52.22	72.50	1.3030	-3	1	4	4	4	11

FASSAITE (AL-AUGITE) – PEACOR, 1967

2THETA	PEAK	D	H	K	L	I(INT)	I(PK)	I(DS)
72.51		1.3024	-4	4	3	2		1
72.53		1.3021	-1	1	4	1		5
73.71	73.74	1.2842	-2	6	2	3	5	7
73.74		1.2838	0	6	2	6		15
73.95	73.96	1.2806	-4	0	4	2	5	5
74.07		1.2789	0	0	4	4		9
74.98	75.10	1.2656	-7	3	1	1	2	3
75.09		1.2640	-4	4	2	1		1
75.16		1.2630	-4	6	1	1		1
75.92	75.92	1.2522	3	5	2	8	5	21
77.19	77.20	1.2348	-1	7	1	2	1	5
77.43	77.44	1.2315	-7	3	2	1	1	1
77.49		1.2307	-4	2	4	1		2
77.60	77.58	1.2292	5	3	1	1	1	2
78.17	78.18	1.2217	-7	1	3	1	1	2
78.77	78.78	1.2139	1	7	1	3	1	7
80.56	80.56	1.1914	-8	2	0	1	0	2
81.56	81.58	1.1793	-8	0	4	1	1	3
81.72	81.72	1.1774	-8	0	2	2	1	6
83.29	83.30	1.1591	-6	0	4	4	2	11
83.37		1.1582	6	4	2	1		2
83.46		1.1572	-6	7	3	1		3
84.31	84.30	1.1477	-1	2	2	1	0	2
85.07	85.08	1.1394	-5	5	3	1	1	2
86.94	86.94	1.1195	-2	5	4	2	1	5
90.00	90.00	1.0893	-8	2	3	3	1	9
91.25	91.28	1.0776	-7	5	3	5	3	15
91.30		1.0771	3	1	4	1		3
91.34		1.0768	4	4	3	4		3
91.68	91.68	1.0737	-2	8	1	4	5	15
91.68		1.0737	7	1	0	6		21
92.26	92.30	1.0684	-6	6	2	2	2	4
92.32		1.0679	7	1	1	1		5
92.59	92.60	1.0655	-5	7	1	4	3	15
93.32	93.32	1.0590	-3	5	4	4	2	13
93.81	93.84	1.0548	-3	1	5	5	1	1
93.86		1.0544	5	7	0	1		3
93.96	93.96	1.0535	8	3	1	1	1	4
94.56	94.56	1.0484	5	8	1	3	1	10
94.78	94.80	1.0466	-2	2	1	1	1	5
96.43	96.54	1.0330	-1	7	5	3	1	4
96.55		1.0320	-9	3	3	3		9
98.15	98.16	1.0194	-9	3	3	3	1	9
98.94	98.94	1.0134	4	0	1	2	1	8
99.60	99.60	1.0084	1	5	4	2	1	6
99.87	99.88	1.0064	4	8	0	2	1	3
100.38	100.38	1.0027	5	5	3	1	1	6
101.08	101.10	0.9976	6	7	3	2	1	9
101.24		0.9965	1	2	5	2		2
102.37	102.38	0.9886	-1	3	5	4	2	17
104.41	104.44	0.9747	-10	0	2	1	1	6
104.67	104.76	0.9731	8	8	2	1	1	2
105.17	105.18	0.9698	-9	3	3	3	1	12
105.20		0.9696	-4	6	4	1		3
105.38		0.9684	-5	3	5	1		3
105.85	105.80	0.9654	-5	7	3	2	1	8
108.24	108.28	0.9506	8	2	2	1	0	2
108.82	108.80	0.9472	4	6	3	1	0	4
109.72	109.74	0.9419	10	0	0	1	1	2
109.76		0.9417	-2	8	3	2		7
112.15	112.50	0.9283	-7	5	4	4	2	18
112.44		0.9267	3	5	4	1		5
112.45		0.9266	2	5	5	3		15
112.56		0.9260	5	7	2	1		2
112.63		0.9257	9	3	1	1		5
112.85		0.9245	7	7	0	2		3
113.25	113.22	0.9224	4	4	4	2	1	7
113.55	113.54	0.9208	7	5	2	2	1	8
114.94	114.90	0.9136	-6	6	4	1	0	6
119.15	119.24	0.8932	7	1	1	1	0	3
119.73	120.00	0.8906	-5	4	0	2	1	3
120.00		0.8894	2	8	3	3		8
120.64	120.60	0.8866	-8	2	5	3	1	13
121.50	121.56	0.8828	-4	0	6	1	1	14
								7

FASSAITE (AL-AUGITE) - PEACOR, 1967

2THETA	PEAK	D	H	K	L	I(INT)	I(PK)	I(DS)
121.55		0.8826	-10	0	4	1		3
121.60		0.8824	-5	9	1	1		5
121.66		0.8821	-3	1	6	1		5
124.42	124.50	0.8707	-6	8	1	2	1	10
124.60		0.8700	-2	10	1	1		6
125.22	125.18	0.8675	10	4	0	2	1	8
126.35	126.32	0.8631	-1	9	3	1	0	5
128.91	129.00	0.8537	-6	0	6	1	0	5
128.94		0.8536	-2	8	4	1		3
129.29	129.24	0.8524	11	1	0	1	0	4
130.37	130.42	0.8486	8	2	3	1	0	4
130.53		0.8481	-11	3	1	1		3
131.22	131.14	0.8457	5	9	1	1	0	3
132.35		0.8420	1	9	3	1		4
132.64		0.8411	0	9	2	3		14
132.68	132.68	0.8409	5	7	3	2	1	9
133.49	133.36	0.8384	-8	6	4	1	1	6
133.94	133.96	0.8370	-4	6	4	2	1	13
134.38		0.8356	-1	3	6	1		4
134.72	134.58	0.8345	5	5	4	1	1	4
135.23	135.22	0.8330	4	10	0	2	1	13
135.70		0.8316	-11	3	3	1		7
137.75	138.12	0.8257	-9	7	1	1	1	6
138.10		0.8248	-9	3	5	4		20
138.49	138.66	0.8237	-6	6	5	1	1	3
138.66		0.8233	-8	8	1	1		7
139.20		0.8218	10	4	2	1		5
141.46	142.02	0.8160	-6	8	2	3	1	3
141.57		0.8157	-4	10	2	3		14
142.06		0.8145	10	0	5	2		16
144.34	144.90	0.8091	-1	7	1	4	1	9
144.88		0.8079	-7	9	1	1		20
145.79	145.74	0.8059	11	1	1	1	1	8
147.48	147.60	0.8023	-12	2	2	1	0	7
149.50	150.40	0.7984	-3	9	3	1	2	7
149.73		0.7979	-5	7	5	6		33
150.12		0.7972	7	9	0	1		4

2THETA	PEAK	D	H	K	L	I(INT)	I(PK)	I(DS)
150.16		0.7971	-6	4	6	1		4
150.35		0.7968	-12	2	1	7		39
150.53		0.7964	9	1	3	1		4
150.66		0.7962	0	4	6	2		11
151.16		0.7953	10	6	0	2		11
151.24		0.7952	-2	10	3	1		4
151.25		0.7951	-11	5	2	3		15
151.55		0.7946	-10	6	3	1		4
151.93	151.96	0.7940	-1	11	1	1	2	7
152.11		0.7936	-3	5	6	11		62
152.21		0.7935	-11	3	4	1		4
155.18	155.10	0.7887	2	2	1	1	1	7
155.23		0.7886	-12	2	6	3		34
158.51	160.14	0.7840	3	11	3	1	0	7
160.03		0.7821	5	3	5	3		18
161.35	160.20	0.7806	1	7	5	2	0	15

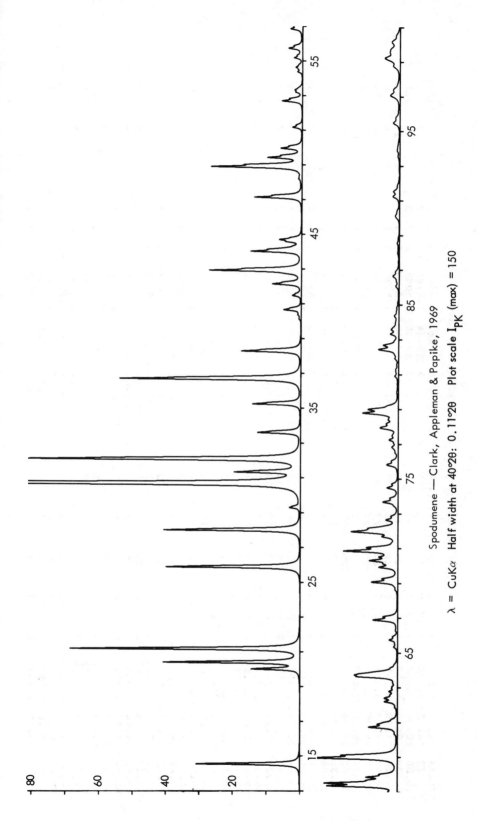

Spodumene — Clark, Appleman & Papike, 1969

$\lambda = CuK\alpha$ Half width at 40°2θ: 0.11°2θ Plot scale I_{PK} (max) = 150

SPODUMENE — CLARK, APPLEMAN, AND PAPIKE, 1969

2THETA	PEAK	D	H	K	L	I(INT)	I(PK)	I(DS)
14.52	14.52	6.095	1	1	0	16	21	15
20.00	20.00	4.437	-2	0	1	8	10	7
20.39	20.40	4.352	-1	1	1	23	27	22
21.17	21.18	4.193	0	2	0	39	46	37
25.87	25.88	3.442	1	1	1	25	27	25
27.99	28.00	3.185	0	2	1	26	27	26
29.28	29.28	3.047	2	2	0	2	2	2
30.65	30.66	2.914	-2	2	1	100	100	100
31.31	31.32	2.854	-3	1	1	12	13	12
32.06	32.06	2.789	3	1	0	56	55	56
33.58	33.58	2.666	1	3	0	9	9	9
35.22	35.22	2.546	-2	0	2	10	10	10
36.64	36.68	2.450	-1	1	2	2	36	2
36.67		2.449	0	2	2	38		39
38.26	38.26	2.350	2	1	1	13	12	13
40.34	40.34	2.234	-3	1	2	1	1	1
40.63	40.64	2.218	4	0	0	4	4	4
41.46	41.46	2.176	-2	2	2	2	2	2
42.12	42.12	2.143	3	1	1	6	6	7
42.89	42.90	2.107	-1	3	1	21	18	22
44.00	44.00	2.056	-3	3	1	12	10	13
44.19	44.10	2.048	-4	2	1	2	7	2
44.56	44.58	2.032	3	3	0	3	3	3
44.67	44.68	2.027	-4	0	2	3	4	4
47.11	47.12	1.9273	0	4	1	12	9	13
48.18	48.18	1.8870	2	0	2	1	1	1
48.87	48.86	1.8622	-2	4	1	23	18	25
49.40	49.40	1.8432	-5	1	1	8	7	9
49.94	49.94	1.8247	-4	2	2	5	5	6
51.15	51.16	1.7841	-3	1	2	2	2	2
51.16	51.16	1.7362	5	1	0	5	4	6
52.67	52.68	1.7208	2	2	2	1	1	2
53.18	53.18	1.7173	2	2	1	1	1	1
53.30	53.30	1.6864	1	2	1	1	2	2
54.36	54.36	1.6790	2	1	1	1	1	2
54.61	54.62	1.6634	-1	3	3	2	2	2
55.17	55.18	1.6480	-3	1	0	4	3	4
55.73	55.74		1	5	0			

2THETA	PEAK	D	H	K	L	I(INT)	I(PK)	I(DS)
56.84	56.84	1.6183	-2	4	2	3	2	3
57.35	57.34	1.6053	-2	2	3	18	14	21
57.52	57.50	1.6009	-3	1	1	10	15	11
57.85	57.84	1.5925	0	4	2	7	6	8
58.95	58.96	1.5654	-5	3	1	25	18	30
59.03		1.5634	-4	4	1	1		1
60.34	60.34	1.5327	1	5	0	8	1	1
60.73	60.74	1.5237	4	4	0	1	6	9
60.84		1.5212	0	0	3			
61.00	60.90	1.5176	-6	3	0	2	4	3
61.87	61.88	1.4983	5	1	2	1	1	1
62.19	62.20	1.4914	1	1	3	3	3	4
62.41	62.40	1.4866	5	1	1	2	2	3
62.77	62.78	1.4789	1	0	3	2	2	3
63.06	63.04	1.4730	4	2	0	1	1	1
63.63	63.64	1.4610	-6	1	1	9	8	11
63.73	63.74	1.4591	-1	3	1			
63.73		1.4590	-5	1	3	3	9	5
63.82		1.4572	3	5	0	4	4	2
65.33	65.34	1.4270	-4	4	2	2	2	1
65.74	65.74	1.4192	-6	2	2	1	1	3
66.88	66.88	1.3977	0	0	4	8	5	10
68.63	68.64	1.3663	4	4	1	1	1	1
69.07	69.06	1.3587	-3	5	2	8	5	11
69.92	69.92	1.3443	4	2	2	4	4	6
69.93		1.3440	0	6	1	1		2
70.30	70.30	1.3379	-2	4	3	8	6	10
70.64	70.50	1.3323	1	3	3	1	4	1
70.85	70.84	1.3289	5	3	1	17	11	21
71.68	71.68	1.3155	2	2	3		3	5
71.97	71.96	1.3109	-7	1	2	13	9	17
72.08		1.3091	-5	3	2	3	3	4
72.64	72.64	1.3004	-2	0	4	4	4	5
73.67	73.68	1.2848	-3	1	0	4	3	6
74.48	74.48	1.2728	-4	0	4	4	2	5
75.77	75.82	1.2543	-5	5	1	1	2	1
75.80		1.2539	2	6	0	1		1

SPODUMENE — CLARK, APPLEMAN, AND PAPIKE, 1969

2THETA	PEAK	D	H	K	L	I(INT)	I(PK)	I(DS)
75.84	76.52	1.2534	7	1	0	2		3
76.51	76.52	1.2441	3	1	3	1	1	1
76.82	76.82	1.2397	6	2	1	1	1	2
77.24	77.24	1.2340	5	1	2	2	2	4
77.59	77.48	1.2293	-6	4	2	3	1	1
77.91	77.92	1.2252	-2	0	2	4	4	1
77.97		1.2243	0	6	0	3		5
78.38	78.40	1.2190	5	5	0	2	2	4
78.46		1.2179	-4	2	4	1		2
78.77	78.78	1.2139	0	6	2	11	7	15
78.95		1.2115	-7	3	1	3	6	5
78.98	78.98	1.2112	-5	5	1	3		
80.90	80.90	1.1872	4	5	0	1		1
81.29	81.28	1.1825	4	7	0	1	1	2
81.90	81.90	1.1752	4	4	2	1	1	1
82.41	82.42	1.1692	3	5	2	8	4	10
82.68	82.66	1.1662	-1	7	1	2	3	2
83.32	83.32	1.1588	-6	0	4	3	2	4
83.58	83.58	1.1559	2	4	3	1	2	1
83.70	83.70	1.1545	7	3	0	1		2
84.36	84.36	1.1472	3	3	3	1	1	2
84.84	84.84	1.1419	1	7	1	1	0	1
86.06	86.06	1.1288	-8	2	1	1	1	2
86.64	86.64	1.1227	-7	3	3	2	1	3
87.96	87.96	1.1092	8	0	0	1	0	1
88.54	88.54	1.1034	6	4	1	1	0	1
90.02	90.12	1.0891	-2	6	3	1	1	1
90.14		1.0880	-4	4	4	1		2
91.22	91.22	1.0778	-7	1	4	3	1	5
91.54	91.52	1.0750	-8	2	1	1	1	2
93.53	93.54	1.0573	0	4	4	1	0	1
93.90	93.90	1.0540	3	7	1	1	1	1
95.47	95.48	1.0408	-7	5	2	3	1	5
97.04	97.08	1.0281	-6	6	2	3	2	4
97.11		1.0275	-3	5	4	2		3
97.43	97.38	1.0250	0	8	1	1	1	1
98.75	98.76	1.0149	-2	8	1	3	1	4
99.16	99.22	1.0118	-5	7	1	2	3	3
99.22		1.0113	-7	5	0	4		7
99.28		1.0108	-5	1	5	1		2
99.75	99.56	1.0074	-3	3	4	3		4
100.19	100.16	1.0041	-2	1	5	2	2	4
101.74	101.74	0.9929	5	7	0	1	2	2
102.47	102.48	0.9879	7	1	2	2	1	3
103.13	103.16	0.9833	2	8	1	3	1	5
103.39	103.42	0.9816	8	2	1	2	1	3
104.66	104.80	0.9731	-9	1	1	1	2	2
104.79		0.9722	5	5	3	5		8
105.78	105.82	0.9659	3	4	4	1	1	2
105.84		0.9655	3	7	3	2		2
107.23	107.24	0.9568	-5	3	5	1	1	2
108.01	108.10	0.9520	-2	6	4	1	2	1
108.05		0.9518	-1	3	5	2		3
108.12		0.9513	-9	3	0	4		7
109.24	109.26	0.9447	-10	0	2	2	1	2
109.87	109.90	0.9411	-4	6	4	2	1	4
110.08	110.28	0.9398	1	7	3	1	1	2
110.29		0.9386	5	7	1	3		5
111.26	111.56	0.9332	5	5	3	3	1	2
111.55		0.9316	-5	7	3	1		4
111.87	111.92	0.9298	-4	4	5	1	1	1
112.30	112.28	0.9275	-2	4	5	2	1	3
113.39	113.48	0.9216	-10	2	0	2	1	3
113.51		0.9210	0	2	6	2		3
114.36	114.34	0.9166	-1	9	1	1	0	1
115.23	115.24	0.9121	-7	5	4	4	1	6
115.87	115.68	0.9089	-10	2	1	1	1	1
117.12	117.14	0.9028	9	1	1	1	0	1
118.22	118.22	0.8976	-2	8	3	4	1	7
119.84	120.00	0.8901	1	7	5	2	1	3
120.02		0.8893	-7	7	2	2		4
120.15		0.8887	3	9	0	1		1
120.46	120.42	0.8874	10	0	0	1	1	2
121.20	121.18	0.8841	-8	0	2	2	1	4

SPODUMENE — CLARK, APPLEMAN, AND PAPIKE, 1969

2THETA	PEAK	D	H	K	L	I(INT)	I(PK)	I(DS)
122.01	121.96	0.8806	8	2	2	2	1	2
123.52	123.54	0.8743	2	2	5	3	1	5
124.00	124.10	0.8724	-6	8	1	1	1	1
124.41	124.46	0.8707	-7	7	0	3	1	5
124.87	125.00	0.8689	-4	0	6	2	2	3
125.04		0.8682	3	5	4	4		8
125.06		0.8681	10	2	0	1		1
126.22	126.64	0.8636	9	3	1	1	1	2
126.56		0.8623	-3	1	6	3		4
126.83		0.8613	-10	4	2	1		2
127.08	127.04	0.8604	4	4	4	2	1	3
128.05	128.10	0.8568	-7	7	3	1	1	1
128.38		0.8556	-7	5	0	2		2
133.41	133.46	0.8386	0	10	0	1	1	4
134.07	134.28	0.8366	5	3	4	2	1	1
134.27		0.8359	2	8	2	2		4
134.35		0.8357	-5	9	3	3		6
135.79	136.06	0.8313	-2	6	5	1	1	2
136.10		0.8304	-8	4	5	2		4
137.45	137.48	0.8266	0	10	1	2	1	3
138.43	138.40	0.8239	-7	1	6	1	1	2
139.32	139.90	0.8215	-9	5	4	1	2	2
139.42		0.8212	-2	10	1	1		1
139.89		0.8200	-9	3	5	8		15
140.42	140.88	0.8186	-1	9	3	1	2	2
140.98		0.8172	10	4	0	5		10
141.35		0.8162	0	0	6	2		4
144.25	144.88	0.8093	10	2	1	2	1	3
144.87		0.8079	-10	4	4	4		7
144.92		0.8078	-11	3	1	2		4
146.23	146.10	0.8050	-7	7	4	3	1	6
147.17	147.26	0.8030	11	1	0	2	1	3
147.32		0.8027	-4	1	6	2		4
148.05		0.8012	0	2	6	1		2
152.04	152.44	0.7938	1	9	3	1	1	1
152.27		0.7934	0	10	2	4		7
152.47		0.7930	5	9	1	11		22

2THETA	PEAK	D	H	K	L	I(INT)	I(PK)	I(DS)
155.12	154.98	0.7888	-5	9	3	4	1	7
156.57	158.24	0.7866	-12	0	5	2	1	4
156.59		0.7866	-3	2	5	1		2
158.07		0.7846	8	2	3	8		16
158.12		0.7845	-9	7	1	3		6
158.18		0.7844	4	10	0	1		3
158.48		0.7840	5	7	3	3		6
159.54		0.7827	-10	6	2	2		3
160.11		0.7820	4	6	4	6		11
160.67		0.7813	-8	8	1	2		5

TABLE 12. CLINOPYROXENES

Variety	Johannsenite	Ureyite	NaIn(SiO$_3$)$_2$	Omphacite
Composition	CaMn(SiO$_3$)$_2$	NaCr(SiO$_3$)$_2$	NaIn(SiO$_3$)$_2$	(Ca,Na)(Mg,Al)(SiO$_3$)$_2$
Source	Venetia, Italy USNM #R3118	Synthetic	Synthetic	Tiburon Pen, California
Reference	Freed & Peacor, 1967	Clark, Appleman & Papike, 1969; 1968	Christensen and Hazell, 1967	Clark & Papike, 1968
Cell Dimensions				
a Å	9.978	9.58	9.916	9.596
b	9.156	8.72	9.132	8.771
c	5.293	5.26	5.371	5.265
α	90	90	90	90
β deg	105.48	107.20	107.00	106.93
γ	90	90	90	90
Space Group	C2/c	C2/c	C2/c	P2
Z	4	4	4	4

TABLE 12. (cont.)

Variety	Johannsenite	Ureyite		Omphacite
Site Occupancy				
M2	Ca 1.0	Na 1.0	Na 1.0	*(see M2 table below)*
M1	Mg 0.02, Fe 0.02, Mn 0.96	Cr 1.0	In 1.0	*(see M1 table below)*
Method	3-D, counter	3-D, counter	3-D, counter	3-D, counter
R & Rating	0.066 (1)	0.04 – 0.06 (1)	0.049 (1)	0.08 (1)
Cleavage and habit		{110} perfect / Variable		
μ	333.5	541.1	471.4	187.9
ASF	0.1064	0.8988×10^{-1}	0.3842	0.1140
Abridging factors	0.3:0:0	0.3:0:0	0.5:0:0	0.5:0:0

Omphacite — M2 site occupancy:

	M2	M2(1)	M2H	M2(1)H	Total
Ca	0.09	0.16	0.24	0.09	(0.58)
Na	0.16	0.09	0.01	0.16	(0.42)

Omphacite — M1 site occupancy:

	M1	M1(1)	M1H	M1(1)H	Total
Mg	0.20			0.20	(0.40)
Fe	0.05			0.05	(0.10)
Fe		0.01	0.04		(0.05)
Al		0.24	0.21		(0.45)

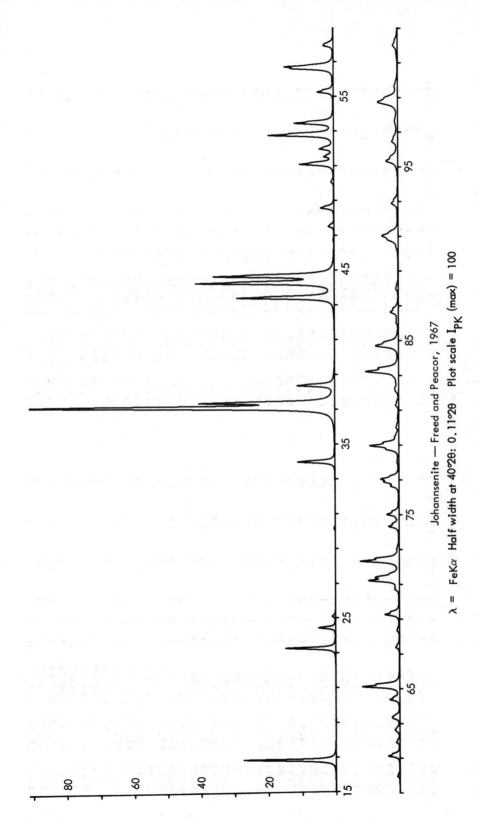

Johannsenite — Freed and Peacor, 1967

$\lambda = FeK\alpha$ Half width at 40°2θ: 0.11°2θ Plot scale I_{PK} (max) = 100

JOHANNSENITE - FREED AND PEACOR, 1967

2THETA	PEAK	D	H	K	L	I(INT)	I(PK)	I(DS)
16.79	16.80	6.631	1	1	0	20	27	9
23.23	23.24	4.808	2	0	0	12	15	7
24.41	24.42	4.578	0	2	1	4	5	3
25.08	25.08	4.459	-1	1	1	1	1	1
30.12	30.12	3.725	1	1	1	0	0	0
33.95	33.96	3.315	-2	2	0	11	11	10
37.03	37.04	3.048	2	2	1	100	100	100
37.32	37.32	3.025	3	1	0	36	40	36
38.33	38.34	2.948	-3	1	1	11	11	12
43.37	43.38	2.620	-1	3	1	32	29	38
44.22	44.22	2.572	2	2	2	45	41	55
44.61	44.62	2.550	0	4	0	39	36	49
47.49	47.50	2.404	4	0	1	2	2	3
48.54	48.54	2.355	3	1	1	5	4	7
50.03	50.04	2.289	-3	3	0	1	1	1
50.98	51.06	2.249	-1	1	2	0	10	1
51.06	51.50	2.246	-3	1	1	12		18
51.50	51.50	2.228	0	2	2	3	3	4
51.95	51.94	2.210	-3	3	1	5	4	7
52.73	52.74	2.180	-3	3	1	24	19	37
53.42	53.42	2.153	-4	2	1	14	12	23
55.23	55.24	2.088	0	4	1	6	5	10
56.58	56.60	2.043	-4	0	2	13	13	23
56.68	56.68	2.039	-4	0	2	10	15	17
57.91	57.92	1.9995	-1	3	2	3	3	6
57.99		1.9969	-2	4	1	1	1	2
59.94	59.94	1.9378	-5	1	1	1	1	2
61.10	61.10	1.9043	3	3	1	5	3	9
61.90	61.90	1.8821	5	1	0	1	1	3
62.52	62.62	1.8653	-4	2	2	1	2	2
62.62		1.8626	2	4	1	2	2	3
63.28	63.36	1.8453	2	4	2	2	2	3
63.35		1.8433	4	2	1	2	2	4
64.36	64.36	1.8176	4	4	0	4	3	8
65.11	65.12	1.7989	1	5	1	16	11	34
67.37	67.38	1.7453	-3	1	2	2	1	4
68.33	68.34	1.7237	-1	5	1	1	1	3
69.25	69.26	1.7035	0	4	2	7	4	15
70.69	70.70	1.6732	-3	3	3	2	1	4
70.82	70.86	1.6705	-1	5	1	0	1	1
71.21	71.22	1.6627	-5	4	0	14	9	32
71.45	71.38	1.6577	-4	4	0	5	7	12
72.32	72.32	1.6405	-2	2	3	18	12	44
73.01	73.00	1.6271	6	0	0	1	1	2
74.31	74.32	1.6027	3	5	0	5	3	12
75.00	75.00	1.5900	-6	0	0	6	4	14
76.62	76.64	1.5614	-6	2	1	3	3	9
76.88	76.88	1.5571	2	6	1	5	5	13
77.03	77.04	1.5545	0	6	2	6	5	15
77.99	78.00	1.5382	-5	3	0	1	1	2
78.74	78.76	1.5260	0	0	3	5	4	13
78.86	78.76	1.5240	-4	4	2	0	9	1
78.96	78.96	1.5225	0	6	2	12		32
79.46	79.46	1.5144	-4	4	3	1	1	4
80.00	80.00	1.5059	-1	5	2	1	1	3
80.53	80.54	1.4976	4	0	1	1	0	1
81.27	81.28	1.4864	-3	3	3	2	1	5
82.09	82.10	1.4741	-6	2	2	1	1	4
82.24	82.24	1.4720	4	2	1	1	1	3
82.61	82.62	1.4666	-5	1	3	1	1	3
83.21	83.22	1.4578	5	5	1	18	10	54
84.66	84.68	1.4374	-3	5	2	9	7	26
84.72		1.4366	1	5	3	6		18
86.03	86.04	1.4189	1	3	3	1	1	4
86.62	86.62	1.4112	2	2	3	5	2	15
86.83	86.82	1.4085	-7	1	1	0	1	1
87.97	87.98	1.3939	-2	4	3	2	1	5
90.34	90.38	1.3649	4	4	3	1	1	5
90.88	91.02	1.3585	-5	1	0	2	1	3
91.03		1.3568	-7	1	2	7	5	7
92.87	92.88	1.3359	-5	3	3	5	2	23
95.30	95.32	1.3098	-2	6	2	2	4	16
95.33		1.3095	-1	6	2	2	2	5
95.75	95.56	1.3052	-6	2	3	1	2	25

JOHANNSENITE - FREED AND PEACOR, 1967

2THETA	PEAK	D	H	K	L	I(INT)	I(PK)	I(DS)
96.45	96.46	1.2980	-3	1	4	4	2	14
96.51		1.2974	-1	1	4	1		2
96.86	96.70	1.2938	-4	6	1	0	1	1
97.09	97.08	1.2915	-7	3	0	0	1	1
97.41	97.66	1.2884	4	6	3	1	1	2
97.50		1.2874	-6	4	2	1	1	2
97.65		1.2860	4	4	4	1		5
98.61	98.74	1.2767	-5	5	2	3	6	3
98.64		1.2764	-4	0	5	9		11
98.73		1.2756	3	5	2	4		34
98.76		1.2752	0	0	4	2		16
99.63	99.62	1.2671	-1	1	7	4	1	7
102.01	102.02	1.2455	-1	7	1	1	1	13
103.61	103.66	1.2317	-7	1	3	0	0	3
103.87	103.90	1.2295	-4	2	4	1	1	3
104.71		1.2285	0	6	4	0		2
105.00	105.00	1.2225	-4	0	2	1	0	1
107.18	107.26	1.2201	-8	0	2	2		4
107.28		1.2028	-8	2	1	3	1	7
107.28	107.26	1.2020	8	0	0	1		11
110.60	110.60	1.1774	6	2	2	0	0	2
111.28	111.28	1.1726	-1	7	2	5		1
113.11	113.12	1.1601	-6	0	4	1	1	20
113.36		1.1584	2	0	0	1		3
114.07	114.08	1.1538	-5	5	3	1	1	6
114.26		1.1525	3	7	1	0		2
116.33	116.34	1.1394	-3	7	2	0	0	2
116.89	116.90	1.1359	-5	3	4	0	0	2
116.93		1.1357	0	6	3	0		1
119.08	119.08	1.1230	2	2	4	2	0	7
120.53	120.64	1.1148	-4	4	4	1	1	5
120.66		1.1140	0	8	4	1		4
120.78		1.1134	2	8	0	0		2
122.89	123.00	1.1021	-2	2	1	10	4	51
123.04		1.1013	-8	3	0	5		23
123.50	123.52	1.0989	-7	0	0	14	6	68
123.67		1.0980	-7	5	2	10		50
124.89	124.92	1.0918	-5	4	1	8	3	42
125.09		1.0909	4	4	3	1		3
125.29	125.30	1.0899	-6	6	2	3	2	16
125.47		1.0890	4	6	1	1		4
126.04	126.04	1.0860	7	1	2	4	1	18
126.09		1.0834	1	1	4	2	1	8
126.63	126.62	1.0816	-7	3	4	3	1	17
127.00	127.00	1.0816	3	5	0	0		5
127.01		1.0755	6	7	2	2	1	2
128.32	128.62	1.0750	8	2	1	1		12
128.44		1.0740	2	8	1	4		20
128.65		1.0661	-3	5	4	7		36
130.46	130.48	1.0658	-1	5	4	1	2	4
130.53		1.0619	5	3	3	5		49
131.45	131.44	1.0542	-5	7	2	3	2	4
133.33	133.44	1.0536	-3	7	1	1	0	4
133.49		1.0501	-3	1	5	1		3
134.38	134.64	1.0491	-1	7	3	5	1	30
134.64		1.0443	-2	8	2	1		4
135.91	135.94	1.0442	0	8	1	2	1	4
135.95		1.0404	-9	3	1	8	1	45
136.99	137.00	1.0371	-3	7	3	1	1	5
137.94	137.50	1.0334	4	8	0	2	1	12
139.02	139.00	1.0298	-2	2	5	6		34
140.09	140.12	1.0296	7	5	2	5	1	3
140.16		1.0272	5	7	1	1		29
140.90	140.82	1.0259	3	3	4	6	1	4
141.32		1.0195	4	0	4	1	1	37
143.41	143.52	1.0186	-4	2	5	6	2	5
143.74	144.32	1.0175	-5	5	4	2		10
144.09		1.0167	1	5	4	6		38
144.39		1.0166	-8	4	3	1		3
144.43		1.0148	-5	1	5	1		4
145.05								2
146.54	146.52	1.0108	6	6	2	9	1	52
147.43		1.0084	9	3	0	1		6
155.30	156.88	0.9909	4	8	1	4	2	22
155.70		0.9902	8	0	2	4		27

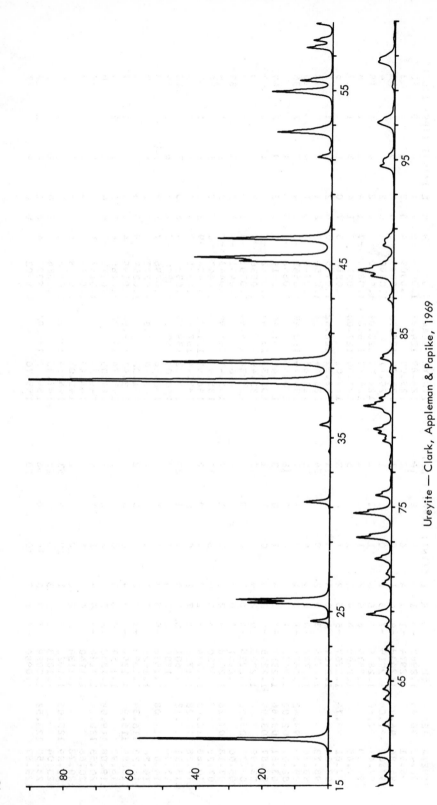

Ureyite — Clark, Appleman & Papike, 1969

λ = FeKα Half width at 40°2θ: 0.11°2θ Plot scale I$_{PK}$ (max) = 100

UREYITE — CLARK, APPLEMAN AND PAPIKE, 1969

2THETA	PEAK	D	H	K	L	I(INT)	I(PK)	I(DS)
17.64	17.64	6.313	2	1	0	43	58	15
24.43	24.44	4.576	-1	0	1	5		
25.46	25.48	4.392	1	1	1	19	25	11
25.65	25.66	4.360	0	2	0	22	28	13
31.28	31.28	3.591	1	1	1	7	8	5
34.19	34.20	3.293	-2	2	1	0	0	0
35.72	35.72	3.157	0	2	1	3	3	0
38.20	38.20	2.959	-2	2	1	100	100	100
39.29	39.30	2.879	3	1	0	51	51	53
40.90	40.90	2.770	1	3	0			1
45.11	45.12	2.523	-1	1	2	27	28	34
45.23	45.32	2.517	-1	3	1	2	42	2
45.32		2.512	0	2	2	40	35	52
46.41	46.42	2.457	3	1	1	38	4	51
51.14	51.14	2.243	-3	1	2	5	1	8
51.56	51.56	2.226	-1	1	0	1	1	1
52.56	52.56	2.186	0	4	2	20	17	32
52.72	52.66	2.180		0	0	0	12	
54.91	54.90	2.100	-3	4	1	20	18	35
55.56	55.56	2.077	-4	3	1	12	10	20
57.46	57.46	2.014	-4	2	2	10	8	18
57.90	57.90	1.9999	0	4	0	7	6	13
58.86	58.86	1.9702	0	0	2	6	5	12
59.53	59.52	1.9499	-1	3	2	2	2	4
60.65	60.66	1.9172	-2	4	1	7	5	14
62.47	62.48	1.8668	-5	1	1	3	2	7
63.94	63.94	1.8282	-4	4	2	3	2	6
64.51	64.52	1.8137	3	2	0	1	1	7
64.88	64.88	1.8046	-3	3	2	2	1	2
65.42	65.42	1.7913	-3	3	1	4	1	4
65.75	65.74	1.7833	5	1	0	1	2	3
66.81	66.82	1.7582	2	4	1	3	2	6
68.11	68.12	1.7285	4	2	1	2	2	5
68.81	68.82	1.7132	-1	5	1	11	7	26
69.07	68.96	1.7075	1	1	3	1	5	2
70.58	70.58	1.6755	-3	1	1	4	3	10
71.16	71.16	1.6636	-3	1	3	2	2	5
72.01	72.02	1.6466	0	4	2	7	5	18
73.23	73.24	1.6229	-2	2	1	16	11	44
74.63	74.64	1.5969	-5	3	1	18	11	49
75.66	75.66	1.5783	4	1	0	8	5	21
76.50	76.50	1.5635	0	4	2	1	1	2
77.22	77.22	1.5512	5	1	1	1	1	3
77.36	77.38	1.5488	5	1	1	1	1	1
77.72	77.72	1.5428	1	3	3	0	1	3
78.79	78.78	1.5253	6	0	0	4	3	13
79.23	79.24	1.5181	0	0	0	6	4	18
79.49	79.48	1.5140	3	5	0	7	6	20
80.79	80.78	1.4937	-1	3	2	14	9	43
81.24	81.24	1.4868	-4	2	2	6	4	18
81.74	81.74	1.4793	-4	4	2	1	1	2
82.78	82.78	1.4641	-3	2	3	2	4	5
83.52	83.54	1.4534	-1	5	2	0		1
83.53		1.4533	0	6	0	6		20
83.71	83.72	1.4507	-5	1	1	2	3	6
84.94	84.94	1.4336	-6	2	2	0	0	1
86.92	86.92	1.4072	4	2	2	3	2	11
88.27	88.28	1.3901	-3	5	1	12	7	45
88.62	88.62	1.3857	5	1	2	19	11	68
90.04	90.04	1.3685	5	2	3	6	3	24
90.16		1.3671	-2	6	1	0		1
90.41	90.38	1.3640	-2	4	3	4	3	15
94.62	94.62	1.3169	-7	3	3	10	4	41
95.03	94.88	1.3126	-5	1	1	5	4	22
96.95	97.12	1.2929	7	2	1	3	5	12
97.00		1.2924	6	2	1	0		2
97.06		1.2918	-3	4	1	4		17
97.14		1.2910	-1	4	3	8		31
97.94	97.92	1.2832	-6	2	4	0	1	2
97.95		1.2831	-4	0	3	0		1
99.20	99.22	1.2711	-1	7	2	2	1	9
100.81	100.62	1.2580	-3	0	6	11	5	50
100.82	100.82	1.2562	0	6	4	5	5	23
101.98	101.98	1.2458	-6	2	2	1	1	5

UREYITE - CLARK, APPLEMAN AND PAPIKE, 1969

2THETA	PEAK	D	H	K	L	I(INT)	I(PK)	I(DS)
102.93	102.94	1.2375	-7	3	1	1	0	3
103.30	103.30	1.2343	1	7	1	1	1	4
104.01	104.04	1.2283	4	4	0	2	1	9
104.20	104.20	1.2268	4	6	2	1	1	5
105.01	105.02	1.2200	-3	5	2	11	4	52
106.27	106.28	1.2099	-1	7	1	3	1	13
106.63	106.56	1.2071	0	2	4	1	1	3
107.24	107.24	1.2023	2	4	3	1	0	3
109.17	109.18	1.1878	1	7	1	4	1	18
110.08	110.06	1.1811	-8	0	2	0	0	2
111.73	111.74	1.1695	5	5	1	1	0	3
112.26	112.24	1.1658	1	5	3	1	0	4
114.33	114.50	1.1520	-8	2	1	3	2	15
114.50		1.1510	-6	0	4	7		36
114.75		1.1493	6	4	1	2		4
115.60	115.60	1.1439	8	0	3	2	1	12
115.72		1.1432	-6	4	3	0		2
118.59	118.76	1.1258	-2	4	4	1		3
118.60		1.1258	-1	7	2	0	0	2
118.82		1.1245	-5	5	3	4		4
123.66	123.70	1.0980	-4	4	4	2	1	11
123.73		1.0977	0	6	3	0		3
124.14	124.12	1.0956	-3	7	2	1	1	5
124.64	124.60	1.0931	2	2	4	1	0	6
125.11	125.04	1.0908	1	7	2	0	0	2
125.58	125.54	1.0884	0	4	4	1	0	4
128.33	128.34	1.0755	-8	2	3	7	1	42
129.33	129.24	1.0710	-7	1	4	2	1	13
131.82	132.24	1.0603	2	8	0	0		3
132.22		1.0587	-7	1	2	9	2	58
133.85	133.92	1.0522	-6	6	1	1	3	4
133.85		1.0522	-2	8	1	11		73
134.46	134.50	1.0498	-6	6	2	8	4	54
134.65		1.0491	3	4	1	6		38
135.20	135.52	1.0470	4	4	3	0	5	3
135.44		1.0461	7	4	0	16		103
135.71		1.0451	-3	5	4	9		61

2THETA	PEAK	D	H	K	L	I(INT)	I(PK)	I(DS)
136.06	136.00	1.0438	-3	1	5	1	5	8
136.11		1.0436	-5	7	1	9		63
139.06	139.08	1.0332	7	1	2	7	1	45
140.09	139.64	1.0298	5	7	0	2	1	12
141.45	142.16	1.0255	-1	1	5	0	2	3
141.81		1.0243	6	1	2	1		5
142.11		1.0234	2	8	1	8		53
142.12		1.0234	2	6	3	0		3
142.91	142.90	1.0210	8	2	1	6	2	39
143.38	144.14	1.0196	-2	2	5	5	3	36
144.17		1.0173	-1	2	3	14		100
145.66	145.66	1.0131	-1	4	3	7	2	51
146.11		1.0119	-4	2	5	2		12
146.19		1.0117	-7	1	4	0		3
148.27	148.22	1.0063	3	7	2	4	1	32
150.95	151.08	0.9999	0	8	1	8	0	7
153.07	153.12	0.9953	-9	3	1	10	1	79
154.15		0.9931	3	3	4	1		5
155.94	157.48	0.9897	-4	8	1	2	1	15
156.34		0.9890	-8	4	4	4		34
157.29		0.9873	1	5	4	8		65
158.62		0.9851	4	0	4	9		68
159.28		0.9840	4	8	0	3		21
162.28	164.00	0.9797	7	3	2	0	1	3
162.58		0.9793	0	2	5	2		9
162.84		0.9789	7	5	1	1		6
164.10		0.9774	5	7	1	13		106

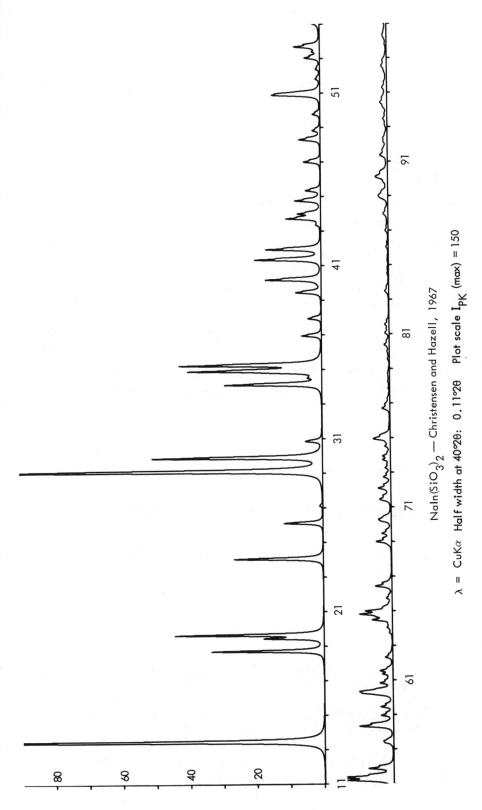

$NaIn(SiO_3)_2$ — Christensen and Hazell, 1967

$\lambda = CuK\alpha$ Half width at $40°2\theta$: $0.11°2\theta$ Plot scale I_{PK} (max) = 150

265

NAIN(SIO3)2 - CHRISTENSEN AND HAZELL, 1967

2THETA	PEAK	D	H	K	L	I(INT)	I(PK)	I(DS)
13.45	13.46	6.578	1	1	0	100	100	100
18.70	18.70	4.741	2	0	0	24	22	33
19.42	19.44	4.566	0	2	0	12	12	17
19.64	19.64	4.517	1	1	1	32	30	46
24.03	24.04	3.701	-1	1	1	21	18	39
26.09	26.10	3.413	1	0	2	10	8	19
27.09	27.10	3.289	2	2	0	1	1	2
29.09	29.10	3.067	-2	2	1	88	69	207
29.89	29.90	2.987	3	1	0	45	34	108
30.82	30.82	2.898	1	3	0	4	3	11
34.10	34.10	2.627	-1	1	2	27	19	77
34.48	34.48	2.599	-2	0	2	2	3	6
34.80	34.90	2.576	-1	2	2	8	27	25
34.91		2.568	0	2	2	32	28	94
35.23	35.24	2.545	2	2	1	39		116
36.92	36.92	2.433	1	3	1	5	4	18
37.92	37.92	2.371	-3	1	2	4	3	13
39.41	39.42	2.284	0	4	0	4	5	15
39.44		2.283	-3	0	1	3		12
40.17	40.16	2.243	1	1	2	17	11	60
41.33	41.34	2.183	-3	3	1	20	13	76
41.93	41.94	2.153	-4	2	1	17	11	64
43.33	43.34	2.086	0	4	1	1		5
43.70	43.70	2.069	-4	0	2	10	7	42
43.98	43.98	2.057	2	0	4	6	5	26
44.74	44.74	2.024	2	4	0	8	5	34
45.33	45.32	1.9990	-2	4	1	5	3	20
46.99	47.00	1.9321	-5	1	2	6	4	25
48.24	48.28	1.8849	-4	2	2	1		6
48.28		1.8833	3	3	1	6		28
48.80	48.80	1.8647	-3	3	2	2	2	11
49.75	49.76	1.8311	2	4	1	2	2	12
50.87	50.94	1.7934	1	5	0	13	10	67
50.95		1.7908	4	2	2	7	9	38
51.80	51.80	1.7634	-5	1	3	1	1	7
52.37	52.38	1.7454	-1	1	3	1	1	7
53.01	53.02	1.7259	3	1	2	5	3	28

2THETA	PEAK	D	H	K	L	I(INT)	I(PK)	I(DS)
53.37	53.38	1.7151	-2	4	2	2	2	11
53.67	53.68	1.7063	0	4	2	9	5	48
55.18	55.18	1.6631	-2	2	3	9	5	85
55.37	55.36	1.6579	1	3	1	12	11	69
55.86	55.86	1.6444	4	0	0	8	5	49
57.34	57.36	1.6055	0	2	3	2	2	14
57.43	57.44	1.6031	3	5	0	6	2	14
58.30	58.32	1.5814	1	1	3	2	7	38
58.33		1.5806	-6	0	1	5		12
58.45	58.46	1.5805	-1	4	2	1		31
58.94	58.94	1.5777	-3	1	4	4	5	4
59.43	59.44	1.5658	-6	2	1	4	2	25
60.13	60.24	1.5538	0	2	3	3	3	27
60.22		1.5376	2	1	3	8	6	17
60.31	60.32	1.5354	-1	3	3	3		52
60.33		1.5333	4	2	0	4		20
60.80	60.80	1.5329	-4	0	3	1		26
60.81		1.5220	0	6	0	4		5
61.14	61.14	1.5220	2	0	6	4	3	24
61.34	61.34	1.5145	-5	3	2	1	1	7
61.54	61.52	1.5101	-1	1	5	3		20
62.30	62.30	1.5057	-3	5	1	3	3	21
63.72	63.72	1.4889	-5	1	3	3	2	19
64.02	64.02	1.4593	0	6	1	1		7
64.04		1.4532	4	2	2	1	1	8
64.46	64.46	1.4528	3	5	1	2		4
64.80	64.80	1.4442	-3	5	2	5		52
64.98	64.98	1.4376	5	1	1	7	7	85
65.00	65.00	1.4336	-5	3	1	4	5	27
65.26	65.16	1.4285	-2	6	1	1	2	10
65.72	65.72	1.4196	1	6	1	3	1	15
66.41	66.42	1.4066	-2	2	1	2		16
66.42		1.4063	2	3	3	2		35
68.98	68.98	1.3603	-7	1	2	6	3	46
69.46	69.46	1.3521	-5	3	3	4	2	30
70.17	70.26	1.3400	1	7	0	3	3	24
70.27		1.3384	6	2	1	4		30

NAIN(SIO3)2 - CHRISTENSEN AND HAZELL, 1967

2THETA	PEAK	D	H	K	L	I(INT)	I(PK)	I(DS)
71.36	71.44	1.3206	-6	2	3	4	2	15
71.44		1.3193	-3	1	0	4		31
71.68	71.64	1.3156	5	5	0	1		6
72.07	72.06	1.3093	0	6	2	5	3	42
72.70	72.72	1.2995	6	4	0	1	1	5
72.71		1.2993	-4	0	4	2		20
73.24	73.24	1.2912	-6	4	2	2	1	15
73.72	73.72	1.2841	0	0	5	3	2	26
73.92	73.92	1.2810	-5	5	2	2	2	8
73.94		1.2808	4	6	0	1		7
74.49	74.50	1.2726	-2	2	4	2	1	18
74.92	74.94	1.2664	3	5	1	6	3	55
75.00		1.2653	-1	7	1	3		23
76.68	76.68	1.2417	-1	7	2	3	2	31
77.84	77.84	1.2261	-4	6	0	1	0	6
78.26	78.26	1.2205	-8	0	2	1	0	7
79.53	79.52	1.2042	-3	7	1	1	0	9
79.97	79.96	1.1986	6	0	2	1		5
80.35	80.36	1.1939	-8	2	1	2	1	21
81.01	81.04	1.1859	4	2	3	1	1	10
81.06		1.1853	8	0	0	2		12
81.53	81.54	1.1796	-6	0	4	2	1	20
83.46	83.46	1.1572	-2	4	4	1	1	11
83.48		1.1570	2	0	4	1		8
84.47	84.48	1.1459	3	7	1	1	0	9
86.01	86.02	1.1293	-4	4	4	1	1	15
86.95	86.98	1.1195	-6	6	1	1		7
86.98		1.1192	0	4	4	1		12
87.55	87.54	1.1134	-4	6	3	1	0	6
87.96	87.94	1.1093	-8	2	3	2	1	25
88.85	89.00	1.1004	-2	8	1	3	2	31
88.94		1.0995	-7	1	1	1		17
89.01		1.0988	-7	5	2	3		37
90.05	90.10	1.0888	-5	5	4	4	2	35
90.13		1.0880	7	1	0	2		48
91.33	91.34	1.0769	-3	5	4	2	2	22
91.35		1.0767	3	1	4	2		25

2THETA	PEAK	D	H	K	L	I(INT)	I(PK)	I(DS)
91.38	91.62	1.0764	-8	4	2	1	1	9
91.81	92.12	1.0725	-1	5	4	1	1	9
92.11	92.12	1.0698	2	8	1	2	1	17
92.30	92.30	1.0681	7	1	2	1	1	22
93.07	93.12	1.0613	6	4	2	1		11
93.17		1.0604	2	6	3	1		8
93.46	93.46	1.0578	8	2	1	2	1	21
94.15	94.18	1.0519	-1	7	3	2	1	20
94.38	94.40	1.0499	5	3	3	3	1	41
94.76	94.68	1.0467	9	1	0	1	1	8
95.26	95.26	1.0426	-2	2	5	2	1	27
95.31		1.0421	-3	7	1	1		8
96.30	96.52	1.0340	-4	8	1	1	1	11
96.45		1.0328	-5	1	5	2		27
96.53		1.0322	-9	1	1	1		10
96.91	96.84	1.0291	-5	5	4	1		7
97.80	97.80	1.0222	-6	6	3	1	0	23
98.24	98.28	1.0188	5	7	1	2	1	19
98.37		1.0178	-1	5	4	1		22
99.09	99.10	1.0123	7	5	0	2	1	13
99.13		1.0120	4	0	4	1		8
99.54	99.44	1.0089	1	9	0	1	1	23
101.14	101.18	0.9972	6	2	3	2	1	30
101.21		0.9967	-1	3	5	2		15
102.39	102.56	0.9884	-10	0	2	1		16
102.42		0.9882	-4	6	3	2		34
102.54		0.9874	-9	3	1	3		28
102.60		0.9869	-5	7	3	2		10
102.93	102.90	0.9847	4	8	1	1	1	17
103.08		0.9837	-5	3	5	1		7
104.62	104.64	0.9734	8	0	2	1	0	19
105.74	105.74	0.9661	3	9	0	1	0	27
106.42	106.42	0.9618	-2	8	3	2	1	8
106.72	106.78	0.9599	-8	6	1	1	1	10
106.80		0.9594	-7	6	1	1	1	14
108.64	108.86	0.9483	-10	0	4	1		39
108.85		0.9470	-7	0	5	3		

NAIN(SIO3)2 – CHRISTENSEN AND HAZELL, 1967

2THETA	PEAK	D	H	K	L	I(INT)	I(PK)	I(DS)
109.63	109.68	0.9424	-8	4	4	1	1	11
109.69		0.9421	-1	3	5	1		17
110.19	110.18	0.9392	-6	8	1	1	1	10
110.37	110.40	0.9382	-9	5	2	1	1	12
110.42		0.9379	3	7	3	1		9
111.41	111.42	0.9323	-6	6	4	1	1	9
111.41		0.9323	-3	5	4	2		33
111.45		0.9321	-3	9	1	1		12
111.72	111.84	0.9306	9	3	2	1	1	19
111.97		0.9293	1	2	5	1		12
112.12	112.10	0.9284	7	2	4	2	1	31
112.44	112.46	0.9267	5	5	0	2	1	27
112.73		0.9251	4	4	0	2		25
115.02	115.06	0.9132	0	10	2	1	0	15
115.12		0.9127	9	5	6	1		11
116.25	116.26	0.9071	-10	4	3	1	0	20
116.71	117.10	0.9048	-7	7	5	1	1	10
116.98		0.9035	-2	8	1	1		24
117.13		0.9027	-8	2	4	2		29
117.14		0.9027	-5	9	0	1		11
117.89	117.94	0.8991	-10	10	0	1	0	9
117.99		0.8986	-10	0	4	1		12
118.40	118.52	0.8967	2	10	0	1	1	10
118.84	119.22	0.8947	5	9	0	2	1	12
119.24		0.8929	-4	0	6	1		20
119.24		0.8928	-7	7	1	1		17
119.72	119.72	0.8907	-3	1	6	1		16
121.07	121.12	0.8847	6	8	1	1	0	21
121.32		0.8836	-5	9	2	1	0	11
122.46	122.42	0.8787	3	9	2	1		13
123.18	123.16	0.8757	10	4	0	2	0	40
123.18		0.8757	-2	4	5	1	1	10
126.77	127.06	0.8616	5	10	2	1	0	16
126.77		0.8615	5	9	1	1		15
127.07		0.8604	0	10	0	1		26
127.65	127.70	0.8582	11	1	2	1	1	16
127.81		0.8577	-11	3	1	1		24

2THETA	PEAK	D	H	K	L	I(INT)	I(PK)	I(DS)
128.26	128.28	0.8561	0	0	6	2	0	12
129.34	129.36	0.8522	4	10	0	2	0	31
130.21	130.30	0.8492	5	7	3	2	1	35
130.40		0.8485	8	3	3	1		39
131.26	130.96	0.8456	-11	3	5	1		22
132.87	132.98	0.8403	-9	3	6	2	0	48
132.97		0.8400	-7	7	1	1	1	28
133.06		0.8397	-9	1	5	1		28
133.18		0.8393	-6	6	5	1		11
133.33		0.8389	-8	8	1	1		24
134.19	134.24	0.8362	-10	4	4	2	1	37
134.42		0.8355	-4	10	2	2		37
134.73		0.8345	-11	1	4	1		16
136.61	136.62	0.8290	-3	7	5	1	0	17
136.62		0.8289	-12	6	2	1		13
137.06	137.26	0.8277	-1	7	6	1		12
139.78	139.86	0.8203	-1	7	5	1	0	22
139.84		0.8201	6	4	4	1	0	18
139.96		0.8198	-11	11	1	1		14
141.20	141.82	0.8166	-7	9	4	1		17
141.22		0.8166	-7	0	2	1	1	33
141.66		0.8155	10	0	2	2		53
141.83		0.8150	-9	7	3	2		37
142.06		0.8145	-10	2	5	3		12
142.69	142.66	0.8130	-5	5	3	1		62
142.85		0.8126	2	8	4	1	1	13
143.04		0.8121	7	9	0	1		21
143.59	143.62	0.8108	5	9	2	1	1	24
144.00		0.8099	-6	4	6	1		31
144.02		0.8098	-10	6	3	1		13
144.89	144.98	0.8079	-8	6	3	1	1	32
144.97		0.8077	-8	4	3	1		12
144.97		0.8077	-11	5	2	2		36
145.08		0.8074	-3	9	4	2		35
145.86	146.20	0.8057	0	10	3	1	1	13
145.93		0.8056	-1	9	4	1		27
146.21		0.8050	-12	2	1	4		93

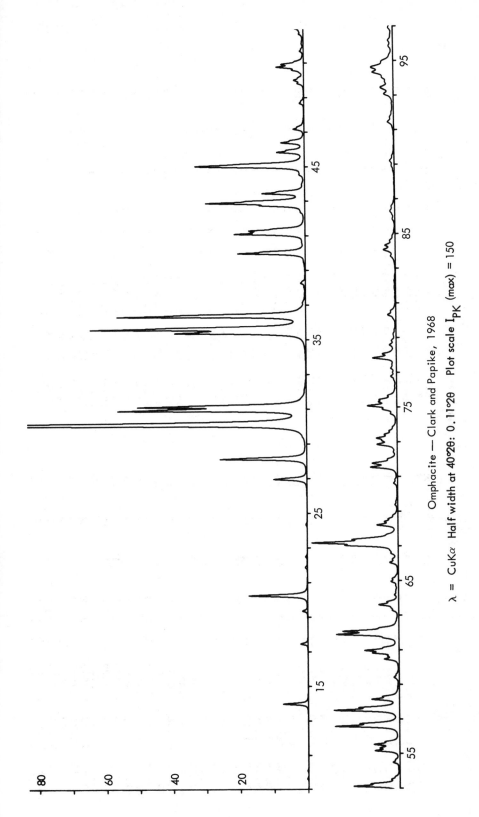

Omphacite — Clark and Papike, 1968

$\lambda = CuK\alpha$ Half width at $40°2\theta$: $0.11°2\theta$ Plot scale I_{PK} (max) = 150

OMPHACITE – CLARK AND PAPIKE, 1968

2THETA	PEAK	D	H	K	L	I(INT)	I(PK)	I(DS)
13.95	13.96	6.342	-1	1	0	4	5	3
17.43	17.44	5.084	-1	0	1	1	1	1
19.32	19.32	4.590	0	1	1	1	1	1
20.17	20.24	4.399	-1	1	1	1	12	1
20.23		4.385	0	0	2	9		7
26.93	26.94	3.307	-2	2	0	6	7	6
28.12	28.12	3.171	0	2	1	16	17	16
30.11	30.10	2.966	-2	2	1	100	100	100
30.92	30.92	2.889	3	1	0	35	38	36
31.14	31.14	2.870	-3	1	1	30	34	30
35.28	35.38	2.542	-2	1	2	2	26	2
35.39		2.534	-1	3	1	24	43	27
35.62	35.62	2.518	0	0	2	42	38	47
36.36	36.36	2.469	2	2	1	40		44
38.24	38.24	2.351	1	3	1	1	1	1
39.98	39.98	2.253	3	1	1	15	13	18
41.00	41.10	2.194	-2	2	2	15	14	18
41.10		2.184	-1	2	2	10	11	12
41.30	41.30	2.114	0	2	2	8	9	10
42.74	42.74	2.106	3	3	0	20	20	25
42.90	42.90	2.081	-3	3	1	9	8	12
43.44	43.44	2.033	-4	2	1	1	1	1
44.52	44.52	2.013	4	0	2	11	22	14
44.99	45.06	2.010	-4	4	0	19	5	24
45.05		1.9783	0	0	2	6	5	8
45.83	45.84	1.9562	0	4	1	5	2	7
46.38	46.38	1.9257	-1	3	2	3	1	3
47.15	47.16	1.8696	-2	4	1	1	1	2
48.66	48.66	1.8295	-5	1	1	1	1	2
49.80	49.80	1.8227	-4	2	2	2	2	3
49.99	50.00	1.8033	3	2	1	2	2	3
50.57	50.58	1.7971	2	4	0	6	5	8
50.76	50.76	1.7911	5	1	2	3	4	4
50.94	50.90	1.7676	1	3	1	1	1	1
51.67	51.66	1.7364	2	4	1	1	1	1
52.67	52.66	1.7230	4	2	0	12	9	18
53.11	53.10		1	5				
53.49	53.48	1.7116	-5	1	2	2	2	2
54.47	54.48	1.6830	3	1	2	1	1	2
55.15	55.16	1.6639	-1	1	3	6	5	10
55.35	55.30	1.6583	-3	5	1	1	4	
55.52	55.52	1.6537	0	4	2			
56.58	56.58	1.6251	-2	2	3	6	5	9
57.51	57.52	1.6011	-5	3	1	17	13	28
58.13	58.14	1.5854	4	5	0	18	13	29
59.39	59.40	1.5549	5	3	0	7	6	12
60.45	60.46	1.5300	6	0	0	1	1	2
60.81	60.82	1.5219	3	5	0	4	3	6
60.96	60.96	1.5185	-6	0	2	6	5	10
61.86	61.90	1.4983	-5	3	2	1	7	1
61.87		1.4977	-4	2	3			
61.90		1.4936	-1	3	3	14	13	25
62.09	62.08	1.4688	4	0	2	7		12
63.25	63.26	1.4618	2	2	0	1	11	2
63.59	63.60	1.4502	0	6	0	5	4	9
64.16	64.16	1.4349	-5	1	3	1	1	2
64.93	64.94	1.4320	-6	2	2	2	2	3
65.08	65.10	1.4138	4	2	1	1	1	1
66.02	66.02	1.4039	0	6	1	2	2	4
66.55	66.56	1.3950	-3	5	2	1	1	2
67.03	67.04	1.3929	2	6	0	12	11	23
67.14	67.20	1.3921	5	0	1	1	17	1
67.19		1.3872	1	5	2	19		35
67.46	67.38	1.3739	1	5	3	3	9	6
68.20	68.20	1.3677	-2	4	3	6	4	12
68.55	68.54	1.3537	-7	1	3	2	2	4
69.36	69.36	1.3331	0	4	1	2	1	2
70.59	70.60	1.3179	-7	3	3	1	1	1
71.53	71.54	1.3137	-5	1	2	8	5	17
71.79	71.76	1.3042	3	3	3	4	5	9
72.40	72.40	1.2983	5	1	1	1	1	1
72.84	72.84	1.2973	6	2	2	4	4	8
72.86		1.2970	7	1	0	2		3

2THETA	PEAK	D	H	K	L	I(INT)	I(PK)	I(DS)
73.21	73.20	1.2917	-3	1	4	6	2	4
73.62	73.62	1.2855	-1	4	4	2	2	2
74.60	74.62	1.2710	-4	0	4	2	1	4
74.86	74.86	1.2672	-2	6	2	3	3	3
75.07	75.08	1.2643	2	6	0	8	6	6
75.42	75.42	1.2592	0	0	4	4	4	17
76.75	76.76	1.2407	-7	3	1	1	1	9
77.22	77.22	1.2344	4	4	2	8	5	3
77.82	77.82	1.2263	4	5	1	1	1	18
78.56	78.56	1.2166	-1	7	1	8	5	4
79.19	79.18	1.2085	-7	1	3	2	1	1
80.30	80.30	1.1945	1	7	1	3	2	6
83.69	83.70	1.1546	-8	2	2	4	1	3
84.06	84.06	1.1504	-6	0	4	4	2	10
84.32	84.32	1.1475	8	0	0	2	2	4
84.58	84.58	1.1447	-6	4	3	1	1	2
85.82	85.82	1.1313	-1	7	2	1	1	1
86.24	86.28	1.1269	-5	5	1	1	1	1
86.27		1.1265	6	4	2	1	1	2
88.93	88.92	1.0996	-4	4	4	1	1	1
89.19	89.20	1.0970	2	2	4	3	1	4
91.44	91.44	1.0759	-8	2	2	4	2	7
93.06	93.06	1.0614	-7	8	1	5	3	11
93.46	93.44	1.0579	-2	1	4	1	2	12
94.00	94.02	1.0531	3	1	4	1	2	3
94.01		1.0531	-6	6	2	3		5
94.15	94.34	1.0519	4	6	3	2	4	3
94.33		1.0504	7	5	0	1		18
94.58	94.62	1.0483	-5	7	1	7	4	10
94.66		1.0475	-3	5	1	4	4	10
95.00	94.90	1.0447	-3	1	1	4	2	3
95.83	95.84	1.0378	7	1	2	1	2	5
96.19	96.16	1.0350	5	7	0	2	1	3
96.90	96.90	1.0292	8	2	1	1	1	6
97.42	97.42	1.0251	2	8	1	2	1	3
97.81	97.82	1.0220	5	3	3	3	2	9
97.92		1.0212	-2	2	5	1		3

2THETA	PEAK	D	H	K	L	I(INT)	I(PK)	I(DS)
98.39	98.36	1.0176	-1	7	3	3	1	6
99.85	99.84	1.0066	-2	4	4	1	0	2
101.04	101.06	0.9979	-9	3	1	2	1	7
101.99	102.28	0.9912	1	5	1	2	1	5
102.26		0.9893	4	8	0	1		2
102.28		0.9891	4	0	4	2		5
103.25	103.26	0.9825	5	7	1	2	1	4
103.63	103.62	0.9799	-1	5	3	1	1	2
104.32	104.32	0.9753	-1	3	5	5	2	14
104.98	104.96	0.9710	-6	2	2	2	1	6
107.23	107.24	0.9568	-10	0	3	3	2	5
107.26		0.9566	-9	3	3	3		10
107.73	107.66	0.9537	-5	7	2	2	2	7
110.97	110.96	0.9348	-10	2	2	1	0	2
112.01	112.04	0.9291	-2	8	3	2	1	6
112.89	112.86	0.9243	8	0	2	4	0	2
113.91	113.94	0.9189	-7	5	4	4	1	14
114.08		0.9180	10	0	0	1		3
115.90	115.94	0.9088	2	2	5	2	1	13
116.10		0.9077	3	5	4	2		3
116.38	116.38	0.9064	5	7	0	1	1	3
116.47		0.9060	7	0	2	1		3
116.86	116.86	0.9041	-6	6	4	1	1	3
117.28	117.36	0.9020	-1	7	4	2	1	2
117.36		0.9017	4	4	4	1		5
117.42		0.9013	9	3	1	2		6
118.15	118.10	0.8979	7	5	2	2	1	11
122.29	122.32	0.8794	-8	2	5	3	1	6
122.85	122.84	0.8771	-0	10	0	2	1	2
122.88		0.8769	0	10	2	1		5
123.39	123.64	0.8748	-4	0	6	4	2	2
123.51		0.8744	-10	4	1	1		14
123.65		0.8738	2	8	3	4		4
123.84		0.8730	-3	0	6	3		6
123.95		0.8726	-10	0	9	1		2
124.99	124.96	0.8684	-5	9	1	1	0	5
128.00	128.02	0.8570	-2	10	1	1	0	4

271

TABLE 13. ORTHO - AND CLINOPYROXENES

Variety	Hypersthene	Ferrosilite	Clinoenstatite	Ferro-Pigeonite	Clinoferrosilite
Composition	$(Fe,Mg)_2Si_2O_6$	$FeSiO_3$	$MgSiO_3$	$(Fe^2,Ca)(Mg,Fe^2)$ Si_2O_6	$FeSiO_3$
Source	Synthetic	Synthetic	Heated Bishopville, S. C., enstatite	Isle of Mull	Synthetic
Reference	Ghose, 1965	Burnham, 1967	Morimoto, Appleman and Evans, 1960	Morimoto and Güven, 1968	Burnham, 1967
Cell Dimensions					
a Å	18.310	18.431	9.620	9.706	9.7085
b	8.927	9.080	8.825	8.950	9.0872
c	5.226	5.238	5.188	5.246	5.2284
α	90	90	90	90	90
β deg	90	90	108.33	108.59	108.432
γ	90	90	90	90	90
Space Group	Pbca	Pbca	P2$_1$/c	P2$_1$/c	P2$_1$/c
Z	8	16	8	4	8

TABLE 13. (cont.)

Variety	Hypersthene	Ferrosilite	Clinoenstatite	Ferro-Pigeonite	Clinoferrosilite
Site Occupancy					
M_1	{Mg 0.85 Fe² 0.15	Fe² 1.0	Mg 1.0	{Mg 0.73 Fe 0.27	Fe² 1.0
M_2	{Mg 0.10 Fe 0.90	Fe² 1.0	Mg 1.0	{Mg 0.05 Fe 0.77 Ca 0.18	Fe² 1.0
Method	3-D, counter & film	3-D, counter	3-D, film	3-D, counter	3-D, counter
R & Rating	0.118 (2)	0.051 (1)	0.13 (av.) (3)	0.036 (1)	0.043 (1)
Cleavage and habit	{210} perfect		Prismatic		{110} good
μ	220.8	241.6	102.1	235.1	241.6
ASF	0.4276×10^{-1}	0.5213×10^{-1}	0.7692×10^{-1}	0.6610×10^{-1}	0.9750×10^{-1}
Abridging factors	1:1.4:0.5	1:1.4:0.5	0.5:0:0	0.5:0:0	0.5:0:0

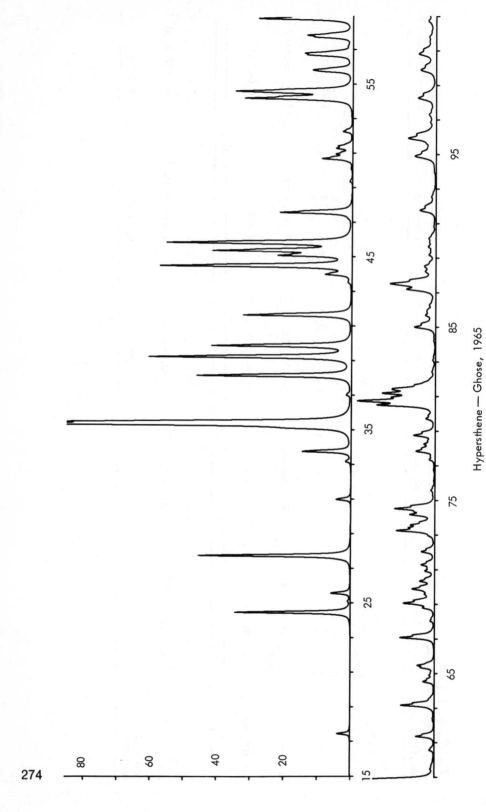

Hypersthene — Ghose, 1965

λ = FeKα Half width at 40°2θ: 0.11°2θ Plot scale I_{PK} (max) = 100

HYPERSTHENE - GHOSE, 1965

2THETA	PEAK	D	H	K	L	I(INT)	I(PK)	I(DS)
17.42	17.44	6.391	2	1	0	3	4	2
24.42	24.42	4.577	4	0	0	32	34	24
25.54	25.54	4.379	1	1	1	6	6	4
27.69	27.70	4.046	2	1	1	44	45	37
30.96	30.96	3.627	3	1	1	5	4	4
33.14	33.14	3.394	0	2	1	2	2	1
33.72	33.72	3.337	1	1	2	16	15	15
35.26	35.28	3.196	4	2	0	100	100	100
35.42	35.42	3.182	2	2	1	74	88	75
36.99	37.00	3.052	6	0	0	1	2	1
38.09	38.10	2.966	3	2	0	52	46	55
39.17	39.18	2.888	6	1	0	69	60	75
39.81	39.82	2.843	5	1	1	47	42	52
41.59	41.60	2.726	4	2	1	38	32	44
43.95	43.96	2.587	1	3	1	8	8	10
44.43	44.44	2.560	1	1	3	71	57	87
45.04	45.04	2.527	6	1	1	23	22	29
45.32	45.32	2.513	2	0	2	47	41	59
45.77	45.78	2.489	5	3	0	30	55	37
45.78		2.489	2	3	1	37		47
47.53	47.54	2.402	3	0	2	28	22	36
49.33	49.34	2.320	3	1	1	1	1	1
50.66	50.66	2.263	7	1	1	11	9	16
50.93	50.92	2.251	4	2	2	5	5	7
51.25	51.26	2.238	1	3	2	6	5	8
51.41	51.38	2.232	0	4	0	3	4	4
52.22	52.22	2.199	4	1	2	4	3	6
54.05	54.14	2.130	6	3	0	4	32	7
54.14		2.127	5	2	2	39		59
54.47	54.56	2.115	3	2	2	17	35	25
54.55		2.112	5	3	1	35		53
55.71	55.78	2.072	7	2	1	9	12	14
55.79		2.069	5	1	2	11		16
56.62	56.68	2.041	8	1	1	5	14	8
56.66		2.040	1	4	2	12		18
56.76	56.76	2.037	8	2	0	9	14	14
57.70	57.72	2.006	4	4	0	13	12	20

2THETA	PEAK	D	H	K	L	I(INT)	I(PK)	I(DS)
57.81	57.82	2.003	2	4	1	11	14	17
58.77	58.78	1.9728	6	3	1	41	28	66
59.45	59.44	1.9522	1	3	2	1	1	1
60.55	60.54	1.9202	5	2	2	2	2	2
61.34	61.34	1.8976	8	0	2	8	6	4
63.15	63.16	1.8486	7	2	1	15	10	14
63.52	63.50	1.8389	6	3	1	1	1	27
64.52	64.52	1.8136	6	2	2	5	3	3
64.88	64.88	1.8045	10	1	0	2	4	9
65.32	65.34	1.7937	5	4	1	6	2	3
65.46	65.46	1.7904	2	5	0	17	4	10
67.06	67.06	1.7524	5	3	2	2	5	8
68.03	68.04	1.7304	8	3	1	5	1	32
68.78	68.78	1.7139	7	2	2	14	4	4
69.05	69.04	1.7079	1	2	3	2	10	9
69.31	69.20	1.7023	10	1	1	3	2	28
69.58	69.58	1.6965	8	1	2	5	6	4
69.86	69.88	1.6905	1	1	4	5	3	5
69.90		1.6898	2	4	2	7	7	11
70.33	70.34	1.6807	2	4	3	5	5	10
70.92	70.92	1.6686	2	2	4	6	3	14
71.27	71.26	1.6615	5	1	1	7	4	9
72.02	72.02	1.6464	3	3	3	9	4	13
73.24	73.24	1.6228	1	0	2	19	12	14
73.57	73.56	1.6165	9	2	3	10	7	40
74.17	74.18	1.6053	9	0	2	11	8	21
74.52	74.52	1.5989	9	3	1	19	12	24
75.57	75.56	1.5799	5	1	3	3	1	41
77.34	77.34	1.5492	6	5	1	3	2	4
77.83	77.84	1.5410	5	4	3	9	6	6
77.90		1.5397	12	0	2	2		21
78.24	78.24	1.5341	9	3	1	5	3	3
78.75	78.76	1.5258	1	2	2	11	7	11
79.71	79.70	1.5105	10	2	3	5	18	26
80.49	80.50	1.4983	1	3	1	29	18	10
80.75	80.74	1.4943	10	6	0	31	24	68
81.17	81.18	1.4878	0	6	0	26	16	61

HYPERSTHENE - GHOSE, 1965

2THETA	PEAK	D	H	K	L	I(INT)	I(PK)	I(DS)
81.45	81.40	1.4836	5	2	3	1	13	3
81.46		1.4835	2	3	3	13		30
81.49		1.4831	6	4	2	1		3
82.41	82.42	1.4694	1	5	2	1	1	3
83.08	83.08	1.4597	3	3	3	1	1	3
84.99	85.00	1.4329	3	5	1	13	6	32
85.68	85.68	1.4237	7	4	2	5	3	12
85.84	85.88	1.4214	10	2	2	1	3	3
86.01		1.4192	7	5	1	3		7
86.29	86.26	1.4156	10	4	0	1	2	3
86.33		1.4150	4	6	0	1		3
87.18	87.20	1.4039	11	0	2	17	9	44
87.51	87.50	1.3996	11	3	1	26	14	67
88.14	88.22	1.3917	12	2	1	1	3	4
88.22		1.3907	5	5	3	4		12
88.53	88.50	1.3868	11	1	3	5	4	14
89.17	89.18	1.3790	7	2	2	4	2	11
89.64	89.64	1.3732	0	4	3	5	2	12
89.96	89.90	1.3694	1	4	3	1	2	3
90.12	90.12	1.3675	5	5	2	4	2	9
90.82	90.82	1.3593	8	5	1	1	1	3
91.74	91.74	1.3486	6	5	3	12	5	33
92.11	91.98	1.3444	13	3	1	2	3	5
92.53	92.54	1.3397	3	4	3	2	1	5
92.57		1.3392	11	2	2	1		2
93.98	93.98	1.3238	8	4	3	2	1	6
94.77	94.88	1.3153	4	4	1	15	6	43
94.89		1.3141	12	3	2	11		31
95.91	95.92	1.3035	12	1	3	5	8	13
95.92		1.3034	7	3	3	6		17
95.94		1.3032	1	0	4	2		4
96.16	96.16	1.3009	13	2	1	2	6	5
96.27		1.2998	9	5	0	3		10
96.84	96.84	1.2940	14	1	4	4	2	13
97.29	97.28	1.2895	1	1	2	14	5	43
98.24	98.26	1.2802	2	6	2	1	4	2
98.45	98.48	1.2783	10	5	0	1		

2THETA	PEAK	D	H	K	L	I(INT)	I(PK)	I(DS)
98.52	98.52	1.2776	3	0	4	1	5	4
99.87	99.88	1.2649	3	6	2	12		37
99.89		1.2647	3	1	4	1		2
100.06		1.2631	2	7	0	1		2
100.80	100.80	1.2563	4	0	4	11	5	33
100.82		1.2561	14	1	1	3		9
100.90		1.2554	7	6	1	1		2
101.07	101.06	1.2539	0	2	3	3	4	5
101.54	101.54	1.2496	10	1	3	3	2	5
101.79	101.78	1.2474	8	6	0	3	1	9
102.18	102.06	1.2440	1	5	3	1	1	10
102.44	102.72	1.2417	10	5	1	6	2	3
102.71	103.12	1.2393	8	5	1	4	1	4
103.17	104.02	1.2354	2	5	0	1		17
103.99		1.2285	4	7	2	1		12
104.04	105.12	1.2280	13	1	2	4	3	3
104.82		1.2216	3	5	3	7		2
105.12		1.2192	5	6	2	2		12
105.14		1.2190	5	2	4	3		21
106.34	106.34	1.2093	4	4	4	7	1	7
108.25	108.38	1.1946	13	2	2	3		9
108.37		1.1937	9	5	2	2	4	6
108.51		1.1927	11	5	3	11		37
109.37	109.36	1.1863	5	2	4	3	1	11
110.19	110.22	1.1803	15	5	3	1	1	4
110.48	110.48	1.1783	15	5	1	3	1	3
111.14	111.14	1.1736	7	7	1	1	1	9
111.82	112.08	1.1688	14	0	4	3	1	3
112.08		1.1671	7	3	1	2		6
113.26	113.26	1.1591	7	6	1	6	2	19
113.28		1.1589	15	2	1	1		6
114.86	114.98	1.1486	6	7	1	2	1	7
114.97		1.1479	16	0	1	1		4
115.53	115.54	1.1444	1	7	0	5	2	18
115.61		1.1438	8	0	4	1		4
117.10	117.10	1.1346	8	0	4	8	2	28
117.76	117.46	1.1307	7	2	4	1	2	3

HYPERSTHENE - GHOSE, 1965

2THETA	PEAK	D	H	K	L	I(INT)	I(PK)	I(DS)
118.30	118.28	1.1275	0	4	4	2	1	7
119.11	119.08	1.1228	2	6	3	1	0	3
119.66	119.62	1.1197	7	7	1	1	0	3
120.33	120.42	1.1159	0	8	0	1	1	2
120.37		1.1157	11	3	3	1		3
120.71	121.12	1.1138	6	3	4	1	1	2
120.96		1.1124	3	6	3	1		3
121.08		1.1118	4	7	2	1		2
121.10		1.1116	12	2	3	3		10
121.54	121.62	1.1092	16	1	0	2	2	7
121.67		1.1085	16	2	2	4		15
121.83		1.1077	2	8	0	1		2
122.15	122.10	1.1059	15	0	2	2	2	8
122.60	122.60	1.1036	11	5	2	5	2	20
122.71		1.1030	14	5	1	1		2
123.34	123.36	1.0997	8	4	4	6	2	24
123.41		1.0993	9	0	4	1		3
123.56		1.0986	10	4	3	1		3
123.61		1.0983	4	6	3	1		3
125.02	125.04	1.0912	9	6	2	14	3	54
125.04		1.0911	9	1	4	1		3
125.35	125.38	1.0895	8	7	1	2	3	7
126.27	126.52	1.0851	11	6	1	1	3	3
126.42		1.0844	16	8	0	4		15
126.47		1.0841	4	4	0	3		12
126.54		1.0838	13	8	2	1		3
126.58		1.0836	2	5	1	5		21
126.95		1.0818	13	8	1	13		53
128.60	128.64	1.0742	3	8	1	5	3	20
128.77		1.0735	15	2	2	3		11
131.16	131.62	1.0631	9	5	3	12	4	48
131.53		1.0615	4	8	1	10		41
131.86		1.0602	12	5	2	1		6
132.08	133.10	1.0593	11	4	3	2	5	6
132.84		1.0562	10	6	2	5		22
132.87		1.0560	10	1	4	23		94
133.12		1.0551	14	5	0			

2THETA	PEAK	D	H	K	L	I(INT)	I(PK)	I(DS)
135.04	135.36	1.0476	17	1	1	1	3	3
135.08		1.0474	2	5	4	8		34
135.34		1.0465	10	0	4	8		32
135.51		1.0458	5	8	1	6	1	24
136.80	136.76	1.0411	16	1	2	4	1	16
137.40	137.38	1.0390	3	5	5	4	2	18
138.12	138.78	1.0364	1	1	2	1		4
138.27		1.0359	14	4	4	1		4
138.67		1.0346	10	2	2	7		2
138.77		1.0342	14	5	1	22		30
140.70	140.74	1.0278	13	3	3	7	5	95
140.73		1.0277	13	0	4	4		30
141.09		1.0266	17	2	1	3		11
141.67	141.30	1.0248	8	7	2	4	4	18
141.73		1.0246	1	8	2	3		13
142.39	142.90	1.0226	2	7	7	1	4	3
142.88		1.0211	11	1	2	19		82
142.91		1.0210	11	6	4	7		29
143.01		1.0207	12	1	3	1		2
143.81	143.60	1.0184	13	5	3	2	3	8
143.83		1.0183	16	2	0	8		33
145.64	146.24	1.0132	5	4	4	1	2	5
145.93		1.0124	4	5	5	1		5
146.08		1.0120	3	1	2	2		11
146.29		1.0114	2	8	5	5		22
146.29		1.0114	8	4	4	3		12
147.94	147.88	1.0071	7	7	1	4	1	17
149.23	149.26	1.0040	4	8	3	1	1	4
149.29		1.0038	3	2	5	3		5
149.63		1.0030	8	3	0	1		4
153.60	154.36	0.9943	17	3	1	17	4	80
153.80		0.9939	11	7	3	1		3
153.99		0.9935	15	1	5	2		9
154.02		0.9934	4	0	1	12		56
154.53		0.9924	12	6	4	16		72
154.58		0.9923	18	1	1	2		8
154.84		0.9918	18	2	0	1		4

HYPERSTHENE — GHOSE, 1965

2THETA	PEAK	D	H	K	L	I(INT)	I(PK)	I(DS)
155.29		0.9909	15	4	2	3		15
155.44		0.9906	5	7	3	2		9
155.98		0.9896	17	1	2	10		45
156.09		0.9894	15	5	1	1		7
156.81	157.94	0.9881	5	8	2	2	5	9
157.63		0.9867	14	3	3	54		253
157.95		0.9862	9	4	4	5		24
157.99		0.9861	2	9	0	3		12
158.65		0.9850	8	8	1	28		132
159.79		0.9832	13	4	3	3		12
160.08		0.9828	6	1	5	15		71
160.42		0.9823	14	6	0	2		11

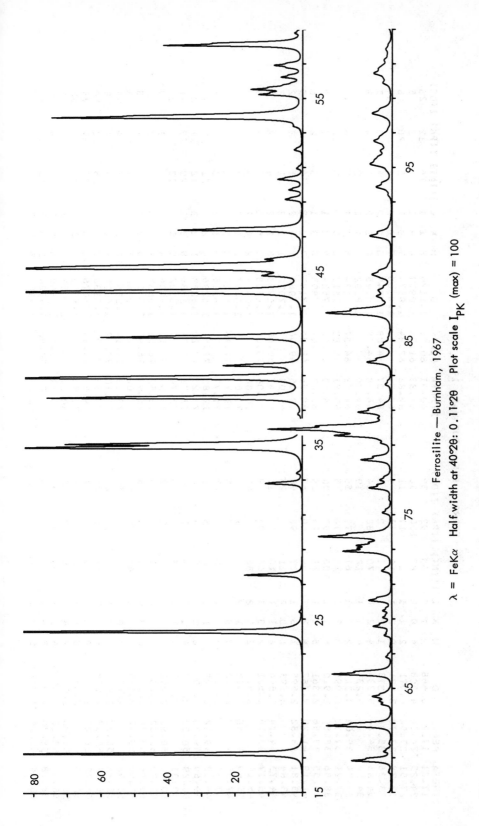

Ferrosilite — Burnham, 1967

λ = FeKα Half width at 40°2θ: 0.11°2θ Plot scale I_{PK} (max) = 100

279

FERROSILITE — BURNHAM, 1967

2THETA	PEAK	D	H	K	L	I(INT)	I(PK)	I(DS)
17.21	17.22	6.468	2	1	0	68	92	39
24.25	24.26	4.608	4	0	0	82	100	56
27.51	27.52	4.071	2	1	1	15	17	11
32.78	32.78	3.431	0	2	0	10	11	9
33.36	33.36	3.373	1	1	2	1	2	1
34.83	34.84	3.234	4	2	0	70	88	64
34.84		3.233	2	2	1	12		10
35.04	35.04	3.215	2	2	1	61	71	55
37.71	37.72	2.995	3	2	1	77	76	75
38.86	38.86	2.910	6	1	0	100	97	100
39.55	39.56	2.861	5	1	1	23	24	24
41.19	41.20	2.752	4	2	1	64	60	67
43.81	43.82	2.595	1	3	0	87	96	98
43.84		2.593	1	0	1	21		23
44.73	44.74	2.544	6	1	1	10	12	11
45.17	45.18	2.521	2	3	1	68	94	79
45.19		2.519	3	1	1	36		42
45.69	45.68	2.493	1	3	2	9	11	10
47.35	47.38	2.410	3	0	1	4	37	5
47.38		2.409	3	1	1	39		47
49.12	49.12	2.329	3	0	2	6	5	8
49.69	49.70	2.304	8	0	0	5	4	6
50.29	50.30	2.278	4	3	1	4	7	5
50.31		2.277	7	1	2	5		6
50.92	50.92	2.252	4	2	1	2	2	3
51.99	52.00	2.209	6	1	1	3	3	4
53.36	53.36	2.156	5	3	2	2		3
53.90	53.92	2.136	5	3	1	45	75	63
53.92		2.135	7	0	2	49		70
55.22	55.22	2.089	5	2	1	15	13	22
55.52	55.52	2.078	5	2	2	17	15	25
56.22	56.22	2.054	8	0	0	5	6	7
56.23		2.054	8	1	1	3		4
56.91	56.90	2.032	6	2	1	8	9	12
58.09	58.10	1.9937	3	4	1	53	41	82
58.78	58.78	1.9725	4	3	1	2	3	4
60.14	60.14	1.9320	5	2	2	1	1	2

2THETA	PEAK	D	H	K	L	I(INT)	I(PK)	I(DS)
60.81	60.82	1.9126	8	2	1	16	12	26
61.33	61.34	1.8979	4	4	1	3	3	5
62.82	62.84	1.8574	4	3	1	5	20	8
62.84		1.8568	7	0	2	22		37
64.07	64.08	1.8249	6	2	2	2	3	7
64.28	64.28	1.8195	4	3	2	2	4	3
64.29		1.8192	7	1	2	2		3
64.51	64.52	1.8134	5	4	0	7	6	12
65.81	65.82	1.7817	2	5	1	25	18	45
66.80	66.80	1.7584	9	2	1	3	2	6
68.03	68.04	1.7303	8	3	1	3	6	15
68.56	68.56	1.7187	7	4	1	8	3	2
69.05	69.06	1.7080	1	1	2	1		7
69.08		1.7072	1	4	2	4	3	9
69.45	69.46	1.6993	8	1	1	5		7
70.04	70.06	1.6868	2	4	2	4	3	4
70.06		1.6864	2	1	3	2		2
70.09		1.6857	3	5	1	1		5
71.71	71.76	1.6526	3	1	3	3	3	42
71.77		1.6515	3	2	3	21		24
72.88	72.88	1.6296	0	2	3	12	15	41
73.21	73.20	1.6233	1	2	3	20	11	26
73.72	73.74	1.6136	3	3	1	13	22	9
73.74		1.6132	9	0	2	4		23
75.09	75.10	1.5884	9	5	1	11	3	8
76.52	76.52	1.5633	6	5	0	4	7	3
76.97	76.98	1.5555	5	5	1	1	4	32
77.02		1.5546	5	5	1	15		8
78.13	78.14	1.5359	12	0	3	4		56
79.10	79.12	1.5201	0	6	2	25	9	73
79.53	79.54	1.5133	10	3	0	33	3	58
79.89	79.90	1.5076	1	3	1	26	17	4
79.91		1.5073	11	2	3	2	37	5
80.02		1.5056	6	1	1	2	4	35
80.51	80.52	1.4980	2	5	3	15	10	4
80.87	80.88	1.4924	2	3	3	2		
81.00		1.4905	5	2	3			

FERROSILITE – BURNHAM, 1967

2THETA	PEAK	D	H	K	L	I(INT)	I(PK)	I(DS)
81.20	81.06	1.4875	1	5	2	3	7	8
82.47	82.48	1.4686	3	3	3	3	2	8
83.75	83.76	1.4502	3	5	2	14	8	33
84.64	84.66	1.4378	4	6	0	3	4	8
84.66		1.4375	7	5	1	2		5
84.68		1.4373	7	4	2	1		3
86.58	86.60	1.4117	11	3	1	24	20	61
86.60		1.4114	11	0	2	16		39
87.55	87.56	1.3992	5	3	2	6	5	16
87.90	87.90	1.3947	11	1	2	6	4	16
88.62		1.3857	7	2	3	4	6	10
88.76	88.80	1.3840	0	4	3	5		14
88.82		1.3833	5	5	2	5		13
91.02	91.02	1.3569	13	3	1	15	7	39
91.34	91.24	1.3532	6	1	1	2	4	1
93.37	93.38	1.3304	13	1	1	1	1	6
93.82	93.86	1.3255	8	2	3	2	5	2
93.86		1.3251	4	4	1	10		28
95.14	95.18	1.3114	12	3	3	7	7	14
95.18		1.3110	7	1	3	5		22
95.25		1.3103	12	1	2	8		5
95.33		1.3095	13	0	1	2		6
95.97	95.94	1.3029	14	0	0	5	4	13
96.42		1.2983	7	5	2	3	6	8
96.52	96.52	1.2973	2	6	2	11		30
97.91		1.2835	2	1	4	2		6
98.11	98.12	1.2815	3	6	2	13	6	39
98.19		1.2807	3	0	4	1		3
99.87	99.88	1.2649	8	0	4	5	3	15
99.92		1.2644	8	6	1	1		4
100.43	100.44	1.2596	14	0	4	13	6	39
100.59	100.70	1.2582	4	0	4	1		4
100.85		1.2559	0	2	2	1		2
100.86		1.2557	10	1	3	1	4	4
100.90		1.2554	5	1	3	3		11
101.22	101.20	1.2525	10	3	2	4	3	11
101.66	101.84	1.2486	8	2	2	1	2	2

2THETA	PEAK	D	H	K	L	I(INT)	I(PK)	I(DS)
101.83	101.83	1.2471	2	5	3	3		8
101.86	101.86	1.2468	13	0	2	1		4
103.26	103.28	1.2346	5	3	2	8	3	26
103.40		1.2335	3	7	1	2		2
103.45		1.2330	3	5	3	1		7
104.40	104.68	1.2250	7	4	3	1	1	2
104.68		1.2227	5	1	4	1		4
105.43	105.46	1.2166	9	3	4	1	1	3
105.78	105.76	1.2138	4	3	4	1	1	3
106.76	106.76	1.2061	9	2	4	12	4	39
107.24	107.02	1.2023	13	5	1	1	3	3
108.31	108.32	1.1941	6	1	4	2	1	7
108.66	108.72	1.1915	5	7	1	1	2	2
108.72		1.1911	5	2	4	3		3
108.76		1.1908	5	1	4	3		9
110.74	110.78	1.1764	14	3	1	3	1	11
112.20	111.20	1.1731	7	6	2	8	3	26
112.70	112.70	1.1628	11	1	1	2	1	4
113.61	113.62	1.1568	15	2	0	5	2	6
114.35	114.36	1.1519	16	0	1	9	3	18
116.48	116.50	1.1385	8	0	2	1	2	32
117.01	116.84	1.1352	7	2	4	1		3
117.16		1.1343	4	4	4	2		3
118.37	118.44	1.1271	11	7	2	6	0	2
119.13	119.14	1.1227	16	3	0	1	1	5
120.21	120.58	1.1166	11	2	3	3	3	9
120.22		1.1165	14	5	0	2		9
120.59		1.1144	15	0	1	6		3
120.88	120.94	1.1128	8	2	4	1	2	21
120.96		1.1124	9	6	2	3		2
122.35	122.58	1.1049	8	8	1	1	4	12
122.47		1.1043	9	5	0	2		3
122.57		1.1037	4	8	2	13		8
122.89	122.94	1.1021	2	0	3	3	4	48
123.03		1.1013	13	5	1	3		12
124.67	124.92	1.0929	13	1	1	1	4	13
124.85		1.0920	16	2	0	4		17

281

FERROSILITE - BURNHAM, 1967

2THETA	PEAK	D	H	K	L	I(INT)	I(PK)	I(DS)
124.94		1.0916	3	8	1	12		44
126.47	126.50	1.0841	5	4	4	2	1	7
127.25	127.74	1.0804	15	2	2	2	4	7
127.68		1.0784	4	8	1	11		42
127.78		1.0780	12	6	0	9		34
127.86		1.0776	12	3	3	1		3
129.04	129.56	1.0723	9	5	3	7	3	6
129.48		1.0703	12	2	3	7		27
129.64		1.0696	13	5	2	4		18
130.03	130.50	1.0679	10	2	2	1	5	5
130.50		1.0659	14	6	0	20		81
131.09	130.98	1.0634	7	7	2	1	4	5
131.37		1.0622	5	8	1	4		17
131.71	131.82	1.0608	10	7	0	3	3	11
131.81		1.0604	1	5	1	1		6
131.85		1.0602	10	1	4	9		35
133.09	133.08	1.0552	2	5	4	10	2	42
135.09	135.32	1.0474	16	1	2	3	1	11
135.29		1.0466	3	5	4	2		9
135.87	135.86	1.0445	14	4	1	4	2	18
135.90		1.0444	14	4	2	1		3
137.18	137.32	1.0398	1	8	2	3	2	11
137.28		1.0394	10	5	3	2		10
137.73	137.84	1.0378	8	7	2	3	2	12
138.73	138.80	1.0343	13	3	3	17	4	72
138.90		1.0338	17	2	1	3		12
139.38	139.38	1.0322	11	6	2	15	5	63
139.49		1.0318	11	0	4	4		19
140.60	140.68	1.0281	12	4	3	2	2	9
140.69		1.0279	13	5	2	4		15
140.89		1.0272	16	4	0	2		9
141.04		1.0268	3	8	2	2		7
141.54	141.40	1.0252	11	1	4	4	2	18
142.50	142.36	1.0222	7	8	1	4	1	17
143.51	143.10	1.0192	1	2	5	2	1	7
143.88	144.08	1.0182	8	8	0	1	1	3
144.11		1.0175	18	1	0	2		8
144.76	145.06	1.0156	4	7	3	1	1	3
145.14		1.0146	2	2	5	4		18
149.28	150.20	1.0038	6	8	4	10	4	47
149.98		1.0022	5	8	2	2		7
150.13		1.0018	17	3	1	14		66
150.17		1.0017	17	0	2	7		30
151.17	151.14	0.9994	8	8	1	20	4	93
151.38		0.9990	15	5	1	2		7
151.48		0.9988	15	5	3	1		4
152.47	152.52	0.9966	4	4	5	9	4	43
152.53		0.9965	12	2	5	14		65
152.90		0.9957	17	0	2	9		43
154.10	154.18	0.9930	14	6	0	1	5	5
154.23		0.9930	14	3	3	43		198
154.60		0.9923	9	4	4	1		7
155.60		0.9903	13	4	3	2		9
155.66		0.9902	0	6	4	4		18
157.90	158.92	0.9863	6	8	2	1	3	4
157.98		0.9861	6	7	3	2		11
158.26		0.9857	6	1	5	5		25
158.65		0.9850	7	5	4	1		6
158.67		0.9850	2	9	1	14		67
159.10		0.9843	9	3	5	19		91
159.45		0.9838	2	5	5	3		13
161.19		0.9812	5	2	1	3		14
161.24		0.9811	18	2	3	9		43

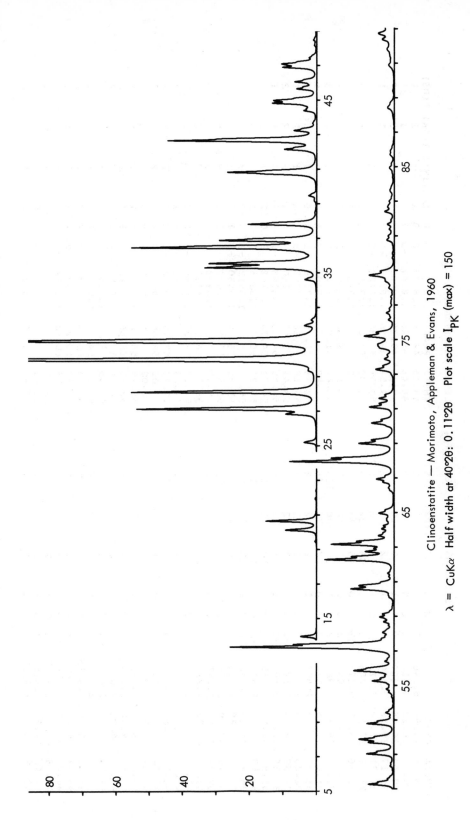

Clinoenstatite — Morimoto, Appleman & Evans, 1960

$\lambda = CuK\alpha$ Half width at $40°2\theta$: $0.11°2\theta$ Plot scale I_{PK} (max) = 150

CLINOENSTATITE – MORIMOTO, APPLEMAN AND EVANS, 1960

2THETA	PEAK	D	H	K	L	I(INT)	I(PK)	I(DS)
13.94	13.96	6.346	1	1	0	3	3	2
20.11	20.12	4.412	0	2	0	5	6	5
20.64	20.64	4.300	0	1	1	9	10	8
25.17	25.18	3.534	-1	1	1	2		2
26.81	26.82	3.322	1	2	0	5	6	5
27.11	27.12	3.286	0	2	1	34	36	34
28.10	28.10	3.173	2	0	0	36	37	35
29.32	29.32	3.044	-2	1	1	1	2	1
29.96	29.96	2.980	2	1	0	100	100	100
31.05	31.06	2.878	-2	2	1	68	77	68
31.93	31.94	2.800	3	1	0	2	2	2
32.25	32.26	2.773	1	3	1	1	1	1
34.59	34.60	2.591	-1	3	1	2	2	2
35.28	35.28	2.542	-1	3	1	23	22	24
35.52	35.52	2.525	0	0	2	10	21	10
35.53		2.524	-2	0	2	12		12
36.46	36.46	2.462	0	2	2	39	37	41
36.87	36.88	2.436	-2	2	1	20	19	21
37.79	37.80	2.378	-2	3	1	15	14	16
39.44	39.44	2.283	4	0	1	2	2	2
40.71	40.82	2.214	3	1	1	8	18	9
40.81		2.209	1	0	2	15		16
42.13	42.12	2.143	-3	1	2	7	6	7
42.64	42.64	2.119	3	3	1	34	30	37
42.71		2.115	3	3	0	1		1
43.21	43.22	2.092	-4	0	1	5	5	5
43.62	43.62	2.073	-2	4	1	1	1	1
44.37	44.38	2.040	-3	2	2	3	3	3
44.79	44.82	2.022	-1	4	0	3		2
44.81		2.021	-4	1	1	8	9	9
44.98	44.96	2.013	0	4	1	7	9	7
45.63	45.64	1.9865	2	0	2	5	4	5
46.04	46.04	1.9697	-4	1	2	5	4	5
46.88	46.88	1.9365	-2	3	1	8	7	9
47.08	47.08	1.9287	2	0	2	7	7	8
47.42	47.42	1.9156	-2	3	0	2	2	1
47.43		1.9151	1	4	1	1		1
48.15	48.16	1.8882	0	3	2	1	1	1
48.46	48.46	1.8767	-5	1	1	1	1	1
49.26	49.26	1.8483	-4	3	1	3	1	7
49.57	49.56	1.8372	-4	2	2	7	5	7
50.31	50.30	1.8122	-3	1	2	1	1	1
51.02	51.02	1.7885	5	1	0	1	1	1
51.02		1.7885	-3	4	1	6	6	7
51.68	51.68	1.7672	2	2	2	1	1	1
51.87	51.88	1.7610	-5	2	1	6	5	6
51.89		1.7606	2	5	0	3	7	3
52.78	52.78	1.7329	-1	1	0	4	5	5
53.18	53.18	1.7208	-5	2	1	7	5	8
53.41	53.40	1.7140	4	1	1	1	2	1
54.59	54.58	1.6798	-1	4	2	2	1	2
54.78	54.78	1.6742	3	0	2	1	2	2
55.24	55.24	1.6615	0	0	2	1		2
55.25		1.6612	-2	4	2	3	4	4
55.55	55.54	1.6530	-1	1	3	1		1
55.84	55.86	1.6449	-3	2	2	9	8	11
55.91		1.6432	3	4	0	3		4
56.39	56.38	1.6303	-5	2	2	2	2	2
57.22	57.24	1.6085	-2	3	1	19	33	23
57.23		1.6082	-1	5	1	27		32
57.70	57.70	1.5962	3	4	0	2	3	2
58.03	58.04	1.5880	5	1	1	2	2	2
58.09		1.5865	4	4	0	1		1
58.45	58.46	1.5775	-6	1	1	1	1	2
58.67	58.66	1.5721	4	3	1	2	2	2
58.94	58.94	1.5656	-4	1	3	2	3	2
58.95		1.5653	3	2	2	1		2
59.13	59.12	1.5611	1	4	2	1	2	1
60.08	60.10	1.5386	0	2	3	1	1	1
60.19	60.24	1.5361	5	1	1	1	1	1
60.53	60.58	1.5282	-3	5	1	3	9	3
60.58		1.5271	-6	0	2	4		5
60.59		1.5269	3	3	0	7		8
60.81	60.76	1.5220	6	0	0	5	7	6

CLINOENSTATITE - MORIMOTO, APPLEMAN AND EVANS, 1960

2THETA	PEAK	D	H	K	L	I(INT)	I(PK)	I(DS)
61.29	61.28	1.5112	1	2	3	1	1	1
61.48	61.48	1.5069	-6	2	1	1	1	1
62.25	62.28	1.4901	-4	4	2	19	14	2
62.28		1.4896	-2	3	3			23
62.73	62.74	1.4798	-1	3	3	7	6	9
63.16	63.16	1.4708	0	0	0	18	13	22
63.65	63.66	1.4607	4	0	2	4	3	4
63.74		1.4587	-1	5	1	1		
64.05	64.06	1.4526	-5	1	3	2	3	3
64.07		1.4521	2	4	2	1		1
64.23	64.22	1.4488	-1	1	3	1	2	2
64.52	64.52	1.4431	-6	2	2	1	2	2
64.62	64.62	1.4411	4	2	2	1	2	2
64.95	64.96	1.4345	0	5	3	4	3	5
65.00		1.4335	0	3	3	1		1
66.26	66.28	1.4093	5	6	0	1	1	1
66.39	66.40	1.4069	-3	4	2	1	1	3
66.74	66.74	1.4004	3	5	1	3	2	2
66.90	66.92	1.3974	-5	2	3	1	4	3
66.92		1.3969	5	5	1	2		
67.91	67.96	1.3790	5	3	3	32	21	42
67.96		1.3781	-5	4	1	1		1
68.24	68.14	1.3732	-2	4	2	6	13	7
68.99	68.98	1.3601	1	4	3	5	7	7
68.99		1.3601	-7	1	1	1		1
69.22	69.18	1.3561	-7	0	2	7	5	9
70.16	70.16	1.3402	-7	1	1	7	5	10
71.08	71.08	1.3251	0	4	2	2	5	3
71.58	71.58	1.3170	-5	3	4	3	3	3
71.59		1.3169	-3	0	3	4		6
73.24	73.26	1.2912	7	1	4	4	4	3
73.29		1.2905	6	0	0	2		1
73.84	73.86	1.2823	-2	2	1	1	2	3
73.87		1.2818	-3	1	4	2		1
74.15	74.08	1.2776	5	1	2	1	1	1
74.49	74.52	1.2727	5	1	2	2	3	3
74.51		1.2723	3	1	3	1		1

2THETA	PEAK	D	H	K	L	I(INT)	I(PK)	I(DS)
74.62	74.62	1.2708	-2	6	2	3	3	4
74.73	74.72	1.2692	5	5	0	2	3	2
75.18	75.18	1.2627	0	6	2	9	6	12
75.22		1.2621	-4	0	4	2		2
76.12	76.12	1.2494	-4	1	4	2	1	2
76.53	76.52	1.2437	-7	3	1	2	2	3
77.06	77.08	1.2364	4	6	0	1	1	1
77.45	77.44	1.2312	0	0	4	1	1	1
77.97	77.96	1.2243	1	6	2	1	1	1
78.35	78.36	1.2194	0	1	4	2	2	3
78.46	78.46	1.2179	-3	5	3	1	2	2
78.67		1.2152	3	7	0	1	5	5
78.71	78.70	1.2147	0	5	3	7		
79.70	79.70	1.2020	-4	6	2	1	0	1
80.74	80.74	1.1892	-4	5	3	2	2	3
81.36	81.36	1.1817	-7	2	2	1	1	1
81.65	81.64	1.1815	3	3	3	1		1
82.38	82.38	1.1781	3	6	2	2	1	2
82.39		1.1695	2	5	2	1	2	2
82.75	82.64	1.1695	-5	2	4	2		3
82.80		1.1653	-3	7	0	1	1	1
82.97	82.98	1.1648	3	1	1	1	1	2
83.56	83.58	1.1628	-7	1	1	1		1
83.67		1.1561	-8	2	2	1		1
84.87	84.92	1.1549	-6	5	0	1	1	1
84.95		1.1415	-6	0	4	1		1
85.66	85.68	1.1407	1	1	4	1		1
88.37	88.40	1.1330	8	2	0	1	1	2
88.38		1.1051	3	6	0	2		2
88.47		1.1050	-8	7	1	2		2
90.78	90.88	1.1041	-7	0	3	3	1	3
90.88		1.0820	-8	2	4	1		3
91.75	91.80	1.0810	-7	0	4	2	1	2
91.83		1.0730	-7	2	4	1		1
91.88		1.0723	2	8	0	1		1
92.38	92.38	1.0719	8	0	3	1		1
		1.0674	-7	7	2	5	2	7

285

CLINOENSTATITE - MORIMOTO, APPLEMAN AND EVANS, 1960

2THETA	PEAK	D	H	K	L	I(INT)	I(PK)	I(DS)
92.74	92.72	1.0642	-2	8	1	5	3	8
92.89		1.0629	-5	4	4	1		1
93.28	93.28	1.0594	-6	6	2	2	2	3
93.48	93.48	1.0577	0	6	0	2	1	2
93.91	93.78	1.0539	-5	7	1	1	1	1
94.48		1.0491	7	5	0	5	1	7
95.15	94.48	1.0435	-8	3	3	2	2	4
95.31	95.16	1.0421	-3	5	3	1	2	2
95.66	95.62	1.0393	3	5	3	1		1
96.08	96.32	1.0358	5	2	3	1	1	1
96.31		1.0339	3	0	4	3	2	2
96.34		1.0337	-9	2	2	3		5
96.55	96.56	1.0320	2	8	1	3	2	4
97.19	97.22	1.0269	3	1	4	2	2	3
97.23		1.0266	-4	5	4	2		4
97.59	97.54	1.0238	-5	7	2	2	1	2
98.08	98.06	1.0200	7	1	2	2	1	2
98.82	98.82	1.0143	8	2	0	2	1	3
98.88		1.0138	8	4	4	1		1
99.42	99.38	1.0098	0	5	3	4	1	2
100.49	100.50	1.0019	5	3	1	4	2	6
100.68		1.0005	-4	8	1	1		2
100.91	100.84	0.9989	-9	3	3	2	2	2
101.87	102.06	0.9920	-9	2	3	1	1	2
102.07		0.9907	-7	6	2	3		4
103.00	102.98	0.9842	-5	2	1	1	1	1
103.54	103.56	0.9805	5	7	1	1	1	2
103.79	103.90	0.9789	0	1	5	1	1	1
104.08	104.12	0.9770	7	5	1	3	1	5
104.55	104.54	0.9738	-2	3	1	1	1	1
105.08	104.96	0.9704	-3	6	4	2	1	3
106.12	106.36	0.9637	-6	5	4	1		1
106.30		0.9626	-1	9	1	1		1
106.37		0.9621	-9	3	3	1	1	2
106.66	106.68	0.9603	-10	0	3	2	1	1
107.06	107.08	0.9578	-4	6	4	1	1	1
107.08		0.9577	-5	7	3	1		1

2THETA	PEAK	D	H	K	L	I(INT)	I(PK)	I(DS)
107.26	107.46	0.9566	-5	8	1	1	1	1
107.57	108.56	0.9547	-10	1	2	1	1	1
108.55	109.36	0.9488	4	8	1	1	0	2
109.35	109.66	0.9441	0	6	4	1	1	1
109.64	110.32	0.9424	8	4	1	1	1	1
110.34	111.14	0.9383	-10	2	2	1	0	1
111.11	111.92	0.9340	0	8	3	3	1	5
111.89		0.9297	-2	3	5	1		1
112.00	112.36	0.9291	-6	3	3	1	1	1
112.32		0.9273	-1	8	5	1		1
112.37		0.9271	-10	1	2	1		5
112.41		0.9269	1	3	5	1		1
112.85	112.82	0.9245	6	5	3	1	1	1
113.34	113.28	0.9219	-7	7	2	1	1	1
114.02	114.02	0.9183	-7	0	0	2	1	3
115.02	115.02	0.9132	10	0	2	1	1	1
115.65	115.50	0.9100	8	3	5	2	1	5
117.22	117.62	0.9023	5	7	2	1	1	5
117.64		0.9003	3	5	4	4		2
119.40	119.40	0.8921	7	5	2	1	1	1
120.41	120.40	0.8876	2	2	5	1		5
120.46		0.8874	-5	8	2	2		2
121.48	121.52	0.8829	0	10	3	3	1	3
121.57		0.8825	-5	0	0	2		1
122.04	122.04	0.8805	0	4	2	1	1	1
123.34	123.80	0.8750	-10	2	1	1	1	5
123.76		0.8733	-10	9	4	3		5
124.32	124.32	0.8711	-5	6	3	3	1	2
124.88	124.88	0.8688	-7	8	1	1	1	1
124.92		0.8687	2	0	1	1		5
125.57	125.64	0.8661	0	9	4	1	1	3
125.67		0.8658	-9	4	4	3		1
126.20	126.18	0.8637	2	3	5	2	1	2
127.07	127.34	0.8604	-3	0	6	4	2	1
127.16		0.8601	-10	5	4	5		1
127.29		0.8596	-3	1	6	1		1
127.29		0.8596	-4	6	2	2		4

286

2THETA	PEAK	D	H	K	L	I(INT)	I(PK)	I(DS)
127.38		0.8593	-8	3	5	5		6
129.18	129.20	0.8527	-2	9	3	4	1	6
129.48		0.8517	6	8	1	1		1
129.71	129.72	0.8509	-1	8	3	1	1	1
131.06	131.20	0.8463	4	5	4	1	1	2
131.18		0.8458	3	6	4	5		8
131.49		0.8448	-2	6	5	1		1
131.81	131.78	0.8438	10	0	0	1		2
131.88		0.8435	-10	5	2	1		2
132.48	132.42	0.8416	-11	2	3	1	1	1
132.53		0.8414	-1	0	3	1		1
133.61	133.64	0.8380	-9	2	6	4	1	7
133.89		0.8371	6	0	5	1		2
134.32	134.28	0.8358	10	2	4	1	1	2
135.58	136.20	0.8320	5	4	4	1	2	1
135.99		0.8308	0	10	1	2		4
136.20		0.8302	5	3	1	8		16
136.49		0.8293	-11	1	3	2		1
138.27	138.26	0.8243	-10	5	3	1	1	4
139.60	139.62	0.8207	-7	0	6	1	1	2
140.74	140.86	0.8178	-11	1	4	1		2
140.94		0.8173	-5	0	6	1		2
141.37	141.48	0.8162	-5	9	3	1	1	2
141.42		0.8161	-5	8	4	1		1
141.78		0.8152	-10	4	4	2		1
142.33	142.28	0.8138	5	7	3	2	1	3
143.31	143.30	0.8115	-8	8	1	1	1	2
143.49		0.8111	10	5	0	3		1
144.12	144.16	0.8096	8	2	3	3		7
144.68		0.8083	-8	8	2	2	1	2
145.31	145.00	0.8070	0	8	6	1		2
145.53		0.8065	4	6	4	1		2
146.25	146.56	0.8049	-10	1	5	1	1	2
146.55		0.8043	-4	4	6	1		3
146.63		0.8041	-10	1	2	2		3
146.98		0.8034	3	8	4	1		1
147.55	147.54	0.8022	3	4	5	1	1	1
148.93	149.70	0.7995	3	7	4	2		4
149.07		0.7992	1	11	0	2	1	3
149.39		0.7986	-2	7	5	1		2
149.50		0.7984	-12	1	2	1		2
149.64		0.7981	-5	4	6	1		2
149.78		0.7978	-7	9	1	1		1
150.67	150.66	0.7962	7	7	2	1		1
150.77		0.7960	-9	4	5	3		5
151.83	152.04	0.7941	10	4	1	1	1	4
152.34		0.7932	8	8	0	1		2
152.42		0.7931	8	3	3	2		4
152.42		0.7931	-9	6	4	1		1
152.91		0.7923	1	1	6	1		2
153.04		0.7921	-9	7	3	1		3
155.37	155.64	0.7884	-5	10	2	2	0	2
155.57		0.7881	-5	7	5	3		3
157.11	157.98	0.7859	-2	10	3	3	1	1
157.33		0.7856	-10	5	4	1		6
157.41		0.7855	11	1	1	1		1
158.09		0.7845	-8	8	3	2		4
158.18		0.7844	-1	10	0	1		2
158.64		0.7838	7	7	1	1		2
160.49	162.34	0.7815	-3	10	3	1	1	2
161.36		0.7806	2	8	4	1		2
161.90		0.7800	10	0	2	12		25
162.08		0.7798	5	9	2	1		3
162.14		0.7797	3	9	3	1		3
162.33		0.7795	-12	2	3	2		3
163.68		0.7781	-12	2	1	1		5
163.72		0.7781	7	1	7	10		20
164.56		0.7773	0	1	3	2		3
164.74		0.7771	0	10	4	2		4
164.90		0.7770	-1	9	1	1	1	1
164.90			11	4	0	1		2

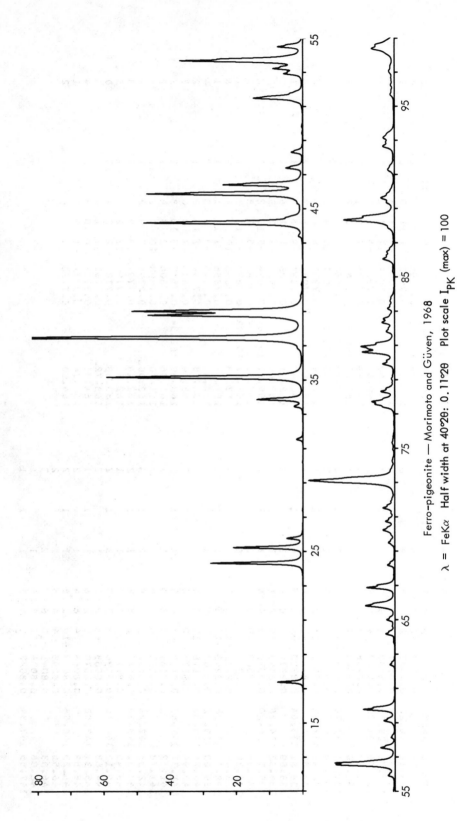

Ferro-pigeonite — Morimoto and Güven, 1968

λ = FeKα Half width at 40°2θ: 0.11°2θ Plot scale I_{PK} (max) = 100

FERRO-PIGEONITE – MORIMOTO AND GÜVEN, 1968

2THETA	PEAK	D	H	K	L	I(INT)	I(PK)	I(DS)
17-36	17-36	6.415	1	1	0	6	8	3
24-30	24-30	4.600	-2	0	1	23	28	16
25-21	25-22	4.436	-1	1	1	18	21	13
25-74	25-74	4.347	0	1	1	4	5	3
31-51	31-52	3.565	1	1	1	2	3	2
33-43	33-44	3.366	-1	1	2	2	3	2
33-84	33-84	3.326	0	2	0	13	14	12
35-13	35-14	3.208	-2	2	1	58	59	54
37-41	37-42	3.018	-2	1	2	100	100	100
38-75	38-76	2.918	-1	2	1	45	47	46
38-98	38-98	2.901	1	1	2	49	52	51
43-37	43-38	2.620	-1	0	3	1	1	1
44-14	44-14	2.576	-1	3	1	52	48	61
44-47	44-46	2.558	0	1	3	3	5	3
45-29	45-30	2.514	-1	1	3	1	2	1
45-83	45-84	2.486	0	2	2	53	47	64
46-38	46-38	2.458	2	2	1	26	24	32
47-37	47-36	2.410	-2	3	1	6	5	7
48-29	48-30	2.367	-1	2	3	6	5	5
51-45	51-46	2.230	3	0	1	15	15	20
51-51		2.228	-1	2	2	4		5
51-75	51-74	2.218	-2	2	2	3	4	4
52-90	52-90	2.173	0	2	2	6	6	9
53-20	53-20	2.162	-1	1	2	9	9	13
53-65	53-66	2.145	-3	3	1	45	37	65
53-83	53-76	2.138	-4	0	3	3	26	5
54-49	54-50	2.114	-3	1	2	9	8	13
55-90	55-90	2.065	-4	0	2	2	8	3
56-49		2.046	-4	2	2	4	18	7
56-53	56-54	2.044	0	4	0	16	18	25
56-64	56-64	2.040	-4	1	2	10	4	15
57-51	57-52	2.012	-2	4	1	5	1	8
58-12	58-12	1.9928	-4	1	1	1	1	7
59-10	59-10	1.9628	0	4	1	4	4	20
59-75	59-76	1.9433	-2	2	2	12	10	4
62-36	62-36	1.8696	-4	3	1	4	3	7
64-15	64-16	1.8228	3	3	1	4		

2THETA	PEAK	D	H	K	L	I(INT)	I(PK)	I(DS)
64-60	64-60	1.8116	-3	4	0	1	1	1
64-97	64-98	1.8022	-5	3	2	3	3	6
65-68	65-80	1.7850	1	3	2	2	9	3
65-78		1.7825	2	2	1	10		19
65-85		1.7808	-5	2	1			4
65-94		1.7787	-1	2	1	1		2
66-86	66-86	1.7570	4	1	0	12	9	23
68-17	68-18	1.7272	-1	5	1	3	2	6
69-35	69-48	1.7014	-1	5	3	1	2	1
69-47		1.6988	-1	1	1	2		4
69-92	69-94	1.6893	0	5	1	1	1	2
70-16	70-22	1.6842	-2	4	1	1	4	2
70-22		1.6830	-3	1	3	4		8
70-72	70-72	1.6727	0	4	2	1	3	8
71-19	71-20	1.6631	3	1	2	5	1	2
71-49	71-50	1.6569	-5	2	2	1	4	10
71-89	71-88	1.6490	-2	2	3	19	26	2
73-00	73-10	1.6274	-1	5	1	1		39
73-08		1.6259	-5	3	1	28		58
73-10		1.6254	-1	2	3	1	1	1
73-71	73-70	1.6138	-1	4	0	1	1	1
74-25	74-24	1.6038	-6	1	3	1	1	1
74-91	74-92	1.5918	-4	0	3	1	1	2
75-31	75-30	1.5845	3	5	0	6	6	13
77-54	77-56	1.5459	-6	0	0	7	7	16
77-69	77-70	1.5432	0	0	1	1	4	15
78-30	78-30	1.5333	-6	2	3	6	1	13
79-04	79-02	1.5212	-2	5	3	3	4	37
79-90	79-90	1.5075	-1	0	3	3	4	34
80-59	80-60	1.4967	-3	0	3	16	10	7
80-92	80-92	1.4917	-1	6	0	15	10	4
81-77	81-80	1.4788	3	3	3	3	3	14
81-83		1.4779	-1	5	2	1		6
82-32	82-34	1.4707	4	2	2	6	4	8
82-56	82-54	1.4672	-2	4	2	3	4	4
83-13	83-14	1.4589	-6	2	2	2	2	6
83-57	83-58	1.4527	0	5	2	2	1	4

FERRO-PIGEONITE - MORIMOTO AND GÜVEN, 1968

2THETA	PEAK	D	H	K	L	I(INT)	I(PK)	I(DS)
85.30	85.30	1.4288	0	6	1	1	1	2
86.02	86.02	1.4190	-3	3	2	1	4	15
86.03		1.4189	5	2	0	6		1
86.42	86.44	1.4138	-3	1	3	1		3
86.46		1.4132	3	5	1	3	2	7
88.27	88.32	1.3900	-5	3	1	1		1
88.31		1.3896	5	3	1	32	15	82
89.33	89.34	1.3770	-2	4	3	4	3	10
89.61	89.60	1.3736	-1	1	3	7	4	17
91.21	91.26	1.3548	-3	3	3	2	2	6
91.29		1.3538	0	2	3	3		7
92.63	92.64	1.3386	-7	1	2	8	4	21
93.12	93.14	1.3331	-5	3	3	6	3	17
93.24		1.3318	0	0	4	2		5
95.66	95.66	1.3061	-3	3	3	3	1	8
96.22	96.22	1.3003	-7	1	0	1	2	6
96.65	96.64	1.2960	-2	1	4	2	1	4
97.01	97.04	1.2924	-3	1	4	1	2	6
97.08		1.2917	6	1	4	2		6
97.44	97.28	1.2881	-2	2	2	1	1	3
98.13	98.38	1.2814	-5	1	2	1	7	6
98.36		1.2791	0	6	2	2		42
98.58	98.58	1.2770	-4	0	4	14	6	16
100.72	100.76	1.2570	-2	2	4	6	1	5
100.83		1.2560	-3	6	0	2		2
101.33	101.32	1.2515	4	6	6	1	1	6
102.17	102.28	1.2441	-1	5	0	3	2	3
102.28		1.2431	0	5	4	2		
103.37	103.40	1.2337	-3	3	5	3	1	10
103.66	103.68	1.2313	0	1	3	2	2	8
104.01	104.12	1.2283	-7	1	1	1	4	4
104.13		1.2273	3	5	4	1		4
106.90	106.92	1.2049	-4	6	2	10	1	32
107.60	107.58	1.1995	-8	0	0	4	1	11
107.93	107.90	1.1970	-4	5	3	1	1	3
109.72	109.78	1.1837	-5	2	2	2	2	6
109.78		1.1833	2	6	2	3		10

2THETA	PEAK	D	H	K	L	I(INT)	I(PK)	I(DS)
110.06	110.06	1.1812	-3	7	1	1	1	4
111.53	111.56	1.1708	7	1	1	2	1	6
111.82	111.88	1.1688	-6	5	2	2	1	3
112.18	112.20	1.1664	-8	0	4	2	1	6
112.52	112.52	1.1641	-6	2	4	3	1	10
113.33	113.32	1.1586	-8	2	1	1	1	5
113.98	113.98	1.1543	-6	0	4	1	0	3
114.44	114.66	1.1513	6	0	4	1	1	3
114.65		1.1499	8	2	0	3		9
115.66	115.66	1.1436	1	4	4	1	0	3
117.82	117.84	1.1304	-2	3	1	1	0	3
119.99	120.06	1.1178	3	7	3	1	1	4
120.09		1.1172	2	6	2	1		4
120.68	120.68	1.1139	8	0	2	1	1	12
120.71		1.1138	1	7	0	2		7
121.60	121.56	1.1089	-7	5	1	1	0	5
124.25	124.48	1.0950	-8	0	4	1	1	9
124.47		1.0939	-7	0	4	3		4
124.50		1.0938	8	8	0	2		5
125.86	126.36	1.0871	2	1	4	2	1	7
126.14		1.0857	-7	4	1	2		5
126.37		1.0846	-7	7	3	1		9
127.15	127.52	1.0809	2	2	4	2	5	5
127.40		1.0798	-7	5	2	10		6
127.55		1.0791	-2	8	1	14		39
128.98	128.98	1.0725	-6	6	2	9	2	55
129.74	129.46	1.0692	7	5	0	6	2	36
132.06	132.08	1.0594	-8	4	0	3	3	12
132.59	132.56	1.0572	6	5	3	14	2	56
133.08	133.08	1.0552	-8	3	3	3	2	3
133.12		1.0551	-3	3	4	3		13
134.42	134.50	1.0500	5	7	0	5	1	20
134.54		1.0495	3	5	3	4		16
135.13	135.58	1.0473	4	6	2	3	2	7
135.55		1.0457	2	1	1	7		13
136.07	136.08	1.0438	-9	2	2	1	2	31
136.15		1.0435	5	2	3	1		3

2THETA	PEAK	D	H	K	L	I(INT)	I(PK)	I(OS)
136.51		1.0421	3	0	4	3		12
137.22	137.16	1.0396	-4	5	4	3	1	14
137.89	137.90	1.0372	-5	7	2	3	1	14
138.49	138.50	1.0351	3	1	4	6	2	25
138.61		1.0348	-4	1	5	1		3
141.01	141.06	1.0269	7	1	2	5	1	20
141.68	141.70	1.0248	-2	8	2	1	1	6
142.32	142.72	1.0228	8	4	0	3	1	12
142.70		1.0216	8	0	1	5		21
142.90		1.0210	0	5	4	1		5
143.47		1.0194	-7	5	3	1		4
144.67	145.40	1.0159	-2	2	5	5	2	21
145.10		1.0147	-7	6	1	1		3
145.37		1.0139	-4	8	1	5		20
146.94	146.98	1.0097	5	3	1	20	3	91
147.65	147.64	1.0079	-9	3	1	9	3	42
147.86		1.0074	-8	4	3	1		5
147.94		1.0072	7	2	2	1		2
149.78	149.82	1.0027	-9	2	3	2	1	11
149.85		1.0025	-7	6	2	3		13
150.32		1.0014	-8	5	1	1		3
151.65	152.54	0.9984	-1	2	5	2	1	7
152.20		0.9972	-1	4	4	2		11
152.58		0.9964	-8	2	4	1		3
152.76		0.9960	-5	2	5	3		12
157.56	158.84	0.9869	1	5	4	1	1	5
158.00		0.9861	7	6	0	1		4
158.20		0.9858	7	5	1	4		20
158.73		0.9849	-6	1	5	1		5
158.88		0.9847	-2	3	5	10		48
160.17		0.9827	-7	4	4	1		3
160.19		0.9826	-3	6	4	4		18

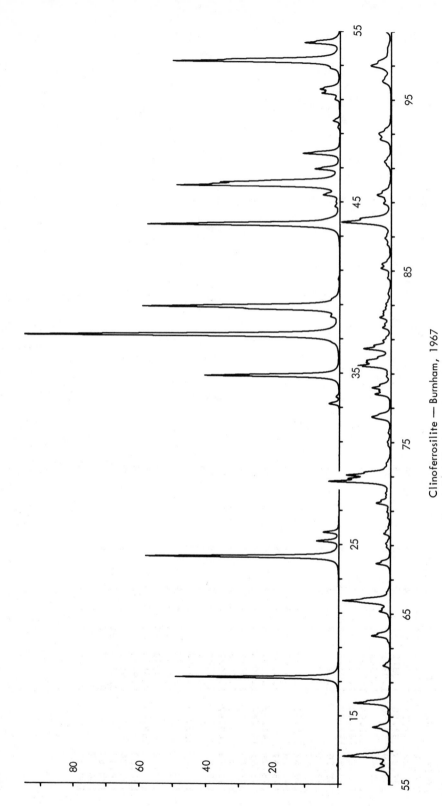

Clinoferrosilite — Burnham, 1967

λ = FeKα Half width at 40°2θ: 0.11°2θ Plot scale I_{PK} (max) = 100

292

CLINOFERROSILITE - BURNHAM, 1967

2THETA	PEAK	D	H	K	L	I(INT)	I(PK)	I(DS)
17.21	17.22	6.469	1	1	0	36		21
24.27	24.28	4.605	2	0	1	49	58	35
25.18	25.18	4.441	-1	1	1	6	7	4
25.69	25.70	4.354	0	1	1	4	5	3
33.19	33.20	3.390	-1	1	1	3	3	3
33.59	33.60	3.350	0	2	0	1	1	1
34.83	34.84	3.234	-2	2	1	40	41	38
37.20	37.20	3.035	-2	2	1	100	100	100
38.28	38.28	2.952	1	2	1	2	3	2
38.73	38.74	2.920	-3	1	1	21	28	22
38.88	38.88	2.909	3	1	0	55	59	58
39.32	39.30	2.877	-1	3	0	1	3	1
43.66	43.66	2.603	-1	1	1	63	58	73
44.71	44.72	2.545	-2	3	0	1	2	2
44.98	44.98	2.531	2	1	2	1	1	1
45.38	45.38	2.509	-1	3	0	5	5	6
45.95	45.96	2.480	0	0	2	53	49	65
46.15	46.14	2.470	-2	2	1	30	36	38
46.91	46.92	2.432	-2	3	1	8	8	10
47.81	47.82	2.389	1	3	1	12	11	16
49.30	49.32	2.321	-3	0	2	1	1	1
49.72	49.72	2.303	4	0	0	2	2	3
50.99	51.00	2.249	-3	1	2	2	2	8
51.35	51.36	2.234	3	1	1	6	6	8
51.59	51.58	2.225	-1	0	2	6	6	2
51.69	51.68	2.220	-2	2	0	1		3
52.81	52.82	2.177	0	2	2	2	3	15
53.23	53.24	2.161	1	1	2	11	50	74
53.23		2.161	-1	3	2	51		1
53.35		2.156	-3	3	0	13	11	19
54.31	54.32	2.121	-4	2	1	2	4	3
55.85	55.88	2.067	-3	2	2	3		5
55.90		2.065	0	4	1	3	3	4
56.23	56.24	2.054	4	2	0	16	14	26
56.68	56.68	2.039	-4	0	2	2		2
56.73		2.037	2	1	0	2	2	3
58.22	58.24	1.9898	-4	1	2			

2THETA	PEAK	D	H	K	L	I(INT)	I(PK)	I(DS)
58.38	58.38	1.9847	-2	4	1	1	5	9
59.16	59.16	1.9610	1	1	1	1		1
59.79	59.80	1.9421	2	0	2	14	11	24
61.97	61.98	1.8802	-4	3	1	3	2	5
63.39	63.40	1.8423	-3	3	2	1	1	1
63.71	63.72	1.8342	3	1	1	8	6	14
63.91	63.84	1.8289	-3	4	1	1	4	1
65.14	65.14	1.7982	2	4	2	4	3	8
65.35	65.28	1.7930	1	3	2	1		1
65.64	65.76	1.7858	2	4	2	8	14	14
65.76		1.7830	-1	5	0	14		26
66.27	66.26	1.7709	-5	2	1	1	1	3
67.92	67.92	1.7328	-5	0	2	6	4	2
68.88	68.88	1.7116	4	1	1	1	1	12
69.12	69.10	1.7064	-1	5	1	1	1	3
69.66	69.66	1.6948	0	5	1	2	2	2
69.67		1.6946	-2	4	3	1		3
70.07	70.06	1.6861	-1	4	2	1		3
70.60	70.60	1.6751	3	0	2	1	1	2
70.95	70.96	1.6679	0	4	3	1	1	3
71.45	71.46	1.6578	-3	1	3	6	4	13
71.84	71.84	1.6500	-3	2	2	1	2	2
72.04	72.02	1.6461	-5	1	1	1	2	2
72.72	72.72	1.6328	1	5	1	28	19	58
73.04	73.08	1.6266	-5	3	1	17	13	2
73.09		1.6256	0	1	3			36
73.78	73.76	1.6126	-2	2	3	1	1	3
74.79	74.82	1.5939	-1	2	3	1	1	1
74.84		1.5931	4	3	1	1		1
75.03	75.02	1.5895	-6	1	3	1	1	2
75.55	75.54	1.5802	-3	3	1	1	1	2
76.38	76.48	1.5656	-4	2	3	4	6	9
76.48		1.5639	-3	5	1	7		16
77.80	77.80	1.5415	3	5	0	8	5	17
78.18	78.18	1.5352	-6	0	0	8	6	19
78.84	78.84	1.5243	-6	2	1	2	2	4

CLINOFERROSILITE – BURNHAM. 1967

2THETA	PEAK	D	H	K	L	I(INT)	I(PK)	I(DS)
79.46	79.46	1.5145	0	6	0	16	10	36
79.54		1.5131	-4	2	3	1		1
79.78	79.76	1.5094	-2	3	3	9	7	20
80.45	80.46	1.4989	-1	9	2	14	8	32
80.93	80.92	1.4916	-3	1	3	3	3	8
81.67	81.68	1.4803		4	2	4	3	10
81.95	81.88	1.4762	2	2	2	5	2	2
82.28	82.28	1.4713	4	4	0	1	3	13
82.61	82.64	1.4665	-5	0	3	3	2	1
82.65		1.4659	0	5	2	3		7
82.89	82.86	1.4624	-1	2	3	1	2	2
83.07	83.08	1.4598	-6	2	2	2	2	5
83.59	83.66	1.4524	-6	1	2	1	1	2
83.68		1.4512	0	3	3	1		3
84.57	84.58	1.4387	-2	6	0	2	1	5
85.14	85.14	1.4309	-3	5	2	5	3	13
85.42	85.38	1.4271	3	2	1	3	3	6
86.53	86.54	1.4123	-5	2	3	2	1	5
86.91	86.90	1.4074	-1	5	2	3	2	8
87.84	87.84	1.3955	5	3	1	31	15	82
88.94	88.94	1.3818	-2	4	3	1	2	11
89.41	89.42	1.3760	-1	2	3	9	4	24
91.17		1.3552	-2	2	3	2		5
91.37	91.38	1.3529	-7	0	2	3	2	9
92.67	92.68	1.3381	-7	1	2	7	4	20
92.79		1.3368	0	4	3	1		2
93.04	93.02	1.3341	-5	2	3	6	4	18
96.03	96.08	1.3023	7	1	0	3	3	9
96.16		1.3015	-3	0	4	4		11
96.76	97.00	1.2948	6	2	1	2	6	5
96.99		1.2925	0	6	2	14		41
97.03		1.2921	-2	1	4	1		2
97.42	97.22	1.2883	-3	1	4	1	4	4
97.69	97.66	1.2856	-3	3	1	1	1	2
97.89	97.96	1.2837	-1	0	4	1	1	2
97.99		1.2827	5	1	2	1		2
98.08		1.2818	3	1	3	1		2

2THETA	PEAK	D	H	K	L	I(INT)	I(PK)	I(DS)
99.05	99.06	1.2725	-4	0	4	5	2	15
99.49	99.32	1.2684	-3	6	2	1	2	3
99.81	99.80	1.2654	-4	6	0	3	2	9
100.37	100.36	1.2602	-4	1	4	1	1	4
100.98	100.98	1.2546	-2	2	4	1	1	3
101.38	101.34	1.2511	-1	5	3	1	1	4
102.60	102.64	1.2403	-6	3	3	1	2	4
102.61		1.2402	-3	5	2	3		7
102.64		1.2400	0	0	4	9		8
103.09	103.08	1.2361	3	5	2	2	4	30
103.98	103.98	1.2286	0	1	4	1	1	6
104.61	104.60	1.2233	-5	0	4	5	1	2
105.52	105.54	1.2159	-4	6	2	1	2	17
107.16	107.18	1.2029	-4	5	3	2	1	3
107.66	107.68	1.1991	-8	0	2	5	1	5
108.29	108.30	1.1943	-8	1	2	2	2	17
110.06		1.1812	-5	2	4	4		8
110.85	110.88	1.1756	-6	5	1	5	1	4
111.27	111.26	1.1727	7	1	1	1	1	5
111.91	111.90	1.1682	-8	2	2	3	1	9
113.05	113.08	1.1605	-6	0	4	2	1	8
113.21		1.1594	-8	2	2	1		3
114.28		1.1524	6	0	0	2		7
114.43	114.44	1.1514	8	0	0	3	1	12
114.47		1.1511	-6	2	4	1		5
115.82		1.1426	1	2	3	1		4
116.17	116.16	1.1404	4	3	2	2	2	6
118.44	118.42	1.1267	-3	6	1	1	1	5
119.11	119.08	1.1228	-8	2	1	1	1	3
120.30	120.30	1.1161	8	0	4	1	0	6
120.98	120.96	1.1123	2	8	0	2	1	7
122.74		1.1028	-7	3	4	2		9
122.98	122.96	1.1016	-2	8	0	1	1	5
124.41	124.46	1.0942	-8	2	1	13	4	52
124.58		1.0934	-8	3	3	3		14
125.09	124.80	1.0908	-7	0	4	3	3	14

CLINOFERROSILITE – BURNHAM, 1967

2THETA	PEAK	D	H	K	L	I(INT)	I(PK)	I(DS)
125.59	125.50	1.0884	0	4	3	1		
125.90		1.0869	-7	4	2	1	3	3
126.25	126.26	1.0852	-7	4	3	10		41
126.70	126.70	1.0831	-7	5	1	3	3	11
127.27	127.28	1.0803	-2	2	4	1		3
127.27		1.0786	-6	6	2	10	3	39
127.64	127.72	1.0781	-5	7	1	1		5
127.75			-6	6	0	6	3	24
130.20		1.0672	-8	4	1	1		4
130.52	130.54	1.0658	7	5	0	13	3	55
131.62		1.0611	5	8	0	2		8
131.98	132.06	1.0597	2	8	1	7	2	31
132.36	132.40	1.0581	-3	5	3	4	2	19
132.92	132.94	1.0559	-8	3	3	5	3	23
133.05		1.0553	4	6	2	4		18
133.25		1.0545	3	5	3	2		8
136.44	136.54	1.0424	-4	5	4	5	2	21
136.80		1.0411	3	0	4	5		20
138.73	138.72	1.0343	3	1	2	6	1	25
139.22	139.26	1.0327	0	8	5	1	1	3
139.62		1.0313	-4	1	1	1		3
140.61		1.0281	-7	1	0	5	2	22
140.97		1.0270	8	4	1	6		7
141.11	141.14	1.0266	-4	8	4	3		28
141.84	141.88	1.0243	0	5	2	4	2	13
142.01		1.0237	8	2	1	6		28
142.24		1.0230	-7	5	3	1		4
145.51	146.30	1.0135	-2	2	5	4	4	19
146.19		1.0117	-6	6	3	1		3
146.22		1.0116	5	3	3	20		94
146.73	146.78	1.0103	-9	3	1	11	4	50
147.12		1.0092	-8	4	3	2		8
147.24		1.0089	-7	6	2	5		22
147.24		1.0089	-7	2	3	1		3
149.90	149.98	1.0024	-9	2	2	4	1	18
150.47		1.0010	5	7	1	1		5
150.54		1.0009	-8	5	2	1		4
154.57	155.52	0.9923	-4	8	2	2	2	2
155.16		0.9912	-7	5	1	5		26
155.59		0.9904	-1	9	1	8		40
155.80		0.9900	1	3	4	1		5
156.70		0.9883	-3	3	5	2		9
157.43		0.9871	-3	6	4	7		36
159.67	159.74	0.9834	-2	3	5	20	1	98
159.72		0.9833	-7	3	4	1		3
160.62		0.9820	-4	4	5	1		3
160.76		0.9818	-6	1	5	1		4
161.68		0.9805	-2	8	2	5		27
161.98		0.9801	-2	9	1	1		5
162.02		0.9800	-5	8	1	3		16
162.60		0.9793	-1	6	4	2		11
163.51		0.9781	-6	5	4	5		24

TABLE 14. PYROXENE-LIKE SILICATES

Variety	α-Wollastonite	Pectolite	Bustamite	Rhodonite
Composition	$CaSiO_3$	$Ca_2NaHSi_3O_9$	$CaMnSi_2O_6$	$(Mn,Ca)SiO_3$
Source	not given	Bergen Hill, N.J.	Franklin, N.J.	Pajsberg, Sweden
Reference	Buerger & Prewitt, 1961	Prewitt, 1968	Peacor & Buerger, 1962	Peacor & Niizeki, 1963
Cell Dimensions				
a Å	7.94	7.988	15.412	7.682
b	7.32	7.040	7.157	11.818
c	7.07	7.025	13.824	6.707
α deg	90	90.52	89.48	92.355
β	95.37	95.18	94.85	93.948
γ	103.43	102.47	102.93	105.665
Space Group	P1̄	P1̄	F1̄	P1̄
Z	6	2	12	10

TABLE 14. (cont.)

Variety	α-Wollastonite	Pectolite	Bustamite	Rhodonite
Site Occupancy				
M_1	Ca 0.120 Mn 0.815 Mg 0.040 Zn 0.015 Fe 0.01			Mn 1.0
M_2	Ca 1.0			Mn 1.0
M_3	Mn 0.935 Mg 0.04 Zn 0.015 Fe 0.01			Mn 0.70 Mg 0.16 Fe 0.14
M_4	Ca 1.0			Ca 0.80 Mn 0.20
M_5				
Method	3-D, counter	3-D, counter	3-D, counter	3-D, counter
R & Rating	0.089 (1)	0.039 (1)	0.095 (1)	0.154 (2)
Cleavage and habit	{100}perfect, {001}, (1̄02) less perfect. Tabular parallel to {100} or {001}	{100}, {001}perfect. Fibrous, elongated to b axis	{100}perfect;{001}. (102)less perfect Granular	{100},{001} perfect; {010} less perfect Tabular parallel to {010}. Elongated to b axis
Comment	Dreierketten - 3 tetrahedra in repeat unit along silicate chain			Fünferketten - 5 tetrahedra in repeat unit along silicate chain
μ	218.4	175.9	331.1	250.7
ASF	0.7933×10^{-1}	0.6401×10^{-1}	0.1531×10^{-1}	0.6255×10^{-1}
Abridging factors	1:1.3:0.6	1:1.3:0.6	1:1.3:0.5	1:1.3:0.6

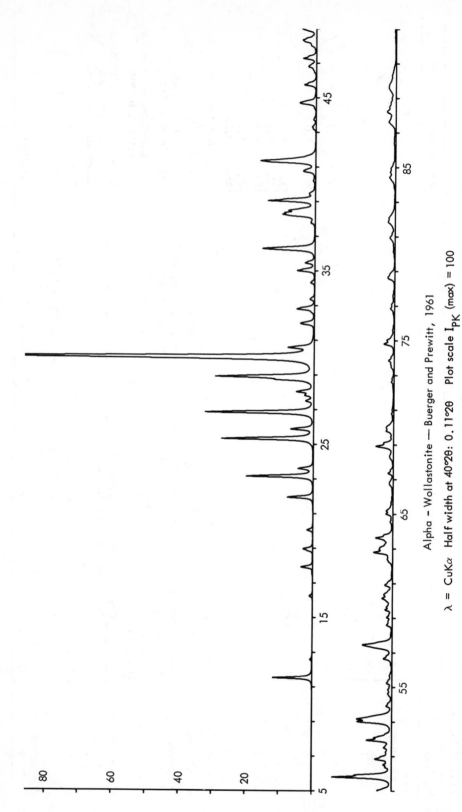

Alpha – Wollastonite — Buerger and Prewitt, 1961

λ = CuKα Half width at 40°2θ: 0.11°2θ Plot scale I$_{PK}$ (max) = 100

ALPHA-WOLLASTONITE - BUERGER AND PREWITT, 1961

2THETA	PEAK	D	H	K	L	I(INT)	I(PK)	I(DS)
11.50	11.50	7.687	1	0	0	16	12	11
16.22	16.24	5.459	-1	0	1	1	1	1
17.87	17.90	4.958	1	0	1	2	4	2
17.91		4.949	-1	1	1	3		3
18.95	18.96	4.680	0	1	1	3	3	3
20.03	20.04	4.429	-1	1	1	3	2	2
21.92	21.92	4.051	-1	1	0	12	8	10
23.12	23.12	3.844	-2	0	0	31	20	27
23.58	23.58	3.770	-1	0	0	6	5	6
25.29	25.30	3.519	-2	0	1	5	27	5
25.29		3.519	0	0	2	39		35
25.85	25.86	3.443	-2	1	1	11	7	10
26.81	26.82	3.323	-1	1	2	55	32	51
27.47	27.48	3.244	2	0	1	4	3	3
27.80	27.82	3.207	-1	2	1	2	3	2
27.81		3.205	0	0	1	2		2
28.01	28.02	3.183	0	1	2	8	5	8
28.71	28.72	3.107	-1	1	2	15	12	14
28.83		3.094	2	1	0	8	29	7
28.88	28.88	3.089	-1	2	2	42		41
29.95	29.95	2.981	-2	2	0	100	100	100
30.00	30.00	2.976	0	0	0	99		99
30.51		2.928	-2	1	1	2	8	2
30.56	30.56	2.923	-1	1	2	10		10
31.92	31.96	2.802	-2	2	1	4	4	4
31.96		2.798	-1	2	2	4		5
32.78	32.78	2.729	-2	0	2	9	5	10
32.86		2.723	1	1	0	1		1
33.34	33.34	2.685	-2	1	1	2	1	2
34.29	34.30	2.613	-3	0	1	2	1	2
34.99	35.00	2.562	3	0	0	10	5	11
35.43	35.44	2.531	-1	2	2	3	3	3
35.44		2.531	0	-2	2	3		3
36.20	36.26	2.479	2	0	2	11	16	12
36.25		2.476	1	-2	2	11		13
36.27		2.475	0	2	2	12		14
37.75	37.76	2.381	3	-1	1	2	1	3

2THETA	PEAK	D	H	K	L	I(INT)	I(PK)	I(DS)
38.16	38.24	2.356	-3	-2	0	10	10	12
38.23	38.32	2.352	-2	2	0	10		12
38.34		2.346	0	0	3	8	9	10
38.44	38.46	2.340	-2	-2	2	3	8	4
38.48		2.338	-1	2	2	3		4
38.50		2.337	3	0	1	5		6
39.02	39.02	2.306	-1	0	3	27	14	33
39.39	39.40	2.286	0	2	2	2	2	2
39.45	39.44	2.282	-3	2	1	2	2	2
40.17	40.18	2.243	-2	-1	3	1	1	1
40.71	40.74	2.214	2	1	2	4	4	5
40.76		2.212	-1	2	1	4		5
41.25	41.32	2.187	-3	-2	1	20	16	25
41.32		2.183	2	2	1	20		25
43.17	43.18	2.094	0	1	3	2	1	2
43.30	43.28	2.088	-3	1	2	2	1	2
43.70	43.70	2.070	-1	3	2	1	1	2
44.64	44.68	2.028	-3	2	2	6	5	9
44.70		2.026	-1	2	2	6		8
45.72	45.72	1.9825	3	0	2	8	4	11
46.61	46.60	1.9470	-3	3	1	1	1	2
46.76	46.80	1.9411	-3	3	1	2	2	3
46.80		1.9395	1	-2	3	1		2
46.82		1.9386	0	2	3	1		2
47.26	47.26	1.9218	2	-1	0	5	4	3
47.26		1.9218	1	0	3	5		8
47.98	48.06	1.8945	3	-2	2	2	2	2
48.05		1.8920	2	2	2	5		7
48.24	48.26	1.8848	-4	2	0	5	4	8
48.33	48.34	1.8817	3	2	0	5	4	2
48.45		1.8772	-2	-3	1	5		7
48.95	49.02	1.8593	-3	2	1	4	4	7
49.03		1.8565	-1	-2	1	1		7
49.78	49.78	1.8300	-1	4	0	40	18	61
50.37	50.38	1.8101	4	0	1	3	2	5
50.51	50.50	1.8052	-2	-2	1	1	3	5
50.81	50.84	1.7954	2	-2	3	5	5	8

ALPHA-WOLLASTONITE - BUERGER AND PREWITT, 1961

2THETA	PEAK	D	H	K	L	I(INT)	I(PK)	I(DS)
50.81		1.7953	2	1	3	1		2
50.85	51.16	1.7940	1	-2	1	5		8
51.15		1.7842	4	-2	1	1	2	2
51.23	51.24	1.7815	3	1	1	2	2	2
51.49	51.50	1.7733	-3	3	2	4	8	6
51.93	51.94	1.7593	0	0	4	17	6	27
52.16	52.06	1.7520	-1	0	4	5	10	7
52.97	53.00	1.7272	-3	2	3	12		20
53.01		1.7258	-2	-2	3	12		20
53.16	53.16	1.7216	-4	2	2	8	11	13
53.23		1.7194	-3	-2	2	8		14
53.89	53.90	1.6999	-4	3	0	2	2	3
53.92		1.6990	0	3	4	1		2
54.49	54.62	1.6824	-1	-3	3	2	1	3
54.61		1.6790	-4	3	1	1		3
54.86	54.86	1.6722	-2	-3	2	2	1	4
55.21	55.22	1.6624	-1	-1	4	2	2	2
55.24		1.6613	-2	0	4	1		5
56.64	56.64	1.6237	-1	-4	2	3	3	5
56.65		1.6234	-1	4	2	3		13
57.35	57.44	1.6051	3	2	2	7	9	12
57.42		1.6033	-2	-4	2	6		11
57.45		1.6026	0	-4	4	6	2	4
58.55	58.58	1.5751	0	4	2	2		4
58.61		1.5738	3	3	0	2		3
58.84	58.74	1.5681	3	-2	4	1	1	3
59.02	59.02	1.5637	1	4	1	1	2	5
59.04		1.5632	0	2	2	1	2	4
59.45	59.46	1.5535	-2	-2	4	2		9
59.47		1.5531	-1	2	4	2	3	4
60.06	60.12	1.5392	-4	-2	1	2		9
60.13		1.5374	5	0	0	5	3	4
60.36	60.32	1.5322	-4	2	3	2		4
60.42		1.5308	-3	-2	2	2	2	4
60.86	60.88	1.5208	-3	0	4	2		4
60.88		1.5202	1	3	0	2		4
60.93		1.5192	-1	-4	2	2		4

2THETA	PEAK	D	H	K	L	I(INT)	I(PK)	I(DS)
62.23	62.24	1.4905	-4	4	0	2	1	4
62.35	62.38	1.4879	2	-4	0	2	2	4
62.75	62.78	1.4794	1	-2	4	8	6	15
62.79		1.4786	-2	4	4	7		15
63.04	63.04	1.4733	3	-4	2	2	4	6
63.07		1.4727	-1	0	1	3		5
63.13		1.4714	5	4	2	2		5
63.49	63.62	1.4639	-4	-2	2	2	5	5
63.57		1.4623	-3	1	4	1		2
63.61		1.4615	-2	-1	2	1		17
63.62		1.4613	-5	0	0	9		5
64.99	65.00	1.4338	0	-4	4	3	2	5
65.01		1.4333	-1	4	3	1		3
65.15	65.18	1.4306	0	-4	4	1	2	2
65.25		1.4287	-1	3	4	1		3
66.68	66.72	1.4015	0	-5	1	1	1	2
66.72		1.4008	-4	-4	2	1		3
66.82	66.82	1.3988	-2	-4	4	4	1	3
67.34	67.34	1.3893	3	0	2	1	2	8
68.27	68.34	1.3726	-4	3	4	4	1	3
68.35		1.3712	-3	5	2	1		2
68.85	68.90	1.3625	5	-2	0	3	5	6
68.90		1.3615	2	0	4	9		20
68.91		1.3615	-1	5	4	3		6
69.43	69.42	1.3526	2	-5	2	1	1	3
69.77	69.78	1.3467	-2	5	2	3	2	6
69.78		1.3466	-5	0	3	2		4
69.89	69.98	1.3448	1	5	0	2	2	4
70.02		1.3425	-4	4	4	1		3
70.08		1.3416	-3	-2	2	1		3
70.77	70.76	1.3301	0	-5	1	1	1	3
72.62	72.60	1.3008	-1	2	5	1	1	2
72.81	72.80	1.2979	-4	5	0	1	1	2
73.45	73.46	1.2881	-3	1	1	1	1	2
73.60	73.62	1.2858	5	4	2	2	1	4
73.83	73.84	1.2824	-6	0	1	3	2	7
73.91		1.2812	-6	0	0	1		2

ALPHA-WOLLASTONITE - BUERGER AND PREWITT, 1961

2THETA	PEAK	D	H	K	L	I(INT)	I(PK)	I(DS)
74.78	74.80	1.2684	1	-4	4	5	3	11
74.80		1.2681	-1	4	4	5		11
74.93		1.2663	4	-5	1	5		1
75.56	75.58	1.2573	5	-1	3	1	1	1
75.61		1.2565	-6	3	0	1		2
76.83	76.84	1.2396	4	0	4	3	1	7
76.93		1.2382	-1	5	3	1		2
77.39	77.40	1.2320	2	0	4	1	1	2
78.52	78.62	1.2172	-5	0	4	1	2	4
78.58		1.2164	-5	2	2	3		8
78.65		1.2154	-4	-2	4	3		7
78.74		1.2143	-2	5	1	1		2
80.89	80.92	1.1873	-3	6	0	2	2	6
80.97		1.1863	0	6	0	2		5
81.65	81.82	1.1782	-6	4	0	2	2	5
81.80		1.1764	-1	0	6	2		6
81.83		1.1761	4	-2	2	1		5
81.91		1.1751	-4	2	5	1		2
81.96		1.1745	-3	-2	5	1		2
81.99		1.1742	0	-6	1	2		5
82.10	82.06	1.1729	5	0	0	1	2	2
82.22		1.1715	6	-4	1	2		4
84.25	84.42	1.1483	6	-4	3	1	2	2
84.29		1.1479	1	5	3	1		4
84.37		1.1470	2	-6	2	1		4
84.41		1.1465	-1	6	2	2		5
84.43		1.1463	4	4	1	3		8
84.75	84.72	1.1428	-6	4	2	3	2	9
84.92	84.92	1.1410	-4	-4	2	1	1	7
87.48	87.62	1.1141	4	-6	1	1	2	4
87.61		1.1128	-2	-2	6	3		4
87.62		1.1127	-1	1	6	1		8
87.62		1.1127	-2	-6	1	3		3
87.73		1.1115	2	-5	0	1		3
88.13	88.22	1.1076	-3	0	6	1	3	5
88.16		1.1072	4	-4	4	2		5
88.20		1.1068	1	-2	6	3		8

2THETA	PEAK	D	H	K	L	I(INT)	I(PK)	I(DS)
88.23		1.1066	0	2	6	3		8
88.27		1.1061	5	0	4	1		3
88.29		1.1059	-2	4	4	2		4
88.68	88.68	1.1021	-7	2	2	2	2	6
88.79		1.1010	-6	-2	2	2		6
89.39	89.60	1.0951	6	1	1	1	2	4
89.44		1.0947	0	5	4	1		3
89.49		1.0942	-3	4	5	1		3
89.53		1.0938	-1	-4	5	1		7
89.58		1.0933	6	-4	2	3		6
89.63		1.0928	2	0	6	2		7
89.77	89.74	1.0915	4	0	4	2	2	7
90.28	90.28	1.0866	-6	0	4	2	1	4
94.04	94.08	1.0528	-5	3	5	1	1	4
94.25	94.46	1.0511	-7	4	1	1	1	3
94.45		1.0493	-5	-4	1	2		5
94.97	94.98	1.0450	7	-2	2	2	1	5
95.09		1.0440	6	-2	2	2		6
96.07	96.20	1.0359	-4	6	3	3	2	6
96.18		1.0350	-1	-6	3	2		4
96.20		1.0349	-5	6	2	1		4
96.25		1.0344	-4	2	6	1		3
96.37		1.0335	-3	-2	6	1		4

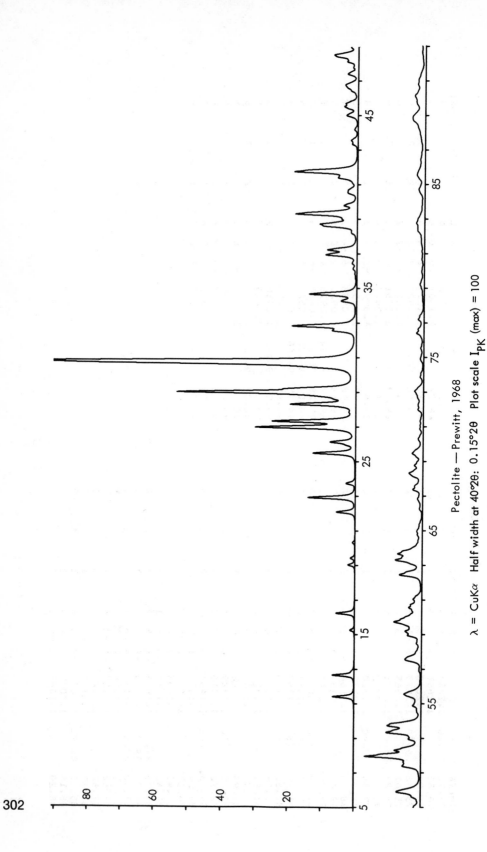

Pectolite — Prewitt, 1968

Half width at 40°2θ: 0.15°2θ   Plot scale I_{PK} (max) = 100

λ = CuKα

302

PECTOLITE - PREWITT, 1968

2THETA	PEAK	D	H	K	L	I(INT)	I(PK)	I(DS)
11.39	11.40	7.765	1	0	0	8	6	6
12.65	12.66	6.993	0	0	1	8	7	6
15.23	15.24	5.811	-1	1	0	2	1	1
16.22	16.24	5.459	1	0	1	7	6	6
19.00	19.02	4.666	-1	1	1	3	2	2
19.39	19.40	4.574	-1	1	1	2	1	1
22.06	22.06	4.027	1	1	0	7	6	7
22.89	22.90	3.882	-2	0	0	19	14	17
23.72	23.72	3.747	-2	1	0	3	3	3
25.45	25.46	3.497	0	1	1	18	13	17
26.10	26.10	3.412	-2	0	1	10	8	9
26.93	26.94	3.307	-1	1	1	43	30	40
27.28	27.30	3.266	2	0	0	35	25	33
28.27	28.28	3.154	0	0	2	28	20	27
28.95	28.96	3.081	-1	1	2	59	53	57
28.97		3.080	0	1	2	18		18
29.17	29.16	3.059	-1	1	2	21	22	20
30.64	30.72	2.916	-1	2	0	98	100	97
30.73		2.907	-1	2	0	3		3
30.74		2.906	-2	2	1	97		97
32.52	32.54	2.751	-1	0	1	9	7	9
32.78	32.80	2.730	-2	2	0	21	19	22
32.80		2.728	-2	2	1	10		10
34.23	34.24	2.617	-3	1	0	7	5	7
34.63	34.64	2.588	3	0	0	23	14	25
36.01	36.10	2.487	0	0	1	1	1	1
36.41	36.42	2.466	2	1	2	15	9	16
36.90	36.92	2.434	-1	1	2	14	9	15
37.19	37.20	2.415	-2	0	2	2	2	2
38.17	38.18	2.356	3	0	0	10	11	11
38.55	38.68	2.333	2	0	3	5		6
38.59		2.331	0	0	3	10		11
38.70	38.70	2.325	-3	2	2	10	4	11
38.97	38.98	2.309	-1	1	2	4		5
39.25	39.26	2.293	-1	2	1	4	18	35
39.36		2.287	-2	2	2	4		5
39.76	39.76	2.265	3	1	0	3	4	4
40.46	40.52	2.228	0	-1	3	2	3	3
40.51		2.225	2	1	2	3		3
40.69	40.60	2.216	-1	-1	1	2	3	3
41.36	41.36	2.181	-3	0	2	8	6	10
41.70	41.70	2.164	3	-2	1	17	19	20
41.70		2.164	2	1	1	16		20
41.83		2.158	-2	0	3	4		5
43.31	43.32	2.087	-2	3	1	3	2	3
43.53	43.52	2.077	-2	-3	1	3	2	4
44.15	44.16	2.050	-2	-1	3	2	1	2
44.94	44.98	2.015	3	-1	2	4	4	3
44.99		2.013	-2	1	2	4		4
45.38	45.50	1.9966	-3	2	2	5	4	5
45.49		1.9923	3	0	0	5		5
45.68	45.66	1.9845	-4	1	0	3	4	6
46.64	46.74	1.9457	1	0	3	3		3
46.73		1.9422	4	0	0	2		2
46.76		1.9412	-1	3	1	1		2
46.81		1.9392	-4	2	0	1		2
47.36	47.38	1.9177	2	0	3	4	3	5
47.38		1.9170	-4	0	2	2		2
47.75	47.76	1.9032	0	2	3	2	2	3
48.08	48.10	1.8906	0	-3	2	5	3	7
48.36	48.48	1.8803	3	-2	2	5		6
48.40		1.8790	-1	1	3	4		5
48.46		1.8769	-1	-2	3	3		3
48.50		1.8754	2	2	2	4		6
48.92	48.92	1.8604	-2	-2	3	3	2	6
49.59	49.60	1.8368	-2	2	3	4	3	4
49.84	49.88	1.8282	-1	-3	1	8	7	5
49.88		1.8265	4	0	0	4		11
49.91		1.8256	3	1	1	3		6
51.32	51.36	1.7789	2	-2	2	4	6	4
51.34		1.7781	3	-2	1	3		5
51.38		1.7768	4	-1	0	2		5
51.59	51.60	1.7700	4	1	1	4	5	5
51.61		1.7693	1	2	3	3		4

PECTOLITE — PREWITT, 1968

2THETA	PEAK	D	H	K	L	I(INT)	I(PK)	I(DS)
51.91	51.92	1.7599	-1	4	0	33	17	47
52.28	52.28	1.7483	0	0	4	12	8	17
52.33		1.7466	1	3	2	1		2
52.93	52.94	1.7284	-3	3	2	6	4	8
53.29	53.30	1.7176	-3	-2	2	15	11	22
53.33		1.7162	-2	-2	3	6		8
53.69	53.70	1.7058	-4	2	2	15	10	22
53.77		1.7032	-1	4	1	2		3
53.81		1.7023	-3	2	3	5		8
54.49	54.50	1.6826	0	1	4	1	1	2
54.86	54.86	1.6721	1	0	4	4	3	6
55.14	55.16	1.6641	4	-1	2	2	2	2
55.35		1.6583	0	-3	3	2		6
55.43	55.52	1.6561	3	0	3	4	5	6
55.52		1.6537	-2	0	4	4		2
55.53		1.6534	-1	-1	4	1		2
55.66		1.6499	-1	1	3	1		2
56.87	56.88	1.6176	-2	3	3	1	1	8
57.47	57.54	1.6022	3	-2	2	5	5	8
57.55		1.6002	2	1	2	5		2
57.66		1.5974	3	2	1	1		3
58.45	58.46	1.5775	-1	-4	2	1	2	9
58.90	58.92	1.5666	-1	4	2	2	4	11
58.92		1.5662	0	4	3	5		5
59.17	59.16	1.5600	0	-4	2	7	5	4
59.53	59.70	1.5516	4	2	0	3	8	10
59.60		1.5499	1	-2	4	3		3
59.71		1.5474	-2	4	2	6		6
59.71		1.5473	-5	0	1	2		2
59.72		1.5471	-5	2	0	4		5
59.85		1.5440	-4	-2	1	1		2
60.35	60.36	1.5323	2	-4	2	3	3	5
60.48	60.54	1.5294	-2	2	2	1	4	7
60.54		1.5280	-3	-2	3	4		
60.71	60.70	1.5241	0	4	2	3	3	5
61.02	61.02	1.5172	-4	2	3	4	3	7
62.41	62.42	1.4866	5	0	1	12	7	20
62.47		1.4855	-1	-4	2	3		4
63.11	63.28	1.4718	-5	0	2	2	8	4
63.27		1.4686	-2	-2	2	13		22
63.33		1.4673	-4	-1	2	2		4
63.63	63.64	1.4610	-1	2	2	13	7	21
63.72		1.4591	-5	2	4	2		3
63.79		1.4578	0	-4	2	2		4
64.74	64.74	1.4386	3	-4	3	2	2	3
65.01	64.94	1.4335	1	4	3	2	2	3
65.34	65.34	1.4269	4	0	2	3	1	3
66.23	66.24	1.4098	0	-3	4	4	1	3
67.32	67.34	1.3897	-2	4	2	3	3	2
67.35		1.3892	3	0	4	4		2
67.63	67.54	1.3841	-2	3	4	6		10
68.11	68.30	1.3754	-2	-4	2	4	3	2
68.18		1.3743	-2	1	4	1		7
68.21		1.3737	5	-1	2	4		2
68.31		1.3720	5	0	2	1		9
68.72	68.74	1.3648	-4	0	4	3	3	3
68.76		1.3640	-4	3	2	2		6
69.25	69.44	1.3556	-3	-3	3	2	3	3
69.42		1.3527	-5	-1	4	4		7
69.46		1.3520	-1	-4	4	1		3
69.56		1.3503	2	-1	3	2		3
70.29	70.28	1.3380	-3	3	4	2	2	4
70.84	70.84	1.3291	-4	-1	2	1	1	3
71.30	71.30	1.3216	-3	5	1	3	1	3
71.97	71.96	1.3110	1	-5	5	1	2	5
72.33	72.20	1.3053	-2	5	2	2	2	3
72.86		1.2971	-2	0	5	2		3
73.02	73.02	1.2946	-6	0	0	4	3	8
73.84	73.84	1.2822	-3	3	4	2	1	4
74.81	74.88	1.2681	3	4	1	1	1	2
74.91		1.2666	-1	-4	1	1		3
75.50	75.50	1.2581	-5	0	5	2	1	3
76.38	76.38	1.2458	-1	-4	4	5	2	10
76.65	76.60	1.2421	4	0	4	1	2	2

PECTOLITE – PREWITT, 1968

2THETA	PEAK	D	H	K	L	I(INT)	I(PK)	I(DS)
77.17	77.18	1.2350	-1	4	4	4	2	9
78.23	78.28	1.2209	4	3	2	1	1	1
78.69	78.70	1.2149	-1	2	0	1	1	2
78.99	79.20	1.2111	-1	-4	4	1	1	2
79.15		1.2091	-3	-3	4	1		2
79.20		1.2084	-2	-5	1	1		2
79.92	79.92	1.1993	-3	4	4	1	1	3
80.62	80.64	1.1906	-5	-2	3	1	1	3
81.12	81.12	1.1846	-6	2	3	1	1	3
81.52	81.68	1.1797	-6	0	3	1	1	2
81.67		1.1780	6	0	2	1		2
81.91	81.94	1.1751	-3	-4	3	1		2
82.25	82.32	1.1711	-1	5	2	1	2	2
82.26		1.1710	-5	-1	4	1		4
82.29		1.1706	-3	-2	5	2		4
82.43	82.44	1.1690	-1	0	6	2		6
82.73	82.74	1.1655	-4	-4	1	3	2	3
82.75		1.1653	-4	-4	1	1	3	2
82.78		1.1649	-5	4	3	1		4
82.90		1.1635	-4	2	5	2		1
83.83	83.84	1.1530	-1	1	6	1	1	3
84.41	84.68	1.1466	-2	0	0	1	2	5
84.54		1.1452	0	6	0	2		6
84.70		1.1433	-3	6	0	2		2
85.27		1.1372	-6	3	1	3		6
85.33	85.52	1.1366	-6	3	3	1	2	6
85.50		1.1347	6	-4	1	1		2
85.52		1.1346	-6	-2	2	1		3
85.66		1.1330	-3	-4	1	1		3
86.31	86.40	1.1261	3	-6	6	1	1	1
86.41		1.1248	-6	0	2	1		3
86.43		1.1234	0	6	1	1		2
86.57		1.1202	-4	3	3	2		4
86.88	86.84	1.1125	3	3	4	1	1	3
87.22	87.94	1.1098	4	-4	3	1	2	3
87.64		1.1097	5	-4	3	1		2
87.90			2	-6	2	1		
87.91								
88.64	88.84	1.1025	-3	0	6	2	4	3
88.67		1.1021	6	-2	3	1		3
88.69		1.1020	6	2	1	4		9
88.76		1.1013	5	2	1	1		3
88.76		1.1006	7	-2	1	4		9
88.83		1.0994	1	-2	6	1		3
88.95		1.0993	-2	2	6	2		5
88.96		1.0979	-2	-6	2	1		2
89.11	89.76	1.0940	0	2	6	1	3	2
89.51		1.0930	4	-4	4	3		6
89.61		1.0918	-5	0	5	3		7
89.73		1.0907	-3	6	2	1		2
89.85		1.0891	-1	5	5	1		2
90.02	90.10	1.0884	2	0	6	2	3	5
90.10		1.0881	5	-5	2	1		2
90.12		1.0880	-5	2	5	1		3
90.13		1.0880	-4	-1	4	1		5
90.14		1.0878	2	-2	4	2		4
90.16			6	-4	2	2		2
90.76	90.78	1.0821	-5	4	5	2	2	5
90.78		1.0820	4	2	2	1		3
90.78		1.0820	-7	0	1	1		3
91.00	91.02	1.0799	-4	-4	3	1	2	3
91.34		1.0768	-6	4	3	3		6
92.28	92.64	1.0683	-1	-6	6	3		2
92.64		1.0650	-1	2	2	2		6
93.09	93.14	1.0610	-4	6	6	6		6
93.16		1.0604	-1	-6	2	1		2
93.94	93.92	1.0537	-4	6	2	1		6
94.45	94.48	1.0493	6	2	2	3		2
94.50		1.0489	-7	-2	2	1		3
94.91	94.92	1.0455	-5	-4	1	2	1	2
95.51	95.54	1.0405	-7	0	3	1	1	5
95.54		1.0402	-7	4	1	2		3
96.82	96.64	1.0299	-5	6	6	1	1	4
97.22	97.22	1.0267	0	3	3	1	1	3
99.35	99.38	1.0103	-1	-6	3	1	1	2
99.37		1.0102	-2	-6	2	1		3

PECTOLITE - PREWITT, 1968

2THETA	PEAK	D	H	K	L	I(INT)	I(PK)	I(DS)
101.02	101.12	0.9981	-6	0	5	1	1	3
101.12		0.9973	-7	-2	1	1		4
101.50	101.42	0.9947	-8	2	1	2		4
103.14	103.32	0.9833	5	-6	2	2	1	5
103.34		0.9819	2	6	2	2		5
104.27	104.28	0.9757	1	-4	6	2	1	5
104.96	105.46	0.9711	0	-6	4	1	1	2
105.46		0.9679	-1	4	6	2		4
107.04	107.12	0.9579	4	-2	6	1	1	4
107.05		0.9579	-8	3	2	1		2
107.51	107.40	0.9550	3	2	6	1	1	3
107.96	108.04	0.9523	0	-7	2	1	1	2
108.10		0.9515	3	-6	4	2		6
108.47		0.9492	8	0	1	1		2
108.50		0.9491	6	-5	3	1		2
109.14	109.14	0.9453	0	6	4	2	1	6
110.33	110.36	0.9384	-2	-4	6	1	0	2

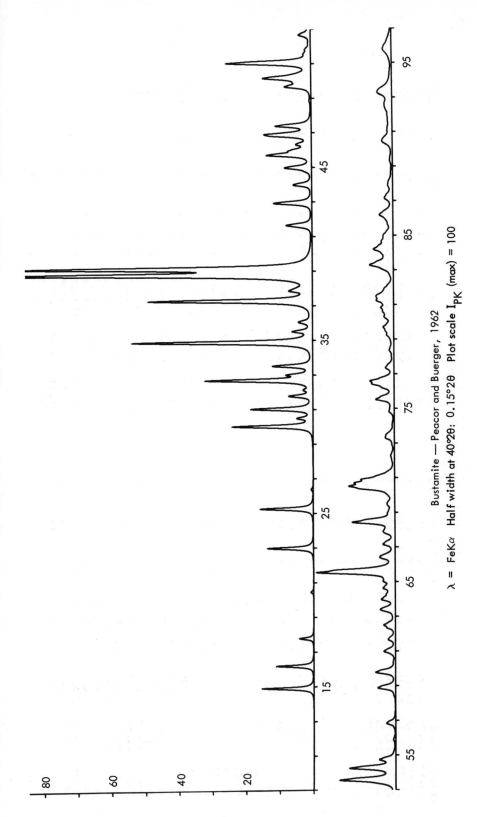

Bustamite — Peacor and Buerger, 1962

λ = FeKα Half width at 40°2θ: 0.15°2θ Plot scale I$_{PK}$ (max) = 100

BUSTAMITE – PEACOR AND BUERGER, 1962

2THETA	PEAK	D	H	K	L	I(INT)	I(PK)	I(DS)
14.86	14.88	7.484	2	0	0	10	14	4
16.16	16.16	6.887	0	0	1	8	11	4
17.75	17.76	6.276	-1	1	0	5	6	2
22.95	22.96	4.866	2	0	1	13	16	7
25.22	25.24	4.433	-3	0	1	13	16	8
29.99	30.00	3.742	-4	1	0	23	25	17
30.46	30.46	3.685	-1	1	3	3	4	2
30.99	31.00	3.623	-3	1	1	18	20	14
31.75	31.76	3.539	1	1	3	7	8	5
32.09	32.10	3.502	3	1	1	2	3	2
32.65	32.66	3.443	1	1	1	31	34	26
32.97	32.98	3.411	-4	0	0	9	9	5
33.48	33.50	3.361	-3	1	3	12	13	10
34.83	34.84	3.234	-2	0	4	52	53	46
35.48	35.48	3.177	4	0	2	3	6	4
36.03	36.04	3.130	-3	0	3	3	4	3
37.23	37.24	3.032	-2	0	4	49	48	47
37.82	37.84	2.987	-5	1	3	4	6	4
37.98		2.975	-3	1	3	1		1
38.71	38.72	2.921	-4	2	0	100	100	100
39.07	39.08	2.895	4	2	1	86	89	87
39.16		2.888	-1	1	1	4		3
41.53	41.62	2.730	-2	2	2	3	8	3
41.62		2.725	-2	2	2	6		7
42.88	42.90	2.648	-4	2	2	13	12	15
43.95	43.96	2.587	-1	1	5	5	5	6
44.94	44.96	2.533	-1	1	5	8	8	10
45.67	45.68	2.495	6	0	0	15	14	18
45.92	45.90	2.463	-2	2	4	5	7	6
46.29	46.30	2.438	0	2	4	4	5	5
46.79	46.88	2.433	0	2	4	11	14	13
46.88		2.410	4	2	4	8		10
47.36	47.36	2.307	4	2	0	12	11	15
49.62	49.64	2.289	3	-1	5	8	8	11
50.04	50.14	2.285	-6	2	0	1		2
50.12		2.284	6	0	2	12	14	16
50.14						3		4

2THETA	PEAK	D	H	K	L	I(INT)	I(PK)	I(DS)
50.22	51.00	2.281	-4	2	6	1	26	2
50.99	51.52	2.249	-2	0	2	31	2	44
51.52	51.78	2.227	-4	-2	2	2	2	2
51.79	52.64	2.217	-6	2	4	1	4	2
52.64	53.54	2.183	2	2	2	4	17	6
53.53	54.22	2.149	4	2	2	20	14	30
54.21	54.72	2.124	-6	-2	2	17	5	26
54.71	56.82	2.107	-4	0	6	3	3	8
56.80	58.86	2.035	-6	-2	4	5	5	5
58.83	59.76	1.9709	-6	0	4	2		8
58.87	60.96	1.9696	0	2	6	8	6	4
59.74	61.64	1.9435	0	2	0	4	4	14
60.95	62.48	1.9087	2	-2	0	3	2	7
61.62	63.36	1.8898	8	0	0	3	4	5
62.31	63.96	1.8709	8	0	0	3		5
62.49	64.24	1.8663	6	-2	4	2		6
63.30	64.60	1.8448	6	2	0	4	5	3
63.36	65.02	1.8431	-8	2	0	5		7
63.95	65.52	1.8279	-3	-3	3	2	4	10
64.26	66.40	1.8200	-6	-2	3	4	3	4
64.60	67.14	1.8115	-8	-2	2	3	4	8
65.00	67.88	1.8016	-2	2	2	33		6
65.51	68.42	1.7891	-7	4	0	3	23	66
65.68	68.94	1.7851	2	3	1	3		6
66.33	69.48	1.7695	8	0	6	1	5	5
66.39	70.50	1.7680	3	3	2	4		6
66.46	70.68	1.7665	4	-2	3	2		3
67.13		1.7507	-2	4	2	2	4	8
67.79		1.7358	8	0	3	2	4	4
67.88		1.7337	-2	4	2	2		5
68.42		1.7217	0	0	8	19	13	40
68.96		1.7098	-2	0	8	1	2	3
69.39		1.7006	-7	3	3	2	2	4
69.49		1.6984	-5	3	5	1		3
70.35		1.6803	-6	2	2	12	14	26
70.49		1.6773	-4	-2	6	10		23
70.71		1.6730	-6	-2	4	8	12	18

BUSTAMITE - PEACOR AND BUERGER, 1962

2THETA	PEAK	D	H	K	L	I(INT)	I(PK)	I(DS)
70.75	70.88	1.6721	4	-4	2	2		3
70.90		1.6689	-8	2	4	8	10	18
73.27	73.28	1.6222	6	0	6	1	3	3
73.27		1.6221	-3	-3	5	3		6
73.55	73.42	1.6169	-4	0	8	1		3
74.78	74.80	1.5940	-2	4	4	2	3	6
75.49	75.52	1.5812	2	-4	4	2	2	5
75.51		1.5809	6	2	4	2	6	16
76.41	76.42	1.5651	8	-2	4	7	7	17
76.59	76.58	1.5621	0	-4	4	7	7	16
76.59	77.34	1.5494	0	4	4	4	3	9
77.33	78.34	1.5325	-4	-4	4	2	1	5
78.34	79.22	1.5183	-4	2	8	2	2	6
79.22	79.52	1.5136	8	2	0	2	3	4
79.51	79.68	1.5109	-2	-2	8	3	3	9
79.68	80.28	1.5027	-10	2	0	1	3	3
80.21		1.5014	-8	-2	2	2		6
80.29		1.4967	-10	2	0	3		9
80.59	80.60	1.4934	-10	0	2	4	3	10
80.81	80.80	1.4874	-6	-2	6	3	4	8
81.20	81.22	1.4871	-8	2	4	4	5	10
81.22		1.4831	-6	4	4	3		8
81.49	81.46	1.4818	-2	-4	4	3	5	7
81.57		1.4604	4	4	0	3		10
83.03	83.28	1.4570	-2	2	8	3	7	29
83.27		1.4474	-8	4	0	10		6
83.95	84.18	1.4440	4	-2	8	2	6	26
84.19		1.4375	10	0	2	9		13
84.65	84.64	1.4325	6	-4	4	5	4	8
85.02	84.96	1.4199	-10	2	4	3	3	4
85.96	86.16	1.4172	-10	0	4	1	4	19
86.16		1.4047	-4	4	4	7	3	12
87.12	87.14	1.3954	0	-4	6	4	3	6
87.84	87.88	1.3774	4	0	10	2	2	3
89.30	89.32	1.3650	-4	-4	4	1	1	6
90.33	90.44	1.3640	-6	0	8	2		15
90.41		1.3624	-8	4	4	5		4
90.55						1		

2THETA	PEAK	D	H	K	L	I(INT)	I(PK)	I(DS)
91.02	91.04	1.3569	-1	-5	1	1	2	4
91.22		1.3546	-1	5	3	1		3
91.63	91.64	1.3499	8	2	4	2	1	6
92.29	92.58	1.3424	4	8	8	2	2	7
92.46		1.3405	-5	5	3	2		6
92.66		1.3383	10	-2	4	2		6
93.22	93.26	1.3321	10	0	4	9	4	28
93.36		1.3305	6	-2	8	3		9
95.73	95.78	1.3054	-10	0	6	4	3	12
95.79		1.3047	-8	2	8	2		5
95.95		1.3031	-6	-2	8	1		4
98.71	99.06	1.2757	-3	-3	5	1	1	4
98.90		1.2739	-1	-3	9	1		2
99.08		1.2723	-4	2	10	2		7
99.72	100.10	1.2663	-2	-2	10	1	2	4
99.87		1.2648	-2	2	10	2		7
100.09		1.2628	6	4	2	1		4
100.13		1.2624	3	5	1	2		8
101.45	101.92	1.2504	10	-4	2	1	4	5
101.81		1.2473	12	0	0	1		5
101.87		1.2467	-2	4	8	6		23
101.96		1.2459	-12	0	2	5		17
102.90	103.28	1.2378	10	2	0	1	3	3
103.27		1.2345	2	-4	8	7		25
105.43	105.46	1.2166	8	0	8	2	1	8
107.19	107.18	1.2027	-6	4	8	1	1	3
108.33	108.30	1.1940	8	-4	6	1	0	3
110.20	110.24	1.1803	-10	2	8	3	2	2
110.20		1.1802	-8	-2	8	3		11
110.22		1.1801	-10	0	8	2		12
112.20	112.82	1.1662	-8	-2	8	1		7
112.74		1.1625	0	6	0	4	2	3
113.64	114.14	1.1566	-6	-4	6	1	2	18
113.84		1.1552	-10	4	6	1		4
114.13		1.1534	8	4	4	2		3
114.17		1.1531	4	2	10	1		10
114.53	114.88	1.1507	-5	5	7	1	2	3

BUSTAMITE - PEACOR AND BUERGER, 1962

2THETA	PEAK	D	H	K	L	I(INT)	I(PK)	I(DS)
114.72		1.1495	-1	5	7	1		5
114.77		1.1492	-2	0	12	3		13
114.99		1.1478	-0	12	0	2		7
115.11		1.1470	-6	6	12	2		7
115.53		1.1443	-6	-2	10	1		2
115.69	115.70	1.1434	-8	2	10	1	2	3
115.80		1.1427	-12	4	0	2		8
115.88		1.1422	12	0	4	1		3
116.05		1.1411	-6	-2	10	1		5
116.16		1.1404	-8	4	8	1		3
118.07		1.1289	-8	6	4	1		4
118.41		1.1269	13	-3	1	1		3
118.43		1.1268	8	4	2	3		12
118.61	118.76	1.1257	-2	6	4	2	2	8
118.84		1.1244	-4	0	12	3		11
118.98		1.1235	-2	-6	4	1		3
119.40		1.1212	5	5	3	1		5
120.05	120.68	1.1174	4	-6	4	3	3	14
120.56		1.1146	12	-4	2	4		19
120.75		1.1136	-8	-4	4	5		23
121.72	121.68	1.1083	-12	4	4	4	2	19
122.36		1.1049	-0	-6	4	1		4
122.44		1.1044	-8	6	0	4		3
122.54	125.00	1.0936	2	6	2	4	3	17
124.54		1.0936	-5	5	1	1		3
124.86		1.0920	4	4	8	1		3
124.94		1.0916	-11	5	3	5		25
125.26		1.0900	10	-2	8	1		3
125.37		1.0894	-4	2	12	2		7
125.57	125.90	1.0885	0	2	12	1	3	6
125.87		1.0870	13	1	1	5		26
125.94		1.0867	10	0	8	1		3
126.29	126.46	1.0850	-2	-2	12	2	3	8
126.56		1.0837	8	-6	2	2		8
126.57		1.0837	2	-2	12	4		18
127.25	127.24	1.0805	7	2	12	7	3	32
127.78	127.72	1.0780	8	-4	8	4	3	21

2THETA	PEAK	D	H	K	L	I(INT)	I(PK)	I(DS)
127.79		1.0780	-6	0	12	1		7
128.31	129.10	1.0756	-9	-3	7	1	3	3
128.49		1.0747	8	4	4	5		26
128.73		1.0737	-3	-3	11	1		3
128.87		1.0730	12	-2	6	1		3
129.02		1.0724	12	2	2	2		11
129.06		1.0722	4	0	12	5		26
129.23		1.0714	-12	-2	4	3		16
129.48		1.0703	-6	4	10	1		6
129.58		1.0699	7	-5	7	1		7
129.76		1.0691	14	0	0	1		3
130.07		1.0678	-14	-2	4	3		15
132.15		1.0590	-12	0	6	1		7
132.20	132.34	1.0588	-10	2	10	4	2	21
132.43		1.0578	-8	-2	10	4		20
132.46		1.0577	-2	6	6	1		5
132.92		1.0558	-8	-4	6	2		8
133.47		1.0537	11	-3	7	1		3
133.54	133.54	1.0534	3	5	7	4	2	22
133.55		1.0533	-12	0	8	3		16
133.59		1.0532	-12	4	6	1		5
133.73		1.0526	-4	-2	12	1		4
134.14		1.0511	2	6	4	1		4
134.36		1.0502	-1	-1	13	1		3
136.04	136.18	1.0439	-5	5	9	1	1	3
136.17		1.0434	9	-3	9	1		4
137.32	137.96	1.0392	-1	5	9	1	1	3
137.37		1.0391	-4	-6	2	1		6
137.82		1.0375	-10	6	0	1		6
138.77	138.52	1.0342	-11	-3	9	2	1	10
140.70	141.84	1.0278	10	4	0	3	2	15
140.73		1.0277	-8	-4	10	3		15
140.81		1.0275	-10	-4	2	4		19
141.24		1.0261	-1	-5	9	1		6
141.87		1.0242	12	2	4	5		28
142.10		1.0235	-4	-4	10	2		10

2THETA	PEAK	D	H	K	L	I(INT)	I(PK)	I(DS)
142.11		1.0234	-5	-5	7	1		3
142.33		1.0228	-12	-2	6	1		5
142.68		1.0217	6	-6	6	1		5
143.11	143.90	1.0204	-14	2	6	1	2	8
143.66		1.0188	-14	4	2	5		30
143.70		1.0187	-14	4	0	1		7
143.77		1.0185	10	-6	2	2		10
144.07		1.0176	14	-2	4	6		31
145.11		1.0147	-5	7	1	1		4
145.78	147.16	1.0128	-9	-1	11	1	3	21
145.99		1.0122	6	4	8	4		6
146.11		1.0119	5	-7	1	1		32
146.34		1.0113	-8	6	6	6		5
146.38		1.0112	15	-1	1	1		5
146.41		1.0111	7	3	9	1		28
146.85		1.0100	-4	-6	4	5		25
147.08		1.0094	-2	-6	6	4		29
147.67		1.0078	-8	2	12	5		31
147.89		1.0073	-10	6	4	5		5
147.97		1.0071	8	4	6	1		3
148.01		1.0070	1	-7	1	1		25
148.62		1.0055	-6	-2	12	4		4
148.75		1.0051	11	-5	5	1		4
153.88	155.82	0.9937	-13	-1	7	1	1	3
154.04		0.9934	15	-3	1	1		21
154.89		0.9917	4	6	4	3		7
155.86		0.9899	-11	5	7	1		9
155.91		0.9898	10	2	8	2		44
157.88	157.74	0.9863	-2	0	14	7	1	3
158.36		0.9855	-4	6	8	1		14
158.40		0.9854	-8	-4	8	2		10
158.79		0.9848	-12	4	8	2		10
159.96		0.9830	12	-2	8	2		36
162.03		0.9800	10	-6	4	6		4
163.24		0.9784	-1	-7	1	1		

311

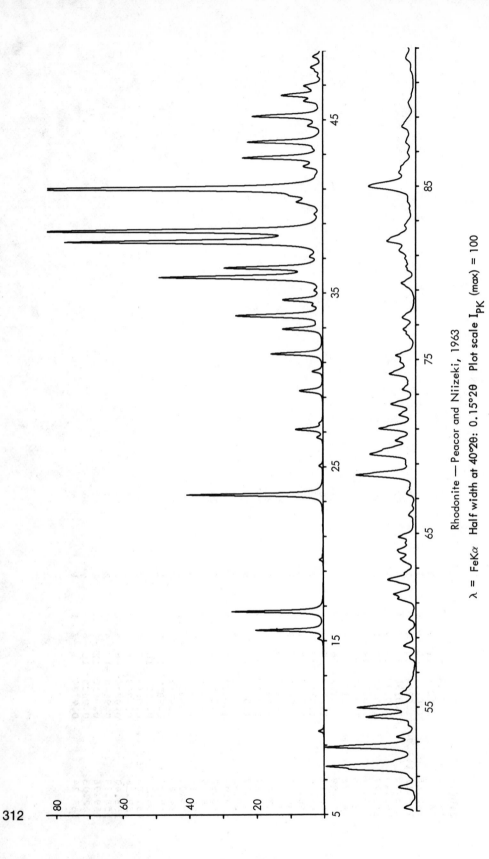

Rhodonite — Peacor and Niizeki, 1963

λ = FeKα Half width at 40°2θ: 0.15°2θ Plot scale I$_{PK}$ (max) = 100

RHODONITE - PEACOR AND NIIZEKI, 1963

2THETA	PEAK	D	H	K	L	I(INT)	I(PK)	I(DS)
9.78	9.78	11.357	0	1	0	1	2	7
15.58	15.60	7.139	-1	1	0	14	20	7
16.67	16.68	6.678	0	0	1	19	27	10
23.37	23.38	4.779	-1	1	1	30	40	19
26.66	26.66	4.199	0	2	1	1		4
27.12	27.14	4.129	-1	2	1	6	8	2
29.31	29.36	3.827	-2	1	0	3	7	2
29.37		3.819	-1	3	0	3		2
30.45	30.46	3.686	-2	0	0	3	3	2
31.47	31.48	3.570	-2	2	0	13	16	10
32.90	32.92	3.418	-2	1	1	10	12	8
33.63	33.68	3.347	-2	0	1	11	26	9
33.66		3.344	1	-3	1	11		4
33.70	33.70	3.339	0	0	2	2	3	9
34.25	34.26	3.288	-1	3	1	7		1
34.56	34.60	3.259	0	-1	2	4	12	6
34.61		3.254	1	1	1	1		3
35.11	35.12	3.209	0	3	2	9	3	1
35.78	35.90	3.151	-1	0	2	40	49	8
35.90		3.141	-2	3	0	13		35
36.41	36.44	3.098	2	-2	0	14	29	11
36.45		3.095	-2	-2	1	2		13
37.06	37.08	3.046	-1	3	0	21	4	2
37.87	37.94	2.983	-1	-1	2	56	77	20
37.94		2.978	1	1	2	83	87	52
38.55	38.56	2.932	-1	4	0	1	3	78
39.51	39.52	2.864	-1	3	1	5	8	5
40.23	40.24	2.815	1	-2	2	100	100	100
40.98	41.00	2.765	2	2	0	3	6	3
42.23	42.30	2.687	1	3	1	2		2
42.29		2.683	-1	-2	2	24	24	25
42.77	42.78	2.655	-1	4	1	1	3	1
43.31	43.34	2.623	-2	1	0	23	22	24
43.69	43.70	2.602	1	1	3	4	5	4
44.62	44.62	2.550	-3	2	0	22	21	24
45.17	45.18	2.520	-3	2	1	5	6	4
46.05	46.06	2.475	2	2	1			6
46.38	46.38	2.458	1	2	2	12	12	14
46.78	46.94	2.438	2	-1	2	2	6	5
46.94		2.430	0	3	1	5		1
47.48	47.50	2.404	-3	2	1	1	2	2
47.79	47.80	2.390	-3	-2	0	2	3	3
48.00	48.02	2.380	-3	-2	1	2	4	3
48.63	48.64	2.351	-1	-3	1	3	3	3
49.16	49.18	2.327	3	1	0	4	4	5
50.34	50.36	2.276	-2	-1	2	5	6	7
51.36	51.42	2.234	0	2	2	1	20	1
51.39		2.233	-2	-2	2	16		20
51.43		2.231	0	-4	2	2		2
51.58	51.58	2.225	-2	5	0	23	27	30
52.68	52.68	2.182	-1	0	3	32	28	41
53.27	53.26	2.159	0	1	3	5	6	7
53.93	53.94	2.135	-1	-1	3	2	2	2
54.40	54.42	2.118	0	-2	3	1		2
54.41		2.117	2	-5	1	16	15	22
54.96	54.96	2.098	3	1	1	20	18	28
55.76	55.78	2.070	-3	2	2	5	5	7
56.42	56.40	2.048	-1	-4	2	1	2	2
57.76	57.78	2.004	-2	-3	3	2	2	3
58.49	58.50	1.9812	-2	1	3	4	4	6
59.25	59.30	1.9581	3	-2	0	1	2	2
59.32		1.9560	-3	-5	0	2		4
60.04	60.04	1.9348	0	0	5	2	2	2
61.08	61.22	1.9049	3	0	0	4	6	6
61.22		1.9011	-3	-3	2	5		7
61.48	61.48	1.8937	-1	5	1	2	7	3
61.49		1.8935	1	5	5	4		6
62.14	62.32	1.8756	1	2	3	2	8	3
62.28		1.8717	-2	-1	3	4		6
62.33		1.8706	-4	3	0	2		5
62.39		1.8688	0	3	3	3		5
62.70	62.68	1.8606	2	-5	0	3	3	4
63.38	63.52	1.8428	4	0	2	2	5	5
63.52		1.8391	2	3	3	2		4

RHODONITE — PEACOR AND NIIZEKI, 1963

2THETA	PEAK	D	H	K	L	I(INT)	I(PK)	I(DS)
63.52	64.02	1.8390	-2	0	3	1		2
63.89		1.8295	-3	-2	2	4	6	7
64.01		1.8265	-2	-6	1	2		3
64.09		1.8244	-2	3	3	1		2
64.44	64.44	1.8155	-4	0	1	2	3	4
64.77		1.8072	0	-4	3	5	5	2
64.78	64.78	1.8070	3	1	1	3		9
66.03	66.04	1.7767	4	-3	1	3	2	5
66.57	66.58	1.7638	2	1	1	1	1	2
67.22	67.26	1.7486	-3	6	0	2	3	4
67.29		1.7471	-1	4	3	1		2
67.62	67.62	1.7397	4	0	1	1		2
68.23	68.36	1.7260	-4	-1	1	5	2	8
68.35		1.7232	-1	6	0	22	18	39
68.96	68.96	1.7099	-4	1	2	8	3	3
69.48	69.54	1.6987	-2	4	3	8	14	15
69.55		1.6971	-3	5	2	11		20
70.32	70.32	1.6810	-3	-3	2	7		13
71.01	71.02	1.6668	0	-1	4	15	6	28
71.79	71.80	1.6509	-1	-1	3	6	11	11
72.31		1.6408	-1	1	4	2	5	3
72.41	72.42	1.6387	1	6	1	8	7	16
73.26	73.26	1.6224	0	7	0	4		9
73.66	73.64	1.6149	-4	-2	1	1	4	3
74.14	74.16	1.6058	-4	-1	1	10	2	19
74.21		1.6045	0	6	2	2	8	4
74.68		1.5959	4	-3	2	1		2
74.83	74.86	1.5931	3	-3	3	4	5	7
75.22	75.22	1.5861	0	-3	3	7	6	15
76.48	76.48	1.5639	-3	-2	3	3	3	5
76.74	76.72	1.5595	4	0	2	4	4	8
77.34		1.5492	-4	6	0	3		7
77.41	77.40	1.5480	-3	-6	2	3		5
79.18	79.38	1.5189	-1	3	4	2	4	7
79.39		1.5155	-4	6	1	5		5
80.14	80.16	1.5038	-2	-7	2	2	3	10
80.36	80.34	1.5003	-4	2	3	2		4
80.62	80.82	1.4963	0	-7	3	1	3	3
80.81		1.4934	4	1	2	3		6
81.16	81.28	1.4880	-1	-7	1	1	5	3
81.25		1.4867	2	5	1	4		9
81.35		1.4851	-5	4	0	2		4
81.78	81.80	1.4787	-3	5	3	11	9	24
81.90		1.4770	-1	0	1	3		6
82.68	82.70	1.4654	5	-3	1	2	3	4
82.74		1.4646	5	-1	1	2		4
83.46	83.48	1.4543	1	-8	1	1	2	3
83.69	83.68	1.4510	-3	1	4	2	2	6
84.08	84.10	1.4455	-1	3	1	2	14	5
84.75		1.4363	-4	-1	3	3		21
84.89	84.98	1.4342	5	-4	1	9		29
84.96		1.4333	-2	2	4	12		17
85.07		1.4319	-2	-6	2	7		9
85.74	85.78	1.4228	-2	8	1	4	5	7
85.82		1.4217	-2	-3	4	3		10
86.13	86.08	1.4177	-5	3	2	4	5	4
86.92	86.92	1.4073	2	-4	4	2	3	8
87.51	87.54	1.3997	0	6	3	3	3	18
88.43	88.44	1.3880	1	-5	4	7	4	4
89.80	89.82	1.3714	0	8	1	2	2	3
89.88		1.3704	1	1	2	1		6
91.72	91.74	1.3489	2	-7	3	2	2	3
92.30		1.3422	2	-5	2	1	3	4
92.33		1.3419	-1	1	2	1		6
92.42	92.42	1.3409	5	-1	2	2		6
93.21	93.32	1.3322	3	-7	2	2	4	11
93.23		1.3319	5	-4	2	4		4
93.36		1.3305	3	0	4	1		4
93.36		1.3305	-3	-6	1	2		4
94.51	94.84	1.3180	-1	1	3	2	3	13
94.84		1.3146	1	-7	3	5		3
95.34	95.66	1.3094	-2	-6	3	1	2	3
95.53		1.3074	4	-7	2	1		14
96.90	96.98	1.2935	-5	3	3	5	4	

RHODONITE - PEACOR AND NIIZEKI, 1963

2THETA	PEAK	D	H	K	L	I(INT)	I(PK)	I(DS)
96.96		1.2928	-2	-7	2	2		5
96.99		1.2926	0	-6	4	1		3
96.99		1.2925	1	-9	1	1		3
97.29		1.2895	-1	-7	3	3		7
98.49	98.52	1.2779	-3	5	4	1	1	3
99.22	99.44	1.2709	4	5	0	2	2	5
99.39		1.2693	2	-6	4	2		6
99.50		1.2682	-6	3	1	1		4
100.03	100.10	1.2633	-4	-1	4	1	2	4
100.37	100.36	1.2602	-6	4	0	1	2	4
100.60		1.2581	-2	-5	4	1		3
100.94	101.32	1.2550	3	-9	1	1	2	3
101.15		1.2531	-1	4	5	1		3
101.32		1.2516	-6	4	1	4		12
102.08	101.96	1.2448	-2	6	4	1	1	2
102.59	102.60	1.2404	1	5	4	5	2	14
104.75	105.08	1.2221	5	-3	3	1	2	4
105.02		1.2200	-3	1	5	1		4
105.05		1.2197	5	3	1	2		7
105.12		1.2191	4	-2	4	1		2
105.30		1.2177	2	8	0	1		3
105.61		1.2152	0	6	4	1		2
107.29	107.60	1.2019	-2	-7	3	1	2	4
107.61		1.1995	-4	5	1	2		5
107.83		1.1978	2	1	5	1		2
108.08		1.1959	4	-7	3	1		3
109.49	109.72	1.1854	5	5	2	1	1	3
109.75		1.1835	-5	-4	4	1		3
113.07	113.22	1.1604	-5	3	4	3	2	11
113.16		1.1598	-1	-7	4	1		2
113.18		1.1596	1	7	4	1		2
114.18	114.14	1.1530	4	5	2	2	1	6
114.19		1.1530	3	-2	5	2		6
115.31	115.28	1.1458	-2	-9	3	2	1	9
118.14	118.16	1.1285	-5	-1	4	2	2	6
118.17		1.1283	0	-9	3	2		6
118.76		1.1248	-6	4	3	1		2
119.83	120.46	1.1187	6	-7	1	2	1	7
120.26		1.1163	0	-8	4	1		4
121.71	122.28	1.1083	0	10	1	3	3	9
122.21		1.1056	6	2	1	1		5
122.22		1.1056	1	8	3	1		3
122.25		1.1054	5	-9	1	1		4
122.30		1.1051	3	8	0	3		12
122.31		1.1051	5	2	3	1		2
122.37		1.1048	2	9	0	1		5
123.45	123.62	1.0991	4	-10	1	7	3	27
123.71		1.0978	-1	-2	6	3		11
123.89		1.0969	1	9	2	1		3
124.19		1.0953	-4	10	1	1		4
124.34		1.0946	-5	-2	4	1		2
125.08	125.92	1.0909	-2	0	6	3	3	10
125.22		1.0902	-3	-6	4	1		2
125.59		1.0884	1	0	6	1		4
125.81		1.0873	-1	0	6	1		5
125.89		1.0869	-5	-4	3	3		10
126.01		1.0864	-6	-3	1	3		9
126.11		1.0858	-7	4	1	1		4
126.59		1.0835	3	-8	4	2		7
126.96	127.48	1.0818	-4	-2	5	3	2	10
127.02		1.0815	5	-1	4	1		3
127.30		1.0802	-5	-5	2	1		5
127.46		1.0795	4	-7	4	1		4
127.70		1.0783	-1	8	4	1		3
127.73		1.0782	-6	1	1	1		6
128.16	128.44	1.0762	1	5	5	3	3	4
128.31		1.0756	3	8	1	3		10
128.37		1.0753	-2	2	6	1		4
128.43		1.0750	-1	-3	6	1		5
128.64		1.0741	-7	0	3	1		3
128.78		1.0734	6	-2	3	2		8
129.08		1.0721	1	1	6	5		20

TABLE 15. SINGLE CHAIN SILICATES

Variety			Alamosite	Proto-enstatite	Ferro-carpholite
Composition	Na_2SiO_3	Li_2SiO_3	$PbSiO_3$	$MgSiO_3$	$FeAl_2(OH)_4Si_2O_6$
Source	Synthetic	Synthetic	Alamos, Sonora, Mexico	Synthetic	Celebes
Reference	McDonald & Cruickshank, 1967	Maksimov, Kharitonov, Ilyukhin & Belov, 1968	Boucher & Peacor, 1968	Smith, 1959	MacGillavry, Korst, Moore & Van der Plas, 1956
Cell Dimensions					
$\underline{a}$ Å	10.48	9.38	11.23	9.25	13.77
$\underline{b}$ Å	6.07	5.40	7.08	8.74	20.18
$\underline{c}$ Å	4.82	4.68	12.26	5.32	5.109
α deg	90	90	90	90	90
β deg	90	90	113.25	90	90
γ deg	90	90	90	90	90
Space Group	$Cmc2_1$	$Cmc2_1$	P2/n (P2/c)	Pbcn	Ccca
Z	4	4	12	8	8

TABLE 15. (cont.)

Variety			Alamosite	Proto-enstatite	Ferro-carpholite
Site Occupancy M_1 M_2 M_3					M_1 $\begin{cases} Fe & 0.80 \\ Mg & 0.20 \end{cases}$ $\left.\begin{matrix} M_2 \\ M_3 \end{matrix}\right]$ $\begin{cases} Al & 0.95 \\ Fe & 0.05 \end{cases}$
Method	3-D, counter	2-D, film	3-D, counter	Powder data	3-D, film
R & Rating	0.035 (1)	0.13 (av.) (3)	0.111 (1)	~ 0.2 (3)	0.14 (2)
Cleavage and habit		A perfect cleavage parallel to c axis. Elongated to c	{010} perfect Fibers parallel to b axis		Fibrous
Comment				Psuedo space group is given B estimated	
μ	78.7	63.2	1152.0	98.6	175.9
ASF	0.8968×10^{-1}	0.9115×10^{-1}	0.4440	0.1013	0.6924×10^{-1}
Abridging factors	0.3:0:0	0.3:0:0	1:0:0	1:1.3:0.5	1:1.3:0.5

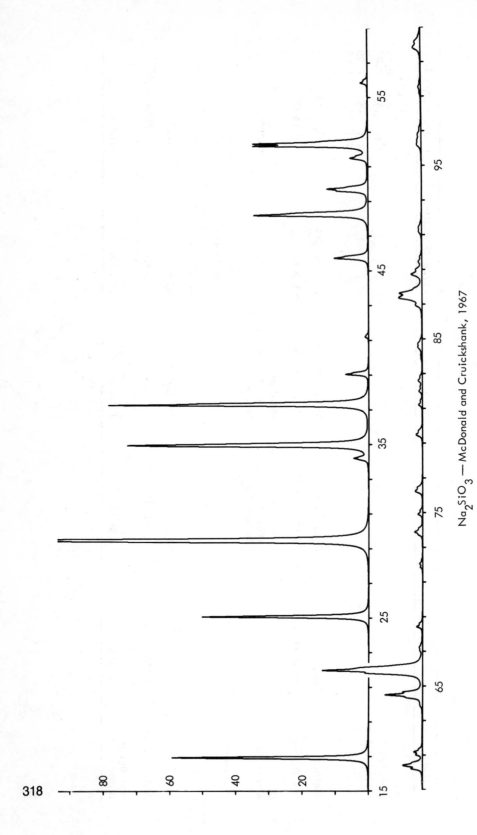

318

Na$_2$SiO$_3$ — McDonald and Cruickshank, 1967

λ = CuKα Half width at 40°2θ: 0.11°2θ Plot scale I$_{PK}$ (max) = 150

NA2SIO3 - MCDONALD AND CRUICKSHANK, 1967

2THETA	PEAK	D	H	K	L	I(INT)	I(PK)	I(DS)
16.86	16.90	5.253	1	1	0	20	40	19
16.91		5.240	2	0	0	25		25
25.05	25.06	3.551	1	1	1	40	34	40
29.40	29.48	3.035	0	2	0	41	100	41
29.48		3.028	3	1	0	100	100	100
34.11	34.20	2.626	2	2	0	2		3
34.19		2.620	4	0	0	3	3	3
34.90	34.96	2.568	0	2	1	38	49	38
34.97		2.564	3	1	1	41		41
37.28	37.28	2.410	0	0	2	74	52	76
39.02	39.02	2.306	2	2	1	7	5	7
41.18	41.18	2.190	1	1	2	1	1	1
45.62	45.72	1.9866	1	3	0	9	7	10
45.71		1.9832	4	2	0	9		10
45.76		1.9812	5	1	0	2		2
48.17	48.20	1.8873	0	2	2	20	23	21
48.22		1.8856	3	1	2	23		24
49.59	49.60	1.8367	1	3	1	8	6	9
49.67	49.70	1.8341	4	2	1	2	8	2
49.71		1.8324	5	1	1	7		8
51.42	51.48	1.7757	4	0	2	5	4	2
51.48		1.7737	2	2	2	2		5
52.20	52.20	1.7509	3	3	0	36	23	39
52.33	52.34	1.7467	6	0	0	17	23	19
55.82	55.82	1.6456	3	3	1	2	1	3
60.18	60.18	1.5364	1	1	3	4	2	4
60.33	60.38	1.5329	1	3	2	4	4	2
60.39		1.5314	4	2	2	2		2
60.44		1.5304	5	1	2	5		5
61.01	61.02	1.5175	6	2	0	0	1	0
61.17	61.16	1.5139	2	4	0	2	2	3
63.80	63.90	1.4576	5	3	0	0	1	1
63.89		1.4557	3	3	2	0		0
64.46	64.46	1.4443	6	2	1	12	8	14
65.70	65.74	1.4200	0	2	3	9	12	10
65.74		1.4192	1	3	3	10		11
65.88	65.88	1.4165	3	3	2	27	20	32
66.00	66.00	1.4143	6	0	2	9		11
67.02	67.04	1.3952	2	4	1	0	1	1
67.11	67.20	1.3936	4	3	1	1	1	0
67.21		1.3917	7	1	1	1	1	1
68.39	68.40	1.3705	4	4	0	3	1	3
71.83	71.84	1.3131	8	0	0	1	1	1
72.03	72.02	1.3100	0	4	2	1	1	1
73.71	73.86	1.2841	6	2	2	2	2	2
73.86		1.2819	1	3	3	3		3
74.88	74.88	1.2670	4	4	1	2	1	1
76.13	76.22	1.2493	1	3	3	4	2	3
76.19		1.2484	4	2	3	3		2
76.23		1.2479	5	5	0	1		1
76.36	76.40	1.2460	5	3	2	3		2
79.39	79.46	1.2059	1	5	0	1	1	1
79.47		1.2050	0	0	4	3		3
81.18	81.18	1.1838	3	1	4	2	1	2
81.96	81.98	1.1745	3	1	4	1	1	1
81.98		1.1743	1	1	4	1		1
82.55	82.58	1.1677	7	3	1	1	1	2
82.61		1.1670	8	4	1	1		1
83.82	83.84	1.1531	4	4	2	0	0	0
84.01	84.04	1.1510	8	0	2	2	0	0
84.40	84.50	1.1467	3	5	0	1	1	1
84.50		1.1455	6	4	0	2		2
84.68	84.68	1.1436	9	0	2	2	1	2
86.90	86.92	1.1200	0	2	4	1	1	1
86.94		1.1196	3	1	4	1		1
87.33	87.34	1.1156	6	5	1	10	5	13
87.44		1.1145	6	4	1	5		6
87.61	87.60	1.1127	0	6	3	5	5	7
88.56	88.70	1.1032	6	4	3	0	2	2
88.71		1.1018	6	2	3	6		8
89.43	89.42	1.0948	4	0	4	2	1	2
91.12	91.18	1.0788	5	5	3	1	1	1
91.16		1.0785	3	5	3	1		1
91.22		1.0779	7	1	3	0		1

319

NA2SIO3 — MCDONALD AND CRUICKSHANK, 1967

2THETA	PEAK	D	H	K	L	I(INT)	I(PK)	I(DS)
91.35	91.38	1.0767	7	3	2	2	1	1
91.41		1.0762	8	2	2	2		1
96.12	96.20	1.0355	3	5	2	0	1	2
96.23		1.0346	6	4	2	2		2
96.41	96.42	1.0332	9	1	2	1	1	1
96.77	96.76	1.0303	1	3	4	1	1	2
96.83		1.0298	4	2	4	1		1
99.17	99.18	1.0117	0	6	0	1	0	1
99.49	99.50	1.0092	9	3	0	1	1	2
101.79	101.80	0.9926	3	3	4	5	2	7
101.89		0.9919	6	0	4	1		2
102.15	102.12	0.9901	0	6	1	1	1	2
106.00	106.24	0.9645	1	5	3	0	0	1
106.26		0.9628	8	2	3	0		1
109.42	109.56	0.9437	0	6	3	1	0	1
109.54		0.9430	7	5	4	0		1
109.57		0.9428	6	4	4	1		1
111.22	111.34	0.9334	3	5	3	4	2	6
111.32		0.9328	0	6	2	3		4
111.34		0.9327	6	4	3	3		2
111.53	111.60	0.9317	9	1	3	3	2	5
111.67		0.9309	9	1	2	2		3
112.99	112.98	0.9238	11	1	1	1	0	2
113.93	113.98	0.9188	10	2	1	1	1	2
113.97		0.9186	3	1	3	2		2
114.01		0.9184	2	6	2	1		1
116.67	116.68	0.9050	2	2	5	0	0	1
122.45	122.54	0.8788	4	6	2	1	0	1
122.60		0.8781	7	5	2	1		2
123.25	123.22	0.8754	6	0	0	0	0	0
123.76	123.74	0.8733	12	0	0	1	0	1
125.39	125.40	0.8668	5	1	5	0	0	0
126.82	126.80	0.8613	6	6	3	1	0	1
128.25	128.26	0.8561	0	6	4	0	0	0
129.52	129.56	0.8515	7	2	4	1	0	1
129.60		0.8513	8	2	5	0		1
131.60	131.74	0.8445	3	3	5	1	0	1

2THETA	PEAK	D	H	K	L	I(INT)	I(PK)	I(DS)
131.79		0.8438	8	4	3	1		1
132.47	132.34	0.8416	3	7	0	0	0	1
136.58	137.02	0.8291	3	7	1	3	1	5
136.99		0.8279	9	2	5	3		5
137.36		0.8268	12	2	1	3		4
138.82	138.82	0.8228	6	6	2	5	1	7
139.47	139.50	0.8211	12	0	3	0	0	1
142.36	142.56	0.8138	0	4	5	1	1	2
142.38		0.8137	1	7	2	1		2
142.48		0.8135	7	5	3	0		1
142.58		0.8132	6	2	5	3		4
142.62		0.8131	11	1	3	1		1
143.05		0.8121	2	4	5	1		3
146.65	146.88	0.8041	5	3	5	5	0	1
146.80		0.8037	3	0	6	5		2
147.00		0.8033	1	1	3	6		1
151.59	151.80	0.7945	2	0	6	3	0	2
151.84		0.7941	1	1	6	6		5
151.87		0.7941	2	0	6	2		2
152.19		0.7935	9	5	2	0		3
164.79	164.92	0.7771	4	4	5	1	0	2

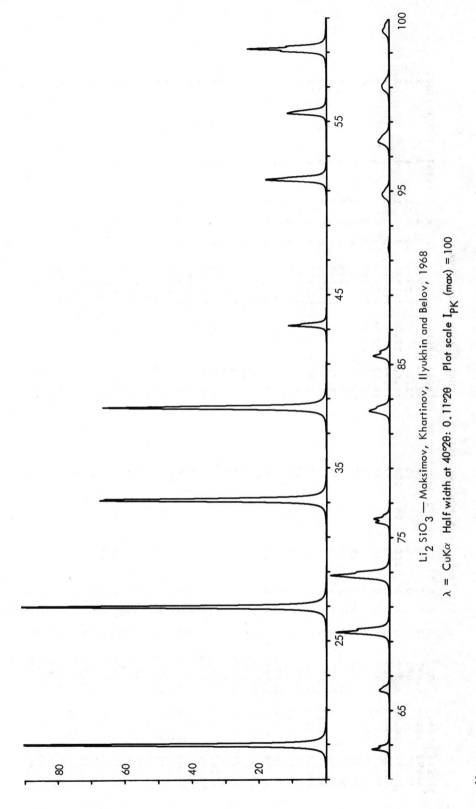

Li$_2$SiO$_3$ — Maksimov, Khartinov, Ilyukhin and Belov, 1968

λ = CuKα Half width at 40°2θ: 0.11°2θ Plot scale I$_{PK}$ (max) = 100

LI2SIO3 - MAKSIMOV, KHARTINOV, ILYUKHIN AND BELOV, 1968

2THETA	PEAK	D	H	K	L	I(INT)	I(PK)	I(DS)
18.91	18.94	4.690	2	0	0	58	100	58
18.95		4.680	1	1	1	40		40
26.92	26.92	3.309	1	1	1	100	98	100
33.08	33.08	2.706	3	1	1	66	68	67
33.15	33.14	2.700	0	2	0	28	66	28
38.35	38.44	2.345	2	0	0	1	67	1
38.39		2.342	3	1	0	26		27
38.44		2.340	0	2	1	46		47
38.46		2.339	0	0	2	15		15
43.17	43.18	2.094	2	1	2	4	11	4
43.19		2.093	1	1	2	2		2
43.19		2.093	2	2	1	10		10
51.53	51.62	1.7721	5	2	0	5	18	5
51.58		1.7705	4	2	0	1		1
51.59		1.7699	3	0	2	14		14
51.65		1.7683	0	2	2	10		11
51.66		1.7677	1	3	0	2		2
55.39	55.44	1.6573	5	1	1	3	12	3
55.42		1.6564	4	0	2	1		1
55.44		1.6559	4	2	1	10		11
55.49		1.6546	2	2	2	2		2
55.52		1.6537	1	3	1	5		5
59.04	59.06	1.5633	6	0	0	15	14	16
59.18	59.18	1.5600	3	3	0	27	24	29
62.73	62.72	1.4799	1	1	3	8	3	9
66.08	66.14	1.4127	4	1	2	1		2
66.12		1.4119	4	2	2	3		3
66.20		1.4105	2	2	0	3		3
69.41	69.50	1.3529	6	2	0	3	16	4
69.49		1.3515	3	1	3	18		20
69.53		1.3507	0	2	3	9		10
69.58		1.3500	6	0	2	1		2
72.67	72.80	1.2999	6	2	1	8	18	9
72.69		1.2997	6	2	1	3		4
72.75		1.2988	5	3	0	3		3
72.80	72.80	1.2980	2	2	3	6		7
72.80		1.2980	3	3	2	16		18

2THETA	PEAK	D	H	K	L	I(INT)	I(PK)	I(DS)
72.84		1.2973	2	4	0	2		2
72.86		1.2971	4	1	1	1		1
75.86	75.86	1.2531	7	1	1	7	4	8
75.97		1.2515	5	3	1	1		1
76.06	76.06	1.2502	8	2	0	5	5	5
82.13		1.1725	6	0	2	2		3
82.24	82.30	1.1712	6	2	3	3	6	4
82.27		1.1709	5	1	3	3		4
82.31		1.1705	4	2	3	5		5
82.35		1.1700	0	4	0	0		0
82.35		1.1700	4	0	3	1		1
82.37		1.1697	1	3	3	1		1
82.40		1.1694	0	4	2	3		3
85.42	85.46	1.1356	5	3	2	4	5	1
85.45		1.1352	2	5	4	2		3
85.47		1.1351	1	3	4	1		1
85.47		1.1350	4	0	1	7		9
85.51		1.1346	2	4	2	1		1
88.58	88.58	1.1031	3	3	3	0	0	0
91.48		1.0755	8	2	0	1		1
91.69	91.70	1.0735	0	2	3	2	0	2
94.58	94.78	1.0483	8	2	4	1	2	1
94.59		1.0482	0	4	2	2		2
94.66		1.0476	8	2	1	1		1
94.74		1.0469	7	0	4	2		2
94.79		1.0465	4	2	4	1		1
94.79		1.0465	2	4	2	2		2
94.87		1.0458	4	5	0	1		2
97.65	97.84	1.0233	9	2	3	2	3	8
97.81		1.0221	6	4	3	1		1
97.85		1.0218	6	2	0	2		3
97.97		1.0208	0	4	0	1		2
97.97		1.0208	3	5	3	3		2
100.90	101.08	0.9989	7	1	3	3	2	4
101.00		0.9982	6	3	1	2		2
101.10		0.9975	2	4	3	1		1
101.12		0.9974	3	5	1	2		3

LI2SIO3 - MAKSIMOV, KHARTINOV, ILYUKHIN AND BELOV, 1968

2THETA	PEAK	D	H	K	L	I(INT)	I(PK)	I(DS)
104.04	104.26	0.9772	8	2	2	1		1
104.11		0.9767	7	3	2	1		2
104.20		0.9761	4	2	4	1		1
104.27		0.9757	1	3	4	2		2
104.33		0.9753	1	5	2	3	2	3
110.47	110.78	0.9376	9	2	2	2		2
110.63		0.9367	6	0	4	1		1
110.69		0.9364	6	4	2	1		2
110.76		0.9360	3	4	3	2		1
110.76		0.9360	4	4	3	2	2	2
110.82		0.9357	3	5	2	3		4
114.11	114.12	0.9178	1	1	5	1	0	1
114.12		0.9178	5	5	1	1		5
117.30	117.32	0.9019	9	3	0	4	1	5
117.70	117.72	0.9000	0	6	0	2	1	2
120.89	121.10	0.8854	8	2	3	3	3	4
120.94		0.8852	8	4	0	1		1
120.97		0.8851	7	3	3	1		1
121.00		0.8850	6	2	4	1		1
121.09		0.8846	3	1	5	8		10
121.14		0.8844	0	2	5	3		4
121.22		0.8840	1	5	3	2		2
124.42	124.80	0.8707	10	0	2	0	1	1
124.44		0.8706	10	2	1	3		3
124.63		0.8698	7	1	4	1		1
124.64		0.8698	8	4	1	1		1
124.76		0.8693	5	3	4	1		1
124.83		0.8691	2	5	5	1		2
124.83		0.8690	5	5	2	1		1
124.96		0.8685	2	6	1	0		1
128.36	128.70	0.8557	9	1	3	6	2	7
128.62		0.8547	6	4	3	4		5
128.77		0.8542	3	5	3	4		6
132.26	132.52	0.8423	11	1	0	1	1	1
132.48		0.8416	9	3	2	6		8
132.70		0.8409	7	5	0	1		1
132.97	132.98	0.8400	0	6	2	4	1	5

2THETA	PEAK	D	H	K	L	I(INT)	I(PK)	I(DS)
136.61	137.20	0.8290	11	1	1	1	1	1
136.73		0.8286	10	2	2	1		2
136.96		0.8280	8	4	2	2		1
137.07		0.8276	5	1	5	3		3
137.08		0.8275	7	5	1	3		3
137.13		0.8272	4	2	5	1		1
137.23		0.8270	1	3	5	0		0
137.29		0.8269	4	6	2	2		2
137.35		0.8269	2	6	2	3		2
147.35	147.38	0.8026	3	3	5	1	0	4
147.35		0.8026	5	5	3	3		2
152.77	153.60	0.7925	11	1	2	2	2	1
153.21		0.7918	8	2	4	4		12
153.36		0.7915	7	5	2	8		7
153.48		0.7913	7	5	2	5		6
153.82		0.7908	4	6	4	4		7
153.84		0.7908	1	5	4	5		1
160.39	161.80	0.7817	12	0	0	1	0	2
161.12		0.7808	9	3	3	2		10
161.86		0.7800	0	0	6	7		1
161.88		0.7800	6	6	0			
162.27		0.7796	0	6	3	1		

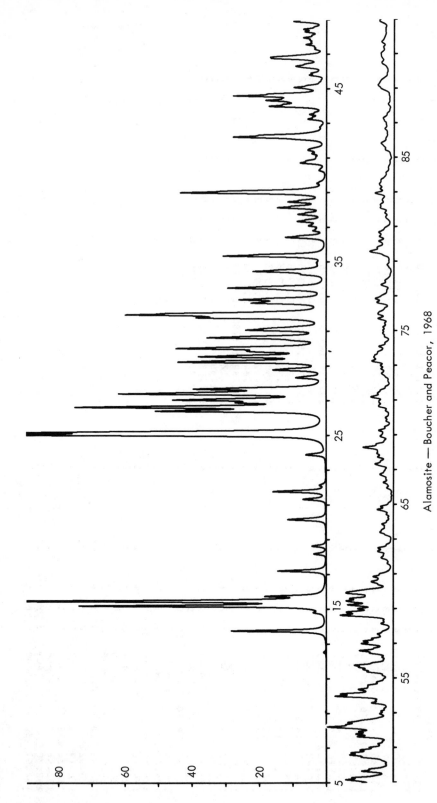

Alamosite — Boucher and Peacor, 1968

λ = CuKα Half width at 40°2θ: 0.11°2θ Plot scale I_{PK} (max) = 100

ALAMOSITE - BOUCHER AND PEACOR, 1968

2THETA	PEAK	D	H	K	L	I(INT)	I(PK)	I(DS)
13.73	13.74	6.446	1	0	1	27	28	11
14.77	14.78	5.994	0	1	1	2	4	1
15.16	15.18	5.838	-1	1	0	71	74	32
15.44	15.46	5.732	1	1	1	98	100	46
15.72	15.72	5.632	0	0	2	15	18	7
17.17	17.18	5.159	-2	0	0	15	15	8
18.15	18.16	4.884	0	1	2	4	4	2
18.60	18.60	4.766	1	1	2	4	4	3
20.13	20.14	4.408	-2	1	1	12	12	8
20.20		4.393	0	2	0	2		1
21.29	21.30	4.169	-2	0	2	7	7	6
21.74	21.74	4.084	-1	1	3	17	16	14
22.09	22.10	4.020	-1	2	1	2	2	2
23.86	23.86	3.726	-2	0	3	7	6	6
25.00	25.02	3.558	2	1	1	100	94	100
25.14	25.14	3.538	-1	1	3	90	91	90
26.37	26.38	3.377	0	2	2	2		2
26.37		3.377	-2	1	2	53	51	58
26.60	26.60	3.348	1	2	0	83	75	92
26.85	26.86	3.317	-3	1	1	21	27	24
27.02	27.02	3.297	-3	1	3	47	46	54
27.37	27.38	3.256	2	0	3	70	62	81
27.65	27.66	3.223	1	0	2	41	40	49
28.30	28.30	3.151	1	2	0	10	9	12
28.75	28.76	3.103	-1	1	2	18	16	23
29.20	29.20	3.055	-1	2	2	53	44	69
29.51	29.52	3.024	-2	0	4	44	38	58
29.78	29.84	2.997	0	2	1	13	24	18
29.83		2.993	-2	2	1	11		15
29.98	29.98	2.978	3	0	3	51	45	70
30.60	30.60	2.919	2	1	0	42	35	59
31.04	31.04	2.879	1	0	4	27	24	39
31.75	31.76	2.816	0	0	2	43	39	64
31.92	31.92	2.801	-4	1	1	67	60	101
32.03	32.00	2.792	-1	1	1	16	49	25
32.60	32.60	2.745	3	3	1	26	23	40
32.79	32.80	2.729	1	2	2	30	26	47

2THETA	PEAK	D	H	K	L	I(INT)	I(PK)	I(DS)
33.36	33.46	2.684	2	2	1	10	29	16
33.47		2.675	-1	2	2	34		55
34.24	34.24	2.617	0	1	4	7	8	12
34.40	34.42	2.604	-4	1	2	3	22	5
34.42		2.603	-2	2	3	24		42
34.80	34.82	2.576	1	2	2	1	3	3
34.93	34.92	2.566	-1	2	1	2	3	3
35.32	35.32	2.539	-3	2	0	41	31	73
36.39	36.40	2.467	-3	2	2	17	12	31
36.77	36.86	2.442	-3	0	4	4	8	8
36.94		2.431	-4	1	1	1		2
37.06	37.06	2.424	2	1	3	6	4	11
37.31	37.32	2.408	4	0	0	11	5	22
37.71	37.72	2.383	-1	2	2	11	9	21
38.10	38.10	2.360	0	3	0	19	9	38
38.43	38.44	2.340	0	1	4	14	15	30
38.96	38.96	2.310	1	3	1	48	43	101
38.98		2.308	0	3	1	14		30
39.03		2.306	-4	1	4	2		5
39.51	39.50	2.279	-1	2	4	3	3	6
40.26	40.26	2.238	3	1	5	7	8	16
40.68	40.70	2.216	-3	3	1	4	6	10
40.70		2.215	-5	0	3	3		7
40.85	40.78	2.207	4	1	2	2	3	4
41.06	41.04	2.197	-4	2	4	4	6	10
41.23	41.22	2.188	-3	2	5	5	3	12
41.42	41.42	2.178	-4	2	1	2	4	6
41.48		2.175	-2	3	3	11		27
42.16	42.18	2.142	-2	0	1	27	28	66
42.19		2.140	-5	1	4	7		4
42.74	42.74	2.114	-5	1	2	6	2	19
43.23	43.24	2.091	1	1	1	21	6	15
43.50	43.50	2.079	0	1	5	4	5	54
43.96	43.98	2.058	3	3	2	5	17	11
44.00		2.056	4	1	3	14		13
44.29	44.32	2.043	-1	3	1	13	18	37
44.32		2.042	-2	0	2			

325

ALAMOSITE - BOUCHER AND PEACOR, 1968

2THETA	PEAK	D	H	K	L	I(INT)	I(PK)	I(DS)
44.38		2.040	-1	0	5	1		3
44.58	44.58	2.031	-1	2	4	39	28	103
44.96	45.06	2.014	-2	2	5	3	10	8
45.04		2.011	-2	3	3	3		9
45.06		2.010	-4	2	4	8		22
45.35	45.34	1.9981	0	3	1	1	2	
45.77	45.76	1.9809	-3	3	2	7	5	18
45.88	45.88	1.9760	4	1	2	2	5	5
46.20	46.28	1.9632	-3	2	5	4	9	12
46.23		1.9622	-2	1	6	2		7
46.28		1.9599	-1	1	5	9		25
46.73	46.76	1.9421	4	1	1	18	17	53
46.81	46.80	1.9391	-3	1	6	13	17	39
47.28	47.28	1.9210	-4	0	6	4	4	11
47.40	47.40	1.9163	-1	1	6	1	4	4
47.54	47.54	1.9108	-3	3	3	7	6	22
47.72	47.68	1.9041	2	0	2	1	5	4
47.80	47.82	1.9012	-3	3	1	2	5	7
47.81		1.9006	0	2	5	2		6
47.94	47.94	1.8959	-5	2	2	7	7	21
48.13	48.06	1.8890	1	3	3	1	4	4
48.29	48.28	1.8832	-5	1	5	8	6	23
48.42	48.42	1.8776	-5	2	2	2	7	6
48.44		1.8774	0	0	6	4		12
48.88	48.88	1.8616	-5	2	1	12	10	38
49.10	49.10	1.8540	-4	3	6	13	14	41
49.13		1.8529	3	2	5	6		18
49.22	49.22	1.8495	3	3	1	1	13	21
49.59	49.60	1.8368	3	1	4	3	9	11
49.61		1.8361	5	0	4	10		32
49.71	49.72	1.8324	-4	2	2	2	7	7
50.02	50.02	1.8220	-6	0	2	2	2	5
50.34	50.42	1.8111	-5	3	4	1	10	5
50.41		1.8088	0	3	4	10		33
50.53	50.54	1.8048	-4	1	2	11	13	35
50.62	50.62	1.8015	-6	1	2	5	13	15
50.67	50.66	1.7999	-3	3	4	7		24

2THETA	PEAK	D	H	K	L	I(INT)	I(PK)	I(DS)
50.34	50.82	1.7945	-4	3	1	9	10	29
50.97	50.96	1.7900	-3	1	4	4	7	13
51.19	51.18	1.7828	5	2	0	5	5	16
51.59	51.60	1.7700	0	4	0	2	11	8
51.60		1.7698	-4	3	3	12		43
51.76	51.76	1.7645	-6	1	4	7	11	25
51.88	51.88	1.7610	2	1	5	4	10	15
51.89		1.7604	4	1	1	4		14
52.11	52.16	1.7535	-6	3	1	11	20	40
52.16		1.7520	-3	0	6	21		73
52.36	52.28	1.7459	-1	2	7	2	14	6
52.40	52.42	1.7445	1	3	0	2	14	45
52.42		1.7439	2	4	3	13		17
52.50		1.7416	-1	3	1	5		16
52.93	52.92	1.7285	-5	1	6	4	4	6
53.07	53.06	1.7242	3	3	2	2	4	6
53.22	53.22	1.7197	-5	0	5	2	3	20
53.53	53.54	1.7104	1	2	5	5	6	15
53.56		1.7094	-5	5	4	4		41
53.90	53.92	1.6995	-2	2	5	11	18	57
53.93		1.6988	-1	3	2	16		4
53.99		1.6970	-4	4	4	1		12
54.05	54.06	1.6951	-3	3	7	3	16	14
54.10		1.6937	-1	0	7	4		7
54.28	54.32	1.6886	-2	1	7	2	10	4
54.28		1.6884	0	4	2	1		14
54.31		1.6878	-4	2	6	4		5
54.33		1.6871	-6	1	1	1		31
54.54	54.46	1.6812	1	5	8	8	8	6
54.57	54.56	1.6802	-1	4	1	1	8	27
54.78	54.72	1.6742	2	0	7	2	5	9
55.14	55.16	1.6641	-2	2	2	3	3	11
55.34	55.34	1.6586	0	4	4	9	8	35
55.46	55.48	1.6555	-4	2	5	7	9	27
55.60	55.60	1.6516	-6	3	1	8	10	31
55.71	55.72	1.6485	-6	2	2	6	12	22

ALAMOSITE — BOUCHER AND PEACOR, 1968

2THETA	PEAK	D	H	K	L	I(INT)	I(PK)	I(DS)
55.71		1.6485	-1	3	7	7		29
56.04	56.02	1.6397	3	2	4	2	3	8
56.18	56.18	1.6358	1	4	2	2	3	10
56.42	56.42	1.6296	0	3	5	8	7	34
56.55	56.56	1.6260	2	1	1	10	10	41
56.90	56.96	1.6168	-4	2	3	1	10	5
56.96		1.6153	-5	0	7	11		44
56.98		1.6149	3	0	5	2		7
57.11	57.10	1.6115	4	0	1	1	9	6
57.26	57.26	1.6076	-6	2	3	4	7	16
57.51	57.52	1.6010	-2	4	3	5	5	21
57.60		1.5988	0	4	1	5		23
58.09	58.10	1.5865	-3	1	6	1	4	6
58.11		1.5860	-6	1	4	1		5
58.53	58.58	1.5756	2	3	4	4	16	17
58.58		1.5745	2	0	6	5		22
58.60		1.5738	3	1	0	14		60
58.67		1.5721	-5	3	4	9		36
58.79	58.76	1.5692	0	1	7	1	14	43
58.98	58.96	1.5647	-2	2	7	10	11	32
59.20	59.20	1.5595	-6	2	2	7	14	77
59.42	59.42	1.5540	5	2	5	18	14	59
59.45		1.5535	5	3	2	14		6
59.53	59.56	1.5514	2	4	0	1	13	21
59.55		1.5510	-2	3	2	5		26
59.84	59.88	1.5443	-7	0	6	6	14	21
59.88		1.5439	1	4	5	5		41
59.94		1.5432	1	0	3	9		13
60.37	60.38	1.5419	-4	2	1	3	5	52
60.48	60.52	1.5320	2	4	7	12	7	20
60.50		1.5294	1	1	4	4		14
60.56		1.5289	1	2	4	3		9
60.82	60.82	1.5276	-2	4	4	4	6	20
60.83		1.5216	1	3	6	6		26
60.93		1.5215	-3	1	3	3		13
		1.5192	-2	0	8	2		8

2THETA	PEAK	D	H	K	L	I(INT)	I(PK)	I(DS)
61.57	61.58	1.5049	-5	3	5	2	2	10
61.86	61.90	1.4986	0	4	4	2	4	10
61.90		1.4976	-3	1	8	3		16
61.97		1.4963	-4	4	2	2		7
62.24	62.28	1.4904	-4	4	1	2	5	12
62.26		1.4898	-4	3	6	3		14
62.31		1.4888	6	0	2	3		14
62.47	62.46	1.4854	-2	1	8	4	5	19
63.36	63.38	1.4667	4	2	4	5	4	23
63.49		1.4640	2	2	1	1		6
63.71	63.68	1.4595	4	4	0	2	3	8
63.83	63.84	1.4569	6	1	2	5	3	23
64.20	64.20	1.4495	-7	2	4	2	2	8
64.40		1.4454	-1	1	6	2		10
64.68	64.68	1.4400	4	3	3	8	5	39
64.87	64.86	1.4362	-6	4	1	2	4	12
65.92	65.92	1.4158	-3	1	5	3	2	15
66.23	66.24	1.4098	4	1	5	5	3	25
66.61	66.70	1.4028	-2	3	7	1	4	7
66.69		1.4014	-1	5	1	4		22
66.74		1.4004	-8	0	4	2		9
67.22	67.30	1.3915	-6	0	8	1	6	8
67.27		1.3906	-4	2	8	1		7
67.31		1.3899	-5	4	2	6		35
67.69	67.70	1.3830	1	5	1	1	5	7
67.71		1.3827	-5	4	3	4		24
67.80		1.3810	0	1	8	2		12
67.93	67.92	1.3787	-1	5	2	3	5	14
68.04	68.04	1.3768	-1	3	7	3	5	16
68.06		1.3763	-8	1	1	1		6
68.26	68.26	1.3738	-4	4	4	4	9	20
68.29		1.3728	-4	4	5	10		55
68.63	68.62	1.3664	6	2	2	2	3	14
68.89	68.90	1.3618	-1	1	8	4	4	11
68.91		1.3615	-3	2	9	1		22
68.99		1.3600	-2	5	2	3		14

ALAMOSITE - BOUCHER AND PEACOR, 1968

2THETA	PEAK	D	H	K	L	I(INT)	I(PK)	I(DS)
69.23	69.24	1.3559	-8	1	2	5	5	26
69.26		1.3555	-5	1	4	5		14
69.77	69.78	1.3468	6	4	3	3	2	19
69.91	69.96	1.3444	1	5	2	1	2	7
69.96		1.3435	5	4	0	1		8
70.36	70.36	1.3370	-3	1	9	1	3	9
70.37		1.3368	-1	4	5	2		10
70.60	70.62	1.3330	-5	3	7	1	5	7
70.61		1.3328	3	2	5	2		11
70.61		1.3327	-4	3	5	2		12
70.69		1.3315	-4	1	9	4		21
70.81	70.84	1.3295	0	3	7	1	4	8
70.94	70.90	1.3274	-5	0	9	3	4	19
71.14	71.16	1.3242	3	2	6	6	7	36
71.17		1.3236	-3	4	5	3		19
71.23		1.3227	-1	1	6	3		21
71.41	71.36	1.3199	-3	5	2	1	5	9
71.53	71.54	1.3180	-7	3	4	3	4	16
71.56		1.3173	-7	3	2	3		16
72.07	72.08	1.3094	3	5	0	1	2	7
72.41	72.50	1.3040	1	8	8	2	4	8
72.47		1.3031	-8	2	3	2		14
72.52		1.3022	-8	2	1	2		13
72.78	72.78	1.2982	7	2	7	3	4	21
72.94	72.96	1.2959	7	2	1	2	4	11
72.96		1.2955	5	4	1	2		11
73.19	73.22	1.2920	-7	3	5	3	7	19
73.22		1.2916	3	5	5	5		34
73.27		1.2908	-7	3	1	3		17
73.34	73.44	1.2898	8	0	0	2	6	10
73.38		1.2892	0	5	5	2		11
73.46		1.2879	4	0	6	4		14
73.56	73.54	1.2865	-7	1	8	4	6	27
73.64		1.2852	-3	3	8	1		7
73.68	73.70	1.2846	-6	4	3	2	5	10
73.72		1.2840	-1	1	9	1		7
73.74		1.2837	7	1	2	1		8

2THETA	PEAK	D	H	K	L	I(INT)	I(PK)	I(DS)
73.83		1.2824	-2	5	4	1		9
74.06	74.06	1.2790	3	4	4	6	5	41
74.62	74.64	1.2707	-3	2	9	3	3	18
74.70	74.82	1.2696	-6	4	4	2	3	10
74.81		1.2680	4	4	3	1		8
74.99	75.00	1.2654	-6	4	1	3	3	19
75.19	75.18	1.2626	-6	0	4	2	3	16
75.35	75.40	1.2602	-6	1	9	1	3	8
75.43		1.2591	6	3	2	1		8
75.65	75.68	1.2559	-5	3	8	3	5	19
75.69		1.2554	-8	2	1	5		34
75.95	75.94	1.2517	-7	3	6	4	5	25
76.06		1.2502	7	3	0	5		34
76.57	76.60	1.2432	-5	1	8	6	5	40
76.63		1.2423	-1	2	8	3		23
76.71		1.2413	4	5	0	1		9
76.82	76.82	1.2397	-2	0	8	5	6	32
76.93		1.2383	-9	0	5	2		11
77.29	77.28	1.2334	0	1	4	2	2	11
77.36		1.2325	-1	5	0	1		8
77.57	77.62	1.2296	-1	1	9	2	4	16
77.64		1.2287	-9	3	4	5		37
78.61	78.76	1.2160	5	3	4	2	2	11
78.75		1.2141	-3	1	5	3		22
79.47	79.54	1.2049	-3	1	5	3	7	20
79.52		1.2044	-8	3	10	1		10
79.52		1.2043	-6	2	4	6		43
79.54		1.2040	-2	0	2	2		12
79.61		1.2032	-8	1	0	5		36
79.82	79.76	1.2006	-7	1	8	2	5	13
79.97	80.14	1.1986	-6	3	8	1	5	10
80.14		1.1966	-2	7	7	7		50
80.43	80.42	1.1930	-5	5	3	2	4	16
80.43		1.1930	3	4	5	1		8
80.54	80.64	1.1916	4	4	4	2	5	16
80.62		1.1907	0	4	7	3		20
80.66		1.1902	6	4	1	2		16

ALAMOSITE - BOUCHER AND PEACOR, 1968

2THETA	PEAK	D	H	K	L	I(INT)	I(PK)	I(DS)
80.66		1.1901	-8	3	5	1		8
81.40	81.52	1.1812	2	2	4	1	3	11
81.50		1.1800	0	0	9	2		18
81.56		1.1793	-9	0	7	1		8
81.77	81.76	1.1767	-9	2	4	2	3	13
82.04	82.06	1.1736	0	6	1	5	4	39
82.09		1.1730	5	4	3	1		9
82.18	82.22	1.1719	-9	2	3	2	4	15
82.23		1.1713	1	1	9	3		25
82.45	82.50	1.1688	-9	2	5	1	3	8
82.53		1.1678	-2	3	5	2		12
82.56		1.1676	5	5	0	1		8
82.60		1.1670	-8	3	1	3		12
82.89	82.90	1.1636	-2	5	6	4	6	25
82.90		1.1635	-6	1	10	1		30
82.94		1.1632	-1	5	5	3		23
83.32	83.44	1.1588	-3	5	6	2	4	8
83.34		1.1585	-3	4	8	1		16
83.37		1.1582	-1	1	6	2		14
83.45		1.1573	4	1	7	2		19
83.48		1.1570	-5	3	9	2		18
83.59	83.62	1.1557	-3	2	10	2	4	13
83.72		1.1542	-8	2	8	2		15
84.07	84.08	1.1503	2	6	0	2	2	18
84.45	84.36	1.1461	8	1	2	1	1	8
84.79	84.76	1.1424	-1	5	9	1	3	10
85.42	85.44	1.1356	5	5	1	5	3	37
85.60	85.78	1.1336	-1	6	3	2	4	12
85.78		1.1318	8	3	0	1		9
85.78		1.1317	9	1	0	4		35
85.81		1.1314	5	3	5	1		9
86.23	86.06	1.1270	-9	2	1	1	3	10
86.35	86.36	1.1257	0	6	3	1	2	11
86.60	86.64	1.1231	-7	1	10	1	3	10
86.65		1.1226	-3	6	2	2		13
86.66		1.1225	-10	1	4	1		9
86.83	86.88	1.1208	-9	1	8	1	2	10

2THETA	PEAK	D	H	K	L	I(INT)	I(PK)	I(DS)
87.09	87.12	1.1181	-6	5	4	3	3	29
87.24	87.24	1.1166	5	4	6	3	3	28
87.25		1.1164	4	4	5	1		9
87.28		1.1161	3	6	0	1		13
88.07	88.06	1.1081	-9	2	2	1	2	12
88.58	88.58	1.1030	-2	3	4	2	2	14
88.95	89.16	1.0994	5	5	7	1	5	11
88.96		1.0993	-2	6	4	1		12
89.01		1.0988	4	2	4	3		28
89.12		1.0977	9	0	1	1		13
89.14		1.0976	-6	0	5	1		11
89.15		1.0975	-8	2	5	2		19
89.15		1.0975	-7	3	8	3		23
89.32	89.82	1.0958	-7	4	9	2	2	15
89.86		1.0907	9	1	4	1		15
90.31	90.44	1.0864	-3	6	7	2	4	10
90.38		1.0857	-3	3	0	1		14
90.42		1.0853	-9	3	10	2		12
90.44		1.0852	-4	0	6	1		27
90.52		1.0844	-8	4	1	3		10
90.53		1.0843	6	3	8	1		15
90.72	90.84	1.0825	3	7	3	2	4	11
90.83		1.0815	-10	0	2	1		16
90.84		1.0814	8	1	3	1		11
90.92	91.00	1.0806	7	2	6	2	4	16
90.99		1.0800	-9	3	3	2		19
91.03		1.0797	-4	6	6	3		26
91.08		1.0791	-3	3	9	1		12
91.40	91.36	1.0763	-4	4	6	2	3	22
91.61	91.68	1.0743	6	0	9	3	4	24
91.68		1.0737	2	6	3	4		34
91.81		1.0725	-2	3	10	1		11
91.85	92.28	1.0721	-10	1	7	1	3	14
92.26		1.0684	-6	5	6	1		11
92.29		1.0681	-10	5	5	1		14
92.44	92.44	1.0668	5	1	7	3	3	26
92.79	92.88	1.0637	4	5	4	1	4	13

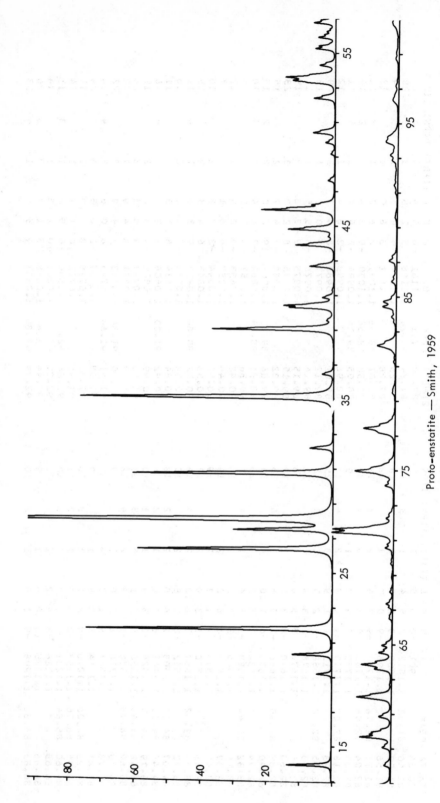

Proto-enstatite — Smith, 1959

λ = CuKα Half width at 40°2θ: 0.11°2θ Plot scale I$_{PK}$ (max) = 150

PROTO-ENSTATITE – SMITH, 1959

2THETA	PEAK	D	H	K	L	I(INT)	I(PK)	I(DS)
13.93	13.94	6.353	1	1	0	8	9	8
19.17	19.18	4.625	2	0	0	3	3	3
20.30	20.32	4.370	0	2	0	8	8	8
21.77	21.78	4.079	0	1	1	50	49	49
26.37	26.38	3.377	1	2	1	43	39	42
27.49	27.50	3.241	2	1	1	21	20	21
28.07	28.10	3.176	2	2	0	18	18	18
28.11		3.172	1	1	1	100	100	100
30.72	30.72	2.908	3	0	0	47	40	48
32.18	32.19	2.779	1	3	0	6	5	6
35.07	35.14	2.556	1	1	2	24	51	25
35.14		2.551	3	1	0	46		48
36.45	36.46	2.463	1	3	1	3	2	3
39.03	39.04	2.306	2	0	2	32	25	34
39.54	39.54	2.277	3	2	0	2	2	1
40.29	40.30	2.237	2	2	1	11	9	11
40.42	40.40	2.230	2	1	2	8	10	8
40.86	40.86	2.207	1	3	1	3	3	3
42.66	42.66	2.118	3	1	1	7	5	7
43.89	43.90	2.061	4	1	0	11	8	12
44.38	44.38	2.039	2	2	2	1	1	1
44.80	44.80	2.021	0	4	0	13	9	14
45.89	45.92	1.9756	2	4	1	2	15	2
45.92		1.9746	0	4	1	19		21
46.21	46.22	1.9626	3	1	2	5	5	6
47.62	47.62	1.9080	4	2	1	2	1	2
49.15	49.16	1.8520	2	4	1	1	1	1
49.81	49.80	1.8292	3	2	2	3	2	3
50.37	50.38	1.8099	5	1	0	6	4	7
50.43		1.8080	2	3	2	1		1
52.38	52.38	1.7452	4	0	1	6	4	7
53.39	53.42	1.7146	1	3	1	7	9	7
53.43		1.7135	4	3	1	2		3
53.50		1.7114	5	1	1	2		2
53.61	53.60	1.7080	4	1	2	11	10	13
54.28	54.28	1.6884	0	4	3	4	3	5
55.26	55.26	1.6610	1	4	2	6	4	7

2THETA	PEAK	D	H	K	L	I(INT)	I(PK)	I(DS)
55.41	55.40	1.6567	3	2	0	2	3	2
55.91	55.90	1.6432	1	2	3	5	3	6
56.23	56.22	1.6345	1	5	1	2	2	3
56.75	56.76	1.6207	4	2	0	7	4	8
58.05	58.02	1.5882	2	4	1	2	1	2
59.05	59.06	1.5630	2	5	0	2	2	3
59.66	59.66	1.5484	4	1	3	10	6	12
59.95	59.94	1.5417	4	3	0	5	4	6
60.81	60.84	1.5218	3	5	1	2	2	3
60.87		1.5206	5	1	0	3		3
61.86	61.96	1.4985	5	2	0	9	7	4
61.96		1.4964	3	0	1	5		11
63.58	63.58	1.4621	3	1	2	9	3	6
63.85	63.84	1.4567	0	6	0	4	6	11
64.53	64.52	1.4429	1	6	1	3	3	5
64.95	64.94	1.4346	5	2	1	1	2	4
66.49	66.50	1.4050	0	6	2	1	1	1
67.14	67.14	1.3930	1	5	1	1	1	2
67.36	67.34	1.3890	3	6	1	1	1	2
68.03	68.02	1.3769	6	0	3	3	1	1
70.55	70.54	1.3338	6	1	0	7	2	4
71.40	71.42	1.3200	2	4	3	13	12	17
71.42		1.3197	6	1	1	7		9
71.49		1.3186	1	0	4	6	8	8
71.62	71.62	1.3165	6	1	0	9	12	12
72.25	72.24	1.3066	7	1	0	3	2	3
74.11	74.12	1.2782	2	0	4	3	2	3
74.15		1.2776	0	6	2	1		1
74.90	74.96	1.2667	5	1	3	7	8	9
74.98		1.2656	1	6	2	9		12
75.04		1.2647	2	1	4	3		4
77.26	77.42	1.2338	1	5	3	3	6	4
77.36		1.2325	4	1	2	2		2
77.43		1.2315	2	6	2	8		11
79.45	79.46	1.2052	1	7	1	1	1	1
79.59	79.60	1.2034	7	3	0	1	1	1
81.24	81.24	1.1831	4	4	3	1	0	1

PROTO-ENSTATITE - SMITH, 1959

2THETA	PEAK	D	H	K	L	I(INT)	I(PK)	I(DS)
82.02	82.18	1.1738	7	3	1	2	4	2
82.11		1.1727	7	1	2	2		1
82.17		1.1720	5	5	2	1		9
82.30		1.1705	2	3	4	6		1
83.54	83.56	1.1562	8	0	0	1	1	1
83.84	83.84	1.1529	4	0	4	1	1	2
84.42	84.42	1.1465	5	5	0	6	3	9
84.73	84.70	1.1430	4	1	4	3	3	5
85.86	85.86	1.1308	3	7	1	2	1	3
87.06	87.06	1.1183	4	6	2	3	2	5
89.54	89.56	1.0937	2	0	6	1	0	1
91.00	90.98	1.0799	5	0	3	1	0	2
91.43	91.42	1.0760	4	7	1	2	1	1
91.89	91.88	1.0717	5	1	4	1	0	2
92.07	92.06	1.0702	0	8	4	1	0	1
92.53	92.56	1.0660	3	4	4	1	0	1
92.86	92.86	1.0631	1	8	0	2	1	4
93.17	93.16	1.0604	8	0	2	1	2	1
93.17		1.0604	6	5	2	2		2
93.89	93.94	1.0541	7	5	0	5	2	7
93.97		1.0534	8	3	1	2		3
94.06		1.0527	8	1	2	2		2
94.19	94.18	1.0516	1	5	4	1	2	1
94.22		1.0513	5	6	0	1		1
96.19	96.34	1.0349	5	7	1	1	1	1
96.30		1.0340	7	5	0	1		1
96.33		1.0338	0	2	5	2		3
96.45		1.0328	5	1	5	1		2
97.13	97.14	1.0274	1	2	3	1	1	1
97.19		1.0270	6	4	2	1		1
98.67	98.74	1.0155	4	7	1	1	1	1
98.76		1.0147	7	0	3	1		2
99.79	99.78	1.0070	6	0	4	1	0	1
100.60	100.84	1.0011	3	5	1	1	1	1
100.69		1.0004	6	1	4	2		3
100.86		0.9992	3	1	5	5		4
101.34	101.32	0.9958	7	3	3	5	2	7

2THETA	PEAK	D	H	K	L	I(INT)	I(PK)	I(DS)
103.14	103.14	0.9832	9	2	1	1	0	1
104.12	104.12	0.9767	1	6	1	3	1	4
105.43	105.44	0.9681	5	4	4	1	0	2
105.99	106.00	0.9645	5	7	1	1	0	1
106.32	106.60	0.9624	4	5	2	1	1	2
106.59		0.9608	2	6	4	2		3
107.26	107.26	0.9566	0	4	5	2		3
107.60	107.64	0.9545	6	7	1	1	1	2
107.85		0.9530	9	1	2	2	1	1
108.09	108.00	0.9515	1	4	5	1	0	1
110.68	110.70	0.9364	9	2	2	1	0	2
111.81	111.82	0.9302	0	8	3	1	0	1
112.75	112.74	0.9250	10	0	0	1	0	2
113.94	113.96	0.9187	5	5	4	2	1	3
115.27	115.30	0.9119	8	2	2	3		3
115.35		0.9115	2	7	3	1		1
116.15	116.16	0.9075	6	7	2	2	1	2
116.44	116.54	0.9061	7	5	0	2	1	2
116.86		0.9040	7	7	3	1		2
118.85	119.02	0.8946	4	7	1	1		8
119.02		0.8938	7	7	4	4	1	2
119.40	119.42	0.8921	5	1	3	1	1	4
120.62	120.66	0.8867	10	0	2	2	1	1
120.89	121.12	0.8855	0	4	6	1		1
121.07		0.8847	8	4	1	1		2
121.23		0.8840	9	4	3	1		1
121.47		0.8829	4	4	5	1		2
122.32	122.32	0.8793	4	9	1	3	1	4
122.59		0.8782	5	3	6	1		2
123.60	123.98	0.8740	1	10	0	1		1
123.94		0.8726	0	0	4	1		2
123.94		0.8726	8	4	4	4	1	7
124.14		0.8718	3	5	5	1	1	3
126.53	126.56	0.8624	0	10	1	2		2
127.22	127.52	0.8598	5	9	0	1		2
127.40		0.8592	0	6	5	1	0	1
127.50		0.8588	2	10	0	1	1	1

PROTO-ENSTATITE - SMITH, 1959

2THETA	PEAK	D	H	K	L	I(INT)	I(PK)	I(DS)
127.53		0.8587	1	10	2	1		2
127.91	128.02	0.8573	8	6	2	1	1	2
128.07		0.8567	10	2	2	2	1	3
129.44	129.46	0.8518	10	4	0	2	0	3
132.63	132.74	0.8411	10	4	1	1	0	2
132.79		0.8406	9	5	2	1		1
133.41	133.40	0.8386	6	3	5	2	1	4
134.28	134.02	0.8359	8	3	4	1	0	1
136.82	137.40	0.8284	6	6	1	1	0	3
136.99		0.8279	4	0	6	1		2
137.35		0.8269	11	1	5	1		3
138.00	137.96	0.8250	7	1	5	2	0	3
141.45	141.48	0.8160	11	2	1	3	1	6
142.74	143.08	0.8128	6	4	5	1	1	1
143.07		0.8121	6	9	1	5		9
143.42		0.8112	10	4	2	1		2
144.42	143.92	0.8089	2	4	6	1	1	2
145.71	145.58	0.8061	10	2	3	1	0	3
147.75	147.76	0.8018	11	0	2	2	0	4
147.76		0.8018	3	10	2	1		1
150.16	150.22	0.7971	7	3	5	5	1	10
150.64		0.7962	5	1	6	1		3
155.21	155.46	0.7886	11	2	2	2	0	4
157.69	159.10	0.7851	6	9	2	3	1	5
157.88		0.7848	9	7	1	1		2
158.62		0.7839	6	7	4	6		11
159.08		0.7833	3	7	5	5		10
160.55	160.66	0.7815	4	10	2	1	1	1
161.08		0.7809	10	6	0	8		16
161.21		0.7807	8	5	4	5		10

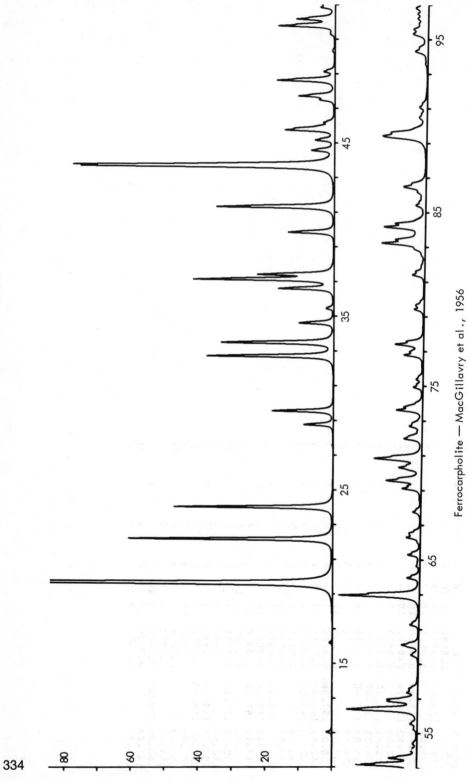

Ferrocarpholite — MacGillavry et al., 1956

λ = FeKα Half width at 40°2θ: 0.11°2θ Plot scale I_{PK} (max) = 200

334

FERROCARPHOLITE - MACGILLAVRY ET AL., 1956

2THETA	PEAK	D	H	K	L	I(INT)	I(PK)	I(DS)
19.60	19.60	5.687	2	2	0	100	100	100
22.12	22.14	5.045	0	4	1	32	30	33
23.98	23.98	4.660	1	1	1	25	24	27
28.73	28.74	3.902	1	3	1	5	4	6
29.51	29.52	3.801	2	2	1	11	9	12
32.66	32.66	3.442	4	0	0	23	19	29
33.42	33.44	3.367	3	1	1	11	17	14
33.45		3.363	0	6	0	11		14
34.57	34.58	3.258	4	2	0	7	5	8
35.41	35.42	3.183	2	2	1	1	1	2
36.56	36.56	3.086	1	5	1	10	8	14
37.08	37.08	3.045	3	3	0	27	21	35
37.36	37.36	3.022	3	3	0	14	12	18
39.80	39.80	2.844	4	2	1	9	7	13
41.27	41.26	2.747	4	4	1	25	18	35
43.60	43.60	2.607	3	5	1	47	38	69
43.70	43.68	2.601	2	6	1	24	39	36
44.54	44.54	2.554	0	0	2	7	4	7
45.13	45.14	2.522	0	8	0	4	3	6
45.71	45.72	2.492	1	1	2	10	7	16
45.86	45.80	2.485	4	4	1	3	6	5
46.15	46.14	2.470	1	7	0	3	2	2
47.45	47.46	2.406	4	0	2	2	2	3
47.68	47.68	2.395	2	6	1	8	5	12
48.58	48.58	2.353	1	3	2	13	9	21
49.09	49.10	2.330	2	2	2	13	2	4
51.74	51.74	2.219	1	7	1	8	8	22
52.14	52.14	2.203	3	7	0	8	6	14
52.81	52.82	2.177	4	6	1	14	9	24
53.15	53.16	2.164	2	8	1	6	4	10
53.55	53.54	2.149	1	5	2	1	1	2
53.99	54.00	2.132	5	5	1	2	1	3
55.52	55.52	2.078	5	0	2	11	11	20
56.31	56.34	2.051	6	2	1	8		15
56.36		2.050	2	6	2	3	5	
56.81	56.82	2.035	4	8	0	3		
56.83		2.034	0	6	2	3		6

2THETA	PEAK	D	H	K	L	I(INT)	I(PK)	I(DS)
56.94	56.94	2.031	1	9	0	3	5	6
57.33	57.34	2.018	0	10	2	2	2	4
57.57	57.56	2.010	4	2	2	2	1	3
59.49	59.50	1.9509	2	6	2	2	1	3
59.98	60.08	1.9365	2	10	0	2	3	4
60.08	60.08	1.9336	4	4	1	3		6
61.25	61.24	1.9003	5	7	1	4	1	4
62.89	62.90	1.8554	7	1	1	21	12	42
63.94	63.94	1.8282	1	8	3	3	2	7
65.27	65.28	1.7949	0	8	2	4	2	7
66.26	66.26	1.7710	7	3	1	2	2	8
66.52	66.52	1.7649	2	8	2	2	1	5
67.74	67.74	1.7368	6	0	2	1	3	4
69.08	69.08	1.7072	5	5	2	5	5	11
69.47	69.54	1.6989	1	1	3	1		3
69.53		1.6976	8	0	0	8		17
70.16	70.28	1.6967	6	2	1	1		3
70.28		1.6842	2	8	1	2	3	4
70.72	70.80	1.6817	0	12	0	4		11
70.80		1.6727	1	9	2	5	7	25
71.95	71.96	1.6710	7	5	1	4		10
72.04		1.6479	4	10	1	5	3	3
72.67	72.68	1.6461	5	9	1	1	1	17
73.59	73.60	1.6336	2	12	0	3		4
73.87	73.76	1.6161	3	11	1	8	4	5
74.91	74.92	1.6110	6	8	0	6	3	5
75.37	75.36	1.5917	3	1	3	2	2	14
76.77	76.78	1.5835	0	10	3	6	3	3
77.06	76.96	1.5590	7	5	1	1	1	21
77.38	77.38	1.5539	7	7	2	9	4	7
79.40	79.40	1.5485	7	7	0	3	1	10
79.68	79.64	1.5155	6	10	0	2		4
81.38	81.38	1.5110	4	12	3	3		10
82.86	82.90	1.4847	3	5	3	4	2	5
83.20	83.20	1.4629	5	11	0	2	6	37
83.48	83.40	1.4580	1	7	3	14	4	6
		1.4539	7	5	2	2		

335

2THETA	PEAK	D	H	K	L	I(INT)	I(PK)	I(DS)
83.83	83.84	1.4490	4	12	1	1	2	3
84.14	84.14	1.4447	5	1	3	14	6	37
84.37	84.34	1.4414	0	14	0	2	5	6
85.39	85.40	1.4274	8	0	2	3	1	9
86.16	86.16	1.4173	3	11	2	2	1	5
86.42		1.4138	6	8	0	1		3
86.45	86.46	1.4134	8	2	1	3	3	8
86.47		1.4131	3	13	1	4		11
89.28	89.38	1.3777	9	5	1	4	7	12
89.33		1.3770	10	0	0	5		15
89.39		1.3763	2	12	2	10		29
89.62	89.58	1.3735	8	4	2	2	5	11
90.47	90.46	1.3633	5	5	3	3	1	6
91.06	91.06	1.3565	6	12	0	6	1	8
94.29	94.30	1.3205	1	13	2	3	2	17
95.30		1.3098	9	1	2	6		18
95.61	95.56	1.3065	9	7	1	2	2	7
95.92	95.88	1.3034	6	10	2	2		5
96.20	96.20	1.3005	4	11	2	2	1	7
96.48	96.48	1.2977	7	11	2	2		7
96.82	96.80	1.2943	5	7	3	2	1	13
97.42	97.58	1.2883	9	3	1	4	2	6
97.58		1.2867	4	14	0	2		10
98.29	98.28	1.2798	7	1	4	3	1	2
99.39	99.40	1.2693	1	1	4	2		6
100.26	100.26	1.2612	0	16	0	1	0	2
100.85	100.86	1.2558	2	0	4	1	1	7
101.54	101.70	1.2497	1	3	4	2	0	3
101.70		1.2482	9	9	2	1	1	4
101.93		1.2462	2	2	4	1		2
103.05	103.06	1.2365	10	2	0	2	1	5
104.21	104.22	1.2266	7	5	1	4	0	3
105.18	105.20	1.2186	2	9	3	1	1	13
105.43	105.44	1.2167	5	1	1	1	1	7
105.80	105.74	1.2137	11	1	1	1		4
105.99		1.2121	1	0	2	1		3
106.43	106.42	1.2086	10	8	0	2	1	7

2THETA	PEAK	D	H	K	L	I(INT)	I(PK)	I(DS)
107.80	107.86	1.1980	6	12	2	3	1	8
107.87		1.1975	4	0	4	2		5
108.00	108.00	1.1965	11	3	1	1		4
109.24	109.30	1.1872	6	14	1	2	1	8
109.42	109.48	1.1859	1	15	2	1	1	2
109.50		1.1853	7	13	1	1		5
110.65	110.68	1.1770	3	5	4	2		6
114.30	114.32	1.1522	3	15	2	2	1	5
114.30		1.1433	1	17	1	4		14
115.71	115.74	1.1402	1	13	3	1	0	2
116.21	116.32	1.1395	12	0	0	1	1	4
116.31		1.1368	5	11	3	4		3
116.75	116.74	1.1328	9	9	2	3	1	16
117.42	117.38	1.1212	3	17	1	1	1	10
119.38	119.50	1.1211	0	18	0	1		3
119.40		1.1203	11	7	1	1		3
119.55		1.1161	6	0	4	5		4
120.30	120.30	1.1160	2	16	2	1	1	18
120.32		1.1130	13	3	3	1		2
120.85	120.70	1.1128	12	3	2	2		2
120.89		1.1087	11	3	4	1	1	3
121.63	121.64	1.1062	9	9	0	4		8
122.10	122.24	1.1053	6	16	0	1		4
122.27		1.1051	8	14	3	1	1	14
122.31		1.0954	9	5	1	1		3
124.18	124.24	1.0930	12	4	2	2		5
124.65	124.74	1.0925	10	8	1	1		2
124.75		1.0883	8	12	4	4	2	8
125.62	125.62	1.0792	0	10	2	1	1	38
127.51	127.52	1.0787	3	9	4	4		15
127.62		1.0732	3	1	2	1	0	3
128.83	128.78	1.0661	5	17	1	2	1	7
130.45	130.60	1.0657	5	17	0	1		4
130.54		1.0654	10	13	4	7		26
130.62		1.0593	6	6	4	1		3
132.08	132.16	1.0590	5	13	3	2	1	7
132.14								

FERROCARPHOLITE - MACGILLAVRY ET AL., 1956

2THETA	PEAK	D	H	K	L	I(INT)	I(PK)	I(DS)
132.24		1.0586	9	7	3	1		5
133.74	133.78	1.0526	1	15	1	1	0	5
134.91	135.26	1.0481	3	17	2	3	1	4
135.26		1.0467	12	0	2	3		12
136.78	137.38	1.0412	12	2	2	2	1	13
137.36		1.0391	10	10	2	4		17
137.98	137.92	1.0369	1	19	1	1	1	4
141.38	141.54	1.0257	8	0	4	4	1	17
141.58		1.0251	13	3	1	2		10
141.63		1.0249	12	4	2	3		11
143.10	143.10	1.0204	8	2	4	3	1	12
144.03	144.00	1.0177	1	1	5	2	1	9
144.85	145.28	1.0154	2	18	2	1	1	4
145.19		1.0144	6	16	2	2		7
145.28		1.0142	3	19	1	10		44
148.31	148.88	1.0062	2	12	4	12	3	56
148.52		1.0057	2	2	5	1		4
148.75		1.0051	8	4	4	7		32
148.85		1.0049	11	9	2	2		8
148.99		1.0045	13	5	1	9		41
149.15		1.0042	7	7	4	1		5
151.67	152.58	0.9983	2	20	0	1	1	3
152.06		0.9975	11	3	3	2		7
152.37		0.9968	7	17	1	2		7
152.91		0.9957	10	14	0	3		12
156.72	158.42	0.9883	6	18	1	1	1	4
156.90		0.9880	1	5	5	2		10
157.71		0.9866	3	3	5	3		15
158.22		0.9857	5	15	3	15		69
159.44		0.9838	4	18	2	1		5
159.44		0.9838	1	13	1	4		18
159.57		0.9836	14	0	0	1		7
159.74		0.9833	10	12	2	2		6
161.25		0.9811	8	6	4	2		7
162.18	162.50	0.9798	2	20	1	3	1	15
162.53		0.9793	9	1	4	2		7
162.78		0.9790	12	10	1	6		30

2THETA	PEAK	D	H	K	L	I(INT)	I(PK)	I(DS)
162.85		0.9789	14	2	0	2		10
163.15		0.9785	11	5	3	5		26
164.16		0.9773	13	1	2	1		5

337

TABLE 16. MISCELLANEOUS CHAIN SILICATES

Variety	Clinohedrite	Euclase	Stokesite	Batistite	Barylite
Composition	$2CaZnSiO_4 \cdot H_2O$	$AlBeSiO_4(OH)$	$CaSnSi_3O_9 \cdot 2H_2O$	$Na_2BaTi_2O_2Si_4O_{12}$	$BaBe_2Si_2O_7$
Source	Franklin, N.J.	Villa Rica, Minas Gerais, Brazil	not given	Central Aldan	Vishnevye Hills, USSR
Reference	Nikitin and Belov, 1963	Mrose & Appleman, 1962	Vorma, 1963	Nikitin & Belov, 1962	Abrashev, Ilyukhin & Belov, 1965
Cell Dimensions					
a Å	5.16	4.763	14.465	10.40	9.79 (9.79)
b Å	15.94	14.290	11.625	13.85	4.63 (11.65)
c Å	5.41	4.618	5.235	8.10	11.65 (4.63)
α deg	90	90	90	90	90
β deg	103.90	100.25	90	90	90
γ deg	90	90	90	90	90
Space Group	Cc	$P2_1/a$	Pnna	Ima2	$Pna2_1$ ($Pn2_1a$)
Z	2	4	4	4	4

TABLE 16. (cont.)

Variety	Clinohedrite	Euclase	Stokesite	Batistite	Barylite
Method	2-D, film	2-D, film	2-D, film	2-D, film	2-D, film
R & Rating	0.10 (av.) (3)	0.13 (av.) (2)	0.10 (av.) (2)	0.174 (av.) (3)	0.115 (av.) (3)
Cleavage and habit	{010} perfect Variable	{010} perfect Prismatic parallel to c axis	Prismatic cleavage		{011},{201} good Variable
Comment	Similar to euclase B estimated	$[Be_2(OH)_2(SiO_4)_2]_n^{-6n}$ chains	$[Si_6O_{12}]_\infty$ chains parallel to b axis	$[Si_4O_{12}]_\infty$ chains parallel to c axis	
μ	198.7	85.2	336.4	401.7	621.7
ASF	0.4022	0.9431×10^{-1}	0.2534	0.2369	0.3421
Abridging factors	1:1:0.5	1:1:0.5	0.5:0:0	0.5:0:0	0.5:0:0

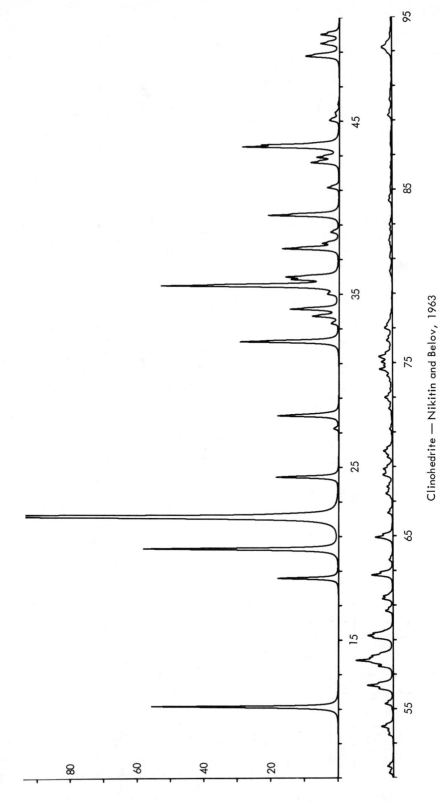

Clinohedrite — Nikitin and Belov, 1963

λ = CuKα Half width at 40°2θ: 0.11°2θ Plot scale I$_{PK}$ (max) = 200

340

CLINOHEDRITE – NIKITIN AND BELOV, 1963

2THETA	PEAK	D	H	K	L	I(INT)	I(PK)	I(DS)
11.09	11.10	7.970	0	2	0	24	28	20
18.55	18.56	4.779	1	1	0	8	9	8
20.23	20.24	4.385	0	2	1	28	29	27
22.08	22.08	4.023	-1	1	1	100	100	100
24.40	24.40	3.645	1	3	0	10	9	10
27.95	27.96	3.189	1	1	1	10	9	11
32.22	32.22	2.776	1	3	1	18	15	21
33.28	33.28	2.689	1	5	0	1	1	2
33.71	33.72	2.657	0	4	0	5	4	6
34.12	34.12	2.626	0	6	0	9	7	11
34.99	35.00	2.562	-1	1	2	2	2	2
35.45	35.46	2.530	-1	5	1	33	27	42
35.82	35.82	2.504	0	2	2	7	7	9
35.98	35.98	2.494	0	2	2	1	8	10
37.58	37.62	2.391	-2	2	1	10	8	13
37.61		2.389	2	0	0	3	3	4
37.92	37.92	2.371	0	6	1	2	3	2
38.56	38.56	2.333	-1	3	2	14	11	19
39.53	39.54	2.278	-1	7	0	2	2	3
41.13	41.14	2.193	0	4	2	3	2	4
42.57	42.60	2.122	-2	0	2	4	4	5
42.60		2.120	2	4	0	19	19	29
42.90	42.90	2.106	2	2	2	4	3	6
43.50	43.50	2.079	-2	7	0	2	14	7
43.62	43.60	2.073	1	2	2	4	12	2
45.03	45.04	2.011	-2	2	2	4	1	7
48.72	48.74	1.8675	0	6	1	1	5	6
48.76		1.8659	1	4	2	3		2
49.41	49.48	1.8431	-2	4	1	3	3	6
49.48		1.8405	-2	4	2	1		2
49.98	50.00	1.8232	-2	6	1	3	3	5
50.01		1.8224	2	6	0	2		2
51.65	51.64	1.7683	-1	5	2	1	1	5
53.95	53.96	1.6980	-1	3	3	3	2	4
55.27	55.28	1.6605	3	1	0	2	1	5
56.13	56.14	1.6372	-2	6	2	3	2	2
56.34	56.34	1.6317	-3	3	1	5	4	9
57.45	57.46	1.6027	0	0	2	3	2	6
57.66	57.78	1.5974	-2	2	3	3	6	6
57.77		1.5946	-2	2	2	4		7
57.79		1.5940	0	10	0	4		7
58.06	58.06	1.5873	0	8	3	4	3	7
59.09	59.10	1.5621	-1	5	0	3	3	8
59.21	59.22	1.5592	2	8	1	2	4	7
59.35	59.36	1.5557	1	9	2	1	3	4
60.64	60.64	1.5257	-3	3	1	1	1	3
61.34	61.34	1.5100	-3	3	1	5	2	5
62.71	62.72	1.4802	-1	7	3	4	3	2
64.90	64.90	1.4356	3	5	1	1	3	5
66.32	66.32	1.4082	-1	11	3	1	1	3
67.44	67.44	1.3876	-1	1	1	1	1	3
67.82	67.82	1.3806	-3	1	3	2	1	3
68.51	68.50	1.3684	-1	11	1	2	2	5
68.84	68.84	1.3626	0	10	2	2	1	5
69.54	69.54	1.3506	3	5	0	3	1	3
69.89	69.88	1.3447	2	0	2	2	1	3
72.33	72.34	1.3052	-3	7	0	3	2	4
72.97	72.98	1.2954	0	2	4	1	1	3
74.24	74.24	1.2764	-1	7	3	1	1	9
74.61	74.60	1.2710	-3	5	2	1	2	6
75.02	75.02	1.2649	-2	10	0	2	2	7
75.34	75.34	1.2605	2	8	2	1	1	3
76.26	76.26	1.2474	3	7	1	1	1	3
76.94	77.00	1.2381	-4	2	4	1	1	4
77.02		1.2370	1	13	1	2		5
84.33	84.34	1.1474	1	13	0	2	1	3
89.27	89.28	1.0963	0	8	4	4	1	5
93.19	93.20	1.0602	-1	8	5	1	2	6
93.24		1.0597	1	5	1	1		4
99.15	99.16	1.0118	-1	13	1	1	1	4
101.12	101.12	0.9973	-3	13	1	1	0	4
102.85	102.88	0.9852	-4	10	0	1	0	2
105.45	105.50	0.9679	-4	6	2	1	0	2
105.86	105.82	0.9653	1	3	5	1	0	2

CLINOHEDRITE - NIKITIN AND BELOV, 1963

2THETA	PEAK	D	H	K	L	I(INT)	I(PK)	I(DS)
106.65	106.76	0.9604	-5	3	3	1	1	4
106.78		0.9595	1	13	3	1		5
107.03	107.02	0.9580	-3	13	1	1	1	3
108.84	108.84	0.9471	-3	13	1	1	0	5
110.23	110.36	0.9390	-5	7	1	1	0	3
110.37		0.9382	1	5	5	1		4
111.30	111.22	0.9329	2	14	2	1		2
111.57	111.58	0.9315	0	16	1	1		2
114.85	114.78	0.9140	-1	15	3	1		3
116.78	117.02	0.9045	5	5	1	1		4
117.03		0.9032	-3	15	1	1		3
119.78	119.78	0.8904	-2	2	6	1		3
121.19	121.16	0.8841	2	10	4	1		3
123.29	123.40	0.8753	0	0	6	1		4
123.39		0.8749	-4	15	3	1		3
124.43	124.60	0.8706	-4	10	1	1	0	2
124.58		0.8700	0	2	6	1		3
124.69		0.8696	4	10	2	1		3
124.96		0.8685	3	11	3	1		4
125.76	125.70	0.8654	3	15	1	1	0	3
127.14	127.14	0.8602	0	14	4	1	0	4
129.29	129.30	0.8523	-6	2	2	1	0	3
129.93	129.96	0.8501	-2	16	3	1		3
131.96	132.12	0.8433	-3	15	3	1		3
132.12		0.8427	-4	14	2	1		3
133.25	133.22	0.8391	0	18	0	1		4
134.61	134.62	0.8349	2	18	0	1	0	3
134.64		0.8348	-6	0	0	1	0	4
135.98	136.12	0.8308	-1	17	4	1		3
136.34		0.8298	2	12	5	1	0	4
137.84	137.84	0.8255	-3	11	2	1		4
140.78	140.98	0.8177	4	12	2	2		9
140.97		0.8172	-2	8	8	1		4
141.76	141.88	0.8152	-4	0	6	1	1	6
141.97		0.8147	-2	18	2	1		6
142.36	142.86	0.8138	4	0	4	1	1	4
142.78		0.8127	1	11	5	1		4

2THETA	PEAK	D	H	K	L	I(INT)	I(PK)	I(DS)
142.86		0.8126	3	13	3	1		4
142.93		0.8124	-1	19	1	1		3
143.49		0.8111	-1	13	5	2		8
146.20	146.38	0.8050	-3	1	5	1	0	3
146.40		0.8046	-5	19	5	3		14
147.97	147.30	0.8014	-0	18	6	1		4
150.27	150.52	0.7969	3	3	5	1		3
150.54		0.7964	6	6	0	1		6
150.82		0.7959	1	15	4	1		3
153.42	154.22	0.7914	6	2	1	1	0	3
154.00		0.7905	5	11	3	1		8
154.07		0.7904	5	11	4	1		5
154.12		0.7903	-2	16	1	1		7
154.99	155.14	0.7890	-5	13	1	2		11
155.99		0.7875	-6	8	2	3		16
159.50	161.88	0.7827	5	3	3	1		6
160.58		0.7814	3	5	5	3		14
160.90		0.7811	-5	7	5	1		4
160.91		0.7811	-2	10	6	2		11
161.01		0.7810	-6	4	4	2		4
161.55		0.7803	1	19	2	1		4
161.87		0.7800	6	4	1	1		4
161.98		0.7799	-4	16	0	1		3
162.22		0.7796	4	16	6	1		7
162.47		0.7794	-4	6	6	1		7
162.75		0.7791	1	7	7	1		7
163.74		0.7781	4	6	4	1		8
163.96		0.7779	2	18	2	2		9
164.88		0.7770	-3	13	5	2		10

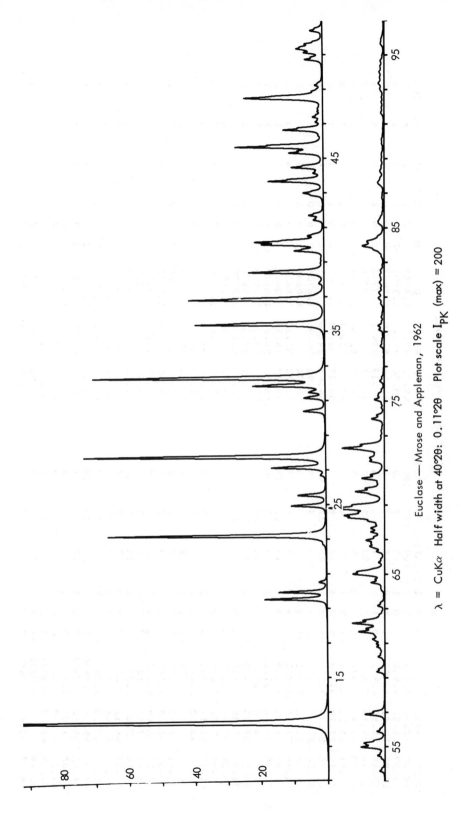

Euclase — Mrose and Appleman, 1962

$\lambda = CuK\alpha$ Half width at $40^\circ 2\theta$: $0.11^\circ 2\theta$ Plot scale I_{PK} (max) = 200

EUCLASE - MROSE AND APPLEMAN, 1962

2THETA	PEAK	D	H	K	L	I(INT)	I(PK)	I(DS)
12.38	12.38	7.145	0	2	0	100	100	100
19.52	19.52	4.544	0	0	1	10	9	10
19.92	19.92	4.454	1	1	0	8	7	8
20.50	20.50	4.331	0	1	1	1	1	1
23.18	23.18	3.834	0	2	1	39	33	40
24.90	24.90	3.572	0	4	0	6	5	7
25.51	25.52	3.489	-1	1	1	5	4	5
27.10	27.10	3.288	0	3	1	10	8	10
27.73	27.74	3.214	-1	0	2	46	37	48
30.36	30.36	2.942	-1	1	2	4	3	4
31.12	31.12	2.871	1	3	1	4	3	4
31.46	31.46	2.841	-1	4	0	3	2	3
31.84	31.84	2.809	0	4	1	14	11	15
32.28	32.28	2.771	1	2	1	48	35	50
35.37	35.38	2.535	-1	4	1	28	20	30
36.80	36.80	2.440	1	5	0	30	21	32
38.38	38.38	2.343	2	0	0	17	11	18
39.63	39.64	2.272	0	0	2	6	4	7
39.99	40.00	2.253	-2	0	1	12	9	14
40.15	40.14	2.244	0	2	2	11	10	12
40.47	40.50	2.227	2	2	0	1	1	1
40.51		2.225	-2	1	2	2	2	2
41.41	41.42	2.179	-1	1	2	2	2	2
41.68	41.68	2.165	0	3	2	3	2	4
42.83	42.84	2.110	0	6	1	2	2	2
42.98	42.96	2.103	2	3	0	3	3	4
43.66	43.66	2.071	0	5	1	13	8	14
44.12	44.12	2.051	-2	3	1	7	5	8
44.45	44.46	2.036	-1	3	1	8	5	8
45.29	45.30	2.001	-1	6	1	21	13	23
45.64	45.64	1.9860	-1	0	1	10	6	11
46.62	46.62	1.9465	2	4	1	2	1	2
47.08	47.08	1.9287	2	2	1	2	2	2
47.38	47.38	1.9172	0	4	2	8	12	9
48.43	48.48	1.8781	2	2	1	14		16
48.48		1.8760	-1	4	1	1	2	1
48.87	48.86	1.8622	0	7	1			

2THETA	PEAK	D	H	K	L	I(INT)	I(PK)	I(DS)
49.21	49.22	1.8499	-1	2	2	2	3	5
50.70	50.70	1.7991	-2	0	2	4	3	5
51.09	51.10	1.7862	-1	8	0	4	3	4
51.42	51.32	1.7786	0	5	1	5	4	6
51.62		1.7756	-1	7	1	3		2
51.62	51.62	1.7691	-2	5	1	2	3	4
52.37	52.38	1.7455	-1	5	2	3	2	3
53.57	53.58	1.7093	-1	3	0	2	1	2
54.47	54.46	1.6831	-2	2	0	2	1	2
54.92	54.96	1.6704	2	6	0	1	4	1
54.96		1.6691	-2	8	1	5		6
55.20	55.20	1.6624	0	6	2	4	3	5
55.88	55.88	1.6440	-1	6	1	2	1	2
56.86	56.86	1.6178	-3	1	1	6	3	7
59.33	59.34	1.5563	-1	8	2	2	1	3
60.21	60.22	1.5356	-2	5	2	2	1	2
60.78	60.78	1.5226	0	7	2	2	1	2
60.96	60.96	1.5185	-2	7	1	1	1	1
61.22	61.22	1.5126	-2	0	2	2	2	3
61.65	61.66	1.5031	-1	7	1	6	4	7
61.90	61.82	1.4978	-1	2	2	1	3	1
62.14	62.14	1.4925	-1	0	2	9	5	10
63.16	63.16	1.4709	2	3	2	2	1	2
64.50	64.50	1.4435	-2	6	3	4	2	5
64.90	65.00	1.4356	-1	4	1	1		1
64.98		1.4339	-3	3	0	6	5	7
65.01	65.12	1.4334	2	4	2	2		3
65.11	65.66	1.4314	3	8	0	4	4	5
65.67		1.4206	2	2	1	1	1	2
66.29	66.30	1.4088	2	7	2	2	2	4
66.60	66.60	1.4030	-3	1	1	3	3	2
66.94	66.94	1.3967	3	7	1	3	3	4
66.98		1.3960	-1	3	1	3		2
67.42	67.42	1.3878	2	4	2	3	2	4
67.67	67.64	1.3834	-3	2	3	2	2	2
68.06	68.06	1.3764	-3	1	1	3	3	4
68.25	68.26	1.3731	-3	5	1	7	6	8

EUCLASE — MROSE AND APPLEMAN, 1962

2THETA	PEAK	D	H	K	L	I(INT)	I(PK)	I(DS)
68.37	68.38	1.3709	3	5	0	8	6	10
68.74	68.80	1.3643	-2	2	3	6	8	7
68.81		1.3632	0	10	1	13		16
69.75	69.76	1.3471	1	2	3	8	4	9
69.94	69.94	1.3440	-1	5	3	1	3	1
70.27	70.28	1.3384	-0	4	3	1	1	1
70.52	70.52	1.3343	-2	3	3	7	3	8
71.51	71.52	1.3181	1	3	3	4	2	5
71.64	71.72	1.3161	2	8	0	2	2	2
71.74		1.3145	2	9	1	1		1
72.14	72.26	1.3082	-3	6	1	2	6	3
72.26		1.3063	3	6	0	6		8
72.27		1.3063	1	8	2	5		7
72.30		1.3057	3	4	1	1		1
73.95	73.96	1.2806	1	4	3	4	2	6
75.06	75.06	1.2643	-3	5	2	3	2	4
75.42	75.42	1.2592	2	5	1	1	1	2
76.15	76.14	1.2491	0	11	1	1	1	2
78.16	78.16	1.2219	-1	11	1	1	1	2
79.10	79.10	1.2097	-0	10	2	2	2	2
79.33	79.34	1.2067	-2	10	0	1	1	2
80.93	80.92	1.1868	-4	0	1	2	2	2
81.26	81.24	1.1829	-3	2	3	1	1	2
81.84	81.84	1.1760	3	8	0	1	1	1
83.93	83.94	1.1519	2	10	1	6	3	8
84.09		1.1501	1	8	2	2		2
84.25	84.20	1.1483	-1	1	4	2	3	3
87.52	87.62	1.1136	3	9	0	1	1	2
87.63		1.1125	-4	2	2	2		2
88.29	88.28	1.1060	-3	5	3	1	0	1
91.63	91.64	1.0741	-4	4	4	1	1	1
93.35	93.34	1.0588	-1	1	4	2	0	2
94.64	94.64	1.0477	-1	12	2	1	1	2
95.64	95.64	1.0395	0	10	3	2	1	2
99.22	99.22	1.0113	-1	2	2	1	1	3
101.00	101.28	0.9982	-3	12	4	1	1	2
101.12		0.9973	1	14	0	1		1
101.26	101.29	0.9963	-2	10	3	1		2
101.29		0.9962	-4	3	0	1		1
101.42		0.9952	2	13	2	1		1
101.73	101.76	0.9930	-2	12	2	1	1	2
101.78		0.9927	0	7	4	1		2
101.83		0.9923	3	2	3	1		1
102.40	102.38	0.9883	-1	11	3	1	1	2
102.72	102.72	0.9861	-4	7	2	1	1	1
104.98	105.14	0.9710	-4	1	4	1	1	2
105.10		0.9702	-3	4	4	1		1
105.16		0.9698	-2	7	4	3		5
105.94	106.00	0.9648	3	4	3	1	1	1
106.02		0.9644	4	2	2	1		2
106.60	106.60	0.9607	4	7	2	2	1	2
106.93	107.16	0.9586	0	8	4	1	1	1
107.17		0.9572	2	13	1	1		2
107.19		0.9570	-3	9	3	2		2
107.76	107.76	0.9536	4	3	4	2	1	3
107.90		0.9527	-4	8	2	1		1
108.23	108.18	0.9507	-3	5	4	3	1	1
108.24		0.9506	-4	9	1	1		1
108.84	108.68	0.9471	3	12	0	1		1
109.08	109.10	0.9456	3	5	3	1	0	1
109.49	109.48	0.9432	-1	11	3	1	0	1
110.40	110.40	0.9380	-2	13	2	3	1	5
110.41		0.9380	-2	8	4	1		1
110.65		0.9366	-4	6	4	2		1
111.94	111.92	0.9294	5	2	0	1	0	1
112.53	112.56	0.9262	-1	14	2	2	1	2
112.77	112.82	0.9249	2	4	4	1		1
112.81		0.9247	-5	1	2	1	1	1
113.52	113.50	0.9209	-1	15	1	2	1	3
113.90	113.92	0.9189	-5	2	2	1	1	1
115.67	115.70	0.9099	2	10	3	2	0	2
116.32	116.32	0.9067	5	4	0	2	1	2
116.44		0.9061	4	10	1	1		1
117.00	116.82	0.9034	-5	5	1	1	1	1

EUCLASE – MROSE AND APPLEMAN, 1962

2THETA	PEAK	D	H	K	L	I(INT)	I(PK)	I(DS)
117.37	117.40	0.9016	0	2	5	1	0	1
117.39		0.9015	-2	1	5	1		1
119.26	119.28	0.8928	0	3	5	1	0	2
120.21		0.8885	2	6	4	1		1
120.44	120.44	0.8857	-2	3	5	2	1	3
120.85		0.8846	5	1	2	3		1
121.10	121.06	0.8815	-4	10	3	1		5
121.82		0.8812	-5	5	1	1		
121.86	121.60	0.8787	-3	11	3	1	1	1
122.45	122.46	0.8783	-1	5	1	1		1
122.55		0.8764	3	11	5	1		1
123.02	123.10	0.8757	0	16	2	1	0	1
123.18		0.8745	-2	1	5	1		2
123.48		0.8723	-1	15	2	1		1
124.02	123.96	0.8668	5	6	0	1	0	1
125.39	125.46	0.8661	-1	16	1	1	0	2
125.57		0.8612	0	5	5	1		3
126.86	126.86	0.8561	5	4	1	2		2
128.23	128.60	0.8561	1	16	4	1	0	1
128.49		0.8552	0	11	0	1		2
128.54		0.8550	-3	14	1	1		
128.68		0.8545	3	14	4	2		2
129.12	129.10	0.8530	-1	15	0	1	1	1
129.13		0.8529	-3	1	2	1		1
129.66	129.70	0.8511	-3	9	5	1	0	2
129.94		0.8501	-1	1	4	1		1
130.58	130.56	0.8479	-1	14	3	1	0	1
131.42	131.54	0.8451	-3	12	2	1	0	2
131.61		0.8444	-2	14	1	1		1
132.20	132.28	0.8425	3	12	4	1	0	1
132.77	132.78	0.8406	-4	12	1	2	1	2
132.83		0.8404	-2	11	4	1		4
134.76	134.78	0.8345	4	11	1	2	0	3
135.42	135.66	0.8325	5	11	3	1	1	2
135.74		0.8315	5	6	1	2		3
137.38	137.24	0.8268	-3	14	2	1	0	2
138.68	138.86	0.8232	-4	7	4	2	0	3

2THETA	PEAK	D	H	K	L	I(INT)	I(PK)	I(DS)
138.92	138.92	0.8225	-5	8	2	1		2
139.12		0.8220	0	12	4	1	0	1
139.87	139.82	0.8200	-3	4	3	1	1	1
140.40	140.54	0.8187	4	12	5	1		3
140.55		0.8183	-4	12	2	2		4
141.19	141.20	0.8166	-5	6	3	1	1	1
141.20		0.8166	-5	9	1	2		3
142.33	142.26	0.8138	-3	15	0	1	0	3
142.52		0.8134	3	15	1	2		2
143.20	143.22	0.8118	2	16	1	1	1	2
143.29		0.8115	1	17	2	1		2
144.15	144.18	0.8095	-2	12	4	1	0	2
144.44		0.8089	-1	15	1	1		1
144.99	145.72	0.8076	5	9	3	1	1	2
145.19		0.8072	0	15	0	1		1
145.54		0.8064	-4	8	3	1		2
145.71		0.8061	-2	10	5	2		3
146.49	146.52	0.8044	4	10	1	1	1	2
146.55		0.8043	-3	6	5	1		1
146.90		0.8035	4	12	1	2		4
147.80	147.88	0.8017	-2	13	0	1	1	1
148.66	149.54	0.8000	5	16	2	1	1	2
148.74		0.7998	-5	7	3	1		2
149.55		0.7983	-1	9	5	7		13
150.62	150.72	0.7963	5	4	2	1	1	1
151.06	152.14	0.7955	2	10	4	1	1	2
151.50		0.7947	5	8	1	2		4
151.52		0.7947	-5	1	4	1		4
151.97		0.7939	0	18	0	1		1
152.01		0.7938	-6	0	1	6		2
153.55	154.72	0.7912	2	17	0	3	2	11
153.69		0.7910	-5	2	4	7		5
154.34		0.7900	0	13	4	5		13
154.57		0.7896	4	3	5	2		10
154.69		0.7894	4	7	3	3		3
154.71		0.7894	-4	0	5	1		7
154.74		0.7893	-3	15	2	4		7

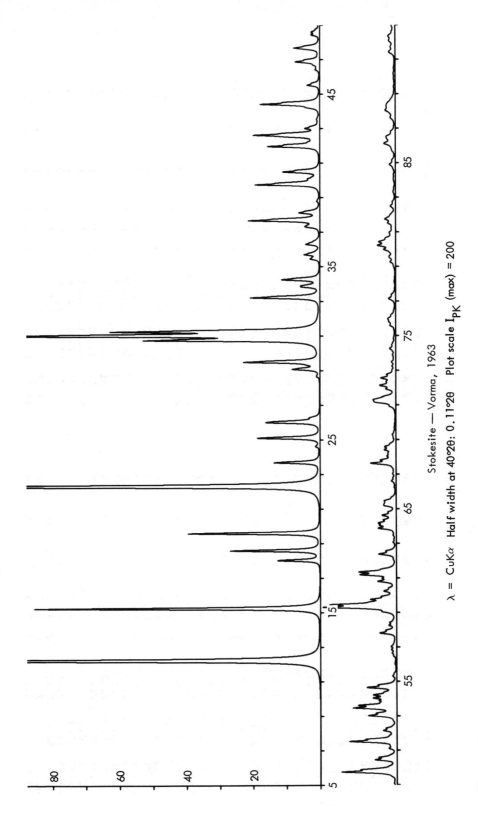

Stokesite — Vorma, 1963

λ = CuKα Half width at 40°2θ: 0.11°2θ Plot scale I$_{PK}$ (max) = 200

STOKESITE – VORMA, 1963

2THETA	PEAK	D	H	K	L	I(INT)	I(PK)	I(DS)
12.23	12.24	7.232	2	0	0	100	100	100
15.23	15.24	5.812	0	2	0	44	43	50
18.00	18.02	4.923	1	0	1	7	6	9
18.57	18.58	4.773	0	2	1	15	13	19
19.58	19.58	4.531	2	1	1	22	20	29
22.30	22.30	3.984	2	2	0	96	83	145
23.66	23.68	3.756	1	1	1	8	7	13
24.60	24.60	3.616	4	0	0	1	1	1
25.09	25.10	3.547	3	0	1	12	9	19
25.99	26.00	3.426	2	2	1	9	8	17
26.07		3.416	2	3	0	2		3
26.25	26.24	3.392	3	1	1	1		2
29.06	29.06	3.070	4	2	0	5	4	10
29.48	29.48	3.027	0	4	1	14	11	29
30.74	30.74	2.906	4	1	1	32	26	67
31.00	31.00	2.882	1	3	1	68	53	143
31.24	31.24	2.861	2	2	2	37	31	78
33.19	33.20	2.697	4	2	1	14	10	32
33.81	33.84	2.649	4	3	0	3	3	7
33.88		2.644	2	4	0	2		4
34.23	34.24	2.617	0	3	2	7	6	17
34.24		2.616	3	3	1	1		2
35.42	35.42	2.532	5	1	1	2	2	5
35.67	35.68	2.515	1	1	2	3	2	8
36.28	36.28	2.474	5	0	1	3	2	7
37.26	37.30	2.411	6	0	0	2	2	4
37.31		2.408	0	1	2	2		4
37.66	37.66	2.387	4	2	2	15	11	40
38.10	38.10	2.360	4	3	1	4	3	10
38.76	38.76	2.321	5	1	1	1	0	1
39.74	39.74	2.266	2	2	2	14	10	39
40.08	40.06	2.248	2	4	2	2	2	5
40.47	40.48	2.227	6	2	1	8	6	24
41.95	41.94	2.152	6	1	1	12	8	36
42.51	42.60	2.125	0	5	1	5		15
42.60		2.120	4	0	2	12	10	37
42.99	42.98	2.102	1	5	1	3	2	9

2THETA	PEAK	D	H	K	L	I(INT)	I(PK)	I(DS)
44.40	44.40	2.039	2	5	1	14	9	44
45.50	45.50	1.9919	4	2	1	3	2	10
46.39	46.40	1.9557	4	5	2	1	1	3
46.85	46.86	1.9375	5	6	0	6	4	20
47.59		1.9091	5	4	0	2		7
47.66	47.66	1.9064	6	3	1	5	4	18
48.42	48.42	1.8782	2	6	1	1	1	7
48.61	48.58	1.8715	4	5	1	1	1	5
49.72	49.72	1.8320	7	0	1	14	8	53
49.93	49.84	1.8249	8	6	0	1	5	5
50.43	50.44	1.8081	1	6	2	5	3	20
50.58	50.56	1.8029	8	0	1	1	1	4
51.08	51.08	1.7866	6	1	0	2	3	6
51.49	51.50	1.7733	2	6	2	13	7	49
51.83	51.82	1.7623	6	4	0	1	1	3
52.13	52.14	1.7530	8	2	1	3	2	12
52.26	52.26	1.7489	0	8	0	1	1	3
52.99	53.00	1.7265	1	2	3	6	4	23
53.02		1.7257	4	4	1	1		4
53.43		1.7135	5	5	0	1		5
53.45	53.44	1.7129	4	6	2	9	7	36
53.46		1.7126	3	6	1	1		4
53.62	53.60	1.7078	6	2	0	4	3	16
53.87	53.88	1.7003	8	1	0	3	3	11
54.02	54.02	1.6961	2	5	2	4	4	15
54.20	54.20	1.6909	1	1	2	2	3	10
54.22		1.6901	5	2	1	2		7
54.63	54.64	1.6785	2	5	3	8	4	32
55.28	55.28	1.6603	1	2	3	3	1	4
56.46	56.46	1.6284	2	2	3	3	1	3
56.64	56.62	1.6236	4	6	1	1	1	2
56.83	56.82	1.6186	2	7	0	1	1	4
57.01	57.00	1.6141	1	4	2	1	1	3
57.79	57.80	1.5941	6	5	1	5	3	21
58.23	58.24	1.5830	0	7	1	2	1	8
58.39	58.38	1.5791	3	3	3	1	1	3
59.02	59.04	1.5637	8	3	1	2	2	11

STOKESITE - VORMA, 1963

2THETA	PEAK	D	H	K	L	I(INT)	I(PK)	I(DS)
59.28	59.28	1.5574	4	6	3	9	11	45
59.29		1.5573	0	6	2	9		44
59.43	59.44	1.5540	2	3	1	7	10	33
59.75	59.74	1.5464	2	7	0	6	4	27
60.23	60.22	1.5352	8	4	2	3	2	13
60.79	60.78	1.5224	2	6	2	4	3	22
61.17	61.18	1.5137	6	4	0	9	6	46
61.33	61.34	1.5102	8	0	2	5	6	25
62.36	62.36	1.4877	8	4	1	5	3	24
63.05	63.04	1.4732	8	1	1	1	0	4
63.44	63.44	1.4650	2	4	3	1	0	3
63.86	63.86	1.4564	4	3	0	5	0	25
64.02	64.04	1.4531	4	0	1	1	3	6
64.17	64.16	1.4501	0	8	0	3	3	18
64.35	64.34	1.4465	4	7	2	1	2	4
64.61	64.60	1.4412	10	0	2	4	2	19
65.17	65.18	1.4303	8	2	2	3	2	14
65.24		1.4288	4	6	3	1		6
65.27		1.4283	3	3	1	1		4
65.46	65.44	1.4247	9	3	0	2	2	10
66.58	66.58	1.4032	2	8	1	1	1	7
66.99	67.00	1.3956	6	1	3	2	4	4
67.62	67.62	1.3843	10	5	1	1	1	43
68.02	68.02	1.3770	8	1	1	8	4	9
68.16	68.18	1.3747	2	8	1	2	1	5
68.40	68.40	1.3704	6	5	3	3	1	15
68.55	68.56	1.3676	2	2	2	2	2	11
69.09	69.10	1.3583	9	5	1	1	0	3
70.90	70.92	1.3280	6	7	1	1	1	8
71.13	71.18	1.3243	8	4	3	3	3	21
71.20		1.3232	6	3	2	2		14
71.28	71.28	1.3219	8	0	1	4	3	24
71.91	71.92	1.3119	10	6	0	2	1	15
72.11	72.12	1.3087	0	3	4	3	2	16
72.15		1.3081	6	0	2	1		7
72.54	72.54	1.3020	4	5	3	4	2	27
73.00	72.98	1.2950	10	4	0	1	1	4
73.23	73.22	1.2915	1	6	3	3	1	4
73.47	73.46	1.2878	2	0	4	1	1	6
73.87	73.88	1.2818	9	5	1	1	1	4
74.21	74.22	1.2768	0	2	3	1	1	6
74.24		1.2763	2	6	2	1		6
75.56	75.58	1.2573	2	2	4	1	1	6
75.79	75.80	1.2541	6	6	1	1	1	6
75.92	75.96	1.2521	0	9	1	1		7
75.98		1.2513	3	6	3	2		7
77.13	77.12	1.2356	2	8	2	1	1	14
79.24	79.24	1.2078	6	5	3	1	1	10
79.55	79.54	1.2039	4	5	3	3	2	6
79.97	79.98	1.1987	10	8	0	2	1	20
80.21	80.22	1.1957	0	4	5	4	3	28
80.40	80.42	1.1933	8	7	1	3	2	23
80.59	80.58	1.1910	1	0	4	1	2	8
80.94	80.94	1.1867	8	4	3	2	1	11
81.47	81.50	1.1803	2	7	3	1	1	4
81.50		1.1800	12	0	2	2		12
81.72	81.72	1.1774	8	6	0	2	2	16
81.82		1.1761	2	4	2	2		6
82.46	82.46	1.1687	9	1	3	1		12
83.29	83.30	1.1591	12	6	1	2	1	6
84.87	84.88	1.1415	10	7	0	1	0	4
86.10	86.10	1.1283	4	2	3	2	1	15
86.50	86.50	1.1242	6	4	4	4	2	29
86.52		1.1240	12	8	1	1		7
87.54	87.62	1.1134	6	0	2	2	1	16
87.63		1.1125	12	4	0	1	1	6
88.03	88.04	1.1086	8	9	1	3		10
88.07		1.1081	0	1	3	3	2	22
88.21	88.30	1.1067	8	8	2	2		6
88.42		1.1047	4	7	0	1	1	12
91.38	91.40	1.0764	3	10	1	1	1	7
91.43		1.0760	6	7	3	3		13
92.05	92.08	1.0703	12	0	2	2	1	14
92.14		1.0695	10	3	4	1		14
			6	4	5	1		5

STOKESITE – VORMA, 1963

2THETA	PEAK	D	H	K	L	I(INT)	I(PK)	I(DS)
92.33	92.34	1.0678	10	7	1	2	1	16
92.94	92.94	1.0624	10	0	2	1	1	9
93.19	93.22	1.0602	8	0	4	1	1	6
93.23		1.0598	10	6	2	1		11
93.95	93.96	1.0536	12	3	2	1		7
94.24	94.24	1.0512	6	10	2	2	1	15
94.56	94.56	1.0485	12	5	0	1	1	10
94.71		1.0471	6	10	0	1		8
95.21	95.20	1.0430	8	8	4	3	1	25
95.63	95.64	1.0395	8	8	2	1	1	8
95.72		1.0388	4	6	8	2		14
96.07	96.04	1.0359	0	11	0	1	1	11
96.40	96.42	1.0332	14	0	0	1	1	8
96.54		1.0321	8	1	5	1		8
96.74	96.74	1.0305	8	9	1	1	1	10
97.10	97.10	1.0277	2	9	3	1	1	10
97.38		1.0255	2	11	1	1		11
97.42	97.42	1.0246	12	4	2	1	1	9
98.16	98.16	1.0193	14	10	0	1	1	12
98.43	98.44	1.0173	14	2	0	1	1	7
99.71	99.74	1.0077	3	2	5	1	1	6
100.15	100.16	1.0044	10	5	3	2	1	17
100.48	100.52	1.0020	8	7	5	1	1	14
100.54		1.0016	4	11	1	1		7
100.61		1.0010	8	3	4	2		16
101.05	100.90	0.9979	2	9	4	1	1	8
101.31	101.32	0.9960	4	4	8	2	1	15
101.33		0.9959	8	1	4	1		8
102.30	102.36	0.9890	12	6	1	2	1	19
102.42		0.9882	8	10	3	2		16
103.94	103.96	0.9778	8	0	5	2	1	19
104.61	104.64	0.9734	0	8	4	1		8
104.76		0.9725	2	3	4	2		19
106.10	106.12	0.9638	2	8	2	1	1	12
106.54	106.56	0.9611	14	0	3	2	1	19
106.58		0.9608	2	3	0	2		15
106.68		0.9602	2	12	0	1		18
106.87	106.90	0.9590	12	7	1	1	1	11
107.18		0.9571	6	1	5	1		6
107.76	107.80	0.9535	6	9	3	2	1	16
107.81		0.9532	12	6	1	1		14
108.05	108.04	0.9518	6	11	1	1	1	16
108.65	108.66	0.9482	14	2	2	1	1	13
108.94	109.04	0.9465	2	5	5	1	1	8
109.10		0.9456	6	6	3	1		7
110.80	110.78	0.9358	4	12	0	1	0	6
111.44	111.46	0.9321	6	3	5	1	0	9
111.83	111.88	0.9300	12	11	4	1		15
111.98		0.9292	10	6	0	1	1	10
112.24	112.24	0.9278	4	1	3	1		6
112.77	112.76	0.9249	10	5	5	1	0	13
113.12	113.16	0.9230	12	8	4	1	0	11
113.60	113.54	0.9205	14	0	2	1	0	8
114.84		0.9141	12	7	2	1		11
115.16		0.9125	14	5	4	1	1	17
115.20	115.22	0.9123	12	9	2	1		12
115.32		0.9117	14	4	3	1		11
116.88	116.94	0.9040	0	11	0	1		7
117.01		0.9033	8	2	3	1	1	10
117.43	117.40	0.9014	2	12	5	1		7
117.63		0.9003	8	0	1	1	1	12
117.95	117.92	0.8989	8	13	0	1		14
119.14	119.12	0.8933	16	2	5	1	1	8
120.40		0.8876	6	3	3	1		7
120.66	120.74	0.8865	6	5	5	1	1	10
120.84		0.8857	14	0	1	1		9
121.63		0.8823	0	7	2	1	0	11
121.81	121.88	0.8815	8	3	5	3	1	10
121.89		0.8811	4	4	1	1		40
122.37	122.36	0.8791	2	12	2	1		13
122.88	122.92	0.8770	4	11	5	3	1	12
122.99		0.8765	12	2	4	1		7
123.36	123.42	0.8750	2	13	1	1	1	15
123.49		0.8745	12	8	2	1		19

STOKESITE - VORMA, 1963

2THETA	PEAK	D	H	K	L	I(INT)	I(PK)	I(DS)
124.83	125.10	0.8690	12	9	1	1	0	8
125.04		0.8682	16	3	1	1		15
125.17		0.8677	10	6	4	1		8
125.81	125.78	0.8652	14	7	0	1	1	14
126.32	126.42	0.8633	16	4	4	1	1	8
126.40		0.8629	2	10	2	1		8
126.43		0.8628	0	2	6	2		31
126.93	126.94	0.8610	14	6	5	2	1	28
127.11		0.8603	4	7	5	1		9
128.06	128.14	0.8568	2	2	6	2	1	27
128.14		0.8565	8	8	4	1		15
128.19		0.8563	10	10	2	2		21
128.68	128.72	0.8545	16	0	0	2	1	7
128.85		0.8539	8	12	0	1		11
129.53	129.54	0.8515	12	7	3	1	0	15
129.92	129.94	0.8502	6	12	2	2	0	21
130.53	130.60	0.8481	12	4	4	1	1	15
131.01	131.22	0.8464	6	11	3	2	1	29
131.17		0.8459	10	1	5	2		30
131.31		0.8454	16	2	4	1		11
131.42		0.8451	4	10	4	2		29
131.67		0.8442	8	5	5	1		14
136.21	136.82	0.8301	2	4	6	1	1	11
136.76		0.8285	10	3	5	3		37
136.79		0.8284	12	5	4	1		15
137.00		0.8279	6	13	1	2		27
139.58	140.20	0.8208	9	5	5	1	1	8
139.72		0.8204	6	0	6	1		18
139.95		0.8198	16	4	0	1		17
140.16		0.8193	16	6	2	1		18
140.80	140.78	0.8176	6	10	4	2	1	32
142.53	143.22	0.8134	0	9	5	1	1	15
142.94		0.8124	6	2	6	1		15
143.18		0.8118	8	12	2	1		18
143.54		0.8110	14	0	4	1		20
144.26	144.62	0.8093	4	14	0	2		24
144.59		0.8085	8	11	3	2		28

2THETA	PEAK	D	H	K	L	I(INT)	I(PK)	I(DS)
144.61	144.61	0.8085	3	14	1	2		12
144.72		0.8083	2	8	5	2		36
145.26	145.26	0.8070	10	8	4	1	1	28
145.64		0.8062	12	6	0	2		11
146.25	146.16	0.8049	10	12	2	2	1	29
147.08	147.12	0.8032	14	2	4	3	1	44
147.85	147.94	0.8016	8	1	1	1		20
148.24		0.8008	16	1	3	1		19
149.45	150.34	0.7985	6	13	2	1	1	11
150.00		0.7974	14	9	1	1		27
150.21		0.7970	12	10	2	4		57
150.35		0.7968	10	10	5	3		42
150.87	151.10	0.7958	0	13	3	1	1	16
151.02		0.7956	0	6	6	2		30
151.12		0.7954	8	7	5	2		31
152.17	152.84	0.7935	4	9	5	1	1	22
152.27		0.7934	15	5	3	1		17
152.81		0.7925	18	1	1	5		87
152.89		0.7923	8	13	1	2		38
153.40	153.90	0.7915	0	14	2	2	1	11
153.67		0.7910	2	13	3	2		28
153.83		0.7908	2	6	6	2		31
154.60		0.7896	6	4	6	5		87
155.21	156.68	0.7886	12	1	5	1	1	17
155.70		0.7879	9	11	3	1		10
156.46		0.7868	2	14	1	4		62
156.56		0.7866	12	9	3	2		32
156.99		0.7860	16	3	3	2		35
157.17		0.7858	8	0	6	1		18
157.27		0.7857	12	11	1	2		29
157.68		0.7851	6	14	0	1		19
157.71		0.7851	16	7	1	2		32
158.66		0.7838	14	7	3	3		38
159.02		0.7833	8	10	4	5		49
163.09	163.00	0.7787	8	2	6	2	0	78
163.16		0.7786	0	12	2	2		42
163.60		0.7782	6	11	4	1		23

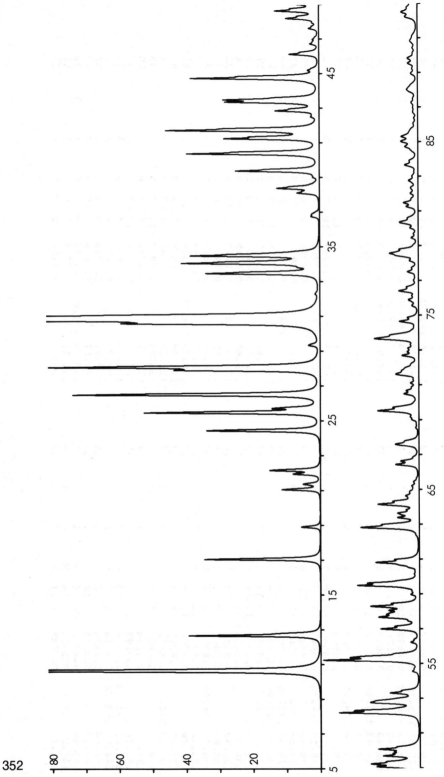

Batisite — Nikitin and Belov, 1962

$\lambda = \text{Cu}K\alpha$ Half width at $40°2\theta$: $0.11°2\theta$ Plot scale I_{PK} (max) = 200

BATISITE - NIKITIN AND BELOV, 1962

2THETA	PEAK	D	H	K	L	I(INT)	I(PK)	I(DS)
10.63	10.64	8.316	1	1	0	73	93	28
12.65	12.66	6.992	0	0	1	15	20	6
17.04	17.04	5.200	2	0	0	15	17	8
18.88	18.88	4.696	1	1	1	2	3	1
21.04	21.04	4.220	2	0	1	5	6	3
21.35	21.36	4.158	2	2	0	2	3	1
21.93	21.94	4.050	0	2	0	3	4	2
22.14	22.14	4.011	3	1	0	7	8	5
24.43	24.44	3.641	0	1	2	17	17	13
25.46	25.46	3.496	1	2	2	26	26	20
25.71	25.70	3.463	0	0	2	6	7	5
26.48	26.48	3.363	3	1	0	37	37	31
27.90	27.90	3.195	2	0	1	19	22	17
28.07	28.08	3.176	2	3	1	39	41	35
29.31	29.32	3.044	1	4	1	1	2	1
30.57	30.58	2.922	1	3	2	25	30	25
30.79	30.80	2.901	3	2	1	100	100	100
30.86		2.895	2	2	1	7		7
31.00	31.00	2.882	3	4	0	39	48	39
33.45	33.46	2.677	2	5	0	19	17	21
33.79	33.80	2.650	1	1	3	2	3	2
34.04	34.04	2.632	0	4	0	22	21	25
34.47	34.46	2.600	4	0	2	21	19	24
36.72	36.72	2.445	1	2	3	1	1	2
36.90	36.82	2.434	4	2	1	1	1	5
38.08	38.08	2.361	2	4	2	4	3	9
38.35	38.36	2.345	3	4	1	7	6	1
38.98	39.00	2.308	0	6	0	1	1	20
39.35	39.36	2.288	3	1	3	15	12	34
40.36	40.36	2.233	1	5	2	24	20	24
41.22	41.22	2.188	4	2	2	17	14	2
41.35	41.32	2.182	2	0	3	1	11	40
41.70	41.70	2.164	4	5	1	27	23	1
42.47	42.48	2.127	3	3	0	1	1	12
42.83	42.82	2.110	2	6	3	8	7	18
43.33		2.086	4	2	2	12	14	7
43.34	43.34	2.086	1	4	3	4		

2THETA	PEAK	D	H	K	L	I(INT)	I(PK)	I(DS)
43.49	43.48	2.079	4	4	0	11	14	17
44.71	44.72	2.025	0	0	2	25	19	40
45.17	45.18	2.005	2	4	2	1	2	2
46.09	46.10	1.9675	6	1	4	6	4	10
46.69	46.70	1.9437	1	7	0	1	1	2
46.95	46.96	1.9335	0	2	4	1		1
47.60	47.60	1.9086	3	5	3	2	1	2
47.93	47.94	1.8964	3	5	0	1	1	3
48.18	48.18	1.8870	2	0	4	6	5	11
48.62	48.62	1.8711	4	6	2	8	7	15
49.04	49.04	1.8559	4	1	3	8	7	14
49.22	49.20	1.8496	5	5	1	6	5	11
49.33	49.34	1.8458	1	1	4	2	6	4
49.67	49.68	1.8340	3	3	2	7	4	13
49.91	49.90	1.8257	4	4	2	3	6	6
50.06	50.06	1.8206	2	7	1	6		12
52.15	52.16	1.7523	1	7	2	16	12	32
52.29	52.28	1.7480	2	2	4	3	10	7
52.70	52.74	1.7355	1	5	3	1	8	2
52.72		1.7348	5	1	4	5		11
52.77		1.7333	6	3	0	4		8
52.84		1.7313	5	3	2	2		5
53.29	53.30	1.7175	3	5	0	6	5	12
54.53	54.52	1.6815	1	5	4	1	1	3
55.17	55.18	1.6633	5	2	4	20	14	44
55.40	55.32	1.6569	0	7	2	7	10	16
56.20	56.20	1.6352	6	0	0	2		4
56.97	56.98	1.6149	6	3	3	8	6	19
57.44	57.46	1.6030	6	4	2	1	2	3
57.65	57.66	1.5976	4	6	1	7	6	16
57.72		1.5959	4	3	2	2		4
57.81	57.80	1.5935	3	7	2	1	5	1
57.91		1.5911	3	6	3	2		5
58.03	58.04	1.5880	1	6	0	4	5	9
58.28	58.28	1.5819	4	7	2	10	7	23
58.95	58.96	1.5654	3	6	3	1	1	3

BATISITE — NIKITIN AND BELOV, 1962

2THETA	PEAK	D	H	K	L	I(INT)	I(PK)	I(DS)
59.47	59.48	1.5529	6	2	2	12	9	29
59.60	59.62	1.5500	6	4	0	4	8	11
60.79	60.80	1.5223	2	9	0	7	7	17
60.80		1.5222	8	8	4	1		3
62.79	62.80	1.4786	3	5	4	11	9	30
62.85		1.4772	7	1	0	4		10
63.36	63.36	1.4666	2	3	5	4	3	12
63.64	63.62	1.4610	6	6	1	4	3	10
64.09	64.14	1.4517	2	9	4	2	6	7
64.14		1.4506	4	4	1	7		19
64.14		1.4506	6	1	4	1		2
64.52	64.62	1.4430	5	4	3	1	2	2
64.62		1.4410	4	8	0	2		6
64.89	64.82	1.4358	3	2	5	1	1	3
65.19	65.18	1.4298	7	2	1	1	1	3
65.44	65.40	1.4250	1	9	2	1	1	2
66.41	66.42	1.4065	3	9	0	6	4	16
66.64	66.60	1.4023	1	7	4	2	3	4
67.52	67.58	1.3861	6	6	0	2	4	5
67.58		1.3850	0	10	0	4		11
67.62		1.3842	5	3	1	1		4
68.52	68.52	1.3682	4	1	4	1	1	2
68.99	69.00	1.3601	4	7	5	1	1	4
69.13	69.16	1.3576	4	8	3	1	1	3
69.49	69.50	1.3514	5	7	2	8	6	23
69.50		1.3513	3	4	5	2		6
69.58		1.3500	0	0	6	1		4
69.79	69.68	1.3463	7	4	1	1	4	2
70.46	70.46	1.3352	3	3	2	3	2	10
70.62	70.64	1.3326	1	1	6	1	2	4
70.84	70.84	1.3287	3	9	2	1	1	2
71.08	71.08	1.3251	0	2	6	2	1	4
71.54	71.60	1.3177	4	3	5	3	3	5
71.60		1.3168	8	5	1	1		11
71.65		1.3159	0	8	4	1		5
71.69		1.3154	1	6	5	1		2
71.94	71.80	1.3114	6	6	2	1	3	4

2THETA	PEAK	D	H	K	L	I(INT)	I(PK)	I(DS)
72.67	72.68	1.3000	8	0	0	4	3	14
73.00	73.08	1.2949	2	9	3	1	2	2
73.08		1.2936	6	1	3	2		6
73.60	73.64	1.2858	1	3	6	3	7	7
73.64		1.2853	5	5	4	8		27
73.72		1.2840	2	2	6	4		12
74.62	74.62	1.2708	2	10	2	5	3	17
75.52	75.54	1.2578	0	1	6	1	1	4
75.88	75.86	1.2528	3	1	6	2	2	7
76.38	76.38	1.2458	7	5	2	5	3	18
76.50		1.2441	0	11	0	1		2
77.02	77.00	1.2371	5	9	0	1	2	9
77.48	77.48	1.2308	6	4	4	3	1	4
78.11	78.12	1.2225	2	4	6	1	1	3
78.12		1.2224	4	10	0	1		2
78.41	78.54	1.2185	2	7	5	3	4	11
78.42		1.2185	8	2	2	1		5
78.53		1.2170	8	4	0	3		18
78.54		1.2168	1	9	4	1		4
78.78	78.76	1.2137	3	3	6	5	3	7
79.44	79.44	1.2054	1	5	6	2	1	13
80.01	80.02	1.1981	4	0	6	2	1	7
80.32	80.32	1.1944	4	11	1	4	3	9
80.39		1.1934	7	1	4	2		10
81.21	81.24	1.1835	3	11	0	1	2	6
81.24		1.1831	5	9	2	3	1	2
81.45	81.46	1.1806	4	2	6	2		3
82.00	82.00	1.1741	8	6	1	3	2	9
82.32	82.32	1.1702	4	10	2	1	1	2
82.34		1.1701	5	7	4	1	1	15
82.73	82.72	1.1656	8	4	2	2		4
83.26	83.26	1.1595	3	9	4	1	1	6
83.64	83.64	1.1552	3	5	6	4	2	7
84.52	84.70	1.1454	6	6	4	1	2	14
84.66		1.1438	0	10	5	2		
84.72		1.1432	1	6	6	1		
85.01	85.00	1.1400	6	7	2	3	3	14

354

BATISITE — NIKITIN AND BELOV, 1962

2THETA	PEAK	D	H	K	L	I(INT)	I(PK)	I(DS)
85.07		1.1393	6	9	1	1		3
85.28	85.28	1.1371	2	6	6	2	3	10
85.39		1.1359	3	11	2	1		5
86.07	86.02	1.1286	5	1	6	1	1	2
86.34	86.34	1.1258	2	1	7	1	1	2
86.68	86.68	1.1223	4	11	0	1	1	4
86.81		1.1210	2	3	4	1		2
87.24	87.24	1.1165	2	10	2	1	1	6
87.88		1.1100	0	12	2	1	2	7
88.00	88.00	1.1088	1	7	6	2		11
88.12		1.1076	5	3	6	2		4
88.91		1.0998	7	5	4	1		5
88.94	88.94	1.0995	8	0	4	3	1	13
89.51	89.52	1.0940	2	9	5	1	1	4
89.83	89.80	1.0909	8	6	2	1	2	5
89.84		1.0909	2	12	0	1		2
90.40	90.40	1.0855	6	10	0	1	1	6
90.77	90.80	1.0820	9	3	0	1	1	4
90.83		1.0815	7	2	5	1		5
90.95		1.0804	9	3	0	1		3
92.21	92.22	1.0689	7	9	0	4	2	20
92.48	92.50	1.0665	9	5	0	3	2	16
92.69		1.0646	0	8	6	2		4
93.03	93.02	1.0616	3	7	0	1	1	8
93.23		1.0598	1	13	0	1		3
93.71	93.74	1.0557	5	9	4	2	1	7
93.89	94.04	1.0541	4	1	7	1	1	3
94.06		1.0527	6	2	6	2		9
94.37	94.36	1.0501	9	5	3	1	1	3
94.59	94.60	1.0482	5	5	6	1	1	6
94.79	94.92	1.0465	4	10	4	1	1	3
94.93		1.0453	6	0	2	2		7
95.19	95.20	1.0431	8	4	4	2	1	8
95.46	95.48	1.0409	5	11	2	2	2	11
95.57		1.0400	10	0	0	1		4
97.47	97.44	1.0247	3	7	7	1	1	3
97.85	97.96	1.0218	3	11	4	1	1	3
97.97		1.0208	4	12	2	3		14
98.34	98.30	1.0180	6	4	1	1	1	3
99.39	99.38	1.0100	1	9	6	1	0	3
99.81	99.80	1.0069	8	8	2	1	1	4
100.06	100.06	1.0051	0	1	8	1	1	6
100.38	100.40	1.0027	9	1	4	1	1	3
100.61	100.60	1.0010	0	12	4	1	1	3
101.19	101.22	0.9968	10	2	2	3	2	16
101.23		0.9965	7	1	6	1		5
101.62	101.56	0.9938	2	0	8	1	1	7
102.95	103.16	0.9846	1	3	8	1	2	6
103.07		0.9838	3	13	0	2		13
103.20		0.9828	2	2	8	2		11
104.14	104.16	0.9765	5	7	6	1	1	5
104.60	104.86	0.9735	7	3	6	1	2	4
104.85		0.9719	2	13	4	4		20
104.86		0.9718	2	14	0	1		6
105.31	105.26	0.9689	0	4	8	2	2	11
105.59	105.60	0.9671	9	7	2	2	1	8
106.54	106.56	0.9610	0	6	6	1	0	5
107.48	107.64	0.9553	2	14	4	2	1	3
107.63		0.9543	6	4	8	1		10
108.17	108.20	0.9510	10	6	4	2	1	7
108.19		0.9510	3	3	8	1		3
108.27		0.9504	3	13	6	1		4
108.85	109.18	0.9470	1	1	8	1	2	8
109.15		0.9453	7	5	6	4		22
109.43	109.44	0.9436	9	9	0	3	2	18
109.45		0.9435	5	5	8	1		9
109.49		0.9433	4	0	0	1		5
110.68	110.98	0.9364	11	1	6	1	1	4
110.83		0.9356	8	0	8	2		5
110.96		0.9348	4	12	2	2		6
110.98		0.9347	6	12	0	1		4
111.00		0.9346	7	11	2	2		5
111.40	111.38	0.9324	7	7	7	1	1	10
112.09	112.22	0.9286	4	11	5	1	1	8

BATISITE – NIKITIN AND BELOV, 1962

2THETA	PEAK	D	H	K	L	I(INT)	I(PK)	I(DS)
112.21		0.9280	8	2	6	1		7
112.34		0.9272	0	6	8	1		4
112.73	112.78	0.9251	10	0	4	1	1	7
112.79		0.9248	8	8	4	1		6
112.83		0.9246	4	14	0	1		5
112.93		0.9240	9	6	0	1		5
113.08		0.9232	10	2	6	2		10
114.24	114.26	0.9172	1	11	6	1	1	7
114.26		0.9171	3	5	8	1		10
114.28		0.9170	10	2	4	2		6
115.59	115.64	0.9103	3	4	8	2		15
115.69		0.9098	5	13	4	1		7
115.97	116.00	0.9084	4	1	8	2	1	11
116.43	116.40	0.9061	4	10	6	1	1	6
116.88	117.04	0.9039	8	3	2	2	1	8
117.09		0.9029	11	3	2	1		13
118.13	118.14	0.8980	1	7	4	1	1	8
118.37	118.34	0.8969	1	15	0	1	1	7
118.76		0.8951	9	11	4	1		4
118.82	118.82	0.8948	11	5	0	1	1	6
119.04		0.8938	10	4	4	1		6
119.17		0.8932	5	3	8	1		6
119.45	119.44	0.8919	7	7	6	2	1	11
119.89	119.88	0.8899	3	11	6	1		10
120.12		0.8889	0	14	4	1		4
122.81		0.8773	2	12	4	2		6
123.07	123.10	0.8762	9	14	6	1	1	27
123.08		0.8761	6	0	8	4		4
123.53	123.64	0.8743	11	5	2	1		8
123.67		0.8737	4	6	8	1		6
123.76		0.8733	3	7	8	1		10
124.01		0.8723	3	7	8	1		8
124.25	124.18	0.8713	9	6	5	2	1	6
124.29		0.8712	7	11	4	1		5
125.13	125.22	0.8679	6	2	8	1	1	4
125.16		0.8677	6	6	6	2		5
125.25		0.8674	6	2	8	2		12

2THETA	PEAK	D	H	K	L	I(INT)	I(PK)	I(DS)
125.85	125.84	0.8650	2	12	6	1	1	7
125.90		0.8649	5	5	8	1		7
126.54	126.54	0.8624	9	6	3	1	1	9
126.67		0.8619	2	8	8	1		8
127.09	127.46	0.8603	6	13	0	3	1	4
127.40		0.8592	6	14	4	1		22
127.52		0.8587	5	13	4	1		9
127.59		0.8585	8	10	8	1		9
130.64	131.00	0.8477	6	4	2	1	1	6
130.70		0.8475	4	0	8	1		9
130.94		0.8467	7	13	2	1		5
130.99		0.8465	0	16	2	1		11
131.65	131.72	0.8443	6	10	6	1	2	9
131.70		0.8441	8	12	2	3		22
131.76		0.8439	5	15	0	1		7
132.30	132.70	0.8422	4	13	5	1	2	4
132.36		0.8420	5	11	6	1		8
132.60		0.8412	12	2	2	1		10
132.63		0.8411	4	14	4	3		20
132.74		0.8407	12	4	0	3		25
132.77		0.8407	9	9	4	3		23
132.82		0.8405	6	14	2	1		4
133.60	133.32	0.8380	1	15	6	1	2	9
133.80		0.8374	15	1	4	1		5
134.41	134.62	0.8355	2	16	2	2	1	15
134.66		0.8347	14	7	2	4		28
135.03	135.16	0.8336	11	13	6	1	1	8
135.19		0.8331	9	11	3	3		20
135.47		0.8323	11	6	3	1		5
135.69		0.8316	10	10	0	1		10
137.29	137.40	0.8270	5	7	8	1	1	11
137.45		0.8266	9	8	6	1		6
138.51	138.60	0.8237	8	8	6	1	1	9
138.65		0.8233	7	3	8	2		17
139.22	139.28	0.8217	3	6	8	2	1	14
140.48	140.58	0.8184	6	2	8	3	1	27

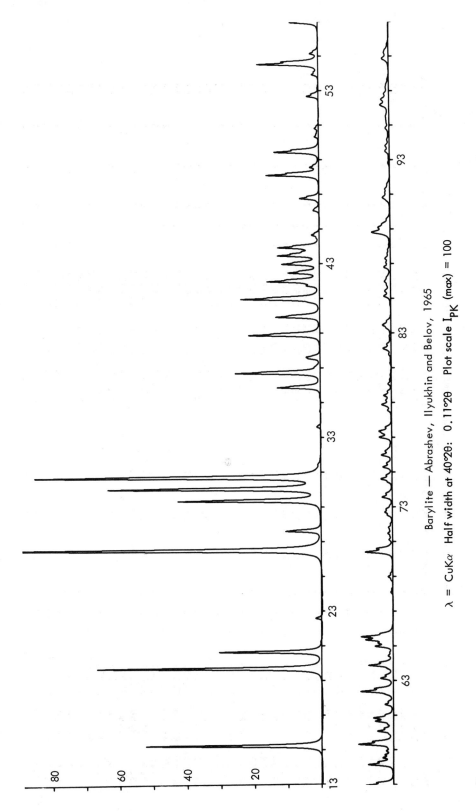

Barylite — Abrashev, Ilyukhin and Belov, 1965

λ = CuKα Half width at 40°2θ: 0.11°2θ Plot scale I_{PK} (max) = 100

357

BARYLITE - ABRASHEV, ILYUKHIN AND BELOV, 1965

2THETA	PEAK	D	H	K	L	I(INT)	I(PK)	I(DS)
15.20	15.20	5.825	0	0	2	44	53	21
19.65	19.66	4.513	2	0	1	60	67	40
20.63	20.64	4.303	1	1	1	28	31	20
22.55	22.56	3.939	0	1	1	2	2	2
26.48	26.48	3.364	2	1	0	100	100	100
27.58	27.58	3.232	2	1	1	100	100	100
29.33	29.34	3.042	2	0	3	44	43	52
30.01	30.02	2.975	0	1	3	68	64	82
30.67	30.68	2.913	2	1	2	88	86	110
30.67		2.912	0	0	4	5		6
33.57	33.58	2.667	3	1	0	2	1	2
35.85	35.84	2.503	4	0	0	15	13	24
36.69	36.68	2.447	4	0	1	30	26	49
37.04	37.02	2.425	3	1	2	1	2	2
37.52	37.60	2.395	4	0	1	3	4	6
37.59		2.391	1	2	1	3		4
38.87	38.88	2.315	0	2	2	26	22	46
39.92	39.92	2.256	4	0	0	16	13	29
39.98		2.253	1	2	2	1		1
40.95	40.96	2.202	2	1	1	29	24	54
41.71	41.72	2.164	4	1	0	4	4	7
41.96	41.96	2.151	0	2	2	19	16	37
42.45	42.46	2.127	4	1	1	11	10	23
42.95	42.96	2.104	2	1	5	13	12	27
43.01		2.101	2	1	2			2
43.44	43.44	2.081	0	1	5	15	13	32
43.68	43.68	2.071	4	0	3	4	6	8
43.92	43.92	2.060	2	1	1	15	13	32
44.46	44.48	2.036	1	1	5	1	1	2
44.64	44.64	2.028	2	2	2	3	3	6
46.04	46.06	1.9695	2	2	2	2	2	5
46.10		1.9671	3	1	0	1		2
46.74	46.74	1.9417	0	1	6	8	6	18
48.10	48.10	1.8902	4	0	4	21	16	52
48.54	48.54	1.8737	4	2	3	3	3	8
49.43	49.44	1.8423	2	2	3	19	14	47
50.30	50.30	1.8123	0	2	4	2	1	5
50.57	50.56	1.8034	5	1	1	2	1	2
50.79	50.78	1.7961	3	2	2	2	1	5
52.65	52.66	1.7369	4	2	4	5	4	15
53.90	53.90	1.6995	2	2	4	3	2	8
54.51	54.52	1.6818	4	2	0	13	19	38
54.52		1.6816	2	1	6	14		40
55.13	55.12	1.6646	4	2	1	3	3	10
56.94	56.94	1.6159	6	0	1	2	3	5
56.94		1.6158	4	2	2	11	9	35
58.13	58.14	1.5855	4	1	5	11	8	36
58.53	58.52	1.5757	2	0	7	4	3	13
58.71	58.70	1.5712	6	0	2	1	3	4
58.92	58.92	1.5662	0	1	7	2	2	7
59.30	59.30	1.5569	2	2	5	15	11	52
60.07	60.06	1.5389	6	1	0	6	5	22
60.65	60.64	1.5256	6	0	6	7	6	24
60.84	60.84	1.5211	4	0	3	6	6	19
61.60	61.60	1.5043	6	1	0	4	3	14
62.36	62.36	1.4879	1	0	6	8	10	28
62.36		1.4877	6	1	2	7		24
63.11	63.12	1.4719	2	3	0	6	4	21
63.86	63.86	1.4565	4	0	8	3	7	12
63.87		1.4562	6	1	6	8		29
64.42	64.42	1.4451	4	0	6	2	2	8
64.97	64.96	1.4342	0	3	6	8	6	32
65.15	65.16	1.4307	6	1	8	1	5	3
65.33	65.34	1.4271	2	3	3	11	8	44
65.52	65.52	1.4235	6	0	8	9	10	34
68.07	68.06	1.3762	4	0	7	1	1	5
68.78	68.80	1.3637	6	1	5	1	1	4
68.96	68.96	1.3606	6	1	5	2	2	9
70.38	70.40	1.3365	2	1	8	3	8	13
70.39		1.3364	0	1	5	12		52
71.09	71.08	1.3250	6	2	1	2	1	8
71.45	71.44	1.3192	4	1	7	2	2	10
71.79	71.80	1.3137	2	3	4	2	1	9
72.50	72.52	1.3026	2	2	7	2	2	11

BARYLITE - ABRASHEV, ILYUKHIN AND BELOV, 1965

2THETA	PEAK	D	H	K	L	I(INT)	I(PK)	I(DS)
72.67	72.68	1.3000	6	2	2	3	2	12
72.84	72.84	1.2973	4	3	1	3	2	2
73.54	73.54	1.2867	0	3	5	5	3	9
73.72	73.74	1.2841	6	1	5	1	2	26
74.59	74.60	1.2713	4	2	6	5	4	3
75.27	75.28	1.2614	4	2	0	5	3	25
75.97	75.98	1.2515	4	0	8	5	3	24
75.98		1.2514	2	0	9	1		26
76.32	76.20	1.2466	0	1	9	1	2	3
76.99	77.00	1.2374	4	3	3	7	4	5
77.35	77.34	1.2326	0	3	8	6	4	34
78.53	78.54	1.2171	8	0	1	2	1	31
78.87	78.86	1.2126	6	2	4	5	3	11
79.38	79.38	1.2060	8	0	6	4	2	28
80.06	80.06	1.1976	8	0	0	2	1	22
80.57	80.56	1.1913	4	3	4	1	0	11
81.25	81.24	1.1830	4	2	7	1	1	4
81.75	81.76	1.1771	8	1	1	5	1	3
82.09	82.10	1.1730	8	3	6	5	3	8
82.59	82.58	1.1672	8	0	3	2	1	27
83.26	83.44	1.1594	8	1	2	2	3	3
83.43		1.1575	0	4	0	2		13
83.43		1.1575	6	2	5	3		10
85.11	85.12	1.1389	0	3	2	5	2	16
85.45	85.40	1.1353	6	3	5	2	2	28
85.78	85.78	1.1318	2	4	2	1	1	11
85.95	85.96	1.1299	6	1	3	3	2	6
86.11		1.1282	4	3	7	1		20
86.78	86.78	1.1212	6	3	0	2	2	4
86.78		1.1212	4	0	1	2		16
87.28	87.28	1.1161	6	3	9	3	1	12
87.80	87.78	1.1108	4	4	2	2	1	21
88.79	88.80	1.1010	6	3	8	6	6	9
88.80		1.1009	2	2	9	2		11
88.80		1.1009	2	1	10	4		37
88.80		1.1008	2	1	10	2		11
88.96		1.0993	6	2	6	1		26
								5

2THETA	PEAK	D	H	K	L	I(INT)	I(PK)	I(DS)
89.29	89.26	1.0961	8	1	4	4	3	26
90.63	90.80	1.0834	8	4	6	1	2	7
90.79		1.0819	8	2	0	1		5
90.79		1.0818	2	4	3	3		29
91.29	91.28	1.0773	8	0	1	1	2	20
91.46		1.0757	0	4	2	1		4
92.79	92.80	1.0637	8	2	8	4	1	10
93.47	93.48	1.0577	6	1	5	1	1	25
93.80	93.78	1.0549	8	1	8	4	1	10
94.15	94.12	1.0519	4	0	10	1	1	7
94.80	94.80	1.0464	4	4	3	2	1	14
94.80		1.0464	6	3	4	1		8
95.30	95.32	1.0422	8	4	1	1	1	5
95.30		1.0422	4	4	7	1		6
95.48	95.48	1.0407	6	2	10	1	1	7
95.49		1.0407	2	3	8	1		6
96.15	96.16	1.0353	8	0	6	1	3	6
96.15		1.0352	2	0	11	5		34
96.16		1.0351	0	1	11	2		15
96.50	96.48	1.0324	4	3	7	3	3	19
96.82	96.80	1.0299	4	4	2	2	2	17
97.16	97.16	1.0272	4	3	9	1	1	6
97.33		1.0258	8	0	4	2		10
98.84	98.84	1.0142	2	4	5	4	2	12
98.84		1.0141	6	0	9	4		31
98.85		1.0141	0	1	6	2		5
99.35	99.18	1.0103	8	0	9	1	2	13
101.56	101.58	0.9942	0	4	6	1	1	12
101.91	101.94	0.9918	6	0	9	1	1	8
102.07		0.9906	6	3	7	1		10
102.75	102.88	0.9859	8	0	7	1	1	11
102.92		0.9848	4	4	4	1		11
104.99	104.98	0.9710	6	3	6	2	1	15
105.84	105.88	0.9655	0	1	7	1	1	10
106.03		0.9643	8	0	2	1		7
107.08	107.10	0.9577	4	2	10	2	1	16
107.41	107.56	0.9557	8	3	1	1	1	9

BARYLITE – ABRASHEV, ILYUKHIN AND BELOV, 1965

2THETA	PEAK	D	H	K	L	I(INT)	I(PK)	I(DS)
107.58		0.9546	10	1	1	3		25
108.14	108.10	0.9513	4	1	11	3	1	25
108.66	108.56	0.9481	6	0	10	3	1	7
108.99	109.18	0.9462	8	3	2	2	2	16
109.17		0.9451	10	1	2	1		7
109.18		0.9451	8	1	6	2		7
109.20		0.9450	2	2	11	3		26
111.67	111.88	0.9309	8	3	3	1	2	9
111.84		0.9300	10	1	3	1		10
111.85		0.9299	6	3	7	2		22
112.04		0.9289	6	1	9	1		8
112.04		0.9289	10	0	10	2		15
112.20		0.9280	4	4	6	2		21
113.48	113.50	0.9211	4	3	9	2	1	20
113.85	113.94	0.9192	6	4	3	2	1	12
114.20	114.20	0.9174	6	3	10	2	1	24
114.96	114.96	0.9135	8	3	4	3	1	28
115.49	115.52	0.9108	10	1	4	3	1	27
115.67		0.9099	2	5	0	2		5
115.68		0.9099	8	2	7	2		20
116.24	116.22	0.9071	0	4	8	2	1	23
116.43		0.9061	0	0	12	2		22
117.20	117.58	0.9024	2	5	3	3	1	5
117.55		0.9007	6	5	2	2		29
117.92	117.98	0.8990	10	4	4	2	1	24
118.11		0.8981	8	2	2	2		19
119.63	119.68	0.8911	6	0	9	2	1	19
120.03	120.20	0.8893	6	2	8	2	1	12
120.22		0.8884	8	3	0	3		32
120.23		0.8884	10	0	11	1		10
120.60	120.74	0.8867	4	3	5	1	1	9
120.79		0.8859	4	1	5	1		12
120.80		0.8858	2	1	7	1		6
120.82		0.8858	0	0	13	1		10
121.80	121.86	0.8815	2	0	13	2	1	17
122.20	122.26	0.8798	0	1	13	1	1	14
122.56	122.70	0.8783	10	2	3	1	1	9

2THETA	PEAK	D	H	K	L	I(INT)	I(PK)	I(DS)
122.78		0.8774	6	2	10	1		14
123.36	123.78	0.8750	6	4	5	2	1	24
123.56		0.8742	10	0	11	1		11
123.77		0.8733	8	3	11	1		9
123.78		0.8733	0	1	10	3		27
124.97	124.96	0.8685	8	2	8	1	1	11
124.98		0.8685	2	5	4	1		14
126.82	126.88	0.8614	10	2	8	5	1	52
127.04		0.8605	0	5	4	2		25
127.25	127.28	0.8598	8	3	6	2	1	20
127.45		0.8590	10	1	10	1		13
130.03	130.08	0.8498	4	4	8	4	1	44
130.04		0.8497	2	4	9	2		21
130.24		0.8490	6	4	6	1		11
130.69	130.68	0.8475	6	3	9	2	1	26
131.34	131.84	0.8453	8	5	9	5	1	52
131.80		0.8438	8	0	10	1		14
132.68	132.70	0.8409	8	1	10	1	1	9
132.72		0.8408	4	4	0	3		17
133.37	133.36	0.8387	8	2	12	3	1	35
135.96	136.24	0.8309	8	3	7	2	2	24
136.20		0.8302	4	5	4	1		13
136.20		0.8302	10	2	11	3		41
136.21		0.8301	8	1	13	3		40
136.22		0.8301	6	2	11	2		24
136.46		0.8294	4	1	10	3		39
136.97	136.90	0.8280	10	2	11	3	2	38
137.41		0.8267	10	1	13	1		11
138.15	138.44	0.8246	2	5	0	4	2	55
138.42		0.8239	2	2	13	3		41
138.45		0.8238	4	3	11	5		64
138.95	139.02	0.8225	8	4	7	5	2	60
139.17		0.8219	6	4	7	1		12
139.44		0.8212	0	4	10	1		17
139.45		0.8211	6	4	10	2		19
139.46		0.8211	6	1	12	2		19
142.33	142.38	0.8138	12	0	1	3	1	38

TABLE 17. AMPHIBOLES

Variety	Tremolite	Hornblende	Cummingtonite	Mn-Cummingtonite	Grunerite
Composition	$Ca_2Mg_5Si_8O_{22}$ $(OH)_2$	$(Ca,Na,K)_3(Mg,Fe,$ $Al,Ti)_5Si_8O_{22}(OH)_2$	$(Mg,Fe^2)_7Si_8O_{22}$ $(OH)_2$	$Mn_2(Mg,Fe)_5Si_8O_{22}$ $(OH)_2$	$(Fe^2,Mg)_7Si_8O_{22}$ $(OH)_2$
Source	Gouverneur Mining Dist., New York; Ross et al., 1968	Kakanui, New Zealand	Seven Islands, Quebec, Canada	Wabush Iron Fm, La-brador, Canada; Klein #2, 1964	Quebec, Canada
Reference	Papike & Clark, 1968b	Papike & Clark, 1967; 1968b	Fischer, 1966; Ghose, 1961	Papike & Clark, 1968b	Ghose & Hellner, 1959
Cell Dimensions					
a̲ Å	9.818	9.870	9.51	9.583	9.56_4
b̲ Å	18.047	18.058	18.19	18.091	18.30_2
c̲	5.275	5.307	5.33	5.315	5.34_8
α	90	90	90	90	90
β deg	104.65	105.20	101.90	102.63	101.833
γ	90	90	90	90	90
Space Group	C2/m	C2/m	C2/m	C2/m	C2/m
Z	2	2	2	2	2

TABLE 17. (cont.)

Variety	Tremolite	Hornblende	Cummingtonite	Mn-Cummingtonite	Grunerite
Site Occupancy	M$_1$ Mg 1.0 M$_2$ Mg 1.0 M$_3$ Mg 1.0 M$_4$ Ca 1.0	M$_1$ | Mg 0.570 M$_2$ \} Fe 0.254 M$_3$ | Ti 0.098 Al 0.078 M$_4$ \{ Ca 0.815 Na 0.130 Fe 0.055 A \{ K 0.39 Na 0.61 T$_1$ \{ Si 0.62 Al 0.38 T$_2$ \{ Si 0.88 Al 0.12	M$_1$ \{ Mg 0.84 Fe^2 0.16 M$_2$ \{ Mg 0.95 Fe^2 0.05 M$_3$ \{ Mg 0.84 Fe^2 0.16 M$_4$ \{ Mg 0.13 Fe^2 0.87	M$_1$ \{ Mg 0.80 Fe 0.20 M$_2$ \{ Mg 0.85 Fe 0.15 M$_3$ \{ Mg 0.88 Fe 0.12 M$_4$ \{ Mn 0.90 Ca 0.09 Mg 0.01	M$_1$ | M$_2$ \} \{Mg$_2$ 0.60 M$_3$ | \{Fe 0.40 M$_4$ Fe^2 1.00
Method	3-D, counter	3-D, counter	3-D, counter	3-D, counter	3-D, counter
R & Rating	0.05 (1)	0.049 (1)	0.102 (2)	0.04 (1)	0.176 (2)
Cleavage and habit		{110} perfect — Prismatic			
Comment					
μ	133.1	226.6	188.9	210.2	210.2
ASF	0.5897×10^{-1}	0.5855×10^{-1}	0.3543×10^{-1}	0.6209×10^{-1}	0.8679×10^{-1}
Abridging factors	1:1:0.5	1:1.3:0.5	1:1:0.5	1:1:0.5	1:1.3:0.5

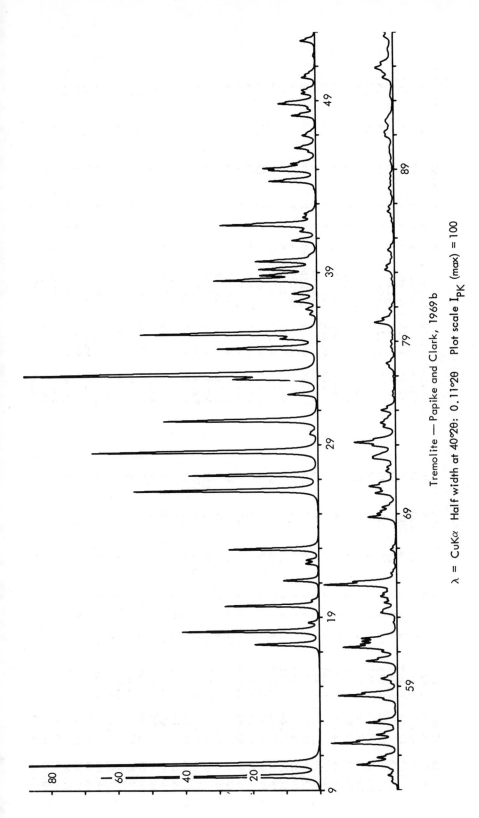

Tremolite — Papike and Clark, 1969 b

λ = CuKα Half width at 40°2θ: 0.11°2θ Plot scale I$_{PK}$ (max) = 100

363

TREMOLITE – PAPIKE AND CLARK, 1968b

2THETA	PEAK	D	H	K	L	I(INT)	I(PK)	I(DS)
9.79	9.80	9.023	0	2	0	48	65	40
10.52	10.52	8.406	1	1	0	74	100	63
17.36	17.44	5.104	0	2	1	6	19	5
17.43		5.082	-1	3	0	13	41	12
18.19	18.20	4.873	1	3	0	34	3	30
18.67	18.68	4.749	-2	0	1	2	28	2
19.66	19.66	4.512	0	4	0	24	2	22
19.97	19.98	4.442	0	2	0		11	1
21.12	21.12	4.203	0	0	1	9	4	8
22.09	22.10	4.021	-2	0	0	3	4	3
22.29	22.30	3.985	1	1	1			
22.94	22.96	3.873	-1	3	1	25	27	23
26.35	26.36	3.380	1	3	1	53	55	50
27.24	27.24	3.271	2	4	0	37	39	35
28.60	28.60	3.119	-1	4	0	66	67	64
29.59		3.016	3	1	0	2	2	2
29.74	29.72	3.002	-3	4	1	1	2	1
30.39	30.40	2.939	-1	5	1	6	46	6
30.40		2.938	-2	2	2	41		40
31.91	31.92	2.802	3	3	0	9	9	9
32.81	32.82	2.727	-3	3	2		25	23
33.09	33.10	2.705	-1	5	2	23	94	100
34.58	34.58	2.591	0	6	1	100	30	33
35.14	35.14	2.552	-2	0	2	32	11	10
35.44	35.44	2.530	0	2	2	9	53	59
36.56	36.56	2.455	-2	2	2	57	3	2
36.86	36.86	2.436	-2	6	1	3	3	3
37.30	37.30	2.408	0	6	1	3	7	6
37.32		2.408	-2	4	2	6		
37.76	37.76	2.380	3	5	1	8	8	8
38.54	38.54	2.334	-3	5	1	34	31	36
38.83	38.84	2.317	-4	2	1	17	17	18
39.18	39.18	2.297	-1	7	0	18	17	19
39.64	39.64	2.272	3	3	2	20	18	22
39.99	39.98	2.253	-2	4	1		2	1
40.85	40.86	2.207	-2	4	2	8	7	9
41.38	41.38	2.180	1	7	1	5	5	5

2THETA	PEAK	D	H	K	L	I(INT)	I(PK)	I(DS)
41.66	41.76	2.166	1	3	2	2	29	2
41.76		2.161	-2	6	1	32	35	35
42.19	42.20	2.140	-3	1	2	4	4	4
42.41	42.42	2.129	2	2	1	3	4	4
44.30	44.30	2.043	3	5	0	17	14	19
44.94	44.94	2.015	1	5	2	15	14	17
45.06	45.06	2.010	3	7	0	10	16	11
45.32	45.32	1.9992	-4	0	1			
46.00	46.00	1.9715	-1	7	1	8	8	9
46.22	46.24	1.9622	3	7	0	2	2	2
46.23		1.9620	-4	2	0	2	6	7
46.95	46.96	1.9336	1	9	0	6	4	5
47.16	47.06	1.9254	4	2	1	2	3	2
48.12	48.12	1.8893	5	1	0	9	7	10
48.52	48.52	1.8748	-4	6	1	2	2	2
48.81	48.82	1.8642	-1	1	6	13	11	15
48.90	48.92	1.8609	-2	9	2	3		4
49.39	49.40	1.8435	-1	7	2	5	4	6
49.60	49.60	1.8363	-4	4	2	4	4	5
50.32	50.32	1.8116	0	10	0	5	4	6
50.53	50.46	1.8047	0	10	1	5	4	6
52.47	52.48	1.7424	-5	1	2	6	3	3
53.38	53.48	1.7148	5	1	2	1		7
53.47		1.7121	-5	1	1	2	2	2
53.68	53.66	1.7061	3	7	2	2		2
53.83	53.84	1.7015	0	10	1	1	3	1
54.44	54.46	1.6839	0	2	3	2		3
54.47		1.6832	-2	8	2	8	12	10
54.59	54.60	1.6797	-1	3	3	8		10
54.68		1.6771	-3	1	3	1		1
54.87	54.86	1.6717	0	2	3	7	6	9
55.71	55.72	1.6485	4	6	1	27	20	35
56.19	56.20	1.6355	1	11	0	8	6	10
56.91	56.90	1.6167	1	9	0	13	9	17
58.23	58.24	1.5831	6	0	0	4	4	6
58.47	58.48	1.5770	-1	5	3	24	18	32
59.15	59.18	1.5605	2	10	1	1	2	2

TREMOLITE – PAPIKE AND CLARK, 1968b

2THETA	PEAK	D	H	K	L	I(INT)	I(PK)	I(DS)
59.20	59.44	1.5593	-6	2	0	1		1
59.40		1.5546	-5	7	1	1	5	2
59.45		1.5535	4	0	2	5		7
60.41	60.48	1.5309	-6	2	2	11	9	1
60.47		1.5297	0	7	2	1		15
60.48		1.5294	5	7	0	1		1
61.00	61.00	1.5175	-1	9	2	5	5	6
61.28	61.28	1.5113	-2	6	3	22	16	29
61.62	61.62	1.5039	0	12	1	11	11	15
61.76	61.78	1.5009	-4	8	2	3	11	4
61.79		1.5001	5	5	1	5		7
62.69	62.68	1.4807	0	6	3	1	1	1
63.23	63.26	1.4693	-2	10	2	2	5	2
63.26		1.4688	4	4	2	4		6
63.35		1.4668	2	2	3	2		3
63.47	63.42	1.4643	3	7	0	5	4	2
63.84	63.84	1.4567	3	11	1	5	4	7
64.19	64.22	1.4498	-1	7	3	3	5	5
64.24		1.4487	-6	4	2	4		5
64.38	64.38	1.4459	-3	11	1	1	4	2
64.83	64.90	1.4369	4	10	0	1	22	45
64.99		1.4338	2	12	0	32		2
65.53	65.54	1.4231	-5	3	3	2	2	3
66.89	66.90	1.3975	-5	9	1	1	2	2
66.91		1.3972	6	6	1	2		2
68.80	68.80	1.3633	5	11	2	13	9	19
69.15	69.14	1.3572	-5	1	5	5	5	8
69.39	69.38	1.3532	7	5	3	5	4	4
70.18	70.18	1.3398	1	11	0	8	6	12
70.25		1.3388	-1	13	1	2		2
70.56	70.60	1.3337	-3	11	2	4	8	6
70.59		1.3332	3	3	3	3		4
70.61		1.3327	5	5	3	6		4
71.40	71.48	1.3199	-1	9	3	1	3	9
71.48		1.3187	-5	9	2	3		3
71.70	71.70	1.3151	-7	3	2	1	3	2

2THETA	PEAK	D	H	K	L	I(INT)	I(PK)	I(DS)
72.20	72.22	1.3072	-7	5	1	9	7	13
72.31	72.30	1.3055	-1	1	4	7	7	10
72.95	72.96	1.2959	6	8	0	1		11
72.96		1.2956	0	12	2	7	12	23
73.14	73.14	1.2928	-2	12	2	15		3
73.26		1.2909	-3	9	3	2		5
74.00	74.04	1.2798	6	6	1	3	4	5
74.06		1.2790	-1	3	4	1		3
74.09		1.2785	5	5	2	2		2
74.27	74.26	1.2759	0	0	4	4	5	6
74.42	74.42	1.2736	-3	3	4	6	3	3
75.00	75.00	1.2652	-4	4	3	1	4	9
75.84	75.84	1.2533	1	9	3	1	1	2
77.40	77.50	1.2320	-7	7	1	2	2	2
77.50	77.72	1.2306	-4	2	3	3		3
77.73	77.44	1.2275	-2	14	1	2	3	2
78.44	78.44	1.2182	-4	4	4	1	1	2
80.10	80.10	1.1971	-5	11	2	10	6	16
80.89	80.88	1.1874	8	0	0	2	2	4
83.14	83.14	1.1608	2	0	4	5	3	9
83.34	83.36	1.1585	-8	4	2	5	2	3
83.73	83.72	1.1541	1	15	1	2	1	2
84.27	84.30	1.1481	4	6	3	3	2	3
84.33		1.1474	2	10	3	2		3
85.98	85.98	1.1296	5	9	3	1	1	6
86.53	86.54	1.1238	7	9	0	3	2	2
86.82	86.80	1.1209	5	13	0	1	2	4
87.55	87.56	1.1133	-5	3	3	3	1	3
87.91	87.90	1.1097	-6	4	4	1	2	3
88.45	88.44	1.1044	-7	7	3	1	2	2
88.75	88.74	1.1014	0	16	1	1	1	3
89.46	89.46	1.0945	3	13	2	1	1	2
89.97	89.98	1.0896	-3	3	4	2	2	3
90.93	90.94	1.0809	5	5	3	3	2	3
90.89		1.0805	4	12	2	1		5
91.12	91.14	1.0788	4	14	1	4	3	7
91.62	91.82	1.0742	-9	1	2	1	2	3

TREMOLITE — PAPIKE AND CLARK, 1968b

2THETA	PEAK	D	H	K	L	I(INT)	I(PK)	I(DS)
91.74		1.0732	-7	1	4	2		2
91.88	93.28	1.0719	-7	3	2	2		2
93.29		1.0593	-9	3	2	1		
94.34	94.34	1.0503	5	11	2	7	1	2
94.72	94.90	1.0471	-2	2	1	5	3	13
94.89		1.0457	-7	11	0	1	5	
94.92		1.0454	-8	6	3	8		14
95.88	96.00	1.0375	-2	14	3	5	1	9
96.74	96.76	1.0305	8	6	1	1	3	3
97.12	97.10	1.0275	-1	17	4	6	3	8
97.66	97.66	1.0232	-1	11	0	4	2	11
98.23	98.20	1.0188	-4	16	5	1	1	7
99.36	99.36	1.0102	-3	5	1	2	1	4
100.04	100.04	1.0052	-8	0	4	3	2	6
100.35	100.36	1.0029	-3	17	4	2	2	3
101.47	101.50	0.9949	-2	6	5	1	1	1
101.84	101.92	0.9923	-7	7	4	1	1	
101.95		0.9915	-2	12	4	2		2
102.83	102.84	0.9854	8	0	2	2	1	4
103.42	103.46	0.9814	-9	5	3	4	4	8
103.47		0.9810	-4	6	5	7		14
103.74	103.76	0.9792	6	6	3	6	5	13
103.78		0.9789	4	16	1	2		3
104.06	104.06	0.9767	9	7	0	1	3	3
104.11		0.9744	-10	0	2	1	2	3
104.46	104.54	0.9739	-5	11	1	1		3
104.53		0.9739	2	14	4	1		
104.54		0.9729	0	12	3	3		6
104.69		0.9682	-4	12	4	1	1	1
105.42	105.64	0.9670	-7	13	2	1		3
105.60		0.9666	0	8	5	2		
105.67		0.9627	8	4	2	1	1	1
106.28	105.96	0.9587	3	9	4	1	1	3
106.91	106.92	0.9499	10	0	7	1	1	3
108.37	108.40	0.9483	-9	9	3	1		2
108.63	108.70	0.9482	-5	17	3	1		8
108.64		0.9478	-2	16	3	2		2

2THETA	PEAK	D	H	K	L	I(INT)	I(PK)	I(DS)
109.26	109.08	0.9446	2	2	5	1	1	2
111.49	111.60	0.9319	8	12	0	1	1	2
111.59		0.9313	-10	6	1	2		5
111.84	111.94	0.9299	-6	6	5	1	1	1
111.89		0.9297	-5	15	3	1		3
112.17		0.9282	-6	16	1	1		1
113.86	114.02	0.9191	-4	16	3	1	1	2
113.90		0.9189	2	12	8	1		4
114.05		0.9182	-8	8	3	2		1
114.21		0.9173	-7	13	11	1		14
115.89	116.08	0.9088	3	11	3	6	3	8
116.03		0.9081	-1	17	6	4		11
116.09		0.9078	-6	16	2	1		2
116.17		0.9074	-7	5	5	5		4
117.65	117.82	0.9002	7	15	0	1	2	5
117.74		0.8998	-9	11	2	2		2
117.87		0.8992	-7	11	4	2		5
118.01		0.8986	-2	16	3	2		2
118.23	118.26	0.8975	-7	15	2	1	1	1
118.41		0.8967	3	3	5	1		2
118.94	119.04	0.8942	5	7	4	3	1	2
119.07		0.8936	4	18	1	1		7
119.28		0.8927	5	17	1	1		1
119.94	119.92	0.8897	10	2	4	1	1	1
119.97		0.8856	6	0	7	1		3
121.90	122.34	0.8811	-7	7	3	3	1	3
122.27		0.8795	-11	3	2	1		7
122.28		0.8782	3	5	5	3		7
122.60	122.86	0.8780	-8	10	4	1	2	4
122.63		0.8766	-3	16	6	2		7
122.96		0.8754	-2	0	0	2		7
123.25		0.8697	10	8	8	1	1	3
124.66	124.66	0.8625	-2	18	1	3	1	0
126.51	126.54	0.8596	11	1	0	1	1	3
127.30	127.32	0.8586	-5	17	9	3		8
127.55		0.8581	9	9	3	3		9
127.70	127.76		3	13	4	1	1	2

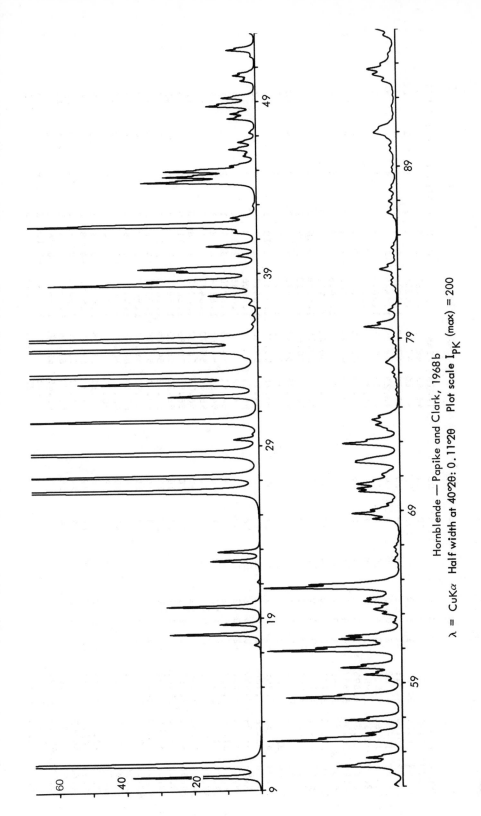

Hornblende — Papike and Clark, 1968 b

λ = CuKα   Half width at 40°2θ: 0.11°2θ   Plot scale I_{PK} (max) = 200

HORNBLENDE — PAPIKE AND CLARK, 1967; 1968b

2THETA	PEAK	D	H	K	L	I(INT)	I(PK)	I(DS)
9.79	9.80	9.029	0	2	0	18	20	11
10.49	10.50	8.425	0	1	1	91	100	55
18.05	18.06	4.911	-1	1	0	14	14	10
18.62	18.62	4.762	-2	0	1	6	6	11
19.65	19.66	4.514	0	4	0	15	14	6
22.29	22.30	3.985	-1	1	1	8	7	6
22.83	22.84	3.893	-1	3	1	7	6	1
26.29		3.387	-1	4	1	1		
26.34	26.34	3.381	1	1	1	68	59	58
27.19	27.20	3.276	2	0	0	44	37	38
28.52	28.52	3.127	2	1	0	75	61	67
29.35	29.36	3.041	-3	1	1	44	36	41
30.42	30.42	2.936	-2	1	1	17	14	16
31.84	31.84	2.808	3	1	0	34	27	33
32.59	32.60	2.745	-3	3	1	100	77	100
33.08	33.08	2.706	-1	3	1	68	52	70
34.54	34.54	2.595	0	5	1	73	56	77
35.14	35.14	2.552	-2	6	2	9	1	1
37.16	37.18	2.417	-2	0	2	9	1	10
37.69	37.70	2.385	-2	5	1	2	7	2
37.75		2.381	3	0	3			
38.34	38.34	2.346	-3	5	1	42	31	48
38.56	38.56	2.333	-4	2	1	17	17	20
39.09	39.10	2.302	-1	7	1	13	12	16
39.27	39.26	2.292	-3	1	2	21	18	25
40.00	40.00	2.252	3	3	1	4	3	4
40.58	40.58	2.221	-2	4	2	10	8	11
41.36	41.36	2.181	1	7	1	4	3	4
41.76	41.76	2.161	2	6	1	48	35	59
41.83		2.158	-3	3	2			
42.23	42.22	2.138	-1	5	2	6	4	8
44.30	44.30	2.043	1	0	2	4	17	5
44.62	44.62	2.029	-4	0	2	25	14	32
44.94	44.94	2.015	3	5	1	19	14	24
45.25	45.24	2.002	3	7	0	19	4	26
45.81	45.82	1.9789	-3	7	1	4	2	6
46.20	46.20	1.9634	1	9	0	6	4	3

2THETA	PEAK	D	H	K	L	I(INT)	I(PK)	I(DS)
46.62	46.62	1.9467	-3	5	2	3	3	5
47.16	47.16	1.9254	4	2	1	1	1	2
47.98	47.98	1.8944	5	1	0	6	4	9
48.28	48.28	1.8835	-4	6	1	5	4	7
48.72	48.72	1.8673	-1	9	1	11	7	15
48.79		1.8648	-5	3	1	1		
48.89	48.86	1.8612	2	4	2	2	6	3
49.19	49.20	1.8508	-4	4	2	4	5	6
49.22		1.8496	-1	7	0	4		
50.19	50.20	1.8162	5	3	0	3	2	5
50.50	50.50	1.8058	0	10	0	4	3	5
50.64	50.62	1.8010	1	9	1	1	3	7
51.98	51.98	1.7578	-5	1	2	7	4	2
53.09	53.10	1.7236	-5	5	1	2	1	10
53.78	53.78	1.7030	0	10	0	8	2	2
54.18	54.20	1.6915	-1	3	3	7	10	13
54.21		1.6906	-2	8	0	1		
54.28		1.6885	2	10	0	2		
54.51	54.50	1.6820	-3	9	1	7	3	11
54.67	54.66	1.6774	0	6	3	34	20	12
55.70	55.70	1.6487	4	8	1	14	5	56
56.09	56.10	1.6382	4	11	0	26	9	12
56.86	56.86	1.6178	1	5	3	9	17	24
58.19	58.20	1.5840	-1	10	1	15	2	45
59.13	59.14	1.5610	2	0	2	1	6	3
59.49	59.48	1.5525	4	0	2	31	9	16
59.92	59.92	1.5424	-6	5	0	5	6	27
60.07	60.08	1.5389	-3	6	3	1	9	2
60.92	60.92	1.5195	-2	9	3	3	6	55
60.94		1.5189	1	9	2	3		
61.11	61.08	1.5151	-3	9	1	14	20	9
61.38	61.40	1.5090	-4	8	2	7	13	3
61.57	61.58	1.5048	0	12	0	2	4	6
61.77	61.76	1.5006	5	1	1	4	9	25
62.49	62.50	1.4849	0	6	3	2	8	12
63.01	63.02	1.4740	-2	10	2	4	4	3
63.22	63.28	1.4695	-1	5	3	2	3	8

HORNBLENDE — PAPIKE AND CLARK, 1967; 1968b

2THETA	PEAK	D	H	K	L	I(INT)	I(PK)	I(DS)
63.29		1.4681	4	4	2	3		6
63.29		1.4680	2	2	3	2		3
63.70	63.76	1.4596	-6	4	2	2	6	5
63.77		1.4582	3	11	0	7		14
63.91	63.92	1.4553	-1	7	1	3	5	6
64.57	64.56	1.4421	-6	6	3	38	21	73
64.73	64.74	1.4388	-4	10	0	3	14	5
66.07	66.08	1.4129	-4	6	3	2	1	3
66.87	66.86	1.3979	6	2	1	1	1	3
68.52	68.52	1.3683	-5	5	3	7	4	15
68.85	68.84	1.3625	5	1	2	13	7	27
69.18	69.04	1.3568	7	10	0	3	5	6
69.40	69.38	1.3530	2	1	2	1	2	3
70.11	70.12	1.3410	10	11	2	10	6	21
70.27	70.30	1.3384	-3	11	3	6	6	12
70.55	70.54	1.3338	2	6	2	9	6	18
70.63		1.3325	5	3	3	9		
71.04	71.06	1.3257	-5	9	2	2	3	5
71.14		1.3242	-1	9	2	2		3
71.70	71.82	1.3152	1	13	3	1	7	3
71.81		1.3134	-7	5	1	4		21
71.91	71.90	1.3118	-1	8	4	7	7	16
72.77	72.92	1.2985	-6	9	0	1	9	1
72.83		1.2975	-3	9	3	1		3
72.84		1.2974	0	12	0	3		7
72.92		1.2962	-2	12	2	15		33
73.66	73.68	1.2850	-1	6	4	2	2	6
73.96	73.98	1.2805	6	6	1	1	4	3
73.97		1.2803	0	0	3	3		8
74.13	74.28	1.2780	5	0	4	4	4	13
74.28		1.2758	-4	8	2	6		3
74.43	74.48	1.2735	-6	10	0	1	3	3
74.53		1.2721	4	12	0	1		3
75.71	75.72	1.2551	1	9	3	1	0	
76.02	76.02	1.2508	0	14	1	1	0	2
76.44	76.44	1.2450	2	14	0	1	0	1
76.76	76.76	1.2405	0	10	3	1	1	3

2THETA	PEAK	D	H	K	L	I(INT)	I(PK)	I(DS)
77.01	77.00	1.2372	-7	7	1	1	1	3
77.53	77.60	1.2302	-4	2	3	1	2	3
77.60		1.2293	-2	14	1	2		6
77.72		1.2277	-4	4	4	1		2
78.16	78.16	1.2218	-8	2	1	1	1	1
78.66	78.74	1.2153	-6	10	1	1	1	2
78.75		1.2142	6	0	2	1		1
78.94	78.94	1.2117	7	3	1	1	1	2
78.95		1.2116	-2	12	2	1		1
79.67	79.68	1.2024	-5	11	2	10	5	26
80.62	80.62	1.1906	8	0	0	3	1	7
81.73	81.74	1.1773	4	10	0	1	0	2
82.10	82.12	1.1728	-6	10	2	2	1	4
82.32	82.34	1.1703	-7	5	1	1	1	2
82.66	82.66	1.1664	-8	4	2	1	1	4
83.00	83.00	1.1624	2	0	4	6	3	15
83.67	83.68	1.1548	-1	15	1	1	1	2
83.92	83.92	1.1520	0	14	2	1	1	4
84.25	84.26	1.1483	2	10	3	1	1	2
86.00	86.04	1.1294	5	9	2	2	1	4
86.24	86.30	1.1269	9	9	1	1	1	2
86.31		1.1261	7	9	3	3		7
86.67	86.58	1.1224	5	13	0	2	1	5
86.95	86.94	1.1195	-6	4	4	1	1	2
87.61	87.60	1.1128	5	3	3	1	1	7
87.62		1.1126	-7	7	3	3		5
88.67	88.68	1.1022	-0	16	1	2	1	4
88.89	88.92	1.1000	-1	9	4	1	1	3
89.27	89.26	1.0963	-1	15	2	1	1	2
89.90	89.90	1.0902	3	1	4	2	1	5
90.20	90.20	1.0874	-2	16	1	1	1	4
90.83		1.0815	-9	1	2	2		4
90.93	90.94	1.0805	4	12	2	3	3	10
91.04		1.0804	5	5	3	3		8
91.07		1.0792	4	14	1	3		10
91.30	91.22	1.0771	-6	12	2	2	3	6
91.90	91.88	1.0717	7	3	2	1	0	2

HORNBLENDE — PAPIKE AND CLARK, 1967, 1968b

2THETA	PEAK	D	H	K	L	I(INT)	I(PK)	I(DS)
92.50	92.50	1.0663	-9	3	2	1	1	3
93.13	93.14	1.0607	-2	16	1	1	0	3
93.96	93.98	1.0535	-8	6	3	5	2	14
94.03		1.0529	-2	2	5	1		
94.35	94.34	1.0502	-5	11	2	7	3	21
94.64	94.66	1.0478	-9	5	1	2	4	5
94.66		1.0476	7	11	0	7		23
95.03	94.96	1.0445	0	10	4	1	2	3
95.19		1.0431	-5	0	1	1		2
95.52	95.52	1.0404	-2	14	3	1	1	4
96.62	96.64	1.0314	8	6	1	4	2	13
97.05	97.04	1.0281	-1	17	1	7	3	21
97.25	97.24	1.0265	-1	11	4	5	3	15
98.09	98.10	1.0199	4	16	0	1	1	2
98.53	98.80	1.0165	-3	5	5	2	2	5
98.79		1.0146	-8	0	4	4		13
98.79		1.0146	-6	14	1	1		2
98.91		1.0136	-9	3	3	1		2
100.15	100.16	1.0044	-3	17	1	2	1	6
102.31	102.48	0.9890	-9	5	3	4	3	12
102.48		0.9878	-4	6	4	8		25
102.83	102.82	0.9853	8	0	2	2	2	8
103.30	103.18	0.9822	-5	11	5	1	1	2
103.52	103.76	0.9807	-10	0	2	1	3	3
103.65		0.9798	-5	11	4	2		6
103.71		0.9794	4	16	1	3		10
103.75		0.9791	9	7	0	5		3
103.81		0.9788	6	6	3	3		17
104.35	104.08	0.9751	0	12	4	3	2	9
104.66	104.66	0.9731	-4	12	4	1	1	3
105.19	105.16	0.9697	0	8	5	2	1	7
105.16		0.9627	8	4	2	1	0	2
106.28	106.30	0.9593	3	9	4	2	0	3
106.82	106.84	0.9533	-9	9	2	1	1	
107.79	107.92	0.9525	10	0	0	2		6
107.94		0.9501	-2	16	3	2		8
108.33	108.32	0.9501	2	3	3	1	1	2
108.99	108.68	0.9462	2	2	5	1	1	3

2THETA	PEAK	D	H	K	L	I(INT)	I(PK)	I(DS)
109.80	109.80	0.9415	0	16	3	1	1	2
110.89	110.94	0.9352	-10	6	1	1		9
110.99		0.9347	-7	3	0	1		3
111.17		0.9337	8	12	1	2		3
111.78	111.74	0.9303	-6	16	1	1	1	6
112.68	112.68	0.9254	-8	8	3	1	1	5
113.26	113.24	0.9223	-4	16	2	1	1	2
113.72	113.72	0.9199	2	12	4	2	1	7
114.61	114.62	0.9153	-7	17	5	5	1	20
115.69	115.76	0.9098	-1	11	3	3	2	11
115.78		0.9094	-3	11	4	6		23
116.61	116.64	0.9053	-7	11	4	2	1	8
116.81		0.9043	-9	1	2	2		8
117.10	117.04	0.9029	0	20	0	1		2
117.36		0.9016	7	15	0	1	1	3
117.87	117.86	0.8992	2	16	3	1		6
118.22	118.20	0.8976	3	3	5	1		4
118.96	118.98	0.8941	5	7	4	3	1	2
118.98		0.8940	4	18	1	1		13
119.18		0.8931	5	17	1	1		3
120.06	120.18	0.8892	6	0	4	1		4
120.25		0.8883	-7	7	5	1	1	5
120.67	120.64	0.8864	-1	17	3	1		2
121.04	121.40	0.8848	-11	3	2	1		2
121.11		0.8845	-8	10	4	2		4
121.36		0.8834	-3	1	6	3	1	9
121.86	122.02	0.8812	-2	0	6	1		10
122.07		0.8803	3	5	6	5		15
122.13		0.8801	-4	0	8	4		3
124.19	124.18	0.8716	-2	18	3	3	1	13
125.87	125.98	0.8650	8	10	2	2	1	3
125.89		0.8649	11	1	0	1		7
126.08	126.48	0.8642	-8	6	5	2		2
126.47		0.8627	-5	17	3	1		18
127.56	127.60	0.8586	3	13	4	4	1	3
127.64		0.8583	7	3	3	1		2
127.72		0.8580	10		6	1		3

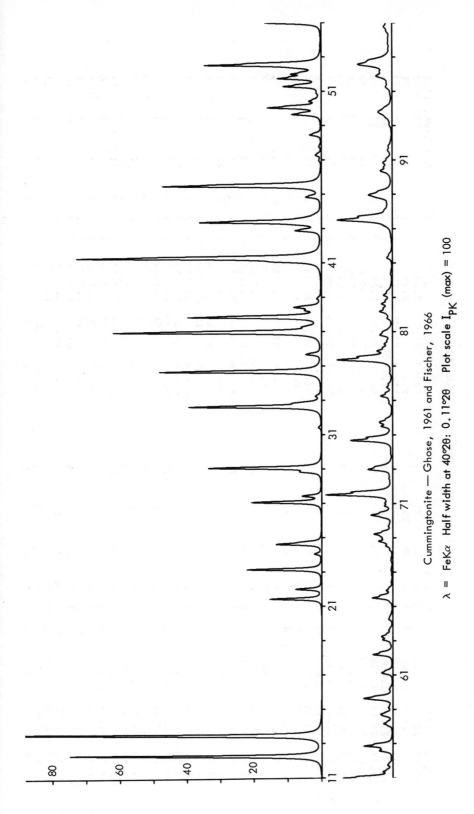

Cummingtonite — Ghose, 1961 and Fischer, 1966

$\lambda = FeK\alpha$ Half width at $40°2\theta$: $0.11°2\theta$ Plot scale I_{PK} (max) = 100

CUMMINGTONITE - GHOSE, 1961 AND FISCHER, 1966

2THETA	PEAK	D	H	K	L	I(INT)	I(PK)	I(DS)
12.22	12.22	9.095	0	2	0	69	75	42
13.42	13.42	8.284	1	1	0	94	100	58
21.39	21.40	5.215	1	3	0	16	15	11
21.97	21.98	5.080	0	0	1	8	8	6
23.11	23.12	4.832	-1	1	1	24	22	17
24.02	24.02	4.653	2	0	0	2	2	2
24.58	24.58	4.547	0	2	1	14	13	11
27.03	27.04	4.142	2	2	0	24	21	18
27.39	27.40	4.088	-1	1	1	6	6	5
28.79	28.80	3.894	-2	0	1	6	7	5
29.02	29.04	3.863	1	3	1	40	34	32
31.38	31.38	3.579	-1	1	1	1	1	1
32.59	32.60	3.450	-2	2	1	49	40	42
32.81	32.80	3.428	0	4	1	7	9	6
33.20	33.20	3.388	1	5	0	2	2	1
34.63	34.64	3.252	2	4	0	62	48	55
35.64	35.64	3.163	2	0	1	6	5	5
36.91	36.92	3.058	3	1	0	82	62	75
37.24	37.24	3.032	0	6	0	4	6	4
37.81	37.82	2.988	2	2	1	53	40	50
38.21	38.22	2.958	-1	2	1	4	5	4
38.39	38.40	2.944	-2	4	1	10	8	9
38.88	38.88	2.908	-3	1	1	1	1	1
41.24	41.24	2.749	-1	5	1	100	73	100
42.85	42.86	2.650	-3	3	1	10	8	10
43.35	43.36	2.621	0	6	1	51	36	53
44.80	44.80	2.540	2	6	0	6	5	7
45.43	45.46	2.507	0	2	2	3	5	3
45.45	45.46	2.506	-2	2	2	67	47	73
47.24	47.24	2.416	-1	2	2	3	3	3
48.42	48.42	2.360	3	3	1	5	3	6
49.57	49.62	2.309	-4	0	1	9	9	1
49.62	49.62	2.307	-1	1	2	12		14
50.02	50.02	2.290	-3	5	1	24	16	28
50.39	50.40	2.274	0	8	0	4	4	5
50.87	50.88	2.254	4	0	1	1	2	1
51.25	51.26	2.238	-4	2	1	17	11	20

2THETA	PEAK	D	H	K	L	I(INT)	I(PK)	I(DS)
51.69	51.72	2.221	-1	3	2	2	13	2
51.72		2.219	3	1	2	17		21
51.97	51.98	2.209	1	7	1	11	9	13
52.50	52.50	2.189	2	6	1	52	35	64
54.95	54.96	2.098	-3	3	2	3	17	30
54.96		2.098	-2	8	1	24		4
55.34	55.34	2.084	0	4	2	3	3	3
55.72	55.72	2.071	4	2	0	2	2	2
56.53	56.56	2.044	2	4	2	4	4	4
56.57		2.043	2	8	0	14		6
56.84	56.84	2.034	3	5	1	2	9	18
58.04	58.16	1.9954	1	7	1	3	3	2
58.15		1.9920	3	3	2	3		5
58.59	58.70	1.9783	0	0	2	2	4	3
58.63		1.9770	1	6	2	4		3
58.69		1.9751	-1	9	0	2		5
59.08	59.08	1.9635	-2	8	1	5	1	2
59.56	59.64	1.9489	-3	1	2	12	9	6
59.63		1.9468	-4	0	2	1		17
60.15	60.14	1.9315	-2	6	2	3		2
61.08	61.10	1.9050	-3	5	2	3	1	4
61.12		1.9037	-4	2	1	11	3	15
62.18	62.18	1.8744	-1	9	1	5		7
63.04	63.04	1.8515	5	1	0	2		3
63.24	63.24	1.8462	2	8	1	2	6	3
63.27		1.8456	-4	6	1	2	3	3
63.60	63.60	1.8370	-1	9	1	11		17
64.23	64.24	1.8209	-4	4	2	3		4
65.48	65.48	1.7897	5	3	0	1	1	2
65.92	65.92	1.7792	-2	2	3	3	1	5
68.27	68.28	1.7251	0	6	3	7	6	11
68.51	68.52	1.7196	0	2	3	4	2	6
68.78	68.78	1.7138	-1	3	3	8		13
69.07	69.20	1.7076	0	8	2	2	2	2
69.20		1.7047	-1	3	3	3	4	6
69.65	69.66	1.6949	4	2	1	8	1	
70.18	70.32	1.6839	-2	8	2	3	7	6

CUMMINGTONITE – GHOSE, 1961 AND FISCHER, 1966

2THETA	PEAK	D	H	K	L	I(INT)	I(PK)	I(DS)
70.31	70.86	1.6812	-5	1	2	11	3	17
70.85	71.02	1.6701	-5	5	1	4		6
71.01	71.50	1.6666	-3	9	1	1	2	2
71.50	71.50	1.6567	-4	6	1	40	20	66
72.44	72.44	1.6381	-4	4	2	1	1	2
72.96	72.96	1.6281	1	11	0	14	8	23
73.07		1.6261	-1	4	0	2		4
74.66	74.66	1.5962	1	8	0	26	13	45
75.48	75.48	1.5815	-5	5	3	2		11
75.74	75.72	1.5769	2	10	0	6	3	10
76.81	76.82	1.5581	2	2	1	6	4	3
77.24	77.24	1.5509	6	0	2	2	1	13
77.75	77.76	1.5423	1	9	2	3	2	6
79.34	79.36	1.5164	-2	6	3	19	17	34
79.37		1.5158	0	12	0	19		34
79.86	79.86	1.5081	5	6	3	7	5	14
80.28	80.28	1.5016	4	5	1	7	3	7
80.78	80.80	1.4938	4	4	2	7	4	13
80.87		1.4925	-4	8	1	2		3
81.77	81.78	1.4788	-4	8	2	3	2	6
82.16	82.16	1.4731	-1	0	2	8	4	15
82.61	82.62	1.4664	-7	1	3	7	3	13
83.11	83.12	1.4592	-3	11	1	6	3	11
84.33	84.34	1.4420	-3	11	0	1	1	2
85.29	85.28	1.4289	-4	10	1	4	2	8
87.50	87.50	1.3999	-6	6	1	44	16	89
88.48	88.50	1.3874	-2	8	3	1	2	3
88.80	88.94	1.3835	-4	6	3	7	7	15
88.94		1.3818	-4	1	2	15		31
90.35	90.36	1.3648	5	6	3	9	4	19
90.82	90.82	1.3592	1	11	2	7	3	14
91.54	91.54	1.3509	5	3	2	3	1	7
91.82	91.78	1.3477	-1	13	1	1	1	2
93.06	93.04	1.3342	-1	9	3	2	1	4
93.78	93.78	1.3273	-5	5	3	3	4	7
93.79		1.3259	-1	1	4	4		8
		1.3258	7	1	0	4		8
93.87		1.3250	-6	6	2	2		4
95.23	95.24	1.3105	0	12	2	5	2	11
95.72	95.88	1.3054	3	5	3	1	5	3
95.87		1.3039	0	0	4	13		29
96.39	96.56	1.2986	-1	3	2	3		3
96.55		1.2970	-2	12	1	27	10	62
96.60		1.2965	-6	8	1	2		4
96.67		1.2958	-5	9	2	3		7
96.82		1.2943	-7	1	3	1		3
97.48	97.46	1.2876	-3	9	3	2	1	3
97.77	97.78	1.2849	-6	2	0	1	1	4
98.14	98.12	1.2813	6	8	0	2	1	3
98.52	98.76	1.2775	1	9	4	1	3	4
98.75		1.2754	-2	4	1	5		12
99.17	99.34	1.2713	-7	5	1	5	3	11
99.31		1.2700	4	12	0	4		9
99.44		1.2688	-7	3	3	3		6
100.22	100.36	1.2616	-2	10	2	2	1	4
100.37		1.2602	4	2	3	3		7
100.74	100.66	1.2568	0	10	3	1	1	3
101.17	101.18	1.2529	-4	0	4	6	2	15
103.97	103.98	1.2286	2	12	2	5	2	12
104.23	104.22	1.2265	6	0	2	2	2	6
107.30	107.28	1.2018	2	14	1	2	1	5
108.06	108.12	1.1960	-4	12	2	3	1	7
108.44	108.44	1.1932	2	10	0	5	2	14
110.04	110.04	1.1814	-5	11	1	17	4	43
110.44	110.34	1.1786	-8	2	2	2	3	5
110.44		1.1733	4	6	1	1	1	3
111.18	111.14	1.1650	8	15	3	3	1	9
112.38	112.46	1.1632	5	0	0	3	1	8
112.65	112.62	1.1429	6	9	2	2	1	4
115.76	115.76	1.1370	6	6	2	2	1	5
116.72	116.76	1.1365	5	3	3	3	1	4
116.80		1.1323	-4	14	0	1		3
117.49	117.08	1.1269	-8	0	4	1	1	3
118.40	118.42	1.1200	3	1	4	7	2	19
119.59	119.66							

CUMMINGTONITE – GHOSE, 1961 AND FISCHER, 1966

2THETA	PEAK	D	H	K	L	I(INT)	I(PK)	I(DS)
119.88	119.92	1.1184	5	13	0	3	2	8
119.95		1.1180	-6	0	4	2		7
121.28	121.32	1.1107	7	9	0	5	1	13
121.54		1.1092	3	13	2	3		8
122.44	122.76	1.1044	2	16	0	1	1	4
122.79		1.1026	5	5	3	2		7
124.39	124.40	1.0943	4	12	2	11	2	33
126.08	126.14	1.0860	8	6	0	2	1	5
126.08		1.0860	4	14	1	2		5
126.14		1.0857	-6	4	4	1		4
126.49	126.48	1.0841	6	12	0	1	1	4
127.58	127.64	1.0789	7	3	3	2	1	6
127.92	127.90	1.0774	-5	13	2	2	1	7
128.95	128.94	1.0727	-3	13	3	1	1	4
131.39	131.40	1.0621	5	11	2	18	3	57
132.75	132.74	1.0566	2	8	4	3	1	8
132.75		1.0565	-6	10	3	1		4
132.78		1.0564	-6	12	2	2		7
134.47	134.52	1.0498	-7	1	4	1	1	3
136.26	136.36	1.0430	-7	10	3	11	3	36
136.37		1.0427	4	16	2	4		14
136.51		1.0421	0	16	2	3		8
136.82	136.84	1.0410	-3	3	3	2	3	8
136.84		1.0409	-9	3	1	1		3
136.90		1.0407	0	14	3	2		5
138.01	138.20	1.0368	1	17	1	23	6	74
138.21		1.0361	-1	11	4	5		17
138.22		1.0361	7	11	0	21		68
138.37		1.0356	8	8	2	5		8
138.38		1.0356	-9	1	2	3		18
140.16	140.08	1.0296	-7	3	4	3	1	11
140.60		1.0281	-4	2	5	1		4
141.74	141.78	1.0246	8	6	1	14	2	47
142.31	142.40	1.0228	-3	11	4	2	2	5
142.45		1.0224	-9	3	2	3		11
142.56		1.0221	4	4	4	1		4
145.07	145.76	1.0148	-3	5	5	14	5	48
145.09		1.0147	-9	5	1	6		20
145.77	145.77	1.0128	-8	6	3	32		107
148.51	148.56	1.0057	-3	17	1	10	2	35
149.24	149.38	1.0039	4	14	2	2	2	7
149.51		1.0033	-6	8	4	5		16
150.26	150.34	1.0015	-1	15	3	4	1	14
150.80		1.00C3	1	11	4	1		14
151.00		0.9998	-3	13	5	1		3
152.06	152.90	0.9975	-9	1	5	3		9
152.77		0.9960	6	14	0	1		4
152.83		0.9958	6	6	3	41	3	144
153.24		0.9950	-8	10	3	4		14
155.82	156.68	0.9899	2	14	3	1		5
156.26		0.9891	8	0	2	13	3	47
156.62		0.9885	0	12	4	42		147
157.86		0.9863	0	6	5	14		50
158.24		0.9857	4	16	1	1		5
158.70		0.9849	-5	15	2	1		5
158.91		0.9846	5	13	2	4		13
160.37	162.42	0.9824	-1	7	5	2	2	8
160.80		0.9817	-4	16	2	6		21
161.06		0.9814	-3	15	3	1		4
161.25		0.9811	3	9	4	5		17
161.49		0.9808	3	17	1	8		28
162.56		0.9793	-4	6	5	24		88
162.90		0.9789	-3	7	5	10		35
162.94		0.9788	-9	7	1	3		12
163.47		0.9781	-2	18	1	3		12
164.84		0.9765	1	5	5	9		31
164.99		0.9763	1	17	2	2		8

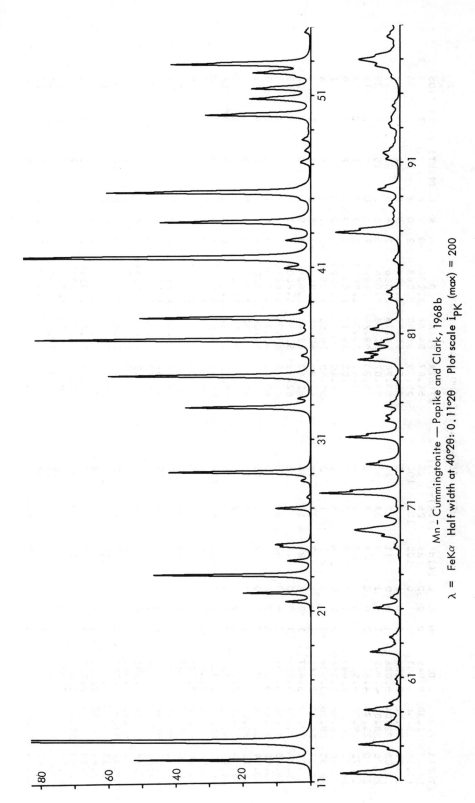

Mn – Cummingtonite — Papike and Clark, 1968b

λ = FeKα   Half width at 40°2θ: 0.11°2θ   Plot scale I_{PK} (max) = 200

375

MN-CUMMINGTONITE - PAPIKE AND CLARK, 1968b

2THETA	PEAK	D	H	K	L	I(INT)	I(PK)	I(DS)
12.29	12.30	9.045	0	2	0	25	26	25
13.38	13.40	8.307	1	1	0	100	100	100
21.51	21.52	5.186	0	0	1	11	10	13
22.02	22.04	5.068	1	3	0	27	23	32
23.06	23.08	4.842	-1	1	1	4	3	5
23.90	23.90	4.676	2	0	0	5	5	6
24.72	24.72	4.523	0	4	0	6	5	7
24.85	24.84	4.499	2	2	0	7	5	8
26.95	26.96	4.154	-2	0	1	2	1	2
28.55	28.56	3.926	-1	3	1	24	21	36
29.04	29.04	3.861	1	3	1	24	19	36
32.81	32.82	3.427	1	5	0	2	2	3
33.34	33.34	3.374	2	4	0	41	30	63
34.65	34.66	3.251	2	0	1	2	1	3
35.82	35.84	3.147	3	1	0	57	41	91
36.74	36.74	3.072	2	2	1	36	25	59
38.01	38.02	2.973	-1	5	1	2	2	3
38.50	38.50	2.936	3	3	1	5	4	9
40.92	40.92	2.769	-1	1	2	74	50	133
41.51	41.52	2.732	3	3	1	3	5	6
42.56	42.56	2.667	-3	5	1	33	22	62
43.28	43.30	2.625	-1	1	2	1	2	2
43.60	43.60	2.607	0	6	1	47	30	90
44.92	44.94	2.534	-2	6	0	1	2	3
45.32	45.32	2.513	-3	5	2	1	1	3
47.08	47.10	2.424	-3	1	1	5	2	4
47.76	47.76	2.391	-2	6	1	1	1	11
48.40	48.40	2.362	-3	5	0	20	16	43
49.82	49.84	2.298	-3	7	1	2		5
49.83		2.298	0	8	0	13	9	28
50.69	50.80	2.261	-4	2	1	14		31
50.79		2.257	-3	1	2	2		5
51.38	51.38	2.233	-1	3	2	6	2	12
52.08	52.10	2.205	1	7	2	8	2	17
52.30	52.30	2.197	-2	4	1	35	6	77
52.31		2.196	1	1	1	3	5	
52.79	52.80	2.177	2	6	3	5	4	21

2THETA	PEAK	D	H	K	L	I(INT)	I(PK)	I(DS)
54.67	54.68	2.108	-3	0	2	1	1	3
55.39	55.40	2.083	2	4	2	15	9	36
55.56	55.50	2.077	-4	0	0	2	6	4
56.78	56.78	2.036	2	8	0	10	2	7
57.07	57.08	2.026	3	5	1	4	6	25
58.23	58.24	1.9895	3	7	0	1	2	10
58.47	58.34	1.9820	-1	5	2	2	2	3
59.02	59.10	1.9652	1	9	2	8	5	5
59.10		1.9629	-4	0	2	3		21
59.50	59.50	1.9508	-3	7	1	8	2	6
60.88	60.88	1.9107	-3	5	0	1	3	3
62.48	62.48	1.8664	-1	9	1	8	5	22
62.71	62.62	1.8603	5	1	0	2	3	5
63.06	63.06	1.8511	-1	7	2	1	3	3
63.33	63.32	1.8440	-4	6	0	2	2	6
65.04	65.04	1.8006	-4	4	3	8	4	21
65.63	65.62	1.7863	5	3	1	3	1	8
68.55	68.56	1.7189	-2	2	3	2	1	5
69.21	69.22	1.7044	0	8	2	5	3	14
69.47	69.58	1.6989	-1	0	3	4	7	12
69.51		1.6981	0	2	2	3		10
69.56		1.6970	-3	7	1	2		5
69.59		1.6964	-5	1	2	7		22
70.32	70.36	1.6809	-2	8	1	2	2	6
70.37		1.6798	-5	6	2	3		9
71.72	71.72	1.6525	4	6	0	25	12	78
73.10	73.12	1.6254	-1	11	0	1	1	5
73.40	73.40	1.6198	1	5	3	11	5	33
74.98	74.98	1.5904	-4	0	4	18	8	57
75.92	75.92	1.5736	2	10	1	3	2	14
76.22	76.20	1.5684	6	0	0	4	2	12
76.79	76.80	1.5585	-1	9	2	5	2	16
78.33	78.34	1.5327	-2	6	3	2	1	6
79.48	79.48	1.5141	1	0	5	13	6	47
79.89	79.90	1.5076	0	12	1	11	5	37
80.39	80.40	1.4998	5	5	1	3	4	12
80.40		1.4998	0	6	3	5		17

MN-CUMMINGTONITE - PAPIKE AND CLARK, 1968b

2THETA	PEAK	D	H	K	L	I(INT)	I(PK)	I(DS)
81.22	81.24	1.4871	-6	0	2	6	4	21
81.28		1.4863	4	4	2	6		17
81.54	81.44	1.4823	-4	8	2	5	3	7
81.58		1.4818	3	7	3	2		7
83.01	83.02	1.4607	-1	1	0	1	2	4
83.44	83.44	1.4546	3	11	1	4	2	14
85.27	85.28	1.4291	-4	10	0	5	1	17
86.90	86.90	1.4075	-6	6	1	27	10	103
87.66	87.66	1.3978	-1	11	2	2		7
88.02	88.00	1.3932	6	0	1	2	1	6
88.47	88.48	1.3876	-4	6	3	2	2	15
89.38	89.38	1.3764	5	1	2	4	3	38
91.16	91.16	1.3553	2	6	3	10	2	24
91.53	91.52	1.3510	1	6	1	6	3	25
92.00	91.98	1.3456	5	3	2	6	1	7
92.40	92.38	1.3411	-1	13	1	2	1	6
93.01	93.04	1.3344	-5	5	1	1	1	13
93.20	93.22	1.3322	-7	11	3	3	2	12
93.61	93.60	1.3278	-3	5	0	3	1	9
94.26	94.26	1.3208	-1	11	2	2	1	14
95.92	95.94	1.3033	-1	1	4	3	1	12
96.59	96.60	1.2966	0	12	2	8	4	34
96.61		1.2964	0	0	4	1		5
96.73	96.96	1.2951	3	5	3	1		6
96.97		1.2928	-6	2	2	17	6	73
97.93	97.92	1.2833	-2	12	0	1	1	5
98.27	98.30	1.2800	6	8	2	2	2	8
98.32		1.2795	-7	0	0	2		18
99.02	99.02	1.2728	-2	5	1	4	1	14
99.38	99.32	1.2694	1	4	4	3	1	6
99.64	99.64	1.2670	4	9	3	2	1	11
100.79	100.80	1.2564	-4	12	0	4	1	20
101.28	101.04	1.2520	4	0	4	2	1	9
101.51	101.50	1.2499	0	2	3	1	1	5
104.65	104.86	1.2230	6	2	2	3	1	9
104.86		1.2212	2	12	3	4		7
108.11	108.14	1.1956	-4	12	2	1	1	13
109.77	109.76	1.1845	2	0	4	4	3	20
111.81	111.82	1.1689	-5	11	2	12	1	59
113.35	113.34	1.1585	8	0	0	2	1	13
118.73	118.72	1.1250	1	15	1	2	0	8
120.23	120.26	1.1165	-6	0	4	2	1	8
120.92	121.04	1.1126	5	13	0	2	0	15
121.07		1.1118	7	9	4	3	1	24
122.73	122.72	1.1029	3	1	4	4	1	7
124.01	124.02	1.0962	3	13	3	4	0	12
125.54	125.54	1.0886	5	5	3	2	1	42
127.04	127.04	1.0814	4	12	2	7	1	7
127.82		1.0778	4	14	1	1	0	8
128.09	128.10	1.0766	-5	13	2	2	1	10
129.46	129.46	1.0704	7	7	3	3		9
132.22		1.0587	-3	13	3	2		10
132.55	132.56	1.0573	-6	12	2	13		80
134.22	134.24	1.0507	5	11	2	7		42
134.59		1.0493	-7	8	4	2		11
135.74	135.72	1.0450	2	8	2	4		23
137.56		1.0384	-9	1	0	1		9
137.98	138.02	1.0369	8	8	0	16	3	99
138.01		1.0368	7	11	0	8		8
138.12		1.0364	-7	7	4	1		8
138.18		1.0362	0	16	2	1		15
139.59	139.70	1.0315	4	10	3	5	3	32
139.64		1.0313	-1	11	4	2		10
139.71		1.0311	1	17	1	17		105
141.50	141.52	1.0253	8	6	1	10	2	64
142.81	143.04	1.0213	-9	5	1	4	3	25
142.90		1.0210	-3	11	4	1		8
143.05		1.0206	-8	6	3	20		128
145.50	145.52	1.0148	4	4	4	1	2	7
145.52		1.0136	5	15	0	4		23
147.90	147.94	1.0072	-3	5	5	10	1	66
149.46	150.08	1.0034	3	15	2	4	1	24

MN-CUMMINGTONITE - PAPIKE AND CLARK, 1968b

2THETA	PEAK	D	H	K	L	I(INT)	I(PK)	I(DS)
150.07		1.0020	-3	17	1	9		61
152.38	152.58	0.9968	-1	15	3	2	1	16
155.34	155.42	0.9908	6	6	3	30	2	209
157.02		0.9878	8	0	2	8		59
159.08	161.24	0.9843	-9	7	1	1	2	9
159.08		0.9843	4	6	4	1		6
159.46		0.9838	2	14	3	2		11
159.92		0.9830	0	12	4	27		194
161.01		0.9814	-8	0	4	9		61
161.08		0.9813	4	16	1	3		25
161.41		0.9809	0	6	5	12		83
162.01		0.9800	5	13	2	2		17
162.22		0.9798	-4	16	2	3		24
162.32		0.9796	-4	6	5	20		143
162.97		0.9788	-3	15	3	1		5
163.58		0.9780	-1	7	5	1		9
163.64		0.9779	-8	8	3	1		8
164.09		0.9774	-3	7	5	5		39

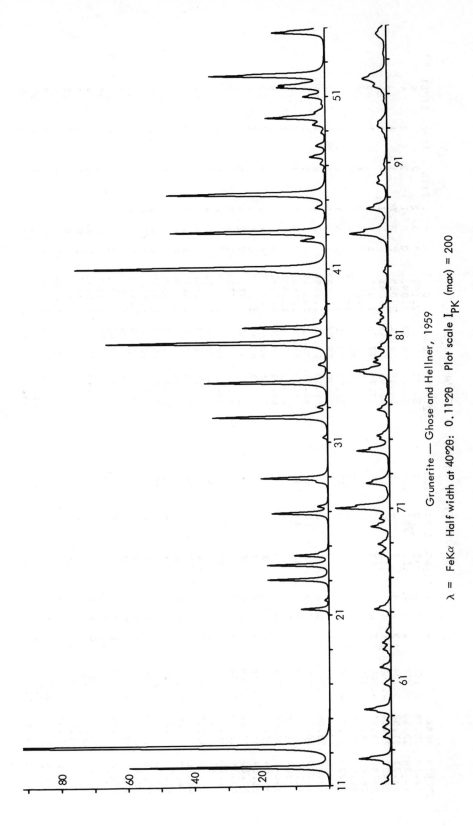

Grunerite — Ghose and Hellner, 1959

λ = FeKα Half width at 40°2θ: 0.11°2θ Plot scale I_{PK} (max) = 200

379

GRUNERITE - GHOSE AND HELLNER, 1959

2THETA	PEAK	D	H	K	L	I(INT)	I(PK)	I(DS)
12.14	12.16	9.151	0	2	0	30	30	29
13.34	13.34	8.334	1	1	0	100	100	100
21.31	21.32	5.234	0	1	1		4	5
23.03	23.04	4.849	-1	0	1	10	9	12
23.87	23.88	4.680	0	4	0	10	9	13
24.43	24.44	4.575	0	2	1	6	5	7
24.60	24.60	4.544	2	0	0		1	1
26.86	26.88	4.167	2	2	0	10	8	13
27.26	27.26	4.108	-1	1	1	2	2	2
28.67	28.68	3.910	-2	0	1	2	2	2
28.89	28.90	3.880	-1	3	1	12	10	17
32.42	32.42	3.468	1	5	1	23	17	33
32.99	33.00	3.409	2	4	0	2	2	3
34.42	34.42	3.272	2	0	1	25	18	38
35.44	35.44	3.180	3	1	0		1	3
36.69	36.70	3.076	3	3	0	47	33	76
37.60	37.60	3.004	-3	2	1	18	13	29
40.78	40.80	2.778	1	7	1	7	8	13
41.00	41.00	2.764	-3	1	1	56	38	98
42.62	42.62	2.663	-1	5	1	5	4	9
43.05	43.10	2.638	-3	3	1	4		7
43.10		2.635	-1	6	0	32	23	59
44.52	44.52	2.556	2	6	0	2	2	4
45.25	45.28	2.516	0	0	2	1		2
45.28		2.514	-2	2	2	36	24	69
47.06	47.06	2.425	-2	6	1	1	1	2
47.47	47.48	2.405	-2	5	1	4	2	8
48.11	48.12	2.375	3	5	0	2	1	4
49.33	49.32	2.320	-1	3	2	3	2	5
49.74	49.74	2.302	-3	5	1	15	9	31
50.06	50.06	2.288	0	8	1	1	2	3
50.97	50.96	2.250	-4	2	1	5	3	11
51.50	51.52	2.228	2	2	2	1	1	3
51.66	51.62	2.222	-3	7	1	6	7	24
52.11	52.18	2.204	-2	4	2	2		12
52.17		2.201	2	6	1	28	17	61
54.68	54.68	2.108	2	0	2	12	8	28
54.71	54.98	2.107	-3	3	2	2		5
55.00	55.22	2.096	0	8	1	2	2	4
56.19	56.22	2.055	2	8	0	1	1	3
56.48	56.48	2.046	3	5	2	9	5	21
57.74	57.76	2.005	1	7	1	1	1	3
57.77		2.004	3	1	1	1		3
58.30	58.32	1.9872	3	9	0	1	1	3
58.71	58.72	1.9746	-2	8	1	1	1	3
59.20	59.22	1.9597	-3	7	1	3	2	7
59.36	59.36	1.9550	-4	0	2	6	4	15
60.83	60.82	1.9119	-4	2	1	3	1	3
61.79	61.80	1.8853	0	9	1	1	1	8
62.83	62.84	1.8571	-2	8	1	2	1	6
63.21	63.20	1.8471	-4	6	1	5	3	13
65.15	65.16	1.7978	-4	4	2	3	3	8
68.39	68.38	1.7225	0	8	2	1	1	3
68.77	68.92	1.7139	2	2	3	3	2	9
68.92		1.7108	-1	13	1	5	3	15
69.94	69.94	1.6889	-5	1	2	5	3	8
70.40	70.40	1.6793	-5	5	1	3	2	54
71.02	71.02	1.6665	-5	4	1	18	8	23
72.44	72.44	1.6381	1	6	0	7	4	3
72.56		1.6359	4	11	0	1		35
74.32	74.32	1.6025	-1	8	3	11	5	9
75.00	75.02	1.5900	4	5	0	3	1	8
75.21	75.20	1.5862	-4	0	1	2	2	12
76.70	76.70	1.5601	2	10	0	4	1	4
77.26	77.26	1.5505	6	0	0	1	2	23
78.79	78.82	1.5252	1	12	2	7	4	27
78.95	78.96	1.5225	-2	6	3	8	5	14
79.45	79.46	1.5145	0	6	1	4	2	8
79.71	79.68	1.5105	5	5	1	3	2	10
80.26	80.26	1.5019	-5	4	2	4	2	14
81.68	81.68	1.4802	-6	0	2	2	1	9
82.19	82.18	1.4727	-1	17	1	2	1	10
82.50	82.48	1.4681	3	11	0	3	1	6
84.68	84.68	1.4372	-4	10	1	2	1	

GRUNERITE - GHOSE AND HELLNER, 1959

2THETA	PEAK	D	H	K	L	I(INT)	I(PK)	I(DS)	2THETA	PEAK	D	H	K	L	I(INT)	I(PK)	I(DS)
86.88	86.88	1.4078	-6	6	1	16	6	60	126.65	126.62	1.0833	-5	13	2	1	0	4
88.32	88.32	1.3894	5	1	1	3	3	21	129.95	129.96	1.0683	5	11	2	5	1	32
88.34		1.3893	-4	6	3	5		13	131.37	131.48	1.0622	-6	12	2	1	1	5
89.81	89.82	1.3712	2	6	3	4	2	17	131.56		1.0614	2	8	4	1		5
90.19	90.18	1.3667	1	11	2	3	1	13	134.87	135.18	1.0482	4	10	3	1		8
90.89	90.88	1.3585	5	11	1	1	0	4	134.90		1.0481	0	16	2	1		7
92.80	92.84	1.3366	-3	3	2	1	1	5	135.13		1.0472	-7	1	4	3		18
93.07	93.12	1.3337	-7	1	0	1	1	6	135.85	136.44	1.0446	-3	3	5	1		4
93.14		1.3329	1	5	3	2		7	136.25		1.0431	1	17	1	6		39
93.36	93.34	1.3305	-5	5	0	3	1		136.46		1.0423	7	11	1	1		40
95.41	95.42	1.3086	-1	1	4	2	2	6	136.62		1.0417	8	8	0	1		3
95.86	95.86	1.3040	0	0	2	5	4	22	136.82		1.0410	-9	1	2	2		13
98.28	98.48	1.2799	-2	12	2	10	1	42	136.88		1.0408	-1	11	4	2		12
98.42		1.2785	-7	5	1	2		8	139.82	139.84	1.0307	8	6	1	5		30
98.50		1.2778	4	12	0	2		9	140.71	140.50	1.0278	-9	3	3	1		7
99.58	99.66	1.2675	-2	10	3	2	0	10	140.86		1.0273	-3	11	2	1		4
99.69		1.2665	-4	2	3	1		3	140.99		1.0269	4	4	4	1	1	4
100.69	100.70	1.2572	-4	0	2	2	1	4	143.80	143.98	1.0184	-3	5	5	9	1	29
103.15	103.18	1.2356	2	12	2	2	1	11	144.03		1.0177	-8	8	3	9		57
103.40	103.40	1.2335	6	0	2	2	1	10	146.17	146.16	1.0118	-3	17	1	3	0	22
107.21	107.22	1.2025	-4	12	2	1	0	6	147.77	147.84	1.0076	-6	8	4	1	0	10
107.78	107.78	1.1982	-2	4	4	2	1	6	148.07		1.0068	-1	15	3	1		4
109.17	109.18	1.1878	-5	11	1	6	1	10	148.81	148.86	1.0050	2	11	4	2	0	14
109.51	109.46	1.1852	-8	2	2	1	1	28	149.71	150.40	1.0028	-9	5	2	1	1	72
111.36	111.64	1.1720	1	15	1	1	1	3	150.37		1.0013	-6	6	3	11		7
111.47		1.1701	7	0	0	1		4	150.59	150.59	1.0008	-8	10	1	1		28
111.64		1.1703	8	0	0	1		3	153.29	154.20	0.9949	-8	0	2	4	1	61
115.67	115.74	1.1435	6	6	2	4	0	6	154.16		0.9931	0	12	4	9		4
118.70	118.74	1.1252	5	13	0	9	1	3	155.59		0.9904	5	13	2	1		33
118.73		1.1250	3	1	4	2		4	155.69		0.9902	-9	0	6	5		8
119.20	119.14	1.1223	-6	0	4	1	1	13	159.02	159.98	0.9844	-9	7	1	1	1	11
120.07	120.10	1.1173	7	9	0	1	0	7	159.27		0.9841	-2	18	2	2		44
120.38	120.38	1.1156	3	13	2	1	0	5	160.02		0.9829	-4	6	5	6		18
121.70	121.68	1.1084	5	5	3	4	1	22	160.27		0.9825	-3	7	5	3		4
123.15	123.16	1.1007	4	12	2	1	0	3	160.59		0.9820	1	17	3	1		21
125.14	125.16	1.0906	6	12	2	1			161.74		0.9804	1	5	5	3		23
126.26	126.28	1.0851	7	3	2	1	0	3	163.98		0.9775	-8	0	4	3		

TABLE 18. AMPHIBOLES

Variety	K-Richterite	Glaucophane	Riebeckite	Tirodite	Protoamphibole
Composition	$(K,Na,Ca)_3(Mg,Fe)_5$ $Si_8O_{22}(OH)_2$	$(Na,Ca)_2Mg,Fe^2,Al,$ $Fe^3)_5Si_8O_{22}(OH)_2$	$(Na,Ca,K)_2(Fe^2,$ $Fe^3,Mn,Al,Li)_5$ $Si_8O_{22}(OH)_2$	$Mn_2Mg_5Si_8O_{22}(OH)_2$	$(Mg,Li)_2Si_8O_{22}$ F_2
Source	West Kimberley, West Australia; Papike, Clark & Hulbner, 1968	West Tiburon, Calif.	Hurricane Mt., New Hampshire	Talcville, N.Y.	Synthetic; Bloss, Shell & Gibbs, 1960
Reference	Papike & Clark, 1968b Papike et al., 1968	Papike & Clark, 1968a	Colville & Gibbs, 1965;1968	Papike & Clark, 1968b	Gibbs, 1964; 1968
Cell Dimensions					
a	10.021	9.541	9.812	9.550	5.288 (9.330)
b Å	18.306	17.740	18.038	18.007	9.330 (17.879)
c	5.289	5.295	5.329	5.298	17.879 (5.288)
α	90	90	90	90	90
β deg	104.98	103.66	103.69	102.65	90
γ	90	90	90	90	90
Space Group	C2/m	C2/m	C2/m	$P2_1/m$	Pnnm (Pnmn)
Z	2	2	2	2	2

TABLE 18. (cont.)

Variety	K-Richerite	Glaucophane	Riebeckite	Tirodite	Protoamphibole
Site Occupancy M_1	{Mg 0.90 / Fe 0.10 (spanning M_1–M_3)	{Mg$_2$ 0.84 / Fe$_2$ 0.16	Fe^2 1.0	Mg 1.0	Mg 1.0
M_2		{Al 0.91 / Fe^3 0.09	Fe^3 1.0	Mg 1.0	Mg 1.0
M_3		{Mg 0.71 / Fe^2 0.29	{Fe^2 0.50 / Li 0.25 / Al 0.25	Mg 1.0	Mg 1.0
M_4	{Ca 0.50 / Na 0.50	{Na 0.98 / Ca 0.02	Na 1.0	{Mn 0.575 / Mg 0.200 / Ca 0.185 / Fe 0.040	{Mg 0.75 / Li 0.25
A	K 1.0				
Method	3–D, counter	3–D, counter	3–D, counter	3–D, counter	3–D, counter
R & Rating	0.076 (1)	0.080 (1)	0.107 (2)	0.072 (1)	0.064 (1)
Cleavage and habit	{110} perfect — Prismatic (spanning K-Richerite, Glaucophane)				{011} Elongated to $\underline{c}$ axis
Comment			A site may be partially filled by K		
μ	146.1	145.1	206.5	185.6	92.1
ASF	0.4910×10^{-1}	0.7868×10^{-1}	0.1567	0.1418×10^{-1}	0.5176×10^{-1}
Abridging factors	1:1:0.5	1:1:0.5	0.5:0:0	1:1:0.5	1:1:0.5

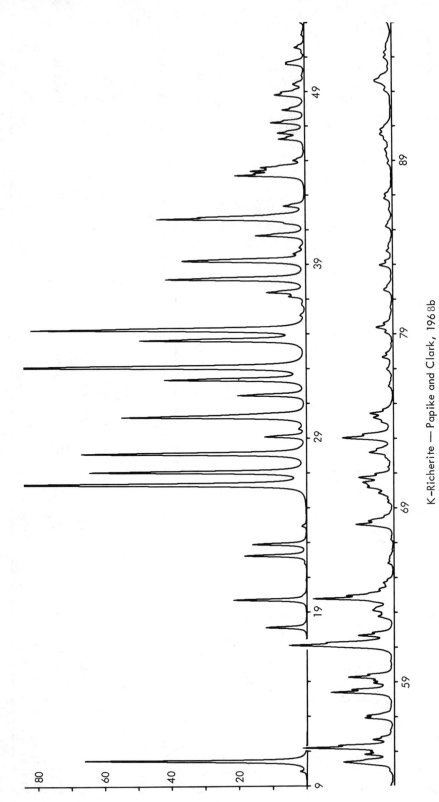

K-Richerite — Papike and Clark, 1968b

λ = CuKα Half width at 40°2θ: 0.11°2θ Plot scale I$_{PK}$ (max) = 100

384

K-RICHTERITE - PAPIKE AND CLARK, 1968b

2THETA	PEAK	D	H	K	L	I(INT)	I(PK)	I(DS)
9.80	9.80	9.018	0	1	0	1	2	1
10.36	10.36	8.530	-1	1	0	48	66	39
18.08	18.08	4.903	-1	1	1	10	12	8
19.67	19.68	4.509	0	2	0	18	22	15
22.21	22.22	4.000	-1	1	1	16	19	14
22.86	22.86	3.887	1	3	1	14	16	12
23.92	23.92	3.717	-2	1	2	1	1	1
26.28	26.28	3.388	1	3	1	84	95	78
26.34		3.381	0	4	0	3		3
27.00	27.00	3.299	2	1	0	59	65	55
28.07	28.08	3.176	3	1	1	63	67	60
29.06	29.06	3.071	3	1	1	11	12	11
29.50	29.50	3.025	-3	1	2	2	2	1
30.17	30.18	2.959	-2	2	2	52	55	50
31.44	31.44	2.843	3	1	1	19	20	19
32.34	32.34	2.767	-3	1	1	41	42	41
33.04	33.04	2.709	-1	1	1	100	100	100
34.59	34.60	2.591	0	0	2	50	50	51
35.10	35.20	2.555	0	0	2	7		7
35.11		2.554	-2	2	2		82	
35.20		2.547	-2	0	2	79		81
36.04	36.04	2.490	-1	1	0	1	2	1
37.12	37.12	2.420	-4	0	0	2	5	2
37.12		2.420	-2	6	1	2		3
37.36	37.36	2.405	3	2	1	12	12	12
38.10	38.12	2.360	-4	2	1	21	42	22
38.13		2.358	-3	5	1	27		29
39.14	39.18	2.299	-1	7	2	24	37	26
39.20		2.296	-3	1	1	22		24
39.60	39.60	2.274	3	1	1	3	4	3
39.95	39.96	2.254	0	8	0	1	2	1
40.55	40.64	2.223	-2	4	2	15	15	17
40.64		2.218	-1	7	1	4		5
41.34	41.36	2.182	2	1	1	48	44	53
41.60	41.60	2.169	2	6	1	10	32	11
41.77	41.70	2.161	-3	3	2	6	7	7
42.34	42.34	2.133	-1	5	2			

2THETA	PEAK	D	H	K	L	I(INT)	I(PK)	I(DS)
44.12	44.12	2.051	2	2	2	23	21	26
44.28	44.22	2.044	0	8	1	1	15	1
44.37	44.58	2.040	-4	0	2	15	16	17
44.59		2.030	3	5	1	12	13	14
44.98	44.98	2.013	3	7	0	3	4	4
46.22	46.22	1.9624	-1	9	1	9	8	10
46.55	46.56	1.9494	4	2	1	2	8	2
46.57		1.9484	-3	5	2	7		8
47.17	47.18	1.9250	5	1	0	12	10	14
47.91	47.92	1.8969	-4	6	1	8	7	9
48.73	48.78	1.8669	2	2	2	4	9	4
48.79		1.8649	4	4	1	8		9
48.97	48.92	1.8584	-4	4	2	4	7	5
49.34	49.40	1.8454	-1	7	1	2	3	2
49.41		1.8429	-1	3	0	3		3
50.56	50.58	1.8036	0	6	1	6	5	7
50.65	50.66	1.8006	0	10	0	4	6	4
51.54	51.54	1.7718	-1	9	1	4	3	4
52.99	53.00	1.7266	-5	1	2	1	1	2
54.29	54.36	1.6883	-2	7	2	6	15	8
54.36		1.6861	-1	8	3	10		13
54.39	54.82	1.6855	-3	3	1	4	9	5
54.81	55.18	1.6735	0	2	3	10	27	13
55.18	55.68	1.6632	4	6	0	35	7	46
55.67	56.92	1.6496	4	8	1	11	9	10
56.91	57.04	1.6166	-6	11	0	4	8	14
57.03	58.38	1.6134	1	0	0	4	8	6
58.38	58.54	1.5793	-1	5	3	25	19	34
58.58	58.92	1.5744	-5	7	1	2	13	3
58.91	59.06	1.5664	4	0	2	6	6	9
59.05	59.26	1.5631	2	10	1	2	6	3
59.26	61.10	1.5580	-6	0	1	17	14	23
60.00		1.5183	9	2	1	6	31	8
61.09	61.24	1.5176	5	5	1	7		10
61.11		1.5155	-2	5	3	34		47
61.23		1.5152	-3	9	2	2		2
		1.5126	-4	8	2	4	21	6

K-RICHTERITE — PAPIKE AND CLARK, 1968b

2THETA	PEAK	D	H	K	L	I(INT)	I(PK)	I(DS)
61.66	61.66	1.5030	0	0	0	14	11	19
62.64	62.74	1.4818	0	6	3	2	5	2
62.74		1.4796	4	2	2	5		7
63.08	63.10	1.4726	-6	4	2	3		4
63.10		1.4720	-2	10	2	3		5
63.18		1.4705	2	2	3	2		7
63.78	63.78	1.4580	-6	0	1	32	24	46
64.36	64.30	1.4462	4	10	0	3	5	4
64.69	64.68	1.4397	-5	3	3	3	3	5
65.81	65.82	1.4178	6	2	1	3	3	3
67.16	67.16	1.3925	-7	3	1	1	1	2
67.92	68.04	1.3789	-7	1	0	3	11	4
68.04		1.3767	5	1	2	16		24
68.33	68.24	1.3715	-5	5	0	6	8	9
69.32	69.32	1.3544	2	10	2	3		3
69.71	69.84	1.3477	7	3	0	1	5	2
69.83		1.3457	5	3	2	5		7
69.90	70.16	1.3446	-6	1	2	8	8	2
70.16		1.3402	-11	2	2	8		13
70.46	70.46	1.3353	2	6	3	8	9	13
70.72	70.74	1.3310	-5	9	2	3	11	4
70.75		1.3305	-7	5	1	11		17
71.34	71.34	1.3210	-1	3	1	2	2	3
71.90	71.92	1.3120	-6	8	0	2	3	2
72.18	72.18	1.3076	-1	1	4	10	7	16
72.31		1.3056	-3	1	3	3		4
72.96	73.02	1.2956	-3	9	3	3		2
72.96		1.2955	6	6	1	3	15	4
72.97		1.2954	0	12	2	5		2
73.05	73.22	1.2945	-2	12	2	18		30
73.36		1.2895	-5	5	4	3	9	2
73.54	73.94	1.2895	5	5	1	3	4	5
73.66		1.2809	-1	3	4	1		3
73.93	74.18	1.2809	-3	3	1	1	6	2
74.05		1.2791	-3	4	1	4		7
74.17		1.2773	0	0	4	4		3
74.21		1.2768	4	12	0	2		2
74.42	74.42	1.2737	-4	0	4	7	7	12
75.98	75.98	1.2513	-7	7	1	2	2	3

2THETA	PEAK	D	H	K	L	I(INT)	I(PK)	I(DS)
76.93	76.96	1.2383	0	10	3	1	2	3
77.00		1.2373	-1	13	2	1		2
77.65	77.66	1.2286	-6	0	0	2	3	3
77.67		1.2283	8	0	0	3		3
79.07	79.08	1.2101	-5	11	2	7	2	5
79.38	79.38	1.2061	2	2	2	7	5	13
81.50	81.52	1.1800	-8	4	2	2	3	6
81.58		1.1790	-6	10	2	7		4
82.94	82.94	1.1631	2	0	4	2	4	13
83.76	83.78	1.1539	1	15	1	1	2	2
83.79		1.1535	4	6	3	2		3
84.04	84.04	1.1489	2	10	0	3		2
85.18	85.18	1.1382	7	1	3	1	1	6
85.29		1.1370	9	9	0	1		2
86.16	86.16	1.1277	5	13	0	3		3
86.78	86.80	1.1213	-6	4	1	1		5
86.82		1.1208	-5	3	1	3	2	3
87.01	87.02	1.1189	-7	7	3	3	2	4
88.64	88.82	1.1008	0	16	1	1	1	4
89.62	89.62	1.0927	0	3	4	3	2	5
90.16	90.26	1.0877	5	5	3	2	2	4
90.23		1.0871	-7	1	4	4		3
90.29		1.0865	2	16	1	1		2
90.50	90.50	1.0846	-7	3	2	1		3
90.51		1.0845	4	12	2	2		6
90.72	90.78	1.0825	4	14	1	1	4	5
90.81		1.0817	-6	12	2	2		3
90.98		1.0801	-9	3	2	1		3
93.05	93.06	1.0615	-8	6	3	4		7
93.14		1.0607	2	16	3	1		2
93.54	93.62	1.0571	7	11	0	6	5	13
93.66		1.0562	5	11	2	8		15
94.45	94.44	1.0494	-2	2	5	2	1	3
94.95	94.96	1.0451	-8	6	1	5		10
95.23	95.26	1.0428	-5	9	4	1	2	2
95.75	95.74	1.0385	-2	14	3	1	2	10
97.16	97.16	1.0271	1	17	1	5	3	10

K-RICHTERITE – PAPIKE AND CLARK, 1968b

2THETA	PEAK	D	H	K	L	I(INT)	I(PK)	I(DS)
97.57	97.56	1.0240	-1	11	4	5	4	11
98.09	98.08	1.0199	-8	0	3	4	3	9
98.68	98.68	1.0154	0	2	5	1	1	2
98.94	98.96	1.0134	-3	5	5	2	2	4
100.16	100.16	1.0043	-3	17	1	2	1	3
100.99	101.02	0.9983	8	0	2	2	2	4
101.01		0.9981	-9	5	5	2		8
101.20		0.9968	-2	6	3	4		1
101.54	101.34	0.9944	-10	0	2	1	2	3
101.77	101.80	0.9927	-2	12	4	1	1	1
101.79		0.9926	-9	7	0	1		2
102.68	102.82	0.9864	6	6	3	8	5	17
102.84		0.9853	-4	6	5	10		21
103.72	103.72	0.9793	-5	11	4	2	2	4
104.00		0.9775	2	10	2	1		1
104.42	104.62	0.9747	8	4	4	1	2	2
104.42		0.9747	2	14	3	1		3
104.63		0.9733	0	12	4	3		7
104.87		0.9717	-4	12	0	1		2
105.44	105.50	0.9680	10	0	5	1	1	2
105.53		0.9675	0	6	3	3		6
106.18	106.20	0.9633	-9	7	1	1	1	1
106.24		0.9629	-9	9	2	1		2
106.59	106.58	0.9607	9	9	4	1		3
107.99	108.06	0.9521	-3	13	4	1	1	1
108.60	108.62	0.9485	-10	6	1	3	2	7
108.61		0.9484	-2	16	2	2		5
109.00	108.98	0.9461	8	12	5	1	1	1
109.61	109.58	0.9425	8	2	3	1	1	3
110.05	110.02	0.9400	0	16	1	1	1	1
110.83	111.16	0.9356	-7	3	5	2	1	3
111.14		0.9338	-6	16	1	2		4
111.98	111.96	0.9292	-4	8	4	2	1	4
113.37	113.38	0.9217	8	16	3	1	1	2
113.73	113.74	0.9198	-2	12	4	2	1	4
114.45	114.46	0.9161	-7	5	5	5	1	11
115.52	115.58	0.9107	3	19	0	1	3	1

2THETA	PEAK	D	H	K	L	I(INT)	I(PK)	I(DS)
115.56	115.56	0.9105	3	11	4	7	3	18
116.02	116.02	0.9082	-1	17	3	3		8
116.18		0.9074	-7	15	1	1		2
116.22		0.9072	-7	11	4	1		3
117.33	117.30	0.9018	0	20	0	1	1	2
117.90	117.96	0.8991	2	16	3	2	1	5
117.98		0.8987	3	3	4	2		4
118.07		0.8983	5	7	4	1		2
118.57	118.62	0.8959	5	17	1	1	2	7
118.65		0.8956	4	18	1	1		2
118.73		0.8952	6	0	4	1		2
119.98	120.10	0.8895	10	8	0	1	1	1
120.07		0.8891	10	10	5	1		2
120.37	120.52	0.8878	-7	7	5	1	1	3
121.84	122.02	0.8814	-8	10	5	5	2	13
122.05		0.8804	-3	5	6	4		10
122.39	122.48	0.8790	11	1	0	2	2	4
122.56		0.8783	-3	19	2	1		1
122.56		0.8783	-2	0	6	3		8
122.77		0.8774	-4	0	6	1		2
123.74	123.72	0.8734	8	10	2	3	1	8
124.55	124.54	0.8701	-2	18	3	1	1	3
125.13	125.10	0.8678	-6	16	5	3	1	6
125.59	125.60	0.8661	-8	6	3	1	1	3
126.43	126.40	0.8628	-8	6	5	2	1	4
127.34	127.48	0.8594	-5	17	3	2		6
127.50		0.8588	3	13	4	1		2
127.96	128.04	0.8571	-6	18	1	1	1	5
128.18		0.8563	3	7	5	1		2
129.12	129.30	0.8530	-7	9	5	1	1	2
129.29		0.8524	10	10	0	4		10
129.52		0.8515	-7	17	1	1		3

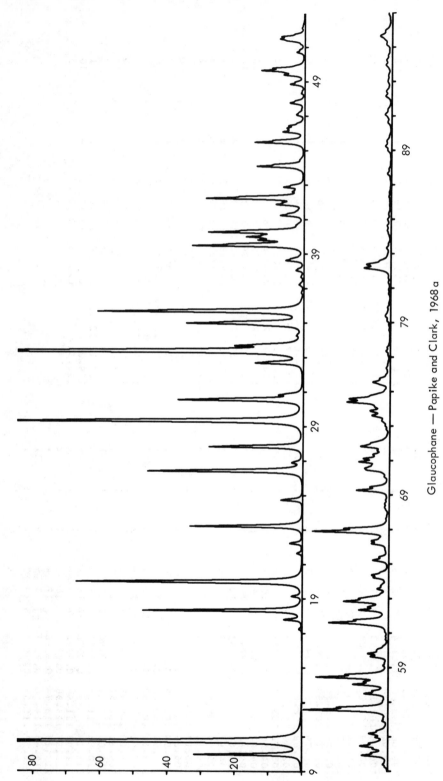

Glaucophane — Papike and Clark, 1968 a

$\lambda = CuK\alpha$ Half width at $40^\circ2\theta$: $0.11^\circ2\theta$ Plot scale I_{PK} (max) = 200

GLAUCOPHANE - PAPIKE AND CLARK, 1968*

2THETA	PEAK	D	H	K	L	I(INT)	I(PK)	I(DS)
9.96	9.98	8.870	0	2	0	100	100	16
10.76	10.76	8.217	1	1	0	100	100	16
17.78	17.78	4.986	1	3	0	3	3	3
18.31	18.32	4.840	-1	1	1	26	24	28
19.13	19.14	4.636	2	0	0	2	2	2
19.93	20.00	4.451	0	4	0	11	34	12
20.00		4.435	0	2	1	33		35
22.20	22.20	4.000	1	3	1	2	2	2
23.19	23.20	3.832	-1	3	1	20	17	22
24.72	24.72	3.599	-2	2	1	4	3	4
26.40	26.40	3.373	1	1	1	30	23	34
26.88	26.88	3.314	2	4	0	2	2	2
27.82	27.82	3.205	3	1	0	18	14	21
29.31	29.32	3.045	3	1	0	63	47	74
30.53	30.54	2.926	2	2	1	25	19	30
30.81	30.80	2.900	-1	1	1	4		4
32.67	32.68	2.739	3	3	0	10	7	12
33.35	33.34	2.685	1	5	1	86	60	106
33.67	33.66	2.660	-3	3	1	11	10	14
34.97	34.98	2.564	0	6	0	23	17	29
35.66	35.66	2.516	-2	0	2	44	31	56
38.02	38.02	2.364	-2	6	1	2		4
38.60	38.60	2.330	3	5	1	3	3	4
39.47	39.48	2.281	-3	5	1	25	17	33
39.79	39.80	2.263	-1	7	1	8	7	11
39.98	39.98	2.253	-4	2	1	10	10	14
40.25	40.26	2.238	-3	4	2	21	14	28
41.22	41.22	2.188	-2	7	1	5	4	7
41.85	41.86	2.157	-1	4	2	6	4	8
42.21	42.20	2.139	-2	6	1	23	15	32
42.60	42.60	2.121	-1	5	2	4	3	6
42.86	42.86	2.108	-3	3	2	11	7	16
44.07	44.08	2.053	2	0	2	12	7	17
45.46	45.46	1.9935	-4	5	1	5	5	7
46.07	46.06	1.9686	3	0	0	3	3	5
46.29	46.28	1.9596	3	7	0	2	2	3
46.44	46.42	1.9538	1	5	2	2	2	3
47.04	47.06	1.9300	-3	7	1	1	1	2
47.72	47.76	1.9040	-3	1	2	1	2	2
47.76		1.9027	4	2	1	2		4
48.84	48.84	1.8631	2	4	2	4	2	5
49.37	49.38	1.8442	5	1	0	5	3	7
49.63	49.64	1.8353	-1	9	1	10	6	15
49.79	49.76	1.8297	-4	6	1	1	5	2
50.69	50.70	1.7993	4	4	1	2	1	3
51.37	51.48	1.7770	-1	9	1	3	4	5
51.47		1.7740	1	9	0	4		6
51.61	51.60	1.7693	5	5	1	3	4	5
53.89	53.90	1.6998	-5	1	2	7	4	11
54.26	54.26	1.6891	5	3	1	3	4	5
54.44	54.44	1.6839	-0	2	2	5	2	8
54.93	54.94	1.6700	-5	5	1	1	4	2
54.95		1.6696	-3	1	3	2	2	3
55.17	55.16	1.6635	-2	8	2	4	3	7
55.89	55.90	1.6436	-3	9	1	1	1	2
56.52	56.52	1.6267	4	6	1	23	13	38
57.46	57.46	1.6023	4	8	1	6	4	10
58.00	58.00	1.5889	-1	11	1	9	5	15
58.41	58.42	1.5785	-4	5	3	20	11	32
59.61	59.62	1.5496	4	0	2	3	3	5
59.80	59.78	1.5452	6	0	1	3	3	8
60.04	59.96	1.5396	2	10	1	1	2	2
61.44	61.56	1.5077	-1	9	3	3	2	5
61.57		1.5050	-2	6	3	16	9	27
62.31	62.30	1.4889	-6	0	2	8	4	14
62.56	62.48	1.4835	0	6	3	1	3	2
62.80	62.80	1.4783	0	12	1	8	7	14
62.80		1.4783	5	5	1	4		6
62.88	62.98	1.4766	2	2	3	2		3
63.09	62.98	1.4722	-4	8	2	2	5	3
63.54	63.54	1.4629	4	4	2	2	2	4
64.19	64.20	1.4498	-2	10	2	4	2	7
64.32	64.34	1.4470	-1	7	3	3	3	5
65.19	65.20	1.4298	3	11	0	5	2	9

GLAUCOPHANE – PAPIKE AND CLARK, 1968a

2THETA	PEAK	D	H	K	L	I(INT)	I(PK)	I(DS)
66.15	66.16	1.4115	-6	4	2	3	3	6
66.16		1.4113	-4	10	1	1		2
66.29	66.30	1.4087	4	10	0	2		4
66.81	66.86	1.3991	-5	3	3	2	3	3
66.86		1.3981	-6	6	1	22	11	41
69.25	69.24	1.3556	5	1	2	11		20
70.03	70.04	1.3423	2	10	2	1	5	2
70.42	70.44	1.3359	2	6	3	5	1	10
70.52	70.52	1.3342	-5	5	3	5	4	9
70.90	70.90	1.3280	1	11	2	8	5	16
71.08	71.10	1.3250	5	3	0	3	4	6
71.35	71.32	1.3208	7	1	3	3	3	5
71.78	71.80	1.3139	-1	9	3	2	4	3
71.80		1.3135	-1	1	4	6		12
71.90		1.3121	-3	11	1	2		4
73.57	73.58	1.2863	0	0	4	2	3	5
73.61		1.2856	-1	3	2	3		5
73.87	73.86	1.2818	0	12	2	4	3	8
74.36	74.36	1.2745	-2	12	2	12	6	24
74.53	74.56	1.2721	-7	5	1	6	6	13
75.41	75.52	1.2595	6	0	4	2	3	4
75.52		1.2578	-4	2	5	5		10
79.37	79.36	1.2062	-2	14	0	1	1	2
82.13	82.14	1.1725	-5	11	2	9	4	18
82.32	82.34	1.1703	2	0	0	4	1	9
83.31	83.30	1.1589	8	0	0	1	1	3
84.97	84.96	1.1404	-6	10	2	1	1	2
87.63	87.62	1.1125	5	3	3	1	1	3
89.21	89.22	1.0969	3	1	4	2	1	4
89.57	89.52	1.0934	-6	14	4	1	1	2
90.58	90.60	1.0839	0	16	1	1	1	2
91.09	91.06	1.0791	5	5	3	2	1	4
92.13	92.14	1.0696	4	12	2	3	1	6
92.96	92.98	1.0622	4	14	1	1	1	3
93.92	93.92	1.0539	-7	1	4	3	1	6
95.15	95.16	1.0435	-9	1	2	2	1	4
95.63	95.64	1.0395	5	11	2	5	2	13

2THETA	PEAK	D	H	K	L	I(INT)	I(PK)	I(DS)
97.62	97.64	1.0235	-7	11	0	6	3	15
98.06	98.06	1.0201	-1	11	4	3	3	8
98.09		1.0199	-8	6	3	4		10
98.72	98.72	1.0151	-9	5	1	1	1	3
99.09	99.20	1.0122	8	1	1	3	2	7
99.21		1.0114	-1	17	5	5		12
99.37		1.0102	-3	15	1	2		5
101.15	101.16	0.9971	-5	15	5	1	1	2
102.91	102.94	0.9848	-3	17	1	2	1	5
102.98		0.9843	-8	0	4	2		5
104.01	104.04	0.9774	-4	6	5	4		12
104.33	104.36	0.9752	6	6	3	4	2	11
104.52		0.9740	8	0	2	2		4
104.79	104.70	0.9723	-5	15	2	2	1	1
104.85		0.9719	0	6	5	2		5
105.55	105.44	0.9673	2	14	3	1	1	2
106.14	106.14	0.9636	4	16	1	2	1	6
106.47	106.50	0.9615	-5	11	4	2	1	3
107.37	107.38	0.9559	-9	5	3	1	1	7
107.63		0.9543	9	7	7	2		2
110.69	110.70	0.9364	-2	16	0	1	0	5
112.36	112.38	0.9271	-10	0	0	2	0	4
112.90	112.82	0.9242	-9	7	3	1	0	2
113.63	113.60	0.9204	-6	6	5	1	0	2
114.16	114.14	0.9176	2	12	4	1	0	2
114.82	114.84	0.9142	-7	3	5	1		2
115.88	116.00	0.9088	-6	16	1	2	1	5
116.01		0.9082	3	11	4	5		14
116.33	116.36	0.9066	-10	6	1	2	1	5
116.65		0.9051	6	16	4	1		2
117.77	118.16	0.8997	-8	8	4	1	1	4
118.14		0.8980	-11	17	3	3		9
118.69	118.68	0.8954	-7	5	4	4	1	12
118.82		0.8948	5	7	3	1		2
119.63	119.62	0.8911	2	16	0	2	1	5
119.74		0.8906	1	6	4	4		2
121.02	121.06	0.8849	5	11	3	3	1	8

GLAUCOPHANE - PAPIKE AND CLARK, 1968a

2THETA	PEAK	D	H	K	L	I(INT)	I(PK)	I(DS)
121.65	121.92	0.8822	7	15	0	1	1	4
121.86		0.8813	-2	0	6	2		5
122.01		0.8806	-3	1	6	2		5
122.30	122.26	0.8794	4	18	1	2	1	6
122.57		0.8783	5	17	1	1		2
122.65		0.8779	-7	15	2	1		2
124.79	124.80	0.8692	-7	7	5	1		3
126.99	127.10	0.8607	-8	10	4	1	0	4
127.57	127.58	0.8586	-2	18	3	2	1	5
128.45	128.40	0.8554	10	8	0	1	0	4
128.89	128.86	0.8538	8	10	2	1	0	2

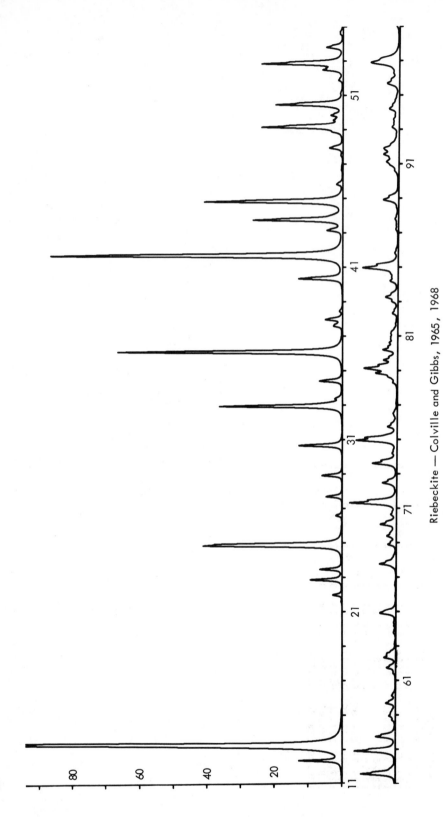

392

Riebeckite — Colville and Gibbs, 1965, 1968

λ = FeKα Half width at 40°2θ: 0.11°2θ Plot scale I$_{PK}$ (max) = 300

RIEBECKITE — COLVILLE AND GIBBS, 1965, 1968

2THETA	PEAK	D	H	K	L	I(INT)	I(PK)	I(DS)
12.32	12.32	9.019	0	2	0	100	100	100
13.19	13.20	8.429	1	1	0	4	4	4
21.94	21.96	5.086	1	3	0	1	1	1
22.83	22.84	4.892	-1	1	1	4	3	4
23.43	23.44	4.767	1	0	0	2	2	3
24.79	24.80	4.509	-2	4	0	14	14	18
24.90	24.88	4.490	0	2	1	11	13	13
26.56	26.56	4.214	0	2	1	2	1	1
27.66	27.66	4.050	2	1	1	2	2	3
28.88	28.88	3.881	-1	3	0	3	2	3
30.63	30.64	3.665	1	1	1	6	4	8
32.89	32.90	3.419	-2	3	1	16	12	24
33.34	33.34	3.374	2	5	0	1	1	1
34.37	34.38	3.276	1	3	1	3	2	5
36.03	36.04	3.130	1	5	0	31	22	49
37.56	37.58	3.006	3	4	0	2	2	2
37.94	37.94	2.978	-3	1	1	2	2	4
40.31	40.32	2.810	-2	2	1	6	4	11
41.62	41.64	2.724	1	3	1	43	29	75
41.70	41.70	2.719	1	5	1	6		10
43.12	43.12	2.634	-3	1	1	2	2	4
43.72	43.72	2.600	-1	6	0	14	9	25
44.78	44.78	2.541	0	0	2	22	14	41
45.80	45.80	2.488	-2	2	2	1	1	2
47.92	47.92	2.385	3	5	0	1	1	2
47.93		2.383	4	0	0	1		2
48.85	48.86	2.341	-3	1	2	13	8	27
49.12	49.12	2.329	-3	5	1	1	1	2
49.53	49.54	2.311	-4	2	1	1	1	2
49.82	49.82	2.298	-1	7	1	2	1	3
50.44	50.44	2.272	-3	3	1	11	7	23
51.86	51.86	2.214	-1	4	2	1	0	1
52.46	52.46	2.190	-2	6	1	3	2	7
52.82	52.82	2.176	2	6	0	14	8	30
53.79	53.78	2.140	-3	3	2	3	2	6
55.53	55.54	2.078	2	0	2	6	3	14
56.86	56.86	2.033	3	5	1	7	4	17

2THETA	PEAK	D	H	K	L	I(INT)	I(PK)	I(DS)
57.71	57.72	2.006	-4	0	2	2	2	9
58.68	58.68	1.9755	1	5	2	1	0	1
58.92	58.92	1.9681	-3	7	1	1	1	2
59.65	59.66	1.9462	4	2	1	2	1	4
60.09	60.10	1.9333	-3	5	2	2	1	1
61.71	61.72	1.8873	2	5	2	2	1	4
62.31	62.32	1.8711	-4	6	1	1	1	6
62.55	62.54	1.8645	-1	9	1	2	1	3
64.85	64.90	1.8053	1	9	0	1	2	5
64.91		1.8038	0	10	0	2		5
67.74	67.74	1.7369	-5	1	2	3	2	9
68.87	68.88	1.7118	-5	5	1	2	1	5
69.37	69.38	1.7010	-1	3	3	2	1	5
70.03	70.04	1.6870	2	10	0	1	2	3
70.05		1.6865	-2	0	2	1		7
70.62	70.62	1.6748	-3	9	1	2	1	2
71.28	71.28	1.6612	4	6	1	10	5	9
72.45	72.46	1.6379	-3	6	0	3	2	9
73.59	73.60	1.6161	1	11	1	9	4	2
74.18	74.18	1.6050	-6	2	1	1	3	28
74.92	74.92	1.5915	-1	5	3	9	2	6
75.07	75.08	1.5889	6	0	0	2	1	6
75.72	75.72	1.5772	4	0	0	1	0	2
76.49	76.50	1.5637	-4	2	3	1	0	3
78.29	78.30	1.5334	-3	5	3	1	1	3
78.62	78.64	1.5279	1	9	2	4	1	3
78.83	78.84	1.5245	-6	0	1	7	2	13
79.11	79.10	1.5200	-2	6	3	3	1	23
79.59	79.58	1.5124	5	5	1	3	1	9
80.18	80.18	1.5032	0	12	0	1	1	11
80.59	80.58	1.4968	0	6	3	1	1	5
80.94	80.94	1.4914	2	2	2	1	1	3
81.11	81.12	1.4888	4	4	1	1	1	2
81.24		1.4869	1	5	3	1	1	3
82.31	82.32	1.4709	-2	2	10	1	1	5
83.00	83.00	1.4609	-1	1	7	1	1	5
83.25	83.24	1.4572	3	11	0	3	1	9

393

RIEBECKITE — COLVILLE AND GIBBS, 1965, 1968

2THETA	PEAK	D	H	K	L	I(INT)	I(PK)	I(DS)
84.17	84.18	1.4442	-6	4	2	2	1	6
84.94	84.94	1.4337	-6	6	1	9	4	34
85.81	85.80	1.4219	-5	3	3	1	0	3
87.18	87.18	1.4039	-4	6	3	1	0	2
87.78	87.78	1.3963	-1	11	2	5	0	2
88.89	88.90	1.3824	5	1	2	1	2	18
90.57	90.60	1.3622	2	10	0	2	0	3
90.92	91.10	1.3580	7	1	2	3	1	6
91.09		1.3561	-5	5	3	1		10
91.53	91.54	1.3510	5	3	2	2	2	6
91.54		1.3510	2	6	3	3		2
91.90	91.88	1.3468	1	11	3	3	2	14
93.11	93.12	1.3332	-3	11	2	2	1	8
93.57	93.56	1.3282	-7	9	0	1	1	2
93.58		1.3281	-1	1	3	1		3
94.19	94.18	1.3216	-1	1	4	3	1	11
95.65	95.66	1.3061	-7	5	1	4	1	17
96.26	96.28	1.2999	0	12	2	4	1	5
96.80	96.86	1.2944	0	0	4	2	3	9
96.84		1.2941	-1	3	4	1		5
96.87		1.2938	-2	12	1	6		25
96.92		1.2933	3	5	2	1		4
99.17	99.26	1.2714	4	12	0	3	1	4
99.26		1.2706	-4	0	4	1		11
103.69	103.74	1.2310	-7	7	2	1	0	3
103.74		1.2305	6	0	2	1		4
104.20	104.02	1.2267	-2	14	1	1	0	3
104.66	104.64	1.2229	-4	4	4	1	0	2
108.20	108.22	1.1950	-5	11	2	5	1	24
108.64	108.52	1.1917	8	0	0	1	1	7
110.17	110.18	1.1805	2	0	4	3	1	12
110.49	110.46	1.1782	7	5	1	1	1	3
112.47	112.48	1.1644	-6	10	0	1	0	5
117.78	117.80	1.1306	5	3	2	1	0	4
118.48	118.48	1.1264	7	9	3	1	0	4
119.15	119.18	1.1226	0	8	0	1	0	3
121.70	121.72	1.1084	3	1	4	2	0	10

2THETA	PEAK	D	H	K	L	I(INT)	I(PK)	I(DS)
122.98	123.02	1.1016	0	16	1	1	0	8
123.12		1.1008	3	13	1	1		3
123.93	123.92	1.0967	5	5	3	1		8
125.64	125.66	1.0881	4	12	2	2	0	8
126.22	126.14	1.0853	-2	16	1	1	0	13
126.91	126.86	1.0820	-4	14	2	3	0	6
129.16	129.20	1.0718	-7	1	4	1	1	6
129.21		1.0715	-9	1	2	2		15
129.47		1.0704	-6	12	2	1		10
132.20	132.22	1.0588	5	11	0	6	1	5
132.53	132.64	1.0574	9	1	2	1	1	37
132.65		1.0569	-7	3	4	1		4
132.70		1.0567	-9	3	2	1		7
133.98	134.00	1.0516	0	10	4	7	0	6
135.01	135.04	1.0477	7	11	0	1	1	40
136.25	136.44	1.0431	-9	5	3	5	1	8
136.25		1.0423	-8	6	1	1		31
137.11	137.00	1.0400	-2	14	3	1	1	4
137.46	137.76	1.0387	-4	10	4	4	1	6
137.79		1.0376	8	6	1	4		25
139.83	139.92	1.0307	-1	11	4	4	1	26
140.41	140.56	1.0288	0	2	5	6	1	8
140.63		1.0281	-1	17	1	1		36
143.36	143.90	1.0197	-4	16	0	4	1	8
143.55		1.0191	-3	5	5	1		5
143.87		1.0182	6	10	2	1		23
144.44		1.0165	-6	14	1	4		8
146.44	146.36	1.0110	-8	0	4	4	0	24
149.67	149.96	1.0029	-8	0	4	4	1	25
150.01		1.0021	-3	17	1	3		5
150.73		1.0005	-4	14	3	1		6
151.73	152.58	0.9982	5	9	3	1	1	4
152.39		0.9968	-8	8	3	3		19
152.48		0.9966	8	8	0	1		8
153.59	154.54	0.9943	8	0	2	11		73
154.46		0.9925	6	6	3	11		11
155.22		0.9911	1	11	4	2		

RIEBECKITE — COLVILLE AND GIBBS, 1965, 1968

2THETA	PEAK	D	H	K	L	I(INT)	I(PK)	I(DS)
155.69		0.9902	-5	15	2	1		7
157.03	157.82	0.9878	2	10	4	1	1	9
157.74		0.9865	-4	6	5	11		73
160.09	162.18	0.9828	-4	16	2	1	1	4
161.06		0.9814	-3	7	5	1		8
161.11		0.9813	4	16	1	5		38
161.42		0.9809	0	12	4	7		45
161.50		0.9807	2	14	3	2		16
161.60		0.9806	-3	15	3	1		9
162.26		0.9797	9	7	0	1		4
162.74		0.9791	0	6	5	6		43
162.79		0.9790	-8	4	4	1		4
163.30		0.9784	-5	5	5	2		14
163.36		0.9783	-9	5	3	10		70

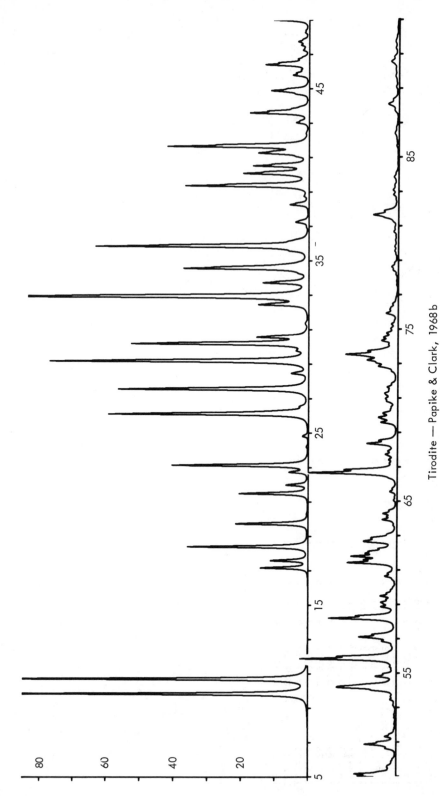

Tirodite — Papike & Clark, 1968b

$\lambda = CuK\alpha$ Half width at $40°2\theta$: $0.11°2\theta$ Plot scale I_{PK} (max) = 100

TIRODITE - PAPIKE AND CLARK, 1968b

2THETA	PEAK	D	H	K	L	I(INT)	I(PK)	I(DS)
9.82	9.82	9.003	0	2	0	65	93	46
10.68	10.68	8.276	1	0	0	67	93	48
17.14	17.14	5.169	0	0	1	11	14	8
17.56	17.56	5.046	1	3	0	9	11	7
18.37	18.37	4.826	-1	1	1	29	36	22
19.70	19.72	4.502	0	4	0	17	22	14
19.79		4.483	0	2	1	1		1
21.46	21.46	4.138	2	0	0	17	21	14
21.95	21.96	4.045	1	1	1	5	7	4
22.70	22.72	3.913	-2	0	1	4	6	4
23.11	23.12	3.846	1	3	1	35	40	30
24.79	24.80	3.589	-2	2	1	2	2	1
26.08	26.08	3.414	1	1	1	53	59	47
26.51	26.50	3.359	1	5	0	3	3	1
27.53	27.54	3.237	0	4	1	53	56	48
28.44	28.44	3.136	2	0	1	4	5	4
29.15	29.16	3.061	3	1	0	74	77	69
29.74	29.74	3.001	0	6	0	2	4	2
30.15	30.16	2.962	2	2	1	49	52	47
30.24		2.953	-2	4	1	4		3
30.54	30.54	2.925	-3	1	1	12	15	1
30.54		2.924	-1	5	1	14		12
32.43	32.44	2.759	3	3	0	14	15	14
32.90	32.90	2.720	-1	5	1	100	100	100
33.69	33.70	2.658	3	1	1	13	14	13
34.53	34.54	2.595	-3	3	1	37	37	38
35.82	35.82	2.505	0	6	1	65	63	69
37.23	37.22	2.413	-2	2	2	3	4	3
38.23	38.24	2.352	1	2	2	6	6	6
38.73	38.74	2.323	-4	0	1	2	2	1
39.33	39.34	2.289	-1	5	0	25	36	28
39.34		2.288	-3	7	1	15		17
40.02	40.06	2.251	0	8	0	2	19	3
40.05		2.249	-4	2	1	18		20
40.49	40.50	2.226	-3	4	2	17	17	20
41.21	41.24	2.189	-2	4	2	7	15	8
41.25		2.187	-1	7	1	10		12
41.62	41.62	2.168	2	6	0	46	42	54
43.01	43.02	2.101	-3	3	2	4	4	5
43.57	43.56	2.076	-2	0	2	20	17	24
43.71	43.68	2.069	4	0	0	1	12	2
43.83	43.83	2.064	0	8	1	2	5	3
44.77	44.86	2.023	2	8	0	2	11	2
44.87		2.018	3	1	1	11		14
45.76	45.76	1.9812	3	5	1	5	5	6
45.92	45.88	1.9745	1	7	0	1	4	2
46.07	46.06	1.9685	4	0	1	1	2	1
46.37	46.38	1.9566	-4	2	2	10	13	12
46.38		1.9562	-1	9	1	5		6
46.50	46.48	1.9511	1	8	0	3	9	1
46.71	46.70	1.9429	-2	8	1	3	3	4
47.22	47.22	1.9231	-4	4	1	1	2	1
47.51	47.52	1.9120	4	2	1	2	3	3
47.73	47.72	1.9039	-3	5	2	3	3	4
48.24	48.24	1.8849	-4	4	2	1	1	2
48.98	48.98	1.8581	-1	9	1	12	11	16
49.10	49.10	1.8537	5	1	0	7	12	10
49.38	49.38	1.8439	-1	7	2	1	3	2
49.49	49.50	1.8402	-4	6	1	1	3	2
49.58		1.8370	-4	0	3	1		2
49.82	49.82	1.8286	2	8	1	2	2	2
50.60	50.60	1.8024	-1	9	1	2	3	2
50.84	50.84	1.7944	3	7	1	11	9	15
51.29	51.28	1.7798	5	3	0	4	3	5
53.43	53.44	1.7134	-2	2	3	2	2	3
54.12	54.14	1.6932	-1	9	1	9	18	13
54.15		1.6924	0	10	1	5		7
54.20		1.6909	-5	1	2	2		3
54.21		1.6906	-3	7	2	10		14
54.78	54.80	1.6742	-2	8	2	2	6	3
54.80		1.6737	-5	1	1	5		7
55.80	55.80	1.6460	4	6	1	2		4
56.83	56.84	1.6187	4	8	0	37	29	56
57.07	57.08	1.6123	1	11	0	13	11	20

TIRODITE - PAPIKE AND CLARK, 1968b

2THETA	PEAK	D	H	K	L	I(INT)	I(PK)	I(DS)
58.16	58.16	1.5848	-1	5	3	26	20	41
58.84	58.84	1.5681	-4	0	2	5	5	8
59.11	59.10	1.5616	2	10	1	4	5	7
59.47	59.46	1.5530	6	0	0	6	5	9
59.82	59.80	1.5448	4	2	2	1	2	2
60.61	60.62	1.5264	-1	9	2	4	4	6
61.40	61.40	1.5087	-2	6	3	19	15	30
61.77	61.76	1.5006	0	12	0	17	14	28
62.05	62.04	1.4943	0	6	3	4	9	7
62.06		1.4942	5	5	1	4		6
62.20	62.22	1.4911	2	2	3	1	6	2
62.62	62.64	1.4822	-6	8	2	8	10	13
62.64		1.4818	4	0	1	1		2
62.69		1.4808	4	4	2	6		11
62.88	62.84	1.4767	-4	8	2	3	8	5
62.91		1.4760	3	7	2	1		2
63.91	63.92	1.4553	-1	11	3	6	4	10
64.26	64.26	1.4482	3	11	0	5	4	9
65.03	65.02	1.4330	-3	11	1	1	1	2
65.53	65.54	1.4232	-4	10	1	3	2	5
66.63	66.64	1.4023	-6	6	1	39	26	69
67.20	67.20	1.3919	-1	12	2	1	2	2
67.40	67.40	1.3882	6	2	1	2	3	3
67.70	67.70	1.3828	-4	6	3	4	4	7
68.33	68.34	1.3715	5	1	2	13	9	24
69.56	69.56	1.3503	2	11	3	7	5	13
69.86	69.86	1.3453	1	5	3	7	6	13
70.13	70.08	1.3408	5	13	2	4	4	7
70.47	70.46	1.3351	-1	13	1	2	2	3
70.79	70.80	1.3299	-5	5	3	3	3	6
70.93	70.94	1.3275	7	1	0	3	3	5
71.18	71.24	1.3236	-1	9	3	2	4	3
71.26		1.3223	-3	11	2	2		3
71.26		1.3221	-2	0	4	2		4
71.62	71.62	1.3165	-1	1	4	4	3	7
72.53	72.54	1.3021	-7	1	2	2	2	3
72.82	72.82	1.2977	0	12	2	6	6	12
72.89	73.04	1.2966	-6	8	1	1	7	5
73.01		1.2948	-5	9	2	3		5
73.17	73.18	1.2923	0	0	4	9	9	18
73.26		1.2909	-6	2	3	1		3
73.38	73.50	1.2892	-1	6	3	1	16	2
73.51		1.2873	-2	12	1	21		41
73.85	73.70	1.2820	-3	9	2	2	9	3
74.11	74.10	1.2783	6	8	0	1	3	2
74.28	74.32	1.2757	-7	3	1	2	5	5
74.33		1.2749	-7	5	1	4		8
74.50	74.52	1.2725	6	6	1	3	4	5
74.77	74.76	1.2686	-2	4	3	3	3	6
75.05	75.00	1.2645	-1	9	3	1	2	4
75.26	75.26	1.2615	4	12	0	2	4	10
75.91	75.90	1.2524	-4	0	4	5	2	4
76.25	76.24	1.2476	-4	2	3	2	2	2
76.44	76.46	1.2450	0	10	3	1	2	3
78.41	78.42	1.2186	6	0	0	1	2	4
78.60	78.60	1.2161	2	12	0	2	2	8
81.46	81.62	1.1805	2	12	2	4	8	24
81.61		1.1786	-5	11	2	11		4
82.08	82.80	1.1648	8	0	3	1	2	3
83.08	83.08	1.1615	4	6	3	2	2	5
83.81	83.82	1.1532	1	15	1	1	1	3
85.62	85.62	1.1335	5	9	2	2	1	3
87.71	87.72	1.1117	5	13	0	1	1	4
88.05	88.06	1.1083	7	9	4	3	3	7
88.08		1.1080	3	13	2	3		7
89.07	89.06	1.0982	3	1	3	1	1	3
89.69	89.70	1.0923	3	13	3	1	1	3
90.55	90.54	1.0841	4	5	4	5	2	11
91.34	91.34	1.0767	4	12	1	2	1	4
91.79	91.72	1.0727	5	14	2	1	1	3
94.01	94.02	1.0531	-7	3	4	7	4	18
94.68	94.68	1.0474	-9	1	2	2	2	9
95.39	95.38	1.0415	-9	1	2	2	1	4
96.45	96.46	1.0328	7	11	0	7	4	19

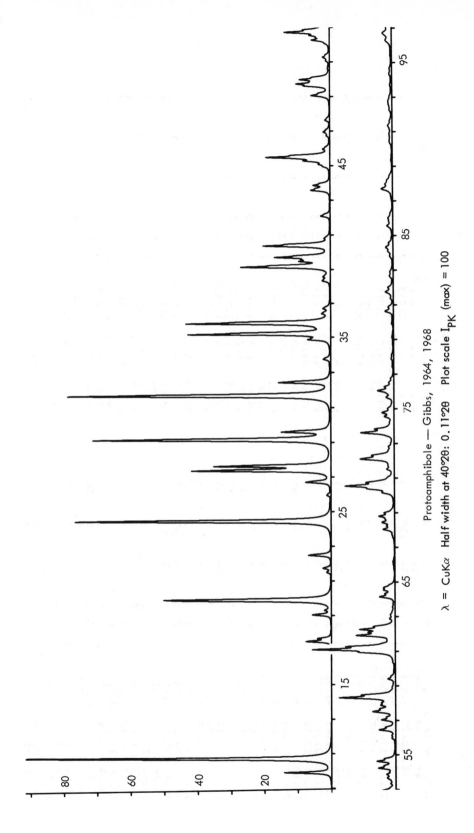

Protoamphibole — Gibbs, 1964, 1968

$\lambda = CuK\alpha$ Half width at $40°2\theta$: $0.11°2\theta$ Plot scale I_{PK} (max) = 100

PROTOAMPHIBOLE - GIBBS, 1964, 1968

2THETA	PEAK	D	H	K	L	I(INT)	I(PK)	I(DS)
9.89	9.90	8.939	0	2	0	13	14	12
10.69	10.70	8.271	0	1	1	91	100	85
17.47	17.48	5.071	0	1	1	7	8	7
17.64	17.64	5.022	1	0	3	4	6	3
19.01	19.02	4.665	0	2	0	6	6	5
19.28	19.28	4.600	1	1	0	1	2	1
19.85	19.90	4.470	1	0	4	38	50	36
19.91		4.455	0	1	1	26		25
21.47	21.48	4.136	1	1	2	1	1	2
21.71	21.72	4.091	0	2	2	3	2	2
22.46	22.46	3.955	1	1	2		7	7
24.42	24.42	3.642	1	1	3	86	76	84
25.93	25.94	3.433	2	1	1	1	1	1
26.67	26.68	3.339	0	1	5	8	8	8
27.35	27.36	3.258	1	2	2	47	42	46
27.61	27.62	3.227	0	2	4	39	35	38
29.12	29.12	3.064	3	1	1	85	71	84
29.58	29.58	3.017	1	2	5	17	15	17
31.66	31.66	2.823	1	1	3	100	78	100
32.44	32.44	2.757	0	3	3	19	16	19
33.78	33.78	2.651	1	3	1	3	2	3
34.91	34.92	2.568	1	1	2	7	7	7
35.25	35.26	2.544	2	1	0	56	43	57
35.87	35.88	2.501	1	2	6	10	43	10
35.88		2.501	1	1	5	47		48
36.44	36.44	2.464	0	3	7	3	3	3
36.72	36.72	2.445	1	2	3	2	2	3
38.32	38.32	2.347	1	3	5	3	3	3
38.56	38.56	2.332	0	1	0	2	2	2
39.13	39.14	2.300	2	2	2	36	27	37
39.46	39.46	2.281	2	2	2	10	9	10
39.57	39.56	2.276	2	0	0	2	9	2
39.70	39.70	2.268	0	1	4	21	17	21
40.32	40.36	2.235	0	1	6	1	20	28
40.35		2.233	1	1	8	27		
40.78	40.78	2.211	2	2	7	2	2	2
42.07	42.08	2.146	2	2	3	4	3	4

2THETA	PEAK	D	H	K	L	I(INT)	I(PK)	I(DS)
43.56	43.56	2.076	2	4	1	8	6	8
43.63		2.073	1	1	5	1		1
43.85	43.86	2.063	1	2	7	7	5	7
44.94	44.94	2.016	0	2	8	2	2	2
45.06	45.06	2.010	1	1	1	2	3	2
45.26	45.26	2.002	2	3	6	3	8	2
45.47	45.48	1.9929	1	3	1	9	8	10
45.94	45.94	1.9738	0	4	6	27	19	29
46.71	46.72	1.9430	0	1	9	2	2	3
46.93	46.92	1.9345	1	1	9	1	1	1
47.61	47.62	1.9083	2	2	5	3	2	3
49.04	49.04	1.8559	2	3	3	2	1	2
49.59	49.60	1.8365	0	5	1	8	6	9
49.71	49.72	1.8326	2	4	4	9	8	10
49.96	49.96	1.8238	1	1	5	12	10	13
51.04	51.04	1.7879	1	4	9	1	1	2
51.26	51.26	1.7808	2	5	0	3	2	3
52.25	52.26	1.7491	1	3	3	7	5	8
52.52	52.52	1.7408	2	4	0	7	6	7
52.71	52.70	1.7350	1	4	1	19	13	21
52.96	52.84	1.7275	1	4	6	2	8	2
53.32	53.32	1.7166	2	2	7	1	2	1
53.57	53.66	1.7092	2	7	0	1	3	3
53.65		1.7068	0	0	8	3		3
53.87	53.86	1.7004	3	1	2	3	3	3
54.22	54.22	1.6903	3	0	3	8	5	9
54.62	54.62	1.6790	2	1	8	8	6	10
55.95	55.96	1.6419	3	2	1	2	1	2
56.44	56.44	1.6289	0	4	4	8	5	9
57.02	57.02	1.6137	2	4	8	6	4	7
57.44	57.50	1.6029	2	2	8	2	7	3
57.51		1.6012	0	1	11	9		11
57.71	57.66	1.5961	1	3	9	1	5	1
57.87	57.88	1.5920	1	2	10	5	5	6
57.98	58.00	1.5892	3	2	3	3	4	3
58.28	58.32	1.5817	2	3	7	1		1
58.31		1.5810	3	0	5	27	17	31

PROTOAMPHIBOLE – GIBBS, 1964, 1968

2THETA	PEAK	D	H	K	L	I(INT)	I(PK)	I(DS)
59.38	59.38	1.5550	0	6	0	4	2	4
60.94	60.96	1.5190	2	5	1	14	11	16
61.11	61.10	1.5152	1	5	6	35	25	41
61.28	61.26	1.5114	3	3	2	4	15	4
61.64	61.64	1.5035	2	2	9	1	3	1
61.91	61.92	1.4974	3	1	6	18	12	21
61.96		1.4963	2	3	8	2		2
62.26	62.26	1.4899	0	0	12	17	11	19
62.37		1.4874	1	3	0	1		1
62.87	62.86	1.4770	2	5	3	2	2	3
63.13	63.12	1.4715	1	6	2	1	1	2
64.14	64.14	1.4507	3	0	7	8	5	10
64.53		1.4429	2	0	4	4	3	5
64.65	64.66	1.4405	0	3	11	4	4	4
68.03	68.04	1.3768	1	6	5	6	4	7
68.39		1.3705	2	4	10	1		2
68.44	68.44	1.3697	1	1	11	5	4	6
68.79	68.78	1.3635	3	3	6	8	5	10
70.15	70.16	1.3404	2	6	0	5	4	7
70.38	70.38	1.3366	2	6	1	9	8	11
70.54	70.54	1.3340	1	6	6	23	15	29
70.83		1.3292	0	7	1	3	9	4
71.49	71.50	1.3185	3	0	9	3	2	3
72.10	72.10	1.3089	1	1	0	14	10	17
72.11		1.3087	3	4	5	5		4
72.17		1.3077	2	4	6	3		6
72.32	72.30	1.3054	4	1	1	1	7	1
72.80	72.80	1.2980	2	0	12	2	2	2
72.80		1.2856	2	1	12	20	10	25
73.61	73.62	1.2856	4	2	0	2	3	6
74.54	74.54	1.2719	3	2	9	2	4	2
74.76	74.76	1.2688	3	2	2	2		2
74.78		1.2684	3	5	1	3		3
75.64	75.66	1.2562	1	1	4	2	2	2
75.68		1.2556	0	4	12	2		2
76.04	76.04	1.2505	2	2	12	10	5	12
76.51	76.50	1.2440	3	1	10	2	2	2
77.87	77.88	1.2256	0	5	11	1	1	2

2THETA	PEAK	D	H	K	L	I(INT)	I(PK)	I(DS)
80.62	80.62	1.1907	2	4	11	5	3	7
81.36	81.38	1.1816	0	7	7	1	1	2
81.74	81.74	1.1771	3	5	6	5	2	7
81.74		1.1771	2	7	3	1		1
82.59	82.66	1.1662	0	8	0	1	1	2
82.67		1.1662	4	4	0	2		3
84.09	84.10	1.1501	4	4	1	4	2	4
84.30	84.30	1.1477	2	4	12	1	2	5
85.55	85.54	1.1342	4	2	4	4	2	5
86.01	86.00	1.1293	4	1	3	1	1	2
87.69	87.70	1.1120	2	5	11	8	3	11
88.17	88.18	1.1071	0	5	13	1	2	2
88.20		1.1068	1	7	9	2		3
90.44	90.44	1.0852	2	8	5	1	1	2
90.87	90.86	1.0811	4	4	13	2	1	2
91.34	91.34	1.0768	2	5	1	3	1	4
92.25	92.26	1.0686	4	3	8	2		2
92.78	92.76	1.0638	1	8	6	2	1	3
94.33	94.34	1.0504	2	8	3	1	1	3
96.19	96.30	1.0349	5	1	3	2	2	7
96.29		1.0341	0	11	11	4		11
96.72	96.70	1.0306	1	7	11	6	3	9
97.40	97.40	1.0253	1	1	17	6	2	6
98.15	98.08	1.0195	4	1	11	1	1	2
99.63	99.78	1.0082	5	2	15	2	2	6
99.77		1.0072	1	2	17	4		3
99.98	100.00	1.0056	4	6	1	2	2	23
100.57	100.58	1.0013	3	7	6	16	5	1
102.01	102.02	0.9910	5	1	6	3	1	5
102.01		0.9910	5	2	5	1		1
102.89	103.14	0.9850	5	2	11	3	2	7
103.13		0.9833	4	4	15	1		1
103.42	103.46	0.9813	3	3	14	5		1
104.52	104.52	0.9740	4	3	11	1	1	2
104.85	104.92	0.9718	5	1	17	1	1	2
104.91		0.9715	4	5	8	2		1
105.00		0.9709	1	6	18	1		1
105.11		0.9702	0	6	14	1		1

PROTOAMPHIBOLE – GIBBS, 1964, 1968

2THETA	PEAK	D	H	K	L	I(INT)	I(PK)	I(DS)
105.54	105.54	0.9674	4	2	12	3	1	4
105.76	105.94	0.9660	3	2	15	1	1	4
105.89		0.9651	2	9	0	1		1
106.11	106.24	0.9637	2	9	6	1	1	1
106.27		0.9627	1	9	8	2		2
106.70	106.66	0.9600	3	7	6	1		4
106.73		0.9599	2	6	13	1		1
107.08	107.12	0.9577	5	4	2	1	1	1
107.17		0.9571	0	7	13	1		1
107.29		0.9564	5	2	7	1		1
108.49	108.50	0.9491	5	3	6	1	0	1
109.47	109.48	0.9434	2	4	4	1	1	1
109.47		0.9434	1	4	17	2		2
109.48		0.9433	1	5	16	1		1
110.25	110.26	0.9388	4	4	11	2	1	3
110.31		0.9385	3	8	5	1		1
110.58	110.60	0.9370	4	6	7	1	1	1
111.29	111.28	0.9330	0	10	0	2	1	3
111.82	111.76	0.9301	5	4	5	1	1	1
112.03		0.9289	5	1	9	1		1
114.37	114.40	0.9165	5	4	6	2	1	3
115.57	115.56	0.9104	4	4	12	1	0	2
116.76	117.06	0.9045	3	5	14	1	2	2
117.04		0.9032	3	0	17	7		10
117.06		0.9031	3	4	18	2		3
117.59	117.54	0.9005	1	4	18	2	2	7
117.96	117.98	0.8988	4	5	11	4	2	2
118.06		0.8983	4	6	9	2		2
119.71	119.70	0.8907	2	6	15	1	0	1
120.06	120.18	0.8892	3	9	2	1	1	2
120.21		0.8885	0	7	15	1		1
121.63	121.90	0.8823	4	4	13	1	1	1
121.84		0.8813	6	0	0	4		7
122.36	122.36	0.8791	5	5	6	9	2	15
122.64		0.8780	0	2	20	1		1
123.02	122.84	0.8764	6	1	1	1	1	2
125.87	125.88	0.8650	1	5	18	5	1	8

2THETA	PEAK	D	H	K	L	I(INT)	I(PK)	I(DS)
126.32	126.42	0.8633	2	10	4	1	1	2
126.74	126.76	0.8616	3	1	18	2	1	3
126.92		0.8610	1	10	8	1		2
127.29	127.30	0.8596	1	6	17	2	1	3
128.23	128.24	0.8561	4	6	11	6	1	10

TABLE 19. DOUBLE CHAIN SILICATES

Variety	Anthophyllite	Vlasovite	Elpidite	Narsarsukite	Xonotlite
Composition	$(Mg,Fe^2)_7Si_8O_{22}(OH)_2$	$Na_4Zr_2Si_8O_{22}$	$Na_2ZrSi_6O_{15} \cdot 3H_2O$	$Na_2TiOSi_4O_{10}$	$Ca_6Si_6O_{17}(OH)_2$
Source	Dillon Complex, Beaverhead, Co., Montana; Rabbitt, 1948	Russia	Lowosero, Kola Peninsula, USSR	Sage Creek, Montana	South Osetie, USSR
Reference	Finger, 1968	Voronkov & Pyatenko, 1962	Neronova & Belov, 1965	Peacor & Buerger, 1962	Mamedov & Belov, 1955
Cell Dimensions					
$\underline{a}$	18.555	10.98	14.40 (7.40)	10.727	7.04A (16.50 kx)
$\underline{b}$ Å	18.014	10.00	7.05 (14.40)	10.727	7.33 (7.32 kx)
$\underline{c}$	5.2803	8.52	7.40 (7.05)	7.948	16.53 (7.03 kx)
α	90	90	90	90	90
β deg	90	100.40	90	90	90
γ	90	90	90	90	90
Space Group	Pnma	C2/c	Pmma (Pbmm)	I4/m	P2/c (P2/a)
Z	4	4	2	4	2

TABLE 19. (cont.)

Variety	Anthophyllite	Vlasovite	Elpidite	Narsarsukite	Xonotlite
Site Occupancy					
M_1	Mg 0.994 / Fe 0.006				
M_2	Mg 0.982 / Fe 0.018				
M_3	Mg 0.915 / Fe 0.085				
M_4	Mg 0.402 / Fe 0.598				
Method	3-D, counter	2-D, film	3-D	3-D, counter	3-D, film
R & Rating	--- (2)	0.167 (av.) (3)	0.17 (av.) (3)	0.080 (1)	0.21 (av.) (3)
Cleavage and habit	{210} perfect, {010} {100} imperfect.	{110} perfect	{110} Fibrous	{100},{110} perfect Tabular	{001} perfect; {100} less so
Comment		$[Si_8O_{22}]_\infty$ chains parallel to [101] B estimated	$[Si_6O_{15}]_\infty$ ribbon	Complex chain	$[Si_2O_8]_\infty$ chains B estimated
μ	189.1	174.6	123.1	146.9	199.4
ASF	0.4300×10^{-1}	0.9205×10^{-1}	0.8429×10^{-1}	0.8230×10^{-1}	0.8288×10^{-1}
Abridging factors	1:1:0.5	1:1:0.5	1:1.2:0.5	1:1:0.5	1:1:0.5

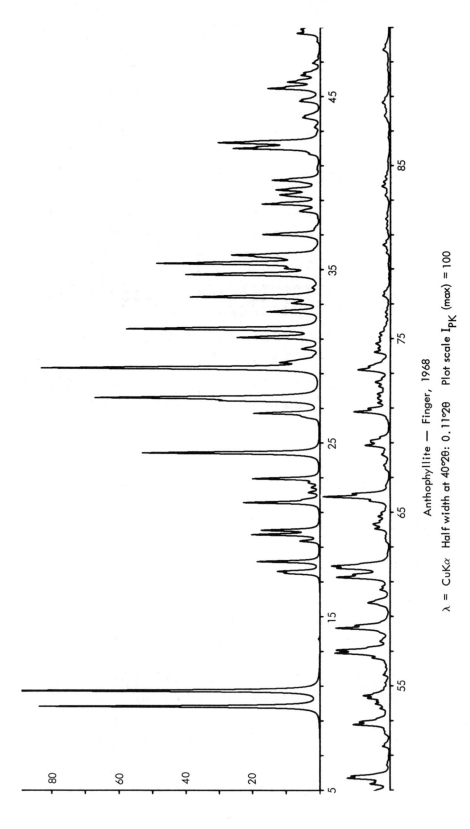

Anthophyllite — Finger, 1968

$\lambda = $ CuKα Half width at 40°2θ: 0.11°2θ Plot scale I$_{PK}$ (max) = 100

405

ANTHOPHYLLITE - FINGER, 1968

2THETA	PEAK	D	H	K	L	I(INT)	I(PK)	I(DS)
9.81	9.82	9.007	0	2	0	74	84	55
10.72	10.72	8.248	1	1	0	91	100	68
17.45	17.48	5.079	1	0	1	6	11	5
17.49		5.067	0	3	0	3		3
17.58	17.58	5.041	2	1	0	11	13	9
18.13	18.14	4.888	1	3	0	19	19	15
19.32	19.34	4.589	0	4	0	6	6	5
19.70	19.70	4.503	2	1	1	20	20	17
19.95	19.96	4.447	0	4	1	17	18	15
21.53	21.54	4.124	2	2	1	24	23	21
21.72	21.70	4.089	4	0	1	4	6	3
22.12	22.12	4.016	2	0	0	3	4	3
22.40	22.40	3.965	0	3	1	2	3	2
22.67	22.68	3.920	3	1	1	3	4	3
22.91	22.92	3.878	1	3	1	22	20	19
24.39	24.40	3.646	3	3	1	60	53	55
26.43		3.370	2	4	1	3	6	3
26.52	26.52	3.358	1	5	0	3		3
26.68	26.68	3.338	3	3	0	22	20	21
27.42	27.42	3.250	4	1	0	28	30	27
27.58	27.58	3.231	4	4	0	75	67	73
29.28	29.28	3.048	6	1	0	100	83	100
29.61	29.62	3.014	4	3	1	10	12	11
30.39	30.40	2.938	1	5	1	6	6	6
31.06	31.06	2.877	5	2	1	30	25	30
31.54	31.54	2.834	2	5	1	73	58	76
32.54	32.54	2.749	6	5	0	19	16	21
33.03	33.04	2.709	5	3	1	9	9	10
33.38	33.38	2.682	3	5	1	50	39	53
33.93	33.94	2.640	6	1	1	2	2	2
34.65	34.68	2.587	1	1	2	3	40	3
34.68		2.584	6	6	0	49		54
35.04	35.04	2.559	6	2	0	9	11	10
35.32	35.32	2.539	2	0	2	62	49	70
35.63	35.72	2.517	5	4	1	1	19	1
35.71		2.512	2	6	1	12		13
35.74		2.510	1	2	2	4		4

2THETA	PEAK	D	H	K	L	I(INT)	I(PK)	I(DS)
35.82	35.82	2.505	4	5	1	24	26	27
36.99	37.00	2.428	3	0	2	23	17	26
38.32	38.34	2.347	6	2	2	5	6	6
38.36		2.344	3	2	2	3		4
38.75	38.76	2.322	5	5	1	23	17	28
39.21	39.30	2.296	1	7	1	6	12	7
39.23		2.295	4	0	2	2		2
39.29		2.291	7	2	1	10		12
39.56	39.58	2.276	4	1	2	7	13	9
39.59		2.275	8	6	1	11		13
40.11	40.14	2.246	2	7	1	17	14	20
40.14		2.245	3	0	1	3		4
41.65	41.66	2.167	5	0	2	35	3	44
41.96	41.96	2.151	4	3	2	5	26	4
42.12	42.06	2.143	3	4	2	5	19	6
42.25	42.30	2.137	5	1	1	11	30	14
42.27		2.136	5	6	2	28		36
42.30		2.135	5	2	1			3
43.20	43.20	2.092	4	7	2	2	2	3
43.69	43.76	2.070	7	1	1	2	5	3
43.76		2.067	8	2	1	6		7
43.87	43.86	2.062	8	4	0	2	5	2
43.95		2.058	1	8	1	2		2
44.70	44.70	2.026	5	3	2	6	6	8
44.71		2.025	2	8	2	2		3
44.79	44.80	2.022	2	8	1	3	6	3
45.43	45.44	1.9946	6	3	1	22	16	30
45.80	45.80	1.9794	7	5	1	11	10	14
45.83		1.9781	6	5	0	1		5
46.18	46.20	1.9641	3	8	1	6	6	2
46.20		1.9631	5	7	0	1		9
46.37	46.34	1.9565	4	2	2	4	5	5
46.90	46.90	1.9354	3	9	1	3	2	4
48.16	48.16	1.8880	6	3	1	1	1	2
48.63	48.64	1.8706	7	0	2	10	7	14
48.87	48.86	1.8622	9	9	1	6	6	8
48.93		1.8598	7	6	2	2		2

ANTHOPHYLLITE - FINGER, 1968

2THETA	PEAK	D	H	K	L	I(INT)	I(PK)	I(DS)
49.33	49.34	1.8457	10	1	0	8	6	11
49.62	49.66	1.8355	8	6	0	2	13	3
49.65		1.8346	2	9	1	8		11
49.67		1.8339	6	4	2	8		11
49.74		1.8315	7	2	1	2		2
49.80	49.78	1.8294	8	5	1	6	12	9
50.41	50.42	1.8086	5	8	0	1	1	2
51.51	51.50	1.7728	10	3	0	4	2	5
52.27	52.28	1.7487	5	6	2	1	2	2
52.73	52.74	1.7344	8	1	2	6	11	10
52.75		1.7337	8	8	1	11		17
52.90	52.90	1.7293	2	0	3	1	9	11
52.96		1.7275	7	4	2	7		11
53.68	53.68	1.7060	1	8	2	4	3	6
53.94	53.94	1.6982	2	2	3	6	4	9
54.07	54.08	1.6947	9	5	1	9	7	2
54.26	54.26	1.6890	0	8	3	7	8	13
54.41	54.42	1.6848	2	8	2	2		11
54.51		1.6821	1	3	3	2	2	3
54.89	54.84	1.6711	5	9	1	1	2	2
55.16	55.16	1.6638	3	8	2	2	2	2
55.62	55.62	1.6511	3	5	2	2	2	3
55.64		1.6505	5	7	2	1		2
56.07	56.06	1.6388	4	0	1	1	1	2
56.59	56.60	1.6249	9	1	3	7	6	11
56.84	56.86	1.6184	9	6	2	5	17	8
56.86		1.6178	9	8	1	20		32
56.95		1.6156	8	8	0	3		5
57.06	57.04	1.6127	2	11	1	14	16	23
57.54	57.56	1.6003	4	10	0	2	3	4
57.59		1.5991	9	2	2	1		2
58.29	58.30	1.5815	0	5	3	26	16	44
58.53	58.44	1.5758	5	5	3	4	11	7
59.75	59.76	1.5462	12	0	0	8	7	13
59.83		1.5445	3	9	2	3	3	5
60.52	60.52	1.5285	9	4	2	4		7
61.19	61.22	1.5134	1	6	3	10	16	18

2THETA	PEAK	D	H	K	L	I(INT)	I(PK)	I(DS)
61.24	61.38	1.5123	10	6	1	19	11	33
61.43	61.74	1.5081	6	2	3	1	16	2
61.74		1.5012	0	12	0	22	18	40
61.86	61.88	1.4986	6	8	3	17		3
61.86		1.4985	2	6	3	1		30
63.32	63.32	1.4675	11	8	1	1	1	3
63.62	63.62	1.4612	9	0	1	1	1	3
64.04	64.04	1.4528	0	7	1	7	5	2
64.31	64.30	1.4472	6	11	0	6	5	14
64.73	64.74	1.4389	7	8	2	2	4	11
64.75		1.4386	10	4	2	3		6
64.98	64.94	1.4339	11	0	1	2	3	4
65.62	65.64	1.4215	11	1	2	8		16
65.85	65.86	1.4171	11	2	1	9	6	17
65.87		1.4167	1	6	1	27	20	50
67.43	67.50	1.3878	11	7	2	1		2
67.50		1.3864	4	3	3	1	1	2
67.68	67.68	1.3832	11	3	0	1	2	2
68.21	68.30	1.3737	8	0	1	1	3	2
68.30		1.3721	12	5	1	3		6
68.34		1.3715	6	8	1	1		2
68.82	68.82	1.3630	6	6	3	13	8	26
69.10	69.10	1.3581	7	5	3	1	6	2
69.13		1.3577	3	11	2	7		13
70.74	70.76	1.3306	12	6	1	9	11	17
70.76		1.3303	12	2	2	13		27
71.29	71.28	1.3218	14	6	0	4		8
71.55	71.56	1.3176	0	1	3	3	4	3
71.60		1.3167	7	9	0	3		3
71.76	71.80	1.3142	8	4	4	4		7
71.82		1.3132	1	9	1	1	4	3
72.22	72.24	1.3069	2	0	4	3		6
72.24		1.3066	0	5	4	3		4
72.47	72.48	1.3030	5	11	3	2	5	6
72.51		1.3025	11	2	2	4		3
73.18	73.18	1.2922	2	12	2	17	10	36

407

ANTHOPHYLLITE - FINGER, 1968

2THETA	PEAK	D	H	K	L	I(INT)	I(PK)	I(DS)
73.26		1.2910	3	0	4	4		8
73.53	73.38	1.2870	13	5	1	2	7	5
74.20	74.20	1.2769	3	12	2	9	5	19
74.59	74.70	1.2712	4	9	3	2	4	5
74.70		1.2697	4	0	4	6		13
75.07	75.08	1.2643	10	2	3	3	3	6
75.10		1.2638	1	4	4	3		4
75.26	75.30	1.2615	9	9	2	2		3
75.35		1.2602	9	12	0	2	3	5
76.26	76.26	1.2475	8	10	3	2	1	5
76.53	76.52	1.2437	2	0	4	2	2	3
77.46	77.46	1.2311	5	12	2	3	2	8
79.96	79.96	1.1989	5	4	2	2	1	4
80.40	80.40	1.1933	8	11	4	4	2	9
81.36	81.36	1.1817	7	0	4	1	1	3
81.91	81.90	1.1751	10	6	3	2	1	5
82.28	82.26	1.1708	7	12	2	1	1	3
83.24	83.26	1.1597	16	0	0	2	2	4
83.79	83.80	1.1535	9	11	2	6	3	14
84.05	84.04	1.1506	3	15	1	1	2	3
84.35	84.34	1.1473	8	0	4	2	2	6
87.86	87.88	1.1103	10	13	0	1	2	4
87.92		1.1096	9	1	1	1		3
88.38	88.38	1.1051	14	9	0	2	2	6
88.62	88.64	1.1026	9	12	2	4	1	11
90.50	90.50	1.0846	8	13	2	1	1	4
91.67	91.70	1.0737	10	1	4	2	2	6
91.70		1.0735	11	11	2	4		10
93.22	93.24	1.0599	16	1	2	1	1	3
94.90	94.88	1.0456	16	3	2	1	1	4
95.35	95.36	1.0418	10	10	3	1	1	4
95.57	95.64	1.0400	13	6	3	1	1	3
95.83	95.84	1.0379	11	1	4	2	1	5
96.26	96.28	1.0344	12	11	2	3	2	10
96.49	96.52	1.0325	2	17	1	2	2	6
96.53		1.0322	11	12	2	2		5

408

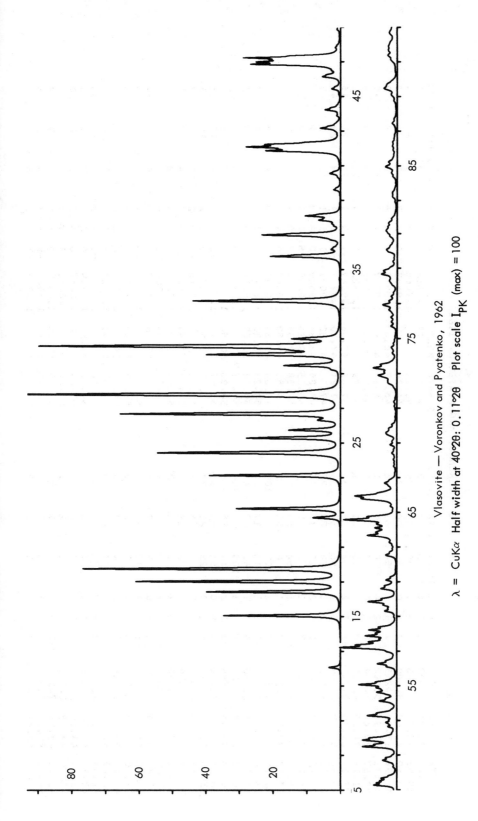

Vlasovite — Voronkov and Pyatenko, 1962

λ = CuKα Half width at 40°2θ: 0.11°2θ Plot scale I_{PK} (max) = 100

VLASOVITE - VORONKOV AND PYATENKO, 1962

2THETA	PEAK	D	H	K	L	I(INT)	I(PK)	I(DS)
12.05	12.06	7.337	-1	1	0	3	3	2
15.03	15.04	5.890	-1	1	1	30	35	25
16.40	16.40	5.400	-2	0	0	35	40	30
16.99	17.00	5.213	0	1	1	54	61	46
17.72	17.74	5.000	0	2	0	70	77	61
20.67	20.68	4.294	0	0	2	8	8	7
21.19	21.20	4.190	-1	1	2	30	31	27
23.10	23.10	3.847	-1	3	0	38	39	35
24.40	24.40	3.644	-2	2	1	56	54	53
25.26	25.26	3.523	2	0	1	28	28	27
25.72	25.72	3.460	-1	2	2	15	15	15
26.29	26.30	3.387	-3	1	1	5	7	5
26.64	26.64	3.344	-3	1	1	68	66	67
27.69	27.76	3.219	2	2	2	8	100	8
27.75		3.211	-1	3	1	100	100	100
29.44	29.44	3.031	-1	3	1	18	17	18
30.06	30.06	2.970	-1	1	1	42	40	44
30.52	30.52	2.926	-3	3	1	100	90	104
30.98	30.98	2.884	-3	1	2	14	15	15
33.15	33.16	2.700	4	0	0	51	44	55
35.74	35.74	2.510	0	1	3	25	21	28
36.15	36.14	2.483	-4	0	1	28	23	32
36.97	36.98	2.429	-3	3	1	2	2	2
37.51	37.52	2.396	0	4	1	2	3	2
37.65	37.64	2.387	-4	2	1	7	11	8
37.84	37.84	2.376	4	2	0	12	2	14
38.08	38.08	2.361	-3	0	3	2	3	3
39.60	39.60	2.274	3	0	1	4	22	5
40.53	40.54	2.224	-4	2	3	27	28	33
41.83	41.84	2.158	-1	1	2	31	23	39
42.05	42.06	2.147	0	1	1	17	22	21
42.22	42.16	2.139	-5	2	4	17	28	21
43.14	43.14	2.095	0	0	1	7	23	9
43.40	43.38	2.083	-1	1	4	1	6	1
43.88	43.88	2.062	-2	4	4	1	2	1
44.21	44.22	2.047	1	3	3	6	5	6
45.42	45.42	1.9951	3	1	3	3	3	4

2THETA	PEAK	D	H	K	L	I(INT)	I(PK)	I(DS)
46.11	46.12	1.9667	5	5	1	3	6	4
46.12		1.9666	1	1	0	3		4
46.20	46.48	1.9633	-3	3	3	1		2
46.47	46.82	1.9523	4	1	2	2	3	2
46.82	46.96	1.9387	-2	2	2	33	27	45
46.93		1.9345	2	4	2	3	25	3
46.99		1.9322	0	5	2	11		15
47.08	47.08	1.9287	-1	5	1	11	21	15
47.21	47.20	1.9237	-2	2	3	29	29	40
47.81	47.82	1.9007	-1	0	3	1	1	2
48.85	48.86	1.8629	0	4	3	1	7	1
49.25	49.26	1.8485	-5	1	0	9	6	13
49.38	49.38	1.8441	-1	0	0	2	5	3
49.50	49.50	1.8397	0	4	2	4	5	6
49.66	49.64	1.8344	-4	4	1	4	1	5
50.29	50.30	1.8126	5	3	0	1	4	2
50.58	50.68	1.8032	-1	5	0	2		3
50.67		1.7999	-6	0	2	4		6
50.83	50.80	1.7947	-1	3	0	2	3	3
51.47	51.46	1.7740	-6	0	4	14	10	21
51.78	51.86	1.7639	-5	3	2	2	10	3
51.86		1.7616	-4	4	2	12		17
51.97	51.96	1.7581	3	3	1	2	8	5
52.63	52.64	1.7375	2	5	3	3	3	5
52.87	52.86	1.7302	-6	2	4	4	3	6
53.25	53.26	1.7187	-4	2	1	4	9	6
53.29		1.7176	6	5	1	9		14
53.48	53.40	1.7120	-4	1	5	4	6	6
54.11	54.10	1.6935	6	2	5	2	5	3
54.50	54.60	1.6822	3	2	1	7	6	11
54.60		1.6792	-1	1	5	2		3
54.87	54.86	1.6719	-6	2	0	7	6	10
55.05	55.04	1.6667	0	6	2	13	12	20
55.07		1.6662	-3	1	4	1		2
55.20	55.20	1.6600	-1	5	2	1	7	2
56.10	56.10	1.6381	-5	3	3	4	3	6
56.27	56.26	1.6335	-1	5	3	6	6	10

410

VLASOVITE - VORONKOV AND PYATENKO, 1962

2THETA	PEAK	D	H	K	L	I(INT)	I(PK)	I(DS)
57.19	57.18	1.6094	4	4	2	24	17	38
57.25		1.6078	6	2	1	1		2
57.33	57.34	1.6057	0	4	4	2	13	4
57.52	57.52	1.6008	-2	4	4	8	9	14
57.85	57.86	1.5925	2	6	0	4	10	6
57.87		1.5921	1	1	5	9		15
58.20	58.20	1.5837	1	5	3	11	9	18
58.40	58.36	1.5788	5	3	2	1	6	2
59.29	59.30	1.5573	5	1	3	5	4	8
59.61	59.64	1.5496	-7	1	1	1	4	3
59.65		1.5486	0	1	2	3		5
59.85	59.84	1.5441	-3	3	3	11	9	19
60.58	60.58	1.5270	4	0	4	7	5	12
60.68		1.5248	7	1	0	1		2
61.09	61.08	1.5157	-2	6	2	5	3	8
62.49	62.52	1.4850	6	2	2	2	3	4
62.53		1.4840	2	4	4	3		5
63.24	63.24	1.4691	-3	3	5	2	2	4
63.54	63.66	1.4630	2	6	2	2	9	3
63.65		1.4607	6	4	0	4		7
63.66		1.4604	4	2	4	9		16
63.92	63.84	1.4551	7	1	1	3	6	6
64.09	64.10	1.4517	-7	3	3	2	7	4
64.09		1.4517	-1	1	5	5		9
64.33	64.32	1.4468	-6	4	2	5	6	9
64.55	64.56	1.4425	-5	1	5	9	16	17
64.56		1.4422	-6	2	2	13		23
64.61		1.4413	-5	5	1	2		4
65.35	65.34	1.4268	3	5	3	2	2	3
65.74	65.78	1.4193	-7	3	1	9	11	16
65.79		1.4182	-4	6	0	8		15
65.89	65.92	1.4163	-2	6	3	1	13	2
65.90		1.4162	1	7	0	1		2
65.93		1.4155	-2	0	6	9		16
66.51	66.54	1.4046	6	4	1	1	3	3
66.54		1.4040	-1	1	6	1		2
66.66	66.66	1.4019	-1	7	1	4	4	7

2THETA	PEAK	D	H	K	L	I(INT)	I(PK)	I(DS)
67.65	67.64	1.3838	-4	6	2	1	1	3
68.41	68.40	1.3702	-5	5	3	2	2	3
68.88	68.88	1.3620	-2	2	2	3	2	6
69.50		1.3514	-1	7	2	3		6
69.58	69.58	1.3500	8	0	0	3	3	5
69.84	69.82	1.3456	7	3	1	2	2	4
71.37	71.36	1.3205	6	4	2	1	1	2
72.39		1.3043	0	6	4	2		3
72.47	72.46	1.3032	4	4	5	2	3	4
72.56	72.56	1.3017	4	7	1	2	3	5
72.79		1.2982	3	5	3	3		6
72.88	72.88	1.2968	-1	5	1	5	3	10
72.89		1.2966	-2	0	6	3		6
73.27		1.2908	-3	7	2	3		6
73.31	73.30	1.2902	-6	4	4	10	7	20
73.46	73.50	1.2879	3	5	5	1	5	2
74.18	74.16	1.2773	4	2	6	2	1	4
74.48	74.48	1.2728	-5	1	1	2	2	5
75.72	75.72	1.2550	8	2	3	1	2	2
75.99	75.96	1.2513	-1	1	7	2	2	5
76.22	76.22	1.2480	7	1	3	3	2	6
76.64	76.64	1.2422	3	5	2	1	2	2
76.94	76.94	1.2381	5	5	3	7	4	14
77.41	77.42	1.2318	-3	2	6	1	2	2
77.43		1.2315	7	4	3	2		4
78.18	78.32	1.2216	-6	5	0	1	2	2
78.70		1.2147	-6	0	6	2		4
78.71	78.72	1.2147	5	6	6	6	5	12
78.87	78.90	1.2126	-1	6	1	3	4	7
79.45	79.46	1.2053	7	1	7	5	1	10
80.04	80.14	1.1978	0	8	2	1	3	4
80.15		1.1964	-5	7	1	3		2
80.90	80.96	1.1872	8	2	3	4	2	6
81.09	81.24	1.1849	7	2	3	2	3	6
81.25		1.1830	-7	5	3	3		2
81.46	81.46	1.1804	-6	2	6	2	3	6
81.67	81.68	1.1780	-5	5	5	3	3	6

VLASOVITE - VORONKOV AND PYATENKO, 1962

2THETA	PEAK	D	H	K	L	I(INT)	I(PK)	I(DS)
81.77		1.1768	7	3	3	1		2
83.20	83.20	1.1601	3	5	5	1	1	3
83.40	83.40	1.1581	4	0	6	1	1	3
84.37	84.38	1.1470	5	3	1	2	1	5
84.56	84.58	1.1449	-9	3	1	2	2	4
84.91	84.94	1.1411	-3	5	6	1	3	3
84.94		1.1408	-1	1	7	5		12
85.90	85.90	1.1305	-5	1	7	1	1	3
86.11	86.12	1.1283	4	2	6	2	1	4
86.33	86.34	1.1259	-4	6	4	1	1	2
87.43	87.44	1.1146	-9	3	3	3	2	7
88.74	88.76	1.1015	1	3	1	1	1	4
89.08	89.08	1.0980	8	4	2	2	2	5
89.10		1.0946	-10	0	0	4		9
89.44	89.44	1.0946	-1	7	5	2	4	4
89.45		1.0946	9	3	1	2		4
89.45		1.0932	-6	1	1	3		7
89.59		1.0926	9	9	6	1		3
89.65		1.0926	-6	4	0	1		3
90.99		1.0800	10	0	6	3	2	7
91.11	91.12	1.0789	-2	5	7	4	3	10
91.34	91.36	1.0768	-5	3	3	2	3	4
91.58	91.66	1.0745	4	8	2	2		5
91.70		1.0734	0	8	4	3	3	8
91.87	91.86	1.0720	-2	0	4	2		4
92.03		1.0705	-9	6	6	2	2	5
92.60	92.66	1.0654	-9	1	5	2		4
92.67		1.0648	7	5	3	2		4
92.77		1.0639	1	9	2	1		4
93.37	93.38	1.0586	5	7	1	2	2	6
93.72	93.70	1.0556	10	2	7	2	2	6
94.18	94.24	1.0516	8	0	0	1	2	4
94.27		1.0508	4	4	4	1		4
94.57	94.56	1.0483	-8	2	6	1	2	4
94.66		1.0476	-7	7	2	1		3
95.32	95.34	1.0421	-4	0	8	1	1	4
95.38		1.0415	-2	2	8	1		6
95.65	95.64	1.0393	7	1	5	1	1	3
96.22	96.24	1.0347	2	8	4	2	2	5
96.27		1.0343	-1	9	3	1		3
96.71	96.68	1.0308	-4	8	4	1	2	3
96.92	96.96	1.0291	8	2	7	1	2	3
97.22	97.22	1.0267	3	3	7	3	2	7
97.62	97.60	1.0236	-7	7	6	1		4
97.64		1.0234	-2	6	5	2		7
98.04	98.00	1.0203	-5	7	5	2	2	5
98.34	98.36	1.0180	-9	5	3	2	2	4
98.39		1.0176	-10	2	4	2		4
99.58	99.60	1.0086	3	7	5	1	1	5
99.87	99.88	1.0064	-1	0	6	1	1	3
101.41	101.62	0.9953	2	5	8	1		2
101.58		0.9941	5	7	6	1		4
101.81	101.98	0.9925	-7	5	4	2	2	2
101.96		0.9914	10	6	1	1		6
102.05		0.9908	-5	9	7	2		4
102.31	102.32	0.9889	-5	5	3	1	1	2
102.55		0.9873	-10	4	6	1		3
102.83	102.86	0.9854	-8	4	2	2	2	5
103.13	103.14	0.9834	10	2	0	1	1	4
103.14		0.9833	2	10	7	1		1
103.27		0.9824	-2	6	6	1		6
103.37		0.9817	-6	4	6	2		6
103.66	103.64	0.9797	-2	4	8	1	1	4
104.20	104.20	0.9761	2	2	8	1		5
104.32		0.9753	3	9	3	1		4
104.72	104.84	0.9727	0	10	2	1	1	4
104.83		0.9720	5	1	1	3		8
105.06	105.08	0.9705	5	1	7	1	1	4
105.24		0.9694	8	4	4	1		2
105.44	105.48	0.9680	-3	9	4	1	1	2
105.56		0.9673	4	8	4	1		4
106.40	106.40	0.9619	-6	4	4	2	1	6
106.75	106.76	0.9597	-10	9	3	1	1	2
107.25	107.30	0.9567	-5	3	1	1	1	3
107.62	107.62	0.9544	-11	3	1	3	1	8

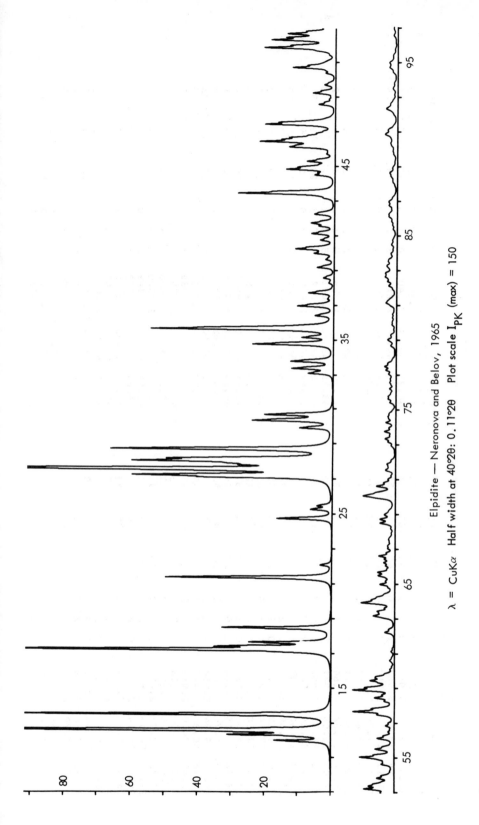

Elpidite — Neronova and Belov, 1965

$\lambda = CuK\alpha$ Half width at $40°2\theta$: $0.11°2\theta$ Plot scale I_{PK} (max) = 150

ELPIDITE — NERONOVA AND BELOV, 1965

2THETA	PEAK	D	H	K	L	I(INT)	I(PK)	I(DS)
11.95	11.96	7.400	0	1	0	9	11	9
12.28	12.30	7.200	2	0	0	16	21	15
12.54	12.56	7.050	0	0	1	91	100	84
13.44	13.44	6.582	1	1	0	60	64	55
17.17	17.18	5.160	2	1	0	69	72	64
17.36	17.36	5.104	1	1	1	17	23	16
17.59	17.60	5.037	0	1	1	14	17	13
18.43	18.44	4.811	2	0	1	22	22	20
21.32	21.32	4.164	2	1	1	35	33	33
22.05	22.06	4.027	3	1	0	2	2	2
24.71	24.72	3.600	4	0	0	12	11	12
25.24	25.24	3.525	0	0	2	4	4	4
25.45	25.46	3.497	3	1	1	3	3	3
27.07	27.08	3.291	2	0	2	24	27	24
27.20	27.20	3.276	2	2	0	36	40	36
27.53	27.54	3.237	2	2	1	100	88	100
27.80	27.80	3.206	4	1	0	12	19	12
28.01	28.02	3.182	0	2	1	37	40	37
28.16	28.16	3.166	2	1	2	29	33	29
28.70	28.70	3.107	2	2	1	52	44	52
29.94	29.94	2.982	1	1	2	7	7	7
30.36	30.36	2.942	4	1	1	18	16	19
30.69	30.70	2.911	2	1	2	16	14	16
33.08	33.08	2.706	5	1	0	6	5	6
33.36	33.36	2.684	3	2	1	10	8	10
33.76	33.76	2.652	3	1	2	10	8	11
34.74	34.74	2.580	4	2	0	20	16	21
35.13	35.14	2.552	0	2	2	7	6	7
35.61	35.62	2.519	4	2	2	46	36	49
35.70	35.70	2.513	1	2	2	1	1	1
36.39	36.40	2.467	0	3	0	4	4	5
36.94	36.94	2.431	3	3	0	9	7	10
37.35	37.36	2.406	2	2	2	1	1	1
37.69	37.70	2.384	4	1	2	6	5	7
38.27	38.28	2.350	1	0	3	1	1	1
38.55	38.56	2.334	2	3	0	3	2	3
39.16	39.16	2.298	1	3	1	4	3	5

2THETA	PEAK	D	H	K	L	I(INT)	I(PK)	I(DS)
39.62	39.62	2.273	5	2	0	2	2	2
39.97	39.98	2.253	3	2	2	5	4	5
40.23	40.24	2.240	0	1	3	10	8	11
40.73	40.74	2.213	1	3	0	3	3	4
41.11	41.10	2.194	3	1	1	6	5	7
41.54	41.54	2.172	6	1	1	4	3	4
41.72	41.72	2.163	5	2	1	5	5	6
42.22	42.22	2.139	2	1	3	5	4	5
42.29		2.135	1	3	2	1		2
43.15	43.16	2.095	2	3	1	2	2	2
43.42	43.42	2.082	5	2	0	28	19	31
44.49	44.48	2.035	3	3	2	5	5	6
44.81	44.80	2.021	6	2	0	13	10	15
44.98	44.92	2.014	2	3	1	4	7	5
45.27	45.28	2.001	6	2	0	7	6	8
46.09	46.10	1.9678	4	3	1	10	9	12
46.15		1.9651	1	2	3	4		4
46.41	46.40	1.9550	4	2	3	20	15	24
46.64	46.64	1.9458	2	2	3	10	9	11
46.89	46.88	1.9361	6	1	1	3	4	4
47.40	47.40	1.9162	2	1	3	19	14	23
47.50	47.52	1.9124	7	1	0	4	11	5
47.62		1.9080	2	3	1	1		2
48.55	48.56	1.8734	5	3	0	5	3	6
48.85	48.86	1.8626	3	3	1	1	1	1
49.21	49.22	1.8500	7	0	0	7	4	8
49.64	49.68	1.8349	4	4	0	1	2	1
49.69		1.8333	2	3	3	2		3
50.35	50.36	1.8106	5	3	0	13	8	16
50.67	50.68	1.8000	8	0	0	2	6	2
50.92	50.80	1.7918	0	4	0	12	14	15
51.83	51.84	1.7625	4	3	2	9		11
51.83		1.7623	2	2	3	18	13	23
52.28	52.28	1.7484	6	0	1	3	10	3
52.42	52.42	1.7441	1	2	3	12	10	15
52.63	52.64	1.7374	2	4	1	1		1
52.66		1.7366	1		4			

ELPIDITE - NERONOVA AND BELOV, 1965

2THETA	PEAK	D	H	K	L	I(INT)	I(PK)	I(DS)
52.95	52.98	1.7276	7	1	2	1	3	1
53.00		1.7262	3	4	0	1		2
53.20	53.20	1.7201	6	3	0	8	6	10
53.39	53.36	1.7145	1	1	4	3	5	5
53.48	53.46	1.7120	0	0	4	4	4	3
53.80	53.82	1.7025	2	1	4	2	3	3
53.83		1.7015	1	0	3	3		4
54.61	54.62	1.6791	6	3	0	3	3	4
55.01	55.02	1.6679	2	0	4	10	7	12
55.44	55.50	1.6558	2	1	3	1	4	2
55.50		1.6543	5	3	2	5		6
56.10	56.12	1.6381	0	4	2	3	3	3
56.12		1.6375	6	1	0	3		4
56.49	56.48	1.6276	4	4	0	1	1	2
56.83	56.84	1.6186	8	2	0	3	2	3
57.43	57.46	1.6031	8	0	2	3	4	1
57.46		1.6024	4	4	1	3		4
57.66	57.66	1.5973	2	0	4	12	8	15
58.23	58.24	1.5830	5	2	1	2	2	3
58.45	58.44	1.5776	4	2	2	2	4	7
58.89	58.90	1.5667	8	0	0	13	8	17
59.01	59.04	1.5639	9	1	0	2	6	3
59.32	59.44	1.5565	5	2	2	3	5	4
59.44		1.5537	4	2	4	6		7
59.68	59.66	1.5479	9	1	1	4	4	5
60.60	60.60	1.5268	7	1	3	1	1	2
61.11	61.12	1.5151	4	4	2	1	1	1
62.21	62.22	1.4910	5	1	2	3	2	5
63.04	63.16	1.4732	8	2	2	2	4	3
63.15		1.4710	6	2	0	6		8
63.43	63.42	1.4652	5	4	0	2	4	3
63.44		1.4649	4	3	4	2		3
63.91	63.92	1.4554	8	2	0	10	7	13
63.98		1.4540	9	1	1	3		3
64.79	64.80	1.4377	0	2	3	1	1	2
64.98	64.98	1.4340	0	3	4	4	3	5
65.45	65.48	1.4249	2	4	3	2	3	3

2THETA	PEAK	D	H	K	L	I(INT)	I(PK)	I(DS)
65.49		1.4241	8	3	2	1		2
65.50	65.66	1.4239	5	4	1	1		2
65.67	65.66	1.4206	6	0	4	3	3	4
66.04	66.04	1.4135	10	1	0	2	2	3
66.18	66.04	1.4109	10	0	1	1		2
66.22	66.22	1.4100	0	0	5	1		2
66.41	66.42	1.4064	2	0	3	2		3
66.59	66.59	1.4031	8	1	4	5	3	6
67.02	67.02	1.3951	6	1	3	2	2	3
67.15	67.18	1.3928	5	2	4	1	2	2
67.53	67.54	1.3859	10	1	1	1	1	2
68.49	68.48	1.3688	0	5	0	5	3	8
68.73	68.70	1.3646	2	1	2	3	3	4
68.99	68.98	1.3601	1	1	5	2	2	3
69.08		1.3585	8	5	2	3		2
69.92	70.04	1.3442	4	3	1	1		5
69.95		1.3437	0	5	0	6		9
70.03		1.3423	9	2	0	3		5
70.06		1.3420	10	5	2	2		8
70.13	70.60	1.3407	2	3	2	5	4	2
70.59		1.3331	10	0	0	1		3
70.64	71.00	1.3322	2	3	4	2	2	3
71.01	71.54	1.3262	4	2	4	3	2	2
71.48	71.88	1.3186	6	2	1	2		4
71.55	72.94	1.3176	9	2	5	1		2
71.89		1.3121	0	2	5	1		2
72.93	73.16	1.2960	1	2	5	2	2	3
73.15	73.32	1.2927	2	2	4	1	1	4
73.32	73.74	1.2901	8	4	0	2	2	6
73.74	74.26	1.2837	4	3	4	3	2	4
74.26	74.60	1.2761	5	3	4	4	2	3
74.60	74.74	1.2711	0	2	4	3	2	2
74.74		1.2690	8	4	4	2	2	2
75.42	75.42	1.2593	8	0	1	2	1	2
75.61	75.62	1.2565	2	4	4	1	2	2
75.76	75.76	1.2545	9	3	2	2	2	3
76.21	76.20	1.2482	5	1	5	2	1	3

ELPIDITE - NERONOVA AND BELOV, 1965

2THETA	PEAK	D	H	K	L	I(INT)	I(PK)	I(DS)
76.56	76.56	1.2433	6	4	3	2	2	3
77.06	77.04	1.2365	8	3	3	2	2	3
77.26	77.28	1.2338	2	5	3	2	2	3
77.47	77.48	1.2310	6	3	4	3	2	4
77.63	77.64	1.2288	1	6	0	1		2
78.64	78.66	1.2156	2	6	0	1	1	2
78.69		1.2149	0	1	1	1		2
79.02	79.00	1.2107	11	1	2	2	2	3
80.03	80.02	1.1979	2	6	1	1	1	1
80.30	80.28	1.1945	3	6	0	1	1	1
80.45	80.48	1.1927	9	4	1	1	1	1
80.50		1.1922	8	2	4	1		1
80.98	81.14	1.1863	6	1	5	3	2	3
81.12		1.1845	12	1	0	1		1
81.25		1.1830	12	0	1	1		2
81.26		1.1828	4	5	3	3		5
82.11	82.12	1.1728	10	3	2	1	2	1
82.37	82.38	1.1698	9	1	4	3	2	4
82.50	82.64	1.1682	12	1	0	2	2	3
82.62		1.1668	5	4	4	1		2
82.63		1.1667	10	2	3	2		3
82.75	82.84	1.1653	0	6	2	1		1
82.85		1.1641	2	6	2	1	2	4
83.18	83.26	1.1603	2	0	6	2		2
83.24		1.1597	0	5	5	2		2
83.30		1.1590	4	3	1	2		3
83.66	83.54	1.1550	6	2	6	2	2	3
84.00	83.98	1.1511	4	6	1	1	1	3
84.49	84.50	1.1457	2	1	6	2	1	3
84.59		1.1446	9	4	2	1		1
84.72	84.74	1.1432	8	5	0	1	1	1
85.38	85.38	1.1360	12	0	2	1	1	2
85.59	85.62	1.1337	5	6	2	1	1	1
85.82	85.84	1.1313	3	6	2	1	2	2
86.09	86.12	1.1285	8	5	1	2	2	3
86.25	86.24	1.1267	6	4	4	2	2	3
86.72	86.74	1.1219	10	4	5	1	2	1

2THETA	PEAK	D	H	K	L	I(INT)	I(PK)	I(DS)
86.74	86.94	1.1216	8	3	4	2	2	4
86.91		1.1199	0	2	6	1	1	1
86.94		1.1196	2	5	4	1		2
88.08	88.16	1.1081	2	4	5	1	1	1
88.12		1.1077	4	6	2	1		1
88.22		1.1066	2	2	6	2		3
88.43	88.46	1.1045	0	1	6	3		5
88.62	88.62	1.1027	10	1	2	1	2	1
89.02	88.90	1.0987	11	3	5	3	1	2
89.71	89.72	1.0921	0	6	3	1	1	2
89.88	89.90	1.0905	6	3	5	1	1	2
90.20	90.34	1.0874	8	5	2	2		3
90.30		1.0865	9	2	2	3	1	6
90.35		1.0860	12	4	2	3		5
90.83	90.84	1.0815	10	5	4	3		2
90.87		1.0811	4	6	6	1		2
91.07	91.10	1.0793	5	2	4	1	2	4
92.16	92.34	1.0693	4	3	4	2	2	5
92.32		1.0679	9	2	6	3		2
92.34		1.0677	10	3	4	1		2
92.46		1.0667	12	6	0	1		3
93.45	93.50	1.0579	1	7	6	1	1	1
93.54		1.0571	0	3	6	1		4
94.44	94.48	1.0495	13	2	2	1		4
94.45		1.0493	6	6	5	2		2
94.68	94.72	1.0474	9	6	3	1	2	2
94.70		1.0472	6	1	4	1		3
94.96	94.98	1.0450	4	4	4	2	2	3
95.00		1.0447	8	7	6	1		1
95.45	95.30	1.0410	2	5	1	2	10	1
96.23	96.24	1.0346	10	3	0	1	0	2
96.54	96.56	1.0321	12	0	3	1		2
96.58		1.0318	14	3	3	1	1	2
96.98	96.96	1.0286	12	5	4	1	1	3
97.21	97.22	1.0268	6	5	5	2	1	2
97.45	97.46	1.0249	0	4	5	2	1	2
97.96	97.96	1.0209	10	5	5	1	1	3

ELPIDITE - NERONOVA AND BELOV, 1965

2THETA	PEAK	D	H	K	L	I(INT)	I(PK)	I(DS)	2THETA	PEAK	D	H	K	L	I(INT)	I(PK)	I(DS)
98.40	98.42	1.0175	4	3	6	1	1	3	108.46	108.46	0.9493	14	3	0	3		2
98.75	98.74	1.0148	6	2	6	1	2	7	109.20	109.20	0.9450	10	5	3	2		4
99.62	99.68	1.0083	14	1	1	1	1	1	109.54	109.60	0.9429	5	1	7	1	1	1
99.66		1.0080	1	6	4	1		1	109.82		0.9413	9	6	2	1		1
100.21	100.22	1.0040	4	7	1	1	1	1	110.15	110.40	0.9394	6	4	5	1	1	1
100.38	100.38	1.0027	2	7	2	1	1	2	110.37		0.9382	4	2	7	2		3
100.66	100.68	1.0007	2	6	1	1	1	2	110.44		0.9378	5	4	6	1		1
101.00	101.04	0.9983	10	1	5	1	1	2	110.63	110.56	0.9367	10	3	0	1	1	3
101.08		0.9976	9	4	4	1		2	111.35	111.58	0.9327	12	5	5	1	1	1
101.22		0.9966	12	1	1	1	1	2	111.45		0.9321	6	6	0	1		1
101.37	101.38	0.9956	1	1	7	1		2	111.60		0.9313	4	7	4	1		2
101.39		0.9954	6	3	6	1		2	112.07	112.06	0.9287	6	0	7	1	1	2
101.59		0.9940	5	6	3	1		1	112.10		0.9286	10	6	1	1		1
101.75	101.76	0.9929	13	1	3	1	1	1	112.68	112.80	0.9254	12	4	3	1	1	2
101.89		0.9919	12	0	0	1		2	112.93		0.9241	12	5	1	1		2
102.09	102.18	0.9905	10	5	2	2	1	1	113.42	113.56	0.9215	6	1	7	1	1	2
102.23		0.9895	1	4	6	1		1	113.57		0.9207	2	6	5	1		1
102.38		0.9885	2	0	7	1		1	113.64		0.9203	12	3	4	1		1
103.04	103.24	0.9839	8	1	6	1	1	2	113.65		0.9202	0	5	6	1		2
103.16		0.9831	12	4	1	1		2	113.87	114.04	0.9191	15	1	2	1	1	1
103.24		0.9826	2	0	4	2		4	114.25		0.9171	0	8	1	1		2
103.42		0.9813	14	2	6	1		1	114.88	114.88	0.9139	8	3	6	1	1	2
103.81	104.00	0.9787	14	1	1	2	1	3	114.89		0.9138	12	0	5	1		2
103.99		0.9775	8	6	2	2		4	115.09		0.9128	2	5	6	1		2
104.41	104.40	0.9748	4	7	4	3	2	6	115.34	115.38	0.9116	8	7	0	1	1	2
104.69	104.74	0.9729	4	6	5	2	2	4	115.57		0.9104	10	0	1	2		4
104.79		0.9722	9	3	6	1		1	116.26	116.30	0.9070	12	1	5	1	1	2
104.93		0.9713	3	4	6	1		3	116.34		0.9066	0	7	4	1		3
105.44	105.46	0.9680	12	2	7	2	1	3	116.96	116.90	0.9036	10	1	6	1	1	3
105.53		0.9674	6	4	0	1		3	117.47	117.48	0.9011	12	5	2	2	1	3
106.22	106.22	0.9631	2	7	7	1	0	2	117.82	117.88	0.8995	2	7	4	1		2
106.85	106.94	0.9591	8	5	4	1	1	2	117.93		0.8989	4	6	5	1		2
106.96		0.9585	6	7	1	1		2	118.86	118.88	0.8946	6	7	3	1	1	2
107.32	107.36	0.9562	4	4	6	1	1	2	119.73	119.74	0.8906	10	5	4	3	1	4
107.68	107.72	0.9540	14	2	2	1	1	3	120.23	120.28	0.8884	14	0	0	3	1	6
107.76		0.9535	5	6	4	1		1	120.32		0.8880	8	5	5	2		4
108.20	108.18	0.9509	8	2	6	2	1	3									

417

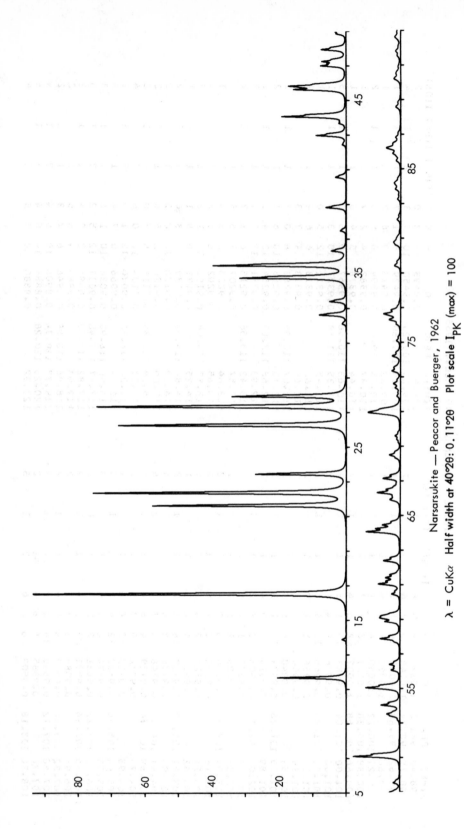

Narsarsukite — Peacor and Buerger, 1962

$\lambda = \text{CuK}\alpha$ Half width at $40°2\theta$: $0.11°2\theta$ Plot scale I_{PK} (max) = 100

418

NARSARSUKITE-PEACOR AND BUERGER, 1962

2THETA	PEAK	D	H	K	L	I(INT)	I(PK)	I(DS)
11.66	11.66	7.585	1	1	0	19	20	19
13.86	13.86	6.386	1	0	1	1	1	1
16.51	16.52	5.363	2	0	0	100	100	100
21.62	21.62	4.107	2	1	1	56	58	59
21.62		4.107	1	2	1	6	6	6
22.35	22.36	3.974	0	0	2	83	76	87
23.44	23.44	3.793	2	2	0	30	27	32
26.25	26.26	3.392	3	1	0	19		21
26.25		3.392	1	3	0	61	68	67
27.33	27.34	3.261	3	0	1	91	74	101
27.92	27.92	3.193	2	0	2	40	34	45
32.10	32.10	2.786	3	2	1	2	1	2
32.61	32.62	2.744	2	2	1	11	8	12
33.38	33.38	2.682	4	0	0	6	5	8
34.74	34.74	2.580	3	1	2	5		6
35.47	35.48	2.528	3	3	0	36	32	44
36.30	36.30	2.473	4	1	1	53	40	65
37.46	37.46	2.399	4	2	0	5	4	6
38.80	38.80	2.319	2	1	3	2	2	3
40.55	40.54	2.223	4	0	2	9	6	11
42.43	42.44	2.129	3	0	3	5	3	6
42.95	42.96	2.104	5	1	0	2	1	2
43.66	43.66	2.071	4	3	1	12	9	17
44.06	44.06	2.054	4	2	2	2	2	2
44.06		2.054	4	2	2	23	19	31
45.62	45.62	1.9870	0	0	4	5	16	7
45.82	45.82	1.9786	3	2	3	22	17	30
45.82		1.9786	2	3	3	20		27
46.99	46.98	1.9322	5	2	1	3	8	4
46.99		1.9322	2	5	1	8		12
47.25	47.24	1.9221	1	1	4	10	7	14
47.93	47.94	1.8963	4	4	0	11	7	16
48.84	48.84	1.8632	0	4	2	4	3	6
48.95	48.96	1.8593	2	5	2	2	3	3
49.50	49.50	1.8397	3	3	4	3	2	4
51.04	51.04	1.7878	0	6	0	23	14	34
53.39	53.40	1.7145	1	4	3	3	3	5
53.50	53.50	1.7114	3	7	0	4	4	7
54.95	54.02	1.6961	6	5	2	9	10	13
55.03	55.04	1.6695	3	3	1	2	3	3
55.03		1.6673	4	4	3	11		16
56.05	56.06	1.6673	0	5	4	5	6	7
56.05		1.6393	5	4	1	1		2
56.38	56.38	1.6393	4	6	1	3	3	5
57.87	57.86	1.6304	5	5	0	1	6	8
57.87		1.5921	2	1	0	5		9
58.86	58.86	1.5677	6	0	0	6	5	6
58.86		1.5677	3	3	1	4		8
59.08	59.08	1.5623	3	6	2	5	6	4
59.18	59.18	1.5600	0	2	0	2	6	11
59.18		1.5600	6	2	2	7		7
60.45	60.44	1.5302	2	4	4	2	3	4
60.45		1.5302	4	2	0	2		3
61.03	61.02	1.5170	2	5	5	4	7	6
61.03		1.5170	5	7	1	2		14
61.39	61.38	1.5089	1	1	0	8	5	12
61.58	61.56	1.5047	2	5	0	7	3	2
62.37	62.38	1.4876	0	4	0	1	5	4
62.37		1.4876	6	6	0	3		12
64.05	64.06	1.4525	4	3	5	7	10	33
64.23	64.22	1.4488	0	7	1	19	8	4
64.45	64.44	1.4445	7	1	4	2	5	13
65.84	65.90	1.4173	1	5	2	8	3	2
65.91		1.4159	5	5	3	1		5
65.91		1.4159	5	4	3	2		3
66.30	66.30	1.4085	7	5	0	11	7	19
66.30		1.4085	3	3	1	2		5
67.13	67.14	1.3932	6	7	4	4	5	7
67.13		1.3932	4	6	2	7		12
68.47	68.48	1.3690	6	3	3	4	3	8
69.58	69.58	1.3499	5	3	4	2	2	3

NARSARSUKITE - PEACOR AND BUERGER, 1962

2THETA	PEAK	D	H	K	L	I(INT)	I(PK)	I(DS)
70.84	70.94	1.3290	0	6	4	5	10	10
70.93		1.3276	7	3	2	10		19
70.93		1.3276	3	7	0	3		5
70.99		1.3265	0	1	3	7		13
71.88	71.88	1.3123	8	1	6	1	1	2
72.35	72.36	1.3049	1	1	0	3	2	6
72.61	72.60	1.3008	2	6	4	5	2	6
73.32	73.32	1.2900	7	2	3	1	3	9
73.48	73.52	1.2877	4	4	5	4	2	2
74.18	74.18	1.2772	3	4	5	1	3	8
74.18		1.2772	6	6	0	3		2
75.08	75.08	1.2642	2	5	6	2		5
76.04	76.04	1.2506	7	5	0	6	1	4
76.30	76.30	1.2470	5	2	5	3	1	11
76.62	76.62	1.2425	2	5	5	7	4	6
76.62		1.2425	1	7	4	4	5	13
79.40	79.42	1.2058	6	4	4	4		7
80.60	80.74	1.1908	8	1	3	2	2	2
80.75		1.1890	3	8	6	1		4
80.75		1.1890	2	4	6	1		2
82.06	82.06	1.1734	5	5	5	3		3
83.25	83.26	1.1596	9	1	4	2	1	5
83.82	83.82	1.1531	1	9	5	3	1	5
84.18	84.18	1.1491	9	3	4	3	2	7
85.45	85.46	1.1352	3	9	2	2	1	6
85.87	85.88	1.1307	6	3	0	1		3
85.87		1.1307	3	6	5	4	2	3
86.19	86.18	1.1274	1	5	6	3		8
86.19		1.1274	5	1	6	6	4	7
86.81	86.80	1.1210	0	7	2	1		13
86.81		1.1210	3	9	0	2	2	3
88.56	88.56	1.1032	0	9	0	4		5
90.18	90.20	1.0876	3	9	3	2	2	9
90.18		1.0876	0	9	6	1		3
90.25		1.0870	7	7	0	3		8
90.61	90.48	1.0836	7	7	0	2	1	4

2THETA	PEAK	D	H	K	L	I(INT)	I(PK)	I(DS)
91.54	91.54	1.0750	3	5	6	2	2	5
91.79	91.80	1.0727	0	10	0	3	2	6
92.61	92.62	1.0653	7	6	3	3	2	8
93.12	93.10	1.0608	3	2	7	2	1	5
93.46	93.48	1.0579	1	10	1	4	1	3
93.65	93.64	1.0562	7	2	4	3	2	9
94.15	94.14	1.0519	10	2	0	2	1	8
94.92	94.96	1.0454	7	7	3	1	1	5
94.98		1.0449	8	5	3	2		3
96.10	96.10	1.0356	0	10	2	1	1	5
98.04	98.04	1.0203	4	7	5	1	1	3
98.04		1.0203	4	3	5	2		5
99.75	99.74	1.0074	5	5	7	3		5
100.26	100.26	1.0035	4	3	6	1		4
101.06	101.08	0.9978	1	10	3	3		8
101.31	101.34	0.9960	6	2	0	3		7
102.15	102.22	0.9900	1	10	6	1		2
102.26		0.9893	9	3	8	3		2
102.87	102.88	0.9851	9	3	4	1	1	8
103.21	103.20	0.9827	3	10	2	1		2
103.21		0.9827	10	5	2	3		3
105.74	105.74	0.9661	4	10	4	1	1	2
105.74		0.9661	2	2	8	2		4
106.54	106.54	0.9611	10	2	2	3		7
107.04	107.00	0.9579	3	10	3	1		4
107.04		0.9579	10	5	0	1		2
107.75	107.88	0.9536	0	10	9	1		4
107.92		0.9525	10	5	1	1		2
107.92		0.9525	2	11	4	1		1
109.37	109.42	0.9439	9	8	0	2		2
109.91	109.92	0.9408	11	3	0	3	1	3
109.91		0.9408	7	6	5	2	1	5
110.25	110.26	0.9389	7	6	4	3	1	9
111.90	112.18	0.9296	10	2	4	2	1	6
112.06		0.9288	9	6	3	1		3
112.17		0.9281	2	8	6	4		10

NARSARSUKITE - PEACOR AND BUERGER, 1962

2THETA	PEAK	D	H	K	L	I(INT)	I(PK)	I(DS)
112.61	112.62	0.9258	6	3	7	1	1	2
112.79		0.9248	8	5	5	2		4
112.79		0.9248	8	8	5	1		3
113.18	113.18	0.9227	5	9	4	1	1	3
113.73	113.74	0.9198	10	6	0	2	1	6
113.73		0.9198	6	10	0	1		3
114.10	114.12	0.9179	2	4	8	2	1	5
114.10		0.9179	4	2	8	2		5
114.56	114.56	0.9155	7	9	2	2	2	7
114.56		0.9155	11	3	2	1		4
114.56		0.9155	3	11	2	1		2
115.19	115.02	0.9123	0	7	7	2	1	5
117.26	117.28	0.9021	2	11	3	2	1	7
118.02	118.06	0.8985	4	9	5	1	1	3
118.05		0.8984	5	1	8	2		5
118.05		0.8984	1	5	8	3		10
118.53	118.52	0.8961	10	6	2	2	1	5
118.53		0.8961	6	10	2	2		4
119.78		0.8904	4	10	4	2		5
120.08	120.12	0.8891	8	4	6	3	1	8
120.37	120.72	0.8878	5	11	0	5		2
120.74		0.8861	1	10	5	1		14
120.93		0.8853	9	3	1	1		2
123.56	123.60	0.8742	5	5	8	2	1	2
123.56		0.8725	3	11	8	1		5
123.96	124.02	0.8721	1	0	4	1	1	3
124.06		0.8721	0	12	2	1		3
125.50	125.54	0.8664	11	5	2	1	0	2
126.20	126.40	0.8637	4	7	7	1	1	2
126.41		0.8629	3	10	5	2		2
126.41		0.8629	10	3	5	1		7
127.90	127.92	0.8573	0	3	9	2	0	2
127.92		0.8573	2	6	8	2		5
128.35	128.42	0.8557	8	8	4	1	0	2
129.87	129.88	0.8503	11	1	4	1	1	4
129.87		0.8503	7	9	4	1		4
129.87		0.8503	3	11	4	1		3
130.53	130.50	0.8480	4	12	0	1	1	7
131.63	131.64	0.8444	9	8	3	2	1	7
131.63		0.8444	12	1	3	1		4
131.63		0.8444	1	12	3	1		3
132.11	132.24	0.8428	9	6	0	1		3
132.54		0.8414	9	9	5	1		3
133.72	133.76	0.8376	10	8	0	1	1	6
133.72		0.8376	6	10	0	2		2
134.66	134.90	0.8347	7	6	4	1	1	7
134.87		0.8341	10	7	3	2		11
135.02		0.8336	0	10	3	3		2
135.02		0.8336	6	8	6	1		4
135.87	135.84	0.8311	1	7	8	1	2	2
135.87		0.8311	7	1	8	1		3
135.87		0.8311	5	5	8	7		23
136.47	136.50	0.8294	4	12	2	3	1	9
137.59	137.56	0.8262	6	4	8	1	1	3
137.59		0.8262	4	6	8	1		3
138.85	139.32	0.8227	13	1	0	1	2	4
138.85		0.8227	1	13	0	1		7
139.08		0.8221	0	9	7	2		6
139.34		0.8214	10	9	5	2		5
139.34		0.8214	2	11	5	6		20
139.60		0.8207	0	13	1	1		2
140.02		0.8196	8	10	2	2		4
142.84	142.88	0.8126	7	6	7	1	1	7
142.84		0.8126	9	2	7	1		2
143.39		0.8113	13	1	1	1		3
145.76	146.02	0.8060	12	2	4	1	1	5
145.76		0.8060	2	12	4	2		8
145.91		0.8056	11	7	2	2		2
145.91		0.8056	13	1	2	2		6
145.91		0.8056	1	13	2	1		2
146.67	146.68	0.8040	13	3	0	9	1	29
146.67		0.8040	3	13	0	2		5
148.89	148.78	0.7995	12	6	0	2	0	6

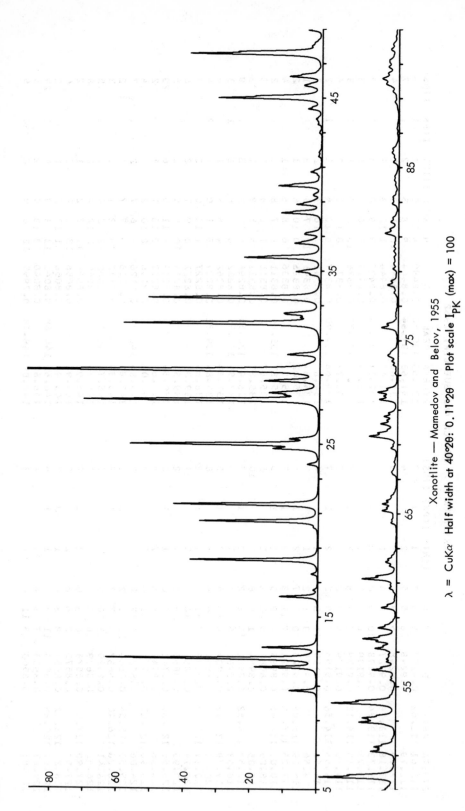

Xonotlite — Mamedov and Belov, 1955

$\lambda = CuK\alpha$ Half width at $40°2\theta$: $0.11°2\theta$ Plot scale I_{PK} (max) = 100

422

XONOTLITE - MAMEDOV AND BELOV, 1955

2THETA	PEAK	D	H	K	L	I(INT)	I(PK)	I(DS)
10.69	10.70	8.265	0	0	2	6	8	4
12.06	12.08	7.330	0	0	0	14	19	10
12.56	12.58	7.040	1	0	0	49	63	37
13.20	13.20	6.701	0	1	1	13	16	9
16.15	16.16	5.484	0	1	1	9	11	7
17.45	17.46	5.077	1	1	0	2	2	1
18.26	18.26	4.854	-1	1	1	30	38	24
18.26		4.854	-1	1	1	3		2
20.51	20.52	4.326	-1	1	2	28	36	24
20.51		4.326	1	1	2	39		33
21.48	21.50	4.132	-1	0	4	3	43	2
23.81	23.82	3.734	1	1	3	11	4	10
24.71	24.72	3.600	0	1	4	15	14	14
24.96	24.96	3.564	-1	0	4	37	56	34
24.96		3.564	-1	0	4	6		6
25.28	25.28	3.520	2	0	0	63	9	61
27.52	27.52	3.239	-2	0	2	5	70	5
27.52		3.239	-2	0	2	4		4
27.81	27.82	3.205	-1	0	4	2	11	1
27.95	27.94	3.190	-1	1	4	11	15	11
28.62	28.62	3.116	-1	2	1	5	17	5
28.62		3.116	-2	1	1	11		11
29.24	29.24	3.052	0	2	1	100	100	100
30.14	30.14	2.962	2	1	2	2	10	2
30.14		2.962	-2	1	2	7		
31.94	31.94	2.800	-2	2	3	60	59	64
32.28	32.28	2.770	-1	1	5	3	6	3
32.47	32.54	2.755	1	2	6	7	11	8
32.53		2.750	0	1	3	5		
33.41	33.42	2.680	2	0	4	53	51	58
33.41		2.680	-2	0	4	3		
34.76	34.76	2.579	0	1	6	7	3	3
34.94	34.94	2.566	1	2	6	3	7	8
35.75	35.76	2.509	-2	2	1	22	23	25
35.75		2.509	-2	2	1	8		
36.57	36.58	2.455	0	2	5	8	8	10

2THETA	PEAK	D	H	K	L	I(INT)	I(PK)	I(DS)
37.09	37.10	2.422	-1	1	6	5	5	5
37.17	37.16	2.417	0	3	0	2	5	2
38.32	38.32	2.347	3	0	3	13	12	16
38.82	38.82	2.318	-1	2	5	7	7	8
38.82		2.318	-1	2	5	1		2
38.99	38.90	2.308	-1	3	0	1	5	1
39.38	39.38	2.286	1	3	2	9	13	2
39.90	39.90	2.257	-3	0	1	5	3	12
39.90		2.257	-3	0	1	3		6
40.70	40.70	2.215	-3	0	2	1	3	3
43.77	43.78	2.066	-1	1	8	5	2	2
44.35	44.36	2.041	-3	0	4	37	4	6
44.98	44.98	2.014	-3	0	5	2	31	51
45.66	45.72	1.9851	-1	2	5	4	5	3
45.72		1.9826	1	0	8	2		5
46.22	46.22	1.9623	-3	2	1	9	10	3
46.22		1.9623	-3	2	1	49		13
47.55	47.56	1.9106	-1	2	7	1	39	70
47.98	47.98	1.8944	-2	1	3	4	3	2
48.21	48.18	1.8859	-2	3	3	1	2	6
48.92	48.92	1.8602	-3	2	5	30	4	45
49.17	49.16	1.8515	3	1	0	7	23	10
49.71	49.72	1.8325	-2	4	8	1	7	2
51.22	51.22	1.7819	-1	4	0	7	6	11
51.22		1.7819	-2	2	8	12		19
51.49	51.48	1.7734	-1	4	0	11	11	17
52.91	52.90	1.7291	-2	2	5	23	10	37
53.16	53.16	1.7214	-4	0	5	3	19	5
54.01	54.02	1.6963	3	2	2	1	2	16
54.01		1.6963	-3	3	2	10	7	5
55.36	55.36	1.6581	-3	0	9	3	3	2
55.95	55.96	1.6420	0	4	0	1	3	16
56.41	56.42	1.6297	-1	4	10	7	6	5
56.57	56.56	1.6254	-2	4	9	8	6	11
57.19	57.20	1.6092	-1	0	2		10	11
57.59	57.60	1.5991	0	2	9			13
57.76	57.76	1.5949	2	4	5			

XONOTLITE - MAMEDOV AND BELOV, 1955

2THETA	PEAK	D	H	K	L	I(INT)	I(PK)	I(DS)
57.76		1.5949	-2	4	2	2		4
58.38	58.38	1.5793	-4	2	1	4	5	7
58.38		1.5793	-4	2	8	3		5
59.56	59.56	1.5508	3	0	3	5	5	10
60.69	60.68	1.5246	-4	2	4	2	2	3
61.22	61.22	1.5126	-2	0	10	15	10	27
61.97	61.98	1.4962	-2	4	6	2	2	4
62.20	62.20	1.4912	-2	0	9	3	3	5
62.34	62.36	1.4881	2	4	1	3	3	2
64.46	64.46	1.4443	3	2	0	2	2	4
65.16	65.16	1.4304	-4	4	5	5	5	10
65.56	65.56	1.4227	-3	2	2	4	4	9
66.00	66.00	1.4141	3	4	0	1	1	2
67.28	67.40	1.3904	0	2	11	2	2	3
68.00	68.00	1.3775	0	0	12	1	1	2
69.38	69.46	1.3533	2	1	0	2	2	4
69.47		1.3519	-1	0	12	9	8	19
69.65	69.66	1.3488	-2	5	1	2	6	2
69.85	69.84	1.3454	3	2	9	2	3	3
70.18	70.18	1.3398	-4	0	8	3	5	6
70.18		1.3328	5	0	4	4		9
70.61	70.62	1.3328	-4	0	0	6	4	12
71.59	71.60	1.3169	4	2	7	1	6	2
71.59		1.3169	-4	2	1	6		13
72.01	72.02	1.3102	-5	5	7	8	6	18
73.58	73.58	1.2861	2	2	4	1	1	3
73.81	73.80	1.2828	2	5	12	2	2	3
74.09	74.14	1.2785	5	2	3	2	3	4
74.16		1.2775	-4	4	8	2		5
75.75	75.74	1.2547	0	4	2	8	5	18
77.74	77.74	1.2274	5	2	10	2	1	4
78.20	78.20	1.2214	-1	4	5	4	3	10
79.14	79.14	1.2092	-3	2	10	2	2	5
80.17	80.16	1.1962	-3	1	11	1	1	3
80.45	80.44	1.1927	0	6	3	1	1	3
81.15	81.18	1.1842	-1	2	13	1	4	3
81.18		1.1838	3	4	8	2		10
81.18	81.84	1.1838	-3	4	8	2		4
81.84	82.90	1.1759	-1	6	3	3	2	7
82.90		1.1635	5	0	8	1	2	3
82.90		1.1635	-5	0	8	2		4
83.30	83.30	1.1590	-2	4	10	2	2	6
83.98	83.98	1.1513	-2	6	1	1	1	3
84.24	84.24	1.1484	5	2	7	1	1	3
84.47	84.48	1.1459	0	6	5	1	1	3
85.30	85.30	1.1369	-2	2	13	3	2	8
85.84	85.84	1.1310	1	6	5	2	2	6
88.78	88.78	1.1011	0	4	12	2	1	5
89.39	89.40	1.0952	-6	2	3	3	2	9
89.82	89.82	1.0910	-4	2	11	2	4	6
89.82		1.0910	-4	2	11	4		12
89.96	90.14	1.0896	2	6	5	2	5	6
90.15		1.0879	-1	4	12	5		13
90.45	90.44	1.0851	0	6	7	2	3	5
90.82	90.82	1.0816	-4	6	8	2		4
90.82		1.0816	-4	4	8	1	3	7
91.04	91.14	1.0795	-6	0	6	3		6
91.22		1.0778	5	4	4	2		9
91.82	91.84	1.0724	-1	6	7	4	3	10
91.88		1.0719	5	0	10	1		3
93.82	93.82	1.0547	3	0	14	1	1	4
94.27	94.28	1.0509	2	4	12	3	2	8
95.13	95.12	1.0437	-1	2	15	4	2	11
95.95	95.94	1.0369	-2	6	7	1	1	4

TABLE 20.   DOUBLE CHAIN ALUMINO-SILICATES

Variety	Sillimanite	2:1 Mullite	1.92:1 Mullite	1.71:1 Mullite
Composition	Al_2SiO_5	$Al_{4.8}Si_{1.2}O_{9.6}$ $(2Al_2O_3 \cdot SiO_2)$	$Al_{4.76}Si_{1.24}O_{9.62}$ $(1.92Al_2O_3 \cdot SiO_2)$	$Al_{4.62}Si_{1.37}O_{9.68}$ $(1.71Al_2O_3 \cdot SiO_2)$
Source	LaBelle Co., Quebec, Harvard No. 97215	Synthetic	Synthetic	Synthetic
Reference	Burnham, 1963a	Sadanaga, Tokonami, & Takéuchi, 1962	Burnham, 1964,1965	Durović, 1962
Cell Dimensions				
a Å	7.4856	7.583	7.584	7.549
b Å	7.6738	7.681	7.693	7.681
c	5.7698	2.8854	2.890	2.884
α	90	90	90	90
β deg	90	90	90	90
γ	90	90	90	90
Space Group	Pbnm	Pbam	Pbam	Pbam
Z	4	1	1	1

TABLE 20. (cont.)

Variety	Sillimanite	2:1 Mullite	1.92:1 Mullite	1.71:1 Mullite

Site Occupancy	Al_1 1.0
	Al_2 1.0
	Si 1.0
	Oa 1.0
	Ob 1.0
	Oc 1.0
	Od 2.0
		Al 2.0 (1.0)
		M_1 {Al 2.0 (0.80)
		     {Si 1.2 (0.20)
		M_2 Al 0.8 (1.0)
		O_1 4.0 (1.0)
		O_2 4.0
		O_3 0.8 (0.40)
		O_4 0.8 (0.20)
			Al 2.0 (1.0)
			M_1 {Al 2.0 (0.81)
			     {Si 1.24 (0.19)
			M_2 Al 0.76 (1.0)
			O_1 4.0 (1.0)
			O_2 4.0 (1.0)
			O_3 0.86 (0.43)
			O_4 0.76 (0.19)
				Al 2.0 (1.0)
				M_1 {Al 2.20 (0.842)
				     {Si 1.42 (0.158)
				M_2 {Al 0.416 (0.416)
				      {Si 0.216 (0.216)
				O_1 4.0 (1.0)
				O_2 4.0 (1.0)
				O_3 1.68 (0.42)
				O_4 0.0 (0.0)

Method	3-D, counter	2-D, film	3-D, counter	2-D, film
R & Rating	0.056 (1)	0.11 (2)	0.068 (1)	0.095 (av.) (2)
Cleavage and habit	{010} perfect Columnar or fibrous parallel to $\underline{c}$ axis	{010} good — Prismatic		
Comment			Defect sillimanite structures	
μ	101.6	99.7	99.8	101.4
ASF	0.1073	0.6706×10^{-1}	0.7068×10^{-1}	0.7710×10^{-1}
Abridging factors	0.5:0:0	0.5:0:0	0.5:0:0	0.5:0:0

426

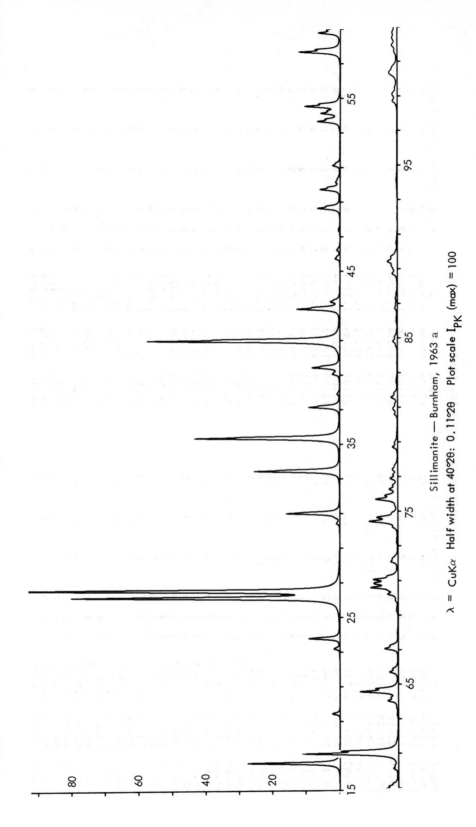

Sillimanite — Burnham, 1963 a

$\lambda = CuK\alpha$ Half width at 40°2θ: 0.11°2θ   Plot scale I_{PK} (max) = 100

SILLIMANITE - BURNHAM, 1963a

2THETA	PEAK	D	H	K	L	I(INT)	I(PK)	I(DS)
16.53	16.54	5.358	1	1	0	24	28	23
19.41	19.42	4.570	1	0	1	2	2	2
23.16	23.16	3.837	0	2	0	2	2	2
23.75	23.76	3.743	2	0	0	9	9	9
26.07	26.08	3.414	1	2	0	79	80	79
26.47	26.48	3.364	2	1	0	100	100	100
30.39	30.40	2.938	1	2	1	1	1	1
30.97	30.98	2.885	0	0	2	17	16	18
33.42	33.42	2.679	2	2	0	29	26	30
35.30	35.30	2.540	1	1	2	51	43	53
37.12	37.12	2.421	1	3	0	11	9	12
37.88	37.88	2.373	3	1	0	1	1	1
39.03	39.04	2.306	0	2	2	2	2	3
39.40	39.40	2.285	2	0	2	10	8	10
40.92	40.92	2.204	2	2	1	71	58	77
42.78	42.78	2.112	2	3	0	16	13	18
43.21	43.22	2.092	3	2	0	3	3	4
45.71	45.72	1.9832	2	3	1	2	2	2
46.20	46.20	1.9632	4	0	0	1	1	1
47.34	47.34	1.9185	0	4	1	1	1	1
48.61	48.62	1.8714	3	1	2	9	7	10
49.71	49.70	1.8326	4	1	0	8	6	9
50.13	50.14	1.8181	4	1	1	2	1	2
50.37	50.26	1.8102	1	1	3	1	1	1
51.09	51.10	1.7861	3	3	1	3	2	3
53.64	53.64	1.7072	4	2	0	9	7	11
54.11	54.12	1.6935	3	2	2	8	6	9
54.51	54.50	1.6820	0	4	2	15	11	18
57.65	57.66	1.5975	2	4	2	18	12	22
58.76	58.76	1.5700	4	3	0	9	6	12
59.08	59.08	1.5623	1	1	4	3	3	4
60.10	60.10	1.5381	4	2	2	2	2	2
60.95	60.96	1.5186	3	3	2	44	29	55
63.63	63.64	1.4611	4	3	1	1	1	1
64.02	64.02	1.4531	4	0	2	5	3	6
64.55	64.56	1.4424	4	4	0	19	12	24
65.70	65.70	1.4200	2	0	4	4	3	5

2THETA	PEAK	D	H	K	L	I(INT)	I(PK)	I(DS)
67.05	67.04	1.3947	5	2	0	7	4	9
67.15		1.3929	1	1	4	1		1
69.25	69.26	1.3557	5	1	1	1	1	1
69.85	69.84	1.3454	3	4	2	1	1	1
70.20	70.20	1.3396	4	4	0	1	3	6
70.58	70.58	1.3333	1	5	2	5	8	17
70.86	70.82	1.3287	2	1	4	13	7	9
71.04	71.04	1.3257	5	3	0	8	8	11
72.07	72.08	1.3094	5	1	2	3	2	5
72.20	72.22	1.3073	3	5	0	1	2	2
73.19	73.18	1.2921	2	3	3	2	2	3
74.06	74.08	1.2790	0	6	0	2	9	20
74.40	74.40	1.2740	2	5	2	15	6	5
74.67	74.62	1.2701	2	2	4	4	7	16
75.32	75.34	1.2607	6	2	0	12	3	6
75.67	75.68	1.2557	5	0	2	4	2	6
76.25	76.26	1.2476	6	0	0	4	3	2
76.87	76.86	1.2391	1	3	4	1	2	3
77.35	77.42	1.2326	3	1	4	2	2	2
77.44		1.2314	6	1	0	3		5
78.68	78.68	1.2150	4	5	0	2	2	1
79.05	78.92	1.2103	2	4	4	4	1	1
80.58	80.58	1.1911	6	3	0	2	1	3
80.88	80.90	1.1875	3	2	4	1	1	1
80.94		1.1867	5	1	4	4		1
80.96		1.1865	0	5	2	3	1	2
81.57	81.56	1.1792	5	3	2	4	2	5
82.41	82.42	1.1692	4	6	0	1	1	1
84.54	84.56	1.1451	0	6	2	1	1	1
84.78	84.78	1.1425	4	0	4	2	1	3
84.79		1.1424	5	2	4	1		1
85.94	85.96	1.1300	4	4	4	1	2	1
86.69	86.72	1.1222	3	3	4	1	0	2
86.77		1.1213	6	3	0	1	1	1
87.29	87.28	1.1160	2	6	2	1	1	2
88.70	88.70	1.1018	6	4	2	4	2	5
89.15	89.16	1.0975	2	5	4	2	2	2

SILLIMANITE – BURNHAM, 1963 a

2THETA	PEAK	D	H	K	L	I(INT)	I(PK)	I(DS)
89.17		1.0973	6	2	2	1		2
89.41	89.42	1.0950	4	2	4	5	3	8
89.68	89.68	1.0924	5	4	2	1	2	2
90.49	90.48	1.0847	1	7	0	1	0	1
94.13	94.16	1.0521	2	7	0	1	0	1
98.69	98.70	1.0153	1	7	2	4	2	7
99.13	99.08	1.0119	3	5	4	3	2	5
100.25	100.34	1.0037	2	7	0	4	3	7
100.39		1.0027	5	1	4	4		7
101.56	101.58	0.9942	7	2	2	2	1	4
101.93	101.94	0.9916	4	6	2	2	2	6
102.39	102.36	0.9884	2	1	2	2	2	4
102.65	102.66	0.9866	7	3	0	3	2	5
103.14	103.14	0.9833	6	4	2	3	2	6
103.39	103.38	0.9816	4	4	0	4	2	6
105.34	105.36	0.9687	3	5	4	1	1	2
106.32	106.38	0.9624	5	3	4	2	0	3
107.20	107.20	0.9570	0	6	4	1	1	3
108.11	108.12	0.9514	1	8	0	2	0	2
108.47	108.50	0.9492	1	6	4	1	1	4
108.93	108.94	0.9465	1	6	0	2	1	4
109.43	109.42	0.9436	6	0	4	4	1	7
110.66	110.78	0.9366	6	1	0	2	1	3
110.81		0.9357	8	0	0	2		3
111.19	111.14	0.9335	7	3	0	1	1	1
111.58	111.56	0.9314	2	0	6	1	1	1
111.98	112.02	0.9292	2	8	0	1	1	1
112.05		0.9288	8	1	0	1		1
112.36	112.64	0.9271	2	6	2	1		1
112.64		0.9256	1	6	6	4	1	6
114.12	114.16	0.9178	6	5	2	4	1	6
114.38		0.9164	4	5	4	1		2
114.41		0.9163	6	6	2	1		1
115.60	115.60	0.9102	0	8	0	2	0	4
119.19	119.20	0.8931	6	6	0	1	1	1
120.93	121.14	0.8853	6	3	4	1	1	1
121.11		0.8845	5	7	0	1		2

2THETA	PEAK	D	H	K	L	I(INT)	I(PK)	I(DS)
121.20		0.8841	8	1	2	1		2
123.66	123.66	0.8737	3	2	6	1	0	2
125.37	125.38	0.8669	1	7	4	1	1	2
127.27	127.28	0.8597	0	4	6	3	1	6
128.46	128.50	0.8553	4	0	6	3	1	5
128.81	128.94	0.8541	4	8	0	1	1	1
128.93		0.8536	1	6	6	1		2
129.95	130.00	0.8501	4	1	6	1	1	4
129.97		0.8500	2	7	4	2		2
130.79	130.92	0.8472	1	9	0	2	2	20
130.93		0.8467	3	3	6	6		4
131.25		0.8456	5	1	2	2		2
133.15	133.16	0.8394	7	5	0	6	1	11
134.63	134.56	0.8348	4	2	6	1	0	2
135.79	135.76	0.8313	2	9	0	6	0	2
138.44	138.44	0.8239	3	7	4	10	1	20
142.12	142.16	0.8144	7	3	4	7	1	14
142.73	142.86	0.8129	9	2	0	2	1	4
142.74		0.8128	1	9	2	1		2
143.00		0.8122	7	6	0	1		2
143.91	143.88	0.8101	1	5	6	6	1	11
144.98	145.04	0.8077	5	8	0	3	1	5
145.11		0.8074	8	4	2	6		12
145.36		0.8068	3	9	0	1		2
145.60	145.94	0.8063	5	6	4	2	1	4
146.38		0.8046	5	5	4	2		3
146.76		0.8038	6	4	0	1		5
149.20	149.30	0.7989	8	5	0	3	1	5
149.25		0.7988	2	5	2	2		1
149.30		0.7987	0	8	4	6		6
150.64	150.62	0.7962	2	9	2	10	1	20
151.40	151.74	0.7949	9	1	2	1	1	1
151.77		0.7942	6	7	0	4		9
153.16	153.22	0.7919	5	2	6	2	1	5
153.27		0.7917	7	2	4	10		20
153.46		0.7914	5	5	4	1		1
153.69		0.7910	4	7	4	1		2

SILLIMANITE - BURNHAM, 1963 a

2THETA	PEAK	D	H	K	L	I(INT)	I(PK)	I(DS)
158.48	160.02	0.7840	7	4	4	1	1	2
159.79		0.7824	9	2	2	10	1	21
160.80		0.7812	4	4	6	3		6
160.84		0.7811	2	8	4	4		9
161.03		0.7809	8	1	4	3		6
161.66		0.7802	4	8	3	1		2
162.80		0.7790	4	6	5	1		1
164.06		0.7778	5	8	2	3		5

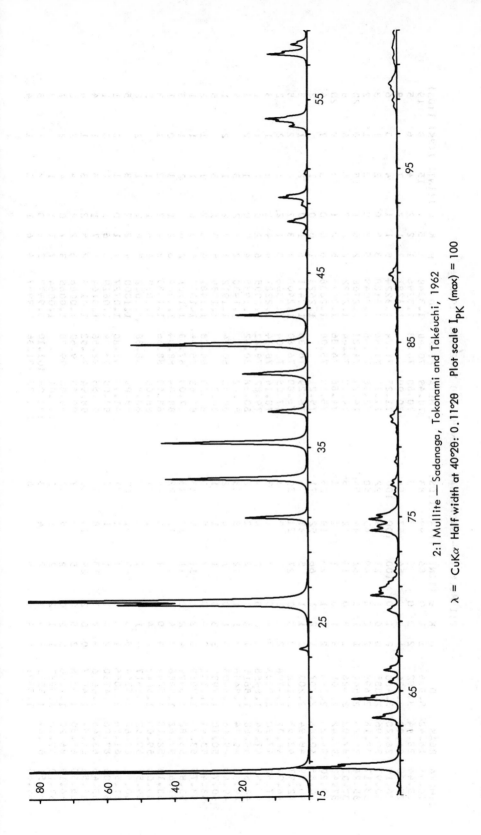

2:1 Mullite — Sadanaga, Tokonami and Takéuchi, 1962

λ = CuKα Half width at 40°2θ: 0.11°2θ Plot scale I_{PK} (max) = 100

2:1 MULLITE – SADANAGA, TOKONAMI AND TAKÉUCHI, 1962

2THETA	PEAK	D	H	K	L	I(INT)	I(PK)	I(DS)
16.41	16.42	5.396	1	1	0	86	99	83
23.44		3.792	2	0	0	3	3	3
25.98	26.00	3.426	1	2	0	54	57	54
26.19	26.20	3.400	2	1	0	100	100	100
30.96	30.97	2.885	0	0	1	21	19	22
33.17	33.18	2.698	2	2	0	49	43	50
35.24		2.544	1	1	1	51	44	54
37.02	37.03	2.426	1	3	0	14	12	15
37.42		2.401	3	1	0	1	1	1
39.20		2.296	2	0	1	24	19	25
40.85	40.86	2.207	1	2	1	66	52	71
40.94	40.99	2.200	2	1	1	5	37	5
42.57	42.58	2.122	2	3	0	27	21	30
46.01	46.02	1.9708	2	2	1	1	0	1
47.30		1.9203	0	4	0	2	1	2
47.94	47.95	1.8958	4	0	0	8	6	9
48.88	48.89	1.8615	1	4	0	2	2	2
49.33	49.34	1.8456	3	1	1	12	8	13
49.46	49.48	1.8405	4	1	0	1	6	2
50.70	50.71	1.7988	3	3	0	7	5	8
53.44		1.7131	2	4	0	11	9	13
53.75	53.76	1.7039	3	2	1	11	12	13
53.88	53.89	1.6999	4	2	0	18	12	22
57.61	57.62	1.5986	0	4	1	7	5	8
58.18		1.5844	4	0	1	1	1	1
59.00		1.5642	1	4	1	1	1	2
59.52		1.5517	4	1	1	1	1	1
60.61	60.62	1.5264	3	3	1	49	31	60
61.54		1.5056	1	5	0	1	1	1
63.05	63.06	1.4730	2	4	1	1	1	2
63.46		1.4646	4	2	1	13	8	17
64.54		1.4427	0	0	2	24	14	30
65.50		1.4238	2	5	0	5	3	7
66.19	66.20	1.4106	1	0	2	8	5	10
67.10		1.3937	1	1	2	2	1	2
69.51		1.3511	3	4	1	2	2	3
69.63	69.64	1.3491	4	4	0	5	4	6
70.49	70.48	1.3348	1	5	1	15	8	19
70.80	70.88	1.3296	1	2	2	4	6	5
70.90		1.3281	2	1	2	7		9
71.25	71.24	1.3224	5	1	1	4	3	6
71.85	71.86	1.3128	3	5	0	2	1	3
72.36	72.36	1.3049	5	3	0	2	1	2
74.21	74.22	1.2768	2	5	1	15	8	20
74.52	74.44	1.2722	2	2	2	6	6	8
74.86	74.86	1.2673	5	2	1	15	9	20
75.10	75.08	1.2638	6	0	0	3	7	4
75.21		1.2623	1	6	0	1		2
76.29	76.30	1.2471	6	1	0	2	1	3
76.80	76.80	1.2400	1	3	2	4	2	5
77.05	77.04	1.2366	3	1	2	1	2	2
78.14	78.14	1.2221	4	4	1	2	1	3
79.82	79.82	1.2005	6	2	0	1	1	1
80.38	80.42	1.1935	2	3	2	2	2	2
80.42		1.1930	3	2	2	2		4
80.58	80.74	1.1912	5	4	0	3	2	1
80.76		1.1889	4	5	0	1		5
84.28	84.28	1.1481	4	0	2	3	1	2
85.63	85.64	1.1333	6	3	0	1	1	1
87.08	87.08	1.1181	2	6	1	1	1	3
88.04	88.04	1.1084	6	2	1	2	1	3
88.53	88.56	1.1035	2	4	2	2	2	3
88.60		1.1029	4	2	2	1		2
88.87	88.88	1.1002	5	4	1	3	1	2
88.89		1.1000	4	2	2	1		5
91.79	91.80	1.0727	7	1	0	1	2	1
93.00	93.04	1.0619	4	6	0	1	1	4
98.55	98.56	1.0164	1	7	1	3	0	4
98.94	98.92	1.0134	7	0	1	3	0	6
99.58	99.60	1.0086	5	2	2	4	1	7
99.86	99.88	1.0065	3	7	0	1	1	2
100.01		1.0054	7	1	1	1		3
101.08	101.10	0.9977	7	3	0	2	1	3
101.35	101.36	0.9957	4	6	1	2	1	4

2:1 MULLITE – SADANAGA, TOKONAMI AND TAKÉUCHI, 1962

2THETA	PEAK	D	H	K	L	I(INT)	I(PK)	I(DS)
101.96	101.96	0.9914	6	4	1	3	1	5
102.15		0.9900	2	7	1	1		5
102.83	102.82	0.9854	4	4	2	3	1	2
104.99	105.00	0.9710	3	5	2	1	0	2
105.48	105.44	0.9677	5	3	2	1	0	2
107.10	107.12	0.9575	0	6	2	2	0	2
108.24	108.30	0.9507	6	0	2	2	1	4
108.35		0.9500	1	6	0	1		2
108.70	108.72	0.9479	8	1	2	2	1	3
108.87		0.9469	1	7	2	2		3
109.45	109.44	0.9435	7	4	0	1	1	1
109.46		0.9435	6	1	2	2		3
109.92	109.86	0.9407	8	0	0	1	1	1
111.42	111.46	0.9323	2	8	0	1	0	2
111.70	111.74	0.9307	2	8	0	1	0	1
112.57	112.60	0.9260	1	2	3	2	1	4
112.84	112.86	0.9245	6	5	1	2	1	3
113.17		0.9228	6	2	2	1		1
115.45	115.46	0.9110	0	8	1	1	0	2
117.83	117.84	0.8994	6	6	0	1	0	2
118.90	118.92	0.8944	8	1	1	1	0	2
119.24	119.58	0.8928	3	3	3	1		1
119.60		0.8912	6	7	2	1		2
120.09	120.08	0.8890	5	5	0	1	0	2
123.29	123.30	0.8753	3	2	3	1	0	2
125.18	125.20	0.8676	1	7	2	1	0	1
126.97	127.20	0.8608	7	1	2	1	1	4
127.19		0.8577	0	4	0	2		2
127.80	127.80	0.8565	4	8	3	1	1	2
128.12		0.8547	4	6	0	1		1
128.63	128.60	0.8500	8	4	2	1	0	1
129.98	130.52	0.8496	5	7	0	1	2	2
130.08		0.8482	3	3	3	8		14
130.50		0.8481	1	9	0	2		3
130.52	131.04	0.8464	7	5	1	4	2	8
131.03			9	1	0	1	0	2
133.75	133.88	0.8375						

2THETA	PEAK	D	H	K	L	I(INT)	I(PK)	I(DS)
133.89	133.89	0.8371	4	2	3	2		4
135.37	135.32	0.8326	2	9	0	1	0	1
137.84	137.86	0.8255	9	7	2	8	1	15
138.76	138.58	0.8230	7	3	0	1	1	2
139.66	139.64	0.8206	8	4	2	3	1	6
141.72	141.76	0.8153	3	4	3	3		5
142.20	142.44	0.8141	8	4	1	1	0	1
142.40		0.8137	1	9	1	1		2
143.42	143.70	0.8112	5	8	0	5		9
143.72		0.8105	1	5	3	1	1	1
144.66	144.62	0.8084	6	5	2	1		3
144.95		0.8077	8	1	3	1		5
145.43		0.8067	2	9	0	2		1
148.67	148.74	0.8000	2	5	1	3	0	13
150.23	150.34	0.7970	6	7	1	7	1	2
150.56		0.7964	1	8	1	1		7
151.39	151.46	0.7949	5	2	2	3	1	14
151.53		0.7947	4	7	3	7		1
152.34	152.84	0.7932	8	0	2	1		12
152.97		0.7922	9	2	1	6		9
153.44		0.7914	4	8	3	4	0	3
159.17	160.00	0.7831	2	8	2	2		6
160.02		0.7821	3	9	1	3		8
163.20		0.7786	4	9	0	4		1
163.59		0.7782						

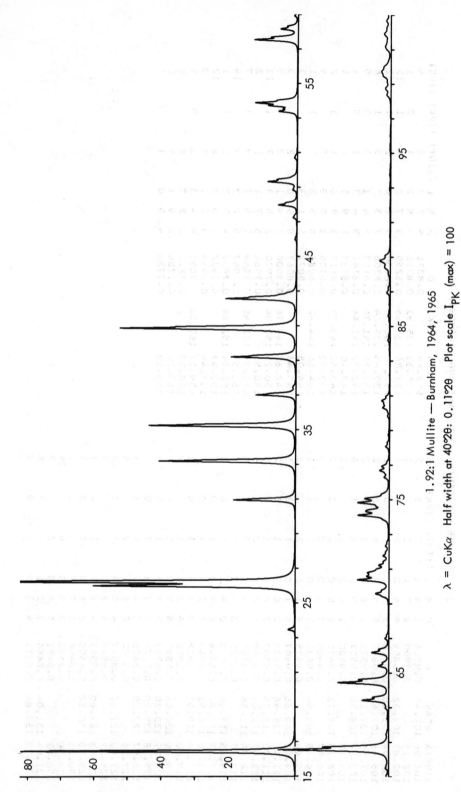

1.92:1 Mullite — Burnham, 1964, 1965

$\lambda = CuK\alpha$ Half width at $40°2\theta$: $0.11°2\theta$ Plot scale I_{PK} (max) = 100

434

2THETA	PEAK	D	H	K	L	I(INT)	I(PK)	I(DS)
16.40	16.40	5.401	1	1	0	74	82	71
23.44	23.44	3.792	2	0	0	2	2	2
25.95	25.96	3.430	1	2	0	58	60	58
26.18	26.18	3.401	2	1	0	100	100	100
30.91	30.92	2.890	0	0	1	21	18	21
33.15	33.14	2.700	2	2	0	48	41	50
35.19	35.20	2.548	1	1	1	53	44	55
36.97	36.98	2.429	1	3	0	14	12	15
37.41	37.42	2.402	3	1	0	1	1	1
39.16	39.16	2.299	2	1	1	24	19	26
40.79	40.80	2.210	2	2	1	67	53	72
40.94	40.88	2.202	0	2	1	4	36	4
42.52	42.52	2.124	2	3	0	27	21	30
45.96	45.96	1.9731	2	2	1	1	1	1
47.22	47.22	1.9233	0	4	0	2	1	2
47.94	47.94	1.8960	4	0	0	8	6	9
48.81	48.82	1.8642	1	1	1	2	2	2
49.29	49.30	1.8471	3	1	1	12	9	14
49.47	49.42	1.8409	4	1	1	2	6	2
50.66	50.66	1.8003	3	3	0	8	6	9
53.37	53.36	1.7152	2	5	0	11	9	12
53.70	53.70	1.7055	3	2	1	13	13	15
53.86	53.86	1.7006	4	2	0	20	13	24
57.51	57.52	1.6011	0	4	1	8	5	9
58.14	58.14	1.5853	4	0	1	1	1	1
58.90	58.90	1.5666	1	1	1	1	1	2
59.48	59.48	1.5527	4	1	1	1	1	2
60.54	60.54	1.5280	3	3	1	52	33	64
61.44	61.44	1.5079	5	1	0	1	1	1
62.34	62.34	1.4882	2	4	0	1	0	1
62.96	62.96	1.4750	4	2	1	1	1	2
63.41	63.40	1.4657	0	5	1	13	8	16
64.42	64.42	1.4450	2	5	0	25	15	32
65.40	65.40	1.4257	5	2	0	5	3	6
66.17	66.16	1.4111	1	1	2	9	5	11
66.98	66.98	1.3959	1	4	0	2	1	2
69.57	69.58	1.3502	4	4	0	6	4	7
70.36	70.36	1.3369	1	5	1	16	9	21
70.68	70.68	1.3317	1	2	2	5	6	6
70.78	70.78	1.3300	2	1	1	8	7	10
71.21	71.20	1.3230	5	1	1	4	3	6
71.75	71.76	1.3143	3	0	0	3	2	4
72.31	72.32	1.3055	5	0	1	2	1	2
74.09	74.08	1.2786	2	2	1	16	9	21
74.39	74.32	1.2741	2	2	2	7	7	10
74.81	74.82	1.2680	5	2	1	16	9	22
75.07	75.04	1.2642	1	6	0	2	7	2
75.09		1.2640	6	0	0	4		5
76.27	76.28	1.2473	1	3	2	3	2	4
76.94	76.66	1.2419	3	1	1	5	3	7
76.94	76.90	1.2382	2	4	2	2	2	2
78.05	78.06	1.2233	4	6	1	3	1	3
78.71	78.72	1.2146	1	2	0	1	0	1
79.80	79.80	1.2008	6	3	0	1	1	2
80.28	80.28	1.1948	2	5	2	3	2	4
80.29		1.1947	4	3	0	1		1
80.45	80.52	1.1927	5	2	1	1	2	1
80.69	80.68	1.1898	3	1	1	4	2	5
84.17	84.16	1.1493	4	0	2	1	1	2
85.32	85.32	1.1367	4	1	2	1	0	1
85.59	85.58	1.1338	6	3	0	3	1	4
86.93	86.92	1.1197	2	6	1	2	1	2
87.99	88.00	1.1089	6	2	2	4	1	4
88.37	88.40	1.1051	2	4	1	5	2	4
88.47		1.1041	4	5	1	4		2
88.77	88.76	1.1012	5	4	1	4	3	7
88.77		1.1011	2	2	0	2		3
91.77	91.78	1.0728	7	1	1	1	0	1
92.84	92.88	1.0633	3	6	0	1	0	1
93.72	93.72	1.0556	2	7	1	1	0	1
98.34	98.36	1.0179	1	7	0	3	1	5
98.74	98.72	1.0149	2	5	2	1	1	5
99.15	99.12	1.0118	5	5	1	1	1	1
99.45	99.48	1.0096	5	2	2	5	2	8

1.92:1 MULLITE - BURNHAM, 1964, 1965

2THETA	PEAK	D	H	K	L	I(INT)	I(PK)	I(DS)
99.68	99.68	1.0079	3	7	0	2	3	9
99.96	99.94	1.0058	7	1	1	2	2	3
101.03	101.16	0.9980	7	3	0	3	2	4
101.18		0.9969	7	6	1	4		6
101.86	101.88	0.9921	4	4	1	5	2	7
101.95		0.9915	2	7	1	2		3
102.66	102.66	0.9866	4	4	2	4	1	7
104.78	104.78	0.9723	3	5	2	2	1	4
105.28	105.34	0.9687	5	3	2	2	1	2
106.86	106.86	0.9591	0	6	0	2	1	3
107.69	107.74	0.9540	1	8	2	1	1	2
108.10	108.12	0.9515	6	6	0	3	2	2
108.12		0.9514	1	0	3	2		5
108.62	108.62	0.9484	6	1	1	2	2	4
108.68		0.9480	8	0	0	2		4
109.33	109.32	0.9442	6	1	2	2	1	4
109.37		0.9440	1	4	2	1		1
109.47		0.9433	7	3	1	1		1
109.90	109.74	0.9409	8	1	0	1	1	1
111.17	111.20	0.9337	2	0	3	2	1	3
111.45	111.48	0.9321	2	8	0	1	1	2
111.87	111.88	0.9298	2	6	2	1	1	1
112.30	112.32	0.9275	1	2	3	3	1	5
112.70	112.72	0.9253	6	5	1	3	1	4
113.03		0.9235	4	2	1	2		2
113.55	113.48	0.9208	6	4	2	1	0	3
115.17	115.18	0.9124	0	5	1	2	0	3
117.04	117.02	0.9032	4	8	1	1	0	3
117.67	117.68	0.9001	7	7	0	2	0	3
118.84	118.88	0.8947	6	1	1	1		2
118.97		0.8941	8	1	3	1		1
119.43	119.40	0.8920	6	3	2	1	1	2
119.88	119.88	0.8900	5	7	0	1	1	2
122.99	122.98	0.8765	3	2	3	1	0	1
126.81	126.84	0.8614	0	4	3	3		6
127.50	127.46	0.8588	4	0	3	2	1	3

2THETA	PEAK	D	H	K	L	I(INT)	I(PK)	I(DS)
127.82	127.82	0.8576	4	8	0	1		2
129.81	130.14	0.8505	5	7	1	2	3	3
129.87		0.8503	8	4	0	1		1
130.14		0.8494	1	9	0	3		5
130.15		0.8494	3	3	3			
130.86	130.78	0.8469	7	5	1	11	2	20
133.54	133.60	0.8382	4	4	3	6	1	11
133.71		0.8377	9	2	0	3		5
137.42	137.44	0.8267	3	7	2	1		2
138.70	138.12	0.8231	9	2	1	11	2	22
139.43	139.42	0.8212	7	3	2	6	1	11
141.55	141.62	0.8157	8	7	1	4	1	8
141.73		0.8153	3	4	3	1		1
141.88		0.8149	6	1	1	1		3
143.03	143.14	0.8122	5	8	0	6	1	12
143.18		0.8118	1	5	3	1		2
144.31	144.04	0.8092	6	5	2	2		4
144.53		0.8087	5	1	3	2		4
145.24	145.22	0.8071	8	5	0	2	1	4
146.42	146.14	0.8045	9	1	1	1	0	2
148.06	148.14	0.8012	2	9	1	3	1	6
149.58	149.68	0.7982	2	5	3	9	1	18
150.13	150.86	0.7972	6	7	1	2	2	4
150.70		0.7961	1	8	2	2		9
150.99		0.7956	5	2	3	5		20
151.74		0.7943	4	7	2	3		2
152.70	152.60	0.7926	8	0	2	8	1	16
153.29		0.7917	9	2	1	6		12
158.35	159.22	0.7842	4	4	3	2	0	3
159.06		0.7833	2	8	2	5		9
160.22	162.42	0.7819	5	8	1	1	0	1
162.13		0.7797	3	9	1	8		16
162.57		0.7792	4	9	0	1		2
164.50		0.7774	8	5	1	10		21

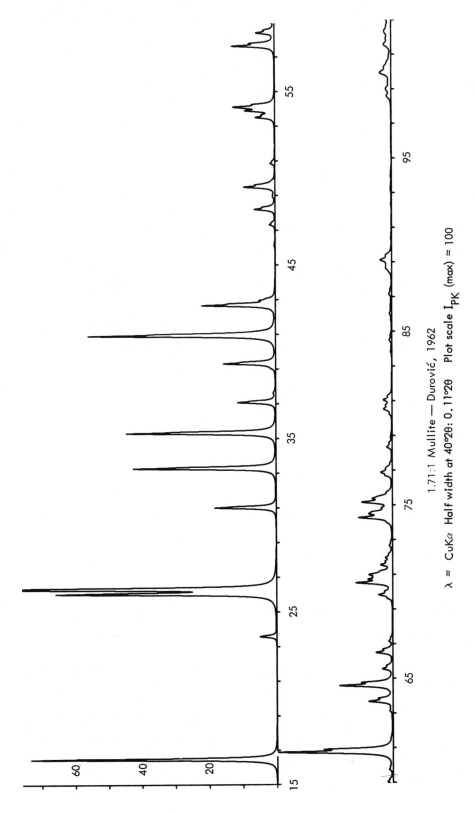

1.71:1 Mullite — Durović, 1962

λ = CuKα Half width at 40°2θ: 0.11°2θ Plot scale I_{PK} (max) = 100

1.71:1 MULLITE – ĎUROVIC, 1962

2THETA	PEAK	D	H	K	L	I(INT)	I(PK)	I(DS)
16.45	16.46	5.384	1	1	0	63	74	60
23.55	23.56	3.775	2	0	0	5	5	5
26.01	26.02	3.423	1	2	0	63	66	63
26.29	26.28	3.388	2	1	0	100	100	100
30.98	30.98	2.884	0	0	1	20	18	20
33.25	33.26	2.692	2	2	0	47	43	49
35.27	35.28	2.542	1	1	1	51	45	53
37.04	37.04	2.425	1	3	0	13	11	14
37.58	37.58	2.391	3	1	0	1	1	1
39.28	39.28	2.292	2	0	1	19	16	20
40.88	40.88	2.206	1	2	1	69	56	74
41.07	40.98	2.196	2	1	1	3	38	4
42.63	42.64	2.119	2	3	0	27	22	30
42.93	42.92	2.105	3	2	0	4	5	5
47.30	47.30	1.9203	0	4	0	2	2	2
48.18	48.18	1.8873	4	0	0	8	6	9
48.90	48.90	1.8610	1	4	0	1	1	1
49.47	49.48	1.8408	3	1	1	13	9	14
49.70	49.60	1.8327	4	1	0	1	6	2
50.83	50.84	1.7947	3	3	0	2	1	2
53.49	53.50	1.7115	2	4	0	8	6	9
53.88	53.88	1.7002	3	2	1	11	9	13
54.10	54.10	1.6938	4	2	0	15	12	17
57.62	57.62	1.5984	0	4	1	19	13	23
58.39	58.38	1.5792	4	0	1	8	5	10
59.02	59.02	1.5637	4	1	1	2	2	2
59.73	59.74	1.5468	1	4	1	1	1	2
60.73	60.74	1.5237	3	4	0	53	34	66
63.11	63.12	1.4718	2	4	1	11	7	14
63.66	63.66	1.4605	4	2	1	26	16	33
64.57	64.58	1.4420	0	5	0	5	3	6
65.55	65.56	1.4229	5	2	1	8	5	10
66.48	66.48	1.4051	1	1	2	2	1	2
67.14	67.14	1.3929	4	4	0	6	4	7
69.81	69.82	1.3460	1	5	1	18	11	24
70.51	70.50	1.3345	1	2	2	5	6	7
70.85	70.84	1.3289						

2THETA	PEAK	D	H	K	L	I(INT)	I(PK)	I(DS)
70.98	70.98	1.3268	2	1	2	9	7	11
71.54	71.54	1.3178	1	5	1	5	4	7
71.95	71.96	1.3112	3	5	0	3	2	4
72.64	72.64	1.3005	5	3	0	1	1	2
74.26	74.26	1.2760	2	2	2	17	10	24
74.60	74.48	1.2711	5	1	0	7	7	10
75.15	75.14	1.2632	2	5	1	16	9	21
75.22			1	6	0			
75.50	75.36	1.2582	6	0	0	4	6	6
76.68	76.70	1.2416	6	1	0	3	2	4
76.85	76.86	1.2394	1	3	2	5	3	6
77.18	77.08	1.2349	3	1	2	1	2	2
78.32	78.32	1.2197	4	4	1	3	2	4
78.89	78.88	1.2123	2	6	0	1	1	1
80.21	80.22	1.1956	6	2	0	3	3	5
80.50	80.50	1.1921	3	2	2	1	1	1
80.56		1.1914	4	5	0	2		2
80.71	80.72	1.1896	3	6	0	1	2	6
81.04	81.04	1.1856	5	3	1	4	3	2
84.48	84.48	1.1458	4	0	2	2	1	2
85.64	85.64	1.1333	1	6	1	1	0	1
86.02	86.02	1.1292	6	3	0	2	1	3
86.51	86.50	1.1241	3	3	2	1	0	2
87.13	87.14	1.1176	6	2	1	2	1	4
88.43	88.62	1.1045	2	4	2	1	2	2
88.61		1.1028	4	5	1	3		4
88.77	88.76	1.1011	5	1	1	1	2	2
89.10	89.10	1.0980	2	1	1	6	3	8
89.14		1.0976	4	2	0	2		3
90.36	90.36	1.0859	5	5	0	1	0	1
93.10	93.12	1.0610	1	7	0	4	0	1
93.94	93.94	1.0537	3	6	1	3	0	1
98.57	98.58	1.0162	2	7	1	3	1	6
99.02	98.98	1.0128	5	2	2	5	2	5
99.88	99.92	1.0064	2	5	0	4	3	8
99.96		1.0058	3	7	1	5		9
101.53	101.56	0.9945	4	6	1	4	2	7

1.71:1 MULLITE - DUROVIĆ, 1962

2THETA	PEAK	D	H	K	L	I(INT)	I(PK)	I(DS)
101.61		0.9939	7	3	0	3		4
102.21	102.34	0.9897	2	7	1	3	2	4
102.36		0.9886	6	4	1	4		7
103.04	103.04	0.9840	4	4	2	4	2	7
105.12	105.12	0.9701	3	5	2	2	1	4
105.79	105.78	0.9658	5	3	2	1	0	2
107.14	107.14	0.9573	0	6	2	2	1	3
107.94	107.96	0.9525	1	8	0	2	1	3
108.39	108.70	0.9497	6	6	2	1	2	2
108.68		0.9480	6	0	2	4		6
108.96	108.98	0.9464	1	1	3	3	2	4
109.43	109.40	0.9436	8	0	0	3	1	5
109.90	109.88	0.9409	6	1	2	3	1	5
110.00		0.9403	7	4	0	1		1
110.12		0.9396	7	3	1	1		2
111.54	111.62	0.9316	2	0	3	1	1	2
111.74	111.70	0.9305	2	8	0	1	1	2
112.21	112.20	0.9280	2	6	2	1	1	1
112.66	112.68	0.9255	1	2	3	3	1	6
112.78		0.9249	5	6	1	1		1
113.27	113.24	0.9223	6	5	1	3	1	5
113.62	113.64	0.9204	6	4	2	1		2
113.99		0.9185	4	5	2	2		2
115.46	115.46	0.9110	0	8	1	1	1	4
117.47	117.48	0.9011	4	7	1	2	0	2
118.27	118.28	0.8973	6	6	0	1	1	4
119.44	119.72	0.8920	3	1	3	2	1	2
119.69		0.8908	8	2	0	2		3
120.08	120.14	0.8890	6	3	1	2	1	3
120.40	120.36	0.8876	5	7	2	2	0	4
123.49	123.50	0.8744	3	1	3	1	0	1
125.24	125.22	0.8674	1	7	2	1	0	6
127.28	127.30	0.8596	0	4	3	3	1	1
127.66	128.14	0.8582	7	1	2	1	1	3
128.10		0.8566	4	0	3	2		1
128.11		0.8566	3	8	1	1		3
128.34		0.8557	4	8	0	2		3

2THETA	PEAK	D	H	K	L	I(INT)	I(PK)	I(DS)
128.80	128.76	0.8541	1	4	3	1	1	1
128.89		0.8538	4	6	2	1		1
130.44	130.72	0.8484	5	7	1	1	3	3
130.53		0.8480	1	9	0	3		6
130.72		0.8474	3	3	3	13		24
130.87		0.8469	8	4	0	1		1
131.74	131.30	0.8440	7	5	1	7	2	14
134.23	134.24	0.8361	4	2	3	3	1	6
134.96	134.92	0.8338	9	1	0	3	1	3
135.43		0.8324	2	0	0	1		1
138.03	138.04	0.8250	3	7	2	12	2	24
140.09	140.52	0.8195	9	2	0	2	1	3
140.53		0.8183	7	1	2	6		12
142.48		0.8136	3	9	1	9		2
142.85	142.88	0.8126	8	4	1	1	1	2
143.86	143.82	0.8102	1	5	3	6	2	12
143.87		0.8102	5	8	0	9		17
144.73	144.70	0.8082	3	9	0	2		4
145.39	145.40	0.8068	6	6	2	1	1	1
145.55		0.8064	8	5	1	1	1	1
146.66	146.56	0.8040	5	1	3	3		6
148.14	148.88	0.8010	8	5	0	3	1	5
148.76		0.7998	9	2	1	1	1	2
150.46	150.56	0.7966	2	5	3	2	1	5
151.37	151.60	0.7949	2	7	1	12	2	24
151.48		0.7947	6	6	1	4		8
152.24	152.18	0.7934	1	8	2	5	2	11
152.49		0.7930	5	7	3	12		24
152.78		0.7925	7	6	1	1		1
154.59	154.70	0.7896	4	7	2	2	1	3
155.46	155.46	0.7883	8	0	2	12	1	24
155.88		0.7876	9	2	1	8		17
157.42		0.7855	7	4	2	3		6
159.86	160.00	0.7823	8	1	2	4	1	8
160.24		0.7818	2	4	3	3		7
161.87		0.7800	5	8	1	1		10

439

TABLE 21. MISCELLANEOUS CHAIN SILICATES

Variety	Sapphirine	Astrophyllite
Composition	$Mg_4Al_8Si_2O_{20}$	$K_{1.8}Na_{1.2}(Fe,Mn)_7Ti_2Si_8(O,OH)_{31}$
Source	Fiskenaesset Harbour, W. Greenland	El Paso Co., Colorado
Reference	Moore, 1968	Woodrow, 1967
Cell Dimensions		
$\underline{a}$ Å	11.26	5.36
$\underline{b}$	14.46	11.76
$\underline{c}$	9.95	21.08
α	90	85.13
β deg	125.33	90
γ	90	103.22
Space Group	$P2_1/a$	$A\bar{1}$
Z	4	2

TABLE 21. (cont.)

Variety	Sapphirine	Astrophyllite
Site Occupancy	M_1 - M_4 Mg M_5 - M_8 Al T_1 - T_4 Al T_5 - T_6 Si	$\begin{matrix} M_1 \\ M_2 \\ M_3 \\ M_4 \end{matrix} \left\{ \begin{matrix} Fe & 0.9 \\ Mn & 0.1 \end{matrix} \right.$
Method	3-D, counter	2-D, film
R & Rating	0.115 (2)	0.10 (2)
Cleavage and habit	None Tabular parallel to {010} (?)	{001} perfect
Comment	No site distribution given. Occupation above is arbitrary [T_6O_{18}]$_\infty$ chain B estimated	Quasi-layer structure
μ	105.1	390.6
ASF	0.4421×10^{-1}	0.8627×10^{-1}
Abridging factors	1:1:0.5	0.5:0:0:0

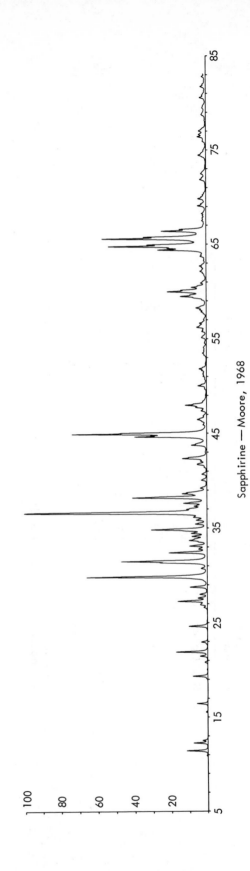

Sapphirine — Moore, 1968

λ = CuKα Half width at 40°2θ: 0.11°2θ Plot scale I_{PK} (max) = 100

442

SAPPHIRINE - MOORE, 1968

2THETA	PEAK	D	H	K	L	I(INT)	I(PK)	I(DS)
11.40	11.40	7.754	0	1	0	11	12	10
12.23	12.24	7.230	0	0	1	7	8	6
12.49	12.50	7.078	0	2	0	2	3	2
16.40	16.42	5.399	0	1	1	6	6	5
19.31	19.32	4.593	2	0	0	8	8	7
20.79	20.80	4.268	-1	2	0	1	2	1
21.42	21.42	4.144	-1	3	0	4	4	4
21.87	21.88	4.060	0	1	2	18	18	17
22.92	22.92	3.877	-1	3	1	4	4	3
24.60	24.62	3.615	-2	2	0	12	10	11
26.59	26.60	3.350	-2	4	0	2	2	2
26.86	26.86	3.316	0	3	2	2	2	3
26.98	26.94	3.302	-3	0	1	4	4	2
27.23	27.24	3.272	-2	2	2	2		16
27.26		3.269	-2	4	1	16	16	
27.57	27.58	3.232	2	1	1	5	5	5
27.96	27.96	3.188	0	1	3	5	4	4
28.73	28.74	3.105	3	3	1	11	10	11
29.80	29.80	2.996	-1	4	2	77	66	74
30.69	30.70	2.910	-1	2	2	3	4	3
31.31	31.32	2.855	2	4	2	26	26	25
31.47	31.46	2.841	-2	0	2	52	47	51
32.38	32.38	2.763	-2	1	3	25	21	25
32.75	32.76	2.732	0	5	1	1		1
32.85	32.82	2.724	0	0	3	7	4	4
33.08	33.08	2.706	-4	2	3	4	10	7
33.08		2.705	0	3	1	4		4
33.51	33.52	2.672	-4	0	3	11	5	4
33.67	33.68	2.660	0	1	3	9	10	11
33.84	33.82	2.647	-4	1	3	9	7	6
34.09	34.10	2.628	-4	1	0	9	9	9
34.42	34.42	2.604	0	4	2	35	31	35
34.81	34.82	2.575	-3	4	2	7	7	7
35.39	35.40	2.534	0	2	3	6	6	6
35.80	35.80	2.506	-4	2	3	7	8	7
36.11	36.12	2.486	-4	2	2	6	6	6
36.54	36.56	2.457	-2	5	2	100	100	100

2THETA	PEAK	D	H	K	L	I(INT)	I(PK)	I(DS)
36.59		2.453	-2	0	4	54		54
36.95	36.94	2.431	-4	4	1	8	11	8
37.05	37.04	2.425	-2	4	2	3	10	3
37.28	37.28	2.410	0	1	0	6	7	6
37.59	37.58	2.391	2	2	2	16	13	16
38.11	38.18	2.359	0	3	3	5	41	5
38.16		2.356	1	4	2	28		28
38.18		2.355	0	5	2	23		24
38.61	38.62	2.330	-4	2	4	13	14	13
38.72	38.72	2.323	-2	0	2	3	11	3
38.78		2.320	-4	3	1	5		5
39.19	39.20	2.297	4	0	1	4	4	4
40.02	40.02	2.251	-1	1	4	4	3	4
40.65	40.66	2.217	-4	2	4	4	3	4
41.39	41.40	2.180	-2	5	3	2	2	2
41.66	41.70	2.166	0	4	1	3	5	3
41.66		2.166	2	2	4	2		2
41.72		2.163	-1	4	4	4		4
42.01	42.02	2.149	-4	4	0	4	4	4
42.28	42.02	2.136	-2	6	1	15	14	15
42.31	42.32	2.134	-5	2	2	3		4
42.46	42.42	2.127	-5	1	4	11	10	12
43.72	43.72	2.069	-1	3	4	6	9	6
43.92	43.82	2.060	-2	0	4			
44.61	44.62	2.029	0	5	4	51	39	54
44.90	44.90	2.017	-4	5	2	99	73	105
44.94		2.015	1	7	0	2		3
45.89	45.88	1.9759	0	5	3	5	2	3
46.22	46.22	1.9626	-4	5	3	5	3	2
46.32		1.9582	-4	5	4	4		3
46.43	46.44	1.9539	0	2	5	5	5	2
46.47		1.9525	-4	4	1	2		4
46.83	46.82	1.9385	-3	4	4	1	2	3
47.36	47.36	1.9177	-5	3	4	3	3	3
47.72	47.72	1.9043	1	6	2	7	6	7
47.96	47.96	1.8953	-5	2	4	15	12	8
48.37	48.36	1.8801	-3	5	4	2	2	17

443

SAPPHIRINE - MOORE, 1968

2THETA	PEAK	D	H	K	L	I(INT)	I(PK)	I(DS)
49.66	49.66	1.8344	-6	0	4	1	1	2
50.00	50.00	1.8226	5	1	0	7	5	8
50.68	50.68	1.7997	0	6	3	1	1	
50.99	50.98	1.7896	-4	6	1	3	2	3
51.22	51.22	1.7820	-4	6	4	2	2	2
51.34	51.34	1.7781	-6	2	4	1	2	1
51.60	51.60	1.7696	0	4	4	3	3	
51.79	51.78	1.7637	1	1	4	5	4	4
53.23	53.24	1.7193	-2	6	0	1	1	6
53.46	53.46	1.7125	3	7	2	2	2	2
53.73	53.74	1.7044	-6	3	6	2	2	3
54.02	54.04	1.6961	-1	8	2	1	2	2
55.10	55.10	1.6653	-4	7	0	3	3	1
55.20		1.6626	-4	4	6	1		4
55.75	55.76	1.6474	-4	0	6	3	3	1
55.95	55.92	1.6419	0	7	3	1	2	1
56.18	56.20	1.6358	-6	4	1	6	5	7
56.24		1.6343	-4	7	4	3		3
56.63	56.62	1.6240	-6	4	2	4	4	4
57.00	57.02	1.6143	-6	3	5	2	2	1
57.65	57.64	1.5977	-6	3	1	1	2	1
57.82	57.82	1.5932	-7	1	4	3	3	3
58.02	58.02	1.5883	-5	2	6	2	4	2
58.19		1.5841	0	2	5	2	6	
58.25	58.24	1.5826	1	9	0	6		7
59.08	59.10	1.5624	-1	8	2	2	3	3
59.29	59.30	1.5572	-2	0	6	8	7	9
59.44	59.44	1.5536	2	6	4	16	14	19
59.50		1.5523	0	1	0	3		4
59.84		1.5442	-6	6	4	5		6
59.95	59.96	1.5417	-6	1	6	32	21	37
60.08	60.10	1.5386	0	5	2	3	14	3
60.41	60.42	1.5311	-1	0	5	8	8	9
60.44		1.5304	0	7	0	4		5
60.88	60.88	1.5204	3	6	4	4	3	4
61.66	61.66	1.5030	0	8	3	3	1	1
61.93	61.92	1.4971	-4	8	3	2	2	3

2THETA	PEAK	D	H	K	L	I(INT)	I(PK)	I(DS)
62.08	62.12	1.4939	0	9	2	2	3	2
62.13		1.4927	-4	8	1	3		3
62.51	62.52	1.4845	-5	6	5	3	3	4
62.68	62.68	1.4810	0	4	5	4	4	5
63.23	63.24	1.4693	-5	7	4	1	1	2
63.70	63.70	1.4597	-6	0	0	4		5
64.37	64.38	1.4460	0	10	0	43	27	52
64.75	64.76	1.4385	-4	5	6	92	53	112
65.55	65.56	1.4229	3	9	2	86	57	106
65.56		1.4227	-7	0	6	13		16
65.61		1.4218	-8	0	4	3		3
66.37	66.36	1.4073	-1	9	2	43	25	53
67.47	67.46	1.3870	0	10	3	2	2	3
67.77	67.76	1.3815	-4	9	3	1	2	2
68.03	68.02	1.3769	-4	9	1	2	2	3
68.22	68.22	1.3734	-1	4	6	8	5	11
69.07	69.06	1.3588	-3	6	6	1	3	1
69.38	69.26	1.3534	-1	7	5	1		1
69.75	69.82	1.3471	0	6	5	2	5	2
69.79		1.3465	-7	4	6	6		7
69.82		1.3458	-2	9	2	3		4
70.22	70.22	1.3393	-1	9	4	6	2	7
71.83	71.84	1.3131	-5	11	0	3	3	5
72.45	72.46	1.3035	-1	9	4	6		2
72.59		1.3013	-5	0	2	4		8
74.41	74.42	1.2739	-4	4	0	1	4	1
74.55		1.2717	-4	3	4	6		3
74.74	74.60	1.2690	-1	10	3	1		3
75.89	75.92	1.2527	-1	6	6	2	3	11
75.96		1.2517	-8	7	4	1	2	2
76.39	76.38	1.2457	-2	10	7	9	5	6
76.62	76.62	1.2425	-7	6	4	2	4	2
76.71		1.2412	-9	4	4	2		5
77.01	77.00	1.2371	-6	8	8	4	4	6
78.66	78.74	1.2153	-8	6	4	1	3	2
78.74		1.2143	-6	1	2	3		5
79.43	79.44	1.2055	-7	1	8	1	2	2

SAPPHIRINE - MOORE, 1968

2THETA	PEAK	D	H	K	L	I(INT)	I(PK)	I(DS)
80.53	80.54	1.1918	1	9	4	6	3	8
80.59		1.1911	-4	11	2	1		2
80.97	80.94	1.1864	-3	1	8	2	2	3
81.70	81.70	1.1776	0	10	4	5	3	7
82.79	82.80	1.1648	-8	0	8	4	2	6
83.59	83.60	1.1557	-3	3	8	1	2	2
84.25	84.26	1.1483	8	0	0	3	2	5
84.57	84.52	1.1448	-3	12	2	2	2	2

445

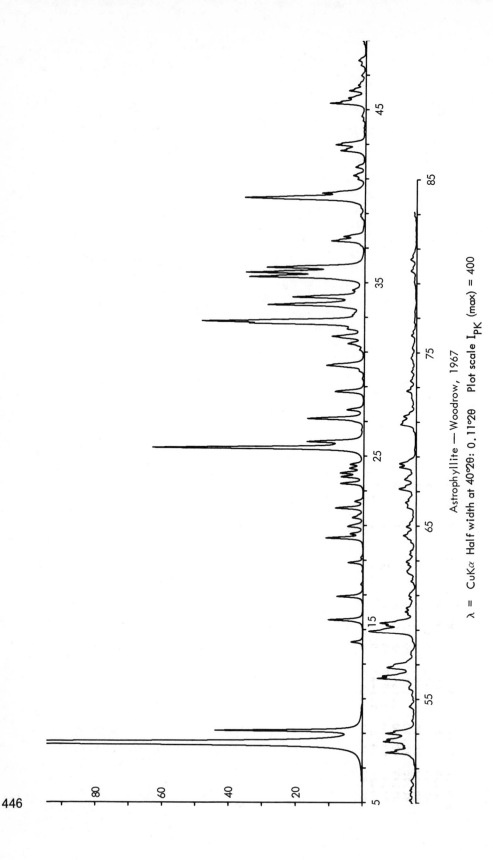

Astrophyllite — Woodrow, 1967

$\lambda = $ CuKα Half width at 40°2θ: 0.11°2θ   Plot scale I$_{PK}$ (max) = 400

446

ASTROPHYLLITE - WOODROW, 1967

2THETA	PEAK	D	H	K	L	I(INT)	I(PK)	I(DS)
8.41	8.42	10.500	0	0	2	100	100	100
8.49		10.410	0	0	1	12		12
9.13	9.14	9.675	0	-1	1	11	11	11
14.24	14.26	6.212	0	0	3	3	1	1
15.53	15.54	5.702	0	0	0	2		4
16.87	16.88	5.250	0	0	4	1	1	3
18.83	18.84	4.710	-1	0	2	3	3	2
20.25	20.26	4.383	-1	-1	0	2	2	6
20.49	20.50	4.331	-1	0	1	1	1	2
20.91	20.92	4.244	-1	1	3	1	3	2
20.92		4.244	0	-2	2	3		2
21.42	21.42	4.145	-1	1	3	1	1	2
21.97	21.98	4.042	-1	-1	4	1	2	5
22.35	22.36	3.974	1	0	2	3	1	2
23.39	23.40	3.799	1	-1	3	2	2	4
23.78	23.78	3.738	0	-2	4	2	2	4
24.00	24.00	3.705	0	0	4	2	2	4
24.27	24.28	3.664	-1	-2	3	1	1	3
24.49	24.50	3.631	-1	1	6	1	1	2
25.43	25.44	3.500	0	0	4	22	16	48
25.80	25.80	3.451	-1	-2	4	5	4	11
27.13	27.14	3.284	1	-2	3	6	4	14
27.64	27.64	3.225	0	-3	5	2	1	4
28.68	28.68	3.110	1	1	6	3	2	7
28.72		3.106	0	-2	7	1		2
30.11	30.20	2.966	-1	0	5	1		4
30.19		2.958	-1	1	5	3	3	8
30.28		2.949	0	1	5	1		3
31.46	31.46	2.841	-1	-1	7	1	1	4
31.80	31.88	2.812	-1	-2	4	1	2	4
31.88		2.804	1	2	6	2		7
32.30	32.30	2.769	-1	3	5	11	1	32
32.61	32.62	2.743	-1	0	1	14	9	43
32.74	32.74	2.733	-1	4	1	1	12	2
33.11	33.10	2.704	0	-3	3	10	1	32
33.68	33.68	2.658	1	-1	3	1	7	2
33.73		2.655	2	3	3			2
34.13	34.14	2.625	0	0	8	8	5	25
34.54	34.54	2.594	1	-3	5	12	1	3
35.30	35.30	2.540	-1	4	3	12	9	40
35.56	35.56	2.523	-1	-3	3	10	9	40
35.85	35.84	2.503	-2	-1	3	1	7	33
35.85		2.503	-1	1	3			3
37.37	37.38	2.404	-1	3	5	4	3	13
37.64	37.64	2.388	-1	-4	4	2	2	8
39.80	39.86	2.263	-2	-1	6	6	9	24
39.87		2.259	-1	-3	5	8		31
39.87		2.259	-1	-4	5	2		9
40.14	40.14	2.244	-1	2	5	4	3	16
40.95	40.96	2.202	-1	4	2	1	1	3
41.16	41.16	2.191	-2	4	0	1	1	5
41.64	41.64	2.167	-1	3	7	3	1	5
42.59	42.60	2.121	-2	4	6	1	1	4
42.94	42.94	2.104	-1	3	0	1	2	11
43.04	43.04	2.100	1	-4	10	3	2	13
45.33		1.9987	0	0	7	1	2	4
45.34	45.34	1.9986	-2	-1	7	1	3	5
45.55	45.44	1.9898	-1	4	6	3	2	15
45.65	45.64	1.9856	-1	4	8	1	1	3
46.07	46.08	1.9686	-1	-3	7	1	1	7
46.38	46.38	1.9560	-1	-5	1	2	1	10
51.91	51.92	1.7599	-2	-1	9	4	1	3
52.23	52.22	1.7500	2	-1	12	1	2	23
52.34	52.36	1.7463	0	0	10	1	1	5
52.54	52.54	1.7403	-1	-2	10	4	1	3
52.67	52.68	1.7363	2	0	10	1	1	21
53.01	53.02	1.7259	-1	-3	9	4	2	4
53.13	53.14	1.7222	-1	-3	11	4	2	21
54.56	54.56	1.6805	3	0	3	1	2	4
55.58	55.60	1.6520	2	0	3	1	0	4
56.20	56.20	1.6354	0	7	1	5	1	3
56.32	56.34	1.6321	1	1	11	2	3	32
56.77	56.82	1.6203	2	-2	10	2	3	10
56.79		1.6198	1	-4	10	1	2	4
56.83		1.6186	-3	3	5			3

ASTROPHYLLITE - WOODROW, 1967

2THETA	PEAK	D	H	K	L	I(INT)	I(PK)	I(DS)
56.84		1.6183	0	7	3	1		8
58.86	58.92	1.5677	-3	5	1	4	4	26
58.92		1.5661	3	2	0	4		27
59.24	59.26	1.5585	3	2	2	2	2	11
59.26		1.5580	3	-5	1	1		9
59.39	59.40	1.5550	2	-1	11	1	3	7
59.39		1.5548	-2	1	11	2		13
59.89	59.90	1.5431	-3	5	3	1	1	5
60.04	60.04	1.5395	-3	-2	2	1	1	6
60.27	60.26	1.5342	-1	4	12	1	1	6
60.79	60.78	1.5223	-1	-3	11	1	0	4
61.80	61.80	1.5000	0	0	14	1	0	4
62.39	62.38	1.4872	0	7	7	1	1	4
62.75	62.76	1.4795	2	-4	10	1	1	4
62.90	62.92	1.4762	0	-7	5	1	1	4
63.35	63.36	1.4670	-2	2	12	1	0	4
64.06	64.06	1.4524	3	2	6	1	1	4
64.34		1.4467	1	4	12	1		5
64.44	64.44	1.4446	1	3	13	1	1	10
64.96	64.94	1.4344	1	-4	12	1	1	9
66.04		1.4136	-3	5	7	1		4
66.06	66.06	1.4131	1	-5	11	1	1	4
66.14		1.4116	1	2	14	1		5
66.35	66.34	1.4076	-3	-2	6	1	1	5
67.08	67.14	1.3941	0	7	9	1	1	6
67.13		1.3932	2	6	2	1		8
67.16		1.3927	-2	8	0	1		10
67.66	67.68	1.3834	-2	1	13	1	1	9
67.74		1.3821	0	-7	7	1		5
68.38	68.38	1.3707	2	6	4	2	1	13
68.39		1.3705	3	2	8	1		6
68.57	68.58	1.3674	3	-8	2	1	1	7
68.66		1.3658	3	-5	7	1		5
70.83	70.84	1.3292	-4	2	2	2	1	15
70.87		1.3285	-2	8	6	1		4
70.97	71.02	1.3269	-3	5	9	1	1	4
71.28	71.32	1.3219	2	-8	4	1	1	5

2THETA	PEAK	D	H	K	L	I(INT)	I(PK)	I(DS)
71.33		1.3210	-2	-6	4	1		7
71.35		1.3207	-3	-2	8	1		5
72.77	72.78	1.2984	-4	-1	4	1	1	6
72.78		1.2899	1	3	15	1		7
73.33	73.36	1.2814	-1	-4	14	1		5
73.90	73.90					1	0	5
79.70	79.70	1.2020	2	6	10	1	0	9
80.35	80.36	1.1940	2	-8	8	1	0	8

TABLE 22. RING SILICATES

Variety	Dioptase	Tourmaline	Benitoite	Kainosite	High Pressure Polymorph
Composition	$Cu_6(Si_6O_{18}) \cdot 6H_2O$	$(Na,K,Ca)Mg_3(Al, Mg,Fe)_6(OH)_3(BO_3)_3 Si_6O_8$	$BaTiSi_3O_9$	$Ca_2(Y,Tr)_2(Si_4O_{12}) CO_3 \cdot H_2O$	$CaSiO_3$
Source	Kirgisenisteppe, USSR	DeKalb, N.Y.	St. Benito, Calif.	Ontario, Canada	Synthetic, 65 kb
Reference	Boll-Dornberger, Heide, Thilo & Thilo, 1955	Burnham & Peacor, Buerger, 1962	Zachariasen, 1930	Rumanova, Volodina & Belov, 1967	Trojer, 1968
Cell Dimensions					
a Å	14.61	15.951	6.61 (6.60 kx)	14.30 (12.93)	6.695
b Å	14.61	15.951	6.61 (6.60 kx)	12.93 (14.30)	9.257
c Å	7.80	7.240	9.73 (9.71 kx)	6.73 (6.73)	6.666
α	90	90	90	90	86.633
β deg	90	90	90	90	76.133
γ	120	120	120	90	70.383
Space Group	R$\bar{3}$	R3m	P$\bar{6}$c2	Pnma (Pmnb)	P$\bar{1}$
Z	3	3	2	4	6

449

TABLE 22. (cont.)

Variety	Dioptase	Tourmaline	Benitoite	Kainosite	High Pressure Polymorph
Site Occupancy		$M_1 \begin{cases} Na & 0.60 \\ Ca & 0.40 \end{cases}$ M_2 Mg 1.0 $M_3 \begin{cases} Al & 0.903 \\ Mg & 0.092 \\ Fe & 0.005 \end{cases}$			
Method	2-D, film	3-D, film - counter.	2-D, film	3-D, film	3-D, --
R & Rating	0.16 (2)	0.069 (1)	-- (3)	0.12 (av.) (2)	0.065 (2)
Cleavage and habit	{10$\bar{1}$1} perfect Short prisms parallel to c axis	{11$\bar{2}$0}{10$\bar{1}$1} very poor. Prismatic parallel to c axis		None Prismatic Parallel to c axis	
Comment	B estimated	Extra O and F in chem. analysis ignored in refinement	B estimated	Tr=trace rare earth elements	[Si_3O_9] rings B estimated
μ	120.4	96.1	560.2	382.3	284.8
ASF	0.3116	0.7715×10^{-1}	0.5798	0.1242	0.6368×10^{-1}
Abridging factors	1:1:0.5	1:1:0.5	0:1:0:0	1:1:0.5	1.5:1:1

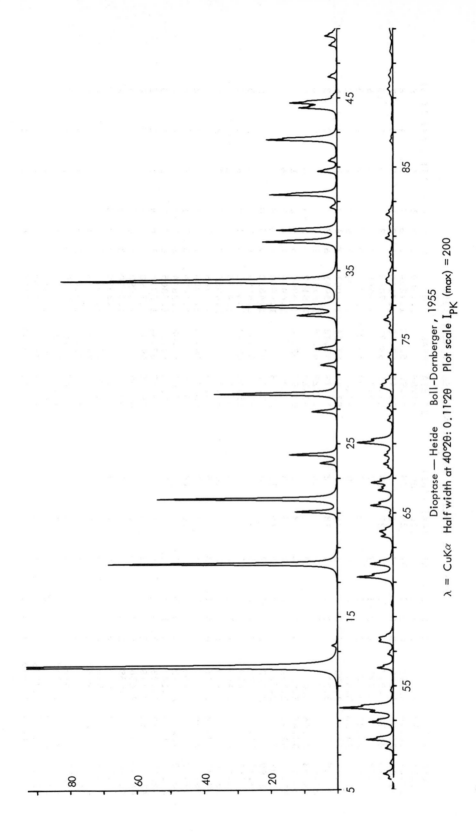

Dioptase — Heide   Boll-Dornberger, 1955

λ = CuKα Half width at 40°2θ: 0.11°2θ Plot scale I$_{PK}$ (max) = 200

DIOPTASE - HEIDE, BOLL-DORNBERGER ET AL., 1955

2THETA	PEAK	D	H	K	L	I(INT)	I(PK)	I(DS)
12.11	12.12	7.305	1	1	0	100	100	100
18.04	18.04	4.913	0	0	3	37	34	38
21.05	21.06	4.218	3	0	0	7	6	7
21.78	21.78	4.077	2	1	1	24	27	25
21.78		4.077	1	2	-1	7		7
23.85	23.86	3.727	0	2	2	3	3	3
24.35	24.36	3.653	2	2	0	9	7	9
26.83	26.84	3.320	2	2	2	5	4	5
27.85	27.86	3.200	1	3	1	1	18	2
27.85		3.200	3	1	-1	22		24
29.53	29.54	3.022	1	2	2	3	2	3
30.47	30.48	2.931	4	0	0	4	3	5
32.40	32.40	2.761	1	4	-1	7	6	8
32.89	32.90	2.720	3	0	4	14	15	15
32.89		2.720	2	3	-1	7		8
34.35	34.36	2.609	3	1	1	5	41	6
34.35		2.609	1	3	-2	53		60
36.66	36.66	2.449	1	1	3	4	11	4
36.66		2.449	1	1	-3	12		14
37.33	37.32	2.407	0	5	-1	13	9	15
38.63	38.64	2.329	2	3	1	1	1	2
39.38	39.38	2.286	2	4	2	15	10	18
40.73	40.74	2.213	3	0	1	4	3	5
41.34	41.34	2.182	5	1	0	1	2	1
42.55	42.56	2.123	5	0	2	2	1	2
42.65	42.64	2.118	2	0	3	16	11	19
44.40	44.40	2.038	4	2	-2	2	8	3
44.40		2.038	2	4	-2	6	6	8
44.69	44.70	2.026	5	2	0	2	7	3
44.69		2.026	2	5	0	8		10
46.19	46.20	1.9635	1	4	2	3	1	3
48.02	48.02	1.8928	1	6	-1	1	1	1
48.56	48.56	1.8730	6	1	-1	3	2	4
49.63	49.64	1.8353	3	4	0	1	1	1
49.89	49.90	1.8263	4	4	-2	2	2	3
51.37	51.36	1.7773	3	3	-3	1	1	2
51.88	51.88	1.7609	7	0	1	1	4	2

2THETA	PEAK	D	H	K	L	I(INT)	I(PK)	I(DS)
51.88	52.90	1.7609	3	5	1	6	4	8
52.90	53.48	1.7294	1	6	-2	6	4	8
53.48	53.72	1.7118	6	2	1	5	8	7
53.73		1.7045	1	3	1	9		13
53.73		1.7045	3	1	-4	4		6
56.03	56.02	1.6399	0	7	-2	2	2	3
56.03		1.6399	3	5	-2	3		2
57.55	57.56	1.6001	2	6	0	2	2	4
57.79	57.78	1.5941	6	3	0	6	2	3
61.29	61.28	1.5112	2	3	4	4	5	8
61.97		1.5112	5	4	-4	1		6
62.05	62.04	1.4961	4	2	-2	4	3	2
62.05		1.4944	4	4	3	2		6
63.63	63.64	1.4944	4	5	-3	3	2	5
63.93	63.92	1.4610	5	8	0	3	3	5
65.41	65.42	1.4549	1	1	-1	4	3	7
65.41		1.4255	3	1	-5	2		3
66.30	66.30	1.4255	7	3	-3	3	2	4
66.30		1.4086	7	1	-3	1		2
66.74	66.74	1.4086	7	3	1	1	3	2
66.74		1.4004	3	7	-1	5		8
66.74		1.4004	8	2	0	1		2
67.61	67.62	1.4004	8	2	0	2	1	3
67.83	67.82	1.3844	1	6	4	1	1	2
68.33	68.34	1.3805	6	3	3	1		2
69.05	69.06	1.3716	3	6	3	3	5	4
69.05		1.3590	6	3	-3	1		2
69.05		1.3590	3	7	-3	3		9
70.34	70.34	1.3590	6	4	-5	5	1	2
72.25	72.24	1.3590	3	7	-2	2	2	5
72.39	72.40	1.3372	2	4	-4	3	2	2
72.67	72.64	1.3065	0	6	-4	1	1	2
74.00	74.00	1.3043	0	1	6	1	1	2
75.67	75.68	1.3000	1	5	-6	6	1	2
76.15	76.16	1.2799	6	5	-2	1	1	5
78.49	78.50	1.2557	10	0	-1	3	1	2
		1.2489	6	6	0	1		2
		1.2175						

452

DIOPTASE - HEIDE, BOLL-DORNBERGER ET AL., 1955

2THETA	PEAK	D	H	K	L	I(INT)	I(PK)	I(DS)
80.88	80.88	1.1875	2	9	2	2	1	3
82.23	82.24	1.1713	7	4	3	1	1	2
86.48	86.48	1.1244	10	2	-1	1	0	2
90.30	90.30	1.0864	4	9	1	1	1	2
91.64	91.66	1.0740	1	8	5	1	0	2
92.98	92.96	1.0620	1	3	7	1	1	2
99.73	99.74	1.0075	6	3	-6	1	1	3
103.16	103.20	0.9831	11	3	-1	1	1	2
103.22		0.9827	0	10	5	1		2
103.99	104.00	0.9776	12	1	2	1	0	2
104.68	104.66	0.9730	7	6	4	1	0	2
105.37	105.36	0.9685	11	0	3	1	1	1
105.79	105.78	0.9658	8	7	-1	1	1	2
105.79		0.9658	8	1	-1	1		2
107.19	107.24	0.9570	7	5	5	1	0	1
108.59	108.58	0.9485	3	5	7	1	1	1
108.59		0.9485	7	0	7	1		1
109.31	109.32	0.9443	8	7	-2	1	0	1
113.04	113.04	0.9235	7	4	-6	1	0	1
113.63	113.60	0.9203	3	12	0	1	0	2
114.14	114.10	0.9177	6	9	4	1	0	2
115.69	115.68	0.9098	5	0	8	1	0	1
118.41	118.30	0.8967	12	1	-4	1	0	2
121.35	121.38	0.8835	3	11	4	1	1	2
121.35		0.8835	11	3	-4	1		2
122.13	122.12	0.8801	1	13	-3	1	1	2
122.13		0.8801	1	13	3	1		1
125.77	125.78	0.8654	8	6	5	1	0	2
126.77	126.76	0.8615	8	8	3	1	0	2
130.64	130.72	0.8477	12	1	5	1	0	2
131.71	131.82	0.8441	5	11	3	1	0	1
131.97		0.8433	2	2	9	1		2
134.08	134.12	0.8365	11	3	5	1	0	3
134.17		0.8362	10	0	7	1		2
135.98	135.96	0.8308	2	9	-7	1	0	2
137.95	137.94	0.8252	9	7	-4	1	0	2
142.12	142.24	0.8143	8	1	-8	1	0	2

2THETA	PEAK	D	H	K	L	I(INT)	I(PK)	I(DS)
142.34	143.08	0.8138	2	11	-6	1		2
143.24		0.8117	9	9	0	1	0	2
146.49	146.52	0.8044	3	7	8	1	0	2
149.93	150.14	0.7975	2	14	3	1		2
150.32		0.7968	5	2	-9	1		2
150.32		0.7968	2	5	9	1		3
151.01	150.88	0.7956	10	2	-7	1	0	3
156.65	157.12	0.7865	7	9	-5	1	0	5
156.99		0.7860	13	3	4	2		3
157.32		0.7856	6	5	-8	1		3
157.32		0.7856	1	9	-8	1		1
158.77		0.7837	4	13	-3	1		1
158.77		0.7837	13	4	3	1		2
159.31		0.7830	4	4	9	1		3

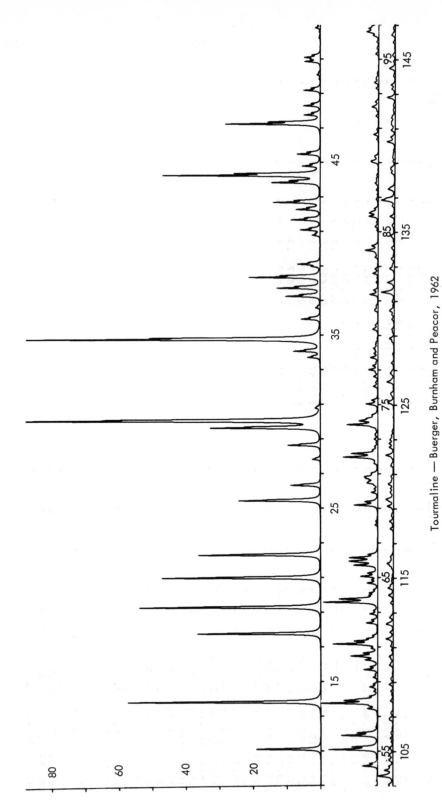

454

Tourmaline — Buerger, Burnham and Peacor, 1962

λ = CuKα Half width at 40°2θ: 0.07°2θ Plot scale I_{PK} (max) = 100

TOURMALINE - BUERGER, BURNHAM AND PEACOR, 1962

2THETA	PEAK	D	H	K	L	I(INT)	I(PK)	I(DS)
11.08	11.08	7.976	1	1	0	15	19	14
13.80	13.80	6.413	1	0	1	44	57	42
17.73	17.74	4.998	0	2	1	31	37	29
19.26	19.26	4.605	3	1	0	46	54	44
20.96	20.96	4.235	2	0	1	41	47	40
22.27	22.28	3.988	0	2	2	33	36	32
25.41	25.42	3.502	2	1	1	23	24	22
26.29	26.30	3.386	1	3	1	8	9	8
27.80	27.80	3.206	2	0	2	2	2	2
28.61	28.62	3.117	4	0	0	2	2	2
29.61	29.60	3.014	4	1	0	9	10	9
30.01	30.02	2.975	1	2	2	33	33	33
30.77	30.78	2.903	2	2	1	100	100	100
33.68	33.68	2.659	3	2	0	1	2	1
34.04	34.04	2.631	3	1	2	4	4	4
34.72	34.72	2.581	0	5	1	8	8	8
35.91	35.90	2.499	0	0	3	90	89	92
36.56	36.56	2.456	2	4	1	6	6	6
37.22	37.22	2.413	0	4	2	1	1	1
37.69	37.70	2.384	3	3	1	11	10	11
38.32	38.32	2.347	5	1	1	14	13	14
38.96	38.96	2.310	1	3	2	22	21	23
39.09	39.08	2.302	6	0	0	6	7	7
40.76	40.76	2.212	5	2	0	2	2	3
41.06	41.06	2.196	5	0	2	6	6	7
41.64	41.64	2.167	4	3	1	9	9	10
42.24	42.24	2.138	3	0	3	4	7	5
42.24	42.24	2.138	0	2	3	3	2	3
42.66	42.66	2.117	2	2	3	15	14	16
43.80	43.80	2.065	2	5	1	16	14	17
44.22	44.22	2.047	1	6	1	51	47	55
44.76	44.77	2.023	1	4	2	6	5	6
45.45	45.46	1.9939	4	4	0	8	7	9
47.20	47.20	1.9238	3	4	1	32	28	35
47.72	47.73	1.9040	7	0	1	2	5	3
47.73	47.73	1.9040	3	5	1	3		3
48.26	48.26	1.8840	1	4	3	6	5	6

2THETA	PEAK	D	H	K	L	I(INT)	I(PK)	I(DS)
49.15	49.16	1.8519	6	2	1	6	5	6
50.05	50.06	1.8208	6	1	2	1	1	1
50.83	50.84	1.7947	1	0	4	6	5	6
51.07	51.06	1.7869	3	3	3	5	5	6
52.79	52.78	1.7327	0	7	2	2	2	2
54.12	54.12	1.6932	6	0	3	6	5	6
55.08	55.08	1.6659	0	6	3	4	15	16
55.08	55.08	1.6659	2	7	0	14	11	16
55.89	55.90	1.6436	4	0	4	14	3	4
57.43	57.44	1.6032	0	5	4	4	17	25
57.75	57.74	1.5951	5	1	0	22	1	1
58.44	58.44	1.5779	8	0	1	1	3	4
58.69	58.68	1.5717	3	2	1	3	3	2
59.24	59.24	1.5585	8	0	0	2	1	6
59.68	59.68	1.5479	4	6	0	5	4	5
60.24	60.24	1.5349	9	7	2	4	4	12
60.47	60.48	1.5296	7	2	2	10	8	2
60.91	60.92	1.5196	7	3	2	2	2	20
61.16	61.16	1.5140	0	5	4	17	13	4
61.47	61.46	1.5072	8	2	0	3	3	5
62.37	62.38	1.4875	2	4	2	4	3	4
62.90	62.90	1.4763	1	8	4	3	2	26
63.57	63.58	1.4622	5	1	1	21	16	5
63.78	63.76	1.4580	1	7	4	4	12	3
64.10	64.10	1.4516	6	4	3	3	3	5
64.67	64.66	1.4401	0	4	5	4	3	8
65.06	65.06	1.4324	7	0	1	7	5	8
65.70	65.70	1.4200	6	4	4	7	7	3
65.70	65.70	1.4200	1	9	0	2		
65.94	65.94	1.4154	4	4	1	9	9	11
66.14	66.14	1.4116	6	3	3	6		7
66.14	66.14	1.4116	3	6	1	2		2
68.02	68.02	1.3770	10	8	0	10	1	13
69.17	69.18	1.3569	3	0	1	1	7	2
69.90	69.90	1.3445	9	1	2		2	2
69.90	69.90	1.3445	5	6	1	1		
70.31	70.32	1.3377	9	2	1	1	1	1

TOURMALINE - BUERGER, BURNHAM AND PEACOR, 1962

2THETA	PEAK	D	H	K	L	I(INT)	I(PK)	I(DS)
70.45	70.46	1.3354	0	4	5	4	3	2
70.54	70.54	1.3339	3	5	4	1	3	5
70.74	70.74	1.3307	1	5	3	3	4	2
70.83	70.82	1.3293	6	6	0	5	4	3
71.58	71.58	1.3170	2	3	5	1	1	6
71.95	71.96	1.3112	10	1	0	15	11	2
72.16	72.16	1.3079	4	3	2	4	1	19
72.57	72.56	1.3016	8	7	1	1	8	5
73.82	73.82	1.2825	5	5	2	13	1	2
74.10	74.04	1.2784	2	0	5	2	9	17
74.19	74.18	1.2771	9	8	3	4	6	3
75.02	75.02	1.2650	1	3	0	5	4	5
76.04	76.04	1.2506	7	5	4	2	1	7
76.60	76.60	1.2427	0	2	5	2	1	3
77.00	77.00	1.2373	11	1	2	4	3	6
77.70	77.70	1.2279	0	11	1	3	2	5
78.23	78.22	1.2209	4	4	2	2	1	3
79.33	79.34	1.2067	6	0	6	2	1	3
80.40	80.42	1.1933	3	1	5	4	2	2
80.49	80.48	1.1922	0	6	4	1	1	4
81.35	81.36	1.1818	6	1	3	3	3	1
82.58	82.58	1.1673	11	1	6	1	2	2
82.58		1.1673	0	3	6	1		8
83.51	83.52	1.1566	10	0	1	2	4	2
83.91	83.92	1.1521	10	3	3	2	1	3
85.27	85.28	1.1372	9	1	2	3	1	3
85.87	85.88	1.1307	6	4	4	3	3	4
85.87		1.1307	1	5	4			
86.06	86.06	1.1288	9	9	3	1	1	1
87.41	87.42	1.1147	3	3	0	2	1	2
87.80	87.80	1.1109	9	10	1	1	2	3
89.02	89.08	1.0988	3	3	6	2		4
89.08		1.0981	10	0	4	3		
90.62	90.62	1.0835	8	0	2	2	2	3
91.69	91.68	1.0736	5	6	2	1	1	2
92.21	92.22	1.0689	6	9	5	2	2	3
92.22		1.0688	0	6	6	1		2

2THETA	PEAK	D	H	K	L	I(INT)	I(PK)	I(DS)
93.63	93.62	1.0564	9	6	0	2	1	2
93.83	93.84	1.0546	12	1	2	1	1	2
95.29	95.28	1.0423	12	2	1	1	1	2
96.36	96.36	1.0335	5	10	1	2	2	3
96.51		1.0323	4	11	4	2	4	3
96.58	96.58	1.0318	0	11	4	4		6
96.76	96.76	1.0303	7	7	3	2	3	3
96.76		1.0303	2	11	3	1		3
97.57	97.58	1.0239	5	6	2	7	5	11
98.13	98.12	1.0196	7	8	1	1	1	2
98.52	98.50	1.0166	11	4	4	1	1	2
99.21	99.20	1.0114	2	12	1	2	1	2
99.75	99.74	1.0073	1	1	6	1	1	2
100.82	100.84	0.9995	0	10	5	1	1	2
100.91	100.90	0.9989	11	1	4	1		1
100.91		0.9989	4	9	4	1		2
101.91	101.92	0.9918	2	9	5	1	1	2
102.48	102.48	0.9878	4	11	2	2	1	3
102.87	102.86	0.9851	7	9	1	2	1	1
103.14	103.14	0.9832	3	6	7	1	1	1
103.97	103.96	0.9777	10	6	1	2	1	3
104.10	104.10	0.9768	7	5	5	2	2	2
104.47	104.46	0.9744	11	13	0	2	1	3
105.07	105.08	0.9704	2	6	3	2	1	5
105.49	105.48	0.9677	9	9	6	3	2	2
105.49		0.9677	6	5	6	2		3
106.34	106.34	0.9623	5	8	4	3	2	3
106.41		0.9619	6	8	7	2		3
107.58	107.56	0.9547	5	5	2	2	1	2
108.02	108.02	0.9520	6	10	6	2	1	3
108.42	108.42	0.9496	4	12	1	1	2	2
108.58	108.56	0.9486	9	9	6	1	1	3
108.58		0.9486	0	2	6	3		1
109.14	109.14	0.9453	13	0	1	1	0	1
109.55	109.56	0.9429	14	1	6	1	1	1
109.71	109.72	0.9420	2	8	7	1	1	2
109.83	109.84	0.9413	4	3		3	2	5

TOURMALINE - BUERGER, BURNHAM AND PEACOR, 1962

2THETA	PEAK	D	H	K	L	I(INT)	I(PK)	I(DS)
109.98	109.96	0.9404	1	13	3	1	1	1
111.84	111.84	0.9300	13	3	1	1	1	2
111.84		0.9300	9	8	1	1		4
112.27	112.26	0.9276	12	3	3	1	1	1
112.58	112.58	0.9260	12	4	2	1	1	2
113.14	113.14	0.9230	9	4	5	1	1	2
113.14		0.9230	9	11	1	1	2	2
113.15		0.9229	7	8	6	1		2
113.43	113.52	0.9214	8	8	3	1		2
113.52		0.9209	15	0	0	3	2	5
114.16	114.16	0.9176	6	11	1	2	1	2
114.40	114.44	0.9164	8	7	4	1	2	2
114.45		0.9161	3	5	7	1		2
115.48	115.48	0.9108	3	10	5	1	1	2
115.58		0.9104	12	2	4	3		1
116.10	116.10	0.9078	12	13	2	1	2	5
116.77	116.76	0.9045	8	9	2	2	2	4
116.97	116.98	0.9035	5	10	1	1	1	2
117.06		0.9031	0	6	8	3		1
119.11	119.18	0.8934	6	6	6	2	2	3
119.19		0.8931	11	4	4	3		5
120.32	120.34	0.8880	5	9	5	2	2	3
120.34		0.8879	10	1	6	1		2
120.34		0.8879	1	15	1	1		2
121.41	121.42	0.8832	8	10	5	1	1	2
122.67	122.68	0.8778	8	1	1	1	1	2
122.83	122.84	0.8772	12	1	5	2	2	1
122.85		0.8771	9	3	6	1		3
123.95	123.96	0.8726	11	7	1	1	1	1
123.95		0.8726	3	14	1	1		1
124.21	124.22	0.8715	7	9	4	1	1	2
124.44	124.44	0.8706	10	7	3	1	2	1
124.44		0.8706	7	10	8	1		1
124.53		0.8702	2	3	4	1		1
125.52	125.52	0.8663	10	6	4	6	4	11
125.52		0.8663	0	14	4	1		2

2THETA	PEAK	D	H	K	L	I(INT)	I(PK)	I(DS)
125.58		0.8661	4	6	7	1		1
126.84	126.90	0.8613	2	13	4	1	1	2
126.90		0.8610	7	3	7	2		4
127.07	127.06	0.8604	15	0	3	1	1	1
127.43	127.44	0.8591	10	8	2	1	1	1
128.08	128.08	0.8567	0	13	5	1	1	1
128.08		0.8567	7	8	6	1		1
128.10		0.8566	8	5	5	1		2
128.79	128.78	0.8541	14	3	2	1	1	1
128.79		0.8541	7	11	2	1		5
129.91	129.90	0.8502	1	5	8	2	2	2
130.85	130.86	0.8470	10	5	5	7	4	12
130.96		0.8466	4	12	4	4		6
132.39	132.46	0.8419	14	1	5	1	1	3
132.46		0.8417	6	5	7	2		3
132.64	132.64	0.8411	13	4	3	1	1	2
132.64		0.8411	4	13	3	1		2
132.75	132.74	0.8407	3	4	8	2	1	3
133.56	133.56	0.8381	14	4	1	1	1	1
133.74	133.74	0.8376	4	11	5	1	1	1
134.23	134.22	0.8361	16	1	0	4	3	7
134.23		0.8361	11	8	0	1		1
134.51	134.50	0.8352	2	15	2	1	1	2
134.51		0.8352	13	5	2	1		3
135.26	135.24	0.8329	12	0	6	1	1	2
135.36	135.38	0.8326	13	3	4	1	1	1
135.43		0.8324	3	8	7	2		2
135.62	135.62	0.8319	9	9	3	1	1	3
135.74	135.72	0.8315	6	1	8	2	1	2
136.80	136.80	0.8284	7	7	6	3	2	6
136.80		0.8284	2	11	6	3		3
137.29	137.30	0.8270	15	3	0	2	1	4
138.49	138.50	0.8237	6	11	0	3	2	6
138.89	138.90	0.8226	5	5	8	1	1	2
140.00	140.00	0.8197	9	7	5	1	1	2
140.42	140.42	0.8186	12	6	3	2	1	3
140.42		0.8186	6	12	3	1		1

TOURMALINE - BUERGER, BURNHAM AND PEACOR, 1962

2THETA	PEAK	D	H	K	L	I(INT)	I(PK)	I(DS)
140.86	140.82	0.8175	9	10	2	1	1	2
141.48	141.48	0.8159	13	6	1	1	1	3
141.69	141.70	0.8154	6	10	5	3	2	3
144.39	144.38	0.8090	12	7	2	2	1	6
145.90	145.90	0.8057	4	5	8	4	2	4
146.47	146.46	0.8044	0	0	9	1	0	8
146.96	146.94	0.8034	2	16	1	1	1	2
148.24	148.24	0.8008	6	13	2	3	2	2
148.46	148.44	0.8004	1	1	9	1	1	6
149.24	149.24	0.7989	12	4	5	5	3	2
149.40	149.40	0.7986	1	15	4	4	3	9
149.93	149.94	0.7976	7	2	8	1	2	8
149.93		0.7976	10	10	0	4		8
151.39	151.40	0.7949	1	14	5	2	1	4
152.82	152.82	0.7924	3	0	9	1	1	2
153.42	153.42	0.7915	4	15	1	1	1	3
153.42		0.7915	9	11	1	1		1
153.90	154.02	0.7907	3	14	4	1	2	1
154.01		0.7905	7	6	7	3		6
154.33	154.32	0.7900	1	16	3	2	3	3
154.33		0.7900	16	1	3	1		2
154.33		0.7900	11	8	3	1		3
154.33		0.7900	8	11	3	1		1
154.99	154.98	0.7890	16	2	2	2	1	4
155.91	155.92	0.7876	12	8	1	1	1	2
156.24	156.26	0.7871	3	13	5	1	1	2
157.12	157.12	0.7859	6	4	8	4	4	9
157.13		0.7858	13	7	0	1		2
158.68	158.68	0.7838	17	1	1	1	1	2
159.42	159.42	0.7828	4	9	7	6	9	11
159.42		0.7828	11	1	7	1		2
159.81	159.82	0.7824	15	3	3	3	9	5
159.81		0.7824	3	15	3	3		6

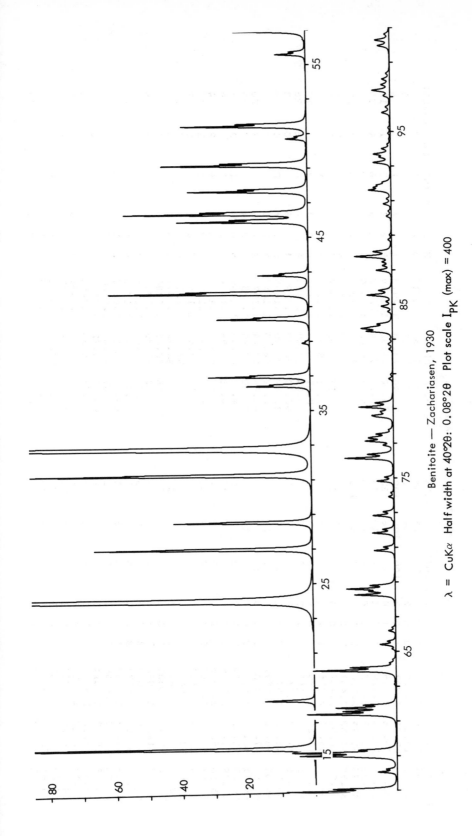

Benitoite — Zachariasen, 1930

$\lambda = CuK\alpha$ Half width at $40°2\theta$: $0.08°2\theta$ Plot scale I_{PK} (max) = 400

BENITOITE – ZACHARIASEN, 1930

2THETA	PEAK	D	H	K	L	I(INT)	I(PK)	I(DS)
15.47	15.48	5.724	1	0	0	23	26	14
18.22	18.22	4.865	0	0	2	3	4	2
23.98	23.98	3.707	1	0	2	100	100	100
26.95	26.96	3.305	1	1	0	17	16	20
28.50	28.50	3.129	1	1	1	11	10	14
31.22	31.22	2.862	2	0	0	24	22	34
32.73	32.74	2.734	2	0	1	70	62	108
36.39	36.38	2.467	1	0	3	6	5	10
36.92	36.92	2.432	2	0	2	9	8	16
38.87	38.88	2.315	1	1	3	1	1	1
40.25	40.24	2.239	2	1	0	8	7	17
41.71	41.72	2.164	2	1	1	19	15	41
42.78	42.78	2.112	2	0	3	5	4	10
45.86	45.86	1.9769	2	1	2	12	10	30
46.30	46.30	1.9591	1	1	4	17	14	44
47.61	47.62	1.9081	3	0	0	11	9	30
49.11	49.10	1.8535	2	0	4	15	11	41
50.67	50.68	1.7999	3	0	2	2	2	6
51.39	51.40	1.7764	2	1	3	13	9	38
55.56	55.56	1.6525	2	2	0	3	2	11
56.91	56.90	1.6167	2	2	1	12	9	42
58.04	58.04	1.5877	3	1	0	2	2	7
58.89	58.98	1.5669	3	1	1	1	1	5
58.98	58.98	1.5647	2	2	2	9	8	35
59.16	59.16	1.5603	1	0	6	7	7	24
61.37	61.38	1.5093	3	1	2	10	10	39
61.73	61.74	1.5013	3	0	4	6	5	25
63.08	63.08	1.4724	2	2	3	1	1	1
63.89	63.88	1.4559	1	1	6	9	9	40
64.33	64.32	1.4469	2	1	5	1	1	2
65.13	65.12	1.4311	4	0	0	0	0	1
65.39	65.38	1.4260	3	1	3	2	2	8
66.17	66.18	1.4109	4	0	1	1	1	4
68.25	68.26	1.3729	4	0	2	5	5	22
68.60	68.60	1.3669	2	2	4	5	5	26
70.81	70.80	1.3295	3	1	4	3	3	13
71.82	71.82	1.3133	3	2	0	3	1	14
72.57	72.58	1.3015	3	2	1	0	0	1
72.82	72.82	1.2976	2	1	6	3	2	13
73.90	73.90	1.2813	1	1	7	1	1	4
74.82	74.82	1.2679	3	2	2	2	0	12
76.14	76.14	1.2492	4	1	0	6	4	35
76.88	76.88	1.2390	4	1	1	1	1	6
77.12	77.12	1.2357	3	0	6	3	2	16
77.28	77.30	1.2335	2	2	5	2	1	12
77.53	77.52	1.2302	3	1	5	1	1	5
78.51	78.58	1.2173	3	2	3	1	2	8
78.59	78.58	1.2162	0	0	8	2	3	12
79.08	79.08	1.2099	4	1	2	4	3	26
80.70	80.70	1.1897	1	0	8	1	0	3
82.39	82.40	1.1695	2	1	7	0	0	3
82.72	82.70	1.1657	4	1	3	3	3	20
83.44	83.44	1.1574	2	2	6	4	2	23
83.60	83.60	1.1556	3	2	4	1	0	3
84.56	84.56	1.1449	5	0	0	1	1	9
84.88	84.88	1.1414	1	1	8	4	2	24
85.52	85.52	1.1345	3	1	6	2	1	15
86.96	86.96	1.1194	2	0	8	5	2	7
87.44	87.44	1.1144	5	0	2	1	1	35
87.76	87.76	1.1112	4	1	4	3	3	3
88.72	88.72	1.1017	3	3	0	0	0	5
90.08	90.08	1.0886	3	2	5	1	1	8
90.80	90.80	1.0818	4	2	0	2	2	23
91.59	91.60	1.0745	4	2	1	6	1	14
91.75	91.74	1.0730	4	0	6	2	1	21
93.19	93.18	1.0602	2	1	8	3	2	19
93.67	93.66	1.0560	4	2	2	2	1	3
94.23	94.22	1.0512	4	1	5	0	0	3
94.87	94.86	1.0458	3	1	7	0	0	11
95.19	95.18	1.0431	3	3	3	1	1	7
96.07	96.08	1.0359	5	0	4	0	0	1
97.04	97.04	1.0281	1	1	9	1	1	7
97.11	97.04	1.0275	5	1	0	0	1	11
97.36	97.36	1.0256	3	0	8	2	1	20

2THETA	PEAK	D	H	K	L	I(INT)	I(PK)	I(DS)
97.76	97.68	1.0224	5	1	1	0	1	1
98.00	98.00	1.0206	3	2	6	1	1	9
99.94	99.94	1.0059	5	1	2	2	1	16
100.27	100.28	1.0035	3	3	4	1	1	13
102.22	102.22	0.9896	4	1	6	2	1	20
102.38	102.38	0.9885	2	2	8	1	1	11
103.69	103.68	0.9795	2	1	9	1	1	11
105.59	105.60	0.9671	3	1	8	0	0	
105.83	105.84	0.9655	1	0	10	1	1	8
106.83	106.84	0.9592	6	0	0	1	1	13
107.67	107.66	0.9541	5	0	4	1	1	5
108.85	108.84	0.9470	4	3	0	1	1	14
109.86	109.86	0.9411	4	3	1	1	1	9
110.63	110.72	0.9367	6	0	2	0	1	4
110.71	110.88	0.9362	6	0	6	3	1	26
110.88		0.9353	5	1	6	0	0	8
111.22	111.20	0.9334	1	1	10	2	0	21
112.00	112.00	0.9291	4	0	7	1	0	5
112.43	112.42	0.9268	4	3	8	0	0	3
112.95	112.94	0.9240	4	3	2	1	0	10
113.46	113.40	0.9212	2	2	10	0	1	3
114.34	114.34	0.9166	5	2	1	1	0	15
115.13	115.16	0.9126	5	2	0	0	1	5
115.40	115.40	0.9113	3	3	6	3	0	27
116.91	116.92	0.9038	4	3	3	3	1	32
117.54	117.54	0.9008	5	2	2	3	1	20
117.72	117.70	0.8999	4	2	6	2	1	24
119.34	119.34	0.8923	3	2	8	2	1	16
119.35	120.28	0.8882	6	0	4	1	1	16
120.27	120.44	0.8874	2	1	10	1	0	30
120.45	121.66	0.8821	5	2	3	0	1	13
121.67	122.70	0.8777	4	3	4	3	1	65
122.70	123.86	0.8730	6	1	0	1	2	20
123.85	124.24	0.8714	4	1	8	6	1	20
124.23	125.00	0.8683	5	1	6	2	1	12
125.01	125.40	0.8668	3	0	10	2	0	
125.40	127.40	0.8592	6	1	2	1	0	

2THETA	PEAK	D	H	K	L	I(INT)	I(PK)	I(DS)
127.79	127.78	0.8578	5	2	4	5	1	55
130.78	130.78	0.8472	4	3	5	0	0	4
132.05	132.06	0.8430	6	1	3	2	0	1
133.46	133.46	0.8385	2	2	10	1	1	24
134.68	134.70	0.8347	3	2	9	0	0	3
135.02	135.02	0.8336	5	0	8	1	1	11
136.39	136.40	0.8296	3	1	10	3	1	37
136.51		0.8293	5	2	5	0		2
137.45	137.56	0.8266	5	1	7	2	0	22
137.57		0.8263	4	4	0	4	1	47
139.00	139.02	0.8223	6	0	6	0	1	40
139.25	139.24	0.8217	6	1	4	1	1	12
140.74	140.76	0.8178	7	0	0	0		10
140.74		0.8178	5	3	0	1		7
140.86		0.8175	4	1	9	1	0	13
141.24	141.24	0.8165	3	3	8	1	1	6
141.89		0.8149	5	3	1	0		32
142.02	142.02	0.8146	4	4	2	2		20
142.28		0.8140	3	0	6	1	0	10
143.59	143.58	0.8108	0	0	12	2	1	34
144.69	144.68	0.8083	4	3	8	2	1	42
145.53	145.54	0.8065	5	3	2	3		11
145.53		0.8065	7	0	0	0		33
146.38	146.42	0.8046	4	0	10	2	1	11
147.25	147.30	0.8028	1	0	12	5	0	77
149.70	149.70	0.7980	5	2	6	1	1	14
152.00	152.00	0.7938	6	3	0	1	0	11
152.51	152.50	0.7930	5	3	3	1	0	41
155.98	155.98	0.7875	6	1	6	3	0	52
157.62	157.58	0.7852	4	4	6	6	1	85
158.92	158.96	0.7835	3	2	10	2	1	33
159.82	160.36	0.7823	2	0	12	3	1	39
160.28		0.7818	1	0	2	2		65
161.74	161.74	0.7801	4	3	10	4	1	17
162.54		0.7793	4	3	7	1		

462

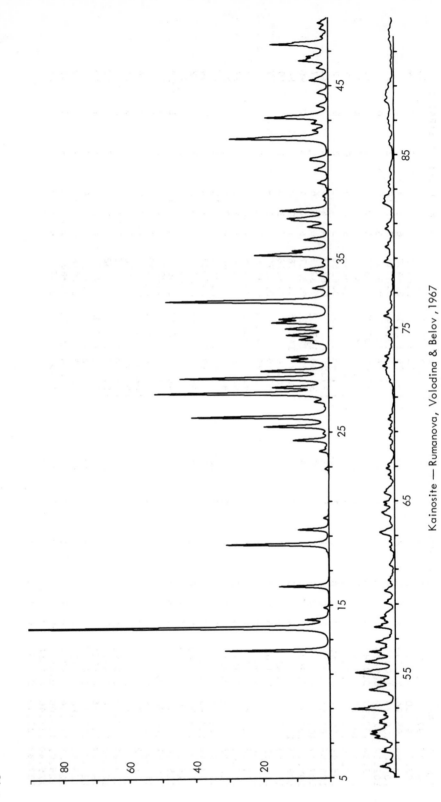

Kainosite — Rumanova, Volodina & Belov, 1967

λ = CuKα Half width at 40°2θ: 0.11°2θ Plot scale I$_{PK}$ (max) = 100

KAINOSITE — RUMANOVA, VOLODINA AND BELOV, 1967

2THETA	PEAK	D	H	K	L	I(INT)	I(PK)	I(DS)
12.37	12.38	7.150	2	0	0	100	31	28
13.69	13.70	6.465	2	2	0	100	100	100
14.14	14.14	6.257	1	1	0	6	7	6
14.83	14.84	5.970	0	1	1	2	2	2
16.07	16.08	5.509	1	1	1	15	15	17
18.49	18.50	4.795	2	2	1	33	31	42
19.35	19.36	4.582	2	1	0	10	9	13
20.01	20.02	4.433	1	1	1	2	2	2
22.84	22.84	3.890	3	0	1	1	1	2
23.87	23.88	3.725	3	1	1	3	3	5
24.50	24.52	3.630	0	1	1	13	11	21
24.88	24.90	3.575	4	0	0	1	2	2
25.29	25.30	3.518	4	1	0	22	19	39
25.83	25.84	3.446	4	1	1	49	41	86
26.72	26.72	3.333	3	2	1	4	4	8
27.20	27.20	3.276	1	0	2	64	52	119
27.54	27.56	3.236	2	3	0	6	17	12
27.57		3.233	4	1	1	13		25
28.08	28.08	3.175	0	4	0	54	44	106
28.51	28.50	3.129	1	2	1	25	20	49
29.09	29.10	3.067	4	2	0	13	11	26
29.31	29.32	3.045	2	1	2	15	13	30
30.13	30.14	2.964	2	4	1	5	5	10
30.32	30.32	2.945	2	0	2	10	9	21
30.57	30.58	2.922	3	3	0	15	13	33
30.94	30.94	2.888	3	4	1	16	13	34
31.30	31.30	2.855	1	4	1	21	17	46
31.51	31.50	2.837	4	2	1	17	15	38
32.48	32.52	2.754	2	2	2	3	49	7
32.51	32.51	2.752	4	3	0	51		117
32.54		2.749	3	0	2	15		34
33.30	33.30	2.689	3	1	2	6	5	14
34.03	34.04	2.632	5	0	1	4	3	9
34.36	34.36	2.608	3	3	2	10	7	24
34.75	34.76	2.579	5	1	1	2	2	4
35.20	35.20	2.547	4	3	1	30	22	78
35.45	35.46	2.530	3	2	2	12	11	32

2THETA	PEAK	D	H	K	L	I(INT)	I(PK)	I(DS)
36.09	36.10	2.487	2	3	2	8	7	4
36.10		2.486	4	4	1	8		22
36.84	36.84	2.438	5	2	1	13	6	22
37.22	37.22	2.414	0	1	2	9	11	35
37.32	37.32	2.407	4	0	0	4	12	24
37.71	37.76	2.383	6	1	1	17	14	11
37.76		2.380	1	5	1	1		48
38.37	38.38	2.344	6	0	0	1	1	4
38.59	38.58	2.331	0	4	2	2	2	5
39.36	39.36	2.287	2	5	1	4	3	13
40.10	40.10	2.247	5	0	1	6	4	17
40.68	40.68	2.216	1	1	3	6	6	18
40.73		2.213	6	0	0	1	1	4
40.79	40.78	2.210	1	3	3	3	5	8
41.88	41.90	2.155	0	6	0	18	29	58
41.91		2.154	3	1	3	28		91
42.40	42.40	2.130	4	2	2	5	5	18
42.78	42.78	2.112	2	1	3	6	5	20
43.11		2.096	1	3	3	5		16
43.14	43.14	2.095	5	0	0	23	19	77
43.80	43.82	2.065	2	6	0	4	3	12
43.84		2.063	6	2	2	2		5
44.55	44.56	2.032	2	2	3	4	3	13
45.29	45.30	2.001	5	1	1	8	6	27
45.49	45.40	1.9922	6	3	1	2	4	6
46.01	46.00	1.9709	1	3	3	4	3	14
46.41	46.42	1.9548	7	0	1	13	9	46
46.66	46.66	1.9448	3	1	2	8	7	29
46.97	46.98	1.9328	5	1	2	1	2	4
47.38	47.38	1.9171	2	3	3	25	17	94
47.83	47.82	1.9002	4	0	3	3	2	10
48.28	48.28	1.8836	5	5	1	3	3	12
48.62	48.62	1.8711	2	6	2	5	3	18
49.33	49.34	1.8456	1	0	3	2	2	10
49.60	49.60	1.8363	6	3	3	7	5	29
49.98	49.98	1.8231	3	2	4	2	2	8
50.46	50.46	1.8070	2	4	5	7	4	27

KAINOSITE - RUMANOVA, VOLODINA AND BELOV, 1967

2THETA	PEAK	D	H	K	L	I(INT)	I(PK)	I(DS)
50.66	50.60	1.8003	1	6	2	2		10
51.02	51.06	1.7884	2	7	0	1	3	6
51.05		1.7875	8	0	0	2		7
51.24	51.28	1.7813	7	1	1	1	6	5
51.27		1.7802	4	3	1	2		7
51.28		1.7799	6	6	1	6		24
51.51	51.50	1.7728	1	3	2	9	8	40
51.67	51.66	1.7676	7	7	1	5	7	22
51.75		1.7651	5	0	3	2		8
51.94	51.92	1.7590	2	6	2	3	4	14
52.14	52.14	1.7525	6	5	0	6	5	25
52.26	52.28	1.7489	5	1	3	3	5	14
52.35		1.7462	7	0	0	1		5
52.59	52.60	1.7387	4	3	3	4	4	19
52.86		1.7305	7	1	2	2	13	8
52.96	52.96	1.7276	8	1	1	20	2	90
53.79	53.80	1.7028	5	2	0	2	2	9
54.02	54.06	1.6960	3	6	1	6	8	26
54.02		1.6960	6	5	3	1		5
54.07		1.6946	0	5	2	8		35
54.48	54.50	1.6828	1	5	1	2	6	8
54.49		1.6825	0	0	4	5		25
54.97	55.06	1.6690	8	2	1	5	12	23
54.98		1.6686	3	7	1	1		7
55.06		1.6664	5	5	0	14		67
55.61	55.70	1.6511	8	3	2	5	9	22
55.70		1.6489	2	5	3	12		59
56.09	56.12	1.6381	4	4	3	2	5	8
56.11		1.6378	2	0	4	4		19
56.26	56.27	1.6335	6	3	0	9	8	45
56.60	56.60	1.6248	2	1	3	3	3	13
56.75	56.84	1.6206	6	1	4	2	6	9
56.84		1.6185	7	3	2	7		36
57.20	57.20	1.6089	1	7	2	2	2	8
57.69	57.68	1.5967	3	5	2	11	6	54
58.05	58.05	1.5876	2	2	4	3	3	16
58.09	58.06	1.5866	3	0	4	1		6

2THETA	PEAK	D	H	K	L	I(INT)	I(PK)	I(DS)
58.20	58.20	1.5838	6	2	2	5	4	23
58.41	58.38	1.5786	8	0	0	2		10
58.50	58.48	1.5765	2	8	0	3	3	13
59.08	59.08	1.5622	1	0	1	3	3	28
59.75	59.78	1.5464	9	1	1	3	3	16
59.81		1.5450	1	0	3	3		16
60.22	60.24	1.5354	9	5	1	1	1	7
60.40	60.40	1.5313	4	3	1	1	2	6
60.41		1.5310	2	1	3	1		7
61.26	61.28	1.5119	4	0	4	1	2	5
61.32		1.5104	7	8	3	2		9
62.14	62.14	1.4925	3	8	1	2	1	12
62.31	62.30	1.4889	3	8	4	2	2	8
62.64	62.64	1.4818	4	2	0	3	2	15
63.07	63.18	1.4727	4	4	3	5	5	26
63.16		1.4708	7	8	2	2		10
63.18		1.4704	8	5	0	4		25
63.79	63.80	1.4579	5	5	3	1		8
64.20	64.32	1.4494	1	8	0	3	2	17
64.31		1.4472	7	2	1	4	4	24
64.74	64.74	1.4387	4	5	2	5	3	29
64.91	64.90	1.4354	4	8	4	3	3	19
65.39	65.46	1.4260	0	2	1	2		9
65.43		1.4253	4	6	3	2		9
65.48		1.4243	3	4	4	3	3	18
65.63	65.62	1.4213	10	0	0	1		6
65.75		1.4189	1	7	3	1		8
66.27	66.28	1.4091	5	5	0	2	1	11
66.49	66.48	1.4050	7	2	1	2	2	15
66.87	66.86	1.3980	9	0	3	4	3	27
67.12	67.08	1.3933	8	8	1	1	1	12
68.01	68.00	1.3773	3	8	4	1	2	7
68.58	68.62	1.3671	5	2	3	1		9
68.63		1.3664	10	3	0	1		8
69.15	69.16	1.3572	8	0	1	3	2	20
69.70	69.70	1.3479	9	6	2	3	2	22
71.84	71.88	1.3129	9	4	2	1	2	9

KAINOSITE - RUMANOVA, VOLODINA AND BELOV, 1967

2THETA	PEAK	D	H	K	L	I(INT)	I(PK)	I(DS)
72.17	72.30	1.3077	10	0	1	1	2	9
72.29		1.3059	1	8	3	3		17
72.41	72.42	1.3039	2	6	2	2		15
72.55	72.54	1.3018	6	6	3	2		17
72.79	72.78	1.2982	5	8	2	5	4	33
73.13	73.12	1.2930	0	10	0	2	2	12
73.31	73.34	1.2901	9	1	3	1	3	8
73.33		1.2898	2	8	3	3		22
74.51	74.48	1.2724	2	0	0	2	2	11
74.75	74.72	1.2690	7	7	2	2	2	14
75.29	75.28	1.2611	5	9	1	1	1	8
75.65	75.66	1.2559	9	5	2	6	3	45
76.39	76.40	1.2456	7	8	1	2	1	12
78.78	78.80	1.2137	6	5	4	6	3	44
78.88		1.2125	5	1	5	1		11
79.28	79.28	1.2074	11	1	2	1	1	10
79.65	79.64	1.2027	1	10	0	4	3	32
79.81	79.86	1.2006	10	3	2	1	1	8
80.12	80.10	1.1968	5	2	5	1		8
80.69	80.68	1.1898	1	5	5	5	3	37
80.90	80.94	1.1872	11	4	1	1	2	9
80.99		1.1862	7	8	2	1		10
81.48	81.48	1.1803	8	8	1	1		9
82.18	82.18	1.1720	5	3	5	4	3	37
82.34	82.42	1.1700	3	10	2	2	3	13
82.46		1.1686	12	1	1	4		10
82.58	82.56	1.1673	11	3	2	1	3	32
83.10	83.14	1.1613	10	0	3	3	2	12
83.17		1.1605	5	10	1	1		9
83.40	83.38	1.1579	0	11	1	1	1	11
83.73	83.66	1.1541	11	1	1	1	1	9
84.07	84.06	1.1504	2	8	4	2		15
84.20	84.18	1.1489	6	8	3	2	1	14
85.56	85.58	1.1341	9	7	2	2	1	20
85.74	85.74	1.1322	12	3	1	2	1	17
86.52	86.52	1.1240	7	0	5	1	1	15
86.84	86.84	1.1207	6	10	1	2	2	15

2THETA	PEAK	D	H	K	L	I(INT)	I(PK)	I(DS)
87.07	87.08	1.1182	1	0	6	3	2	25
87.46	87.32	1.1142	5	9	3	1	2	13
88.07	88.14	1.1082	4	8	4	2	3	18
88.16		1.1072	8	5	4	2		28
88.53	88.48	1.1036	7	6	3	2		19
88.61		1.1027	6	7	4	1		12
88.70	88.68	1.1019	1	5	6	2	2	14
88.71		1.1018	5	5	5	2		10
89.10	89.08	1.0979	11	1	2	3	2	25
90.14	90.14	1.0880	3	1	6	1	1	12
90.49	90.44	1.0847	1	7	5	1	1	10
91.16	91.16	1.0784	7	10	1	2	1	15
91.66	91.68	1.0738	9	8	2	2		19
91.73		1.0732	2	5	6	1		11
92.25	92.24	1.0686	12	2	1	3		25
93.39	93.46	1.0584	3	3	6	3		25
93.48		1.0577	10	8	1	3		16
93.63	93.62	1.0564	10	1	3	3	2	29
95.23	95.26	1.0428	12	6	0	1	1	11
96.16	96.18	1.0352	8	10	1	1	1	15
97.18	97.18	1.0270	0	0	5	5	1	24
97.41	97.44	1.0252	10	0	4	2	1	19
97.89	97.94	1.0214	14	0	0	1		14
97.95		1.0210	8	7	4	2		18
98.75	98.80	1.0149	5	3	6	1		11
98.75		1.0148	2	10	4	1		13
98.82		1.0143	9	2	5	2		19
98.88		1.0138	6	10	3	1		11
100.52	100.52	1.0017	11	8	1	2	1	17
101.15	101.12	0.9971	11	6	0	1	1	12
101.65	101.86	0.9939	14	3	0	1	1	8
101.80		0.9936	4	8	5	1		7
101.88		0.9926	1	6	6	1		9
103.29	103.32	0.9823	9	10	1	2	1	18
103.31		0.9821	8	11	0	1		6
103.38		0.9816	1	13	1	1		10

465

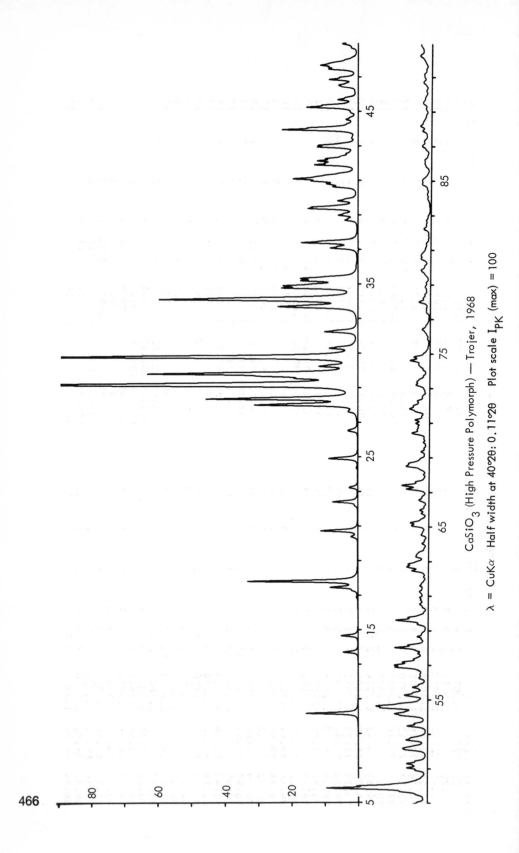

466

CaSiO₃ (High Pressure Polymorph) — Trojer, 1968

$\lambda = CuK\alpha$ Half width at 40°2θ: 0.11°2θ   Plot scale I_{PK} (max) = 100

CASIO3 (HIGH PRESSURE POLYMORPH) — TROJER, 1968

2THETA	PEAK	D	H	K	L	I(INT)	I(PK)	I(DS)
10.14	10.14	8.717	0	1	0	11	16	6
13.67	13.68	6.470	0	0	1	3	5	2
14.63	14.64	6.048	1	0	1	4	5	2
17.43	17.44	5.084	1	0	1	6	8	4
17.79	17.80	4.983	1	1	1	27	33	17
20.70	20.70	4.287	1	1	0	9	11	7
22.37	22.38	3.971	1	2	1	7	8	5
23.21	23.22	3.830	1	-1	1	2	3	2
24.88	24.88	3.576	0	2	1	8	9	7
26.48	26.52	3.363	-1	-2	1	2	3	2
27.59	27.60	3.230	2	1	1	28	31	26
27.99	28.00	3.185	1	0	2	42	45	39
28.34	28.34	3.147	1	1	2	61	99	57
29.10	29.18	3.066	2	0	0	16		15
29.14		3.061	0	-1	2	44		42
29.19		3.057	2	1	0	11	19	10
29.51	29.52	3.024	2	2	1	55	63	54
29.79	29.78	2.997	2	2	1	9	12	9
30.21	30.22	2.955	0	1	2	100	100	100
30.74	30.74	2.906	1	3	1	7	9	7
31.26	31.26	2.858	1	2	1	10	10	10
32.22	32.22	2.776	-1	2	1	10	24	10
33.62		2.664	-1	3	1	5		6
33.67		2.660	-2	-1	1	11		12
33.68	33.68	2.659	-1	1	0	7		8
34.04	34.08	2.632	-2	2	0	22	59	25
34.09		2.628	-1	-1	2	13		14
34.09		2.628	0	2	2	12		13
34.09		2.628	-2	2	2	6		7
34.09		2.627	0	3	1	21		24
34.75	34.76	2.579	2	3	0	16	23	18
34.90	34.88	2.569	0	0	3	14	22	16
35.17	35.18	2.549	-2	0	1	2	17	2
35.28	35.30	2.542	2	0	2	12	17	14
35.32		2.539	-1	-2	2	8	8	10
37.05	37.06	2.424	-1	1	2	8		10
37.37	37.38	2.404	-1	2	2	17	17	21

2THETA	PEAK	D	H	K	L	I(INT)	I(PK)	I(DS)
38.67	38.68	2.326	1	3	2	4	4	5
38.96	38.96	2.310	2	4	1	6	6	8
39.35	39.36	2.288	2	-1	1	14	15	19
39.44	39.44	2.283	-1	-3	1	6	14	7
39.80	39.80	2.263	-2	3	1	6	6	8
40.61	40.62	2.220	3	1	1	7	8	10
40.79	40.78	2.211	1	0	3	7	10	9
40.95	40.96	2.202	2	-2	0	11	14	14
41.06	41.08	2.196	2	1	1	9	20	13
41.10		2.194	3	3	1	7		10
41.88	41.88	2.155	3	0	0	14	13	19
42.11	42.10	2.155	1	2	0	8	12	11
42.12		2.144	3	2	2	4		5
42.26	42.22	2.144	2	4	0	4	10	6
42.55	42.54	2.137	-1	0	2	13	4	3
43.02	42.92	2.123	1	-1	3	5	12	18
43.92	43.02	2.105	0	-1	2	27	23	8
44.13	43.92	2.101	-1	1	2	3	15	39
44.51	44.02	2.060	2	4	1	3		2
45.18	44.52	2.051	0	2	2	2	15	5
45.23	45.22	2.034	3	1	2	16		4
45.36		2.005	-1	-3	2	2	11	24
45.66	45.34	2.003	2	0	3	7	6	2
46.41	45.66	1.9974	-1	2	0	4	6	10
46.49	46.50	1.9853	2	4	1	4		7
46.83		1.9548	0	-2	3	10		6
47.44	46.84	1.9515	-2	4	0	9	9	16
47.53	47.44	1.9382	3	3	1	3	9	14
47.66	47.56	1.9150	-3	1	1	4	9	5
47.67	47.66	1.9114	-3	-1	1	4	11	7
47.73		1.9066	-1	3	3	4		7
48.66		1.9062	-1	3	2	2		6
48.91	48.68	1.9037	2	1	3	5	2	3
49.65	48.90	1.8695	-2	-1	1	6	5	8
49.85	49.66	1.8608	-2	-3	2	19	9	9
49.85	49.86	1.8346	-2	3	0	15	30	32
		1.8277	0	-4	2			25

CASIO3 (HIGH PRESSURE POLYMORPH) - TROJER, 1968

2THETA	PEAK	D	H	K	L	I(INT)	I(PK)	I(DS)
49.90	49.98	1.8261	-1	-4	2	2	21	3
49.99		1.8229	3	2	0	2		4
50.10	50.90	1.8192	2	3	3	5		8
50.90		1.7926	0	5	0	3	4	6
51.04	51.04	1.7878	3	1	2	3	6	6
51.05		1.7876	1	2	1	2		3
51.28	51.28	1.7797	2	5	1	7	6	12
51.29	52.16	1.7542	-1	5	1	6	7	11
52.09		1.7517	1	5	1	5		9
52.17	52.66	1.7370	-1	2	3	7	7	13
52.65		1.7343	3	4	3	2		4
52.73	53.46	1.7124	0	3	3	8	6	14
53.46	54.18	1.6945	3	0	3	7	10	14
54.07		1.6919	-2	3	1	7		13
54.16		1.6905	-3	0	1	2		3
54.21	54.54	1.6817	-2	-4	2	14	14	26
54.52	54.62	1.6796	-2	-5	2	2	16	4
54.59		1.6788	-2	2	2	10		19
54.62	55.26	1.6628	3	0	4	2	7	5
55.19		1.6609	3	3	2	7		14
55.26	55.72	1.6484	2	5	2	5	5	9
55.71	56.24	1.6346	4	3	1	2	3	3
56.23	56.98	1.6177	3	5	0	10	10	21
56.87	57.16	1.6149	4	2	2	7	10	13
56.98	57.34	1.6096	2	1	4	6	8	12
57.18	58.00	1.6032	4	1	0	11	4	22
57.43		1.5888	3	-1	3	2		3
58.00	58.50	1.5766	4	0	3	3	10	6
58.49	58.64	1.5729	-1	-5	2	3	2	6
58.64	59.36	1.5556	1	5	2	3	3	6
59.36	59.60	1.5502	0	-5	2	7		16
59.60		1.5499	-2	4	0	5	10	10
59.60	60.76	1.5230	2	6	0	2	2	4
60.76	61.24	1.5120	4	0	2	2	3	6
61.25	61.58	1.5048	0	2	4	3	3	6
61.58	62.46	1.4858	-1	3	4	5	5	12
62.45	62.68	1.4809	-1	0	4	4	6	8

2THETA	PEAK	D	H	K	L	I(INT)	I(PK)	I(DS)
62.72	62.82	1.4801	-1	-6	1	2		4
62.83	63.26	1.4777	-2	-4	3	3	6	6
63.26	63.38	1.4688	-4	-3	1	4	4	8
63.38	64.04	1.4663	4	1	3	2	4	5
64.03	64.52	1.4529	0	6	0	2	2	5
64.51	65.06	1.4432	3	0	5	2	2	6
65.04		1.4328	4	5	1	5	5	13
65.10	65.22	1.4316	-1	-1	4	2		4
65.23		1.4291	1	-2	2	2	6	4
65.24		1.4288	2	-2	2	2		4
65.75	65.76	1.4190	-4	0	1	2	2	4
65.97	65.96	1.4149	-2	2	3	2	3	5
66.54	66.54	1.4041	1	-3	5	3	3	8
67.41	67.18	1.3923	4	5	2	9	7	23
68.48	67.40	1.3881	2	4	5	8	8	19
68.57	68.50	1.3690	-3	-2	3	7	7	17
68.66		1.3673	-1	-6	2	3		9
69.33	68.66	1.3659	1	6	2	3		6
69.86	69.32	1.3543	-1	5	4	2	6	8
70.37	69.86	1.3452	4	-2	3	3	2	5
70.42	70.42	1.3368	-3	-3	1	3	5	8
70.43		1.3359	-4	1	4	3		4
70.79	70.78	1.3357	-3	0	3	3		9
70.97	70.98	1.3299	-4	-2	2	2	3	4
71.18	71.18	1.3270	-2	-1	4	3	4	9
71.78	71.82	1.3235	3	-3	3	4	4	11
71.82		1.3139	0	-4	4	2	6	5
72.28	72.26	1.3132	-4	-3	3	2		6
72.59	72.48	1.3061	1	-1	5	5	4	6
73.16	73.18	1.3013	2	6	3	3	2	4
73.19		1.2924	4	3	4	2		6
73.23		1.2921	2	-5	0	2		5
73.34	73.36	1.2915	-1	1	5	2		4
73.50		1.2898	-1	2	4	2	4	5
74.06	74.08	1.2874	-1	-4	4	2		6
74.21	74.36	1.2790	-2	5	1	2	2	5
		1.2767	-3	3	3	2	3	4

CASIO3 (HIGH PRESSURE POLYMORPH) - TROJER, 1968

2THETA	PEAK	D	H	K	L	I(INT)	I(PK)	I(DS)
74.60	74.60	1.2711	-2	-7	1	2	6	5
74.61		1.2709	-4	0	4	6		16
74.80	74.80	1.2682	-1	3	4	3	5	9
76.39	76.40	1.2457	4	4	1	2	2	6
77.98	77.98	1.2242	4	-3	1	2	3	7
77.98		1.2242	-4	-6	1	3		8
78.42	78.42	1.2184	0	-6	3	2	2	5
79.77	79.84	1.2012	-4	3	0	2	3	6
79.86		1.2001	0	-5	4	2		7
80.46	80.50	1.1926	5	0	3	2		6
80.52		1.1919	0	6	3	2	3	7
80.66	80.68	1.1902	3	-1	5	2		6
80.82		1.1882	-2	-7	2	2	3	5
82.22	82.22	1.1711	-2	6	0	2	2	6
85.97	85.98	1.1297	3	8	0	2	2	7
86.98	87.00	1.1192	-1	-8	1	2	3	8
87.33	87.28	1.1156	-5	-2	2	2	2	5
87.89	87.90	1.1099	-2	8	2	4	3	14
88.19	88.18	1.1070	-2	-2	5	2		6
88.36		1.1053	6	2	2	2	3	7
89.57	89.50	1.0934	1	-1	6	2	3	6
90.66	90.70	1.0831	6	3	0	2		6
90.68		1.0828	1	5	5	2		6
91.17	91.16	1.0783	0	0	6	3	2	11
92.22	92.24	1.0688	3	2	6	2	4	9
92.29		1.0681	6	5	4	2		6
92.73	92.86	1.0642	6	4	3	2	3	8
92.88		1.0629	4	8	2	4		14
93.13	93.12	1.0607	2	7	4	3	4	11
93.64	93.62	1.0563	6	1	0	3	2	10
94.00	93.96	1.0532	1	3	6	2	2	6
95.55	95.58	1.0402	0	-8	2	2	1	6
96.11	96.12	1.0356	3	-1	6	2	1	7
96.61	96.62	1.0315	4	7	4	4	1	6
98.22	98.20	1.0189	0	-3	6	2	2	15
98.93	98.94	1.0135	1	-8	1	2	1	6
100.77	100.92	0.9999	5	-3	3	2	3	8

2THETA	PEAK	D	H	K	L	I(INT)	I(PK)	I(DS)
100.78	100.78	0.9998	1	9	1	2		4
100.92		0.9988	3	-4	5	2		7
101.12		0.9974	-2	-5	5	1		5
102.45	102.46	0.9880	-1	6	4	1	1	6
104.60	104.66	0.9735	3	5	6	2	2	9
104.72		0.9727	-5	0	3	1		6
104.96	104.96	0.9712	-4	-5	4	1	2	5
107.86	107.84	0.9529	6	7	3	2	1	7
108.47	108.38	0.9492	5	2	6	1	1	5

TABLE 23. MUSCOVITE MICAS

Variety	Muscovite 2M$_1$ (1)	Muscovite 2M$_1$ (2)	Muscovite 3T	Paragonite	Phengite
Composition	$K_2Al_4(Si_3Al)_2O_{20}(OH)_4$	$(K,Na)_2Al_4(Si_3Al)_2O_{20}(OH)_4$	$(K,Na,Ca,Ba)_2(Al,Fe^{2+},Fe^{3+},Mg,Ti)_4(Si_{3.11}Al_{0.89})_2O_{20}[(OH)F]_4$	$(K,Na)_2Al_4(Si_3Al)_2O_{20}(OH)_4$	$(K,Na,Ba,Ca)_{1.94}(Al,Ti,Fe^{3+},Fe^{2+},Mg)_{4.16}(Si_{3.39}Al_{0.61})_2O_{20}[(OH)_{3.84}O_{0.16}]$
Source	Spotted Tiger Mine, Central Australia	Alpe Sponde, Switzerland	Snokomish Co., Washington	Alpe Sponde, Switzerland	Tiburon, Calif.
Reference	Radoslovich, 1960; Gatineau, 1963	Burnham and Radoslovich, 1965;1968	Güven & Burnham, 1967	Burnham and Radoslovich, 1965;1968	Güven, 1968
Cell Dimensions					
a Å	5.189	5.1740	5.1963	5.1342	5.2112
b Å	8.996	8.9762	5.1963	8.9072	9.0383
c Å	20.096	19.875	29.9705	19.376	19.9473
α	90	90	90	90	90
β deg	95.18	95.63	90	94.58	95.77
γ	90	90	120	90	90
Space Group	C2/c	C2/c	P3$_1$12 (P3$_2$12)	C2/c	C2/c
Z	2	2	1.5	1	2

TABLE 23. (cont.)

Variety	Muscovite 2M₁(1)	Muscovite 2M₁(2)	Muscovite 3T	Paragonite	Phengite
Site Occupancy	$T_1 \begin{cases} Si & 0.75 \\ Al & 0.25 \end{cases}$ $T_2 \begin{cases} Si & 0.75 \\ Al & 0.25 \end{cases}$	$M_1 \begin{cases} Na & 0.358 \\ K & 0.642 \end{cases}$ $T_1 \begin{cases} Si & 0.75 \\ Al & 0.25 \end{cases}$ $T_2 \begin{cases} Si & 0.75 \\ Al & 0.25 \end{cases}$	$M_1 \begin{cases} K & 0.90 \\ Na & 0.06 \\ Ca & 0.01 \\ Ba & 0.01 \end{cases}$ $Al_1 \begin{cases} Al & 0.830 \\ Fe & 0.038 \\ Fe & 0.038 \\ Mg & 0.086 \\ Ti & 0.010 \end{cases}$ $T_1 \begin{cases} Si & 0.55 \\ Al & 0.45 \end{cases}$ $T_2\ Si\ 1.0$	$M_1 \begin{cases} K & 0.15 \\ Na & 0.85 \end{cases}$ $T_1 \begin{cases} Si & 0.75 \\ Al & 0.25 \end{cases}$ $T_2 \begin{cases} Si & 0.75 \\ Al & 0.25 \end{cases}$	$M_1 \begin{cases} K & 0.87 \\ Na & 0.07 \\ Ba & 0.01 \\ Ca & 0.02 \end{cases}$ $Al_1 \begin{cases} Al & 0.715 \\ Fe^2 & 0.045 \\ Fe^3 & 0.025 \\ Mg & 0.250 \\ Ti & 0.005 \end{cases}$ $T_1 \begin{cases} Si & 0.82 \\ Al & 0.18 \end{cases}$ $T_2 \begin{cases} Si & 0.82 \\ Al & 0.18 \end{cases}$
Method	3-D, film	3-D, counter	3-D, counter	3-D, counter	3-D, counter
R & Rating	0.16 (2)	0.038 (1)	0.064 (1)	0.038 (1)	0.038 (1)
Cleavage and habit		{001} perfect Tabular parallel to {001}			
Comment	Reported Fe^3 ignored.				
μ	120.3	107.6	127.2	91.4	133.2
ASF	0.4963×10^{-1}	0.3230×10^{-1}	0.5097×10^{-1}	0.3801×10^{-1}	0.3947×10^{-1}
Abridging factors	1:1:0.5	1:1:0.5	1:1:0.5	1:1:0.5	1:1:0.5

471

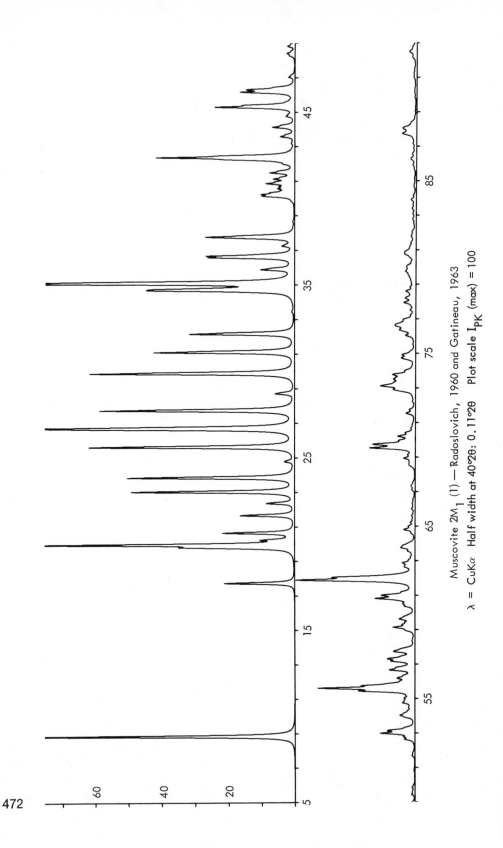

472

Muscovite 2M₁ (1) — Radoslovich, 1960 and Gatineau, 1963

λ = CuKα Half width at 40°2θ: 0.11°2θ Plot scale I$_{PK}$ (max) = 100

MUSCOVITE 2M1 — RADOSLOVICH, 1960 AND GATINEAU, 1963

2THETA	PEAK	D	H	K	L	I(INT)	I(PK)	I(DS)
8.83	8.84	10.007	0	0	2	58	87	50
17.71	17.72	5.004	0	0	4	16	21	14
19.72	19.80	4.497	0	2	0	4	35	3
19.80		4.481	-1	1	1	20		18
19.95	19.96	4.448	1	1	0	59	78	53
20.22	20.22	4.388	0	2	1	5	10	5
20.63	20.64	4.301	1	1	1	17	22	15
21.64	21.66	4.102	0	2	2	13	16	12
22.34	22.36	3.975	1	1	2	7	8	6
23.01	23.02	3.862	-1	1	3	41	49	38
23.84	23.84	3.729	0	2	3	42	50	39
24.77	24.78	3.592	1	1	3	2	3	2
25.60	25.60	3.476	-1	1	4	55	62	51
26.63	26.70	3.345	0	4	0	40	83	38
26.70		3.336	0	0	6	46		43
27.72	27.72	3.215	-1	1	5	53	59	50
28.69	28.70	3.109	-1	3	1	57	6	5
29.85	29.86	2.990	0	2	5	40	62	55
31.07	31.08	2.876	-1	1	6	31	42	38
32.15	32.14	2.782	-1	3	3	26	31	30
34.64	34.68	2.587	-1	1	7	14	44	26
34.69		2.584	1	1	6	10		14
35.04	35.06	2.581	1	3	0	36	100	10
35.06		2.558	-2	0	2	64		36
35.86	35.88	2.557	-1	3	1	5	10	64
35.88		2.502	1	1	0	5		5
36.57	36.58	2.501	-1	3	3	23	26	5
36.66	36.66	2.455	-1	2	2	11	27	24
37.23	37.24	2.449	-2	0	7	3	4	11
37.68	37.76	2.413	0	2	4	8	27	3
37.75		2.385	-1	3	3	22		9
40.11	40.12	2.381	-2	2	1	8	10	23
40.22	40.22	2.246	2	2	0	4	10	9
40.32	40.32	2.240	0	4	5	4	8	4
40.55	40.54	2.235	-1	3	8	4	5	4
40.68	40.66	2.223	2	0	4	2	5	2

2THETA	PEAK	D	H	K	L	I(INT)	I(PK)	I(DS)
40.84	40.84	2.208	2	2	1	8	9	8
41.10	41.10	2.194	0	4	2	7	7	6
41.46	41.46	2.176	-2	2	3	8	8	8
41.97	41.98	2.151	-2	2	2	2	4	2
42.34	42.34	2.133	-1	3	5	33	41	35
42.38		2.131	0	4	4	3		3
43.57	43.58	2.075	2	2	2	5	4	5
44.11	44.12	2.051	0	4	5	7	7	8
44.74	44.74	2.024	-2	2	4	2	3	3
45.27	45.26	2.001	2	0	5	27	24	30
46.14	46.14	1.9658	-1	3	7	18	17	19
46.31	46.28	1.9589	0	2	6	9	15	10
47.01	47.02	1.9312	-1	3	3	2	2	2
48.25	48.26	1.8844	-2	1	7	1	2	2
48.40	48.40	1.8791	-1	3	8	2	2	2
48.80	48.80	1.8646	0	4	6	3	2	3
49.41	49.42	1.8428	-1	3	8	1	1	1
51.67	51.68	1.7676	0	6	0	1	1	1
52.61	52.62	1.7380	-2	4	1	3	3	4
52.75	52.76	1.7338	-2	1	8	1		1
52.96	52.96	1.7273	-1	1	11	12	10	14
53.16	53.12	1.7214	-3	3	9	5	8	6
54.01	54.02	1.6963	-2	4	0	3	5	3
54.73	54.74	1.6757	-1	5	3	3	4	4
55.01	55.00	1.6678	-3	3	1	2	3	2
55.42	55.44	1.6564	-2	4	3	13	17	15
55.60	55.60	1.6564	-2	2	10	5		6
55.86	55.74	1.6516	1	3	9	27	29	33
56.16	56.18	1.6444	-3	1	4	4	16	5
56.22		1.6363	-2	4	4	4	5	5
56.64	56.64	1.6349	0	5	3	2		2
56.73		1.6235	-1	5	4	9	8	10
57.15	57.16	1.6212	-1	3	1	7	7	2
57.25	57.30	1.6104	-3	5	5	3	8	9
57.31		1.6077	0	8	8	3		3
57.70	57.72	1.6064	-2	1	3	6	6	4
		1.5962	-2	4	5			7

MUSCOVITE 2M1 – RADOSLOVICH, 1960 AND GATINEAU, 1963

2THETA	PEAK	D	H	K	L	I(INT)	I(PK)	I(DS)
57.77		1.5944	1	5	4	3		4
58.93	58.94	1.5659	-3	1	6	3	3	3
59.12	59.12	1.5613	-3	1	4	7	7	8
59.41	59.40	1.5543	-2	2	10	2	3	3
59.61	59.60	1.5496	-2	0	6	4	5	5
59.70		1.5475	-1	5	5	2		2
60.35	60.36	1.5325	-1	5	6	3	3	4
60.47	60.50	1.5296	2	4	5	2		2
60.69	60.80	1.5246	-1	1	12	2	12	17
60.79		1.5223	-1	3	11	14		
61.01	60.98	1.5174	2	0	10	6	10	8
61.83	61.86	1.4992	0	6	0	16	36	21
61.88		1.4982	-3	3	1	34		43
61.97		1.4961	-1	5	6	3		3
62.02	62.02	1.4950	0	6	1	1	25	2
62.60	62.62	1.4826	0	6	1	1	3	2
62.86	62.86	1.4809	3	1	2	2		2
63.21	63.20	1.4771	-2	4	6	3	4	4
63.55	63.56	1.4699	-2	0	11	1	1	1
63.74	63.74	1.4628	-1	3	12	2	2	2
63.85	63.84	1.4588	0	1	11	4	4	5
64.58	64.60	1.4566	1	5	13	3	1	4
64.79	64.78	1.4419	0	2	7	1	3	1
65.20	65.20	1.4378	0	2	10	4	1	5
65.58	65.58	1.4296	0	0	14	1	1	2
65.99	65.98	1.4222	1	1	13	1	1	2
66.33	66.32	1.4145	0	4	11	1	1	1
67.09	67.08	1.4080	-3	1	9	1	1	1
67.36	67.34	1.3940	-2	4	14	1		2
68.50	68.56	1.3890	-3	0	9	1	2	2
68.92	68.92	1.3686	-2	3	7	4	3	5
69.52	69.52	1.3612	-1	2	11	19	14	26
69.74	69.74	1.3510	3	1	13	7	13	12
70.19	70.20	1.3470	0	1	12	7	5	9
70.65	70.64	1.3396	1	4	12	1	1	1
71.52	71.52	1.3321	1	1	14	1	3	6
		1.3180	-2	2	13	4		

2THETA	PEAK	D	H	K	L	I(INT)	I(PK)	I(DS)
72.78	72.90	1.2983	1	3	13	1	7	2
72.81		1.2978	-2	6	1	2		2
72.88		1.2967	-3	2	0	4		6
72.91		1.2963	-4	0	2	3		4
73.10	73.10	1.2935	-2	6	0	9	11	12
73.20		1.2920	-4	0	9	5		7
73.49	73.50	1.2875	-3	6	8	5	7	7
73.59		1.2860	0	6	7	4		6
73.77	73.76	1.2833	3	3	0	6	7	8
74.65	74.72	1.2704	0	4	2	2	4	3
74.72		1.2693	+2	0	4	5		7
74.89	74.90	1.2669	0	0	16	3	4	4
76.01	76.04	1.2509	0	2	14	3	3	2
76.05		1.2504	-2	1	1	1		5
76.34	76.36	1.2463	-1	7	1	4	5	2
76.40		1.2456	-4	2	6	1		3
76.57	76.60	1.2431	-4	0	9	2	6	2
76.58		1.2431	0	6	4	1		5
76.68	76.66	1.2417	2	6	3	4	6	4
76.97	76.92	1.2377	-3	5	3	3	4	2
77.49	77.50	1.2306	-1	7	2	2	2	4
77.92	77.92	1.2249	3	5	2	3	4	4
77.93		1.2249	2	2	13	3		4
78.22	78.18	1.2210	2	1	3	3	3	6
78.59	78.58	1.2162	-4	2	5	4	3	2
79.02	79.02	1.2107	-3	5	4	1	1	3
79.55	79.58	1.2039	-1	7	11	2	2	3
79.74	79.74	1.2015	-3	0	10	1	3	3
79.87		1.1999	0	6	9	2		2
80.01	80.00	1.1982	-4	2	4	1	3	3
80.08		1.1973	3	3	2	2		5
80.72	80.74	1.1894	3	5	15	3	3	2
80.83	80.84	1.1880	-2	1	11	1	3	3
80.87		1.1876	3	1	5	2		3
81.38	81.38	1.1815	4	2	4	1	2	2
81.76	81.76	1.1769	-4	3	7	1	2	2
82.76	82.76	1.1652	1	3	15	1	1	2

MUSCOVITE 2M1 - RADOSLOVICH, 1960 AND GATINEAU, 1963

2THETA	PEAK	D	H	K	L	I(INT)	I(PK)	I(DS)
83.90	83.90	1.1523	-1	7	7	2	2	4
85.89	85.90	1.1305	-2	6	8	1	1	2
86.27	86.26	1.1266	-2	4	14	3	2	4
87.73	87.74	1.1115	-2	6	10	7	4	11
87.86		1.1102	-1	5	14	1		2
88.09	88.04	1.1079	-1	1	4	4	4	6
92.18	92.18	1.0692	-4	0	12	4	1	3
92.48	92.48	1.0665	-2	6	10	2	2	6
93.60	93.62	1.0566	-2	0	18	3	1	3
93.88	93.88	1.0542	-3	3	17	2	2	7
96.03	96.06	1.0362	-3	3	15	4	1	3
96.23	96.26	1.0346	0	6	14	1	1	2
96.50	96.50	1.0324	3	7	13	2	1	3
99.29	99.38	1.0108	-3	3	5	1	1	3
100.10	100.10	1.0047	-4	0	14	2	1	3
100.48	100.48	1.0020	2	6	12	3	2	5
101.42	101.44	0.9952	-5	1	7	1	1	1
101.91	101.92	0.9918	-1	3	19	3	1	6
102.24	102.24	0.9895	2	0	18	1	1	2
103.49	103.78	0.9809	-1	9	2	1	2	1
103.54		0.9805	-4	6	2	1		2
103.61		0.9801	-5	3	1	1		2
103.73		0.9793	-1	3	3	2		2
103.79		0.9789	-5	3	3	2		3
103.82		0.9787	-4	6	0	1		2
105.37	105.44	0.9685	-5	3	5	2	3	4
105.38		0.9684	1	9	3	3		5
105.50		0.9676	4	6	12	3		5
105.52		0.9675	3	5	12	1		2
105.83	105.76	0.9656	-2	8	8	3	2	1
106.63	106.94	0.9605	0	6	16	3	3	6
106.89		0.9589	1	7	13	3		5
106.95		0.9585	3	3	15	3		6
107.20		0.9569	-4	6	6	1		2
107.22		0.9568	1	9	5	1		2
108.16	108.14	0.9511	-4	2	15	1	1	2
109.82	109.86	0.9414	2	8	8	1	1	1

2THETA	PEAK	D	H	K	L	I(INT)	I(PK)	I(DS)
110.06	110.12	0.9399	5	1	7	1	1	1
111.63	111.68	0.9311	5	3	5	1	1	2
113.28	113.28	0.9222	-4	2	16	2	1	3
114.11	114.12	0.9178	-3	7	11	1	1	1
114.13		0.9177	1	0	21	1		2
115.70	115.76	0.9098	0	0	22	1	1	1
115.73		0.9096	3	5	14	1		2
116.01	116.08	0.9082	-1	9	8	1	1	2
116.05		0.9080	0	8	13	1		2
116.23		0.9071	4	6	7	1		2
117.29	117.32	0.9020	5	1	9	5	1	2
118.13	118.16	0.8980	4	4	12	5	3	2
119.40	119.60	0.8921	1	9	8	2		9
119.65		0.8910	4	6	8	1		10
120.93	120.92	0.8853	0	10	4	2	1	1
121.35	121.38	0.8835	2	8	11	1	1	3
122.57	122.60	0.8783	-3	7	13	2	1	4
123.04	123.06	0.8763	-1	7	22	1		1
124.27	124.62	0.8713	-2	8	13	1		2
124.47		0.8705	-4	6	12	1		2
124.61	124.84	0.8699	-1	9	11	1	1	2
124.84		0.8690	-4	4	16	1		3
125.77	125.88	0.8654	-3	9	1	3	2	2
125.91		0.8648	-6	0	2	2		4
126.21		0.8637	-2	6	18	2	2	4
127.84	127.60	0.8576	-1	9	11	1	1	2
128.62	129.08	0.8548	-5	5	9	1	1	1
129.02		0.8533	2	2	21	1		2

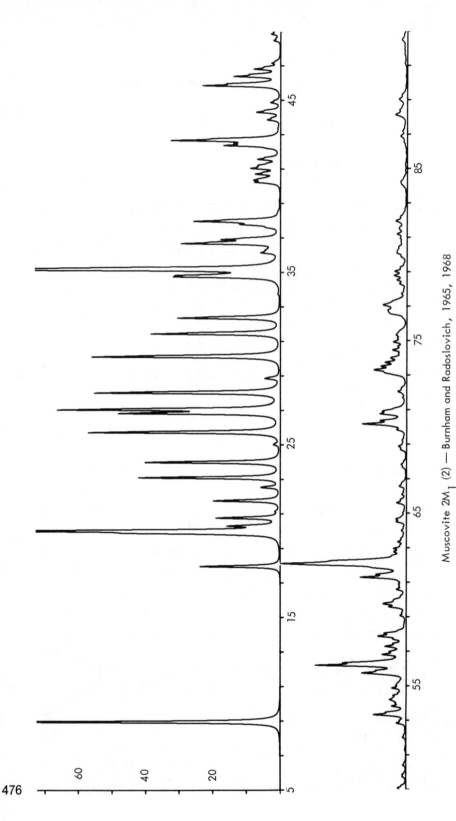

476

Muscovite 2M$_1$ (2) — Burnham and Radoslovich, 1965, 1968

λ = CuKα Half width at 4°2θ: 0.11°2θ Plot scale I$_{PK}$ (max) = 100

MUSCOVITE 2M1 - BURNHAM AND RADOSLOVICH, 1965, 1968

2THETA	PEAK	D	H	K	L	I(INT)	I(PK)	I(DS)
8.93	8.94	9.890	0	0	2	97	99	94
17.92	17.92	4.945	0	0	4	27	24	27
19.99	20.00	4.438	-1	1	1	100	92	100
20.27	20.28	4.377	1	1	0	14	16	14
20.74	20.74	4.280	-1	1	2	22	19	22
21.73	21.74	4.087	0	2	1	24	20	24
22.50	22.50	3.948	1	1	2	7	6	7
23.07	23.08	3.852	-1	1	3	52	42	53
23.96	23.96	3.710	0	2	3	51	40	52
24.98	25.00	3.561	1	1	3	2	2	2
25.70	25.70	3.463	-1	1	4	73	57	75
26.80	26.80	3.323	0	2	4	59	48	61
27.02	27.02	3.297	2	0	0	83	67	86
28.00	28.00	3.184	1	1	4	74	56	77
28.83	28.84	3.094	-1	1	5	6	5	6
30.09	30.08	2.968	0	2	5	79	56	83
31.42	31.42	2.845	1	1	5	54	39	57
32.33	32.34	2.767	-1	1	6	44	31	46
34.64	34.72	2.587	-1	3	1	4	32	5
34.81	34.80	2.581	-1	3	0	37		40
35.12	35.18	2.575	-2	0	2	17	100	18
35.14		2.553	-1	3	2	55		60
35.17		2.551	-1	3	1	12		13
36.12	36.12	2.549	-1	1	7	96	6	104
36.30	36.30	2.485	-1	3	2	7	5	8
36.66	36.66	2.473	0	0	8	3	30	3
36.88		2.449	-2	3	3	43	18	47
36.89	36.88	2.435	-1	0	8	22	5	24
37.58		2.391	-2	2	6	4		4
37.75	37.60	2.381	-1	2	7	14	12	16
37.94	37.76	2.369	0	3	3	37	26	41
40.22	37.94	2.360	-2	0	8	10	8	11
40.35	40.22	2.240	-2	2	6	5	8	6
40.42	40.34	2.233	0	4	2	4		5
40.69		2.230	-1	3	2	11	8	12
41.02	40.70	2.215	2	3	1	9	9	10
41.03	41.02	2.198	2	0	0	4		4

2THETA	PEAK	D	H	K	L	I(INT)	I(PK)	I(DS)
41.22	41.20	2.188	0	0	4	7	7	8
41.55	41.54	2.172	-2	0	2	10	7	11
42.35	42.36	2.132	-2	0	3	24	17	27
42.52	42.46	2.124	0	4	0	49	14	5
42.65	42.64	2.118	1	3	2	6	33	56
43.85	43.86	2.063	2	0	2	11	4	7
44.29	44.30	2.043	0	4	0	5	7	13
44.85	44.86	2.019	2	2	0	38	3	5
45.83	45.84	1.9780	0	0	10	1	23	44
45.94		1.9738	-1	3	0	23		2
46.37	46.38	1.9564	-2	2	4	12	14	26
46.80	46.80	1.9396	2	0	6	3	8	14
47.14	47.14	1.9261	-2	2	0	3	3	4
48.45	48.46	1.8771	-1	3	7	4	2	4
48.84	48.84	1.8633	0	4	4	4	3	5
49.07	49.06	1.8551	-1	3	6	2	3	4
49.70	49.70	1.8328	-1	3	8	2	1	2
50.37	50.38	1.8100	0	4	7	5	1	2
51.99	52.00	1.7573	0	4	8	1	3	2
52.82	52.82	1.7317	-2	2	8	17	10	6
53.23	53.32	1.7192	-1	1	11	7		2
53.31		1.7170	-1	1	9	1	4	21
53.80	53.80	1.7025	-2	3	3	1	5	8
54.06	54.18	1.6948	-2	0	8	5		1
54.10		1.6936	-1	5	1	2		2
54.17		1.6917	-1	4	1	3		7
54.27		1.6889	-3	1	0	1	4	2
54.42	54.42	1.6844	-1	5	2	6		2
54.43		1.6843	2	2	7	3	4	3
54.85	54.86	1.6722	-3	1	2	6	3	8
55.13	55.12	1.6645	1	5	1	2	13	3
55.66	55.72	1.6500	-2	4	2	19		8
55.73		1.6481	-2	4	2	2	27	24
56.13	56.18	1.6372	-2	0	10	7		2
56.17		1.6362	-2	2	9	41		8
56.18		1.6359	1	3	9	5	19	51
56.29	56.32	1.6329	-2	4	4			7

MUSCOVITE 2M1 — BURNHAM AND RADOSLOVICH, 1965, 1968

2THETA	PEAK	D	H	K	L	I(INT)	I(PK)	I(DS)
56.43	56.80	1.6292	-1	5	3	3	7	4
56.79	56.94	1.6196	-1	5	4	12	5	15
57.01		1.6139	-2	4	3	2	3	3
57.26	57.26	1.6074	-3	1	5	10	6	13
57.67	57.68	1.5971	3	1	3	5	4	6
57.84	57.86	1.5927	-2	4	5	9	9	12
57.88		1.5918	2	2	5	9		6
58.03	58.02	1.5880	1	5	4	5	7	6
59.06	59.06	1.5629	-3	1	6	3	2	4
59.55	59.56	1.5510	1	1	4	9	5	11
59.71	59.74	1.5473	0	2	12	1	7	1
59.72		1.5470	-2	2	10	3		4
59.77		1.5458	-2	4	6	6		7
60.01	59.90	1.5404	-1	5	5	3	4	4
60.56	60.56	1.5276	-1	1	5	5	3	6
60.87	60.86	1.5207	2	0	6	3	2	3
61.27	61.27	1.5116	-1	3	11	26	14	35
61.54	61.44	1.5057	-1	1	12	2	9	3
61.98	62.06	1.4960	0	6	0	27	38	35
62.06		1.4941	-3	3	1	56		74
62.17		1.4918	0	6	1	2		3
62.75	62.76	1.4794	-3	3	3	1	4	2
62.76		1.4792	0	6	2	2		3
62.94	62.94	1.4754	3	1	3	4	4	5
63.32	63.32	1.4674	2	4	6	4		6
63.97	63.98	1.4540	-2	0	12	2	3	3
64.48	64.48	1.4438	1	3	11	5	1	7
64.62	64.64	1.4410	0	2	13	3	3	4
65.00	65.00	1.4335	1	5	7	3	3	2
65.58	65.58	1.4223	2	2	10	6	1	8
66.07	66.08	1.4129	0	0	14	3	3	3
66.52	66.54	1.4045	1	1	13	2	2	2
66.54		1.4041	-3	1	9	2		2
66.58		1.4033	0	4	11	2		2
67.63	67.66	1.3841	-2	4	9	2	1	2
67.80	67.80	1.3811	-1	1	14	2	2	3
68.67	68.68	1.3657	-3	3	7	3	2	5

2THETA	PEAK	D	H	K	L	I(INT)	I(PK)	I(DS)
68.86	68.86	1.3623	0	6	6	3	2	4
69.22	69.22	1.3562	3	3	5	2	1	2
69.80	69.82	1.3462	-2	2	11	5	3	7
70.14	70.14	1.3405	-1	0	12	29	13	40
70.74	70.74	1.3306	-2	4	12	14	8	20
70.87	70.88	1.3285	0	4	12	9	7	12
71.68	71.68	1.3155	-2	2	14	2	1	2
72.02	72.02	1.3100	-2	2	13	5	3	8
73.14	73.28	1.2928	-4	0	2	4	10	6
73.27		1.2907	-2	6	2	14		21
73.50	73.50	1.2873	4	0	0	7	8	9
73.70	73.72	1.2844	-3	1	3	3	7	4
73.72		1.2841	-3	3	9	6		9
73.99	73.96	1.2800	0	6	8	6	5	8
74.43	74.42	1.2736	3	3	7	8	4	12
74.90	74.90	1.2667	-2	6	1	7	4	11
75.00		1.2652	0	2	15	1		1
75.30	75.30	1.2610	-1	0	2	5	3	7
75.41		1.2594	-2	4	13	3		5
76.53	76.66	1.2437	-2	2	14	7	6	10
76.63		1.2424	-4	2	1	2		3
76.64		1.2423	-4	2	1	1		2
76.67	76.74	1.2418	-3	5	2	2	6	3
76.99	77.06	1.2408	-4	0	9	2	7	3
77.04		1.2374	-4	2	9	3		5
77.05		1.2367	0	6	4	1		2
77.08		1.2367	2	0	16	2		3
77.16		1.2363	0	0	3	6		9
77.25		1.2351	-3	5	2	4		5
77.69	77.70	1.2280	-1	7	3	4	2	7
78.27	78.28	1.2204	3	1	3	4	2	3
78.49	78.50	1.2175	1	7	4	1	3	5
78.77	78.76	1.2140	-4	2	9	5	4	6
79.01	79.00	1.2108	-2	2	13	4	4	8
79.20	79.22	1.2083	-3	5	5	2	2	3

MUSCOVITE 2M1 – BURNHAM AND RADOSLOVICH, 1965, 1968

2THETA	PEAK	D	H	K	L	I(INT)	I(PK)	I(DS)
79.86	79.88	1.2000	-1	7	4	3	2	4
80.06	80.08	1.1976	-3	3	11	4	3	6
80.18	80.18	1.1961	-4	2	3	1	3	2
80.18		1.1960	-4	2	6	1		2
80.19		1.1956	0	4	14	1		2
80.21		1.1932	0	4	4	2		2
80.41	80.42	1.1891	-3	6	10	2	3	4
80.75	80.74	1.1870	3	5	6	1	2	3
80.92	80.92	1.1838	3	6	9	5	3	4
81.18	81.18	1.1800	-2	2	4	4	2	7
81.50	81.50	1.1791	0	2	15	2		4
81.58		1.1751	-1	7	5	3		4
81.91	81.92	1.1748	3	1	11	2	3	5
81.94		1.1747	-4	2	4	3		3
81.95			-4	2	7	3		4
84.04	84.20	1.1507	-4	2	5	1	2	2
84.11		1.1499	-1	1	17	1		2
84.21		1.1488	-1	1	7	3		5
86.53	86.54	1.1236	-2	6	8	4	1	5
86.87	86.88	1.1203	-2	4	14	1	2	2
86.88		1.1202	-4	4	4	1		6
87.23	87.16	1.1167	-4	4	0	1	1	2
87.93	88.12	1.1096	-4	4	4	4		2
88.11		1.1077	-2	6	10	10	4	16
89.01	89.00	1.0988	4	0	8	5	3	9
91.12	91.10	1.0788	3	5	8	1	1	2
91.65	91.66	1.0739	-2	5	15	2		2
92.51	92.48	1.0662	-4	4	12	3	1	3
93.31	93.32	1.0591	2	0	10	6	2	6
93.51		1.0574	-1	6	15	1		10
94.58	94.58	1.0482	-2	5	18	3	1	2
95.30	95.30	1.0423	0	0	17	6		5
96.63	96.64	1.0314	-3	3	15	3	2	10
97.16	97.14	1.0272	0	6	14	3	1	6
97.83	97.84	1.0219	-3	3	13	3	1	4
99.56	99.62	1.0087	-3	7	5	2	1	6
99.91	99.94	1.0062	3	3	3	3		3
100.06		1.0051	-2	8	5	1		2
100.56	100.56	1.0014	-4	0	14	2	1	4
101.04	100.94	0.9979	2	2	17	1	1	2
101.56	101.56	0.9943	2	6	12	5	2	9
101.56		0.9933	-5	1	7	1		1
101.69		0.9890	0	0	20	1		2
102.30	102.28	0.9869	0	4	18	1	1	1
102.60	102.64	0.9826	-1	3	19	6	1	10
103.24	103.24	0.9782	-4	6	1	1	2	1
103.89	104.12	0.9782	-1	6	2	6	3	3
103.89		0.9771	-4	2	9	1		2
104.06		0.9771	-1	6	1	2		4
104.08		0.9769	-5	0	18	3		5
104.15		0.9764	-2	3	3	2		3
104.25		0.9758	-5	0	0	4		7
105.68	105.76	0.9665	-4	6	3	4	2	8
105.77		0.9659	1	9	3	5		9
106.05	106.06	0.9642	-2	4	2	1		2
106.22		0.9631	-1	8	8	3	3	5
107.17	107.24	0.9571	-1	8	4	2	1	3
107.22		0.9569	-3	3	17	2		3
107.51	107.60	0.9551	-4	6	6	4	2	8
107.58		0.9546	-1	9	5	2		3
107.85	107.92	0.9530	0	6	16	4	2	8
107.96		0.9523	1	1	13	1		3
108.22	108.22	0.9508	5	3	15	2	2	3
108.65	108.64	0.9482	3	3	15	5	2	8
108.73		0.9477	-4	0	7	4		3
110.60	110.62	0.9369	2	8	8	1	2	1
111.19	111.18	0.9336	5	1	14	1	0	1
111.54	111.62	0.9316	2	6	8	1	1	1
111.57		0.9315	-4	6	7	1	1	1
111.74		0.9305	-1	9	7	1		3
111.83		0.9300	-3	3	13	1		1
112.57	112.54	0.9260	5	5	5	2	1	5
113.11	113.04	0.9231	1	7	14	1	0	3
113.95	113.98	0.9187	-4	2	16	3	1	1
114.18		0.9175	3	1	17	1	1	1

MUSCOVITE 2M1 - BURNHAM AND RADOSLOVICH, 1965, 1968

2THETA PEAK	2THETA	D	H	K	L	I(INT)	I(PK)	I(DS)
114.44	114.43	0.9162	-1	9	8	1	1	1
	114.61	0.9153	-3	7	11	1		1
	114.73	0.9146	4	4	11	1		2
117.40	117.07	0.9030	0	8	13	1	1	1
	117.28	0.9020	4	6	7	1		2
	117.43	0.9013	3	5	14	1		1
117.88	117.89	0.8991	0	0	22	2	1	4
	118.04	0.8984	2	8	10	1		1
118.72	118.76	0.8950	5	1	9	1	1	3
119.58	119.56	0.8914	-5	3	11	9	2	18
	119.80	0.8903	-4	2	17	1		1
	119.82	0.8902	4	4	12	2		3
120.22	119.85	0.8901	-4	4	15	1	2	1
	120.24	0.8883	1	9	9	8		15
	120.25	0.8883	-2	8	12	1		1
120.82	120.85	0.8856	4	6	8	8	2	16
121.34	121.21	0.8841	-1	9	10	1	1	3
	121.41	0.8832	0	10	4	1		2
122.52	122.57	0.8782	2	8	11	1	0	2
123.26	123.25	0.8754	-3	7	13	2	1	4
123.78	123.74	0.8734	3	5	15	1	1	2
	123.78	0.8732	-5	5	7	1		1
124.52	124.30	0.8711	1	9	10	1	1	1
	124.49	0.8704	3	7	11	3		6
	124.68	0.8696	-5	1	14	1		1
125.06	125.02	0.8683	-4	6	12	2	1	5
	125.10	0.8680	-2	8	13	1		1
	125.25	0.8674	-1	9	11	1		2
125.54	125.42	0.8667	1	9	11	2	1	5
	125.60	0.8660	-4	4	13	1		1
	125.64	0.8659	-4	4	16	2		4
126.34	126.29	0.8634	-3	9	1	5	2	11
	126.45	0.8628	-4	2	18	1		1
	126.56	0.8623	3	9	0	1		2
	126.57	0.8623	-6	0	2	3		6
	126.80	0.8614	5	3	9	1		3
127.62	127.59	0.8585	-2	6	18	3	1	7

2THETA PEAK	2THETA	D	H	K	L	I(INT)	I(PK)	I(DS)
	127.66	0.8582	6	0	0	1		2
	127.70	0.8581	-1	1	23	2		4
128.14	128.57	0.8549	-3	7	14	1	1	1
129.00	129.02	0.8533	-1	9	11	1	1	2
	129.08	0.8531	-5	5	9	1		2

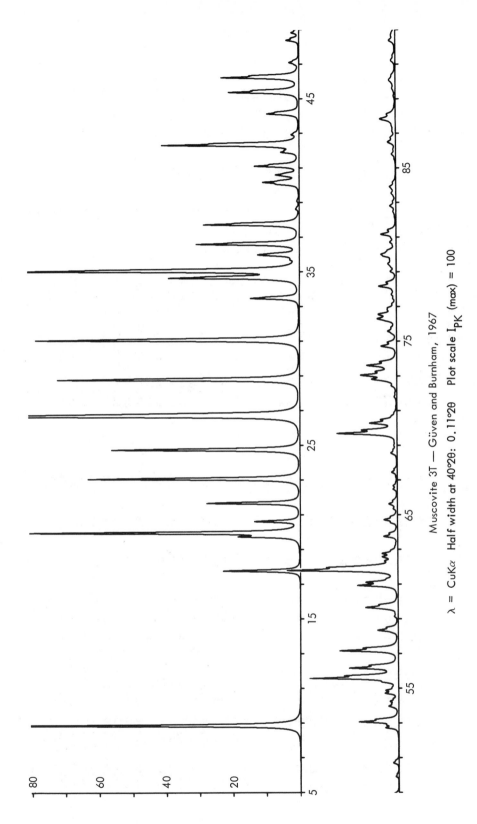

Muscovite 3T — Güven and Burnham, 1967

λ = CuKα Half width at 40°2θ: 0.11°2θ Plot scale I$_{PK}$ (max) = 100

MUSCOVITE 3T – GUVEN AND BURNHAM, 1967

2THETA	PEAK	D	H	K	L	I(INT)	I(PK)	I(DS)
8.84	8.86	9.990	0	0	3	77	100	65
17.74	17.74	4.995	0	0	6	20	23	17
19.71	19.72	4.500	1	0	0	13	19	11
19.93	19.94	4.450	1	0	1	71	83	63
20.59	20.60	4.310	1	0	-2	12	14	10
21.64	21.64	4.103	1	0	2	25	28	23
23.03	23.04	3.858	1	0	-4	59	63	53
24.72	24.72	3.599	1	0	4	54	56	49
26.64	26.64	3.343	0	0	9	84	93	78
26.75	26.70	3.330	1	0	-6	44	86	41
28.76	28.76	3.102	1	0	7	74	72	70
31.03	31.04	2.879	1	0	8	83	79	80
33.45	33.44	2.677	1	0	9	16	15	16
34.62	34.62	2.588	-1	1	-1	4	39	4
35.02	35.02	2.560	1	1	2	37	90	37
35.68	35.68	2.515	-1	1	3	99	3	99
35.93	35.96	2.498	1	1	0	1	12	1
35.97		2.494	0	0	12	6		6
36.57	36.58	2.455	1	0	10	9		9
37.70	37.70	2.384	-1	1	4	34	31	35
38.60	38.60	2.331	1	1	5	33	29	33
40.04	40.06	2.250	1	0	11	1	1	1
40.16	40.16	2.244	2	0	0	6	17	7
40.51	40.58	2.225	2	0	-2	9	11	10
40.58		2.221	2	0	1	3	7	3
41.08	41.08	2.195	-1	1	7	6	13	6
41.88	41.88	2.155	2	0	-4	16	5	17
42.30	42.30	2.135	1	1	8	50	41	53
42.89	42.90	2.107	2	0	5	2	2	2
44.10	44.10	2.052	2	0	6	11	10	12
45.35	45.36	1.9980	0	0	15	27	21	29
46.20	46.20	1.9632	-1	1	10	29	23	33
47.07	47.08	1.9289	2	0	-10	3	3	3
48.37	48.36	1.8803	1	1	11	3	4	4
48.37		1.8803	-1	1	11	2		2
48.80	48.80	1.8644	2	0	9	3	2	3
49.90	49.90	1.8261	1	0	15	1	1	1
50.71	50.68	1.7994	2	0	10	2	2	2
52.71	52.72	1.7349	1	1	11	2	4	2
53.06	53.06	1.7244	-1	1	-13	5	12	6
53.95	53.96	1.6982	2	1	1	15		18
54.23	54.22	1.6900	2	1	2	3	3	3
54.69	54.70	1.6768	2	1	3	2	3	2
54.69		1.6768	-2	1	-10	2		4
54.87	54.86	1.6717	2	0	12	2	4	3
55.34	55.36	1.6587	2	1	4	2	5	3
55.34		1.6587	-2	1	-1	2		2
55.58	55.58	1.6522	1	1	14	36	27	44
56.16	56.16	1.6363	-1	1	5	5	15	6
56.16		1.6363	-2	1	-1	15		19
57.15	57.16	1.6103	-2	1	5	3	18	4
57.16		1.6101	2	0	13	11		13
58.32	58.32	1.5807	-1	1	6	11		14
58.32		1.5807	2	1	7	5	6	6
59.11	59.12	1.5616	-2	1	7	3	1	4
59.55	59.64	1.5509	1	0	18	1	10	1
59.65		1.5487	2	0	14	3		3
60.92	60.92	1.5194	2	1	8	3		3
61.13	61.10	1.5147	-1	1	8	10	12	13
61.13		1.5147	-2	1	-16	17	10	22
61.79	61.80	1.5000	1	1	9	2		3
61.88		1.4982	3	0	0	47	33	61
62.32	62.32	1.4886	3	0	1	7		9
62.56	62.56	1.4834	3	0	19	3		4
62.76	62.74	1.4793	2	1	3	5	5	6
63.74	63.74	1.4588	3	0	10	4	5	5
64.53	64.54	1.4428	2	1	17	6	4	8
64.69	64.70	1.4396	2	1	11	2	2	3
65.33	65.32	1.4272	2	0	16	5	4	6
65.61	65.60	1.4218	0	0	21	4	3	5
66.45	66.44	1.4058	-2	1	-12	1	1	2

MUSCOVITE 3T - GÜVEN AND BURNHAM, 1967

2THETA	PEAK	D	H	K	L	I(INT)	I(PK)	I(DS)
67.43	67.42	1.3877	2	0	17	2	2	3
68.55	68.54	1.3677	3		9	2	2	3
68.97	68.98	1.3604	-1	0	21	1	1	
69.68	69.68	1.3483	-1	-1	19	30	18	42
70.27	70.26	1.3384	2	0	18	13	8	18
71.77	71.78	1.3141	3	0	11	2	1	1
72.42	72.42	1.3039	1	0	22	2	2	3
72.73	72.80	1.2991	2	2	0	2	2	2
72.79		1.2981	1	1	20	1		8
72.81		1.2979	2	2	1	6		9
72.81		1.2979	-2	-2	1	2		
73.05	73.04	1.2942	-2	-2	2	12	11	17
73.22	73.24	1.2916	2	0	19	2	7	3
73.59	73.60	1.2859	3	0	12	15	9	21
74.70	74.70	1.2696	-2	-2	5	8	5	12
75.42	75.56	1.2592	2	1	16	1	3	
75.56		1.2573	3	0	13	2		3
76.17	76.28	1.2488	0	0	24	3	6	4
76.28		1.2472	2	0	20	3		4
76.29		1.2470	3	-1	1	1		7
76.54	76.54	1.2431	-3	-2	2	4	6	7
76.58		1.2385	-2	-2	7	5	3	4
76.91	76.90	1.2285	-3	-1	3	3	2	2
77.66	77.68	1.2219	3	0	14	9	5	13
78.15	78.16	1.2109	3	-1	5	1	2	2
79.00	79.00	1.2109	-3	-1	6	2		3
79.00		1.2065	-1	-1	22	3		2
79.35	79.44	1.2052	2	0	21	3	3	5
79.45		1.1996	2	0	15	4		7
79.90	80.00	1.1982	3	1	7	3	5	5
80.00		1.1982	-3	-1	7	3		5
80.00		1.1898	-2	-1	18	3	2	4
80.68	80.68	1.1841	3	-1	8	6	5	10
81.16	81.16	1.1841	-3	-1	8	2		4
81.16		1.1648	-1	-1	23	2	2	4
82.78	82.78	1.1648	1	1	23	2	1	2
82.80		1.1566	2	1	19	1		
83.51	83.52							
83.51	83.51	1.1566	-2	-1	19	1		2
83.90	83.90	1.1522	-3	-1	10	3	2	5
85.78	85.78	1.1318	-2	-2	13	4	1	3
86.48	86.48	1.1244	-2	-1	20	1	2	7
87.10	87.12	1.1180	4	0	3	1	1	2
87.63	87.82	1.1126	4	0	4	1	5	2
87.82		1.1106	-2	-2	14	10		16
90.08	90.08	1.0885	-1	-1	25	1	2	2
90.13		1.0881	4	0	7	1		4
91.18	91.20	1.0782	3	1	14	2	1	8
92.37	92.38	1.0675	2	2	16	5	2	11
92.84	92.66	1.0633	-2	-1	22	6	2	7
93.94	93.94	1.0537	1	0	26	4	1	3
96.31	96.30	1.0340	3	0	21	6		6
99.25	99.26	1.0110	3	-2	6	3		9
100.37	100.36	1.0028	-2	0	19	5	2	2
101.38	101.40	0.9955	2	0	27	1	1	8
101.41		0.9953	3	2	8	4		3
102.21	102.22	0.9897	-1	-1	28	4		5
103.40	103.42	0.9815	-4	-1	1	2	2	1
103.40		0.9815	4	1	1	3		6
103.63	103.66	0.9799	-4	-1	2	3		14
103.63		0.9799	-4	1	2	1		2
104.57	104.56	0.9737	4	1	4	7		2
105.28	105.28	0.9691	4	-1	5	5		3
105.85	105.64	0.9654	-3	-1	11	1	2	2
106.14	106.16	0.9636	4	0	6	1	1	19
106.75	106.78	0.9597	3	0	24	10	4	8
106.86		0.9590	3	1	20	4		4
107.17	107.14	0.9572	-4	0	7	2	3	2
108.62	108.62	0.9484	-1	0	17	1	1	1
109.15	109.10	0.9452	3	0	31	1	0	2
109.66	109.68	0.9422	-3	-2	13	1	1	2
109.66		0.9422	3	2	13	1		2
110.03	110.08	0.9401	-3	-2	22	1	1	1
110.14		0.9395	2	2	21	1		2

483

MUSCOVITE 3T – GÜVEN AND BURNHAM, 1967

2THETA	PEAK	D	H	K	L	I(INT)	I(PK)	I(DS)
110.67	110.60	0.9365	3	0	25	1	1	2
111.26		0.9332	-4	-1	10	1	1	3
111.85	111.24	0.9299	-3	-2	14	1	1	3
111.82		0.9238	-4	-1	11	1	0	2
112.97	112.98	0.9203	3	1	22	2	1	4
113.65	113.64	0.9169	1	0	32	1	1	2
114.29		0.9159	4	0	19	2		3
114.48	114.46	0.9139	4	1	12	1		1
114.88	114.88	0.9082	0	0	33	1	1	2
116.01	116.04	0.9042	-3	-2	16	1	0	1
116.84	116.96	0.9035	4	1	13	3	1	6
116.98		0.9013	3	1	23	1		1
117.42	117.44	0.8997	4	0	20	1	1	3
117.77		0.8964	5	0	3	1		1
118.47	118.48	0.8926	4	1	14	14	3	29
119.30	119.30	0.8926	-4	-1	14			1
119.30		0.8909	-3	-2	17	1		1
119.67	119.76	0.8858	5	0	6	2	2	2
120.82	120.82	0.8835	4	0	21	1	1	3
121.34		0.8828	3	1	24	1	1	1
121.51	121.40	0.8813	-4	-1	15	1		1
121.85	121.88	0.8813	4	1	15	1		2
121.85		0.8811	1	1	32	2	1	5
121.90		0.8774	-3	-2	18	2		4
122.77	122.76	0.8751	5	0	8	1	1	2
122.77		0.8697	-4	-1	16	2	1	5
123.32	123.28	0.8675	4	0	22	1	2	3
124.65	124.68	0.8660	3	3	0	4		9
125.23	125.72	0.8657	3	3	1	1		2
125.59		0.8657	-3	-3	1	1		2
125.68		0.8650	1	0	34	4	2	2
125.68		0.8647	2	0	32	1		2
125.85		0.8646	3	1	25	1		1
125.95	126.12	0.8638	3	2	19	2		2
125.97		0.8638	-3	-2	19	1		2
126.17		0.8628	-3	-3	3	1	2	2
126.17								
126.43	126.50							
126.59	126.59	0.8622	-2	-2	26	3		6
126.80	126.80	0.8614	2	1	30	1		2
127.75	127.70	0.8579	4	1	17	2	1	4
128.66	128.60	0.8546	5	0	11	1	1	3
128.92	129.98	0.8502	3	2	20	1	1	2
129.93		0.8501	-4	-2	1	1		2
130.23	130.24	0.8491	4	2	1	1		2
130.73	131.28	0.8474	4	2	3	1		2
131.18		0.8459	-4	-1	18	1		2
131.18		0.8459	4	1	18	1		2
131.43		0.8450	2	0	33	3	2	6
132.30	132.34	0.8422	4	-2	5	1		2
132.35		0.8420	-4	-2	5	3		7
132.35		0.8420	5	1	0	1		1
132.82	132.90	0.8405	2	1	31	3	1	5
133.48	133.48	0.8384	5	0	13	2		3
133.48		0.8384	4	-2	6	2		4
133.48		0.8384	-4	-2	6	1		2
134.09	134.14	0.8365	3	-2	21	3	1	6
134.09		0.8365	-3	-2	21	1		3
134.66	135.00	0.8348	-1	-1	34	1		3
134.86		0.8341	4	2	7	8		17
135.02		0.8337	-4	-1	19	1		3
135.02		0.8337	4	1	19	1		2
135.40		0.8325	0	0	36	2		4
135.75	135.64	0.8315	3	0	30	2	2	4
136.45	136.42	0.8294	3	1	27	1	1	1
136.45		0.8294	-3	-1	27	1		2
136.48		0.8293	-4	-2	8	2		2
136.48		0.8293	4	2	8	2		4
137.63	137.68	0.8261	2	2	28	2	1	5
137.89		0.8254	3	3	11	1		2
137.89		0.8254	-3	-3	11	1		2
139.36	139.64	0.8214	4	1	20	2		4
139.36		0.8214	-4	-1	20	2	1	5
139.59		0.8207	2	0	34	2		4
139.72		0.8204	2	1	32	1		2

MUSCOVITE 3T - GUVEN AND BURNHAM, 1967

2THETA	PEAK	D	H	K	L	I(INT)	I(PK)	I(DS)
139.72		0.8204	-2	-1	32	1		2
139.74		0.8204	4	0	25	1		3
140.55	140.52	0.8183	3	3	12	2	2	5
140.55		0.8183	-3	-3	12	6		14
142.88	143.52	0.8125	-3	-1	28	1	1	3
143.17		0.8118	4	2	11	1		3
143.42		0.8112	5	0	16	2		4
143.63		0.8107	-3	-3	13	4		9
145.25	145.16	0.8071	-5	-1	2	1	1	3
145.25		0.8071	5	1	2	1		1
146.15	146.22	0.8051	4	0	26	1	1	1
146.18		0.8051	-4	-2	12	1		2
146.88	146.98	0.8036	5	1	4	3	1	6
147.24		0.8028	-3	-3	14	1		1
147.84	148.10	0.8016	5	0	17	1	1	3
148.07		0.8011	-2	-1	33	1		2
148.15		0.8010	5	1	5	3		7
148.50		0.8003	2	0	35	1		2
149.76	149.82	0.7979	-5	-1	6	2	1	4
149.76		0.7979	5	1	6	5		12
150.44	150.58	0.7966	-4	-1	22	4	1	10
150.77		0.7960	-3	-1	29	1		1
151.55	151.64	0.7946	-3	-3	15	1	1	2
151.78		0.7942	5	1	7	5		11
151.78		0.7942	-5	-1	7	1		2
153.23	153.18	0.7918	5	0	18	8	1	19
154.09	154.14	0.7904	-4	-2	14	1	1	2
154.28		0.7901	-5	-1	8	4		9
159.59	161.76	0.7826	-2	-1	34	3	1	8
159.87		0.7823	-3	-2	25	1		2
161.16		0.7808	2	0	36	3		9
161.92	161.88	0.7799	3	1	30	1	1	2
161.92		0.7799	-3	-1	30	3		8
165.00	164.90	0.7769	3	0	33	27	1	69

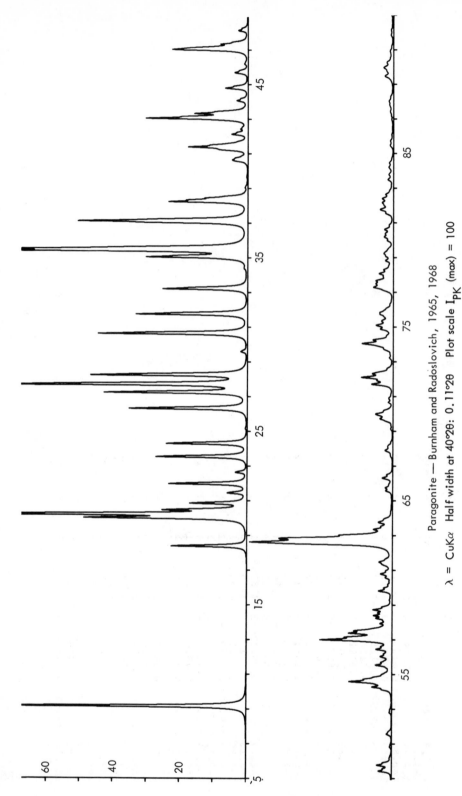

486

Paragonite — Burnham and Radoslovich, 1965, 1968

$\lambda = CuK\alpha$ Half width at $40°2\theta$: $0.11°2\theta$ Plot scale I_{PK} (max) = 100

PARAGONITE — BURNHAM AND RADOSLOVICH, 1965, 1968

2THETA	PEAK	D	H	K	L	I(INT)	I(PK)	I(DS)
9.15	9.16	9.656	0	0	2	76	86	74
18.36	18.36	4.828	0	0	4	23	23	23
19.92	20.00	4.454	0	2	0	4	49	4
19.99		4.437	0	2	1	42		42
20.20	20.20	4.392	-1	1	1	100	100	100
20.45	20.46	4.340	0	2	2	21	25	21
20.83	20.84	4.260	-1	1	2	16	17	16
21.43	21.44	4.143	1	1	1	6	6	6
21.96	21.96	4.044	-1	1	2	25	23	25
22.61	22.62	3.929	0	2	2	3	4	
23.52	23.52	3.779	-1	1	3	30	27	30
24.28	24.28	3.663	0	2	3	27	24	27
26.29	26.30	3.387	-1	2	3	41	35	42
27.22	27.22	3.274	0	2	4	49	43	50
27.69	27.70	3.219	-1	0	4	89	77	92
28.23	28.24	3.159	1	1	6	56	47	57
29.55	29.56	3.020	-1	2	4	2	2	2
30.61	30.62	2.918	0	1	5	55	45	57
31.73	31.74	2.817	-1	1	5	42	33	43
33.19	33.20	2.697	1	1	5	32	25	34
35.03	35.04	2.559	-1	3	6	30	30	32
35.04		2.559	-1	3	0	6	6	7
35.42	35.42	2.532	-1	0	1	82	72	87
35.53	35.52	2.524	2	0	2	51	79	54
35.55		2.523	1	1	6	9		9
37.03	37.12	2.425	-2	0	1	31	51	33
37.12		2.420	1	1	7	4		5
37.13		2.419	-2	0	3	45		48
38.22	38.22	2.353	-1	3	7	31	24	33
38.34	38.32	2.345	1	3	7	3	18	4
38.43	38.42	2.340	1	3	1	11	13	12
40.57		2.222	-2	0	4	3		4
40.63	40.62	2.219	-1	2	1	4	5	4
41.39	41.38	2.180	2	3	5	21	18	23
42.10	42.10	2.144	-1	2	3	6	5	7
42.40	42.40	2.130	-2	2	2	2	2	2
42.94	43.04	2.104	0	4	3	3	31	4
43.04	43.04	2.100	-1	3	5	41		45
43.32	43.32	2.087	-2	0	6	20	16	22
44.05	44.06	2.054	-2	2	3	4	3	5
44.78	44.78	2.022	0	4	5	10	7	11
45.67	45.68	1.9847	-2	2	4	1	4	6
46.17	46.16	1.9645	2	2	4	5	1	1
46.96	47.02	1.9332	0	0	9	1	23	1
47.01		1.9313	0	0	10	1		1
47.08		1.9287	2	0	6	28		32
47.33	47.32	1.9189	-1	3	7	10	8	11
48.11	48.12	1.8895	-2	2	6	9	3	10
49.41	49.42	1.8430	2	2	7	4	4	5
49.75	49.74	1.8313	-2	1	6	6	5	7
51.53	51.54	1.7719	0	4	8	4		3
52.78	52.78	1.7328	-2	0	10	2	2	5
54.10	54.12	1.6936	0	4	7	1	1	2
54.25	54.26	1.6893	0	2	8	4	3	1
54.49	54.56	1.6824	0	0	0	1		4
54.54		1.6811	-2	1	8	7	6	8
54.56		1.6805	0	4	1	2		2
54.59		1.6798	-2	1	9	2	13	2
54.82	54.70	1.6732	1	4	0	12		14
55.08	54.96	1.6659	-3	1	2	5	9	6
55.52	55.52	1.6536	2	1	1	1	3	2
55.77	55.66	1.6468	-3	1	3	1	5	1
56.01	56.02	1.6403	2	1	3	7	3	8
56.43	56.44	1.6291	3	1	1	5	4	5
56.88	56.98	1.6173	1	5	2	7	5	8
56.98		1.6149	1	3	3	3	22	3
57.04		1.6132	-2	4	9	32		38
57.19	57.12	1.6094	0	0	4	3		3
57.36	57.36	1.6050	-2	0	12	1		2
57.37		1.6047	-1	5	10	15	15	18
57.48	57.50	1.6020	-1	5	3	8	13	2
57.58		1.5993	-2	2	9	2	12	9
57.91	57.92	1.5910	3	1	3	4	4	5

PARAGONITE – BURNHAM AND RADOSLOVICH, 1965, 1968

2THETA	PEAK	D	H	K	L	I(INT)	I(PK)	I(DS)
58.21	58.22	1.5835	-3	1	5	6	5	8
58.37	58.38	1.5795	2	2	8	5	6	5
58.52	58.52	1.5758	1	5	4	3	6	3
58.71	58.71	1.5711	-2	1	6	8	4	9
59.79	59.80	1.5453	3	1	4	6	4	8
60.16	60.16	1.5368	-3	1	6	1	1	2
60.56	60.56	1.5277	1	5	5	3	2	4
60.78	60.78	1.5226	-2	5	6	5	4	6
61.34		1.5099	-2	2	10	2	4	3
61.45	61.46	1.5075	-1	5	6	4	4	5
62.49		1.4850	2	0	10	12		14
62.51	62.60	1.4845	0	6	0	25	43	31
62.60		1.4827	-3	3	1	53		65
62.72		1.4802	0	6	1	2		2
62.84	62.80	1.4775	-1	3	11	36	34	44
63.33	63.34	1.4673	0	6	2	2	6	2
63.34		1.4670	3	1	1	5		7
63.73	63.74	1.4590	0	4	10	5	4	6
63.79		1.4579	2	4	6	3		3
65.55	65.56	1.4228	1	3	11	11	2	4
65.71	65.72	1.4198	-1	5	7	7	3	1
66.26	66.28	1.4093	0	2	13	2	3	2
66.29		1.4087	2	0	10	4		5
67.88	67.88	1.3795	0	1	14	3	2	4
68.17	68.08	1.3744	-3	1	9	2	2	2
69.11	69.10	1.3581	-2	3	9	1	1	1
69.57	69.70	1.3501	0	3	5	3	4	3
69.69		1.3481	-1	6	6	5		6
69.92	69.96	1.3442	0	6	14	1	5	2
69.98		1.3433	-1	3	7	7		9
70.65	70.64	1.3322	-3	3	11	2	1	3
70.90	70.88	1.3280	2	1	8	11	1	16
71.70	71.70	1.3154	3	2	12	8	6	23
72.08	72.08	1.3091	-1	0	13	12	10	5
72.39	72.30	1.3044	3	3	13	18	6	2
73.20	73.20	1.2919	0	4	12	4	1	2
73.68	73.72	1.2847	-2	1	14	2	3	2
73.72		1.2841	-2	0	0	2		3
73.87	74.02	1.2818	-4	2	2	3	9	4
74.02		1.2796	-2	6	0	13		17
74.03		1.2794	-2	0	2	4		5
74.24	74.22	1.2764	-2	1	13	2	6	3
74.50	74.44	1.2726	3	3	9	1	2	2
74.87	74.88	1.2672	3	6	7	6	4	8
75.05	75.08	1.2646	0	6	8	3	5	3
75.09		1.2640	-1	1	13	3		5
75.35	75.32	1.2603	-1	1	15	2	4	2
75.39		1.2597	-3	3	9	5		7
75.71	75.72	1.2552	-3	0	2	5	4	6
75.82	75.80	1.2536	-2	6	4	1	4	1
76.28	76.28	1.2472	-2	4	11	2	1	3
77.11	77.26	1.2358	0	4	13	6	6	8
77.26		1.2339	-1	7	1	2		3
77.30		1.2332	-1	1	2	2		2
77.41	77.48	1.2318	-4	2	2	1	5	2
77.45		1.2312	-3	5	2	1		2
77.57		1.2296	-2	2	0	6		8
77.65	77.64	1.2286	4	2	4	4	5	5
78.05	78.04	1.2233	-3	6	3	3	4	4
78.05		1.2232	-4	0	6	1		1
78.23	78.26	1.2209	0	2	9	4	3	5
78.81	78.82	1.2133	3	5	2	1	2	2
79.07	79.08	1.2101	-1	2	14	4	2	4
79.20	79.22	1.2084	-2	7	3	1	3	2
79.22		1.2081	-1	4	2	3		3
79.30		1.2071	4	0	16	1		4
79.95	79.96	1.1989	-4	2	2	2	2	5
80.16	80.16	1.1964	-2	2	13	3	3	2
80.34	80.34	1.1941	-2	2	12	3	3	2
80.34		1.1941	-2	1	5	1		2
80.60	80.60	1.1909	3	5	7	1	3	4
80.61		1.1908	-3	2	4	3		4
81.51	81.54	1.1799	-1	9	3	3	2	4
81.54		1.1796	-4	2	6	1		2

PARAGONITE — BURNHAM AND RADOSLOVICH, 1965, 1968

2THETA	PEAK	D	H	K	L	I(INT)	I(PK)	I(DS)
81.71	81.74	1.1775	3	5	4	4	4	5
81.75		1.1770	0	6	10	4		5
82.12	82.14	1.1727	0	4	14	3	3	4
82.15		1.1723	-3	3	11	1		2
82.34	82.36	1.1701	4	2	4	4	3	6
82.38		1.1696	1	7	5	2		3
82.62	82.60	1.1668	3	1	11	2	2	2
83.48	83.48	1.1570	-4	2	7	3	2	5
83.82	83.76	1.1532	-2	6	8	1	1	2
84.19	84.18	1.1490	-1	2	15	3	1	4
85.44	85.44	1.1354	-1	7	7	2	1	3
87.38	87.40	1.1151	2	6	8	1	1	2
87.79	87.80	1.1109	-4	4	2	2	1	2
87.99	87.98	1.1089	-3	1	14	1	1	2
89.42	89.46	1.0949	-2	4	14	2		3
89.46		1.0945	4	0	8	4	3	6
89.95	89.94	1.0898	-2	6	10	7		10
90.90	90.90	1.0809	-1	5	14	1	1	2
91.82	91.82	1.0724	1	8	5	2	1	2
94.39	94.40	1.0499	-2	6	8	6	2	8
94.49		1.0490	-2	4	10	1		2
95.17	95.17	1.0433	-4	0	15	5	2	7
96.08	96.08	1.0358	-1	5	15	1	1	2
97.60	97.60	1.0237	-1	3	17	5	2	7
98.18	98.16	1.0192	-2	0	18	2	1	3
98.95	98.96	1.0133	3	6	13	4	2	5
99.32	99.30	1.0105	0	6	14	3	1	5
99.84	99.84	1.0066	-3	3	15	3	1	5
100.76	100.72	1.0000	-3	7	3	1	1	1
101.02	101.02	0.9981	-3	7	5	1	1	1
101.45	101.40	0.9950	-1	2	8	1	1	1
101.78	101.78	0.9927	-5	1	5	5	1	8
102.96	102.98	0.9845	2	6	6	1	2	1
103.08		0.9837	2	2	12	1		1
103.89	103.88	0.9782	-4	0	14	1	1	2
105.18	105.18	0.9697	1	9	1	1	2	2
105.27		0.9691	4	6	0	1		1
105.48	105.48	0.9678	-5	3	3	5		7
105.81	105.82	0.9656	0	0	20	1	2	2
106.36	106.38	0.9622	2	0	18	2	1	4
106.46		0.9616	-4	6	4	1		1
106.80	106.94	0.9595	-2	6	14	1	4	1
106.93		0.9586	-1	9	3	4		6
106.95		0.9585	4	6	2	4		7
106.95		0.9585	-1	3	19	5		7
107.27	107.32	0.9565	-2	5	16	2	3	3
107.36		0.9559	-5	3	15	1		1
107.45		0.9554	-4	2	14	1		3
108.97	109.02	0.9463	5	3	5	2	1	3
109.06		0.9458	-1	6	6	2		3
109.35	109.36	0.9440	-4	7	13	2	1	4
109.93	109.96	0.9407	-1	7	15	3	1	4
110.20	110.22	0.9392	3	3	15	2	1	3
110.66	110.62	0.9365	0	3	17	1	1	3
111.29	111.28	0.9330	-3	3	19	1	1	1
111.42		0.9323	1	1	7	1		2
111.80	111.82	0.9302	5	1	7	1	1	1
111.90		0.9296	-2	0	8	1		1
112.10		0.9285	-2	3	20	2		3
113.28	113.28	0.9222	5	3	11	1	1	1
115.22	115.36	0.9122	-1	7	14	1		1
115.39		0.9113	1	7	9	1		1
115.97	115.94	0.9084	-5	3	17	1	1	1
116.16		0.9075	3	1	17	1		1
116.78	116.76	0.9044	3	7	11	2	0	2
117.49	117.50	0.9010	-3	7	16	2		3
118.38	118.36	0.8967	-4	2	16	1	0	4
118.41		0.8934	1	9	5	1	1	1
119.11	118.86	0.8916	3	5	14	2	1	2
119.50	119.58	0.8910	5	1	9	2	1	2
119.66		0.8908	0	8	13	1		1
119.69		0.8894	2	8	10	1		1
120.01	120.00	0.8883	-5	5	1	1	1	1
120.26			1	1	21	1		1

PARAGONITE - BURNHAM AND RADOSLOVICH, 1965, 1968

2THETA	PEAK	D	H	K	L	I(INT)	I(PK)	I(DS)
121.02	121.04	0.8849	4	4	12	1	1	2
121.93	122.06	0.8810	4	6	8	1	2	12
122.13		0.8801	1	9	9	7		11
122.66	122.60	0.8779	0	0	22	2	2	3
123.08	123.08	0.8761	-5	3	11	7	2	12
123.13		0.8759	0	10	4	1		1
123.37		0.8749	-2	8	12	1		1
123.74	123.56	0.8734	-1	9	10	1	1	1
124.02		0.8723	-5	1	13	1		1
124.27	124.26	0.8713	-4	6	11	1	1	1
124.47		0.8705	2	8	11	1		1
124.92	124.82	0.8687	-4	2	17	1	1	1
125.77	126.16	0.8653	3	5	15	1	1	1
126.12		0.8640	3	7	11	2		4
126.49		0.8626	-5	5	7	1		1
127.05	127.04	0.8605	4	4	13	1	1	1
127.07		0.8604	-3	7	13	1		1
128.07	128.32	0.8567	-3	9	1	5	2	9
128.34		0.8558	-3	9	2	1		9
128.36		0.8557	-1	9	11	3		6
128.36		0.8557	-6	0	2	2		4
128.94	128.92	0.8536	-4	6	12	5	2	8
129.13		0.8529	6	0	0	2		3

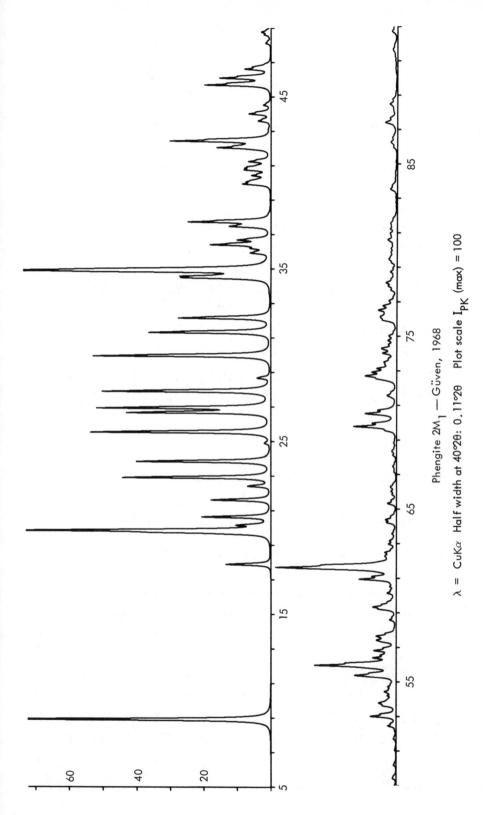

Phengite 2M$_1$ — Güven, 1968

λ = CuKα Half width at 4°2θ: 0.11°2θ Plot scale I$_{PK}$ (max) = 100

PHENGITE 2M1 – GÜVEN, 1968

2THETA	PEAK	D	H	K	L	I(INT)	I(PK)	I(DS)
8.90	8.92	9.923	0	0	2	100	100	100
17.86	17.86	4.962	0	0	4	15	13	16
19.84	19.84	4.471	-1	1	1	83	79	88
20.13	20.14	4.406	0	2	1	9	11	9
20.61	20.62	4.306	1	1	2	23	21	25
21.59	21.60	4.113	0	2	2	21	18	22
22.38	22.60	3.968	-1	1	3	8	7	8
22.90	22.90	3.880	1	1	3	54	44	58
23.82	23.84	3.732	0	2	3	50	40	55
24.87	24.88	3.577	-1	1	4	2	2	2
25.52	25.52	3.487	1	1	4	69	54	76
26.66	26.66	3.341	0	0	6	53	43	59
26.93	26.94	3.308	0	2	5	64	52	72
27.89	27.90	3.196	1	1	5	66	50	75
28.64	28.64	3.114	-1	1	6	5	4	6
29.94	29.94	2.982	2	2	1	71	53	82
31.30	31.30	2.855	1	1	6	50	36	59
32.14	32.14	2.783	1	3	1	38	27	44
34.40	34.48	2.605	-1	3	0	30	27	36
34.47		2.600	-1	1	3	4		5
34.57	34.56	2.592	-2	0	2	51		62
34.85	34.94	2.572	-2	0	4	86	87	104
34.93		2.566	-1	3	1	15		18
35.02		2.560	1	1	6	11		14
35.91	35.92	2.499	-1	3	7	8	6	9
36.18	36.18	2.481	0	0	8	6	6	8
36.39	36.40	2.467	-1	3	3	25	18	30
36.67	36.66	2.449	2	0	2	13	10	16
37.41	37.46	2.402	-2	2	7	4	12	4
37.46		2.399	1	3	4	14		17
37.71	37.72	2.383	-2	0	3	35	25	44
39.86	39.92	2.260	0	4	0	1	9	2
39.92		2.257	-2	2	1	10		13
40.06	40.04	2.249	2	2	1	4	7	5
40.13		2.245	0	4	1	5		6
40.41	40.42	2.230	-1	3	5	7	6	9
40.74	40.74	2.213	2	2	1	10	8	13

2THETA	PEAK	D	H	K	L	I(INT)	I(PK)	I(DS)
40.83	40.84	2.208	2	0	4	3	8	4
40.93	40.92	2.203	-2	0	2	6	7	8
41.22	41.22	2.188	0	4	2	9	7	12
42.03	42.04	2.148	-2	0	6	22	16	28
42.23	42.12	2.138	0	4	3	4	11	5
42.42	42.42	2.129	1	3	5	46	30	60
43.59	43.60	2.075	2	2	5	6	4	8
44.00	44.00	2.056	0	4	4	10	7	13
44.50	44.50	2.034	-2	2	5	3	2	4
45.67	45.68	1.9846	0	2	10	30	20	40
45.69		1.9841	-1	2	7	2		2
46.07	46.08	1.9684	-2	3	6	24	15	32
46.59	46.60	1.9476	-2	0	8	12	8	16
46.79	46.72	1.9399	0	2	7	3	6	4
48.11	48.12	1.8897	-1	3	6	3	2	3
48.60	48.60	1.8717	0	4	7	3	2	4
48.77	48.76	1.8658	-1	3	8	3	3	5
49.39	49.40	1.8436	0	2	10	2	1	2
50.16	50.16	1.8171	-1	4	2	1	1	2
51.69	51.68	1.7670	0	2	8	4	1	6
52.44	52.44	1.7434	-2	1	11	1	3	2
52.96	52.98	1.7273	-1	1	9	12	8	18
52.99		1.7266	-1	3	8	5		7
53.59	53.62	1.7086	-1	0	8	1	4	2
53.65		1.7069	1	5	0	1		2
53.70	53.76	1.7054	-1	5	1	4	5	6
53.77		1.7034	-2	4	0	1		2
53.84		1.7014	-3	1	1	2		3
54.18	54.18	1.6915	-1	2	7	5	3	7
54.41	54.40	1.6848	-3	1	3	3	4	4
54.70	54.70	1.6765	-2	4	3	6	3	9
55.27	55.36	1.6606	-2	0	10	18	12	27
55.35		1.6584	-2	4	4	5		7
55.85	55.92	1.6448	-2	4	9	37	24	56
55.93		1.6427	1	3	3	3		4
56.04		1.6397	-1	5	3	11		16
56.37	56.38	1.6307	-1	5	4	11	7	16

PHENGITE 2M1 – GÜVEN, 1968

2THETA	PEAK	D	H	K	L	I(INT)	I(PK)	I(DS)
56.63	56.52	1.6239	-2	4	3	2	5	3
56.79	56.80	1.6197	-3	1	1	10	7	15
57.30	57.40	1.6065	3	1	3	5	6	7
57.39		1.6042	-2	4	1	8		12
57.63	57.62	1.5982	2	5	6	4	6	6
57.64		1.5979	1	5	4	5		7
58.57	58.56	1.5747	-3	1	6	3	2	5
59.19	59.20	1.5596	3	1	9	8	6	13
59.31	59.32	1.5569	-2	2	10	5	7	5
59.31		1.5568	-2	2	6	3		9
59.46	59.46	1.5531	0	2	12	1	4	4
59.61	59.60	1.5496	1	5	5	3	3	7
60.12	60.12	1.5377	-1	5	6	4	3	3
60.49	60.50	1.5291	2	4	3	2		32
60.91	60.92	1.5195	-1	3	11	20	11	36
61.50	61.58	1.5064	0	6	1	22	36	75
61.58		1.5047	-3	3	10	47		14
61.59		1.5046	2	0	10	9		3
61.70		1.5021	0	6	1	2		3
62.20	62.26	1.4911	4	4	10	1	4	3
62.24		1.4903	-3	3	3	2		4
62.29		1.4893	0	6	2	3		5
62.48	62.46	1.4852	3	3	1	3	3	5
62.95	62.96	1.4752	-2	4	6	4	2	6
63.14	63.14	1.4712	2	0	11	1	2	2
63.56	63.56	1.4625	-2	3	12	2	1	4
64.21	64.22	1.4493	1	3	11	5	3	9
64.36	64.36	1.4463	0	5	11	3	3	6
64.54	64.54	1.4414	1	5	13	1	2	2
65.31	65.30	1.4275	2	2	7	5	3	8
65.82	65.82	1.4176	0	0	10	2	1	4
66.00	66.00	1.4143	-3	4	14	2	2	3
66.23	66.26	1.4099	0	4	9	2	2	3
66.29		1.4087	1	1	11	2		3
67.12	67.14	1.3933	1	4	13	1	1	2
67.46	67.46	1.3871	-2	4	9	1	1	3
68.09	68.10	1.3758	-1	1	14	2	1	4
68.37	68.34	1.3709	0	6	6	2	1	3
69.52	69.54	1.3509	2	2	11	5	5	8
69.75	69.74	1.3471	-1	3	13	26	13	45
70.49	70.50	1.3348	2	0	12	12	9	21
70.50		1.3346	0	4	12	8		15
71.43	71.56	1.3194	1	1	14	1		2
71.55		1.3176	1	1	13	5	3	9
72.41	72.54	1.3040	-2	2	13	2		4
72.51		1.3025	-2	6	1	5	6	9
72.53		1.3022	2	6	0	2		6
72.65	72.68	1.3004	-4	0	15	3		3
72.68		1.2999	-1	1	15	1	10	23
72.92	72.90	1.2962	-2	6	2	13		11
73.11	73.10	1.2933	3	3	0	6	8	14
73.39	73.32	1.2890	-3	3	9	8	6	3
73.48	73.48	1.2876	-1	0	13	2	4	11
74.00	74.00	1.2800	0	6	8	6	4	15
74.28	74.26	1.2757	3	3	7	8	5	13
74.74	74.74	1.2690	-2	6	4	7	3	8
75.02	74.98	1.2650	0	4	2	4	2	5
75.91	75.94	1.2523	-1	7	13	5	4	9
75.99		1.2513	-4	2	1	1		3
76.03		1.2507	-3	5	2	3		2
76.06		1.2503	-4	0	6	3		6
76.13		1.2493	-2	2	14	2		4
76.37		1.2460	4	2	0	1	6	2
76.48		1.2444	-3	5	3	5		10
76.50	76.50	1.2441	0	6	9	3		6
76.52		1.2439	0	0	16	2		3
76.77	76.74	1.2404	-1	7	3	4	4	7
77.06	77.02	1.2365	1	5	3	2	2	4
77.66	77.68	1.2284	1	3	3	3	3	6
77.87		1.2256	-4	2	5	3	3	6
78.07	78.08	1.2230	-3	2	5	3	3	9
78.52	78.52	1.2172	0	2	13	5	1	3
78.70	78.70	1.2148	1	2	13	1	2	7
79.24	79.26	1.2078	1	7	4	2	2	4

PHENGITE 2M1 – GÜVEN, 1968

2THETA	PEAK	D	H	K	L	I(INT)	I(PK)	I(DS)
79.40	79.46	1.2059	-3	3	11	3	2	5
79.46		1.2050	-4	2	6	1		2
79.59		1.2035	-4	2	3	1		2
79.87	79.86	1.1999	0	6	10	2	2	3
80.04	80.06	1.1977	-3	5	6	2	2	4
80.48		1.1924	3	3	9	2		4
80.58	80.58	1.1911	3	5	4	4	3	8
80.95		1.1866	1	7	5	2		4
80.96	80.96	1.1864	1	7	5	2	2	3
81.20	81.20	1.1836	-2	2	15	2	2	4
81.35	81.38	1.1818	-4	2	7	2	2	3
81.54	81.54	1.1796	3	1	11	3	2	5
81.83	81.78	1.1761	-2	4	11	1	1	2
83.45		1.1572	-4	2	13	1		2
83.53	83.52	1.1564	-1	2	5	3	2	6
83.68	83.70	1.1547	-1	1	7	1	1	2
85.95	85.96	1.1299	-1	1	17	4	2	7
86.25	86.24	1.1268	-2	6	8	1	2	3
86.48	86.50	1.1243	-2	4	14	4	2	20
87.38	87.38	1.1150	4	4	0	10	4	3
88.01	88.04	1.1087	-2	6	10	2	1	11
88.47	88.46	1.1042	-1	5	14	5	2	3
90.99	91.00	1.0800	-4	0	8	1	1	6
91.65	91.64	1.0740	-2	4	15	3	1	11
92.70	92.70	1.0645	-4	6	12	5	2	2
92.88		1.0629	-2	6	10	1		7
93.97	93.96	1.0535	-1	5	15	3	1	13
94.87	94.88	1.0458	-2	0	18	6	2	7
95.82	95.84	1.0379	1	3	17	3	1	5
96.51	96.34	1.0324	-3	3	15	3	1	6
97.33	97.34	1.0259	0	6	14	2	1	5
98.60	98.62	1.0160	3	3	13	3	1	3
99.03	99.04	1.0127	3	7	5	2	1	3
99.11		1.0121	-2	7	3	1		3
99.61	99.58	1.0084	-4	8	14	2	1	5
100.89	100.90	0.9990	0	6	12	4	1	10
101.83	101.86	0.9923	0	0	20	1	1	2

2THETA	PEAK	D	H	K	L	I(INT)	I(PK)	I(DS)
102.61	102.66	0.9869	-1	3	19	5	2	11
102.79	103.06	0.9857	-1	9	1	1	3	2
102.86		0.9851	-4	6	1	1		3
103.04		0.9850	-4	6	2	1		1
103.06		0.9839	-5	3	1	2		2
103.09		0.9838	-1	9	1	3		4
103.25		0.9836	-5	3	0	3		6
104.57		0.9825	4	3	5	4		4
104.76	104.76	0.9737	-5	3	3	1	2	8
105.05	105.06	0.9725	1	9	3	1	2	9
105.18		0.9705	4	6	2	4		10
106.12		0.9697	-2	8	8	1		2
106.28	106.32	0.9637	3	5	12	4	2	9
106.38		0.9627	-3	3	17	2		4
106.52		0.9621	-4	6	6	4		3
107.10	107.14	0.9612	-1	9	5	2	2	11
107.17		0.9575	0	6	16	5		8
107.22		0.9572	1	7	13	3		3
107.64	107.54	0.9568	5	3	3	1	2	4
108.07	108.06	0.9543	-4	8	2	5	2	12
109.64	109.62	0.9517	3	3	15	3	0	2
110.29	110.34	0.9424	2	8	8	2	1	2
110.35		0.9386	5	1	7	1		2
110.61	110.76	0.9383	-4	6	8	6	1	2
110.78		0.9368	-1	9	7	8		2
110.80		0.9358	2	6	14	7		1
111.10	111.08	0.9341	-5	1	11	13	1	4
111.56	111.54	0.9315	3	5	13	7	1	4
112.26	112.24	0.9277	5	1	5	5	1	4
112.78	112.78	0.9249	1	7	14	6	0	6
113.26	113.22	0.9223	-4	2	16	2	1	1
113.39		0.9216	-3	7	11	1		2
115.84	116.04	0.9090	-1	0	21	1	1	3
116.02		0.9081	0	8	13	1		2
116.21		0.9072	4	6	7	4		2
116.63	116.60	0.9052	3	5	14	1	1	2

PHENGITE 2M1 - GÜVEN, 1968

2THETA	PEAK	D	H	K	L	I(INT)	I(PK)	I(DS)
117.26	117.28	0.9021	0	0	22	2	1	5
117.80	118.12	0.8996	5	1	9	1	2	5
118.11		0.8981	-5	3	11	9		23
118.93	119.02	0.8943	-4	4	12	2	2	4
118.97		0.8941	-2	8	12	1		2
119.05		0.8937	-1	9	9	7		18
119.74	119.72	0.8906	4	6	8	7	2	19
119.92		0.8898	-1	6	10	1		2
120.05		0.8892	0	10	4	1		4
120.12		0.8889	0	10	5	1		1
121.45	121.88	0.8830	2	8	11	2	1	2
121.85		0.8813	-3	7	13	1		5
122.21	122.30	0.8798	-5	5	7	1	1	1
123.11	123.46	0.8760	-5	1	14	1	1	2
123.37		0.8749	3	7	11	2		7
123.48		0.8745	-4	6	12	2		5
123.71	124.06	0.8735	-2	9	13	1	1	2
124.03		0.8722	-1	8	11	2		5
124.12		0.8719	0	4	21	1		2
124.16		0.8717	-4	4	16	2		4
124.59	124.76	0.8700	1	1	22	1	2	2
124.63		0.8698	4	1	13	1		1
124.74		0.8694	-3	9	1	5		12
124.98		0.8685	-6	0	2	3		8
124.99		0.8684	-4	2	18	1		1
125.02		0.8683	3	9	0	1		2
126.09	126.30	0.8641	6	0	0	1	1	2
126.30		0.8633	-2	6	18	3		9
126.35		0.8631	-5	3	13	1		1
126.80	126.82	0.8614	-1	1	23	1	1	4
127.02		0.8606	-3	7	14	1		2
127.32		0.8595	-5	5	9	1		2
128.54	129.02	0.8550	4	6	10	1	1	3
128.74		0.8543	4	0	16	2		4
128.98		0.8534	2	10	0	1		4
129.00		0.8534	-4	8	0	1		2
129.48		0.8517	4	8	0	1		3

TABLE 24. MISCELLANEOUS MICAS

Variety	Celadonite	Phlogopite	Fluorphlogopite	Ferri-annite
Composition	$K_{1.6}(Mg_{1.4}Fe_{2.8})$ $(Si_{3.6}Al_{0.4})_2O_{20}$ $(OH)_4$	$(K,Mn)_2Mg_6(Si_3Fe)_2$ $O_{20}(OH)_4$	$(K,Na)_2(Mg)_6(Si_3Al)_2$ $O_{20}[F(OH)]_4$	$K_2Fe_6^2(Si_3Fe)_2O_{20}$ $(OH)_4$
Source	Pobuzhé, USSR Malkina collection No. 43-B	Sweden	Synthetic	Synthetic
Reference	Zvyagin, 1957	Steinfink, 1962b	McCauley, 1968	Donnay, Morimoto, Takeda & Donnay, 1964
Cell Dimensions				
$\underline{a}$ Å	5.21 (5.20 kx)	5.36	5.308	5.430
$\underline{b}$ Å	9.02 (9.00 kx)	9.29	9.183	9.404
$\underline{c}$	10.27 (10.25 kx)	10.41	10.139	10.341
α	90	90	90	90
β deg	100.02	100.0	100.07	100.07
γ	90	90	90	90
Space Group	C2/m	C2/m	C2/m	C2/m
Z	1	1	1	1

TABLE 24. (cont.)

Variety	Celadonite	Phlogopite	Fluorphlogopite	Ferri-annite
Site Occupancy	M_1 K 0.80 M_2 Mg 0.10 M_3 {Mg 0.30, Fe 0.70} T_1 {Si 0.90, Al 0.10}	M_1 {K 0.90, Mn 0.10} T_1 {Si 0.75, Fe 0.25}	T_1 {Si 0.75, Al 0.25}	T_1 {Si 0.75, Al 0.25}
Method	Electron diffrac.	3-D, film	3-D, counter	3-D, film
R & Rating	--- (3)	0.13 (2)	0.064 (1)	0.093 (2)
Cleavage and habit		{001} perfect Platy		
Comment	Related to glauconite B estimated		Minor Na & (OH) ignored in powder pattern calculation.	
μ	198.7	217.7	118.3	251.8
ASF	0.1961	0.2964×10^{-1}	0.5883×10^{-1}	0.2438
Abridging factors	1:1:0.5	1:1:0.5	1:1:0.5	0.5:0:0

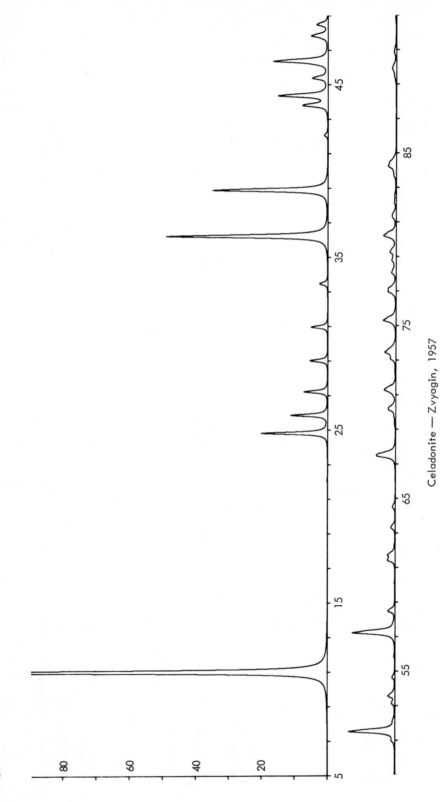

Celadonite — Zvyagin, 1957

λ = FeKα Half width at 40°2θ: 0.15°2θ Plot scale I_{PK} (max) = 200

498

CELADONITE - ZVYAGIN, 1957

2THETA	PEAK	D	H	K	L	I(INT)	I(PK)	I(DS)
10.99	11.00	10.112	0	0	1	10C	100	100
24.79	24.80	4.509	0	2	0	12	10	15
25.83	25.84	4.331	-1	1	1	7	6	8
27.19	27.20	4.118	0	1	1	4	4	6
28.98	29.00	3.868	-1	1	2	3	3	5
30.93	30.94	3.630	-1	1	1	3	2	5
33.43	33.44	3.365	0	2	2	2	1	2
36.19	36.20	3.117	1	1	1	34	24	54
38.85	38.86	2.911	-1	1	3	26	18	42
43.76	43.78	2.597	-1	2	1	5	4	9
44.30	44.34	2.568	-1	0	3	2	8	3
44.34	44.34	2.565	-2	0	0	10		19
45.35	45.36	2.511	1	3	0	4	2	7
46.34	46.34	2.460	1	1	3	13	8	25
47.82	47.82	2.389	2	0	1	4	3	7
48.46	48.46	2.359	-1	1	4	3	2	5
51.48	51.50	2.229	1	3	1	12	7	26
53.53	53.54	2.150	-1	3	3	4	1	4
57.19	57.20	2.022	0	0	5	12	7	28
58.45	58.46	1.9827	-2	0	4	2	1	5
61.39	61.40	1.8963	-1	3	4	2	1	5
61.69	61.68	1.8879	2	0	3	2	1	5
63.28	63.28	1.8453	0	2	5	1	1	3
67.44	67.46	1.7438	-2	0	5	5	3	14
67.54		1.7414	0	2	3	2		5
70.11	70.14	1.6853	1	0	5	2	1	6
71.23	71.28	1.6622	1	3	4	1	2	4
71.30		1.6608	-2	0	5	2		6
73.05	73.06	1.6264	-2	2	5	2	1	5
73.43	73.44	1.6192	-3	1	3	4	2	11
75.28	75.28	1.5850	-1	1	5	4	1	13
76.95	76.96	1.5558	1	5	3	2	1	8
77.14	77.12	1.5525	-1	5	2	1	1	3
78.03	78.04	1.5377	2	4	0	2	1	5
78.76	78.76	1.5257	-2	2	6	1	1	4
79.23	79.24	1.5181	3	1	2	2	1	7
80.19	80.20	1.5030	0	6	0	4	2	13

2THETA	PEAK	D	H	K	L	I(INT)	I(PK)	I(DS)
84.21	84.22	1.4437	-3	3	3	3	1	11
89.86	89.86	1.3706	3	3	2	3	1	7
90.83	90.84	1.3592	-1	3	6	1	0	4
95.05	95.04	1.3124	-4	0	7	2	1	6
96.14	96.40	1.3012	-4	0	1	2	2	9
96.16		1.3009	-2	6	0	2		10
96.38		1.2986	-4	0	2	1		5
96.57		1.2968	2	6	0	2	1	10
105.39	105.52	1.2170	1	5	5	3		14
105.53		1.2158	-3	3	3	2		11
106.72	106.76	1.2063	0	6	5	4	1	16
109.10	109.14	1.1882	-1	7	3	1		6
110.98	111.44	1.1747	-3	5	4	2	0	5
111.48		1.1712	3	5	2	2	0	8
116.48	116.56	1.1385	-2	6	5	2		12
119.67	119.72	1.1196	-4	0	6	1		6
120.60	120.58	1.1144	-2	6	6	2	0	11
128.47	129.40	1.0748	-2	6	6	2	0	11
129.35		1.0709	-1	5	7	2	1	10
129.43		1.0705	4	0	4	2		13
131.34	131.38	1.0623	2	0	8	1	0	6
133.04	133.18	1.0554	-4	0	7	1	0	6
134.37	134.58	1.0502	2	6	5	1	0	8
135.82	136.00	1.0447	-3	3	8	3	0	16
143.74	144.02	1.0186	-3	1	9	1	0	7
146.38	146.40	1.0112	0	10	0	1	0	7
147.45		1.0084	-5	1	4	1		7
149.79	149.84	1.0026	-2	0	10	3	0	18
156.82	159.76	0.9881	3	7	2	1	0	7
157.16		0.9875	1	5	8	1		7
158.90		0.9846	0	8	6	1		52
159.46		0.9837	-4	6	1	8		6
159.67		0.9834	-1	9	0	1		
160.18		0.9827	-4	6	2	3	0	20
160.83		0.9817	1	5	5	1		5
163.01		0.9787	-2	2	10	4		24
164.22		0.9772	-5	3	3	1		6

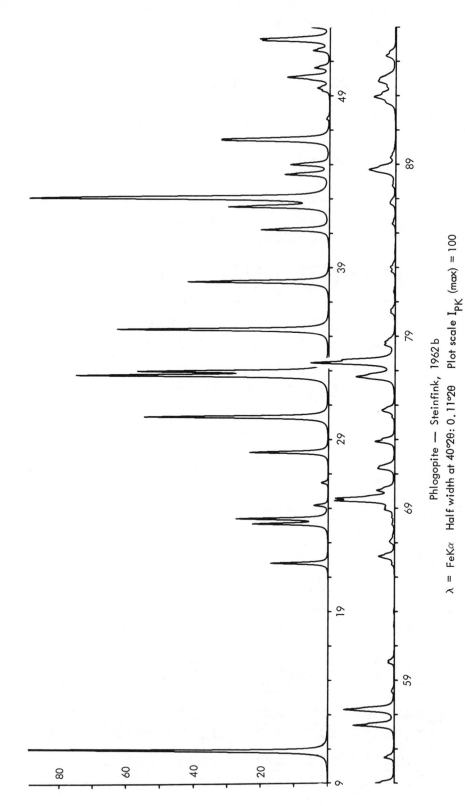

Phlogopite — Steinfink, 1962 b

$\lambda = FeK\alpha$ Half width at $40°2\theta$: $0.11°2\theta$ Plot scale I_{PK} (max) = 100

PHLOGOPITE – STEINFINK, 1962b

2THETA	PEAK	D	H	K	L	I(INT)	I(PK)	I(DS)
10.84	10.84	10.252	0	0	1	95	100	48
21.77	21.78	5.126	0	0	2	19	17	12
24.06	24.06	4.645	0	2	0	25	23	16
24.35	24.36	4.589	1	1	0	31	27	20
25.15	25.16	4.447	-1	1	1	5	4	3
26.45	26.46	4.231	0	2	2	2	2	2
28.22	28.22	3.971	1	1	2	29	24	21
30.26	30.26	3.709	-1	1	2	70	55	53
32.67	32.68	3.442	0	2	3	97	75	77
32.91	32.92	3.417	1	1	3	69	57	55
35.35	35.36	3.188	-1	1	3	87	63	73
38.13	38.14	2.964	-1	1	4	60	42	54
41.18	41.18	2.753	0	2	4	30	20	29
42.50	42.50	2.671	1	1	4	29	30	29
42.51		2.670	-2	0	1	14		14
42.99	43.00	2.642	-1	3	1	100	89	100
43.03		2.639	2	0	0	49		49
44.38	44.40	2.563	0	0	5	11	13	12
44.40		2.562	1	3	3	9		9
44.95	44.96	2.532	-2	0	2	6	12	6
44.97		2.531	-1	3	2	12		13
46.37	46.38	2.459	-1	1	5	42	32	45
46.44	46.44	2.455	-2	0	3	21	32	23
47.64	47.64	2.397	-1	3	3	1		3
49.26	49.28	2.322	0	4	0	3	2	3
49.43	49.42	2.315	-2	2	3	5	4	5
50.01		2.290	-2	2	1	7		8
50.06	50.06	2.288	-1	3	4	14	13	16
50.60	50.60	2.265	0	4	2	7	5	8
51.62	51.62	2.223	-2	2	4	8	5	9
52.20	52.20	2.200	-1	3	5	29	20	36
52.30	52.30	2.196	2	0	6	15	21	19
52.97	52.98	2.171	0	4	2	9	6	11
54.46	54.46	2.115	-2	2	4	5	3	6
56.23		2.054	-2	2	3	3		4
56.34	56.34	2.050	0	4	5	20	12	26
57.19	57.26	2.023	-2	0	4	10	15	13
57.26		2.020	-1	3	4	20		27
59.99	60.00	1.9362	-1	3	3	3	2	4
60.11	60.12	1.9328	-2	0	3	1	2	2
65.70	65.70	1.7844	-2	0	5	1	1	2
66.08	66.18	1.7754	-1	3	1	4	5	6
66.18		1.7731	-1	3	1	7		11
66.97	67.06	1.7545	-3	1	1	1	2	2
67.06		1.7524	-2	4	2	1		2
68.05	68.06	1.7299	-3	1	0	1	1	2
68.45	68.46	1.7210	0	4	1	1	1	2
68.90	68.90	1.7112	-1	5	1	4	3	7
69.02	69.04	1.7086	0	5	2	1		2
69.41	69.42	1.7002	-1	5	0	33	18	54
69.54	69.56	1.6973	-2	0	6	16	18	26
69.96		1.6884	-1	5	2	2		13
70.02	70.02	1.6872	-2	0	2	8	5	15
71.29	71.30	1.6610	-3	1	3	9	5	7
71.39	71.40	1.6589	-3	1	1	4	5	8
72.83		1.6307	-2	4	3	5		15
72.87	72.86	1.6298	-1	5	4	9	6	2
74.26	74.28	1.6036	0	2	6	1	1	8
74.60	74.68	1.5973	-1	5	3	4	4	10
74.69		1.5958	-2	4	1	6		16
76.50	76.62	1.5635	-2	0	6	9	12	5
76.58		1.5621	-3	1	5	3		32
76.62		1.5615	-1	1	4	17		4
76.73		1.5595	0	6	5	2		37
77.39	77.42	1.5483	0	6	0	20	25	75
77.44		1.5476	-3	3	1	40		4
78.44	78.48	1.5310	0	3	2	2	3	4
78.46		1.5306	-3	3	0	2		5
78.51		1.5298	-1	3	0	2		4
78.79	78.64	1.5276	-2	4	4	2	3	3
80.35	80.36	1.5253	-1	4	1	2	3	3
80.51	80.52	1.5004	-1	0	3	3	1	2
81.57	81.58	1.4980	-2	2	6	1	1	
		1.4818						2

501

PHLOGOPITE - STEINFINK, 1962b

2THETA	PEAK	D	H	K	L	I(INT)	I(PK)	I(DS)
82.74	82.74	1.4645	0	0	7	3	1	6
83.02	83.00	1.4606	-1	1	7	3	2	5
85.52	85.52	1.4257	2	2	1	3	1	6
86.66	86.70	1.4107	-3	3	4	2	2	4
86.68		1.4103	0	6	3	2		3
86.80		1.4088	3	3	2	1		3
87.74	87.74	1.3968	0	2	7	2	1	5
88.53	88.66	1.3868	-2	0	7	9	8	20
88.67		1.3851	-1	3	6	18		39
89.39	89.38	1.3763	0	4	6	4	2	9
92.69	92.92	1.3380	1	1	7	1	7	3
92.91		1.3355	2	6	0	11		26
92.94		1.3352	-4	0	2	6		13
93.51	93.52	1.3289	-2	2	7	5	3	11
93.79	93.82	1.3258	-3	3	5	6	5	15
93.84		1.3253	0	6	4	7		15
93.98	93.96	1.3237	3	3	3	7	5	16
95.29	95.32	1.3098	-4	0	3	3	3	8
95.31		1.3096	-2	4	0	7		16
97.99	98.16	1.2827	-2	6	3	5	4	13
98.11		1.2815	0	0	8	3		7
98.17		1.2810	4	0	1	2		6
98.17		1.2809	2	6	2	4		10
99.38	99.38	1.2694	-4	4	0	2	1	5
100.28	100.30	1.2610	-1	7	2	3	1	8
101.56	101.58	1.2495	-3	5	5	4	1	11
101.66		1.2486	3	5	1	2		3
103.07	103.20	1.2363	-3	7	2	5	3	5
103.10		1.2360	1	0	6	2		4
103.15		1.2356	0	6	5	2		5
103.23		1.2349	4	2	1	4		11
103.32		1.2342	3	3	4	2		5
104.81	104.82	1.2217	-1	7	3	3	1	7
104.82		1.2216	-4	4	3	2		5
106.81	106.84	1.2057	-3	2	4	1	1	4
106.96		1.2045	3	5	3	1		3
107.73	107.74	1.1985	-2	2	8	1	0	3

2THETA	PEAK	D	H	K	L	I(INT)	I(PK)	I(DS)
108.91	108.94	1.1896	-1	5	6	2	1	5
113.55	113.74	1.1571	-1	5	7	1	1	4
113.74		1.1559	-2	2	6	2		6
115.43	115.60	1.1450	-4	0	6	6	3	18
115.61		1.1439	2	0	4	12		36
117.45	117.46	1.1326	-3	2	7	1	0	4
118.65	118.68	1.1254	0	8	2	2	0	6
119.31	119.06	1.1217	-3	1	4	1	1	3
123.25	123.28	1.1002	-2	4	6	9	2	28
123.63	123.62	1.0982	4	0	0	4	2	13
124.61	124.56	1.0932	-1	5	7	2	1	5
126.78	126.84	1.0827	-1	3	9	9	2	29
127.06		1.0814	-2	8	0	5		15
130.79	131.20	1.0647	-3	3	8	5	2	16
130.95		1.0640	0	6	7	5		17
131.23		1.0628	3	3	6	5		16
133.59	133.58	1.0532	2	8	1	2	0	4
134.47	134.52	1.0497	2	2	8	1	0	7
136.18	136.30	1.0433	-3	1	7	1	0	4
136.32		1.0428	3	7	1	8		7
139.12	139.12	1.0330	-2	6	7	7	1	27
139.68	139.66	1.0312	4	0	5	4	1	14
141.09	141.44	1.0266	2	8	2	2	1	6
141.36		1.0257	5	1	1	2		4
141.54		1.0252	0	0	10	2		7
144.25	144.62	1.0171	-1	1	10	5	2	20
144.61		1.0161	-2	0	10	10		38
145.70	146.42	1.0130	1	9	0	2	2	8
145.85		1.0126	-4	6	1	2		9
145.88		1.0125	-5	3	2	2		9
146.29		1.0114	-1	9	1	6		23
146.40		1.0111	-4	6	2	6		23
146.57		1.0107	-5	3	3	6		23
150.85	151.04	1.0002	-1	1	9	10	2	36
150.92		1.0000	-4	6	3	9		36
151.26		0.9992	5	3	0	10		39
156.82	157.46	0.9881	-4	2	8	1	2	4

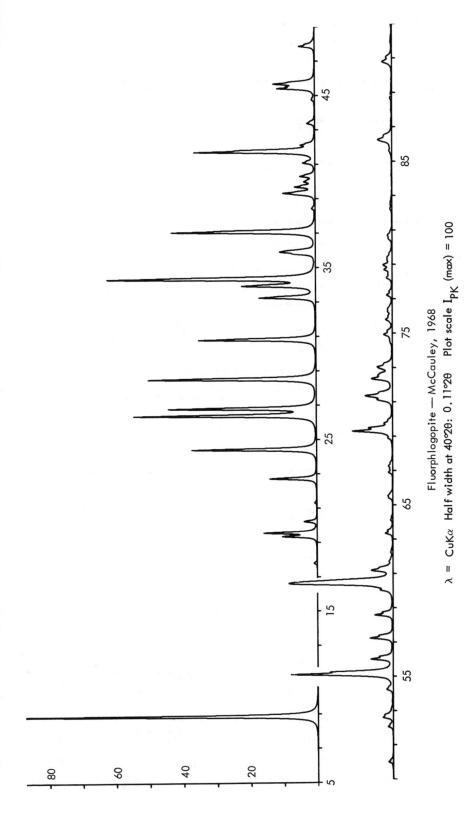

Fluorphlogopite — McCauley, 1968

$\lambda = \text{CuK}\alpha$ Half width at $40°2\theta$: $0.11°2\theta$ Plot scale I_{PK} (max) = 100

FLUORPHLOGOPITE - MCCAULEY, 1968

2THETA	PEAK	D	H	K	L	I(INT)	I(PK)	I(DS)
8.85	8.86	9.983	0	0	1	100	100	100
17.75	17.76	4.991	0	0	2	1	1	1
19.31	19.32	4.591	1	1	0	11	11	12
19.53	19.54	4.542	0	2	0	18	16	19
20.19	20.20	4.394	1	1	1	4	4	5
22.69	22.70	3.916	-1	1	1	18	14	19
24.39	24.40	3.647	1	1	1	48	37	51
26.35	26.36	3.379	-1	1	2	71	54	77
26.77	26.78	3.328	0	0	3	58	44	63
28.48	28.48	3.131	1	1	2	69	50	76
30.77	30.78	2.903	-1	1	3	49	35	55
33.22	33.22	2.694	0	2	3	24	17	28
33.88	33.90	2.644	1	3	0	11	22	13
33.91		2.641	-2	0	0	21	62	25
34.29	34.32	2.613	-1	3	1	32		37
34.31		2.611	1	3	1	64		74
35.76	35.78	2.509	-2	0	2	8	7	9
35.86	35.88	2.502	1	3	1	4	11	4
35.90		2.499	0	2	4	8		9
35.95		2.496	-1	3	2	4		4
37.02	37.04	2.426	-1	1	4	22	43	26
37.04		2.425	2	0	1	45		53
38.39	38.40	2.343	-1	3	2	1	1	2
39.21	39.30	2.296	-1	1	4	6	10	7
39.29		2.291	0	4	0	11		13
39.65	39.66	2.271	-2	2	1	9	6	11
39.93	39.96	2.256	-2	2	3	8	3	9
39.97		2.254	-1	3	2	3		4
40.27	40.28	2.237	0	4	1	7	5	8
41.05	41.06	2.197	-2	2	2	5	4	6
41.69	41.70	2.165	-1	3	2	20	36	24
41.70		2.164	0	0	5	39		48
42.09	42.08	2.145	-2	2	4	5	4	6
43.34	43.34	2.086	0	2	2	4	2	5
44.72	44.72	2.025	-2	2	2	1	1	2
45.39	45.38	1.9966	0	0	6	18	11	23
45.63	45.66	1.9866	-2	0	4	7	13	9
45.66		1.9851	1	3	3	14		18
47.83	47.84	1.9001	-2	0	3	3	5	4
47.84		1.8998	-1	3	4	5		7
49.98	49.98	1.8233	-2	2	4	2	1	3
52.04	52.04	1.7557	-2	2	5	3	2	4
52.55	52.58	1.7400	0	0	6	2	3	2
52.58		1.7389	-3	1	4	3		5
52.63		1.7374	-1	3	4	1		2
52.76	52.74	1.7334	0	4	3	1	3	2
54.20	54.22	1.6908	-1	5	1	2	2	2
54.24		1.6896	0	4	4	1		2
55.02		1.6675	-1	5	2	5		7
55.04		1.6671	0	0	5	2		3
55.10	55.10	1.6652	-1	3	5	18	30	24
55.11		1.6651	2	4	0	36		49
55.15		1.6638	1	3	4	1		2
55.99	56.00	1.6410	-3	1	3	8	7	11
56.00		1.6406	-1	3	5	4		6
57.20	57.22	1.6090	-2	2	5	5	7	7
57.23		1.6083	-1	3	2	9		12
58.56	58.56	1.5750	-3	0	6	6	5	9
58.57		1.5747	-1	3	3	5		6
60.01	60.02	1.5403	-2	2	6	2	4	3
60.03		1.5399	-1	5	2	3		4
60.37	60.42	1.5318	-3	1	6	38	31	55
60.43		1.5305	-2	0	7	19		28
60.46		1.5299	-1	6	0	9		13
60.50		1.5291	-1	3	1	18		26
61.15	61.16	1.5142	-3	3	6	4	6	6
61.16		1.5140	-3	3	0	5		7
61.21		1.5128	0	6	4	4		6
61.69	61.70	1.5023	-2	4	4	1	1	1
62.22	62.22	1.4908	-1	1	6	1	1	2
63.32	63.32	1.4675	2	0	5	3	2	2
63.32		1.4675	-1	3	6	1		4
64.10	64.10	1.4514	-2	2	6	1	1	2
65.38	65.40	1.4261	-2	0	7	2	1	3

FLUORPHLOGOPITE - MCCAULEY, 1968

2THETA	PEAK	D	H	K	L	I(INT)	I(PK)	I(DS)
65.50	65.50	1.4239	-1	1	7	2	1	3
66.87	66.88	1.3979	2	1	5	3	1	4
66.88	68.90	1.3619	0	2	7	2	1	3
69.26	69.28	1.3554	-2	2	0	9	12	14
69.29		1.3548	1	0	7	19		29
69.74	69.74	1.3472	0	4	1	4	3	6
69.74	71.12	1.3257	-4	0	6	3	5	5
71.05		1.3246	-2	6	1	6		10
71.11	71.34	1.3219	-4	0	2	11	8	9
71.28		1.3207	2	6	0	4		17
71.36		1.3057	-3	3	5	4		7
72.30	72.34	1.3054	3	3	3	4	6	7
72.32		1.3051	3	1	3	1		2
72.34		1.3047	0	6	6	4		7
72.37		1.3040	-1	3	4	1		2
72.41	72.54	1.3000	-2	2	7	4	4	7
72.67	73.02	1.2957	-4	0	3	4	5	6
72.95		1.2945	2	6	1	7		12
73.03	74.90	1.2673	4	0	1	2	3	3
74.86		1.2665	-2	6	3	4		7
74.92	75.78	1.2544	0	2	8	4	2	6
75.77	76.24	1.2479	0	0	2	2	1	4
76.23	76.38	1.2457	-1	7	2	2	1	3
76.39	77.20	1.2347	-3	5	3	3	2	5
77.19		1.2346	3	5	1	1		2
77.20	78.20	1.2216	4	2	1	2	2	5
78.18		1.2206	-1	7	2	3		3
78.26	78.68	1.2155	-3	3	6	2	2	3
78.64		1.2152	3	3	4	2		3
78.67		1.2147	3	6	5	2		3
78.71	78.96	1.2114	0	4	7	3	3	5
78.96		1.2109	-1	7	4	1		2
79.00	79.40	1.2069	-1	3	7	2	2	4
79.32		1.2057	-4	1	3	1		2
79.41	80.52	1.1930	-4	0	4	2	1	3
80.43		1.1919	2	6	3	2		3
80.51	80.68	1.1899	3	5	2	1		2

2THETA	PEAK	D	H	K	L	I(INT)	I(PK)	I(DS)
82.23	82.22	1.1713	-2	2	8	1	1	2
85.58	85.58	1.1338	-2	4	6	1	1	2
86.15	86.24	1.1278	-4	0	6	5	4	8
86.24		1.1269	-2	6	4	9		17
88.64	88.66	1.1025	-3	1	8	1	1	6
90.74	90.76	1.0824	-4	0	4	3	3	6
90.77		1.0820	-2	6	6	7		12
93.61	93.62	1.0566	-1	3	9	4	2	7
93.61		1.0565	2	0	8	2		3
95.09	95.14	1.0439	-3	3	8	2	2	5
95.13		1.0436	3	3	6	2		5
95.16		1.0434	0	7	1	3		5
96.67	96.68	1.0310	3	0	7	2	1	2
98.73	98.76	1.0150	4	2	5	2	1	3
98.77		1.0147	-2	6	7	1		6
100.65	100.74	1.0007	-5	1	2	2		3
100.70		1.0004	-4	6	6	1		4
100.79		0.9998	-1	3	9	2		4
100.93		0.9987	0	0	10	1		2
100.99		0.9983	0	0	10	2		4
101.85	101.90	0.9922	-2	1	3	4	2	7
101.89		0.9919	-1	3	0	4		7
102.28	102.30	0.9892	5	3	3	3	3	7
102.32		0.9889	-4	0	6	3		7
102.41		0.9883	-1	9	3	3		7
104.13	104.22	0.9766	-5	3	4	1	0	1
104.20		0.9761	4	6	1	1		1
104.28		0.9756	-1	9	2	1		1
105.33	105.54	0.9687	5	3	1	2	2	1
105.36		0.9685	-4	4	4	1		1
105.46		0.9679	-1	9	6	1		6
105.50		0.9676	-3	3	8	3		6
105.55		0.9673	3	3	7	3		6
105.58		0.9671	0	4	8	3		1
108.23	108.32	0.9507	4	6	2	1	1	2
108.29		0.9503	-3	5	8	1		1
108.31		0.9502	-1	9	3	1		1

FLUORPHLOGOPITE - MCCAULEY, 1968

2THETA	PEAK	D	H	K	L	I(INT)	I(PK)	I(DS)
108.37		0.9499	-2	8	8	1		1
109.22	109.18	0.9448	0	8	6	1	0	
111.77	111.82	0.9303	4	2	6	1	0	2
112.14	112.16	0.9283	4	4	5	1	0	2
116.05	116.10	0.9080	5	3	3	6	3	13
116.07		0.9079	-4	6	6	6		13
116.15		0.9075	0	0	11	1		3
116.17		0.9074	-1	9	5	6		13
117.99	117.98	0.8987	-5	5	4	1	0	1
118.94	118.96	0.8942	0	5	7	1	0	3
121.07	121.26	0.8847	-6	8	2	3	2	6
121.18		0.8842	-5	3	7	2		5
121.25		0.8839	-3	9	1	6		13
121.29		0.8837	4	4	4	2		5
121.37		0.8834	-1	9	5	2		2
121.87	121.72	0.8812	-6	0	3	1	2	3
122.05		0.8804	3	9	0	1		2
124.72	124.70	0.8695	2	6	8	1	0	3
126.63	126.66	0.8621	5	1	5	1	0	1
126.75		0.8616	2	8	6	1		2
130.71	131.20	0.8474	-5	7	6	1	2	2
131.04		0.8463	4	6	5	5		12
131.18		0.8459	4	0	8	5		12
131.27		0.8456	1	9	6	5		12
131.75		0.8440	0	4	11	1		2
134.10	134.20	0.8365	-3	3	11	3	1	6
134.17		0.8362	3	9	3	3		6
134.20		0.8361	0	6	10	3		6
134.29		0.8359	1	3	11	1		2
134.93	135.12	0.8339	-6	0	6	1	1	3
134.96		0.8338	-5	9	9	1		1
134.99		0.8337	6	0	2	1		1
135.16		0.8332	-3	9	5	1		3
135.17		0.8332	-4	6	8	1		2
135.17		0.8332	5	3	3	1		2
135.19		0.8331	3	3	3	3		7
135.31		0.8328	-1	9	7	1		2

2THETA	PEAK	D	H	K	L	I(INT)	I(PK)	I(DS)
135.37		0.8325	4	0	8	1		2
135.39		0.8325	-2	6	10	1		3
135.61		0.8319	0	0	12	1		2
138.91	138.90	0.8226	-2	2	12	1	0	4
142.06	142.12	0.8145	-6	4	4	2	1	1
142.10		0.8144	6	4	0	5		4
142.11		0.8144	-1	3	12	2		12
142.13		0.8143	-4	0	11	1		6
142.44		0.8136	2	4	10	1		2
143.80	143.92	0.8103	-5	7	4	1	0	4
144.05		0.8098	1	11	2	1		2
144.42	144.92	0.8089	-6	0	7	2	1	2
144.75		0.8082	3	9	4	1		4
145.01		0.8076	-5	1	9	1		3
145.20		0.8072	4	6	6	1		3
145.31		0.8069	1	9	7	2		6
147.75	147.72	0.8018	2	8	11	1	0	2
149.88	149.72	0.7976	0	8	9	1	0	2
151.85	151.74	0.7941	-5	5	8	1	0	2
159.12	161.28	0.7832	5	5	5	3	1	3
160.00		0.7821	0	4	12	9		8
161.06		0.7809	-3	3	12	9		24
161.09		0.7809	-2	6	11	1		2
161.24		0.7807	3	3	10	9		23
161.31		0.7806	0	6	11	9		24
161.39		0.7805	3	9	5	1		2
163.01		0.7788	-4	8	7	1		2
164.92		0.7770	-5	7	6	1		3

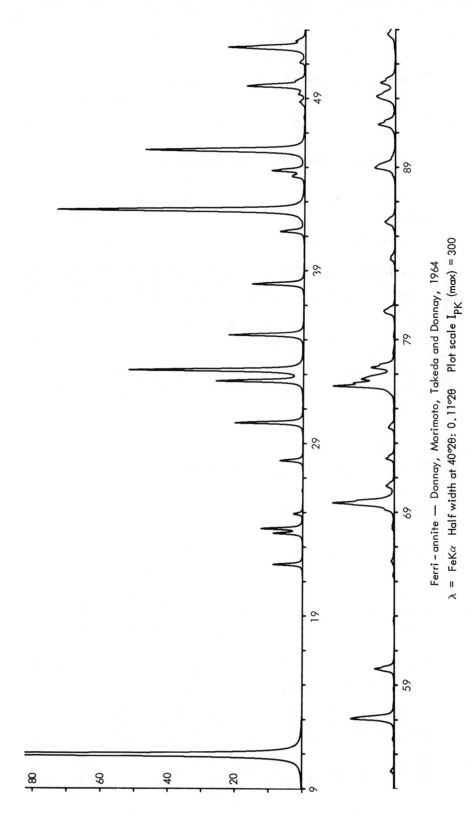

Ferri – annite — Donnay, Morimoto, Takeda and Donnay, 1964

λ = FeKα Half width at 40°2θ: 0.11°2θ Plot scale I$_{PK}$ (max) = 300

FERRI-ANNITE — DONNAY, MORIMOTO, TAKEDA, AND DONNAY, 1964

2THETA	PEAK	D	H	K	L	I(INT) 100	I(PK) 100	I(DS) 100
10.93	10.94	10.163	0	0	1	100	100	100
21.96	21.98	5.082	0	0	2	3	3	5
23.77	23.78	4.700	0	2	0	3	3	5
24.04	24.04	4.648	1	1	0	5	4	7
24.89	24.90	4.491	-1	1	1	1	1	2
27.98	27.98	4.005	1	1	1	3	2	5
30.16	30.16	3.721	-1	1	2	9	7	15
32.58	32.60	3.450	0	2	2	12	9	21
33.21	33.22	3.388	1	1	2	25	17	44
35.24	35.24	3.198	-1	1	3	11	7	20
38.20	38.20	2.959	0	0	4	8	5	16
41.25	41.26	2.748	-1	1	4	4	2	8
42.45	42.48	2.674	0	2	3	13	24	31
42.49		2.671	-1	3	1	28		63
44.43	44.46	2.560	0	2	4	5	3	13
44.79	44.83	2.541	1	3	1	9	16	23
45.91	45.98	2.482	0	0	5	19		48
45.98		2.479	-1	3	2	1	1	2
48.79	48.78	2.344	-2	2	1	1	1	3
49.23	49.24	2.324	2	0	2	7	6	20
49.68	49.70	2.304	2	2	1	4		10
49.73		2.302	1	3	3	1	1	3
50.02	50.00	2.290	-2	0	4	1	1	3
51.07	51.08	2.246	0	4	0	5	8	15
51.87	51.96	2.213	-2	2	4	10		29
51.96		2.210	2	2	2	1	1	3
52.34	52.34	2.195	0	4	2	1	0	2
53.98	53.98	2.133	0	4	3	6	4	20
57.02	57.04	2.028	1	3	5	3		10
57.09		2.026	-2	2	6	2	2	5
59.82	59.92	1.9413	-1	5	2	3	6	10
59.92		1.9383	-2	0	8	2		21
69.40	69.52	1.7003	-1	3	8	10	4	41
69.52		1.6978	0	1	7	1	1	3
69.70	69.66	1.6939	0	6	1	1		3
70.43	70.52	1.6786	3	1	3	1		5
70.52		1.6768	-3	1	3			

2THETA	PEAK	D	H	K	L	I(INT)	I(PK)	I(DS)
72.08	72.10	1.6453	1	1	5	1	1	5
72.12		1.6445	-2	4	3	1		3
73.86	73.90	1.6111	-2	4	2	1	1	4
73.93		1.6098	-1	5	1	1		3
76.28	76.30	1.5673	-3	1	3	10	6	45
76.32		1.5667	0	6	5	5		22
76.69	76.70	1.5602	-1	5	5	5	3	25
76.80		1.5585	-2	0	6	3		13
77.33	77.36	1.5493	3	3	1	2	2	9
77.37		1.5486	-3	3	3	2		8
77.39		1.5484	0	6	0	2		9
80.53		1.4977	-1	5	5	1		4
80.66	80.66	1.4957	3	3	0	2	1	7
83.63	83.64	1.4519	3	3	2	1		6
85.70	85.82	1.4233	0	6	3	1	0	6
85.80		1.4220	-3	3	6	1	1	6
85.83		1.4216	-1	5	4	1		6
88.93	88.96	1.3818	-2	0	6	5	2	28
89.05		1.3804	-4	0	7	2		14
91.45	91.46	1.3519	2	6	2	2	2	11
91.47		1.3517	3	6	4	4		21
92.97	93.10	1.3348	0	0	6	2	2	14
93.08		1.3335	-3	3	6	2		14
93.13		1.3330	-3	3	6	3		13
93.88	93.90	1.3248	-1	6	9	3		16
93.91		1.3246	-2	2	7	2		9
93.92		1.3244	-4	0	1	2	1	4
96.58		1.2967	-2	2	3	1		6
96.72	96.72	1.2952	4	0	6	2	1	14
98.27	98.32	1.2800	-2	2	8	2	0	5
99.26	99.26	1.2704	-2	8	3	1	0	6
99.27		1.2626	0	5	7	1	0	4
100.11	100.10	1.2500	-3	5	2	1	0	4
100.50	101.54	1.2352	4	2	0	1	0	3
103.20	103.40	1.2347	0	0	7	1		3
103.25		1.2192	1	4	7	1	0	4
105.12	105.16	1.2192	2	6	3	1	1	6
114.31	114.38	1.1521	-3	6	4	4		31

508

FERRI-ANNITE — DONNAY, MORIMOTO, TAKEDA, AND DONNAY, 1964

2THETA	PEAK	D	H	K	L	I(INT)	I(PK)	I(DS)
114.47		1.1511	-4	0	6	2		16
122.03	122.36	1.1066	4	0	4	2	1	13
122.35		1.1049	-2	6	6	3		27
127.94	128.18	1.0772	2	0	8	1	1	12
128.18		1.0761	-1	3	9	3		24
130.49	130.82	1.0659	3	3	6	2	1	22
130.73		1.0649	0	6	7	2		21
130.88		1.0643	-3	3	8	2		21
137.87	138.34	1.0373	4	0	5	2	1	15
138.33		1.0357	-2	6	7	3		29
141.95	142.08	1.0239	-5	3	1	2	1	24
142.08		1.0235	-4	6	2	3		24
142.15		1.0233	-1	9	1	3		25
144.38	144.52	1.0167	-5	3	3	1	0	5
144.52		1.0163	0	0	10	1		11
146.02	146.38	1.0122	5	3	0	4	1	37
146.27		1.0115	-4	6	3	4		38
146.31		1.0114	-1	9	2	4		36
146.49		1.0109	1	3	9	4		42
146.82		1.0100	-2	0	10	2		21
151.41	151.62	0.9989	4	6	1	2	0	21
151.64		0.9984	1	9	2	2		22
151.64		0.9984	-5	3	4	2		23
157.10	157.62	0.9876	3	3	7	4	1	46
157.62		0.9868	0	6	8	5		49
157.96		0.9862	-3	3	9	5		51

TABLE 25. CHLORITES

Variety	Corundophillite	Prochlorite	Ripidolite
Composition	$(Mg,Al,Fe^2,Cr)_3(Si, Al)_4 O_{10}(OH)_2(Mg,Al Fe^2,Fe^3,Cr)_3(OH)_6$	$(Mg,Fe,Al)_7(Si,Al)_4 O_{10}(OH)_8$	$(Mg,Fe,Al)_7(Si, Al)_4 O_{10}(OH)_8$
Source	Mochako District, Kenya	Washington, D. C., USNM 45875	Tazawa Mine, Japan
Reference	Steinfink, 1958a	Steinfink, 1958b, 1961, 1962a	Shirozu & Bailey, 1965
Cell Dimensions			
$\underline{a}$ Å	5.34	5.37	5.390
$\underline{b}$ Å	9.27	9.30	9.336
$\underline{c}$	14.36	14.25	14.166
α	90	90	90
β deg	97.36	96.28	90.0
γ	90	90	90
Space Group	C1	C2	C2/m
Z	2	2	2

TABLE 25. (cont.)

Variety	Corundophillite		Prochlorite		Ripidolite	
Site Occupancy						
	M_1, M_2 {Mg Fe Al	0.833 0.067 0.100	M_1 Mg	1.0	M_1 {Fe Mg Al	0.67 0.16 0.16
	M_3, M_4 {Mg Fe Al	0.833 0.067 0.100	M_2 {Mg Fe	0.75 0.25	M_2 {Fe Mg Al	0.67 0.16 0.16
			M_3 Fe	0.90		
	M_5, M_6 {Mg Fe Al	0.833 0.067 0.100	M_4 {Al Fe	0.75 0.25	M_3 {Fe Mg Al	0.472 0.264 0.264
			M_5 {Al Fe	0.75 0.25		
	T_1 {Si T_2 {Al	0.625 0.375	M_6 Mg	0.75	M_4 {Fe Mg Al	0.472 0.264 0.264
			T_1 {Si Al	0.30 0.70	T_1 {Si Al	0.67 0.33
			T_2 {Si Al	0.80 0.20		
Method	2-D, film		2-D, film		3-D, film	
R & Rating	0.13 (av.) (2)		0.13 (av.) (3)		0.087 (1)	
Cleavage and habit	Tabular		{001} perfect ——————— Scaly			
Comment	Minor Cr ignored					
μ	148.4		156.4		167.6	
ASF	0.1096×10^{-1}		0.6444×10^{-1}		0.1202	
Abridging factors	1:1:0.5		1:1:0.5		1:1:0.5	

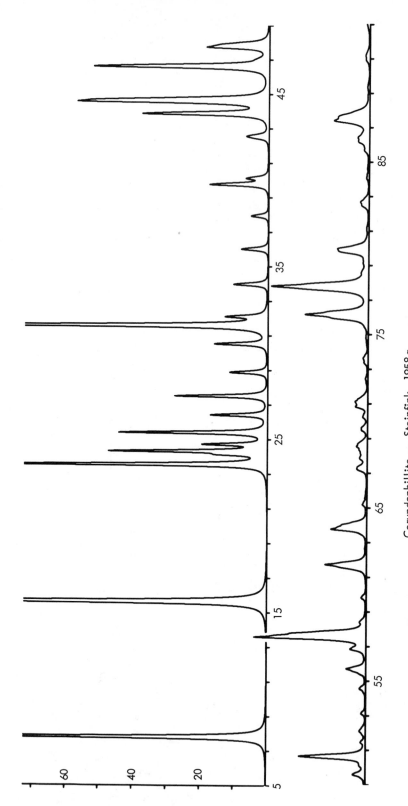

Corundophillite — Steinfink, 1958a

λ = FeKα Half width at 40°2θ: 0.15°2θ Plot scale I_{PK} (max) = 200

CORUNDOPHILLITE – STEINFINK, 1958 d

2THETA	PEAK	D	H	K	L	I(INT)	I(PK)	I(DS)
7.79	7.80	14.242	0	0	1	61	64	58
15.63	15.64	7.121	0	0	2	100	100	100
23.53	23.54	4.747	0	0	3	48	44	52
24.30	24.32	4.598	1	1	0	14	24	16
24.30		4.598	-1	1	1	11		12
24.70	24.72	4.526	1	1	-1	7	10	8
24.70		4.526	-1	1	2	3		3
25.37	25.38	4.407	0	2	1	10	22	11
25.37		4.407	-2	0	1	15		16
26.39	26.40	4.240	1	1	1	9	9	10
27.48	27.50	4.075	-1	1	3	12	14	13
27.48		4.075	1	1	-2	4		5
28.86	28.86	3.885	0	2	2	7	6	8
30.49	30.50	3.681	-2	0	2	9	8	9
30.49		3.681	1	1	2	2		2
31.55	31.56	3.560	0	0	4	68	53	80
32.08	32.08	3.503	-1	1	4	7	6	8
33.94	33.96	3.316	1	1	-3	6	5	8
35.99	36.00	3.133	0	2	3	5	4	6
37.91	37.92	2.980	-2	0	3	4	3	5
39.73	39.74	2.848	0	0	5	13	9	16
40.10	40.10	2.824	0	2	4	3	3	4
40.10		2.824	-2	0	4	1		1
42.43	42.54	2.675	-1	3	1	2	3	3
42.53		2.669	1	3	0	2		3
42.59		2.665	-3	1	1	1		1
43.84	43.84	2.593	1	3	-1	10	19	13
43.84		2.593	-1	3	2	9		13
43.84		2.593	-2	0	2	9		12
44.55	44.56	2.554	1	3	1	18	28	25
44.55		2.554	-3	1	2	18		25
44.55		2.552	-1	3	3	2		3
44.70		2.552	1	1	-3	18		25
44.70		2.545	-2	0	5	14		20
46.55	46.60	2.450	-2	0	5	18	26	20
46.59		2.448	1	3	2	14		20
46.59		2.448	-3	1	3	14		20
47.71	47.72	2.393	-1	3	3	8	9	11
47.71	47.71	2.393	-1	3	3	6		9
49.57	49.56	2.309	-1	-1	6	1	2	2
50.52	50.60	2.268	-2	0	6	6	10	9
50.60		2.265	1	-3	3	6		9
50.60		2.265	-1	1	6	6		9
55.57	55.70	2.077	-2	2	0	2	3	3
55.69		2.072	0	3	3	2		3
55.69		2.072	-1	3	3	2		3
56.82	56.84	2.035	0	0	7	3	2	5
57.47	57.48	2.013	-1	3	5	14	17	22
57.47		2.013	-1	3	5	14		23
61.54	61.70	1.8920	-2	0	7	4	6	7
61.70		1.8877	1	3	3	5		8
61.70		1.8877	-1	3	4	4		7
63.74	63.74	1.8334	-1	3	6	5	5	8
63.74		1.8334	-1	3	6	5		9
68.34	68.34	1.7235	-2	0	7	1	1	2
68.52	68.52	1.7194	-1	3	7	1	2	2
70.81	70.82	1.6709	-1	1	6	2	2	4
70.81		1.6709	-1	3	6	2		3
71.11	71.10	1.6647	-2	0	6	2	2	4
75.92		1.5737	-2	0	8	2		17
76.14	76.14	1.5699	-1	3	7	8	9	17
76.14		1.5699	-1	3	7	8		17
77.59	77.74	1.5450	0	6	0	12	14	26
77.75		1.5424	-3	3	1	13		26
77.75		1.5424	-3	3	1	12		26
79.75	79.94	1.5099	0	6	2	3	5	6
79.75		1.5099	-6	0	2	3		6
79.83		1.5086	0	6	2	2		5
79.83		1.5086	-3	3	3	2		5
79.98		1.5063	3	3	1	2		5
79.98		1.5063	-3	3	1	2		4
85.64	85.66	1.4242	3	3	-3	1	1	5
86.15	86.18	1.4173	0	0	10	1	2	4
86.16		1.4172	-6	0	4	1		2
86.45	86.40	1.4134	-1	3	3	1	2	2

CORUNDOPHILLITE - STEINFINK, 1958a

2THETA	PEAK	D	H	K	L	I(INT)	I(PK)	I(DS)
86.45		1.4134	3	-3	3	1		2
87.37	87.40	1.4014	-1	3	9	7	5	17
87.37		1.4014	-1	-3	9	8		17
87.73	87.58	1.3969	-2	0	8	7	5	17
93.66	93.72	1.3272	-2	-6	2	3	2	6
93.66		1.3272	-2	6	2	3		7
93.96	93.92	1.3240	4	0	0	1	2	7
94.26		1.3207	2	-6	1	1		3
95.58	95.62	1.3068	-2	-6	3	2	1	4
95.58		1.3068	-2	6	3	2		4
95.95	95.86	1.3030	4	0	1	2		4
96.59	96.64	1.2966	2	-6	2	3	3	7
96.59		1.2966	2	6	2	2		6
96.60		1.2964	2	-6	4	2		6
104.19	104.52	1.2268	-4	0	11	2	2	5
104.41		1.2250	-2	0	6	1		3
104.52		1.2241	-4	0	10	2		5
104.52		1.2241	1	-3	10	2		5
107.97	108.04	1.1967	1	3	6	2	1	5
107.97		1.1967	-2	-6	6	2		3
108.59		1.1921	-2	6	4	2		4
109.30	109.60	1.1868	4	0	12	2	1	4
110.09	110.28	1.1810	0	0	7	1		4
110.30		1.1795	-4	0	5	1		3
110.30		1.1795	2	6	5	2		4
116.81	117.26	1.1364	2	-6	11	1	1	4
116.81		1.1364	1	-3	11	1		4
117.17		1.1342	1	3	8	1		3
117.47		1.1324	-4	0	6	1		4
117.47		1.1324	2	-6	6	1		3
122.81	122.88	1.1025	-2	6	8	1	1	7
122.81		1.1025	-2	-6	6	2		8
123.67		1.0980	-2	6	11	3		8
134.87	135.28	1.0482	-3	-3	1	2	1	6
134.87		1.0482	-3	3	10	2		7
135.15		1.0472	0	6	10	2		5
135.15		1.0472	0	-6	10	2		5
135.88	135.88	1.0445	3	3	9	2		7
135.88		1.0445	3	-3	9	2		6
137.59	138.18	1.0383	-4	0	10	4	1	13
138.20		1.0362	-2	6	8	4		12
138.20		1.0362	2	-6	8	2		13
144.19	144.30	1.0173	0	0	14	2	0	8
148.09	148.82	1.0068	1	9	1	2	1	7
148.09		1.0068	1	-9	1	2		5
148.66		1.0053	-4	6	0	2		7
148.66		1.0053	-4	-6	0	2		6
148.68		1.0053	-5	-3	3	1		5
148.68		1.0053	-5	3	3	1		6
152.00	153.84	0.9976	-1	-9	2	2	1	8
152.00		0.9976	1	9	2	2		8
152.52		0.9965	-5	3	4	2		9
152.52		0.9965	-5	-3	4	3		9
152.72		0.9961	4	6	1	2		9
152.72		0.9961	4	-6	1	2		8
153.80		0.9938	-1	-9	3	2		6
153.80		0.9938	1	9	3	2		7
154.01		0.9934	-3	3	12	1		3
154.01		0.9934	-3	-3	12	1		3
154.15		0.9931	-3	-3	12	1		5
154.17		0.9931	-2	0	14	2		6
154.17		0.9931	-2	0	14	2		7
154.56		0.9923	-4	6	4	1		3
154.56		0.9923	0	6	4	1		3
155.08		0.9913	0	-6	11	1		2
155.13		0.9912	-4	0	11	1		5
155.13		0.9912	1	3	11	1		5
155.19		0.9911	5	-3	1	2		8
155.19		0.9911	5	3	1	2		8
155.85		0.9899	3	-3	10	1		3
155.85		0.9899	3	3	10	1		3

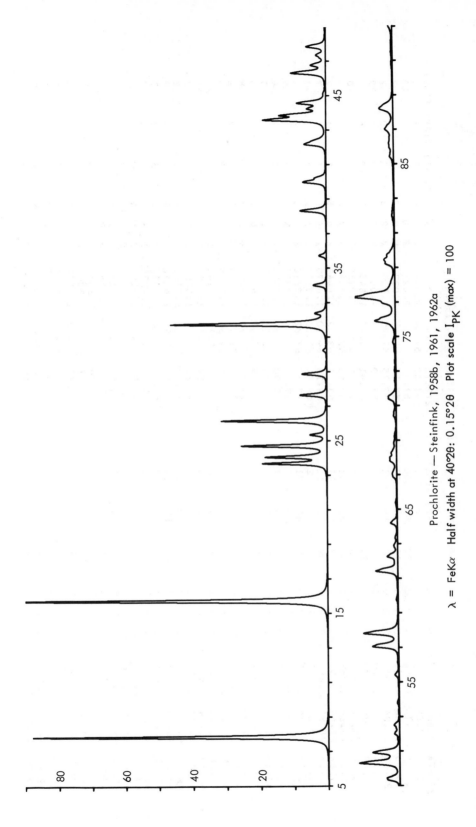

Prochlorite — Steinfink, 1958b, 1961, 1962a

λ = FeKα   Half width at 40°2θ: 0.15°2θ   Plot scale I_{PK} (max) = 100

PROCHLORITE – STEINFINK, 1958b, 1961, 1962a

2THETA	PEAK	D	H	K	L	I(INT)	I(PK)	I(DS)
7.84	7.84	14.164	0	0	1	84	88	80
15.71	15.72	7.082	0	0	2	100	100	100
23.66	23.68	4.721	0	0	3	21	19	23
24.03	24.04	4.650	-1	0	1	19	19	21
24.14		4.629	-1	1	1	2		2
24.68	24.70	4.529	-1	1	1	29	26	32
25.31	25.32	4.418	0	2	1	5	5	6
26.13	26.14	4.282	-1	1	2	36	31	41
27.62	27.64	4.055	-1	1	2	10	8	11
28.84	28.86	3.887	0	2	1	9	8	11
30.19	30.20	3.717	-1	1	3	1	1	2
31.73	31.74	3.541	0	2	4	59	46	71
32.35	32.36	3.474	0	0	3	4	4	5
33.98	33.98	3.313	-1	1	3	5	4	7
35.69	35.70	3.159	0	2	3	3	2	4
38.30	38.30	2.951	-1	1	3	11	8	15
39.96	39.98	2.833	-1	1	4	10	7	13
40.19	40.18	2.817	0	2	5	4	4	5
42.15	42.16	2.692	0	2	1	9	7	13
43.58	43.58	2.608	-1	1	4	27	19	38
43.84	43.84	2.593	-2	0	4	18	14	25
44.22	44.22	2.572	-2	0	1	6	6	9
44.54	44.56	2.554	-1	3	2	13	9	18
46.29	46.30	2.463	-1	3	0	16	10	24
46.73	46.74	2.441	-2	0	2	6	5	9
47.17	47.32	2.419	0	2	3	2	3	2
47.32		2.412	-1	3	2	3		5
47.82	47.84	2.388	-1	1	5	9	6	14
49.31	49.34	2.320	-1	3	1	3	4	5
49.33		2.320	-2	0	5	1		2
49.44	49.42	2.315	-1	3	3	2		4
50.29	50.30	2.278	-1	3	1	19	12	30
50.89	50.90	2.253	-2	0	3	12	8	18
50.95		2.251	2	2	0	1		2
51.98	51.98	2.209	0	4	1	2	2	4
52.49	52.50	2.189	-1	1	6	2	2	3
55.38	55.38	2.083	-1	3	4	2	1	3

2THETA	PEAK	D	H	K	L	I(INT)	I(PK)	I(DS)
57.03	57.04	2.028	-2	2	4	5	8	9
57.04		2.027	0	0	7	8		13
57.16		2.023	-2	0	3	1		2
57.75	57.80	2.005	-1	2	5	17	10	30
57.80		2.003	0	3	5	12		22
61.41	61.42	1.8958	-1	3	5	6	6	10
62.24	62.24	1.8728	-2	0	4	1	3	2
62.78	62.88	1.8584	0	2	5	4	1	7
63.31	63.32	1.8445	-2	0	4	1	1	3
64.18	64.20	1.8220	2	0	5	3	2	5
66.84	67.00	1.7575	-1	3	6	2	1	4
67.72	67.82	1.7372	-3	1	1	1	2	3
67.80		1.7355	-2	2	6	3		3
68.00	67.98	1.7310	-1	5	1	2	2	3
68.26	68.26	1.7253	-2	4	2	1	1	8
70.39	70.40	1.6794	-1	3	6	4	2	3
71.37	71.38	1.6595	-2	0	7	1	1	2
71.51	71.52	1.6565	-1	5	6	2	2	28
74.65	74.66	1.5965	-1	3	4	13	6	21
75.90	75.90	1.5740	0	6	7	10		39
77.29	77.30	1.5500	-1	3	0	18	12	12
77.29		1.5500	-1	3	1	5		9
79.20	79.22	1.5186	0	6	2	4	3	7
79.48	79.44	1.5142	-3	3	2	3	3	6
79.76	79.70	1.5097	-3	3	3	2	2	4
85.40	85.42	1.4274	0	0	10	4	1	4
85.96	85.98	1.4199	0	6	0	3	1	16
86.22	86.48	1.4164	-3	0	9	2	1	32
86.51		1.4126	-2	3	2	2	2	17
87.00	87.00	1.4062	-1	6	1	7	1	4
88.20	88.22	1.3910	-1	3	7	13	4	5
93.37	93.38	1.3304	-2	6	4	7	2	8
94.78	94.80	1.3152	-4	0	4	1	1	6
95.84	95.88	1.3042	-2	3	5	2	2	3
95.86		1.3040	-2	6	2	3		
96.61	96.66	1.2964	-4	6	4	2	1	
96.68		1.2957	0	6	6	1		3

516

PROCHLORITE – STEINFINK, 1958ℓ, 1961, 1962α

2THETA	PEAK	D	H	K	L	I(INT)	I(PK)	I(DS)
97.51	97.60	1.2874	-3	3	7	1	1	3
103.65	103.70	1.2314	2	6	1	1	0	3
104.42	104.48	1.2249	1	3	10	1	1	4
106.77	106.78	1.2059	4	0	4	1	0	4
108.32	108.36	1.1941	-2	6	6	3	1	10
110.18	110.24	1.1804	0	0	12	1	0	4
115.14	115.24	1.1468	-2	6	7	1	0	4
116.39	116.72	1.1390	2	6	6	2	1	5
116.78		1.1366	1	3	11	2		7
118.47	118.56	1.1265	-4	0	8	1	1	5
118.55		1.1260	-2	0	12	1		3
121.25	121.30	1.1108	4	6	6	2	0	8
123.65	123.72	1.0981	-2	3	8	4	1	12
133.63	133.78	1.0530	3	3	9	3	1	12
136.69	137.40	1.0415	2	6	8	7	1	25
137.58		1.0383	-3	0	11	5		19
137.58		1.0383	-2	0	12	1		4
140.25	140.32	1.0293	-1	3	13	2	1	8
140.31		1.0291	-4	0	10	3		12
146.17	147.06	1.0117	0	0	14	2	1	9
146.34		1.0113	4	6	0	3		13
146.68		1.0104	1	9	1	4		15
147.36		1.0086	-5	3	3	6		23
149.70	150.84	1.0028	4	6	1	3	1	12
149.93		1.0023	4	9	6	1		5
150.34		1.0013	1	3	2	2		9
151.14		0.9995	5	3	1	3		9
151.68		0.9983	-5	3	4	1		12
152.09		0.9974	3	3	10	2		9
152.70		0.9961	-1	9	3	3		8

517

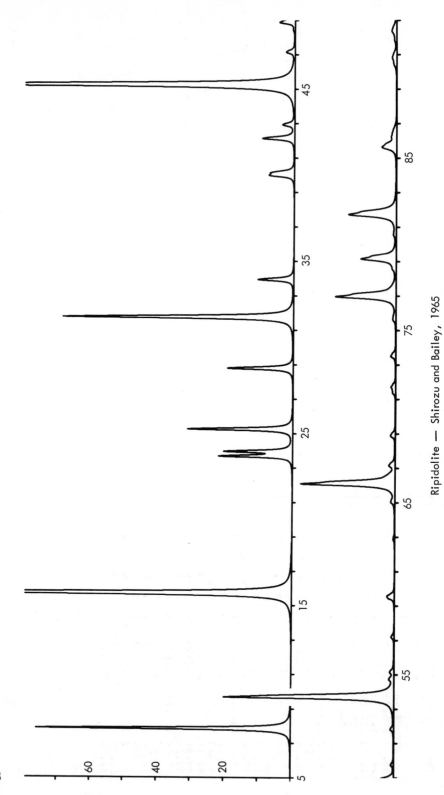

Ripidolite — Shirozu and Bailey, 1965

λ = FeKα Half width at 40°2θ: 0.15°2θ Plot scale I$_{PK}$ (max) = 200

518

RIPIDOLITE - SHIROZU AND BAILEY, 1965

2THETA	PEAK	D	H	K	L	I(INT)	I(PK)	I(DS)
7.84	7.84	14.166	0	0	1	36	38	34
15.71	15.72	7.083	0	0	2	100	100	100
23.66	23.68	4.722	0	0	3	12	11	13
23.94	23.94	4.668	0	1	0	7	10	8
23.94		4.668	1	1	0	4		4
25.22	25.24	4.433	1	1	1	11	16	12
25.22		4.433	-1	1	1	6		7
28.76	28.78	3.898	0	2	1	6	10	7
28.76		3.898	-1	1	2	6		7
31.74	31.74	3.541	0	2	2	44	34	54
33.92	33.92	3.320	1	1	3	5	5	3
39.96	39.98	2.833	-1	1	3	5	4	6
40.13	40.12	2.821	-1	3	1	5	4	7
42.10	42.10	2.695	2	0	0	3	5	4
45.20	45.20	2.519	-1	3	2	1	57	2
45.20		2.519	-1	3	2	59		89
45.20		2.519	0	3	3	30		46
47.11	47.12	2.422	-1	1	5	1	1	2
48.86	48.86	2.341	-1	3	3	1	2	2
48.86		2.341	-1	3	3	1		2
53.66	53.66	2.145	-1	3	4	8	25	13
53.66		2.145	-1	3	4	22		37
53.66		2.145	-2	0	4	11		20
66.06	66.06	1.7759	-1	3	6	4	14	6
66.06		1.7759	-1	3	6	16		33
66.06		1.7759	-2	0	6	8		17
73.48	73.48	1.6183	-1	3	7	1	1	3
76.94	76.94	1.5560	0	6	0	1	9	3
76.94		1.5560	0	6	2	7		17
79.13	79.14	1.5198	0	6	2	14	5	34
79.13		1.5197	-3	3	2	4		10
79.13		1.5197	-3	3	2	4		11
81.70	81.72	1.4799	1	3	8	9	7	22
81.70	81.70	1.4799	-1	3	3	3		9
81.70		1.4799	-2	0	8	2		5
81.70		1.4799	-2	0	8	4		11
85.61	85.62	1.4246	0	6	4	2	2	6
85.61		1.4245	3	3	4	2		6
85.61		1.4245	-3	3	4	2		6
86.21	86.20	1.4166	0	0	10	7	1	4
93.98	94.00	1.3238	2	6	2	3	3	21
93.98		1.3238	-4	0	2	1		10
96.33	96.36	1.2992	0	6	6	2	1	4
96.33		1.2992	-3	3	6	2		5
96.33		1.2992	2	3	6	1		4
100.45	100.50	1.2594	-2	3	4	2	1	5
100.45		1.2594	-2	6	4	2		5
100.46		1.2594	-4	0	4	1		3
101.06	101.06	1.2539	-1	6	10	3	1	10
101.06		1.2539	-2	6	10	2		5
111.61	111.64	1.1703	2	0	6	3	2	19
111.61		1.1703	-4	0	6	2		9
127.07	127.18	1.0813	-1	3	12	3	1	6
129.03	129.10	1.0723	-2	6	8	4	1	5
129.03		1.0723	-2	6	8	4		21
129.03		1.0723	4	0	8	3		12
135.06	135.16	1.0475	-3	3	10	1	1	14
135.06		1.0475	1	3	10	1		16
135.06		1.0475	-3	3	10	1		14
143.71	144.44	1.0186	4	0	0	5	1	5
143.72		1.0186	5	3	6	5		5
143.72		1.0186	0	0	6	3		6
146.13	147.60	1.0119	0	0	14	6	1	6
147.50		1.0083	-1	9	2	6		25
147.50		1.0083	-4	3	2	2		26
147.51		1.0082	5	2	2	7		28
154.68	155.06	0.9921	0	6	11	1	0	3
154.69		0.9921	3	3	11	1		4
154.69		0.9921	-3	3	11	1		4
162.83	163.28	0.9790	-1	9	4	7	1	31

519

RIPIDOLITE - SHIROZU AND BAILEY, 1965

2THETA	PEAK	D	H	K	L	I(INT)	I(PK)	I(DS)
162.83		0.9790	-1	9	4	6		26
162.84		0.9789	4	6	4	7		33
162.84		0.9789	-4	6	4	6		25
162.85		0.9789	5	3	4	7		31
162.85		0.9789	-5	3	4	7		30
164.99	163.32	0.9764	2	6	10	1	1	6
164.99		0.9764	-2	6	10	6		29
165.00		0.9763	4	0	10	4		17

TABLE 26. SEPTECHLORITES

Variety	Amesite (2 layer)	Cronstedtite (1 layer)	Al-Serpentine
Composition	$(Mg_2Al)(SiAl)O_5(OH)_4$	$Fe_3^2(Fe^3,Si)O_4(OH)_5$	$(Mg,Al,Fe,Ca,K)_{2.91}(Si,Al)_2O_5(OH)_4$
Source	Saranovskoye Dist. USSR, USNM 103312	Wheal Maudlin, England	Lake Superior, Minn.
Reference	Steinfink and Brunton, 1956	Steadman & Nuttall, 1963	Jahanbagloo & Zoltai, 1968
Cell Dimensions			
$\underline{a}$ Å	5.31	5.490	5.294
$\underline{b}$ Å	5.31	5.490	5.294
$\underline{c}$	14.04	7.085	63.99
α	90	90	90
β deg	90	90	90
γ	120	120	120
Space Group	$P6_3$	$P31m$	$P3_1$
Z	2	1	9

TABLE 26. (cont.)

Variety	Amesite (2 layer)	Cronstedtite (1 layer)	Al-Serpentine
Site Occupancy		$T_1 \begin{cases} Si & 0.5 \\ Fe & 0.5 \end{cases}$	$M_1 \begin{cases} Mg & 0.67 \\ Al & 0.20 \end{cases}$ $M_2 \begin{cases} Mg & 0.67 \\ Al & 0.20 \end{cases}$ $M_3 \begin{cases} Mg & 0.67 \\ Al & 0.20 \end{cases}$ $T \begin{cases} Si & 0.70 \\ Al & 0.30 \end{cases}$
Method	2-D, film	2-D, film	3-D, film, counter
R & Rating	0.14 (av.) (3)	0.15 (av.) (3)	0.102 (2)
Cleavage and habit	{0001} Hexagonal prisms		Platy
Comments		B estimated	Minor Fe,Ca, and K ignored
μ	148.5	190.0	71.4
ASF	0.5285×10^{-1}	0.1929	0.8395×10^{-1}
Abridging factors	1:1:0.5	0.5:0:0	0.5:0:0:0

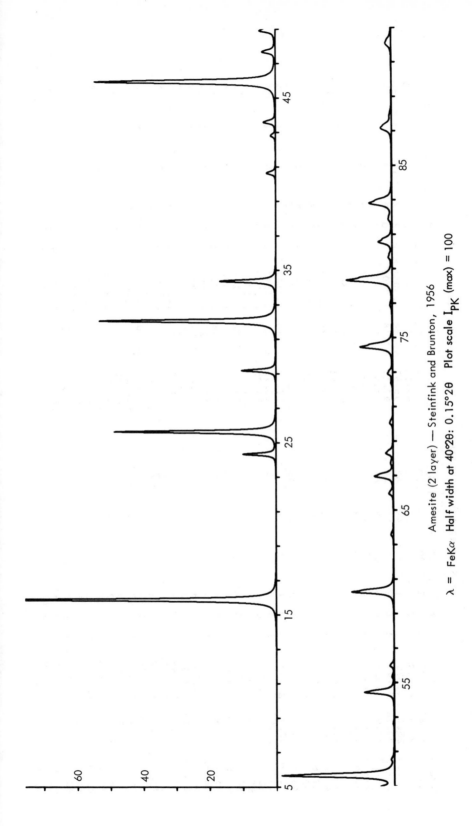

Amesite (2 layer) — Steinfink and Brunton, 1956

$\lambda = FeK\alpha$ Half width at $40°2\theta$: $0.15°2\theta$ Plot scale I_{PK} (max) = 100

AMESITE (2 LAYER) — STEINFINK AND BRUNTON, 1956

2THETA	PEAK	D	H	K	L	I(INT)	I(PK)	I(DS)
15.85	15.86	7.020	0	0	2	100	100	100
24.30	24.32	4.599	1	1	0	11	10	12
25.59	25.60	4.370	1	1	1	56	49	62
29.15	29.16	3.847	1	0	2	13	10	14
32.02	32.02	3.510	0	0	4	69	53	81
34.33	34.34	3.280	1	1	3	22	17	27
40.60	40.60	2.790	1	1	3	4	3	5
42.76	42.78	2.655	1	0	1	2	2	3
43.56	43.58	2.609	1	1	2	6	4	8
45.88	45.90	2.483	1	1	5	86	54	121
47.65	47.66	2.397	1	1	6	6	4	9
48.87	48.88	2.340	0	0	3	7	5	10
49.56	49.58	2.309	1	1	4	56	34	83
50.50	50.50	2.269	2	0	0	1	1	2
54.41	54.42	2.117	1	2	1	16	9	25
55.31	55.32	2.086	1	0	6	1	1	4
55.95	55.96	2.064	2	1	3	2	1	4
60.23	60.24	1.9292	1	1	5	24	13	40
63.54	63.54	1.8385	1	0	7	2	1	3
65.93	—	1.7790	2	0	2	3	1	5
66.93	66.94	1.7555	1	1	6	11	6	20
66.95	—	1.7550	0	0	8	1	1	2
67.69	67.70	1.7381	2	1	1	1	1	2
68.27	68.28	1.7249	2	2	0	4	2	7
70.02	70.02	1.6872	1	1	1	1	1	2
72.90	72.90	1.6294	2	2	2	2	1	4
74.44	74.44	1.6004	2	1	3	3	2	5
76.85	76.86	1.5576	1	2	7	23	10	45
78.32	78.32	1.5329	3	1	2	1	1	1
79.65	79.66	1.5115	2	0	4	34	14	70
80.54	80.54	1.4976	2	0	7	2	1	5
81.84	81.86	1.4779	2	1	1	10	4	22
82.78	82.78	1.4641	1	2	5	5	1	2
87.11	87.14	1.4048	3	1	8	19	7	41
87.17	—	1.4040	0	0	10	9	3	20
92.06	92.08	1.3450	1	1	9	6	2	15

2THETA	PEAK	D	H	K	L	I(INT)	I(PK)	I(DS)
93.64	93.66	1.3275	2	2	0	1	1	3
94.18	94.18	1.3216	2	2	1	1	1	3
95.82	95.84	1.3044	2	2	2	12	3	30
97.15	97.18	1.2909	2	0	9	3	1	4
98.04	98.08	1.2822	3	0	6	4	1	8
98.57	98.58	1.2771	2	2	3	2	1	10
99.30	99.28	1.2702	3	1	1	4	1	5
100.96	100.98	1.2549	1	3	10	5	0	4
102.51	102.54	1.2411	3	1	3	2	1	14
103.74	103.76	1.2305	2	2	5	5	1	4
107.52	107.62	1.2001	1	2	6	1	1	14
107.71	—	1.1987	0	0	8	1	1	3
111.65	111.72	1.1700	3	1	12	4	0	4
112.94	113.08	1.1612	2	0	11	2	1	3
113.93	114.00	1.1546	3	1	5	5	1	15
114.59	—	1.1545	1	0	8	4	—	3
120.32	120.32	1.1503	2	1	11	2	0	11
121.95	122.00	1.1160	2	0	7	7	1	5
129.41	129.48	1.1070	2	1	12	3	1	20
132.21	132.28	1.0707	1	2	8	9	—	10
138.44	138.54	1.0587	3	0	9	11	1	29
146.46	146.62	1.0353	2	0	10	7	1	37
149.69	150.62	1.0110	3	0	14	3	1	23
150.51	—	1.0029	0	4	1	3	1	9
150.51	—	1.0009	1	1	1	3	—	10
152.09	154.20	1.0009	4	4	7	1	1	10
154.02	—	0.9974	4	1	2	12	—	3
154.02	—	0.9934	4	4	5	12	—	42
157.13	—	0.9934	3	2	5	1	—	42
157.13	—	0.9876	2	1	9	1	—	3
157.23	—	0.9876	1	3	9	1	—	3
157.23	—	0.9874	3	1	9	1	—	2
161.18	161.02	0.9874	1	4	3	5	0	18
161.18	—	0.9812	4	1	3	5	—	18
163.97	—	0.9775	2	0	13	4	—	16

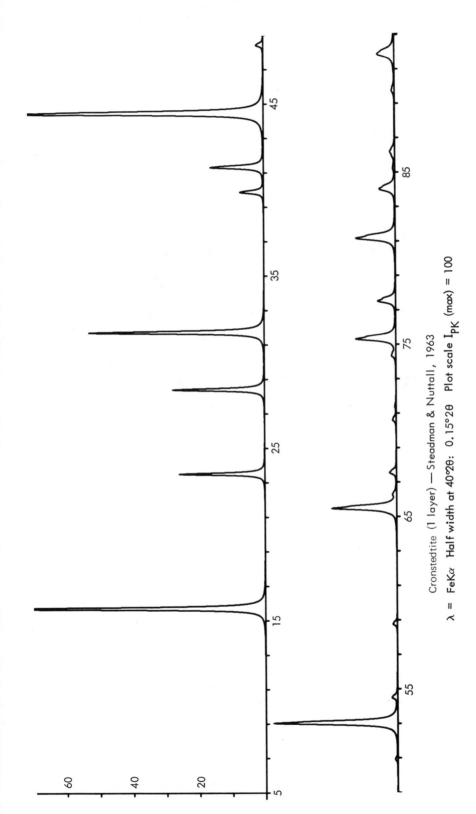

Cronstedtite (1 layer) — Steadman & Nuttall, 1963

λ = FeKα Half width at 40°2θ: 0.15°2θ Plot scale I$_{PK}$ (max) = 100

CRONSTEDTITE (1 LAYER) - STEADMAN AND NUTTALL, 1963

2THETA	PEAK	D	H	K	L	I(INT)	I(PK)	I(DS)
15.71	15.72	7.085	0	0	1	80	100	48
23.49	23.50	4.754	1	0	0	23	26	16
28.39	28.40	3.948	1	0	0	27	28	20
31.72	31.72	3.542	0	0	2	54	53	43
39.85	39.86	2.841	1	0	1	8	7	8
41.30	41.30	2.745	1	0	1	19	16	18
44.44	44.46	2.560	1	0	1	90	80	90
44.44		2.560	-1	-1	1	9		9
48.39	48.40	2.362	-1	0	1	3	2	3
50.87	50.88	2.254	2	0	0	3	1	1
52.99	53.00	2.170	2	0	1	27	37	31
52.99		2.170	-1	-1	2	25		29
54.47	54.48	2.115	-1	-1	0	2	2	3
58.73	58.74	1.9740	1	0	2	2	1	3
65.46	65.46	1.7903	-1	-1	3	28	20	40
65.46		1.7903	1	0	3	4		5
66.25	66.26	1.7712	0	0	4	1	1	2
67.52	67.52	1.7419	2	-1	1	2	2	2
67.52		1.7419	-2	-1	1	2		3
70.59	70.60	1.6754	2	0	3	2	1	3
71.35	71.36	1.6598	1	0	4	1	0	1
74.31	74.32	1.6026	-2	-1	2	1	1	1
74.31		1.6026	-2	-1	0	1		2
75.29	75.30	1.5848	3	0	1	23	12	37
77.49	77.50	1.5466	-1	0	4	10	5	17
81.14	81.16	1.4883	1	-1	4	5	12	9
81.14		1.4883	-1	1	4	19		35
84.00	84.00	1.4467	3	0	2	11	5	20
85.20	85.20	1.4301	0	-1	4	1	1	1
85.92	86.16	1.4203	-2	-1	3	1	2	3
86.18		1.4170	0	0	5	3		5
89.70	89.72	1.3725	2	2	0	2	1	4
91.84	91.86	1.3474	-2	2	1	3	5	7
91.84		1.3474	2	-2	1	12		23
94.71	94.72	1.3160	3	0	3	6	2	13
96.61	96.62	1.2964	3	-1	1	1	1	2
96.61		1.2964	-3	-1	1	1		1
98.29	98.32	1.2798	-2	2	2	4	2	9
98.29		1.2798	-2	-2	2	2		4
100.49	100.50	1.2591	-1	-1	5	1	1	3
100.49		1.2591	-1	1	5	3		7
103.12	103.16	1.2358	3	-1	5	1	0	2
105.36	105.40	1.2172	2	0	5	1	0	3
109.32	109.36	1.1867	-2	2	3	2	2	4
109.32		1.1867	-2	-2	3	9		21
114.44	114.48	1.1513	3	1	4	1	0	1
126.31	126.40	1.0849	2	2	4	10	3	27
126.31		1.0849	-2	-2	4	2		6
126.35		1.0847	-1	-1	6	5		13
126.35		1.0847	-2	1	6	4		13
132.50	132.86	1.0576	3	0	5	17	2	3
132.80		1.0563	0	0	6	1		51
136.42	136.62	1.0425	-3	-2	5	2	0	2
137.81	137.92	1.0375	-4	-1	2	1	0	5
141.10	141.20	1.0266	-4	-1	0	5	3	15
141.10		1.0266	4	0	1	20		63
146.03	146.12	1.0121	0	0	7	3	0	9
152.91	153.12	0.9957	-4	-1	2	14	1	45
152.91		0.9957	-4	1	2	7		23
155.66		0.9902	3	-2	3	1		3
155.66		0.9902	-3	-2	3	1		3
155.81		0.9900	-3	0	7	1		2
157.48	157.72	0.9870	4	0	4	2	1	6
157.56		0.9868	2	-1	6	1		2
157.56		0.9868	-2	-1	6	1		2
158.15		0.9859	-2	-2	5	3		11
158.15		0.9859	2	2	5	7		23

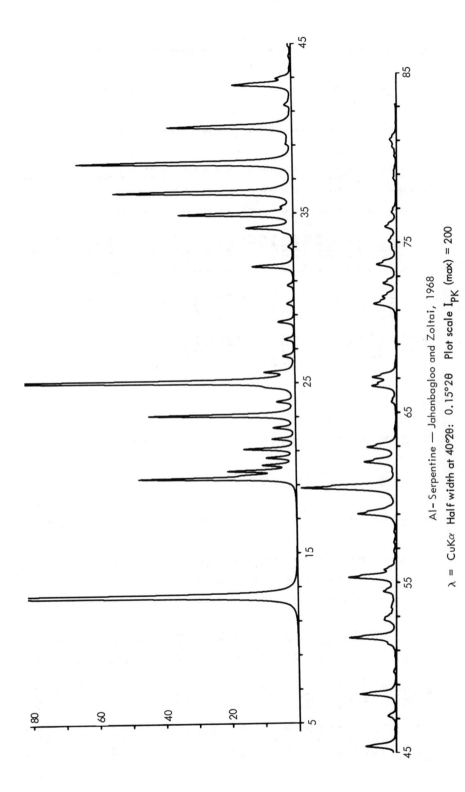

Al – Serpentine — Jahanbagloo and Zoltai, 1968

λ = CuKα   Half width at 40°2θ: 0.15°2θ   Plot scale I_{PK} (max) = 200

527

AL-SERPENTINE – JAHANBAGLOO AND ZOLTAI, 1968

2THETA	PEAK	D	H	K	L	I(INT)	I(PK)	I(DS)
12.44	12.44	7.110	0	0	9	100	100	100
19.34	19.36	4.586	1	0	0	21	24	21
19.39		4.574	0	1	1	5		5
19.79	19.80	4.483	1	0	3	5	10	5
19.79		4.483	0	1	4	5		5
20.13	20.14	4.408	1	0	4	5	5	5
20.56	20.56	4.317	1	0	5	4	4	4
21.07	21.08	4.213	0	1	6	4	8	4
21.07		4.213	1	0	6	4		4
21.66	21.68	4.099	1	0	7	4	4	4
22.33	22.34	3.978	0	1	8	3	3	3
23.06	23.06	3.854	1	0	9	13	22	13
23.06		3.854	0	1	9	13		13
23.85	23.86	3.727	0	0	10	64	3	3
25.03	25.04	3.555	0	1	12	2	54	65
25.60	25.60	3.477	1	0	12	2	5	2
25.60		3.477	0	1	12	2		2
26.54	26.56	3.355	0	1	13	2	2	2
27.53	27.54	3.237	1	0	14	2	2	2
28.55	28.56	3.123	1	0	15	2	2	2
28.55		3.123	0	1	15	2		2
29.61	29.62	3.014	0	0	16	1	1	1
30.70	30.72	2.909	0	1	17	1	1	1
31.82	31.84	2.810	1	0	18	4	6	4
31.82		2.810	1	1	18	4		4
32.97	32.98	2.714	0	0	19	1	1	1
34.10	34.10	2.627	1	1	3	5	7	6
34.10		2.627	1	1	-3	5		3
34.14		2.624	1	0	20	3		1
34.89	34.90	2.570	1	1	-6	20	17	20
34.89		2.570	1	1	6	4		5
35.34	35.34	2.538	0	2	1	1	2	1
35.34		2.538	1	1	21	1		1
36.17	36.18	2.481	1	1	-9	35	27	37
36.17		2.481	1	1	9	3		3
37.79	37.92	2.379	0	2	23	1	33	1
37.91		2.371	1	1	12	39		41

2THETA	PEAK	D	H	K	L	I(INT)	I(PK)	I(DS)
37.91		2.371	0	1	-12	1		1
37.93		2.370	0	0	27	7	19	8
40.05	40.06	2.250	2	1	15	28	1	29
41.34	41.34	2.182	0	2	0	1		1
41.34		2.182	2	1	9	1		1
42.54	42.54	2.123	1	1	18	13	9	13
42.54		2.123	1	1	-18	1		1
42.92	42.92	2.105	0	2	27	1		1
42.92		2.105	1	2	-27	1		1
45.34	45.34	1.9985	1	1	21	5	2	6
45.34		1.9985	2	0	21	2		2
47.13	47.12	1.9268	1	2	18	1		1
47.13		1.9268	0	2	-18	8	6	8
48.41	48.42	1.8787	1	2	24	2		2
48.41		1.8787	1	1	-24	2		2
51.36	51.38	1.7775	0	3	36	12	1	13
51.72	51.72	1.7658	1	1	27	1	7	1
51.72		1.7658	1	1	-27	1		1
52.77	52.78	1.7332	2	0	0	1	2	1
52.77		1.7332	1	2	9	1		1
54.44	54.44	1.6839	2	1	9	1	2	1
54.44		1.6839	1	2	-9	1		1
54.44		1.6839	2	1	-9	1		1
55.26	55.26	1.6610	1	1	30	12	7	13
55.26		1.6610	1	1	-30	1		1
55.73	55.72	1.6479	2	0	27	1	2	1
55.73		1.6479	0	2	27	1		1
58.99	59.00	1.5644	1	1	33	8	6	9
59.26	59.14	1.5579	2	1	18	3	4	3
59.26		1.5579	1	2	-18	1		1
60.52	60.52	1.5285	3	0	0	1		1
62.05	62.06	1.4944	0	3	9	28	14	31
62.05		1.4944	3	0	-9	5	5	5
62.92	62.92	1.4757	1	1	36	3	5	4
62.92		1.4757	1	1	-36	5		6

AL-SERPENTINE - JAHANBAGLOO AND ZOLTAI, 1968

2THETA	PEAK	D	H	K	L	I(INT)	I(PK)	I(DS)
65.59	65.60	1.4220	0	0	45	1	1	2
66.50	66.54	1.4048	0	2	36	1	4	1
66.50		1.4048	2	0	36	1		1
66.53		1.4042	3	0	18	3		4
66.53		1.4042	0	3	18	3		4
67.05	67.04	1.3947	1	1	-39	6	4	7
67.05		1.3947	1	1	39	1		1
71.16	71.36	1.3237	2	2	0	1	3	1
71.32		1.3212	2	2	3	1		1
71.32		1.3212	2	2	-3	2		2
71.36		1.3205	1	1	-42	5		6
71.79	71.80	1.3137	2	2	6	1	2	1
71.79		1.3137	2	2	-6	3		4
72.58	72.58	1.3014	2	2	9	1	2	1
72.58		1.3014	2	2	-9	4		4
73.67	73.68	1.2848	2	2	12	1	3	1
73.67		1.2848	2	2	-12	3		3
73.69		1.2845	0	0	27	2		2
73.69		1.2845	3	3	27	2		2
75.07	75.08	1.2643	2	2	15	1	1	1
75.07		1.2643	2	2	-15	1		1
75.88	75.90	1.2527	1	1	-45	3	2	3
76.76	76.76	1.2405	2	2	18	1	1	1
78.75	78.76	1.2141	2	2	21	1	1	1
78.75		1.2141	2	2	-21	1		1
80.62	80.62	1.1907	1	1	-48	1	1	1
81.03	81.04	1.1857	2	2	-24	1	1	2
81.08		1.1850	0	0	54	1		1
83.59	83.58	1.1557	2	2	-27	2	1	3

TABLE 27. CLAYS

Variety	Kaolinite	Dickite	Nacrite	Mg-Vermiculite
Composition	$Al_2Si_2O_5(OH)_4$	$Al_2Si_2O_5(OH)_4$	$Al_2Si_2O_5(OH)_4$	$(Mg,Fe^3,Al)_3(Si,Al)_4O_{10}(OH)_2(Mg_{0.48},K)(H_2O)_{4.72}$
Source	not given	Schuylkill, Penn.	Pikes Peak, Colo. USNM #83593	Llano, Texas
Reference	Drits & Kashaev, 1960	Newnham, 1961	Threadgold, 1963	Shirozu & Bailey, 1966
Cell Dimensions				
a̲ Å	5.14	5.150	8.909	5.349
b̲ Å	8.93	8.940	5.146	9.255
c̲	7.37	14.424	14.606	28.89
α	91.80	90	90	90
β deg	104.50	96.73	100.25	97.12
γ	90.00	90	90	90
Space Group	C1	Cc	Cc	C2/c
Z	2	4	4	4

TABLE 27. (cont.)

Variety	Kaolinite	Dickite	Nacrite	Mg-Vermiculite
Site Occupancy				M_1 — Mg 0.95
				M_2 — Al 0.05
				M_3
				M_4 Mg 0.41
				T_1 {Si 0.715 / Al 0.285
				T_2 {Si 0.715 / Al 0.285
Method	2-D, film	3-D, film	3-D, film	3-D, film
R & Rating	0.15 (3)	0.075 (2)	0.103 (2)	0.091 (1)
Cleavage and habit	{001} perfect Finely divided	{001} good Tabular	{001} good Tabular	Platy
Comment				Minor Fe and K ignored
μ	78.5	78.5	78.3	61.5
ASF	0.1842×10^{-1}	0.7767×10^{-1}	0.6618×10^{-1}	0.4959
Abridging factors	1:1:0.5	1:1:0.5	1:1:0.5	0.1:0:0

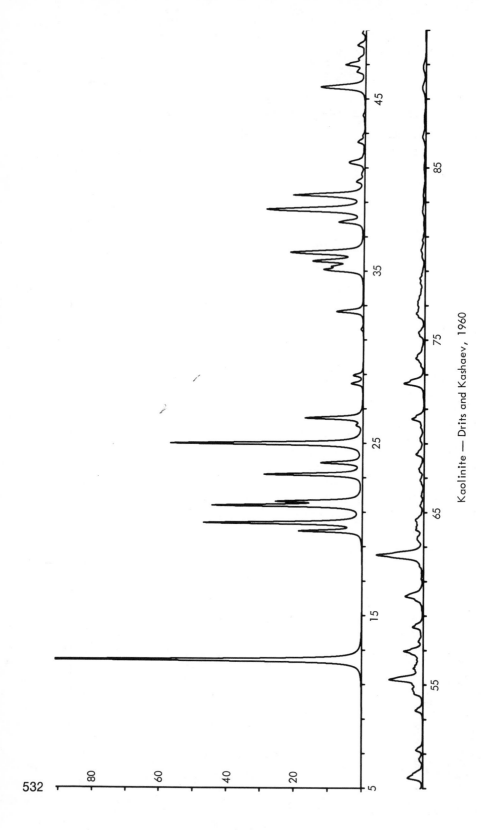

Kaolinite — Drits and Kashaev, 1960

λ = CuKα Half width at 40°2θ: 0.15°2θ Plot scale I_{PK} (max) = 100

532

KAOLINITE - DRITS AND KASHAEV, 1960

2THETA	PEAK	D	H	K	L	I(INT)	I(PK)	I(DS)
12.40	12.40	7.131	0	0	1	100	100	100
19.88	19.88	4.463	0	2	0	50	19	19
20.34	20.36	4.361	-1	1	0	50	47	50
21.35	21.36	4.158	-1	1	1	46	44	47
21.59	21.60	4.112	-1	-1	1	24	26	25
23.15	23.16	3.839	0	0	2	33	29	33
23.84	23.86	3.729	0	-2	1	14	12	14
24.95	24.96	3.566	0	0	2	66	57	68
26.01	26.02	3.423	1	1	0	2	2	2
26.43	26.44	3.369	-1	1	2	20	17	21
28.44	28.44	3.136	-1	-1	2	4	4	4
28.91	28.92	3.086	1	2	2	4	3	4
32.62	32.62	2.743	0	2	2	11	8	11
34.98	35.06	2.563	-2	0	2	9	12	9
35.07		2.557	-1	3	1	8		8
35.24	35.24	2.544	-1	-3	1	9	10	9
35.53	35.56	2.524	-1	3	1	15	15	16
35.59		2.520	-1	-1	2	6		6
35.99	36.06	2.493	-1	3	0	16	22	17
36.07		2.488	-2	0	2	17		18
36.15		2.483	1	0	3	2		2
37.81	37.82	2.377	0	0	3	10	8	11
38.48		2.338	-1	3	0	24		26
38.56	38.56	2.333	0	-3	2	22	29	24
38.68		2.326	-2	0	2	2		2
39.26		2.293	-1	1	3	3		4
39.38	39.38	2.286	-1	-1	3	28	21	30
40.17	40.18	2.243	-1	-3	2	3	3	3
41.21		2.188	-1	3	1	4		4
41.28	41.28	2.185	-1	-1	2	3	5	3
41.37		2.181	-2	0	3	11		12
42.46	42.48	2.127	0	-2	3	3	2	4
44.00	44.00	2.056	-2	2	2	1	1	1
45.59		1.9880	-1	-3	2	11		12
45.66	45.66	1.9850	-2	0	3	13	13	14
46.55	46.56	1.9494	2	2	1	4	3	4
46.95	46.96	1.9336	1	3	2	9	6	10
47.31	47.30	1.9197	0	-4	3	3	2	3
48.10	48.10	1.8902	-1	-3	4	4	3	4
48.80	48.80	1.8645	0	4	2	2	1	2
49.56	49.58	1.8379	-1	3	0	5	5	6
49.61		1.8360	-2	0	2	3		3
49.85	49.84	1.8277	-2	2	2	3	3	4
50.67	50.68	1.7999	0	0	4	1	1	1
51.19	51.20	1.7829	-2	0	2	4	2	5
53.49	53.50	1.7116	-2	-4	1	2	2	2
54.44	54.48	1.6839	-3	1	1	1	3	1
54.49		1.6825	-2	4	1	1		1
54.63		1.6784	-1	-5	0	2		2
54.72	54.74	1.6759	-1	4	1	3	3	3
54.76	54.76	1.6748	-2	-1	0	10	10	11
55.01	55.02	1.6679	-1	-5	1	1	3	1
55.25	55.30	1.6611	1	3	3	8	10	9
55.30		1.6597	-1	0	4	2		2
55.32		1.6593	1	4	2	3		3
55.80	55.80	1.6462	-2	0	4	2	2	2
55.95	55.96	1.6420	-3	-1	1	3	3	3
56.12	56.10	1.6374	0	2	4	1	1	2
56.28	56.26	1.6332	-3	0	2	1	1	1
56.46	56.46	1.6284	3	-1	1	3	2	3
56.78		1.6199	1	-5	1	8		9
56.93	56.94	1.6159	-1	3	3	1	6	1
57.49	57.50	1.6017	-2	1	3	6	2	7
58.34	58.34	1.5804	-1	-3	4	4	3	5
60.10	60.12	1.5382	-1	-1	2	4	6	5
60.14		1.5372	2	0	4	2		2
60.20		1.5358	-3	-1	3	11		13
61.86	61.88	1.4985	-3	0	1	11	1	12
62.37		1.4875	-2	6	3	1		1
62.49		1.4849	0	6	3	1		1
62.53	62.52	1.4842	-2	-3	3	3	14	13
63.29		1.4681	-3	2	1	1		1
63.43	63.44	1.4651	-2	-2	3	1	2	1
63.56	63.54	1.4624	-3	-3	2	1	2	2

KAOLINITE – DRITS AND KASHAEV, 1960

2THETA	PEAK	D	H	K	L	I(INT)	I(PK)	I(DS)
63.99	64.03	1.4538	-1	5	2	1	2	2
63.99		1.4537	-3	3	2	2	2	2
64.15	64.16	1.4504	-1	-5	3	1		1
64.33	64.34	1.4469	0	6	0	1	2	2
64.48	64.56	1.4438	3	3	1	1	2	2
64.58		1.4419	2	0	0	1	3	2
65.37	65.38	1.4263	0	2	3	3		3
65.94	65.94	1.4154	0	-4	5	2	1	2
66.64	66.66	1.4022	-2	4	4	1	1	1
66.89	66.86	1.3976	-2	0	2	1	1	1
67.36	67.38	1.3890	0	-6	5	2	1	2
67.52	67.52	1.3861	-3	1	2	2	3	2
68.34	68.36	1.3713	0	3	3	2		2
68.37		1.3709	-3	3	4	1		2
68.38		1.3707	-3	3	3	1	1	2
69.30	69.30	1.3548	-3	3	1	1	1	2
69.92	69.92	1.3442	-1	-2	5	1	1	1
70.41	70.40	1.3361	-1	-3	5	8	4	10
72.44	72.46	1.3035	-1	3	3	8	6	9
72.48		1.3029	-2	0	4	7		8
73.48	73.50	1.2876	-2	-6	1	2	1	2
73.73	73.92	1.2839	-2	6	1	1	3	2
73.90		1.2814	-2	6	0	1		1
73.92		1.2811	-4	0	0	2		3
74.52	74.50	1.2722	-2	6	2	1	1	1
75.22	75.26	1.2621	-2	-6	2	3	2	3
76.32	76.36	1.2466	-2	6	2	3	3	4
76.51	76.52	1.2440	4	0	0	3	2	4
76.84	76.72	1.2395	2	-6	1	1	2	1
76.98	77.04	1.2376	-4	1	0	1		2
77.04		1.2368	-4	0	3	1		1
77.23	77.22	1.2342	-1	-7	1	2	2	2
77.93	77.90	1.2249	-3	-5	2	1	1	1
78.58	78.56	1.2163	-3	5	0	1	2	2
79.79	79.82	1.2009	-3	4	6	1	1	1
80.79	80.78	1.1886	0	0	6	1	1	2
82.33	82.42	1.1702	-2	4	5	1	1	1

2THETA	PEAK	D	H	K	L	I(INT)	I(PK)	I(DS)
86.75	86.76	1.1215	-2	-6	4	2	1	3
89.48	89.64	1.0942	-2	6	2	2	1	2
89.66		1.0926	4	0	2	2		2
90.65	90.86	1.0831	2	-6	3	1	1	1
90.86		1.0812	-4	0	3	1		1
93.75	93.78	1.0553	2	6	4	2	1	2
96.96	97.00	1.0287	3	-3	4	1	1	2
99.10	99.08	1.0122	3	3	4	1	1	2
101.59		0.9940	2	-6	6	2		3
101.80	101.80	0.9925	-4	0	7	1	1	1
103.42	103.46	0.9814	-1	3	6	1		1
103.45		0.9811	2	0	6	1		1
104.66	104.86	0.9731	-1	-2	4	1	1	1
104.77		0.9724	-4	1	1	1		1
104.84		0.9719	-5	7	4	1		2
105.63	105.64	0.9668	2	6	4	2	1	2
106.24	106.10	0.9629	1	-9	1	1	1	2
106.64	106.66	0.9604	-5	-3	3	1	1	2
107.01	107.08	0.9581	-4	6	0	2	1	2
107.16		0.9572	-5	9	3	1		2
108.02	108.26	0.9520	5	9	0	2	1	2
108.26		0.9505	4	6	0	2		1
109.43	109.40	0.9436	0	-6	6	1	1	1
109.59		0.9427	-3	-3	7	1		1
112.20	112.24	0.9280	3	-3	5	1	1	2
112.23		0.9279	-3	3	7	1		2
114.82	114.90	0.9142	0	6	6	1	0	2
114.96		0.9135	3	3	5	2		3
118.56	119.12	0.8960	1	-9	3	2	1	3
119.01		0.8939	-5	-3	5	1		2
119.55		0.8914	0	0	8	1		2
120.35	120.48	0.8878	4	-6	6	2	1	2
120.54		0.8870	-5	3	5	2		2
120.92		0.8853	-4	9	7	1		1
123.45	123.76	0.8746	3	9	3	1	1	2
123.73		0.8735	4	6	2	1		2
127.59	128.12	0.8585	-3	9	1	1	1	2

KAOLINITE – DRITS AND KASHAEV, 1960

2THETA	PEAK	D	H	K	L	I(INT)	I(PK)	I(DS)
127.71		0.8580	-3	-9	1	1		2
128.13		0.8565	-6	0	2	2		2
128.95		0.8536	0	-6	7	2		3
132.26	133.12	0.8423	-2	10	0	1	1	1
132.54		0.8414	-3	-9	3	1		1
132.93		0.8401	3	-3	6	1		1
134.70	134.84	0.8346	-1	-9	5	1	1	2
134.92		0.8340	-4	8	0	1		1
135.05		0.8336	-4	-6	6	2		3
136.37	136.42	0.8297	-4	0	8	1	1	1
136.48		0.8294	6	0	0	1		2
137.09	137.04	0.8276	3	3	6	1	1	1
137.32		0.8269	0	-8	6	1		1
137.92		0.8253	1	9	4	2		1
140.71	141.12	0.8179	-4	6	6	2	1	3
140.89		0.8174	5	-3	3	1		3
141.10		0.8169	-3	-9	4	2		3
141.15		0.8167	1	-3	8	2		1
141.26		0.8165	-2	0	9	2		3
143.56	144.18	0.8109	-1	9	5	2	1	3
144.16		0.8095	5	3	3	2		4
144.53		0.8087	0	-10	4	1		1
148.10	149.48	0.8011	6	0	1	1	1	1
148.67		0.8000	5	5	2	1		1
149.39		0.7986	-5	7	7	1		1
149.67		0.7980	1	-9	5	1		1
150.46		0.7966	-5	-3	7	1		2
150.72		0.7961	1	-11	1	1		1
150.95		0.7957	-5	-7	1	1		1
151.16		0.7953	-6	4	3	1		2

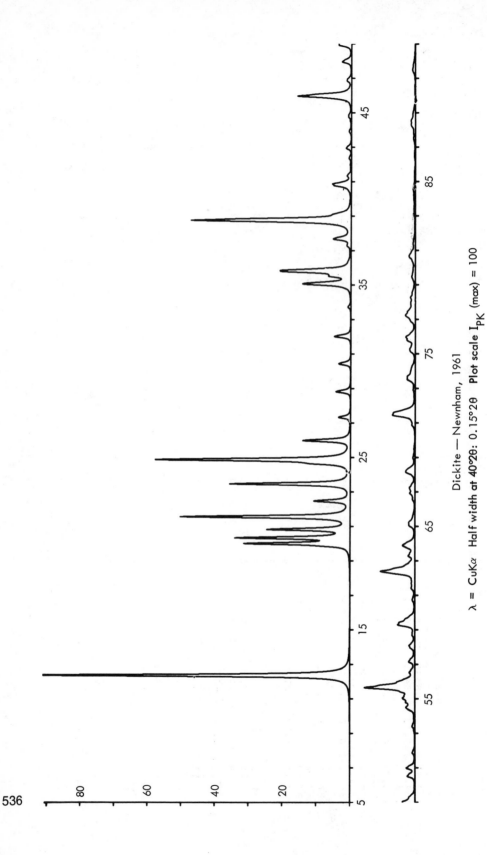

Dickite — Newnham, 1961 Plot scale I_{PK} (max) = 100

λ = CuKα Half width at 40°2θ: 0.15°2θ

536

2THETA	PEAK	D	H	K	L	I(INT)	I(PK)	I(DS)
12.35	12.36	7.162	0	0	2	100	100	100
19.98	20.00	4.439	-1	1	0	32	31	32
20.31	20.32	4.368	-1	1	1	35	34	36
20.80	20.80	4.267	0	2	1	26	25	26
21.53	21.54	4.123	1	1	1	56	50	56
22.44	22.46	3.958	-1	1	2	12	11	12
23.44	23.44	3.792	0	0	4	41	36	42
24.84	24.86	3.581	0	2	3	69	58	70
25.96	25.96	3.430	-1	1	3	17	14	17
27.31	27.32	3.263	0	2	3	4	4	4
28.79	28.80	3.098	-1	1	3	6	4	6
30.41	30.42	2.937	-1	1	4	5	4	5
32.00	32.00	2.795	0	2	4	7	5	7
33.69	33.70	2.658	-1	1	4	1	1	1
35.01	35.04	2.561	-1	3	1	13	14	13
35.06		2.557	-2	0	0	8		9
35.49	35.50	2.527	-1	1	5	23	21	25
35.76	35.82	2.509	1	3	1	11		12
35.84		2.503	-2	0	2	1	1	1
37.25	37.26	2.412	0	0	6	7	5	8
37.64	37.66	2.387	-1	3	3	49	48	53
38.71	38.72	2.324	-2	0	3	22		23
38.72		2.212	-2	0	3	6	6	6
40.75	40.84	2.207	1	3	0	4		4
40.85		2.184	-2	2	0	2	2	2
41.30	41.30	2.106	-2	2	1	2	2	3
42.91	42.92	2.062	2	2	1	1	1	1
43.87	43.88	2.024	0	4	3	1	1	1
44.73	44.74	1.9790	-2	2	4	3		3
45.81	45.94	1.9746	1	3	4	9	16	9
45.92		1.9738	-1	3	5	16		18
45.94		1.9388	2	2	3	2	1	2
46.82	46.82	1.8960	0	4	4	5	3	5
47.94	47.94	1.8612	1	3	5	5	4	6
48.89	48.92	1.8571	-2	0	6	3	2	3
49.01	49.00	1.8480	-2	2	5	2	2	2
49.26	49.24							
50.49	50.50	1.8062	2	2	2	4	3	4
50.96	50.96	1.7906	0	0	8	4	2	4
51.84	51.84	1.7622	0	4	5	1	1	2
53.35	53.38	1.7149	-2	2	6	1	1	2
54.35	54.48	1.6865	-2	4	1	2	3	1
54.48	54.78	1.6829	3	1	0	2		2
54.77		1.6747	2	1	5	2		2
54.79	55.02	1.6741	1	5	1	2	4	2
54.98		1.6688	-2	2	2	2		2
55.04	55.64	1.6671	-1	5	6	3		4
55.61		1.6513	-2	0	7	9	15	10
55.65		1.6502	-1	3	7	18		21
56.51	56.50	1.6270	-2	4	3	2	2	2
57.12	57.14	1.6110	-1	5	2	1	2	2
57.13		1.6108	1	5	2	2		2
57.93	58.06	1.5906	2	4	3	2		2
58.06		1.5872	-2	2	7	2	2	2
58.66	58.68	1.5723	-1	5	1	1	1	1
59.26	59.28	1.5579	-1	3	7	9	5	10
59.39	59.40	1.5549	-2	4	8	3	5	3
60.71	60.70	1.5242	-3	1	5	1	1	1
61.37	61.38	1.5093	-3	1	7	2	1	2
61.68	61.68	1.5025	-1	5	0	1	1	1
62.26	62.38	1.4900	-2	6	1	7	10	9
62.37		1.4875	-3	3	8	15		18
63.27	63.28	1.4686	-2	2	1	3	2	4
63.74	63.88	1.4588	0	6	2	2	4	3
63.83		1.4569	3	3	3	2		2
63.88		1.4560	-3	1	1	4		5
65.00	65.06	1.4336	-2	0	7	2		2
65.06		1.4325	0	4	10	1	2	2
65.27	65.18	1.4283	-2	6	6	1	1	1
66.54	66.56	1.4040	-1	5	8	3	2	3
66.90	66.90	1.3974	0	6	9	2	2	3
67.25	67.22	1.3910	-1	3	6	2	2	3
68.10	68.16	1.3757	-2	0	3	2	3	3
68.17		1.3745	3	3	3	3		3

DICKITE — NEWNHAM, 1961

2THETA	PEAK	D	H	K	L	I(INT)	I(PK)	I(DS)
68.25	68.96	1.3729	-3	3	5	2		1
68.95		1.3607	-2	2	9	1	1	1
70.83	70.86	1.3291	2	2	8	1	1	2
71.43	71.46	1.3195	-1	3	3	1	7	2
71.57	71.56	1.3173	-2	0	13	13	7	15
71.50		1.2874	2	0	10	7	7	8
73.52	73.52	1.2851	-4	0	0	4	3	5
73.65	73.66	1.2634	2	0	3	2	2	2
75.13	75.26	1.2616	-3	3	5	1	2	1
75.26		1.2543	2	0	2	2		2
75.77	75.80	1.2515	-4	0	4	4	3	6
75.97	75.98	1.2349	-2	6	2	3	3	4
77.18	77.22	1.2340	-4	0	6	2	3	3
77.25		1.2339	3	5	0	4		2
77.26		1.2315	1	7	1	1		2
77.43		1.2243	-4	4	3	1	1	2
77.97	77.96	1.2205	-3	5	1	2	1	2
78.26	78.24	1.1907	-1	3	11	1	2	2
80.61	80.64	1.1894	2	6	4	2		2
80.72		1.1062	-4	6	6	4		4
88.26	88.28	1.1035	-4	0	8	3		2
88.53	88.52	1.0759	-4	0	10	2	1	2
91.44	91.46	1.0758	-2	6	8	1	1	2
91.45		1.0326	0	3	10	2	1	2
96.47	96.52	1.0326	-3	3	6	1		4
96.68		1.0310	-1	3	11	2		2
101.86	102.02	0.9921	-2	3	13	2	1	2
102.04		0.9909	-1	0	14	1		1
102.55	102.58	0.9873	-4	0	8	2	1	3
102.61		0.9869	-2	6	10	1		2
104.74	104.74	0.9726	-5	3	1	2	1	2
106.77	107.00	0.9596	-1	1	9	3		3
106.98		0.9583	-4	6	4	2		2
107.00		0.9582	5	5	3	1		1
108.14	108.36	0.9512	-1	3	17	1	1	1
108.25		0.9506	-1	7	10	1		1
108.29		0.9504	4	6	2	1		1

H	K	L	I(INT)	I(PK)	I(DS)	D	2THETA	PEAK
-5	3	5	1		1	0.9491	108.50	108.50
3	3	11	1	1	1	0.9317	111.52	111.74
0	3	12	2		1	0.9316	111.54	
-3	3	13	1		2	0.9303	111.78	
-3	7	6	1	0	1	0.9166	114.34	114.70
3	5	7	1		1	0.9147	114.72	
-5	5	12	1	1	1	0.8968	118.39	118.72
-3	0	16	2		2	0.8953	118.71	
-1	9	7	3	1	3	0.8880	120.31	120.50
5	3	5	2		4	0.8873	120.47	
-4	6	8	2		3	0.8868	120.59	
-3	9	7	1	0	1	0.8727	123.91	124.26
4	9	6	1		2	0.8723	124.02	
-3	0	2	1	1	1	0.8598	127.24	127.48
-6	0	12	1		2	0.8583	127.63	
3	3	13	1		1	0.8575	127.87	
0	6	14	1	1	2	0.8436	131.85	132.38
-6	2	1	1		1	0.8435	131.90	
-4	6	10	1		1	0.8416	132.48	
3	9	5	1		1	0.8398	133.03	
-4	0	12	1		2	0.8387	133.37	
-2	10	10	1	1	1	0.8371	133.90	133.98
-3	9	3	1		1	0.8366	134.05	
6	0	2	1		1	0.8362	134.19	
4	6	12	1		1	0.8358	134.30	
-2	6	14	1		1	0.8350	134.56	
-4	8	5	1	1	1	0.8256	137.79	138.90
1	9	9	1		1	0.8251	137.99	
-4	6	14	1		1	0.8247	138.12	
-5	3	11	2		3	0.8233	138.63	
0	10	7	2		3	0.8230	138.75	
2	6	8	2		4	0.8214	139.34	
-5	3	11	3		5	0.8192	140.17	
2	10	16	2	1	2	0.8157	141.56	141.86
-1	3	17	3		3	0.8152	141.78	
-1	9	10	1		1	0.8146	142.02	
3	9	5	1	0	2	0.8092	144.32	144.70

DICKITE - NEWNHAM, 1961

2THETA	PEAK	D	H	K	L	I(INT)	I(PK)	I(DS)
144.53		0.8087	-3	9	7	1		2
145.17	145.08	0.8072	-6	0	8	1	0	2
145.82		0.8058	5	5	6	1		2
148.73	149.04	0.7998	0	10	8	1	0	1
148.94		0.7994	-5	7	3	1		1
149.02		0.7993	-1	11	2	1		1
150.52		0.7965	-6	4	0	1		2
150.69		0.7962	-6	4	4	1		1
153.34	155.08	0.7916	-5	5	10	1	0	2
154.79		0.7893	3	5	13	1		2
154.94		0.7890	-1	9	11	2		4
155.08		0.7888	5	3	9	1		1
155.61		0.7880	-4	6	12	1		2
156.16		0.7872	1	3	17	1		1
156.69		0.7865	-2	0	18	1		1
157.13		0.7858	5	5	7	1		1
159.92		0.7822	6	4	2	1		1

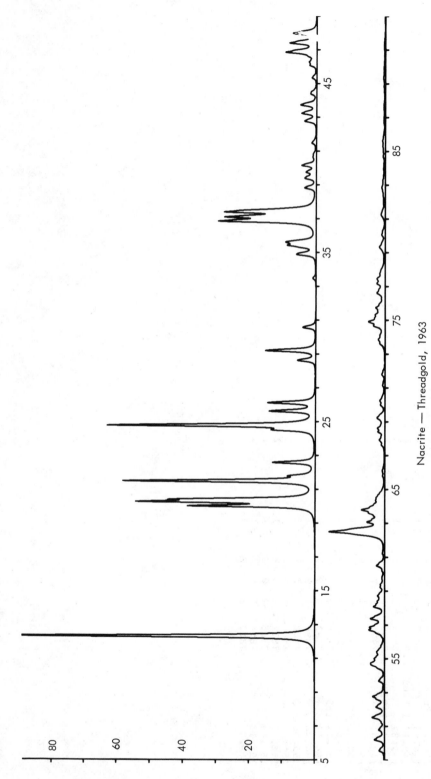

540

Nacrite — Threadgold, 1963

λ = CuKα Half width at 40°2θ: 0.15°2θ Plot scale I$_{PK}$ (max) = 100

NACRITE – THREADGOLD, 1963

2THETA	PEAK	D	H	K	L	I(INT)	I(PK)	I(DS)
12.31	12.32	7.186	0	0	2	100	100	100
19.99	20.00	4.438	1	1	0	38	39	38
20.24	20.26	4.383	-2	0	0	50	54	51
20.39	20.38	4.352	-1	1	1	37	45	38
21.46	21.48	4.137	1	1	1	64	58	65
21.77	21.78	4.079	-2	0	2	6	9	6
22.56	22.56	3.938	-1	1	2	14	13	15
24.48	24.50	3.633	-1	1	2	11	13	12
24.76	24.76	3.593	1	2	1	74	63	76
25.60	25.60	3.477	-2	0	2	16	14	16
26.09	26.10	3.413	-1	1	3	17	15	18
28.60	28.60	3.119	-1	1	3	7	6	7
29.17	29.18	3.059	-2	0	4	19	15	20
30.54	30.56	2.924	-1	1	4	5	4	5
33.44	33.46	2.677	-1	1	4	1	1	1
34.84	34.86	2.573	-3	0	1	2		2
34.86		2.571	2	1	4	6	6	6
34.96		2.564	3	0	0	1		1
35.29		2.541	0	1	4	4		4
35.41	35.42	2.533	-3	0	1	8	9	9
35.57		2.522	-1	1	2	5		9
35.62	35.60	2.518	3	1	2	5	9	5
36.83	36.84	2.438	-3	0	2	40	30	42
37.08	37.08	2.422	0	2	2	33	28	35
37.37	37.38	2.404	-3	1	2	35	28	37
37.51		2.395	0	0	5	10		10
38.79	38.80	2.319	3	1	1	5	4	5
39.36	39.38	2.287	3	1	2	5	4	5
39.73	39.74	2.267	0	2	3	4	3	4
40.13	40.14	2.245	-3	1	4	6	4	7
41.46	41.46	2.176	-2	1	2	5	1	2
42.74	42.76	2.114	3	1	3	5	4	6
43.21	43.22	2.092	0	2	4	7	5	7
43.70	43.72	2.069	-3	0	5	6	5	7
43.73		2.068	2	1	2	1		2
44.38	44.38	2.039	-4	0	4	3		3
45.32	45.32	1.9995	4	0	2	2		2
46.06	46.08	1.9689	-2	2	4	3	2	3
46.27	46.26	1.9604	-2	0	6	2	2	3
46.83	46.84	1.9382	3	1	4	15	10	16
47.38	47.38	1.9172	0	2	5	13	8	14
47.95	47.96	1.8956	-3	1	6	12	7	13
49.54	49.54	1.8384	-2	2	4	5	3	3
50.19	50.20	1.8163	0	2	8	3	2	6
50.77	50.78	1.7966	0	0	5	6	3	7
51.51	51.52	1.7726	4	1	5	4	2	5
52.12	52.12	1.7533	-3	2	6	3	4	4
52.59		1.7386	-2	2	4	4	3	4
52.76	52.76	1.7336	-1	3	7	3	3	3
53.67	53.68	1.7063	4	2	6	4	4	4
54.63	54.70	1.6786	-4	1	1	2	2	2
54.72		1.6760	4	0	0	3	2	2
54.99	54.98	1.6685	5	1	3	4	5	5
55.30	55.32	1.6597	-5	1	6	2	5	7
55.40	55.40	1.6569	5	2	1	6	4	4
56.69	56.70	1.6223	-1	3	3	4		3
56.86	56.84	1.6180	0	1	7	3		5
57.34		1.6055	-3	1	2	4		7
57.36	57.36	1.6049	-2	2	7	6	3	7
58.06	58.06	1.5873	4	2	2	6	3	6
58.37	58.36	1.5797	1	3	3	2	2	2
58.40		1.5788	-4	2	6	1		1
58.70	58.70	1.5715	3	0	7	3	3	4
59.20	59.20	1.5594	-4	1	2	2	2	2
60.49	60.50	1.5293	-3	3	3	4	1	5
62.33		1.4884	-3	0	6	9	2	11
62.48		1.4852	1	2	8	18	17	21
62.52	62.50	1.4843	-6	2	6	10		12
63.05	63.06	1.4730	0	2	8	8	5	10
63.42	63.60	1.4654	-4	2	6	8	5	5
63.57		1.4622	-2	2	8	3		3
63.77	63.78	1.4582	-3	1	9	2	7	3
63.80		1.4576	-3	3	3	8		9
64.36	64.34	1.4463	-6	0	4	2	3	2

NACRITE - THREADGOLD, 1963

2THETA	PEAK	D	H	K	L	I(INT)	I(PK)	I(DS)
64.49	64.48	1.4436	2	2	7	2	3	3
64.64		1.4406	4	2	4	1	1	1
65.78	65.78	1.4185	5	1	4	1	1	1
66.10	66.10	1.4123	-5	1	7	1	1	1
67.60	67.62	1.3845	6	0	2	2	1	2
67.92	67.92	1.3789	3	3	3	3	2	3
68.40	68.40	1.3704	-3	1	8	4	2	5
68.63	68.60	1.3664	-3	3	5	2	2	3
69.17	69.22	1.3569	0	2	9	3	2	3
69.26		1.3554	-2	2	2	2	1	2
69.98	69.98	1.3433	-3	1	10	3	1	3
70.26	70.24	1.3386	2	2	8	1	2	3
71.14	71.14	1.3242	-1	3	7	1	1	3
71.78	71.78	1.3139	-4	0	10	2	1	2
73.61	73.62	1.2857	-6	2	2	2	2	2
73.85	73.84	1.2821	-4	0	8	2	2	2
74.12	74.08	1.2781	-6	2	3	1	2	2
74.63	74.66	1.2706	-6	2	0	5	4	7
74.70		1.2696	3	3	5	2		2
74.90	74.90	1.2667	3	1	9	3	5	4
74.92		1.2664	0	4	2	5		6
75.31	75.30	1.2608	-6	2	4	2	3	6
75.72	75.72	1.2551	-3	3	7	3	3	3
75.74		1.2548	0	2	10	3		4
76.60	76.60	1.2428	-3	1	11	3	3	4
76.62		1.2425	0	4	3	2		2
77.14	77.18	1.2355	-7	1	1	2	3	2
77.19		1.2348	-5	3	2	1		2
77.44	77.44	1.2313	-7	1	3	2	2	2
77.50		1.2306	-7	1	1	1		3
77.78	77.74	1.2269	-2	4	4	1	2	1
79.29	79.32	1.2073	-5	3	2	1	2	2
79.34		1.2066	2	4	2	2		3
80.04	80.06	1.1977	0	0	12	2	1	3
81.20	81.24	1.1836	6	2	3	1	1	2
81.32		1.1822	-5	3	5	1		1
82.78	82.82	1.1650	0	2	11	1	1	1

2THETA	PEAK	D	H	K	L	I(INT)	I(PK)	I(DS)
82.86	82.86	1.1640	-6	2	7	1	1	2
83.25	83.22	1.1596	-2	2	10	1	1	1
84.68	84.68	1.1436	6	0	4	2	1	2
85.62	85.60	1.1334	0	4	6	1	1	2
89.91	89.90	1.0902	6	2	7	1	1	2
93.58	93.58	1.0568	6	0	2	1	1	2
94.85	94.86	1.0460	0	4	8	2	1	3
95.62	95.62	1.0396	-3	3	3	1	1	2
97.28	97.30	1.0263	-3	3	11	1	1	1
100.89	100.82	0.9990	-7	3	1	1	1	1
102.05	102.02	0.9907	-6	2	11	1	0	2
105.72	105.86	0.9662	-9	4	4	1	1	1
105.83		0.9656	-9	4	0	1		1
105.96		0.9647	-3	5	1	1		1
106.30	106.32	0.9626	-3	5	3	1	1	1
106.51	106.54	0.9612	-6	4	4	1	1	1
106.75		0.9609	-6	1	13	1		1
106.57		0.9598	-9	1	2	1		1
107.76	107.60	0.9535	0	2	14	1	1	1
108.41	108.80	0.9496	-6	4	5	1	1	1
108.73		0.9477	-9	3	1	1		1
108.98		0.9462	-3	1	15	1		1
109.35		0.9441	6	0	10	1		2
110.35	110.36	0.9383	3	3	11	1	1	2
112.41	112.52	0.9269	-6	3	2	1	1	2
112.47		0.9265	-3	3	13	1		2
113.03	112.96	0.9235	-3	5	5	1	1	2
113.58	113.90	0.9206	0	0	14	1	1	2
113.85		0.9192	-4	4	10	1		2
113.93		0.9188	-3	3	6	1		1
114.45	114.36	0.9161	-7	3	9	1	1	2
114.88		0.9139	7	3	5	2		2
115.95	116.38	0.9085	9	1	3	1	1	1
116.38		0.9064	-6	2	13	1		1
116.42		0.9062	8	2	0	1		1
118.06	118.16	0.8983	0	0	16	2	1	3
120.52	120.66	0.8871	9	1	4	1	0	1

NACRITE – THREADGOLD, 1963

2THETA	PEAK	D	H	K	L	I(INT)	I(PK)	I(DS)
121.12	121.08	0.8845	6	4	5	1	0	1
121.75	121.78	0.8817	3	5	6	1	0	
122.01		0.8806	-1	5	8	1		1
123.11	123.28	0.8760	-3	5	8	1	0	1
123.85	123.82	0.8730	-6	4	9	1	0	1
124.62	124.62	0.8699	-9	1	10	1	0	1
127.26	127.88	0.8597	7	3	7	1	1	1
127.66		0.8582	3	5	7	1		2
127.81		0.8577	0	6	0	1		3
127.95		0.8572	-9	3	3	2		3
128.82		0.8540	3	1	15	1		1
129.06	129.38	0.8532	-4	4	12	1	1	1
129.34		0.8522	-3	5	9	1		2
129.50		0.8516	0	6	2	1		2
130.06	130.18	0.8497	3	3	13	1	1	2
130.24		0.8490	-6	4	10	1		2
130.28		0.8489	-9	3	5	1		2
130.52		0.8481	0	2	16	1		1
131.19		0.8458	-9	4	10	1		2
132.25	133.08	0.8423	4	6	0	1	1	1
132.44		0.8417	2	6	3	1		1
132.77		0.8407	-10	2	2	1		1
133.14		0.8395	-3	4	15	1		1
133.44		0.8385	0	3	13	1		1
133.61		0.8380	9	3	1	1		1
134.02		0.8367	6	4	7	2		2
137.14	137.74	0.8275	-3	5	10	1	0	1
137.71		0.8259	-5	5	8	1		1
138.29		0.8242	-6	4	11	1		1
139.58	139.48	0.8208	1	3	15	1	0	1
140.34	140.38	0.8188	10	0	4	1	1	1
144.44	145.06	0.8089	5	5	6	2		2
144.59		0.8085	3	1	16	1		1
145.07		0.8075	0	6	6	2		3
145.15		0.8073	-5	6	9	2		1
145.35		0.8069	-10	2	8	1		1
146.13		0.8052	-5	3	15	1		1

2THETA	PEAK	D	H	K	L	I(INT)	I(PK)	I(DS)
146.16		0.8051	8	0	10	1		1
147.43	149.06	0.8024	0	4	14	1	1	2
147.52		0.8023	-3	5	11	1		1
147.95		0.8014	-9	3	9	2		4
148.90		0.7995	-4	6	2	1		1
149.02		0.7993	-11	1	4	1		1
149.11		0.7991	-8	4	8	1		1
149.15		0.7990	-6	4	12	1		2
150.92		0.7957	-9	1	13	1		2
151.02		0.7956	-7	5	4	1		2
153.95	155.22	0.7906	-4	6	4	1	0	2
154.72		0.7894	9	1	8	1		2
154.81		0.7892	5	5	7	1		1
155.49		0.7882	4	6	2	1		2
155.50		0.7882	7	1	12	1		1
155.91		0.7876	11	1	0	1		1
155.92		0.7876	-5	5	10	1		2
159.08		0.7833	3	5	10	2		4

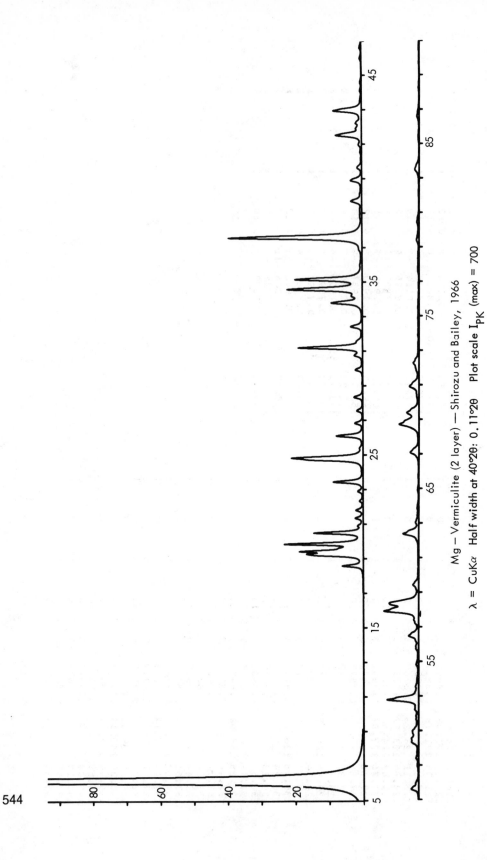

Mg – Vermiculite (2 layer) — Shirozu and Bailey, 1966

λ = CuKα Half width at 40°2θ: 0.11°2θ Plot scale I$_{PK}$ (max) = 700

544

MG-VERMICULITE (2 LAYER) - SHIROZU AND BAILEY, 1966

2THETA	PEAK	D	H	K	L	I(INT)	I(PK)	I(DS)
6.16	6.16	14.334	0	0	2	100	100	100
18.55	18.56	4.778	0	0	6	1	1	1
19.16	19.26	4.627	0	2	0	1	2	1
19.18		4.625	-1	1	1	0		0
19.26		4.604	1	1	0	2		2
19.41	19.42	4.568	0	1	1	2	3	2
19.84	19.84	4.471	1	1	1	4	3	4
20.47	20.48	4.334	-1	1	3	2	2	2
20.87	20.88	4.253	1	1	2	0	0	0
21.32	21.32	4.165	0	2	3	0	0	0
21.77	21.78	4.078	-1	1	4	0	0	0
22.30	22.30	3.984	1	1	3	0	0	0
22.86	22.86	3.888	0	2	4	0	0	0
23.42	23.44	3.795	-1	1	5	1	1	2
24.82	24.82	3.601	0	2	5	2	3	3
24.83		3.583	0	0	8	3		1
26.07	26.08	3.415	0	1	5	0	0	0
26.80	26.80	3.324	-1	2	6	0	0	0
27.52	27.54	3.238	1	2	1	0	0	0
28.30	28.32	3.150	-1	1	7	0	0	0
29.88	29.88	2.988	-1	1	6	0	0	0
30.71	30.72	2.909	1	1	8	4	3	4
31.17	31.18	2.867	0	0	10	0	0	0
31.55	31.54	2.833	0	2	8	1	0	1
32.38	32.40	2.762	-1	1	9	1	1	1
33.74	33.78	2.654	-2	0	0	1		1
33.77		2.652	-1	1	2	0		0
34.14	34.16	2.624	0	3	2	2	0	0
34.52	34.56	2.596	-2	0	9	3	3	2
34.56		2.593	1	3	4	3		3
35.11	35.14	2.553	-2	1	2	3	3	1
35.13		2.552	-1	1	4	3		3
37.54	37.56	2.394	-2	0	4	6	6	6
37.55		2.393	-1	3	6	6		6
37.62		2.389	0	3	12	0		0
39.64	39.68	2.272	-2	0	8	0	0	0
39.66		2.271	0	2	11	0		0
39.68	39.68	2.270	1	3	6	0	0	0
40.86	40.88	2.207	2	0	6	0	0	0
40.86		2.206	-1	3	8	0		0
41.58	41.60	2.170	1	1	11	0	0	0
41.64		2.167	-2	2	6	0		0
42.91	42.92	2.106	-2	2	7	1	0	1
43.46	43.50	2.080	-2	0	10	1	1	1
43.50		2.078	1	3	8	0		0
43.52		2.078	-1	3	3	0		0
43.89	43.88	2.061	2	1	5	0	0	0
44.19	44.20	2.048	0	0	14	1	1	1
44.91	44.92	2.016	2	0	8	1	0	0
44.92		2.016	-1	3	10	0		0
44.96		2.014	0	4	7	1		0
49.57	49.58	1.8373	-1	3	12	0	0	0
49.57		1.8373	2	0	10	0		0
49.91	49.90	1.8255	-2	2	11	0	0	0
51.36	51.38	1.7773	2	2	9	0	0	0
52.30	52.30	1.7476	-3	1	3	0	0	0
52.52	52.52	1.7409	-2	4	3	0	0	0
52.54		1.7402	1	5	1	0		0
52.82	52.84	1.7318	-1	5	3	0	0	0
52.98	52.96	1.7269	1	3	12	0	0	0
54.18	54.18	1.6915	2	4	14	2	1	2
54.74	54.76	1.6754	-1	3	12	1		1
54.75		1.6753	-2	0	12	0		0
54.77		1.6747	-3	1	7	0		0
54.78		1.6742	3	1	3	0	0	0
57.70	57.70	1.5963	0	4	13	0	0	0
58.42	58.44	1.5785	-2	0	16	1	0	1
58.46		1.5774	3	1	14	0		0
59.20	59.22	1.5594	2	4	7	0	0	0
59.86	59.38	1.5549	-2	2	15	2	2	2
59.86	59.88	1.5437	-3	3	2	1		1
59.91		1.5425	0	6	0	1		1
60.24	60.34	1.5349	-3	3	4	1	1	1
60.25		1.5348	3	3	0	1		1

MG-VERMICULITE (2 LAYER) - SHIROZU AND BAILEY, 1966

2THETA	PEAK	D	H	K	L	I(INT)	I(PK)	I(DS)
100.26		1.0036	-4	6	6	0		0
100.34		1.0030	-1	9	4	0		0
101.32	101.70	0.9960	-3	3	24	0	0	0
101.38		0.9955	3	3	20	0		0
101.39		0.9954	0	6	22	0		0
101.70		0.9933	5	3	2	0		0
101.72		0.9931	-4	6	8	0		0
101.81		0.9925	-1	9	6	0		0

2THETA	PEAK	D	H	K	L	I(INT)	I(PK)	I(DS)
60.30		1.5336	0	6	2	1		1
60.36		1.5321	-1	6	16	1		1
60.37		1.5320	-2	3	14	0		1
61.37	61.40	1.5092	-3	3	6	0	0	0
61.39		1.5090	3	3	2	0		0
61.43		1.5080	-0	6	4	0		0
64.32	64.36	1.4470	-2	0	18	0	1	1
64.36		1.4462	1	3	16	1		1
65.01	65.00	1.4334	-0	0	20	0	0	0
69.03	69.06	1.3594	-3	3	12	0	0	0
69.06		1.3589	3	3	8	0		0
69.09		1.3583	0	6	10	0		0
70.49	70.68	1.3347	-4	0	4	0	1	0
70.56		1.3336	-2	6	0	0		0
70.65		1.3321	-2	0	20	1		1
70.69		1.3314	2	3	18	1		1
70.83		1.3291	-1	2	19	0		0
71.31	71.36	1.3215	-4	0	6	0	1	0
71.38		1.3203	2	6	2	1		1
72.81	72.90	1.2979	-4	0	8	0	0	0
72.89		1.2967	2	6	4	0		0
74.21	74.24	1.2767	4	0	4	0	0	0
74.26		1.2761	-2	6	8	0		0
81.29	81.38	1.1825	-4	0	14	0	0	0
81.31		1.1814	2	6	10	0		0
85.40	85.48	1.1358	-4	0	16	0	0	0
85.50		1.1348	2	6	12	0		0
88.54	88.58	1.1034	-2	6	16	0	0	0
88.56		1.1032	4	0	14	0		0
93.74	93.76	1.0555	-4	0	14	0	0	0
93.75		1.0553	-2	6	18	0		0
94.30	94.34	1.0506	-3	3	22	0	0	0
94.36		1.0502	3	3	18	0		0
94.37		1.0500	0	6	20	0		0
95.31	95.32	1.0421	-1	3	26	0	0	0
99.64	99.68	1.0082	-2	6	20	0	0	0
100.23	100.28	1.0038	5	3	0	0	0	0

TABLE 28. WATER-FREE SHEET SILICATES

Variety	Haradaite	Hardystonite	Gehlenite	Gillespite	Sanbornite
Composition	$SrVSi_2O_7$	$Ca_2ZnSi_2O_7$	$Ca_2Al_2SiO_7$	$BaFeSi_4O_{10}$	$BaSi_2O_5$
Source	Iwate, Japan	Franklin, N.J.	Synthetic	Incline, Calif.	Incline, Mariposa Co., Cal.
Reference	Takéuchi & Joswig, 1967	Warren & Trautz, 1930	Raaz, 1930	Pabst, 1943	Douglass, 1958
Cell Dimensions					
$\underline{a}$ Å	5.33 (7.06)	7.85 (7.83 kx)	7.70 (7.69 kx)	7.495	13.53 (4.63)
$\underline{b}$ Å	14.64	7.85 (7.83 kx)	7.70 (7.69 kx)	7.495	7.69 (7.69)
$\underline{c}$ Å	7.06 (5.33)	5.00 (4.99 kx)	5.11 (5.10 kx)	16.050	4.63 (13.53)
α deg	90	90	90	90	90
β deg	90	90	90	90	90
γ deg	90	90	90	90	90
Space Group	Cmcm (Amam)	$P\bar{4}2_1m$	$P\bar{4}2_1m$	P4/nnc	Pmna (Pcmn)
Z	4	2	2	4	4

TABLE 28. (cont.)

Variety	Haradaite	Hardystonite	Gehlenite	Gillespite	Sanbornite
Method	3-D, film	2-D, film	2-D, film	2-D, film	2-D, film
R & Rating	0.086 (2)	--- (3)	--- (2)	--- (3)	0.10 (av.) (3)
Cleavage and habit	{010} perfect; {100} & {001} distinct. Platy.	{001} mod, {110} poor. Equidimensional	{001} good, {110} less perfect. Prismatic and tabular	{001} good; {100} poor Platy	{100} perfect; {001} less so, {010} poor. Platy
Comment		Melilite structure B estimated	Akermanite group B estimated	B estimated	B estimated
μ	232.3	213.5	206.9	494.8	673.8
ASF	0.2432	0.2248	0.2033	0.3104	0.4164
Abridging factors	1:1:0.5	1:1:0.5	1:0:0	1:1:0.5	1:1:0.5

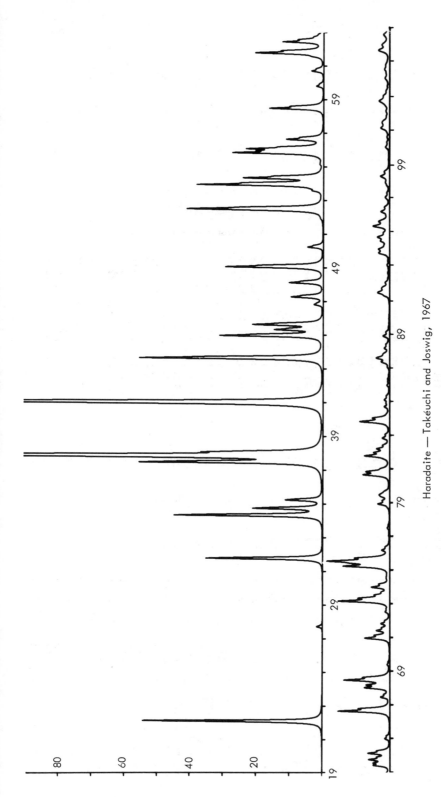

Haradaite — Takéuchi and Joswig, 1967   Plot scale I_{PK} (max) = 200

λ = FeKα Half width at 40°2θ: 0.11°2θ

HARADAITE — TAKÉUCHI AND JOSWIG, 1967

2THETA	PEAK	D	H	K	L	I(INT)	I(PK)	I(DS)
12.08	12.08	7.320	0	2	0	22	27	15
21.74	21.74	4.085	0	1	1	16	18	14
24.30	24.30	3.660	1	0	0	21	22	19
24.72	24.72	3.599	0	3	0	10	11	9
25.22	25.22	3.530	0	4	1	5	6	5
27.44	27.44	3.249	1	1	2	27	28	26
27.80	27.80	3.207	1	3	1	100	100	100
28.04	28.04	3.180	0	2	2	12	18	12
30.97	30.98	2.885	1	2	2	84	77	91
33.60	33.60	2.665	2	0	0	31	28	35
34.93	34.94	2.566	1	5	0	17	16	21
35.29	35.30	2.541	0	4	2	7	11	7
35.59	35.60	2.520	1	3	2	12	11	14
36.80	36.80	2.440	2	6	0	2	1	2
37.25	37.26	2.412	0	5	1	6	5	7
38.10	38.10	2.360	2	2	1	6	15	8
39.02	39.02	2.306	0	6	1	18	15	23
40.22	40.22	2.240	0	2	3	3	2	4
42.40	42.46	2.130	0	1	3	6	21	8
42.46	42.46	2.127	2	0	3	22	21	31
43.56	43.58	2.076	1	5	2	1	2	2
43.90	43.90	2.061	2	4	1	24	19	35
44.31	44.32	2.042	2	2	1	15	12	22
45.80	45.80	1.9795	0	4	3	17	14	26
46.04	46.04	1.9697	1	3	3	13	12	20
46.61	46.62	1.9469	1	7	0	7	6	11
48.46	48.46	1.8769	1	7	1	11	8	18
49.78	49.78	1.8300	2	8	0	1	1	2
50.68	50.68	1.7996	2	6	0	2	2	4
51.75	51.76	1.7650	0	0	4	14	10	25
52.42	52.42	1.7439	2	6	1	8	6	15
52.73	52.72	1.7345	1	5	3	3	2	2
53.38	53.38	1.7149	3	2	3	3	3	5
53.51	53.52	1.7111	2	1	1	4	3	2
53.72	53.72	1.7048	1	7	2	4	3	7
54.09	54.10	1.6939	0	6	3	4	3	8
54.95	54.96	1.6695	3	3	0	1	1	2

2THETA	PEAK	D	H	K	L	I(INT)	I(PK)	I(DS)
56.60	56.60	1.6247	3	3	1	10	8	20
56.60	56.60	1.6247	0	8	2	1	1	2
57.42	57.42	1.6033	2	6	2	2	1	4
57.98	57.98	1.5891	2	4	3	5	4	10
58.16	58.16	1.5847	1	3	4	2	4	4
58.44	58.44	1.5778	3	1	2	10	7	20
60.92	60.94	1.5194	1	9	1	3	3	6
60.94	60.94	1.5189	3	5	0	3	1	5
61.38	61.38	1.5092	3	3	2	2	2	4
61.79	61.78	1.5001	1	7	3	2	1	5
63.13	63.13	1.4715	2	0	4	12	8	26
63.49	63.48	1.4640	0	10	0	3	2	9
63.96	63.96	1.4543	1	5	4	4	3	9
65.20	65.20	1.4295	2	6	3	11	7	24
65.51	65.50	1.4237	1	9	2	14	10	31
66.16	66.16	1.4113	3	1	3	3	1	7
68.90	68.90	1.3616	3	3	3	3	2	4
69.34	69.44	1.3540	3	7	0	2	2	4
69.44	69.44	1.3523	0	10	2	6		
70.63	70.64	1.3325	4	0	1	3	4	16
70.79	70.90	1.3298	3	7	5	3	4	7
71.74	71.74	1.3145	1	1	4	6	2	16
72.18	72.18	1.3076	1	9	3	3	1	7
72.81	72.82	1.2978	1	9	0	2	1	5
73.78	73.78	1.2831	2	10	0	8	5	21
75.07	75.08	1.2642	3	7	2	1	1	3
77.33	77.34	1.2329	4	4	1	2	1	5
77.55	77.56	1.2300	2	2	5	2	2	6
77.62		1.2289	4	2	5	1		3
78.14	78.14	1.2221	0	6	5	1	1	4
78.86	78.86	1.2127	1	11	2	3	1	4
81.42	81.44	1.1810	2	4	5	2	2	8
83.76	83.78	1.1538	4	6	1	1		7
83.83		1.1531	0	12	2	1		3
83.98	84.00	1.1513	3	5	4	2	2	5
84.51	84.52	1.1455	1	7	6	2	1	6
84.73	84.74	1.1430	1	1	5	2	2	8

HARADAITE – TAKÉUCHI AND JOSWIG, 1967

2THETA	PEAK	D	H	K	L	I(INT)	I(PK)	I(DS)
85.39	85.38	1.1359	3	9	2	5	3	17
86.25	86.24	1.1268	0	10	4	5	1	9
88.34	88.34	1.1054	4	4	3	3	1	10
90.07	90.08	1.0887	1	13	1	2	1	5
91.20		1.0781	3	3	5	3	1	10
91.61	91.50	1.0743	3	7	4	1	1	4
92.82	92.82	1.0635	4	0	4	4	2	12
94.16	94.18	1.0518	1	13	2	4	1	4
94.69	94.70	1.0473	4	6	3	1	1	5
95.83	95.84	1.0379	2	10	4	2	2	16
96.77	96.76	1.0303	5	1	2	5	1	6
98.33	98.34	1.0180	5	3	2	2	1	6
100.91	100.92	0.9989	5	1	3	1	0	2
101.05		0.9979	1	13	3	1		3
102.34	102.34	0.9887	1	1	7	1	0	3
102.82	102.80	0.9854	4	10	0	1	0	5
103.80	104.34	0.9788	3	1	6	1	1	5
104.02		0.9773	3	7	5	1		6
104.78	104.80	0.9723	0	4	7	1		4
105.07	105.08	0.9704	3	11	3	1	0	2
108.13	108.44	0.9513	1	15	1	1	0	2
108.39		0.9497	5	7	0	1	1	2
108.49		0.9491	4	10	2	1		4
110.19	110.30	0.9392	2	12	4	1	1	4
110.32		0.9385	1	9	6	3		11
110.33		0.9384	2	14	2	1		5
111.46	111.44	0.9321	0	6	7	1	0	4
114.00	114.04	0.9184	3	13	3	2	0	4
114.97	114.94	0.9134	2	4	2	1	0	5
116.17	116.18	0.9074	0	16	1	1	0	4
117.56	117.76	0.9007	4	6	5	1		5
117.80		0.8995	2	14	3	1		4
120.24	120.24	0.8883	6	0	0	1	0	4
121.09	121.14	0.8846	5	9	1	1	0	2
121.57	121.70	0.8825	0	0	8	1	1	3
121.69		0.8820	4	0	6	1		4
121.72		0.8819	3	13	3	1		3

2THETA	PEAK	D	H	K	L	I(INT)	I(PK)	I(DS)
122.20	122.18	0.8798	2	6	7	3	1	12
123.19	123.20	0.8757	4	6	2	1	0	3
123.23		0.8755	3	1	7	1		3
124.11	124.20	0.8719	4	12	2	1	1	5
124.30		0.8712	5	3	4	1		5
124.93	124.88	0.8687	1	13	5	1	1	7
125.29	125.28	0.8672	2	10	6	1	1	6
126.00	126.00	0.8645	5	9	2	3	1	15
126.79	126.79	0.8615	6	0	2	1	1	4
127.07	127.08	0.8604	4	10	4	3		16
127.46	127.48	0.8590	2	16	1	2	1	10
128.03	128.02	0.8569	6	4	1	1	1	6
128.15		0.8564	4	8	5	1		5
128.39		0.8556	6	2	2	1		4
129.16	129.10	0.8528	0	16	3	2	1	9
130.19	129.80	0.8492	3	15	1	1	0	3
131.05	130.94	0.8463	1	9	7	1	0	4
132.82	132.94	0.8405	3	1	8	3	1	5
132.95		0.8401	3	7	6	1		14
133.56	133.62	0.8381	5	3	5	3		11
133.68		0.8378	2	0	8	3		16
134.14		0.8364	5	7	4	1		6
136.61	136.58	0.8290	6	6	1	1	0	4
141.00	141.06	0.8171	4	14	1	1	0	4
143.00	143.02	0.8122	2	16	3	3	0	15
143.75	143.84	0.8105	6	4	3	1		6
144.03		0.8098	5	11	2	2		4
145.21	145.10	0.8072	2	12	6	2	0	9
146.55	146.70	0.8043	1	13	6	1	0	5
146.78		0.8038	7	1	8	1		7
147.82	147.78	0.8017	4	12	4	1	0	6
148.06		0.8012	4	14	2	1		3
151.42	152.18	0.7948	1	11	7	1	1	8
151.95		0.7939	1	15	5	1		8
152.19		0.7935	6	0	4	4		22
154.53	155.50	0.7897	3	11	6	1	1	5
155.05		0.7889	3	13	5	2		14

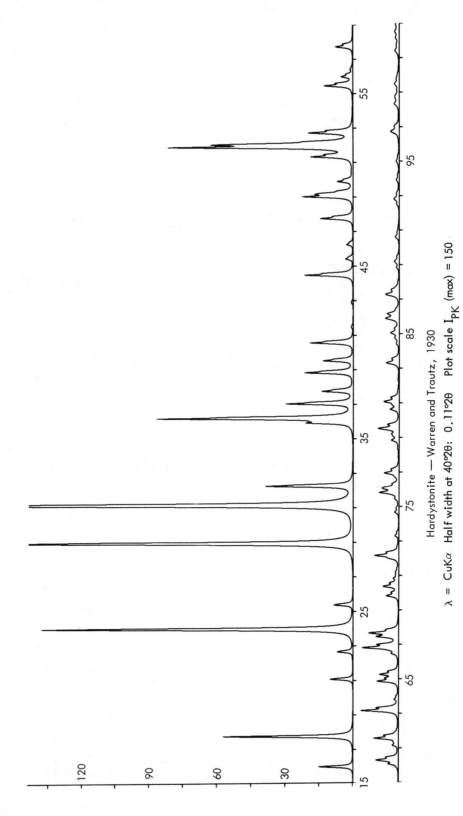

Hardystonite — Warren and Trautz, 1930

λ = CuKα Half width at 40°2θ: 0.11°2θ Plot scale I$_{PK}$ (max) = 150

HARDYSTONITE – WARREN AND TRAUTZ, 1930

2THETA	PEAK	D	H	K	L	I(INT)	I(PK)	I(DS)
15.95	15.96	5.551	1	1	0	5	7	4
17.72	17.74	5.000	0	0	1	21	25	16
21.05	21.06	4.217	1	0	1	4	5	3
22.63	22.64	3.925	2	0	0	3	3	2
23.93	23.94	3.715	2	1	0	54	61	46
25.35	25.36	3.511	1	1	1	3	4	3
28.89	28.90	3.087	2	1	1	71	73	67
31.10	31.10	2.873	2	2	0	100	100	100
32.22	32.22	2.775	2	2	1	17	17	18
35.89	35.90	2.500	0	0	2	8	9	9
36.15	36.16	2.482	3	1	1	41	38	45
37.01	37.02	2.427	2	2	1	14	13	16
37.73	37.74	2.382	1	0	2	7	6	8
38.81	38.82	2.318	3	0	1	11	10	13
39.50	39.50	2.279	2	1	2	6	6	8
40.54	40.54	2.223	3	1	1	11	9	12
44.45	44.46	2.036	3	2	1	2	2	15
45.39	45.40	1.9962	4	0	1	2	1	2
46.22	46.22	1.9625	4	1	0	12	6	2
47.72	47.73	1.9039	2	2	2	5	10	11
49.00	49.00	1.8575	3	3	0	4	8	17
49.20	49.14	1.8503	4	1	1	10	3	7
49.88	49.88	1.8268	4	0	1	45	8	5
51.30	51.30	1.7793	3	1	1	15	36	15
51.86	51.86	1.7615	4	2	2	11	28	69
52.06	52.00	1.7553	4	2	0	8	9	23
52.70	52.70	1.7353	3	3	1	3	6	17
55.43	55.44	1.6562	4	2	1	5	2	13
55.96	55.96	1.6419	3	1	2	2	3	5
57.70	57.70	1.5963	5	1	3	6	2	8
60.04	60.04	1.5395	2	0	3	1	5	4
60.28	60.28	1.5341	4	1	1	7	1	10
61.13	61.14	1.5147	2	1	1	11	7	3
61.54	61.54	1.5056	5	1	3	1	5	13
63.13	63.14	1.4713	5	1	1	1	8	21
63.79	63.80	1.4577	5	2	0	1	1	2
64.85	64.84	1.4366	4	2	2	6	4	12

2THETA	PEAK	D	H	K	L	I(INT)	I(PK)	I(DS)
65.24	65.24	1.4288	2	2	3	5	4	10
66.45	66.46	1.4057	3	0	1	2	2	3
66.79	66.78	1.3994	5	2	1	11	7	22
67.43	67.44	1.3877	4	1	0	6	5	12
67.65	67.64	1.3837	3	3	3	6	6	12
69.80	69.80	1.3463	5	1	1	4	3	8
70.34	70.34	1.3372	4	4	0	5	3	10
70.81	70.80	1.3296	4	3	2	11	5	10
72.13	72.14	1.3083	6	0	0	7	5	15
73.29	73.30	1.2905	6	1	0	1	1	3
74.65	74.66	1.2704	4	0	3	1	1	3
75.42	75.42	1.2593	5	2	2	1	1	3
75.79	75.80	1.2540	4	1	3	6	4	15
76.08	76.04	1.2500	0	0	4	2	3	6
76.92	76.92	1.2383	3	3	3	5	3	11
78.82	78.82	1.2133	4	4	2	2	2	6
79.18	79.18	1.2086	6	1	2	7	4	17
79.50	79.50	1.2046	5	5	1	3	2	7
80.62	80.62	1.1907	5	3	2	5	3	13
81.06	81.06	1.1853	6	0	2	5	3	13
83.28	83.28	1.1592	4	3	3	1	1	3
84.75	84.76	1.1428	2	2	4	3	2	7
85.03	85.04	1.1397	5	1	3	6	3	15
85.86	85.86	1.1309	3	0	4	2	2	5
86.14	86.12	1.1279	5	5	1	5	2	16
87.25	87.24	1.1164	3	4	3	6	3	5
89.17	89.16	1.0972	5	2	4	2	1	3
90.59	90.58	1.0838	5	4	1	1	1	6
92.79	92.80	1.0637	6	4	1	2	1	4
93.90	93.90	1.0540	7	2	1	1	1	4
94.98	94.98	1.0449	4	1	4	5	2	14
96.71	96.72	1.0308	7	3	0	3	1	8
98.31	98.32	1.0182	7	3	1	4	1	8
99.46	99.46	1.0095	6	2	3	2	1	8
101.38	101.38	0.9955	6	2	2	3	1	6
102.15	102.16	0.9901	7	2	2	2	1	6
102.51	102.50	0.9876	5	4	3	1	1	5

HARDYSTONITE — WARREN AND TRAUTZ, 1930

2THETA	PEAK	D	H	K	L	I(INT)	I(PK)	I(DS)
103.01	102.96	0.9842	1	1	5	1	1	5
103.93	103.92	0.9779	4	3	4	1	1	2
104.57	104.58	0.9737	7	4	0	1	0	3
105.07	105.28	0.9704	5	1	4	1	1	3
105.28		0.9690	2	0	5	2		6
106.25	106.26	0.9629	8	0	1	3	1	11
106.43		0.9617	2	1	5	1		3
107.40	107.42	0.9557	8	1	1	2	1	5
107.40		0.9557	7	4	1	1		3
107.86	107.86	0.9529	7	3	2	4	2	13
108.02		0.9520	8	2	0	1		4
109.91	109.92	0.9408	2	2	5	1	0	3
111.76	112.08	0.9304	7	0	3	1		2
112.06		0.9288	4	4	4	2		7
112.73	112.94	0.9251	6	6	1	1	1	5
112.95		0.9240	7	1	3	1		5
112.95		0.9240	5	5	3	3		10
114.46	114.46	0.9160	6	3	4	3	1	11
115.37	115.36	0.9114	6	4	2	2	1	6
116.20	116.20	0.9073	7	4	2	1	1	4
116.91	116.90	0.9038	6	0	4	4	1	14
118.19	118.18	0.8977	6	5	1	2	1	9
119.65	119.96	0.8910	4	0	5	1		6
119.95		0.8896	8	2	2	2		8
121.98	122.22	0.8808	6	3	2	2	1	2
122.22		0.8797	3	3	5	3		10
122.96	122.76	0.8766	7	2	3	1	1	5
124.87	125.22	0.8689	4	6	2	4	1	6
125.19		0.8676	6	1	4	1		17
125.38		0.8669	9	5	0	2		4
127.93	127.94	0.8572	7	2	5	1	0	7
129.55	129.52	0.8515	9	0	3	4	0	5
131.26	131.28	0.8456	8	1	3	3	1	17
132.75	132.76	0.8407	8	4	1	3		12
132.75		0.8407	7	2	1	1	1	12
133.17	133.38	0.8394	9	1	5	3		3
133.41		0.8386	5	1	5	3		13

2THETA	PEAK	D	H	K	L	I(INT)	I(PK)	I(DS)
135.13	135.12	0.8333	0	0	6	1	0	3
136.71	136.92	0.8287	1	0	4	1	0	3
136.91		0.8281	8	2	2	1		5
137.44	137.46	0.8266	8	2	3	1	0	5
138.16	138.20	0.8246	5	1	5	1	1	6
138.35		0.8241	1	1	6	1		4
139.57	139.66	0.8208	8	5	1	3	1	13
140.25	140.28	0.8190	9	1	2	5	1	24
141.26	141.26	0.8165	7	2	4	2	2	9
141.30		0.8164	9	3	1	7		33
143.39	143.56	0.8113	4	4	1	2		3
143.60		0.8108	2	1	6	1	0	11
145.75	145.82	0.8060	9	2	2	1	0	3
146.39	146.46	0.8046	8	3	3	1	1	5
148.44	148.46	0.8004	7	5	2	6	1	27
149.62	149.54	0.7981	2	2	6	2	1	8
150.20	151.20	0.7970	9	4	0	1		5
151.19		0.7953	7	3	4	8		37
151.88		0.7940	3	0	6	1		5
154.32	154.34	0.7900	3	1	6	3	0	16
160.87	163.14	0.7811	10	1	0	3	1	16
163.10		0.7787	6	2	5	15		76
163.53		0.7783	3	2	6	2		8

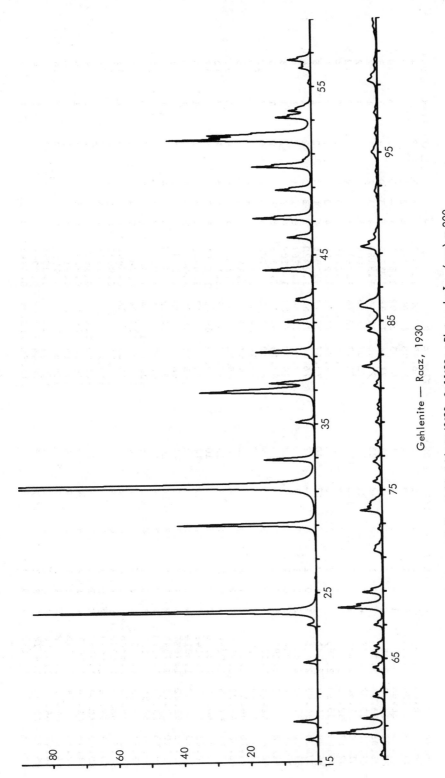

Gehlenite — Raaz, 1930

λ = CuKα Half width at 40°2θ: 0.11°2θ Plot scale I_{PK} (max) = 200

GEHLENITE – RAAZ, 1930

2THETA	PEAK	D	H	K	L	I(INT)	I(PK)	I(DS)
16.27	16.28	5.445	1	1	0	1	1	1
17.34	17.34	5.110	0	0	1	3	4	2
20.85	20.86	4.258	1	0	1	2	2	1
23.08	23.08	3.850	2	0	0		1	1
23.86	23.86	3.726	1	1	1	38	43	33
29.01	29.02	3.075	2	0	1	20	21	19
31.30	31.30	2.856	2	1	1	100	100	100
32.87	32.87	2.722	2	2	0			
35.09	35.10	2.555	0	0	2	8	3	3
36.88	36.88	2.435	3	1	1	3	17	20
37.04	36.98	2.425	1	2	0	18	14	7
37.40	37.40	2.403	2	0	2			8
39.25	39.24	2.294	3	1	1	6	9	12
41.02	41.02	2.198	3	2	0	7	4	6
42.28	42.28	2.136	3	2	2	10	3	4
42.40	42.40	2.129	2	0	2	5		2
44.10	44.10	2.052	2	1	1	3	8	6
46.02	46.02	1.9704	3	2	0	3	4	15
47.17	47.18	1.9250	4	0	0	11	9	10
48.84	48.84	1.8630	2	2	2	7	6	17
50.23	50.23	1.8149	3	0	3	11	9	
50.63	50.62	1.8014	4	1	1	9		2
51.82	51.82	1.7627	3	1	2	1	22	41
52.10	52.10	1.7541	4	2	0	27	16	26
53.15	53.16	1.7218	4	2	1	17	5	11
53.54	53.54	1.7102	3	3	1		3	
53.77	53.68	1.7033	0	0	3	4	2	2
56.08	56.08	1.6386	3	2	0	1	2	4
56.56	56.56	1.6256	2	1	2	2	3	7
59.27	59.27	1.5577	2	1	3	4	1	2
60.60	60.60	1.5268	4	3	1	1	9	21
61.44	61.44	1.5077	5	2	2	12	4	9
62.74	62.74	1.4796	2	2	3	5	2	4
64.26	64.26	1.4482	5	2	1	2	1	2
64.47	64.46	1.4440	2	2	2	1	2	2
65.29	65.30	1.4278	4	2	3	3	1	5
65.74	65.74	1.4192	3	0	3	2	2	4
66.99	67.00	1.3957	3	1	3	2	1	4
68.03	68.02	1.3770	5	2	1	11	7	22
68.93	68.92	1.3612	5	4	0	5	3	10
71.36	71.38	1.3205	4	3	0	2	1	3
71.46	71.44	1.3189	5	1	2	1	1	2
72.67	72.68	1.3000	4	0	3	5	3	12
73.77	73.76	1.2833	6	0	3	2	2	
74.16	74.16	1.2775	0	0	1	4	1	4
74.29	74.28	1.2756	4	1	3	2	1	5
75.47	75.48	1.2585	4	3	3	2	1	4
76.66	76.66	1.2420	3	1	1	2	3	4
77.64	77.64	1.2287	6	2	2	2	2	3
79.00	79.00	1.2109	4	2	2	1	1	5
81.14	81.14	1.1843	6	0	1	2	1	12
82.30	82.30	1.1706	5	1	4	5	3	9
84.39	84.40	1.1468	6	3	3	3	2	11
84.67	84.66	1.1437	3	3	3	4	3	10
85.82	85.86	1.1313	5	3	1	4	3	17
85.95	85.92	1.1300	5	1	3	6	1	3
89.39	89.40	1.0951	5	4	2	1	1	3
90.13	90.12	1.0880	3	4	4	1	1	
95.01	95.00	1.0447	7	0	0	1	2	12
96.09	96.10	1.0357	4	0	5	4	1	3
97.31	97.32	1.0259	7	1	3	1	1	5
99.25	99.26	1.0111	1	2	4	4	0	13
100.14	100.14	1.0045	5	2	2	2	1	14
103.27	103.24	0.9824	7	3	1	4	2	4
110.03	110.14	0.9401	8	1	4	5		4
110.26	110.26	0.9388	7	4	0	3		10
110.26		0.9388	4	4	5	2		7
111.56	111.56	0.9315	5	3	5	4	0	5
114.05	114.14	0.9182	5	5	3	4	1	13
114.18		0.9175	6	0	4	3		14
116.59	116.62	0.9054	7	2	1	2	1	4
118.01	118.02	0.8985	4	1	5	1	1	4
118.44	118.48	0.8965	3	1	2	1	1	7
118.86	118.86	0.8946	8	1	2			5

GEHLENITE – RAAZ, 1930

2THETA	PEAK	D	H	K	L	I(INT)	I(PK)	I(DS)
119.75	119.74	0.8905	3	3	5	2	1	7
128.51	128.54	0.8551	6	6	2	3	1	13
131.02	131.00	0.8464	5	1	5	1	0	5
135.05	135.22	0.8336	7	0	4	1	1	5
135.22		0.8331	7	4	3	2		8
135.22		0.8331	8	1	1	2		9
137.39	137.40	0.8268	2	1	6	3	1	13
138.29	138.18	0.8242	9	2	1	1	0	5
140.34	140.32	0.8188	8	2	3	2	0	9
141.98	141.98	0.8147	7	2	4	1	0	4
143.24	143.22	0.8117	9	3	0	2	0	10
145.37	145.70	0.8068	9	1	2	4	1	18
145.75		0.8060	8	5	1	8		35
147.84	147.80	0.8016	9	3	1	3	1	13
151.99	152.70	0.7938	7	6	2	1	1	5
152.60		0.7928	7	3	4	8		38
152.86		0.7924	7	5	3	3		15
153.64		0.7911	3	2	6	1		5
159.48	159.70	0.7828	6	2	5	6	0	27
163.05	163.68	0.7788	5	4	5	4	0	19
164.00		0.7778	7	7	0	3		14
164.35		0.7775	8	5	2	4		20

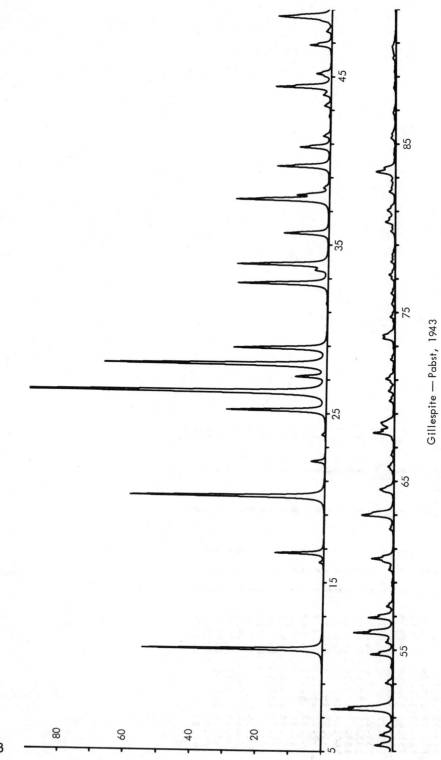

Gillespite — Pabst, 1943

λ = CuKα Half width at 40°2θ: 0.11°2θ Plot scale I_{PK} (max) = 100

GILLESPITE – PABST, 1943

2THETA	PEAK	D	H	K	L	I(INT)	I(PK)	I(DS)
11.02	11.02	8.025	0	0	2	42	55	18
16.17	16.18	5.478	1	0	0		1	1
16.71	16.72	5.300	1	0	1	12	15	7
20.06	20.06	4.422	1	0	2	52	59	38
22.13	22.14	4.012	0	0	4	4	4	3
23.72	23.72	3.748	1	0	3	1	1	1
25.15	25.16	3.537	1	1	0	29	30	27
26.22	26.22	3.396	2	0	0	100	100	100
27.15	27.16	3.281	2	1	1	9	10	9
27.86	27.86	3.199	1	1	1	69	67	75
28.84	28.84	3.093	2	1	4	29	28	33
32.67	32.68	2.739	2	0	1	30	27	40
33.47	33.48	2.675	0	0	6	3	4	5
33.80	33.80	2.650	2	2	2	30	28	42
35.65	35.66	2.516	2	2	2	15	14	23
37.63	37.64	2.388	1	1	6	30	28	49
37.68		2.385	3	0	2	3		5
37.93	37.92	2.370	3	1	0	10	10	16
38.36	38.36	2.345	3	1	1	2	2	3
39.61	39.62	2.273	3	1	2	19	16	33
40.77	40.78	2.211	2	2	4	11	9	20
41.44	41.44	2.177	2	1	6	2	2	5
43.23	43.24	2.091	2	2	1	3	2	5
43.88	43.88	2.062	3	2	4	21	17	44
44.35	44.36	2.041	3	1	5	2	2	4
45.01	45.02	2.012	3	2	2	5	5	10
45.15	45.16	2.006	0	0	8	8	7	18
46.84	46.84	1.9380	1	1	8	2	2	4
47.65	47.66	1.9067	3	0	5	8	6	19
48.47	48.54	1.8763	1	1	8	16	16	38
48.54		1.8738	4	0	4	7		17
49.33	49.34	1.8458	3	1	0	5	6	12
49.94	49.94	1.8247	4	0	1	3	4	6
50.48	50.48	1.8063	3	1	6	25	19	65
51.47	51.48	1.7740	4	0	0	3	13	8
51.63	51.60	1.7687	4	1	8	3	2	
53.03	53.04	1.7253	3	2	2			

2THETA	PEAK	D	H	K	L	I(INT)	I(PK)	I(DS)
54.72	54.72	1.6759	4	2	0	10	7	27
55.44	55.44	1.6558	4	1	4	3	2	8
56.01	56.00	1.6405	4	2	2	17	12	50
56.90	56.90	1.6168	3	3	8	11	8	32
57.57	57.58	1.5995	4	2	2	2	2	9
59.75	59.74	1.5465	4	2	4	2	2	7
60.19	60.26	1.5361	1	1	10	3	3	5
60.25		1.5347	1	0	0			5
60.40	60.40	1.5313	5	1	6	8	7	25
62.14	62.14	1.4925	3	3	1	2	1	6
62.94	62.96	1.4754	4	0	10	12	10	40
63.00		1.4741	0	0	10	4		15
64.38	64.50	1.4458	3	3	6	2	4	5
64.49		1.4436	5	1	2	4		13
65.69	65.70	1.4202	3	2	8	5	1	17
65.84	65.84	1.4174	4	1	6	1	2	5
66.53	66.54	1.4042	5	2	3	2	1	8
67.85	67.84	1.3802	5	0	4	1	6	6
68.26	68.36	1.3728	5	1	10	10	4	40
68.35		1.3713	2	2	0	2		9
68.45	68.46	1.3694	5	0	8	3	4	13
69.75	69.76	1.3471	4	1	8	4	2	14
70.32	70.32	1.3375	4	0	8	2	2	9
70.84	70.86	1.3290	3	1	12	3	1	10
71.03	71.04	1.3258	3	3	10	1	3	6
71.09		1.3249	4	4	0	3		11
72.20	72.20	1.3072	4	4	2	2	1	6
73.44	73.44	1.2882	5	1	6	6	4	8
73.58	73.62	1.2862	5	2	8	2	4	24
73.63		1.2854	1	3	0	1		10
74.73	74.72	1.2692	4	2	10	2	1	6
75.50	75.50	1.2581	5	4	4	1	1	10
76.14	76.12	1.2492	6	0	1	1	1	7
76.99	77.00	1.2374	6	1	0	1	1	6
77.22	77.22	1.2343	5	0	3	2	2	9
77.99	77.98	1.2241	5	3	4	2	1	11

GILLESPITE - PABST, 1943

2THETA	PEAK	D	H	K	L	I(INT)	I(PK)	I(DS)
79.80	79.80	1.2008	5	0	8	2	1	8
80.34	80.34	1.1940	2	2	12	5	3	25
81.02	81.04	1.1857	5	1	8	3	2	15
81.08		1.1851	6	2	0	1		7
82.15	82.14	1.1723	6	2	2	4	2	19
82.79	82.80	1.1648	3	1	12	1	1	7
83.29	83.32	1.1592	4	2	10	6	6	32
83.34		1.1586	5	3	6	7		37
85.33	85.34	1.1365	6	2	4	3	1	16
86.85	86.82	1.1205	1	1	14	1	1	6
88.32	88.32	1.1056	4	4	8	1	1	8
90.07	90.08	1.0886	4	0	12	2	1	12
90.56	90.58	1.0840	5	1	10	1	1	7
90.74	90.74	1.0823	5	3	8	2	1	11
94.28	94.28	1.0508	7	1	2	1	1	8
94.92	94.92	1.0454	2	2	12	3	1	19
95.65	95.64	1.0394	4	4	0	3	1	20
96.55	96.68	1.0320	3	1	14	2	1	14
96.71		1.0308	6	4	2	2		13
97.46	97.46	1.0248	7	1	4	2	1	11
97.46		1.0248	5	5	4	2		14
97.85	97.80	1.0218	4	4	10	1	1	7
98.03		1.0204	6	2	8	1		8
102.77	102.82	0.9858	6	0	10	2	2	10
102.79		0.9856	1	1	16	1		11
102.82	102.82	0.9854	7	1	6	3		25
103.93	104.08	0.9779	5	5	6	1		4
104.10		0.9768	4	0	14	1	1	8
105.29	105.30	0.9690	7	3	2	1		8
105.32		0.9688	2	0	16	1		9
106.44	106.44	0.9617	6	4	6	1	0	4
107.39	107.40	0.9558	3	3	14	3	1	26
107.79	107.78	0.9534	7	3	4	4	2	28
108.98	109.00	0.9462	6	2	10	2	1	13
109.83	109.82	0.9413	4	4	12	1	0	5
110.54	110.56	0.9372	7	1	8	2	1	14

2THETA	PEAK	D	H	K	L	I(INT)	I(PK)	I(DS)
110.54		0.9372	5	5	8	1		7
110.60		0.9369	8	0	0	1		7
111.73	111.74	0.9306	8	0	2	1	0	10
112.42	112.44	0.9268	5	3	12	3	1	9
112.98	113.02	0.9238	3	1	16	2	1	16
113.01		0.9236	7	3	6	2		14
113.15		0.9229	6	4	8	1		9
115.07	115.08	0.9129	6	0	12	1	1	11
115.52	115.52	0.9106	4	3	14	2	1	5
115.87	115.86	0.9089	8	2	0	1	1	16
116.87	116.98	0.9040	5	1	14	2	1	10
117.05		0.9031	8	2	2	2		12
120.54	120.58	0.8870	6	2	12	2	1	21
120.67		0.8864	8	2	4	1		5
121.02	121.12	0.8849	5	2	14	2	1	8
121.11		0.8845	7	1	10	1		5
121.11		0.8845	5	5	10	1		7
121.33		0.8836	7	3	8	2		19
122.32	122.36	0.8793	1	1	18	1	0	11
122.63	122.62	0.8780	6	6	2	1	1	12
123.99	124.02	0.8724	4	4	10	2	1	16
124.01		0.8723	3	3	16	1		22
125.23	125.52	0.8674	2	0	18	1	0	5
125.36		0.8669	4	4	14	1		7
125.56		0.8662	7	5	2	2		10
126.48	126.54	0.8626	6	6	4	1		12
127.03	127.02	0.8606	8	2	6	1	0	14
128.39	128.40	0.8556	5	3	14	2	1	23
129.55	129.54	0.8514	7	5	4	2	0	17
130.22	130.22	0.8489	8	0	8	1	0	7
131.41	131.56	0.8451	2	2	18	1	0	7
131.55		0.8446	6	0	14	1		8
131.90	131.90	0.8435	7	3	8	1	0	8
133.29	133.60	0.8390	8	3	10	1	0	11
133.62		0.8380	8	4	0	1		11
134.72	135.08	0.8346	3	1	18	3		27
135.09		0.8334	8	4	2	3	1	30

GILLESPITE - PABST, 1943

2THETA	PEAK	D	H	K	L	I(INT)	I(PK)	I(DS)
136.01	136.00	0.8307	7	1	12	1	1	11
136.75	136.80	0.8286	5	1	16	4	2	44
136.79		0.8284	7	5	6	6		60
136.98		0.8279	8	2	8	2		22
138.39	138.46	0.8240	6	2	14	1	1	10
138.63		0.8233	9	1	2	1		13
139.61	139.64	0.8207	6	4	12	5	1	56
139.78		0.8203	8	4	4	2		17
143.68	143.74	0.8106	9	1	4	4	1	48
144.34	144.46	0.8091	8	0	10	2	1	25
144.47		0.8088	7	6	2	1		7
144.65		0.8084	6	6	8	1		9
146.57	146.56	0.8042	7	1	13	1	0	7
147.40	147.46	0.8025	0	0	20	1	0	13
148.77	148.96	0.7998	4	4	16	1	0	17
149.08		0.7992	7	5	8	2		27
150.76	150.58	0.7960	3	3	18	2	0	20
152.67	153.90	0.7927	7	3	12	1	1	9
153.76		0.7909	8	2	10	2		25
153.81		0.7908	5	3	16	2		25
153.88		0.7907	9	1	6	4		52
154.30		0.7900	9	3	0	1		12
154.48		0.7898	5	4	15	1		8
163.53	163.42	0.7783	7	1	14	6	0	71
163.53		0.7783	5	5	14	1		13

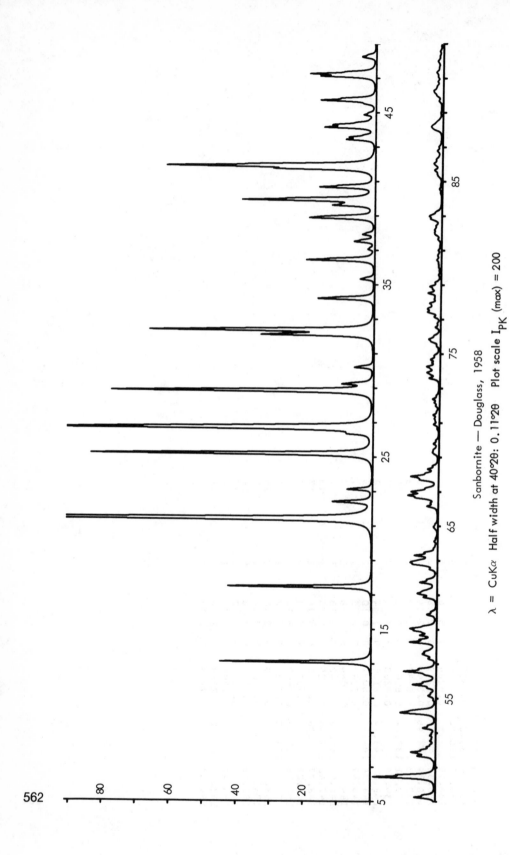

Sanbornite — Douglass, 1958

$\lambda = $ CuKα Half width at 40°2θ: 0.11°2θ Plot scale I$_{PK}$ (max) = 200

562

SANBORNITE - DOUGLASS, 1958

2THETA	PEAK	D	H	K	L	I(INT)	I(PK)	I(DS)
13.08	13.08	6.765	2	0	0	20	22	15
17.44	17.46	5.079	2	1	1	20	21	15
21.39	21.40	4.150	0	1	1	100	100	100
22.39	22.43	3.968	1	1	0	6	6	6
23.11	23.12	3.845	0	2	0	4	4	6
25.15	25.16	3.538	2	1	1	45	42	56
26.33	26.34	3.382	4	0	0	3	4	3
26.64	26.64	3.343	2	0	1	53	50	72
28.81	28.82	3.096	4	1	0	44	39	68
29.22	29.22	3.054	3	1	1	4	5	7
30.18	30.18	2.959	1	2	1	3	3	5
32.06	32.06	2.789	4	0	1	18	17	33
32.33	32.34	2.767	2	2	0	39	33	71
34.17	34.18	2.622	2	2	1	10	8	20
35.31	35.32	2.540	4	1	1	3	2	5
36.42	36.42	2.465	4	2	0	13	10	27
37.04	37.04	2.425	0	0	2	3	3	2
37.49	37.49	2.397	1	1	2	4	3	9
37.90	37.90	2.372	2	3	0	2	2	5
38.85	38.86	2.316	5	0	1	10	10	25
38.91		2.313	2	1	2	4		8
39.59	39.60	2.274	1	3	1	7	6	18
39.90	39.90	2.258	4	2	2	25	20	63
40.65	40.66	2.218	2	1	1	11	8	27
41.70	41.72	2.164	6	0	2	14	15	38
41.72		2.163	3	0	1	2		4
41.87	41.86	2.156	2	3	2	37	31	99
43.42	43.42	2.082	3	1	2	5	4	15
43.58	43.54	2.075	0	2	2	3		7
44.12	44.12	2.051	6	3	1	10	8	29
44.30	44.26	2.043	4	0	1	5	6	14
44.86	44.86	2.019	5	2	2	2	2	6
45.69	45.70	1.9839	2	2	0	11	8	35
47.08	47.08	1.9285	4	1	1	11	10	34
47.24	47.22	1.9225	0	4	1	9	2	28
48.17		1.8873	4	3	1	1		4
48.23	48.22	1.8852	3	2	2	2		7

2THETA	PEAK	D	H	K	L	I(INT)	I(PK)	I(DS)
49.23	49.24	1.8493	2	4	0	5	3	16
50.39	50.40	1.8094	6	2	1	14	9	51
51.63	51.64	1.7688	4	2	2	4	3	14
51.86	51.86	1.7616	1	3	2	5	4	18
52.16	52.16	1.7522	7	1	1	1	1	6
52.52	52.52	1.7410	5	3	1	1	1	4
52.83	52.82	1.7315	2	4	1	3	2	4
53.26	53.26	1.7185	3	3	2	3	2	11
54.12	54.14	1.6931	6	3	0	6	5	26
54.19		1.6912	8	0	0	3		14
55.12	55.14	1.6647	3	4	1	1	1	5
55.59	55.60	1.6518	8	1	0	5	4	6
55.78	55.78	1.6467	5	2	1	7	5	20
56.54	56.54	1.6262	6	2	2	2	1	29
57.28	57.28	1.6070	0	1	3	1	2	7
57.50		1.6013	6	3	1	2		5
57.56	57.56	1.5997	8	0	1	6	4	9
58.24	58.24	1.5829	4	3	2	3	3	29
58.64	58.64	1.5730	4	1	1	4	4	16
58.92	58.92	1.5662	8	1	3	2	2	18
59.03	59.04	1.5635	2	3	3	3	2	19
59.85	59.84	1.5440	3	3	3	3	3	11
60.85	60.86	1.5210	7	0	2	2	2	15
61.07	61.08	1.5160	0	4	2	1	3	16
61.17		1.5138	3	3	1	5		8
61.71	61.72	1.5018	5	2	3	3	3	19
62.48	62.48	1.4852	2	4	2	5	1	8
62.76	62.78	1.4793	8	2	1	4	4	23
62.87	62.88	1.4770	6	2	3	5	4	19
62.97		1.4748	6	5	0	4		7
63.28	63.28	1.4682	8	1	1	5	4	28
66.14	66.14	1.4117	0	0	0	3	2	15
66.62	66.64	1.4025	8	5	1	5	4	25
66.75	66.76	1.4001	6	3	0	5	5	29
67.00	66.98	1.3956	6	0	2	5	4	21
67.05		1.3946	8	3	3	4		8
67.66	67.68	1.3834	0	3	3	3	3	17

SANBORNITE - DOUGLASS, 1958

2THETA	PEAK	D	H	K	L	I(INT)	I(PK)	I(DS)
67.76	67.86	1.3817	5	2	3	2	4	11
67.87		1.3797	4	2	3	5		26
68.30	68.30	1.3722	8	1	2	3	2	15
69.26	69.26	1.3554	2	3	2	1	1	8
69.76	69.77	1.3468	4	3	1	1	1	8
70.90	70.90	1.3281	6	0	3	1	1	9
71.44	71.44	1.3193	5	2	3	1	1	7
72.36	72.36	1.3048	10	0	1	1	1	9
73.56	73.56	1.2864	10	1	1	3	2	20
73.88	73.90	1.2817	0	6	0	1	2	10
73.91		1.2812	2	5	2	1		10
74.24	74.24	1.2763	10	2	0	3	2	21
74.63	74.64	1.2706	6	4	0	2	1	10
74.69		1.2698	8	0	3	1		8
75.70	75.82	1.2553	6	2	2	2	2	11
75.83		1.2534	3	5	1	2		15
76.52	76.52	1.2439	7	4	1	1	1	11
77.12	77.12	1.2357	1	3	3	1	1	11
77.40	77.41	1.2318	5	3	3	2	2	19
77.57	77.58	1.2297	8	4	1	3	2	15
77.92	77.92	1.2250	8	3	2	2	2	18
78.50	78.50	1.2174	4	5	0	2	2	18
78.88	78.90	1.2125	2	0	4	1	2	10
78.91		1.2121	0	1	4	2		13
82.02	82.03	1.1737	0	2	4	1	1	10
82.81	82.82	1.1646	2	6	1	3	2	20
82.97	82.98	1.1628	10	3	1	1	1	23
85.62	85.63	1.1334	10	2	2	2	1	16
86.00	86.02	1.1294	6	5	2	2	1	14
86.76	86.76	1.1214	2	6	2	1	1	10
88.01	88.18	1.1087	8	5	1	2	2	15
88.11		1.1077	2	4	2	1		12
88.18		1.1070	1	0	3	2		15
89.64	89.66	1.0927	6	4	1	1	1	12
90.13	90.24	1.0881	12	1	0	2	1	17
90.78	90.78	1.0869	6	6	0	1		18
		1.0819	12	2	0	1	1	12

2THETA	PEAK	D	H	K	L	I(INT)	I(PK)	I(DS)
91.34	91.34	1.0768	5	2	1	2	1	17
93.32	93.32	1.0591	2	7	1	2	1	23
93.75	93.66	1.0553	4	3	4	1	1	11
96.18	96.18	1.0350	10	1	3	1	1	10
99.37	99.42	1.0102	12	3	1	1	1	14
101.46	101.46	0.9950	10	5	1	2	1	12
102.06	102.06	0.9907	12	2	2	2	1	19
102.51	102.48	0.9876	6	7	0	1	1	12
102.93	102.90	0.9847	11	0	3	1	1	7
103.12		0.9834	0	0	5	1		9
104.99	105.02	0.9710	7	5	3	1	1	16
105.32	105.38	0.9688	5	4	4	1	1	11
105.39		0.9683	4	6	3	1		8
106.03	106.00	0.9642	8	2	1	1		7
106.39	106.42	0.9620	4	7	2	1	0	7
106.81	106.82	0.9594	1	1	4	1	0	8
107.39	107.38	0.9558	3	5	5	1	1	10
107.89	108.02	0.9527	1	2	5	1	0	10
107.98		0.9522	2	5	4	1	1	7
108.06		0.9517	2	8	0	1		13
109.04	109.02	0.9459	9	1	4	1	0	9
109.84	109.88	0.9412	1	1	4	1	1	10
109.94		0.9406	5	5	4	1		8
111.03	111.04	0.9344	2	8	1	1	0	8
111.05		0.9344	3	6	5	1		12
111.75	111.74	0.9305	10	6	0	1		9
113.45	113.70	0.9213	8	7	0	1	0	9
113.55		0.9208	14	2	1	1	1	11
113.65		0.9203	0	3	5	1		8
113.74		0.9198	5	1	5	1		15
113.86		0.9192	4	2	5	1		10
114.82	114.82	0.9142	7	4	3	1	1	12
115.00		0.9133	0	7	3	1		11
115.89	115.86	0.9088	4	8	1	1	0	9
116.72	116.76	0.9047	12	4	0	1	1	13
116.81		0.9043	14	3	0	1		11
117.35	117.20	0.9017	3	3	5	1	0	8

SANBORNITE - DOUGLASS, 1958

2THETA	PEAK	D	H	K	L	I(INT)	I(PK)	I(DS)
118.76	118.96	0.8950	11	5	2	1	1	12
118.94		0.8942	12	5	1	1		20
119.13		0.8934	9	3	4	1		10
120.36	120.36	0.8878	2	8	2	1	0	15
121.07	121.04	0.8847	6	5	4	1	0	12
121.96	121.94	0.8808	2	6	4	1	0	8
123.01	123.16	0.8764	11	4	3	1	1	12
123.09		0.8761	14	2	2	1		10
123.23		0.8755	1	6	5	1		19
124.18	124.44	0.8716	3	6	1	1		9
124.27		0.8713	9	6	4	1	1	13
124.46		0.8705	10	3	5	1		14
124.49		0.8704	6	0	2	1		23
125.78	125.84	0.8653	5	8	1	2	1	17
125.84		0.8651	11	7	3	1		16
126.10	126.32	0.8641	10	1	3	1	1	18
126.39		0.8630	8	5	2	1		18
127.58	127.54	0.8585	10	7	4	1	0	9
127.96	127.96	0.8571	13	3	3	1	0	11
129.81	129.90	0.8505	14	2	1	1	0	14
129.90		0.8502	5	4	2	1		11
130.26	130.48	0.8490	14	8	0	1	0	12
130.82	130.86	0.8471	15	3	2	1	0	11
130.97		0.8465	12	2	6	1		13
131.72	131.66	0.8441	5	6	4	1	0	16
132.38	132.86	0.8419	0	9	1	1		11
132.80		0.8406	16	1	0	1		12
132.86		0.8404	10	7	1	2		27
133.37	133.40	0.8387	9	6	3	1	0	9
135.09	135.14	0.8335	16	0	1	1	0	12
135.68	135.88	0.8317	13	7	3	1	1	11
135.95		0.8309	7	3	5	2		29
136.80	136.80	0.8284	4	9	0	1		21
136.89		0.8282	1	8	3	1	1	18
138.20	138.38	0.8245	9	0	5	1		10
138.25		0.8244	11	3	4	1		16
138.40		0.8239	8	8	1	1		25
138.42		0.8239	2	4	5	1		13
138.62		0.8234	10	4	4	1		13
141.30	141.92	0.8164	3	5	5	1	1	21
141.64		0.8155	1	1	6	2		30
141.88		0.8149	14	4	2	2		13
142.04		0.8145	16	2	1	2		31
142.63		0.8131	12	2	4	1		12
143.88	143.88	0.8102	9	5	1	1	0	16
145.17	145.72	0.8072	14	5	1	2	1	31
145.19		0.8072	7	6	4	1		21
145.67		0.8062	9	2	5	1		16
145.82		0.8058	4	8	3	1		25
146.71	146.82	0.8039	3	1	6	1	1	12
146.87		0.8036	8	7	3	1		20
147.13		0.8031	16	3	0	2		28
147.59	148.08	0.8021	1	2	6	1	1	13
147.83		0.8016	2	9	2	2		16
148.32		0.8006	12	6	2	2		37
149.16	149.10	0.7990	6	9	0	1	0	18
150.23	151.08	0.7970	4	7	4	1	1	26
150.97		0.7956	12	5	3	2		35
151.01		0.7956	16	1	2	1		30
152.15	152.22	0.7936	5	5	5	2	1	45
152.72		0.7926	10	1	5	1		12
152.92	153.44	0.7923	14	4	3	1	1	20
153.42		0.7915	8	8	3	1		14
153.52		0.7913	12	3	2	1		11
153.65		0.7911	3	2	6	2		32
153.72		0.7910	13	1	4	1		21
156.62	157.92	0.7866	15	5	1	2	0	33
156.86		0.7862	5	0	6	2		36
157.56		0.7853	4	9	2	2		47
157.84		0.7849	9	3	5	1		14
158.80		0.7836	10	8	0	2		46
160.84	161.24	0.7811	1	1	6	2	0	47
161.09		0.7809	11	6	3	2		38
161.37		0.7805	9	7	3	1		25

TABLE 29. MISCELLANEOUS SHEET SILICATES

Variety		Dalyite	Hodgkinsonite	Sepiolite	Chloritoid
Composition	β-$Na_2Si_2O_5$	$K_{1.7}Na_{0.3}ZrSi_6O_{15}$	$Zn_2MnSiO_4(OH)_2$	$H_6Mg_8Si_{12}O_{30}(OH)_{10}$	$(Fe^2,Mg)_2AlAl_3Si_2O_{10}(OH)_4$
Source	Synthetic	Ascension Island	Franklin, N. J.	Ampandrandava, Madagascar	Point des Chats, Ile de Groix, Moribhan, France
Reference	Pant, 1968	Fleet, 1965	Rentzeperis, 1963	Brauner & Preisinger, 1956	Harrison & Brindley, 1957
Cell Dimensions					
$\underline{a}$	12.329	7.731	8.171	5.28 (13.4)	9.52
$\underline{b}$ Å	4.848	7.730	5.316	13.4 (26.8)	5.47
$\underline{c}$	8.133	6.912	11.761	26.8 (5.28)	18.19
α	90	106.23	90	90	90
β deg	104.24	111.45	95.25	90	101.65
γ	90	100.00	90	90	90
Space Group	P2$_1$/a	P$\bar{1}$	P2$_1$/a	Pnna (Pncn)	C2/c
Z	4	1	4	2	4

TABLE 29. (cont.)

Variety		Dalyite	Hodgkinsonite	Sepiolite	Chloritoid
Site Occupancy					M1a Al 0.75, Fe^3 0.25 M1b Fe^2 0.75, Al 0.15 M2a Al 1.0 M2b Al 1.0 $(OH)_1$ 0.85 $(OH)_2$ 0.85
Method	3-D, counter	2-D, film	3-D, counter	(?), film	2-D, film
R & Rating	0.0427 (1)	0.11 (av.) (2)	0.067 (1)	--- (3)	0.23 (av.) (3)
Cleavage and habit	{100}perfect, {001} imperfect. Platy, elongated parallel to b axis	$\{\bar{1}01\}$&$\{010\}$	{001}perfect Commonly elongated to c axis	Mica-like scales and fibers	{001}perfect Sometimes platy.
Comment				B estimated	B estimated
μ	80.2	171.2	343.9	67.0	206.0
ASF	0.558×10^{-1}	0.7896×10^{-1}	0.1493	0.3064	0.9457×10^{-1}
Abridging factors	1:1:0.5	1:1:0.5	1:1:0.5	0.2:0:0	1:1:0.5

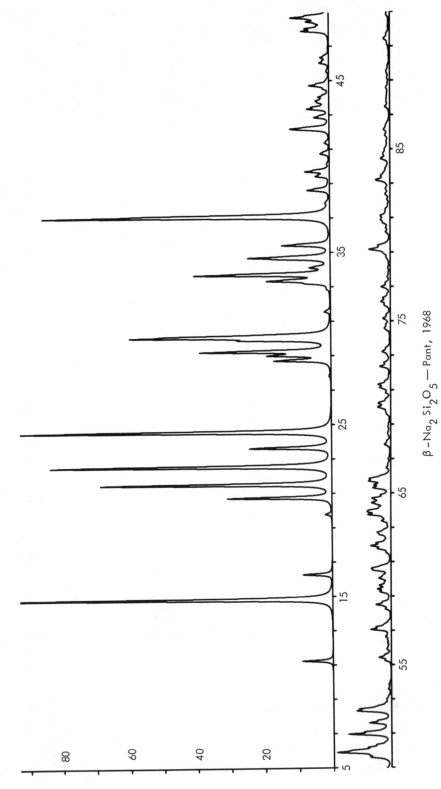

β-Na₂Si₂O₅ — Pant, 1968

λ = CuKα Half width at 40°2θ: 0.11°2θ Plot scale I_PK (max) = 100

568

BETA NA2SI2O5 - PANT, 1968

2THETA	PEAK	D	H	K	L	I(INT)	I(PK)	I(DS)
11.21	11.22	7.883	0	0	1	7	9	7
14.81	14.82	5.975	2	0	0	83	100	78
16.25	16.26	5.451	-2	0	1	7	9	7
19.75	19.76	4.492	1	1	0	2	2	2
20.73	20.74	4.282	2	1	0	28	31	27
21.50	21.50	4.130	0	0	2	64	69	61
22.54	22.54	3.942	0	1	1	79	84	75
23.61	23.62	3.765	2	1	1	21	24	21
23.66		3.757	-1	1	1	2		2
24.55	24.56	3.623	-2	1	1	91	93	87
28.69	28.70	3.109	-1	1	2	17	17	16
28.99	29.00	3.078	1	1	0	17	19	17
29.22	29.22	3.054	-3	0	1	39	39	38
29.88	29.90	2.988	4	0	0	23	28	23
30.05	30.06	2.971	2	0	2	57	60	55
30.15	30.12	2.961	-2	1	1	18	51	17
31.52	31.52	2.836	1	1	1	2	2	2
33.17	33.18	2.698	3	1	2	6	8	6
33.33	33.34	2.686	-3	0	3	19	19	19
33.69	33.68	2.658	-2	0	3	45	40	45
34.09	34.10	2.628	0	0	3	5	6	5
34.60	34.68	2.590	4	0	1	14	24	14
34.69		2.584	-4	1	1	19		18
35.40	35.40	2.533	2	1	2	16	14	16
37.06	37.06	2.424	0	2	3	100	85	100
38.59	38.60	2.331	-2	1	1	8	7	8
39.40	39.40	2.285	4	1	1	5	4	5
39.68	39.68	2.269	-4	0	3	8	7	9
40.70	40.72	2.215	-2	2	1	2	3	2
40.73		2.213	1	0	3	2		2
41.34	41.34	2.182	1	1	3	2	1	2
42.12	42.18	2.144	5	1	0	6	12	6
42.17		2.141	4	0	2	11		11
42.83	42.84	2.109	2	2	1	5	5	6
43.33	43.34	2.086	-5	1	2	8	7	8
43.47	43.44	2.080	-1	2	2	2	5	2
43.72	43.72	2.069	-3	2	1	3	4	3

2THETA	PEAK	D	H	K	L	I(INT)	I(PK)	I(DS)
43.81	43.80	2.065	0	2	0	3	4	3
44.03	44.02	2.055	-6	1	1	2	3	2
44.50	44.52	2.034	-2	2	2	2	2	2
44.70	44.70	2.026	-2	0	4	7	6	7
46.01	46.00	1.9708	4	0	2	4	3	4
46.32	46.32	1.9584	-4	2	1	3	3	3
47.87	47.88	1.8985	-6	1	2	10	8	11
48.05	48.02	1.8919	2	2	1	5	7	5
48.31		1.8823	-4	1	2	3		3
48.42	48.42	1.8783	-1	2	2	6	7	6
48.65		1.8701	-6	0	4	5		5
48.66	48.66	1.8695	3	1	4	5	11	5
48.67		1.8691	6	1	2	5		5
49.58	49.58	1.8370	0	2	3	6	6	7
49.68	49.72	1.8337	-3	1	1	2	7	2
49.86	49.88	1.8272	-6	0	4	18	16	19
49.91		1.8257	3	1	3	2		3
49.94		1.8246	6	3	1	3		4
50.17	50.16	1.8169	-3	1	0	7	7	7
50.33	50.30	1.8113	-6	0	3	1	5	1
50.94	50.94	1.7912	-4	2	2	17	13	18
51.59	51.60	1.7700	-2	2	3	8	7	9
52.29	52.30	1.7481	2	0	4	5	11	5
52.40	52.40	1.7447	-4	1	4	13	10	13
52.97	52.96	1.7271	6	2	1	1	2	1
53.55	53.56	1.7098	-6	1	1	1	1	1
55.22	55.24	1.6619	-7	1	3	2	2	2
55.42	55.42	1.6566	0	3	3	5	4	5
55.93	55.92	1.6426	0	2	4	1	1	1
56.23	56.24	1.6344	-6	1	3	1	1	1
56.62	56.62	1.6241	-5	1	3	2	2	2
57.04	57.04	1.6133	-2	0	4	9	6	9
57.15	57.18	1.6102	-7	0	0	2	5	2
58.23	58.24	1.5831	0	1	1	7	5	7
58.49	58.48	1.5766	0	3	5	6	5	4
59.24	59.24	1.5584	-5	2	3	4	4	4
59.41	59.40	1.5544	-2	2	4	3	4	3

BETA NA2SI2O5 - PANT, 1968

2THETA	PEAK	D	H	K	L	I(INT)	I(PK)	I(DS)
59.62	59.62	1.5493	-2	1	1	4	4	4
60.02	60.02	1.5400	6	1	2	4	4	4
60.49	60.50	1.5292	0	2	2	6	6	7
60.59	60.60	1.5268	-8	0	1	5	6	6
61.51	61.52	1.5062	5	1	3	2	2	2
61.75	61.76	1.5010	-1	1	2	5	4	5
61.91	61.92	1.4975	3	3	0	6	6	7
62.08	62.08	1.4938	8	0	0	3	5	3
62.56	62.58	1.4835	-2	3	2	2	4	4
62.65	62.70	1.4816	1	1	4	1	4	1
62.69		1.4806	-4	2	4	3		4
63.37	63.36	1.4664	-8	1	1	3	3	3
63.73	63.74	1.4591	6	2	1	9	7	10
63.86	63.90	1.4563	-8	0	2	3	7	3
63.98		1.4539	-6	1	3	6		7
64.33	64.42	1.4468	-3	2	1	3	6	4
64.43		1.4449	-3	3	1	6		7
64.69	64.68	1.4396	2	0	5	7	6	8
65.27	65.30	1.4283	-4	3	1	4	5	5
65.31		1.4275	8	1	0	2		3
65.32		1.4272	6	0	3	2		2
65.47	65.46	1.4244	-7	2	1	2	5	3
65.67	65.68	1.4205	4	1	4	6	7	6
65.72		1.4196	2	3	2	3		3
65.81	65.82	1.4178	8	2	0	5	7	6
66.45	66.44	1.4058	2	0	3	2	1	2
67.81	67.82	1.3809	-2	3	3	3	2	3
68.47	68.46	1.3691	6	1	3	1	1	1
68.84	68.84	1.3627	-8	0	3	1	1	1
69.99	69.98	1.3430	-6	2	4	5	4	6
70.25	70.22	1.3387	5	3	0	3	4	4
70.83	70.84	1.3292	-4	3	6	1	2	2
71.13	71.12	1.3244	-5	2	2	4	3	5
71.29	71.30	1.3217	0	1	5	4	4	4
72.35	72.34	1.3050	-2	1	5	5	3	6
73.20	73.20	1.2919	-8	2	2	4	3	4
74.46	74.56	1.2731	-5	0	5	1	2	2

2THETA	PEAK	D	H	K	L	I(INT)	I(PK)	I(DS)
74.56	74.56	1.2717	8	2	0	2	2	3
75.13	75.14	1.2634	-1	3	4	4	2	4
75.85	75.92	1.2532	-6	3	2	1		1
75.92		1.2522	0	3	3	2		3
76.11	76.12	1.2496	3	3	3	3	3	1
76.13		1.2493	-3	3	4	1		2
76.32	76.32	1.2467	-6	0	6	2	2	2
76.97	76.96	1.2378	2	1	5	5	3	6
78.69	78.70	1.2150	6	1	5	10	6	13
79.15	79.16	1.2090	-8	1	4	1		1
79.27		1.2075	-6	3	5	2		3
80.15	80.16	1.1964	-1	4	3	2	1	3
80.71	80.72	1.1896	-10	1	1	3	2	3
80.85	80.88	1.1878	2	4	0	1	3	2
80.87		1.1876	1	4	1	2		1
81.11	81.10	1.1847	2	1	6	3	3	4
81.51	81.50	1.1799	8	0	3	1	1	2
82.04	82.04	1.1736	7	3	0	1	1	3
82.38	82.38	1.1696	-10	1	0	2	2	3
83.19	83.20	1.1603	-10	0	4	7	4	9
83.60	83.44	1.1556	-10	1	4	1	3	2
83.44	83.44	1.1464	8	0	1	4	3	5
84.95	84.94	1.1407	10	0	3	1	2	2
85.42	85.42	1.1355	3	4	1	2	1	3
85.81	85.80	1.1314	5	3	3	2	2	3
86.13	86.12	1.1280	-2	3	7	2	2	3
87.46	87.46	1.1143	-8	3	1	3	1	3
88.02	88.00	1.1087	-6	2	6	1	1	1
88.54	88.56	1.1035	-8	0	7	1	1	1
88.61		1.1028	-2	4	3	1		1
89.12	89.12	1.0978	4	4	4	2	1	1
89.54	89.52	1.0937	4	3	4	2	1	2
91.12	91.14	1.0788	4	1	6	3	2	4
91.43	91.42	1.0760	-6	1	7	3	1	1
92.80	92.84	1.0636	-10	1	5	2	2	4
93.19	93.14	1.0602	-8	2	7	2	2	2
94.69	94.72	1.0474	10	1	2	1	1	2

BETA NA2SI2O5 - PANT, 1968

2THETA	PEAK	D	H	K	L	I(INT)	I(PK)	I(DS)
95.19	95.20	1.0431	-10	2	4	2	1	1
95.42	95.46	1.0413	-5	4	3	2	1	1
96.50	96.52	1.0324	0	4	4	2	2	2
98.85	98.80	1.0141	5	4	2	1	1	3
99.62	99.66	1.0083	-6	4	1	2	1	2
99.84	99.78	1.0067	-8	1	7	1	1	2
100.16	100.14	1.0043	-6	2	7	1	1	2
100.87	100.98	0.9991	10	0	3	2		2
100.99		0.9983	-7	4	1	1	1	1
101.29	101.30	0.9962	-12	1	3	1	1	1
101.67	101.66	0.9935	9	3	1	1	1	1
102.11	102.04	0.9903	-9	3	4	1	1	1
102.41	102.42	0.9883	7	4	0	1	1	1
102.45		0.9880	-8	3	8	2		2
102.83	102.78	0.9854	-2	0	6	1		1
104.44	104.44	0.9745	-12	3	4	1	1	1
104.78	104.90	0.9723	-7	1	5	1	1	1
104.87		0.9717	8	4	7	1		1
104.89		0.9716	2	0	4	1		2
104.92		0.9714	3	2	4	1		1
105.21	105.24	0.9695	-3	4	5	2	1	1
105.48		0.9678	10	1	8	2		3
105.81	105.80	0.9656	4	3	0	2	1	3
106.57	106.58	0.9608	5	1	7	1	1	3
107.01	107.06	0.9581	5	3	5	2	1	3
107.12		0.9575	-1	3	5	2	1	2
107.86	107.86	0.9529	-3	5	2	1	1	1
109.54	109.72	0.9430	-2	3	7	1	1	2
109.55		0.9429	-5	3	5	1	2	2
109.66		0.9423	-12	2	1	2		2
109.69		0.9421	-11	2	5	1		2
109.75		0.9417	4	4	4	1		3
110.07	110.10	0.9399	-12	2	2	2	2	1
110.11		0.9397	-11	5	3	1		1
110.20		0.9391	-4	2	1	1		1
110.30		0.9386	-2	5	5	1		1
110.32		0.9385	-12	2	2	1		1

2THETA	PEAK	D	H	K	L	I(INT)	I(PK)	I(DS)
112.00	112.14	0.9291	3	5	1	1	1	1
112.09		0.9286	-3	5	2	1		1
112.21		0.9279	-7	4	2	1		2
112.57	112.50	0.9260	-5	3	8	1	1	1
114.25	114.26	0.9171	-11	1	1	1	1	2
114.85	115.06	0.9141	-9	2	7	1	1	1
115.05		0.9130	4	3	6	1		1
115.09		0.9128	0	2	8	1		2
115.39		0.9113	-6	3	7	1		1
116.21	116.22	0.9072	1	3	7	1	0	1
117.16	117.30	0.9026	-4	0	9	1	0	1
117.32		0.9018	8	2	5	1		1
118.03	117.96	0.8985	5	5	0	1	1	1
118.70	119.02	0.8953	-9	3	6	1	1	2
118.96		0.8941	-5	5	2	1		1
119.03		0.8938	-9	4	3	1		1
119.51	119.50	0.8916	-11	2	4	1	1	1
119.67		0.8909	6	0	7	1		1
119.68		0.8908	0	4	6	1		1
121.40	121.40	0.8832	4	5	2	1	1	2
121.43		0.8831	-13	2	2	1		2
122.73	122.80	0.8776	-1	4	6	1	1	1
123.45	123.84	0.8746	-1	5	4	1	1	1
123.76		0.8733	7	3	5	1		2
123.86		0.8729	-10	0	8	1		2
124.12	124.18	0.8719	-14	1	0	1	1	1
124.13		0.8718	6	4	4	1		1
124.17		0.8717	-6	1	9	1		1
124.37		0.8709	3	5	3	1		1
124.55		0.8702	8	0	6	1		1
124.59		0.8700	0	5	4	1		1
126.26	126.28	0.8635	12	2	0	1	0	1
127.31	127.36	0.8595	13	0	3	1	1	1
127.42		0.8591	-10	1	8	1		2
129.14	129.12	0.8529	9	2	5	1	0	1

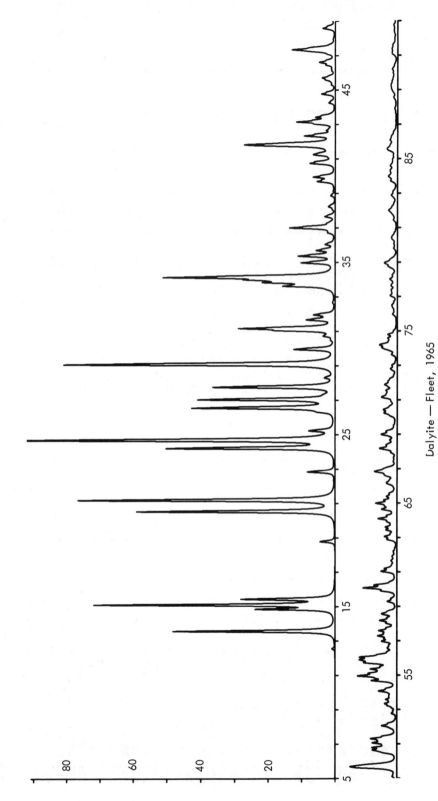

Dalyite — Fleet, 1965

λ = CuKα Half width at 40°2θ: 0.11°2θ Plot scale I$_{PK}$ (max) = 100

572

DALYITE – FLEET, 1965

2THETA	PEAK	D	H	K	L	I(INT)	I(PK)	I(DS)
13.56	13.56	6.525	1	0	0	41	48	35
14.84	14.84	5.966	0	0	1	18	24	16
15.09	15.10	5.865	-1	1	0	61	72	54
15.37	15.44	5.759	-1	0	1	3	28	3
15.43		5.736	0	1	1	21		19
18.76	18.76	4.726	-1	1	1	4	4	4
20.53	20.54	4.323	-1	1	1	55	59	52
21.18	21.20	4.190	1	1	0	74	77	70
22.81	22.82	3.894	0	1	1	8	8	8
24.20	24.20	3.674	-1	1	1	50	50	50
24.67	24.68	3.605	0	2	0	92	100	92
24.69		3.603	-1	0	2	8		8
25.21	25.22	3.530	0	0	2	7	7	7
26.54	26.54	3.356	-1	1	2	43	43	44
27.02	27.02	3.297	-1	-1	2	42	41	44
27.75	27.76	3.212	0	1	2	39	36	40
28.28	28.28	3.153	-1	1	1	3	3	3
29.06	29.06	3.070	-1	1	2	89	81	95
29.93	29.94	2.983	-2	0	0	13	13	14
30.70	30.70	2.910	-1	2	0	3	4	3
31.03	31.04	2.880	-2	2	0	17	19	18
31.16	31.16	2.868	0	-2	1	26	29	29
31.65	31.66	2.825	-1	1	2	9	9	10
31.84	31.94	2.808	-1	-1	2	1	7	2
31.94		2.799	-2	1	2	6		7
33.60	33.60	2.665	2	1	1	16	16	18
33.81	33.82	2.649	0	-2	2	20	22	23
33.98	33.98	2.636	2	1	0	21	28	24
34.12	34.12	2.625	-2	1	1	52	51	60
34.95	34.96	2.565	-2	-1	2	11	10	13
35.35	35.36	2.537	0	-3	1	13	11	15
35.68	35.68	2.514	1	1	3	6	6	7
36.02	36.02	2.491	1	1	3	3	3	4
36.70	36.70	2.447	2	0	1	2	2	2
36.99	36.98	2.428	-3	1	1	17	14	20
37.25	37.24	2.412	-3	0	1	2	3	2
37.63	37.64	2.388	-2	-2	1	4	3	4
38.21	38.22	2.353	0	3	0	2	2	3
38.84	38.84	2.316	0	-3	2	2	1	2
39.68	39.68	2.270	-2	-3	1	6	6	8
39.94	39.94	2.256	-1	-3	1	8	7	10
40.73	40.74	2.213	-1	-1	2	9	8	11
40.82	40.82	2.208	-1	0	3	2	6	2
41.22	41.24	2.188	-1	-2	3	5	7	7
41.26		2.186	-1	2	3	3		4
41.75	41.82	2.161	-2	2	1	15	27	20
41.79		2.159	-3	1	1	1		2
41.81		2.158	-1	3	2	21		28
42.32	42.32	2.134	1	-1	3	11	9	14
42.64	42.64	2.119	0	-1	1	2	3	3
43.01	43.02	2.101	-1	-3	3	6	7	8
43.14	43.14	2.095	2	-2	0	11	12	15
43.41	43.40	2.083	-2	3	1	6	6	8
44.24	44.24	2.045	-2	-2	3	2	2	3
44.75	44.76	2.023	1	2	5	5	4	7
45.29	45.30	2.000	1	-1	1	1	1	2
45.58	45.58	1.9886	0	0	3	3	3	4
45.69	45.70	1.9839	2	-1	2	3	4	5
46.44	46.44	1.9537	0	3	2	5	3	4
46.60	46.60	1.9472	0	2	1	1	5	7
47.23	47.34	1.9229	1	-1	4	5	13	2
47.25		1.9221	-2	-3	1	1		7
47.30		1.9199	-3	2	0	5		2
47.34		1.9187	3	1	3	12		17
48.58	48.58	1.8724	-3	-1	1	4	4	6
49.08	49.08	1.8547	1	-2	3	3	1	2
49.58	49.66	1.8370	2	0	1	11	14	16
49.65		1.8347	-3	-2	1	7		11
49.66		1.8342	1	2	3	5		7
49.73		1.8317	-4	1	1	2		3
50.52	50.62	1.8049	3	0	0	3	7	5
50.59		1.8025	-2	4	0	1		2
50.62		1.8016	-4	2	2	5		7
50.77	50.76	1.7966	2	-4	1	5	7	8

DALYITE – FLEET, 1965

2THETA	PEAK	D	H	K	L	I(INT)	I(PK)	I(DS)
51.04	51.04	1.7879	-2	-3	3	8	7	12
51.11		1.7856	3	-3	1	1		2
51.29	51.28	1.7797	-3	-2	3	9	8	14
52.04	52.04	1.7559	-1	-4	1	5	5	8
52.75	52.76	1.7338	-1	-1	3	5	2	3
53.19	53.22	1.7204	-1	-3	2	2	3	4
53.31	53.32	1.7169	-4	0	1	3	4	5
53.56	53.56	1.7094	-1	-1	0	4	3	6
53.80	53.80	1.7024	-2	-2	3	2	2	3
54.04		1.6954	-1	-1	4	5	5	7
54.07	54.06	1.6947	2	-2	2	2		4
54.66	54.66	1.6776	-3	3	3	8	8	13
54.93	54.94	1.6700	-4	-2	2	11	12	18
54.97		1.6689	1	0	3	5	9	8
55.11	55.08	1.6651	0	-4	2	1		2
55.34	55.34	1.6588	-2	0	0	8	9	13
55.68		1.6493	-1	4	1	6	8	10
55.71	55.72	1.6484	-2	-4	2	1		2
55.73		1.6481	-3	2	3	5		8
55.79		1.6464	-1	2	1	3		4
55.85	55.88	1.6446	-4	-1	1	1		2
55.90		1.6434	-1	-4	3	6	11	10
56.01	56.02	1.6405	-3	-4	0	4		7
56.04		1.6395	3	2	0	7	11	11
56.35	56.34	1.6314	-4	0	0	2	3	4
56.56	56.56	1.6258	-2	1	3	2	3	3
56.59		1.6249	-4	-1	2	1		2
56.94		1.6158	-1	1	4	3	5	6
56.96	56.96	1.6152	-3	-3	2	2		4
57.25	57.26	1.6077	-4	3	1	6	6	10
57.31		1.6061	0	-2	4	1		2
57.55	57.56	1.6001	-1	-3	4	2	5	3
57.57		1.5996	-3	0	1	5		8
58.00	58.00	1.5889	-3	4	3	4	3	6
58.15		1.5850	1	-4	1	3	5	5
58.16	58.16	1.5848	1	4	0	2		3

2THETA	PEAK	D	H	K	L	I(INT)	I(PK)	I(DS)
58.50	58.54	1.5764	-2	-4	3	3	5	4
58.55		1.5750	3	-1	2	2		3
58.57		1.5748	-3	-4	1	2		3
59.27	59.28	1.5577	1	-3	1	4	3	7
59.88	59.90	1.5433	-1	-5	3	6	5	10
60.06	60.06	1.5392	2	-3	1	12	10	20
60.24	60.22	1.5348	-4	-2	2	3	7	6
61.00	61.02	1.5176	-1	-2	1	3	5	5
61.02		1.5171	-2	-4	2	1		7
61.17	61.18	1.5139	1	-3	1	2	4	2
61.65	61.64	1.5032	2	-1	1	2	2	3
61.90	61.90	1.4977	-1	-5	0	1		3
62.84	62.90	1.4774	2	-1	2	3	4	6
62.89		1.4764	-1	0	4	3		6
63.05	63.04	1.4732	-1	-1	1	4	4	8
63.43	63.42	1.4653	-4	-1	5	3	3	5
63.58	63.58	1.4621	-4	-1	1	3	4	4
63.89	63.92	1.4557	-2	-3	3	2	3	4
63.92		1.4551	-2	2	4	2		3
64.08	64.08	1.4519	-1	-1	2	2	6	9
64.08		1.4518	-5	-2	5	5		6
64.61	64.64	1.4413	2	0	3	3	5	6
64.64		1.4406	2	-5	3	3		7
64.98	64.98	1.4340	0	-2	4	4	6	11
64.98		1.4340	-3	-4	6	6		
65.34	65.38	1.4270	3	0	1	2	3	2
65.39		1.4260	-4	2	2	1		4
65.54	65.54	1.4230	-2	2	1	3	2	3
66.12	66.14	1.4119	-1	0	4	2	3	2
66.20		1.4104	-5	0	5	1		5
66.55	66.64	1.4039	-2	-4	4	2	4	5
66.62		1.4026	-1	-3	3	2		5
66.81	66.82	1.3990	-1	-5	6	6	7	11
66.89		1.3976	-3	0	4	4		7
67.39	67.40	1.3884	3	-4	1	1	1	2
67.92	67.94	1.3788	1	-5	1	2	2	4
68.10	68.16	1.3757	-3	-4	2	2	5	3

DALYITE - FLEET, 1965

2THETA	PEAK	D	H	K	L	I(INT)	I(PK)	I(DS)
68.17		1.3744	4	-4	1	1		13
68.62	68.64	1.3665	-5	3	1	7	2	2
68.67		1.3657	-5	-1	3	1		2
68.95	68.98	1.3608	0	1	3	1	5	6
69.00		1.3600	-1	-2	4	3		9
69.14	69.16	1.3574	-3	-1	5	5	3	2
69.50	69.50	1.3513	-4	4	2	1	1	2
70.17		1.3400	1	-2	3	1	4	3
70.22	70.22	1.3392	-4	1	4	2		7
70.40	70.40	1.3363	1	-2	2	3		5
70.50		1.3346	-3	5	3	3	4	3
71.13		1.3244	4	-2	2	1		11
71.21	71.14	1.3230	-2	0	5	5	5	6
71.52	71.52	1.3181	-2	4	5	3		5
71.56		1.3174	-3	0	5	3	3	4
71.86	71.86	1.3126	-4	2	4	2		8
71.90		1.3121	-2	-5	1	4	4	2
72.21	72.20	1.3071	-2	1	5	1		4
72.70	72.78	1.2995	1	5	0	1	2	2
72.79		1.2982	0	3	3	3	2	2
73.61	73.62	1.2857	0	-2	5	3		3
73.93	74.02	1.2810	1	-6	1	3	2	5
74.02		1.2797	0	4	1	2	5	7
74.03		1.2795	-5	3	5	3		4
74.17	74.18	1.2774	-3	-3	1	4		6
74.57	74.56	1.2714	-4	-1	5	3	5	9
74.78	74.78	1.2685	0	-6	2	3	3	7
76.48	76.48	1.2445	2	-1	1	2	3	5
77.02	77.04	1.2371	-2	-1	5	2	2	4
77.06		1.2366	5	-3	1	2		4
77.38	77.38	1.2322	-5	-2	4	2	2	4
77.58		1.2295	3	3	1	1		4
77.62	77.62	1.2282	-5	3	2	2		2
77.68		1.2239	-6	1	1	1	2	3
78.00	78.00	1.2155	-1	1	4	1		6
78.65	78.76	1.2142	-2	3	4	3	3	4
78.75		1.2123	-6	2	2	2		6
78.89	78.94		-1	-6	2	3	4	6

2THETA	PEAK	D	H	K	L	I(INT)	I(PK)	I(DS)
78.96	78.96	1.2114	-5	-2	1	3		6
79.80	79.80	1.2008	-6	2	1	1	1	3
80.31	80.34	1.1945	3	-5	3	2	2	4
80.43	80.44	1.1930	-3	-4	5	2	2	3
81.43	81.42	1.1808	4	-3	0	1	3	3
81.98	81.98	1.1744	-5	-1	5	2	1	4
81.98		1.1743	-1	-6	1	1	3	3
82.77	82.86	1.1651	-1	1	3	1		2
82.87		1.1640	-3	-4	4	3	2	6
83.71	83.72	1.1543	1	1	5	1		3
84.01	83.96	1.1510	-6	0	4	3	3	3
84.35	84.32	1.1472	1	-5	5	1	2	3
84.55	84.58	1.1450	-2	-5	4	1	2	3
84.60		1.1445	-5	4	3	1		2
84.71		1.1432	4	-2	3	1		3
84.96	84.94	1.1406	-5	4	6	1	2	3
85.14	85.16	1.1386	-3	-1	0	2	1	3
85.49	85.54	1.1348	-4	3	6	2	3	4
85.55		1.1342	-2	-3	2	2		4
85.63		1.1334	-1	6	4	2		4
86.06	86.06	1.1288	-6	0	0	1	2	6
86.22		1.1271	0	4	3	2		4
88.53	88.54	1.1035	-4	4	1	1		4
88.91	89.06	1.0998	-1	6	6	1	2	4
89.05		1.0984	-3	5	3	2		4
89.20	89.22	1.0970	-5	-1	1	1	1	4
89.31		1.0959	5	1	1	1	2	3
89.50	89.48	1.0941	5	-4	2	1		3
89.95	89.96	1.0897	-1	0	1	1		3
90.18	90.18	1.0876	5	-1	0	2	2	4
91.08	91.12	1.0791	-4	6	5	1	1	3
91.37	91.36	1.0765	0	-7	3	1	2	3
92.41	92.42	1.0671	-5	-4	3	1	2	3
92.74	92.74	1.0642	-1	-6	5	1	1	3
93.18	93.20	1.0603	-2	7	0	1	1	3
93.90	93.96	1.0541	-4	-5	2	1	1	3

DALYITE - FLEET, 1965

2THETA	PEAK	D	H	K	L	I(INT)	I(PK)	I(DS)
93.98		1.0534	-6	-1	5	1		4
94.77	94.76	1.0466	-1	7	0	1		3
94.85		1.0460	3	5	0	1	2	3
95.20	95.20	1.0431	-1	-7	2	1		3
95.52	95.72	1.0404	-5	-1	4	1	2	3
95.66		1.0393	-2	-1	5	1	2	3
95.73		1.0387	-6	5	1	2		5
96.30	96.30	1.0341	0	-7	4	2	3	4
96.32		1.0339	-1	-5	6	1		5
96.36		1.0336	-6	-2	5	1		3
96.72	96.64	1.0306	5	-6	1	1	2	4
97.04	97.02	1.0281	-1	6	2	1	1	3
100.17	100.18	1.0042	-7	-1	3	2	1	5
100.62	100.68	1.0009	-3	7	1	1	1	4
100.72		1.0002	2	-2	7	2		5
101.29	101.30	0.9962	4	-4	4	1	1	2
101.31		0.9960	-5	-1	6	1		3
101.51	101.54	0.9946	-1	4	6	1	1	2
101.55		0.9943	0	0	6	1		3
101.94	101.94	0.9915	5	-2	3	1	1	3
101.96		0.9914	-1	3	5	1		3
102.16	102.20	0.9900	-7	-1	7	1	1	2
102.58	102.64	0.9871	-3	-2	2	1	1	2
102.65		0.9866	-3	2	7	1		4
102.89	102.92	0.9849	-7	4	6	1	1	3
103.41	103.40	0.9814	-7	2	0	1	1	4
103.77	103.86	0.9790	3	-3	5	1	2	3
103.87		0.9783	-6	0	6	2		4
104.00	103.98	0.9774	-3	5	4	1	1	3
104.01		0.9774	5	1	2	1		2
104.04		0.9772	-2	2	6	1		2
106.01	106.26	0.9644	-2	-5	3	1	2	3
106.18		0.9633	5	-5	7	1		2
106.21		0.9631	-2	-1	6	2		5
106.27		0.9628	-6	6	2	2		5
106.48	106.54	0.9615	-2	-8	2	1	2	2
106.53		0.9611	-7	3	4	1		2
106.54		0.9610	-4	-6	3	1		2
106.54		0.9610	-4	-6	4	1		2
106.78		0.9596	-5	-4	6	1		2
106.80		0.9595	-2	-4	7	1		3
107.07	107.10	0.9578	-4	4	5	1	2	2
107.08		0.9577	-1	-3	7	1		2
107.22		0.9568	-7	-2	3	1		2
107.29		0.9564	-2	6	3	1		4
107.37		0.9559	1	-8	1	1		2
108.27	108.32	0.9504	-6	-3	1	1	1	3
108.47		0.9493	2	5	2	1		2
108.65	108.68	0.9482	0	-8	2	1	1	2
108.72		0.9478	-6	-4	4	1		3
109.19	109.16	0.9450	-1	-4	7	1	1	2
109.46	109.46	0.9434	-2	-7	1	2	1	5
110.07	110.10	0.9399	6	-1	2	1	1	4

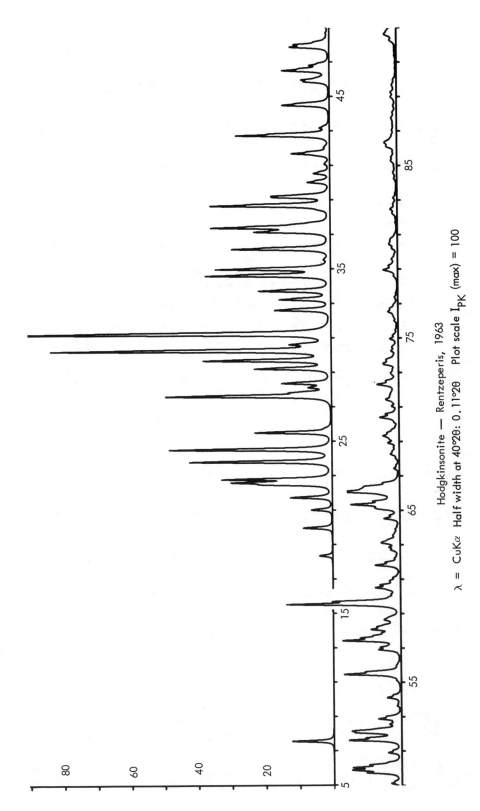

Hodgkinsonite — Rentzeperis, 1963

$\lambda = $ CuKα Half width at 40°2θ: 0.11°2θ Plot scale I_{PK} (max) = 100

HODGKINSONITE – RENTZEPERIS, 1963

2THETA	PEAK	D	H	K	L	I(INT)	I(PK)	I(DS)
7.54	7.54	11.712	0	0	1	9	12	3
18.31	18.32	4.841	0	1	1	3	4	2
19.93	19.94	4.450	-1	1	0	7	9	5
20.98	20.98	4.231	-1	1	1	5	6	4
21.70	21.70	4.093	-1	1	1	10	12	7
22.45	22.46	3.957	-2	0	1	15	22	11
22.57	22.58	3.936	0	0	2	21	30	15
22.76	22.76	3.904	0	1	2	26	33	19
23.78	23.78	3.739	2	0	1	38	42	28
24.49	24.50	3.632	-1	1	2	44	48	34
25.47	25.48	3.494	-2	1	0	21	23	17
27.59	27.60	3.231	2	1	0	48	49	42
27.80	27.80	3.206	-1	1	2	8	13	7
28.09	28.10	3.174	2	0	1	5	7	5
28.34	28.34	3.147	-2	1	1	13	15	12
29.18	29.18	3.058	2	1	1	21	23	20
29.65	29.66	3.010	-1	0	3	36	38	35
30.22	30.22	2.955	-2	0	1	82	83	79
30.50	30.58	2.928	0	0	2	2	12	12
30.59		2.920	-2	1	2	8		8
31.20	31.20	2.865	1	1	3	100	100	100
32.58	32.58	2.746	2	1	3	17	17	18
33.20	33.20	2.696	-1	0	3	15	15	16
33.69	33.70	2.658	0	2	0	22	21	24
34.57	34.58	2.592	0	1	4	38	37	43
34.96	34.96	2.565	1	2	1	35	34	40
35.50	35.50	2.527	-2	2	0	1	2	2
36.12	36.12	2.484	0	2	2	31	29	37
37.11	37.12	2.420	-1	2	1	22	23	27
37.35	37.36	2.405	0	1	3	7	35	9
37.36		2.405	-3	1	2	29		36
38.63	38.64	2.329	3	1	1	40	35	51
39.12	39.14	2.300	-3	1	1	15	17	20
39.20		2.296	-1	2	2	9	18	11
39.99	40.00	2.253	-2	1	4	7	7	10
40.50	40.50	2.225	2	2	0	5	5	7
41.05	41.04	2.197	0	2	3	2	2	2
41.66	41.66	2.166	2	2	1	11	11	16
42.00	42.00	2.149	-1	2	3	11	2	2
42.70	42.70	2.116	-2	2	2	21	28	31
42.71		2.115	-2	0	5	11		16
43.14	43.14	2.095	2	1	4	2	3	3
44.46	44.48	2.036	-4	0	1	13	14	20
44.50		2.034	-4	0	0	5		7
45.82	45.84	1.9785	-4	0	2	7	8	12
45.93	45.94	1.9740	2	0	1	5	8	7
46.43	46.48	1.9541	2	0	5	3	14	5
46.48		1.9519	0	0	6	14		23
46.78	46.78	1.9403	-1	1	0	6	6	9
47.84	47.86	1.8998	4	2	0	10	12	16
47.88		1.8984	3	2	1	5	11	9
48.02	47.98	1.8932	-3	2	2	6	11	10
49.09	49.09	1.8542	-4	1	6	11	11	6
49.72	49.72	1.8323	0	1	6	12	15	20
49.89	49.88	1.8262	-2	0	4	10	15	22
50.03	50.02	1.8214	-1	1	4	6	10	17
50.20	50.18	1.8159	-2	2	2	5	5	10
50.88	50.88	1.7931	3	1	4	20	16	38
51.63	51.64	1.7688	-4	1	3	5	14	9
52.05	52.08	1.7555	1	1	6	9	15	17
52.08		1.7547	-3	1	1	9		19
52.55	52.16	1.7521	0	3	0	10	15	6
52.86	52.56	1.7399	4	2	3	3	4	16
53.61	52.86	1.7304	2	2	1	8	7	3
54.09	53.62	1.7080	-1	2	2	1	1	10
55.47	54.10	1.6939	-2	2	5	21	4	44
55.52	55.48	1.6552	4	1	3	3	17	5
55.97		1.6536	4	2	4	1		2
56.88	55.96	1.6416	-3	3	4	4	2	9
56.89	56.90	1.6172	-2	3	5	3	7	7
57.03	57.04	1.6171	3	3	3	5	6	5
57.43	57.42	1.6136	0	3	4	3	17	49
57.72	57.70	1.6032	4	0	4	23	9	14
		1.5959	0	1	7	7		

HODGKINSONITE - RENTZEPERIS. 1963

2THETA	PEAK	D	H	K	L	I(INT)	I(PK)	I(DS)
58.07	58.08	1.5871	-4	2	2	10	9	22
58.16		1.5847	-4	2	1	2		5
58.24	58.22	1.5827	-3	1	6	1	7	3
58.34		1.5804	-2	3	2	3		6
58.63	58.62	1.5733	0	2	6	4	5	9
58.70		1.5714	1	2	3	2		5
59.56	59.56	1.5509	2	3	2	47	34	107
60.12	60.12	1.5376	-5	1	2	1	3	3
60.37	60.50	1.5318	-2	1	7	2	8	5
60.50		1.5290	4	2	1	8		19
60.67	60.66	1.5252	5	1	1	3	6	6
60.91	60.90	1.5197	-2	3	3	3	3	7
61.80	61.80	1.4998	2	0	7	11	8	25
62.21	62.20	1.4910	-5	1	1	2	2	4
62.68	62.70	1.4808	2	3	3	2	3	6
62.83	62.84	1.4778	1	1	4	1	3	3
63.09	63.10	1.4722	5	1	2	6	6	14
63.17		1.4707	4	0	5	4		9
63.56	63.54	1.4626	3	3	1	2	2	6
64.52	64.52	1.4431	-2	3	4	6	5	16
65.14	65.14	1.4323	2	2	3	5	8	14
65.15		1.4307	3	2	2	6		14
65.35	65.34	1.4268	-5	1	1	17	15	44
66.03	66.08	1.4136	-1	2	4	2	17	6
66.06		1.4132	0	3	7	14		36
66.13		1.4118	-1	1	8	12		32
66.40	66.38	1.4067	-3	2	6	11	11	28
67.42	67.42	1.3879	5	2	0	2	2	6
68.00	68.00	1.3774	-4	1	5	1	1	3
68.26	68.26	1.3728	2	2	2	3	3	8
68.66	68.66	1.3658	5	1	1	1	2	4
68.91	68.92	1.3614	-6	0	0	4	3	11
68.92		1.3561	6	1	1	2		6
69.22	69.32	1.3561	6	0	1	5	5	15
69.32		1.3544	-4	1	7	1		
69.47	69.50	1.3518	-5	0	5	1	4	4
69.66	69.66	1.3486	-6	0	2	1	2	4
70.10	70.12	1.3412	-5	2	3	1		4

2THETA	PEAK	D	H	K	L	I(INT)	I(PK)	I(DS)
70.38	70.38	1.3366	-4	3	1	6	6	8
70.41		1.3361	4	0	1	3		3
70.58	70.58	1.3333	6	1	7	1	4	8
71.04	71.04	1.3258	3	4	1	3	2	8
71.36	71.44	1.3205	0	4	2	1	4	5
71.40		1.3200	-4	2	3	2		5
71.46		1.3193	3	0	6	1		3
71.77	71.76	1.3190	-6	1	3	1	2	5
72.23	72.30	1.3140	4	1	0	2	6	4
72.31		1.3068	-1	4	1	1		26
72.86	72.96	1.3056	-3	1	6	9	2	4
72.97		1.2971	6	2	1	1		6
73.45	73.50	1.2954	-4	0	2	2	5	13
73.53		1.2881	1	3	5	4		13
73.73	73.72	1.2868	-1	4	2	1	4	3
73.82		1.2840	-1	2	8	1		3
74.22	74.24	1.2826	5	0	3	2	2	3
74.27		1.2766	-6	1	3	1		3
74.94	74.94	1.2759	-1	4	9	4	3	12
75.98	75.98	1.2662	-1	2	8	2	1	5
76.50	76.52	1.2514	-1	4	8	3	3	11
76.64	76.70	1.2441	-4	3	4	1		4
76.69		1.2422	-2	1	2	2	1	5
77.36	77.36	1.2415	-1	3	3	3	4	5
78.57	78.58	1.2325	0	2	9	6	2	21
78.94	78.94	1.2165	-6	1	7	2	2	6
79.57	79.58	1.2117	-1	4	4	2	2	8
80.20	80.20	1.2036	-1	4	8	2	3	6
80.39	80.44	1.1958	2	4	0	1	2	5
80.77	80.74	1.1934	3	4	4	1	2	5
82.16	82.24	1.1889	4	3	4	4	2	5
82.25		1.1721	-2	3	4	2		8
82.45	82.46	1.1712	0	0	10	1	2	4
82.83	82.72	1.1687	0	2	9	2	1	6
83.12	83.12	1.1644	6	2	2	2	1	4
84.22	84.24	1.1610	6	1	4	2	1	4
		1.1486	-2	1	9	1	1	4

HODGKINSONITE - RENTZEPERIS, 1963

2THETA PEAK	PEAK	D	H	K	L	I(INT)	I(PK)	I(CS)
84.57	84.58	1.1448	-4	0	9	1	1	4
85.22	85.26	1.1378	-6	1	6	1	2	5
85.49	85.48	1.1348	-4	3	6	1	2	4
86.07	86.10	1.1286	0	3	8	3	3	10
86.17		1.1276	-2	1	10	2		7
86.30	86.32	1.1263	-4	2	8	3		13
86.80	86.78	1.1210	-3	4	4	2		6
87.01	87.04	1.1189	-1	1	10	1		4
87.44	87.44	1.1145	-4	1	8	1		4
87.77	87.74	1.1111	-6	2	5	2	1	7
88.10	88.06	1.1079	-2	3	8	1	1	4
88.48	88.46	1.1041	-6	0	7	1	1	4
88.93	88.96	1.0996	2	2	9	2	2	7
88.99		1.0990	2	1	10	1		4
89.24	89.24	1.0966	7	1	2	1	2	6
89.64	89.64	1.0928	5	1	7	2	1	7
92.27	92.28	1.0683	2	3	8	3	2	11
92.60	92.62	1.0654	6	3	1	3	4	12
92.63		1.0651	7	0	3	2		8
92.68		1.0647	0	0	11	1		6
92.80		1.0636	-7	2	2	3		14
94.23	94.24	1.0512	-1	2	10	3	1	6
94.79	94.80	1.0465	3	0	5	3	2	13
94.84		1.0461	0	5	2	1		5
95.60	95.62	1.0397	-1	4	7	1	2	6
95.61		1.0397	-6	2	5	5		6
95.94	95.94	1.0369	-3	4	6	4	3	22
96.27	96.26	1.0343	-2	1	9	2	3	16
96.88	96.92	1.0294	5	4	0	2	2	8
96.97		1.0287	4	5	0	2		9
97.75	97.80	1.0225	2	1	8	2	3	10
97.81		1.0217	5	1	6	1		6
97.86		1.0217	-7	1	0	1		6
98.04	98.08	1.0203	3	4	1	1	2	6
99.50	99.52	1.0092	5	5	2	2	1	9
99.80	99.84	1.0070	-8	0	3	2	3	9
99.85		1.0066	-6	6	5	4		17

2THETA	PEAK	D	H	K	L	I(INT)	I(PK)	I(DS)
100.07	100.08	1.0050	-3	1	11	1	2	11
100.69	100.74	1.0004	3	4	6	3	2	12
100.84		0.9994	0	5	4	2		10
102.15	102.40	0.9900	2	1	11	1	3	5
102.25		0.9894	-8	1	3	1		3
102.29		0.9891	2	5	3	2		11
102.37		0.9886	-2	2	9	1		8
102.47		0.9879	-5	2	1	1		7
102.49		0.9877	8	1	1	3		7
103.41	103.42	0.9814	-7	5	5	2	2	14
103.44		0.9812	-5	4	3	1		8
103.96	103.98	0.9777	-2	1	7	2	2	9
103.98		0.9776	-2	1	10	1		7
104.06		0.9770	-4	5	0	1		5
104.38	104.34	0.9749	7	2	10	2	2	8
104.93	105.18	0.9713	1	1	4	1	1	7
105.17		0.9698	1	4	8	1		4
106.23	106.22	0.9630	8	0	3	2	1	5
106.56	106.54	0.9609	4	1	10	1	1	8
107.92	107.96	0.9526	-6	5	5	1	1	7
108.18	108.26	0.9510	6	3	1	1	1	4
108.98	108.98	0.9462	-6	4	11	2	1	5
109.11		0.9455	3	1	6	1		4
109.49	109.68	0.9433	7	2	8	1		8
109.66		0.9423	2	4	5	1		3
109.76		0.9417	4	5	0	1		5
110.43	110.46	0.9378	-8	2	3	1	1	7
110.57		0.9371	-8	0	6	1		5
110.72		0.9362	-6	3	7	1		3
111.11	111.14	0.9340	2	4	3	1		3
111.17		0.9337	0	5	10	1		4
111.53	111.54	0.9317	-3	1	6	1	1	4
111.94	111.94	0.9294	-6	2	12	1	1	7
112.44	112.76	0.9267	7	5	10	1	2	5
112.71		0.9252	8	2	3	1		3
112.77		0.9249	-4	5	2	4		7
114.26	114.60	0.9170	-6	4	9	1	2	24
								6

HODGKINSONITE - RENTZEPERIS, 1963

2THETA	PEAK	D	H	K	L	I(INT)	I(PK)	I(DS)
114.30		0.9169	-5	1	11	3		4
114.44		0.9162	0	2	12	3		15
114.62		0.9152	6	3	6	3		18
116.21	116.56	0.9072	-4	5	3	1	2	8
116.50		0.9058	-2	3	11	2		9
116.58		0.9054	4	3	9	4		21
118.18	118.32	0.8978	-7	2	8	1	2	7
118.27		0.8973	0	5	7	3		17
118.44		0.8966	6	2	8	1		4
118.51		0.8962	-7	1	9	3		17
119.53	119.58	0.8915	-3	2	12	1	1	3
119.58		0.8913	9	1	10	3		19
119.76		0.8905	8	1	5	1		3
120.26	120.04	0.8882	0	6	1	1	1	3
120.77	120.80	0.8860	0	6	13	3	1	17
120.89		0.8855	-2	5	7	1		5
121.02		0.8849	-8	3	1	1		6
121.52	121.48	0.8827	9	1	2	1	1	7
121.54		0.8826	-8	3	2	1		5
122.08	122.12	0.8803	3	1	12	2	1	4
122.51	122.88	0.8785	-5	2	11	2	1	9
122.87		0.8770	6	3	7	2		11
122.87		0.8770	-7	4	1	1		6
123.49	123.50	0.8744	-8	3	1	2	2	11
123.55		0.8742	-7	4	2	4		26
124.02	124.00	0.8723	-1	6	2	1	1	3
125.20	125.30	0.8676	-3	1	13	1	1	6
125.28		0.8673	3	4	10	2		11
125.62	125.80	0.8660	-5	4	8	2	1	6
125.70		0.8656	-6	1	11	1		4
126.20	126.62	0.8637	-5	5	4	1	1	8
126.36		0.8631	-5	2	0	1		6
126.58		0.8622	8	3	13	2		12
126.62		0.8621	2	6	2	1		5
126.99	126.93	0.8607	-4	2	11	1	1	6
127.09		0.8603	-1	5	8	1		5
128.19	128.64	0.8563	-1	2	13	1	1	8

2THETA	PEAK	D	H	K	L	I(INT)	I(PK)	I(DS)
128.38		0.8556	4	3	10	1	1	5
128.57		0.8549	5	4	7	1	1	5
128.74		0.8543	-7	1	10	1	1	5

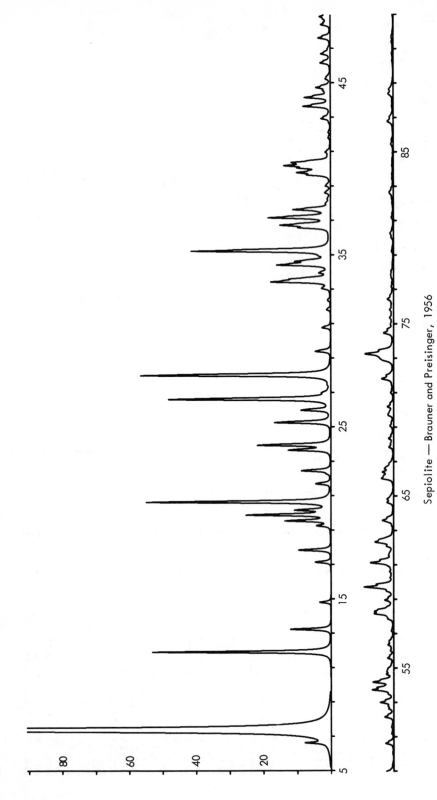

Sepiolite — Brauner and Preisinger, 1956

$\lambda = CuK\alpha$ Half width at $40°2\theta$: $0.11°2\theta$ Plot scale I_{PK} (max) = 700

582

SEPIOLITE - BRAUNER AND PREISINGER, 1956

2THETA	PEAK	D	H	K	L	I(INT)	I(PK)	I(DS)
6.59	6.60	13.400	0	1	1	100	100	100
7.37	7.38	11.985	0	2	0	100	100	100
11.90	11.90	7.433	0	1	3	8	8	8
13.20	13.22	6.700	0	2	2	2	2	2
14.77	14.78	5.993	1	0	0	1	0	1
17.10	17.10	5.180	0	1	5	1	1	1
17.81	17.82	4.977	0	4	0	2	1	2
19.23	19.24	4.612	1	1	0	1	1	1
19.51	19.52	4.545	0	3	3	2	2	2
19.86	19.86	4.467	1	1	1	2	2	2
20.14	20.14	4.406	0	0	6	2	2	2
20.62	20.62	4.305	1	3	1	9	8	9
21.67	21.68	4.098	1	2	2	1	1	1
22.42	22.42	3.962	0	3	5	1	1	1
23.63	23.64	3.761	1	3	2	2	2	2
23.92	23.92	3.716	0	6	0	4	3	4
25.23	25.24	3.526	0	2	6	3	3	3
25.94	25.94	3.431	0	4	5	2	2	2
26.59	26.60	3.350	0	4	4	9	7	10
26.96	26.96	3.305	1	3	3	0	0	0
27.98	27.98	3.186	1	3	3	11	8	11
29.36	29.36	3.039	1	2	7	1	1	1
30.73	30.74	2.907	0	7	1	1	0	1
31.78	31.78	2.813	1	3	6	0	0	0
33.02	33.02	2.710	1	4	6	0	0	0
33.41	33.42	2.680	0	4	6	4	3	4
33.58	33.56	2.667	2	0	1	2	2	2
33.93	33.92	2.640	2	6	0	1	1	1
34.38	34.38	2.606	1	2	8	3	2	3
34.60	34.60	2.590	2	1	0	2	1	2
34.92	34.94	2.567	0	5	3	0	1	0
35.21	35.22	2.546	1	2	9	9	6	9
36.55	36.56	2.456	2	2	0	2	2	2
36.71	36.70	2.446	2	1	9	3	3	3
37.14	37.14	2.419	2	2	8	4	3	4
37.61	37.60	2.390	1	3	5	2	2	2
38.57	38.58	2.332	2	1	5	0	0	0
38.98	38.98	2.309	1	5	3	0	0	0
39.62	39.62	2.273	2	3	6	1	1	1
39.77	39.76	2.265	2	0	1	2	1	2
40.02	40.02	2.251	1	5	4	0		0
40.02		2.251	1	2	10	1		1
40.17	40.16	2.243	1	3	9	2	2	2
40.35	40.34	2.233	0	6	0	2	2	2
40.35		2.233	0	0	12	1		1
40.94	40.94	2.203	2	3	3	0	0	0
42.88	42.88	2.107	1	4	10	1	1	1
43.61	43.62	2.074	2	3	0	2	0	2
44.12	44.12	2.051	0	4	9	2	1	2
44.42	44.42	2.038	1	6	4	0	0	0
44.53	44.54	2.033	1	1	13	0	1	0
44.70	44.70	2.026	2	6	3	1	0	1
45.20	45.20	2.004	1	5	6	0	1	0
46.12	46.12	1.9663	1	6	3	1	1	1
46.65	46.64	1.9455	1	5	8	1	1	1
47.58	47.58	1.9094	0	7	1	1	0	1
48.35	48.36	1.8807	2	4	6	1	0	1
48.82	48.82	1.8638	1	5	9	1	1	1
50.59	50.58	1.8028	0	7	5	1	0	1
52.13	52.14	1.7529	1	6	8	0	0	0
52.61	52.62	1.7381	1	2	14	1	1	1
52.98	52.98	1.7268	3	1	3	0	0	0
53.47	53.48	1.7122	0	7	7	1	1	1
53.71	53.72	1.7051	2	6	0	0	0	0
53.83	53.84	1.7016	2	6	1	1	1	1
54.15	54.16	1.6924	1	6	9	0	1	0
54.18		1.6914	2	1	12	1	0	1
54.61	54.62	1.6791	2	1	15	0	0	0
54.86	54.86	1.6722	3	2	3	0	0	0
55.57	55.58	1.6524	2	2	12	0	0	0
55.99	56.00	1.6409	1	2	15	0	0	0
56.33	56.34	1.6317	2	6	10	0	0	0
57.83	57.84	1.5929	1	6	6	0	0	0
58.14	58.14	1.5854	2	1	8	1	1	1

SEPIOLITE - BRAUNER AND PREISINGER, 1956

2THETA	PEAK	D	H	K	L	I(INT)	I(PK)	I(DS)
58.25	58.26	1.5826	1	3	15	1	1	1
58.38	58.38	1.5792	0	6	12	0	1	1
58.94	58.94	1.5657	0	1	17	1	0	1
59.46	59.48	1.5531	1	8	4	0	0	0
59.70	59.70	1.5476	3	1	8	1	1	2
59.72	59.72	1.5472	2	7	1	1	0	1
60.24	60.24	1.5349	3	4	3	0	0	0
61.11	61.12	1.5151	3	0	9	2	1	2
61.54	61.54	1.5055	3	1	9	0	0	0
62.07	62.08	1.4941	1	7	10	0	0	1
62.31	62.30	1.4889	0	0	18	1	1	1
62.41		1.4866	0	9	1	1	1	1
63.57	63.58	1.4623	3	1	10	1	0	1
64.61	64.62	1.4413	1	8	8	1	1	1
65.99	66.00	1.4143	2	8	0	1	1	1
66.10		1.4124	2	8	1	0	0	0
66.38	66.38	1.4071	1	8	9	1	1	1
66.61	66.58	1.4028	0	1	19	0	0	0
67.83	67.82	1.3805	3	4	9	0	0	0
68.33	68.32	1.3716	1	0	10	0	0	0
69.06	69.06	1.3588	0	5	17	0	0	0
69.67	69.70	1.3484	2	8	6	0	0	0
69.76		1.3470	3	5	8	0	0	0
70.17	70.18	1.3400	0	10	0	0	0	0
70.98	70.98	1.3267	2	1	17	0	0	1
71.80	71.80	1.3136	4	1	0	1	1	0
72.87	72.98	1.2969	2	9	0	0	0	0
72.97		1.2953	2	0	18	0	0	0
73.09		1.2936	2	3	17	0	0	1
73.24	73.24	1.2913	4	2	1	1	1	2
73.27		1.2908	1	1	18	1		1
74.45	74.46	1.2732	2	2	18	1	0	1
74.96	74.96	1.2659	4	3	0	0	0	0
75.35	75.34	1.2603	4	1	6	0	0	0
75.80	75.80	1.2538	3	0	15	0	0	0
76.41	76.42	1.2454	2	3	18	0	0	0

2THETA	PEAK	D	H	K	L	I(INT)	I(PK)	I(DS)
76.41		1.2454	2	9	6	0	0	0
79.99	80.00	1.1985	2	3	19	0	0	0
80.25	80.24	1.1952	3	3	21	0	0	0
81.15	81.16	1.1842	4	5	0	1	0	0
85.44	85.44	1.1353	4	6	1	0	0	1
86.76	86.76	1.1215	2	6	18	1	0	1
88.37	88.36	1.1052	2	7	17	1	0	0
91.38	91.38	1.0764	3	9	8	0	0	0
94.05	94.06	1.0528	2	8	17	0	0	0
95.19	95.18	1.0431	2	7	19	0	0	0
96.06	96.06	1.0360	4	8	1	0	0	0

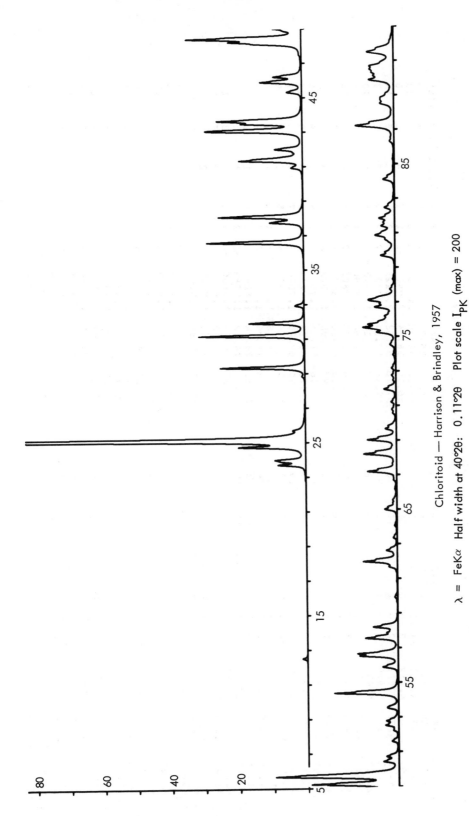

Chloritoid — Harrison & Brindley, 1957

λ = FeKα Half width at 40°2θ: 0.11°2θ Plot scale I$_{PK}$ (max) = 200

CHLORITOID - HARRISON AND BRINDLEY, 1957

2THETA	PEAK	D	H	K	L	I(INT)	I(PK)	I(CS)
23.68	23.68	4.718	-1	1	0	4	4	4
23.87	23.96	4.681	-1	1	1	3	5	2
23.97		4.662	-2	1	0	3		3
24.72	24.72	4.523	-2	0	2	8	10	8
25.11	25.12	4.454	0	0	4	92	100	92
25.13		4.450	-1	0	1	8		8
29.32	29.32	3.825	-2	0	4	13	13	14
31.16	31.16	3.604	2	0	4	17	16	19
31.90	31.90	3.523	-1	1	3	9	8	10
32.91	32.92	3.418	-1	1	4	1	1	2
36.57	36.58	3.085	-1	1	4	17	14	21
37.72	37.72	2.994	-1	1	5	6	5	7
38.05	38.06	2.969	0	0	6	15	13	19
40.90	40.90	2.771	-2	0	6	2	2	3
41.31	41.32	2.744	0	1	1	11	9	16
41.96	41.98	2.703	-3	1	0	3	4	4
41.98		2.702	0	0	3	2		3
43.01	43.04	2.641	3	2	6	13	14	19
43.06		2.637	-3	1	2	8		11
43.46	43.46	2.615	0	1	1	10	9	14
43.63	43.64	2.605	3	1	4	14	13	20
45.29	45.30	2.514	-3	2	3	3	2	4
45.87	45.88	2.484	0	1	2	8	6	12
46.18	46.18	2.468	3	1	2	5	4	8
48.09	48.20	2.376	-4	0	5	7	11	11
48.21		2.370	-2	2	2	8		13
48.39	48.38	2.362	-3	2	5	20	17	33
49.08	49.08	2.331	0	1	4	16	13	27
49.52	49.52	2.311	3	1	3	23	18	38
49.59		2.308	2	2	1	3		6
50.40	50.40	2.274	-2	2	3	3	2	5
50.69	50.70	2.261	-4	2	2	2	1	4
51.58	51.58	2.225	-2	2	8	1	2	2
52.50	52.50	2.189	-2	0	4	3	2	4
52.75	52.76	2.179	-2	2	3	2	2	4
53.49	53.52	2.151	4	0	2	1		3
54.35	54.36	2.119	2	2	3	14	10	25

2THETA	PEAK	D	H	K	L	I(INT)	I(PK)	I(DS)
55.86	55.86	2.067	-2	2	5	3		6
56.50	56.50	2.045	-4	0	6	8	6	15
56.65	56.64	2.040	-3	1	7	4	6	8
57.53	57.52	2.012	0	2	4	7	5	14
57.84	57.84	2.002	2	2	4	2	2	3
58.17	58.18	1.9913	3	1	8	5	4	10
61.66	61.66	1.8890	-3	2	8	2	1	3
61.96	61.96	1.8805	2	2	7	8	5	16
62.60	62.60	1.8632	0	0	8	2	2	5
64.97	64.98	1.8022	-4	1	9	3		6
67.16	67.16	1.7500	-3	2	8	7		16
68.19	68.18	1.7269	0	2	7	8		19
69.00	69.00	1.7090	3	0	5	7		17
69.63	69.64	1.6955	-4	2	6	2		4
71.18	71.18	1.6633	-5	0	7	1		3
71.93	71.94	1.6482	-1	2	9	3		8
73.15	73.14	1.6245	-2	0	2	1		3
74.51	74.52	1.5989	-6	0	0	1		3
75.22	75.24	1.5860	-3	1	5	5	3	11
75.52	75.52	1.5808	-3	0	8	8	5	19
75.71	75.70	1.5775	-4	1	4	2		6
75.86	75.84	1.5748	1	3	10	2		4
76.18	76.18	1.5691	-6	0	5	2		5
76.69	76.70	1.5602	-3	3	4	5	2	13
77.06	77.10	1.5540	6	0	3	4		10
77.13		1.5527	3	3	0	5		12
79.67	79.68	1.5111	-3	1	1	4	2	11
80.85	80.86	1.4928	0	2	11	4	3	12
80.95		1.4913	-1	0	10	1		12
81.39	81.40	1.4846	0	1	7	3	2	9
81.83	81.84	1.4779	4	0	12	4	2	11
82.45	82.46	1.4688	3	1	9	3	1	8
84.09	84.10	1.4454	-6	0	8	1	2	10
86.29	86.30	1.4156	-2	2	8	8	1	4
87.19	87.20	1.4038	-3	3	11	6	6	25
87.21		1.4035	-2	3	7	6		17
88.49	88.50	1.3873	3	3	5	4	2	13

CHLORITOID — HARRISON AND BRINDLEY, 1957

2THETA	PEAK	D	H	K	L	I(INT)	I(PK)	I(DS)
88.65	88.72	1.3854	-4	0	12	1	2	4
88.85	88.82	1.3829	-6	2	4	3	2	8
89.87	89.88	1.3705	-6	2	3	8	4	26
90.29	90.30	1.3654	-6	2	1	5	3	15
90.46	90.48	1.3635	0	4	1	4	4	13
90.67	90.66	1.3610	-6	2	4	5	4	15
91.47	91.50	1.3517	0	4	2	7	4	22
91.52		1.3511	6	2	0	3		11
94.31	94.42	1.3203	-6	2	6	2	2	5
94.40		1.3192	-3	1	13	2		6
94.45		1.3187	-2	2	12	1		4
95.78	95.80	1.3048	0	2	12	3	1	9
96.02	96.02	1.3024	6	2	2	1	1	4
96.75	96.76	1.2949	4	0	10	2	2	13
96.98	96.98	1.2927	3	1	11	2	1	6
100.70	100.70	1.2572	-6	2	8	2	1	8
102.40	102.48	1.2421	0	4	6	2	2	6
102.48		1.2413	-2	2	13	4		13
103.32	103.34	1.2341	6	2	4	1	1	5
104.38	104.34	1.2252	0	2	13	3	1	4
110.10	110.12	1.1810	-6	2	10	2	1	9
112.33	112.34	1.1653	0	4	8	3	1	12
113.77	113.80	1.1557	6	2	6	2	1	8
118.43	118.48	1.1267	8	0	2	1	0	4
120.77	120.78	1.1135	0	0	16	4	1	16
123.28	123.32	1.1000	-6	2	12	3	1	14
125.30	125.36	1.0898	-3	3	13	2	1	10
125.39		1.0894	-4	4	8	4		7
126.34	126.34	1.0848	0	4	10	3	1	19
128.50	128.54	1.0747	3	3	11	4	1	12
128.55		1.0745	6	2	8	2		17
131.63	131.96	1.0611	-8	2	7	2	1	7
131.99		1.0596	-6	2	13	1		9
134.43	134.50	1.0499	1	5	4	1	0	6
135.77	135.80	1.0449	0	4	11	4	0	7
136.94	136.96	1.0406	-3	1	17	4	1	19
138.66	138.74	1.0346	6	2	9	1	1	6
138.70	138.77	1.0344	-9	1	5	2		8
138.77		1.0342	-3	5	5	1		7
139.65	139.68	1.0313	0	2	16	3	1	17
139.75		1.03C9	-6	4	4	2		11
140.53	140.48	1.0284	-3	5	3	2	1	10
140.58		1.0282	-1	5	6	1		7
140.68		1.0279	-9	1	1	1		7
141.09	141.12	1.0266	6	0	0	1	1	6
141.20		1.0262	3	5	1	2		11
142.20	142.22	1.0231	3	5	15	4	1	18
142.41		1.0225	-8	0	12	1		6
142.43		1.0224	-8	2	3	1		7
143.24	142.94	1.0200	-6	2	14	1	1	5
143.97	144.18	1.0179	9	1	0	1	1	6
144.18		1.0173	8	0	6	2		11
144.19		1.0173	-9	1	7	4		19
144.21		1.0172	-7	3	9	2		10
145.76	145.76	1.0129	-6	4	6	3	1	23
147.04	147.10	1.0095	-3	5	5	3	1	17
148.23	148.82	1.0064	-6	0	16	6	2	30
148.64		1.0054	3	5	3	4		20
148.83		1.0049	9	1	1	5		26
148.92		1.0047	6	4	2	4		23
149.43		1.0035	-4	4	11	2		11
151.78	151.70	0.9981	-3	3	15	7	1	36
152.92		0.9957	-3	5	6	1		4
152.92		0.9957	6	2	10	2		10
157.47	159.24	0.9870	-3	1	18	1		5
158.95		0.9846	3	3	13	12	1	64
159.61		0.9835	-6	4	8	4		19
160.67		0.9819	-8	2	11	2		12
160.91		0.9816	-6	0	15	1		4
162.45		0.9795	-3	5	7	4		21
162.62		0.9792	6	0	12	7		37

TABLE 30. MISCELLANEOUS SHEET SILICATES

Variety	Datolite	Apophyllite	Prehnite	Leucophanite	Meliphanite
Composition	$CaBSiO_4(OH)$	$(K,Na)Ca_4Si_8O_{20}$ $F \cdot 8H_2O$	$Ca_2(Al,Fe)_2Si_3O_{10}$ $(OH)_2$	$CaNaBeSi_2O_6F$	$Ca(Na,Ca)$ $BeSi_2O_6F$
Source	Bergen Hill, N.J.	Northern Michigan	Radautal, Hartz-burg, Germany	Brevig, Norway	Brevig, Nor-way
Reference	Ito & Mori, 1953; Pank & Cruickshank, 1967	Colville and Anderson, 1968	Preisinger, 1965	Cannillo, Giuseppetti & Tazzoli, 1967	DalNegro, Rossi and Ungaretti, 1967
Cell Dimensions					
a ⦧Å	4.84	8.963	18.48 (4.627)	7.401	10.516
b ⦧Å	7.60	8.963	5.490 (5.490)	7.420	10.516
c ⦧Å	9.62	15.804	4.627 (18.48)	9.939	9.887
α ⦧deg	90	90	90	90	90
β ⦧deg	90.15	90	90	90	90
γ ⦧deg	90	90	90	90	90
Space Group	$P2_1/c$	$P4/mnc$	$Pma2$ $(P2cm)$	$P2_12_12_1$	$I\bar{4}$
Z	4	2	2	4	8

TABLE 30. (cont.)

Variety	Datolite	Apophyllite	Prehnite	Leucophanite	Meliphanite
Site Occupancy		$\{$ K 0.84, Na 0.16			$\{$ Na 0.63, Ca 0.37
Method	3-D, film		2-D, film	3-D, film	3-D, film
R & Rating	0.128 (2)	0.05 (1)	0.09 (av.) (2)	0.094 (2)	0.104 (1)
Cleavage and habit	None Variable, usually prismatic	{001} perfect, {110} poor. Variable	{100} good, {011} poor Tabular		{001} Tabular
Comment	Alternating sheets SiO_4 and BO_3		Minor Fe^3 ignored. B estimated		
μ	178.3	141.1	160.8	145.8	170.6
ASF	0.5096×10^{-1}	0.8094×10^{-1}	0.6358×10^{-1}	0.6467×10^{-1}	0.1321
Abridging factors	1:1:0.5	1:1:0.5	1:0:0	1:0:0	1:0:0

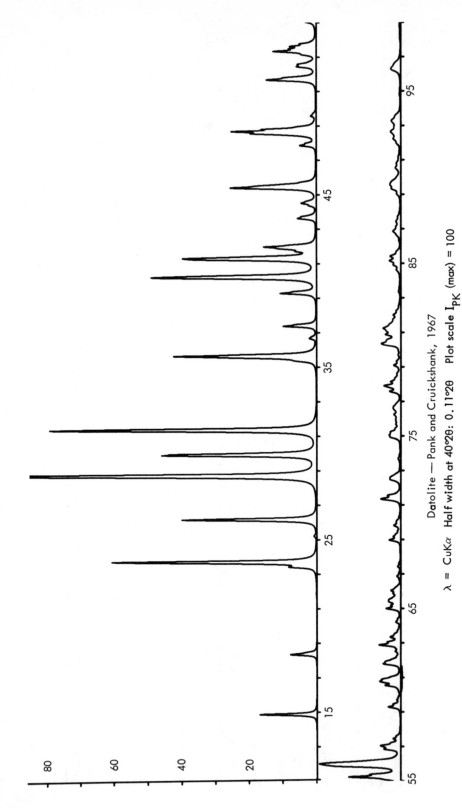

Datolite — Pank and Cruickshank, 1967

λ = CuKα Half width at 40°2θ: 0.11°2θ Plot scale I$_{PK}$ (max) = 100

590

DATOLITE - PANT AND CRUICKSHANK, 1967

2THETA	PEAK	D	H	K	L	I(INT)	I(PK)	I(DS)
14.84	14.86	5.964	0	1	1	18	17	15
18.31	18.32	4.840	1	0	0	8	8	7
23.39	23.40	3.800	0	2	0	8	8	7
23.64	23.68	3.761	-1	1	1	27	61	25
23.67		3.755	1	1	1	52		48
26.06	26.14	3.416	-1	0	2	2	40	2
26.13		3.407	1	0	2	50		48
28.62	28.68	3.116	-1	1	2	100	100	100
28.69		3.109	1	1	2	59		59
29.87	29.88	2.989	1	2	0	55	46	56
29.94	29.94	2.982	0	2	1	20	44	20
31.30	31.32	2.856	-1	2	1	74	79	78
31.33		2.853	1	2	1	43		45
35.53	35.62	2.524	-1	1	3	23	43	26
35.61		2.519	1	1	3	47		53
36.65	36.66	2.450	0	3	1	15		3
37.36	37.36	2.405	0	0	4	16	2	17
39.26	39.26	2.293	1	3	1	33	10	19
40.14	40.18	2.244	0	1	4	21	11	40
40.16		2.244	-2	1	2	35	49	25
40.20		2.241	0	2	4	41		42
41.22	41.28	2.188	-1	2	3	6	40	51
41.26		2.186	-1	3	1	28		7
41.28		2.185	1	2	3	6		35
41.70	41.70	2.164	-2	0	2	11	6	7
41.86	41.96	2.156	-1	0	4	16	16	13
41.96		2.152	1	1	4	8		21
43.60	43.60	2.074	-1	1	4	2	6	10
43.69	43.70	2.070	1	2	2	5	5	3
44.34	44.36	2.041	1	2	2	1	5	3
44.48	44.48	2.035	2	2	2	23		7
44.55		2.032	-1	2	4	24		2
45.36	45.38	1.9977	-2	2	1	8	26	30
45.40		1.9959	-2	2	0	4		32
47.83	47.84	1.9000	0	4	2		5	11
48.44	48.54	1.8775	2	2	2		20	5
48.50	48.50	1.8752	-1	2	4	6		8
48.53		1.8743	-2	1	3	20	25	28
48.65	48.64	1.8699	-2	1	1	26	17	36
48.78	48.78	1.8652	0	0	5	2		3
48.82		1.8640	0	4	0	9		12
49.56	49.56	1.8377	1	3	2	20	2	3
51.64	51.64	1.7686	-1	4	2	9	15	30
51.68		1.7671	0	4	2	8		13
52.41	52.42	1.7442	-1	1	5	3	6	13
52.49		1.7419	-1	4	1	1		4
52.58	52.56	1.7392	-1	4	1	2		30
53.32	53.32	1.7165	0	2	5	20	6	17
53.61	53.60	1.7081	-2	0	4	11	13	2
53.76	53.74	1.7036	0	0	4	1	8	39
55.21	55.20	1.6624	2	0	4	25	5	28
55.90	55.96	1.6434	2	3	2	18	16	21
55.95		1.6419	-1	3	4	13	24	18
56.03		1.6399	0	3	4	12		12
56.82	56.82	1.6190	-1	3	4	2		11
56.91	57.02	1.6166	-1	2	5	7	5	5
57.04		1.6133	3	0	0	3	6	11
57.42	57.42	1.6033	0	2	4	7	2	3
59.26	59.26	1.5579	-2	2	3	2	4	13
59.68	59.68	1.5480	-1	4	3	8	1	13
60.53	60.52	1.5284	3	0	2	8	5	11
60.75	60.76	1.5232	-1	0	6	6	6	14
60.86	60.86	1.5208	-3	0	0	8	6	7
61.76	61.78	1.5007	-1	3	2	4	5	4
61.87	61.88	1.4984	3	1	4	2	2	16
62.49	62.48	1.4850	3	0	2	9	6	5
62.85	62.86	1.4772	0	2	6	3		6
62.86		1.4771	-2	4	1	4		2
63.29	63.32	1.4682	-3	2	1	1	3	3
63.34		1.4671	3	2	5	3		
63.69	63.68	1.4599	1	3	0	1	1	6
64.17	64.18	1.4502	1	5	2	1	2	2
64.21		1.4494	0	5	2	2		3

DATOLITE — PANT AND CRUICKSHANK, 1967

2THETA	PEAK	D	H	K	L	I(INT)	I(PK)	I(DS)
64.97	64.98	1.4341	-1	5	1	7	4	12
64.99		1.4338	-1	5	1	2		3
65.30	65.38	1.4278	-2	4	2	1	4	2
65.36		1.4265	-1	4	2	4		7
65.41		1.4255	-1	4	2	2		3
65.48	65.48	1.4242	1	3	4	3	4	6
65.83	65.84	1.4174	-3	1	3	3	2	5
65.98	66.02	1.4145	3	1	3	1	3	2
66.02		1.4138	-1	1	3	2		4
66.12		1.4119	-1	2	2	2		2
66.84	66.84	1.3985	2	5	1	1	1	3
67.40	67.40	1.3881	1	3	2	2	1	3
68.94	68.94	1.3608	3	4	0	7	3	14
68.95		1.3554	-2	2	3	1		2
69.14	69.14	1.3463	3	4	0	4	2	7
69.26		1.3414	-3	0	3	1	1	2
69.80	69.80	1.3217	-1	5	1	1	1	11
70.00	70.00	1.3210	-3	1	4	6	6	
70.09		1.3209						
71.29	71.32	1.3039	3			8	3	15
71.32		1.3033	-1	1	7	4		
71.34		1.3014	-1	5	5	3		8
72.42	72.46	1.2849	0	6	0	4		6
72.45	72.56	1.2667	-2	6	2	3	3	3
72.58		1.2649						
73.66	73.66	1.2622	0	2	4	6	1	11
74.90	74.92	1.2595	-3	2	6	5	3	10
75.02	75.02	1.2558	-1	2	1	3	4	5
75.20	75.20	1.2537	2	2	6	1	3	2
75.40	75.40	1.2517	0	6	1	2	2	4
75.66	75.66	1.2479	-3	3	3	1	2	3
75.82		1.2438	1	3	3	1	2	3
75.94	75.94	1.2430	-2	2	7	3		7
75.96		1.2423						
76.22	76.22	1.2414	-1	5	4	2	2	4
76.52	76.60	1.2298	3	4	0	2		16
76.63		1.2253	0	4	6	10	5	22
77.56	77.56							
77.88	77.88							

2THETA	PEAK	D	H	K	L	I(INT)	I(PK)	I(DS)
78.33	78.30	1.2196	3	4	1	1	2	3
79.07	79.08	1.2100	0	0	0	1	2	11
79.66	79.66	1.2025	2	5	8	5	1	3
80.26	80.34	1.1951	3	1	3	6	6	13
80.27		1.1949				2		4
80.35		1.1939	-4	0	5	9		20
80.45		1.1927	2	1	5	1		2
80.86	80.88	1.1877	0	1	8	7	5	15
80.90		1.1873						
81.23	81.22	1.1832	3	3	2	2	5	17
81.35		1.1818	-2	3	4	7		5
81.75	81.74	1.1770	-3	1	7	2	2	7
81.99	81.98	1.1742	3	2	5	2	1	2
82.23	82.22	1.1714	-1	5	5	5	1	2
83.34	83.36	1.1585	-1	5	1	1	1	4
84.60	84.60	1.1444	-4	2	2	2	2	9
84.90	84.90	1.1412	-2	2	7	4		10
85.12	85.14	1.1388	-3	0	6	3	3	6
85.13		1.1387						
85.40	85.40	1.1358	3	3	0	5	3	12
86.68	86.84	1.1222	2	6	0	3	2	4
86.83		1.1207	0	6	4	1		7
86.85		1.1206						3
87.40	87.40	1.1148	-2	6	2	2		3
89.01	89.02	1.0989	-3	5	1	3	1	6
89.32	89.32	1.0958	-3	1	5	3	1	6
89.50	89.54	1.0941	-1	6	4	6	3	14
89.68		1.0923	1	4	6	5		12
89.73		1.0918	3	4	0	2		5
90.49	90.52	1.0847	1	3	7	2	2	4
90.50		1.0846	4	1	1	1		4
90.55		1.0841	-4	3	3	2		3
91.95	91.98	1.0712	-2	1	4	3		5
92.18	92.20	1.0691	2	5	5	3	1	8
92.19		1.0690	1	1	4	2	2	7
92.38	92.62	1.0673	-4	2	8	2		5
92.61		1.0653	-4	2		4	3	11

2THETA	PEAK	D	H	K	L	I(INT)	I(PK)	I(DS)
92.63		1.0651	2	1	8	2		6
92.73		1.0642	4	3	2	2		5
93.16	93.26	1.0605	-1	3	8	3	3	9
93.28		1.0594	1	7	0	4		11
93.28		1.0594	1	1	8	1		3
94.80	94.80	1.0464	-3	5	3	1	0	3
96.22	96.26	1.0347	-1	1	7	3	3	8
96.23		1.0346	1	7	2	2		4
96.25		1.0345	-1	1	7	2		6
96.32		1.0339	1	6	5	1		4
96.37		1.0335	-1	6	9	2		4
96.40		1.0333	1	4	5	2		5
101.04	101.06	0.9979	4	5	2	1	1	4
101.04		0.9979	-1	3	7	2		6
101.44	101.44	0.9951	-4	4	4	2	1	5
101.60		0.9940	1	6	8	2		5
102.04	102.20	0.9908	0	6	6	1		2
102.14		0.9902	3	1	1	1	2	3
102.15		0.9901	-2	3	8	2		6
102.23		0.9896	-4	7	5	6		16
102.46	102.48	0.9879	0	2	4	1	2	3
104.33	104.56	0.9753	4	6	5	2	1	5
104.54		0.9739	3	6	6	2		7
104.63		0.9733	-1	1	9	1		3
105.03	105.02	0.9707	-1	6	6	1	1	2
105.15		0.9699	-2	1	5	2		2
105.31	105.30	0.9688	-2	6	9	1	1	6
105.39	106.36	0.9620	2	0	10	2	0	2
106.39	106.70	0.9602	0	1	0	1	0	3
106.67	107.40	0.9569	5	1	6	1	1	3
107.20		0.9557	-1	1	1	2		3
107.40		0.9485	5	0	2	1		3
108.60	108.64	0.9478	-2	2	9	1	1	2
108.72		0.9431	-1	0	10	1		2
109.52	109.68	0.9423	-4	5	1	1		4
109.65		0.9421	-5	1	2	3		10

2THETA	PEAK	D	H	K	L	I(INT)	I(PK)	I(DS)
110.03	110.10	0.9401	-2	5	7	1	1	4
110.26	110.26	0.9388	2	5	7	2	1	7
110.40		0.9380	5	5	0	1		4
110.55	~110.66	0.9372	-4	2	6	1	1	2
110.61		0.9368	-1	1	10	2		6
110.78		0.9359	1	1	10	1		4
110.94	110.98	0.9350	4	2	6	2	1	5
111.14	111.14	0.9338	-5	2	1	2	1	5
111.21		0.9334	3	2	8	1		2
111.22		0.9334	5	2	1	1		2
111.37		0.9326	0	2	10	1		3
111.98	112.22	0.9292	-4	5	2	2	1	4
112.22		0.9279	-1	8	1	1		7
112.23		0.9278	1	8	8	1		2
112.57	112.60	0.9260	-1	5	8	1		3
114.41	114.62	0.9163	-2	7	4	1		3
114.55		0.9156	2	7	4	1		4
114.63		0.9152	-1	4	9	1		3
114.64		0.9151	-1	8	2	1		3
114.68		0.9149	-1	6	7	1		3
114.80		0.9143	1	6	7	1		2
116.17	116.12	0.9074	4	5	3	1	1	4
116.98	117.30	0.9035	-4	3	6	1	1	12
117.27		0.9021	-3	3	8	4		3
117.68	117.80	0.9001	3	3	8	1	2	16
117.84		0.8993	0	3	10	5		2
118.30	118.30	0.8972	5	0	4	1	1	3
118.35		0.8969	-3	7	1	1		2
118.41		0.8967	3	7	1	2		7
118.71	118.68	0.8953	-1	8	3	1	1	2
119.30	119.16	0.8926	-5	1	9	1	0	3
120.76	121.14	0.8861	-3	7	2	3	2	11
120.86		0.8856	3	7	2	2		6
120.97		0.8851	-1	3	10	1		2
121.09		0.8846	-2	8	0	2		8
121.16		0.8843	3	6	5	1		3
121.20		0.8841						

DATOLITE - PANT AND CRUICKSHANK, 1967

2THETA	PEAK	D	H	K	L	I(INT)	I(PK)	I(DS)
121.27		0.8838	1	3	10	1		2
123.37	123.42	0.8749	4	6	0	3	1	10
123.51		0.8743	0	6	9	2		6
124.21	124.02	0.8715	-4	6	1	1	1	2
124.29		0.8712	-4	6	1	1		4
124.35		0.8709	-2	2	10	1		2
124.91	124.80	0.8687	-2	6	7	1		2
126.69	127.16	0.8618	-4	4	6	2	1	7
126.98		0.8607	-1	5	9	1	2	4
127.15		0.8601	4	4	6	5		18
127.16		0.8601	-4	5	6	1		7
127.47		0.8589	3	4	9	2		5
127.65	127.60	0.8583	0	4	10	3	1	9
129.32	129.44	0.8523	0	2	11	1	1	5
129.41		0.8519	0	7	7	1		2
129.44		0.8518	0	8	5	3		9
129.44		0.8518	4	0	8	3		9
129.93	130.30	0.8501	-4	5	5	1	1	4
130.34		0.8487	-4	1	8	4		14
130.82	131.24	0.8471	-5	3	4	1	2	7
130.89		0.8468	-3	6	6	2		7
131.23		0.8457	-5	3	4	5		17
131.26	131.82	0.8456	3	6	6	5		19
131.74		0.8440	-5	2	5	1	1	3
131.76		0.8439	-3	7	4	1		3
132.01		0.8431	3	7	4	1		2
132.25		0.8423	2	3	10	3		9
133.06	133.14	0.8397	-1	2	11	2	1	9
133.29		0.8390	1	2	11	1		5
133.35		0.8388	-1	7	2	1		2
133.36		0.8388	1	8	5	1	1	3
135.85	136.38	0.8312	-2	4	8	1		4
136.17		0.8302	-2	8	4	1		3
136.35		0.8297	2	8	4	1		2
136.38		0.8296	-5	0	6	2		8
137.05	137.16	0.8277	5	0	6	1	1	5
137.23		0.8272	-3	0	10	4		15
138.11	138.00	0.8247	-5	1	6	2		6
138.79	138.90	0.8229	-2	5	9	1	1	5
138.85		0.8227	-4	6	4	3		11
139.21	139.70	0.8218	1	5	9	2	1	7
139.69		0.8205	3	1	10	5		19
139.79		0.8202	-4	4	7	1		3
139.90		0.8200	2	4	8	1		4
140.46	140.46	0.8185	-2	6	8	1	1	3
141.03	141.58	0.8170	2	1	11	1	1	2
141.20		0.8166	0	6	3	1		3
141.25		0.8165	-5	3	5	1		2
141.36		0.8162	-3	7	0	1		3
141.48		0.8159	-4	3	6	4		14
141.54		0.8158	-3	8	1	1		9
141.62		0.8156	3	8	1	2		7
141.89		0.8149	-3	5	8	2		2
142.07		0.8145	-3	5	6	2		7
142.38	142.38	0.8137	-5	5	1	1	1	4
142.49		0.8134	3	5	8	1		4
142.73		0.8129	0	5	10	1		8
142.89		0.8125	-5	4	4	2		2
145.10	145.92	0.8074	-4	3	8	3	2	13
145.17		0.8072	-3	6	7	1		4
145.36		0.8068	3	6	7	2		7
145.44		0.8067	6	0	2	1		12
145.71		0.8061	-1	8	6	3		6
145.74		0.8060	3	6	7	1		5
145.76		0.8060	3	4	9	1		11
145.81		0.8058	-1	7	8	3		4
145.85		0.8058	1	6	6	2		8
145.87		0.8053	-1	8	3	3		10
146.06		0.8051	4	7	1	1		3
146.14		0.8051	9	7	1	2	1	10
146.35		0.8051	2	2	11	1		3
146.44		0.8045	-2	2	5	2		7
146.78		0.8038	-2	8	1	1		6
149.85	150.40	0.7977	-4	1	9	2	1	7

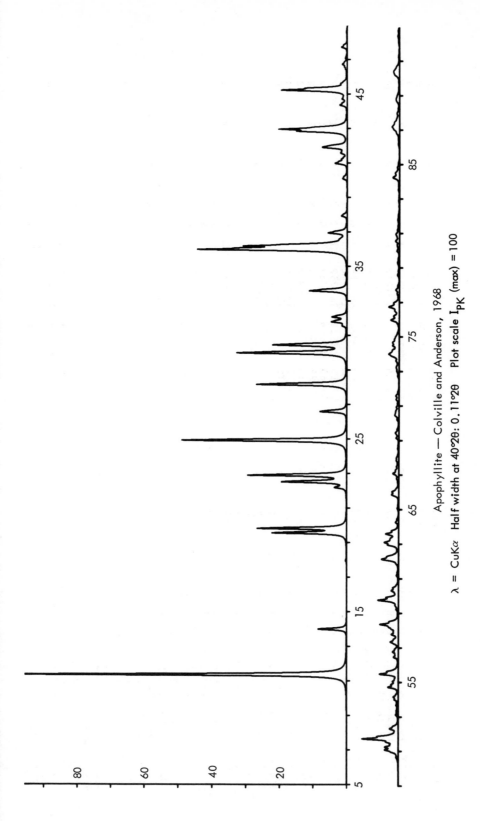

Apophyllite — Colville and Anderson, 1968

λ = CuKα Half width at 40°2θ: 0.11°2θ Plot scale I$_{PK}$ (max) = 100

APOPHYLLITE - COLVILLE AND ANDERSON, 1968

2THETA	PEAK	D	H	K	L	I(INT)	I(PK)	I(DS)
11.34	11.34	7.796	1	1	0	100	100	100
13.96	13.96	6.338	1	1	1	9	8	9
19.53	19.54	4.542	1	1	2	24	22	25
19.79	19.80	4.481	2	0	0	29	27	30
22.16	22.16	4.008	2	0	1	4	4	4
22.48	22.48	3.951	0	0	4	23	19	24
22.79	22.88	3.898	2	1	1	1		2
22.87		3.885	1	1	2	34	29	37
24.89	24.90	3.575	2	1	1	61	49	67
26.56	26.56	3.353	1	1	2	10	8	11
28.14	28.14	3.169	2	1	4	35	27	39
29.95	29.96	2.981	2	0	0	44	33	51
30.36	30.42	2.941	2	2	5	5	22	6
30.42		2.936	3	0	2	25		29
31.54	31.54	2.834	3	1	1	2	1	2
31.77	31.78	2.814	3	1	0	6	5	7
32.05	32.06	2.790	3	1	4	6	4	7
33.56	33.56	2.668	3	1	1	16	11	19
35.95	35.96	2.496	3	1	2	62	44	77
36.16	36.16	2.482	2	1	3	36	31	45
36.56	36.54	2.456	1	1	5	8	2	2
36.92	36.92	2.432	3	2	1	18	6	10
37.91	37.92	2.371	2	1	6	2	2	3
40.07	40.08	2.248	3	2	2	2	1	3
40.96	40.96	2.201	3	2	3	5	4	7
41.20	41.20	2.189	1	0	6	2	2	2
41.56	41.56	2.171	3	0	5	2	2	3
41.87	41.90	2.156	4	1	2	7	7	10
41.91		2.154	4	1	1	5		7
42.82	42.82	2.110	3	2	5	20	15	26
42.95	42.94	2.104	3	2	4	20	20	27
44.35	44.34	2.041	3	3	2	3	2	4
44.70	44.70	2.026	2	2	6	2	2	2
45.20	45.20	2.004	2	2	0	31	19	43
45.59	45.58	1.9883	4	2	1	2	2	3
46.72	46.72	1.9427	4	2	2	2	2	3
47.71	47.72	1.9046	4	1	4	3	2	4

2THETA	PEAK	D	H	K	L	I(INT)	I(PK)	I(DS)
50.94	51.06	1.7911	4	1	5	3	4	4
51.05		1.7874	4	2	4	5		7
51.24	51.24	1.7812	5	0	1	2		3
51.24		1.7812	4	3	1	2	4	4
51.53	51.54	1.7720	2	1	8	4		4
51.72	51.72	1.7659	1	1	7	17	11	6
52.28	52.30	1.7482	3	3	2	2		25
52.32		1.7470	5	0	1	1	3	4
53.10	53.10	1.7232	1	0	9	2	1	3
53.65	53.66	1.7067	4	0	6	1	1	2
53.99	54.00	1.6970	4	3	3	1	1	2
54.14	54.14	1.6926	4	1	5	2	2	2
54.70	54.70	1.6766	4	1	6	3	3	4
54.70		1.6764	2	2	8	3		4
55.03	55.02	1.6674	5	1	3	3		3
55.46	55.46	1.6552	5	1	2	2	2	16
56.75	56.46	1.6287	5	2	1	10	6	4
57.23	56.76	1.6207	3	1	8	2	2	5
57.32	57.32	1.6084	2	1	9	3	2	2
57.75	57.76	1.6060	5	2	4	1		6
58.34	58.34	1.5950	4	0	2	4	2	6
59.74	59.74	1.5804	3	0	6	8	6	13
60.29	60.30	1.5466	3	2	8	11	6	18
60.46	60.46	1.5338	5	2	4	2	2	2
61.17	61.16	1.5299	5	1	1	1	2	2
62.08	62.10	1.5139	3	0	9	1	1	2
62.13		1.4938	6	0	0	6	5	9
62.93	62.94	1.4927	3	1	1	5		8
63.07	63.08	1.4756	5	3	3	5	3	9
63.34	63.24	1.4727	5	2	5	5	4	8
63.58	63.58	1.4671	6	1	1	2	3	4
63.59		1.4621	5	1	6	2		3
64.25	64.24	1.4620	4	2	8	5		8
65.85	65.84	1.4485	6	2	1	2	1	3
66.00	66.00	1.4172	2	2	2	4	2	6
66.77	66.76	1.4143	2	2	10	2	2	3
		1.3958	5	4	0	1	1	2

2THETA	PEAK	D	H	K	L	I(INT)	I(PK)	I(DS)
67.07	67.06	1.3943	5	4	1	3	2	6
67.47	67.46	1.3870	5	1	7	1	1	2
67.82	67.82	1.3806	6	1	1	1	1	2
69.13	69.12	1.3577	4	4	4	2	1	3
69.43	69.42	1.3525	2	2	11	1	1	3
70.19	70.20	1.3397	5	2	7	1	1	3
70.44	70.44	1.3355	6	1	5	1	1	2
70.93	70.92	1.3276	5	3	6	1	1	3
71.35	71.36	1.3208	4	2	9	2		3
71.59	71.58	1.3170	0	0	12	1	1	2
73.59	73.60	1.2860	6	1	1	2	1	3
73.89	73.89	1.2815	3	1	11	3	3	6
74.00	74.00	1.2799	5	4	5	5		10
74.48	74.48	1.2729	5	2	8	3	2	6
74.63	74.64	1.2706	5	3	7	2	2	5
75.97	75.96	1.2516	7	1	2	3	3	6
75.97		1.2516	5	5	2	2		4
76.50	76.50	1.2442	7	0	3	2		
76.73	76.74	1.2410	4	2	10	6	3	12
77.70	77.70	1.2278	6	4	2	2	1	3
78.57	78.58	1.2165	7	2	1	1	1	3
79.98	79.98	1.1986	4	1	11	1	1	3
80.32	80.30	1.1944	3	1	12	1	1	2
80.54	80.54	1.1916	6	3	6	2	1	3
80.94	80.94	1.1868	7	3	5	2	1	5
84.23	84.24	1.1486	4	4	10	5	2	11
87.00	87.16	1.1189	3	1	13	3	2	6
87.17		1.1173	7	2	6	4		10
87.35		1.1153	8	0	2	1		2
87.95	87.94	1.1093	6	5	1	1	1	2
88.68	88.80	1.1020	4	4	12	1	1	3
88.83		1.1006	5	2	11	1		3
90.18	90.38	1.0876	2	2	14	3	2	6
90.29		1.0866	2	1	14	2		4
90.39		1.0856	6	0	10	3		7
93.78	93.78	1.0551	6	2	10	1	1	3
95.13	95.14	1.0436	7	3	7	3	1	8

2THETA	PEAK	D	H	K	L	I(INT)	I(PK)	I(DS)
95.58	95.58	1.0399	8	3	2	2	1	4
95.64		1.0394	4	2	13	1		3
96.97	97.04	1.0287	6	1	11	1	1	3
97.07		1.0278	3	2	14	1		4
100.39	100.42	1.0026	5	4	11	2	1	5
101.60	101.64	0.9939	9	0	1	1	0	2
101.83	101.84	0.9923	6	5	5	1	1	3
102.23	102.20	0.9896	7	5	5	1	0	1
102.97	102.94	0.9844	8	4	6	1	0	2
103.56	103.56	0.9804	6	6	6	1	0	2
103.97	103.96	0.9776	7	3	9	1	0	2
104.44	104.44	0.9746	8	0	8	1	0	3
104.71	104.74	0.9728	9	1	3	1	1	3
105.08	105.10	0.9703	9	2	1	1	0	1
105.31	105.30	0.9688	8	1	8	1	0	2
105.93	105.92	0.9649	7	4	8	1		2
105.93		0.9649	9	2	2	1		3
107.35	107.40	0.9560	7	6	3	1	1	2
107.48		0.9552	8	4	5	1		2
107.76	107.76	0.9535	5	5	13	3	1	8
108.62	108.62	0.9484	8	5	1	1	0	2
109.01	109.02	0.9460	7	5	7	1	0	2
109.48	109.44	0.9433	8	5	2	1	0	2
112.47	112.48	0.9265	9	1	6	2	1	4
112.47		0.9265	8	3	8	3		8
113.18	112.88	0.9227	6	2	13	1	1	2
115.25	115.26	0.9120	9	2	6	1	0	1
117.86	117.86	0.8993	4	1	16	1	1	2
118.41	118.48	0.8967	6	5	11	1	0	2
118.53		0.8961	8	1	14	1	0	3
119.74	119.76	0.8906	8	6	2	1		2
119.82		0.8902	5	2	15	1	1	4
120.77	120.78	0.8860	4	2	16	1	0	4
121.38	121.36	0.8833	3	1	17	1	0	7
121.78	121.84	0.8816	7	0	13	1	0	3
122.34	122.48	0.8792	8	1	11	1	1	8

APOPHYLLITE - COLVILLE AND ANDERSON, 1968

2THETA	PEAK	D	H	K	L	I(INT)	I(PK)	I(DS)
122.58		0.8782	6	6	10	1		4
123.59	123.56	0.8740	8	3	10	1	0	2
124.02	124.06	0.8723	7	6	8	1	0	2
125.37	125.36	0.8669	10	2	3	1	0	2
127.81	127.84	0.8577	2	1	18	2	1	6
128.21		0.8562	7	7	6	1		3
128.97	128.96	0.8535	10	3	2	1	0	2
129.01		0.8534	8	0	12	1		4
129.81	129.70	0.8505	9	2	9	1	0	2
130.91	131.32	0.8468	10	2	5	1	1	3
131.27		0.8456	7	3	13	3		9
131.52		0.8447	10	1	6	2		5
133.19	133.20	0.8393	10	5	5	1	0	3
135.00	135.52	0.8337	10	2	6	1	0	2
135.51		0.8322	10	4	0	2		7
136.99	137.06	0.8279	3	2	18	3	1	8
137.08		0.8276	10	4	2	1		2
137.13		0.8275	9	6	1	1		4
138.38	138.34	0.8240	6	6	12	1	0	2
138.41		0.8239	6	0	16	1		3
138.72		0.8231	7	8	8	1		3
139.15	139.12	0.8219	8	4	11	1	0	2
140.46	140.46	0.8185	9	0	11	2	0	7
142.13	142.22	0.8143	10	4	4	1	0	5
142.22		0.8141	4	1	18	2		5
142.73		0.8129	10	1	9	1		3
143.78	143.66	0.8104	11	1	8	1	0	3
144.84	144.74	0.8080	9	4	9	1	0	2
146.13	146.26	0.8052	9	2	11	1	0	2
146.13		0.8052	7	6	11	1		3
146.32		0.8048	10	4	5	1		2
147.15	147.88	0.8030	8	7	6	1	1	3
147.64		0.8020	11	1	3	1		3
147.81		0.8017	10	5	0	2		6
147.87		0.8015	9	6	5	2		8
148.32		0.8006	11	2	1	1		4
149.62	149.62	0.7981	3	1	19	6	1	20

2THETA	PEAK	D	H	K	L	I(INT)	I(PK)	I(DS)
150.42	150.70	0.7966	9	5	8	1	1	3
151.74	152.90	0.7943	6	3	16	4	1	13
152.17		0.7935	10	4	6	2		5
152.75		0.7926	11	2	3	1		4
152.80		0.7925	8	5	11	3		10
152.95		0.7922	8	8	0	1		3
153.02		0.7921	8	1	14	1		2
153.62	153.98	0.7911	7	5	13	1	1	5
154.05		0.79C4	9	6	6	3		10
154.20		0.7902	0	0	20	1		3
154.71		0.7894	9	3	11	2		8
155.45		0.7883	8	1	5	1		2
157.03	157.08	0.7860	11	5	5	3	1	10
157.26		0.7857	11	2	4	1		2
157.26		0.7857	10	5	4	1		3
157.30		0.7856	7	3	10	1		5
157.76		0.7850	7	3	15	1		5
162.19	163.72	0.7798	8	6	10	2	0	6
162.19		0.7796	10	0	10	1		3
162.69		0.7791	7	1	16	2		7
162.69		0.7791	5	5	16	1		3
164.34		0.7775	9	7	3	4		15
164.81		0.7771	10	5	5	3		10
164.81		0.7771	11	2	5	3		9

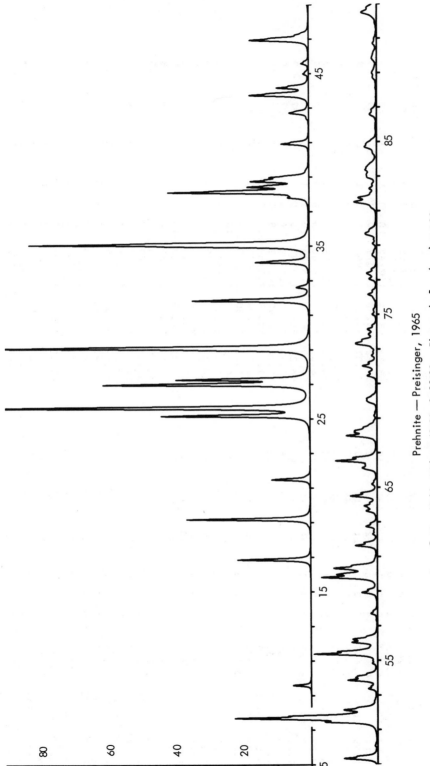

Prehnite — Preisinger, 1965

λ = CuKα Half width at 40°2θ: 0.11°2θ Plot scale I$_{PK}$ (max) = 100

PREHNITE - PREISINGER, 1965

2THETA	PEAK	D	H	K	L	I(INT)	I(PK)	I(DS)
9.56	9.58	9.240	2	0	0	4	5	4
16.83	16.84	5.263	1	1	0	19	22	17
19.17	19.18	4.627	0	0	1	32	37	30
19.19		4.620	4	0	0			2
21.46	21.46	4.137	2	0	1	11	12	10
25.15	25.16	3.538	0	1	1	43	44	42
25.61	25.62	3.475	1	1	1	100	100	100
26.96	26.96	3.304	2	1	1	62	61	63
27.25	27.26	3.269	4	0	1	38	40	39
29.08	29.08	3.068	3	1	1	94	95	98
31.83	31.84	2.809	4	1	0	39	35	42
32.59	32.63	2.745	0	2	0	4	4	5
34.04	34.04	2.631	2	2	0	18	16	20
35.08	35.08	2.556	5	1	1	78	83	89
37.78	37.78	2.379	7	1	0	6	7	7
38.08	38.10	2.361	0	0	2	22	42	26
38.10		2.360	4	2	0	28		34
38.41	38.40	2.342	1	1	2	19	19	23
38.73	38.74	2.323	6	0	0	19	17	23
38.89	38.82	2.313	0	2	1	3	13	3
38.96	38.96	2.310	8	0	0	10	12	12
40.90	40.90	2.204	1	2	1	10	8	13
42.65		2.118	1	1	2	3		4
42.70	42.70	2.116	7	1	1	5	6	7
43.72	43.74	2.069	1	0	2	19	18	24
43.76		2.067	7	0	2	6		7
44.16	44.16	2.049	8	1	0	12	10	15
44.95	44.96	2.015	6	3	0	2	2	3
45.55	45.56	1.9896	3	2	2	3	2	4
46.89	46.92	1.9358	5	1	2	10	18	13
46.93		1.9342	4	1	1	16		21
47.22	47.20	1.8480	8	1	1	4	4	5
49.27	49.30	1.8467	9	0	0	2	10	3
49.30		1.7759	10	0	0	12		17
51.41	51.42	1.7690	5	1	2	17	16	24
51.62	51.62	1.7674	9	2	2	52	42	75
51.67			8	2	0	5		7

2THETA	PEAK	D	H	K	L	I(INT)	I(PK)	I(DS)
52.09	52.12	1.7542	3	3	0	4	10	6
52.13		1.7530	3	1	2	8		12
53.34	53.34	1.7182	10	0	1	3	3	5
53.82	53.82	1.7017	10	0	1	12	9	18
54.07	54.06	1.6946	1	3	1	6	6	9
54.80	54.80	1.6736	2	3	1	2	2	2
55.34	55.34	1.6586	7	0	2	27	19	41
56.01	56.02	1.6403	3	3	1	7	8	10
56.03		1.6400	3	3	1	1		5
56.10		1.6380	5	1	1	6		2
56.22	56.20	1.6347	10	0	2	3	7	9
57.68	57.68	1.5969	9	3	0	7	2	4
58.90	58.90	1.5667	4	3	1	24	5	11
59.77	59.78	1.5458	8	1	2	7	16	39
60.02	59.94	1.5400	5	3	1	8	12	12
60.28	60.30	1.5340	12	0	0	12	13	13
60.33		1.5330	6	2	2	10		19
61.61	61.62	1.5040	10	0	0	2	7	17
62.28	62.28	1.4895	7	3	1	5	1	3
62.72	62.72	1.48C1	6	3	1	5	3	8
63.62	63.62	1.4612	1	3	3	5	3	8
63.92	63.92	1.4552	12	2	1	7	4	9
64.48	64.48	1.4439	10	0	2	6	8	13
64.50		1.4435	10	1	3	2		10
65.16	65.16	1.4303	3	0	3	3	2	3
66.03	66.12	1.4136	7	3	1	5	4	6
66.12		1.4120	4	1	3	20		8
66.52	66.52	1.4045	12	1	2	15	12	35
67.98	67.98	1.3778	8	2	3	8	9	27
68.28	68.26	1.3725	5	4	0	2	7	14
68.38		1.37C6	0	3	3	2		3
69.13	69.14	1.3576	2	4	0	3	1	2
69.90	70.00	1.3446	0	3	3	3	3	4
69.99		1.3431	12	2	1	2		6
70.11	70.10	1.3411	1	2	3	1		4
71.43	71.44	1.3195	11	1	1	3	3	2
71.46		1.3191	13	1	1	3	3	6

600

PREHNITE - PREISINGER, 1965

2THETA	PEAK	D	H	K	L	I(INT)	I(PK)	I(DS)
71.66	71.64	1.3158	0	4	1	1	2	2
72.01	72.02	1.3102	9	3	1	7	4	14
72.61	72.60	1.3009	6	3	2	2	2	4
73.25	73.30	1.2911	4	2	1	6	6	12
73.33		1.2898	12	2	3	7		13
73.54	73.48	1.2868	3	4	1	1	4	2
73.81	73.84	1.2827	8	0	1	1	3	3
73.86		1.2820	12	2	3	2		4
74.13	74.12	1.2779	10	2	2	2	3	7
74.98	74.98	1.2655	4	4	1	4	1	4
76.15	76.14	1.2491	8	1	3	2	3	8
77.04	77.04	1.2367	14	1	1	4	2	7
77.37	77.36	1.2323	6	2	3	3	3	10
78.37	78.44	1.2191	8	3	2	5	1	3
78.47		1.2178	13	1	1	2		3
79.07	79.08	1.2100	6	4	3	1		4
79.61	79.66	1.2032	9	1	0	2	1	8
79.66		1.2021	15	1	0	4		9
81.47	81.50	1.1804	0	4	2	6	7	12
81.50		1.1799	8	4	0	7		14
81.76	81.74	1.1769	1	3	3	4	6	7
82.27	82.26	1.1709	15	2	2	4	2	8
82.92	82.92	1.1635	15	1	1	4	3	9
83.03	83.06	1.1621	8	2	1	2	3	5
83.08		1.1615	12	2	3	2		4
83.36	83.36	1.1583	3	3	3	4	3	8
83.50	83.48	1.1567	0	0	4	4	3	8
84.67	84.68	1.1437	4	4	2	6	4	12
84.84	84.84	1.1419	12	3	1	4	3	8
86.56	86.56	1.1235	5	4	3	4	2	8
88.66	88.72	1.1022	6	4	2	1	1	3
88.75		1.1014	6	3	3	1		3
88.95	88.96	1.0994	4	1	4	1	1	3
89.69	89.82	1.0922	15	2	1	1	1	3
89.82		1.0910	13	3	2	2		4
90.68	90.68	1.0829	0	0	4	3	1	6
92.27	92.50	1.0683	0	5	1	1	5	3

2THETA	PEAK	D	H	K	L	I(INT)	I(PK)	I(DS)
92.45		1.0667	15	1	2	4		9
92.47		1.0666	5	5	1	3	1	6
92.49		1.0664	17	1	0	1	1	3
92.54		1.0660	0	2	4	5		11
93.45	93.44	1.0579	14	2	1	2	2	5
94.07	94.08	1.0526	3	4	4	3	1	8
95.73	95.84	1.0387	4	2	4	3	1	8
95.88		1.0375	16	0	4	2	2	8
96.27	96.22	1.0343	8	2	4	4	2	5
97.20	97.26	1.0268	13	1	3	3	2	5
97.27		1.0263	5	5	1	1		4
97.73	97.76	1.0227	9	3	0	3	2	8
97.82		1.0220	15	1	4	1		5
98.54	98.56	1.0164	8	5	0	1	1	7
98.89	99.00	1.0138	7	1	4	1	2	4
99.01		1.0129	12	5	2	5		12
99.48	99.34	1.0093	6	4	1	3	1	3
100.70	100.70	1.0004	12	4	2	3	1	8
101.04	101.04	0.9979	15	0	4	2	1	5
101.48	101.48	0.9948	10	4	2	5	2	13
103.55	103.56	0.9805	10	0	4	2	1	4
105.38	105.42	0.9684	17	2	4	1	1	6
105.46		0.9679	8	3	0	3		3
107.41	107.38	0.9557	16	4	1	3	1	8
108.11	108.12	0.9514	14	4	0	2	1	6
108.66	108.66	0.9481	15	5	1	3	1	4
108.73		0.9477	9	3	4	1		5
109.14	109.12	0.9453	5	0	2	2	1	4
110.33	110.34	0.9384	18	3	2	4	2	12
110.96	110.96	0.9348	15	3	2	2	1	5
112.40	112.10	0.9286	7	0	5	4	2	4
112.68	112.84	0.9254	0	0	2	2	1	7
112.85		0.9245	16	2	4	1		13
113.06	113.02	0.9234	10	3	4	2	1	7
114.29	114.30	0.9169	7	4	4	4	1	13
115.37	115.36	0.9114	1	5	5	1	1	4
115.54		0.9105	2	6	0	1		4

PREHNITE - PREISINGER, 1965

2THETA	PEAK	D	H	K	L	I(INT)	I(PK)	I(DS)
116.12	116.44	0.9076	13	3	3	1	1	3
116.18		0.9074	4	0	5	1		3
116.44		0.9061	20	0	1	3		10
117.14	116.86	0.9027	3	1	5	1	1	4
118.22	118.40	0.8976	3	6	0	2	1	5
118.43		0.8966	10	4	3	3		8
118.99	118.94	0.8940	20	1	1	1	1	3
120.79	120.82	0.8859	5	1	5	4	1	12
120.95		0.8852	3	5	3	1		3
121.11		0.8845	0	4	4	2		5
122.68	122.70	0.8778	12	2	1	1	1	4
123.85	123.84	0.8730	2	2	5	1	1	4
124.74	124.80	0.8694	5	5	3	6	2	19
124.86		0.8689	21	4	1	1		9
125.08		0.8680	16	3	2	2		4
125.46	125.30	0.8666	17	2	3	2	2	5
127.06	127.16	0.8604	20	2	3	2	1	6
127.45	127.44	0.8590	6	5	1	1	1	4
129.41	129.70	0.8519	15	3	3	3	2	10
129.71		0.8509	0	6	3	5		15
134.94	135.04	0.8339	9	1	2	3	2	11
135.06		0.8335	15	1	5	6		21
136.49	136.48	0.8293	14	2	4	2	1	7
137.48	137.68	0.8265	22	0	1	1	2	4
137.65		0.8260	8	4	4	8		26
138.02		0.8250	1	3	5	3		6
139.82	139.90	0.8201	6	6	2	4	2	12
139.86		0.8200	9	6	3	4		14
139.88		0.8200	10	6	0	4		15
140.26		0.8190	20	2	2	2		8
140.46	140.44	0.8185	3	1	5	4	2	13
142.50	142.46	0.8134	21	1	2	2	0	7
145.10	145.80	0.8074	10	6	1	3	2	9
145.22		0.8071	15	5	1	4		15
145.77		0.8059	5	3	7	1		27
146.13		0.8052	19	3	2	7		5
149.47	149.64	0.7984	8	6	2	2	1	27

2THETA	PEAK	D	H	K	L	I(INT)	I(PK)	I(DS)
149.78		0.7978	10	4	4	5		17
152.24	153.30	0.7934	2	5	4	1	1	4
152.36		0.7932	12	0	5	8		28
152.94		0.7922	10	2	2	5		18
153.45		0.7914	22	2	1	7		26
154.60		0.7896	22	0	2	2		6
157.71	158.22	0.7851	12	1	5	1	1	4
157.91		0.7848	4	5	1	1		4
158.12		0.7845	20	1	3	4		16
158.30		0.7843	16	5	1	2		7
158.48		0.7840	17	1	1	4		14

602

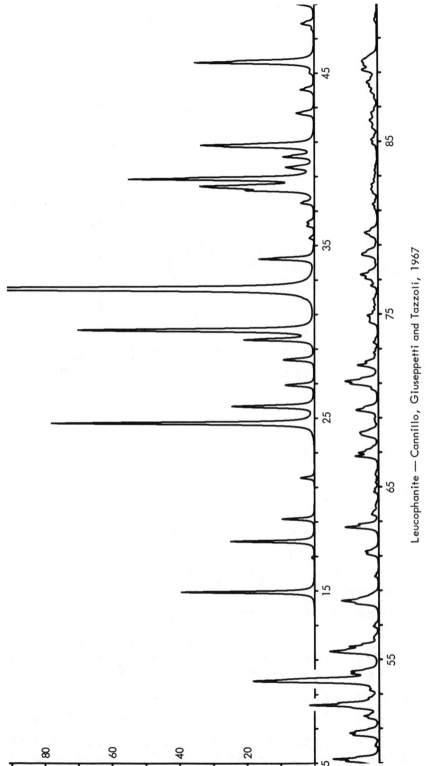

Leucophanite — Cannillo, Giuseppetti and Tazzoli, 1967

$\lambda = CuK\alpha$ Half width at $40°2\theta$: $0.11°2\theta$ Plot scale I_{PK} (max) = 200

603

LEUCOPHANITE - CANNILLO, GIUSEPPETTI AND TAZZOLI, 1967

2THETA	PEAK	D	H	K	L	I(INT)	I(PK)	I(DS)
14.89	14.90	5.946	0	1	1	21	20	18
14.91		5.936	1	0	0	7		6
17.83	17.84	4.969	0	1	1	18	12	15
19.13	19.14	4.635	1	1	0	7	5	6
21.50	21.52	4.129	1	0	1	2	2	1
21.52		4.126	1	1	0	2		1
24.67	24.68	3.606	1	0	2	63	39	57
25.61	25.66	3.476	0	2	1	10	12	9
25.67		3.468	2	0	1	12		11
26.86	26.90	3.317	1	2	1	4	4	4
26.90		3.312	2	0	0	4		4
28.34	28.36	3.146	1	2	1	5	5	5
28.38		3.142	2	1	0	3		3
29.50	29.52	3.025	1	1	3	10	10	9
29.51		3.024	2	1	2	8		8
30.03		2.973	0	0	3	37	35	36
30.08	30.08	2.968	1	2	0	32		31
32.43		2.759	0	2	2	95	100	95
32.46	32.46	2.756	2	2	2	100		100
34.19	34.20	2.620	1	2	2	15	8	15
35.40	35.40	2.533	2	2	0	1	1	1
36.12	36.12	2.485	2	0	4	2	1	2
36.32	36.32	2.471	0	2	3	3	1	2
37.44	37.44	2.400	0	3	1	8	2	4
38.16		2.356	0	1	4	8		8
38.17	38.18	2.356	0	3	0	9	10	9
38.34		2.346	0	3	2	15		16
38.37		2.344	1	0	4	4	17	4
38.40		2.342	1	2	0	4		4
38.42	38.42	2.341	2	1	3	13		13
38.82	38.82	2.318	3	0	4	53	28	57
39.43		2.283	2	2	0	3	4	4
39.51	39.52	2.279	2	2	2	5		6
40.13	40.14	2.245	3	1	1	9	5	10
40.71	40.80	2.214	1	3	4	20	17	22
40.80		2.210	3	0	3	20		22
42.58	42.66	2.121	1	3	2	2	3	3

2THETA	PEAK	D	H	K	L	I(INT)	I(PK)	I(DS)
42.66	44.02	2.118	3	1	2	4	2	4
44.03		2.055	2	2	4	4		5
45.58	45.60	1.9886	1	1	4	19	18	22
45.60		1.9875	2	1	4	20		24
47.83	47.86	1.9001	3	1	2	2	2	2
47.87		1.8985	1	3	2	2		3
48.97	49.08	1.8586	0	4	1	2		3
49.07		1.8550	4	0	0	8	5	10
49.20	49.20	1.8503	0	4	1	10	7	12
49.97	49.98	1.8235	4	1	1	1	1	1
50.58	50.60	1.8029	0	0	4	7	4	9
50.69	50.70	1.7993	2	4	0	4	4	5
50.81	50.80	1.7953	1	1	4	3	4	3
51.57	51.58	1.7706	4	1	1	3	2	4
51.70	51.70	1.7667	1	1	4	3	2	3
52.32	52.34	1.7471	2	1	0	3		4
52.33		1.7467	3	1	3	3	10	3
52.36		1.7459	3	3	0	16		21
52.75	52.74	1.7340	4	2	2	2	2	2
53.20	53.20	1.7203	1	3	3	2	1	2
53.69	53.72	1.7057	3	4	4	28	19	36
53.75		1.7039	1	4	2	23		30
54.16	54.16	1.6919	2	1	2	7	4	9
54.28	54.28	1.6885	4	0	0	5	4	6
55.35	55.44	1.6583	0	6	6	7	7	10
55.42		1.6565	2	0	0	4		5
55.45		1.6558	4	2	2	9		12
55.73	55.74	1.6478	3	3	3	8	4	11
56.91	56.90	1.6167	0	6	4	1	1	1
58.18	58.22	1.5842	2	3	4	3	4	3
58.21		1.5836	3	2	5	3		3
58.22		1.5833	1	4	3	4		4
58.30	58.38	1.5812	1	1	6	7	6	3
58.37		1.5795	4	3	6	5		7
58.42		1.5784	4	1	3	4		5
59.80	59.80	1.5451	4	3	3	1	1	2
61.05	61.10	1.5165	1	3	5	1	2	2

LEUCOPHANITE - CANNILLO, GIUSEPPETTI AND TAZZOLI, 1967

2THETA	PEAK	D	H	K	L	I(INT)	I(PK)	I(DS)
61.11		1.5152	3	1	5	3		4
61.23	61.24	1.5126	0	2	6	1	2	2
61.25		1.5119	2	0	6	1		2
62.63	62.64	1.4819	1	2	6	5	5	8
62.65		1.4815	2	1	6	7		9
63.37	63.38	1.4664	3	3	1	2	1	3
64.69	64.70	1.4397	1	5	1	2	1	3
66.75	66.76	1.4001	2	2	6	9	3	14
66.95	66.94	1.3964	1	5	2	2	3	3
67.98	68.12	1.3779	4	2	4	1	3	4
68.00		1.3774	2	5	0	2		2
68.06		1.3763	0	3	6	2		3
68.12		1.3752	3	0	0	2		3
68.15		1.3748	5	2	3	1		3
69.38	69.42	1.3533	3	2	3	3	3	3
69.39		1.3531	1	3	6	3		5
69.45		1.3522	1	0	3	4		6
69.49		1.3514	5	0	6	1		2
70.64	70.66	1.3322	1	5	3	2	1	3
70.94	70.96	1.3273	2	5	2	9	4	14
71.08	71.10	1.3250	5	2	2	9	5	15
71.89	72.02	1.3121	3	3	5	8	3	13
72.03		1.3100	4	4	0	2		3
72.75	72.74	1.2988	4	1	1	1	1	3
73.36	73.34	1.2895	3	2	6	1	1	2
74.56	74.54	1.2717	3	5	0	3	1	2
74.90	74.90	1.2667	4	4	2	1	2	6
75.38	75.38	1.2599	5	1	4	4	1	2
75.84	75.84	1.2534	5	1	0	1	1	3
76.63	76.64	1.2424	0	6	8	3	1	5
77.05	77.06	1.2367	0	6	0	4	2	7
77.13		1.2356	4	0	6	1		2
77.23	77.28	1.2342	6	0	0	3	3	2
77.28		1.2335	1	4	6	3		6
78.40	78.42	1.2187	4	4	6	5	2	8
78.50		1.2174	4	1	0	4		7
79.63	79.70	1.2030	5	2	4	2	2	4

2THETA	PEAK	D	H	K	L	I(INT)	I(PK)	I(DS)
79.71		1.2019	3	3	6	3		6
81.12	81.12	1.1846	1	6	2	3	1	6
81.34	81.34	1.1819	6	1	2	2	1	3
82.00	82.02	1.1741	1	5	5	2	1	3
82.18	82.22	1.1720	2	4	3	1	1	2
82.48	82.48	1.1684	4	3	7	1	1	2
83.32	83.30	1.1588	2	6	4	1	0	2
84.87	84.88	1.1415	6	2	2	2	1	4
85.07	85.08	1.1393	3	5	4	2	1	4
85.75	85.78	1.1320	4	4	2	2	1	3
86.18	86.20	1.1275	5	2	2	2	1	3
86.24		1.1269	2	1	8	2		4
86.65	86.64	1.1226	2	3	7	3	1	5
87.52	87.52	1.1137	0	6	8	1	1	2
87.86	87.88	1.1102	3	0	0	2	1	3
87.92		1.1096	0	6	8	1		3
88.17	88.18	1.1071	6	0	4	2	1	3
88.40	88.42	1.1048	0	0	8	4	2	8
88.41		1.1048	3	1	4	2		8
89.11	89.14	1.0979	1	6	6	4	2	2
89.16		1.0974	3	1	4	4		6
89.41	89.44	1.0949	1	6	5	1		2
89.59	89.62	1.0932	6	1	6	3		7
89.63		1.0929	4	3	6	1		3
89.74		1.0928	6	7	2	1		3
92.05	92.04	1.0917	5	4	0	3	0	5
99.37	99.40	1.0703	2	1	0	1	1	7
100.96	100.96	1.0102	2	2	6	1	1	3
101.63	101.60	0.9982	4	2	2	1	1	3
104.15	104.18	0.9937	1	7	8	2	0	3
104.54	104.80	0.9765	3	7	0	3	1	5
104.80		0.9739	7	3	0	1		7
107.37	107.38	0.9722	6	2	0	3	0	3
108.59	108.54	0.9559	5	4	6	1	1	3
111.96	111.96	0.9485	2	2	10	1	0	3
116.31	116.66	0.9293	2	1	4	2	1	3
		0.9067	3	7	4	1		3

LEUCOPHANITE — CANNILLO, GIUSEPPETTI AND TAZZOLI, 1967

2THETA	PEAK	D	H	K	L	I(INT)	I(PK)	I(DS)
116.59		0.9054	7	3	4	2		5
117.40	117.08	0.9015	4	4	8	2	1	4
120.85	120.90	0.8856	5	5	6	2	1	5
123.00	123.24	0.8765	0	6	8	2	0	4
123.27		0.8753	6	0	8	1		3
126.16	126.14	0.8638	3	3	10	2	0	5
130.41	130.28	0.8485	7	2	5	1	0	3
137.41	137.46	0.8267	7	2	7	1	0	4
138.41	138.46	0.8239	6	6	4	3	1	9
146.12	146.98	0.8052	7	0	8	1	1	4
146.34		0.8047	2	9	0	1		4
146.45		0.8045	1	8	6	3		7
146.62		0.8041	4	7	6	1		4
146.67		0.8040	2	6	9	1		3
146.88		0.8036	1	2	12	1		4
146.92		0.8035	2	1	12	1		4
147.01		0.8033	7	4	6	1		3
147.18		0.8030	8	1	6	3		8
153.95	154.30	0.7906	2	8	6	2	0	5
154.23		0.7901	7	5	5	2		6
154.50		0.7897	2	2	12	1		3
154.83		0.7892	8	2	6	2		6
154.96		0.7890	5	6	7	2		6
159.51	160.86	0.7827	5	3	10	1	0	3
159.59		0.7826	8	5	1	1		3
160.70		0.7813	3	6	9	2		5
162.30		0.7795	3	9	1	1		4
163.65		0.7782	1	9	4	5		16
163.87		0.7779	9	3	1	2		7

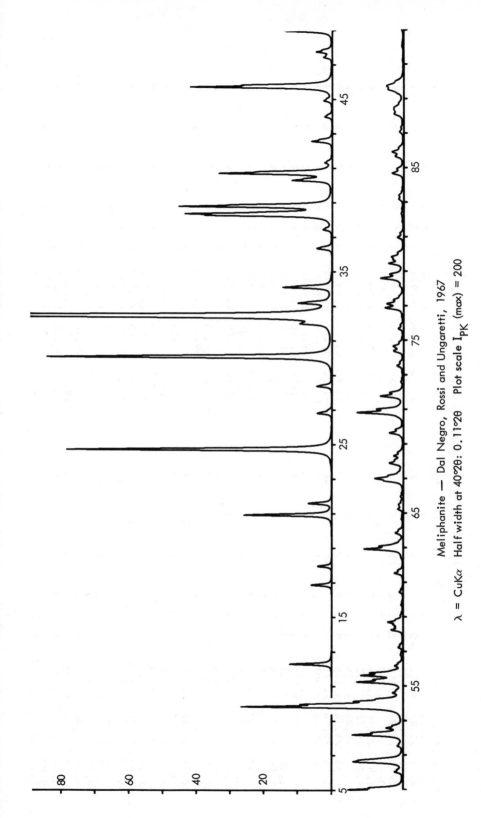

Meliphanite — Dal Negro, Rossi and Ungaretti, 1967

λ = CuKα Half width at 40°2θ: 0.11°2θ Plot scale I$_{PK}$ (max) = 200

MELIPHANITE — DAL NEGRO, ROSSI AND UNGARETTI, 1967

2THETA	PEAK	D	H	K	L	I(INT)	I(PK)	I(DS)
12.28	12.28	7.203	0	1	1	5	6	4
16.85	16.86	5.258	0	0	0	2	3	2
17.93	17.94	4.943	2	2	0	2	2	1
20.90	20.90	4.247	1	0	1	10	13	8
20.90		4.247	1	1	1	1		1
21.57	21.58	4.117	1	1	1	3	4	2
24.70	24.70	3.602	2	1	2	35	39	31
26.79	26.80	3.325	3	2	0	1		
28.35	28.36	3.145	0	1	3	3	2	1
30.05	30.06	2.971	2	2	2	2	2	2
31.96	31.98	2.797	3	1	2	42	42	40
31.96		2.797	2	2	1	1	5	1
32.42	32.42	2.759	3	1	2	2		
32.42		2.759	1	3	2	49	100	49
33.16	33.16	2.699	2	2	2	51		51
33.16		2.699	1	3	2	2	5	2
34.07	34.08	2.629	0	2	4	7	7	8
36.31	36.34	2.472	0	4	0	1	2	1
37.42	37.42	2.401	0	3	3	1	1	1
38.24	38.26	2.351	4	2	0	9	19	10
38.24		2.351	2	4	2	8		
38.34	38.34	2.346	1	1	4	12	22	13
38.76	38.76	2.321	0	4	0	24	23	27
40.28	40.28	2.237	0	4	2	6	6	7
40.68	40.68	2.216	3	1	4	19	17	21
42.54	42.54	2.123	4	2	1	2	3	2
42.54		2.123	2	2	4	2		
45.69	45.70	1.9838	2	4	2	12	21	15
45.69		1.9838	1	3	4	13		16
47.41	47.42	1.9158	3	2	1	1	1	2
47.74	47.74	1.9034	5	1	2	1	2	2
47.74		1.9034	5	1	0	2		2
47.74		1.9034	3	5	0	1		
48.96	48.96	1.8590	4	4	0	10	8	13
50.57	50.58	1.8035	5	3	0	4	7	5
50.58		1.8035	3	0	4	4		5
50.65		1.8008	0	4	0	9	8	13
52.14	52.14	1.7527	0	6	0			

2THETA	PEAK	D	H	K	L	I(INT)	I(PK)	I(DS)
52.55	52.54	1.7400	4	4	2	3	3	4
53.76	53.76	1.7037	4	2	4	16	24	22
53.76		1.7037	2	4	4	14		20
54.08	54.08	1.6943	5	3	2	3	7	5
54.08		1.6943	3	5	2	3		6
55.19	55.20	1.6627	6	2	0	4	7	6
55.19		1.6627	0	6	2	8		11
55.58	55.58	1.6519	0	6	2	2	6	3
55.74	55.74	1.6478	5	1	4	1	5	2
58.21	58.22	1.5835	4	4	4	1	2	2
58.21		1.5835	0	0	6	1		3
58.66	58.66	1.5724	2	2	6	2		3
61.50	61.50	1.5065	3	1	6	4	2	7
62.89	62.90	1.4785	1	3	6	2	1	7
62.89		1.4765	6	4	2	4	6	3
66.83	66.84	1.3987	4	2	6	2		2
66.96	66.96	1.3962	6	4	4	1	3	8
67.88	67.88	1.3796	0	2	4	5		2
67.88		1.3796	6	2	6	1		2
68.29	68.30	1.3723	2	3	4	2	4	3
69.61	69.62	1.3495	4	4	6	1	2	3
69.61		1.3495	4	2	6	6		3
70.78	70.78	1.3299	7	3	2	5	7	10
70.78		1.3299	0	7	0	5		9
71.74	71.74	1.3145	0	8	0	2		10
74.65	74.62	1.2704	6	0	2	4	3	3
76.85	76.86	1.2393	6	0	8	1	2	9
77.31	77.10	1.2359	8	0	0	1	2	3
77.31	77.32	1.2331	6	0	8	3	2	2
78.57	78.56	1.2165	4	4	6	3	3	6
78.57		1.2165	5	3	6	3		6
79.43	79.44	1.2055	7	4	4	2	2	3
79.43		1.2055	3	7	4	2		4
79.82	79.82	1.2005	0	6	6	3	2	5
84.66	84.66	1.1438	8	2	2	2	2	4
84.66		1.1438	4	8	2	2		3

MELIPHANITE - DAL NEGRO, ROSSI AND UNGARETTI, 1967

2THETA	PEAK	D	H	K	L	I(INT)	I(PK)	I(DS)
85.89	85.90	1.1305	9	1	2	1	2	2
85.89		1.1305	1	1	2	1		2
87.05	87.06	1.1185	0	9	2	1		3
88.10	88.10	1.1079	6	4	8	1	1	3
88.28	88.38	1.1060	3	6	8	2	1	5
89.51	89.54	1.0940	2	3	8	1	2	3
89.51		1.0940	4	4	8	2		4
89.71	89.72	1.0921	6	2	6	2		4
89.71		1.0921	4	6	6	2	3	4
104.16	104.16	0.9764	10	4	0	2		5
104.16		0.9764	4	10	0	2	1	5
104.88	104.86	0.9717	0	2	10	1		3
107.44	107.44	0.9555	2	2	10	1	1	3
114.96	114.98	0.9135	8	8	2	1	1	3
116.03	116.06	0.9081	10	0	4	1	0	3
116.03		0.9081	4	10	4	1	1	3
117.62	117.62	0.9004	0	8	8	1		3
120.66	120.64	0.8865	0	10	6	2	0	5
123.33	123.44	0.8751	6	6	8	2	1	6
123.86	123.84	0.8729	4	4	10	1	1	4
126.88	126.86	0.8611	0	6	10	2	1	7
137.67	137.70	0.8260	12	0	4	3	1	10
144.14	144.16	0.8096	8	8	6	1	0	5
146.29	146.32	0.8048	11	3	6	1	1	4
146.29		0.8048	9	7	6	2		8
146.29		0.8048	7	9	6	2		7
146.29		0.8048	3	11	6	1		4
148.79	148.82	0.7997	11	3	12	1	1	5
148.79		0.7997	3	11	12	1		5
162.54	162.90	0.7793	8	10	4	3	0	11
162.54		0.7793	10	8	4	3		11
163.43		0.7784	13	3	2	3		10
163.43		0.7784	3	13	2	3		10

TABLE 31. BRITTLE MICAS AND DOUBLE SHEET SILICATES

Variety	Xanthophyllite	Margarite	α-$BaAl_2Si_2O_8$	$CaAl_2Si_2O_5$
Composition	$Ca(Mg,Al)_3(Si,Al)_4$ $O_{10}(OH)_2$	$(Ca,Na)(Al,Li,Fe)_2$ $(Al,Si)_4O_{10}(OH)_2$		
Source	Chichibu, Japan	Chester, Mass.	Synthetic	Synthetic
Reference	Takéuchi, 1966 Takéuchi & Sadanaga, 1966	Takéuchi, Kawada & Sadanaga, 1968 Takéuchi, 1966	Takéuchi, 1958	Takéuchi & Donnay, 1959
Cell Dimensions				
a Å	5.194	5.123	5.293	5.10
b Å	9.003	8.886	5.293	5.10
c Å	9.802	19.221	7.790	14.72
α	90	90	90	90
β deg	100.10	95.50	90	90
γ	90	90	120	120
Space Group	C2/m	C2/c	C6/mmm	$P6_3$/mcm
Z	2	4	1	2

TABLE 31. (cont.)

Variety	Xanthophyllite	Margarite		
Site Occupancy	M₁ Ca 1.0 M₂ {Mg 0.50, Al 0.50} M₃ Mg 1.00 T {Si 0.25, Al 0.75}	M₁ {Ca 0.88, Na 0.12} M₂ {Al 0.88, Li 0.06, Fe 0.06} T₁ {Si 0.50, Al 0.50} T₂ {Si 0.50, Al 0.50}		
Method	3-D, film	3-D, film	2-D, film	2-D, film
R & Rating	0.108 (2)	0.140 (2)	0.12 (av.) (2)	0.138 (2)
Cleavage and habit	{001} perfect — Tabular parallel to {001}—		{0001} perfect Platy	{0001} perfect
Comment			True cell is probably orthorhombic B estimated Double Sheets	
μ	129.4	143.8	459.6	139.9
ASF	0.8990×10^{-1}	0.4365×10^{-1}	0.5496	0.1198
Abridging factors	0.5:0:0	1:1:0.5	0.1:0:0	0.1:0:0

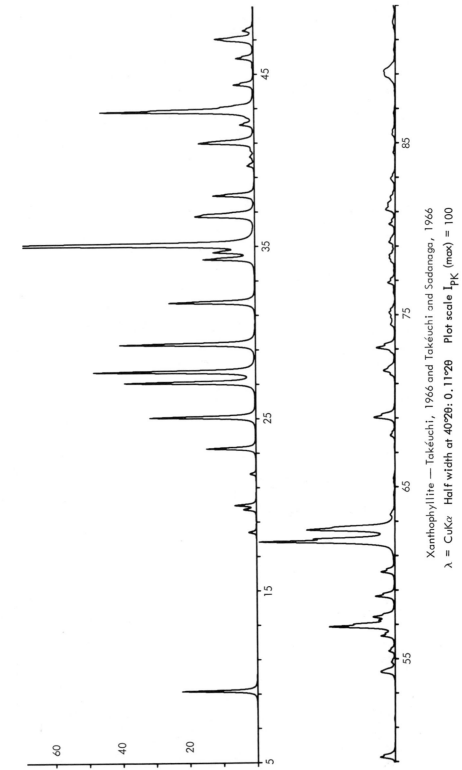

Xanthophyllite — Takéuchi, 1966 and Takéuchi and Sadanaga, 1966

λ = CuKα Half width at 40°2θ: 0.11°2θ Plot scale I$_{PK}$ (max) = 100

XANTHOPHYLLITE - TAKÉUCHI, 1966 AND TAKÉUCHI AND SADANAGA, 1966

2THETA	PEAK	D	H	K	L	I(INT)	I(PK)	I(DS)
9.16	9.16	9.650	0	0	1	15	22	13
18.37	18.38	4.825	0	0	2	2	2	2
19.70	19.72	4.501	0	1	0	3	2	2
19.95	19.96	4.446	0	1	1	5	6	3
21.77	21.78	4.079	0	1	1	2	2	4
23.25	23.26	3.823	1	1	0	2	2	1
25.06	25.06	3.551	-1	1	1	12	15	11
27.07	27.08	3.291	0	0	2	27	32	25
27.71	27.72	3.217	0	2	0	35	39	32
29.29	29.30	3.046	-1	1	2	43	48	41
31.72	31.72	2.818	-1	1	3	37	40	35
34.23	34.24	2.617	0	2	3	25	26	24
34.63	34.64	2.588	1	1	1	15	16	14
34.65		2.586	-2	0	1	8	12	8
35.06	35.06	2.557	-1	3	1	62	100	2
35.07		2.557	2	0	1	38		62
36.71	36.74	2.446	-1	3	1	12	18	38
36.75		2.443	-2	0	2	7		12
37.92	37.92	2.371	-1	0	2	5	12	7
37.93		2.370	1	3	1	8		5
39.66	39.66	2.271	-1	1	4	2	2	9
40.18	40.18	2.243	-2	2	2	1	1	2
40.54	40.54	2.223	-2	2	1	14	17	1
40.97	40.98	2.201	-1	1	2	6		15
41.03		2.198	-2	3	2	4		6
42.04	42.04	2.147	-2	0	2	17	4	4
42.80	42.82	2.111	-1	3	2	35	46	18
42.82		2.110	-2	0	3	7		37
43.09	43.08	2.098	0	4	2	6	10	7
44.37	44.38	2.040	-2	2	2	6	6	7
45.90	45.90	1.9752	-1	3	2	7	5	7
46.93	47.04	1.9344	-2	1	3	3	12	7
46.99		1.9319	0	0	5	2		8
47.04		1.9300	-2	2	2	4		4
47.53	47.52	1.9114	2	2	3		3	4
49.22	49.24	1.8497	2	0	5		5	2
49.25		1.8485	-1	3	4	4		4

2THETA	PEAK	D	H	K	L	I(INT)	I(PK)	I(DS)
49.38	49.36	1.8441	0	4	3	1	4	1
54.18	54.22	1.6915	-2	3	1	1	4	5
54.25		1.6894	-3	0	5	3		3
54.79	54.78	1.6740	-1	1	2	1	1	2
55.43	55.44	1.6563	-2	4	1	2	2	2
56.31	56.32	1.6323	0	2	5	1	4	5
56.32		1.6321	-2	4	2	4		2
56.83	56.86	1.6187	0	0	6	1	20	11
56.87		1.6175	-1	3	5	9		23
57.23	57.22	1.6083	0	3	3	18	5	5
57.39	57.42	1.6042	-3	1	6	4	7	2
57.44		1.6029	-1	1	3	2		7
58.61	58.62	1.5736	-2	5	1	6	6	8
58.66		1.5726	-1	4	3	2		2
60.03	60.04	1.5398	-1	5	3	4	4	5
60.05		1.5394	-3	1	4	2		2
61.69	61.80	1.5024	-3	6	0	2	40	24
61.77		1.5005	0	3	3	19		48
61.81		1.4997	-3	3	5	37		37
62.47	62.48	1.4853	-2	0	6	28	26	18
62.55		1.4837	-1	5	3	13		1
63.34	63.36	1.4670	-1	2	6	1	2	1
63.39		1.4660	0	4	0	1		1
67.94	67.98	1.3786	-1	1	7	1	2	2
68.00		1.3775	-1	6	3	2		5
69.00	69.02	1.3598	0	3	3	4	6	5
69.01		1.3598	-3	3	2	4		5
69.07		1.3586	-1	3	4	2		5
71.51	71.52	1.3182	0	2	7	3	2	5
71.73	71.74	1.3148	-1	3	6	3	4	3
71.73		1.3135	-2	0	7	2		3
71.81		1.2941	-4	0	0	7		10
73.05	73.06	1.2932	-3	3	3	2		3
73.12		1.2743	0	4	4	1		1
74.38	74.40	1.2742	-3	6	5	1	1	1
74.39		1.2730	-3	3	5	1		1
74.47		1.2679	-2	6	1	1		2
74.82	74.84							

XANTHOPHYLLITE - TAKÉUCHI, 1956 AND TAKÉUCHI AND SADANAGA, 1966

2THETA	PEAK	D	H	K	L	I(INT)	I(PK)	I(DS)
74.90		1.2667	-4	0	3	1		1
75.07	75.12	1.2642	-1	3	7	1		1
75.13		1.2634	-1	1	7	1		2
75.31	75.30	1.2609	-2	1	7	1	2	2
76.85	76.86	1.2393	4	0	1	1	1	2
76.86		1.2393	-2	6	3	2	2	3
78.07	78.10	1.2231	2	6	2	2		2
78.17		1.2217	-4	0	4	1	1	1
78.28	78.30	1.2203	-1	7	2	1	2	2
78.46	78.46	1.2179	-2	2	6	2	2	3
78.93	78.94	1.2118	-1	1	8	2	2	2
79.23	79.22	1.2080	-3	5	3	2	2	3
79.37	79.36	1.2063	0	0	8	1	1	2
80.24	80.28	1.1953	1	7	2	1	2	1
80.28		1.1949	4	2	1	2		4
81.08	81.12	1.1851	-2	3	4	2	3	1
81.10		1.1848	3	3	5	2		2
81.12		1.1846	0	6	6	2		3
81.20		1.1836	-3	3	3	2		3
81.49	81.48	1.1801	-1	7	6	2	2	3
81.58		1.1790	-4	2	7	1		1
81.87	81.84	1.1756	0	4	4	2	2	3
82.75	82.92	1.1653	2	6	7	2	1	1
82.87		1.1639	3	5	3	1		2
82.94		1.1631	-3	7	4	1		2
84.42	84.42	1.1465	1	1	3	1	1	1
84.44		1.1463	-4	2	7	1		1
85.45	85.46	1.1353	-2	2	2	1	1	2
86.72	86.72	1.1219	-2	6	8	1	1	2
88.85	88.98	1.1004	2	6	5	4	3	7
88.99		1.0990	-4	0	4	2		4
89.00		1.0990	2	2	7	1		2
89.15	89.16	1.0974	3	3	5	2	4	3
89.18		1.0972	0	3	6	1		2
89.28		1.0962	-3	3	7	1		2
89.31		1.0960	0	8	2	1		1
90.39	90.38	1.0856	4	4	1	1	1	2

2THETA	PEAK	D	H	K	L	I(INT)	I(PK)	I(DS)
91.67	91.70	1.0737	-4	4	4	1	1	2
91.92	91.98	1.0715	-3	1	6	1	1	1
92.07		1.0701	-1	1	8	1		2
92.96	92.96	1.0622	0	0	8	1	1	1
93.72	93.76	1.0556	4	0	4	1	4	6
93.79		1.0550	-2	6	6	3		10
95.44	95.48	1.0411	-1	5	5	1	1	2
95.52		1.0404	-2	4	7	1		1
97.58	97.66	1.0239	2	0	8	1	1	2
97.66		1.0232	-1	3	9	2		4
98.67	98.70	1.0155	-3	3	6	1	1	2
98.71		1.0152	0	6	0	1		1
98.81		1.0144	-3	3	8	1		1
99.10	99.08	1.0122	-2	4	7	2	1	1
99.17		1.0117	-1	5	8	1		1
99.47	99.44	1.0094	-2	7	3	1		1
100.52	100.54	1.0017	-2	8	3	2		2
102.40	102.40	0.9883	4	6	7	2	1	3
103.42	103.68	0.9814	-4	6	1	3	2	5
103.62		0.9800	-1	6	1	1		1
103.68		0.9796	-4	6	1	2		1
103.70		0.9795	-5	3	2	3		1
105.38	105.44	0.9684	-1	6	0	1	1	2
105.44		0.9680	5	0	10	1		3
105.46		0.9679	-1	6	3	1		2
105.91	105.86	0.9650	-4	0	10	1	2	2
106.57	106.60	0.9609	-1	3	9	2	1	1
106.68		0.9602	-2	0	10	1		2
107.40	107.44	0.9557	5	6	2	1	1	1
107.43		0.9555	-5	3	0	4		2
107.54		0.9549	3	3	4	2		2
109.98	110.06	0.9404	0	9	7	1	1	1
110.03		0.9401	-3	6	8	1		1
110.15		0.9394	-1	3	9	1		2
111.53	111.52	0.9317	-1	5	10	1	1	2
112.80	112.74	0.9248	-3	5	8	1	0	1
116.47	116.50	0.9059	4	2	6	1	0	1

XANTHOPHYLLITE — TAKÉUCHI, 1966 AND TAKÉUCHI AND SADANAGA, 1966

2THETA	PEAK	D	H	K	L	I(INT)	I(PK)	I(DS)
118.16	118.12	0.8979	-5	3	6	1	0	1
120.49	120.54	0.8872	5	3	3	3	2	6
120.50		0.8872	-1	9	5	3		6
120.62		0.8866	-4	6	6	3		7
122.39	122.42	0.8790	1	3	10	1	0	1
124.14	124.18	0.8718	0	8	7	1	0	1
125.57	125.64	0.8662	-3	9	1	5	2	11
125.69		0.8657	-6	0	2	2		5
126.29	126.28	0.8634	1	9	5	4	3	9
126.29		0.8634	-4	6	4	4		9
126.54		0.8624	-5	3	7	3		8
128.42	128.74	0.8555	4	4	6	1	1	1
128.64		0.8547	2	0	10	1		2
128.76		0.8542	-1	3	11	2		3
128.85		0.8539	-5	1	8	1		1
131.22	131.44	0.8457	-2	6	8	1	0	3
131.52		0.8447	-4	0	10	1		2
132.46	132.68	0.8417	2	10	1	1	1	1
132.71		0.8408	2	8	6	1		2
132.73		0.8408	5	1	5	1		2
133.92	134.08	0.8370	-6	2	4	1	1	3
134.09		0.8365	-3	9	2	1		2
134.18		0.8362	-3	9	4	1		2
135.19	135.10	0.8331	4	8	1	2	1	4
137.41	137.72	0.8267	3	7	6	1	1	2
137.63		0.8261	-3	7	8	1		2
137.76		0.8257	4	6	5	1		2
137.77		0.8257	-1	9	6	1		2
138.10		0.8248	-5	3	8	1		3
142.53	143.24	0.8134	-1	1	12	1	2	2
143.03		0.8121	-2	8	8	1		3
143.13		0.8119	3	6	9	3		7
143.24		0.8116	0	6	10	3		7
143.29		0.8115	-5	1	9	1		2
143.48		0.8111	-3	3	11	3		7
144.23	144.14	0.8094	-4	0	8	1	1	1
144.49		0.8088	-2	6	10	2		6

2THETA	PEAK	D	H	K	L	I(INT)	I(PK)	I(DS)
149.41	149.44	0.7985	5	1	6	1	0	3
149.41		0.7985	2	8	7	1		1
150.21	150.42	0.7970	-6	4	0	1	0	1
150.38		0.7967	-6	4	4	1		3
154.06	155.28	0.7904	3	9	4	2	1	4
154.17		0.7902	6	0	3	1		1
154.30		0.7900	-3	9	6	1		3
154.66		0.7895	-6	0	7	1		2
154.97		0.7890	-1	11	3	1		2
155.05		0.7889	5	7	1	1		2
155.34		0.7884	2	0	11	2		5
155.61		0.7880	-1	3	12	4		10

615

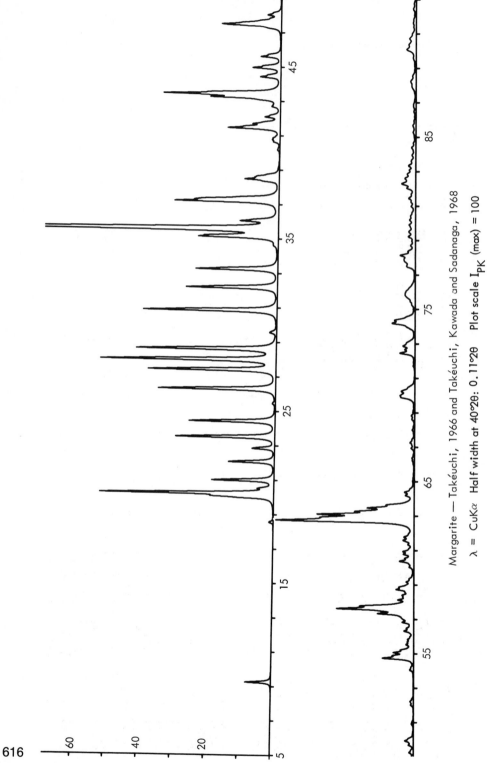

Margarite — Takéuchi, 1966 and Takéuchi, Kawada and Sadanaga, 1968

λ = CuKα   Half width at 40°2θ: 0.11°2θ   Plot scale I_{PK} (max) = 100

616

MARGARITE – TAKEUCHI, 1966 AND TAKEUCHI, KAWADA AND SADANAGA, 1968

2THETA	PEAK	D	H	K	L	I(INT)	I(PK)	I(DS)
9.24	9.24	9.566	0	0	2	8	8	6
18.53	18.54	4.783	0	0	4	2	1	1
20.06	20.06	4.423	-1	1	1	14	19	12
20.21	20.22	4.390	0	1	1	58	52	50
20.50	20.50	4.328	1	1	1	4	5	3
20.97	20.98	4.233	-1	1	2	22	19	19
22.04	22.04	4.030	0	2	2	17	14	14
22.82	22.82	3.893	1	1	2	8	7	7
23.49	23.50	3.785	-1	1	3	37	30	33
24.40	24.40	3.646	0	2	3	32	26	28
26.26	26.26	3.391	-1	1	4	45	35	41
27.37	27.38	3.255	0	2	4	49	38	45
27.96	27.96	3.189	-1	1	6	69	52	63
28.58	28.58	3.121	0	2	5	56	42	51
29.55	29.54	3.021	-1	1	5	3	2	3
30.81	30.82	2.899	2	2	5	55	40	53
32.15	32.16	2.782	1	1	6	39	27	37
33.21	33.22	2.695	0	2	6	35	25	34
34.59	34.60	2.591	2	2	6	1	2	1
35.00	35.10	2.561	-1	3	0	3	24	3
35.09		2.555	1	3	1	27		27
35.17	35.16	2.550	-2	0	2	10	22	10
35.53	35.56	2.525	0	2	3	56	100	56
35.56		2.523	-1	3	1	100		100
36.03	36.04	2.490	1	3	2	14	12	14
37.16	37.16	2.417	-1	3	3	39	31	39
37.17		2.417	0	2	7	6		6
37.33	37.26	2.407	-1	3	4	22	25	23
38.47	38.46	2.338	1	1	8	12	10	13
38.56	38.56	2.328	0	4	1	7	7	7
38.64		2.243	1	3	5	7		7
40.17	40.16	2.218	0	4	3	1	1	1
40.64	40.64	2.211	-2	2	0	3	1	1
40.77	40.76	2.207	0	0	8	2	2	2
40.86	40.86	2.181	4	1	1	2	2	2
41.36	41.44	2.177	-1	3	8	2	2	3
41.43		2.176	1	5	1	20	16	21
41.46			2	2	1	3		4
41.67	41.68	2.165	2	0	4	7	8	8
41.70		2.164	0	4	2	4		4
42.08	42.08	2.145	-2	2	3	7	5	7
42.69	42.70	2.116	2	2	6	3	3	4
43.22	43.22	2.092	-2	0	2	27	21	29
43.41	43.42	2.083	1	3	5	52	35	57
44.42	44.42	2.038	2	2	0	10	6	11
44.95	44.96	2.015	0	4	4	14	9	15
45.60	45.60	1.9876	-2	2	1	10	6	11
47.42	47.48	1.9155	-1	1	10	23	18	26
47.48		1.9133	0	2	10	7		7
48.04	48.04	1.8924	-2	2	6	6	4	7
49.23	49.24	1.8492	2	2	5	2	1	2
49.66	49.66	1.8342	-1	3	7	4	2	3
49.91	49.92	1.8256	3	1	8	2	1	5
51.99	52.00	1.7572	0	2	10	2	1	2
52.22	52.20	1.7503	2	2	6	2	1	3
54.04	54.04	1.6954	2	2	8	3	9	3
54.72	54.72	1.6759	-1	3	9	13	6	17
54.76		1.6749	1	3	0	3		4
55.03	55.06	1.6672	2	4	1	1	6	2
55.06		1.6664	1	5	1	7		9
55.52	55.52	1.6537	-3	1	8	3	2	4
55.82	55.82	1.6457	-2	2	3	2	2	3
56.31	56.30	1.6325	2	4	2	5	3	6
56.81	56.82	1.6191	3	1	2	5	4	7
57.06	57.06	1.6128	-2	0	10	16	11	4
57.32	57.32	1.6060	0	4	10	8	23	21
57.58	57.62	1.5994	-1	5	4	34		11
57.62		1.5984	1	3	9	2		44
57.73	57.76	1.5955	-1	1	12	4	16	2
57.78		1.5944	0	0	12	7		5
58.11	58.12	1.5859	-3	3	1	4	5	9
58.39	58.38	1.5791	3	3	4	8	4	6
58.71	58.72	1.5711	-2	4	5	3	5	11
58.84	58.84	1.5682	1	5	2	4	5	4
59.16	59.16	1.5603	2	2	8	4	2	5

617

MARGARITE - TAKÉUCHI, 1966 AND TAKÉUCHI, KAWADA AND SADANAGA, 1968

2THETA	PEAK	D	H	K	L	I(INT)	I(PK)	I(DS)
60.03	60.04	1.5398	-3	1	6	2	1	3
60.37	60.36	1.5320	3	1	4	8	4	11
60.77	60.78	1.5228	-2	4	6	5	3	7
60.92	60.92	1.5194	-1	5	5	3	3	4
61.32	61.32	1.5104	-2	1	10	3	2	3
61.58	61.58	1.5048	-1	5	6	5	3	7
61.76	61.76	1.5001	1	5	5	2	3	3
62.68	62.74	1.4810	0	6	0	30	41	42
62.75		1.4794	-3	3	1	64		88
63.09	63.10	1.4722	-1	1	11	48	29	67
63.20		1.4699	-1	4	7	2		2
63.38	63.26	1.4662	-2	5	6	2	18	3
63.47	63.46	1.4644	2	0	10	21	14	30
64.19	64.18	1.4497	0	4	10	4	3	6
64.38	64.38	1.4458	0	2	13	3	3	4
66.92	66.92	1.3971	0	2	13	2	1	3
67.26	67.26	1.3908	2	2	10	2	1	4
68.00	68.00	1.3775	-3	1	9	1	1	2
68.61	68.62	1.3666	0	0	14	2	1	3
68.83	68.82	1.3629	0	1	13	1	1	3
69.87	69.88	1.3451	-3	3	6	7	4	10
69.98	69.98	1.3432	0	6	6	6	4	9
70.24	70.22	1.3389	3	1	5	5	4	8
70.33		1.3374	-1	1	14	2		3
72.45	72.44	1.3035	-1	3	13	10	5	16
72.86	72.86	1.2971	-2	0	12	6	4	9
72.98		1.2953	0	4	12	2		4
74.17	74.18	1.2774	-2	6	4	11	7	18
74.25		1.2761	2	1	14	2		4
74.33	74.34	1.2751	-2	2	13	2	6	2
74.34		1.2749	4	0	0	5		8
75.26	75.26	1.2616	-3	3	9	1	1	2
75.43	75.46	1.2591	-2	6	8	2	2	3
75.68	75.82	1.2557	0	0	14	1	3	2
75.74		1.2547	-3	3	7	3		4
75.82		1.2536	-1	1	15	3		4
75.92	75.92	1.2522	-2	6	4	3	3	4

2THETA	PEAK	D	H	K	L	I(INT)	I(PK)	I(DS)
76.05	76.04	1.2503	1	3	13	2	3	4
77.46	77.48	1.2311	-1	7	1	4	3	6
77.56		1.2298	-4	2	2	1		2
77.78	77.80	1.2269	0	4	13	2	3	4
77.85		1.2260	0	2	15	2	3	3
77.89	77.90	1.2254	-4	0	6	1		2
77.93		1.2249	-4	2	3	2		4
78.11	78.12	1.2225	2	6	6	4	5	7
78.14		1.2220	-3	5	3	3		5
78.71	78.70	1.2147	-1	7	3	1	1	2
79.18		1.2087	-2	6	6	1	2	2
79.24	79.22	1.2083	3	5	14	2		2
79.52	79.50	1.2079	1	7	3	2	2	4
79.90	79.90	1.2043	-4	0	5	3	2	3
80.25	80.18	1.1996	0	5	16	2	2	5
80.35	80.42	1.1958	-3	5	5	1	2	4
80.98	80.98	1.1939	1	7	4	3		2
81.22	81.22	1.1863	4	2	3	1	2	5
82.01	81.44	1.1834	2	5	13	3	1	2
82.02	82.02	1.1808	2	5	13	1	1	5
82.25		1.1739	-3	6	11	4		2
82.31	82.28	1.1738	-3	6	10	3	3	7
82.61	82.58	1.1711	0	5	4	4		6
82.80	82.84	1.1704	3	1	9	4		8
82.86		1.1669	3	7	5	2		7
83.07	83.06	1.1647	1	4	14	1		4
83.32	83.32	1.1640	0	4	4	2		2
83.68	83.60	1.1616	4	2	7	3		3
83.96	83.94	1.1587	-4	0	16	1	2	6
84.38	84.38	1.1547	0	2	15	1	2	2
85.29	85.28	1.1515	3	2	5	3	1	5
85.67	85.66	1.1469	4	1	7	1	1	2
87.89		1.1370	-1	7	7	1	1	3
87.98	87.98	1.1329	-3	1	14	1	1	2
		1.1100	-4	4	2	1		3
88.18	88.22	1.1090	-4	6	8	2		2
		1.1070	2	6	8	2	2	3

2THETA	PEAK	D	H	K	L	I(INT)	I(PK)	I(DS)
88.27		1.1061	0	4	15	1		2
89.60	89.60	1.0931	-2	4	14	2	1	3
90.06	90.08	1.0887	-2	6	10	7	3	13
90.16		1.0878	-3	3	13	2		4
90.44	90.34	1.0851	0	4	12	1	2	2
90.70	90.70	1.0827	4	0	8	4	2	8
91.38	91.38	1.0765	-1	5	14	1	1	3
91.59	91.62	1.0745	4	4	3	1	1	2
91.62		1.0667	0	8	5	2	2	3
92.45	92.44	1.0582	4	4	4	1	1	3
93.42	93.44	1.0560	-4	4	7	1	1	2
93.67	93.70	1.0470	-2	4	15	1	2	3
94.72	94.88	1.0458	-4	0	12	6		11
94.87		1.0413	-4	6	10	9		17
95.41	95.40	1.0313	2	5	15	1	3	2
96.64	96.64	1.0163	-1	0	18	2	1	5
98.56	98.56	1.0126	-2	3	17	4	1	9
99.05	99.06	1.0118	1	3	14	2	2	4
99.15		1.0069	3	3	15	2		4
99.81	99.82	1.0043	-3	6	14	2	1	4
100.16	100.16	1.0008	0	3	13	2	1	4
100.64	100.64	0.9996	3	8	3	2	1	1
100.80		0.9977	2	1	2	1	1	1
101.07	101.10	0.9973	5	7	5	1		3
101.13		0.9956	-3	7	3	1	1	2
101.37	101.44	0.9936	3	8	5	1		2
101.65	101.66	0.9933	-2	8	6	1		2
101.69		0.9887	-5	1	16	1		2
102.35	102.34	0.9810	-1	5	7	1	0	2
103.48	103.54	0.9802	-5	1	14	1	1	3
103.58		0.9768	-4	0	17	1		1
104.10	104.26	0.9758	-3	1	12	3	1	7
104.25		0.9684	2	6	2	1	3	2
105.39	105.68	0.9675	-4	5	1	1		1
105.52		0.9672	-5	9	1	1		6
105.57		0.9665	1	6	3	5		10
105.68		0.9662	4	6	0	3		6
105.73								

2THETA	PEAK	D	H	K	L	I(INT)	I(PK)	I(DS)
107.25	107.40	0.9566	0	0	20	2	2	5
107.38		0.9558	-5	3	5	1		3
107.40		0.9557	1	9	3	2		4
107.43		0.9556	-1	7	13	2		1
107.62	107.82	0.9544	4	6	2	2	2	4
107.86		0.9529	-1	3	19	4		9
108.33	108.28	0.9501	-2	8	8	1	1	2
108.46		0.9493	-2	0	18	2		5
108.89	108.86	0.9468	-3	5	14	2	1	5
109.38	109.40	0.9439	-4	6	6	2	1	5
109.40		0.9438	-1	9	5	2		4
109.89	109.82	0.9410	5	3	3	2	1	4
111.11	111.32	0.9340	-1	7	13	2	1	4
111.34		0.9328	-3	3	17	1		3
111.77	111.74	0.9304	0	6	16	2	1	4
112.34	112.32	0.9272	3	3	15	2	1	5
112.70	112.76	0.9253	-2	0	20	1		2
112.89		0.9243	2	3	19	2		5
113.29	113.30	0.9222	1	3	8	1	1	2
113.34		0.9219	5	1	19	2		2
113.82	113.76	0.9193	-4	6	7	1	1	1
113.91		0.9189	-1	9	8	1		2
114.54	114.52	0.9156	5	3	7	1		2
114.82	114.92	0.9142	-5	1	5	1	0	1
114.95		0.9135	-2	8	11	1	0	1
116.74	116.76	0.9046	-1	7	10	1		2
117.54	117.54	0.9008	-3	7	14	1	0	2
118.09	118.12	0.8982	-4	2	11	1	0	4
118.20		0.8977	3	2	16	2	1	2
118.45	118.54	0.8965	-1	7	9	1		1
121.00	121.12	0.8850	2	8	15	1	1	1
121.09		0.8846	5	1	10	1	1	2
121.31	121.50	0.8837	3	5	7	1		3
121.55		0.8826	5	1	14	1		3
122.62	122.72	0.8780	1	9	9	1	1	2
122.68		0.8778	-5	3	21	6		15
123.12	123.18	0.8760	-1	9	11	6	2	15

619

MARGARITE - TAKEUCHI, 1966 AND TAKEUCHI, KAWADA AND SADANAGA, 1968

2THETA	PEAK	D	H	K	L	I(INT)	I(PK)	I(DS)
123.31		0.8752	4	4	12	7		3
123.58	123.60	0.8741	4	6	8	7	2	17
123.68		0.8737	0	10	4	1		3
124.15	124.10	0.8717	-1	3	21	1	2	3
124.28		0.8712	-5	5	6	1		2
124.31		0.8711	-1	9	10	1		2
126.00	125.98	0.8645	-2	8	11	1	0	2
127.15	127.22	0.8601	-3	7	13	1	0	2
128.36	128.84	0.8557	-3	5	15	1	3	2
128.62		0.8547	-3	9	1	8		19
128.75		0.8543	-4	6	12	7		18
128.82		0.8540	-3	9	2	1		1
128.88		0.8538	-6	0	0	4		11
128.89		0.8538	-3	9	0	1		2
128.90		0.8537	-5	1	14	1		2
129.01		0.8534	-1	9	11	6		16
129.16	130.56	0.8528	-2	8	13	1		1
130.45		0.8483	-2	0	22	1	1	3
130.63		0.8477	-4	4	16	2		4
131.25	131.14	0.8456	-1	3	21	2	1	5
133.19	133.64	0.8393	-3	7	14	1	1	3
133.26		0.8391	2	10	0	2		5
133.62		0.8380	-2	6	18	4		10
133.72		0.8377	0	8	15	1		2
133.78		0.8375	4	7	0	1	1	2
134.24	134.26	0.8360	3	7	12	1		2
134.32		0.8358	-6	2	4	2	1	5
135.04	135.12	0.8336	2	10	2	2		3
135.16		0.8332	-4	10	16	1		3
135.26		0.8329	-5	1	15	2	1	3
135.93	136.02	0.8310	-2	10	4	2		4
136.06		0.8306	6	2	1	1		3
136.20		0.8301	-1	1	23	1		3
136.34		0.8298	-4	8	2	1		3
136.57		0.8291	-4	8	5	2		5
137.35	137.54	0.8268	-6	2	6	1	1	3
137.54		0.8263	-3	9	7	2		5
137.58		0.8262	-6	0	8	1		3
137.69	137.69	0.8259	-2	6	17	1		1
137.99	138.06	0.8251	-2	10	5	1	1	4
138.05		0.8249	6	2	2	1		1
138.06		0.8249	3	9	5	2		4
138.50	138.78	0.8237	4	8	3	1		1
138.63		0.8233	6	0	4	1		1
138.82		0.8228	2	8	13	2		5
139.19	139.32	0.8218	5	5	7	1	1	5
139.32		0.8215	1	7	17	1		3
139.68	139.66	0.8205	5	1	12	2		5
139.79		0.8202	-6	2	7	1		3
140.48	140.78	0.8184	-3	7	15	3	1	9
140.71		0.8179	6	2	3	1		5
140.71		0.8178	-2	10	6	3		3
140.79		0.8176	0	10	23	2		7
140.88		0.8174	-4	6	14	3		3
141.29	141.40	0.8164	-1	9	13	3	1	8
141.30		0.8164	-5	5	11	1		3
141.35		0.8162	4	8	4	2		2
141.76		0.8152	-4	8	7	1		7
141.79		0.8151	3	7	13	2		2
142.15	142.80	0.8143	2	10	5	1		1
142.79		0.8127	-5	3	11	5		12
143.04		0.8121	-1	5	16	2		6
143.48		0.8111	-2	8	15	2		5
145.45	145.96	0.8066	-3	5	19	1	2	2
145.78		0.8059	0	10	10	2		6
145.79		0.8059	0	10	10	3		3
145.84		0.8058	-6	0	10	3		8
145.95		0.8055	-3	3	9	6		18
146.09		0.8052	2	10	21	2		6
146.69	146.90	0.8040	-3	9	6	3	2	8
146.89		0.8036	0	9	7	1		17
147.08		0.8032	6	2	20	6		2
147.16		0.8030	-6	0	20	3		10
147.49		0.8023	6	0	6	3		8

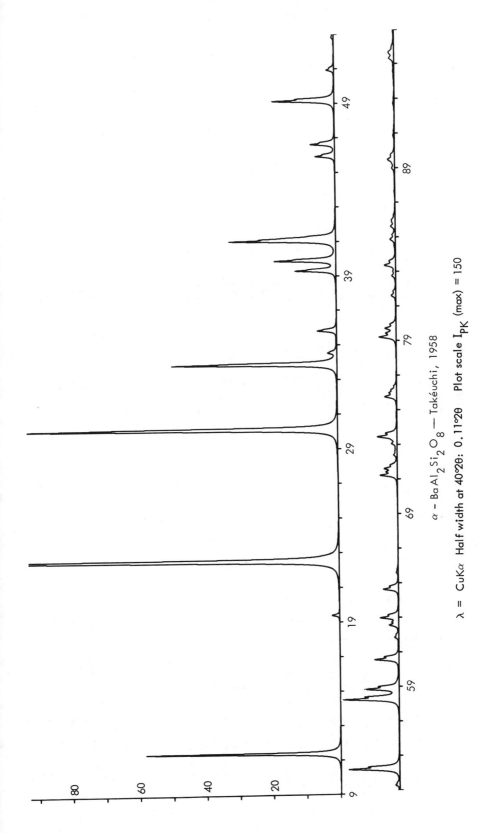

α – Ba Al$_2$ Si$_2$ O$_8$ — Takéuchi, 1958

λ = CuKα Half width at 40°2θ: 0.11°2θ Plot scale I$_{PK}$ (max) = 150

621

ALPHA BAAL2SI2O8 - TAKEUCHI, 1958

2THETA	PEAK	D	H	K	L	I(INT)	I(PK)	I(DS)
11.35	11.36	7.790	0	1	0	32	39	17
19.35	19.36	4.584	0	1	1	1	2	1
22.49	22.50	3.951	1	1	0	100	100	100
30.08	30.08	2.968	1	2	0	70	63	101
33.84	33.84	2.647	0	1	3	39	33	65
34.51	34.52	2.597	1	1	1	3	2	3
35.80	35.80	2.506	1	0	1	5	4	8
39.28	39.28	2.292	2	0	0	10	8	21
39.87	39.86	2.259	2	0	1	16	12	33
41.01	41.02	2.199	2	1	0	27	21	58
41.20	41.12	2.189	1	2	1	8	15	17
45.90	45.90	1.9753	2	1	2	5	4	14
46.60	46.60	1.9475	0	3	1	7	5	17
49.11	49.12	1.8535	1	0	3	18	12	50
50.90	50.90	1.7924	2	1	1	2	2	6
52.79	52.80	1.7325	1	1	0	1	1	2
53.26	53.26	1.7183	1	0	3	1	1	4
54.19	54.18	1.6912	2	2	0	15	10	50
58.23	58.24	1.5830	1	2	2	18	11	64
58.82	58.82	1.5686	2	2	1	11	7	39
60.54	60.54	1.5280	3	0	0	8	5	30
61.82	61.82	1.4994	3	0	1	1	1	5
62.53	62.54	1.4841	2	2	1	3	2	12
62.95	62.96	1.4751	1	1	3	6	4	24
64.61	64.62	1.4412	2	2	2	5	3	22
65.57	65.58	1.4224	3	2	3	1	3	3
71.20	71.20	1.3233	3	0	3	6	3	30
71.59	71.60	1.3169	2	2	1	4	3	21
72.37	72.38	1.3046	2	2	2	1	1	6
72.78	72.78	1.2983	0	2	4	1	0	2
73.03	73.04	1.2944	2	1	4	2	1	8
73.42	73.42	1.2885	3	0	1	7	4	38
74.58	74.58	1.2713	3	1	4	5	0	1
75.74	75.74	1.2547	1	0	6	2	1	25
76.14	76.12	1.2492	3	2	2	7	3	9
79.18	79.18	1.2086	1	1	3	7	3	39
79.69	79.70	1.2021	3	0	4	4	2	24

2THETA	PEAK	D	H	K	L	I(INT)	I(PK)	I(DS)
81.58	81.58	1.1790	2	2	3	2	1	12
82.72	82.72	1.1656	1	1	6	2	1	10
83.35	83.34	1.1585	2	1	5	5	2	31
84.46	84.48	1.1460	4	0	3	0	0	3
84.84	84.84	1.1418	3	1	3	2	1	15
85.59	85.60	1.1338	4	0	1	2	1	13
85.97	85.94	1.1297	2	1	6	1	1	8
87.60	87.60	1.1129	2	0	7	1	0	2
88.95	88.96	1.0994	4	0	2	1	1	9
89.46	89.46	1.0945	2	1	4	4	1	25
90.84	90.84	1.0814	1	1	7	1	1	5
92.69	92.70	1.0646	3	0	3	1	0	2
94.18	94.20	1.0516	3	2	0	0	0	3
94.56	94.56	1.0484	4	1	3	3	1	21
95.31	95.32	1.0422	3	2	1	3	1	15
95.69	95.66	1.0390	1	1	6	2	1	25
97.33	97.32	1.0259	3	2	7	3	1	24
98.69	98.70	1.0153	2	2	5	3	0	1
99.58	99.56	1.0086	5	1	6	3	1	26
100.71	100.72	1.0003	4	1	0	1	0	6
101.86	101.86	0.9921	3	1	6	1	1	8
102.25	102.26	0.9894	3	1	5	3	1	24
102.88	102.88	0.9850	3	1	2	1	0	9
104.41	104.42	0.9747	4	1	3	2	1	4
105.31	105.32	0.9688	2	0	8	0	0	10
107.93	107.94	0.9525	4	1	7	1	1	10
110.70	110.74	0.9363	2	1	3	3	1	33
111.22	111.22	0.9334	4	4	3	0	0	22
113.10	113.10	0.9231	5	0	5	1	0	2
114.32	114.34	0.9168	1	0	8	0	0	8
114.89	114.90	0.9139	5	1	6	3	1	12
115.55	115.56	0.9105	5	1	1	1	1	16
115.98	115.98	0.9084	3	6	3	2	0	23
117.80	117.80	0.8996	2	0	7	1	0	7
118.51	118.28	0.8962	5	0	2	2	0	12
119.34	119.38	0.8924	5	0	2	1	1	16
119.92	119.92	0.8898	4	1	4	4	1	42

ALPHA BAAL2SI2O8 - TAKEUCHI, 1958

2THETA	PEAK	D	H	K	L	I(INT)	I(PK)	I(DS)
121.65	121.66	0.8822	3	3	0	1	0	13
122.98	123.00	0.8766	3	3	1	0	0	4
124.18	124.18	0.8716	3	2	5	4	1	40
125.53	125.56	0.8663	4	2	0	0	0	4
126.00	126.04	0.8645	5	0	3	0	0	5
126.92	126.96	0.8610	4	2	1	2	1	28
127.08		0.8604	3	3	2	0		4
127.41	127.44	0.8592	4	0	6	1	0	10
129.81		0.8505	1	1	9	2	1	18
130.29	130.28	0.8489	2	1	8	2		26
131.25	131.22	0.8456	4	2	2	3	1	30
133.81	133.86	0.8374	3	1	7	1	0	17
134.48	134.48	0.8353	3	3	3	2	0	26
136.44	136.44	0.8295	5	0	4	1	0	8
139.22	139.36	0.8217	4	2	3	1	0	15
139.43		0.8212	3	0	8	1	0	11
140.37	140.36	0.8187	5	1	1	3	0	34
140.98	141.00	0.8172	3	2	6	2	0	27
144.07	144.08	0.8097	2	0	9	3	0	41
145.98	146.00	0.8055	5	1	2	5	1	64
146.89	146.88	0.8036	3	3	4	3	1	35
149.50	149.44	0.7984	4	0	7	0	0	4
152.85	154.14	0.7924	4	1	6	3	0	44
153.39		0.7915	4	2	4	2		26
154.24		0.7901	5	0	5	4		51
157.91	158.02	0.7848	5	1	3	3	0	46
158.29		0.7843	2	2	8	1		16

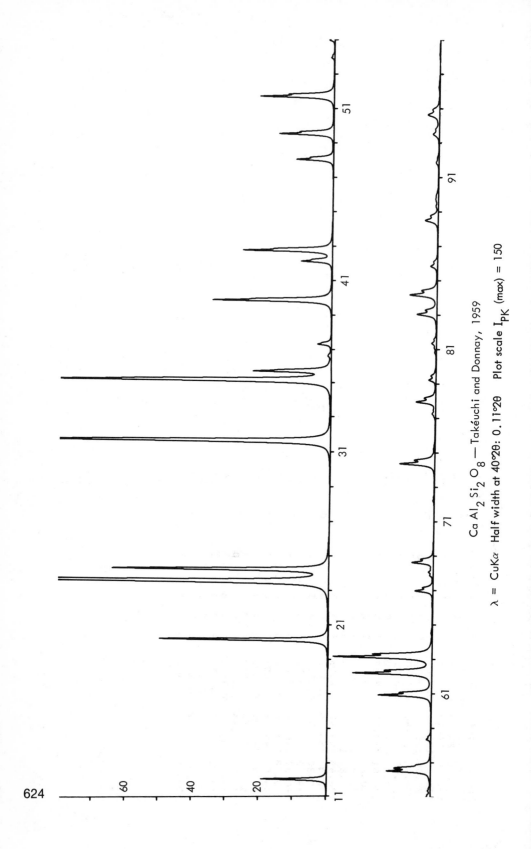

624

Ca Al$_2$ Si$_2$ O$_8$ — Takéuchi and Donnay, 1959

λ = CuKα Half width at 40°2θ: 0.11°2θ Plot scale I$_{PK}$ (max) = 150

CAAL2SI2O8 — TAKÉUCHI AND DONNAY, 1959

2THETA	PEAK	D	H	K	L	I(INT)	I(PK)	I(DS)
12.01	12.02	7.360	0	0	2	11	13	10
20.09	20.10	4.417	1	0	0	32	33	31
23.47	23.48	3.787	1	0	2	100	100	100
24.16	24.16	3.680	0	0	4	44	43	44
31.62	31.62	2.827	1	1	0	75	66	82
35.16	35.16	2.550	1	1	2	69	58	78
35.70	35.70	2.513	1	0	4	18	15	20
36.60	36.60	2.453	0	1	6	1	1	1
37.29	37.28	2.409	1	1	3	3	3	4
39.80	39.80	2.263	1	1	6	30	23	36
42.10	42.10	2.145	1	1	0	8	6	9
42.71	42.72	2.115	2	0	2	23	18	28
48.00	48.00	1.8936	2	0	4	10	7	13
49.49	49.50	1.8400	0	0	8	15	11	20
51.66	51.66	1.7680	1	0	6	21	14	28
53.94	53.94	1.6985	1	0	8	2	1	1
54.96	54.96	1.6694	2	1	0	2	1	2
55.34	55.34	1.6587	2	1	1	1	1	2
56.47	56.48	1.6280	2	1	2	13	9	18
56.69	56.64	1.6224	1	1	7	6	7	9
58.34	58.34	1.5804	2	1	3	1	1	2
60.88	60.88	1.5203	2	1	4	17	11	26
62.16	62.16	1.4921	1	1	8	25	16	39
63.09	63.10	1.4722	1	1	0	33	20	50
64.49	64.50	1.4436	3	0	2	0	0	0
66.03	66.04	1.4136	3	0	0	0	0	0
66.95	66.96	1.3965	3	0	10	6	4	10
67.85	67.86	1.3802	2	0	6	1	1	1
68.04	68.04	1.3767	1	1	9	1	1	2
68.60	68.60	1.3669	2	1	4	8	4	13
72.19	72.20	1.3075	3	1	7	7	0	1
74.33	74.34	1.2750	2	1	0	6	7	13
74.34	74.34	1.2748	2	1	8	0	0	10
74.66	74.54	1.2702	2	2	3	6	4	10
75.20	75.20	1.2624	2	0	0	0	0	1
77.24	77.26	1.2340	2	1	1	1	1	2
77.92	77.94	1.2250	3	1	0	1	4	2
77.93	78.16	1.2248	2	0	10	6	2	12
78.24	78.24	1.2208	3	1	1	0	2	0
79.20	79.20	1.2084	3	1	2	3	1	5
79.49	79.44	1.2047	2	2	4	1	1	1
80.79	80.80	1.1885	3	1	3	1	0	1
81.09	81.08	1.1849	1	1	11	0	0	1
81.34	81.34	1.1819	1	0	12	2	1	3
83.01	83.02	1.1623	3	1	4	9	4	17
84.14	84.14	1.1496	3	0	8	12	5	24
85.82	85.82	1.1313	2	2	6	3	1	6
88.34	88.48	1.1054	1	1	12	0	3	0
88.47		1.1042	4	0	0	0		1
88.48		1.1041	2	1	10	6	0	12
89.31	89.30	1.0960	3	0	6	1	1	2
89.72	89.72	1.0920	4	0	2	1	0	1
89.90	89.90	1.0903	2	2	7	1	1	3
91.83	91.82	1.0723	3	0	12	2	1	3
93.39	93.48	1.0585	3	1	7	3	1	7
93.49		1.0576	4	0	4	7	2	14
94.61	94.62	1.0480	2	2	8	1	0	1
96.20	96.20	1.0349	1	1	13	0	0	1
98.12	98.12	1.0197	2	2	9	0	1	0
99.99	100.22	1.0056	2	2	2	2		5
100.23		1.0038	3	1	12	2	1	5
102.38	102.38	0.9885	3	2	14	2	1	5
104.08	104.10	0.9769	3	1	4	6	2	13
104.82	104.82	0.9720	1	1	14	6	2	14
106.10	106.10	0.9638	4	1	0	6		4
106.11		0.9637	2	1	10	2	1	0
106.43	106.46	0.9618	4	1	8	0	1	0
108.89	109.08	0.9468	4	0	1	0	0	0
109.06		0.9457	3	1	3	3		6
109.78	109.78	0.9416	3	1	10	1	1	2
111.41	111.40	0.9324	3	1	4	0	0	0
113.11	113.12	0.9231	4	2	11	1	0	0
118.33	118.34	0.8971	4	1	6	0	1	10
121.39	121.40	0.8833	4	0	10	1	0	3

CAAL2SI2O8 - TAKÉUCHI AND DONNAY, 1959

2THETA	PEAK	D	H	K	L	I(INT)	I(PK)	I(DS)
122.86	122.90	0.8771	5	0	2	1	0	3
123.07		0.8762	4	1	1	0		1
125.40	125.76	0.8668	3	1	12	2	1	5
125.76		0.8654	1	1	16	3		9
127.47	127.52	0.8589	5	0	4	1	0	2
128.37		0.8556	3	0	14	2	2	6
128.89	128.90	0.8538	4	1	8	11		30
129.96	129.92	0.8500	3	3	0	5	1	15
130.95	130.52	0.8466	2	2	13	0	1	1
134.68	134.70	0.8347	4	2	0	0	1	1
134.69		0.8346	3	2	10	7		20
135.12		0.8333	4	2	1	0		1
135.74	135.34	0.8315	3	1	13	0	1	1
136.13	136.54	0.8304	4	1	9	0	0	0
136.47		0.8294	4	2	2	2		5
136.88	136.84	0.8282	3	3	4	2	0	5
138.80	138.86	0.8229	4	2	3	1	0	2
139.62	139.66	0.8207	4	0	12	1	0	3
139.92		0.8199	1	1	17	0		1
140.74	140.48	0.8178	0	0	18	0	0	1
142.26	142.28	0.8140	4	2	4	6	1	18
143.44	143.38	0.8112	2	2	14	7	1	19
145.59	145.62	0.8063	4	1	10	4	1	12
145.86		0.8057	2	1	16	0		0
146.63	146.60	0.8041	1	0	18	4	1	11
147.08		0.8032	3	3	6	1		3
154.19	155.20	0.7902	4	2	6	1	0	3
155.17		0.7887	5	1	2	3		10
159.21	161.56	0.7831	5	1	3	0	0	1
160.05		0.7821	4	1	11	0		1
160.78		0.7812	3	2	12	6		19
161.68	161.70	0.7802	3	0	16	2	0	6
163.09		0.7787	1	1	18	6		20

TABLE 32. PLAGIOCLASE FELDSPARS

Variety	Low Albite	High Albite	Oligoclase	Bytownite	Transitional Anorthite
Composition	$Ab_{98.5}An_{0.5}Or_1$	$Ab_{97.7}An_{0.7}Or_{1.6}$	$Ab_{69}An_{29}Or_2$	$Ab_{20}An_{80}$	$Ab_{0.9}An_{99.1}$
Source	Ramona, Calif.	Amelia, Va. (inverted from low form)	Mitchell Co., N.C.	St. Louis Co., Minnesota	Miyaké, Japan;
Reference	Ribbe, Megaw and Taylor, 1969;Ferguson, et al., 1958	Ribbe, Megaw and Taylor, 1969;Ferguson, et al., 1958	Colville & Ribbe, 1966,1968	Fleet, Chandrasekhar & Megaw, 1966	Megaw, Kempster & Radoslovich, 1962; Megaw & Ribbe, 1962
Cell Dimensions					
a Å	8.138	8.149	8.169	8.178	8.1815
b Å	12.789	12.880	12.851	12.870	12.8733
c Å	7.156	7.106	7.124	14.187	14.776
α deg	94.33	93.37	93.63	93.50	93.21
β deg	116.57	116.30	116.40	115.90	115.835
γ deg	87.65	90.28	89.46	90.63	91.137
Space Group	C1̄	C1̄	C1̄	I1̄	I1̄ (P1̄)
Z	4	4	4	8	8

TABLE 32. (cont.)

Variety	Low Albite	High Albite	Oligoclase	Bytownite	Transitional Anorthite
Site (%Si) Occupancy					
$T_1(0)$ (0000)	0	72	35	100	100
(00i0)	0	72	35	100	100
(0z00)				6	0
(0zi0)				0	0
$T_1(m)$ (m000)	100	75	81	13	0
(m0i0)	100	75	81	13	0
(mz00)				100	100
(mzi0)				100	100
$T_2(0)$ (0000)	100	78	79	30	0
(00i0)	100	78	79	13	0
(0z00)				100	100
(0zi0)				100	100
$T_2(m)$ (m000)	100	75	81	90	100
(m0i0)	100	75	81		100
(mz00)				13	0
(mzi0)				13	0
Method	3-D, film	3-D, film	3-D, counter	3-D, film	3-D, film
R & Rating	0.068 (1)	0.082 (1)	0.069 (1)	0.118 (2)	0.109 (2)
Cleavage and habit	Sometimes platy parallel to {010}.	{001},{010} perfect —————— Variable			
Comment		Minor K ignored in calculations. Ab = NaAlSi$_3$O$_8$ An = CaAl$_2$Si$_2$O$_8$ Or = KAlSi$_3$O$_8$	Na/Ca assumed random	Na/Ca assumed random	Cell dimensions for synthetic An100ANS-26; Stewart, 1967, p.50
μ	86.3	85.6	101.7	84.0	139.4
ASF	0.5020×10^{-1}	0.4621×10^{-1}	0.4533×10^{-1}	0.2487×10^{-1}	0.1825
Abridging factors	0.5:0:0	0.5:0:0	0.5:0:0	0.5:0:0	0.5:0:0

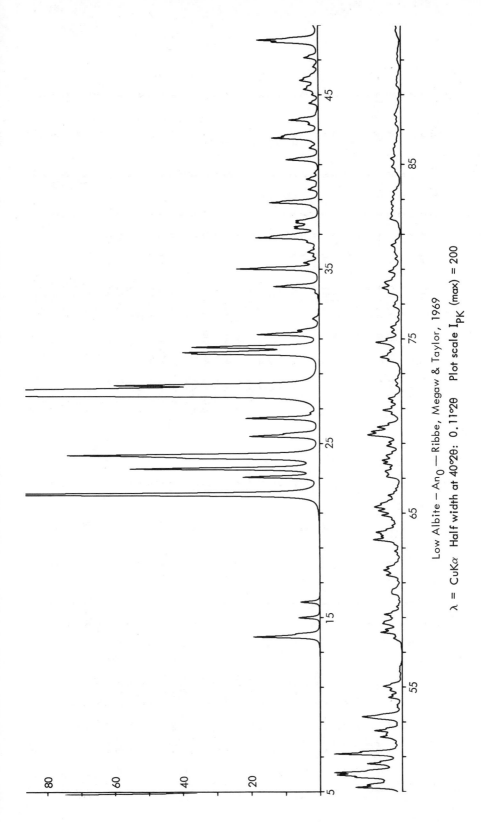

Low Albite — An$_0$ — Ribbe, Megaw & Taylor, 1969

λ = CuKα Half width at 40°2θ: 0.11°2θ Plot scale I$_{PK}$ (max) = 200

LOW ALBITE - AN 0 - RIBBE, MEGAW AND TAYLOR, 1969

2THETA	PEAK	D	H	K	L	I(INT)	I(PK)	I(DS)
13.85	13.88	6.387	0	0	1	7	10	7
13.88		6.376	0	2	0	4		4
13.95		6.343	1	1	0	3		3
14.05	14.04	6.299	-1	-1	1	4	4	3
14.97	14.98	5.912	-1	1	1	4	3	4
15.87	15.88	5.581	-1	1	1	4	3	4
22.05	22.06	4.027	-2	0	1	93	67	92
23.06	23.06	3.854	-1	-1	1	15	11	15
23.54	23.54	3.777	1	1	1	39	28	39
24.24	24.32	3.668	-1	1	1	16	37	16
24.31		3.658	-1	-3	0	41		41
25.39	25.40	3.505	1	3	0	15	10	14
25.55	25.54	3.483	-1	-1	2	6	6	5
26.42	26.42	3.370	-2	-2	1	16	11	16
26.74	26.74	3.332	-1	1	2	1	1	1
27.74	27.74	3.214	-2	2	0	72	55	72
27.91	27.96	3.194	0	0	2	100	100	100
27.96		3.188	0	0	2	66		66
28.31	28.32	3.150	-2	4	0	39	30	39
30.11	30.12	2.965	-1	3	1	25	19	25
30.21	30.20	2.956	-2	-2	2	16	20	16
30.47	30.50	2.931	0	-2	2	12	19	12
30.50		2.928	0	-4	1	18		19
31.23	31.24	2.862	-1	3	1	15	9	15
31.47	31.46	2.840	0	-3	1	4	3	4
32.14	32.14	2.782	0	4	1	1	1	1
33.61	33.62	2.664	-3	-1	1	1	1	1
33.97	33.98	2.637	-1	3	2	11	7	11
35.00	35.00	2.562	-2	-4	1	21	12	21
35.35	35.34	2.537	-3	-1	2	3	2	3
35.71	35.72	2.512	-3	1	1	1	1	1
35.95	35.96	2.496	-2	-2	2	3	2	3
36.20	36.20	2.479	-3	-2	1	1	1	1
36.51	36.52	2.459	-2	2	2	2	2	2
36.78	36.78	2.442	-2	4	1	15	9	16
36.95	36.86	2.431	-1	-5	1	6	7	6
37.32	37.32	2.408	2	4	0	5	4	5
37.43	37.42	2.401	-1	5	0	3	4	3
37.63	37.64	2.388	-2	4	1	3	3	3
37.65		2.387	-3	1	0	2		2
37.77	37.76	2.380	-3	-3	1	3	3	3
38.80	38.80	2.319	-1	-1	2	12	7	13
38.88		2.315	-1	3	1	2		2
39.56	39.56	2.276	-1	-3	2	3	2	3
40.14	40.14	2.245	1	1	2	3	2	3
40.48	40.48	2.226	0	2	2	3	1	3
41.24	41.26	2.187	-2	2	2	1	5	1
41.26		2.186	-2	-1	2	5		5
41.27		2.186	0	1	3	2		2
41.93	41.94	2.153	-1	-3	3	12	7	13
42.50	42.50	2.125	-2	0	6	1	1	1
42.62	42.60	2.119	1	3	1	5	7	5
42.69	42.68	2.116	-3	-3	1	5	5	5
42.73		2.114	-1	3	3	2		2
43.28	43.28	2.088	-3	-1	2	8	4	8
43.54	43.54	2.077	-2	4	1	3	1	3
44.50	44.50	2.034	-4	0	0	2	1	2
44.98	44.98	2.014	-1	5	2	1	2	1
45.20	45.30	2.004	-1	0	3	3		3
45.29		2.001	-4	0	3	1		1
45.33		1.9989	-2	0	3	2	2	2
45.49	45.46	1.9922	-1	3	3	3	3	3
45.74	45.82	1.9817	-0	2	2	1	1	1
45.81		1.9791	0	6	1	4		4
45.97	45.94	1.9724	3	-1	1	2	2	2
46.02		1.9706	-3	-3	3	1		1
46.59	46.58	1.9479	-4	2	2	3	1	3
47.12	47.12	1.9270	-2	-2	2	4	2	4
48.01	48.02	1.8935	-4	-2	1	12	7	13
48.12	48.14	1.8894	-3	-5	2	3	9	3
48.15		1.8883	-1	2	3	9		10
49.24	49.24	1.8491	-4	0	0	13	7	14
49.44	49.36	1.8418	2	6	0	4	5	5
49.79	49.82	1.8299	1	-1	3	2	7	2

630

LOW ALBITE - AN 0 - RIBBE, MEGAW AND TAYLOR, 1969

2THETA	PEAK	D	H	K	L	I(INT)	I(PK)	I(DS)
49.82	49.94	1.8288	-2	6	0	8		9
49.92		1.8253	0	-4	3	5	10	5
49.96		1.8240	0	-6	2	7		8
50.09	50.08	1.8196	4	0	0	12	10	13
50.59	50.60	1.8027	-1	0	3	9	5	10
51.16	51.16	1.7838	-2	2	0	20	10	22
51.57	51.56	1.7706	-1	7	0	1	1	1
51.79	51.78	1.7638	-1	7	0	1	1	1
52.11	52.14	1.7535	4	2	0	2	3	2
52.15		1.7525	-2	-2	4	4		4
52.34	52.28	1.7464	-4	2	3	1	2	1
52.36		1.7460	-4	2	0	1		1
52.50	52.50	1.7417	-4	-4	2	7	4	8
52.37	52.86	1.7302	-1	-1	2	1	1	1
53.19	53.30	1.7204	0	4	3	1	6	1
53.23		1.7193	0	6	2	3		3
53.27		1.7181	1	-3	3	3		3
53.30		1.7171	-4	-4	1	2		2
53.83	53.84	1.7014	-2	6	2	4		4
53.96	53.96	1.6977	-3	-5	3	1	1	1
54.40	54.40	1.6852	-2	-1	4	1	1	1
54.56	54.54	1.6804	-1	1	1	4	2	4
54.78	54.80	1.6744	-1	-7	2	1	2	1
54.83		1.6730	-4	4	1	2		2
55.04	55.04	1.6670	2	4	2	4	3	4
55.08		1.6658	-4	4	2	2		2
55.90	55.90	1.6433	-1	-3	0	1	0	1
57.68	57.80	1.5968	0	0	4	1	1	1
57.79		1.5940	0	8	0	2		2
58.13	58.14	1.5855	-3	5	3	4	3	4
58.14		1.5852	-4	-2	4	3		3
58.34	58.30	1.5802	3	-5	1	2	2	2
58.57	58.64	1.5747	4	0	0	1	3	1
58.64		1.5728	0	-2	4	4		4
58.65		1.5728	-5	-1	1	1		1
58.75	58.74	1.5704	0	-8	1	2	3	3

2THETA	PEAK	D	H	K	L	I(INT)	I(PK)	I(DS)
59.03	59.04	1.5635	-1	3	4	1	2	2
59.03		1.5634	-5	1	1	2		2
59.18	59.18	1.5600	3	5	0	3	3	4
59.20		1.5594	2	0	3	1		2
59.62	59.62	1.5494	-5	-3	3	1	1	1
60.01	60.00	1.5403	-5	1	2	1	1	2
60.35	60.38	1.5323	2	-3	2	2	2	2
60.45	60.44	1.5302	-4	2	3	2	2	2
60.56	60.52	1.5276	-3	-7	2	1	2	1
60.62		1.5261	0	2	4	1		1
61.13	61.14	1.5147	-1	-7	2	2	1	1
61.39	61.44	1.5090	-2	1	4	4	2	2
61.44		1.5078	-2	-8	1	5		4
61.72	61.72	1.5017	-5	-3	3	1	3	6
61.84	61.88	1.4991	-4	-6	2	1	2	1
61.88		1.4981	-5	3	3	1		1
61.93		1.4970	3	-3	2	1		1
62.11	62.10	1.4931	-5	-3	1	1	1	1
62.60	62.60	1.4826	2	-6	1	1	1	1
62.80	62.80	1.4783	-4	-6	1	1	1	1
63.18	63.24	1.4705	-2	-8	2	1	1	1
63.23		1.4693	-5	1	1	1		1
63.40	63.46	1.4657	0	-4	4	4	4	9
63.47		1.4645	2	8	0	7		1
63.48		1.4641	0	-8	2	2		5
63.77	63.78	1.4582	0	6	3	4	3	7
63.89	63.90	1.4558	-2	8	0	6	4	5
64.12	64.06	1.4511	-1	8	0	1	2	2
64.27	64.28	1.4480	-3	7	0	2	1	1
64.61	64.62	1.4413	-4	-6	3	2	2	2
64.70		1.4396	-3	-7	3	1		1
64.87	64.86	1.4362	-4	6	1	5	3	6
65.09	65.12	1.4317	-1	-1	4	1	3	1
65.14		1.4309	2	6	2	5		6
65.33	65.34	1.4270	-4	-6	2	4	4	5
65.42		1.4254	2	-4	3	4		4
65.54	65.52	1.4230	-5	1	1	2	3	3

LOW ALBITE — AN 0 — RIBBE, MEGAW AND TAYLOR, 1969

2THETA	PEAK	D	H	K	L	I(INT)	I(PK)	I(DS)
65.87	65.88	1.4168	4	-4	1	1	2	1
66.01	66.00	1.4141	1	4	1	4		2
66.38	66.38	1.4070	4	4	4	2	2	1
67.07	67.08	1.3943	-5	-3	4	4	1	2
67.14		1.3929	-1	9	0	2		4
67.38	67.40	1.3887	-1	9	0	2	2	3
67.42		1.3878	-4	6	1	3		2
67.64	67.62	1.3839	2	-8	1	1	2	3
67.82	67.84	1.3806	5	3	3	1	3	1
67.84		1.3804	1	-3	4	4		1
68.00	68.02	1.3774	2	-4	3	3	3	5
68.04		1.3767	-4	6	0	1		3
68.21	68.30	1.3737	-5	3	0	4	2	1
68.31		1.3720	-4	0	5	1		4
68.80	68.84	1.3633	-1	9	1	4	2	4
68.85		1.3624	-4	-2	5	1		5
69.07	69.06	1.3587	2	8	1	4	2	1
69.41	69.48	1.3529	-1	-9	3	1	5	5
69.44		1.3523	-6	0	2	5		2
69.49		1.3515	1	-9	2	6		7
69.54		1.3507	-1	-9	1	3		7
69.67	69.66	1.3484	-1	0	5	3	4	4
69.93	69.92	1.3441	4	0	2	7	7	8
70.55	70.66	1.3338	-2	-4	5	1	2	1
70.66		1.3319	-6	-2	2	2		1
70.86		1.3286	-5	5	1	4		3
70.90	70.90	1.3280	-1	-3	5	4	3	1
71.33	71.36	1.3211	-4	-2	2	1		5
71.35		1.3208	4	-2	2	1		1
71.37		1.3205	3	-1	3	1		1
71.51	71.54	1.3182	-3	3	5	1	2	1
71.54		1.3177	0	3	4	1		1
71.58		1.3171	0	-8	3	1		1
72.30	72.30	1.3058	-5	5	3	2	1	2
72.62	72.62	1.3007	-6	2	3	2	1	2
73.68		1.2847	3	-3	3	1		1
73.72	73.72	1.2840	-1	9	2	2	2	3
73.74		1.2837	-3	-5	5	1		1
73.85	73.92	1.2821	-6	-2	1	1	3	1
73.93		1.2809	-4	-8	2	5		6
74.07		1.2789	-6	0	4	1		2
74.32	74.32	1.2752	0	10	0	2	2	2
74.63	74.78	1.2706	4	-6	1	1	4	1
74.75		1.2699	-5	1	5	1		5
74.77		1.2688	-2	0	4	2		4
74.78		1.2686	-6	-2	4	1		4
74.79		1.2685	0	-8	2	5		5
75.31	75.38	1.2683	-6	-1	2	1		2
75.40		1.2608	-6	-4	3	1		3
75.65	75.66	1.2596	-5	-3	5	2	2	2
75.68		1.2560	2	-2	2	1		1
76.23	76.24	1.2555	-4	-4	4	2		3
76.82	76.86	1.2479	-4	6	2	1	1	2
76.87		1.2398	-4	6	4	1		2
76.91		1.2391	0	8	3	1	1	2
77.59	77.66	1.2386	4	4	2	2		1
77.66		1.2293	4	4	2	2	2	2
77.74	77.90	1.2285	-6	8	2	2		2
77.91		1.2274	-4	-9	5	5	3	6
78.09	78.22	1.2252	-2	8	2	2		5
78.24		1.2227	-3	-9	3	3		1
78.31		1.2208	-4	8	2	4		1
78.65	78.64	1.2198	1	5	0	1		5
78.84	78.84	1.2155	-2	-10	0	1		1
78.97		1.2130	6	0	0	4	2	5
79.56	79.58	1.2113	0	-10	0	1		1
79.60		1.2038	4	8	2	1		1
79.97	79.96	1.2033	-3	-5	5	1	1	2
80.10	80.10	1.1987	3	-5	3	1	1	2
80.34	80.36	1.1970	-6	-4	3	1	1	1
80.39		1.1941	-4	-6	5	1		1
81.42	81.44	1.1935	6	2	0	2	0	2
81.84	81.88	1.1810	-6	0	5	1	2	1
		1.1759	-3	1	6	3		4

LOW ALBITE – AN 0 – RIBBE, MEGAW AND TAYLOR, 1969

2THETA	PEAK	D	H	K	L	I(INT)	I(PK)	I(DS)
81.90		1.1752	-1	9	3	1		1
81.97		1.1744	-4	8	4	1		2
82.16	82.14	1.1722	-2	0	6	1	1	1
82.29		1.1707	-1	5	5	1	1	1
82.43	82.44	1.1690	-5	-7	4	1		2
82.49		1.1683	-3	-3	6	1	1	2
82.69	82.70	1.1660	0	-8	3	2	1	2
82.85	82.84	1.1642	3	5	3	2	1	3
83.09	83.12	1.1614	1	1	5	1	1	1
83.18		1.1604	0	4	5	1		1
83.43	83.42	1.1575	1	-7	4	2	1	2
84.58	84.60	1.1447	-3	9	3	2	1	2
84.74	84.86	1.1429	2	10	1	2	1	2
84.85		1.1418	-2	6	5	2	1	3
84.96		1.1405	-3	-9	1	2		3
85.30	85.32	1.1368	6	4	0	2	1	3
85.34		1.1364	-1	-9	4	1		2
85.71		1.1325	-6	-4	0	2		3
85.88	85.86	1.1307	-1	-11	2	1	1	3
86.15	86.14	1.1278	-3	3	6	1	1	1
86.57	86.40	1.1235	-5	1	6	1	1	1
86.78	86.78	1.1213	-5	1	5	1	1	1
87.27	87.32	1.1162	-5	-3	6	1	1	1
87.32		1.1157	-5	3	1	1		1
87.35		1.1154	4	8	1	2		3
87.83	87.84	1.1105	-6	6	3	1	1	1
87.96		1.1092	-1	-3	6	1		1
88.76	88.74	1.1013	-2	-8	5	1	0	1
89.48	89.54	1.0942	4	-4	3	1	1	1
89.54		1.0937	0	8	4	1		1
89.59		1.0932	-4	8	4	1		2
90.01	90.00	1.0892	1	-5	5	1	1	1
90.83	91.02	1.0815	-3	7	7	1	1	1
91.03		1.0796	-3	-11	5	1		1
91.37	91.34	1.0765	-2	-10	4	1	1	1
92.65		1.0650	-7	1	5	2		2
92.71	92.68	1.0644	0	10	3	1	1	1

2THETA	PEAK	D	H	K	L	I(INT)	I(PK)	I(DS)
92.97	92.98	1.0621	-7	-5	2	1	1	1
93.19	93.22	1.0602	2	-2	5	1	1	1
93.89	94.10	1.0541	-4	10	2	1	1	3
94.06		1.0527	-6	-8	3	2		2
94.18		1.0516	5	-5	2	1		1
94.95	95.08	1.0451	6	-4	1	1	1	2
95.54	95.70	1.0402	-2	-12	1	1	1	2
95.68		1.0391	-7	5	5	1		1
95.73		1.0387	5	5	2	2		1
96.54	96.52	1.0321	-3	-7	0	1	1	2
98.15	98.16	1.0195	5	9	7	1	1	1
98.50	98.50	1.0168	-4	-2	3	1	1	1
98.57		1.0162	-8	0	0	1	1	3
99.23	99.26	1.0112	-5	9	3	1		1
99.43	99.56	1.0097	-8	-2	3	2	1	3
100.18	100.20	1.0041	-6	6	5	1	1	2
100.44	100.46	1.0022	-2	10	4	1	1	2
100.46		1.0021	-8	-2	4	1		1
100.83	100.80	0.9994	-3	-11	4	1		4
101.32	101.30	0.9959	4	0	4	1	1	4
101.76	101.72	0.9928	-5	-7	6	1	1	6
102.71	102.72	0.9862	4	-2	4	1	1	4
103.49	103.46	0.9809	6	-2	2	1		2
104.92	104.94	0.9714	-6	-8	5	1		5
105.54	105.56	0.9674	-8	-2	5	1		5
105.79	106.00	0.9658	-3	9	5	1	1	4
106.29	106.28	0.9644	-4	10	4	1	1	6
106.54	106.72	0.9626	-6	0	7	1	1	1
106.78		0.9611	-8	0	1	1		5
107.13	107.14	0.9596	1	7	5	1		2
107.15		0.9574	-4	-12	2	1	1	3
107.25		0.9573	1	11	3	1		3
107.65	107.66	0.9567	-8	4	3	1	1	7
107.70		0.9542	-5	3	7	1		2
107.99	108.00	0.9539	3	-11	2	2	1	3
		0.9521	-1	1	7	1	1	1

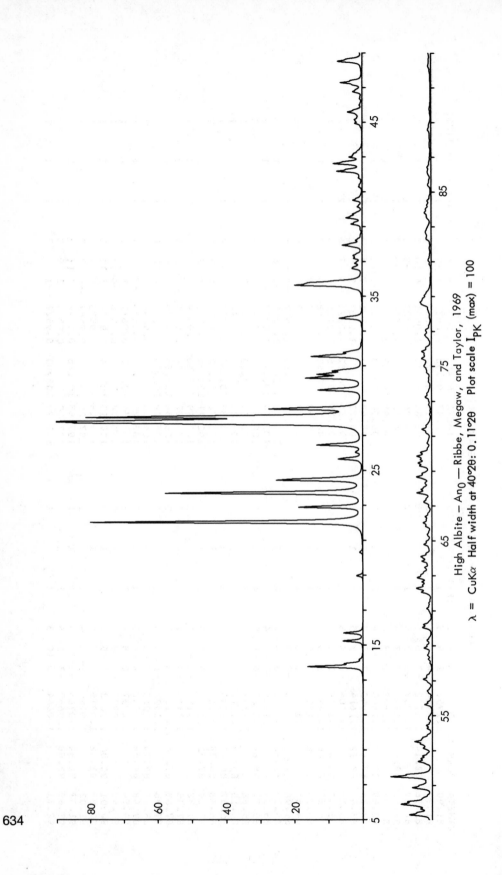

High Albite – An_0 – Ribbe, Megaw, and Taylor, 1969
λ = CuKα Half width at 40°2θ: 0.11°2θ Plot scale I_{PK} (max) = 100

634

HIGH ALBITE – AN 0 – RIBBE, MEGAW AND TAYLOR, 1969

2THETA	PEAK	D	H	K	L	I(INT)	I(PK)	I(DS)
13.77	13.76	6.445	-1	1	0	9	16	9
13.92	13.92	6.425	0	2	0	10		10
13.92	13.92	6.356	0	1	1	4	6	4
14.15	14.14	6.255	-1	0	1	1	1	1
15.21	15.22	5.822	-1	1	1	6	6	6
15.68	15.68	5.648	-1	-1	1	6	6	6
18.94	18.96	4.681	0	2	1	2	2	2
20.29	20.30	4.372	0	0	1	1	1	1
22.03	22.04	4.032	-2	0	1	100	80	98
22.88	22.88	3.884	-1	-1	1	23	19	22
23.69	23.70	3.752	-1	3	0	45	58	44
23.71		3.749	-1	1	1	30		29
24.44	24.44	3.639	1	3	0	30	25	30
25.64	25.64	3.472	-1	-1	2	9	7	9
26.01	26.02	3.422	-2	-2	1	2	2	2
26.44	26.44	3.368	-1	1	2	17	13	17
27.75	27.76	3.212	0	4	0	68	85	68
27.83	27.82	3.203	-2	0	2	83	100	83
28.05	28.06	3.178	0	0	2	100	81	100
28.52	28.52	3.127	2	2	0	35	28	35
29.59	29.60	3.016	1	-3	1	18	13	18
30.28	30.28	2.949	0	-4	1	22	17	22
30.49	30.50	2.929	0	2	2	16	13	16
30.69	30.68	2.911	-2	-2	2	10	9	10
31.53	31.54	2.835	-1	-3	1	21	15	22
31.76	31.76	2.815	-1	1	2	6	6	6
33.72	33.72	2.656	-1	3	2	10	7	10
35.62	35.64	2.518	-2	-4	1	24	20	25
35.65		2.516	-3	-1	1	3		3
35.66		2.515	-1	1	1	3		3
36.58	36.60	2.455	-2	4	0	3	3	3
36.63		2.451	-1	5	1	1		1
37.16	36.88	2.435	2	2	1	3	2	3
37.33	37.16	2.417	-1	-5	0	3	3	3
37.83	37.32	2.407	-3	1	0	3	3	3
37.92	37.92	2.376	3	1	0	3	6	3
		2.371	2	4	0	7		7

2THETA	PEAK	D	H	K	L	I(INT)	I(PK)	I(DS)
38.19	38.20	2.354	-1	5	1	2	2	3
38.80	38.80	2.319	-1	-1	3	1	1	1
39.13	39.12	2.300	-3	3	1	5	3	5
39.49	39.50	2.280	-3	-3	1	7	5	8
39.71	39.70	2.268	-1	1	3	3	3	4
40.12	40.12	2.245	-1	-3	2	2	2	2
40.49	40.50	2.226	-2	4	1	2	2	2
40.51		2.225	-1	-5	1	2		2
41.26	41.26	2.186	0	4	2	2	1	2
41.42	41.38	2.178	-3	3	2	1	1	1
41.77	41.78	2.160	-2	2	3	3	1	1
42.15	42.16	2.142	-1	-5	2	1	8	1
42.16		2.142	0	6	0	11		11
42.59	42.60	2.121	2	-4	1	12	9	13
42.64		2.119	0	-6	1	1		1
42.97	42.98	2.103	-1	5	2	5	4	6
43.18	43.16	2.093	-3	1	3	3	3	3
43.36	43.28	2.085	-3	3	0	3	2	3
44.92	44.92	2.016	-4	0	2	2	2	3
45.08	45.06	2.010	-2	4	1	1	1	1
45.14	45.14	2.007	-4	2	0	3	2	3
45.57	45.58	1.9889	0	-6	1	5	5	6
45.61		1.9874	-1	0	1	1		1
45.97	45.98	1.9724	3	-1	3	1	1	1
46.34	46.34	1.9574	0	2	3	3	3	3
46.74	46.74	1.9418	3	-1	1	4	3	4
46.80		1.9396	-2	-2	2	1		1
47.28	47.28	1.9210	-4	4	0	9	6	10
48.48	48.50	1.8761	-2	2	6	6		7
48.52		1.8747	-2	6	2	6		7
48.54		1.8740	-3	5	0	7		7
49.30	49.30	1.8467	-4	0	3	11	7	12
49.36		1.8448	-3	-3	2	1	1	1
49.64	49.64	1.8349	0	-6	2	4	3	4
49.84	49.92	1.8279	1	-1	3	3	3	3
49.88		1.8267	0	-4	3	3		3
49.92		1.8253	4	0	0	11		12

635

HIGH ALBITE - AN O - RIBBE, MEGAW AND TAYLOR, 1969

2THETA	PEAK	D	H	K	L	I(INT)	I(PK)	I(DS)
50.09	50.06	1.8197	2	6	0	5	7	5
50.82	50.86	1.7952	-1	7	0	3	6	3
50.86		1.7937	-1	1	3	8		9
51.09	50.98	1.7862	-4	-2	3	1	4	1
51.20	51.22	1.7827	-1	-7	1	1	2	1
51.42	51.50	1.7754	3	3	1	1	12	1
51.49		1.7733	-2	0	4	21		23
51.53		1.7721	-4	2	0	1		1
52.32	52.36	1.7472	-1	7	1	1	4	1
52.34		1.7464	1	5	2	1		1
52.37		1.7455	2	-4	2	4		5
52.55	52.52	1.7399	4	2	0	3	3	3
52.68	52.68	1.7359	-2	-2	4	3	3	3
53.06	53.12	1.7245	-1	-3	3	1	3	2
53.13		1.7224	0	6	2	3		3
53.13		1.7222	2	-6	1	1		1
53.35	53.38	1.7157	0	4	3	2	4	2
53.38		1.7150	-4	4	1	3		3
53.50	53.50	1.7112	-4	-4	2	6	5	7
53.59		1.7087	1	-7	1	1		1
53.74	53.72	1.7041	-4	4	2	3	3	4
54.24	54.26	1.6896	-1	1	4	1	2	1
54.26		1.6891	-4	-1	4	2		2
54.43	54.42	1.6841	-2	2	4	2	2	2
54.79	54.80	1.6739	-3	-5	3	1	1	1
54.89	54.90	1.6711	-1	-7	2	1	1	1
55.62	55.62	1.6510	2	1	2	4	2	5
55.94	55.92	1.6423	-1	3	3	1	1	1
56.29	56.30	1.6330	-1	-3	4	1	1	1
56.29		1.6329	2	6	1	1		1
56.39	56.40	1.6301	1	7	1	1	2	1
56.40		1.6300	-3	-3	4	1		1
57.07	57.08	1.6124	-3	5	3	3	3	4
57.31	57.24	1.6062	0	8	0	2	2	2
57.99	58.00	1.5889	0	0	4	1	1	1
58.23	58.22	1.5831	0	-8	1	2	1	2
58.43	58.46	1.5782	4	0	1	1	1	1
58.47	58.47	1.5771	-5	1	1	1		2
58.80	58.86	1.5691	-5	-1	1	1		2
58.85		1.5679	-4	-2	4	2	2	2
58.86		1.5675	0	-2	4	1		3
59.15	59.02	1.5606	-1	3	4	1		2
59.26	59.30	1.5580	-5	1	3	1	2	2
59.32		1.5566	2	0	3	1	2	2
59.93	60.04	1.5421	3	-3	2	2		1
60.02		1.5401	-4	2	4	2	3	1
60.05		1.5395	3	5	1	3		3
60.06		1.5392	-2	6	3	1		1
60.20	60.20	1.5359	-2	-2	5	1	3	1
60.22		1.5353	1	-7	2	1		1
61.11	61.12	1.5151	-2	-6	4	2	2	3
61.43	61.42	1.5081	2	-6	2	1	1	2
61.94	62.04	1.4967	-5	3	1	1	3	1
62.04		1.4946	-2	-8	0	5		6
62.28	62.26	1.4895	-2	8	0	6	5	7
62.30		1.4890	-1	-7	3	1		1
62.53	62.44	1.4842	3	3	2	4		1
62.63	62.64	1.4820	-5	-3	3	5	3	5
62.87	62.88	1.4768	-4	6	1	1	4	6
62.89		1.4765	-5	-3	1	1		1
62.98		1.4746	0	-8	2	2		2
62.99		1.4744	-1	-5	4	1		1
63.46	63.46	1.4646	-4	6	2	2	2	3
63.47		1.4644	0	-6	4	1		1
63.81	63.82	1.4573	0	6	3	5	3	6
63.91		1.4554	-4	-4	4	1		1
64.07	64.08	1.4521	-4	6	0	1		1
64.09		1.4516	2	8	0	5		5
64.49	64.50	1.4437	4	-4	1	1	1	1
64.78	64.78	1.4380	2	-4	3	4	2	2
64.91	64.94	1.4354	-5	-1	1	1	1	1
65.26	65.34	1.4285	-1	-1	4	1	2	1
65.35		1.4268	-5	1	4	2		2
65.76	65.82	1.4189	-3	-7	3	1		1

HIGH ALBITE - AN O - RIBBE, MEGAW AND TAYLOR, 1969

2THETA	PEAK	D	H	K	L	I(INT)	I(PK)	I(DS)
65.83		1.4176	2	6	2	`3		3
65.93	66.02	1.4155	-4	-6	3	1	3	1
66.02		1.4138	-4	6	0	1		1
66.03		1.4136	2	-8	1	2		2
66.20	66.20	1.4105	-1	9	0	2		1
66.37	66.36	1.4073	-2	-6	4	1	2	1
66.92	66.94	1.3971	-1	1	4	2	2	2
66.95		1.3965	-5	3	0	1		1
67.19	67.20	1.3921	-3	1	0	3	1	1
67.26		1.3908	1	4	5	2		4
67.72	67.74	1.3824	4	4	1	2	2	2
67.73		1.3822	-1	9	3	3		2
67.75		1.3819	-4	6	4	5	4	3
67.93	67.94	1.3787	1	-3	4	1		6
68.06	68.04	1.3764	-5	5	2	3	3	1
68.32	68.30	1.3717	-5	-3	4	2	2	4
68.56	68.66	1.3676	1	-9	1	1	2	3
68.62		1.3664	5	3	0	2		3
68.66		1.3658	2	4	3	2	5	2
68.67		1.3656	4	6	0	5		6
69.29	69.30	1.3550	-4	0	5	6		8
69.43	69.44	1.3524	-6	0	2	5	4	6
69.65	69.64	1.3488	-1	-9	2	2	4	2
69.74		1.3472	-4	2	5	2	3	2
69.84	69.84	1.3455	2	8	1	7	4	8
70.08	70.06	1.3416	4	0	0	3		3
70.38	70.26	1.3366	-1	-1	5	3	3	1
70.57	70.58	1.3334	-5	-5	1	2	1	2
70.68		1.3317	-5	5	3	1	1	1
71.14	71.20	1.3241	4	-2	2	2		2
71.20		1.3232	0	-8	3	3	2	2
71.26		1.3222	-6	-2	4	2		1
71.38	71.38	1.3203	-2	-4	5	1	3	4
71.41		1.3198	-1	-3	5	3		1
71.64	71.58	1.3162	3	1	3	1	2	2
72.25	72.24	1.3065	-4	-6	4	1	1	1

2THETA	PEAK	D	H	K	L	I(INT)	I(PK)	I(DS)
72.79	72.80	1.2982	-1	9	2	1	1	2
73.02	73.02	1.2945	-3	-3	3	1	1	1
73.17	73.22	1.2924	-2	-8	3	1	2	1
73.23		1.2914	2	-8	2	3		4
73.38	73.42	1.2891	-5	5	0	1	1	1
74.08	74.10	1.2787	-4	-4	4	1	1	1
74.11		1.2783	-6	0	4	1		1
74.60	74.64	1.2711	0	0	5	1	1	1
74.62		1.2708	0	3	5	1		1
74.67		1.2700	-5	1	5	1		1
74.95		1.2660	4	-4	2	1		4
74.98	74.98	1.2656	2	0	4	2	2	2
75.06		1.2645	-4	8	1	1		2
75.13		1.2635	0	-2	5	2		1
75.39	75.44	1.2597	-4	6	2	1		5
75.43		1.2591	-4	-8	2	5		1
75.51		1.2581	-6	-2	4	1		6
75.53		1.2577	2	-2	4	1		1
75.83	75.84	1.2534	-4	8	2	3	3	4
75.87		1.2529	-6	4	2	2		2
76.66		1.2419	-6	-4	3	2		2
76.78	76.78	1.2404	-5	-3	5	2	2	2
76.81		1.2399	0	8	3	1		2
77.07	77.06	1.2364	3	0	4	1	2	1
77.10		1.2360	-4	4	5	2		2
77.80	77.82	1.2266	-4	-8	3	1	1	1
78.10		1.2227	-2	10	1	1	1	1
78.24	78.24	1.2208	-6	4	1	1		2
78.48	78.64	1.2176	-6	4	2	1	3	2
78.54		1.2168	4	4	0	4		5
78.65		1.2155	6	0	0	4		6
78.70		1.2147	2	8	5	1		1
78.80		1.2135	-5	3	3	1		1
79.04	79.04	1.2104	3	-5	3	2	2	3
79.19		1.2086	-3	5	5	1		1
79.60	79.58	1.2033	-2	-10	0	1	0	1
80.04	80.04	1.1977	-6	2	8	1	0	1

637

HIGH ALBITE - AN 0 - RIBBE, MEGAW AND TAYLOR, 1969

2THETA	PEAK	D	H	K	L	I(INT)	I(PK)	I(DS)
80.83	80.84	1.1880	6	2	0	1	1	1
81.05	81.08	1.1854	-4	8	3	1	1	1
81.15		1.1842	-1	9	1	1		1
81.76	81.76	1.1769	-4	-6	5	1		1
82.26	82.30	1.1710	-3	1	6	2	1	1
82.33		1.1702	0	-8	4	2	2	3
82.40		1.1693	-1	5	5	2		2
82.64	82.66	1.1665	-3	9	3	2		2
82.74		1.1654	1	-7	4	1		2
82.76		1.1652	-2	0	6	1	2	2
83.45	83.58	1.1573	-3	-3	6	1		1
83.48		1.1570	-2	8	4	1		1
83.54		1.1562	-4	-8	4	1		1
83.57		1.1560	2	-8	3	1		1
83.63		1.1553	0	4	5	2		3
83.79	83.80	1.1535	3	5	3	2	2	3
83.81		1.1532	4	-8	1	1		1
83.98	83.98	1.1513	-6	4	0	2	2	3
84.25	84.24	1.1483	-5	-7	4	1	1	2
84.48	84.48	1.1458	-2	6	5	2	1	2
85.38	85.40	1.1360	-6	6	3	1	1	2
85.42		1.1356	-1	-9	4	1		2
85.99	86.12	1.1295	-1	-11	2	1		2
86.12		1.1281	-3	-9	4	2	2	2
86.42	86.40	1.1250	6	4	0	1	1	2
86.97	86.96	1.1192	-5	1	6	1	0	1
87.58	87.60	1.1131	-4	8	0	1	0	1
88.11	88.10	1.1077	6	0	1	1	1	1
88.56	88.58	1.1033	-5	-3	6	1	1	1
88.58		1.1031	1	-3	6	1		1
89.77	89.78	1.0915	-1	-5	5	1	1	1
91.42	91.42	1.0760	-5	9	2	1	0	1
92.23	92.24	1.0687	5	-5	2	1	1	1
92.68	92.60	1.0647	1	9	0	1	1	1
92.90	93.00	1.0628	2	0	5	1	1	1
93.03		1.0616	6	-4	1	1		1
93.24	93.26	1.0598	-7	5	3	1	1	2

2THETA	PEAK	D	H	K	L	I(INT)	I(PK)	I(DS)
93.30	94.00	1.0592	2	6	4	1	0	1
94.00	94.00	1.0532	3	-5	4	1		1
94.65	94.78	1.0477	-7	-5	2	1	0	1
94.80		1.0464	-7	-3	5	1		1
95.26	95.26	1.0425	-7	1	0	1		1
95.49	95.68	1.0406	-7	-5	4	1		1
95.67		1.0392	-5	9	0	1	0	1
96.04	96.00	1.0361	-2	-12	3	1	1	1
96.41	96.38	1.0331	-6	-8	3	1	1	1
97.00	97.02	1.0284	-6	8	3	1	1	1
97.05		1.0280	5	5	2	1		1
97.84	97.88	1.0218	-3	-7	6	1	1	1
98.08	98.12	1.0200	-6	6	5	2	1	2
98.27		1.0185	-8	0	3	1		1
98.84	98.84	1.0142	2	-12	1	1	0	1
99.06	99.16	1.0125	-2	-10	4	1	1	1
99.18		1.0116	-4	0	7	1		1
99.51	99.54	1.0091	-6	-8	4	1	1	1
100.27	100.28	1.0035	5	9	0	1	0	1
101.16	101.24	0.9971	-1	-11	4	1	1	1
101.23		0.9965	-8	-2	4	1		1
101.27		0.9963	4	-2	4	1		1
103.01	103.02	0.9841	-4	10	4	1	0	1
103.96	103.98	0.9777	-3	9	5	1	1	2
104.73	104.90	0.9726	-8	4	3	1	1	1
105.80	105.82	0.9657	-8	0	0	1	1	1
108.11	108.12	0.9514	-4	12	1	1	0	1
108.76	108.80	0.9475	-4	-12	2	1	0	1
108.83		0.9471	-1	1	7	1		1
109.18	109.18	0.9450	3	9	3	1	1	1
110.36	110.48	0.9382	-3	-9	6	1	1	1
110.51		0.9374	4	4	4	1	0	1

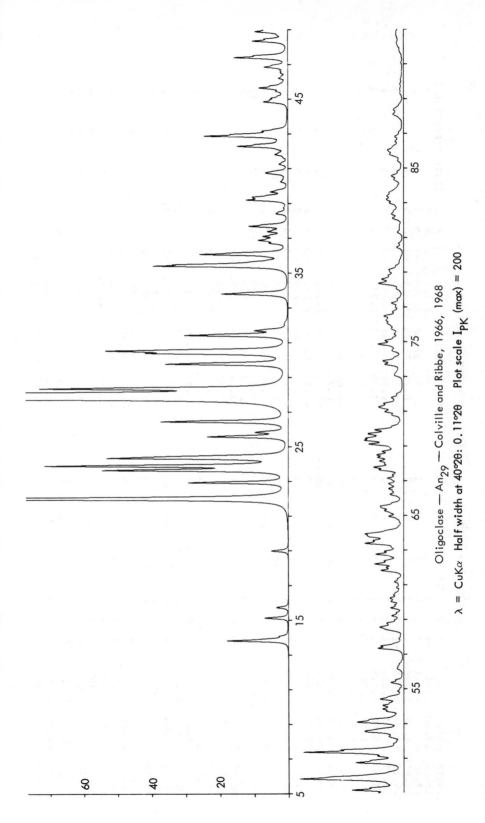

Oligoclase — An$_{29}$ — Colville and Ribbe, 1966, 1968

λ = CuKα · Half width at 40°2θ: 0.11°2θ Plot scale I$_{PK}$ (max) = 200

639

OLIGOCLASE - AN 29 - COLVILLE AND RIBBE, 1966, 1968

2THETA	PEAK	D	H	K	L	I(INT)	I(PK)	I(DS)
13.80	13.80	6.412	-1	1	0	7	6	6
13.80		6.411	0	2	0			6
14.05	14.04	6.298	-1	-1	0	2		2
15.12	15.12	5.856	-1	1	1	1	1	1
15.71	15.72	5.634	-1	-1	1	4	2	3
18.97	18.98	4.675	0	2	1	2	1	2
21.97	21.98	4.042	-2	0	1	3	2	
22.91	22.92	3.879	-1	1	2	91	53	89
23.62	23.62	3.763	-1	-1	2	16	10	16
23.87	23.88	3.725	1	3	0	30	18	30
24.32	24.32	3.657	-1	3	0	40	24	40
25.55	25.56	3.483	-1	-1	2	29	18	28
25.56		3.482	-1	3	0	13	8	13
25.81	25.82	3.449	-2	-2	1	1		1
26.42	26.42	3.371	-1	-1	2	5	3	5
27.76	27.80	3.211	-2	0	2	22	12	22
27.80		3.206	-2	2	0	87	100	87
27.81		3.206	-2	4	0	46		46
28.00	28.00	3.184	0	0	2	67	62	67
28.32	28.32	3.149	2	2	0	100	24	100
29.74	29.74	3.001	-1	-3	1	40	12	40
30.35	30.36	2.942	0	-4	1	23	14	23
30.49	30.50	2.929	-2	-2	2	16	18	24
30.50		2.928	-1	2	1	23		17
31.40	31.40	2.847	-1	-3	2	13	10	14
31.67	31.68	2.823	-1	3	1	20		20
33.78	33.78	2.651	-1	-3	2	5	3	6
35.38	35.38	2.535	-1	3	2	13	7	13
35.46	35.46	2.529	-2	-4	1	26	13	27
35.47		2.524	-3	1	2	5	11	5
35.53		2.489	-2	-1	1	3		3
36.05	36.06	2.447	-2	4	1	18	9	18
36.69	36.70	2.436	-2	4	1	3	2	3
36.86	36.86	2.422	-1	5	0	3	3	3
36.86		2.405	-1	-5	1	2		2
37.08	37.08	2.405	-3	1	0	4	3	5
37.36	37.36	2.386	-3	-1	0	3	2	3
37.66	37.66		3	1	0	3	4	3

2THETA	PEAK	D	H	K	L	I(INT)	I(PK)	I(DS)
37.66	37.66	2.386	2	4	0	5		6
38.38	38.38	2.343	-1	-1	3	2	1	2
38.70	38.70	2.325	-1	-3	1	1	0	1
39.17	39.18	2.298	-3	3	1	8	4	9
39.35	39.34	2.288	-3	3	1	6	4	6
39.65	39.64	2.271	-1	-1	3	3	2	4
40.22	40.22	2.240	-1	-3	3	1	1	2
40.43	40.42	2.229	-3	-1	2	2	0	1
40.68	40.74	2.216	-2	-5	2	2	2	2
40.74		2.213	0	4	1			4
41.23	41.22	2.188	-3	3	2	3	1	2
41.61	41.62	2.169	-2	3	3	2	1	1
41.78	41.78	2.160	-3	-5	0	3	1	3
42.25	42.26	2.137	0	2	0	1	5	1
42.25		2.137	1	-4	1	2		
42.84	42.84	2.109	-2	3	1	10	8	11
42.85		2.109	3	-1	3	7		7
43.05	42.94	2.099	-1	-6	0	12	6	13
43.14	43.14	2.095	-4	0	3	1	3	2
44.80	44.82	2.021	2	4	1	4	2	4
44.82		2.020	-1	5	2	3		4
44.84		2.011	-4	0	1	1		1
45.04	45.04	2.001	-1	-3	3	3	2	3
45.29	45.30	1.9947	3	-1	1	1	1	1
45.43	45.44	1.9870	0	6	1	3	1	2
45.62	45.62	1.9868	-3	-3	2	2	3	5
45.62		1.9642	-2	-2	2	2		2
46.18	46.18	1.9521	-4	-2	4	4	2	4
46.48	46.48	1.9393	-4	-2	2	2	1	3
46.81	46.82	1.9381	-3	-5	2	3	3	4
46.84		1.9173	-2	6	1	2		3
47.37	47.38	1.8817	-3	5	1	12	5	14
48.33	48.32	1.8693	-2	-5	1	8	3	9
48.67	48.68	1.8623	-3	5	1	1	1	1
48.86	48.86	1.8603	-2	6	3	6	3	7
48.92		1.8515	-3	5	3	1		1
49.17	49.16		-4	0	3	12	5	13

OLIGOCLASE - AN 29 - COLVILLE AND RIBBE, 1966, 1968

2THETA	PEAK	D	H	K	L	I(INT)	I(PK)	I(DS)
49.75	49.82	1.8312	0	-6	2	5	10	5
49.80		1.8294	1	-1	3	2		2
49.82		1.8289	4	0	0	13		15
49.82		1.8286	2	6	0	5		2
49.90		1.8260	0	-4	3	3		6
50.74	50.74	1.7978	-1	1	3	10	5	4
50.78		1.7963	-4	-2	3	2		11
51.10	51.12	1.7857	-1	7	0	2	3	3
51.12		1.7852	3	3	1	1		3
51.15		1.7843	-1	-7	1	1		1
51.36	51.36	1.7773	-2	0	4	24	10	28
51.85	51.84	1.7618	-4	2	3	3	1	1
52.18	52.26	1.7514	1	5	2	1	1	2
52.26		1.7491	-4	2	0	2		2
52.50	52.56	1.7414	-2	-2	4	4	4	4
52.52		1.7408	-2	6	2	2		2
52.56		1.7395	-1	7	1	1		1
52.58		1.7391	2	-4	2	5		5
53.06	53.08	1.7243	-4	-4	2	8	5	9
53.11		1.7230	-1	-3	3	2		2
53.12		1.7225	0	6	3	3		3
53.27	53.20	1.7182	0	4	4	2	3	2
53.70	53.70	1.7053	-4	4	1	3	2	4
53.82	53.84	1.7020	-4	-4	1	2	2	3
53.89		1.7000	-1	-7	1	1		1
54.04	54.04	1.6954	-4	4	2	3	2	4
54.14		1.6926	-1	1	4	1		1
54.39	54.40	1.6854	-2	1	4	4	2	5
54.45		1.6838	-2	-2	4	2		2
54.86	54.86	1.6721	-3	-5	3	1	1	1
55.35	55.36	1.6583	-2	-7	2	3	1	4
56.12	56.16	1.6375	-3	4	4	1		1
56.17		1.6360	-1	-3	3	1		1
56.43	56.42	1.6292	-4	-4	3	1	0	1
57.30	57.32	1.6066	3	-5	1	3	3	2
57.33		1.6057	-3	5	3	5		6
57.44	57.46	1.6028	0	8	0	2	2	2

2THETA	PEAK	D	H	K	L	I(INT)	I(PK)	I(DS)
57.88	57.88	1.5918	0	0	4	1	1	1
58.45	58.54	1.5777	-5	1	1	2	0	2
58.54		1.5754	-4	-2	4	3	2	3
58.55		1.5739	-1	-1	4	1		1
58.60		1.5739	-3	3	4	1		1
58.79	58.78	1.5693	0	-2	4	3	2	3
59.20	59.20	1.5593	-5	0	3	2	1	2
59.21		1.5591	2	0	5	1		1
59.66	59.66	1.5486	3	5	1	3	1	3
60.02	60.04	1.5400	-4	-3	1	2	1	1
60.09		1.5385	1	5	3	1		2
60.22	60.20	1.5353	2	-2	3	1		1
60.40	60.40	1.5312	-5	-3	3	1	1	2
60.50		1.5289	1	-7	2	1		1
60.82	60.84	1.5216	0	2	4	5	2	6
61.16	61.16	1.5141	-2	-6	4	1	1	1
61.76	61.80	1.5007	2	-6	2	1	2	7
61.81		1.4996	-2	-8	1	5		1
62.16	62.18	1.4921	-5	3	1	1	3	1
62.18		1.4916	-5	3	3	5		5
62.22		1.4908	3	3	2	1		1
62.44	62.34	1.4860	-5	-3	1	1	2	1
62.61	62.74	1.4824	-4	-6	0	6	3	8
62.75		1.4795	-2	8	0	2		2
62.90	62.92	1.4762	-1	-5	4	3	2	4
63.14	63.16	1.4711	0	-8	2	1	1	1
63.38	63.38	1.4662	-4	6	1	6	4	7
63.38		1.4662	-5	-3	3	3		4
63.45		1.4647	0	-4	4	1		2
63.49		1.4640	-4	-4	4	1		1
63.54	63.54	1.4630	0	6	3	1	3	7
63.75		1.4585	2	8	2	5		6
63.84	63.84	1.4569	0	6	0	5	4	6
63.93	63.92	1.4549	-4	2	4	3	4	4
64.93	64.94	1.4349	2	-4	3	3	1	1
65.18		1.4301	-1	-1	1	1	2	2
65.26	65.26	1.4285	-5	1	4	4	2	5

OLIGOCLASE - AN 29 - COLVILLE AND RIBBE, 1966, 1968

2THETA	PEAK	D	H	K	L	I(INT)	I(PK)	I(DS)
65.38	65.50	1.4261	-3	-7	3	1	2	2
65.43		1.4253	-4	-6	3	1		2
65.52		1.4234	-2	-6	2	4		5
66.16	66.20	1.4113	-2	-6	4	1	1	2
66.21		1.4103	1	1	1	2		3
66.51	66.52	1.4047	2	-8	1	1	1	2
66.54		1.4041	-4	6	0	1		2
66.55		1.4040	-1	9	0	1		1
66.82	66.82	1.3989	4	4	1	2	1	3
66.82		1.3988	-3	1	5	1		1
67.13	67.14	1.3932	-5	3	0	1	2	2
67.14		1.3930	1	9	0	3		4
67.61	67.62	1.3844	-1	5	3	5	2	7
67.76	67.76	1.3817	-1	-3	4	5	3	7
68.03	68.12	1.3769	-1	9	1	1	2	1
68.11		1.3754	5	3	0	1		3
68.12		1.3754	4	6	0	2		3
68.15		1.3748	-4	6	3	2		4
68.34	68.34	1.3714	-2	4	3	3	3	3
68.47	68.48	1.3690	-4	0	5	5	3	7
68.69	68.68	1.3652	-1	-9	1	2	2	3
68.88	68.90	1.3620	-2	2	5	1	1	1
69.06	69.10	1.3589	-5	-5	3	1	3	1
69.10		1.3582	-6	0	2	8		10
69.33	69.32	1.3542	-5	5	1	4	4	1
69.33		1.3542	-4	-2	5	4		5
69.43	69.44	1.3524	-1	-9	2	6	4	7
69.45		1.3521	2	8	1	2		2
69.71	69.70	1.3478	4	0	2	7	4	9
69.78		1.3467	-5	-5	1	1		5
69.93	69.92	1.3441	-1	-1	5	4	3	4
70.74	70.78	1.3306	4	-2	2	2		1
70.78		1.3300	-6	-2	2	2		2
70.95	70.96	1.3272	-5	5	5	3	2	4
71.04		1.3258	-2	-4	3	1		1
71.15	71.24	1.3240	-4	2	5	1	2	1
71.23		1.3226	-1	-3	5	4		5

2THETA	PEAK	D	H	K	L	I(INT)	I(PK)	I(DS)
71.29	71.44	1.3217	0	-8	3	2		2
71.43		1.3195	3	1	3	1	2	1
71.46		1.3190	-6	-6	4	1		2
71.75	71.78	1.3144	-4	-6	4	1	1	3
71.80		1.3137	-6	2	3	2		3
73.04	73.06	1.2944	-1	9	3	2	1	1
73.14		1.2928	3	-3	3	1		5
73.66	73.68	1.2850	-4	-4	4	1		1
73.69		1.2845	2	-8	5	4	2	2
73.83	73.88	1.2824	-5	2	0	1		5
73.84		1.2822	2	10	0	1		1
73.88		1.2816	-6	-10	4	2		2
74.43	74.48	1.2735	-3	-5	5	1	1	2
74.44		1.2734	0	0	5	1		7
74.50		1.2726	-1	3	5	1		7
74.54		1.2719	-5	-1	5	1		1
74.83	74.84	1.2677	2	0	4	2	3	3
74.85		1.2675	-4	-8	2	5		3
75.01	75.04	1.2651	0	-2	5	1	2	4
75.42	75.44	1.2592	-3	9	1	1	1	3
75.51		1.2580	2	-2	4	1		1
75.73	75.74	1.2549	-4	-6	6	1	1	2
75.74		1.2548	-4	8	1	1		2
75.88	76.04	1.2527	-6	-4	2	1	2	2
76.04		1.2506	-6	-4	3	3		4
76.19	76.20	1.2484	-6	4	2	3	2	4
76.31		1.2467	-5	-3	5	1		2
76.47	76.46	1.2445	-4	8	2	4	2	6
76.78	76.70	1.2403	0	0	8	1	1	1
76.95	76.98	1.2380	0	6	4	1	1	3
77.20	77.20	1.2346	-4	-4	5	2	1	1
77.23		1.2342	-4	8	3	1		2
78.02	78.04	1.2236	4	4	2	4	1	4
78.32	78.34	1.2198	2	8	0	3	2	6
78.36		1.2192	-6	0	0	1		5
78.57	78.58	1.2164	-6	-4	1	1		2
78.63		1.2157	-2	10	1	1		1

OLIGOCLASE - AN 29 - COLVILLE AND RIBBE, 1966, 1968

2THETA	PEAK	D	H	K	L	I(INT)	I(PK)	I(DS)
78.76	78.78	1.2140	-5	3	5	2		2
78.81		1.2133	-3	-9	3	2		2
78.96	79.00	1.2115	-1	9	2	1	1	2
78.99		1.2111	-2	-10	2	1		1
79.09	79.12	1.2098	3	-5	3	1	1	1
79.14		1.2091	-3	5	5	3		4
79.56	79.54	1.2039	-5	7	1	1	0	1
79.67	79.66	1.2024	-6	2	0	1	0	2
80.41	80.42	1.1932	6	2	0	1	1	1
80.42		1.1931	4	8	0	1		2
81.26		1.1829	-4	-6	5	1		2
81.33	81.28	1.1821	-1	9	3	1	1	1
81.71	81.70	1.1774	-6	4	4	1	0	5
82.09	82.08	1.1730	-3	1	6	3	1	5
82.32	82.32	1.1704	-1	5	5	2	1	2
82.46		1.1687	0	-8	4	1		1
82.94	82.98	1.1631	0	10	2	2	1	2
82.95		1.1629	1	-7	4	4		2
82.98		1.1627	-4	-8	4	1		2
83.13	83.18	1.1610	-3	-3	6	1	2	3
83.17		1.1605	-3	9	3	2		1
83.35	83.38	1.1585	-3	5	5	2	2	2
83.43		1.1575	0	4	6	2		3
83.55	83.56	1.1562	-5	-7	4	2	1	3
83.71		1.1544	-2	8	4	1		1
84.32	84.32	1.1476	-6	4	0	2	1	4
84.51	84.54	1.1455	4	-8	1	1	1	4
84.53		1.1452	-2	6	5	2		4
85.42	85.42	1.1356	-1	-9	4	1	1	2
85.73	85.78	1.1322	-3	-9	4	1	1	3
85.78		1.1317	-6	4	0	2		1
85.92	86.02	1.1303	-1	6	6	2	2	3
86.03		1.1291	-1	-7	2	1		2
86.05		1.1288	-1	-11	4	1		1
86.54	86.54	1.1237	-7	1	4	1	1	2
86.79	86.80	1.1212	-5	1	6	2	1	1

2THETA	PEAK	D	H	K	L	I(INT)	I(PK)	I(DS)
87.91	88.08	1.1097	6	0	1	1	1	1
88.07		1.1082	-5	-3	6	1		3
88.09		1.1079	-4	8	4	2		2
88.24		1.1065	3	-7	3	1		1
88.39	88.36	1.1050	-1	-3	6	1	1	2
89.27	89.30	1.0963	-2	-8	4	1	0	1
89.52	89.54	1.0939	0	8	4	1	0	1
89.83	89.82	1.0909	1	-5	5	1	0	1
90.05	90.08	1.0888	-3	7	5	1	1	1
92.02	92.16	1.0706	-4	-8	5	1		2
92.17		1.0693	-7	1	5	1		1
92.29		1.0682	-5	9	2	1		1
92.64	92.66	1.0650	5	-5	5	1	1	1
92.70		1.0645	2	0	5	1		1
92.79		1.0637	-1	9	4	1		2
92.90		1.0628	2	6	4	1		2
93.70	93.82	1.0558	-7	-5	3	1	1	2
93.78		1.0551	-7	5	2	1		1
93.85		1.0544	-7	-3	5	2		3
94.13	94.14	1.0521	-7	1	4	1	1	1
94.24		1.0511	3	-5	4	1		2
94.62	94.62	1.0479	-5	-5	6	2	1	2
94.66		1.0475	-6	2	6	1		2
95.44	95.46	1.0411	-6	-8	3	2	1	2
95.83	95.80	1.0379	-2	-12	1	1	1	1
95.96		1.0368	-3	-9	5	1		1
96.37	96.36	1.0335	5	5	0	1	0	1
96.61	96.66	1.0316	-5	9	1	1	1	1
96.77		1.0303	-6	-4	6	1		1
96.91	96.90	1.0292	-7	5	4	1	1	1
97.43	97.42	1.0250	-3	-7	6	2	1	3
97.82	97.94	1.0220	-6	8	3	1	1	1
97.95		1.0210	-8	0	3	1		2
98.52	98.52	1.0166	-6	6	5	3	1	3
98.53		1.0165	-6	-8	4	1		1
98.87	98.84	1.0139	-4	0	7	1	1	2
99.23	99.40	1.0112	-4	-2	7	1	1	1

OLIGOCLASE - AN 29 - COLVILLE AND RIBBE, 1966, 1968

2THETA	PEAK	D	H	K	L	I(INT)	I(PK)	I(DS)
99.37		1.0102	5	9	0	1		1
99.41		1.0099	-2	10	4	1		2
99.45		1.0096	-8	-2	3	1		1
99.60		1.0084	2	-12	1	1		1
100.60	100.64	1.0011	-8	-2	4	2	1	3
100.71		1.0003	4	0	4	1		1
100.78		0.9998	-3	-3	7	1		1
101.25	101.24	0.9964	4	-2	4	1	1	2
101.66	101.66	0.9935	-3	-11	4	1	1	1
101.66		0.9935	6	-2	2	1		1
101.72		0.9931	-8	-2	2	1		1
102.68	102.66	0.9864	-5	-7	6	1	0	1
104.12	104.34	0.9767	-8	0	5	1	1	2
104.35		0.9752	-8	-4	3	1		1
104.35		0.9751	-3	9	5	2		3
105.06	105.08	0.9705	-8	4	3	1	1	1
105.09		0.9703	-6	-8	5	1		2
105.48	105.46	0.9678	-8	0	1	1	1	1
106.69	106.96	0.9601	7	1	1	1	1	1
106.77		0.9596	-6	-2	7	1		2
107.00		0.9582	-5	3	7	1		2
108.01	108.04	0.9520	-4	-12	2	1	0	2
108.35	108.50	0.9500	-4	4	7	1	1	1
108.52		0.9489	3	9	3	1		2
108.53		0.9489	-1	1	7	1		1
109.23	109.26	0.9448	-7	-5	6	1	0	1
109.28		0.9445	-4	12	1	1		2
109.70	109.88	0.9420	-7	-7	5	1	1	1
109.90		0.9409	4	4	4	1		2
109.90		0.9409	-3	-9	6	1		2
110.22	110.24	0.9391	-6	6	6	1	0	2

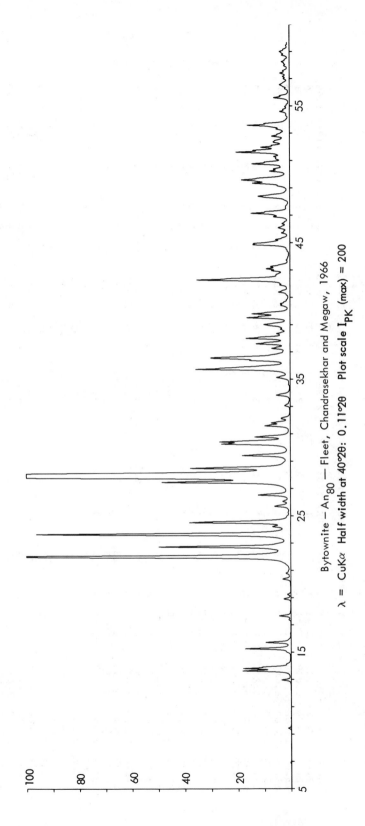

Bytownite – An$_{80}$ – Fleet, Chandrasekhar and Megaw, 1966

λ = CuKα Half width at 40°2θ: 0.11°2θ Plot scale I$_{PK}$ (max) = 200

BYTOWNITE – AN 80 – FLEET, CHANDRASEKHAR, AND MEGAW, 1966

2THETA	PEAK	D	H	K	L	I(INT)	I(PK)	I(DS)
9.41	9.42	9.388	0	-1	1	1	1	1
12.93	12.94	6.839	-1	-1	1	2	2	2
13.62	13.62	6.498	-1	1	0	9	9	9
13.79	13.80	6.417	0	0	2	9	9	8
13.90	13.90	6.364	0	2	0	4	6	4
14.12	14.12	6.265	-1	1	2	1	1	0
15.24	15.24	5.809	-1	-1	2	10	10	9
15.70	15.70	5.641	-1	1	1	5	5	5
17.63	17.64	5.026	-1	-2	1	2	2	2
18.89	18.90	4.694	-1	0	-2	2	1	2
19.21	19.22	4.615	-1	-2	2	1	1	1
20.34	20.36	4.361	0	-2	2	2	2	2
20.77	20.78	4.274	-1	-2	1	1	1	1
21.98	21.98	4.041	-2	0	2	100	81	100
22.71	22.72	3.913	-1	0	3	31	25	31
23.59	23.64	3.767	-1	-3	0	29	48	29
23.64		3.761	1	1	2	37		37
24.20	24.20	3.675	2	0	0	3	3	3
24.49	24.50	3.631	-1	-3	2	23	19	23
24.58		3.618	-1	3	2	5		5
25.67	25.68	3.468	-1	-1	4	3	3	3
25.69		3.465	-1	3	1	1		1
26.04	26.04	3.419	-2	2	1	1	1	1
26.50	26.50	3.360	-1	1	4	8	6	8
27.43	27.44	3.249	-2	2	0	29	24	30
27.88	27.88	3.197	-2	0	4	86	100	88
28.02	28.02	3.182	2	0	0	86	83	88
28.47	28.48	3.132	2	2	0	23	19	23
29.39	29.40	3.036	-1	3	2	12	9	12
30.23	30.24	2.953	0	-4	2	17	13	17
30.40	30.40	2.938	-2	0	4	16	13	16
30.76	30.76	2.905	-2	2	4	9	7	9
31.56	31.56	2.832	-1	3	2	7	5	7
31.78	31.78	2.813	-1	-3	4	4	4	4
32.07		2.788	2	2	1	1	2	1
32.12	32.12	2.784	2	2	1	1	1	1
33.03	33.04	2.709	2	0	2	2	1	2
33.81	33.82	2.649	-1	-3	4	4	3	4
35.01	35.10	2.561	-3	-1	1	1	3	1
35.11		2.554	-2	-1	2	3		3
35.70	35.70	2.513	-2	2	2	20	18	20
35.70		2.513	-2	-4	0	6		7
36.35	36.36	2.469	-2	4	1	8	8	9
36.50	36.54	2.459	-1	1	5	13	15	13
36.56		2.456	-1	5	0	12		13
36.99	36.98	2.428	3	-1	2	1	1	1
37.23	37.24	2.413	-1	-5	0	5	3	5
37.57	37.58	2.392	-1	5	0	9	6	10
37.63		2.388	3	1	0	1		1
37.97	37.96	2.368	2	4	0	13	8	13
38.23	38.32	2.352	-1	-5	2	1	3	1
38.32		2.347	-1	5	2	2		3
38.57	38.66	2.332	-2	0	6	1	2	1
38.65		2.327	-3	-1	2	1		1
38.91	39.00	2.312	-3	-3	4	3	5	4
38.98		2.309	-2	2	6	1		1
39.00		2.308	-3	-4	2	4		4
39.49	39.50	2.280	-3	3	2	12	8	13
39.78	39.78	2.264	-1	1	6	10	7	11
40.39	40.40	2.231	-2	-4	4	2	2	2
40.53	40.52	2.224	-1	-5	4	2	2	2
40.65	40.62	2.217	3	-1	4	1	1	1
41.36	41.36	2.181	-3	-3	6	3	2	3
41.88	41.88	2.155	-1	1	6	2	2	2
42.21	42.26	2.139	-2	-1	6	6	18	6
42.26		2.137	0	-4	0	24		26
42.58	42.58	2.121	2	0	6	1	3	1
42.65	42.64	2.118	-1	3	4	2	3	2
42.93	42.94	2.105	-2	-1	6	4	4	5
43.08	43.06	2.098	-1	5	6	5	5	5
43.25	43.24	2.090	-3	-1	6	4	4	4
43.70	43.70	2.069	0	-4	5	1	1	1
44.82	44.84	2.020	-4	0	4	3	7	3
44.83		2.020	-1	5	4	7		8

BYTOWNITE – AN 80 – FLEET, CHANDRASEKHAR, AND MEGAW, 1966

2THETA	PEAK	D	H	K	L	I(INT)	I(PK)	I(DS)
44.91	44.92	2.017	-4	0	2	5	7	5
45.22	45.22	2.004	3	-1	2	3	3	3
45.69	45.68	1.9841	0	6	2	3	2	2
46.00	46.08	1.9714	0	2	6	1	2	2
46.07		1.9684	3	1	2	2		1
46.37	46.38	1.9563	-2	-2	4	4	2	2
46.87	46.88	1.9368	2	-4	6	4	3	5
47.12	47.12	1.9271	-4	-2	2	2	7	3
47.12		1.9270	-4	2	4	9		10
48.27	48.36	1.8837	-2	6	0	5	6	5
48.32		1.8819	-3	5	2	5		2
48.36		1.8803	2	2	2	6		7
49.13	49.14	1.8528	-3	-5	2	2	2	2
49.33	49.32	1.8459	-4	0	6	11	7	12
49.53	49.56	1.8388	0	-6	4	2	9	2
49.57		1.8375	4	0	0	11		12
49.62		1.8357	-1	-1	6	2		3
50.21	50.20	1.8156	2	6	0	6	4	7
50.40	50.34	1.8089	-2	-6	4	2	3	2
50.72	50.74	1.7984	-3	5	4	1	7	1
50.73		1.7980	1	1	6	8		9
50.76		1.7969	-1	7	0	4		4
51.08	51.12	1.7866	-4	0	0	2	3	2
51.12		1.7851	-4	-5	6	1		1
51.16		1.7841	-4	1	6	1		1
51.28	51.28	1.7800	-1	-7	2	2	3	2
51.59	51.60	1.7700	-2	0	8	18	10	20
51.78	51.72	1.7640	-4	-4	6	8	7	9
51.95	51.96	1.7586	2	-4	4	10	5	9
52.32	52.32	1.7483	-2	6	4	1	3	2
52.32		1.7470	4	2	0	2		3
52.37		1.7455	-1	7	2	1		1
52.44		1.7434	-2	-2	8	1		1
52.75	52.76	1.7338	1	5	8	3	3	3
52.76		1.7336	1	-3	6	2		2
52.80		1.7324	2	-6	2	1		1
53.28	53.28	1.7178	0	6	4	3	3	3
53.45	53.56	1.7127	0	0	6	1	8	1
53.55		1.7097	-4	-4	4	7		8
53.56		1.7096	-4	4	4	7		7
54.30	54.30	1.6879	-1	1	8	6	1	1
54.57	54.58	1.6801	-2	4	8	3	2	3
55.59	54.58	1.6519	2	4	4	5	3	6
56.24	56.24	1.6343	-1	-3	8	1	1	1
56.52	56.54	1.6267	-3	-3	8	1	1	2
56.55		1.6259	1	7	2	1	2	1
56.97	56.98	1.6150	-4	-4	6	6	1	5
57.13	57.14	1.6108	-3	5	0	1	1	2
57.39	57.38	1.6043	0	8	0	3	2	3
57.61	57.54	1.5985	-4	0	8	1	1	1
57.91	57.92	1.5910	0	0	8	1	1	2
58.21	58.22	1.5834	0	-8	8	2	2	2
58.70	58.72	1.5714	0	-2	8	2	1	2
58.98	58.98	1.5647	-4	-2	6	2	2	2
59.00		1.5642	2	0	6	6		1
59.44	59.42	1.5537	-3	-1	9	1	1	1
59.78	59.80	1.5455	2	-2	6	1	1	1
59.95	60.04	1.5416	1	-7	4	1	1	1
60.03		1.5399	3	5	2	1		1
60.11	60.20	1.5378	-4	2	8	8	1	2
60.20		1.5358	-2	-6	6	1		1
60.99	60.98	1.5179	-2	-6	4	1	1	1
61.29	61.30	1.5112	-2	4	8	3	2	3

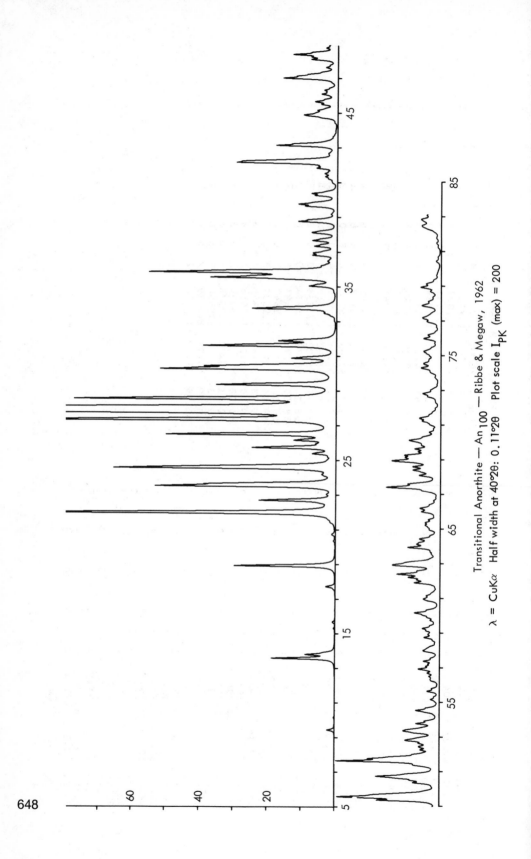

Transitional Anorthite — An$_{100}$ — Ribbe & Megaw, 1962

λ = CuKα Half width at 40°2θ: 0.11°2θ Plot scale I$_{PK}$ (max) = 200

648

2THETA	PEAK	D	H	K	L	I(INT)	I(PK)	I(DS)
9.42	9.42	9.382	0	-1	1	1	1	1
13.57	13.58	6.522	-1	1	0	10	9	9
13.78	13.80	6.419	-1	0	0	4	4	4
17.70	17.70	5.006	-1	-2	1	2	1	1
18.90	18.90	4.691	0	-2	1	17	15	16
20.71	20.72	4.286	-1	0	2	1	1	1
21.98	21.98	4.041	1	-2	0	58	50	55
22.68	22.68	3.918	-1	-1	2	13	11	12
23.01	23.02	3.861	-1	1	2	30	27	29
23.51	23.52	3.780	-1	-2	0	17	19	16
23.64	23.64	3.760	-1	1	2	1	2	1
24.10		3.690	-1	2	3			
24.18	24.18	3.677	-2	0	0	38	33	37
24.56	24.56	3.622	-1	3	0	1	1	1
25.17	25.18	3.535	-2	1	3	4	4	4
25.36	25.36	3.509	-1	-1	3	15	12	14
25.71	25.72	3.462	-1	-1	1	2	3	2
25.93	25.92	3.433	-2	2	2	6	6	6
26.13	26.14	3.407	-2	-2	2	31	25	30
26.48	26.48	3.363	-1	-1	4	64	50	63
27.33	27.34	3.261	-2	2	0	100	100	100
27.89	27.90	3.196	-2	0	4	92	97	92
28.02	28.00	3.182	0	0	4	48	39	48
28.54	28.54	3.125	2	2	0	23	18	23
29.34	29.34	3.042	1	-3	2	33	26	34
30.25	30.26	2.952	0	-4	4	21	19	21
30.42	30.42	2.936	0	-2	2	8	7	8
30.86	30.86	2.895	-2	-2	4	26	19	27
31.59	31.60	2.830	-1	-3	3	10	9	11
31.87	31.88	2.806	-1	1	4	2	2	2
33.46	33.46	2.676	-3	1	2	17	12	18
33.73	33.74	2.655	2	-2	2	5	4	6
35.03	35.02	2.560	-1	-1	4	7	19	7
35.44	35.54	2.530	1	-1	4	21	27	23
35.54		2.524	-2	4	2	1		1
35.82	35.84	2.505	-3	1	4	37		40
35.85		2.503	-2	-4	2			
36.46	36.20	2.479	-2	4	0	2	2	2
36.82	36.46	2.462	-1	5	1	1	1	1
36.96	36.82	2.440	-2	2	2	5	4	5
37.33	36.94	2.430	-3	1	0	3	4	4
37.64	37.32	2.407	-1	-1	2	5	3	5
37.66	37.66	2.388	-1	-5	0	2		2
38.08		2.387	3	1	0	3		3
38.12	38.12	2.361	2	1	0	3	4	4
38.75	38.74	2.359	-1	5	2	8	6	9
38.85	38.84	2.322	-3	3	2	1		1
39.16	39.16	2.316	-1	-1	6	1	1	1
39.63	39.64	2.299	-1	-3	4	6	5	7
39.76	39.76	2.272	-2	-3	2	5	6	6
39.83		2.265	-3	1	6	1		1
40.25	40.26	2.261	-1	-1	4	4	3	5
40.37	40.36	2.238	-1	-3	2	3	4	4
40.96	40.98	2.232	-1	-5	2	1	1	1
41.21	41.22	2.201	-2	4	4	1	1	2
41.33	41.32	2.189	-3	-3	4	2	2	1
41.50	41.50	2.183	-3	-3	4	4	2	2
41.81	41.82	2.174	0	4	0	19	3	4
42.13	42.14	2.159	-3	3	6	9	15	22
42.20	42.20	2.143	-1	-2	2	1	14	11
42.27		2.140	-2	-4	4	14		
42.65	42.64	2.136	-2	-6	0	3	1	1
43.12	43.12	2.118	-4	-1	4	5	9	16
43.21	43.22	2.096	1	-5	4	3	7	2
44.82	44.88	2.092	1	5	2	3	5	3
44.89		2.021	-3	1	6	3		6
45.15	45.14	2.018	-4	0	4	4	3	3
45.49	45.50	2.007	-4	0	2	1	2	4
45.65	45.64	1.9923	3	-1	2	11	3	5
45.97	46.08	1.9854	-1	-3	6		3	1
46.08		1.9724	0	6	2			
46.31	46.30	1.9683	0	2	6		2	3
46.99	47.00	1.9589	3	-1	2		8	14
		1.9320	-4	2	4			

649

TRANSITIONAL ANORTHITE – AN 100 – RIBBE AND MEGAW, 1962

2THETA	PEAK	D	H	K	L	I(INT)	I(PK)	I(DS)
47.03		1.9306	-2	-4	6	1		1
47.04	47.22	1.9300	-3	-3	6	2		2
47.23	47.62	1.9227	-4	-2	4	2	4	6
47.63		1.9077	-4	-2	2	2	2	3
48.09	48.10	1.8905	-3	5	2	2	4	5
48.09		1.8902	-2	6	0	4		2
48.26		1.8843	-2	-6	2	2	4	5
48.38	48.24	1.8798	2	2	4	9	7	11
48.75	48.38	1.8662	3	-3	2	2	2	2
48.76	48.76	1.8458	-3	-5	2	1	9	1
49.33	49.34	1.8455	-4	0	6	2		14
49.34		1.8449	-2	4	6	11		3
49.36		1.8387	-4	0	4	18	15	23
49.53	49.54	1.8377	0	-6	0	3		3
49.56		1.8363	2	-1	6	4		4
49.60		1.8110	1	-1	0	5	4	7
50.34	50.34	1.8061	3	5	4	2	4	2
50.49	50.50	1.8004	-1	7	0	4	9	5
50.66		1.7983	-1	1	6	10	4	13
50.72		1.7799	-4	-2	6	2		1
51.28	51.38	1.7777	3	3	2	2		3
51.35		1.7767	-1	-7	2	2		2
51.38	51.62	1.7694	-2	0	8	25	15	32
51.61		1.7676	2	2	6	2		1
51.67	51.74	1.7617	-2	-4	4	8	11	10
51.85	51.98	1.7543	-2	6	4	1	4	2
52.09	52.24	1.7492	-1	7	2	3	3	4
52.25	52.44	1.7450	4	2	0	1		1
52.39		1.7430	1	5	4	3		4
52.44		1.7370	2	-6	2	2	3	3
52.45	52.64	1.7343	1	-3	6	1	5	1
52.65	52.80	1.7327	-4	4	2	4		6
52.74		1.7309	-2	-2	8	3		4
52.79		1.7192	0	6	4	4	5	5
52.85	53.34	1.7165	-4	4	4	4		6
53.23		1.7160	1	-7	2	1		1
53.33		1.7139	0	4	6	3		4
53.77	53.76	1.7034	-4	-4	4	5	3	7
54.42	54.52	1.6845	-4	-4	2	2	3	2
54.52		1.6817	-2	2	8	5		6
55.04	55.16	1.6670	-1	-7	1	1	1	1
55.16		1.6638	-3	-5	6	2		2
55.63	55.64	1.6506	2	4	4	3	2	4
56.24	56.32	1.6342	3	-5	0	1	1	2
56.32		1.6321	-1	3	8	1		2
56.50	56.48	1.6275	2	6	2	1	2	2
56.69	56.68	1.6224	-3	-3	8	5	2	3
56.92	56.92	1.6164	-3	5	6	2	3	7
57.20	57.20	1.6090	-4	-4	6	2	2	2
57.37	57.36	1.6047	0	8	0	3	2	4
57.93	57.96	1.5905	4	0	2	2	2	3
57.98		1.5892	-5	-1	2	2		3
58.54	58.58	1.5755	-5	3	8	1	1	1
58.58		1.5745	-3	3	8	1		1
58.73	58.74	1.5708	0	-2	8	2		3
58.97	58.90	1.5648	2	0	6	1	2	1
59.21	59.12	1.5614	-4	-2	8	2	2	2
59.26	59.26	1.5592	-1	3	8	1	2	1
59.73		1.5579	3	-3	4	3	3	4
59.74	59.74	1.5468	2	-2	6	2	1	2
59.90	59.90	1.5429	1	-7	4	1	2	2
60.15	60.16	1.5370	3	5	4	4	4	5
60.16		1.5368	-5	3	4	1		2
60.86	60.94	1.5208	2	-6	4	2	2	3
60.94		1.5190	0	2	8	1		3
60.96		1.5186	-5	-3	4	2		3
61.16	61.14	1.5139	-2	4	8	2	2	2
61.26		1.5117	-5	3	2	1		1
61.88	61.88	1.4980	-2	8	0	6	4	9
62.02	62.06	1.4951	2	2	6	1	3	1
62.19	62.38	1.4913	-4	6	4	6	5	9
62.37		1.4876	3	3	4	2	6	3
62.39		1.4871	-2	-8	2	5		8
62.45		1.4858	-3	7	4	1		1

TRANSITIONAL ANORTHITE - AN 100 - RIBBE AND MEGAW, 1962

2THETA	PEAK	D	H	K	L	I(INT)	I(PK)	I(DS)
62.86	62.90	1.4772	-5	-3	6	4	7	6
62.86		1.4771	-5	-3	2	1		6
62.89		1.4766	-5	3	6	3		2
62.91		1.4761	0	-8	4	2		5
62.95		1.4753	-4	6	4	6		3
63.54	63.54	1.4628	-4	-6	4	1	1	8
63.65	63.72	1.4608	-4	4	2	1	2	2
63.93	63.92	1.4550	0	6	6	8	5	1
64.20	64.10	1.4494	2	-4	0	4	3	13
64.43	64.40	1.4449	2	8	0	3	3	5
64.93	64.96	1.4350	3	7	0	2	2	5
64.98		1.4339	-1	-1	8	1		2
65.06	65.20	1.4323	-5	-1	8	1	2	1
65.19		1.4297	-4	6	0	2		3
65.34	65.34	1.4270	-5	1	8	2	2	4
65.53	65.52	1.4233	2	-8	2	1	2	2
65.97	65.98	1.4148	2	6	4	4	3	6
66.09	66.18	1.4126	-5	3	0	1	3	1
66.19		1.4107	1	1	8	2		2
66.19		1.4106	-3	-7	6	1		2
66.61	66.60	1.4028	-2	-6	8	1	1	1
67.38	67.40	1.3885	-4	6	6	4	8	6
67.39		1.3884	1	-3	8	7		11
67.44		1.3876	1	9	0	5		7
67.66	67.58	1.3835	-1	9	2	1	5	1
68.05	68.08	1.3765	-5	5	2	1	3	1
68.08		1.3760	1	-9	0	5		7
68.39	68.38	1.3706	-5	-3	8	2	4	7
68.39		1.3705	5	3	0	4		3
68.58	68.58	1.3673	2	4	6	4	4	6
68.58		1.3620	-4	0	10	4	4	6
68.88	68.92	1.3618	4	6	0	3	7	4
68.89		1.3610	-6	0	4	9		13
68.94		1.3556	-4	0	4	8	5	13
69.25	69.24	1.3495	-1	-9	4	6	4	9
69.61	69.60	1.3434	-4	-2	10	2	5	3
69.97	70.10	1.3421	2	8	2	2		3
70.04								

2THETA	PEAK	D	H	K	L	I(INT)	I(PK)	I(DS)
70.10		1.3413	-5	5	6	2		4
70.11		1.3410	-1	-1	10	3		5
70.56	70.56	1.3336	3	5	4	1		2
70.56		1.3335	-5	-5	2	1	2	1
71.09	71.22	1.3250	-6	-2	4	1		2
71.21		1.3230	0	-6	8	1		2
71.22		1.3229	3	1	6	2		2
71.36	71.38	1.3206	-6	2	6	1	3	4
71.38		1.3203	-1	-3	10	3		1
71.46		1.3190	-3	-3	10	1		1
72.28	72.30	1.3060	3	3	4	3		4
72.62	72.64	1.3007	2	-8	8	1		2
72.75	72.80	1.2987	-4	-6	2	3	3	1
72.78		1.2983	-6	2	4	1		3
72.80		1.2979	-1	9	2	2		2
73.10	72.98	1.2933	-2	-8	6	1	2	1
73.74	73.74	1.2838	0	10	0	1		6
74.00	74.10	1.2798	4	-4	4	1		2
74.09		1.2785	-6	0	8	1		2
74.24	74.26	1.2763	0	-10	8	3	3	1
74.26		1.2761	-3	9	2	1		5
74.30		1.2755	-4	8	2	1		2
74.51	74.50	1.2724	-2	-4	10	2	3	2
74.58		1.2714	-2	0	4	2		4
75.09	75.24	1.2639	-6	4	4	2	3	4
75.18		1.2626	-4	6	8	1		5
75.24		1.2618	-4	8	4	3		2
75.69	75.70	1.2554	-6	-2	8	1	1	1
75.94	75.98	1.2519	-1	-9	6	1	3	7
75.97		1.2514	-4	-8	4	4		2
76.14	76.18	1.2491	-3	-9	4	1	2	2
76.93	76.94	1.2383	-6	-4	6	3	2	5
76.97		1.2377	0	8	6	1		1
77.10	77.16	1.2359	-4	4	10	2	3	4
77.18		1.2349	0	6	8	1		1
77.18		1.2348	-5	-3	10	1		2
77.27		1.2336	-6	4	2	1		2

TRANSITIONAL ANORTHITE - AN 100 - RIBBE AND MEGAW, 1962

2THETA	PEAK	D	H	K	L	I(INT)	I(PK)	I(DS)
77.81	77.86	1.2265	-2	10	2	1	3	2
77.86		1.2258	-6	0	0	4		7
77.93		1.2249	-5	7	2	1		2
77.93		1.2248	3	-5	6	1		1
78.31	78.32	1.2199	4	4	4	1	1	2
78.39		1.2189	-4	-8	6	1		2
78.67	78.68	1.2151	-6	2	0	2	2	3
78.68		1.2151	-5	3	10	1		1
78.91	78.92	1.2121	2	8	4	5	3	9
79.08	79.10	1.2098	-3	5	10	2	2	3
79.73	79.60	1.2017	-3	-9	6	1	1	1
80.41	80.40	1.1933	6	2	0	1	1	3
81.23	81.24	1.1833	-1	9	6	1	1	1
81.45	81.46	1.1806	4	8	0	1	1	1
81.75	81.74	1.1771	-6	0	10	1	1	1
82.30	82.44	1.1706	-4	-6	10	1	3	2
82.39		1.1695	-3	9	6	3		5
82.46		1.1687	-3	1	12	2		3
82.50		1.1682	-1	5	10	1		2
82.73	82.82	1.1655	4	-8	2	1	3	2
82.83		1.1644	-6	4	0	4		7
82.85		1.1642	2	-8	6	1		2
83.05	83.06	1.1619	0	10	4	1	2	2
83.50	83.52	1.1567	-2	8	8	1	1	2
83.65	83.72	1.1550	0	4	10	1	2	2
83.72		1.1542	3	5	6	2		4
84.56	84.58	1.1449	-2	6	10	3	2	5
84.61		1.1443	-6	6	6	2		3
84.90	84.86	1.1412	-5	-7	8	1	1	3
85.37		1.1361	-6	-6	4	1	1	1
85.48	85.48	1.1349	-1	-9	8	2		3

TABLE 33. PLAGIOCLASE AND OTHER RELATED SILICATES

Variety	Anorthite	Celsian (Barium-Feldspar)	Paracelsian	Reedmergnerite
Composition	Ab_0An_{100}	$(Ba_{0.84}K_{0.18})Al_2Si_2O_8$	$\alpha-BaAl_2Si_2O_8$	$NaBSi_3O_8$
Source	Monte Somma, Italy	Broken Hill, N.S.W. Australia	not given	Duchesne Co., Utah
Reference	Megaw, Kempster & Radoslovich, 1962	Newnham & Megaw, 1960	Bakakin & Belov, 1961	Appleman & Clark, 1965
Cell Dimensions				
a Å	8.1768	8.627	9.08	7.833
b Å	12.8768	13.045	9.58	12.360
c Å	14.1690	14.408	8.58	6.803
α	93.17	90	90	93.31
β deg	115.85	115.217	~90	116.35
γ	91.22	90	90	92.05
Space Group	P$\bar{1}$	I2/c	P2$_1$/a	C$\bar{1}$
Z	8	8	4	4

TABLE 33. (cont.)

Variety			Anorthite	Celsian (Barium-Feldspar)	Paracelsian	Reedmergnerite
Site (% Si) Occupancy						
$T_1(0)$		(0000)	100	$B_1(0)$ 79		B 100
		(00i0)	100			
		(0z00)	0			
		(0zi0)	0			
$T_1(m)$		(m000)	0	$B_1(z)$ 23		100
		(m0i0)	100			
		(mz00)	100			
		(mzi0)	0			
$T_2(0)$		(0000)	0	$B_2(0)$ 27		100
		(00i0)	100			
		(0z00)	100			
		(0zi0)	0			
$T_2(m)$		(m000)	100	$B_2(z)$ 82		100
		(m0i0)	100			
		(mz00)	0			
		(mzi0)	0			
Method			3-D, film	3-D, film	3-D (?), film	3-D, film
R & Rating			0.111 (2)	0.069 (1)	--- (3)	0.109 (2)
Cleavage and habit			{001}{010}perfect Variable	{001}perfect{010}good Variable	{110}	{001} perfect Stubby, prismatic elongated to [$\bar{1}$10]
Comment				Site occupancy from Taylor, 1962	Similar but not isoptypic with danburite B estimated	Isotypic with low albite
μ			139.7	390.1	465.8	82.4
ASF			0.4578×10^{-1}	0.1594	0.2263	0.4678×10^{-1}
Abridging factors			0.5:0:0	0.5:0:0	1:0:0	0.5:0:0

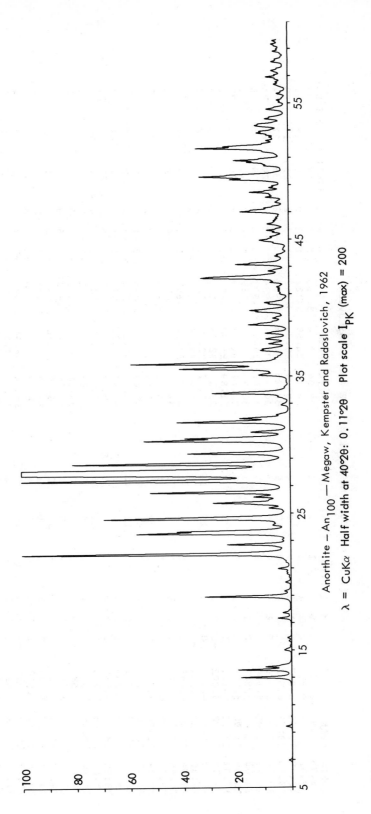

Anorthite — An$_{100}$ — Megaw, Kempster and Radoslovich, 1962

λ = CuKα Half width at 40°2θ: 0.11°2θ Plot scale I$_{PK}$ (max) = 200

ANORTHITE – AN 100 – MEGAW, KEMPSTER AND RADOSLOVICH, 1962

2THETA	PEAK	D	H	K	L	I(INT)	I(PK)	I(DS)
6.94	6.94	12.718	0	0	1	1	1	0
9.42	9.42	9.379	0	-1	1	1	1	1
12.99	13.00	6.807	-1	-1	0	9	9	8
13.56	13.58	6.522	-1	1	0	10	10	9
13.78	13.78	6.420	0	-2	0	4	5	4
14.99	15.00	5.904	0	0	2	1	1	1
15.08		5.869	0	-2	1	1	1	1
15.89	15.90	5.573	0	2	1	1	1	1
17.31	17.32	5.118	-1	-1	1	3	3	3
17.72	17.72	5.002	-1	1	1	2	2	2
18.91	18.92	4.690	-1	-2	1	17	16	16
19.30	19.30	4.594	-1	-1	2	1	1	1
19.94	19.94	4.448	-1	1	2	1	1	1
20.56	20.56	4.316	-1	-1	3	1	1	1
20.70	20.72	4.286	-1	1	2	1	1	1
20.94	20.94	4.239	0	0	3	2	2	2
21.99	22.00	4.039	-2	0	2	58	52	55
22.69	22.70	3.916	-1	-1	2	13	12	12
23.03	23.04	3.859	-1	1	3	1	1	1
23.50	23.50	3.783	-1	3	0	30	29	29
23.67	23.66	3.756	-1	2	2	17	21	16
24.10	24.10	3.690	-1	0	3	3	3	1
24.10		3.689	-2	0	3	2		2
24.29	24.30	3.661	0	-2	3	1	3	1
24.57	24.58	3.620	1	-3	0	38	35	37
25.17	25.18	3.535	-2	1	3	1		1
25.35	25.36	3.510	-1	3	2	4	4	4
25.73	25.74	3.459	-1	-1	4	15	14	15
25.92	25.92	3.434	-2	2	2	2	4	2
25.98		3.426	0	2	3	1		1
26.16	26.16	3.403	-2	-2	2	6	7	6
26.49	26.50	3.362	-1	1	4	31	26	31
27.32	27.32	3.261	-2	2	0	64	53	64
27.39		3.253	1	3	1	1		1
27.72	27.78	3.215	-2	-2	3	1		1
27.77		3.210	0	4	0	53	58	53
27.91	27.92	3.194	-2	0	4	100	100	100

2THETA	PEAK	D	H	K	L	I(INT)	I(PK)	I(DS)
28.04	28.04	3.180	0	0	4	93	100	93
28.57	28.58	3.122	2	0	0	49	41	49
29.34	29.34	3.042	1	-3	2	23	19	23
30.25	30.26	2.952	0	-1	4	33	27	34
30.44	30.44	2.934	0	-2	4	20	20	21
30.89	30.90	2.893	-2	2	1	8	7	8
30.93		2.889	1	3	1	1		1
31.61	31.62	2.828	-1	-3	3	26	21	27
31.90	31.90	2.804	-1	3	1	11	9	11
32.16	32.16	2.781	2	-1	4	2	3	2
32.87	32.88	2.723	-2	-1	5	2	2	2
33.47	33.48	2.675	-3	3	4	2	2	2
33.73	33.74	2.655	0	1	5	18	14	19
33.76		2.653	-1	3	3	1		1
34.20	34.20	2.620	0	-4	3	1	1	2
35.04	35.06	2.559	-2	-3	2	5	6	6
35.08		2.556	2	-2	2	1		2
35.26	35.26	2.543	-3	1	5	2	2	2
35.47		2.529	-2	-1	4	7	4	7
35.52	35.52	2.525	2	4	2	21	21	23
35.83		2.504	-3	3	5	1		1
35.88	35.88	2.501	-2	-4	4	37	30	40
36.31	36.30	2.472	-1	1	5	1		2
36.44	36.42	2.463	-2	5	1	1	3	1
36.80	36.86	2.440	-1	-5	1	1	2	1
36.84		2.437	2	-1	5	1		1
36.87		2.436	-2	-2	2	5	5	5
36.98	36.96	2.429	-1	3	0	1		1
37.34	37.34	2.406	-1	-1	5	3	5	4
37.65		2.387	-1	5	2	5	4	5
37.70	37.70	2.384	-1	1	0	3	4	2
38.10	38.10	2.360	3	-1	1	3	4	3
38.11		2.359	-2	4	2	9		4
38.73	38.74	2.323	-3	3	3	9	8	3
38.75		2.322	-3	-1	5	1		10
38.98	38.96	2.309	0	-2	5	2	3	2

ANORTHITE - AN 100 - MEGAW, KEMPSTER AND RADOSLOVICH, 1962

2THETA	PEAK	D	H	K	L	I(INT)	I(PK)	I(DS)
39.19	39.22	2.297	-2	-4	4	1	3	1
39.22		2.295	-1	-5	3	3		3
39.48	39.50	2.280	-3	-1	1	1	2	1
39.67	39.68	2.270	-3	-3	2	6	6	7
39.69		2.269	-3	-3	1	1		1
39.78	39.78	2.264	-1	1	6	5	7	6
39.85		2.260	-1	-3	4	1		1
40.24	40.24	2.239	1	-5	2	4	4	5
40.36	40.36	2.233	-2	4	4	3	5	4
40.95	41.00	2.202	-3	-3	1	1	2	1
41.00		2.199	-3	1	4	1		1
41.14	41.18	2.192	2	-4	1	1	2	2
41.19		2.190	-3	3	4	1		1
41.34	41.32	2.182	0	4	4	1	2	1
41.50	41.50	2.174	-3	0	3	1	2	2
41.82	41.82	2.158	-2	-4	6	4	4	4
42.13	42.14	2.143	2	-4	2	19	16	22
42.19		2.140	0	-6	0	10		11
42.65	42.66	2.118	-2	-4	5	3	4	4
42.69		2.116	1	3	4	1		
43.13	43.14	2.095	1	5	2	14	10	16
43.22	43.24	2.091	-3	-1	6	2	7	2
43.75	43.76	2.067	0	-4	5	3	3	3
43.88	43.88	2.061	-1	-5	3	1	2	1
44.22	44.22	2.047	-3	-3	5	1	1	1
44.65	44.72	2.028	3	-3	1	1	3	1
44.71		2.025	-1	-3	4	3		3
44.84	44.92	2.019	-4	0	4	3	5	3
44.92		2.016	-4	0	2	3		6
45.18		2.005	3	-1	2	3		6
45.50	45.50	1.9919	-1	-3	6	3	3	4
45.65	45.64	1.9857	0	6	2	4	4	4
46.00	46.10	1.9712	1	-3	5	2	4	5
46.07		1.9683	1	1	6	1		1
46.08		1.9682	0	-6	3	1		2
46.12		1.9664	3	1	2	3		4
46.34	46.34	1.9578	2	-2	4	3	3	3
46.44	46.44	1.9536	-2	-2	7	1	3	1
46.57	46.58	1.9486	-4	0	5	1	2	1
46.61		1.9468	2	-4	7	1		1
46.91	47.00	1.9353	-1	1	7	1	9	1
46.99		1.9319	-4	2	4	11		14
47.06		1.9293	-2	-4	6	1		1
47.08		1.9284	-3	-3	6	1		2
47.28	47.28	1.9209	-4	-2	5	5	5	6
47.39	47.38	1.9165	-2	6	1	1	3	1
47.68	47.68	1.9059	-4	-2	2	2	2	3
47.73		1.9039	-3	-1	7	2		1
48.06	48.06	1.8916	-3	5	2	2	4	2
48.06		1.8914	-2	6	0	4		5
48.43	48.42	1.8781	2	2	4	9	7	12
48.77	48.78	1.8658	-3	-3	5	2	4	3
48.78		1.8653	-4	-2	5	1		2
48.91	48.90	1.8607	-2	-6	3	1		1
49.35	49.36	1.8451	-2	4	6	2		3
49.36		1.8446	-4	0	6	11	11	14
49.37		1.8442	-3	-5	2	1		1
49.48	49.58	1.8406	-1	-3	7	1	17	1
49.56		1.8376	0	-6	4	3		4
49.57		1.8373	4	0	0	18		23
49.64		1.8349	-1	-1	6	4		4
50.37	50.38	1.8100	2	-6	0	5	5	7
50.62	50.62	1.8016	-2	-6	4	5	8	6
50.63		1.8013	-1	7	0	4		6
50.76	50.76	1.7969	-1	1	6	10	10	13
51.33	51.42	1.7785	-4	-2	6	1	5	1
51.39		1.7765	-1	-7	2	2		2
51.41		1.7759	3	3	2	2		3
51.45		1.7744	-2	-4	1	1		1
51.64	51.64	1.7684	-2	0	8	25	17	33
51.67		1.7674	-4	2	6	2		1
51.78	51.78	1.7613	2	-4	4	8	12	10
51.87		1.7552	-2	6	4	1	5	2
52.06	52.00	1.7552	2	6	5	4		2
52.15	52.22	1.7523	-1	3	7	1	4	2

ANORTHITE - AN 100 - MEGAW, KEMPSTER AND RADOSLOVICH, 1962

2THETA	PEAK	D	H	K	L	I(INT)	I(PK)	I(DS)
52.22		1.7501	-1	7	2	3		4
52.25		1.7493	-2	-4	7	1		1
52.44	52.48	1.7433	4	2	0	1	3	1
52.48		1.7422	1	5	4	3		4
52.63	52.64	1.7376	-1	-6	2	3	4	3
52.72	52.78	1.7349	-1	-7	3	1	6	2
52.77		1.7333	-1	-3	6	1		1
52.77		1.7331	-4	4	2	4		6
52.89	52.88	1.7297	-2	-2	8	3	5	4
53.24	53.32	1.7192	0	6	4	4	6	5
53.31		1.7171	-4	4	4	4		6
53.32		1.7166	1	-7	2	1		1
53.35		1.7157	-2	-6	5	1		1
53.43	53.42	1.7132	0	4	6	6	6	4
53.82	53.82	1.7017	-4	-4	4	6	4	7
53.93	53.94	1.6988	0	-6	5	2	3	2
54.48	54.54	1.6828	-4	-4	2	2	4	3
54.54		1.6812	-2	2	8	5		7
55.05	55.04	1.6668	-4	-4	5	1		1
55.05		1.6666	-1	-7	4	1		1
55.20	55.20	1.6625	-3	-5	6	2	2	2
55.68	55.68	1.6493	2	4	4	3	2	4
56.23	56.34	1.6344	-5	5	2	2	3	2
56.36		1.6317	2	-6	3	1		2
56.53	56.52	1.6310	-1	-3	8	1		2
56.74	56.74	1.6265	2	6	2	2	2	2
56.89	56.90	1.6211	-3	-3	8	2	2	3
57.26	57.36	1.6170	-3	5	6	5	4	7
57.36		1.6076	-4	-4	6	2	3	2
57.52	57.50	1.6051	0	8	0	3		4
57.98	58.00	1.6009	-5	-1	5	1	2	1
58.01		1.5892	4	0	2	2	3	3
58.21	58.16	1.5884	-5	2	2	2		3
58.58	58.60	1.5835	0	-8	2	1	2	2
58.59		1.5744	-3	3	8	1	1	1
58.77	58.78	1.5742	-5	-1	2	1		1
		1.5697	0	-2	8	2	3	4

2THETA	PEAK	D	H	K	L	I(INT)	I(PK)	I(DS)
59.03	58.94	1.5635	2	0	6	1	2	1
59.16	59.18	1.5603	-4	-2	8	1	3	3
59.23	59.30	1.5586	-1	3	8	1	3	1
59.29		1.5572	3	-3	4	3		4
59.56	59.56	1.5509	-3	-1	9	3	3	2
59.77	59.76	1.5458	3	-2	6	2	2	1
59.89	59.90	1.5431	-1	-3	7	1	3	2
59.89		1.5431	1	-7	4	2		3
60.03	60.04	1.5398	-4	2	8	2	3	3
60.16	60.20	1.5368	-5	3	5	1	4	6
60.21		1.5356	3	5	2	4		1
60.29		1.5337	-3	-1	9	1		2
60.98	61.00	1.5180	0	2	8	1	4	3
61.00		1.5177	-1	-1	9	2		1
61.02		1.5171	-5	-3	4	2		3
61.17	61.16	1.5138	-2	4	8	1	3	2
61.27		1.5115	-5	3	2	1		1

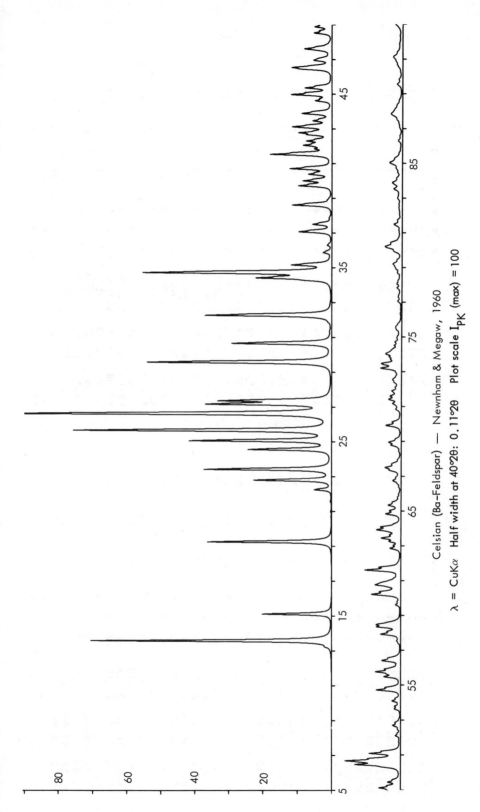

Celsian (Ba-Feldspar) — Newnham & Megaw, 1960

λ = CuKα Half width at 40°2θ: 0.11°2θ Plot scale I$_{PK}$ (max) = 100

CELSIAN (BA-FELDSPAR) — NEWNHAM AND MEGAW, 1960

2THETA	PEAK	D	H	K	L	I(INT)	I(PK)	I(DS)
13.21	13.24	6.698	1	1	0	1	2	0
13.56	13.58	6.522	0	0	2	39	71	21
13.57		6.517	0	2	0	20		11
15.10	15.10	5.864	-1	1	1	17	20	10
19.24	19.24	4.610	0	2	1	32	36	23
22.22	22.22	3.997	1	1	2	5	5	4
22.77	22.78	3.902	2	0	0	21	23	18
23.40	23.40	3.799	-1	3	0	35	37	31
24.54	24.54	3.625	1	3	0	23	24	21
25.04	25.04	3.553	-2	2	1	40	42	38
25.64	25.64	3.472	-1	1	2	74	76	71
26.59	26.60	3.349	2	0	2	100	100	100
27.15	27.16	3.282	-2	0	2	35	37	36
27.34	27.34	3.259	0	0	4	30	33	31
29.55	29.56	3.021	1	3	2	58	54	66
30.63	30.64	2.916	0	4	2	29	29	34
30.63		2.915	0	0	4	3	3	4
32.24	32.24	2.774	-1	3	4	41	37	52
34.39	34.38	2.606	-3	1	4	21	22	29
34.68	34.70	2.584	-2	4	2	43	55	59
34.71		2.582	1	1	1	22		30
35.14	35.14	2.551	3	1	0	12	12	17
35.85	35.86	2.502	2	4	0	3	3	4
36.27	36.28	2.474	-1	5	0	1	1	2
37.05	37.04	2.425	-3	5	0	11	10	17
37.47	37.48	2.398	-3	3	2	6	6	10
37.82	37.82	2.377	-2	0	6	1	1	1
38.58	38.58	2.332	-1	1	6	14	12	22
38.90	38.88	2.313	-2	4	4	1		1
39.69	39.70	2.269	-3	3	4	11	10	18
39.98	39.98	2.253	-1	4	6	9	8	15
40.36	40.36	2.233	-2	2	0	3	7	6
40.36		2.233	3	3	0			7
40.68	40.68	2.216	-1	5	4	14	12	24
41.50	41.50	2.174	0	6	0	21	18	37
41.53		2.172	0	0	6	1		1
41.82	41.82	2.158	-3	1	4	4	5	8
42.03	42.02	2.148	-2	4	2	9	8	15
42.28	42.28	2.136	-4	0	2	8	8	14
42.64	42.64	2.118	-4	0	4	7	7	13
42.76	42.76	2.113	-1	5	4	7	10	13
43.04	43.10	2.100	-2	0	2	3	12	5
43.10		2.097	3	1	2	12		22
43.45	43.46	2.081	-1	3	6	6	6	11
43.89	43.86	2.062	0	6	2	8	9	15
44.60	44.60	2.061	0	2	6	3		6
44.95	44.96	2.036	-4	2	2	5	5	10
45.34	45.34	2.015	-4	2	4	15	12	29
46.42	46.50	1.9986	-3	3	6	9	8	18
46.50		1.9546	-4	2	0	9		18
46.93	46.94	1.9512	-2	6	2	10	12	19
46.99		1.9344	-3	5	3	6		11
47.28	47.28	1.9319	-2	5	6	2	5	4
47.51	47.60	1.9207	2	0	2	1		1
47.59		1.9121	-4	0	6	3	1	6
48.51	48.52	1.9089	3	3	6	9	8	18
48.86	48.86	1.8751	-1	5	4	7	5	14
49.10	49.10	1.8624	-3	5	6	6	5	13
49.43	49.44	1.8538	3	1	7	7	6	16
50.41	50.42	1.8422	0	6	0	4	4	10
50.43		1.8086	0	4	4	4	14	8
50.64	50.64	1.8080	-2	8	6	11	16	25
51.07	51.08	1.8009	2	0	8	19	9	42
52.09	52.10	1.7868	-4	4	2	12	2	27
52.46	52.46	1.7542	-1	5	5	2	1	5
52.68	52.68	1.7427	-3	6	6	1	2	2
52.89	52.82	1.7360	-2	2	8	3	2	7
53.62	53.62	1.7296	-2	6	2	1	2	2
53.77	53.76	1.7076	-5	1	4	2	2	6
54.08	54.08	1.7034	-1	7	4	1	2	8
54.70	54.70	1.6944	-3	1	0	3	3	8
54.78		1.6765	-3	5	6	10	7	25
55.47	55.48	1.6744	-4	4	0	1		1
		1.6552	-1	7	4	7	5	17

CELSIAN (BA-FELDSPAR) - NEWNHAM AND MEGAW, 1960

2THETA	PEAK	D	H	K	L	I(INT)	I(PK)	I(DS)
55.75	55.74	1.6474	3	2	2	10	10	26
55.98	55.90	1.6411	-4	0	8	1	5	4
56.38	56.40	1.6306	0	8	0	5	6	12
56.39		1.6302	-3	3	8	2		5
56.42		1.6294	0	0	8	1		2
56.44		1.6289	-5	1	6	1	1	3
57.39	57.50	1.6042	-2	6	6	1		2
57.50		1.6014	-5	3	4	7	6	18
57.89	57.90	1.5915	-4	3	8	2		3
57.93		1.5905	3	3	4	9	7	24
58.32	58.32	1.5808	0	2	8	5	7	14
58.43	58.44	1.5781	2	2	6	1		2
58.49		1.5765	-2	4	8	1		6
58.55		1.5751	-5	3	2	2		4
59.60	59.60	1.5499	5	1	0	4	2	11
60.16	60.20	1.5368	0	6	6	10	9	28
60.20		1.5359	-5	3	6	6		5
60.32	60.34	1.5331	1	3	6	2	6	7
60.62	60.74	1.5261	-3	7	4	2	8	25
60.73		1.5237	-4	6	2	9		2
60.81		1.5218	-2	8	2	3		9
60.83		1.5214	1	7	4	16	11	50
61.59	61.58	1.5046	2	8	0	5	3	15
62.83	62.84	1.4778	-5	1	8	4		14
63.24	63.24	1.4692	-4	4	0	4	4	2
63.39	63.44	1.4659	1	4	8	2	6	7
63.44		1.4649	-1	7	8	4		12
63.44		1.4649	-3	5	8	1	7	2
63.77	63.90	1.4582	0	5	8	9		5
63.80		1.4576	2	4	6	3		30
63.91		1.4554	4	4	0	6	6	11
64.07	64.08	1.4522	-6	0	4	1	5	21
64.82	64.82	1.4372	-4	6	6	6		2
64.89		1.4358	-4	6	6	1	4	3
65.01	65.00	1.4334	-2	0	10	5	4	17
65.32	65.32	1.4274	4	4	4	5		7
65.43		1.4251	1	9	0	1		3
65.76	65.76	1.4188	-3	3	6	1	1	3
66.37	66.38	1.4073	-5	7	8	1	1	6
66.71	66.72	1.4010	-3	7	2	3	2	2
66.76		1.4000	-2	2	10	1		9
67.33	67.34	1.3895	-5	5	6	4	4	3
67.47	67.48	1.3869	-6	1	8	5	5	15
67.66	67.66	1.3820	-2	6	2	2	3	18
67.74		1.3701	5	1	2	1	1	7
68.41	68.42	1.3640	-6	9	10	6		2
68.77	68.78	1.3612	-4	2	4	2		14
68.92	68.94	1.3447	-1	2	10	4	3	13
69.89	69.90	1.3446	-2	9	4	6	4	23
70.86	70.84	1.3296	-6	8	6	1		4
71.16	71.16	1.3286	5	0	8	1		5
71.41	71.40	1.3238	-1	7	8	3	3	13
71.57	71.58	1.3198	3	0	2	4	4	16
71.88	71.88	1.3172	-2	4	10	3	3	10
72.27	72.40	1.3122	-5	4	0	4	3	14
72.38		1.3063	0	10	10	1	2	2
72.52	72.52	1.3045	0	6	0	2		8
72.62		1.3023	2	6	6	2	3	9
72.93	72.94	1.3008	-6	0	6	1		2
73.18	73.18	1.2960	-4	4	6	1	2	3
73.18		1.2922	-5	8	4	8	6	32
73.47	73.44	1.2921	-5	5	8	1		4
73.83	73.84	1.2879	-6	8	2	7	6	25
73.98	74.00	1.2824	-4	4	4	4	4	15
74.11		1.2802	0	4	10	5	5	18
74.49	74.48	1.2782	-3	9	2	2	2	6
74.68	74.68	1.2727	3	9	0	2	2	8
74.94	74.92	1.2699	-3	9	4	2	1	6
75.03		1.2662	-5	7	0	1		3
75.40	75.52	1.2649	-5	7	4	1		4
75.51		1.2595	3	5	4	2	2	7
75.59		1.2579	-5	3	10	2		7
		1.2568	-5	3	10	1		4
77.01	77.04	1.2372	2	10	0	1	1	3

CELSIAN (BA-FELDSPAR) - NEWNHAM AND MEGAW, 1960

2THETA	PEAK	D	H	K	L	I(INT)	I(PK)	I(DS)
77.06	77.28	1.2365	-5	9	6	1	1	5
77.40	77.80	1.2319	5	7	6	1	1	3
77.80	77.80	1.2266	5	5	2	2	1	9
78.04	78.04	1.2234	-1	5	10	1	1	4
78.66	78.68	1.2154	-5	6	10	1	1	6
78.72		1.2145	5	1	4	1		3
79.04	79.18	1.2104	0	4	10	2	3	7
79.17		1.2088	-2	8	8	2		8
79.21		1.2082	6	4	0	2		5
79.87	79.96	1.1999	1	1	10	2	3	8
79.95		1.1989	-6	6	4	2		9
80.08	80.18	1.1973	3	9	2	1	5	3
80.13		1.1967	-2	6	10	2		10
80.13		1.1966	5	7	0	1		4
80.18		1.1960	4	2	6	2		7
80.22		1.1956	-3	1	12	3		15
80.41	80.40	1.1932	4	6	4	2	3	7
80.96	80.98	1.1865	2	10	2	1	1	2
81.13	81.14	1.1844	-6	6	6	1	1	3
81.44	81.46	1.1807	-7	3	6	2	2	10
81.48		1.1802	-7	1	8	2		8
82.01	82.00	1.1740	-6	6	2	1	1	4
82.40	82.40	1.1694	3	1	2	1	1	4
82.60	82.62	1.1670	-1	11	2	1	1	3
82.90	83.00	1.1636	-6	2	2	1		3
83.00		1.1625	-5	1	8	2		7
83.11		1.1617	-1	3	10	1		2
83.50	83.50	1.1567	-4	8	8	4	2	17
83.55		1.1562	1	7	8	1		2
83.87	83.86	1.1526	0	8	8	4	3	20
84.45	84.46	1.1461	-5	10	2	2	1	8
84.61	84.70	1.1444	-7	3	12	1	2	4
84.70		1.1434	-7	3	8	1		6
84.89	85.02	1.1413	-1	11	2	1	4	6
85.00		1.1401	1	1	2	2		9
85.02		1.1399	-4	4	12	4		18
85.12		1.1389	-6	4	10	2		7

2THETA	PEAK	D	H	K	L	I(INT)	I(PK)	I(DS)
86.27	86.28	1.1266	2	6	8	2	1	8
86.28		1.1264	-1	11	6	1		3
87.06	87.06	1.1184	0	10	6	1	1	3
87.71	87.82	1.1118	6	4	2	3	3	6
87.79		1.1109	-7	5	0	3		13
87.87		1.1101	-7	5	6	1		15
87.98		1.1090	-5	9	4	1		6
89.50	89.52	1.0941	-6	0	12	3	1	3
89.51		1.0939	1	5	10	1		5
90.51	90.68	1.0845	4	10	0	2	1	6
90.66		1.0831	-3	11	4	1		10
91.01	91.10	1.0798	-7	5	2	1	3	6
91.09		1.0791	3	11	0	1		7
91.10		1.0790	-6	2	12	2		7
91.14		1.0786	-7	3	10	1		3
91.19		1.0782	-6	8	4	1		6
91.83	91.84	1.0723	0	12	2	1	1	6
92.31	92.32	1.0680	-8	0	4	1	1	3
92.97	93.10	1.0622	5	9	0	2	2	11
93.02		1.0617	4	6	6	1		6
93.11		1.0609	-6	6	10	3		14
93.22	93.60	1.0599	-6	0	2	1	2	6
93.62	93.94	1.0565	-2	10	8	1	2	12
93.91		1.0539	-8	2	4	1		8
94.02		1.0530	-2	2	2	2		10
94.55	94.56	1.0485	6	2	4	2	1	10
95.25	95.26	1.0426	-3	11	6	2	1	10
95.81	96.06	1.0380	-5	9	8	1	1	5
96.00		1.0365	4	2	8	1		6
96.12		1.0355	3	11	2	1		7
97.40	97.66	1.0253	7	5	0	1	1	3
97.62		1.0236	-8	4	6	1		6
97.68		1.0231	-3	1	14	1		4
98.29	98.32	1.0183	0	10	8	1	1	5
98.39		1.0176	2	10	6	1		6
98.70	98.68	1.0152	1	1	12	1	1	4
99.14	99.12	1.0119	-1	7	10	1	1	5

CELSIAN (BA-FELDSPAR) - NEWNHAM AND MEGAW, 1960

2THETA	PEAK	D	H	K	L	I(INT)	I(PK)	I(DS)
99.49	99.50	1.0093	-3	7	12	2	1	11
99.62		1.0083	-5	1	14	1		3
100.24	100.28	1.0037	-8	2	10	1	1	8
100.30		1.0033	3	1	10	1		3
100.32		1.0032	5	9	2	1		3
100.75	100.86	1.0000	-7	7	8	1	1	6
100.85		0.9993	4	4	8	1		4
100.92		0.9988	-3	3	14	1		9
101.68	101.94	0.9934	3	7	8	1	1	4
101.95		0.9915	1	3	12	1		5
102.88	102.90	0.9851	-5	3	14	1	0	4
103.57	103.58	0.9804	3	3	10	2	1	11
103.77		0.9790	-5	7	12	1		5
104.22	104.28	0.9760	1	13	2	1	1	7
104.33		0.9753	-1	7	12	1		7
104.80	104.80	0.9722	0	12	6	1	1	9
104.84		0.9719	-1	1	14	1		6
105.16	105.24	0.9698	-8	4	10	1	1	6
105.32		0.9688	-4	12	2	2		12
105.66	105.74	0.9666	-1	13	4	1	1	9
105.77		0.9659	-6	10	4	1		6
105.95		0.9648	-2	10	10	1		6
106.24	106.24	0.9629	4	10	4	1	1	9
106.26		0.9628	-2	4	14	1		7
106.50		0.9613	-5	11	6	1		5
107.23	107.28	0.9568	-7	7	0	1	1	8
107.35		0.9560	-8	0	12	1		6
108.43	108.44	0.9495	-9	1	8	2	1	10
108.79	108.82	0.9474	-2	0	12	1	1	4
109.03		0.9459	-8	2	12	1		5
109.54	109.56	0.9430	-5	5	14	2	1	15
109.78	109.76	0.9416	5	1	8	2	1	11

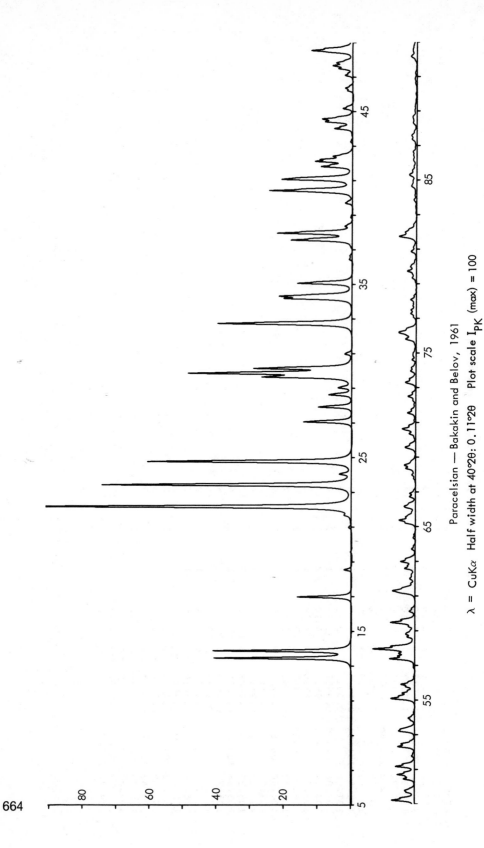

Paracelsian — Bakakin and Belov, 1961

$\lambda = CuK\alpha$ Half width at $40°2\theta$: $0.11°2\theta$ Plot scale I_{PK} (max) = 100

664

PARACELSIAN – BAKAKIN AND BELOV, 1961

2THETA	PEAK	D	H	K	L	I(INT)	I(PK)	I(DS)
13.42	13.44	6.590	1	1	0	36	41	22
13.84	13.44	6.391	0	1	1	36	41	23
16.95	16.96	5.226	-1	1	1	7	16	5
16.96		5.226	1	2	1	8		6
18.51	18.52	4.790	0	2	0	2	2	2
21.64	21.66	4.103	2	1	0	1	2	1
22.13	22.14	4.013	-2	0	1	52	100	52
22.13		4.013	1	2	1	48		48
23.40	23.40	3.799	-1	1	1	36	75	38
23.40		3.799	1	2	1	40		43
24.02	24.02	3.701	-2	1	1	1	4	1
24.02		3.701	-2	1	1	2		2
24.74	24.74	3.595	-1	1	2	32	61	37
24.74		3.595	1	1	2	33		37
27.04	27.04	3.295	-2	2	0	16	15	20
27.90	27.90	3.196	0	2	2	11	10	15
28.60	28.60	3.118	-2	0	2	4	7	6
28.60		3.118	0	2	1	4		5
29.00	29.00	3.076	-2	2	1	2	4	3
29.00		3.076	2	2	1	2		3
29.61	29.62	3.014	-1	2	2	11	27	15
29.61		3.012	-1	2	2	11		16
29.63		3.012	1	3	0	6		9
29.83	29.84	2.993	0	3	1	53	49	78
30.11	30.12	2.965	2	1	2	16	30	23
30.11		2.965	-2	1	2	15		23
30.96	30.96	2.886	3	1	0	3	3	4
32.65	32.72	2.740	0	1	3	8	40	13
32.71		2.735	-3	1	1	21		35
32.71		2.735	1	1	3	21		34
34.15	34.16	2.624	-3	1	1	11	21	19
34.15		2.624	-1	1	3	12		21
34.29	34.28	2.613	-2	2	2	4	22	7
34.29		2.613	2	2	2	4		7
34.30		2.612	-2	3	0	12		22
35.04	35.04	2.559	3	3	0	20	17	36
37.52	37.52	2.395	0	4	0	12	19	24
37.53		2.395	-3	1	2	5		10
37.53		2.395	3	1	2	6		11
37.92	37.92	2.370	-1	2	3	15	23	30
37.92		2.370	1	2	3	13		26
38.33	38.34	2.346	-2	1	3	1	3	2
38.33		2.346	1	1	3	1		2
39.67	39.68	2.270	0	4	1	3	3	6
40.30	40.40	2.236	-1	4	1	3	25	7
40.30		2.236	1	2	1	3		6
40.39		2.231	-2	4	0	14		30
41.04	41.04	2.198	-3	2	2	14	21	32
41.04		2.198	3	2	2	11		25
41.05		2.197	3	3	0	10		23
41.10		2.194	-4	0	1	3		6
41.10		2.194	0	0	1	4		9
41.78	41.78	2.160	-2	2	3	4	10	9
41.78		2.160	2	2	3	6		13
42.09	42.10	2.145	0	3	3	6	12	14
42.21	42.20	2.139	-2	0	4	14	10	32
42.21		2.139	0	1	1	2		5
42.39	42.38	2.130	-4	1	1	2	6	5
43.23	43.22	2.091	0	3	2	6	1	14
43.99	44.00	2.057	0	4	2	1	6	3
43.99		2.057	2	4	1	4		9
44.37	44.40	2.040	-1	1	4	4	9	9
44.37		2.040	1	1	4	4		10
44.42		2.038	-1	4	2	2		9
44.42		2.038	1	2	2	2		4
44.56	44.56	2.031	-3	3	1	5	9	12
44.56		2.031	1	3	3	4		10
45.15	45.16	2.006	-3	2	2	2	3	5
45.15		2.006	4	2	2	2		6
46.19	46.20	1.9638	-4	1	2	1	2	4
46.19		1.9638	1	1	2	1		3
46.34	46.32	1.9577	-4	2	4	2	3	6
47.08	47.08	1.9287	2	3	3	2	3	5

PARACELSIAN – BAKAKIN AND BELOV, 1961

2THETA	PEAK	D	H	K	L	I(INT)	I(PK)	I(DS)
47.08	47.48	1.9287	-2	3	3	2		4
47.47		1.9137	-1	2	4	3	5	7
47.47		1.9137	-1	3	4	3		8
47.65	47.64	1.9069		3	3	3	6	9
47.65		1.9009	-3	2	3	3		9
47.81	47.80	1.9009	-2	1	4	1	5	3
47.85		1.8994	-2	4	2	1		4
47.85		1.8994	-2	4	2	1		3
48.42	48.52	1.8781	3	4	0	6	13	18
48.52		1.8747	1	5	0	13		38
48.65	48.64	1.8699	0	5	1	3	9	8
49.19	49.20	1.8506	-4	2	2	3	7	8
49.19		1.8506	-4	2	2	3		9
49.20		1.8502	4	3	0	4		13
49.65	49.66	1.8347	3	4	1	1	3	4
49.65		1.8347	-3	4	1	2		5
49.74	49.74	1.8315	-1	5	1	1	3	4
49.74		1.8315	-1	5	1	1		4
50.41	50.42	1.8086	-4	3	1	2	4	7
50.41		1.8086	-4	4	1	3		8
50.68	50.68	1.7998	-1	4	3	4	6	12
50.68		1.7998	-1	4	3	4		11
51.15	51.16	1.7842	5	1	0	9	6	27
51.74	51.74	1.7652	2	5	0	2	6	5
52.28	52.30	1.7482	-4	1	3	3	5	10
52.28		1.7482	-4	1	3	4		12
52.48	52.44	1.7421	-3	3	3	1		4
52.48		1.7421	-3	3	3	1	5	4
53.19	53.20	1.7205	-3	4	2	3	5	11
53.19		1.7205	-3	4	2	3		9
53.28		1.7179	-1	5	2	1		5
53.28		1.7179	-1	5	2	1	1	5
53.92	53.92	1.6989	-4	4	2	2	2	4
53.92		1.6989	-4	3	2	1		4
55.04	55.06	1.6669	-4	2	3	3	8	11
55.04		1.6669	-4	2	3	3		10
55.08		1.6658	5	2	1	3		11
55.08	55.38	1.6658	-5	2	1	3		11
55.38		1.6577	-2	3	4	3	4	9
55.38		1.6577	-2	3	2	2		8
55.75	55.76	1.6474	-5	1	1	1	3	4
55.75		1.6474	-5	1	1	1		4
55.88	55.88	1.6438	-3	3	2	3	5	9
55.88		1.6438	-3	3	2	2		7
57.35	57.36	1.6052	-2	0	5	7	8	24
57.64	57.66	1.6052	-2	0	4	5		20
57.69		1.5978	0	4	4	3	5	12
57.92	57.92	1.5967	0	6	0	3		11
57.92		1.5908	-3	5	1	7	13	26
57.93		1.5908	3	5	1	8		29
57.93		1.5905	-1	2	5	2		8
58.61	58.62	1.5905	-1	1	4	3	3	9
58.61		1.5737	-1	2	5	2		6
58.85	58.78	1.5737	-1	4	4	2		7
58.85		1.5679	-1	5	3	1	3	4
59.21	59.22	1.5679	-1	5	3	1		5
59.21		1.5591	-4	0	4	1	2	4
59.45	59.44	1.5591	-4	4	3	1		5
59.45		1.5535	-4	3	3	5	8	21
61.19	61.28	1.5535	-4	3	3	6		23
61.27		1.5133	6	0	0	3	7	12
61.96	61.96	1.5116	0	6	2	8		32
62.56	62.56	1.4964	0	0	6	2	2	10
62.56		1.4835	-2	6	1	2	4	9
62.96	62.96	1.4835	-2	6	1	2		10
62.96		1.4750	3	1	5	3		14
63.68	63.68	1.4750	-3	1	5	3	5	15
63.68		1.4601	1	5	2	1		4
64.32	64.32	1.4601	-5	2	3	1	1	5
65.18	65.20	1.4471	5	0	0	2		7
65.34	65.34	1.4300	0	4	6	5	4	21
65.34		1.4269	4	1	1	2	5	11
65.64	65.52	1.4269	-5	4	1	2		9
65.64		1.4212	2	6	2	1	3	5

PARACELSIAN – BAKAKIN AND BELOV, 1961

2THETA	PEAK	D	H	K	L	I(INT)	I(PK)	I(DS)
65.64		1.4212	-2	6	2	2		5
66.06	66.12	1.4130	-3	4	4	1		7
66.06		1.4130	-3	4	4	1	4	8
66.14		1.4116	-1	5	4	2		8
66.14		1.4116	-1	5	4	2		7
66.70	66.70	1.4010	-4	3	4	2	2	9
66.70		1.4010	-4	3	4	1		7
66.89	66.90	1.3975	-1	1	6	1		5
68.32	68.32	1.3717	-5	1	4	1	2	11
68.32		1.3717	-5	1	4	2	3	12
68.48	68.52	1.3689	-4	0	5	1		5
68.48		1.3689	-4	0	5	1	4	5
69.49	69.54	1.3515	0	7	1	2		10
69.55		1.3505	6	3	1	2		9
69.55		1.3505	-6	3	1	2		9
70.32	70.32	1.3376	-6	0	3	2	2	8
70.32		1.3376	-6	0	3	1		8
70.57	70.60	1.3334	-2	4	5	1		6
70.57		1.3334	-2	4	5	2	4	6
70.61		1.3327	-2	6	3	2		12
70.61		1.3327	-2	6	3	3		14
71.52	71.52	1.3180	5	5	0	5		26
72.48	72.48	1.3029	-6	5	2	2	3	10
72.48		1.3029	-6	3	2	2	3	10
73.28	73.28	1.2906	-1	7	2	2	4	12
73.28		1.2906	-1	7	0	2		11
73.94	73.94	1.2808	0	6	4	2	2	12
74.11	74.12	1.2783	0	5	5	1	2	6
74.59	74.58	1.2712	7	1	1	1	2	7
74.59		1.2712	-7	1	1	1		7
75.77	75.78	1.2543	-2	3	6	3	4	16
75.77		1.2543	-2	3	6	2		13
76.12	76.16	1.2494	-4	6	2	3		15
76.12		1.2494	-4	6	2	3	6	16
76.20		1.2483	-3	2	6	3		19
76.20		1.2483	-3	2	6	3		17
77.44	77.44	1.2313	-7	1	2	1	2	6

2THETA	PEAK	D	H	K	L	I(INT)	I(PK)	I(DS)
78.05	78.06	1.2233	-1	7	3	2	2	9
78.05		1.2233	-1	7	3	2		9
79.31	79.32	1.2070	-5	2	5	1		8
79.31		1.2070	-5	2	5	1	2	7
79.71	79.72	1.2019	-7	2	2	2		15
79.71		1.2019	-7	2	2	3	3	16
80.84	80.84	1.1880	-4	6	3	2	2	11
80.84		1.1880	-4	5	3	2		10
81.70	81.72	1.1776	-3	5	5	5	6	28
81.72		1.1776	-3	5	5	4		27
81.72		1.1774	-1	8	1	1		8
81.72		1.1774	-1	8	1	1		8
81.83		1.1760	-6	1	8	1	1	7
83.82	83.82	1.1531	-6	3	4	1		7
83.82		1.1531	-6	3	3	1		7
84.62	84.62	1.1443	0	3	7	2	2	10
85.22	85.26	1.1377	3	4	6	1	3	8
85.22		1.1377	-3	4	6	1		7
85.29		1.1370	-1	5	6	2		11
85.29		1.1370	-1	5	6	2		14
86.61	86.62	1.1230	-1	5	4	1	2	7
86.61		1.1230	-5	5	0	3		9
87.53	87.54	1.1135	3	8	5	1	2	17
88.26	88.28	1.1062	5	4	5	1	2	9
88.26		1.1062	-5	4	5	1		8
88.66	88.64	1.1023	-7	4	2	1	2	7
88.66		1.1023	-7	4	5	1		8
92.09	92.14	1.0700	0	7	5	1	2	9
92.15		1.0695	6	3	5	1		10
92.15		1.0695	-6	3	5	1		10
93.63	93.62	1.0563	0	9	0	1	1	11
95.61	95.66	1.0397	-1	2	8	2	2	10
97.39	97.36	1.0254	-6	8	1	1	1	8
97.84	97.80	1.0218	-4	6	7	1	2	8
99.99	100.00	1.0056	-3	5	3	1	1	9
99.99		1.0056	-7	5	3	1	2	12

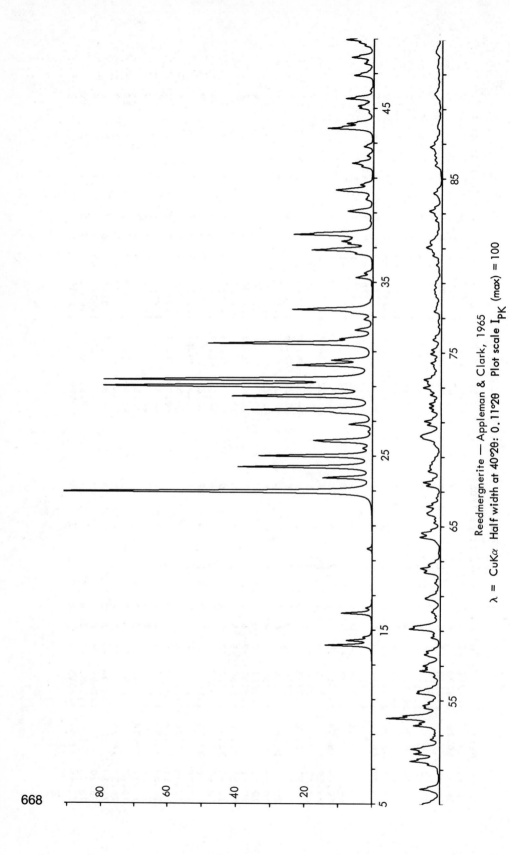

668

Reedmergnerite — Appleman & Clark, 1965

λ = CuKα Half width at 40°2θ: 0.11°2θ Plot scale I$_{PK}$ (max) = 100

REEDMERGNERITE - APPLEMAN AND CLARK, 1965

2THETA	PEAK	D	H	K	L	I(INT)	I(PK)	I(DS)
14.10	14.12	6.275	-1	1	0	12	14	12
14.38	14.38	6.155	0	1	1	6	8	6
14.57	14.58	6.076	0	0	1	2	3	2
15.96	15.96	5.550	-1	-1	1	8	9	8
16.25	16.26	5.449	-1	1	1	1	2	1
19.65	19.66	4.513	0	-1	1	1	2	1
22.94	22.94	3.874	-2	0	1	1	1	1
23.74	23.74	3.745	1	1	1	100	100	100
24.36	24.36	3.651	-1	-1	1	14	15	14
24.99	25.00	3.560	1	3	0	40	40	40
25.42	25.42	3.501	2	0	1	35	34	35
25.81	25.88	3.449	0	0	2	2	3	2
25.89		3.439	-1	-3	1	12	17	13
26.37	26.38	3.377	-1	3	0	10	1	10
26.80	26.80	3.324	-1	-1	2	7	7	7
27.63	27.64	3.225	-1	-1	2	39	38	39
28.42	28.42	3.137	-1	2	2	45	41	46
28.99	29.04	3.078	-2	2	0	41	79	42
29.03		3.073	-2	0	2	53		54
29.38	29.38	3.038	0	0	2	85	79	87
30.19	30.20	2.957	2	2	0	25	24	26
30.49	30.48	2.930	-1	-3	1	12	12	13
31.46	31.46	2.841	0	-4	1	55	48	56
31.73	31.72	2.818	0	-2	2	8	10	6
32.23	32.24	2.775	-2	2	2	6	5	6
32.84	32.84	2.725	-2	2	2	3	3	3
33.24		2.695	-2	2	2	3	3	5
33.41	33.42	2.680	-1	-3	1	27	6	28
33.42		2.544	-1	3	3	6	24	6
35.25	35.26	2.526	-1	-1	2	3	5	3
35.51		2.452	-3	3	1	5	3	5
35.52	35.52	2.438	-2	-1	2	21	18	22
36.62	36.62	2.415	-1	-1	1	7	7	7
36.84	36.84	2.406	-3	-1	2	8	9	9
37.19	37.20	2.394	-2	-4	0	3	6	4
37.35	37.34	2.383	-2	-4	1	28	23	29
37.53	37.52	2.374	-1	5	0	4	17	5
37.72	37.72							
37.86	37.82							

2THETA	PEAK	D	H	K	L	I(INT)	I(PK)	I(DS)
38.52	38.52	2.335	-2	1	1	3	1	1
39.00	39.06	2.307	-1	2	2	3	7	3
39.06		2.304	-1	5	1	7		8
39.71	39.76	2.268	-1	5	1	1	2	1
39.76		2.265	-3	3	0	2		2
40.22	40.28	2.240	-2	2	1	6	11	7
40.29		2.237	-3	1	1	10		10
40.59	40.58	2.221	-1	1	3	4	4	4
41.56	41.62	2.171	-1	3	2	1	4	1
41.62		2.168	-1	3	1	4		4
41.82	41.82	2.158	-3	3	1	7	6	8
42.25	42.24	2.137	-2	2	3	2	2	2
42.77	42.78	2.112	-3	2	2	4	3	4
43.82	43.82	2.064	-2	4	0	18	13	19
44.10	44.10	2.052	0	6	0	9	8	10
44.71	44.72	2.025	0	0	3	1	1	1
44.98	44.98	2.013	-3	1	3	5	4	5
45.12	45.12	2.007	-1	3	1	3	5	3
45.53	45.54	1.9904	-1	5	1	11	8	12
45.99	46.00	1.9716	3	3	0	1	1	1
46.87	46.86	1.9369	-4	0	2	8	6	8
47.55	47.56	1.9105	-2	0	1	3	3	4
47.90	47.90	1.8973	-2	4	1	1	7	1
47.91		1.8972	0	6	1	8		9
48.58	48.58	1.8725	2	2	2	3	3	3
48.76	48.76	1.8659	3	1	1	4	4	4
48.92	48.92	1.8601	-4	2	2	9	8	10
49.13	49.04	1.8529	-2	4	1	1	5	1
49.88	49.88	1.8267	-3	5	1	6	6	7
49.91		1.8257	-2	2	0	1		1
51.27	51.28	1.7802	-4	0	3	10	4	5
51.49	51.50	1.7734	-2	4	2	2	9	12
51.54		1.7718	0	6	2	1		2
51.62	51.62	1.7690	0	4	3	10	7	1
51.92	51.92	1.7595	0	1	3	2	8	11
52.15	52.20	1.7523	0	1	3	10	9	2
52.21		1.7506	4	0	0	9		11

REEDMERGNERITE — APPLEMAN AND CLARK, 1965

2THETA	PEAK	D	H	K	L	I(INT)	I(PK)	I(DS)
52.75	52.74	1.7340	-1	7	0	3	3	2
53.05	53.06	1.7247	-2	-6	2	2	2	2
53.55	53.54	1.7099	1	1	3	10	8	11
53.71	53.70	1.7052	-4	-2	1	1	1	1
53.85		1.7011	-1	-7	1	1	6	6
53.96	53.96	1.6977	-2	0	4	22	16	25
54.15		1.6922	2	-4	2	4	11	4
54.55	54.56	1.6808	3	3	1	1	5	2
54.56		1.6806	-1	7	1	3		4
54.63		1.6786	-1	-7	0	2		2
54.71	54.70	1.6761	2	-6	1	3	5	3
54.91	54.88	1.6708	-4	4	1	3	3	3
55.40	55.42	1.6569	-4	4	2	9	7	10
55.49	55.52	1.6545	1	-7	1	1	7	7
55.50		1.6543	4	2	0	2		2
55.52		1.6536	2	-1	5	2		2
55.81	55.80	1.6457	-1	-1	4	1	2	1
56.35	56.36	1.6313	0	4	3	2	2	2
56.70	56.70	1.6221	-4	-4	2	9	6	10
56.99	56.98	1.6146	-1	1	4	6	5	6
57.06		1.6126	-2	2	4	2		2
57.54	57.64	1.6004	-1	-7	2	3	5	3
57.63		1.5980	-4	-4	1	5		6
57.84	57.82	1.5928	-3	-5	3	4	4	4
58.61	58.60	1.5738	3	-5	2	3	3	3
59.09	59.10	1.5620	2	4	2	12	9	14
59.19	59.24	1.5596	-5	1	2	4	8	5
59.20		1.5595	5	1	3	1		1
59.28		1.5575	1	3	4	1		1
59.29		1.5573	-3	-3	3	2	2	2
59.76	59.76	1.5460	5	1	7	1	2	1
59.93	59.94	1.5420	2	6	0	1	2	2
60.07	60.08	1.5388	0	8	0	3	3	3
60.17		1.5365	-4	0	4	1		1
60.82	60.86	1.5218	0	-8	1	4	4	4
60.88		1.5202	-5	1	1	4		5
61.25	61.24	1.5120	4	0	1	1	1	2

2THETA	PEAK	D	H	K	L	I(INT)	I(PK)	I(DS)
61.74	61.74	1.5012	-5	1	3	2	5	2
62.32	62.34	1.4886	1	-5	3	3	5	3
62.33		1.4883	2	0	3	2		3
62.39		1.4871	1	-7	2	3		4
62.59	62.58	1.4828	-4	2	4	6	6	8
62.86	62.78	1.4772	2	-2	3	3	4	3
63.45	63.46	1.4648	-2	-6	2	1	1	1
64.31	64.38	1.4472	0	2	2	2	6	3
64.37		1.4461	-5	-3	2	1		1
64.38		1.4458	-2	0	0	7		8
64.64	64.62	1.4406	-4	6	1	1	6	8
65.18	65.22	1.4300	-3	-7	1	1	2	1
65.22		1.4293	-1	-7	3	2		3
65.67	65.76	1.4206	0	-8	2	4	4	5
65.77		1.4187	-2	-8	1	4		5
66.15	66.14	1.4114	-5	-3	3	6	4	8
66.64	66.64	1.4023	-5	-3	1	2	1	2
67.33	67.34	1.3895	2	-4	3	6	4	8
67.37		1.3887	-2	-8	2	1		1
67.55	67.54	1.3855	0	6	3	5	5	7
68.13	68.18	1.3751	-5	-1	4	2	4	2
68.17		1.3744	2	8	0	3		3
68.21		1.3737	-2	-8	1	1		1
68.36	68.38	1.3711	-4	-6	1	1	3	1
68.45		1.3695	1	5	3	3		3
68.52	68.50	1.3683	1	-1	3	1		1
68.66	68.68	1.3658	1	7	2	3	3	4
68.73		1.3645	3	-5	1	2		2
68.79		1.3635	-2	8	2	1		1
68.97	68.96	1.3604	3	7	0	1	2	1
68.99		1.3601	-1	9	1	1		1
69.59	69.74	1.3498	-3	-7	3	3	2	3
69.67		1.3485	-3	5	4	4		4
69.73		1.3474	-2	-6	4	1		2
69.98	70.08	1.3433	-4	-6	3	2	5	2
70.05		1.3420	-5	5	2	4		5
70.07		1.3417	-1	-9	3	2		3

REEDMERGNERITE - APPLEMAN AND CLARK, 1965

2THETA	PEAK	D	H	K	L	I(INT)	I(PK)	I(DS)
70.17	70.18	1.3400	-4	6	3	5	6	6
70.29		1.3381	-1	3	5	3		4
70.84	70.84	1.3290	-1	-3	4	4	4	5
70.86		1.3287	-1	9	1	2		2
71.04	71.06	1.3258	1	-9	1	1		2
71.06		1.3254	-1	9	0	3	5	4
71.17		1.3236	-1	5	4	4		1
71.64	71.64	1.3162	4	4	1	4	3	5
71.83	71.84	1.3131	-5	-3	4	3	4	4
72.07	72.06	1.3093	-1	0	5	4	3	5
72.57	72.58	1.3015	-6	0	0	4	3	5
72.86	72.98	1.2971	-2	-8	3	1	5	1
72.97		1.2953	-1	-9	2	4		5
72.99		1.2951	2	4	3	3		3
73.02		1.2946	-5	5	3	2		2
73.29	73.28	1.2905	-4	-2	5	3	4	4
73.40	73.44	1.2888	4	6	0	2	5	5
73.45		1.2881	-4	0	2	6		7
73.87	73.84	1.2818	-1	-1	5	2	2	2
74.09	74.10	1.2786	3	-1	3	2	2	2
74.21	74.20	1.2768	0	-8	3	2	2	3
74.40	74.40	1.2740	2	8	1	4	3	5
74.69	74.62	1.2697	-6	2	2	1	2	1
74.89	74.92	1.2668	-2	-4	3	2	3	2
74.94		1.2661	-1	-3	5	4		4
75.17	75.16	1.2628	-6	-2	2	2	3	3
75.29		1.2611	3	5	1	1		1
75.73	75.74	1.2549	-5	5	0	2	3	3
75.74		1.2547	-6	0	1	1		1
75.77		1.2543	2	-8	2	1		2
75.88		1.2528	-4	0	1	1		1
76.49	76.50	1.2443	3	7	1	1	1	1
76.82	76.84	1.2397	4	2	2	1	1	1
77.36	77.36	1.2325	-4	8	1	4	2	5
77.47	77.47	1.2310	0	10	0	1		1
77.71	77.58	1.2278	-6	0	4	1	2	1
77.84	77.84	1.2260	4	-4	2	1	1	2

2THETA	PEAK	D	H	K	L	I(INT)	I(PK)	I(DS)
78.20	78.22	1.2213	-4	-4	5	2	2	2
78.27		1.2204	-5	1	5	1		1
78.39	78.40	1.2189	-4	8	2	2	2	2
78.54	78.60	1.2169	-6	4	1	2	2	1
78.61		1.2160	-6	-2	1	1		1
78.68		1.2151	0	0	5	1		1
78.83	78.86	1.2131	-3	-5	5	1	2	1
78.88		1.2125	-1	3	5	1		2
78.98		1.2112	0	-2	5	1		2
79.08	79.06	1.2099	2	0	4	1		1
79.60	79.62	1.2033	-1	-9	3	1	2	1
80.03	80.04	1.1979	-3	9	3	2	2	3
80.57	80.58	1.1913	-4	-8	2	4	3	5
80.69		1.1897	-5	7	2	1		1
80.76	80.82	1.1889	-4	-4	5	1		1
80.91	81.04	1.1871	-2	10	0	1	5	1
81.03		1.1857	-6	-4	1	5		6
81.08		1.1850	-5	-3	5	3		4
81.27	81.30	1.1827	-5	7	1	1		1
81.39		1.1813	-6	-4	4	1		1
81.43		1.1808	-6	-4	3	3		4
81.93	81.94	1.1749	-2	-8	4	1	1	1
81.94		1.1748	3	-5	3	1		1
82.33	82.32	1.1702	-5	0	5	3	1	1
82.60	82.60	1.1671	6	0	0	1	2	4
82.89	82.90	1.1637	-4	-8	3	3	2	1
82.92		1.1633	0	-4	5	1		1
83.09	83.12	1.1613	-6	2	0	1	4	3
83.11		1.1611	-3	5	5	1		1
83.14		1.1608	-4	8	3	4		2
84.16	84.16	1.1494	2	8	2	2	3	2
85.71	85.76	1.1325	6	2	0	3		5
85.78		1.1318	-1	9	3	2		3
85.80		1.1315	-6	0	3	1		1
86.10	86.08	1.1283	0	-8	4	2	2	3
86.31	86.32	1.1261	1	-7	4	3	2	3
86.63	86.80	1.1228	-4	-6	5	1	4	2

2THETA	PEAK	D	H	K	L	I(INT)	I(PK)	I(DS)
86.79		1.1212	2	-8	3	1		1
86.79		1.1211	-3	1	6	5		7
86.89		1.1201	-6	6	2	1		1
87.06	87.06	1.1183	4	8	0	1	2	1
87.19		1.1170	-6	4	5	1		2
87.40	87.38	1.1148	-1	5	3	1		2
87.92	87.94	1.1096	0	10	1	2	2	3
88.05	88.04	1.1083	-2	8	4	2	1	2
88.10		1.1078	-3	-3	6	1		1
88.92	88.90	1.0998	-7	-1	2	1	1	1
89.35	89.34	1.0956	-2	6	5	1	1	2
89.70	89.86	1.0922	3	5	3	2	2	2
89.85		1.0907	-5	-7	4	3		5
89.94		1.0898	-5	5	5	1		1
90.40	90.40	1.0855	-7	1	4	1	1	1
90.41		1.0854	-2	-6	4	1		1
90.72	90.74	1.0825	-7	-1	4	1	1	1
90.79		1.0819	-1	-11	2	1		1
90.91		1.0807	-5	-1	6	1		1
91.19	91.18	1.0782	3	-7	3	1	1	1
91.39	91.42	1.0763	-3	-9	4	2	1	3
91.49		1.0754	-1	-1	6	1		1
91.53		1.0750	-4	8	4	1		1
91.62		1.0742	-5	1	6	1		2
91.85	91.82	1.0722	3	-9	2	1	1	1
92.18	92.28	1.0691	-7	-3	3	1	2	2
92.26		1.0685	4	-4	3	3		4
92.39		1.0673	6	4	0	2		3
93.03	93.06	1.0616	3	1	1	1	1	2
93.17		1.0604	-2	10	4	1		1
93.44	93.46	1.0580	-2	10	3	1	2	1
93.50		1.0575	5	-7	1	1		1
93.50		1.0575	-1	-3	6	1		1
93.79	93.78	1.0550	-5	-3	6	1	1	1
94.22	94.20	1.0513	-1	-5	5	1	1	2
94.48	94.54	1.0491	-2	-8	5	1	1	2
94.61		1.0480	-5	9	2	1		2
95.43	95.46	1.0412	-4	10	0	1	1	1
95.93	95.96	1.0370	-5	3	6	1	1	1
96.04		1.0361	5	-5	2	1		1
96.32	96.32	1.0339	-3	4	4	3	1	4
96.90	96.98	1.0292	-7	5	3	2	1	3
97.03		1.0282	6	-4	1	2		2
97.37	97.38	1.0255	0	-12	1	1		2
97.45		1.0249	2	8	3	1		2
98.43	98.62	1.0173	3	-5	4	1		1
98.61		1.0159	2	0	5	1		1
98.62		1.0158	-1	9	4	1		1
99.00	98.94	1.0129	-5	-9	2	1	1	1
99.43	99.40	1.0097	0	6	5	1	1	2
100.39	100.40	1.0027	-7	1	0	1	1	1
100.64	100.66	1.0008	-6	8	3	1	1	2
100.82		0.9995	-7	-3	5	1		1
101.16	101.14	0.9971	0	-12	4	1	1	1
101.42	101.46	0.9952	2	-8	1	1	1	2
102.02	101.98	0.9910	-5	-9	5	2	1	2
102.65	102.72	0.9866	-6	6	6	2		2
102.75		0.9859	-2	-12	1	2		2
103.19	103.22	0.9829	1	11	2	1	1	1
103.28		0.9823	2	-12	1	1		1
103.86	103.90	0.9784	-8	0	3	2		3
103.89		0.9782	-3	-7	6	1		2
103.94		0.9779	-6	-8	3	2		3
104.59	104.54	0.9736	5	5	2	1	1	2
106.85	107.06	0.9591	-8	2	2	1	1	1
107.06		0.9578	5	1	3	1		1
107.48	107.56	0.9552	4	0	4	1	1	1
107.58		0.9546	-3	-3	7	1		1
107.81	107.84	0.9532	-8	-2	4	1	1	1
107.82		0.9531	-7	1	6	1		1
107.94		0.9525	-2	6	6	1		2
108.32	108.28	0.9501	-4	10	4	1	1	1
108.61	108.90	0.9484	-1	-7	6	2	1	2
108.84		0.9471	-1	13	0	1		1

REEDMERGNERITE - APPLEMAN AND CLARK, 1965

2THETA	PEAK	D	H	K	L	I(INT)	I(PK)	I(DS)
108.91		0.9466	-5	1	7	1		1
110.71	110.74	0.9362	4	-8	3	1	1	1
110.76		0.9360	-5	-7	6	2		3
111.30	111.26	0.9329	-6	2	2	1	1	1
111.35		0.9327	-1	13	1	1		1
111.35		0.9327	-8	4	2	1		1
111.57		0.9314	4	2	4	1		1
112.42	112.54	0.9268	-8	0	1	1	1	1
112.49		0.9265	-4	-8	6	1		1
112.52		0.9263	2	12	1	1		2
112.81	112.90	0.9247	-4	12	1	1		2
112.97		0.9238	-2	-4	7	1		1
114.15	114.16	0.9176	0	8	5	1	1	1
114.15		0.9176	-8	-4	4	1		2
116.68	116.70	0.9049	-1	1	7	1	1	1
116.74		0.9047	-5	11	3	1		1
117.30	117.24	0.9019	-3	13	1	1	1	1
118.08	118.10	0.8982	-3	-9	6	1	1	2
118.16		0.8979	-4	-12	2	1		1
119.00	119.10	0.8939	-1	-13	3	1	1	2
119.15		0.8932	3	9	3	1		2
120.30	120.64	0.8881	5	9	3	1	1	2
120.53		0.8870	-4	12	3	1		2
120.69		0.8864	7	3	1	1		1
120.71		0.8862	0	12	3	1		1
120.94		0.8852	-3	13	2	1		1
121.11	121.08	0.8845	0	-12	4	1	1	1
121.37		0.8834	3	-9	4	1		1
121.60	121.90	0.8824	2	-12	3	1	1	1
121.79		0.8816	-2	10	5	1		1
121.83		0.8814	-7	9	2	1		1
121.88		0.8812	-3	5	7	1		1
121.99		0.8807	-6	-10	3	1		2
122.48	122.38	0.8786	-7	9	3	1	1	1
122.60		0.8781	-1	-5	7	1		1
123.18	123.16	0.8757	-5	5	7	1	1	1
124.13	124.08	0.8718	-6	10	4	1	1	1

2THETA	PEAK	D	H	K	L	I(INT)	I(PK)	I(DS)
124.75	124.74	0.8694	0	-2	7	1	1	2
126.03	126.50	0.8644	-8	6	1	1	1	2
126.17		0.8638	0	-14	2	1		1
126.48		0.8626	-7	-3	7	3		5
127.26	127.10	0.8597	4	12	0	1	1	1
127.29		0.8596	5	-1	4	1		2
127.84	127.86	0.8576	-8	4	0	1	1	1
127.91		0.8573	1	5	6	1		1
128.86	128.88	0.8539	0	-4	7	1	1	2
129.01		0.8533	-5	-7	7	1		2

TABLE 34. ALKALI FELDSPARS

Variety	Max. Microcline (granitic)	Intermediate Microcline	Adularia	Orthoclase	High Sanidine
Composition	$Or_{98}Ab_2$	Or_{100}	$Or_{89}Ab_9Cs_2$	$Or_{90}Ab_8An_1$	$Or_{91}Ab_6An_3$
Source	Pellotsalo, Russia Helsinki #SM4709	Kodarma, Bihar India (Spencer U)	St. Gotthard, Switz. (Spencer B)	Mogok, Upper Burma (Spencer C)	Mogok, Upper Burma (Spencer C)
Reference	Brown & Bailey, 1964	Bailey & Taylor, 1955	Colville & Ribbe, 1968	Colville & Ribbe, 1968; Jones and Taylor, 1961	Ribbe, 1963; Cole, Sorum & Kennard, 1949
Cell Dimensions					
a	8.560	8.5784	8.554	8.561	8.5642
b Å	12.964	12.9600	12.970	12.996	13.0300
c	7.215	7.2112	7.207	7.192	7.1749
α	90.605	90.30 (89.70)	90	90	90
β deg	115.833	115.967	116.007	116.01	115.994
γ	87.70	89.125 (90.875)	90	90	90
Space Group	$C\bar{1}$	$C\bar{1}$	C2/m ($P2_1/a$)	C2/m ($P2_1/a$)	C2/m
Z	4	4	4	4	4

TABLE 34. (cont.)

Variety	Max. Microcline (granitic)	Intermediate Microcline	Adularia	Orthoclase	High Sanidine
Site (% Si) Occupancy					
T_1 (o)	6	31	61	65	75
T_1 (m)	97	73	61	65	75
T_2 (o)	99	97	89	85	75
T_2 (m)	98	99	89	85	75
Method	3-D, film	3-D, film	3-D, counter	3-D, film	3-D, film
R & Rating	0.104 (1)	0.138 (2)	0.055 (1)	0.054 (1)	0.099 (1)
Cleavage and habit	{001} and {010} perfect Variable				
Comment	Minor Na, Ca, and Cs ignored in calculations for all K-feldspars. Site occupancy from Brown & Bailey (1964); calculations based on Al = 1.0 per formula unit				
μ	124.4	124.6	124.2	124.1	123.8
ASF	0.4544×10^{-1}	0.5394×10^{-1}	0.5745×10^{-1}	0.5736×10^{-1}	0.5607×10^{-1}
Abridging factors	0.5:0:0	0.5:0:0	0.5:0:0	0.5:0:0	0.5:0:0

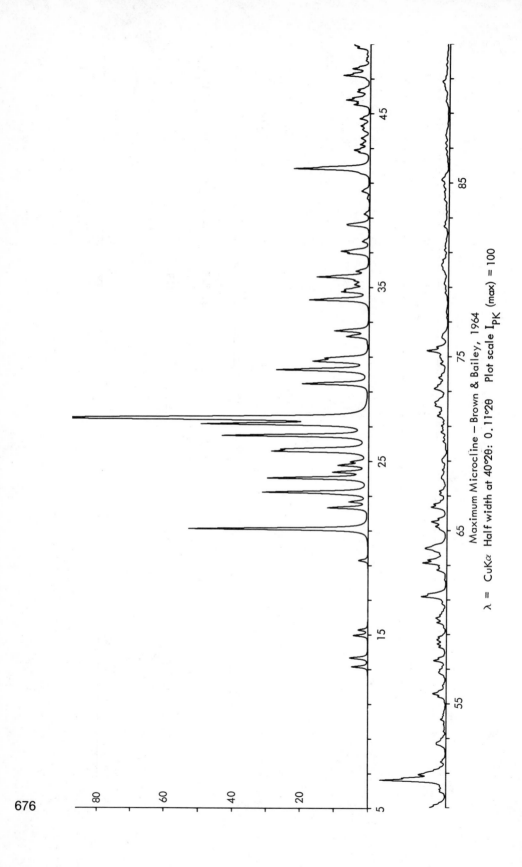

Maximum Microcline — Brown & Bailey, 1964

λ = CuKα Half width at 40°2θ: 0.11°2θ Plot scale I$_{PK}$ (max) = 100

MAXIMUM MICROCLINE - BROWN AND BAILEY, 1964

2THETA	PEAK	D	H	K	L	I(INT)	I(PK)	I(DS)
13.13	13.14	6.736	1	1	0	6	5	6
13.60	13.66	6.506	-1	1	0	1	5	5
13.62		6.494	0	0	1	3		3
13.66		6.477	0	1	0	4		4
14.94	14.96	5.923	-1	-1	1	6	4	5
15.26	15.28	5.800	-1	1	1	4	3	3
19.26	19.28	4.603	-1	0	1	4	3	4
21.07	21.08	4.213	-2	0	1	76	53	73
22.29	22.30	3.984	1	1	1	17	12	16
22.64	22.64	3.924	-1	1	1	8		7
23.20	23.20	3.831	-1	1	0	44	31	43
24.00	24.02	3.704	-1	3	1	46	30	44
24.35	24.36	3.652	-1	-3	1	15	11	14
24.74	24.76	3.595	-1	2	1	13	11	13
24.94	24.94	3.566	-1	-1	1	6	9	6
25.55	25.56	3.484	-1	1	2	40	29	39
25.64		3.471	-2	2	1	2		2
25.68	25.68	3.466	1	1	1	29	26	28
26.44	26.44	3.368	2	2	0	67	43	67
27.11	27.12	3.286	-2	2	0	73	49	73
27.39	27.46	3.253	-2	0	2	61	100	61
27.45		3.247	2	0	0	100		100
27.52		3.238	0	0	0	42		42
29.43	29.44	3.033	-1	3	1	33	20	33
30.15	30.22	2.961	-2	-2	2	6	27	6
30.23		2.954	1	3	1	42		43
30.68	30.74	2.912	0	2	2	8	17	8
30.73		2.907	0	4	1	19		19
30.80		2.900	-2	2	2	3		3
30.93	30.92	2.889	0	-4	1	13	13	13
32.15	32.16	2.782	-1	-3	2	11	7	11
32.43	32.48	2.758	-3	1	1	1	10	1
32.47		2.755	-1	3	2	16		17
34.24	34.24	2.616	-3	-1	1	9	18	10
34.24		2.616	-2	-4	1	21		22
34.28		2.614	2	2	2	1		1
34.72	34.72	2.581	-3	1	2	13	8	14

2THETA	PEAK	D	H	K	L	I(INT)	I(PK)	I(DS)
34.88	34.86	2.570	1	1	2	8	7	9
35.12	35.14	2.553	2	-2	2	3	4	4
35.15		2.551	-1	-1	3	3		3
35.36	35.36	2.536	3	1	1	5	4	5
35.48	35.58	2.528	-2	4	0	2	15	2
35.58		2.521	-2	4	1	26		27
35.91	35.92	2.499	-3	1	0	6	4	6
36.93	36.94	2.432	-1	-5	1	7	6	8
36.94		2.431	-2	4	0	3		3
37.07	37.06	2.423	-3	-3	1	11	8	12
37.62	37.62	2.389	-1	5	1	4	2	4
38.53	38.60	2.334	-3	3	1	6	7	6
38.56		2.333	-1	-1	3	4		5
38.61		2.330	-3	-1	3	6		6
39.21	39.20	2.296	-1	3	3	3	2	3
40.10	40.12	2.247	-3	-3	0	1	1	1
40.13		2.245	-2	-2	2	1		1
40.49	40.52	2.226	3	3	1	2	2	2
40.52		2.224	-1	5	5	2		2
40.53		2.224	-2	2	2	1		2
41.69	41.80	2.165	0	0	3	1	22	1
41.79		2.160	2	0	1	2		14
41.81		2.159	0	6	1	12		35
42.75	42.76	2.113	-4	0	2	6	3	6
42.90	42.88	2.106	-4	0	1	5	5	7
43.21	43.20	2.092	2	-4	1	5	3	6
43.48	43.58	2.079	3	1	2	2	3	2
43.56		2.076	2	0	3	3		3
43.59		2.074	-1	-3	1	1		1
43.98	44.06	2.057	3	-1	2	2	2	2
44.06		2.053	0	-6	1	3		3
44.28	44.28	2.044	0	6	1	5	3	5
44.69	44.70	2.026	-4	-2	2	6	3	7
45.49	45.50	1.9920	2	2	2	9	5	10
45.76	45.76	1.9810	-4	2	2	13	7	15
45.93	45.88	1.9742	-3	-3	3	3	5	3
46.23	46.24	1.9619	2	-2	2	7	4	8

MAXIMUM MICROCLINE - BROWN AND BAILEY, 1964

2THETA	PEAK	D	H	K	L	I(INT)	I(PK)	I(DS)
46.38	46.36	1.9562	-3	-5	1	3	4	3
47.18	47.18	1.9246	4	0	0	16	8	19
47.54	47.58	1.9110	3	3	3	5	5	
47.58		1.9096	-4	0	3	9		11
48.04	48.04	1.8922	-2	6	1	2	1	3
48.18	48.18	1.8862	-3	-5	2	1	1	1
48.40	48.40	1.8790	-3	5	1	1	1	2
48.76	48.78	1.8660	1	1	3	6	4	7
48.78		1.8651	4	2	0	1		2
48.92	48.92	1.8602	3	5	0	3	5	3
48.94		1.8596	3	-3	1	1		2
48.99	49.00	1.8576	1	5	2	2		2
49.00		1.8573	-1	-1	3	4	5	5
49.15	49.14	1.8520	-2	6	0	3	4	3
49.27	49.26	1.8480	-4	-2	3	1	2	1
49.90	49.92	1.8261	-2	-6	2	2	2	2
49.92		1.8253	-4	2	0	2	2	2
50.03	50.04	1.8216	-1	-5	2	1		1
50.20	50.20	1.8158	-4	2	3	1	2	1
50.21		1.8156	1	7	0	1		1
50.49	50.56	1.8060	0	4	3	5	20	6
50.55		1.8042	0	6	2	5		7
50.56		1.8037	-4	-4	1	4		5
50.57		1.8034	-2	0	4	29		36
50.75	50.68	1.7974	-4	-4	2	8	13	3
50.88	50.90	1.7932	0	-4	3	8	8	9
50.93		1.7914	0	-6	2	8		10
51.18	51.16	1.7832	-2	6	2	4	4	5
51.63	51.62	1.7688	2	4	4	3	2	3
52.63	52.66	1.7375	-4	4	1	4	3	5
52.69		1.7357	-4	4	2	1		2
52.71		1.7350	1	3	3	1		1
52.74	52.72	1.7343	-1	-1	4	2	3	2
52.97	52.96	1.7272	2	-4	2	2	2	3
53.40	53.40	1.7143	1	-3	3	1	1	2
53.65	53.66	1.7069	1	7	1	1	1	1
53.78	53.78	1.7032	-5	-1	2	1	1	1

2THETA	PEAK	D	H	K	L	I(INT)	I(PK)	I(DS)
54.07	54.08	1.6945	-3	-5	3	3	2	2
54.38	54.38	1.6857	-5	1	2	1	1	2
54.60	54.52	1.6795	3	1	2	1	1	1
55.10	55.10	1.6654	-5	-1	1	1	1	1
55.39	55.40	1.6573	3	-5	1	5	3	6
55.54	55.56	1.6531	-1	-7	2	3	3	4
55.59		1.6517	-3	5	3	5		6
56.03	56.04	1.6398	1	7	2	2	1	2
56.73	56.80	1.6214	-1	3	4	1	2	1
56.74		1.6211	4	1	1	1		1
56.75		1.6208	-3	3	4	1		1
56.81		1.6191	-3	0	0	3		4
56.95	56.94	1.6155	-5	1	3	2	2	2
57.14	57.12	1.6106	2	0	3	1	1	1
57.46	57.48	1.6024	-4	-2	4	4	4	6
57.48		1.6020	3	-1	5	4		6
58.24	58.24	1.5827	-4	2	4	4	2	5
58.45	58.44	1.5776	0	-2	4	4	3	5
58.68	58.68	1.5719	0	-2	4	5	3	6
58.83	58.84	1.5682	0	-8	1	2	2	2
59.39	59.40	1.5549	2	-2	3	1	1	2
59.67	59.68	1.5482	-5	-3	2	3	3	8
59.90	59.94	1.5428	-3	-7	1	1	3	1
59.94		1.5419	-4	-6	1	6		8
60.15	60.10	1.5369	-5	3	1	1	2	1
60.26	60.26	1.5345	0	6	3	1	1	1
60.27		1.5341	5	5	1	1		1
60.72	60.78	1.5241	1	7	2	6	1	1
60.77		1.5227	0	-6	3	1		3
61.06	61.06	1.5163	2	6	2	2	8	1
61.16	61.16	1.5140	2	8	0	16		22
61.27		1.5116	-5	3	3	4		6
62.14	62.14	1.4924	-2	8	1	1	1	1
62.63	62.72	1.4819	-5	-1	4	1	3	2
62.71		1.4802	-4	6	0	5		7
63.11	63.12	1.4719	-2	8	0	13	7	17
63.13		1.4715	-5	1	4	4		5

MAXIMUM MICROCLINE - BROWN AND BAILEY, 1964

2THETA	PEAK	D	H	K	L	I(INT)	I(PK)	I(DS)
63.26	63.28	1.4687	5	3	0	2	6	3
63.32		1.4675	-2	-8	2	2		2
63.41		1.4657	4	6	0	2		2
63.62	63.62	1.4612	-1	-7	3	2	2	3
63.88		1.4560	-1	7	3	2		3
63.91	63.94	1.4553	2	-4	4	2	6	1
63.92		1.4551	2	4	3	7		10
63.98		1.4539	-4	-6	3	2		3
64.00		1.4535	0	8	2	2		3
64.01		1.4534	-5	-5	2	2		2
64.01		1.4532	-1	-1	4	2		2
64.96		1.4344	-2	0	5	1		1
65.00	65.08	1.4336	3	5	2	1	2	1
65.05		1.4325	-5	3	0	2		3
65.09		1.4319	-3	-1	5	1		1
65.21	65.22	1.4295	2	-4	3	5	3	8
65.49	65.48	1.4241	-6	0	2	9	5	13
66.26	66.32	1.4093	-4	6	0	1	5	2
66.32		1.4082	4	0	6	9		13
66.33		1.4079	-4	-6	3	1		1
66.43		1.4062	-5	-5	3	1		2
66.50	66.50	1.4047	-1	9	0	2	4	3
66.69	66.68	1.4012	-5	5	0	2	2	3
67.05	67.06	1.3946	-4	4	5	4	2	5
67.12		1.3934	-1	-3	4	1		2
67.25	67.24	1.3909	-2	-6	4	3	2	4
67.96	68.00	1.3781	-6	0	0	1	2	2
68.01		1.3773	-2	6	4	2		3
68.10		1.3757	-1	1	5	3		3
68.47	68.48	1.3690	-4	-2	5	1	2	4
68.90	68.94	1.3617	-6	2	3	2	2	1
68.93		1.3610	-5	5	5	2		3
69.04	69.04	1.3593	-6	-2	1	3	2	4
69.13		1.3576	-4	2	5	1		1
69.64	69.66	1.3490	-2	-8	3	1	1	1
69.75	69.74	1.3472	5	5	0	1	1	1
70.09	70.10	1.3413	-1	-9	2	3	1	5

2THETA	PEAK	D	H	K	L	I(INT)	I(PK)	I(DS)
70.36	70.32	1.3369	-6	2	1	1	1	2
70.65	70.64	1.3322	-1	9	1	1	2	7
70.77		1.3301	-1	7	3	1		1
71.01	71.00	1.3263	-6	0	4	4	2	5
71.18	71.20	1.3235	-6	-4	2	1	1	4
71.53	71.58	1.3178	-1	3	5	3	3	7
71.59		1.3170	-1	-3	5	5		2
71.74	71.76	1.3145	-2	-4	5	2	2	3
71.94	71.94	1.3113	-4	-8	1	2	4	3
72.12	72.16	1.3085	-2	4	5	1		2
72.15		1.3081	-4	-8	2	5		8
72.22		1.3069	4	4	2	1		2
72.29	72.34	1.3059	2	0	4	4	4	6
72.35		1.3049	2	6	3	1		1
72.40		1.3042	-5	1	1	2		4
72.97	73.12	1.2953	0	10	0	2	4	3
73.11		1.2932	2	8	2	6		10
73.14		1.2927	0	-6	4	1		2
73.17		1.2923	5	-3	1	1		2
73.21		1.2918	0	-8	3	1		2
73.30	73.30	1.2904	-4	-4	5	1	3	2
73.52	73.50	1.2871	-6	-4	1	2	2	4
73.68	73.76	1.2846	-6	4	2	1	3	1
73.74		1.2837	-3	-9	2	2		4
73.78		1.2831	-6	0	0	3		4
73.91	73.92	1.2811	-5	-7	2	1	2	2
74.55	74.58	1.2718	-1	-7	4	1	2	1
74.57		1.2715	-1	0	5	1		2
74.57		1.2715	-4	-2	5	3		4
74.80	74.80	1.2682	-6	2	2	1	2	1
74.91	75.00	1.2665	-5	-3	5	3	2	1
74.96		1.2658	3	7	2	1		1
75.02		1.2650	-3	-5	3	3		4
75.31	75.32	1.2608	-2	8	2	8	6	12
75.32		1.2607	-4	8	2	9		13
75.96		1.2516	-2	-10	1	1		1
76.05	76.08	1.2504	-3	1	5	2	2	3

2THETA	PEAK	D	H	K	L	I(INT)	I(PK)	I(DS)
76.09		1.2498	-6	4	1	2		3
76.13		1.2493	-6	2	0	1		3
76.19		1.2484	-5	3	5	2		2
77.23	77.22	1.2342	3	5	1	1	0	1
77.54	77.54	1.2300	5	5	1	1	1	2
78.24	78.32	1.2208	-1	5	5	1	0	1
78.64	78.66	1.2156	-6	8	0	1	1	2
78.68		1.2151	-6	0	2	1		1
78.96	78.90	1.2115	-6	-6	5	1		2
79.13	79.20	1.2093	6	4	0	1	1	1
79.18		1.2086	0	4	5	2		3
79.26		1.2076	3	-5	3	1		2
79.87	80.02	1.1999	0	-10	6	2	2	2
80.00		1.1982	-3	-1	5	1		3
80.01		1.1982	-2	-6	6	2		2
80.18	80.28	1.1960	-3	1	3	3	2	3
80.30		1.1945	-2	10	6	2		4
80.32		1.1943	1	1	5	2		4
80.41		1.1932	5	-5	1	1		1
80.55	80.54	1.1915	2	10	0	1	2	2
80.56		1.1913	-2	6	5	1		2
81.08	81.12	1.1851	-2	0	6	1	1	2
81.13		1.1844	-7	-3	3	2		4
81.33	81.34	1.1821	-7	-3	2	1	1	1
81.49	81.48	1.1801	4	8	1	1	1	2
81.57		1.1791	-7	-1	4	1		1
81.75	81.74	1.1771	-6	-4	0	3	1	5
82.48	82.48	1.1684	-4	-8	4	1	1	1
82.58		1.1672	-6	6	2	1		1
82.70	82.72	1.1659	6	0	1	1	1	1
83.11	83.16	1.1612	-3	-3	6	3		3
83.18		1.1603	-7	3	3	1		1
83.47	83.46	1.1571	-6	6	3	1	1	1
83.67	83.68	1.1549	2	2	3	1	1	1
84.02	84.04	1.1509	0	8	4	3	2	3
84.07		1.1503	-5	-1	6	1		2
84.18	84.28	1.1491	-6	-4	5	1	1	1
84.29		1.1479	-6	-6	4	1		1
84.46	84.46	1.1460	-2	-10	6	2		2
84.50		1.1456	-3	-7	3	1		1
84.71	84.82	1.1433	0	-8	5	1		2
84.82		1.1420	-5	7	4	3	2	4
84.86		1.1416	-4	8	4	1		2
85.14	85.16	1.1386	1	1	6	3	2	2
85.16		1.1384	4	4	3	1		4
85.25		1.1374	2	6	4	3		5
86.21	86.20	1.1272	-5	-9	2	1	1	1
86.60	86.62	1.1231	-7	-5	3	1	1	2
86.97	86.96	1.1193	-4	-4	4	2	1	4
87.28	87.26	1.1161	-1	-9	2	1	1	4
87.69	87.72	1.1119	3	9	1	2	1	2
87.73		1.1115	-1	9	2	1		2
87.76		1.1112	6	4	1	1		1
87.86		1.1102	5	5	4	1		1
88.34	88.38	1.1054	7	1	2	1	1	1
88.94	88.94	1.0995	1	5	0	1	1	3
89.63	89.68	1.0928	-7	-1	5	2	1	3
89.69		1.0922	3	-11	1	1		2
89.89	89.92	1.0903	3	11	0	1	1	1
90.28	90.38	1.0866	-7	5	3	1		2
90.36		1.0859	-6	6	0	2	1	4
90.52	90.56	1.0843	-6	-6	0	1		1
90.54		1.0842	-7	-5	1	1		1
90.55		1.0841	6	-4	1	1	2	2
90.79	90.80	1.0818	1	-8	5	3		5
90.83		1.0815	-6	-8	2	2	2	3
91.00		1.0799	1	11	2	1		2
91.28	91.30	1.0773	-5	9	2	1	1	1
91.32		1.0770	5	-5	5	1		1
91.43		1.0760	-2	-8	3	1		2
91.50		1.0754	7	3	0	1		1
91.78	91.76	1.0728	-6	-6	5	1	1	3

MAXIMUM MICROCLINE - BROWN AND BAILEY, 1964

2THETA	PEAK	D	H	K	L	I(INT)	I(PK)	I(DS)
91.78		1.0728	5	9	0	1		2
92.12	92.10	1.0697	3	5	4	1	1	2
92.19		1.0691	-8	0	3	1		1
92.52	92.50	1.0661	0	12	2	1	1	1
92.90	92.86	1.0628	-3	11	3	1	1	2
93.02		1.0617	-8	-2	4	1		1
93.55	93.72	1.0571	-2	-10	4	1	1	1
93.69		1.0558	-7	3	1	1		1
93.74		1.0554	-2	-12	0	1		1
93.84		1.0546	-3	11	0	1		2
94.18	94.16	1.0516	-7	5	1	1	1	1
94.23		1.0512	3	-5	4	1		1
94.54	94.72	1.0486	-3	-11	3	1		1
94.71		1.0471	-2	10	4	2		3
94.91	94.90	1.0455	-6	6	5	2	1	3
95.40	95.22	1.0414	-6	8	3	1	1	2
96.04	96.04	1.0362	-2	12	2	1	1	2
96.64	96.66	1.0313	-5	9	0	1	1	3
96.89	96.94	1.0293	-4	0	7	1	1	2
97.00		1.0284	6	-2	2	1		2
98.17	98.26	1.0193	1	-9	4	1	1	1
98.26		1.0186	5	-3	3	1		1
98.27		1.0185	-4	-2	7	1		1
98.28		1.0184	-8	0	1	2		3
99.02	99.08	1.0128	-3	-7	6	2	1	1
99.06		1.0125	-8	0	5	1		2
99.31	99.34	1.0106	-3	9	5	1	1	2
99.96	99.94	1.0058	-8	-2	7	1	0	2
101.08	101.08	0.9977	-3	3	6	1	1	2
102.46	102.46	0.9879	-5	-7	8	1	1	2
102.70	102.70	0.9862	-6	-8	5	1	1	2
103.98	104.22	0.9776	-4	-12	1	1	1	2
104.23		0.9759	3	3	5	1	1	1
104.53	104.54	0.9739	3	-7	3	1	1	1
105.28	105.58	0.9690	4	10	2	1	1	1
105.62		0.9669	3	-3	5	1		2
106.02	106.28	0.9643	-2	-10	5	1	1	1

2THETA	PEAK	D	H	K	L	I(INT)	I(PK)	I(DS)
106.23		0.9630	0	-12	3	1		2
106.32		0.9624	-2	-4	1	1		2
107.24	107.28	0.9567	-8	4	5	1	0	1
107.55	107.58	0.9548	-8	0	6	1	0	1
107.92	107.96	0.9526	-3	-13	1	1	0	2
108.44	108.42	0.9494	-5	-5	7	1	0	2
109.06	109.08	0.9458	-4	12	1	1	1	2
109.44	109.44	0.9435	-3	-13	2	1	0	2
109.95	110.02	0.9406	-2	0	6	1	1	3
110.00		0.9403	-9	1	4	1		3
110.14	110.36	0.9395	-6	10	2	1	1	3
110.33		0.9384	4	-10	2	1		3
110.37		0.9382	-5	5	5	2		3

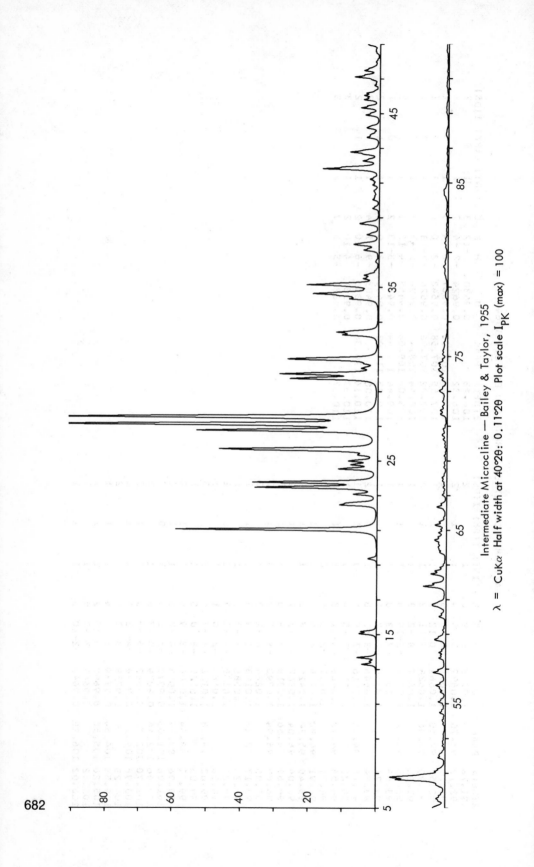

Intermediate Microcline — Bailey & Taylor, 1955

$\lambda = $ CuKα Half width at 40°2θ: 0.11°2θ Plot scale I$_{PK}$ (max) = 100

682

INTERMEDIATE MICROCLINE - BAILEY AND TAYLOR, 1955

2THETA	PEAK	D	H	K	L	I(INT)	I(PK)	I(DS)
13.26	13.28	6.669	-1	1	0	3	4	3
13.43	13.44	6.585	-1	0	0	1	2	1
13.65	13.66	6.483	1	0	0	2	6	2
13.65		6.479	0	2	0	3		3
15.02	15.04	5.893	-1	-1	1	4	5	4
15.15	15.16	5.842	-1	1	1	3	5	3
19.34	19.34	4.587	-1	0	1	2	3	2
21.01	21.02	4.225	-2	0	1	55	59	52
22.43	22.44	3.961	1	1	1	9	11	9
22.55	22.54	3.940	1	0	1	6	10	6
23.05	23.06	3.856	2	0	0	6	7	6
23.44	23.44	3.792	0	3	0	34	36	33
23.73	23.74	3.746	1	3	0	34	37	34
24.51	24.52	3.629	-1	3	0	11	12	10
24.75	24.76	3.593	-1	-3	1	6	8	6
24.97	24.98	3.563	-2	2	1	7	9	7
25.31	25.32	3.515	-2	-2	1	4	6	4
25.59	25.66	3.477	-1	-1	2	30	46	29
25.66		3.468	-1	1	2	25		25
26.71	26.72	3.335	-2	2	0	52	53	52
27.06	27.08	3.293	-2	2	0	50	100	50
27.08		3.290	-2	0	0	54		54
27.49	27.50	3.242	0	0	2	74	97	74
27.51		3.240	0	4	0	26		26
29.70	29.70	3.005	1	3	1	27	26	27
29.97	29.98	2.979	-1	-3	2	30	29	30
30.31	30.32	2.946	-2	2	2	4	5	4
30.58	30.60	2.921	-2	-2	2	3	4	3
30.80	30.84	2.901	-2	2	2	4	27	5
30.81		2.900	0	4	1	12		13
30.84		2.897	0	2	2	4		4
30.85		2.896	0	-4	1	11		11
32.17	32.24	2.780	-3	1	1	10	11	10
32.24		2.774	-1	-3	2	11		12
32.40	32.40	2.761	-1	3	2	9	12	9
34.30	34.30	2.612	-3	-1	1	9	9	19
34.55	34.60	2.593	2	2	1	1		1

2THETA	PEAK	D	H	K	L	I(INT)	I(PK)	I(DS)
34.61		2.589	-2	-4	1	16		17
34.85	34.84	2.572	2	-1	1	2	5	2
35.02	35.12	2.560	1	-1	1	5	21	5
35.11		2.554	1	-1	2	3		4
35.12		2.553	-2	4	1	18		19
35.47	35.48	2.528	3	1	0	4	5	4
35.68	35.68	2.514	-1	-3	1	4	4	4
35.92	35.90	2.498	-2	4	0	1	1	1
36.45	36.46	2.463	-2	-3	0	1	1	1
37.11	37.12	2.420	-1	5	1	5	5	5
37.45	37.44	2.403	-1	5	1	3	7	4
38.00	38.00	2.399	-3	-3	1	6		7
38.61	38.64	2.366	-3	3	1	4	4	5
38.64		2.330	-1	-1	3	3	6	5
39.52	39.52	2.328	-3	-3	3	4		3
40.02	40.02	2.278	-3	-3	1	2	2	2
40.21	40.22	2.251	-2	-2	3	2	2	2
40.25		2.241	1	-3	2	1	1	1
40.48		2.239	1	-3	0	1		1
40.54	40.52	2.227	3	3	0	1	2	1
40.86	40.86	2.223	1	-5	1	2		2
41.20	41.20	2.207	1	-5	1	2	3	3
41.79	41.80	2.189	0	6	0	21	17	24
42.25	42.24	2.160	0	-4	1	6	5	6
42.74	42.76	2.137	2	4	1	4	9	4
42.77		2.114	-4	0	2	5		5
43.60	43.62	2.112	-1	-3	2	2	4	3
43.61		2.074	-1	3	3	3		3
43.63		2.074	3	-1	1	2		1
43.80	43.72	2.073	3	-1	1	2		2
44.14		2.065	0	-2	1	3	3	2
44.18	44.16	2.050	0	-6	1	3	4	3
44.90	44.90	2.048	-4	-2	2	5		3
45.31	45.32	2.017	-4	2	2	7	4	6
45.78	45.78	1.9997	-3	-1	2	6	5	8
46.03	46.02	1.9803	2	-2	2	5	5	7
		1.9702	2	-2	2		5	6

2THETA	PEAK	D	H	K	L	I(INT)	I(PK)	I(DS)
46.17	46.16	1.9643	-3	-3	3	2	4	2
46.60	46.60	1.9474	-3	3	3	2	2	2
47.10	47.10	1.9279	4	0	0	9	7	10
47.46	47.46	1.9139	-4	0	3	5	5	6
47.52		1.9116	-2	6	1	1		1
47.97	47.96	1.8947	3	3	1	1	1	1
48.47	48.48	1.8765	-3	1	0	1	1	1
48.57	48.58	1.8728	-2	-6	2	1	1	1
48.68	48.68	1.8688	-3	-5	2	1	1	1
48.93	49.00	1.8600	1	1	3	3	4	4
49.00		1.8574	1	-1	2	3		3
49.38	49.40	1.8441	1	5	1	3	2	2
49.44		1.8420	-4	-2	3	1		1
49.48	49.52	1.8406	4	2	0	2	3	2
49.54		1.8384	3	5	0	2		1
49.70	49.68	1.8327	1	-5	2	1	2	1
50.24	50.32	1.8144	-2	-6	2	1	3	1
50.31		1.8120	-3	5	0	2		2
50.60	50.60	1.8024	-2	0	4	17	15	21
50.71	50.74	1.7991	0	2	3	4	16	5
50.77		1.7987	0	6	2	3		4
50.77		1.7967	-2	-6	2	2		2
50.78		1.7964	0	-4	3	4		5
50.79		1.7960	0	-6	2	4		5
51.10	51.10	1.7857	-4	-4	1	3	3	3
51.23	51.22	1.7816	-4	4	2	1	2	1
51.88	51.88	1.7608	-4	4	1	3	2	3
52.11	52.10	1.7537	2	2	2	1	2	1
52.56	52.56	1.7397	2	4	1	1	1	1
52.80	52.80	1.7322	-1	-4	4	1	1	1
53.23	53.22	1.7193	-1	3	3	1	1	1
53.80	53.80	1.7025	1	-3	2	1	1	1
54.03	54.02	1.6958	-5	1	1	2	1	2
54.47	54.46	1.6832	-5	1	3	3	2	3
55.09	55.10	1.6656	3	-5	3	2	2	2
55.65	55.66	1.6501	-1	-7	2	2	2	2
55.90	55.90	1.6434	-1	7	2	1	2	2

2THETA	PEAK	D	H	K	L	I(INT)	I(PK)	I(DS)
56.04	56.04	1.6395	3	5	1	3	3	4
56.78	56.78	1.6199	3	-5	0	3	4	4
56.79		1.6198	4	-8	0	2		3
57.61	57.62	1.5986	-4	-2	4	3	2	4
57.93	57.94	1.5905	-4	2	4	3	2	3
58.64	58.68	1.5729	0	2	4	2	3	3
58.69		1.5717	0	0	2	3		4
59.24	59.22	1.5585	2	-2	3	1	1	1
60.00	60.00	1.5405	-5	-3	3	3	2	4
60.61	60.70	1.5264	0	-6	3	1	4	1
60.63		1.5261	-5	3	3	3		4
60.71		1.5242	-5	-1	5	3		4
61.74	61.74	1.5011	-2	8	0	7	7	10
61.75		1.5010	-4	-8	1	3		4
62.45	62.46	1.4857	-2	8	0	7	5	9
62.64	62.62	1.4818	-5	-1	4	2	3	2
62.84	62.82	1.4776	-5	1	5	2	2	2
63.67	63.72	1.4602	-1	-7	3	1	2	1
63.72		1.4592	5	3	0	2		2
63.86	63.88	1.4564	-1	7	3	1	2	2
63.98	64.00	1.4539	1	1	3	1	2	2
64.05		1.4525	-1	-1	4	2		2
64.24	64.24	1.4487	4	6	0	1	2	2
64.38	64.42	1.4459	-5	3	3	1	4	1
64.43		1.4449	2	-4	2	2		2
64.59	64.58	1.4417	-4	-6	1	3	3	3
64.85	64.84	1.4366	-4	4	3	1	2	1
65.28	65.28	1.4281	-2	-4	2	3	3	4
65.28		1.4280	-1	0	0	2		6
66.32	66.32	1.4081	-4	6	2	4	3	6
67.01	67.02	1.3954	-4	0	5	1	1	1
67.04		1.3948	-5	0	3	5		6
67.46	67.46	1.3872	-5	-2	4	1	1	1
67.83	67.84	1.3805	-2	-6	4	1	1	2
68.02	68.02	1.3771	-3	5	5	1	1	2
68.61	68.62	1.3666	-1	-7	2	1	1	2
68.90	68.88	1.3617	-4	2	5	1	1	1

INTERMEDIATE MICROCLINE – BAILEY AND TAYLOR, 1955

2THETA	PEAK	D	H	K	L	I(INT)	I(PK)	I(DS)
69.27	69.28	1.3552	-6	-2	1	1	1	2
69.76	69.76	1.3468	-6	2	1	1	1	1
70.21	70.20	1.3394	-1	-9	2	2	1	2
70.48	70.48	1.3348	-1	9	2	2	2	3
70.81	70.80	1.3296	-6	0	4	2	2	2
71.64	71.66	1.3161	-1	-3	5	2	2	3
71.69		1.3154	-1	3	5	2		2
71.88	71.88	1.3123	-2	-1	4	1	2	1
71.90		1.3120	5	3	1	1		1
72.10	72.10	1.3088	-2	4	5	1	1	1
72.41	72.40	1.3040	-2	0	4	1	1	2
72.53	72.52	1.3021	5	-3	1	1	1	1
72.94	73.00	1.2958	0	10	0	1	2	2
73.02		1.2947	-4	-8	2	2		3
73.02		1.2945	2	6	3	2		1
73.61	73.64	1.2857	-4	-4	5	1	2	1
73.64		1.2853	6	0	0	1		1
73.85	73.86	1.2821	2	8	2	3	2	4
74.15	74.22	1.2777	-6	-4	1	1		3
74.16		1.2775	-4	4	5	1		2
74.24		1.2763	-4	8	2	3	2	5
74.44	74.44	1.2734	-3	-9	2	1	2	2
74.59	74.58	1.2712	2	-8	2	3	3	4
75.10	75.10	1.2638	-6	4	1	1	1	1
75.18		1.2627	-5	-3	5	1		1
75.30	75.32	1.2609	-3	5	5	1	1	1
75.44		1.2590	-3	9	2	1		1
75.71	75.76	1.2552	-5	3	5	1	1	1
75.78		1.2542	-3	5	5	1		1
78.50	78.48	1.2174	-6	0	5	1	1	1
79.76	79.76	1.2013	-2	10	2	1	1	1
80.10	80.16	1.1970	-3	-1	6	1	1	2
80.19		1.1959	-3	1	6	1		1
81.50	81.50	1.1800	-7	-3	3	1	1	1
82.28	82.28	1.1707	-7	3	3	1	1	1
83.17	83.18	1.1604	-4	-8	4	1	1	2
84.27	84.30	1.1481	-4	8	4	1	1	2

2THETA	PEAK	D	H	K	L	I(INT)	I(PK)	I(DS)
84.27	84.40	1.1481	-5	1	6	1	1	1
84.40		1.1466	0	8	4	1	1	2
84.57	84.54	1.1448	0	-8	4	1	1	2
85.97	85.98	1.1297	4	4	3	1	1	2
86.66	86.66	1.1224	4	-4	3	1	1	2
87.75	87.76	1.1114	-7	-5	3	1	1	1
88.99	89.04	1.0990	7	1	0	1	1	1
89.04		1.0985	-7	5	3	1	1	1
89.26	89.28	1.0964	-7	1	0	1	0	1
90.20	90.22	1.0874	1	-5	5	1	1	1
90.53	90.54	1.0843	-1	5	5	1	1	2
92.52	92.58	1.0661	-6	-6	5	1	1	1
93.78	93.80	1.0551	-6	6	6	1	1	1

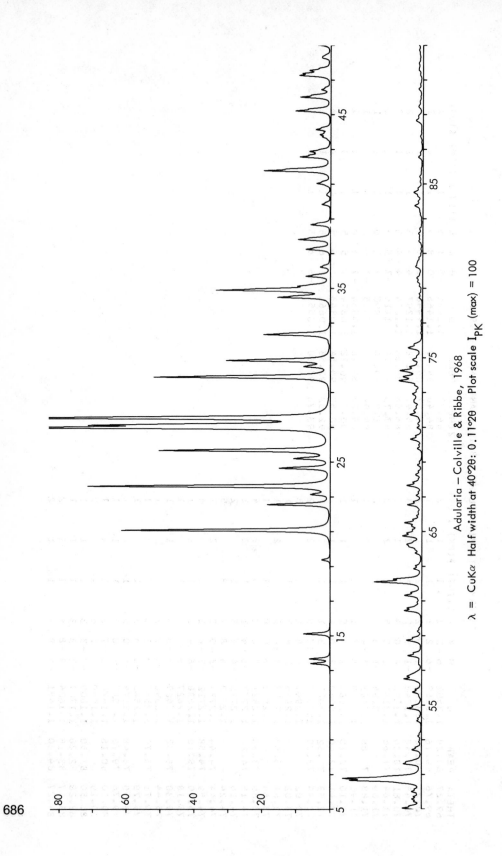

Adularia – Colville & Ribbe, 1968

λ = CuKα Half width at 40°2θ: 0.11°2θ Plot scale I$_{PK}$ (max) = 100

686

ADULARIA - COLVILLE AND RIBBE, 1968

2THETA	PEAK	D	H	K	L	I(INT)	I(PK)	I(DS)
13.38	13.38	6.613	1	1	0	5	6	4
13.64	13.66	6.485	0	2	0	5	6	3
13.66	13.66	6.477	1	0	1	2	1	1
15.10	15.10	5.863	-1	1	1	7	8	6
19.35	19.36	4.583	0	2	1	2	2	2
21.06	21.06	4.215	-2	0	1	59	62	57
22.53	22.54	3.943	2	0	1	18	19	17
23.12	23.12	3.844	1	3	0	5	6	5
23.59	23.60	3.768	2	2	0	72	72	70
24.62	24.64	3.612	-1	3	1	15	15	15
25.18	25.18	3.534	-2	2	1	10	11	10
25.64	25.64	3.471	-1	1	2	52	50	51
26.94	26.94	3.307	2	2	0	100	100	100
27.11	27.10	3.287	-2	0	2	60	71	60
27.48	27.52	3.242	-2	0	2	30	100	30
27.52	27.52	3.239	0	0	2	78		78
29.86	29.86	2.990	1	3	1	57	52	58
30.46	30.46	2.932	-2	2	2	7	8	8
30.81	30.82	2.899	2	4	0	25	31	26
30.83		2.897	0	0	2	10		10
32.32	32.32	2.768	-1	2	2	22	20	23
34.46	34.46	2.600	-3	1	1	17	16	19
34.88	34.88	2.570	-2	4	1	37	34	40
35.12	35.12	2.553	1	1	2	8	10	8
35.68	35.68	2.514	3	1	0	8	7	9
36.21	36.22	2.478	2	4	0	3	3	4
37.23	37.24	2.413	-1	5	1	9	7	9
37.79	37.80	2.378	-3	3	1	12	9	13
38.65	38.64	2.328	-1	3	3	7	6	8
39.82	39.82	2.262	-3	3	2	3	3	3
40.33	40.34	2.234	-2	2	3	3	1	1
40.40	40.40	2.231	1	3	1	1	1	1
41.03	41.02	2.198	1	5	1	5	4	5
41.75	41.76	2.162	0	6	0	25	20	28
41.80		2.159	0	0	3	1		1
42.54	42.54	2.123	2	4	1	11	9	13
42.77	42.78	2.112	-4	0	1	3	5	4
42.88	42.88	2.107	-4	0	2	5	6	6
43.70	43.70	2.070	2	0	2	2	2	3
43.84	43.84	2.063	3	1	1	3	3	3
44.13	44.12	2.050	0	6	1	5	4	6
45.20	45.20	2.004	-4	2	2	14	10	16
45.99	46.00	1.9716	2	2	2	12	9	14
46.42	46.42	1.9544	-3	3	2	3	3	4
47.21	47.26	1.9234	-2	6	0	1	1	1
47.25		1.9220	-2	0	3	11		13
47.36	47.36	1.9178	4	0	1	3	8	4
47.55	47.54	1.9108	-3	5	1	6	6	8
48.26	48.28	1.8842	-2	6	1	2	3	3
48.33		1.8816	3	1	1	2		2
49.03	49.04	1.8563	-1	5	2	7	6	9
49.06		1.8551	-3	5	1	1		1
49.42	49.42	1.8427	4	2	0	2	2	3
49.56	49.56	1.8377	1	5	2	2	3	2
49.70	49.70	1.8328	-4	2	3	1	1	2
49.99	49.98	1.8230	3	5	0	3	3	4
50.63	50.64	1.8014	-2	6	2	23	21	28
50.73	50.74	1.7980	0	0	4	9	23	12
50.76		1.7971	0	4	3	10		12
51.59	51.60	1.7699	-4	4	1	7	5	8
51.69	51.70	1.7670	-4	4	2	3	5	3
52.40	52.40	1.7445	-2	4	3	4	3	5
52.85	52.84	1.7309	-1	1	4	1	1	1
53.19	53.18	1.7206	1	3	3	1	1	2
54.06	54.06	1.6948	-5	1	2	2	1	2
54.80	54.80	1.6739	-3	5	2	6	4	8
55.74	55.74	1.6477	-1	7	1	4	3	5
56.50	56.50	1.6274	-3	5	3	7	5	9
56.73	56.64	1.6237	-5	1	3	1	4	2
57.33	57.36	1.6057	2	0	3	3	1	3
57.41		1.6038	0	8	0	1		1
57.83	57.84	1.5930	-4	2	4	6	4	8
58.65	58.72	1.5727	0	8	1	1	5	1

ADULARIA - COLVILLE AND RIBBE, 1968

2THETA	PEAK	D	H	K	L	I(INT)	I(PK)	I(DS)
58.72		1.5711	0	2	4	6		8
59.23	59.24	1.5586	-2	2	3	1	1	2
60.43	60.44	1.5306	-5	3	3	8	5	10
60.56	60.58	1.5276	0	6	0	2	4	3
60.90	60.90	1.5199	1	5	3	1	1	1
60.93		1.5193	-3	7	2	1		1
61.30	61.30	1.5108	-4	6	1	8	5	10
62.08	62.08	1.4938	2	8	0	22	14	31
62.84	62.84	1.4775	-5	0	4	4	2	5
63.00	63.00	1.4742	4	4	1	1	2	1
63.74	63.74	1.4588	-1	7	3	3	2	4
64.10	64.12	1.4514	1	1	4	4	4	6
64.19	64.24	1.4497	0	8	2	4	5	2
64.24		1.4487	-1	5	4	1		1
64.24		1.4487	5	3	0	3		4
64.73	64.74	1.4389	2	4	3	9	6	12
64.86	64.90	1.4363	4	6	0	2	5	3
65.06	65.06	1.4316	-4	6	3	2	3	3
65.25		1.4287	-3	1	5	1		1
65.32	65.30	1.4274	-5	5	2	2	3	3
65.49	65.48	1.4240	-6	0	2	7	5	10
65.88	65.88	1.4164	1	9	0	2	2	2
66.14	66.10	1.4115	3	5	2	1	1	1
66.52	66.52	1.4045	4	0	2	7	4	10
67.07	67.06	1.3943	-4	4	5	3	2	4
67.63	67.64	1.3841	-5	5	3	2	5	3
67.64		1.3839	-2	6	4	4		5
67.65		1.3838	1	3	4	2		2
68.02	68.00	1.3771	-6	0	1	1	1	1
68.25	68.26	1.3731	6	2	3	1	2	2
68.27		1.3727	-1	1	5	1		2
68.81	68.82	1.3632	-4	1	5	2	1	2
69.75	69.76	1.3471	6	2	1	2	1	3
70.18	70.30	1.3399	-2	8	3	1	4	4
70.29		1.3380	-1	9	2	6		1
70.96	70.96	1.3271	-6	0	4	2	2	9
71.23	71.18	1.3227	5	5	0	1	1	2

2THETA	PEAK	D	H	K	L	I(INT)	I(PK)	I(DS)
71.71	71.72	1.3150	-1	3	5	6	4	9
72.02	71.92	1.3102	-2	4	5	2	3	3
72.16	72.16	1.3080	-5	1	5	2	2	2
72.42	72.44	1.3037	-6	3	2	1	3	1
72.43		1.3037	5	3	1	2		3
72.54	72.54	1.3020	0	0	4	3		4
72.86	72.90	1.2970	0	10	0	2	3	3
72.90		1.2964	0	8	3	2	3	3
72.93		1.2960	0	6	4	1		2
73.39	73.42	1.2890	2	6	3	1	2	1
73.40		1.2888	-4	8	2	1		2
73.58	73.66	1.2861	-4	1	1	1		2
73.66		1.2850	-4	8	0	11	7	16
73.90	73.88	1.2813	6	0	2	2	6	3
73.93		1.2809	-4	4	5	3		4
74.24	74.24	1.2763	-2	8	2	11	7	16
74.84	74.86	1.2676	-6	2	1	3	3	3
74.92		1.2665	-3	9	4	2		3
75.52	75.56	1.2578	-5	3	2	3	4	5
75.56		1.2573	-3	5	5	3		5
75.58		1.2570	6	2	0	2		2
75.62		1.2564	-5	7	2	1	1	1
76.40	76.40	1.2456	3	7	2	1	1	1
78.40	78.44	1.2186	-1	5	5	1	1	2
78.44		1.2181	3	5	3	1		1
78.63	78.64	1.2157	-6	0	5	1	1	2
79.11	79.12	1.2096	5	5	1	1	1	3
79.35	79.36	1.2065	-2	10	2	2	2	2
79.54	79.56	1.2040	-10	4	0	2	2	2
79.63		1.2030	0	10	5	1		2
80.20	80.22	1.1958	-3	1	6	3	2	5
80.35		1.1940	-2	6	5	1		2
80.74	80.72	1.1892	-6	0	2	1	1	2
81.24	81.24	1.1831	-2	8	6	1	1	2
81.65	81.68	1.1782	-4	2	3	1		1
81.67		1.1780	-6	6	3	1	1	1
81.69		1.1777	4	6	1	1	1	1

688

ADULARIA - COLVILLE AND RIBBE, 1968

2THETA	PEAK	D	H	K	L	I(INT)	I(PK)	I(DS)
81.71	81.88	1.1775	2	10	1	1	1	1
81.85	82.12	1.1758	-7	1	4	1	1	2
82.12	82.38	1.1726	-7	3	3	2	1	3
82.40	82.90	1.1693	-7	3	2	1	1	2
82.90	82.90	1.1636	6	0	1	1	1	1
83.23	83.22	1.1598	-5	7	4	1	1	1
83.73	83.74	1.1541	-4	8	4	4	2	7
84.37	84.50	1.1471	1	7	4	4	3	1
84.49		1.1457	0	8	4	4		8
85.11	85.12	1.1389	-2	10	3	3	1	1
85.41	85.40	1.1357	-1	1	6	1	1	1
85.85	85.84	1.1310	-6	6	4	1	1	1
86.52	86.52	1.1239	4	4	3	3	1	6
87.36	87.36	1.1153	2	6	4	2		1
87.73	87.72	1.1116	-1	9	4	3		3
88.61	88.64	1.1027	-7	1	3	3		6
88.66		1.1022	-6	6	0			1
88.71		1.1018	-5	9	2	1		2
89.39	89.44	1.0952	6	4	1	2	1	3
89.47		1.0943	7	1	0	1		3
90.02	90.00	1.0891	5	5	2	2	1	1
90.46	90.46	1.0850	5	5	6	2	1	6
90.97	90.74	1.0802	-6	2	6	2		1
91.26	91.28	1.0775	-3	11	2	1	1	2
91.52	91.60	1.0751	4	11	0	1	1	1
91.63		1.0741	1	11	2	1	1	2
91.71		1.0734	-2	8	5	1	1	2
91.96	91.94	1.0711	3	11	1	1	1	2
92.41	92.52	1.0671	-7	5	1	1	1	1
92.52		1.0661	0	12	1	1	1	2
92.66	92.70	1.0649	0	2	6	1	1	2
92.71		1.0645	7	3	3	1	1	3
93.01	93.02	1.0617	-6	8	3	2	1	2
93.26	93.28	1.0596	-6	6	5	2	1	4
93.46		1.0579	3	5	4	1	1	2
93.79	93.76	1.0550	-8	2	3	2	1	4
94.07	94.12	1.0526	-2	10	4	2	1	3

2THETA	PEAK	D	H	K	L	I(INT)	I(PK)	I(DS)
94.20	94.20	1.0515	5	9	1	2	1	4
94.73	94.72	1.0470	-2	12	2	1	1	2
95.78	95.80	1.0383	-3	11	3	1	1	2
96.60	96.58	1.0317	-6	2	2	1	1	2
97.00	97.00	1.0284	-4	0	7	1	1	2
97.67	97.64	1.0232	5	3	4	1	1	1
97.83	97.82	1.0219	-4	2	4	1	1	1
98.37	98.38	1.0177	-3	9	5	1	1	2
98.98	98.98	1.0131	-8	0	5	1		3
99.69	99.70	1.0078	-3	7	6	1		2
100.62	100.62	1.0010	-8	2	5	1		2
101.02	101.02	0.9981	-3	3	7	1		2
103.82	103.80	0.9787	-5	7	6	1	1	1
105.02	105.10	0.9708	-6	2	7	1	1	2
105.15		0.9699	1	13	1	1		1
105.35	105.42	0.9686	3	3	5	1	1	2
105.47		0.9679	-1	1	7	1		1
105.60	105.66	0.9670	-8	4	5	1	1	2
105.68		0.9665	0	12	3	1		2
106.36	106.40	0.9622	-4	12	7	1	1	3
106.65	106.76	0.9604	-2	4	2	1	1	2
106.89		0.9589	-6	10	2	1		2
107.87	107.86	0.9529	4	10	5	1	1	3
109.43	109.46	0.9436	-5	5	7	2	1	4
109.48		0.9433	-9	1	4	1		2
110.44	110.42	0.9378	2	0	6	1	1	1
111.13	111.12	0.9339	-3	13	2	1		1
112.30	112.32	0.9275	3	3	5	1	1	2
112.33		0.9273	-9	9	3	1		2
113.21	113.24	0.9226	-3	9	6	2	1	3
113.57	113.86	0.9206	-8	6	1	1	1	1
113.83		0.9193	-1	13	3	1	1	2
113.91		0.9189	2	10	7	1		1
114.15	114.16	0.9176	-7	0	7	1	1	3
114.46		0.9160	0	2	7	1		1
114.99	115.00	0.9133	-7	9	7	2	1	4
115.31	115.34	0.9117	-2	6	7	1	1	2

ADULARIA - COLVILLE AND RIBBE, 1968

2THETA	PEAK	D	H	K	L	I(INT)	I(PK)	I(DS)
115.33		0.9116	-3	11	5	1		2
115.84	115.80	0.9091	-5	3	4	1	1	1
116.11		0.9077	-6	8	6	1		1
116.70	116.74	0.9048	-2	14	1	1	1	2
116.93	117.10	0.9037	7	3	2	1	1	1
117.09		0.9030	3	9	4	1		3
117.57	117.64	0.9006	2	14	0	1	1	2
117.63		0.9004	3	13	1	1		3
117.68		0.9001	-5	9	6	1		3
118.01	118.06	0.8986	0	8	6	2	1	3
118.42	118.38	0.8966	2	12	3	1	1	3
119.00	118.92	0.8939	-6	2	7	1	1	1
119.50	119.52	0.8917	-2	14	2	1	1	2
119.91	119.88	0.8898	0	4	7	2	1	4
121.00	121.02	0.8850	-8	8	2	2	1	3
122.00	122.02	0.8807	-9	3	3	1		7
122.55	122.56	0.8783	6	8	2	3	1	1
123.58	124.00	0.8741	-2	10	6	1	1	2
123.99		0.8724	4	4	5	1		2
124.02		0.8723	4	8	4	1		3
124.70	124.62	0.8696	-9	5	5	1	1	3
126.30	126.38	0.8633	-2	14	3	1	1	5
126.38		0.8630	8	4	1	2		2
126.45		0.8628	-2	12	5	1		2
128.68	128.76	0.8545	-7	1	8	1	0	1
129.10	129.42	0.8530	-7	13	3	1	1	2
129.42		0.8519	-5	13	1	2		5
129.89	129.98	0.8502	-6	12	2	1	1	2
130.00		0.8499	-9	5	1	1	1	1
130.42	130.62	0.8484	-4	14	1	1		2
130.52		0.8481	-4	14	2	1		4
130.76		0.8473	-7	11	3	2		3
132.06	132.06	0.8430	-10	0	5	1	0	1
132.07		0.8429	-1	1	8	1		2
132.65	133.06	0.8410	5	11	2	1	1	4
133.03		0.8398	-6	8	7	2		3
133.24		0.8391	1	11	5	1		

2THETA	PEAK	D	H	K	L	I(INT)	I(PK)	I(DS)
134.07	133.64	0.8365	-7	9	6	2		2
134.73	134.66	0.8345	-4	14	0	1	0	1
136.59	136.82	0.8290	-1	3	8	1	1	2
136.69		0.8287	-5	9	7	2		1
136.76		0.8285	1	5	5	2		5
137.09	137.40	0.8276	8	0	2	1	1	2
137.25		0.8272	8	6	0	1		1
137.42		0.8267	8	8	1	1		4
138.01	138.44	0.8250	-8	10	3	1	1	1
138.43		0.8239	-10	4	3	1		2
138.44		0.8238	1	13	4	1		2
138.71		0.8231	3	5	6	1		3
143.37	143.72	0.8113	9	5	0	1	1	2
143.68		0.8106	0	16	0	3		7
144.56	144.56	0.8086	-1	13	5	3	1	8
144.95		0.8077	-6	12	5	1		2
145.77	145.64	0.8059	-4	10	7	2	1	4
145.82		0.8058	-1	5	8	1		2
147.07	147.34	0.8032	-1	1	8	1	1	2
147.38		0.8026	2	12	3	2		5
147.43		0.8024	2	14	3	2		4
148.59	148.46	0.8001	7	5	1	1	1	4
149.17	149.34	0.7990	3	15	1	2		2
149.29		0.7988	-4	0	9	1		4
150.36	150.62	0.7968	4	2	6	1		2
150.51		0.7965	-8	4	8	1		2
150.64		0.7962	2	4	7	1		2
150.68		0.7960	-9	5	7	3		8
151.48	151.66	0.7947	-2	16	1	1		2
151.52		0.7947	-5	13	5	1		2
152.27		0.7934	-2	10	9	1		2
152.77	152.82	0.7925	-10	6	6	3	1	6
153.14		0.7919	6	12	1	2		5
153.85	153.64	0.7907	4	8	2	1	1	2
153.87		0.7907	-9	9	4	2		4
155.98	156.64	0.7875	-6	2	9	2	1	4

690

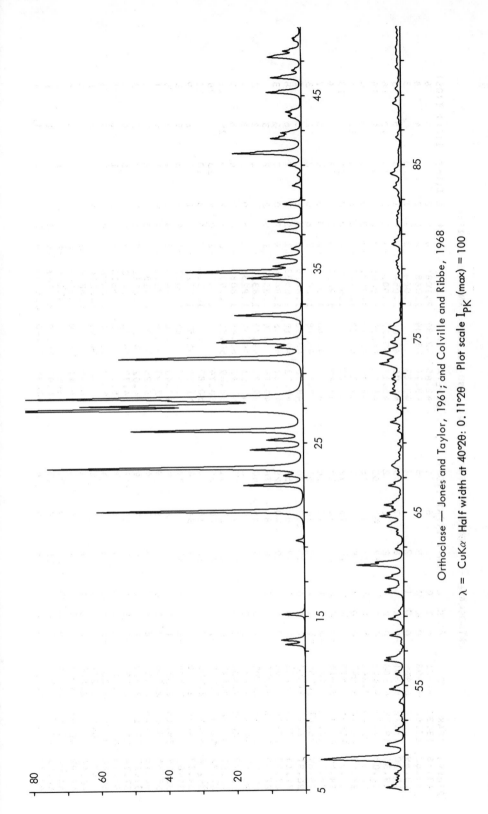

Orthoclase — Jones and Taylor, 1961; and Colville and Ribbe, 1968

λ = CuKα Half width at 40°2θ: 0.11°2θ Plot scale I_{PK} (max) = 100

ORTHOCLASE — — JONES AND TAYLOR, 1961 AND COLVILLE AND RIBBE, 1968

2THETA	PEAK	D	H	K	L	I(INT)	I(PK)	I(DS)
13.36	13.38	6.621	1	1	0	5	6	4
13.62	13.62	6.498	0	2	0	5	7	5
13.69		6.464	0	1	1	2		2
15.10	15.12	5.861	-1	1	1	6	7	5
19.35	19.36	4.583	0	2	1	2	2	2
21.05	21.06	4.217	-2	0	1	57	61	55
22.54	22.54	3.941	1	1	1	17	18	16
23.10	23.10	3.847	2	0	0	5	6	5
23.55	23.56	3.775	-1	3	0	74	75	72
24.60	24.60	3.616	-1	1	2	15	16	15
25.15	25.16	3.537	-2	2	1	10	11	10
25.69	25.70	3.465	-1	1	2	52	51	51
26.91	26.92	3.310	-2	2	0	100	100	100
27.14	27.14	3.283	-2	0	2	59	66	59
27.43	27.44	3.249	0	4	0	29	40	29
27.58	27.58	3.232	0	0	2	78	82	78
29.84	29.84	2.991	-1	3	1	59	54	60
30.48	30.48	2.930	-2	2	1	7	8	8
30.77	30.78	2.903	0	4	1	25	25	25
30.87	30.86	2.894	2	2	0	10	24	11
32.33	32.34	2.767	-1	3	2	22	20	23
34.46	34.46	2.600	-3	1	2	18	16	19
34.77	34.84	2.578	-2	2	1	4		4
34.83		2.574	-2	4	1	37	34	39
35.17	35.16	2.550	1	1	2	8	9	9
35.65	35.66	2.516	3	1	0	8	8	9
36.16	36.16	2.482	2	4	0	3	3	4
37.17	37.18	2.417	-1	5	1	8	7	9
37.74	37.74	2.381	-3	1	1	12	10	13
38.73	38.74	2.323	-1	3	3	7	6	8
39.80	39.80	2.263	-3	3	0	3	3	4
40.39	40.40	2.231	-2	3	2	1	1	1
40.42		2.229	1	1	3	1	1	1
40.98	40.98	2.201	1	5	1	4	4	4
41.66	41.66	2.166	0	6	0	25	20	29
42.51	42.52	2.125	2	4	1	11	9	13
42.73	42.72	2.114	-4	0	1	4	5	4
42.85	42.84	2.108	-4	0	2	5	6	6
43.74	43.84	2.068	-2	0	1	2	4	3
43.83		2.064	3	1	1	3		3
44.05	44.06	2.054	0	6	1	5	5	6
45.17	45.18	2.006	-4	2	2	13	10	16
46.02	46.02	1.9705	-2	2	2	12	9	14
46.44	46.44	1.9536	-3	2	3	3	3	4
47.13	47.22	1.9267	-2	6	1	1		1
47.21		1.9235	4	0	0	3	10	13
47.29		1.9207	-3	5	1	3		4
47.56	47.56	1.9101	-4	0	1	11		7
48.17	48.18	1.8874	2	6	0	6	5	7
48.30	48.30	1.8825	3	1	3	2	2	2
49.12	49.12	1.8533	1	3	0	7	3	8
49.37	49.36	1.8444	4	2	0	2	6	3
49.54	49.54	1.8383	1	5	2	2	3	2
49.71	49.70	1.8326	-4	2	0	1	3	2
49.91	49.92	1.8256	3	5	2	3	2	4
50.43	50.44	1.8080	-2	6	2	4	3	5
50.69	50.74	1.7993	0	6	0	9	5	11
50.80		1.7976	-2	4	4	23	25	28
51.52	51.52	1.7956	0	4	3	10		12
51.63	51.64	1.7722	-4	4	1	7		9
52.40	52.40	1.7687	-2	4	2	3	6	3
52.97	52.98	1.7445	1	4	4	4	5	5
53.25	53.26	1.7271	-1	1	3	1	3	1
54.02	54.02	1.7187	-1	3	3	2	1	2
54.78	54.78	1.6961	-5	1	2	2	2	3
55.01	54.92	1.6742	3	5	2	6	4	8
55.67	55.68	1.6678	-1	7	2	3	3	1
56.42	56.44	1.6496	-3	3	4	3	3	4
56.44		1.6294	-2	5	0	1	6	1
56.61	56.60	1.6289	0	8	0	7	5	9
56.62		1.6245	-5	1	3	2		3
56.91	56.90	1.6241	-1	3	4	1	1	1
57.38	57.40	1.6167	4	2	1	1	1	1
		1.6045						

692

ORTHOCLASE - - JONES AND TAYLOR, 1961 AND COLVILLE AND RIBBE, 1968

2THETA	PEAK	D	H	K	L	I(INT)	I(PK)	I(DS)
57.41		1.6037	2	0	3	1		1
57.89	57.88	1.5916	-4	2	4	1	4	8
58.54	58.56	1.5755	0	8	1	6	1	1
58.84	58.84	1.5681	0	2	4	6	5	9
59.21	59.30	1.5591	-5	3	1	1	1	1
59.30		1.5570	2	3	1	2		2
60.40	60.40	1.5312	-5	3	3	8	6	11
60.56	60.56	1.5275	0	6	3	2	5	3
60.94	60.94	1.5193	1	5	3	5	2	1
61.21	61.20	1.5130	-4	6	0	1	6	11
61.95	61.96	1.4965	-2	8	0	8	14	30
62.87	62.88	1.4768	-5	4	1	3	2	5
62.95		1.4752	4	4	0	1		1
63.71	63.72	1.4594	-1	7	3	3	2	4
63.88	63.88	1.4560	-2	8	2	1	1	2
64.10	64.22	1.4514	0	8	2	1	5	5
64.17		1.4500	5	3	0	3		2
64.23		1.4489	1	1	4	4		4
64.30		1.4474	-1	5	5	4		6
64.76	64.76	1.4382	4	4	0	2	7	3
64.77		1.4381	2	4	3	9		12
65.05	64.94	1.4326	-4	6	3	2	5	3
65.23	65.24	1.4290	-5	5	2	7	3	3
65.42	65.42	1.4253	-6	0	2	7	6	9
65.60	65.60	1.4192	1	9	0	2	3	3
65.74	65.74	1.4119	3	5	0	1	1	1
66.12	66.12	1.4042	4	0	5	7	4	10
66.53	66.54	1.3922	-4	0	5	2	4	3
67.18	67.18	1.3851	-5	5	3	2	2	3
67.57	67.68	1.3833	-2	6	4	3	3	5
67.67		1.3818	1	3	4	1		2
68.20	68.20	1.3739	-6	2	3	1	1	2
68.44	68.42	1.3697	-1	2	5	1	1	2
68.92	68.92	1.3613	-4	2	1	2	2	3
69.67	69.68	1.3485	-6	2	3	3	2	4
70.11	70.18	1.3410	-2	8	1	1	4	1
70.18		1.3400	-1	9	2	6		9

2THETA	PEAK	D	H	K	L	I(INT)	I(PK)	I(DS)
70.96	70.96	1.3270	-6	0	4	2	2	4
71.14	71.16	1.3241	5	5	0	1	2	2
71.50	71.50	1.3184	1	7	3	1	1	1
71.86	71.86	1.3126	-1	3	5	6	4	9
72.14	72.08	1.3082	-2	4	5	2	3	3
72.24	72.24	1.3066	-1	5	4	1	2	2
72.33		1.3052	-5	1	2	1		1
72.38	72.36	1.3044	-6	3	1	2	3	3
72.66	72.66	1.3001	5	0	4	3		4
72.70		1.2996	2	0	8	2		2
72.86	72.88	1.2971	0	10	3	2		3
72.98		1.2952	0	6	4	1		2
73.39	73.54	1.2890	4	4	2	1	7	2
73.40		1.2889	2	6	1	1		1
73.44		1.2882	-4	8	2	1		2
73.53		1.2868	-4	8	0	10		15
73.84	73.74	1.2823	6	0	3	2	4	3
74.02	74.16	1.2796	-4	4	5	3	7	4
74.16		1.2774	-2	8	4	17		15
74.74	74.76	1.2690	-6	2	1	3	3	5
74.78		1.2684	-3	0	9	2		3
74.82		1.2679	0	2	5	1		2
75.50	75.62	1.2581	-5	5	7	1		1
75.51		1.2581	-6	6	2	1		2
75.60		1.2568	-5	3	5	3		5
75.65		1.2560	-3	5	5	4		6
76.34	76.34	1.2464	3	7	2	1	1	1
78.46	78.48	1.2179	3	5	5	1		2
78.53		1.2170	-1	5	5	1		1
78.69	78.70	1.2149	-6	0	5	1		1
79.03	79.04	1.2105	-2	5	1	1		1
79.20	79.22	1.2084	-2	10	2	2	1	2
79.41	79.42	1.2058	-2	10	0	1	2	3
79.77	79.66	1.2011	0	4	5	1	2	1
80.40	80.42	1.1934	-3	1	6	3	1	5
80.44		1.1929	-2	6	5	1		2
80.45		1.1928	-6	4	0	1		1

693

ORTHOCLASE -- JONES AND TAYLOR, 1961 AND COLVILLE AND RIBBE, 1968

2THETA	PEAK	D	H	K	L	I(INT)	I(PK)	I(DS)
80.62	80.64	1.1906	-6	6	2	1	2	2
80.89	80.84	1.1873	1	1	5	1	1	2
81.55	81.68	1.1794	-2	10	1	1	1	1
81.57		1.1791	-6	6	3	1		1
81.64		1.1783	4	2	6	1		1
81.69		1.1777	4	2	3	1	1	1
81.82	81.82	1.1762	-7	1	4	1		1
82.04	82.04	1.1735	-7	3	3	2	1	3
82.30	82.28	1.1705	-7	3	2	1	1	1
83.18	83.22	1.1604	-5	7	4	1	1	1
83.25		1.1596	4	8	1	1	1	1
83.68	83.68	1.1547	-4	8	4	4	2	7
84.40	84.48	1.1467	1	7	4	1	3	1
84.42		1.1464	-5	1	6	2		3
84.49		1.1456	0	8	4	4		7
85.00	84.98	1.1402	-2	10	3	1	1	1
85.20	85.22	1.1380	-6	4	5	1	1	1
85.64	85.66	1.1332	-1	1	6	1	1	2
85.80	85.80	1.1316	-6	6	4	1	1	1
86.55	86.56	1.1236	4	4	3	3	1	6
87.42	87.42	1.1147	2	6	4	1	1	1
87.70	87.70	1.1119	-1	9	4	1	1	3
88.51	88.52	1.1038	-7	5	3	3	2	6
88.54		1.1034	-5	9	2	1		2
88.54		1.1034	6	6	0	1		1
89.31	89.36	1.0959	6	4	1	1	1	2
89.38		1.0952	7	1	0	2		3
89.98	89.98	1.0895	5	5	2	1	1	1
90.59	90.60	1.0837	-3	5	5	3		6
91.06	91.06	1.0793	-3	11	1	1	1	2
91.08		1.0792	-6	2	6	1		1
91.33	91.38	1.0768	4	10	0	1		1
91.47		1.0756	1	11	2	1	1	1
91.75	91.74	1.0731	3	11	0	1	1	2
92.28	92.28	1.0683	-7	5	1	1	1	1
92.30		1.0681	0	12	1	1		2
92.60	92.60	1.0654	7	3	0	1	1	2

2THETA	PEAK	D	H	K	L	I(INT)	I(PK)	I(DS)
92.88	92.88	1.0630	-6	8	3	3	1	3
92.90		1.0628	-0	2	8	1		1
93.25	93.24	1.0596	-6	6	5	2	1	4
93.53	93.56	1.0572	3	5	4	1	1	1
93.69		1.0559	-8	2	3	1		1
94.00	94.00	1.0532	-2	10	4	2	2	3
94.02		1.0530	5	9	0	2		4
94.50	94.32	1.0490	-2	12	1	1		2
95.61	95.60	1.0397	-3	11	3	1	0	2
96.57	96.56	1.0319	-6	0	2	1	1	2
97.27	97.30	1.0263	-4	0	7	1	1	1
97.69	97.66	1.0230	5	3	3	1	1	3
97.94	97.96	1.0211	4	2	5	1	1	2
98.37	98.34	1.0177	-3	9	5	1	1	5
99.81	99.82	1.0069	-3	7	6	1	0	6
100.60	100.60	1.0011	-8	2	5	2	1	5
101.30	101.28	0.9961	-3	3	7	1	1	7
103.89	103.88	0.9782	-5	7	6	1	0	6
104.68	104.84	0.9729	-6	8	5	1	1	8
105.50	105.54	0.9676	0	12	3	2	1	3
105.52		0.9675	3	3	5	1		5
105.56		0.9673	-8	4	5	1		5
105.80	105.94	0.9657	-1	1	7	1	1	7
106.09		0.9639	-4	12	1	1		1
106.66	106.48	0.9603	-6	10	2	1	1	2
106.94	106.92	0.9586	-2	4	7	1	1	7
107.71	107.70	0.9538	4	10	4	3	1	4
109.37	109.62	0.9439	-9	1	4	2	1	4
109.63		0.9424	-5	5	7	2		7
110.72	110.80	0.9362	2	0	6	1	0	6
110.84		0.9355	-3	13	3	1		3
112.18	112.40	0.9281	-9	3	3	2	1	3
112.40		0.9269	5	1	4	1		4
113.28	113.32	0.9222	-3	9	6	1	1	6
113.38		0.9217	-8	6	1	1		1
113.59	113.76	0.9206	-1	13	3	2	1	3
113.86		0.9191	2	10	4	1		4

ORTHOCLASE - - JONES AND TAYLOR, 1961 AND COLVILLE AND RIBBE, 1968

2THETA	PEAK	D	H	K	L	I(INT)	I(PK)	I(DS)
114.31	114.32	0.9168	-7	3	7	1	1	3
114.82	114.80	0.9142	-7	9	4	2	1	4
114.82		0.9142	0	2	7	1		1
115.25	115.26	0.9120	-3	11	5	1	1	2
115.58	115.60	0.9103	-2	6	7	1	1	2
115.93	116.16	0.9086	5	3	4	1	1	1
116.13		0.9076	-6	8	6	1		1
116.34		0.9066	-2	14	1	1		2
117.07	117.28	0.9031	3	9	4	1	1	3
117.20		0.9024	2	14	0	1		1
117.32		0.9019	3	13	1	1		2
117.70	117.74	0.9000	-5	9	6	1	1	3
118.17	118.20	0.8978	0	8	6	1	1	3
118.23		0.8975	2	12	3	1		3
119.07	119.10	0.8936	-8	4	3	1	1	1
119.15		0.8933	-2	14	2	1		1
119.16		0.8932	-6	6	7	1		2
120.27	120.26	0.8882	0	4	7	1	1	4
120.75	120.74	0.8861	-8	8	2	2	1	3
121.80	121.84	0.8815	-9	8	1	2	1	3
122.41	122.40	0.8789	6	8	2	3	1	7
124.03	124.16	0.8722	4	8	4	1	1	2
124.17		0.8717	4	4	5	1		2
124.58	124.58	0.8700	-9	5	5	1	1	3
125.96	126.26	0.8646	-2	14	3	1	1	3
126.23		0.8636	8	4	1	2		4
126.34		0.8632	-2	12	5	1		1
127.26	127.24	0.8597	-8	8	5	1	0	2
128.94	129.12	0.8536	-2	8	7	1	1	2
129.06		0.8532	-5	13	3	2		4
129.42	129.64	0.8519	-7	1	8	1	1	1
129.50		0.8516	-6	12	1	1		2
129.74		0.8508	-9	5	1	1		2
129.97		0.8500	-4	14	1	1		1
130.08	130.38	0.8496	-4	14	2	1		2
130.42		0.8484	-7	11	3	2	1	4
130.88		0.8469	2	14	2	1		2

2THETA	PEAK	D	H	K	L	I(INT)	I(PK)	I(DS)
132.38	132.70	0.8419	5	11	2	1	0	1
132.64		0.8411	-1	1	8	1		2
133.16	133.20	0.8394	-6	8	7	2	1	4
133.22		0.8392	1	11	5	1		3
134.01	133.94	0.8367	-7	9	6	1	1	2
134.24		0.8360	4	14	0	1		1
136.99	137.16	0.8279	8	6	1	1	1	2
137.02		0.8278	8	8	0	1		1
137.10		0.8276	8	8	2	2		3
137.21		0.8273	1	5	7	1		5
137.63	137.84	0.8261	-8	10	3	1	1	2
138.13		0.8247	-10	4	3	1		1
138.20		0.8245	1	13	4	1		3
139.04	138.96	0.8222	3	5	6	1	1	1
139.58		0.8208	-8	0	8	1		2
140.47	140.38	0.8185	-10	0	6	1	0	6
142.99	143.02	0.8122	0	16	0	2	1	2
143.06		0.8121	9	5	0	1		7
144.37	144.42	0.8091	-1	13	5	3	1	4
144.64		0.8084	-6	12	5	1		2
145.23	145.22	0.8071	5	13	1	2	1	4
146.00		0.8055	-4	10	7	1		2
146.97	147.06	0.8034	2	14	3	1	1	4
147.05		0.8032	4	12	3	2		4
147.78	148.00	0.8017	-1	15	8	1	1	2
148.53	148.32	0.8002	-7	5	8	2	1	4
149.91	150.90	0.7976	-2	16	1	1	1	2
150.22		0.7970	-4	0	9	3		4
150.76		0.7960	-9	5	7	3		8
150.87		0.7958	4	2	6	1		8
150.89		0.7958	-8	4	8	1		3
151.08		0.7954	-5	13	5	1		1
151.31		0.7950	2	4	7	1		1
151.82	152.14	0.7941	-2	10	7	1	1	3
152.28		0.7934	-10	6	3	3		6
152.53		0.7929	6	12	1	2		4
153.14		0.7919	-6	0	9	1		2

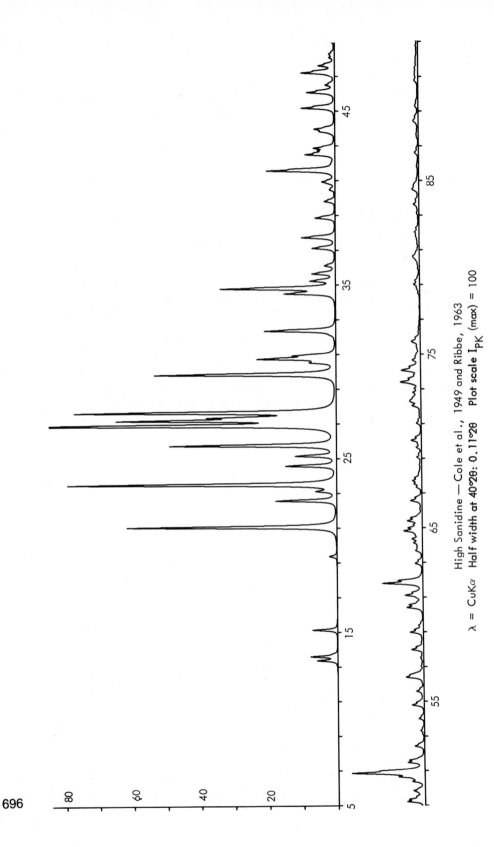

High Sanidine — Cole et al., 1949 and Ribbe, 1963

λ = CuKα Half width at 40°2θ: 0.11°2θ Plot scale I$_{PK}$ (max) = 100

696

HIGH SANIDINE - COLE ET AL., 1949 AND RIBBE, 1963

2THETA	PEAK	D	H	K	L	I(INT)	I(PK)	I(DS)
13.35	13.36	6.628	1	1	0	5	6	4
13.58	13.58	6.515	0	2	0	6	8	6
15.11	15.12	5.857	-1	1	1	6	7	6
19.35	19.36	4.583	0	1	1	2	2	2
21.05	21.06	4.217	-2	0	1	57	62	55
22.56	22.56	3.939	1	1	1	17	18	16
23.09	23.10	3.849	2	0	0	5	6	4
23.50	23.50	3.783	-1	3	0	76	79	74
24.56	24.56	3.621	-1	1	1	14	15	14
25.13	25.14	3.540	-2	2	1	11	12	11
25.74	25.74	3.458	-1	1	2	49	49	48
26.88	26.88	3.314	-2	2	0	100	100	100
27.18	27.18	3.278	-2	0	2	59	65	59
27.35	27.36	3.257	0	2	0	28	38	28
27.64	27.64	3.225	-2	0	1	77	77	77
29.82	29.82	2.994	-2	3	1	57	53	58
30.50	30.50	2.928	0	2	2	7	8	7
30.72	30.72	2.908	0	4	1	24	23	24
30.92	30.92	2.890	0	2	2	11	13	11
32.34	32.34	2.766	-1	3	1	23	21	24
34.47	34.48	2.599	-3	1	2	16	15	17
34.76	34.78	2.578	-2	2	1	4		4
34.77		2.578	-2	4	1	34	34	37
35.22	35.22	2.546	1	1	0	8	7	8
35.63	35.64	2.518	3	1	1	8	7	8
36.09	36.10	2.487	2	4	0	3	3	4
37.09	37.10	2.422	-1	5	1	8	7	9
37.70	37.70	2.384	-3	3	1	12	10	13
38.82	38.82	2.318	-1	1	3	7	6	8
39.78	39.78	2.264	-3	3	2	4	3	4
40.44	40.46	2.228	-1	3	2	1	2	1
40.46		2.228	-2	2	3	1		1
40.81	40.90	2.204	1	3	0	2		2
40.91		2.204	-1	5	1	3	4	4
41.55	41.56	2.172	0	6	0	25	20	29
42.47	42.48	2.127	2	4	1	11	9	12
42.70	42.70	2.116	-4	0	1	4	5	5

2THETA	PEAK	D	H	K	L	I(INT)	I(PK)	I(DS)
42.85	42.84	2.109	-4	0	2	5	6	6
43.78	43.82	2.066	-2	2	1	2	5	3
43.83		2.064	3	1	1	3		3
43.96	43.94	2.058	0	6	1	5	6	6
45.16	45.16	2.006	-4	2	2	13	10	15
46.05	46.06	1.9693	-2	2	2	11	8	13
46.47	46.48	1.9523	-3	3	0	3	3	4
47.19	47.20	1.9245	4	0	0	10	10	12
47.20		1.9238	-3	1	2	3		3
47.60	47.60	1.9086	-4	0	3	6	5	7
48.06	48.06	1.8914	2	6	0	2	2	2
48.28	48.28	1.8834	3	3	1	2	2	2
49.20	49.20	1.8502	1	1	3	7	5	8
49.33	49.34	1.8456	4	2	0	2	2	2
49.52	49.50	1.8392	1	5	2	2	3	2
49.74	49.84	1.8316	-4	3	1	1		2
49.83		1.8284	3	5	0	3	3	4
50.36	50.36	1.8105	-2	2	2	4	4	5
50.63	50.64	1.8013	0	6	2	8	8	10
50.85	50.86	1.7942	0	4	3	9	22	11
50.87		1.7933	-2	0	4	21		26
51.46	51.46	1.7742	-4	4	1	6	5	7
51.59	51.58	1.7701	-4	2	2	2	5	3
52.40	52.40	1.7446	2	4	4	4	3	5
53.11	53.12	1.7230	-1	1	4	1	1	1
53.31	53.30	1.7169	1	3	3	1	1	2
54.00	54.00	1.6966	-5	1	2	2	2	2
54.77	54.78	1.6746	-3	5	1	5	4	7
55.04	54.92	1.6669	3	1	2	1	2	1
55.58	55.58	1.6522	-1	7	1	3	2	4
56.38	56.38	1.6305	3	5	0	7	5	9
56.45		1.6287	0	8	0	2		2
56.51	56.52	1.6271	-3	4	4	1	4	1
56.64		1.6237	-5	1	3	3		1
57.02	57.02	1.6138	-1	3	3	3	1	1
57.49	57.50	1.6016	2	0	2	4	1	1
57.97	57.96	1.5896	-4	2	4	5	4	7

HIGH SANIDINE - COLE ET AL.,1949 AND RIBBE, 1963

2THETA	PEAK	D	H	K	L	I(INT)	I(PK)	I(DS)
58.39	58.38	1.5792	0	8	1	1		1
58.96	58.96	1.5651	0	2	4	6	4	8
59.37	59.36	1.5553	2	2	3	1	1	2
60.40	60.40	1.5313	-5	3	3	7	5	9
60.55	60.56	1.5278	0	6	3	2	4	3
61.10	61.10	1.5154	-4	8	1	7	5	10
61.80	61.80	1.5000	2	8	0	19	12	26
62.91	62.94	1.4762	4	1	4	1	2	4
62.94		1.4755	-5	7	4	3		4
63.67	63.68	1.4603	-1	7	3	3	2	4
63.75		1.4586	-2	8	2	1		1
63.99	64.12	1.4538	0	8	0	3	3	1
64.12		1.4511	5	3	0	3		4
64.36	64.36	1.4463	-1	1	4	1	3	4
64.37		1.4461	-1	5	4	7		10
64.81	64.82	1.4373	-4	4	3	2	6	2
65.00	65.00	1.4336	-4	6	3	2	4	4
65.16	65.16	1.4304	-5	5	2	2	3	2
65.39	65.40	1.4259	-6	0	2	5	4	8
65.55	65.56	1.4228	-3	9	0	1	4	2
66.55	66.56	1.4038	4	0	2	6	4	8
67.33	67.34	1.3895	-4	0	5	2	2	3
67.53	67.54	1.3858	-5	5	3	2	2	3
67.70	67.70	1.3828	-2	6	4	3	3	4
67.86	67.88	1.3799	-1	3	4	1	2	1
67.89		1.3793	-6	0	1	1		1
68.19	68.18	1.3740	-6	2	3	1	1	1
68.62	68.62	1.3664	-1	1	5	3	1	1
69.06	69.06	1.3589	-4	2	5	2	1	3
69.61	69.62	1.3494	-6	2	1	2	2	3
70.02	70.02	1.3425	-1	9	2	5	3	7
70.03		1.3424	-1	0	3	2		1
71.01	71.02	1.3263	-6	0	3	2	1	3
71.05		1.3255	5	1	0	1		1
71.46	71.46	1.3189	1	7	3	1	1	1
72.03	72.02	1.3100	-1	3	5	4	3	6

2THETA	PEAK	D	H	K	L	I(INT)	I(PK)	I(DS)
72.28	72.30	1.3060	-2	4	5	2	3	3
72.35		1.3050	-5	5	1	2		3
72.37		1.3047	-5	1	5	1		2
72.48	72.48	1.3030	0	10	0	1	3	2
72.79	72.78	1.2982	0	8	3	1	2	2
72.79		1.2981	2	0	4	2		3
73.03	73.00	1.2945	0	6	4	1	2	1
73.29	73.40	1.2906	-4	8	1	1	6	1
73.38		1.2892	4	4	2	1		1
73.39		1.2890	-4	2	6	1		12
73.39		1.2890	6	6	3	8		2
73.79	73.78	1.2830	2	0	2	2	2	3
74.06	74.06	1.2790	-4	8	2	8	6	13
74.12		1.2781	-3	4	5	2		3
74.62	74.64	1.2708	-6	9	2	2	3	3
74.66		1.2701	-6	4	1	2		3
75.00	74.86	1.2653	0	0	5	1	2	2
75.38	75.44	1.2598	-5	7	0	1	1	1
75.45		1.2588	6	6	5	2		4
75.70	75.72	1.2553	-5	3	5	2	3	4
75.75		1.2545	-3	5	5	2		4
76.26	76.26	1.2475	3	7	2	1	1	1
78.48	78.48	1.2177	3	0	5	1	1	1
78.79	78.80	1.2136	-6	0	3	1	1	1
78.97	79.00	1.2114	5	5	5	1	2	2
79.01		1.2109	-2	10	2	2		2
79.22	79.22	1.2081	0	10	0	1	1	2
79.93	79.92	1.1992	-6	4	5	1	1	2
80.51	80.62	1.1919	-6	6	2	1	2	2
80.53		1.1918	-2	6	5	1		2
80.63		1.1905	-3	1	6	2		4
81.08	81.06	1.1850	1	11	5	1	1	1
81.34	81.34	1.1819	2	10	1	1	1	1
81.59	81.60	1.1789	4	6	2	1	1	1
81.75	81.82	1.1770	4	2	3	1	1	1
81.84		1.1760	-7	1	4	2		1
82.00	82.00	1.1740	-7	3	3	2	1	3

698

2THETA	PEAK	D	H	K	L	I(INT)	I(PK)	I(DS)
82.24	82.24	1.1713	-7	3	2	1	1	1
83.11	83.12	1.1612	4	8	1	1	1	1
83.14		1.1609	-5	7	1	1		1
83.62	83.62	1.1553	-4	8	4	3	2	5
84.48	84.48	1.1458	0	8	4	3	2	6
84.61		1.1443	1	6	6	1		2
84.84	84.72	1.1419	-2	10	3	1	2	1
85.89	85.78	1.1305	-1	1	6	1	1	1
86.58	86.58	1.1233	4	4	3	3	1	5
87.47	87.64	1.1142	2	6	4	1	1	1
87.64		1.1125	-1	9	2	1		2
88.36	88.42	1.1053	-5	9	2	1	2	2
88.42		1.1046	6	6	0	2		4
88.44		1.1045	-7	5	3	2		2
89.25	89.30	1.0965	6	4	1	1	1	2
89.32		1.0958	7	1	0	1		2
90.73	90.76	1.0824	1	5	5	2	1	4
90.82		1.0816	-3	11	2	1		1
91.10	91.08	1.0790	4	10	0	1	1	1
91.24	91.22	1.0777	1	11	0	1	1	1
91.48	91.48	1.0755	3	11	2	1	1	2
92.00	92.02	1.0708	0	12	1	1	1	1
92.17		1.0693	-7	5	5	1		1
92.53	92.52	1.0661	7	3	0	1	1	1
92.74	92.76	1.0642	-6	8	3	1	1	2
93.28	93.28	1.0594	-6	6	5	2	1	3
93.82	93.86	1.0547	5	9	0	1	1	2
93.89		1.0541	-2	10	4	1		2
94.19	94.16	1.0515	-2	12	1	1	1	2
95.40	95.40	1.0414	-3	11	3	1	0	1
96.57	96.58	1.0319	6	2	2	1	0	1
98.35	98.36	1.0179	-3	7	5	1	0	1
99.94	99.94	1.0059	-8	2	5	1	0	2
100.64	100.66	1.0008	3	3	5	1	0	1
105.70	105.72	0.9664	-4	12	1	1		2
105.75		0.9660	-6	10	2	2		1
106.41	106.40	0.9619	-6	10	2	2		1

2THETA	PEAK	D	H	K	L	I(INT)	I(PK)	I(DS)
107.25	107.54	0.9566	-2	4	7	1	0	1
107.52		0.9550	4	10	2	1		2
109.34	109.34	0.9441	-9	1	4	1	0	2
109.88	109.84	0.9410	-5	5	7	1	0	2
112.53	112.54	0.9262	5	1	4	1	0	1
113.27	113.30	0.9223	-1	13	3	1	1	2
113.34		0.9219	-3	9	6	1		2
114.55	114.62	0.9156	-7	3	7	1	1	2
114.65		0.9150	-7	9	4	1		2
117.01	116.94	0.9033	3	9	4	1	0	1
117.97	118.30	0.8988	2	12	3	1	1	2
118.32		0.8971	0	8	6	1		1
120.51	120.58	0.8871	-8	8	2	1	0	2
120.65		0.8865	-0	0	7	1		2
121.66	121.74	0.8821	-9	3	1	1		2
122.25	122.24	0.8796	6	8	2	2	0	4
124.37	124.50	0.8709	4	4	5	1	1	1
124.55		0.8702	-9	9	5	1		1
125.52	125.50	0.8663	-2	14	3	1	0	3
126.12	126.12	0.8640	8	4	1	1	0	1
127.16	126.72	0.8601	-8	8	5	1	0	1
129.54	129.56	0.8515	-9	5	2	1	1	1
129.55		0.8515	-4	14	4	1		3
130.06	130.04	0.8497	-7	11	3	1	1	1
130.38		0.8486	2	14	2	1		2
133.33	133.28	0.8389	-6	8	7	1	0	2
136.79	136.92	0.8285	8	8	0	1	0	2
137.68	137.72	0.8259	1	8	5	1	0	1
139.39	139.40	0.8213	3	5	6	1	0	3
140.57	140.52	0.8182	-10	0	6	1	0	1
142.11	142.10	0.8144	0	16	0	1	0	3
144.06	144.18	0.8098	-1	13	5	1	0	3
144.59		0.8086	5	11	1	1		2
146.34	146.50	0.8047	2	14	3	1	0	2
146.61		0.8041	4	12	3	1		2
148.50	148.52	0.8003	7	5	3	1		2
151.01	151.94	0.7956	-9	5	7	1	1	4

TABLE 35. ZEOLITES

Variety	Paulingite	Tugtupite	Brewsterite	Yugawaralite	Laumontite
Composition	$(K,Ca,Na,Ba)_{1.23}$ $(Si,Al)_7 O_{14} \cdot 7.34$ (H_2O) (IDEAL)	$Na_8 Al_2 Be_2 Si_8 O_{24}$ $(Cl,S)_2$	$(Sr_{0.58}Ba_{0.30}$ $Ca_{0.15})Al_2Si_6O_{16} \cdot$ $4.9H_2O$	$CaAl_2Si_6O_{16} \cdot 4H_2O$	$CaAl_2Si_4O_{12} \cdot$ $4H_2O$
Source	Wenatchee, Wash.	South Greenland	Strontian, Argyll Scotland	Kanagawa Prefecture, Japan	Insel Mull, G. B.
Reference	Gordon, Samson & Kamb, 1966	Danø, 1966	Perrotta & Smith, 1964	Leimer & Slaughter, 1969	Bartl & Fischer, 1967
Cell Dimensions					
$\underline{a}$ Å	35.093	8.583	6.772	6.73	14.75 (7.57)
$\underline{b}$ Å	35.093	8.583	17.51	13.95	13.10 (14.75)
$\underline{c}$	35.093	8.817	7.744	10.03	7.57 (13.10)
α	90	90	90	90	90 (90)
β deg	90	90	94.30	111.50	112.0 (90)
γ	90	90	90	90	90 (112.0)
Space Group	Im3m	I4̄	P2$_1$/m	Pc	Cm (Am)
Z	96	1	2	2	4

TABLE 35. (cont.)

Variety	Paulingite	Tugtupite	Brewsterite	Yugawaralite	Laumontite
Site Occupancy					
M_1,	K 0.576 / Ca 0.304				
M_2, & M_3	Na 0.110 / Ba 0.010				
T (all)	Si 0.774 / Al 0.226				
Method	3-D, counter	3-D, film	3-D, counter, film	3-D, counter	3-D, film
R & Rating	0.14 (2)	0.089 (1)	0.11 (2)	0.14 (2)	0.126 (2)
Cleavage and habit	None Dodecahedrons	{110}&{101} good	{010} perfect Flattened parallel {010}	{010} imperfect	{010}&{110} perfect. Fibrous parallel to c axis
Comment	Not all H_2O cations located. {110} reflections at d = 17.547 omitted in plot.				
μ	82.0	82.7	138.2	86.9	93.6
ASF	0.4451	0.1724	0.4852×10^{-1}	0.5338×10^{-1}	0.1323
Abridging factors	1:1:0.5	1:1:0.5	1:0:0	1:1:0.5	1:1:0.5

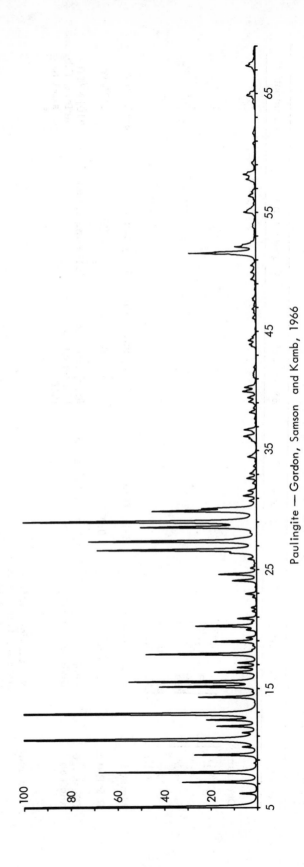

Paulingite — Gordon, Samson and Kamb, 1966

λ = CuKα Half width at 40°2θ: 0.11°2θ Plot scale I$_{PK}$ (max) = 200

2THETA	PEAK	D	H	K	L	I(INT)	I(PK)	I(DS)
5.03	5.04	17.546	2	0	0	79	81	79
6.16	6.16	14.327	2	1	1	4	4	4
7.12	7.12	12.407	2	2	0	16	16	16
7.96	7.96	11.097	3	1	0	34	34	34
9.42	9.42	9.379	3	2	1	14	14	14
10.07	10.08	8.773	4	0	0	3	3	3
10.69	10.70	8.271	3	3	0	100	100	100
10.69		8.271	4	1	1			
11.27	11.28	7.847	4	2	0	4		4
11.82	11.82	7.482	3	3	2	3	3	3
12.35	12.36	7.163	4	2	2	9	9	9
12.85	12.86	6.882	5	1	0	3	3	3
14.26	14.28	6.204	4	4	0	85	83	85
15.13	15.14	5.849	6	0	1	14	13	14
15.55	15.56	5.693	6	1	1	23	21	23
16.36	16.36	5.415	5	4	2	2	28	2
16.74	16.76	5.290	6	2	1	28		28
17.12	17.12	5.174	6	3	1	10	9	10
17.86	17.86	4.963	7	1	0	5	4	5
18.91	18.92	4.689	6	4	2	27	24	27
19.25	19.26	4.608	7	3	0	11	9	11
19.90	19.92	4.457	7	3	1	2	2	2
20.23	20.24	4.387	8	0	0	2	2	2
20.86	20.86	4.256	8	2	0	16	13	16
21.47	21.48	4.136	8	2	2	5	4	5
22.64	22.66	3.924	8	4	0	1	1	1
22.93	22.94	3.875	8	4	1	1	1	1
24.04	24.04	3.699	9	3	0	3	2	3
24.57	24.58	3.620	9	3	2	6	5	6
25.87	25.88	3.441	10	2	0	10	8	11
26.37	26.38	3.377	10	2	2	2	2	2
26.62	26.62	3.346	10	3	1	6	6	6
27.35	27.36	3.258	10	4	0	45	34	46
28.53	28.54	3.126	11	2	1	49	36	50
28.53		3.126	10	5	1	15	25	15
						19		20

2THETA	PEAK	D	H	K	L	I(INT)	I(PK)	I(DS)
28.99	28.98	3.078	9	7	0	2	54	2
28.99	28.98	3.078	11	3	1	73		76
29.88	29.88	2.987	11	4	1	31	22	32
30.10	30.10	2.966	10	6	2	14	12	14
30.76	30.76	2.904	11	5	0	2	2	2
31.19	31.20	2.865	11	5	2	4	3	3
31.61	31.62	2.828	12	4	1	2	2	1
32.24	32.24	2.774	12	4	2	1	1	1
32.65	32.66	2.740	10	8	1	3	2	3
33.06	33.06	2.707	10	9	0	2	1	2
34.45	34.46	2.601	10	0	1	2	2	2
35.79	35.79	2.507	14	2	0	1	1	1
35.98	35.98	2.494	10	1	1	1		2
36.17	36.16	2.481	10	10	0	3	3	3
36.72	36.72	2.445	14	3	1	1		2
36.72		2.445	14	4	1	2	3	3
38.18	38.18	2.355	11	9	2	2	2	2
39.06	39.06	2.304	14	5	0	3	3	3
39.41	39.42	2.284	14	6	2	4	3	3
39.93	39.94	2.256	11	11	0	4	3	4
40.27	40.28	2.237	14	7	1	2	1	2
43.74	43.74	2.068	12	12	0	3	2	3
44.21	44.22	2.047	12	2	1	1	1	1
44.53	44.52	2.033	17	3	0	1	1	1
46.85	46.84	1.9377	17	3	0	2	1	2
49.37	49.36	1.8444	18	2	0	1	1	1
50.52	50.52	1.8050	19	4	1	26	14	29
51.52	51.52	1.7725	19	14	0	7	4	8
52.08	52.08	1.7546	14	14	0	5	3	6
54.96	54.96	1.6692	20	0	1	1	1	2
55.23	55.22	1.6517	21	1	0	1	1	3
56.30	56.30	1.6327	18	11	1	3	1	3
56.70	56.70	1.6222	19	10	1	1	1	1
57.75	57.76	1.5951	18	12	0	3	2	4
58.14	58.14	1.5853	22	7	0	4	3	5
59.04	59.04	1.5632	21	10	2	1	1	2
64.80	64.80	1.4375	20	14	0	3	2	4

PAULINGITE - GORDON, SAMSON & KAMB, 1966

2THETA	PEAK	D	H	K	L	I(INT)	I(PK)	I(DS)
67.10	67.10	1.3937	25	3	0	1	1	1
67.34	67.34	1.3893	21	14	1	4	2	5
69.59	69.60	1.3497	24	10	0	3	1	3
69.83	69.82	1.3458	22	14	0	4	2	5
70.88	70.88	1.3283	24	11	1	2	1	2
73.67	73.66	1.2848	25	11	0	2	1	3
75.84	75.84	1.2533	28	0	0	1	1	2
78.67	78.68	1.2152	28	7	1	2	1	3
79.46	79.46	1.2051	28	8	0	1	1	1
86.80	86.80	1.1210	28	14	0	2	1	3
107.39	107.62	0.9558	32	18	0	1	0	1
107.62		0.9544	34	14	0	1		2
115.95	115.92	0.9085	36	14	0	1	0	1
127.24	127.26	0.8598	35	21	0	2	0	3
149.70	150.14	0.7980	42	13	1	1	0	2
150.36		0.7967	34	28	0	1		1
156.00	156.62	0.7875	32	31	1	1	0	1
156.54		0.7867	42	15	1	1		1
159.49	159.72	0.7827	35	28	1	3	0	6
161.13		0.7808	38	24	0	1		2
162.92		0.7789	45	2	1	1		1

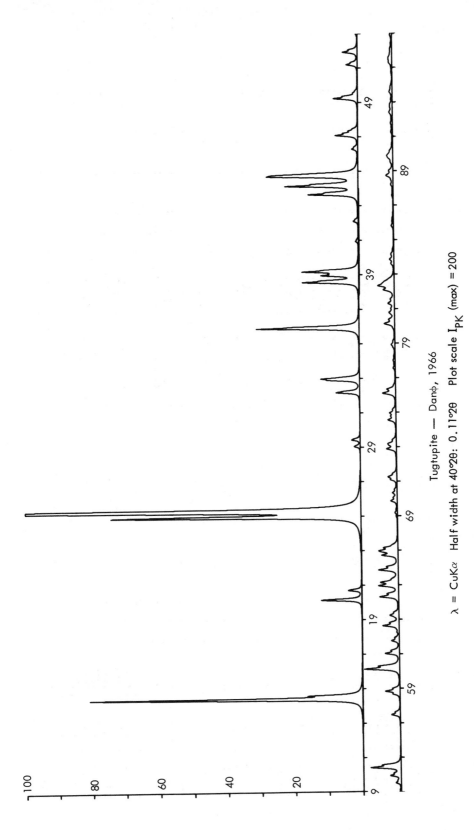

Tugtupite — Danφ, 1966

$\lambda = CuK\alpha$ Half width at $40°2\theta$: $0.11°2\theta$ Plot scale I_{PK} (max) $= 200$

TUGTUPITE - DANØ, 1966

2THETA	PEAK	D	H	K	L	I(INT)	I(PK)	I(DS)
14.39	14.40	6.150	0	1	1	33	40	32
14.58	14.58	6.069	1	1	0	5	8	4
20.12	20.14	4.408	0	0	2	5	6	5
20.68	20.68	4.291	0	1	2	2	2	2
24.94	24.94	3.567	2	1	1	35	37	35
25.28	25.28	3.519	1	1	2	48	100	48
25.28		3.519	2	1	0	52		52
29.41	29.42	3.035	2	0	0	1	1	1
32.16	32.16	2.781	0	3	1	4	4	4
32.88	32.90	2.721	0	0	3	5	5	5
32.97	32.90	2.714	3	1	0	3	6	3
35.90	35.90	2.500	2	2	2	18	15	18
38.55	38.56	2.334	2	1	3	7	8	8
38.55		2.334	3	1	2	3		3
38.93	38.94	2.311	1	3	2	3	6	3
38.93		2.311	3	2	1	3		3
39.16	39.16	2.298	3	2	2	6	8	6
39.16		2.298	2	3	1	6		7
43.65	43.66	2.072	1	1	4	9	7	10
44.14	44.14	2.050	0	3	3	13	11	14
44.69	44.70	2.026	0	1	1	7	14	7
44.69		2.026	4	4	2	7		7
44.76		2.023	1	3	0	8		9
46.26	46.26	1.9607	0	4	1	4	1	1
47.06	47.06	1.9293	0	2	3	4	3	5
49.21	49.22	1.8498	3	4	2	4	4	4
49.21		1.8498	2	3	3	1		1
51.18	51.18	1.7834	2	3	4	2	2	2
51.92	51.92	1.7597	4	2	2	3	2	3
53.93	53.92	1.6987	1	4	2	5	2	2
54.40	54.40	1.6850	4	3	3	5	4	6
57.46	57.46	1.6024	1	1	5	2	2	2
58.66	58.66	1.5725	1	5	2	1	1	1
58.83	58.82	1.5684	5	2	1	1	2	2
58.83		1.5684	2	5	1	1		1
60.13	60.12	1.5375	0	4	4	8	5	9
61.02	61.02	1.5173	4	4	0	3	2	4
61.74	61.74	1.5012	0	3	5	2	2	2
62.62	62.62	1.4823	3	4	3	1	1	2
62.62		1.4823	0	5	3	1		2
63.22	63.22	1.4695	4	0	2	2		2
64.30	64.30	1.4474	2	4	2	2	3	2
64.30		1.4474	4	2	0	2		3
64.94	64.94	1.4347	4	0	5	4		5
65.16	65.14	1.4305	0	6	0	2	3	2
65.86	65.86	1.4170	3	2	5	2	3	3
65.86		1.4170	2	5	3	2		3
66.70	66.70	1.4011	2	5	1	2		4
66.96	66.96	1.3962	5	3	2	3	3	2
67.12	67.14	1.3933	6	1	1	1	2	2
69.84	69.84	1.3455	1	4	5	1	1	2
71.07	71.08	1.3252	5	4	1	1	1	2
71.24	71.24	1.3226	2	2	6	2	1	2
72.86	72.86	1.2970	2	2	2	2	1	2
73.18	73.16	1.2923	1	3	6	1	1	2
74.53	74.54	1.2720	6	1	6	1	1	1
74.53		1.2720	6	6	3	1		1
74.93	74.94	1.2662	3	4	4	4	2	4
76.09	76.10	1.2498	4	1	7	2	2	2
80.12	80.12	1.1968	2	2	6	1		2
80.12		1.1968	1	2	1	1		2
80.76	80.76	1.1889	3	3	6	1	1	1
81.30	81.30	1.1824	2	6	5	1	1	2
82.08	82.08	1.1731	6	6	3	2	2	1
82.08		1.1731	3	1	3	2		2
82.32	82.32	1.1703	1	7	2	2	2	3
82.47		1.1686	7	2	1	2		2
88.71	88.72	1.1017	6	1	5	2	1	3
89.87	89.86	1.0905	6	5	1	1	1	1
95.27	95.30	1.0424	0	8	2	2	0	1
101.17	101.16	0.9970	6	2	6	1	1	1
101.17		0.9970	2	6	6	1		1
102.42	102.42	0.9882	5	6	7	1	0	1
102.75	102.76	0.9859	6	6	2	1	0	1

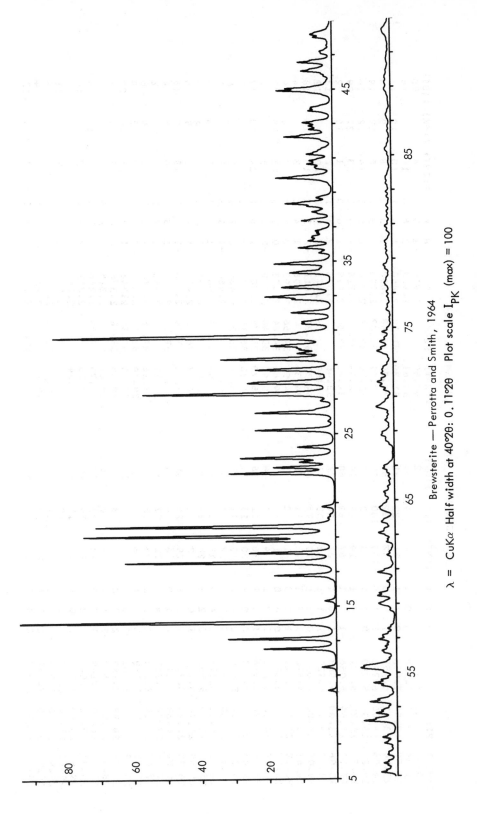

Brewsterite — Perrotta and Smith, 1964

λ = CuKα Half width at 40°2θ: 0.11°2θ Plot scale I$_{PK}$ (max) = 100

BREWSTERITE – PERROTTA AND SMITH, 1964

2THETA PEAK	D	H	K	L	I(INT)	I(PK)	I(DS)
10.10	8.755	0	2	0	2	3	2
11.46	7.718	0	0	1	4	4	4
12.54	7.063	0	1	1	21	22	21
13.12	6.751	1	0	0	32	32	32
14.06	6.299	1	1	0	100	100	100
15.30	5.790	0	2	1	3	3	3
16.78	5.281	1	2	0	19	18	19
17.54	5.056	-1	0	1	67	63	68
18.08	4.903	-1	1	1	27	26	27
18.78	4.721	1	1	1	32	33	33
19.06	4.655	0	3	1	79	75	82
19.62	4.522	-1	2	1	76	71	79
20.10	4.415	1	3	0	1	2	1
20.28	4.377	0	4	1	1	2	1
20.76	4.278	1	2	1	4	4	4
22.70	3.916	-1	3	1	36	32	38
23.04	3.859	-1	0	2	20	18	21
23.34	3.808	0	0	2	10	11	11
23.60	3.769	0	4	1	32	28	35
24.22	3.673	-1	4	1	13	11	14
25.20	3.531	0	2	2	28	24	30
25.70	3.464	-1	1	2	2	2	2
26.20	3.398	1	1	2	29	24	31
26.88	3.314	-1	4	1	4	4	4
27.30	3.265	1	4	1	71	57	78
27.94	3.193	-1	5	1	24	26	27
27.95	3.189	1	2	2	6		7
28.02	3.181	0	0	2	10		11
28.31	3.149	-2	2	0	1	3	2
28.50	3.130	-2	2	1	4	6	5
28.70	3.109	-1	1	2	28	23	32
29.32	3.044	2	2	0	42	34	48
29.64	3.011	0	4	2	11	11	13
29.86	2.990	-2	0	2	9	10	10
29.96	2.979	-1	1	3	6	11	7
30.10	2.967	2	3	0	20	19	23
30.57	2.922	2	0	0	60	84	69

2THETA	PEAK	D	H	K	L	I(INT)	I(PK)	I(DS)
30.60		2.919	-1	5	1	44		50
30.61		2.918	0	6	1	14		16
31.36	31.38	2.850	1	5	1	10	10	11
31.50	31.48	2.838	-1	3	2	8		9
32.01	32.02	2.793	2	1	1	16	9	19
32.78	32.78	2.730	-2	3	1	14	13	16
32.94	32.94	2.716	0	4	2	25	12	29
33.42	33.48	2.679	-1	6	1	1	21	2
33.46		2.676	2	4	1	11	15	13
33.49		2.673	2	3	1	10		12
34.31	34.32	2.611	2	4	0	15	13	18
34.36		2.608	-2	4	1	3		4
34.56	34.56	2.593	0	0	3	23	4	28
34.84	34.84	2.573	-2	0	2	5	18	6
35.48	35.48	2.528	-1	1	3	14	4	17
35.78	35.78	2.508	2	6	1	9	10	11
36.18	36.18	2.481	1	1	3	5	8	6
36.37	36.38	2.468	2	0	2	1	6	1
36.40		2.466	-1	7	1	3		4
36.45	36.44	2.463	1	5	2	5	6	6
36.63	36.62	2.451	-2	0	3	5	5	6
36.77	36.76	2.442	-1	1	3	13	9	16
37.34	37.34	2.406	2	3	2	4	5	4
37.75	37.76	2.381	3	2	1	1		2
37.77		2.380	-1	1	3	5		6
37.86	37.86	2.374	0	7	2	7	6	9
38.19	38.20	2.355	-1	2	3	17	8	22
38.34	38.34	2.346	1	5	0	3	14	4
38.65	38.68	2.328	-2	6	2	3		4
38.69		2.325	1	6	2	2	5	3
39.74	39.84	2.266	1	2	3	12	17	16
39.83		2.261	-2	4	2	2		2
39.84		2.261	2	7	1	2		2
39.85		2.260	-1	3	3	9		11
40.38	40.46	2.232	3	0	1	4	5	6
40.45		2.228	1	7	2	3		4
40.64	40.64	2.218	2	3	0	9	8	12

BREWSTERITE - PERROTTA AND SMITH, 1964

2THETA	PEAK	D	H	K	L	I(INT)	I(PK)	I(DS)
40.84	40.84	2.208	2	6	0	8	7	10
41.21	41.22	2.189	0	8	0	10	8	13
41.39	41.30	2.179	3	1	0	1		2
41.44	41.42	2.177	1	3	3	3	4	4
41.57	41.54	2.171	-1	6	2	2	3	2
41.97	42.02	2.151	-2	4	1	1	4	2
42.01		2.149	-1	2	3	2		3
42.22	42.22	2.139	2	4	2	14	14	18
42.22		2.138	-3	2	1	7		9
42.65	42.64	2.118	3	0	3	8	7	11
42.84	42.84	2.109	-2	1	2	3	7	3
42.85		2.108	-2	5	2	5		7
43.04	43.04	2.100	3	3	0	10	9	14
43.74	43.74	2.068	-3	4	1	9	7	11
43.85	43.86	2.063	3	3	3	2	6	3
43.94	43.94	2.059	3	5	1	6	6	8
44.91	44.92	2.016	-1	5	3	22	17	29
45.05	45.04	2.011	-3	0	2	8	14	11
45.75	45.76	1.9816	1	7	0	3		4
45.88	45.90	1.9760	2	0	3	2	3	2
46.05	46.04	1.9695	-3	4	1	13	9	17
46.12		1.9664	-2	7	1	2		2
46.55	46.56	1.9492	1	5	1	16	10	21
47.05	47.10	1.9299	0	6	3	2	3	3
47.05		1.9296	0	0	4	1		2
47.11		1.9275	2	2	3	2		3
47.20	47.18	1.9240	2	7	1	3	4	4
47.53	47.54	1.9112	-2	0	3	2	2	2
47.80	47.82	1.9010	-3	3	2	1	2	1
48.02	48.02	1.8932	3	5	0	3	6	4
48.02		1.8931	-1	0	4	5		7
48.19	48.18	1.8865	0	9	1	6		9
48.28		1.8835	3	0	2	3		4
49.20	49.20	1.8504	-1	2	4	6	5	8
49.20		1.8503	-1	8	2	2		2
49.46	49.44	1.8413	3	2	2	2	2	2
49.60	49.60	1.8365	2	8	0	2	3	3

2THETA	PEAK	D	H	K	L	I(INT)	I(PK)	I(DS)
49.72	49.72	1.8320	0	3	4	4	4	6
50.19	50.22	1.8160	-2	7	2	2	3	3
50.23		1.8149	-1	8	2	2		2
50.38	50.34	1.8098	1	1	4	1	3	2
50.65	50.64	1.8008	-1	0	4	2	2	3
50.87	50.90	1.7935	0	7	3	4	4	6
50.90		1.7925	3	3	0	3		4
51.22	51.22	1.7820	1	6	2	4	4	6
51.24		1.7815	-1	2	4	2		3
52.17	52.20	1.7516	0	1	3	5	10	7
52.19		1.7510	-3	0	0	9		13
52.20		1.7508	2	7	2	1		2
52.43	52.32	1.7437	-3	5	2	3	6	4
52.63	52.64	1.7376	-1	4	4	5	6	8
52.64		1.7373	0	9	0	1		2
52.64		1.7371	1	3	4	3		5
52.87	52.78	1.7301	3	4	2	13	4	19
53.29	53.30	1.7175	-2	6	3	3	8	4
53.91	53.44	1.7113	1	7	3	1	6	2
53.92	53.92	1.6991	-2	2	4	6	1	9
54.23	54.24	1.6900	0	5	4	4	5	6
54.39	54.38	1.6855	-3	3	3	3	7	5
54.40		1.6852	-2	8	2	2		2
54.57	54.54	1.6802	1	4	4	1	5	2
54.58		1.6800	-4	0	1	2		3
54.75	54.74	1.6752	-4	1	1	2	3	3
55.02	55.04	1.6676	0	8	3	2	4	3
55.04		1.6671	0	10	1	7		10
55.22	55.28	1.6620	-1	0	4	8	11	11
55.27		1.6605	-2	9	1	4		6
55.30		1.6598	2	2	0	5		8
55.39	55.38	1.6572	-4	2	1	1	10	2
55.50		1.6542	-3	7	1	2		2
55.83	55.82	1.6453	-4	8	3	4	2	6
56.13	56.16	1.6371	-1	2	3	2	4	3
56.24	56.26	1.6342	3	0	3	3	4	3
56.25		1.6341	2	9	1	2		2

709

BREWSTERITE - PERROTTA AND SMITH, 1964

2THETA	PEAK	D	H	K	L	I(INT)	I(PK)	I(DS)
56.51	56.50	1.6271	3	1	3	4	3	6
56.64	56.66	1.6236	4	0	1	1	3	2
56.91	56.92	1.6166	2	7	4	2	5	4
56.91		1.6165	3	7	1	5		7
57.16	57.14	1.6102	-4	3	1	4	4	6
57.54	57.54	1.6005	-1	8	1	4	2	6
58.57	58.60	1.5747	4	8	3	1	2	2
58.61		1.5737	3	4	0	1		3
58.88	59.00	1.5672	-3	3	3	2		5
58.91		1.5663	-2	7	2	3	5	5
58.99		1.5645	4	9	1	2		3
59.00		1.5642	2	4	4	3		5
59.45	59.46	1.5535	-3	3	1	8	5	13
59.57	59.60	1.5506	2	8	3	1	5	2
59.62		1.5493	1	7	0	2		3
59.87	59.86	1.5436	0	11	5	1	2	4
60.12	60.12	1.5377	0	0	5	2	2	2
60.25	60.26	1.5348	-4	3	2	2	3	3
60.28		1.5340	-2	10	1	8		3
60.80	60.80	1.5221	1	8	2	1	5	13
61.46	61.48	1.5074	-3	6	5	4	1	2
61.80	61.80	1.4999	4	1	3	4	3	6
62.55	62.56	1.4837	4	2	1	4	3	7
63.06	63.10	1.4730	4	5	2	2	5	6
63.11		1.4718	3	9	3	2		3
63.13		1.4715	0	11	0	2		3
63.68	63.72	1.4601	1	2	2	2	4	4
63.72		1.4592	0	12	5	2		4
63.73		1.4589	-3	0	1	2		3
63.79		1.4577	-4	3	2	1		2
64.35	64.46	1.4464	-1	11	2	1	4	5
64.40		1.4454	-1	2	5	3		5
64.46		1.4443	-1	4	3	3		4
64.55	64.56	1.4424	-4	2	3	4	5	6
64.69		1.4397	-3	7	3	1		2
64.94	64.94	1.4348	-2	9	3	1	2	2
65.22	65.30	1.4293	1	11	2	3	3	4

2THETA	PEAK	D	H	K	L	I(INT)	I(PK)	I(DS)
65.30	65.44	1.4277	3	8	2	2	3	4
65.45		1.4248	2	10	1	3	2	4
65.76	65.76	1.4188	4	6	1	1		2
65.78		1.4185	-4	3	3	1		2
66.09	66.08	1.4125	0	5	2	2	2	4
66.37	66.36	1.4073	2	11	2	2	2	4
66.58	66.60	1.4033	1	10	1	1		2
66.65		1.4020	-1	5	3	3		3
66.80	66.80	1.3992	-1	8	5	2	3	5
66.81		1.3991	4	7	0	1		2
66.94	66.96	1.3967	-4	6	2	2	3	4
68.21	68.24	1.3737	-3	3	4	2	1	3
68.42	68.42	1.3700	0	9	3	2	2	4
68.53	68.52	1.3681	3	7	5	3	2	5
68.73	68.74	1.3645	0	6	2	2	4	4
68.80		1.3633	-2	11	1	1		2
68.89	68.90	1.3619	-1	7	2	2		3
68.94		1.3609	4	1	3	1	2	3
69.28	69.30	1.3550	-1	6	3	1		3
69.51	69.60	1.3512	-2	10	5	1	2	13
69.62		1.3493	-2	2	3	2	3	2
70.04	70.04	1.3423	-4	7	4	2		6
70.33	70.40	1.3375	-1	11	4	4	5	7
70.38		1.3365	4	8	0	3		6
70.41		1.3361	-2	5	5	3		3
70.47		1.3350	2	11	5	2		5
70.80	70.78	1.3297	2	3	2	3	2	7
71.37	71.38	1.3205	-3	10	5	2	3	5
71.37		1.3204	-4	0	4	2		4
71.57	71.60	1.3173	-1	11	3	1		3
71.61		1.3166	-4	1	4	3	4	3
71.64		1.3162	-3	3	5	3		2
71.68		1.3155	0	11	0	1		3
71.83	71.84	1.3131	2	12	2	2	5	3
71.85		1.3127	-5	3	5	2		6
71.99	72.02	1.3107	3	5	1	3	4	3
72.05		1.3097	5	1	2	1		2

BREWSTERITE – PERROTTA AND SMITH, 1964

2THETA	PEAK	D	H	K	L	I(INT)	I(PK)	I(DS)
72.32	72.34	1.3054	-3	9	3	1	4	3
72.33		1.3052	-3	2	1	1		3
72.34		1.3052	-1	4	5	2		3
72.47	72.54	1.3031	4	4	3	1	3	3
72.56		1.3017	-5	1	2	2		3
72.70	72.68	1.2995	3	11	0	2	2	4
73.48	73.62	1.2876	-5	4	1	2	5	4
73.56		1.2864	0	0	6	2		3
73.63		1.2855	-1	10	4	7		13
73.80	73.82	1.2829	0	1	6	1		2
74.13	74.16	1.2780	-1	1	6	3	4	6
74.18		1.2772	-2	12	2	5	4	9
74.72	74.72	1.2692	0	12	3	5		9
75.38	75.38	1.2598	5	5	0	1	3	2
76.01	76.02	1.2510	2	13	0	2	1	3
76.10		1.2497	-3	8	4	2	2	2
76.33	76.32	1.2465	-2	9	4	1		2
76.78	76.78	1.2404	4	8	2	2	2	5
77.06	77.06	1.2365	4	6	3	2	2	4
77.44	77.56	1.2314	-2	10	4	2	2	4
77.56		1.2297	5	5	1	1		2
77.58		1.2295	2	13	1	1		3
77.87	77.80	1.2256	4	0	4	1	2	2
78.37	78.40	1.2191	1	1	6	3	2	3
78.40		1.2186	3	2	5	2		3
78.54	78.60	1.2169	-3	12	1	1	3	3
78.61		1.2159	-4	8	3	2		4
79.27	79.28	1.2075	-5	3	3	2		2
79.34		1.2066	-2	3	6	1		2
79.53	79.54	1.2041	3	3	1	1	2	3
79.73	79.74	1.2016	3	12	1	2	2	4
79.91	79.98	1.1995	4	3	4	1		2
80.02		1.1981	5	3	2	2		3
80.62	80.62	1.1906	4	9	2	2	2	3
80.84	80.86	1.1879	-5	4	3	1	2	2
80.39		1.1873	3	8	4	1		3
81.46	81.44	1.1805	2	13	2	1	2	3

2THETA	PEAK	D	H	K	L	I(INT)	I(PK)	I(DS)
81.74	81.72	1.1771	0	6	6	1	2	2
82.05	82.04	1.1735	-4	2	2	1	2	3
82.55	82.58	1.1677	-4	4	7	1	2	3
82.59		1.1671	1	14	2	1		2
82.86	82.84	1.1641	-5	5	3	3		3
83.34	83.34	1.1586	2	8	5	1	2	6
83.72	83.70	1.1542	0	15	1	1	2	2
84.57	84.58	1.1448	-3	2	6	1	1	2
85.03	85.02	1.1398	-1	15	1	1	1	3
85.23	85.34	1.1377	-5	2	4	2	2	3
85.39		1.1359	-2	6	6	1		4
86.40	86.38	1.1252	6	0	0	1	1	3
87.16	87.32	1.1173	0	15	2	1	1	2
87.32		1.1157	1	7	6	2		4
88.75	88.84	1.1014	5	5	3	1		3
88.84		1.1004	-6	1	2	1	2	3
89.24	89.22	1.0966	-3	5	6	1		3
89.31		1.0959	-2	15	1	1		2
90.54	90.56	1.0842	-1	11	5	1	1	5
90.96	90.94	1.0803	3	1	6	1	1	2
92.17	92.32	1.0692	5	0	4	2	2	4
92.30		1.0681	1	16	0	2	1	5
94.05	94.08	1.0528	4	13	0	1	2	3
95.65	95.64	1.0394	3	13	3	2	1	3

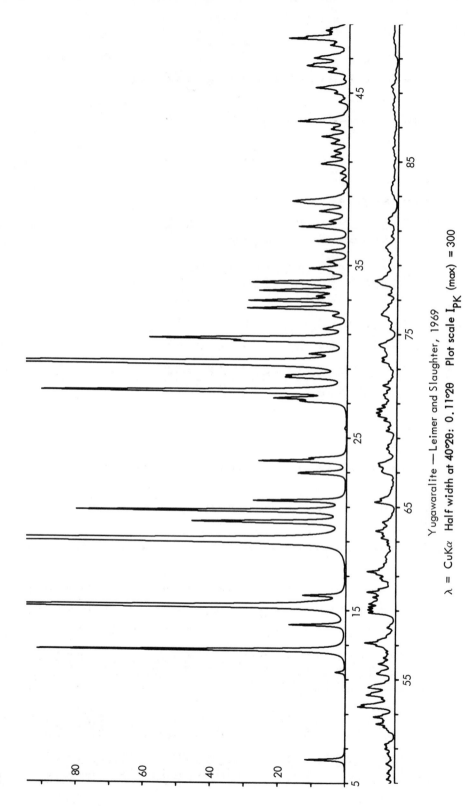

712

Yugawaralite — Leimer and Slaughter, 1969

λ = CuKα Half width at 40°2θ: 0.11°2θ Plot scale I$_{PK}$ (max) = 300

YUGAWARALITE — LEIMER AND SLAUGHTER, 1969

2THETA	PEAK	D	H	K	L	I(INT)	I(PK)	I(DS)
6.33	6.34	13.950	0	1	0	4	4	4
11.40	11.40	7.757	0	1	1	4	4	1
12.68	12.68	6.975	0	2	0	35	31	35
14.13	14.14	6.262	-1	0	1	6	6	6
15.23	15.24	5.814	-1	1	1	100	85	100
15.85	15.86	5.587	0	0	1	4	4	4
19.00	19.08	4.666	0	2	2	35	100	35
19.03		4.660	-1	2	0	48		48
19.09		4.645	-1	1	2	76		76
20.13	20.14	4.407	-1	2	1	16	15	16
20.75	20.76	4.277	0	1	2	33	27	34
21.33	21.34	4.162	0	3	1	11	9	12
22.91	22.92	3.878	0	2	2	6	5	6
22.98		3.866	-1	1	2	2		2
23.63	23.64	3.761	-1	3	0	11	9	11
23.81	23.82	3.733	-1	3	1	11	4	4
27.05	27.12	3.294	0	3	2	4	5	4
27.11		3.286	-1	1	1	4		3
27.25		3.269	-2	0	2	3		6
27.28	27.26	3.267	-1	1	2	6	7	6
27.67	27.68	3.222	-1	0	3	3	30	13
27.69		3.219	1	1	2	12		30
28.43	28.48	3.136	-1	0	2	29	6	3
28.48		3.131	1	1	0	3		6
28.61	28.60	3.117	-2	1	2	5	6	6
29.21	29.30	3.055	-2	1	1	6	65	2
29.29		3.047	1	2	0	2		96
29.39		3.036	1	4	1	93		15
29.84	29.84	2.991	0	1	3	14	4	15
30.56	30.56	2.923	-1	2	2	14	11	28
30.73	30.74	2.907	-2	2	2	27	20	3
31.29	31.30	2.856	-2	2	2	3	2	1
31.46	31.46	2.841	0	2	3	1	1	1
32.01		2.793	0	4	2	1	2	1
32.06	32.06	2.789	0	4	2	1		1
32.47	32.48	2.755	-1	4	1	15	10	16
32.48			1	4	1	15	10	15
32.91	32.92	2.719	-2	1	3	15	10	15

2THETA	PEAK	D	H	K	L	I(INT)	I(PK)	I(DS)
33.19	33.18	2.697	-1	3	3	3	3	4
33.49	33.50	2.673	0	5	1	13	9	14
33.96	33.98	2.638	0	1	1	8	9	9
33.99		2.635	-2	3	2	6		7
34.51	34.52	2.597	-2	2	0	2	2	2
34.79	34.78	2.576	-2	3	3	5	4	5
35.06	35.06	2.557	-1	5	1	1	1	1
35.18	35.18	2.548	1	5	0	2	2	3
35.78	35.78	2.507	-1	0	4	2		2
35.79		2.507	-1	1	4	1		1
36.37	36.38	2.468	-2	1	4	5	3	6
37.22	37.22	2.414	-1	5	3	7	5	7
37.53	37.54	2.395	-1	4	2	2	2	2
38.01	38.10	2.365	0	2	4	2	3	1
38.10		2.360	-1	0	4	4		4
38.56	38.58	2.333	-1	4	0	5		6
38.61		2.330	0	4	1	2	4	2
38.68	38.68	2.326	2	0	4	3	5	3
38.74		2.323	2	4	3	2	2	2
38.76		2.321	-2	3	1	1	1	1
39.93	39.92	2.256	0	6	2	1	1	1
40.32	40.32	2.235	0	3	4	3	3	4
40.85	40.86	2.207	-3	0	1	3	3	3
41.28	41.28	2.185	-1	6	1	2	2	2
41.62	41.62	2.168	-1	3	3	3	3	3
42.04	42.04	2.147	-1	5	1	3	3	3
42.44	42.44	2.128	-2	3	1	2	3	3
42.87	42.88	2.108	-1	1	2	8	8	9
43.31	43.32	2.087	-3	0	0	1	1	2
44.46	44.46	2.036	-1	4	4	3	2	2
44.96	44.96	2.014	-3	3	2	2	1	2
45.18	45.28	2.005	-1	4	1	1	3	1
45.27		2.001	-3	3	1	4		5
46.20	46.20	1.9634	-1	0	4	3	2	4
46.51	46.56	1.9509	-3	0	4	4	4	5
46.56		1.9489	-3	7	1	1		5
46.68	46.66	1.9443	-1	1	4	4		2

YUGAWARALITE — LEIMER AND SLAUGHTER, 1969

2THETA	PEAK	D	H	K	L	I(INT)	I(PK)	I(DS)
46.99	47.00	1.9321	-3	1	4	5	4	5
47.07		1.9289	-2	1	5	1		2
47.76	47.76	1.9027	-1	6	3	3	2	3
48.10	48.14	1.8900	1	2	4	2	6	2
48.14		1.8887	-2	4	2	9		11
48.49	48.46	1.8758	-2	2	5	1	2	1
48.62	48.62	1.8712	-3	2	1	2	2	2
48.74	48.74	1.8666	-2	4	0	2	2	2
49.21	49.22	1.8499	2	6	5	1	1	1
49.56	49.56	1.8377	-1	3	5	1	1	1
50.10	50.10	1.8191	1	4	3	1	1	1
50.98	50.98	1.7897	0	5	4	1	1	1
51.24	51.24	1.7815	-2	6	3	1	1	2
52.43	52.42	1.7437	0	8	0	3	2	3
52.68	52.70	1.7359	-3	5	1	1	1	1
52.83	52.82	1.7314	-3	1	5	3	2	3
53.39	53.40	1.7145	-2	7	1	2	4	2
53.41		1.7141	0	8	1	5		6
53.51	53.52	1.7109	1	4	4	2	3	3
53.87	53.88	1.7004	-2	4	5	2	1	2
54.09	54.10	1.6941	-3	5	3	4	3	4
54.18		1.6914	-2	7	2	2		2
54.50	54.52	1.6824	-1	8	1	1	3	2
54.52		1.6816	-4	0	0	3		3
54.58		1.6798	1	8	0	2		2
55.29	55.34	1.6601	-2	6	3	2	2	2
55.35		1.6584	-1	0	6	2		3
55.91	55.90	1.6431	-2	6	4	2	2	3
56.95	56.96	1.6156	3	6	2	2	2	3
57.12	57.12	1.6112	-3	6	1	4	3	5
57.37	57.26	1.6046	-3	1	4	1	2	2
58.00	58.00	1.5886	-4	4	4	1	1	2
58.84	58.86	1.5682	3	5	2	3	3	3
58.87		1.5673	3	2	4	2		2
59.03	59.02	1.5635	-2	5	3	2	3	2
59.19	59.18	1.5596	-2	3	6	2	3	4
59.37	59.36	1.5553	0	0	6	2	3	3
59.46	59.78	1.5532	3	6	0	3	2	3
59.77		1.5458	0	1	6	5	3	6
60.07	60.06	1.5389	-3	1	6	2	1	2
60.74	60.76	1.5234	2	8	0	1	2	2
60.86	60.88	1.5209	0	3	4	3	2	3
61.10	61.12	1.5153	1	7	4	1	2	3
61.23	61.24	1.5124	-2	7	4	2	2	4
61.24		1.5121	-4	3	5	1		1
62.80	62.80	1.4784	-1	4	5	1	1	2
63.18	63.20	1.4703	-1	1	0	2	2	2
63.56	63.60	1.4626	2	8	4	1	2	3
63.62		1.4613	-4	1	1	3		2
64.00	64.12	1.4535	4	4	4	1	1	2
64.12		1.4512	-4	1	3	2		1
65.21	65.28	1.4295	4	5	0	4	2	5
65.28		1.4282	-4	3	0	1		1
65.57	65.46	1.4225	-2	1	7	1		2
67.35	67.36	1.3891	-4	9	0	2	2	3
67.51	67.52	1.3862	-3	8	6	1	2	2
68.38	68.38	1.3707	-3	8	1	2	1	2
69.41	69.40	1.3529	-2	9	3	1	1	2
70.02	70.02	1.3425	2	9	0	2	2	2
70.28	70.26	1.3382	3	8	6	2	2	3
70.42	70.44	1.3358	1	3	6	2	2	2
70.67	70.66	1.3318	3	6	6	3	3	3
70.87	70.88	1.3285	-4	5	5	2	2	2
71.31	71.32	1.3214	-4	5	3	1	1	2
71.42	71.50	1.3196	3	9	3	2	1	1
72.03	72.04	1.3100	1	9	0	2	2	2
74.27		1.2758	2	10	0	2		1
74.38	74.38	1.2742	2	5	7	1	1	2
75.29	75.30	1.2611	-1	5	2	2	2	2
76.10	76.12	1.2497	-5	4	8	2	2	2
76.17		1.2487	-2	1	4	2		2
76.70	76.70	1.2414	-4	5	6	2	1	2
77.97		1.2244	4	3	3	1	2	2
78.10	78.08	1.2227	-3	7	6	2		2

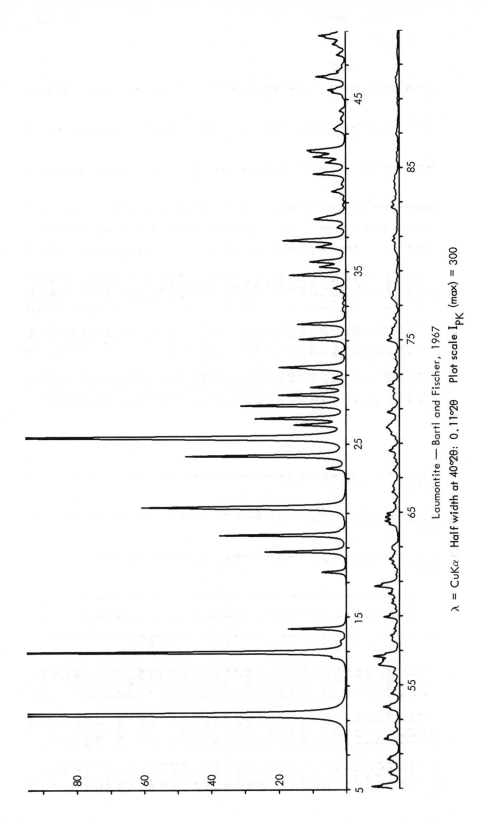

Laumontite — Bartl and Fischer, 1967

λ = CuKα Half width at 40°2θ: 0.11°2θ Plot scale I_{PK} (max) = 300

LAUMONTITE – BARTL AND FISCHER, 1967

2THETA	PEAK	D	H	K	L	I(INT)	I(PK)	I(DS)
9.34	9.34	9.460	1	1	0	100	100	100
12.94	12.94	6.838	-2	0	0	39	38	39
14.29	14.30	6.193	-2	0	1	6	2	3
17.56	17.56	5.046	1	1	1	3	2	3
18.74	18.76	4.730	-2	2	0	9	8	9
19.71	19.72	4.500	2	2	0	14	13	15
21.25		4.178	-2	0	2	12	20	12
21.34	21.34	4.160	-1	3	0	19	20	20
23.57	23.58	3.772	-1	3	1	2	2	2
24.29	24.30	3.661	-4	0	1	20	16	20
25.36	25.36	3.509	0	2	2	46	40	48
26.09	26.10	3.412	-1	3	2	6		6
26.46	26.46	3.365	-3	1	0	11	9	12
27.21	27.22	3.275	0	4	0	14	10	14
27.83	27.84	3.203	-3	3	0	8	7	8
28.28	28.28	3.153	3	3	0	4	4	5
28.81	28.82	3.096	-4	0	2	1	1	1
29.44	29.44	3.031	-4	2	0	8	7	8
31.05	31.06	2.878	-5	1	1	6	5	7
31.94	31.94	2.799	-4	1	2	7	5	7
34.02	34.02	2.633	3	3	1	2	1	2
34.78	34.78	2.577	2	4	1	8	6	9
35.26	35.26	2.543	-1	3	3	3	3	4
35.55	35.56	2.523	2	2	0	3	4	3
35.57		2.522	-2	0	4	2		2
36.42	36.42	2.465	-2	4	2	4	3	4
36.74	36.80	2.444	-5	3	1	2	6	2
36.79		2.441	-4	4	0	8		9
38.01	38.04	2.363	1	5	0	3	3	4
38.05		2.363	-6	2	0	2		2
39.64	39.64	2.272	3	6	0	2	1	2
40.65	40.64	2.218	-6	2	0	5	3	6
41.32	41.32	2.183	0	6	0	3	3	3
41.65	41.66	2.167	-3	3	3	5	3	5
41.93	41.94	2.153	6	2	0	4	4	4
42.01	42.02	2.149	-1	5	2	3	4	3
43.18	43.20	2.094	-3	5	0	1	1	1

2THETA	PEAK	D	H	K	L	I(INT)	I(PK)	I(DS)
45.39	45.40	1.9964	2	0	3	2	2	2
45.54	45.54	1.9901	4	2	1	2	2	2
46.31	46.30	1.9588	-5	5	1	4	3	4
46.62	46.62	1.9466	-1	3	3	1	1	1
47.58	47.58	1.9096	-2	1	3	1	1	2
48.21	48.20	1.8860	-3	6	0	1	1	2
48.57	48.58	1.8729	-3	1	2	1	2	2
48.71	48.70	1.8679	-7	1	3	2	3	3
49.10	49.10	1.8538	0	6	1	4	3	5
49.91	49.92	1.8256	-8	0	2	1	1	1
50.68	50.68	1.7997	-4	2	1	1	1	1
51.87	51.86	1.7611	-4	4	4	2	2	2
52.08	52.02	1.7547	0	4	0	1	1	1
53.56	53.58	1.7095	8	0	0	1	1	1
53.73	53.72	1.7046	-2	4	3	2	2	2
54.49	54.50	1.6825	-6	2	1	1	1	1
56.12	56.16	1.6375	0	8	0	1	2	1
56.15		1.6366	-7	5	1	2		2
56.21		1.6351	-7	1	4	2		2
56.45	56.44	1.6287	-7	4	2	2	2	2
56.66	56.66	1.6231	-7	5	1	1	3	1
56.66		1.6230	-9	1	2	3		3
57.85	57.84	1.5925	2	8	0	1	2	1
58.92	58.92	1.5662	5	5	0	3	1	4
59.74	59.74	1.5467	0	4	4	1	1	1
60.69	60.70	1.5246	2	8	1	5	3	5
60.74		1.5236	-8	4	3	1	1	1
62.28	62.28	1.4894	-4	0	5	1	1	1
64.30	64.30	1.4475	-1	8	1	2	1	2
64.48	64.48	1.4439	-1	1	5	2	1	2
64.71	64.70	1.4393	-9	3	4	1	1	1
64.92	64.90	1.4351	9	1	0	1	1	1
66.16	66.16	1.4112	-7	9	5	1	1	1
68.55	68.56	1.3677	0	6	2	2	2	2
69.89	69.90	1.3448	-6	8	0	2	2	2
70.00		1.3429	-3	9	0	1		1
70.79	70.78	1.3299	6	8	3	2	1	3

LAUMONTITE - BARTL AND FISCHER,1967

2THETA	PEAK	D	H	K	L	I(INT)	I(PK)	I(DS)
72.33	72.34	1.3053	-5	9	1	1	1	1
73.13	73.12	1.2930	7	5	2	1	1	2
74.94	74.96	1.2661	2	8	3	2	1	3
75.27	75.20	1.2614	4	4	4	1	1	2
80.74	80.74	1.1891	9	3	2	1	1	2
82.69	82.68	1.1660	-6	4	6	1	1	2
104.18	104.18	0.9762	13	5	0	1	0	1
107.75	107.76	0.9536	-5	13	1	1	0	1

TABLE 36. ZEOLITES

	Gismondite	Phillipsite	Epistilbite	Heulandite	Na–Type A (dehydrated zeolite)
Composition	$CaAl_2Si_2O_8 \cdot 4H_2O$	$(K,Na)Al_5Si_{11}O_{32} \cdot 10H_2O$	$(Ca_{2.59}Na_{1.06}K_{0.10})(Al_{6.29}Si_{17.71})O_{48} \cdot 15.74\,H_2O$	$Al_2Si_7O_{18} \cdot 6H_2O$	$Na_{12}(AlO_2)_{12}(SiO_2)_{12}$
Source	Hohenberg, Germany	Pacific Ocean	Teigerhorn, Iceland	Giebelsbach, near Fresch, Wallis, Switz.	Synthetic
Reference	Fischer, 1963	Steinfink, 1962c	Perrotta, 1967	Merkle & Slaughter, 1968	Smith & Dowell, 1968
Cell Dimensions					
a Å	10.02	14.252 (9.965)	9.08	17.73	12.28
b Å	10.62	14.252 (14.252)	17.74	17.82	12.28
c	9.84	9.965 (14.252)	10.25	7.43	12.28
α	90	90	90	90	90
β deg	92.41	90	124.54	116.33	90
γ	90	90	90	90	90
Space Group	$P2_1/c$	Amna2 (B2mb)	C2/m	Cm	Pm3m
Z	4	2	1	4	1 (?)

TABLE 36. (cont.)

Variety	Gismondite	Phillipsite	Epistilbite	Heulandite	Na–Type A (dehydrated zeolite)
Site Occupancy				T_1 $Si_{0.75}Al_{0.25}$ T_2 $Si_{0.25}Al_{0.75}$ T_3 Si T_4 $Si_{0.60}Al_{0.40}$ T_5 Si T_6 Si T_7 $Si_{0.75}Al_{0.25}$ T_8 Si T_9 $Si_{0.25}Al_{0.75}$	
Method	3-D, film	3-D, film	3-D, film	3-D, counter	3-D, counter
R & Rating	0.15 (2)	0.14 (3)	0.16 (2)	0.11 (1)	0.11 (2)
Cleavage and habit	Octahedrons	Neither platy nor fibrous	{010} Platy & Prismatic	{010} perfect Platy	
Comment	B estimated				
μ	93.1	61.1	79.9	85.9	71.0
ASF	0.3269×10^{-1}	0.5198×10^{-1}	0.4395×10^{-1}	0.3526×10^{-1}	0.5522×10^{-1}
Abridging factors	1:1:0.5:0.5	1:1:0.5:0.5:0.5	1:1:0.5	1:1:0.5	0.5:0:0

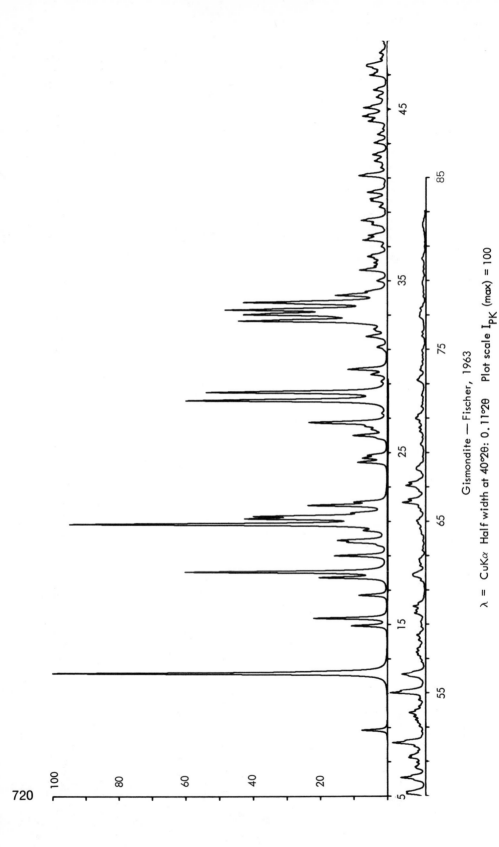

Gismondite — Fischer, 1963

λ = CuKα Half width at 40°2θ: 0.11°2θ Plot scale I$_{PK}$ (max) = 100

720

GISMONDITE - FISCHER, 1963

2THETA	PEAK	D	H	K	L	I(INT)	I(PK)	I(DS)
8.83	8.84	10.011	1	0	0	7	8	7
12.14	12.14	7.285	1	1	0	93	100	91
14.90	14.90	5.941	-1	1	1	10	11	10
15.35	15.36	5.769	1	1	1	22	22	21
16.68	16.68	5.310	0	0	2	8	9	8
17.70	17.72	5.006	0	2	0	20	20	19
18.03	18.04	4.916	0	0	2	61	60	60
18.98	18.98	4.672	0	2	1	16	16	15
19.76	19.78	4.488	-1	0	1	10	12	10
19.89	19.88	4.461	0	1	2	13	15	13
20.44	20.46	4.340	1	2	1	5	5	13
20.80	20.80	4.267	-1	1	2	100	95	100
21.13	21.14	4.202	-1	1	1	39	43	39
21.27	21.28	4.174	-1	2	1	36	40	36
21.48	21.48	4.134	-1	1	1	7	11	7
21.90	21.92	4.054	2	1	1	25	24	25
22.10	22.10	4.018	1	0	2	9	10	9
24.42	24.42	3.642	1	2	2	10	9	10
24.66	24.66	3.607	-2	2	0	7	8	8
24.82	24.82	3.584	-1	0	2	6	6	6
25.97	25.98	3.428	2	2	2	11	10	11
26.33	26.34	3.382	1	2	1	4	5	4
26.50	26.50	3.361	1	3	0	7	7	5
26.69		3.337	0	1	1	16	23	16
26.74	26.74	3.331	1	3	1	14		15
27.26	27.26	3.269	3	1	2	1	2	1
28.00	28.00	3.184	1	3	0	72	60	74
28.48	28.48	3.131	0	1	1	64	54	66
29.12	29.12	3.065	-3	1	3	1	1	1
29.52	29.52	3.024	-1	1	3	5	5	5
29.82	29.82	2.994	3	1	1	14	12	15
31.11	31.10	2.873	3	3	2	3	3	4
31.64	31.74	2.825	3	0	0	7	6	7
31.74	31.74	2.817	-3	2	3	3		2
32.07	32.08	2.789	0	0	3	3	4	3
32.18	32.16	2.779	-1	1	2	2	4	2
32.61	32.64	2.743	1	3	3	4	44	5
32.63	33.00	2.742	-3	2	1	50	43	53
33.00	33.00	2.712	-1	1	2	45		48
33.05		2.708	-3	0	2	4		5
33.10		2.704	-2	1	3	5		5
33.28	33.28	2.690	3	2	1	57	48	59
33.64	33.72	2.662	1	2	2	22	43	23
33.73		2.655	0	4	0	40		43
34.14	34.14	2.624	3	1	2	18	15	19
34.36	34.36	2.608	1	1	3	6	7	6
34.98	34.98	2.563	0	4	1	3	3	4
35.62	35.62	2.518	2	3	2	11	8	11
35.85	35.84	2.503	-2	0	1	4	4	4
36.04	36.04	2.490	-1	1	1	4	4	4
36.24		2.477	-1	3	3	1		1
36.28	36.28	2.474	-2	2	2	3		3
36.41	36.40	2.465	2	3	2	6	6	6
36.53	36.50	2.458	0	2	4	2	5	2
36.99	36.98	2.428	3	3	0	2	2	2
37.36	37.36	2.405	0	4	3	9	7	10
37.53	37.46	2.395	3	1	1	2	6	3
37.64	37.64	2.388	-1	1	4	6	6	6
38.25	38.26	2.351	-1	4	1	5	6	5
38.34	38.34	2.345	4	4	0	1	5	1
38.50	38.50	2.336	2	0	2	9	8	10
38.59		2.331	-3	3	3	2		2
39.28	39.28	2.292	-2	1	1	3		3
39.65	39.66	2.271	-2	4	4	6	2	6
39.78	39.76	2.264	4	3	0	3	5	4
40.15	40.16	2.244	-2	1	4	8	5	9
41.08	41.12	2.196	-2	0	4	1	6	1
41.09		2.195	-1	1	4	3		3
41.12		2.193	0	2	2	8		9
41.22		2.188	-4	2	1	1		1
41.58	41.58	2.170	2	0	4	2	2	2
41.79	41.80	2.160	-1	2	4	1	2	1
41.97	41.96	2.151	3	3	2	5	4	5
42.33	42.34	2.133	-2	3	4	6	4	6

GISMONDITE — FISCHER, 1963

2THETA	PEAK	D	H	K	L	I(INT)	I(PK)	I(DS)
43.02	43.02	2.101	2	4	2	6	4	6
43.52	43.52	2.078	1	5	0	4	3	4
43.85	43.84	2.063	0	4	3	1	1	2
44.28	44.28	2.044	-3	4	1	6	6	7
44.57		2.044	4	4	0	2		2
44.66	44.58	2.031	-1	2	3	9	7	10
44.95	44.68	2.027	-4	2	2	2	6	2
45.07	45.06	2.015	1	3	1	1	7	2
45.18		2.010	-1	0	3	9		10
45.25	45.18	2.002	5	0	0	2		2
45.68	45.68	1.9845	-3	1	4	4	5	4
46.09	46.10	1.9676	5	1	0	5	3	6
46.96	46.96	1.9334	0	1	5	7	4	8
47.20	47.20	1.9238	-2	5	1	2	5	2
47.36	47.36	1.9177	4	1	2	4	3	4
47.46	47.50	1.9141	5	1	1	1	4	1
47.48		1.9132	-1	1	5	3		3
47.52		1.9117	-2	1	5	1	6	1
47.55		1.9105	-4	3	2	3		3
47.60		1.9086	3	1	4	3		3
47.92	47.92	1.8968	-1	2	3	2	2	2
48.27	48.28	1.8838	-1	1	5	3	2	3
49.10	49.12	1.8539	-5	2	1	6	6	7
49.11		1.8534	-5	1	2	2		2
49.20	49.20	1.8502	2	3	4	4	6	5
49.89	49.90	1.8263	-1	2	5	3	4	4
50.04	50.04	1.8212	4	2	0	9	8	11
50.56	50.66	1.8036	0	4	4	3	3	3
50.65		1.8007	1	2	5	3		3
51.21	50.92	1.7918	3	5	0	3	3	3
51.32	51.22	1.7824	0	5	3	7	5	8
51.59	51.32	1.7789	2	6	0	2	5	3
51.86	51.58	1.7700	0	3	5	1	2	1
52.04	51.86	1.7615	-4	3	3	3		3
52.08	52.08	1.7558	3	5	4	3	10	3
		1.7545	-3	3	4	13		15
52.74	52.74	1.7342	3	4	3	1	1	1

2THETA	PEAK	D	H	K	L	I(INT)	I(PK)	I(DS)
53.04	53.04	1.7251	-4	4	2	3	2	3
53.28	53.28	1.7177	4	0	4	1	2	2
53.41	53.42	1.7140	1	6	1	2	3	3
53.60	53.60	1.7084	2	3	5	2	4	2
53.62		1.7078	4	3	4	4		2
53.83	53.84	1.7014	3	4	3	6	5	7
54.20	54.18	1.6909	4	4	2	2	2	2
54.57	54.58	1.6803	2	4	4	3	2	3
54.98	54.98	1.6687	2	6	0	4	11	5
54.99		1.6685	6	0	0	12		15
55.04		1.6670	0	5	3	1		2
55.10		1.6653	-3	2	5	2		2
55.96	56.08	1.6417	0	6	2	5	7	6
56.06		1.6390	-1	0	6	2		2
56.08		1.6385	0	1	6	7		9
56.80	56.80	1.6194	0	4	6	1	1	2
57.05	57.02	1.6130	-4	1	4	1	2	1
57.19	57.20	1.6092	1	1	6	2	2	2
57.36	57.36	1.6049	-4	5	1	3	3	3
57.58	57.56	1.5994	5	2	3	2	3	3
58.05	58.06	1.5875	3	3	5	2	2	2
58.35	58.34	1.5801	0	4	5	2	3	3
58.36		1.5799	1	5	4	1		2
58.84	58.84	1.5681	-5	1	3	1	1	2
59.64	59.64	1.5490	-3	6	1	5	3	5
59.87	59.80	1.5436	-1	6	3	1	3	2
60.04	60.04	1.5395	1	6	3	5	4	6
60.27	60.20	1.5342	6	3	1	1	2	1
61.54	61.54	1.5056	3	3	5	1	2	2
61.78	61.82	1.5004	-6	1	0	2	4	3
61.79		1.5000	0	7	1	1		2
61.82		1.4994	0	0	6	2		2
61.89	61.90	1.4978	-6	1	1	3	4	4
62.40	62.40	1.4870	0	3	6	3	2	3
62.66	62.66	1.4814	-3	1	6	1	2	2
62.77	62.76	1.4791	-1	3	6	2	2	2
63.30	63.30	1.4678	-4	5	3	2	1	2

GISMONDITE – FISCHER, 1963

2THETA	PEAK	D	H	K	L	I(INT)	I(PK)	I(DS)
64.25	64.26	1.4485	6	1	3	2	1	2
64.57	64.56	1.4420	-4	3	5	2	2	2
64.86	64.86	1.4363	4	5	3	2	2	3
65.00	65.02	1.4336	3	1	6	2	3	3
65.58	65.60	1.4222	-3	6	3	1	2	2
65.80	65.82	1.4180	5	4	4	2	2	2
66.08	66.08	1.4127	6	2	0	12	7	15
66.24	66.26	1.4097	6	2	3	1	5	2
66.73	66.74	1.4005	3	6	3	2	2	3
67.06	67.06	1.3944	0	4	1	9	6	11
67.24	67.24	1.3912	6	4	1	1	5	1
67.97	68.02	1.3780	-6	3	3	2	3	3
68.04		1.3768	-3	3	6	2		3
68.12	68.12	1.3753	0	7	1	2	3	3
68.74	68.74	1.3644	-7	2	1	1	1	2
69.41	69.48	1.3528	-1	2	7	1	2	1
69.50		1.3514	6	3	3	2		3
69.87	69.88	1.3450	6	4	2	3	1	2
70.22	70.22	1.3393	3	3	6	1	2	4
70.43	70.42	1.3357	-3	7	2	2	2	2
70.72	70.74	1.3309	2	4	4	2	2	2
70.93	70.96	1.3275	0	8	0	2	3	2
70.95		1.3271	2	0	7	2		2
71.02		1.3260	7	3	0	2		3
71.17	71.16	1.3236	3	7	2	2	2	2
71.68	71.68	1.3156	0	8	1	2	1	2
72.32	72.32	1.3055	0	3	7	2	1	2
73.12	73.20	1.2931	-7	2	3	2	3	2
73.18		1.2921	-1	5	6	2		2
73.22		1.2916	-7	2	3	2		3
73.78	73.78	1.2831	-2	8	0	3	2	4
74.26	74.26	1.2760	-3	2	7	1	1	2
74.92	74.92	1.2663	3	1	7	1	1	1
75.76	75.76	1.2544	7	2	3	2	3	3
76.88	76.88	1.2390	-4	1	7	2	2	2
77.01	77.02	1.2371	-4	4	6	1	2	1
77.37	77.36	1.2323	-6	5	3	3	2	3
78.90	78.88	1.2122	3	6	5	1	1	1
79.52	79.52	1.2044	-6	5	3	1	1	2
80.30	80.28	1.1946	-6	0	1	1	1	2
82.32	82.32	1.1703	6	6	2	1	1	1
84.54	84.56	1.1452	6	0	6	2	1	2

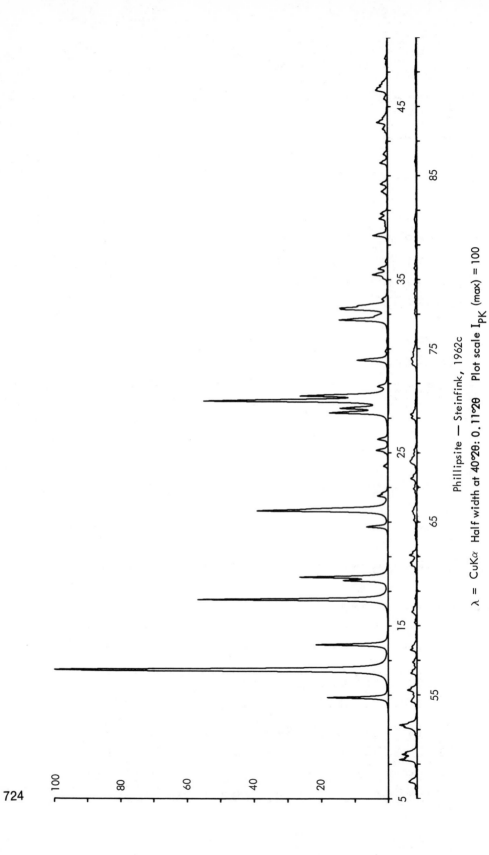

Phillipsite — Steinfink, 1962c

λ = CuKα   Half width at 40°2θ: 0.11°2θ   Plot scale I_{PK} (max) = 100

724

PHILLIPSITE - STEINFINK, 1962C

2THETA	PEAK	D	H	K	L	I(INT)	I(PK)	I(DS)
10.82	10.84	8.167	0	1	1	23	18	23
12.41	12.44	7.126	0	2	0	56	100	56
12.41		7.126	2	0	0	44		44
12.48		7.086	1	1	1	76		76
13.88	13.90	6.374	1	2	0	28	22	28
16.50	16.50	5.369	2	0	1	76	57	76
17.59	17.60	5.039	2	1	1	16	14	16
17.79	17.80	4.982	0	0	2	35	26	35
20.69	20.70	4.288	0	3	1	9	9	9
21.62	21.62	4.106	3	1	1	4		4
21.62		4.106	1	3	1	51	39	52
22.47	22.48	3.953	3	2	0	4	3	4
22.63	22.64	3.925	1	1	2	2	2	2
24.20	24.20	3.674	2	3	1	2	1	2
25.11	25.12	3.543	2	2	2	5	4	5
25.75	25.76	3.457	1	4	0	5	3	5
27.28	27.28	3.266	4	1	1	27	18	28
27.55	27.56	3.235	0	1	3	20	14	21
27.97	27.98	3.187	2	4	0	39	55	40
27.97		3.187	4	2	1	39		39
28.01		3.183	3	3	1	12		12
28.26	28.26	3.155	1	2	3	39	26	39
28.81	28.80	3.097	3	1	2	4	3	4
30.32	30.32	2.946	2	1	3	15	9	16
32.65	32.66	2.740	4	3	1	6	15	6
32.65		2.740	0	5	1	19		20
32.87	32.86	2.722	0	3	3	4	4	4
33.28	33.28	2.691	5	1	1	11	14	12
33.26		2.691	1	5	1	10		10
33.35	33.34	2.685	4	2	2	4		4
33.35		2.685	2	4	2	4	14	5
33.48	33.48	2.674	3	1	3	2	9	2
33.48		2.674	1	3	3	8		9
33.84	33.84	2.647	5	2	0	2	2	2
35.06	35.06	2.558	2	5	1	2	2	2
35.26	35.26	2.543	2	3	3	8	5	8
35.60	35.60	2.519	4	4	0	5	3	5

2THETA	PEAK	D	H	K	L	I(INT)	I(PK)	I(DS)
36.02	36.02	2.491	0	0	4	1		1
37.52	37.52	2.395	4	1	3	8	4	8
38.07	38.06	2.362	3	3	3	1	1	1
38.48	38.48	2.337	5	2	2	5	3	5
38.78	38.78	2.320	2	2	4	4	2	4
39.97	40.08	2.253	6	0	0	1	2	1
40.07		2.248	4	4	2	3		3
40.49	40.50	2.226	5	4	0	4	2	4
41.72	41.72	2.163	3	5	2	4	2	4
42.22	42.22	2.139	1	5	3	2	1	2
43.52	43.52	2.078	6	3	1	1	1	1
43.69	43.70	2.070	2	5	3	3	2	3
44.07	44.06	2.053	6	2	2	2	3	2
44.07		2.053	0	6	2	5		5
44.33	44.32	2.042	0	4	4	2		2
45.43	45.44	1.9948	7	1	1	2	1	2
45.90	45.94	1.9755	1	1	5	2	4	3
45.94		1.9738	1	3	5	4		4
46.07	46.04	1.9687	5	3	3	2	3	2
46.21	46.20	1.9627	2	4	4	3	2	3
46.40	46.34	1.9551	1	5	5	1	1	1
47.44	47.44	1.9146	6	1	1	1	1	1
47.77	47.78	1.9022	2	6	2	2	3	2
49.94	49.94	1.8248	5	3	0	4		5
50.00		1.8227	3	3	5	1		1
51.23	51.24	1.7815	8	0	0	7	5	7
51.55	51.54	1.7715	4	4	4	8	9	9
51.66	51.66	1.7677	6	4	1	1	1	1
53.10	53.12	1.7231	7	1	1	2	3	2
53.10		1.7231	5	5	3	2		3
53.24	53.24	1.7231	1	7	1	2	4	2
53.24		1.7191	6	0	0	4	4	4
53.24		1.7191	6	6	0	4		4
54.59	54.60	1.6796	3	8	0	3	2	4
55.00	55.00	1.6681	0	8	0	1	1	2
55.26	55.26	1.6608	0	0	6	6	3	6

PHILLIPSITE – STEINFINK, 1962

2THETA	PEAK	D	H	K	L	I(INT)	I(PK)	I(DS)
56.24	56.24	1.6343	5	7	1	4	2	4
56.68	56.68	1.6227	1	3	5	1	1	1
56.88	56.88	1.6175	0	2	6	2	1	2
57.58	57.58	1.5993	6	5	3	4	2	5
59.15	59.16	1.5605	8	1	1	1	1	1
59.77	59.78	1.5458	9	2	0	2	2	2
60.05	60.04	1.5393	7	2	4	1	1	2
60.56	60.56	1.5276	2	2	1	2	1	2
62.09	62.10	1.4936	8	5	1	1	1	1
62.46	62.48	1.4855	9	3	1	3	2	4
62.60	62.62	1.4826	5	7	3	3	2	3
63.06	63.06	1.4728	4	4	6	3	2	3
65.08	65.08	1.4321	2	9	1	2	1	2
65.58	65.58	1.4223	1	9	3	2	1	2
66.89	66.88	1.3975	2	10	0	4	2	5
67.49	67.50	1.3866	4	4	6	4	2	5
68.36	68.48	1.3711	9	5	1	2	2	2
68.49		1.3688	9	3	3	3	2	4
68.93	68.92	1.3611	0	6	6	1	1	1
69.14	69.14	1.3575	1	3	7	1	1	1
70.99	70.98	1.3266	4	9	3	4	2	5
71.19	71.20	1.3233	4	10	0	2	2	2
71.19		1.3233	10	4	0	1		1
74.06	74.08	1.2789	4	10	2	1	1	1
74.06		1.2789	10	4	2	1		1
74.14		1.2778	9	5	3	2		2
74.43	74.40	1.2736	4	3	7	1	2	1
74.43		1.2736	0	5	7	2		2
74.78	74.78	1.2685	5	1	7	1	1	1
75.39	75.38	1.2597	8	8	0	1	1	1
77.02	77.02	1.2371	8	6	4	1	1	1
77.67	77.68	1.2283	5	6	6	1	1	2
77.67		1.2283	11	6	3	1	1	2
79.64	79.64	1.2028	3	8	8	1	1	2
81.76	81.76	1.1769	9	2	6	1	0	1
85.80	85.80	1.1315	9	11	3	1	0	2
106.71	106.68	0.9600				1	0	1
112.40	112.42	0.9269	13	7	3	1	0	1
112.45		0.9266	4	4	10	1	1	1
118.33	118.34	0.8971	15	3	3	1	0	1
121.02	121.02	0.8849	4	15	3	1	1	1

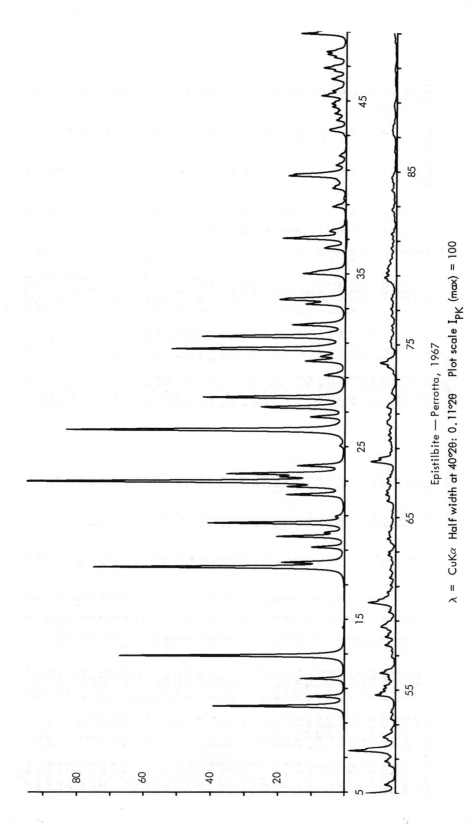

Epistilbite — Perrotta, 1967

λ = CuKα Half width at 40°2θ: 0.11°2θ Plot scale I_{PK} (max) = 100

EPISTILBITE — PERROTTA, 1967

2THETA	PEAK	D	H	K	L	I(INT)	I(PK)	I(DS)
9.96	9.98	8.870	0	0	1	33	39	32
10.51	10.52	8.407	0	1	0	9	11	9
11.56	11.56	7.651	-1	1	1	11	13	11
12.88	12.88	6.867	1	1	0	58	67	57
18.00	18.00	4.923	-1	1	2	69	75	68
18.27	18.28	4.851	-1	3	0	15	19	15
19.15	19.16	4.631	1	1	1	9	10	9
19.78	19.78	4.485	0	0	2	19	20	19
20.00	20.00	4.435	-2	0	1	4	6	4
20.55	20.56	4.318	-1	0	2	38	41	38
20.93	20.94	4.240	-1	2	1	2	3	2
22.19	22.20	4.002	-2	0	2	16	18	16
22.65	22.66	3.923	0	4	1	14	17	14
22.94	22.94	3.873	-1	3	2	100	100	100
23.23	23.24	3.826	-2	2	2	13	20	13
23.40	23.40	3.798	0	2	2	32	35	32
23.88	23.88	3.723	2	0	0	13	14	13
25.02	25.02	3.557	1	3	1	1	2	1
25.93	25.94	3.433	-2	2	0	87	83	88
26.69	26.70	3.337	0	3	1	10	11	10
27.22	27.28	3.273	-1	5	1	18	25	18
27.28		3.266	-1	1	3	13		13
27.83	27.84	3.203	-1	5	0	45	42	46
29.11	29.12	3.065	-2	4	2	7	6	7
29.93	29.94	2.983	-3	1	1	13	12	13
30.20	30.20	2.957	-3	3	2	4	8	4
30.20		2.957	1	3	2	3		3
30.62	30.62	2.917	-1	5	2	57	52	59
31.34	31.34	2.852	-2	4	3	47	43	48
32.03	32.04	2.792	-3	3	1	17	16	17
33.23	33.24	2.694	-3	3	3	13	12	13
33.47	33.50	2.675	1	3	3	16	20	16
33.51		2.672	0	2	4	7		7
33.58		2.666	-2	4	0	5		5
34.99	35.00	2.562	-2	2	4	15	13	15
36.47	36.48	2.462	-2	2	4	7	6	7
36.52		2.458	3	1	0	2		2

2THETA	PEAK	D	H	K	L	I(INT)	I(PK)	I(DS)
37.03	37.04	2.425	-1	5	3	13	19	14
37.03		2.425	-2	6	2	9		10
37.32	37.32	2.408	-3	1	4	2	3	2
37.45	37.44	2.400	2	1	1	4	5	4
37.54	37.54	2.394	-1	4	0	2	4	2
38.86	38.86	2.315	2	6	3	5	4	5
39.93	39.94	2.256	-4	0	3	5	4	5
40.08	40.04	2.248	-3	3	4	2	4	2
40.29	40.30	2.237	-1	1	4	1	2	1
40.63	40.64	2.219	-1	7	3	13	17	13
40.65		2.217	-2	6	0	8		8
40.74	40.74	2.213	1	5	3	5	16	6
40.78		2.211	-2	6	3	3		2
41.26	41.26	2.186	-4	2	2	3	3	4
41.80	41.80	2.159	-4	2	3	2	2	3
43.24	43.32	2.091	2	2	2	5	5	5
43.31		2.087	-1	3	1	4		4
43.87	43.86	2.062	-4	0	4	4	3	4
44.25	44.26	2.045	0	2	5	4	4	5
44.64	44.64	2.028	-2	0	5	5	4	4
44.64		2.020	-3	1	3	3		5
44.83	44.82	2.015	-1	7	1	3	4	5
44.94	44.94	2.005	-3	3	4	1	4	1
45.19	45.28	2.001	-1	4	2	1	8	2
45.28		1.9894	-4	2	4	9		9
45.56	45.56	1.9650	-2	8	2	2	4	5
46.16	46.24	1.9617	-2	6	1	2	5	2
46.25		1.9613	3	1	2	1		2
46.71	46.72	1.9429	-3	7	2	4	4	4
46.88	46.88	1.9364	-2	6	4	4	7	4
46.90		1.9357	1	7	2	2		2
47.49	47.50	1.9128	-4	4	4	4	4	7
47.69	47.68	1.9055	1	9	0	4	6	5
47.69		1.9052	2	8	4	2		2
47.85	47.84	1.8992	0	4	4	4		4
48.88	48.88	1.8617	4	0	0	19	14	20
49.50	49.50	1.8397	-1	9	2	4	3	4

728

2THETA	PEAK	D	H	K	L	I(INT)	I(PK)	I(DS)
50.60	50.60	1.8023	1	9	1	2	2	3
50.85	50.86	1.7940	1	5	1	2	3	4
51.47	51.48	1.7740	0	10	0	11	14	12
51.49		1.7734	3	7	0	10		11
52.10	52.10	1.7540	-3	7	4	2	2	2
52.27	52.26	1.7487	-1	7	4	3	4	3
52.33		1.7466	-5	1	2	1		1
52.69	52.68	1.7358	0	10	1	1	1	1
53.32	53.32	1.7166	4	4	0	2	1	2
53.70	53.70	1.7051	-4	6	1	2	1	2
53.71		1.6768	-2	8	4	2	6	7
54.69	54.70	1.6762	-1	7	3	2		3
54.71		1.6684	-4	0	6	2	4	2
54.99	55.02	1.6670	-1	5	5	3	3	3
55.04		1.6616	-2	0	6	3	3	3
55.22	55.22	1.6516	-3	0	2	2	2	2
55.23		1.6472	1	9	2	1	2	1
55.60	55.60	1.6412	-3	3	6	6	5	7
55.76	55.76	1.6169	-3	9	3	1	1	1
55.98	55.98	1.6075	-3	7	1	1	3	2
56.90	56.92	1.5991	-5	5	4	4		5
57.26	57.26	1.5871	1	5	4	4	3	5
57.59	57.60	1.5732	-5	5	2	4	4	4
57.60		1.5710	2	8	2	2		3
58.06	58.06	1.5522	-5	1	6	3	2	3
58.07		1.5413	-1	1	6	3	8	12
58.63	58.64	1.5391	-3	5	6	1		3
58.72		1.5324	-4	8	4	2	5	3
59.50	59.50	1.5254	0	8	4	2	3	3
59.97	60.06	1.5163	1	11	1	2	2	2
60.06		1.5068	-1	3	6	2	2	2
60.22	60.22	1.4968	-1	3	0	1	2	1
60.35		1.4842	5	1	3	1		2
60.66	60.66	1.4783	1	9	3	1	2	1
61.06	61.06	1.4749	-5	7	3	1	2	1
61.48	61.48	1.4628	-5	5	4	2	2	2
61.49		1.4586	-2	10	4	3	2	3
61.94	61.94							
62.53	62.54							
62.80	62.80							
62.96	62.96							
63.54	63.54							
63.75	63.74							

2THETA	PEAK	D	H	K	L	I(INT)	I(PK)	I(DS)
63.99	63.98	1.4536	-1	7	4	1	2	2
64.13	64.14	1.4509	-5	11	3	2	2	2
64.46	64.46	1.4446	-6	3	0	1	1	2
64.99	65.00	1.4338	-6	2	2	1	1	2
65.94	65.94	1.4155	-3	4	5	2	3	2
65.95		1.4153	4	2	5	4		5
66.30	66.30	1.4085	-3	11	1	1		1
66.45	66.48	1.4057	-3	1	7	7	3	3
66.56		1.4036	-5	1	6	3		2
66.98	66.98	1.3959	-6	2	6	6	3	3
67.23	67.22	1.3913	-4	10	4	2	3	3
68.19	68.22	1.3740	2	12	0	5	8	6
68.23		1.3733	5	0	0	9		10
68.48	68.42	1.3690	3	5	3	2	5	2
68.75	68.72	1.3643	-6	6	4	1	2	2
69.25	69.26	1.3556	0	10	2	1	1	1
69.44	69.44	1.3524	3	11	0	2	2	2
69.76	69.76	1.3469	-3	7	5	1		2
69.77		1.3467	-6	4	6	1	1	1
70.09	69.96	1.3413	-1	11	6	2	2	2
70.41	70.42	1.3360	-1	11	3	2	3	3
70.53	70.54	1.3341	0	6	6	2	2	2
70.94	70.94	1.3274	-6	8	6	3	3	3
71.37	71.38	1.3205	-5	7	6	4		5
71.48		1.3187	-1	7	4	1		1
72.18	72.18	1.3075	0	12	3	1	1	1
73.70	73.72	1.2843	4	2	0	3		3
73.91	73.94	1.2812	1	0	8	3	5	4
73.96		1.2805	-2	2	12	4		4
74.81	74.94	1.2680	-4	2	8	4		2
74.94		1.2661	0	6	6	1	1	1
75.84	75.84	1.2533	3	3	4	1	1	2
76.09	76.08	1.2498	-6	8	4	1	1	2
76.65	76.66	1.2421	1	13	2	1	2	1
77.06	77.06	1.2365	-4	12	3	2	2	2
77.62	77.62	1.2289	-3	13	3	2	2	2
78.70	78.72	1.2148	-7	5	4	2	3	3

729

EPISTILBITE – PERROTTA, 1967

2THETA	PEAK	D	H	K	L	I(INT)	I(PK)	I(DS)
78.86	78.86	1.2127	-2	10	6	2	3	3
78.87		1.2127	-4	12	4	2		2
79.14	79.12	1.2092	0	12	4	2	3	3
79.38	79.38	1.2061	3	5	4	2	2	2
79.54	79.52	1.2041	-6	8	2	1	2	2
79.88	79.86	1.1998	4	8	2	2	1	3
80.25	80.22	1.1952	6	4	0	1	2	2
80.57	80.56	1.1912	-7	5	6	2	2	2
81.39	81.38	1.1813	1	5	6	2	1	3
82.81	82.80	1.1647	-1	13	3	1	2	2
83.89	83.92	1.1524	-1	15	2	2	1	2
83.95		1.1517	-7	7	4	1	1	2
84.41	84.42	1.1465	-7	5	7	2	1	2
84.62	84.64	1.1443	3	7	4	1	1	2
86.14	86.14	1.1279	-8	0	6	1	1	2
87.15	87.16	1.1174	2	0	6	2	1	2
87.37	87.38	1.1151	-6	10	2	1	1	2
92.51	92.52	1.0662	-3	13	6	1	1	2
97.87	97.84	1.0216	-6	12	6	1	1	2
100.02	99.98	1.0053	-8	8	6	1	1	1
100.96	101.02	0.9985	-2	10	8	1	1	1
101.05		0.9979	-2	8	6	1	1	1
101.55	101.42	0.9943	-7	11	3	1	1	1
103.48	103.48	0.9809	7	7	0	1	1	2
104.77	104.78	0.9724	-3	15	6	1	2	1
106.39	106.38	0.9620	3	17	0	1	0	2
106.48		0.9614	-6	6	10	1	1	1
107.66	107.68	0.9541	1	3	8	1	1	1
107.67		0.9541	-7	5	10	1	1	1
108.05	108.10	0.9518	-8	10	6	1	1	2
108.32	108.36	0.9502	-5	7	10	1	1	1
108.37		0.9498	-3	5	10	1		1
108.37		0.9498	-9	5	4	1		1
109.11	109.14	0.9455	2	10	6	2	1	2
109.36	109.38	0.9440	5	5	4	1	1	1
109.90	109.80	0.9409	-9	5	8	1	1	1

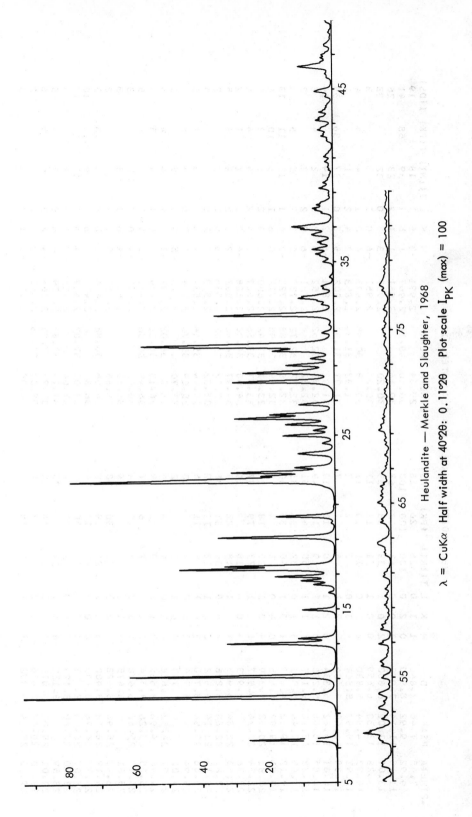

Heulandite — Merkle and Slaughter, 1968

λ = CuKα Half width at 40°2θ: 0.11°2θ Plot scale I$_{PK}$ (max) = 100

HEULANDITE - MERKLE AND SLAUGHTER, 1968

2THETA	PEAK	D	H	K	L	I(INT)	I(PK)	I(DS)
7.45	7.46	11.860	0	1	0	26	26	26
9.92	9.92	8.910	1	2	0	100	100	100
11.13	11.14	7.945	-2	0	0	63	62	63
13.01	13.02	6.800	0	0	1	33	33	33
13.28	13.30	6.659	2	0	0	11	11	11
14.93	14.94	5.930	0	2	1	11	11	11
15.91	15.92	5.564	-1	3	0	2	2	2
16.38	16.40	5.406	-2	3	0	6	7	7
16.61	16.62	5.334	0	1	1	10	10	10
16.85	16.86	5.256	-3	1	1	19	18	19
17.28	17.28	5.127	-1	1	1	50	47	51
17.45	17.46	5.077	3	1	0	32	33	33
19.12	19.12	4.639	-1	1	1	40	35	40
19.91	19.92	4.455	-4	0	1	1	1	1
20.31	20.32	4.369	-1	3	1	20	18	21
22.34	22.34	3.977	-4	3	0	75	79	77
22.36		3.973	2	2	1	18		19
22.65	22.64	3.923	2	4	0	18	22	19
22.87	22.88	3.886	-2	2	1	35	31	36
23.13	23.14	3.842	2	2	1	12	12	12
23.86		3.726	-2	4	1	6	11	6
23.94	23.94	3.714	-2	0	2	9		10
24.01		3.703	0	4	1	2		2
24.51	24.52	3.628	4	2	0	4	4	4
24.96	24.96	3.565	-3	1	2	19	15	19
25.25	25.26	3.524	-1	1	2	7	6	7
25.59	25.62	3.478	-5	1	1	4	15	4
		3.475	1	5	0	14		14
25.97	25.98	3.428	-2	2	2	37	30	38
26.19	26.18	3.400	-4	0	2	21	19	22
26.75	26.76	3.330	0	0	2	13	11	14
28.07	28.06	3.177	-4	2	1	41	30	42
28.50	28.60	3.129	5	1	0	2	27	2
28.59		3.119	-4	4	1	34		35
29.01	29.02	3.075	-1	3	2	17	14	18
29.40	29.40	3.036	-5	1	2	7	7	8
29.83	29.84	2.992	3	3	1	4	18	4
29.84		2.991	-3	5	1	18		19
30.10	30.18	2.967	1	5	1	39	58	41
30.16		2.961	1	1	2	23		25
30.20		2.957	-3	5	0	32		33
30.26		2.951	-6	0	1	3		3
30.88	30.88	2.893	-4	0	2	4	4	4
31.33	31.34	2.853	-2	4	2	1	1	1
31.91	31.92	2.802	-5	3	0	37	36	39
31.92		2.801	-6	2	1	14		14
32.51	32.52	2.752	4	2	1	2	4	2
32.52		2.751	-6	0	2	2		2
32.72	32.72	2.735	-5	3	2	5	5	5
32.88	32.88	2.722	-2	6	1	13	11	13
32.99	32.96	2.712	0	3	2	6	10	6
33.41	33.42	2.680	1	6	1	3	3	4
33.57	33.56	2.667	0	4	2	4	4	4
34.08	34.08	2.628	-6	2	2	1	1	2
34.97	34.98	2.563	2	2	2	4	3	5
35.22	35.32	2.546	-3	5	2	4	4	2
35.32		2.539	6	0	0	5		4
35.44	35.44	2.531	-1	7	1	4	5	6
35.69	35.70	2.514	-5	1	2	5	6	4
35.70		2.513	-7	1	1	4		5
36.01	36.02	2.492	-7	1	0	5	4	4
36.13	36.12	2.484	3	5	1	4	5	4
36.49	36.56	2.460	-6	4	0	5	6	5
36.55		2.456	-4	0	3	2		3
36.55		2.456	-4	6	1	3		2
36.65		2.450	5	1	0	1		2
36.86	36.88	2.436	-2	2	3	3	10	3
36.87		2.436	2	6	1	9		10
37.01	37.02	2.427	4	1	2	11	12	12
37.04		2.425	-7	1	2	4		5
37.79	37.80	2.379	4	6	0	4	4	4
37.96	37.96	2.368	-4	2	3	6	5	5
38.07		2.362	3	1	2	2	6	6
38.08	38.06	2.361	-5	1	3	3		3

HEULANDITE - MERKLE AND SLAUGHTER, 1968

2THETA PEAK	2THETA	D	H	K	L	I(INT)	I(PK)	I(DS)
38.26	38.27	2.350	-2	2	3	6	6	6
38.36	38.43	2.340	-6	4	2	3	5	3
38.80	38.79	2.319	-2	2	1	3	4	1
38.94	38.95	2.310	-3	6	1	1	1	1
39.24	39.15	2.299	1	7	1	1	4	1
	39.23	2.294	3	7	0	1		2
	39.23	2.294	3	4	0	2		3
39.56	39.55	2.277	6	4	0	4	3	4
39.80	39.79	2.263	-7	1	2	2	2	2
40.00	40.00	2.252	-7	1	0	1	1	1
40.28	40.28	2.237	-4	6	0	4	1	2
40.46	40.46	2.227	0	8	0	2	2	2
40.76	40.77	2.211	3	3	2	3	2	3
41.06	41.05	2.197	-6	2	3	4	3	4
41.16	41.23	2.188	-8	0	1	1	1	1
41.96	41.96	2.151	-4	4	3	2	2	2
42.24	42.24	2.137	-2	4	0	5	5	5
42.36	42.42	2.129	4	0	2	2	3	2
42.60	42.60	2.121	-7	3	0	5	4	5
42.70	42.77	2.112	-7	1	1	7	4	2
43.16	43.16	2.094	6	2	1	4	5	8
43.28	43.34	2.086	-3	7	2	6	4	4
43.64	43.64	2.073	-4	6	1	2	4	6
43.86	43.87	2.062	1	1	3	1	2	2
44.00	44.01	2.056	-7	5	1	2	2	1
44.54	44.52	2.033	-3	5	3	1		2
	44.55	2.032	5	5	1	7	5	8
44.84	44.82	2.020	-6	4	3	3		3
44.88	44.88	2.018	-7	5	0	2		2
45.86	45.87	1.9767	-6	4	1	2	2	3
46.30	46.19	1.9637	-8	4	0	2		2
	46.22	1.9624	-5	7	2	2	10	3
	46.24	1.9615	-8	4	1	2		2
46.54	46.30	1.9593	-1	5	3	10	4	12
46.80	46.54	1.9491	-9	1	2	4	4	4
46.96	46.80	1.9396	6	4	1	4	3	4
	46.98	1.9323	-8	2	3	3		4

2THETA PEAK	2THETA	D	H	K	L	I(INT)	I(PK)	I(DS)
47.50	47.48	1.9132	-1	9	1	2	2	2
47.56	47.56	1.9102	-2	8	2	2	2	2
48.26	48.26	1.8841	-2	2	2	1	1	2
48.88	48.87	1.8620	-9	3	2	2	2	2
49.02	49.02	1.8568	-4	0	4	2	3	3
	49.08	1.8547	3	9	0	2		2
49.84	49.82	1.8288	-3	1	4	4	3	4
	49.88	1.8267	-7	5	3	2		3
50.02	50.03	1.8216	5	3	2	2		2
	50.10	1.8191	-6	0	4	2		5
51.22	51.22	1.7820	0	10	0	4	3	2
51.36	51.34	1.7780	0	6	3	2	4	2
51.46	51.44	1.7748	-3	7	3	1	4	2
	51.47	1.7739	5	7	1	1		2
	51.54	1.7717	-10	0	0	2		2
51.76	51.75	1.7650	4	8	1	3	9	4
	51.77	1.7645	-7	7	2	10		11
	51.85	1.7618	-2	2	4	1		2
52.58	52.57	1.7394	-5	7	3	2	3	2
	52.58	1.7391	-3	9	1	2		3
53.18	53.16	1.7214	0	10	1	4	4	5
53.24	53.24	1.7190	3	9	3	4	4	4
53.58	53.58	1.7090	3	5	3	3	3	3
54.02	53.93	1.6986	-9	5	0	2	4	2
	54.01	1.6964	-10	0	3	3		4
	54.08	1.6943	7	7	0	3		3
54.94	54.94	1.6699	-8	2	4	5	3	6
55.62	55.02	1.6676	-1	3	4	1		1
	55.62	1.6511	8	6	1	3	3	5
55.88	55.65	1.6500	-4	10	0	4		1
	55.88	1.6439	-2	8	3	3	4	4
	55.88	1.6438	2	10	1	1		1
	55.95	1.6420	-5	5	4	1		1
56.26	56.25	1.6339	-3	3	5	2	2	2
56.56	56.55	1.6259	-1	5	4	2	1	2
56.90	56.89	1.6170	4	10	0	6	3	8
57.74	57.73	1.5956	3	3	3	5	3	6

HEULANDITE – MERKLE AND SLAUGHTER, 1968

2THETA	PEAK	D	H	K	L	I(INT)	I(PK)	I(DS)
57.78	57.98	1.5942	8	4	1	1		2
57.99		1.5891	10	0	0	2	3	2
58.00		1.5888	-6	8	3	2		3
58.27	58.16	1.5821	9	5	0	3	2	1
58.69	58.68	1.5718	7	1	2	1	1	1
58.99	59.00	1.5644	10	2	0	1	2	2
59.14	59.14	1.5608	8	8	1	2	2	3
59.65	59.64	1.5487	-6	8	2	2	2	3
60.06	60.06	1.5391	4	8	1	1	2	1
60.06		1.5390	-9	7	2	1		2
60.74	60.78	1.5235	5	1	3	1		2
60.77		1.5228	3	9	0	1		2
61.01	60.96	1.5173	4	10	1	2	2	2
62.00	62.00	1.4955	9	3	1	2	2	2
62.16	62.16	1.4922	7	9	0	1	2	2
62.35	62.36	1.4879	-9	7	3	2	2	2
62.49	62.62	1.4850	0	12	0	2	4	4
62.61		1.4825	8	8	0	4		5
62.69	62.70	1.4808	5	3	3	2	4	2
62.72		1.4802	8	6	1	1		2
63.89	63.90	1.4557	-12	2	2	3	3	3
64.06	64.08	1.4522	0	6	4	1	3	4
64.13		1.4508	9	7	0	1		2
64.82	64.82	1.4371	-11	5	3	2	3	3
64.99		1.4338	-5	11	2	2	4	3
65.04	65.02	1.4328	1	9	3	4		4
65.12		1.4311	-11	1	1	1		2
65.40	65.42	1.4257	11	1	2	2	3	3
65.42		1.4253	-12	0	0	2		2
65.69	65.68	1.4202	-2	0	5	3	3	3
65.82		1.4178	9	5	1	1		3
65.96	65.96	1.4149	-11	3	4	4	3	5
66.15	66.14	1.4114	-3	3	5	1	2	1
66.44	66.44	1.4060	-4	12	1	1	1	2
66.67	66.72	1.4016	-6	2	3	2	3	2
66.73		1.4005	-4	4	5	1		1
66.78		1.3997	-9	1	5	1		1

2THETA	PEAK	D	H	K	L	I(INT)	I(PK)	I(DS)
67.11	67.12	1.3935	10	2	1	2	2	2
67.76	67.76	1.3816	-8	10	1	3	2	3
68.36	68.36	1.3711	-5	5	5	3	2	3
69.17	68.60	1.3665	4	10	1	2	2	2
69.25	69.18	1.3569	-7	5	5	1	3	5
69.49	69.50	1.3556	-3	11	3	2		2
69.52		1.3514	-8	8	4	1		1
69.85	69.86	1.3509	-7	11	5	2	3	3
70.10	70.08	1.3455	-3	5	2	2		2
70.38	70.28	1.3412	7	7	3	2	2	2
70.99	70.98	1.3365	-11	6	1	1	2	1
71.31	71.32	1.3266	-6	6	5	2	1	2
71.38		1.3215	-12	6	3	2	2	1
71.98	71.98	1.3203	-13	3	3	3	1	2
72.76	72.76	1.3108	5	7	2	2	1	4
73.58	73.58	1.2985	2	12	1	3	1	2
74.10	74.10	1.2861	-1	5	1	1	2	2
74.62	74.62	1.2784	-5	13	3	3	3	3
74.83	74.84	1.2708	-4	12	1	1	2	2
74.99	74.98	1.2677	11	3	3	3	2	4
75.46	75.46	1.2654	-2	10	1	2	2	2
76.16	76.16	1.2587	5	13	0	2	2	4
77.23	77.24	1.2489	-9	11	2	3	2	1
77.24		1.2343	0	12	3	1		2
77.53	77.56	1.2340	-7	9	2	2	3	3
77.58		1.2301	-12	8	1	2		2
77.64		1.2294	-13	5	2	2		2
78.09	78.08	1.2287	-8	12	1	1		2
82.05	82.06	1.2227	6	12	1	2	2	2
82.75	82.76	1.1735	-11	9	4	2	1	2
85.62	85.60	1.1653	-10	10	4	1	1	2
		1.1334	9	3	3	1		2

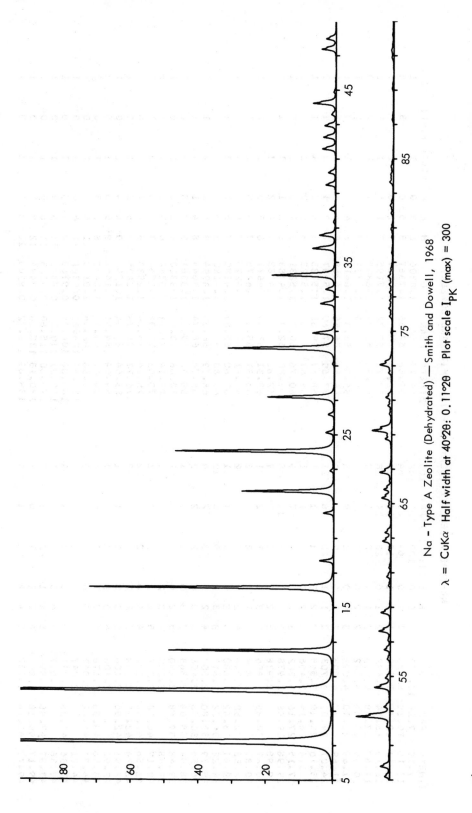

Na – Type A Zeolite (Dehydrated) — Smith and Dowell, 1968

λ = CuKα Half width at 40°2θ: 0.11°2θ Plot scale I$_{PK}$ (max) = 300

735

NA-TYPE A ZEOLITE (DEHYDRATED) – SMITH AND DOWELL, 1968

2THETA	PEAK	D	H	K	L	I(INT)	I(PK)	I(DS)
7.19	7.20	12.280	1	0	0	100	100	100
10.18	10.18	8.683	1	1	0	64	61	64
12.47	12.48	7.090	1	1	1	17	16	18
16.13	16.14	5.492	2	1	0	28	24	28
17.68	17.68	5.013	2	1	1	2	1	1
20.44	20.44	4.342	2	2	0	1	1	1
21.69	21.70	4.093	3	0	0	3	9	3
21.69		4.093	2	2	1	9		9
24.01	24.02	3.703	3	1	1	20	16	21
25.10	25.10	3.545	2	2	2	1	1	1
26.14	26.14	3.406	3	2	0	9	7	9
27.15	27.16	3.282	3	2	1	8	11	8
29.98	29.98	2.978	4	1	0	7		7
29.98		2.978	3	2	2	1		1
30.87	30.88	2.894	4	1	1	2	2	2
30.87		2.894	3	3	0	1		1
32.58	32.58	2.746	4	2	0	2		2
33.41	33.42	2.680	4	2	1	1		1
34.22	34.22	2.618	3	3	2	9	6	10
34.22		2.618	3	3	2	3	2	4
35.79	35.80	2.507	4	2	2	2	2	2
36.56	36.56	2.456	4	3	0	1	1	1
38.04	38.04	2.363	5	1	1	1		1
39.48	39.48	2.280	5	2	0	1		1
39.48		2.280	4	3	2	2	1	1
40.18	40.18	2.242	5	2	1	1	1	1
41.56	41.56	2.171	4	4	0	2	1	2
42.24	42.24	2.138	5	2	2	2		2
42.91	42.90	2.106	5	3	0	1		1
42.91		2.106	4	3	3	1		1
43.56	43.56	2.076	5	3	1	3	0	3
44.22	44.22	2.047	6	0	0	1	2	1
44.21	44.21	2.047	4	4	2	1		1
47.36	47.36	1.9178	5	4	0	2	1	2
47.36		1.9178	4	4	3	1		1
47.97	47.98	1.8948	5	4	1	2	0	1
49.17	49.18	1.8513	6	2	2	1	0	1
49.77	49.76	1.8306	5	4	2	1	1	1
49.77	49.77	1.8306	6	3	0	1	3	1
52.66	52.66	1.7367	5	5	0	4		4
52.66		1.7367	7	1	0	2		2
53.22	53.22	1.7195	7	1	1	1	1	1
54.34	54.34	1.6868	7	2	0	3	2	1
54.89	54.90	1.6711	7	2	1	1	0	3
55.99	56.00	1.6410	6	4	2	1	0	1
56.53	56.54	1.6265	7	2	2	1	1	1
57.60	57.60	1.5987	7	3	1	2	1	2
58.67	58.66	1.5723	6	5	0	1	1	1
58.67		1.5723	6	4	3	1		1
60.76	60.76	1.5231	6	5	2	1	0	1
62.80	62.80	1.4783	8	2	1	1	1	1
62.80		1.4783	7	4	2	1		1
63.31	63.30	1.4677	6	5	3	1	1	1
65.31	65.32	1.4275	7	5	0	1	1	1
65.31		1.4275	7	4	3	1		1
65.80	65.80	1.4180	7	5	1	2	1	3
65.80		1.4180	5	5	5	4		5
66.79	66.80	1.3994	8	3	2	1	1	1
68.74	68.74	1.3644	8	4	1	1	1	1
69.22	69.22	1.3561	8	3	3	1	2	2
70.66	70.66	1.3320	9	2	0	1	1	1
71.14	71.14	1.3242	9	2	1	1	1	1
73.03	73.04	1.2944	7	5	4	2	1	2
75.84	75.84	1.2533	8	4	4	1	0	1
76.77	76.76	1.2405	7	7	0	1	0	1
83.64	83.64	1.1552	9	4	4	1	0	1
84.54	84.54	1.1451	9	5	3	1	0	1
94.47	94.48	1.0492	11	4	0	2	2	2
98.56	98.56	1.0163	11	5	0	1	0	3
103.61	103.62	0.9801	10	6	0	1	0	1
104.08	104.04	0.9769	11	6	0	1	0	1
110.21	110.20	0.9391	11	6	1	1	0	1
115.60	115.60	0.9103	11	7	1	2	0	2
130.13	130.16	0.8494	12	8	1	1	0	1
139.69	139.72	0.8205	12	8	4	1	0	1

NA-TYPE A ZEOLITE (DEHYDRATED) - SMITH AND DOWELL, 1968

2THETA	PEAK	D	H	K	L	I(INT)	I(PK)	I(DS)
143.31	143.34	0.8115	12	9	2	1	0	1
143.31		0.8115	12	7	6	1		1
153.68	154.78	0.7910	15	4	0	1	0	1
154.72		0.7894	11	11	0	1		2
154.72		0.7894	13	8	3	3		5
163.59	163.30	0.7782	14	7	2	1	0	1

TABLE 37. SiO_2 POLYMORPHS

Variety	α-Quartz	High-Quartz (600°C)	High-Tridymite (220°C)	Low-Cristobalite	Coesite
Composition	SiO_2	SiO_2	SiO_2	SiO_2	SiO_2
Source	Natural, not given	Several	Heated Steinbach meteorite	Ellora, Hyderobad India	Synthetic
Reference	Zachariasen and Plettinger, 1965	Young, 1962	Dollase, 1967	Dollase, 1965	Araki & Zoltai, 1969
Cell Dimensions					
a Å	4.9133	5.01	8.74	4.9733	7.173
b Å	4.9133	5.01	5.04	4.9733	12.328
c	5.4048	5.47	8.24	6.9262	7.175
α deg	90	90	90	90	90
β deg	90	90	90	90	120
γ	120	120	90	90	90
Space Group	$P3_12$	$P6_42$	$C222_1$	$P4_12_12$	$C2/c$
Z	3	6	4	4	16

TABLE 37. (cont.)

Variety	α-Quartz	High-Quartz (600°C)	High-Tridymite (220°C)	Low-Cristobalite	Coesite
Method	3-D, counter	3-D, counter	3-D, counter	3-D, counter	3-D, counter
R & Rating	0.018 (1)	0.075 (2)	0.086 (1)	0.044 (1)	0.076 (1)
Cleavage and habit	None Variable	$\{10\bar{1}1\}$ good Holohedral	$\{0001\}$ prob. Flattened on $\{0001\}$	None Octahedral but can be fibrous	Unknown Flattened on $\{010\}$, & elongated on c axis
Comment	Cell dimensions from Swanson et. al. (1954) p. 24	Combined data from several crystals		Cell dimensions from Frondel (1962) p. 276	
μ	91.3	91.3	91.6	81.6	100.3
ASF	0.3270	0.3209	0.6471×10^{-2}	0.3338	0.2074×10^{-1}
Abridging factors	0.1:0:0	0.1:0:0	0.5:0:0	0.1:0:0	0.2:0:0

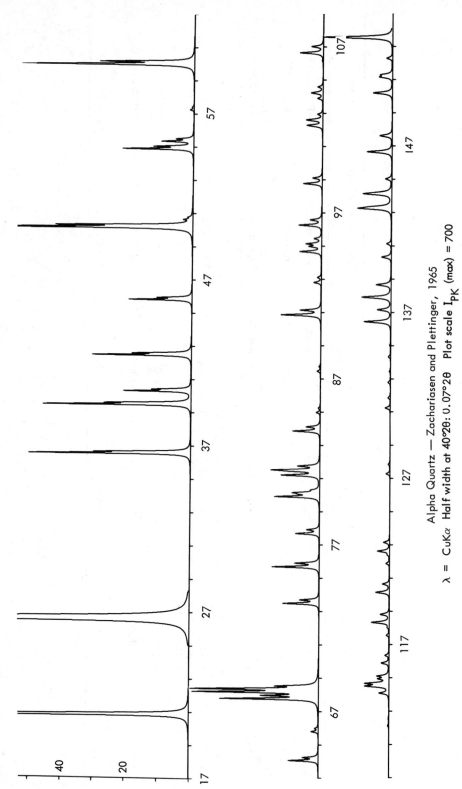

Alpha Quartz — Zachariasen and Plettinger, 1965

λ = CuKα   Half width at 40°2θ: 0.07°2θ   Plot scale I_{PK} (max) = 700

740

ALPHA QUARTZ - ZACHARIASEN AND PLETTINGER, 1965

2THETA	PEAK	D	H	K	L	I(INT)	I(PK)	I(DS)
20.86	20.86	4.255	1	0	0	18	19	18
26.64	26.64	3.343	1	0	1	29	100	29
26.64		3.343	0	1	1	70		70
36.55	36.54	2.457	1	1	0	8	7	8
39.47	39.46	2.281	1	0	2	6	7	6
39.47		2.281	0	1	2	2		2
40.29	40.30	2.236	1	1	1	3	3	4
42.45	42.46	2.128	2	0	0	5	4	6
42.45		2.128	0	2	0	2		3
45.79	45.80	1.9797	2	0	1	2	3	1
45.79		1.9797	0	2	1	1		2
50.14	50.14	1.8178	1	1	2	14	11	16
50.62	50.62	1.8016	0	0	3	0	0	0
54.87	54.88	1.6716	2	0	2	1	3	1
54.87		1.6716	0	2	2	3		4
55.33	55.32	1.6590	1	0	3	2	1	2
57.23	57.24	1.6083	2	1	0	2	0	0
59.96	59.96	1.5415	2	1	1	5	8	6
59.96		1.5415	1	2	1	5		6
64.04	64.04	1.4528	1	1	3	2	0	2
65.78	65.78	1.4183	3	0	0	1	1	1
67.74	67.74	1.3820	2	1	2	2	4	2
68.14	68.14	1.3749	2	0	3	5		5
68.14		1.3749	0	2	3	6	6	7
68.31	68.32	1.3719	3	0	1	2		2
73.47	73.46	1.2878	1	0	4	5	6	6
73.47		1.2878	0	1	4	2		2
75.66	75.66	1.2559	3	0	2	2	2	1
77.67	77.66	1.2283	2	2	0	2	1	3
79.88	79.88	1.1998	2	1	3	2	1	2
79.88		1.1998	1	2	3	3	2	3
80.04	80.12	1.1978	2	2	1	1	1	1
81.17	81.18	1.1839	3	1	2	3	2	4
81.49	81.48	1.1801	3	1	1	3	2	4
83.83	83.84	1.1530	1	1	1	1	1	1
83.83		1.1530	3	1	3	2		2
84.95	84.96	1.1406	2	0	4	0	0	0
87.44	87.44	1.1144	3	1	2	0	0	0
90.83	90.82	1.0815	3	1	0	2	2	3
90.83		1.0815	1	3	0	1	1	1
92.79	92.78	1.0638	3	0	3	1		2
94.65	94.64	1.0477	2	1	4	1	0	1
94.65		1.0477	1	2	4	0		2
95.12	95.12	1.0437	4	0	0	1	1	1
95.12		1.0437	0	4	0	1		2
96.24	96.24	1.0345	2	2	3	2	1	0
98.74	98.74	1.0149	3	2	1	2	1	3
102.19		1.0149	2	3	1	0		2
102.19	102.24	0.9898	4	1	1	1	1	0
102.24		0.9894	1	4	1	0		2
102.57	102.58	0.9872	3	0	4	3	0	1
102.57		0.9872	0	3	4	0		0
103.87	103.88	0.9783	2	2	4	0	0	0
103.87		0.9783	4	1	2	1		1
104.19	104.20	0.9762	1	4	2	2	1	1
106.61	106.60	0.9606	3	3	0	3	1	3
112.10	112.10	0.9285	4	2	0	0	0	1
114.06	114.06	0.9181	3	3	2	2	0	1
114.06		0.9181	2	1	5	2		2
114.47	114.48	0.9160	1	2	5	4	1	6
114.47		0.9160	4	0	4	3		7
114.64	114.64	0.9151	4	2	2	4	1	3
114.64		0.9151	2	4	2	1		2
115.87	115.88	0.9089	5	0	1	2	0	1
117.53	117.54	0.9008	0	5	1	1		0
118.31	118.30	0.8971	1	1	6	1	1	1
118.31		0.8971	3	0	5	2		3
120.12	120.12	0.8888	0	3	5	2	1	2
120.12		0.8888	2	2	5	3	2	3
121.86	121.86	0.8813	3	3	3	1	1	1
122.60	122.60	0.8781	1	4	4	3	2	4
122.60		0.8781	4	1	4	1	1	4
127.25	127.26	0.8597	3	1	6	1	0	1
131.22	131.22	0.8457	3	0	6	1	0	1
132.77	132.76	0.8407	0	5	4	2	0	0

ALPHA QUARTZ - ZACHARIASEN AND PLETTINGER, 1965

2THETA	PEAK	D	H	K	L	I(INT)	I(PK)	I(DS)
134.31	134.30	0.8358	0	4	4		0	0
136.42	136.42	0.8295	0	2	6	2	1	4
136.42		0.8295	2	0	6	0		1
137.89	137.88	0.8254	4	1	3	0	1	1
140.31	140.30	0.8189	1	4	3	3		5
143.22	143.22	0.8117	3	3	0	1	0	2
143.31		0.8115	5	0	2	3		5
144.11	144.10	0.8096	2	2	5	1	1	2
146.62	146.62	0.8041	2	2	1	1		2
150.16	150.16	0.7971	3	3	0	3	1	5
150.16		0.7971	4	2	5	1		2
151.12	151.12	0.7954	1	3	5	0	1	1
151.12		0.7913	3	1	5	1		1
153.52	153.52	0.7913	4	2	1	1	3	2
153.52		0.7859	2	4	1	5		10
157.08	157.08	0.7859	3	2	4	2	2	4
157.08		0.7859	2	1	6	0		1
158.75	158.74	0.7837	3	3	2	0	0	0

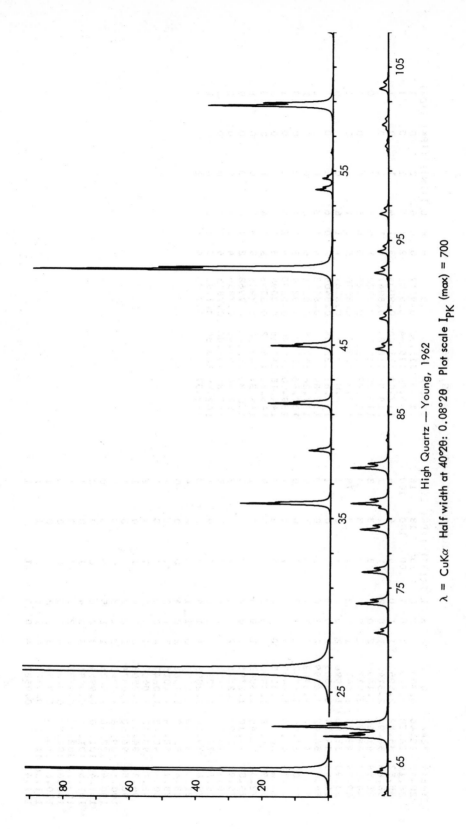

High Quartz — Young, 1962

λ = CuKα Half width at 40°2θ: 0.08°2θ Plot scale I_{PK} (max) = 700

HIGH QUARTZ (600C) – YOUNG, 1962

2THETA	PEAK	D	H	K	L	I(INT)	I(PK)	I(DS)
20.45	20.46	4.339	1	0	0	16	18	16
26.19	26.20	3.399	1	0	1	100	100	100
35.82	35.82	2.505	1	1	0	4	4	4
38.89	38.90	2.314	1	0	2	1	1	1
41.59	41.60	2.169	2	0	0	3	3	3
44.91	44.92	2.017	2	0	1	3	3	4
49.29	49.28	1.8473	1	1	2	17	13	18
53.90	53.90	1.6996	2	0	2	1	1	1
54.55	54.54	1.6809	1	0	3	1	0	1
56.03	56.02	1.6399	2	1	0	0	0	0
58.73	58.72	1.5708	2	1	1	8	5	9
63.00	63.00	1.4742	1	1	3	0	0	0
64.36	64.36	1.4463	3	0	0	1	1	1
66.41	66.42	1.4065	2	0	3	4	3	5
66.99	66.98	1.3958	2	1	2	7	5	9
72.39	72.40	1.3043	1	0	4	1	1	1
74.09	74.10	1.2785	3	0	2	2	1	3
75.90	75.90	1.2525	2	2	0	2	1	2
78.35	78.36	1.2193	2	2	1	2	1	3
79.60	79.60	1.2034	3	1	0	1	0	1
79.84	79.84	1.2003	1	1	4	2	2	3
81.90	81.90	1.1753	3	1	1	3	2	4
83.49	83.50	1.1568	2	0	4	0	0	0
88.74	88.74	1.1015	3	1	2	1	1	2
90.48	90.48	1.0847	4	0	0	1	0	1
93.12	93.12	1.0608	1	0	5	1	1	2
94.34	94.34	1.0503	2	2	3	1	0	1
96.50	96.50	1.0324	2	1	4	1	0	1
100.16	100.16	1.0043	3	1	3	0	0	0
100.40	100.40	1.0026	1	1	5	0	0	0
101.40	101.40	0.9954	3	2	0	1	0	1
101.64	101.64	0.9936	3	0	4	1	0	1
103.72	103.72	0.9793	3	2	1	1	0	0
108.88	108.88	0.9468	4	1	0	0	0	1
110.87	110.88	0.9354	3	2	2	1	0	1
111.30	111.32	0.9329	4	1	1	1	1	1
111.43		0.9322	4	0	3	1	1	1
115.32	115.32	0.9117	0	0	6	0	0	1
115.64	115.64	0.9101	2	1	5	1	1	1
117.00	117.00	0.9034	3	1	4	1	0	1
118.83	118.84	0.8947	1	1	2	0	0	0
119.39	119.34	0.8922	3	3	0	0	0	1
123.97	123.96	0.8725	4	0	5	0	0	0
127.98	128.00	0.8570	5	0	1	0	0	1
128.08		0.8567	1	1	6	1	0	1
132.83	132.84	0.8405	2	0	6	0	0	0
132.89		0.8403	4	1	3	1	0	1
137.26	137.26	0.8271	5	0	2	0	0	0
137.86	137.88	0.8254	3	1	1	0	0	1
139.90	139.90	0.8200	4	2	0	1	1	1
144.18	144.18	0.8095	3	3	2	2	2	3
146.32	146.32	0.8048	3	2	4	1	1	3
149.37	149.38	0.7986	2	1	6	0	0	1
150.33	150.40	0.7968	4	1	2	0	0	0
157.44	157.46	0.7854	4	2	2	0	0	0
158.86	158.86	0.7835	5	1	0	0	0	1
162.55	163.36	0.7793	5	1	1	0	0	1
163.37		0.7784	4	1	4	2	0	4

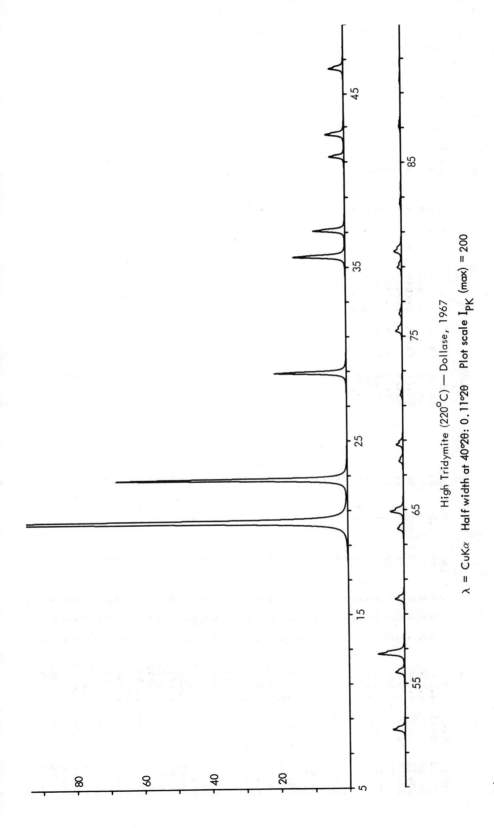

High Tridymite (220°C) — Dollase, 1967

$\lambda = CuK\alpha$ Half width at $40°2\theta$: $0.11°2\theta$ Plot scale I_{PK} (max) = 200

HIGH TRIDYMITE (220C) - DOLLASE, 1967

2THETA	PEAK	D	H	K	L	I(INT)	I(PK)	I(DS)
20.30	20.32	4.370	2	0	0	21	100	21
20.30		4.370	0	0	1	42		42
20.32		4.366	1	1	0	38		38
22.73	22.74	3.909	2	0	1	11	34	11
22.75		3.906	1	1	1	25		25
28.87	28.88	3.090	2	0	2	4	11	4
28.88		3.089	1	1	2	9		9
35.56	35.58	2.522	3	1	0	7	8	7
35.60		2.520	0	0	3	4		4
37.05	37.06	2.424	2	1	0	2	5	2
37.07		2.423	2	1	1	3		3
41.28	41.28	2.185	0	0	4	2	2	4
41.29		2.185	3	1	2	1		3
42.61	42.64	2.120	4	0	1	1	3	1
42.65		2.118	2	2	1	3		3
46.43	46.44	1.9540	2	1	4	2	2	2
46.46		1.9529	2	2	2	1		1
52.29	52.32	1.7480	4	0	3	2	2	2
52.32		1.7470	1	2	3	1		1
55.60	55.62	1.6515	5	1	0	2	1	2
55.62		1.6509	4	2	0	1		1
55.66		1.6498	1	3	0	1		1
56.67	56.68	1.6230	2	0	5	1	4	2
56.67		1.6228	5	1	1	1		1
56.67		1.6228	1	1	5	3		3
56.70		1.6222	4	2	1	1		1
56.73		1.6212	1	3	1	1		2
59.81	59.84	1.5449	5	1	2	1	1	1
59.84		1.5443	4	2	2	1		1
59.84		1.5443	2	2	4	1		1
59.87		1.5435	1	3	2	1		1
63.85	63.90	1.4567	6	0	0	1	1	1
63.91		1.4554	3	3	0	1		1
64.84	64.88	1.4367	5	1	3	1	2	1
64.86		1.4363	4	2	3	1		1
64.90		1.4356	3	3	1	1		1
64.90		1.4356	1	3	3	1		1

2THETA	PEAK	D	H	K	L	I(INT)	I(PK)	I(DS)
68.71	68.72	1.3650	4	0	5	1	1	1
68.74		1.3645	2	2	5	1		2
75.27	75.28	1.2614	3	1	6	1	1	1
75.29		1.2611	6	2	2	1		1
76.29	76.28	1.2471	0	4	1	1	0	1
79.83	79.86	1.2004	5	1	5	1	1	1
79.85		1.2002	4	2	5	1		1
79.88		1.1998	1	3	5	1		1

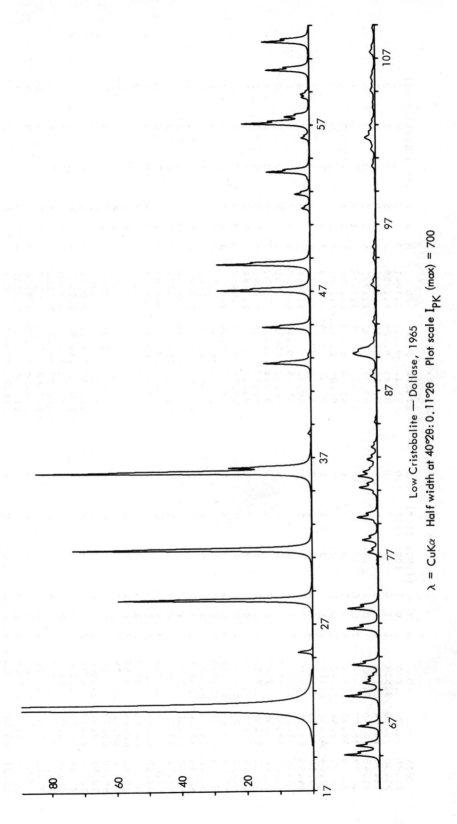

Low Cristobalite — Dollase, 1965

λ = CuKα   Half width at 40°2θ: 0.11°2θ   Plot scale I_{PK} (max) = 700

747

LOW CRISTOBALITE - DOLLASE, 1965

2THETA	PEAK	D	H	K	L	I(INT)	I(PK)	I(DS)
21.98	21.98	4.040	1	0	1	100	100	100
25.30	25.30	3.517	1	1	0	1	1	1
28.44	28.44	3.136	1	1	1	9	9	9
31.45	31.46	2.842	1	0	2	12	11	12
36.09	36.10	2.487	2	0	0	15	12	15
36.38	36.38	2.467	2	0	1	4	4	4
38.43	38.44	2.340	1	1	2	0	0	0
42.66	42.66	2.118	2	1	1	3	2	3
43.16	43.16	2.094	2	0	2	0	0	0
44.83	44.84	2.020	2	1	2	3	2	3
47.04	47.04	1.9300	1	1	3	6	4	6
48.61	48.62	1.8714	2	2	0	6	4	6
51.96	51.96	1.7583	2	2	1	1	0	1
52.83	52.82	1.7315	0	0	4	1	1	1
54.16	54.16	1.6919	2	1	3	3	2	3
56.20	56.20	1.6353	1	0	4	1	1	1
57.08	57.08	1.6122	3	0	1	4	3	4
57.48	57.48	1.6018	2	2	2	1	1	1
58.65	58.66	1.5727	3	1	0	0	0	0
58.85	58.84	1.5678	3	1	1	0	0	0
60.30	60.30	1.5337	2	2	2	3	2	3
62.01	62.02	1.4953	3	1	2	3	2	4
65.08	65.08	1.4320	2	1	4	2	1	3
65.65	65.64	1.4210	2	2	3	2	1	2
66.82	66.82	1.3988	2	2	3	3	2	3
68.63	68.64	1.3663	3	1	3	1	1	1
69.41	69.42	1.3528	2	2	3	1	1	1
69.78	69.78	1.3466	3	0	5	2	1	3
70.51	70.52	1.3344	3	0	3	2	1	3
72.68	72.68	1.2998	3	1	2	3	1	3
73.89	73.90	1.2814	3	2	0	3	1	3
76.56	76.56	1.2433	4	2	0	0	0	0
77.26	77.26	1.2337	2	2	4	1	0	1
78.01	78.02	1.2238	2	0	5	0	1	0
79.06	79.08	1.2101	2	1	0	2	0	2
79.37	79.38	1.2062	4	1	1	0	1	0
80.81	80.82	1.1883	4	1	1	0	0	0
81.16	81.16	1.1841	3	2	3	2	1	2
81.85	81.86	1.1758	2	1	5	2	1	2
82.16	82.10	1.1722	3	3	0	0	0	0
82.33	82.32	1.1702	3	0	2	0	0	0
82.85	82.84	1.1642	3	1	4	1	0	1
83.58	83.58	1.1558	3	3	1	1	0	1
85.09	85.10	1.1391	4	1	2	0	1	0
87.85	87.86	1.1103	4	3	1	1	0	1
89.10	89.22	1.0980	4	3	2	1	0	1
89.22		1.0968	4	2	1	2	1	2
89.44	89.40	1.0947	2	1	6	1	0	1
90.12	90.12	1.0881	2	0	3	0	0	0
91.11	91.12	1.0789	4	2	5	0	0	0
92.19	92.18	1.0691	4	2	5	0	0	0
93.35	93.36	1.0588	3	2	3	1	1	1
94.94	94.94	1.0452	4	3	2	0	0	0
95.63	95.64	1.0395	4	2	2	0	0	0
100.49	100.50	1.0019	4	3	5	0	0	0
101.50	101.50	0.9947	4	1	3	1	0	1
102.20	102.20	0.9897	4	3	0	2	1	1
102.95	102.94	0.9846	3	1	4	1	1	1
104.01	104.02	0.9774	5	1	1	0	0	2
104.32	104.34	0.9753	3	2	5	0	0	1
105.03	105.06	0.9707	5	2	0	0	0	0
105.07		0.9704	2	5	4	1	0	1
105.79	105.80	0.9658	5	0	1	0	0	0
105.92		0.9650	4	3	2	0	0	1
107.35	107.36	0.9560	3	2	6	1	0	0
107.35		0.9560	4	3	2	0	0	1
108.80	108.80	0.9473	5	2	2	0	0	0
110.81	110.82	0.9357	4	1	6	0	0	1
111.73	111.74	0.9306	2	4	4	0	0	0
112.70	112.70	0.9253	4	0	6	0	0	1
113.82	113.82	0.9193	5	2	5	0	0	0
114.58	114.96	0.9154	4	1	7	1	0	1
114.96		0.9135	5	3	3	0	0	0
115.72	115.40	0.9097	4	1	5	0	0	0

LOW CRISTOBALITE - DOLLASE, 1965

2THETA	PEAK	D	H	K	L	I(INT)	I(PK)	I(DS)
116.86	116.86	0.9040	2	1	7	1	0	1
118.03	118.04	0.8985	5	1	3	0	0	1
119.35	119.36	0.8923	5	2	2	1	0	1
120.94	120.94	0.8853	3	2	6	1	0	1
124.05	124.06	0.8722	4	4	1	1	0	0
125.30	125.32	0.8672	4	2	5	1	0	1
126.52	126.54	0.8625	5	0	4	0	0	0
126.52		0.8625	4	3	3	0		1
126.57		0.8623	2	2	7	0		1
127.87	127.86	0.8575	5	2	3	0	0	0
129.13	129.32	0.8529	5	3	0	1	0	1
129.35		0.8521	4	4	2	1		1
130.02	129.86	0.8498	5	1	4	0	0	0
130.98	130.96	0.8465	5	3	1	0	0	0
133.77	133.78	0.8375	3	1	7	0	0	1
138.75	138.76	0.8230	6	0	1	1	0	1
140.80	140.82	0.8176	2	0	8	0	0	1
140.81		0.8176	6	1	0	1		2
141.91	141.58	0.8149	5	2	4	0	0	0
143.11	143.06	0.8120	6	1	1	1	0	0
144.86	145.38	0.8080	4	3	5	1	0	1
145.37		0.8068	2	1	8	2		3
146.68	146.24	0.8040	3	2	7	0	0	1
148.20	148.22	0.8009	4	2	6	1	0	1
149.96	149.96	0.7975	5	1	5	2	0	4
150.92	150.92	0.7957	6	1	2	0	0	0
160.68	160.98	0.7813	6	2	1	2	0	4
161.74		0.7801	6	0	3	1		1

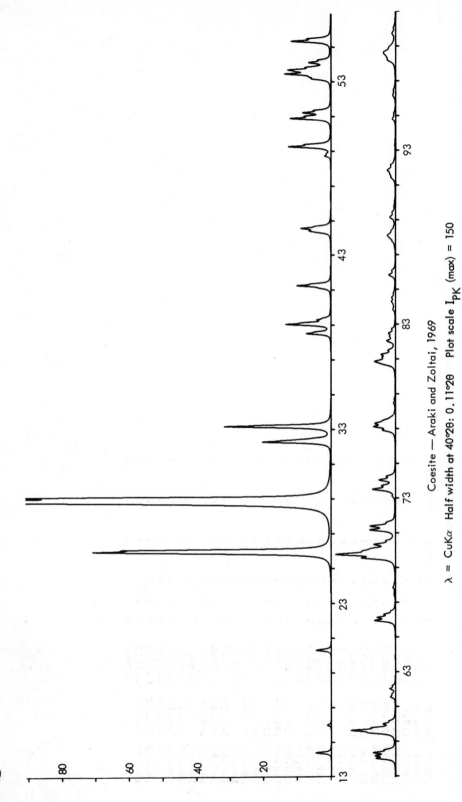

Coesite — Araki and Zoltai, 1969

$\lambda = CuK\alpha$ Half width at $40°2\theta$: $0.11°2\theta$ Plot scale I_{PK} (max) = 150

COESITE — ARAKI AND ZOLTAI, 1969

2THETA	PEAK	D	H	K	L	I(INT)	I(PK)	I(DS)
14.36	14.36	6.164	0	2	0	2	2	2
15.96	15.96	5.548	-1	1	1	0	2	2
15.96		5.548	1	1	0	0	1	0
20.28	20.28	4.376	0	2	1	2		2
25.84	25.86	3.445	-1	1	2	3	2	3
25.85		3.444	1	1	1	35	35	34
25.97	25.96	3.427	-1	3	1	7		7
25.98		3.427	1	3	0	19	31	19
28.71	28.72	3.107	0	0	2	43		43
28.71		3.106	-2	0	2	50	100	50
28.72		3.106	2	0	0	8		8
28.76	28.76	3.100	-2	2	1	42		42
28.95	28.94	3.082	0	4	0	50	99	50
32.24	32.24	2.774	-2	2	2	4	52	4
32.24		2.774	2	2	0	1	10	1
32.24		2.774	-2	2	2	8		8
33.12	33.12	2.703	-1	3	2	5		5
33.12		2.702	1	3	1	16	16	16
38.48	38.48	2.337	-2	4	1	5		5
39.01	39.02	2.307	-1	1	3	2	4	2
39.01		2.307	1	1	2	4	7	4
39.02		2.307	-3	1	2	1		1
39.02		2.306	3	1	1	3		3
39.28	39.28	2.292	-3	3	1	2		2
41.22	41.22	2.188	-1	5	1	1	2	1
41.23		2.188	1	5	0	2	5	2
41.23		2.188	-2	4	2	4		4
44.31	44.40	2.042	-2	4	2	0	3	0
44.32		2.042	0	0	3	1		1
44.39		2.039	-3	1	3	1		1
44.39		2.039	3	1	2	0		0
44.40		2.039	-3	3	2	2		2
44.55	44.54	2.032	-1	5	2	5	4	6
48.70	48.72	1.8680	-2	4	3	0	1	0
48.71		1.8678	2	4	1	1		1
49.23	49.24	1.8494	-3	3	3	4		4
49.23		1.8492	3	3	0	6	6	6

2THETA	PEAK	D	H	K	L	I(INT)	I(PK)	I(DS)
50.86	50.86	1.7937	-2	0	4	9	6	11
51.19	51.20	1.7828	-2	6	1	6	4	7
53.14	53.16	1.7221	-2	2	2	2	3	2
53.15		1.7219	-4	2	2	2		2
53.42	53.42	1.7138	0	6	2	3	7	3
53.42		1.7137	-2	6	0	1		1
53.42		1.7136	-2	6	4	6		7
53.65	53.66	1.7068	-1	1	3	0	6	0
53.66		1.7067	-1	1	3	7		8
53.66		1.7067	1	5	1	0		0
53.67		1.7064	-3	1	4	0		1
53.86	53.80	1.7006	-1	5	3	0	4	1
54.08	54.08	1.6944	-1	7	1	1	3	1
54.08		1.6944	1	7	0	4		4
55.31	55.32	1.6594	-4	2	3	2	6	2
55.32		1.6593	-4	2	1	9		11
57.98	57.98	1.5893	-1	3	4	0	3	0
57.98		1.5892	1	3	3	1		1
57.98		1.5892	-3	3	3	3		4
57.99		1.5890	3	3	1	0		0
58.12	58.14	1.5857	3	5	0	0	3	1
58.32	58.32	1.5809	-1	7	2	2	3	2
58.32		1.5809	1	7	1	2		2
59.45	59.48	1.5534	-4	0	4	1	4	3
59.46		1.5532	0	0	4	2		1
59.47		1.5530	-4	4	0	2		3
59.58	59.60	1.5503	-2	4	4	2	7	2
59.59		1.5501	-4	4	2	2		3
59.60		1.5500	-2	6	1	2		5
59.75	59.74	1.5463	-4	6	0	4	4	2
59.98	59.98	1.5410	-2	8	0	2	2	0
61.51	61.52	1.5063	0	2	4	0	1	3
61.61		1.5059	4	0	3	3		0
61.61		1.5039	-4	4	1	0		1
61.99	62.00	1.4957	0	8	1	1	1	2
65.90	65.92	1.4161	-3	1	5	2	3	1
65.92		1.4158	3	1	2	2		3

751

COESITE – ARAKI AND ZOLTAI, 1969

2THETA	PEAK	D	H	K	L	I(INT)	I(PK)	I(DS)
65.92		1.4158	-2	8	1	1		1
65.92		1.4157	-5	1	3	1		1
65.92		1.4157	-5	1	2	1		1
66.09	66.10	1.4125	-1	5	3	1	3	2
66.28	66.28	1.4090	-1	7	3	1	2	1
66.28		1.4090	-1	7	2	1		1
66.28		1.4089	-3	7	2	0		1
67.47	67.46	1.3870	-4	4	4	0	0	0
67.83	67.84	1.3805	0	8	2	0	0	0
67.83		1.3804	2	8	0	0		0
69.51	69.52	1.3512	2	6	2	3	5	4
69.52		1.3511	-4	6	2	1		6
69.71	69.72	1.3478	-1	1	5	6	9	8
69.71		1.3478	1	1	4	1		1
69.72		1.3476	-5	1	4	2		3
69.73		1.3474	-5	3	1	0		0
69.77		1.3468	-3	3	5	1		1
69.78		1.3466	3	3	2	2		3
69.79		1.3464	-5	3	2	3		4
70.31	70.30	1.3376	1	9	0	4	3	5
71.14	71.14	1.3242	-4	2	5	4	4	6
71.15		1.3240	4	2	1	3		4
71.38	71.36	1.3202	-4	6	3	3	4	4
71.39		1.3202	-4	6	1	0		1
73.08	73.08	1.2938	-2	4	5	0	1	1
73.08		1.2937	2	4	3	0		0
73.43	73.48	1.2884	-2	8	3	1	3	1
73.43		1.2883	2	8	1	1		3
73.48		1.2877	-1	3	5	2		3
73.48		1.2876	1	3	4	1		1
73.49		1.2874	-5	3	4	3		4
74.01	74.02	1.2797	-1	9	2	5	3	6
76.87	76.88	1.2391	0	6	4	1	2	1
76.87		1.2390	-4	6	4	0		0
76.88		1.2389	4	6	0	2		3
77.07	77.08	1.2364	0	8	3	2	3	3
77.07		1.2363	-5	1	5	1		1
77.09		1.2361	5	1	0	0		1
77.24	77.26	1.2341	-3	5	5	0	3	0
77.25		1.2339	-5	5	2	2		3
77.26		1.2338	-5	2	2	1		1
77.33		1.2328	0	10	0	0		1
77.41		1.2318	-1	7	4	1		1
77.42		1.2317	1	7	3	1		1
78.44	78.44	1.2182	0	2	5	1	0	1
80.73	80.84	1.1892	5	3	0	2	3	3
80.82		1.1881	-1	3	5	2		2
80.83		1.1881	-1	3	4	2		3
80.84		1.1879	-5	3	4	2		2
81.23	81.22	1.1832	-3	9	2	0	2	3
81.24		1.1832	-3	9	1	0		2
81.98	82.02	1.1743	-2	0	6	1	1	1
82.00		1.1741	-4	0	2	0		0
82.01		1.1739	-6	0	2	1		2
82.04		1.1736	-6	2	3	1		1
82.44	82.44	1.1689	-2	8	4	1	1	1
82.45		1.1688	2	8	2	1		1
82.45		1.1687	-4	8	2	0		1
83.87	83.84	1.1526	0	4	5	0	0	1
84.26	84.28	1.1482	-3	3	6	0	1	0
84.28		1.1481	3	3	3	1		1
84.48	84.50	1.1458	-4	3	3	1	1	1
84.79	84.74	1.1424	3	3	0	0	0	0
85.79	85.80	1.1317	-4	10	1	1	1	3
85.80		1.1315	4	6	5	1		1
87.92	88.08	1.1096	-5	6	5	0		1
87.93		1.1095	5	5	5	1		1
88.08		1.1080	-3	7	0	1		2
88.09	88.08	1.1079	3	7	2	0	1	1
88.10		1.1078	-5	7	2	1		1
89.10	89.12	1.0980	-6	2	6	0		1
89.11		1.0978	-6	2	1	1		2
89.17		1.0973	-2	4	4	1		0
89.19		1.0971	-6	4	4	0		0

COESITE – ARAKI AND ZOLTAI, 1969

2THETA	PEAK	D	H	K	L	I(INT)	I(PK)	I(DS)
89.76	89.76	1.0915	-2	10	3	0		1
91.43	91.46	1.0760	-3	5	6	1	1	2
91.60	91.82	1.0744	-1	7	5	1		1
91.62		1.0743	-5	7	4	0		1
91.83		1.0723	-1	9	4	1		2
92.12	92.10	1.0697	1	11	1	1	1	1
93.29	93.28	1.0594	0	10	3	0	0	1
94.79	94.80	1.0465	-2	8	5	1		2
94.80		1.0464	2	8	3	0		0
96.14	96.14	1.0353	-6	0	3	0		2
96.39	96.42	1.0333	-6	6	0	0		1
98.15	98.20	1.0195	2	6	4	0		1
98.16		1.0194	-4	6	2	0		1
98.17		1.0193	-6	6	1	1		2
98.33	98.58	1.0181	-6	6	2	1	2	1
98.37		1.0177	-3	11	1	1		1
98.50		1.0167	-7	1	3	1		1
98.51		1.0167	-1	5	6	0		1
98.60		1.0160	1	5	5	0		0
98.61		1.0159	-2	10	4	1		1
98.69		1.0159	-2	10	2	1		2
98.70		1.0153	-4	10	2	1		1
98.69		1.0152	-5	7	0	1	1	1
99.69	99.70	1.0078	5	7	7	1		1
99.71		1.0076	-4	2	5	1		2
101.94	101.98	0.9915	-7	1	3	1	1	2
101.98		0.9913	3	3	4	0		1
102.00		0.9911	-7	3	3	0		1
102.27	102.30	0.9892	3	7	3	2	1	3
102.50		0.9876	-2	12	1	1		1
103.29	103.36	0.9822	2	2	5	1	1	1
103.37		0.9817	0	2	6	1		1
103.41		0.9814	6	4	0	0		1
103.54		0.9805	-6	6	5	0		1
103.55		0.9805	-6	6	1	0		1

2THETA	PEAK	D	H	K	L	I(INT)	I(PK)	I(DS)
104.31	104.30	0.9754	-2	12	2	0	0	0
104.32		0.9754	2	12	2	0		1
105.54	105.60	0.9674	0	8	5	1	1	2
105.60		0.9670	-5	3	7	1		1
105.83	105.84	0.9656	4	10	0	1		1
109.22	109.28	0.9448	-1	8	7	1	1	2
109.26		0.9446	-6	6	3	1		2
109.27		0.9445	-7	1	7	0		0
109.40		0.9437	-3	5	4	1		1
109.42		0.9436	3	1	5	0		0
109.59	109.62	0.9427	1	7	1	1	1	1
109.62		0.9425	5	7	5	0		0
109.85		0.9412	2	12	1	1		1
110.50	110.66	0.9374	1	13	0	1		1
110.67		0.9365	-6	2	7	1		2
110.69		0.9364	6	2	1	0		1
111.38	111.08	0.9325	-2	10	5	1		0
111.39		0.9324	2	10	3	0		1
113.06	113.20	0.9234	-1	3	7	1	0	1
113.09		0.9232	-7	3	6	0		1
113.11		0.9231	-7	3	1	0		0
113.19		0.9227	-5	5	5	1		1
113.22		0.9225	-7	5	7	1		2
113.63	113.62	0.9203	0	12	3	1	0	2
114.31	114.28	0.9168	1	13	2	1	0	2
114.72	114.72	0.9147	-4	6	7	0	0	1
114.74		0.9146	4	6	3	0		1
116.92	116.96	0.9038	-6	8	5	1	0	1
116.93		0.9037	-6	8	1	2		4
117.53	117.52	0.9008	-3	9	6	1	1	1
118.39	118.40	0.8967	4	0	4	1		1
118.42		0.8966	-8	0	4	1		1
118.67		0.8955	0	8	6	0		1
119.55	119.56	0.8914	2	2	5	1	0	1
119.55		0.8914	2	6	2	1	1	1
119.55		0.8914	-4	12	2	0		1
120.91	121.08	0.8854	-5	1	8	1	0	1
120.94		0.8853	-6	1	7	0	1	1

COESITE - ARAKI AND ZOLTAI, 1969

2THETA	PEAK	D	H	K	L	I(INT)	I(PK)	I(DS)
120.96		0.8851	7	1	0	0		1
121.11		0.8845	-1	5	7	1		1
121.11		0.8845	1	5	6	0		2
121.14		0.8844	-7	5	6	0		1
121.33		0.8836	3	7	4	1		1
121.91	121.86	0.8810	-3	11	5	1	0	1
122.32	122.34	0.8793	1	13	2	0	0	1
122.32		0.8793	-3	13	2	0		1
125.20	125.58	0.8676	3	3	5	0		1
125.54		0.8663	-5	7	7	2	0	3
125.84		0.8651	5	1	1	0		1
126.16	126.10	0.8639	-1	11	5	0		0
126.19		0.8637	-5	11	1	0		0
126.78	126.84	0.8615	6	0	2	1	1	1
126.80		0.8614	-8	0	2	1		3
126.93		0.8609	-8	4	4	0		1
127.07		0.8604	-6	6	7	1		1
127.30	127.32	0.8595	0	8	6	0	1	1
127.66		0.8582	-6	10	3	1		2
129.55	129.88	0.8514	-4	8	7	1	0	1
129.91		0.8502	6	10	4	0		2
129.92		0.8501	-6	10	2	1		1
130.79	130.50	0.8472	2	14	0	0		1
134.28	134.52	0.8359	-2	8	7	1	0	2
134.30		0.8359	-2	8	5	2	1	4
134.52		0.8352	-5	8	8	0		1
134.53		0.8351	3	5	5	0		1
134.56		0.8351	5	5	3	0		1
134.58		0.8350	3	7	0	0		1
134.81		0.8343	-7	7	1	1		1
135.09	134.94	0.8335	-2	12	2	1	1	1
135.50		0.8322	-5	11	5	1		1
135.97	136.30	0.8308	-1	13	4	0	0	1
135.97		0.8308	1	13	3	0		1
135.98		0.8308	-3	13	1	1		1
136.31		0.8298	2	4	6	0		1
136.36		0.8297	-8	4	6	1		1
136.38		0.8296	-8	4	4	1	0	1
138.19	138.12	0.8245	-2	2	2	1		2
139.13	139.42	0.8220	-4	14	3	0		1
139.45		0.8211	-4	6	8	0	0	0
140.35	140.78	0.8188	4	12	1	1		1
140.36		0.8187	-7	9	1	0		1
140.78		0.8177	3	11	4	3		5
141.63	141.56	0.8156	-8	2	3	1	0	1
141.94		0.8148	-1	15	1	0		0
141.95		0.8148	1	15	0	0		0
143.79	143.88	0.8104	0	14	3	0		1
146.21	146.26	0.8050	5	9	7	0		1
146.25		0.8049	-7	9	2	3		6
146.27		0.8049	5	9	9	3		5
151.94	152.90	0.7939	-5	1	2	1	1	3
152.01		0.7938	5	1	9	1		2
152.06		0.7937	-9	1	4	2		4
152.55		0.7929	-6	10	5	0		1
152.71		0.7926	3	7	5	2		6
152.73		0.7926	-7	7	7	1		5
152.75		0.7925	5	7	7	0		7
152.78		0.7918	7	7	5	1		5
153.20	153.96	0.7908	0	12	0	2	1	1
153.82		0.7907	1	11	5	1		1
153.88		0.7905	5	11	5	2		5
154.03		0.7904	-2	14	1	3		4
154.05		0.7904	-4	14	2	3		2
154.06		0.7904	4	2	5	4		5
155.00		0.7889	-7	1	0	4		8
160.71	161.32	0.7813	-3	9	6	3	0	6
160.77		0.7812	-9	1	6	1		2
160.88		0.7811	-9	1	3	0		1
160.92		0.7811	9	3	6	0		1
161.07		0.7809	9	3	3	1		2
161.08		0.7809	-9	3	5	1		2
161.24		0.7807	-1	5	8	3		6

TABLE 38. SiO$_2$ POLYMORPHS AND FRAMEWORK SILICATES

Variety	Keatite	Stishovite	β-Spodumene	Nepheline	Kalsilite
Composition	SiO$_2$	SiO$_2$	LiAlSi$_2$O$_6$	(K$_{0.5}$Na$_{1.5}$)Al$_2$Si$_2$O$_4$ (IDEAL)	KAlSiO$_4$
Source	Synthetic	Synthetic	Synthetic	Mt. Summa, Vesuvius	Mt. Nyiragongo, Congo
Reference	Shropshire, Keat & Vaughan, 1959	Preisinger, 1962	Li and Peacor, 1968	Hahn & Buerger, 1955	Perrotta & Smith, 1965
Cell Dimensions					
a Å	7.456	4.179	7.541	10.01	5.161
b Å	7.456	4.179	7.541	10.01	5.161
c	8.604	2.6649	9.156	8.405	8.693
α	90	90	90	90	90
β deg	90	90	90	90	90
γ	90	90	90	120	120
Space Group	P4$_1$2$_1$2	P4$_2$/mnm	P4$_3$2$_1$2	P6$_3$	P6$_3$
Z	12	2	4	4	2

TABLE 38. (cont.)

Variety	Keatite	Stishovite	β-Spodumene	Nepheline	Kalsilite
Site Occupancy				K site 30% vacant	
Method	2-D, film	Powder, counter	3-D, counter	3-D, counter	3-D, counter
R & Rating	0.115 (av.) (2)	0.078 (2)	0.062 (1)	0.180 (2)	0.059 (1)
Cleavage and habit	Unknown	Rutile structure	Keatite structure	{10$\bar{1}$0}&{0001} poor Tends to prisms parallel $\underline{c}$ axis	{10$\bar{1}$0}&{0001} poor Acicular
Comment	B estimated				
μ	87.6	147.5	62.8	97.3	152.6
ASF	0.2261	0.2128	0.2150	0.7126×10^{-1}	0.1683
Abridging factors	0.5:0:0	0:0:0	0.2:0:0	1:1:0.5	0.5:0:0

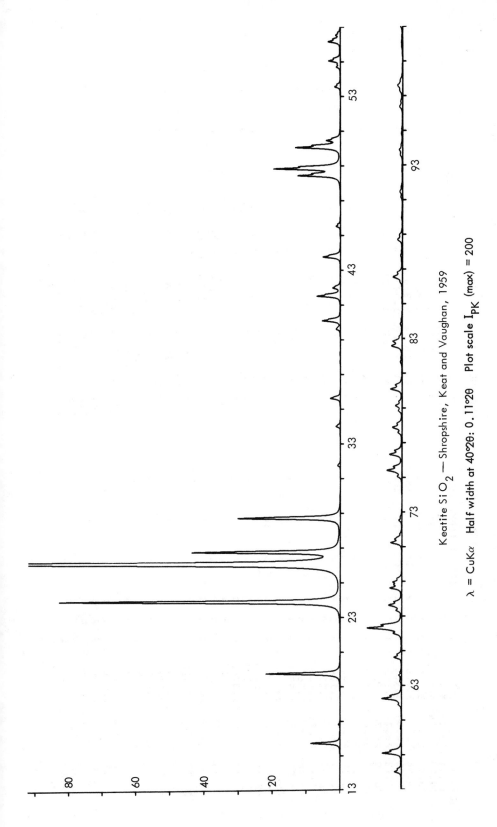

Keatite SiO_2 — Shropshire, Keat and Vaughan, 1959

$\lambda = CuK\alpha$ Half width at $40°2\theta$: $0.11°2\theta$ Plot scale I_{PK} (max) = 200

KEATITE SIO2 — SHROPSHIRE, KEAT AND VAUGHAN, 1959

2THETA	PEAK	D	H	K	L	I(INT)	I(PK)	I(DS)
15.71	15.72	5.635	1	0	1	4	4	3
19.73	19.74	4.495	1	1	0	10	11	9
23.85	23.86	3.728	2	0	0	1	41	1
23.86		3.726	2	1	0	38		38
26.03	26.04	3.421	1	1	2	100	100	100
26.72	26.72	3.333	2	0	1	21	22	16
28.69	28.70	3.109	1	1	3	16	15	2
35.60	35.60	2.519	2	1	2	2	1	1
39.60	39.60	2.274	2	2	0	1	1	3
40.08	40.08	2.248	2	2	2	3	3	3
41.49	41.50	2.174	0	0	4	1		4
41.97	41.96	2.151	3	1	2	2	1	1
43.74	43.76	2.068	1	0	4	1	3	2
43.76		2.067	1	1	4	1		1
45.50	45.50	1.9916	3	1	3	1	1	1
48.42	48.42	1.8782	3	0	3	8	6	9
48.82	48.82	1.8640	4	0	0	10	10	11
48.82		1.8638	3	2	2	3		3
50.04	50.04	1.8213	3	1	1	8	7	9
50.42	50.44	1.8083	4	0	1	2	2	1
50.44		1.8075	2	2	4	1		2
53.53	53.54	1.7104	4	0	2	2	1	1
54.67	54.68	1.6774	3	2	1	1	1	1
55.04	55.04	1.6671	4	2	0	2	2	3
56.15	56.14	1.6368	4	2	2	3	2	3
56.52	56.52	1.6269	3	3	1	1	1	1
57.99	57.98	1.5891	3	1	4	1	1	2
59.08	59.08	1.5624	2	0	5	1	1	5
61.87	61.88	1.4985	3	3	3	3	3	1
62.22	62.22	1.4907	3	2	4	1	0	5
63.57	63.58	1.4622	5	1	1	4	3	1
64.59	64.60	1.4416	5	1	0	1	0	1
64.60		1.4416	4	2	3	1		1
65.97	65.98	1.4148	3	0	5	2	1	2
66.28	66.28	1.4090	5	0	4	5	5	6
66.29		1.4087	4	4	0	2		2
66.32		1.4082	1	0	6	1		1

2THETA	PEAK	D	H	K	L	I(INT)	I(PK)	I(DS)
67.30	67.30	1.3900	3	1	5	1	1	1
67.61	67.60	1.3845	5	1	1	3	2	3
68.59	68.60	1.3670	5	2	1	3	2	3
68.94	68.94	1.3609	3	3	4	2	1	2
71.21	71.22	1.3230	4	3	3	2	2	3
72.49	72.50	1.3027	5	3	1	1	0	1
75.03	75.04	1.2648	5	3	1	4	2	5
75.39	75.40	1.2597	2	2	6	3	2	4
76.30	76.30	1.2469	5	3	3	1	1	1
77.55	77.56	1.2299	6	0	1	1	1	2
77.86	77.86	1.2258	6	1	0	2		1
77.87		1.2257	5	3	2	1		1
78.80	78.82	1.2135	6	1	1	1	1	2
79.13	79.12	1.2093	5	4	1	3	2	4
80.07	80.08	1.1974	5	2	4	2	2	2
82.53	82.56	1.1679	4	3	5	1		2
82.58		1.1673	5	0	7	1	1	1
82.84	82.82	1.1642	5	2	4	1	0	3
83.81	83.80	1.1533	2	4	2	2	1	1
86.52	86.52	1.1240	5	4	4	1		1
86.53		1.1238	4	4	1	1	1	3
88.65	88.68	1.1023	6	3	0	1		1
88.71		1.1018	3	0	8	1	0	2
91.48	91.44	1.0755	0	0	8	1	0	1
93.55	93.58	1.0571	7	0	1	1	0	1
93.85	93.86	1.0544	5	5	0	1	0	1
96.35	96.36	1.0336	5	3	6	1	0	1
97.26	97.28	1.0264	5	2	5	1	0	1
97.54	97.56	1.0242	7	0	3	1	1	1
100.96	100.96	0.9985	7	3	2	1	0	1
107.58	107.62	0.9546	3	6	4	1	1	2
107.66		0.9542	6	2	8	1	0	1
108.89	108.90	0.9468	5	5	4	1		1
112.56	112.82	0.9260	2	0	9	1	0	2
112.81		0.9247	7	2	4	1		6
113.79	113.82	0.9195	7	4	1	1	1	2
113.85		0.9192	5	2	7	1		1

KEATITE SIO2 - SHROPSHIRE, KEAT AND VAUGHAN, 1959

2THETA	PEAK	D	H	K	L	I(INT)	I(PK)	I(DS)
116.88	116.90	0.9039	5	4	6	1	1	1
122.12	122.14	0.8802	7	4	3	1	0	1
126.55	126.56	0.8624	7	5	1	1	0	2
140.68	140.70	0.8179	2	2	10	1	0	2
142.01	142.50	0.8146	8	1	5	1	0	2
142.47		0.8135	9	0	2	1		2
144.59	144.60	0.8086	7	3	6	1	0	2
146.13	146.18	0.8052	9	2	1	1	0	1
146.23		0.8050	7	0	7	1		2
148.41	148.46	0.8005	8	4	3	1	0	1
151.45	151.52	0.7948	7	6	2	1	0	1
154.28	154.22	0.7901	5	4	8	1	0	3
156.53	156.52	0.7867	5	2	9	2	0	2
159.56	159.66	0.7827	9	3	1	1	0	4
163.44	164.08	0.7784	9	2	3	2	0	2
163.51		0.7783	8	3	5	1		2
164.67		0.7772	8	1	6	3		5

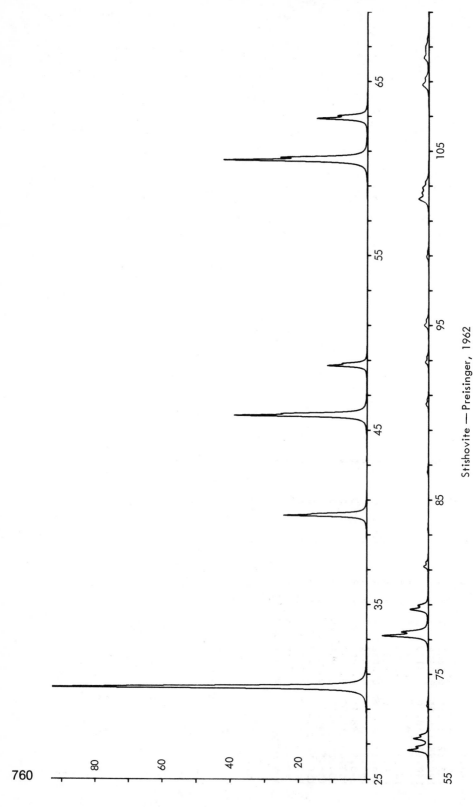

Stishovite — Preisinger, 1962

λ = CuKα Half width at 40°2θ: 0.11°2θ Plot scale I_{PK} (max) = 100

STISHOVITE – PREISINGER, 1962

2THETA	PEAK	D	H	K	L	I(INT)	I(PK)	I(DS)
30.22	30.22	2.955	1	1	0	100	100	100
40.10	40.10	2.247	1	0	1	28	25	31
43.26	43.26	2.089	2	0	0	0	0	0
45.81	45.82	1.9790	1	1	1	48	39	58
48.68	48.68	1.8689	2	1	0	15	12	19
60.45	60.46	1.5301	2	1	1	61	42	89
62.84	62.84	1.4775	2	2	0	22	15	33
70.63	70.64	1.3324	0	0	2	10	6	16
71.30	71.30	1.3215	3	1	0	7	4	12
73.18	73.18	1.2922	2	2	1	1	1	2
77.21	77.20	1.2345	3	0	1	24	13	42
78.71	78.72	1.2147	1	1	2	10	5	17
81.17	81.18	1.1839	3	1	1	3	1	5
83.30	83.30	1.1590	3	2	0	0	0	1
86.57	86.56	1.1235	2	0	2	1	0	2
90.46	90.46	1.0849	2	1	2	2	1	4
92.89	92.88	1.0629	3	2	1	3	1	5
95.00	95.00	1.0447	4	0	0	4	1	7
98.92	98.92	1.0136	4	1	0	2	1	4
102.23	102.24	0.9895	2	2	2	9	3	20
102.88	102.88	0.9850	3	3	0	5	2	10
108.79	108.80	0.9474	4	1	1	7	2	16
110.35	110.36	0.9383	3	1	2	5	1	12
111.03	111.00	0.9345	4	2	0	3	1	6
112.96	112.96	0.9239	3	3	1	1	0	1
121.73	121.74	0.8818	4	2	1	1	0	3
123.48	123.48	0.8745	3	2	2	0	0	1
124.87	124.88	0.8689	1	0	3	2	0	6
129.76	129.78	0.8507	1	1	3	2	0	5
134.31	134.32	0.8358	4	3	0	1	0	2
139.06	139.08	0.8222	4	0	2	6	1	18
140.04	139.98	0.8196	5	1	0	7	1	19
145.42	145.46	0.8067	4	1	2	4	1	13
147.51	147.52	0.8023	2	1	3	14	2	43
149.96	149.96	0.7975	4	3	1	0		1
149.96	149.96	0.7975	5	0	1	15		45
153.04	153.06	0.7921	3	3	2	13	1	41

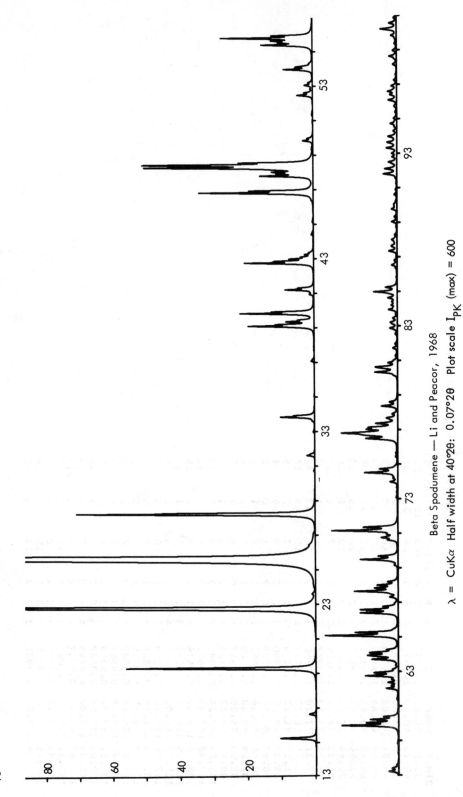

Beta Spodumene — Li and Peacor, 1968

λ = CuKα Half width at 40°2θ: 0.07°2θ Plot scale I$_{PK}$ (max) = 600

762

BETA SPODUMENE - LI AND PEACOR, 1968

2THETA	PEAK	D	H	K	L	I(INT)	I(PK)	I(DS)
15.21	15.22	5.821	1	1	0	2	2	2
16.61	16.62	5.332	1	1	1			
19.25	19.24	4.608	1	1	0	9	9	9
22.70	22.70	3.913	2	0	2	31	33	31
25.53	25.52	3.486	2	1	1	100	100	100
28.17	28.18	3.165	2	1	2	12	12	12
31.60	31.60	2.829	1	1	3	0	0	0
33.81	33.82	2.649	1	1	3	2	2	2
39.00	39.06	2.308	3	1	2	0	3	0
39.33	39.32	2.289	2	0	4	4	1	4
39.80	39.80	2.263	2	1	3	1	4	1
40.92	40.92	2.203	3	0	0	4	0	4
41.18	41.18	2.190	3	0	1	0	1	0
42.72	42.72	2.115	3	1	2	2	3	2
42.96	42.96	2.103	2	1	4	4	1	4
43.22	43.22	2.091	3	2	0	0	0	0
46.78	46.78	1.9403	3	1	3	7	6	7
47.77	47.76	1.9024	3	2	2	3	3	3
47.99	48.00	1.8939	2	1	4	2	2	2
48.23	48.24	1.8853	4	0	0	10	8	11
48.38	48.38	1.8791	3	1	3	7	8	8
49.81	49.82	1.8290	4	0	2	1	1	1
52.44	52.44	1.7432	4	2	0	0	0	0
53.03	53.04	1.7253	3	2	3	2	1	2
53.94	53.94	1.6984	3	1	4	2	1	1
54.14	54.08	1.6924	4	2	2	4	3	0
55.35	55.36	1.6583	4	2	1	3	3	3
55.40		1.6569	3	3	2	1	1	1
55.61	55.60	1.6514	2	1	4	2	2	2
55.76	55.76	1.6472	2	0	5	5	5	5
57.19	57.20	1.6093	3	2	5	0	1	0
59.85	59.84	1.5440	3	2	4	3	3	4
60.20	60.20	1.5359	3	3	3	1	1	2
61.99	62.00	1.4957	1	0	6	1	1	1
62.34	62.34	1.4881	4	3	1			
62.72	62.72	1.4801	3	0	5	2	2	2

2THETA	PEAK	D	H	K	L	I(INT)	I(PK)	I(DS)
62.92	62.90	1.4759	5	2	3	1	1	1
63.68	63.68	1.4600	3	1	1	1	1	2
63.92	63.92	1.4552	4	1	4	2	2	2
64.06	64.06	1.4524	3	0	5	1	1	1
65.06	65.06	1.4325	4	0	2	5	4	5
65.24	65.24	1.4289	4	4	4	0	0	0
66.37	66.36	1.4073	5	1	2	3	2	3
66.55	66.56	1.4039	5	3	0	1	2	1
66.74	66.74	1.4003	5	2	6	0	1	0
67.29	67.28	1.3903	2	1	1	1	0	1
67.62	67.62	1.3842	5	2	5	3	2	3
67.98	67.98	1.3777	3	3	2	1	1	2
69.13	69.14	1.3576	5	2	5	0	0	0
69.45	69.46	1.3521	4	3	3	1	2	1
69.45	69.45	1.3521	4	0	3	2	1	2
70.59	70.60	1.3331	5	1	3	1	1	1
70.73	70.72	1.3309	1	1	0	1	1	1
71.12	71.10	1.3244	2	2	6	5	3	6
73.95	73.96	1.2806	1	3	1	1	0	1
74.48	74.48	1.2728	5	3	2	2	2	3
76.43	76.46	1.2452	6	0	3	1		1
76.47		1.2446	5	3	1	2	2	4
76.77	76.78	1.2404	4	2	5	3	3	2
76.82		1.2397	6	1	0	1		4
77.12	77.10	1.2358	2	0	7	1	1	1
77.34	77.34	1.2328	3	2	6	2	1	2
77.65	77.66	1.2285	6	1	1	1	1	1
78.34	78.34	1.2195	2	1	7	1	1	1
80.30	80.30	1.1945	5	2	4	2	1	2
80.61	80.60	1.1908	5	3	0	1	0	2
80.99	80.98	1.1861	4	0	6	0	0	0
81.69	81.70	1.1777	5	4	5	0	0	0
82.85	82.84	1.1642	4	3	4	0	0	1
83.19	83.18	1.1603	3	0	7	0	0	1
83.93	83.92	1.1520	4	4	4	1	0	1
84.39	84.38	1.1468	3	1	7	0	0	0

BETA SPODUMENE - LI AND PEACOR, 1968

2THETA	PEAK	D	H	K	L	I(INT)	I(PK)	I(DS)
84.60	84.60	1.1445	0	4	8	1	0	1
84.96	84.96	1.1406	5	4	2	2	1	2
86.50	86.50	1.1241	6	3	0	0	0	1
87.31	87.32	1.1158	3	3	1	1	1	2
87.65	87.60	1.1124	5	2	5	0	0	1
87.98	87.98	1.1090	3	4	7	0	0	0
89.02	89.02	1.0987	5	4	3	0	0	0
89.75	89.74	1.0917	6	3	2	0	0	1
90.58	90.58	1.0838	2	1	8	1	0	1
91.79	91.78	1.0727	5	0	6	0	1	0
92.10	92.08	1.0699	7	1	1	1	1	1
92.48	92.48	1.0665	1	0	0	0	0	0
92.48	92.48	1.0665	0	1	3	0	0	0
92.98	92.98	1.0620	6	5	6	0	0	0
93.29	93.30	1.0593	5	1	1	0	0	1
93.63	93.62	1.0564	5	5	5	1	1	1
93.96	93.96	1.0535	3	3	7	1	0	1
94.70	94.70	1.0472	5	4	4	1	0	1
96.08	96.08	1.0358	7	2	0	0	0	0
96.37	96.38	1.0335	4	7	7	1	0	1
96.58	96.58	1.0318	3	1	8	0	0	0
97.23	97.22	1.0266	6	1	5	0	0	0
98.14	98.14	1.0195	6	4	2	1	0	1
98.62	98.62	1.0159	7	0	3	0	0	0
99.83	99.82	1.0068	3	1	8	0	0	0
100.20	100.20	1.0040	5	2	5	2	1	2
102.08	102.08	0.9905	4	4	7	0	0	0
102.43	102.44	0.9882	5	3	3	0	0	0
102.64	102.64	0.9866	2	0	9	0	0	0
103.29	103.30	0.9822	5	1	7	1	1	1
103.65	103.66	0.9798	4	0	8	0	0	0
103.87	103.86	0.9783	7	3	2	0	0	0
105.48	105.48	0.9678	7	1	4	1	0	1
105.65	105.64	0.9667	5	5	4	1	1	1
105.65	105.65	0.9667	5	5	0	0	0	0
105.83	105.84	0.9655	6	0	1	0	0	0
106.68	106.68	0.9602	6	5	1	0	0	0

2THETA	PEAK	D	H	K	L	I(INT)	I(PK)	I(DS)
107.38	107.38	0.9559	5	2	7	1	1	1
109.41	109.42	0.9437	7	2	4	1	1	1
109.73	109.74	0.9419	7	2	3	1	0	1
110.13	110.14	0.9395	6	2	6	0	0	0
110.80	110.80	0.9357	3	1	9	0	0	0
111.41	111.40	0.9323	5	4	6	0	0	0
111.74	111.74	0.9305	8	1	1	1	1	1
112.10	112.14	0.9285	7	4	1	1	0	1
113.40	113.40	0.9216	7	0	5	1	0	1
113.59	113.58	0.9206	6	1	5	0	0	1
114.39	114.38	0.9164	8	2	3	1	0	1
114.69	114.70	0.9148	3	2	9	1	0	1
115.31	115.30	0.9117	4	3	8	0	0	0
116.65	116.64	0.9051	6	3	6	0	0	0
117.37	117.38	0.9016	7	2	5	1	0	1
118.93	118.92	0.8943	7	4	3	1	0	1
119.95	119.94	0.8896	6	5	4	0	0	0
121.31	121.32	0.8836	2	1	10	1	0	1
122.50	122.50	0.8785	8	1	1	1	1	1
123.93	123.94	0.8726	7	5	9	0	0	0
124.32	124.32	0.8711	4	2	5	0	0	0
124.34		0.8710	7	3	1	1	1	1
125.44	125.46	0.8666	8	2	9	0	0	0
125.61	125.62	0.8660	2	0	5	0	0	1
126.48	126.48	0.8626	6	4	5	0	0	1
126.92	126.92	0.8610	3	5	2	0	0	1
127.10	127.10	0.8603	5	0	10	0	0	0
127.97	127.98	0.8571	6	6	8	0	0	0
128.80	128.80	0.8541	6	5	5	0	1	1
129.03	129.02	0.8533	6	2	6	0	0	1
130.19	130.20	0.8492	8	4	3	0	0	1
130.59	130.58	0.8479	8	3	4	0	0	1
132.18	132.18	0.8426	7	5	1	0	0	1
133.11	133.12	0.8396	8	4	9	0	0	0
133.55	133.56	0.8382	5	1	9	1	1	1
134.77	134.78	0.8344	9	0	1	0	0	0

764

BETA SPODUMENE – LI AND PEACOR, 1968

2THETA	PEAK	D	H	K	L	I(INT)	I(PK)	I(DS)
135.25	135.24	0.8330	8	1	5	1	1	1
135.25		0.8330	7	4	4	1		1
135.72	135.72	0.8316	7	0	7	1	1	1
136.04	136.02	0.8306	7	3	6	1	0	1
136.48	136.44	0.8293	9	1	1	0	0	0
137.18	137.18	0.8273	11	0	11	0	0	0
137.46	137.46	0.8265	7	1	7	0	0	0
138.31	138.30	0.8242	9	0	2	1	1	1
138.73	138.72	0.8231	5	2	9	1	1	2
139.58	139.50	0.8208	5	4	8	1	0	2
140.14	140.14	0.8193	9	1	2	0	0	1
140.40	140.38	0.8186	7	5	4	0	0	0
140.60	140.60	0.8181	8	2	5	0	0	1
140.68		0.8179	9	2	0	0		0
141.98	141.98	0.8147	9	2	1	1	1	1
142.28	142.28	0.8140	3	3	10	1	0	1
142.76	142.80	0.8128	2	0	11	1	1	1
142.81		0.8127	8	4	3	0		0
144.51	144.52	0.8087	4	4	9	0	0	0
144.78	144.78	0.8081	2	1	11	1	1	1
146.12	146.12	0.8052	9	2	2	0	0	0
146.12		0.8052	7	6	2	1		1
146.97	147.06	0.8034	9	1	3	0	0	0
147.67	147.66	0.8020	8	0	6	0	0	1
148.86	148.86	0.7996	5	3	9	1	1	1
148.99		0.7993	8	1	0	1		3
149.98	149.98	0.7975	7	4	6	2	1	3
149.98		0.7975	8	4	5	0		2
151.29	151.28	0.7951	8	3	1	1	1	2
153.14	153.14	0.7919	9	3	5	0	0	1
153.89	153.88	0.7907	7	5	5	2	1	2
154.29	154.28	0.7901	5	2	11	1	1	2
154.66	154.66	0.7895	7	3	7	2	1	3
156.71	156.72	0.7864	6	1	9	1	1	1
157.12	157.12	0.7859	3	1	11	1	0	1
158.19	158.18	0.7844	8	2	6	0	0	0
159.15	159.16	0.7832	9	3	2	1	1	2

2THETA	PEAK	D	H	K	L	I(INT)	I(PK)	I(DS)
159.56	159.56	0.7827	4	3	1	2	4	2
159.56		0.7827	5	0	10	1		2

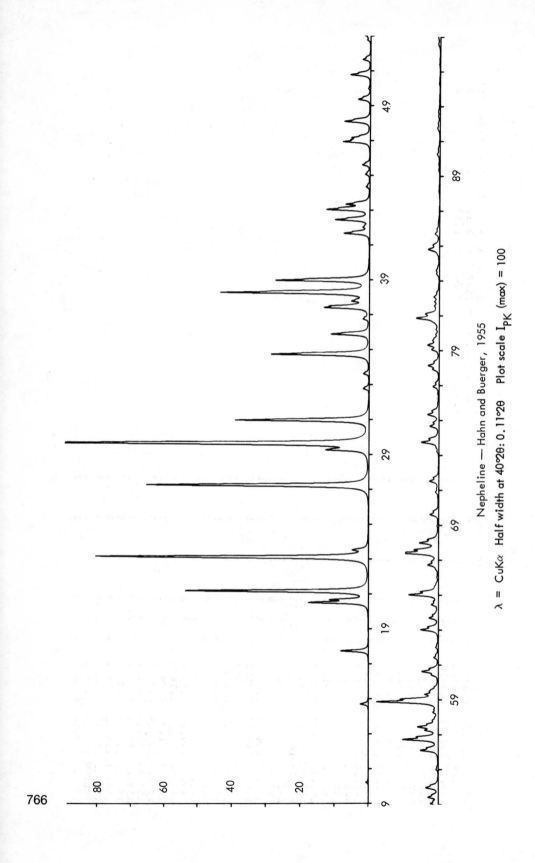

Nepheline — Hahn and Buerger, 1955

λ = CuKα Half width at 40°2θ: 0.11°2θ Plot scale I_{PK} (max) = 100

766

NEPHELINE - HAHN AND BUERGER, 1955

2THETA	PEAK	D	H	K	L	I(INT)	I(PK)	I(DS)
14.67	14.68	6.034	1	1	0	2	2	2
17.71	17.72	5.005	1	1	0	7	8	6
20.47	20.48	4.334	2	1	0	14	18	14
20.64	20.64	4.300	1	0	0	8	12	8
21.12	21.12	4.202	2	0	1	47	54	45
23.07	23.08	3.852	2	0	2	73	81	71
23.51	23.50	3.782	1	1	0	3	5	3
27.19	27.20	3.277	2	1	2	55	65	54
29.23		3.053	2	2	0	2		8
29.23	29.24	3.053	1	1	1	8	13	2
29.58	29.58	3.017	3	0	1	100	100	100
30.92	30.92	2.890	3	0	0	40	39	40
32.74	32.74	2.733	3	1	0	2	2	2
33.59	33.60	2.666	1	1	2	2	2	2
34.68		2.584	2	1	2	28	29	28
34.69	34.69	2.584	2	2	1	4		4
35.85	35.86	2.502	2	2	2	12	11	13
36.73	36.74	2.445	2	1	1	2	2	2
37.37	37.38	2.404	3	1	3	3	13	3
37.37		2.404	3	0	0			
37.46	37.46	2.398	2	2	1	10	12	11
37.75	37.76	2.381	2	2	2	4	5	5
38.22	38.22	2.353	2	0	3	48	44	51
38.93	38.94	2.312	3	1	1	19	28	20
38.93		2.312	3	3	1	13		13
41.64	41.64	2.167	4	0	0	9	8	9
41.98	41.98	2.150	2	2	1	2	2	2
42.41	42.42	2.129	2	1	2	6	10	6
42.41		2.129	1	2	3	7		7
43.01	43.02	2.101	0	0	4	14	13	15
43.07		2.099	4	1	1	1		1
43.32	43.32	2.087	3	2	2	5	7	6
43.32		2.087	1	3	0	2		2
44.32	44.32	2.042	1	0	4	1	1	1
45.03	45.04	2.011	3	3	0	2	1	2
45.57	45.58	1.9888	3	2	0	2	2	3

2THETA	PEAK	D	H	K	L	I(INT)	I(PK)	I(DS)
46.91	46.90	1.9353	3	2	1	2	8	2
46.91		1.9353	2	3	1	7		8
47.14	47.14	1.9262	4	1	1	5	5	6
48.05	48.08	1.8917	4	0	2	3	8	4
48.08		1.8908	2	1	4	6		7
49.34	49.34	1.8455	4	4	1	1	4	2
49.34		1.8455	4	1	1	3		4
50.74	50.74	1.7977	3	4	2	8	6	8
51.63	51.64	1.7688	1	2	2	2	2	2
52.75	52.76	1.7338	5	4	4	1	1	1
53.04	53.04	1.7250	4	0	0	2	2	3
53.40	53.40	1.7142	3	1	2	4	3	4
53.90	53.94	1.6994	3	0	3	1	4	1
53.95		1.6980	5	0	1	4		5
56.09	56.10	1.6383	3	4	0	3	5	4
56.09		1.6383	4	2	2	4		4
56.71	56.72	1.6217	2	3	3	11	11	12
57.24	57.24	1.6080	2	4	1	4	4	4
57.45	57.44	1.6027	5	0	0	6	6	8
57.81	57.80	1.5935	1	1	5	2	2	2
58.26	58.26	1.5822	3	1	4	2	2	2
58.85	58.86	1.5678	4	1	3	8	18	10
58.85		1.5678	1	4	3	2		3
58.87		1.5673	5	0	5	2		19
59.30	59.30	1.5570	1	5	0	16	4	5
59.57	59.46	1.5506	3	3	2	2	3	2
60.41	60.42	1.5309	5	5	1	2	2	2
60.61	60.60	1.5264	2	1	2	6	5	7
61.40	61.40	1.5086	4	0	3	2	1	2
62.99	63.00	1.4743	5	0	3	8	5	9
63.68	63.68	1.4600	1	5	2	3	3	4
64.43	64.44	1.4448	6	0	0	2	2	2
64.45		1.4444	2	3	4	3		1
65.01	65.00	1.4334	3	3	3	13	9	17
66.71	66.72	1.4008	0	0	6	4	3	5
67.40	67.40	1.3881	5	2	0	7	12	9

NEPHELINE – HAHN AND BUERGER, 1955

2THETA	PEAK	D	H	K	L	I(INT)	I(PK)	I(DS)
67.40		1.3881	2	5	0	12		15
67.99	67.98	1.3777	3	1	5	5	6	6
67.99		1.3777	1	3	5	3		4
69.60	69.60	1.3497	4	3	2	1	3	1
69.60		1.3497	3	4	2	3		3
70.89	70.88	1.3283	4	0	5	1	1	2
71.52	71.52	1.3181	2	5	2	4	3	5
73.73	73.74	1.2838	3	2	5	2	5	3
73.73		1.2838	2	3	5	6		7
74.66	74.66	1.2703	4	3	3	3	3	4
74.66		1.2703	1	4	3	1		2
75.33	75.32	1.2605	3	0	6	5	3	6
76.92	76.92	1.2384	5	2	0	2	2	2
76.92		1.2384	3	5	0	1		1
77.91	77.92	1.2252	3	2	6	2	2	2
78.12	78.12	1.2224	2	2	6	4	3	6
79.04	79.04	1.2104	3	1	5	1	2	2
79.04		1.2104	1	3	6	2		3
79.32	79.30	1.2069	5	0	5	4	3	6
79.69	79.68	1.2022	5	6	0	2	2	2
80.84	80.84	1.1879	7	3	2	1	7	1
80.84		1.1879	3	5	2	9		12
80.84		1.1879	3	3	5	2		3
81.15	81.08	1.1841	4	0	5	1	4	2
81.54	81.54	1.1795	2	4	5	2	1	2
81.80	81.80	1.1765	4	4	3	5	2	3
82.07	82.06	1.1732	2	4	6	2	3	7
84.78	84.78	1.1425	4	4	3	5	3	2
84.80		1.1423	1	5	0	2		3
85.69		1.1327	7	0	3	2	1	1
90.24		1.0871	3	4	5	1	1	2
91.74		1.0731	4	5	2	1	1	2
96.56		1.0319	5	4	3	2	1	3
97.18		1.0270	2	7	2	2	1	4
97.94		1.0211	2	0	8	3	1	2
100.24	100.26	1.0037	4	4	5	1	1	2
101.16	101.16	0.9971	7	0	5	1	1	2

2THETA	PEAK	D	H	K	L	I(INT)	I(PK)	I(DS)
101.16	101.16	0.9971	3	5	5	1	2	2
102.07	102.08	0.9907	7	2	2	2	1	3
102.07		0.9907	2	7	3	2		3
102.74	102.72	0.9860	5	2	6	3	2	4
102.74		0.9860	2	5	6	2		3
105.50	105.48	0.9677	4	5	0	2	1	2
106.20	106.24	0.9632	9	0	0	2	1	3
106.27		0.9627	1	3	8	1		2
115.20	115.18	0.9123	7	6	1	1	1	2
118.54	118.52	0.8961	6	2	5	2	0	3
128.31	128.34	0.8559	4	4	5	1	1	2
131.24	131.26	0.8457	2	3	8	1	1	3
133.69	133.78	0.8377	5	5	2	1		2
133.69		0.8377	9	2	2	2		4
133.91		0.8371	8	1	6	1	0	2
139.23	139.20	0.8217	3	3	9	1	1	4
141.89	141.94	0.8149	3	0	10	2	0	3
145.26	145.26	0.8071	6	0	8	2	1	8
147.92	148.08	0.8014	3	9	3	4		2
148.07		0.8012	7	6	3	1		4
148.92	148.96	0.7995	6	6	3	2	1	4
150.36	150.70	0.7968	2	2	10	4	1	7
150.72		0.7961	5	5	3	5		9
150.72		0.7961	7	0	6	1		3
152.08	151.98	0.7937	9	1	3	3	1	3
152.24		0.7934	1	3	10	1		3
153.51	153.48	0.7913	2	9	0	2		4
153.64		0.7911	2	2	6	6	1	9
154.66	154.64	0.7895	10	1	0	5		9
154.66		0.7895	3	0	9	3		6
156.14		0.7873	3	9	2	1		2
162.08	162.02	0.7798	8	3	5	4	0	7
162.08		0.7798	3	8	5	3		6

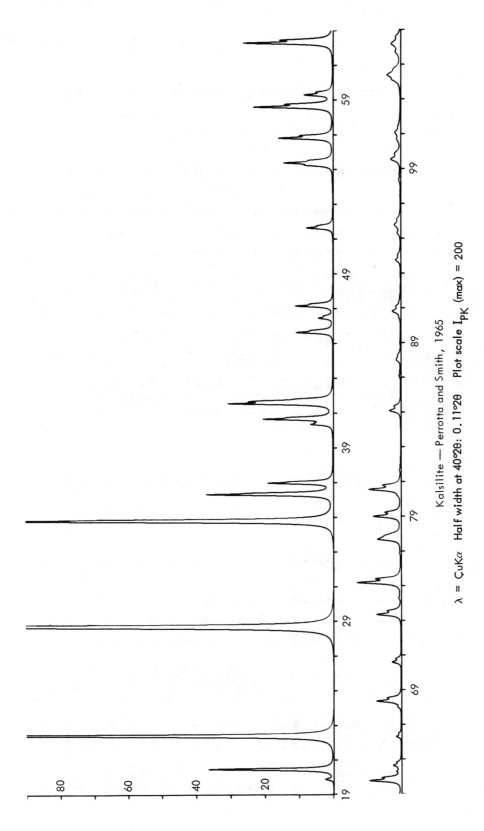

Kalsilite — Perrotta and Smith, 1965

λ = CuKα   Half width at 40°2θ: 0.11°2θ   Plot scale I$_{PK}$ (max) = 200

KALSILITE - PERROTTA AND SMITH, 1965

2THETA	PEAK	D	H	K	L	I(INT)	I(PK)	I(DS)
19.85	19.86	4.469	1	0	0	1	1	1
20.42	20.42	4.345	0	0	1	16	18	15
22.35	22.36	3.974	1	0	1	56	63	52
28.63	28.64	3.115	1	1	0	100	100	100
34.74	34.74	2.580	1	1	1	57	53	62
36.29	36.30	2.473	2	0	0	20	19	23
36.95	36.96	2.431	0	0	2	11	10	12
40.33	40.34	2.234	2	0	1	3	3	3
40.63	40.64	2.218	1	0	2	12	10	14
41.53	41.54	2.172	2	1	0	17	15	20
41.70	41.64	2.164	1	1	2	8	12	9
45.62	45.62	1.9870	2	1	1	7		8
46.44	46.44	1.9538	3	0	0	3	2	3
47.13	47.14	1.9266	2	0	2	7	5	9
51.62	51.62	1.7692	2	2	0	5	4	7
55.23	55.24	1.6618	3	1	0	4	4	6
55.36	55.36	1.6580	2	1	2	4	7	5
55.36		1.6580	1	1	3	4		5
56.79	56.78	1.6198	2	0	5	11	8	16
58.59	58.58	1.5742	2	1	2	7	12	14
58.59		1.5742	2	1	2	6		10
59.28	59.28	1.5576	3	0	4	19	13	30
62.28	62.28	1.4896	1	1	0	4	5	6
63.73	63.72	1.4591	2	1	3	3		5
63.73		1.4591	1	2	3	2		3
64.60	64.60	1.4414	1	1	5	1	1	2
66.27	66.28	1.4091	3	0	2	6	4	10
68.32	68.32	1.3718	2	1	5	1	1	2
70.57	70.58	1.3334	2	2	0	6	4	11
73.32	73.32	1.2900	1	1	6	6	6	20
75.16	75.16	1.2629	3	0	4	11	3	10
77.65	77.66	1.2285	3	1	1	5		3
77.77	77.77	1.2270	0	3	5	4		7
78.97	78.98	1.2113	2	1	2	4	4	7
78.97		1.2113	1	2	5	4		8
80.52	80.52	1.1919	3	1	2	4	5	8
80.52		1.1919	1	3	2	4		5
85.06	85.06	1.1395	3	1	3	2	2	3
85.06		1.1395	3	3	1	2		4
87.96	87.98	1.1092	2	2	4	1	1	3
90.78	90.78	1.0820	4	0	2	3	1	6
93.73	93.74	1.0555	1	0	8	2	1	4
95.29	95.30	1.0423	4	0	3	1	1	3
95.76	95.76	1.0384	3	0	6	2	1	5
99.51	99.52	1.0091	1	3	5	2	1	5
99.51		1.0091	3	3	0	2		5
101.06	101.04	0.9978	3	2	2	1	1	1
101.06		0.9978	2	3	0	1		3
104.08	104.36	0.9769	2	0	8	3	2	7
104.35		0.9751	4	1	0	2		5
104.35		0.9751	1	4	0	3		6
105.69	105.70	0.9664	2	2	3	1	1	3
105.69		0.9664	3	2	3	3		3
106.18	106.16	0.9633	2	4	6	6		9
110.09	110.10	0.9398	5	0	4	5	1	4
112.36	112.36	0.9271	2	4	2	4	1	3
112.36		0.9271	3	3	4	2		2
114.93	114.94	0.9136	1	2	8	8	1	4
114.93		0.9136	2	1	8	4		5
119.95	119.98	0.8896	4	4	1	4	0	2
119.95		0.8896	4	1	5	5		2
121.45	121.46	0.8830	2	3	5	2	1	4
121.45		0.8830	3	2	5	5		5
122.85	122.86	0.8771	3	3	7	7		8
127.18	127.20	0.8600	4	3	0	3	0	3
128.82	128.82	0.8540	5	0	3	1	1	3
132.80	132.82	0.8405	2	4	1	1	0	2
132.80		0.8405	4	2	1	1		1
136.60	136.62	0.8290	2	4	2	3	0	9
136.60		0.8290	4	2	2	3		9
137.47	137.36	0.8265	5	1	0	2	1	6
138.55	138.54	0.8235	1	1	10	7	1	22
141.09	141.10	0.8169	3	1	8	2	1	5
141.09		0.8169	1	3	8	2		5

KALSILITE - PERROTTA AND SMITH, 1965

2THETA	PEAK	D	H	K	L	I(INT)	I(PK)	I(DS)
144.44	144.44	0.8089	1	4	6	5	1	15
144.44		0.8089	4	1	6	5		17
148.84	148.92	0.7996	3	3	4	2	0	6
151.44	151.46	0.7948	5	0	5	3	0	9
154.01	154.84	0.7905	2	3	7	1	1	2
154.01		0.7905	3	2	7	1		
154.81		0.7892	5	1	2	4		13
154.81		0.7892	1	5	2	3		10
163.01	163.50	0.7788	4	0	8	4	0	14
163.88	163.54	0.7779	1	0	11	4	0	13

TABLE 39. MISCELLANEOUS FRAMEWORK SILICATES

Variety	Natrolite	Pollucite	Analcime (disordered)	Hydrosodalite	Sodalite
Composition	$Na_2Al_2Si_3O_{10} \cdot 2H_2O$	$Cs_{12}Na_4Al_{16}Si_{32}O_{96} \cdot 4H_2O$	$NaAlSi_2O_6 \cdot H_2O$	$Na_6Al_6Si_{5.5}(H_4)_{0.5}O_{24} \cdot 0.4NaCl \cdot 0.7NaOH$	$Na_8Si_6Al_6O_{29}Cl_2$
Source	Aussig, Bohemia	Rumford, Maine	Cyclopean Islands, Greece	Synthetic	Bolivia
Reference	Meier, 1960	Beger & Buerger, 1967	Knowles, Rinaldi & Smith, 1965	Bukin & Makarov, 1967	Löns & Schulz, 1967
Cell Dimensions					
a Å	18.30	13.69	13.73	8.887	8.870
b	18.63	13.69	13.73	8.887	8.870
c	6.60	13.69	13.73	8.887	8.870
α	90	90	90	90	90
β deg	90	90	90	90	90
γ	90	90	90	90	90
Space Group	Fdd2	Ia3d	Ia3d	P$\bar{4}$3n	P$\bar{4}$3n
Z	8	1	16	1	1

TABLE 39. (cont.)

Variety	Natrolite	Pollucite	Analcime (disordered)	Hydrosodalite	Sodalite
Method	3-D, film	3-D, counter	3-D, counter	3-D, neutron diff.	3-D, film
R & Rating	0.083 (1)	0.055 (1)	0.10 (2)	0.075 (1)	0.082 (1)
Cleavage and habit	(110) & (1$\bar{1}$0) Fibrous, elongated to $\underline{c}$ axis	None Equidimensional	None (?) Equidimensional	Dodecahedrons (?)	{110} poor Equidimensional
Comment		Analcime group	Minor Ca and K substitution for Na ignored.		
μ	67.6	383.7	68.7	69.6	85.13
ASF	0.5924×10^{-1}	0.4466	0.1302	0.1630	0.2221
Abridging factors	1:1:0.5:0.5	1:1:0.5	1:1:0.5	0.1:0:0	0.1:0:0

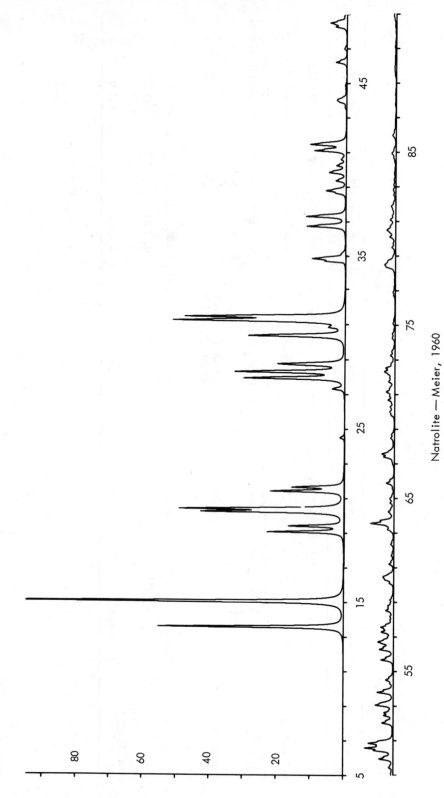

774

Natrolite — Meier, 1960

λ = CuKα Half width at 40°2θ: 0.11°2θ Plot scale I$_{PK}$ (max) = 100

NATROLITE – MEIER, 1960

2THETA	PEAK	D	H	K	L	I(INT)	I(PK)	I(DS)
13.55	13.56	6.528	2	2	0	54	55	54
15.03	15.04	5.890	1	1	1	100	100	100
19.04	19.04	4.657	0	4	0	24	23	24
19.39	19.40	4.575	4	0	0	17	16	17
20.21	20.22	4.391	1	3	1	42	43	42
20.37	20.38	4.356	3	3	0	48	49	49
21.39	21.40	4.151	2	4	0	23	22	24
21.62	21.62	4.106	4	2	0	16	16	16
24.48	24.48	3.633	3	3	1	2	1	2
27.30	27.30	3.264	4	4	0	4	4	4
27.90	27.90	3.195	1	5	1	36	30	37
28.27	28.26	3.155	5	1	1	40	33	41
28.67	28.74	3.111	0	2	2	12	20	12
28.73		3.104	2	2	2	15		16
30.32	30.32	2.945	2	6	0	35	29	36
30.37		2.940	2	2	2	3		3
30.82	30.82	2.899	6	2	0	4	5	4
31.20	31.20	2.865	3	5	1	63	51	64
31.42	31.42	2.845	5	3	1	56	48	57
34.70	34.70	2.583	2	4	2	7	6	7
34.85	34.84	2.572	4	2	2	11	10	12
36.71	36.72	2.446	1	7	1	16	12	17
37.28	37.28	2.410	7	1	1	16	12	17
38.63	38.64	2.329	0	8	0	4	3	4
38.77	38.76	2.321	4	4	2	7	6	7
39.34	39.36	2.288	3	7	0	2	3	3
39.35		2.287	8	0	2	3	3	3
39.83	39.84	2.261	0	6	2	6	5	7
40.23	40.22	2.240	6	0	0	4	3	4
40.57	40.58	2.221	8	2	2	13	10	14
41.08	41.08	2.195	2	6	2	11	11	12
41.43	41.44	2.178	6	2	2	5		5
41.46		2.176	6	6	0	2		3
43.90	44.00	2.061	1	3	3	2	3	3
43.98		2.057	3	3	3	2		2
44.07		2.053	8	4	0	2		2
46.19	46.20	1.9637	1	9	1	2	3	2

2THETA	PEAK	D	H	K	L	I(INT)	I(PK)	I(DS)
46.20		1.9634	3	3	3	4	1	4
46.97	46.98	1.9327	9	1	1	1	1	1
48.25	48.26	1.8844	5	5	3	5	3	5
48.48	48.48	1.8760	8	1	2	7	5	7
49.41	49.42	1.8428	2	2	0	2	1	2
49.91	49.92	1.8255	3	10	3	7	4	7
50.40	50.40	1.8092	5	5	3	8	6	8
50.54	50.54	1.8042	10	3	0	9	9	10
50.80	50.82	1.7957	7	1	1	6	8	7
50.83		1.7947	4	7	1	5		6
52.01	52.02	1.7568	8	4	2	5	3	6
52.44	52.44	1.7433	5	9	1	4	3	4
52.61	52.60	1.7382	4	10	0	2	3	2
53.03	53.04	1.7254	9	5	1	8	5	9
53.11		1.7230	10	0	0	2		2
53.77	53.78	1.7032	10	1	1	8	5	9
54.28	54.28	1.6884	5	7	3	1	1	1
54.49	54.50	1.6824	0	0	3	3	2	4
55.66	55.66	1.6500	3	11	4	6	4	7
56.25	56.26	1.6339	10	7	1	4	4	5
56.26		1.6337	7	3	3	2		2
56.60	56.68	1.6247	9	1	3	3	5	4
56.69		1.6223	11	8	2	6		7
57.00	57.00	1.6143	10	6	1	2	2	2
57.23	57.26	1.6082	2	1	1	2	4	2
57.24		1.6072	6	0	4	4		5
57.54	57.54	1.6004	10	2	2	5	4	5
57.57		1.5997	7	0	4	1		1
57.96	57.96	1.5899	9	1	0	1	1	1
58.18	58.18	1.5842	11	6	0	1	1	2
58.49	58.50	1.5766	5	9	1	2	2	2
58.52		1.5759	7	7	1	2		3
58.79	58.78	1.5694	9	1	1	2	2	2
59.12	59.12	1.5614	11	5	3	2	1	2
60.08	60.10	1.5386	7	5	3	2	1	2
60.28	60.32	1.5340	5	7	3	1		1
60.31		1.5333	2	4	4	2	3	3

NATROLITE - MEIER, 1960

2THETA	PEAK	D	H	K	L	I(INT)	I(PK)	I(DS)
60.41	60.42	1.5310	4	2	4	3	3	3
60.50		1.5290	4	10	2	2		3
61.18	61.18	1.5135	10	4	2	3		2
61.68	61.68	1.5025	1	9	3	1	1	1
62.32	62.32	1.4885	4	1	0	2	1	2
63.19	63.20	1.4702	4	12	0	2	2	3
63.51	63.54	1.4635	3	9	3	3	7	3
63.55		1.4628	8	8	2	11		12
64.08	64.22	1.4519	9	3	1	1		1
64.21		1.4493	12	6	0	1		1
64.73	64.72	1.4389	2	4	4	2	2	3
64.98	64.96	1.4339	6	2	1	2	1	2
65.87	65.86	1.4168	2	9	1	5	2	5
67.38	67.40	1.3885	2	12	2	3	3	3
67.39		1.3883	4	6	4	3		3
67.55	67.56	1.3855	6	4	4	3	4	4
68.46	68.46	1.3693	12	2	2	1	1	1
69.80	69.80	1.3463	2	8	2	2	1	2
69.98		1.3429	4	12	4	1	2	1
70.27	70.26	1.3384	1	11	3	1	1	1
70.28		1.3382	8	0	4	4		4
70.66	70.66	1.3320	2	8	4	4	2	2
70.97	70.98	1.3270	12	4	2	1	3	3
71.11	71.12	1.3246	8	2	4	3		3
71.17		1.3236	11	1	3	1		4
71.46	71.32	1.3190	10	8	0	2	2	2
71.73	71.74	1.3147	6	6	4	1		1
72.16	72.30	1.3080	5	13	1	3	2	3
72.28		1.3060	7	3	3	2	3	3
72.31		1.3055	10	10	0	3		4
72.52	72.50	1.3023	9	7	3	3	3	2
73.24	73.24	1.2914	13	5	1	2	1	3
75.06	75.08	1.2644	12	6	2	2	1	2
76.70	76.72	1.2414	1	5	5	1		1
78.32	78.46	1.2197	6	14	0	3	3	3
78.37		1.2191	3	5	5	2		2
78.46		1.2179	1	15	1	3		4
78.48	78.68	1.2176	5	3	5	2	2	2
78.76		1.2140	10	10	5	1	1	1
79.49	79.48	1.2047	14	6	2	3	2	3
80.09	80.10	1.1972	15	1	0	3		3
80.47	80.48	1.1925	4	10	4	4	3	5
80.54		1.1916	4	14	2	2		2
81.07	81.08	1.1851	10	4	2	3	2	4
81.84	81.84	1.1759	14	4	2	1	1	3
82.51	82.52	1.1681	11	11	1	2	1	2
83.80	83.80	1.1534	9	13	1	3	1	3
84.92	84.92	1.1409	5	13	3	2	1	3
85.95	85.96	1.1299	13	5	1	4	0	1
91.08	91.08	1.0792	1	17	1	6	1	1
92.84	92.88	1.0633	2	4	6	4	1	2
93.19	93.16	1.0602	6	12	4	1	1	2
93.96	94.00	1.0535	12	6	2	4	0	1
97.59	97.62	1.0238	10	10	4	3	0	1
100.84	100.84	0.9994	7	15	3	2	0	1
102.12	102.14	0.9903	15	7	3	1	0	1
103.34	103.64	0.9819	2	18	0	1	1	1
103.61		0.9801	6	18	2	1		1
103.78		0.9790	15	11	1	1		1
105.70	105.72	0.9664	18	2	2	2		2
107.20	107.20	0.9570	14	12	2	2		2
107.45	107.50	0.9554	17	3	3	3		3
108.81	108.82	0.9472	0	10	6	1	0	2
109.57	109.58	0.9428	10	0	4	6	1	1
110.05	110.04	0.9400	16	0	0	4	0	2
110.33	110.30	0.9384	5	13	5	5	1	1
111.42	111.44	0.9323	13	8	5	5	0	2
115.23	115.24	0.9121	8	8	6	6	1	1
115.26		0.9120	5	17	1	5		2
115.89	115.72	0.9088	17	7	7	7	1	1
116.88	116.98	0.9040	3	5	3	3	0	1
117.01		0.9033	5	3	5	7		1
120.30	120.34	0.8881	12	2	2	6	1	1
122.22	122.28	0.8797	4	20	2	2	0	1

776

NATROLITE – MEIER, 1960

2THETA	PEAK	D	H	K	L	I(INT)	I(PK)	I(DS)
		0.7816	16	4	6	1	1	1
160.46								

2THETA	PEAK	D	H	K	L	I(INT)	I(PK)	I(DS)
122.68	122.70	0.8778	20	2	1	1	0	1
123.10	123.18	0.8761	17	9	3	1	0	1
123.26		0.8754	9	19	1	1		1
125.52	125.56	0.8663	20	4	2	1	0	1
125.68		0.8657	19	9	1	1		1
126.91	126.92	0.8610	6	12	6	1	0	1
130.76	130.76	0.8473	7	19	3	1	0	1
131.45	131.40	0.8450	11	17	3	1	0	1
131.98	132.06	0.8432	2	22	0	1	0	1
132.74	132.64	0.8408	14	16	2	1	0	1
133.64	133.70	0.8379	11	19	1	1	0	1
133.77		0.8375	19	7	3	1		1
135.02	135.04	0.8336	4	14	6	1	0	2
136.05	135.72	0.8306	19	11	1	1		1
136.77	136.82	0.8285	22	2	0	1		1
136.87		0.8282	14	4	6	1		1
138.02	138.04	0.8250	0	0	8	2	0	2
138.76	138.76	0.8230	1	11	7	2	0	2
140.03	140.32	0.8196	9	19	3	1		1
140.08		0.8195	11	1	7	1		2
140.38		0.8187	8	18	4	1		2
141.47	141.44	0.8159	16	16	0	2	0	3
141.69		0.8154	3	21	3	1		1
143.24	143.30	0.8117	19	9	3	1		3
143.38		0.8113	18	8	4	2		3
145.12	145.12	0.8074	4	22	2	1	0	2
146.98	147.02	0.8034	21	3	3	1	0	2
147.42		0.8025	7	17	5	1		1
150.75	150.82	0.7960	17	7	5	1	0	2
153.11	153.24	0.7920	15	11	5	1	0	1
153.37		0.7915	14	8	6	1		1
153.98	154.06	0.7905	9	9	7	1	0	1
155.66	155.84	0.7880	14	18	2	1	0	1
157.57	158.18	0.7852	5	23	1	1		1
157.76		0.7850	6	4	8	1		1
157.97		0.7847	18	14	2	1	0	1
158.57	158.22	0.7839	6	20	4	2		3

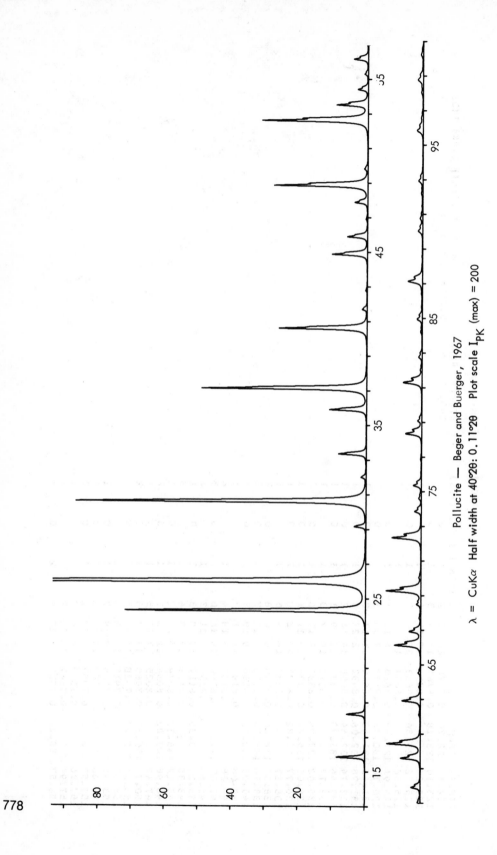

Pollucite — Beger and Buerger, 1967

$\lambda = CuK\alpha$ Half width at $40°2\theta$: $0.11°2\theta$ Plot scale I_{PK} (max) = 200

POLLUCITE – BEGER AND BUERGER, 1967

2THETA	PEAK	D	H	K	L	I(INT)	I(PK)	I(DS)
15.84	15.86	5.589	2	1	1	4	4	2
18.31	18.32	4.840	2	2	0	2	3	2
24.31	24.32	3.659	3	2	1	36	36	33
26.01	26.02	3.422	4	0	0	100	100	100
29.15	29.16	3.061	4	2	0	2	2	2
30.60	30.60	2.919	3	3	2	47	43	57
33.34	33.34	2.685	4	3	1	5	4	6
35.90	35.90	2.499	5	2	1	6	6	9
37.12	37.12	2.420	4	4	0	29	25	44
40.59	40.60	2.221	5	3	2	15	13	26
44.87	44.86	2.018	6	3	1	4	3	14
45.88	45.88	1.9760	4	4	4	3	2	8
47.87	47.88	1.8985	6	4	0	3	2	5
48.84	48.84	1.8630	5	5	2	8	14	17
48.84		1.8630	7	2	1	2		5
52.59	52.60	1.7386	6	3	3	8	16	19
52.59		1.7386	6	5	1	1		3
53.50	53.50	1.7112	7	3	0	21	5	51
54.40	54.40	1.6851	7	4	1	7	2	16
56.16	56.16	1.6363	6	5	3	2	2	5
57.89	57.90	1.5914	8	2	1	3	1	8
59.59	59.60	1.5501	7	5	0	2	3	5
60.43	60.44	1.5306	8	4	1	4	5	13
62.90	62.90	1.4762	7	6	1	8	3	23
66.12	66.12	1.4120	8	4	6	4	4	11
66.91	66.90	1.3972	8	6	3	6	2	22
68.48	68.48	1.3690	7	7	4	3	1	10
69.25	69.26	1.3555	10	1	0	1	5	4
69.25		1.3555	9	5	1	5		19
72.33	72.32	1.3053	9	5	2	4	4	14
72.33		1.3053	7	6	3	5		4
72.33		1.3053	10	3	1	2		19
73.84	73.84	1.2822	8	7	1	2	1	7
75.35	75.36	1.2603	10	5	3	4	1	7
78.33	78.34	1.2196	10	5	1	1	2	9
79.07	79.08	1.2100	8	8	0	1	1	5

2THETA	PEAK	D	H	K	L	I(INT)	I(PK)	I(DS)
81.28	81.28	1.1826	11	3	1	1	3	6
81.28		1.1826	10	5	2	2		11
81.28		1.1826	7	7	6	2		8
84.93	84.94	1.1408	8	8	4	1	1	7
87.12	87.12	1.1178	11	5	1	2	2	8
87.12		1.1178	10	7	1	1		7
90.02	90.02	1.0891	10	5	5	2	1	11
92.92	92.92	1.0626	9	7	3	1	1	7
95.83	95.84	1.0378	11	7	2	2	1	9
98.76	98.76	1.0148	11	6	5	1	1	6
102.45	102.46	0.9880	8	8	8	1	0	3
103.94	103.94	0.9779	12	6	4	2	0	3
104.69	104.70	0.9729	13	5	1	1	1	5
104.69		0.9729	14	1	9	1		8
107.71	107.72	0.9538	9	9	6	1	1	4
107.71		0.9538	11	9	2	2		7
109.24	109.24	0.9447	14	3	11	1	0	12
110.79	110.78	0.9358	11	8	5	1	0	15
113.92	113.94	0.9188	14	3	3	1	1	3
114.72	114.70	0.9147	13	7	4	1		5
117.14	117.14	0.9027	12	8	3	1	0	11
120.45	120.46	0.8874	14	5	6	1	0	6
123.88	123.88	0.8728	11	5	5	1	1	8
123.88		0.8728	14	10	1	1		8
123.88		0.8728	11	7	3	2		5
127.45	127.46	0.8590	14	11	7	1	1	5
127.45		0.8590	15	5	3	1		6
129.31	129.30	0.8523	13	9	9	2	0	17
131.21	131.20	0.8458	13	9	5	1	0	5
135.19	135.22	0.8331	13	10	1	2	0	14
136.23	136.22	0.8301	12	8	8	2	0	17

POLLUCITE - BEGER AND BUERGER, 1967

2THETA	PEAK	D	H	K	L	I(INT)	I(PK)	I(DS)
139.47	139.46	0.8211	14	9	1	1	0	5
139.47		0.8211	15	7	2	1		9
144.17	144.18	0.8095	15	6	5	2	0	15
144.17		0.8095	13	9	6	1		13
146.73	146.72	0.8039	15	8	1	1		9
148.07	148.10	0.8011	16	6	0	1	0	11
149.47	149.48	0.7984	17	2	1	1	1	13
149.47		0.7984	13	10	5	2		17
149.47		0.7984	13	11	2	3		32
155.79	155.86	0.7878	17	3	2	3	0	29
155.79		0.7878	14	9	5	3		28
157.62		0.7852	12	12	4	2		18
161.81	164.00	0.7801	16	6	4	2	0	21
164.30		0.7775	15	9	2	4		45
164.30		0.7775	15	7	6	1		9

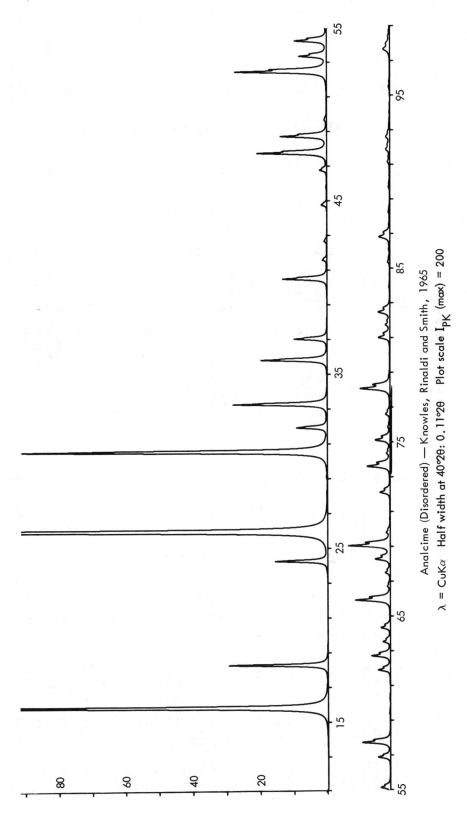

Analcime (Disordered) — Knowles, Rinaldi and Smith, 1965

λ = CuKα Half width at 40°2θ: 0.11°2θ Plot scale I$_{PK}$ (max) = 200

ANALCIME (DISORDERED) - KNOWLES, RINALDI AND SMITH, 1965

2THETA	PEAK	D	H	K	L	I(INT)	I(PK)	I(DS)
15.80	15.80	5.605	2	1	1	66		65
18.26	18.26	4.854	2	2	0	13	15	13
24.23	24.24	3.669	3	2	1	8	8	8
25.94	25.94	3.432	4	0	0	100	100	100
30.51	30.52	2.927	3	3	2	50	46	51
31.90	31.90	2.803	4	2	2	5	5	5
33.24	33.24	2.693	4	3	1	16	14	16
35.79	35.80	2.507	5	2	1	12	10	12
37.01	37.00	2.427	4	4	0		7	6
40.46	40.46	2.227	6	1	1	3	1	4
40.46	40.46	2.227	5	3	2	5		5
44.73	44.74	2.024	6	3	1	1	1	1
46.74	46.74	1.9417	4	4	4	1	1	1
47.72	47.72	1.9040	5	4	3	14	10	15
48.69	48.70	1.8684	6	4	0	2	7	2
48.69		1.8684	7	2	1	7		8
52.43	52.44	1.7437	6	5	1	6		6
52.43		1.7437	8	0	0	14	14	15
53.33	53.34	1.7162	8	1	1	6	4	6
54.23	54.22	1.6900	7	4	1	7	5	7
55.11	55.12	1.6650	8	2	0	2	1	2
56.85	56.86	1.6181	6	6	0	6	2	2
57.71	57.72	1.5961	8	3	1	6	4	6
61.88	61.88	1.4981	8	4	2	3	2	3
62.70	62.70	1.4805	9	2	1	4	3	4
63.51	63.50	1.4636	6	6	4	2	1	2
64.31	64.32	1.4473	7	5	4	2	1	2
65.90	65.90	1.4161	9	3	2	6	5	7
65.90		1.4161	7	6	3	3		3
68.25	68.26	1.3730	8	6	0	4	2	4
69.02	69.02	1.3595	7	7	2	7	6	8
69.02		1.3595	10	1	1	4		4
72.08	72.08	1.3091	10	3	1	2	1	3
73.59	73.60	1.2859	8	7	1	6	3	7
75.09	75.10	1.2639	9	6	1	2	2	2
75.09		1.2639	10	3	3	2		2
78.06	78.06	1.2232	11	2	1	1	4	2
78.06	78.06	1.2232	9	6	3	1		4
78.06		1.2232	10	5	1	3		4
80.99	81.00	1.1861	10	7	1	2	2	2
80.99		1.1861	7	7	6	1		2
82.45	82.46	1.1688	8	7	5	2	2	2
82.45		1.1688	11	5	0	2		2
86.80	86.80	1.1210	10	6	5	3	2	3
97.64	97.64	1.0234	12	6	0	3	1	3
99.83	99.84	1.0067	11	8	3	2	0	1
101.30	101.30	0.9961	10	9	1	1	0	1
102.77	102.78	0.9858	9	8	7	1	0	1
103.51	103.52	0.9807	12	6	4	1	1	2
104.26	104.26	0.9757	14	1	1	1	1	1
107.26	107.26	0.9566	13	6	1	1	1	1
107.26		0.9566	14	3	3	1		1
107.26		0.9566	14	4	1	2		2
108.77	108.78	0.9475	13	5	4	1	0	1
108.77		0.9475	11	8	5	1		1
108.77		0.9475	13	8	2	2		2
110.30	110.30	0.9386	14	3	6	1	0	1
111.07	111.06	0.9342	12	4	1	1	0	1
113.41	113.42	0.9215	14	5	1	1	0	1
116.60	116.60	0.9053	11	10	3	1	0	1
118.22	118.22	0.8976	13	8	7	1	0	1
121.55	121.56	0.8826	15	4	1	1	0	1
123.26	123.24	0.8754	14	5	5	1	0	1
126.78	126.80	0.8615	13	9	2	2	0	2
128.60	128.62	0.8548	13	8	8	1	0	1
129.53	129.52	0.8515	16	2	0	1	0	2
130.48	130.46	0.8482	10	9	9	2	0	1
132.40	132.40	0.8418	11	9	8	1	0	2
134.39	134.40	0.8356	11	10	7	2	0	2
135.40	135.28	0.8325	16	4	0	1	0	1
137.50	137.50	0.8264	16	4	2	2	0	2
143.15	143.18	0.8119	14	9	3	1	0	2
144.37	144.18	0.8090	16	4	4	1	0	1
145.63	145.62	0.8063	15	8	1	3	0	4

ANALCIME (DISORDERED) - KNOWLES, RINALDI AND SMITH, 1965

2THETA	PEAK	D	H	K	L	I(INT)	I(PK)	I(DS)
146.92	146.86	0.8035	16	6	0	3	0	5
148.27	148.22	0.8007	13	10	5	1	1	2
148.27		0.8007	13	11	2	3		4
151.13	150.98	0.7954	15	8	3	1	0	1
154.28	154.32	0.7901	14	9	5	1	0	2
154.28		0.7901	17	3	2	1		2
157.83	159.78	0.7849	17	4	1	2	0	3
159.83		0.7823	16	6	4	4		6
162.04		0.7798	15	9	2	3		4

783

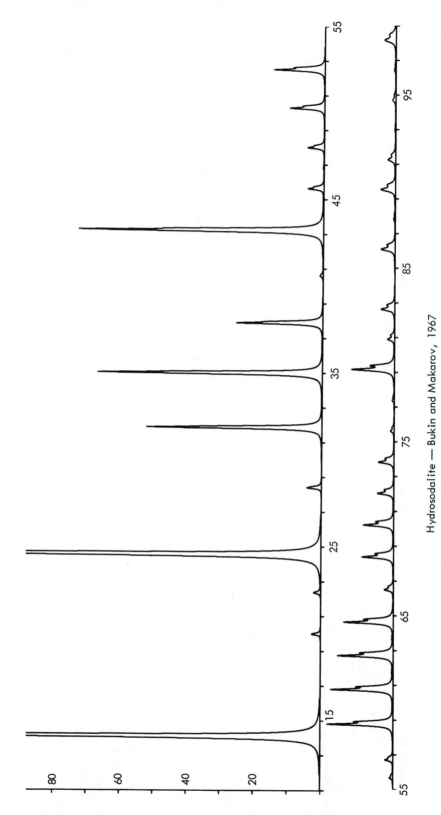

Hydrosodalite — Bukin and Makarov, 1967

$\lambda = CuK\alpha$ Half width at $40°2\theta$: $0.11°2\theta$ Plot scale I_{PK} (max) = 300

784

HYDROSODALITE – BUKIN AND MAKAROV, 1967

2THETA	PEAK	D	H	K	L	I(INT)	I(PK)	I(DS)
14.08	14.08	6.284	1	1	0	100	100	100
19.96	19.98	4.443	2	0	0	1	1	1
22.35	22.36	3.974	2	1	1	1	1	1
24.51	24.52	3.628	2	2	0	89	77	91
28.38	28.38	3.142	2	2	2	2	1	2
31.81	31.82	2.810	3	1	0	23	17	23
34.94	34.94	2.565	3	2	1	31	22	32
37.85	37.84	2.375	3	3	0	12	9	13
40.57	40.58	2.222	4	0	0	0	0	1
43.15	43.16	2.095	4	1	1	14	24	15
43.15		2.095	3	3	2	23		25
45.61	45.62	1.9872	4	2	0	2	2	3
46.80	46.80	1.9393	4	2	2	0	0	0
47.97	47.98	1.8947	3	3	2	3	2	3
50.25	50.26	1.8141	4	3	2	6	3	6
52.46	52.46	1.7429	4	4	2	8	5	9
55.65	55.64	1.6503	5	3	2	0	0	0
55.65		1.6503	4	3	2	0		1
56.68	56.68	1.6225	5	2	1	12	1	2
58.72	58.72	1.5710	4	4	0	6	7	13
60.71	60.72	1.5241	5	4	3	5	6	7
61.70	61.70	1.5022	4	3	1	0	0	0
62.67	62.68	1.4812	6	0	2	3	6	3
62.67		1.4812	4	1	2	8		9
63.63	63.64	1.4610	6	1	1	0	0	0
64.59	64.60	1.4417	6	1	1	0	5	11
64.59		1.4417	5	3	2	9		
66.48	66.48	1.4052	6	2	0	2	1	2
68.35	68.34	1.3713	5	4	1	6	3	8
70.19	70.18	1.3398	6	3	1	6	3	7
72.01	72.00	1.3103	4	4	4	4	2	4
73.81	73.80	1.2827	7	1	0	3	0	4
75.59	75.60	1.2568	5	5	0	0	0	0
75.59		1.2568	6	4	1	0		1
77.36	77.36	1.2324	6	4	0	0	0	1
79.12	79.12	1.2094	7	2	1	5	4	6
79.12	79.12	1.2094	5	5	2	2		2
79.12		1.2094	6	3	3	3		4
80.87	80.88	1.1876	6	4	2	2	1	2
82.61	82.62	1.1669	7	3	0	4	1	4
86.07	86.08	1.1287	6	5	1	3	1	3
86.07		1.1287	8	0	0	1		1
87.80	87.80	1.1109	7	4	1	2	0	2
89.52	89.52	1.0939	8	1	1	0	1	2
89.52		1.0939	5	5	4	1		1
89.52		1.0939	8	2	0	4		3
91.24	91.24	1.0777	6	4	4	2	1	2
91.24		1.0777	6	5	3	0		0
92.96	92.96	1.0622	8	2	2	3	0	1
94.69	94.70	1.0473	8	3	1	2	2	0
96.42	96.42	1.0331	7	5	0	1	3	0
96.42		1.0331	9	1	0	0		0
98.15	98.16	1.0194	6	6	2	4	1	5
99.90	99.90	1.0063	7	6	1	1	0	2
101.65	101.66	0.9936	8	4	0	5	1	7
103.41	103.42	0.9814	9	1	1	0	1	0
103.41		0.9814	8	3	2	3		2
105.19	105.20	0.9697	8	4	2	2	0	1
106.98	106.98	0.9583	8	6	1	1	1	0
106.98		0.9583	7	2	1	0		2
108.79	108.80	0.9474	9	2	1	1	0	1
110.62	110.62	0.9368	6	6	5	1	0	1
110.62		0.9368	6	6	5	2		2
114.35	114.36	0.9166	8	4	4	1	0	1
114.35		0.9166	7	6	3	3		1
116.25	116.26	0.9070	8	6	4	2	0	3
118.19	118.20	0.8977	9	4	4	1	0	2
118.19		0.8977	7	7	1	0		0
122.17	122.18	0.8799	7	7	0	1	0	1
122.17		0.8799	10	1	0	1		0
124.23	124.24	0.8714	10	2	2	0	0	1
126.34	126.34	0.8632	9	4	1	3	0	1

HYDROSODALITE - BUKIN AND MAKAROV, 1967

2THETA	PEAK	D	H	K	L	I(INT)	I(PK)	I(DS)
126.34		0.8632	9	5	0	1		2
128.50	128.52	0.8552	10	2	2	1	1	4
128.50		0.8552	6	6	6	3		1
130.74	130.74	0.8473	10	3	1	1	1	1
130.74		0.8473	9	5	2	4		6
130.74		0.8473	7	6	5	1		2
135.46	135.46	0.8323	8	7	1	2	1	3
135.46		0.8323	7	7	4	1		3
135.46		0.8323	8	5	5	2		2
137.97	137.98	0.8251	10	4	0	2	1	3
137.97		0.8251	8	6	6	3		5
140.61	140.62	0.8181	9	6	1	2	0	3
140.61		0.8181	10	4	3	0		0
143.40	143.42	0.8113	10	4	2	1	0	1
146.40	146.42	0.8046	9	5	4	0	0	1
146.40		0.8046	8	7	3	1		1
153.25	153.30	0.7917	11	2	1	1	0	1
153.25		0.7917	9	6	3	3		4
153.25		0.7917	10	5	1	1		2
157.38	157.44	0.7855	8	8	0	4	0	7
162.38	162.36	0.7794	11	3	0	2	0	3
162.38		0.7794	9	7	0	0		1

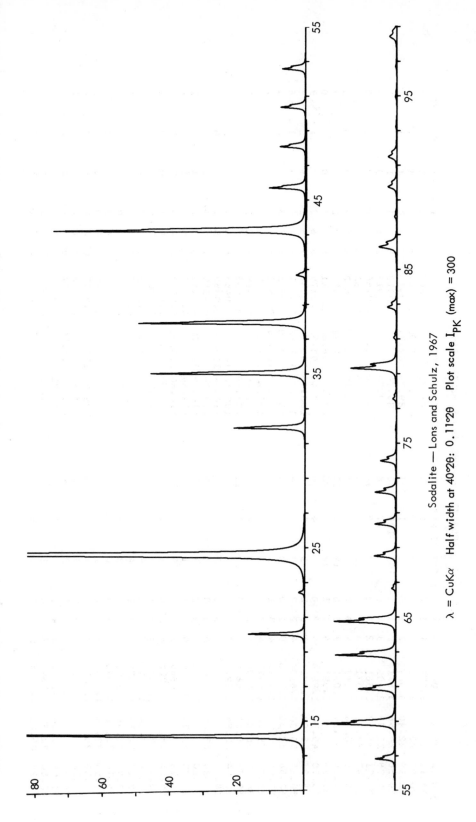

Sodalite — Lons and Schulz, 1967

$\lambda = CuK\alpha$ Half width at 40°2θ: 0.11°2θ   Plot scale I_{PK} (max) = 300

SODALITE - LONS AND SCHULZ, 1967

2THETA	PEAK	D	H	K	L	I(INT)	I(PK)	I(DS)
14.11	14.12	6.272	1	1	0	30	36	29
20.00	20.00	4.435	2	0	0	5	6	5
22.39	22.40	3.967	2	1	0	1	1	1
24.56	24.56	3.621	2	1	1	100	100	100
31.88	31.88	2.805	3	1	0	8	7	8
35.01	35.02	2.561	2	2	2	18	15	18
37.92	37.92	2.371	3	2	1	20	17	21
40.65	40.66	2.217	4	0	0	1	1	1
43.24	43.24	2.091	4	1	1	14	25	15
43.24		2.091	3	3	0	18		20
45.70	45.70	1.9834	4	2	0	5	4	4
48.07	48.08	1.8911	3	3	2	3	3	4
50.35	50.36	1.8106	4	2	2	3	2	0
52.56	52.56	1.7396	5	1	0	0	2	0
52.56		1.7396	4	3	1	3		4
55.76	55.76	1.6471	5	2	0	3		0
55.76		1.6471	4	3	2	3		4
56.80	56.80	1.6194	5	2	1	11	7	13
58.84	58.84	1.5680	4	4	0	3	4	4
60.84	60.84	1.5212	5	3	0	2	6	3
60.84		1.5212	4	3	3	3		3
62.80	62.80	1.4783	6	0	0	7	6	8
62.80		1.4783	4	4	2	1		1
64.73	64.74	1.4389	6	1	1	9	6	11
64.74		1.4389	5	3	2	1		1
66.62	66.62	1.4025	6	2	0	4	2	5
66.63		1.4025	5	4	1	4	2	5
68.50	68.50	1.3687	6	2	2	4	2	5
70.34	70.34	1.3372	6	3	1	3	2	4
72.16	72.16	1.3078	4	4	4	0	0	0
72.17		1.3078	5	4	3	1	0	1
73.97	73.98	1.2803	4	4	3	4	0	4
75.76	75.76	1.2544	5	4	0	1		1
77.54	77.54	1.2300	4	2	1	3		3
79.30	79.30	1.2071	5	5	2	1	5	1
79.30		1.2071	6	3	3	1		2
81.06	81.06	1.1853	6	4	2	3	0	4
82.80	82.80	1.1647	7	3	0	2	1	1
86.28	86.28	1.1265	6	5	1	3	2	3

2THETA	PEAK	D	H	K	L	I(INT)	I(PK)	I(DS)
86.28	86.28	1.1265	7	3	0	1	0	1
88.01	88.00	1.1087	8	0	0	1	1	1
89.74	89.74	1.0918	8	1	1	1		1
89.74		1.0918	7	4	1	0		0
89.74		1.0918	5	5	4	1		3
91.46	91.46	1.0756	8	2	0	0	1	1
91.46		1.0756	6	4	4	2		0
93.19	93.20	1.0602	6	5	3	0	0	0
94.93	94.92	1.0453	8	2	2	0	0	0
96.66	96.66	1.0311	7	5	1	0	1	3
98.41	98.42	1.0175	6	6	2	2	0	2
100.16	100.16	1.0043	7	5	2	1	1	6
101.92	101.92	0.9917	8	4	0	4	0	1
103.69	103.70	0.9795	9	1	0	1	0	1
105.48	105.48	0.9678	8	4	2	1	1	2
107.28	107.28	0.9565	8	3	3	1	1	1
107.28		0.9565	7	6	1	1		1
107.28		0.9565	9	2	1	0		2
109.10	109.10	0.9455	6	6	4	0	0	2
110.94	110.94	0.9350	8	5	1	1	0	0
110.94		0.9350	7	5	4	1		1
114.69	114.70	0.9149	9	3	2	1	0	1
114.69		0.9149	7	6	3	1		1
116.60	116.62	0.9053	8	4	4	2	0	3
118.55	118.56	0.8960	9	4	1	0	0	1
118.55		0.8960	7	7	0	0		0
122.57	122.58	0.8783	7	7	2	1	0	5
122.57		0.8783	10	1	1	0		5
124.64	124.64	0.8698	10	2	2	1	0	4
126.77	126.78	0.8615	9	4	3	0	0	0
126.77		0.8615	9	5	0	0		0
128.96	128.98	0.8535	10	3	2	0	0	2
128.96		0.8535	9	6	0	1		1
131.22	131.24	0.8457	10	3	1	1	1	2
131.22		0.8457	9	5	2	3		5
131.22		0.8457	7	7	5	1		2
136.00	136.00	0.8308	8	7	1	1	0	2

SODALITE – LONS AND SCHULZ, 1967

2THETA	PEAK	D	H	K	L	I(INT)	I(PK)	I(DS)
136.00		0.8308	7	7	4	1		2
136.00		0.8308	8	5	5	1		1
138.54	138.56	0.8236	10	4	0	2	1	3
138.54		0.8236	8	6	4	3		5
141.23	141.24	0.8165	9	6	1	2	0	3
141.23		0.8165	10	3	3	0		0
144.08	144.06	0.8097	10	4	2	0	0	0
147.13	147.16	0.8031	8	7	3	0	0	0
154.19	154.24	0.7902	11	2	1	1	0	1
154.19		0.7902	9	6	3	2		4
154.19		0.7902	10	5	1	1		2
158.50	158.56	0.7840	8	8	0	3	0	6

TABLE 40. MISCELLANEOUS FRAMEWORK SILICATES

Variety	Cancrinite	Danburite	Petalite	Marialite-Scapolite	Hemimorphite
Composition	$(Na,Ca)_6Al_6Si_6O_{24}$ $(Na,Ca)CO_3 \cdot 2H_2O$ $(IDEAL)^3$	$CaB_2Si_2O_8$	$LiAlS_{14}O_{10}$	$(Na,Ca,K)_8(Si,Al)_{24}$ $O_{48}[Cl,CO_3]_2$ (IDEAL)	$Zn_4S_{12}O_7(OH)_2 \cdot$ H_2O
Source	Litchfield, USA	not given	Elba	Gooderham, Ontario	Glencrieff Mine, Wanlockhead
Reference	Jarchow, 1965	Johansson, 1959	Zemann-Hedlik & Zemann, 1955; Liebau, 1961	Papike & Zoltai, 1965	McDonald & Cruickshank, 1967
Cell Dimensions					
$\underline{a}$	12.75	8.04	7.62 (11.76)	12.060	8.370
$\underline{b}$ Å	12.75	7.74	5.14	12.060	10.719
$\underline{c}$	5.14	8.77	11.76 (7.62)	7.572	5.120
α	90	90	90	90	90
β deg	90	90	112.4	90	90
γ	120	90	90	90	90
Space Group	$P6_3$	Pnma	Pc (Pa)	I4/m	Imm2
Z	1	4	2	1	2

TABLE 40. (cont.)

Variety	Cancrinite	Danburite	Petalite	Marialite-Scapolite	Hemimorphite
Site Occupancy	M_1|M_2| {Ca 0.733, Na 0.122}; C 0.5, O_5 0.5, H_2O 0.33			M {Na 0.736, Ca 0.186, K 0.054}; T_1 Si 0.50; T_2 {Si 0.542, Al 0.458}; Cl 1.0	
Method	3-D, film	3-D, film	2-D, film	3-D, counter	3-D, counter & film
R & Rating	0.089 (1)	0.095 (2)	0.096 (2)	0.12 (2)	0.055 (1)
Cleavage and habit	{10$\bar{1}$0} perfect Prismatic	{100} very poor Prismatic parallel to $\underline{a}$ axis	{100} perfect, {201} easy. Tabular parallel to {100}	{100} & {110} perfect Prismatic	{110} perfect, {101} less so. Tabular parallel {010}; also prismatic
Comment	Minor Fe ignored			Minor Mg,Fe ignored. Cl & CO_3 not differentiated	
μ	83.1	145.7	91.10	103.6	153.2
ASF	0.6504×10^{-1}	0.4864×10^{-1}	0.8398×10^{-1}	0.8545×10^{-1}	0.2375
Abridging factors	1:0:0	1:0:0	0.5:0:0	1:0:0	1:1:0.5

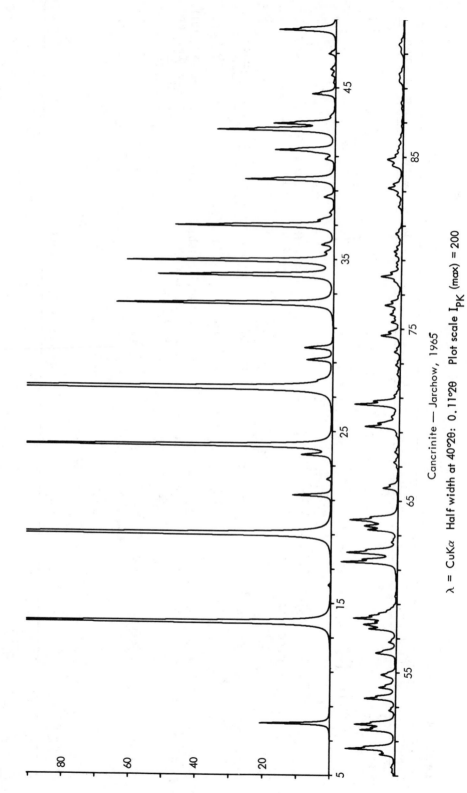

Cancrinite — Jarchow, 1965

$\lambda = CuK\alpha$ Half width at $40°2\theta$: $0.11°2\theta$ Plot scale I_{PK} (max) = 200

CANCRINITE - JARCHOW, 1965

2THETA	PEAK	D	H	K	L	I(INT)	I(PK)	I(DS)
8.00	8.00	11.042	1	0	0	9	10	9
13.88	13.88	6.375	1	1	0	75	81	75
19.03	19.04	4.660	1	0	1	100	100	100
21.27	21.28	4.173	2	1	0	6	6	6
23.63	23.64	3.762	2	0	1	5	5	5
24.16	24.16	3.681	3	0	0	59	54	59
27.51	27.52	3.240	1	2	1	54	89	55
27.51		3.240	2	1	1	51		52
29.13	29.14	3.062	1	3	0	1	4	1
29.13		3.062	3	1	0	3		3
29.83	29.84	2.993	3	0	1	5	4	5
32.40	32.40	2.760	4	0	0	41	32	42
34.06	34.06	2.631	3	1	1	12	26	13
34.05		2.631	1	0	3	21		22
34.88	34.88	2.570	0	0	2	39	31	41
35.84	35.84	2.503	3	0	2	2	2	2
36.93	36.94	2.432	4	0	0	31	24	33
37.29	37.28	2.410	1	4	0	2		2
38.61	38.62	2.330	2	0	2	2	2	2
39.63	39.64	2.272	2	2	0	10	13	11
39.63		2.272	3	1	1	8		8
40.83	40.82	2.208	5	0	0	2	1	2
41.35	41.34	2.182	4	1	1	5	9	5
41.35		2.182	1	4	0	6		7
42.50	42.50	2.125	3	3	0	25	18	27
42.88	42.88	2.107	3	0	2	12	9	13
44.62	44.62	2.029	2	4	0	5	4	5
46.95	46.96	1.9335	4	0	0	1	1	1
48.35	48.34	1.8810	4	3	0	13	8	14
50.22	50.22	1.8153	3	4	0	2	2	2
50.22		1.8153	3	2	2	1		1
50.55	50.54	1.8041	2	3	0	7	7	7
50.55		1.8041	2	3	2	5		6
51.65	51.66	1.7681	5	5	0	3	5	3
51.65		1.7681	2	1	2	5		6
51.98	51.98	1.7578	4	1	2	4	6	4
51.98		1.7578	1	4	2	5		6
52.79	52.80	1.7326	6	0	1	1	1	2
53.49	53.48	1.7117	4	3	1	5	5	6
53.49		1.7117	1	1	1	3		3
54.12	54.12	1.6931	3	0	3	4	3	4
54.76	54.86	1.6749	5	0	2	1	2	2
54.86		1.6720	2	5	0	2		2
56.11	56.14	1.6377	5	1	1	3	3	3
56.16		1.6363	2	3	1	3		3
57.55	57.56	1.6002	6	1	0	3	4	4
57.55		1.6002	1	4	1	3		3
57.80	57.80	1.5938	4	4	0	7	5	8
58.15	58.16	1.5850	2	1	3	5	6	6
58.15		1.5850	3	1	2	5		5
58.46	58.44	1.5774	5	3	0	2	3	2
61.43	61.44	1.5080	0	1	3	6	8	7
61.43		1.5080	5	3	1	6		7
61.43		1.5080	3	5	0	3		3
61.97	61.98	1.4963	6	0	2	8	8	8
62.01		1.4952	1	3	3	2		2
62.01		1.4952	3	3	1	4		4
63.32	63.32	1.4675	6	2	0	4	5	4
63.32		1.4675	2	6	0	4		4
63.56	63.54	1.4625	7	1	0	2	5	5
63.56		1.4625	4	0	3	4		4
63.85	63.88	1.4567	2	5	2	2	7	2
63.89		1.4557	5	0	3	10		12
65.74	65.74	1.4192	3	2	3	2	2	3
65.74		1.4192	3	3	2	2		2
67.84	67.84	1.3802	8	0	0	1	1	1
69.32	69.32	1.3544	4	4	2	9	5	11
69.64	69.52	1.3490	7	2	0	1	3	1
70.60	70.60	1.3330	8	0	1	13	7	16
73.66	73.66	1.2850	0	0	4	4	1	2
74.60	74.60	1.2711	7	1	2	2	3	2
75.39	75.40	1.2597	1	1	4	3	1	3
75.84	75.84	1.2533	1	8	0	2	2	3

CANCRINITE – JARCHOW, 1965

2THETA	PEAK	D	H	K	L	I(INT)	I(PK)	I(DS)
76.37	76.36	1.2460	4	3	3	3	2	4
76.37		1.2460	4	4	0	2		2
77.78	77.78	1.2269	9	0	2	1	1	2
78.04	78.04	1.2234	6	3	2	2	3	3
78.04		1.2234	3	6	2	3		4
78.61	78.62	1.2160	8	0	2	2	1	2
80.53	80.52	1.1918	2	2	4	2	1	2
82.78	82.78	1.1650	4	2	4	2	1	2
83.17	83.18	1.1605	7	0	3	2	2	3
83.17		1.1605	5	3	3	1		2
84.46	84.52	1.1460	2	3	4	1		2
84.46		1.1460	3	2	4	1		2
84.55		1.1450	4	7	0	1		2
84.85	84.84	1.1417	6	2	3	2	2	3
84.85		1.1417	2	6	3	1		3
91.55	91.54	1.0748	8	0	3	2	1	3
94.86	94.86	1.0459	4	7	2	1	1	2
96.07	96.04	1.0359	5	7	1	1	1	2
97.75	97.76	1.0225	8	4	1	3	2	4
97.75		1.0225	4	8	1	1		4
100.70	100.70	1.0003	4	4	4	3	1	4
102.86	102.86	0.9852	11	0	1	3	1	3
105.64	105.66	0.9667	8	4	0	2	1	3
105.64		0.9667	4	8	2	1		2
106.17	106.18	0.9634	4	0	5	2		3
119.62	119.62	0.8911	8	4	3	2	1	3
119.62		0.8911	4	8	3	2	1	3
120.96	120.96	0.8851	9	5	1	1	0	2
122.17	122.22	0.8799	10	3	0	2	0	2
125.53	125.56	0.8663	12	0	2	2	1	3
126.85	126.88	0.8612	3	5	5	1	0	2
133.60	133.64	0.8380	8	7	1	1	0	2
137.19	137.26	0.8273	2	2	6	1	0	4
138.22	138.22	0.8245	8	0	5	3	0	4
143.75	143.82	0.8105	11	4	1	1	1	2
143.75		0.8105	4	11	1	1		
147.08	147.44	0.8032	12	3	0	1	0	2

2THETA	PEAK	D	H	K	L	I(INT)	I(PK)	I(DS)
150.29	150.66	0.7969	8	8	0	4	1	6
150.90		0.7958	5	9	3	1		3
150.90		0.7958	9	5	3	2		3
156.71	157.74	0.7864	1	1	6	1		2
160.25	160.18	0.7818	4	11	2	1	0	3
162.26	162.60	0.7796	14	0	1	1	0	2
163.87		0.7779	1	13	2	3		5
163.87		0.7779	13	1	2	2		3

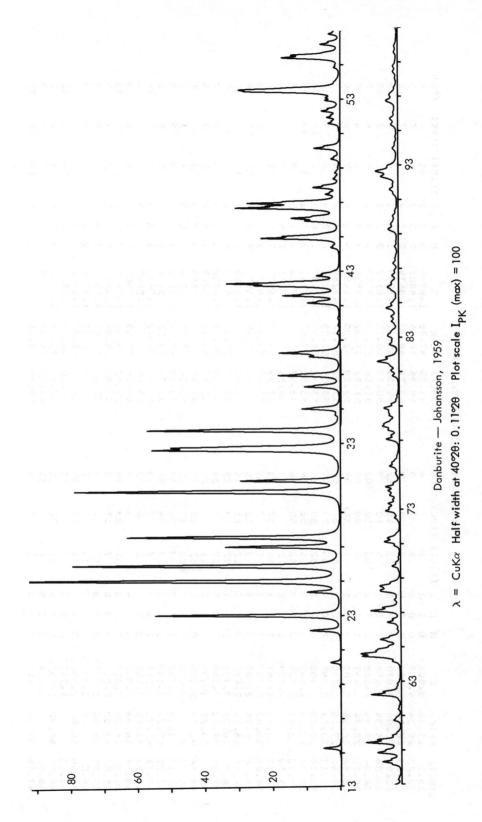

Danburite — Johansson, 1959

λ = CuKα   Half width at 40°2θ: 0.11°2θ   Plot scale I$_{PK}$ (max) = 100

DANBURITE - JOHANSSON, 1959

2THETA	PEAK	D	H	K	L	I(INT)	I(PK)	I(DS)
15.25	15.26	5.803	0	1	1	5	5	4
20.23	20.24	4.385	2	0	0	5	3	3
22.09	22.10	4.020	2	0	0	9	9	8
22.96	22.96	3.870	0	2	0	53	55	52
24.34	24.34	3.654	2	1	0	9	10	9
24.94	24.94	3.568	1	0	1	100	100	100
25.83	25.84	3.447	2	1	1	83	80	84
26.96	26.96	3.305	1	2	1	35	34	35
27.50	27.50	3.240	1	1	1	67	63	69
30.13	30.14	2.963	2	0	2	86	79	91
30.79	30.80	2.902	2	2	0	32	30	35
32.56	32.56	2.747	1	0	3	59	56	64
32.72	32.72	2.735	0	2	2	42	50	46
32.79		2.729	1	2	2	2		2
33.70	33.70	2.657	2	2	1	66	57	73
34.61	34.62	2.589	1	3	0	1	2	1
34.98	34.98	2.563	3	1	1	13	12	15
36.26	36.26	2.475	0	3	1	12	11	14
36.91	36.92	2.433	3	1	1	22	18	25
38.01	38.02	2.366	1	3	1	10	9	11
38.22	38.22	2.353	2	2	2	21	18	24
41.14	41.14	2.192	0	0	4	12	10	14
41.56	41.56	2.171	2	3	0	21	17	25
42.13	42.14	2.143	1	3	1	28	26	34
42.24	42.24	2.137	3	2	1	28	34	34
42.71	42.72	2.115	1	0	4	8	7	10
42.82	42.87	2.108	2	3	1	2	6	3
44.36	44.36	2.040	1	1	4	3	3	4
44.88	44.89	2.018	2	2	0	31	23	39
45.00	45.07	2.010	4	0	0	7	17	8
45.90	45.90	1.9755	3	3	1	12	11	15
46.06	46.06	1.9687	3	2	2	14	15	18
46.64	46.64	1.9458	2	3	3	31	31	40
		1.9455	4	1	0	9		11
46.91	46.92	1.9350	0	4	0	25	28	32
46.93		1.9344	0	3	3	8		11
47.46	47.46	1.9141	3	1	3	5	4	6

2THETA	PEAK	D	H	K	L	I(INT)	I(PK)	I(DS)
47.85	47.86	1.8993	4	1	1	11	8	14
48.35	48.36	1.8807	1	3	1	2	2	3
48.71	48.70	1.8679	2	1	4	2	2	2
49.51	49.52	1.8394	1	4	1	4	3	5
50.13	50.12	1.8183	3	3	1	11	8	15
51.58	51.58	1.7703	0	4	2	5	4	7
51.92	51.92	1.7595	4	2	2	7	6	7
52.29	52.30	1.7480	2	2	1	5	5	10
52.91	52.92	1.7289	1	0	4	2	4	7
53.09	53.08	1.7234	3	4	5	32	29	44
53.42	53.44	1.7137	1	3	5	16	30	22
53.50	53.52	1.7113	3	4	5	9		13
53.52		1.7106	0	2	1	2		2
53.54		1.7101	2	3	1	1		1
53.99	53.98	1.6970	3	0	4	2	2	3
54.82	54.82	1.6732	1	1	5	17	2	24
55.38	55.40	1.6576	3	1	4	11	18	16
55.43		1.6563	4	0	3	7		10
55.57	55.54	1.6523	4	2	2	8	15	12
56.18	56.18	1.6358	1	3	2	2	6	4
56.77	56.78	1.6202	4	2	2	3	3	4
56.79		1.6196	2	4	3	1		2
57.26	57.26	1.6076	4	0	5	3	1	5
58.13	58.14	1.5856	2	3	1	6	3	9
58.29	58.28	1.5816	5	0	0	9	6	14
58.82	58.82	1.5685	3	3	3	2	7	3
58.89		1.5669	1	2	5	2		4
59.16	59.16	1.5603	4	3	1	15	2	23
59.42	59.42	1.5541	3	2	4	3	11	3
59.61	59.58	1.5496	5	1	1	2	7	3
59.84	59.82	1.5443	3	4	1	3	3	4
60.70	60.70	1.5244	0	5	2	3	2	5
61.35	61.36	1.5097	5	0	4	1	1	2
62.20	62.20	1.4911	4	3	4	14	9	22
62.65	62.64	1.4816	4	0	4	1	1	2
64.13	64.14	1.4508	0	4	4	2	3	3
64.44	64.44	1.4446	2	5	0	16	11	25

DANBURITE - JOHANSSON, 1959

2THETA	PEAK	D	H	K	L	I(INT)	I(PK)	I(DS)
64.58	64.60	1.4419	3	1	5	10	12	16
64.77	64.76	1.4381	1	0	6	2	8	3
64.86	64.84	1.4362	1	5	2	7	7	11
65.30	65.42	1.4277	1	4	4	3	6	5
65.42		1.4254	2	1	1	8		12
66.02	66.02	1.4139	1	1	6	1	1	2
66.41	66.42	1.4065	5	2	0	6	4	10
67.08	67.10	1.3940	4	3	3	1	9	
67.10		1.3938	4	4	0	14		23
67.73	67.72	1.3823	3	0	3	2	2	3
68.21	68.26	1.3737	2	4	6	7	6	11
68.29		1.3723	3	2	5	4		7
68.30		1.3721	2	0	5	1		2
68.53	68.52	1.3680	0	5	3	5	6	8
68.57		1.3674	0	2	6	3		5
69.67	69.68	1.3484	5	3	1	2	3	3
69.69		1.3480	1	2	4	3		4
70.87	70.88	1.3285	4	4	2	2	2	3
71.08	71.10	1.3251	4	5	1	2	5	4
71.11		1.3246	6	0	0	4		8
71.30	71.32	1.3216	6	1	0	2	4	3
71.38		1.3204	5	3	2	2		4
72.47	72.48	1.3030	5	0	4	2	2	4
72.89	72.90	1.2967	5	2	0	7	4	12
73.02	73.08	1.2945	2	2	6	5	3	3
73.32	73.32	1.2900	0	6	0	2	3	9
73.78	73.80	1.2832	3	0	6	6	3	3
73.80		1.2829	1	4	5	2		3
74.07	74.06	1.2788	5	1	4	3	3	6
74.29	74.28	1.2757	3	3	3	1	4	9
75.07	75.08	1.2643	6	1	2	5	1	3
75.49	75.50	1.2583	4	4	3	1	3	9
75.64	75.68	1.2561	1	3	6	5	2	3
77.06	77.06	1.2366	5	3	4	6	4	10
77.58	77.60	1.2295	5	2	6	4	3	8
77.67		1.2283	2	6	0	1	1	3
77.81	77.82	1.2264	4	4	5	5	5	9

2THETA	PEAK	D	H	K	L	I(INT)	I(PK)	I(DS)
77.95	77.94	1.2246	5	5	1	5	5	9
78.42	78.44	1.2185	3	6	3	3	3	6
78.57	78.58	1.2165	6	2	1	2	3	3
78.57		1.2164	4	6	1	1		2
78.71	78.66	1.2146	2	5	5	1		3
78.88	78.88	1.2125	5	1	6	7	5	13
79.36	79.36	1.2063	2	5	4	1	1	2
80.74	80.74	1.1892	6	3	0	3	2	7
81.06	81.04	1.1853	5	0	5	3	6	6
81.41	81.58	1.1811	4	5	2	2		3
81.58		1.1791	1	2	7	11	3	21
82.54	82.56	1.1677	0	6	3	5		9
82.66		1.1663	6	4	1	1		3
83.04	83.06	1.1619	0	2	6	4	3	9
83.16		1.1606	1	5	5	1		7
83.72	83.90	1.1542	3	4	6	4	2	7
83.90		1.1523	1	6	1	2		3
84.22	84.18	1.1487	3	5	5	8	4	15
84.68	84.68	1.1436	7	4	3	1	2	3
85.12	84.94	1.1388	4	2	1	3	3	5
85.72	85.88	1.1324	7	1	6	5		10
85.89		1.1306	3	6	3	2		5
86.26	86.16	1.1267	5	3	1	3	2	5
86.56	86.56	1.1236	7	3	1	3	2	6
87.28	87.28	1.1161	6	5	1	3	2	10
88.25	88.26	1.1063	1	5	5	1	1	3
89.20	89.26	1.0970	5	5	1	8	3	5
89.28		1.0962	0	7	1	1		4
89.66	89.54	1.0925	7	2	1	2	2	9
90.02	90.02	1.0891	2	3	7	7	2	25
91.31	91.34	1.0771	5	3	5	5	2	21
91.57	91.58	1.0747	4	3	6	2	1	3
92.31	92.32	1.0679	7	2	2	4	2	9
92.64	92.62	1.0650	3	5	5	1	5	25
92.67		1.0648	6	0	6	1	7	21
93.95	93.96	1.0536	1	5	6	3		3
94.62	94.62	1.0479	2	1	8	1	1	3

DANBURITE – JOHANSSON, 1959

2THETA	PEAK	D	H	K	L	I(INT)	I(PK)	I(DS)
95.33	95.34	1.0420	5	5	3	3	2	7
95.34		1.0419	7	3	3	3		3
95.71	95.68	1.0389	3	3	7	4	2	8
96.28	96.30	1.0342	0	7	3	3	1	2
96.69	96.72	1.0309	6	4	3	4	4	6
96.72		1.0306	1	6	5	5		12
96.75		1.0304	7	2	3	3		5
97.19	97.06	1.0270	3	6	2	2	2	6
98.01	98.02	1.0205	7	3	1	3	1	6
98.41	98.40	1.0174	7	0	4	4	1	3
98.69	98.74	1.0153	7	4	1	3	1	6
98.78		1.0146	3	0	8	1		2
99.55	99.56	1.0088	4	7	0	2		3
101.39	101.40	0.9955	6	0	6	2	1	5
101.59	101.58	0.9940	3	7	2	2	1	3
102.51	102.56	0.9876	5	5	4	2	1	4
102.57		0.9871	7	3	3	1		2
103.17	103.16	0.9830	6	3	7	2	1	5
103.63	103.64	0.9799	4	1	4	3	2	3
103.65		0.9798	1	7	6	7		7
104.72	104.70	0.9727	6	1	0	2	1	4
106.24	106.16	0.9629	8	2	1	1	1	3
107.18	107.20	0.9571	4	7	2	1	1	3
107.51	107.56	0.9551	6	1	6	1		3
107.54		0.9549	4	1	8	1		3
107.71		0.9538	0	8	1	1		3
108.94	109.26	0.9465	2	5	8	3	2	3
108.94		0.9460	7	3	7	1		3
109.02		0.9460	4	8	4	1		3
109.23		0.9448	0	8	2	3		7
110.35	110.34	0.9383	1	1	8	1	1	3
111.12	111.32	0.9340	4	2	2	2	2	5
111.31		0.9329	6	4	8	3		7
111.62		0.9312	8	6	5	1		4
112.91	113.20	0.9242	6	6	1	1	1	3
113.13		0.9230	8	2	3	1		3
113.72	113.74	0.9199	2	2	9	1	1	3

2THETA	PEAK	D	H	K	L	I(INT)	I(PK)	I(DS)
113.76	113.76	0.9197	6	5	4	1		4
113.77		0.9196	4	7	3	2	1	5
114.21	114.20	0.9173	7	6	1	1	1	6
114.63	114.68	0.9152	6	0	7	1		3
114.76		0.9145	5	6	3	1		4
115.14	115.14	0.9126	8	3	3	1		3
117.34	117.62	0.9017	4	3	8	3	1	3
117.61		0.9005	7	5	0	1		7
119.47	119.72	0.8918	8	3	3	1	1	3
119.61		0.8912	3	2	9	1		4
119.74		0.8906	6	3	7	3		9
120.20	120.08	0.8890	2	8	4	2	1	5
122.28	122.22	0.8798	1	7	9	2	1	6
122.97	122.84	0.8795	7	5	4	2		5
123.78	123.82	0.8766	2	8	6	2	1	4
124.27	124.28	0.8733	5	1	8	1	1	7
124.63	124.64	0.8713	3	7	5	2	1	5
125.22	125.24	0.8698	4	7	3	2	1	6
125.51		0.8675	1	8	1	4		10
126.82	126.82	0.8664	2	7	10	2	1	5
128.50	128.62	0.8613	7	6	6	2	1	5
129.15	129.24	0.8552	4	2	9	2	1	5
129.36		0.8529	6	7	0	2		6
129.65	129.76	0.8521	3	9	7	1		6
129.90		0.8511	1	5	1	2		3
132.47	132.86	0.8502	7	1	4	2	1	5
132.55		0.8416	5	7	7	2		6
132.82		0.8413	8	8	4	2		5
133.27	133.18	0.8405	3	9	4	2	1	6
133.82		0.8393	1	2	2	2		6
134.82	135.16	0.8391	8	5	3	3	1	8
135.07		0.8342	3	9	10	2		5
135.24		0.8335	5	2	5	1		5
135.35		0.8330	7	5	7	1		6
136.62	136.64	0.8327	4	6	2	2	2	28
137.90	137.36	0.8289	6	3	9	2	1	6
		0.8253	5	8	1	2		7

DANBURITE - JOHANSSON, 1959

2THETA	PEAK	D	H	K	L	I(INT)	I(PK)	I(DS)
138.89	138.74	0.8226	9	1	4	3	1	8
140.70	140.92	0.8179	2	6	8	1	1	4
140.92		0.8173	4	5	8	6		18
141.72	141.78	0.8153	3	9	1	2	1	5
142.02		0.8146	5	8	2	3		9
144.91	145.64	0.8078	2	5	9	3	2	8
144.94		0.8078	3	8	5	1		3
145.00		0.8076	9	4	1	2		6
145.05		0.8075	4	7	6	2		7
145.39		0.8068	0	8	6	5		15
145.73		0.8060	6	3	8	9		27
146.48		0.8044	7	3	7	5		14
149.24	149.28	0.7989	7	6	4	10	1	31
149.95	150.10	0.7975	3	6	8	3	1	9
150.56		0.7964	10	1	1	1		5
151.40	151.36	0.7949	1	4	10	4	1	12
152.26	152.86	0.7934	1	0	11	2	1	7
152.40		0.7931	3	3	10	2		5
152.44		0.7931	0	1	11	2		5
152.60		0.7928	8	6	0	2		5
153.08		0.7920	3	6	7	7		22
153.81		0.7908	10	0	2	2		6
154.59	155.68	0.7896	8	6	1	3	2	8
155.57		0.7881	6	0	9	14		46
155.78		0.7878	9	3	4	2		7
156.18		0.7872	10	2	0	3		9
156.51		0.7867	10	1	2	2		6
157.61		0.7852	2	9	4	1		4
161.72	161.52	0.7802	8	6	2	4	1	14
162.13		0.7797	9	2	5	8		25
163.69		0.7781	4	9	2	1		4
164.65		0.7772	1	2	11	12		39

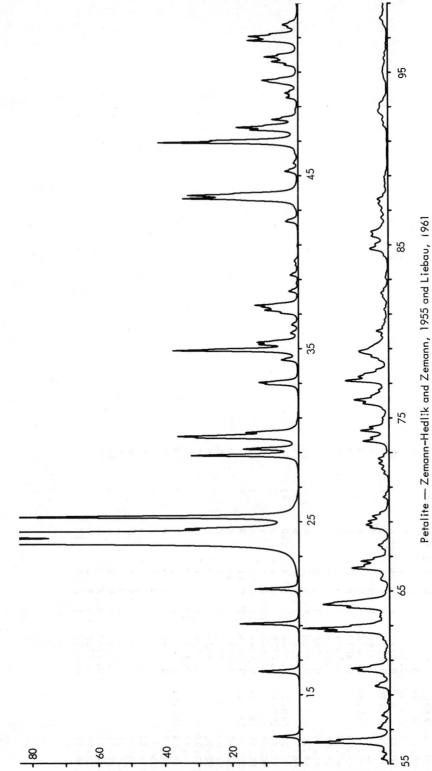

Petalite — Zemann-Hedlik and Zemann, 1955 and Liebau, 1961

λ = CuKα Half width at 40°2θ: 0.11°2θ Plot scale I_{PK} (max) = 500

PETALITE — ZEMANN-HEDLIK AND ZEMANN, 1955 AND LIEBAU, 1961

2THETA	PEAK	D	H	K	L	I(INT)	I(PK)	I(DS)
12.55	12.56	7.045	1	0	0	2	2	2
16.29	16.34	5.436	0	0	2	2	2	2
16.35		5.416	-1	0	1	2		2
19.08	19.08	4.647	0	1	1	6	4	6
21.10	21.10	4.207	-1	1	1	5	3	5
23.80	23.84	3.735	0	1	2	97	100	97
23.84		3.728	-1	1	2	100		100
24.17	24.18	3.679	-1	0	2	79	52	79
24.29	24.30	3.661	-2	0	2	78	54	78
24.58	24.58	3.618	1	1	1	7	7	7
25.26	25.26	3.523	-2	0	3	32	17	32
28.81	28.80	3.097	-1	1	3	13	6	13
29.19	29.20	3.057	-2	1	1	6	3	6
29.84	29.86	2.992	-1	1	2	10	6	10
29.94	29.94	2.982	-2	1	2	9	7	9
30.15	30.14	2.962	-2	0	1	5	3	5
32.92	32.94	2.718	0	0	4	4	2	4
33.05	33.04	2.708	-2	0	4	3	1	3
34.33	34.34	2.610	2	0	1	2	1	2
34.88	34.88	2.570	0	2	2	16	7	17
35.21	35.22	2.546	2	0	0	2	2	2
35.36	35.36	2.536	-3	0	2	4	2	4
35.87	35.88	2.501	0	2	1	1	0	1
36.35	36.36	2.469	-1	1	3	1	1	1
37.04	37.04	2.425	-1	2	2	2	2	2
37.21	37.21	2.414	1	2	0	4	2	4
37.39	37.40	2.403	0	1	4	3	2	3
37.50	37.50	2.396	-2	1	4	3	3	4
38.29	38.30	2.348	3	0	1	1	1	1
39.24	39.24	2.294	-1	2	1	1	1	1
42.33	42.34	2.133	-1	1	5	2	1	2
43.55	43.66	2.076	2	2	0	4	7	5
43.67		2.071	1	1	4	13		14
43.86	43.86	2.062	-3	1	1	13	7	14
45.24	45.24	2.003	0	1	5	2	1	2
45.72	45.72	1.9831	-2	1	3	1	0	1
46.92	46.92	1.9349	-1	2	4	21	8	23

2THETA	PEAK	D	H	K	L	I(INT)	I(PK)	I(DS)
47.64	47.64	1.9074	3	0	2	6	3	7
47.79	47.78	1.9014	-4	0	2	6	4	7
48.23	48.24	1.8851	-3	1	5	4	2	4
49.51	49.52	1.8395	2	0	4	2	1	2
49.78	49.78	1.8303	0	2	2	2	1	2
50.41	50.52	1.8088	2	2	0	3	2	3
50.51		1.8055	-2	0	6	1		1
50.52		1.8052	-3	2	3	2		3
51.40	51.40	1.7762	-1	1	5	2	1	2
51.60	51.60	1.7699	1	1	5	3	2	3
51.60		1.7697	-3	2	3	1		1
51.87	51.86	1.7613	-4	1	3	5	2	5
52.76	52.82	1.7336	3	2	0	1	3	1
52.81		1.7319	-1	1	4	4		8
53.07	53.06	1.7242	-4	1	4	7	3	8
53.58	53.58	1.7090	0	1	6	1	1	1
53.71	53.74	1.7050	-2	2	5	1	1	1
53.77		1.7035	-3	1	6	1		1
55.29	55.30	1.6600	0	2	5	2	1	2
55.70	55.70	1.6488	2	2	3	3	0	8
56.25	56.26	1.6341	0	3	1	7	5	9
56.27		1.6336	-1	3	2	1		1
57.05	57.08	1.6130	-1	0	6	1	1	1
57.09		1.6119	3	3	3	1		1
57.35	57.34	1.6052	-4	0	1	1	1	1
57.80	57.80	1.5939	-2	1	7	1		2
59.46	59.50	1.5532	-1	3	2	2	2	2
59.52		1.5518	-2	3	2	1		2
60.07	60.08	1.5390	3	1	6	1		1
60.36	60.38	1.5322	-4	1	1	1	2	1
60.38		1.5316	-3	2	2	1		1
60.52	60.52	1.5285	-4	2	2	4	2	5
62.68	62.68	1.4810	0	2	6	10	4	12
62.85	62.84	1.4774	-3	1	3	10	5	12
63.50	63.50	1.4637	4	3	1	3	1	3
64.03	64.04	1.4530	4	1	2	2	3	3
64.05		1.4526	3	1	4	4		4

PETALITE - ZEMANN-HEDLIK AND ZEMANN, 1955 AND LIEBAU, 1961

2THETA	PEAK	D	H	K	L	I(INT)	I(PK)	I(DS)
64.19	64.22	1.4496	-5	1	2	2	4	3
64.20		1.4494	0	3	4	1		4
64.28		1.4479	-2	3	4	3		4
64.35		1.4465	-5	1	4	4		4
65.62	65.64	1.4215	2	3	2	1	1	1
65.71	65.70	1.4197	-3	3	2	1	1	1
66.28	66.32	1.4090	5	0	0	3	2	4
66.34		1.4079	2	3	6	4		5
66.64	66.72	1.4022	-3	3	3	1		1
66.72		1.4006	-5	0	3	4	2	5
67.47	67.48	1.3870	-1	1	8	1	0	1
67.61	67.64	1.3844	-3	1	8	1	1	1
67.67		1.3834	-1	3	5	1		1
68.65	68.66	1.3660	-1	3	4	2	1	3
68.80	68.82	1.3634	-3	3	3	2	1	3
68.91		1.3615	-4	2	6	1		1
69.05	69.04	1.3591	0	0	8	2	1	3
69.34	69.32	1.3541	-4	0	8	2	1	3
70.42	70.42	1.3360	3	3	1	1	0	1
71.78	71.78	1.3139	0	1	8	1	0	1
72.06	72.06	1.3094	-4	1	8	1	0	1
72.37	72.36	1.3047	3	2	4	2	1	2
72.65	72.64	1.3003	-5	2	4	1	1	2
73.66	73.66	1.2850	0	4	0	4	2	6
73.81	73.86	1.2827	-5	1	7	1	1	1
74.26	74.26	1.2760	-2	2	8	5	2	7
74.46	74.46	1.2732	-2	2	8	1	1	1
74.80	74.80	1.2681	-6	0	4	1	0	1
74.97	74.98	1.2657	-1	4	1	1	0	1
75.81	75.84	1.2537	-2	3	4	3		4
75.87		1.2529	5	0	2	2		2
76.02	76.04	1.2508	-4	3	4	3	2	4
76.06		1.2502	-6	0	2	2		2
76.37	76.24	1.2459	-1	4	1	1	1	1
76.60	76.60	1.2428	-3	3	6	1	1	1
77.11	77.16	1.2358	4	1	4	1	3	1
77.13		1.2355	5	2	0	5		7
77.19	77.19	1.2347	-2	2	6	3	2	5
77.46	77.38	1.2312	-6	1	1	1	1	1
77.55	77.54	1.2299	-5	2	2	3	2	4
78.33	78.34	1.2197	2	1	7	1	1	1
78.51	78.54	1.2173	5	1	2	2	1	3
78.70	78.84	1.2148	-6	1	2	2	2	3
78.79		1.2136	1	4	8	1		1
78.83		1.2131	1	1	4	1		1
78.88		1.2125	-2	4	2	2		2
78.99		1.2111	0	4	1	1		1
79.75	79.76	1.2014	0	2	8	1	0	1
80.00	80.00	1.1983	-2	3	7	1	1	1
80.02		1.1980	-4	2	8	1		2
80.43	80.42	1.1930	-2	4	3	1	0	1
81.25	81.26	1.1830	2	4	1	1	0	1
81.97	81.98	1.1744	-4	3	6	1	0	1
82.23	82.22	1.1714	-4	3	6	2	1	3
84.73	84.76	1.1431	-2	3	10	1		2
84.74		1.1429	-3	1	4	2		2
84.82		1.1421	-3	1	10	2		2
85.49	85.52	1.1349	4	3	2	2	1	2
85.51		1.1347	3	3	4	2		2
85.64	85.76	1.1332	-5	3	2	2	1	2
85.78		1.1318	-5	3	4	2		2
86.58	86.58	1.1233	-1	2	8	2	0	2
86.99	86.98	1.1191	-5	2	8	2	0	2
87.39	87.38	1.1150	-4	1	10	1	1	2
87.65	87.64	1.1124	-1	1	10	2	0	2
88.39	88.38	1.1050	2	1	8	1	0	1
88.93	88.92	1.0996	-6	1	8	1	0	1
91.89	91.92	1.0717	-1	4	6	1		1
91.94		1.0713	4	4	6	1		1
92.25	92.24	1.0685	5	1	4	1	0	1
92.66	92.76	1.0649	-7	1	4	1	1	1
92.79		1.0637	0	1	10	1		1
93.21	93.16	1.0600	-5	1	10	1		2
94.59	94.76	1.0482	0	4	6	1	0	1

PETALITE — ZEMANN-HEDLIK AND ZEMANN, 1955 AND LIEBAU, 1961

2THETA	PEAK	D	H	K	L	I(INT)	I(PK)	I(DS)
94.74		1.0469	-3	4	6	1		1
95.01	95.02	1.0446	-6	1	2	1	0	1
95.23	95.22	1.0428	-7	1	2	1	0	1
95.80	95.80	1.0381	4	4	0	1	0	1
97.83	97.84	1.0219	4	3	4	1	0	1
98.10	98.14	1.0198	3	0	8	1	0	1
98.17		1.0193	-6	3	4	1		1
98.78	98.78	1.0146	-7	0	8	1	0	1
99.21	99.40	1.0113	5	3	2	1	1	1
99.38		1.0101	0	5	2	1		1
99.40		1.0100	-1	1	2	1		1
99.40		1.0099	-6	3	2	1		3
99.87	99.86	1.0064	7	0	0	2	1	3
100.06		1.0050	-4	3	9	2		3
101.94	101.94	0.9916	4	2	6	2	0	3
102.51	102.50	0.9876	-7	2	6	2	0	3
102.83	102.84	0.9853	6	2	2	1	0	1
103.06		0.9838	-7	0	12	1		2
103.62	103.62	0.9800	-3	2	3	2		1
105.51	105.54	0.9676	-2	3	10	1	0	1
105.60		0.9670	-3	3	10	1		2
108.44		0.9495	5	4	0	2		4
108.50	108.50	0.9491	2	4	6	2		4
108.87	108.86	0.9469	-5	4	6	3	1	1
110.55	110.60	0.9371	7	2	0	1	0	2
111.16	111.18	0.9337	0	4	8	2	0	2
111.45	111.44	0.9321	-4	4	8	2	0	2
112.36	112.34	0.9271	2	1	10	1	0	1
113.18	113.18	0.9227	-2	1	10	1	0	1
113.49		0.9211	5	3	4	1		1
113.93	113.52	0.9188	-7	3	4	1	0	1
114.08		0.9180	0	3	10	1		1
114.53	113.96	0.9157	-5	3	10	1	0	1
114.54		0.9156	-3	2	12	1		1
114.95	114.50	0.9135	7	1	2	1	0	1
115.23	114.94	0.9121	-8	1	2	2		1
116.78	117.08	0.9044	-7	3	2	1	0	1

2THETA	PEAK	D	H	K	L	I(INT)	I(PK)	I(DS)
116.78		0.9044	4	4	4	1		1
117.00		0.9034	-4	5	3	1		1
117.16		0.9026	-6	4	4	1	0	1
118.33	118.50	0.8971	5	4	2	1		1
118.54	118.94	0.8961	-6	4	2	1	0	1
118.96		0.8941	6	1	4	1	0	1
119.36	119.40	0.8923	0	1	12	1		1
119.51		0.8916	-8	2	4	1		1
120.06	120.04	0.8891	-6	1	12	1	0	3
120.48	120.50	0.8873	-1	2	12	2	0	3
120.95	120.96	0.8852	-5	2	12	2	0	3
128.09	128.16	0.8567	0	6	0	2	0	1
128.17		0.8564	3	1	10	1		2
129.79	129.92	0.8506	4	5	2	1		1
129.81		0.8505	3	5	4	1		1
129.98		0.8499	-5	5	2	1		1
130.16		0.8493	-5	5	4	1		1
130.99	131.04	0.8465	-1	6	12	1		1
131.09		0.8461	-1	1	2	1		2
132.07	131.70	0.8429	-7	1	12	1	0	1
135.03	135.48	0.8336	7	1	4	1	0	2
135.42		0.8324	5	0	8	2		3
135.44		0.8324	2	6	0	1		2
135.80		0.8313	-9	1	4	4		3
136.79	136.98	0.8285	-9	0	8	1	0	1
137.09		0.8276	-2	0	14	1		2
137.61	137.62	0.8261	-5	3	14	1	0	2
137.71		0.8259	2	3	10	1		2
139.39	139.80	0.8213	8	0	2	1	0	1
139.83		0.8201	-9	0	2	1		1
141.36	141.32	0.8162	7	3	2	1	0	1
141.77		0.8152	-8	3	2	1		1
150.03	150.44	0.7974	-3	2	14	1	0	2
150.20		0.7970	0	6	5	1		2
150.27		0.7969	-4	2	14	1		2
150.61		0.7963	-7	4	8	2		4
151.11	151.88	0.7954	2	5	7	1	1	1

PETALITE - ZEMANN-HEDLIK AND ZEMANN, 1955 AND LIEBAU, 1961

2THETA	PEAK	D	H	K	L	I(INT)	I(PK)	I(DS)
151.49		0.7947	5	5	2	1		1
151.67		0.7944	-6	5	5	1		2
151.88		0.7940	-6	5	2	1		2
151.90		0.7940	6	4	3	1		1
152.08		0.7937	1	5	8	1		2
152.88	153.00	0.7923	7	4	0	5	1	8
152.95		0.7922	-5	5	8	1		2
153.12		0.7919	5	2	8	5		9
153.80		0.7908	-8	3	9	1		1
154.61	154.34	0.7896	4	1	10	3	1	6
155.30		0.7885	-9	2	8	5		10
155.79		0.7878	-2	2	14	2		4
159.84	160.76	0.7823	8	2	2	1	0	1
160.56		0.7815	3	6	2	3		5
160.69		0.7813	-9	2	2	1		1
160.92		0.7811	-4	6	2	3		6
162.58		0.7792	-3	4	12	5		9
164.19		0.7776	-4	6	1	1		1

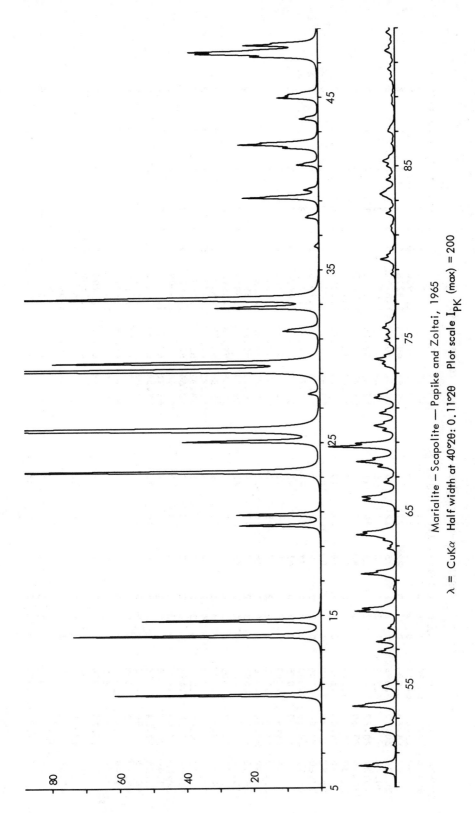

Marialite — Scapolite — Papike and Zoltai, 1965

λ = CuKα Half width at 40°2θ: 0.11°2θ Plot scale I_{PK} (max) = 200

MARIALITE-SCAPOLITE — PAPIKE AND ZOLTAI, 1965

2THETA	PEAK	D	H	K	L	I(INT)	I(PK)	I(DS)
10.36	10.38	8.528	1	1	0	24	31	23
13.80	13.80	6.413	0	1	1	30	37	29
14.68	14.68	6.030	0	0	2	22	27	21
20.20	20.20	4.393	2	1	1	5	12	5
20.20		4.393	1	2	1	6		6
20.81	20.82	4.264	3	1	0	11	13	11
23.30	23.30	3.814	1	1	2	40	49	39
23.30		3.551	0	3	1	7		7
25.06	25.06	3.460	1	3	1	19	21	19
25.72	25.72	3.206	0	2	2	100	100	100
27.80	27.80	3.060	2	3	1	77	74	78
29.16	29.16	3.015	0	4	0	41	40	42
29.60	29.60	2.843	3	3	0	5	5	6
31.44	31.44	2.831	2	2	2	5	5	2
31.58	31.52	2.728	1	4	1	16	16	17
32.80	32.80	2.687	3	1	2	3	5	3
33.32	33.32	2.687	3	3	2	45	47	47
33.32		2.365	4	1	0	2	2	2
38.01	38.02	2.298	3	5	1	9	11	9
39.16	39.16	2.298	4	3	1	4		5
39.16		2.273	3	4	2	2	2	3
39.61	39.62	2.196	4	3	1	3	3	3
41.06	41.06	2.148	5	2	1	2	5	2
42.04	42.04	2.148	4	4	1	14	12	15
42.04		2.138	0	4	3	2	9	2
42.24	42.24	2.132	4	2	3	3	3	4
42.36	42.34	2.068	3	5	2	5	6	6
43.73	43.74	2.015	3	2	3	2		2
44.95	44.96	2.015	2	3	3	2		2
44.95		2.006	5	1	2	2	4	10
45.16	45.16	2.006	5	5	1	9	10	2
45.16		1.9180	6	1	1	2	2	16
47.36	47.36	1.9180	1	6	1	14	18	4
47.36		1.9109	1	4	3	3		13
47.54	47.54	1.9109	4	1	3	11	19	
47.54		1.9069	6	2	0			
47.65	47.66	1.9069						

2THETA	PEAK	D	H	K	L	I(INT)	I(PK)	I(DS)
47.65	47.65	1.9069	2	6	0	2		3
48.02	48.02	1.8930	0	0	4	13	11	15
49.85	49.86	1.8278	4	5	1	2	2	3
50.22	50.22	1.8151	5	3	2	5	6	6
50.22		1.8151	3	5	2	6		2
52.25	52.26	1.7492	6	1	1	2	4	4
52.25		1.7492	3	6	1	3		3
52.43	52.40	1.7438	3	5	3	3		4
53.69	53.70	1.7055	7	7	0	6	4	7
53.78		1.7055	1	7	1	3	6	2
53.69		1.7030	2	6	2	2		2
54.75	54.76	1.6752	5	5	1	1	2	2
54.85	54.84	1.6724	6	4	1	4	3	5
56.84	56.84	1.6183	2	6	4	4	3	5
57.43	57.44	1.6032	0	7	1	2	1	2
58.22	58.22	1.5836	7	3	0	2	6	3
59.21	59.22	1.5591	6	1	1	7	5	8
59.21		1.5591	1	6	6	1	5	2
59.38	59.38	1.5550	5	4	1	7	1	9
61.36	61.36	1.5095	5	3	1	1	1	2
62.82	62.82	1.4779	4	1	4	4	2	2
63.32	63.32	1.4675	7	7	1	4	6	4
63.64	63.64	1.4609	3	3	2	3		5
63.64		1.4609	1	7	5	3	5	4
63.78	63.80	1.4580	0	1	3	3	5	4
65.54	65.64	1.4230	6	6	0	6		8
65.63		1.4213	0	3	5	4	6	6
65.85	65.82	1.4172	0	5	3	2	2	3
66.65	66.66	1.4019	7	7	2	4	3	5
67.58	67.58	1.3849	2	7	1	9	6	11
67.88	67.88	1.3796	2	3	5	10	10	13
68.75	68.76	1.3642	8	2	2	7		9
68.75		1.3642	2	8	0	1		2
69.68	69.68	1.3483	8	4	1	1	2	2
69.68		1.3483	4	8	0	4		2
69.97	69.96	1.3434	6	2	4	4	3	6
70.67	70.68	1.3318	1	9	0	3	2	5

MARIALITE-SCAPOLITE – PAPIKE AND ZOLTAI, 1965

2THETA	PEAK	D	H	K	L	I(INT)	I(PK)	I(DS)
71.43	71.58	1.3195	0	9	1	1	3	3
71.57		1.3172	5	6	3	4		6
71.73	71.74	1.3147	5	7	2	1	3	2
73.39	73.40	1.2890	9	2	1	2	2	1
73.82	73.82	1.2826	4	3	5	1	3	5
74.87	74.88	1.2671	1	7	4	3	2	2
75.23	75.22	1.2620	0	0	6	1	1	3
75.63	75.62	1.2563	9	1	2	2	2	2
75.63		1.2563	1	9	2	1		2
79.59	79.58	1.2035	6	1	5	3	2	5
80.02	80.00	1.1981	1	3	6	1	1	3
82.23	82.24	1.1714	5	7	0	2	1	3
83.24	83.36	1.1597	7	2	2	2	2	3
83.37		1.1582	3	6	5	1		2
83.37		1.1582	6	3	5	1		2
84.18	84.18	1.1491	8	6	2	2	1	3
84.97	84.98	1.1404	5	8	5	1	1	2
85.25	85.26	1.1374	0	7	5	2	2	2
85.33		1.1366	6	9	4	2		3
86.99	87.00	1.1190	5	1	2	2	1	2
90.59	90.58	1.0838	10	0	3	2	1	3
91.67	91.66	1.0738	4	10	7	1	1	4
92.73	92.72	1.0642	4	7	5	2	1	4
93.14	93.08	1.0606	1	2	7	1	1	2
96.21	96.28	1.0348	7	8	3	1	1	2
98.79	98.78	1.0146	4	7	7	1	1	2
99.99	100.04	1.0056	0	11	3	1	1	2
101.89	101.90	0.9919	2	1	3	1	1	2
102.17	102.18	0.9899	9	2	5	2	1	3
108.42	108.40	0.9496	1	6	7	1	1	4
110.40	110.38	0.9380	5	4	7	1	1	2
114.45	114.44	0.9161	0	7	7	1	1	2
116.06	116.10	0.9080	7	8	5	1	1	2
116.15		0.9075	6	10	4	1		2
118.16	118.08	0.8979	9	6	5	1	1	3
120.78	120.78	0.8860	5	6	7	1	0	2
134.45	134.46	0.8354	8	9	5	2	1	5

2THETA	PEAK	D	H	K	L	I(INT)	I(PK)	I(DS)
139.54	139.94	0.8209	7	12	3	1	0	2
139.96		0.8198	3	12	5	1		2
141.48	141.46	0.8159	2	1	9	1		3
142.51	142.50	0.8134	1	14	3	2	0	4
144.32	144.34	0.8091	1	4	2	2	1	4
144.32		0.8091	14	4	2	3		5
146.94	146.96	0.8035	4	14	7	3		6
151.55	151.60	0.7946	10	1	8	3	1	6
151.55		0.7946	2	8	8	3		5
151.68		0.7944	9	2	8	1		2
157.16	158.32	0.7858	8	13	1	2	1	3
157.90		0.7848	15	1	2	1		3
158.16		0.7845	5	5	7	1		3
158.24		0.7843	13	7	2	4		8
159.38		0.7829	7	7	8	3		6

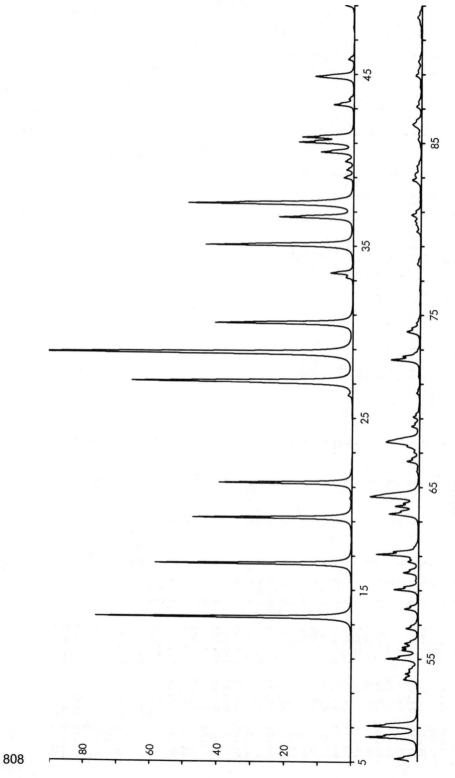

808

Hemimorphite — McDonald and Cruickshank, 1967

$\lambda = CuK\alpha$ Half width at $40°2\theta$: $0.11°2\theta$ Plot scale I_{PK} (max) = 100

HEMIMORPHITE - MCDONALD AND CRUICKSHANK, 1967

2THETA	PEAK	D	H	K	L	I(INT)	I(PK)	I(DS)
13.41	13.42	6.597	1	0	0	60	76	51
16.53	16.54	5.359	0	2	0	48	58	42
19.19	19.20	4.620	0	0	1	40	47	36
21.21	21.22	4.185	2	1	0	35	39	32
27.11	27.12	3.286	1	3	0	58	66	57
28.76	28.76	3.102	2	1	1	100	100	100
30.48	30.48	2.930	0	3	1	42	41	43
33.15	33.16	2.700	3	1	0	7	2	2
33.41	33.42	2.680	0	0	2	2	2	8
35.02	35.02	2.560	0	4	0	48	44	52
36.65	36.66	2.450	3	0	1	25	22	27
37.44	37.44	2.400	2	3	1	55	49	61
38.96	38.96	2.310	0	2	2	3	3	4
39.42	39.42	2.284	1	4	1	2	2	2
39.91	39.92	2.257	2	4	0	3	3	3
40.45	40.46	2.228	3	2	1	11	10	13
41.01	41.02	2.199	3	3	0	19	16	22
41.31	41.30	2.184	2	0	2	17	15	20
43.20	43.20	2.092	2	0	0	7	6	9
43.54	43.54	2.077	1	5	0	2	2	2
44.77		2.022	2	2	2	6	12	8
44.84	44.84	2.020	1	2	3	10		13
45.85	45.86	1.9775	0	5	2	2	2	3
48.99	49.00	1.8577	3	1	1	3	3	4
49.18	49.18	1.8511	3	4	1	8	7	10
50.43	50.42	1.8081	2	5	0	20	15	27
51.04		1.7879	0	6	1		15	5
51.08	51.08	1.7865	4	3	0	17	15	22
53.79	53.80	1.7028	4	3	1	5	4	7
53.89	53.92	1.6999	3	5	2	4	4	2
54.13	54.12	1.6929	2	4	3	4	4	6
54.39	54.38	1.6854	0	4	2	3	3	4
55.00	55.00	1.6681	3	3	0	13	10	18
55.51	55.52	1.6540	4	1	0	6	5	8
55.68	55.68	1.6493	4	6	0	3	5	4
55.91	55.90	1.6431	2	6	2	4	4	6
56.77	56.78	1.6201	4	0	2	5	4	7

2THETA	PEAK	D	H	K	L	I(INT)	I(PK)	I(DS)
57.91	57.90	1.5911	5	0	1	6	4	8
59.03	59.04	1.5634	2	1	3	10	7	15
60.02	60.02	1.5400	0	3	3	6	5	9
60.66	60.66	1.5253	5	2	1	4	3	6
61.08	61.08	1.5159	5	3	0	18	13	28
61.51	61.50	1.5063	1	7	0	1	1	2
63.34	63.44	1.4671	0	6	1	3	9	5
63.44	63.88	1.4650	0	3	2	11		18
63.88		1.4559	2	3	3	10	7	15
64.41	64.42	1.4453	2	6	0	18	15	28
64.50		1.4435	3	2	1	10		16
66.49	66.50	1.4050	3	0	3	6	4	9
67.03	67.04	1.3950	6	2	0	2	2	3
67.34	67.36	1.3892	5	4	1	4	4	7
67.50	67.62	1.3864	4	7	2	2	10	3
67.61		1.3845	2	6	1	11		19
67.70	68.52	1.3828	5	4	1	7		12
68.53	69.08	1.3681	4	6	0	4	2	6
69.07	70.46	1.3587	0	4	3	3	2	5
70.46	72.38	1.3352	5	3	2	1	1	2
72.39	74.02	1.3044	3	0	4	15	9	26
73.99	74.52	1.2800	2	5	3	3	4	5
74.04		1.2793	2	6	4	4		8
74.53	77.98	1.2720	4	6	3	1	1	3
77.99	79.86	1.2240	5	0	1	1	1	2
79.85	80.26	1.2001	1	5	1	2	1	4
80.26	80.46	1.1951	5	6	0	2	2	4
80.45	80.82	1.1927	7	0	1	2	2	3
80.82	82.84	1.1882	0	9	1	4	3	8
82.83	83.18	1.1644	0	0	4	5	3	9
83.21	83.64	1.1600	7	2	1	3	2	5
83.65	85.22	1.1550	5	5	0	2	1	4
85.21	86.08	1.1378	5	7	1	2	1	4
85.96	87.10	1.1299	2	6	3	5	2	7
86.08	88.92	1.1286	2	9	1	3	3	10
87.11		1.1179	2	6	3	3		7
88.92		1.0997	2	7	3	4	2	8

HEMIMORPHITE - MCDONALD AND CRUICKSHANK, 1967

2THETA	PEAK	D	H	K	L	I(INT)	I(PK)	I(DS)
89.73	89.74	1.0919	4	0	4	1	1	3
92.32	92.32	1.0679	7	4	1	3	1	6
95.51	95.52	1.0405	0	6	0	1	1	6
95.97	95.96	1.0368	7	3	2	3	1	6
96.35	96.32	1.0337	5	7	2	3	2	7
98.02	98.04	1.0204	8	1	1	2	1	3
99.42	99.42	1.0098	2	6	4	2	1	4
99.79	99.76	1.0071	3	0	2	2	1	5
101.69	101.70	0.9933	5	6	3	2	1	5
102.10	102.08	0.9904	2	1	5	1	1	3
102.34	102.36	0.9887	0	10	0	1	1	3
102.84	102.86	0.9853	8	3	1	2	1	6
102.98		0.9844	0	5	5	1		3
103.32	103.32	0.9820	3	10	1	3	2	7
103.73	103.92	0.9793	7	0	3	2	3	6
103.92		0.9780	5	3	4	6		14
104.12		0.9767	0	3	3	3		2
104.30	104.28	0.9755	7	6	1	4	3	10
104.31		0.9754	1	7	4	1		2
105.07	105.06	0.9704	5	9	0	2	1	6
105.61	105.42	0.9669	7	5	2	1	1	1
106.18	106.50	0.9633	7	2	3	1	1	2
106.35		0.9622	2	10	2	1		2
106.50		0.9613	3	0	5	4		4
107.00	106.98	0.9582	2	3	0	3	1	9
107.68	107.36	0.9540	4	10	0	1	1	2
108.16	108.16	0.9511	2	9	3	3	1	7
108.99	108.98	0.9462	3	2	5	1	0	2
111.53	111.54	0.9317	4	6	4	1	0	3
112.50	112.50	0.9264	3	1	4	3	0	8
113.74	113.74	0.9198	7	4	3	4	1	10
116.17	116.18	0.9074	5	9	2	1	1	4
116.70	116.64	0.9048	9	4	5	1	0	3
117.70	117.68	0.9000	9	3	0	1	0	2
119.00	119.12	0.8940	4	10	2	2		2
119.15		0.8932	0	12	0	1		1
119.46	119.52	0.8919	5	8	3	1	0	

2THETA	PEAK	D	H	K	L	I(INT)	I(PK)	I(DS)
120.09	120.08	0.8890	5	10	1	1	1	4
120.11		0.8889	8	1	3	1		4
123.70	123.72	0.8736	2	12	0	1	1	2
123.71		0.8735	5	0	5	1		4
124.28	124.28	0.8712	9	0	2	1	1	2
124.37		0.8709	7	1	4	1		2
125.29	125.76	0.8672	1	9	4	3	1	9
125.75		0.8654	8	3	3	3		9
126.34	126.32	0.8632	3	10	3	2	1	6
126.61		0.8622	5	2	5	1		2
127.52	127.52	0.8587	7	6	3	4	1	13
129.01	129.54	0.8533	0	0	6	1	1	4
129.44		0.8518	8	7	1	2		7
129.56		0.8514	8	6	2	1		3
130.23	130.26	0.8491	9	3	2	2	1	6
130.32		0.8488	5	7	4	2		6
130.82	130.90	0.8471	5	7	3	3	1	9
130.98		0.8465	3	6	5	2		6
131.79	131.66	0.8438	7	9	0	1	1	4
131.93		0.8434	0	12	2	2		5
132.13		0.8427	0	2	6	1		2
133.22	133.22	0.8392	3	12	1	2	1	6
133.93	133.94	0.8370	10	0	0	2	1	5
134.21		0.8361	2	0	6	1		2
134.86	134.84	0.8341	2	7	5	2	1	6
135.49	135.50	0.8322	3	9	4	2	1	6
136.08		0.8305	5	4	5	1		2
137.39	137.48	0.8268	2	12	2	3	1	8
137.68		0.8259	10	1	1	1		4
138.53	138.18	0.8236	0	3	6	1	1	2
139.20	139.22	0.8218	0	10	4	2	1	7
139.30		0.8215	4	12	0	1		2
142.09	142.50	0.8144	9	6	1	1	0	3
142.24		0.8140	4	3	6	1		4
142.63		0.8131	0	13	1	1		6
144.32	144.24	0.8091	7	5	4	1	0	3
145.56	146.32	0.8064	2	10	4	1	1	2

HEMIMORPHITE - MCDONALD AND CRUICKSHANK, 1967

2THETA	PEAK	D	H	K	L	I(INT)	I(PK)	I(DS)
146.30		0.8048	10	3	1	4		12
147.94	147.98	0.8014	7	9	2	4	1	11
149.13	149.16	0.7991	2	13	1	5	1	15
149.20		0.7989	10	4	0	2		7
149.72		0.7980	5	10	3	1		4
151.02	151.02	0.7956	10	0	2	5	1	16
151.03		0.7955	3	3	6	1		5
151.40		0.7949	9	7	0	2		8
154.22	155.30	0.7902	4	0	6	1	1	4
154.77		0.7893	6	11	1	1		2
155.22		0.7886	7	10	1	5		17
156.36		0.7869	10	2	2	2		5
157.82	157.84	0.7849	8	8	2	1	1	3
157.94		0.7847	5	6	5	5		18
159.91		0.7822	4	12	2	1		4
160.36		0.7817	4	2	6	1		2
160.61		0.7814	1	13	2	1		4
162.91	163.44	0.7789	5	12	1	6	0	20
164.06		0.7778	7	0	5	5		16
164.96		0.7769	8	9	1	5		15

TABLE 41. MISCELLANEOUS FRAMEWORK SILICATES

	Larsenite	Low Cordierite	Anhydrous Beryl	Milarite	Osumilite
Composition	$PbZnSiO_4$	$(Li,Na,Ca,K)_{0.27}(Mg,Fe,Mn)_2Si_5Al_4O_{18}(H_2O)_{0.48}$	$Be_2Al_3Si_6O_{18}$	$K_2Ca_4Be_4Al_2Si_{24}O_{60} \cdot H_2O$	$(K,Na,Ca)(Mg,Fe,Mn)_2(Si_{10}Al_5)O_{30} \cdot H_2O$
Source	from foundry slag	Guilford, Conn.	Synthetic	Val Giuf, Grisons, Switzerland	Sakkabira, Japan
Reference	Prewitt, Kirchner & Preisinger, 1967	Gibbs, 1966	Gibbs, Breck & Meagher, 1968	Ito, Morimoto & Sadanaga, 1952	Brown & Gibbs, 1969
Cell Dimensions					
$\underline{a}$ Å	8.244	17.083	9.212	10.54	10.155
$\underline{b}$ Å	18.963	9.780	9.212	10.54	10.155
$\underline{c}$ Å	5.06	9.335	9.187	13.96	14.284
α deg	90	90	90	90	90
β deg	90	90	90	90	90
γ deg	90	90	120	120	120
Space Group	$Pna2_1$	Cccm	P6/mcc	C6/mcc	P6/mcc
Z	8	4	2	1	2

TABLE 41. (cont.)

Variety	Larsenite	Low Cordierite	Anhydrous Beryl	Milarite	Osumilite
Site Occupancy		$M_1 \begin{cases} Na & 0.12 \\ Li & 0.12 \end{cases}$ M_2 Al$_1$ 1.0 M_3 Al$_2$ 1.0 $M_4 \begin{cases} Mg & 0.76 \\ Fe & 0.24 \end{cases}$		$T_1 \begin{cases} Si & 0.9 \\ Be & 0.1 \end{cases}$ $T_2 \begin{cases} Si & 0.4 \\ Be & 0.27 \\ Al & 0.33 \end{cases}$	$M_1 \begin{cases} K & 0.71 \\ Na & 0.24 \\ Ca & 0.05 \end{cases}$ $M_2 \begin{cases} Mg & 0.46 \\ Fe & 0.46 \\ Mn & 0.08 \end{cases}$ $T_1 \begin{cases} Si & 0.850 \\ Al & 0.150 \end{cases}$ $T_2 \begin{cases} Al & 0.876 \\ Fe & 0.124 \end{cases}$
Method	3-D, counter	3-D, counter	3-D, counter	3-D, film	3-D, counter
R & Rating	0.055 (1)	0.097 (1)	0.048 (1)	not given (3)	0.066 (1)
Cleavage and habit	Prismatic cleavage (?) Prismatic and sometimes tabular parallel to {010}	{100}distinct,{001} & {010}poor.Tabular parallel to {010};also short prisms elongated to c axis	None Prismatic	None Prismatic	None Prismatic or tabular
Comment	Not an olivine structure, but related	Minor Ca,K&Mn ignored in calculation. H$_2$O not located		B estimated	
μ	918.3	117.0	84.8	113.8	149.2
ASF	0.7590	0.1173	0.1838	0.7209×10^{-1}	0.8982×10^{-1}
Abridging factors	1:1:0.5	1:1:0.5	1:1:0.5	1:1:0.5	1:1:0.5

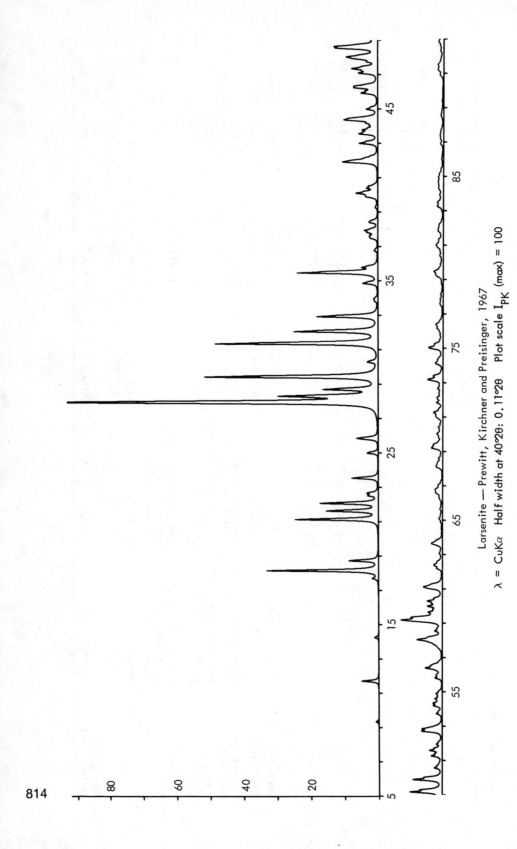

814

Larsenite — Prewitt, Kirchner and Preisinger, 1967

λ = CuKα   Half width at 40°2θ: 0.11°2θ   Plot scale I_{PK} (max) = 100

LARSENITE - PREWITT, KIRCHNER AND PREISINGER, 1967

2THETA	PEAK	D	H	K	L	I(INT)	I(PK)	I(DS)
9.32	9.32	9.481	0	2	0	1	1	0
11.69	11.70	7.560	1	1	0	6	5	2
14.22	14.24	6.221	1	2	0	1	1	1
17.67	17.68	5.016	1	3	1	2	2	1
18.13	18.14	4.889	0	4	0	40	33	21
18.70	18.70	4.741	0	1	1	10	9	6
21.11	21.12	4.205	1	4	0	31	25	20
21.54	21.60	4.122	2	1	0	2	15	1
21.61		4.110	1	4	1	18	17	12
22.05	22.06	4.028	2	0	0	22	17	15
22.49	22.50	3.950	0	3	1	3	3	2
22.63	22.64	3.926	1	2	1	4	3	3
23.51	23.52	3.780	2	2	0	10	8	8
24.97	24.98	3.562	1	2	1	4	3	4
25.78	25.84	3.453	2	3	0	3	6	3
25.84		3.445	1	5	0	6		5
27.89	27.94	3.196	2	0	1	55	100	55
27.94		3.190	1	4	1	100		100
28.21	28.30	3.160	0	6	0	12	30	12
28.29		3.151	2	1	1	31		32
28.67	28.68	3.111	1	4	0	21	16	22
29.41	29.42	3.035	2	5	1	75	51	82
29.47		3.028	1	6	1	2		3
30.26	30.26	2.951	2	3	0	4	3	5
31.34	31.34	2.852	1	5	1	71	48	85
31.38		2.848	3	1	1	3		4
32.04	32.04	2.791	2	5	0	37	25	46
32.90	32.90	2.720	3	0	0	28	18	36
33.80	33.80	2.650	2	4	0	1	1	1
33.93	33.92	2.639	3	2	0	1	1	2
34.83	34.84	2.574	1	7	0	7	24	10
35.45	35.46	2.530	0	0	2	38	5	56
35.77	35.76	2.508	2	6	0	6	1	9
36.74	36.74	2.444	2	5	1	1	2	2
37.45	37.52	2.399	1	1	2	2	2	4
37.51		2.396	3	1	1	2		3
37.63	37.62	2.388	0	7	1	2	2	2

2THETA	PEAK	D	H	K	L	I(INT)	I(PK)	I(DS)
37.81	37.82	2.377	3	0	0	4	4	7
37.93	37.92	2.370	1	8	1	4	4	6
38.37	38.38	2.344	1	2	2	3	3	5
38.43		2.340	3	2	1	2		3
39.24	39.24	2.294	1	1	1	1	1	1
39.87	39.90	2.259	3	1	1	4	4	2
39.93		2.256	2	6	1	2		7
40.09	40.10	2.247	3	3	0	10	6	4
40.37	40.38	2.232	2	5	0	5	3	17
40.50	40.48	2.225	3	0	2	2	10	8
41.86	41.90	2.156	1	4	2	3	5	3
41.90		2.154	1	8	1	13		5
41.95		2.152	3	2	1	4		26
42.98	42.98	2.103	2	7	0	9	6	8
43.53	43.54	2.077	1	9	0	9	5	19
43.77	43.76	2.066	2	8	0	6	3	18
44.03	44.04	2.055	4	2	0	3	4	13
44.16	44.16	2.049	1	9	0	10	3	5
44.34	44.36	2.041	2	3	2	3	7	6
44.35		2.041	2	5	2	4		22
44.38		2.039	3	5	0	6		5
44.44		2.037	1	0	0	8		9
44.97	44.98	2.014	4	6	0	10	6	8
45.91	45.90	1.9751	0	9	0	6	8	12
46.21	46.22	1.9628	2	4	2	9	3	18
46.29	46.30	1.9595	4	3	0	3	9	22
47.06	47.06	1.9292	3	6	2	10	13	14
47.28	47.32	1.9207	1	10	1	4	13	22
47.33		1.9188	3	6	1	9		8
47.73	47.74	1.9038	0	8	1	10	10	24
47.93	48.02	1.8963	2	8	0	18	7	9
48.02		1.8931	1	9	1	12		21
48.52	48.54	1.8745	2	2	1	17	9	25
48.61	48.62	1.8712	4	1	0	1	13	45
49.14	49.14	1.8524	3	10	0	16	13	30
49.27	49.26	1.8480	1	0	0	7	10	43
49.86	49.86	1.8273	4	3	1	16	9	42

LARSENITE - PREWITT, KIRCHNER AND PREISINGER, 1967

2THETA	PEAK	D	H	K	L	I(INT)	I(PK)	I(DS)
50.54	50.54	1.8042	1	7	2	4	2	11
51.24	51.24	1.7812	2	6	2	6	3	15
51.57	51.58	1.7706	4	4	1	7	4	18
51.94	51.94	1.7591	2	9	1	3	2	8
52.68	52.70	1.7359	1	10	1	11	6	30
52.79	52.82	1.7325	3	4	2	4	6	13
52.88		1.7298	0	8	2	3		8
53.71	53.72	1.7050	4	5	1	3	2	10
54.17	54.18	1.6916	3	8	1	4	3	13
54.32	54.32	1.6874	1	11	0	2	3	7
54.58	54.58	1.6800	0	1	3	3	2	10
54.86	54.86	1.6721	3	9	0	3	2	8
55.80	55.80	1.6462	1	1	3	4	2	12
56.02	55.96	1.6403	4	7	0	1	2	4
56.25	56.38	1.6339	4	6	1	1	5	24
56.37		1.6308	2	10	1	8		8
56.61	56.52	1.6244	5	5	0	2	4	5
57.93	58.02	1.5904	2	11	1	1	8	26
58.00		1.5887	1	9	2	8		18
58.05		1.5876	3	9	1	5		13
58.53	58.52	1.5757	4	2	2	4	2	24
59.08	59.16	1.5624	5	1	1	7	12	17
59.13		1.5610	2	0	3	5		7
59.14		1.5607	3	10	0	2		31
59.16		1.5604	1	4	3	9		18
59.16		1.5603	4	7	1	5	9	5
59.29	59.30	1.5573	5	4	0	1		13
59.35		1.5558	2	1	3	4		4
59.37		1.5553	4	8	0	1		22
59.63	59.62	1.5492	4	3	2	6	4	25
59.97	59.98	1.5411	0	5	3	7	4	28
60.28	60.28	1.5341	3	7	1	8	4	19
61.01	61.10	1.5174	0	10	2	5	6	28
61.09		1.5155	2	3	3	8		4
61.15		1.5142	4	4	2	1		4
62.15	62.18	1.4923	1	10	2	1	2	8
62.19		1.4914	3	10	1	2		

2THETA	PEAK	D	H	K	L	I(INT)	I(PK)	I(DS)
62.33	62.38	1.4884	5	4	1	2	3	7
62.41		1.4867	4	8	0	3		11
63.04	63.04	1.4733	3	9	2	2	1	6
63.49		1.4639	3	6	0	1		5
63.59	63.62	1.4618	5	6	0	4	3	17
63.67		1.4603	0	11	0	3		13
65.09	65.08	1.4318	3	7	3	2	1	6
65.63	65.64	1.4213	1	5	2	1	1	6
66.31	66.32	1.4084	5	7	0	2	1	6
66.55	66.58	1.4038	1	11	2	2	2	9
66.59		1.4031	3	11	1	1		5
66.67		1.4016	0	13	1	2		5
66.78	66.76	1.3996	2	9	3	6	2	9
67.03	67.00	1.3949	3	9	2	2	1	8
68.06	68.12	1.3763	3	7	2	2	2	8
68.10		1.3756	4	4	3	1		5
68.13		1.3751	2	13	0	1		5
68.19		1.3740	6	0	0	1		6
68.40	68.36	1.3704	6	1	0	1	2	7
68.59	68.60	1.3669	5	5	2	2	1	27
69.18	69.18	1.3569	5	2	1	6	3	11
69.25		1.3556	1	8	3	6		11
69.79	69.82	1.3465	2	11	2	1	1	6
70.01	70.00	1.3427	6	3	0	2	2	11
70.38	70.38	1.3366	1	14	1	3	1	12
70.97	71.08	1.3270	2	13	0	1	2	7
71.01		1.3262	5	4	2	2		8
71.09		1.3250	4	8	2	1		7
71.26	71.24	1.3223	3	12	1	4	3	17
71.82	71.82	1.3132	6	1	1	4	2	19
72.12	72.10	1.3085	3	6	3	3	2	14
72.43	72.34	1.3037	2	8	3	1	1	6
72.65	72.66	1.3002	1	9	3	2	2	12
73.12	73.16	1.2931	4	2	2	3	5	17
73.18		1.2922	1	14	1	8		41
74.11	74.10	1.2783	4	3	3	4	2	22
74.20		1.2770	6	4	1	1		7

LARSENITE - PREWITT, KIRCHNER AND PREISINGER, 1967

2THETA	PEAK	D	H	K	L	I(INT)	I(PK)	I(DS)
74.97	75.02	1.2657	5	6	2	4	4	23
75.02		1.2650	0	0	4	4		14
75.04		1.2648	3	11	2	5		24
75.48	75.48	1.2585	4	4	0	1	1	6
76.11	76.16	1.2496	1	15	1	1	2	6
76.14		1.2491	1	13	1	1		8
76.29	76.36	1.2471	2	14	1	1		6
76.38		1.2458	1	10	3	3	2	16
77.50	77.60	1.2306	5	1	2	1	1	7
77.61		1.2291	3	8	3	2		9
79.15	79.28	1.2090	1	4	4	1	2	7
79.22		1.2082	2	13	0	2		11
79.27		1.2074	6	1	2	2		10
79.32		1.2069	2	1	4	1		7
79.45	79.44	1.2052	2	10	3	2	3	12
79.47		1.2050	6	1	2	1		6
79.89	79.88	1.1996	2	2	4	1	1	8
80.39	80.38	1.1935	5	2	2	1	1	8
80.88	81.00	1.1875	3	9	3	2	2	9
81.00		1.1860	6	3	2	4		21
81.35	81.24	1.1818	1	14	2	2	1	12
81.77	81.84	1.1768	5	3	3	2	2	11
81.84		1.1759	4	7	3	1		9
81.97		1.1744	0	6	4	1		8
83.23	83.26	1.1598	5	11	1	3	2	21
83.30		1.1590	4	13	1	1		9
83.46	83.46	1.1572	6	8	1	2	1	14
83.91	83.90	1.1522	2	5	4	2	1	15
84.27	84.36	1.1481	3	13	1	2	2	12
84.37		1.1470	3	1	4	2		13
84.50	84.54	1.1455	3	10	3	1	2	7
84.70	84.68	1.1434	4	8	3	1	2	7
84.93	84.94	1.1409	2	12	0	2	2	16
85.12	85.12	1.1388	7	2	1	2	1	11
85.45	85.40	1.1353	1	7	4	1	1	7
85.99	86.00	1.1295	2	6	4	1	1	7
86.86	86.88	1.1204	1	15	2	1	1	9

2THETA	PEAK	D	H	K	L	I(INT)	I(PK)	I(DS)
86.90	86.90	1.1200	3	15	1	1		9
87.40	87.40	1.1149	7	4	2	1	1	15
88.60	88.56	1.1028	6	7	1	2	1	7
89.39	89.38	1.0952	3	14	2	3	1	8
90.87	90.90	1.0811	5	7	1	1	1	21
91.18	91.20	1.0782	7	6	1	3		12
91.50	91.48	1.0753	1	9	4	2	2	15
92.32	92.32	1.0679	5	13	1	2	2	16
92.83	92.84	1.0633	3	12	3	2	1	16
92.90		1.0628	4	3	3	2		13
93.37	93.40	1.0586	6	2	3	2	1	18
93.46		1.0579	3	7	1	2		14
93.91	93.94	1.0540	4	15	1	2	1	12
94.66	94.66	1.0475	1	14	3	4	1	27
95.37	95.56	1.0416	7	4	2	2		8
95.57		1.0400	5	12	0	4	1	30
96.36	96.36	1.0335	3	17	1	2	1	15
98.27	98.28	1.0185	6	10	2	2	1	13
99.42	99.42	1.0098	8	0	1	2	1	17
99.74	99.74	1.0074	7	9	1	2	1	16
100.68	100.70	1.0005	2	18	0	2	1	14
100.85		0.9993	0	3	5	1		5
101.02	101.00	0.9981	5	3	4	2	1	6
101.53	101.54	0.9945	8	5	0	2	1	13
101.69		0.9933	7	7	2	1		4
101.96	101.96	0.9914	4	15	1	1	1	7
102.16		0.9900	2	11	4	1		5
103.20	103.42	0.9828	2	5	5	1	1	7
103.22		0.9827	3	10	0	4		5
103.23		0.9826	1	4	5	1		11
103.26		0.9824	2	15	1	1		9
103.40		0.9815	2	1	5	1		6
103.42		0.9814	7	10	1	1		5
104.65	104.76	0.9732	5	11	3	3	2	11
104.72		0.9727	4	13	3	2		18
104.75		0.9725	0	18	2	2		18

817

LARSENITE - PREWITT, KIRCHNER AND PREISINGER, 1967

2THETA	PEAK	D	H	K	L	I(INT)	I(PK)	I(DS)
104.78		0.9723	1	19	1	2		14
104.88		0.9717	6	8	3	2		14
104.96		0.9711	2	3	5	1		13
105.82	105.82	0.9656	3	18	1	1	0	9
106.33	106.36	0.9623	5	16	0	1	1	6
106.41		0.9619	8	6	1	1		8
106.60	106.68	0.9607	7	2	3	1	1	7
107.23	107.26	0.9568	3	17	2	3	1	27
107.26		0.9566	5	16	4	2		20
107.33		0.9561	3	11	4	2		17
107.52	107.62	0.9550	7	11	1	1	1	5
107.82		0.9532	1	11	2	1		10
108.42	108.44	0.9496	8	2	2	1	1	10
108.46		0.9493	1	13	4	1		8
108.46		0.9493	3	15	3	1		7
108.68		0.9480	0	7	5	1		5
108.99	108.98	0.9462	7	4	5	1	1	11
109.20	109.20	0.9449	3	2	5	1	1	10
109.31		0.9443	7	12	0	1		5
109.74	109.70	0.9418	5	7	5	1	1	8
109.86		0.9411	1	5	4	1		6
110.32	110.28	0.9385	2	6	5	1	0	6
110.80	110.76	0.9358	7	5	3	1	0	5
111.62	111.78	0.9312	3	4	5	1	1	6
111.65		0.9310	2	13	4	1		9
111.72		0.9306	6	0	5	4		7
111.77		0.9304	7	10	2	1		11
111.92		0.9295	6	1	4	2		6
112.15	112.14	0.9283	7	12	1	1	1	16
112.57	112.64	0.9260	1	20	2	2	1	16
112.66		0.9255	8	5	2	3		33
112.78		0.9248	1	8	5	1		6
112.90	113.00	0.9242	5	8	4	1	1	8
113.04		0.9235	7	6	3	1		11
113.21		0.9226	1	19	2	1		8
113.94	113.96	0.9187	1	14	4	1	1	9
114.28	114.30	0.9170	5	13	3	1	1	15

2THETA	PEAK	D	H	K	L	I(INT)	I(PK)	I(DS)
114.77	114.74	0.9144	3	16	3	1	1	8
115.74	116.10	0.9096	7	7	3	1	1	6
115.75		0.9095	3	6	5	1		13
115.88		0.9089	5	17	1	2		26
116.03		0.9081	4	15	3	1		11
116.08		0.9079	2	8	5	1		8
116.14		0.9076	2	17	3	1		5
116.32		0.9067	1	9	5	1		7
117.81	117.84	0.8995	5	16	2	1	1	12
117.84		0.8994	9	4	0	1		9
117.88		0.8992	4	3	5	2		16
120.02	120.14	0.8893	5	14	3	1	1	8
120.09		0.8890	1	15	5	1		15
120.15		0.8887	7	14	0	1		10
120.37	120.36	0.8877	5	18	1	2	1	10
120.40		0.8876	1	10	5	2		18
120.92	120.94	0.8853	8	8	2	1	1	8
121.07		0.8847	7	12	2	1		12
121.31	121.30	0.8836	6	16	1	2	1	18
121.80	121.78	0.8815	3	10	5	1	1	10
122.12	122.28	0.8802	6	7	4	1		8
122.31		0.8794	8	0	3	2		19
122.67	122.74	0.8778	7	9	3	2	1	18
123.05	123.22	0.8762	3	14	4	1	1	7
123.27		0.8753	7	14	1	1		15
123.79	123.80	0.8732	2	18	3	1	1	16
123.91		0.8727	0	11	5	1		6
123.95		0.8726	2	10	5	1		9
124.85	124.88	0.8690	2	21	1	2	1	20
125.53	125.48	0.8663	6	8	4	1		8
125.66		0.8658	3	9	5	1		9
125.96		0.8646	8	4	4	1		7
126.75	126.90	0.8616	5	1	5	1	1	15
126.84		0.8613	4	7	5	1		7
126.89		0.8611	7	1	4	1		12
127.05		0.8605	7	10	3	1		7
127.07		0.8604	9	1	2	1		8

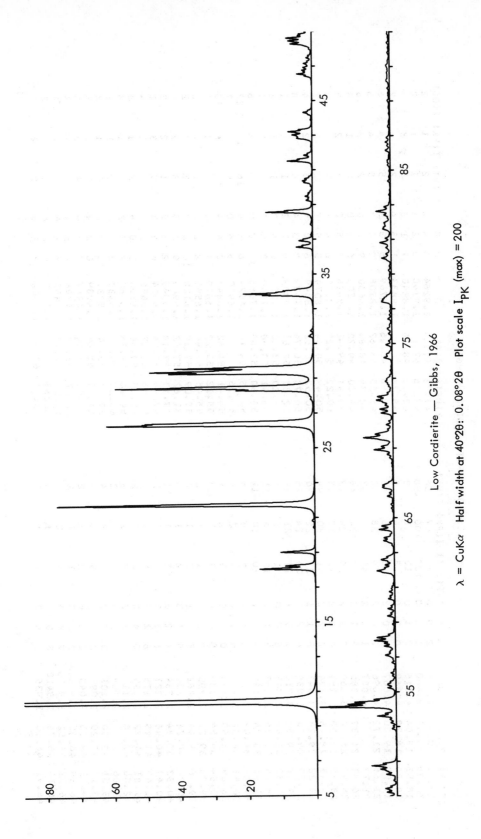

Low Cordierite — Gibbs, 1966

λ = CuKα   Half width at 40°2θ: 0.08°2θ   Plot scale I_{PK} (max) = 200

LOW CORDIERITE – GIBBS, 1966

2THETA	PEAK	D	H	K	L	I(INT)	I(PK)	I(DS)
10.35	10.36	8.541	2	0	0	58	67	58
10.45	10.46	8.460	1	1	0	100	100	100
18.03	18.04	4.916	3	1	0	10	8	10
18.20	18.20	4.869	0	0	2	5	4	5
19.00	19.00	4.667	0	2	0	6	5	6
21.68	21.72	4.096	2	0	2	9	38	10
21.73		4.087	1	1	2	45		47
26.31	26.32	3.385	3	1	2	42	31	45
26.43	26.42	3.369	0	2	2	25	25	27
28.30	28.30	3.151	4	2	0	13	12	15
28.45	28.46	3.134	2	0	1	41	30	44
29.28	29.28	3.047	1	1	1	27	22	30
29.39	29.38	3.036	5	1	0	24	24	26
29.58	29.58	3.018	4	2	2	27	21	29
31.39	31.40	2.847	1	3	1	1	1	1
33.76	33.76	2.653	6	0	0	10	8	11
33.86	33.86	2.645	5	1	2	10	10	11
34.02	34.02	2.633	4	2	2	5	4	5
36.53	36.52	2.458	1	3	2	4	2	4
36.89	36.90	2.434	6	2	0	3	2	3
38.54	38.54	2.334	0	0	4	11	7	13
39.23	39.22	2.295	0	1	1	1	1	2
39.46	39.46	2.282	7	3	3	2	1	2
40.24	40.24	2.239	5	1	1	1	1	2
40.33	40.34	2.234	4	3	3	2	1	1
40.47	40.46	2.227	1	1	3	1	1	1
41.49	41.48	2.175	6	4	2	6	4	7
41.81	41.82	2.159	0	1	2	2	1	4
42.80	42.80	2.111	7	4	1	3	2	3
42.86		2.108	0	1	1	3		3
43.01	43.00	2.101	3	3	4	3	3	4
43.19	43.20	2.093	5	4	2	5	4	6
44.29	44.30	2.043	2	2	4	2	1	2
46.39	46.39	1.9556	8	2	0	4	1	4
46.52	46.52	1.9506	7	3	2	2	2	2
46.74	46.74	1.9418	8	0	2	2	1	4
46.91	46.90	1.9351	1	5	0	1	1	2

2THETA	PEAK	D	H	K	L	I(INT)	I(PK)	I(DS)
47.13	47.14	1.9264	4	4	2	2	1	2
47.46	47.46	1.9140	8	1	1	1	1	2
48.26	48.28	1.8840	7	1	3	3	3	4
48.28		1.8833	1	3	4	2		2
48.46	48.46	1.8769	5	3	2	5	4	6
48.63	48.62	1.8708	2	4	3	5	2	6
48.84	48.84	1.8630	9	1	0	2	2	3
49.20	49.20	1.8503	6	4	1	3	1	3
49.41	49.42	1.8428	3	5	4	3	2	3
50.22	50.22	1.8150	6	0	1	2	1	2
50.43	50.44	1.8079	3	5	4	3	2	4
50.52	50.52	1.8049	6	3	0	3	2	8
50.73	50.74	1.7979	3	1	4	6	4	1
52.87	52.86	1.7303	9	3	2	1	1	2
53.20	53.20	1.7201	6	1	2	1	1	1
53.41	53.40	1.7140	3	6	0	1		6
53.60	53.60	1.7083	10	0	4	4	3	
54.15	54.16	1.6924	6	5	0		11	22
54.16		1.6920	5	1	4	16		7
54.41	54.42	1.6847	0	7	4	5	6	13
55.22	55.22	1.6619	8	0	3	10	1	3
55.45	55.44	1.6557	9	1	0	2	1	3
57.09	57.10	1.6120	10	2	3	1	1	2
57.62	57.62	1.5984	9	1	3	2	2	4
57.77	57.78	1.5945	2	6	3	3	3	7
57.94	57.94	1.5904	6	4	3	1	1	4
58.13	58.12	1.5856	4	5	3	5	3	3
59.35	59.34	1.5558	3	0	6	3	3	3
59.76	59.76	1.5460	6	3	2	2	1	2
60.33	60.32	1.5330	9	6	2	2	1	2
60.73	60.74	1.5237	10	2	4	1	1	3
61.84	61.86	1.4989	8	2	4	2	2	5
61.95	61.94	1.4967	7	4	4	4	3	3
62.27	62.28	1.4896	1	5	6	2	1	4
62.57	62.58	1.4833	3	1	6	3	2	2
62.63		1.4820	2	0	6	1		2
63.67	63.68	1.4602	2	2	6	2	2	3

LOW CORDIERITE - GIBBS, 1966

2THETA	PEAK	D	H	K	L	I(INT)	I(PK)	I(DS)
63.83	63.84	1.4570	11	1	2	1	1	2
64.31	64.32	1.4472	7	5	2	2	2	6
64.53	64.50	1.4428	4	6	2	4	2	3
66.22	66.22	1.4100	6	6	0	2	1	3
66.69	66.74	1.4012	5	1	6	2	2	3
66.75		1.4001	4	2	6	2		3
66.85	66.86	1.3983	1	3	6	1	2	2
68.63	68.62	1.3664	12	0	0	1	1	2
68.90	68.90	1.3617	12	2	0	5	3	8
69.50	69.60	1.3514	3	7	0	1	4	2
69.59		1.3498	6	2	2	8		12
69.87	69.80	1.3451	9	6	1	2	3	3
70.33	70.32	1.3375	3	5	4	4		5
71.00	71.00	1.3263	10	8	2	2	2	3
71.23	71.22	1.3226	8	0	4	2	2	3
71.61	71.62	1.3165	2	4	4	3	2	5
71.74	71.74	1.3146	6	2	6	1	2	2
71.76		1.3142	6	4	5	1		2
71.96	71.94	1.3110	0	4	6	3	2	2
72.66	72.66	1.3002	7	1	6	1	1	2
72.94	72.94	1.2958	2	4	3	3	1	4
74.17	74.16	1.2774	11	1	3	3	1	3
74.91	74.92	1.2665	1	7	0	2	1	2
75.55	75.54	1.2575	8	4	6	3	1	3
75.84	75.84	1.2533	4	1	7	2	1	3
77.37	77.40	1.2323	5	1	7	2	1	2
77.43		1.2316	4	3	7	3		3
77.52	77.52	1.2303	1	3	7	3	1	2
79.90	79.90	1.1995	10	4	4	3	1	6
80.51	80.50	1.1920	1	7	4	2	2	4
80.80	80.80	1.1884	12	2	2	1	1	3
81.57	81.58	1.1791	12	4	2	1	1	2
81.68	81.68	1.1779	0	8	2	2	1	3
81.94	81.94	1.1746	9	5	4	4	2	7
82.39	82.40	1.1695	3	7	4	1	1	2
82.61	82.62	1.1669	0	8	8	4	2	7
86.11	86.12	1.1282	9	3	6	1	1	2

2THETA	PEAK	D	H	K	L	I(INT)	I(PK)	I(DS)
86.60	86.60	1.1231	0	6	6	1	1	3
93.50	93.50	1.0575	8	8	0	2	1	3
93.89	93.88	1.0541	6	2	8	1	1	2
94.34	94.34	1.0503	12	0	6	1	1	3
94.99	95.00	1.0448	6	6	6	2	0	3
100.86	100.86	0.9993	1	5	8	1	1	1
106.02	106.02	0.9644	12	4	6	2		1
106.14		0.9635	10	0	8	1		1
106.43	106.40	0.9617	9	7	5	1	1	1
106.61	106.60	0.9606	5	5	8	1	1	2
106.92	106.92	0.9587	8	8	6	1	1	2
107.54	107.56	0.9549	17	3	7	1		1
107.62		0.9544	12	2	7	1	1	2
107.99	108.00	0.9522	16	2	8	1	1	2
108.02		0.9520	9	9	7	1		1
108.29	108.32	0.9503	13	7	1	1		2
108.48	108.44	0.9492	3	7	7	1		2
109.15	109.18	0.9452	10	4	1	1		1
109.27		0.9446	4	10	5	1	1	1
109.73	109.74	0.9419	14	2	8	1		2
109.77		0.9417	13	7	2	1		1
110.78	110.80	0.9359	1	1	9	1		1
110.98	110.96	0.9347	1	9	5	1		1
112.22	112.22	0.9279	3	1	10	1		2
114.25	114.26	0.9171	18	2	2	2	1	2
114.95	114.96	0.9135	2	2	10	1		2
115.34	115.34	0.9116	12	8	2	1	1	1
116.15	116.14	0.9075	6	4	9	1		3
116.71	116.88	0.9048	6	10	2	1		1
116.88		0.9040	10	4	8	1		1
117.92	117.92	0.8990	19	1	0	1	1	1
118.57	118.56	0.8959	2	10	4	1		2
118.71		0.8953	3	5	6	1		2
119.04	119.06	0.8938	16	6	0	1	0	2
119.35	119.38	0.8923	2	10	5	1	0	2
119.43		0.8920	16	6	6	2		2
120.46	120.48	0.8873	12	2	8	1	0	2

LOW CORDIERITE - GIBBS, 1966

2THETA	PEAK	D	H	K	L	I(INT)	I(PK)	I(DS)
120.91	120.92	0.8854	9	5	8	1	0	2
121.21	121.22	0.8841	1	11	0	1	0	1
121.41	121.40	0.8832	3	7	8	1	0	1
124.83	124.84	0.8690	13	1	8	1	0	1
125.82	125.84	0.8652	18	2	4	1	0	2
127.17	127.18	0.8600	12	8	4	1	0	3
127.99	127.98	0.8570	6	10	4	1	0	2
128.91	128.90	0.8537	17	3	5	1	0	2
129.41	129.42	0.8519	19	3	2	1	0	3
130.58	130.60	0.8479	15	7	3	1	0	2
131.63	131.66	0.8444	14	4	7	1	0	1
132.08	132.28	0.8429	5	11	2	1	0	2
132.31		0.8421	3	11	3	1		1
133.14	133.18	0.8395	7	7	7	1	0	1
133.22		0.8392	1	9	1	1		1
134.28	134.28	0.8359	19	1	4	1	1	1
134.67	134.68	0.8347	7	9	6	1		2
134.67		0.8347	19	3	3	1		3
135.03	135.06	0.8336	11	3	9	1	0	3
135.59	135.58	0.8320	13	9	4	1	0	1
136.33	136.30	0.8298	11	10	0	1	0	1
136.83	136.78	0.8283	8	6	9	1	0	1
137.53	137.52	0.8264	14	5	7	1	0	2
138.17	138.20	0.8246	9	9	0	2	0	5
138.82	138.88	0.8228	3	8	6	1	0	2
140.16	140.16	0.8193	18	0	7	1	0	2
141.48	141.44	0.8159	13	7	6	1	0	3
142.23	142.24	0.8141	0	12	0	1	0	2
143.31	143.30	0.8115	10	2	10	1	0	2
143.76	143.74	0.8104	4	10	6	1	0	3
144.91	144.98	0.8078	10	2	1	1	0	2
145.01		0.8076	21	1	1	1		3
147.16	147.18	0.8030	20	4	0	1	0	2
147.50	147.56	0.8023	9	11	0	1	0	2
147.59		0.8021	20	0	4	1		4
148.98	149.04	0.7994	9	11	1	1	1	5
149.05		0.7992	18	2	6	2		

2THETA	PEAK	D	H	K	L	I(INT)	I(PK)	I(DS)
149.24		0.7989	7	1	11	1		2
149.78	150.06	0.7978	2	4	11	1	0	2
150.01		0.7974	11	5	10	1	1	2
150.89	150.92	0.7958	7	5	10	3	1	7
151.13	151.20	0.7954	10	10	6	2		2
151.22		0.7952	12	8	6	2		4
151.30		0.7951	4	6	10	1		4
151.77	152.04	0.7942	20	4	2	1	1	4
152.60	152.56	0.7928	6	10	6	3	0	1
153.66	153.76	0.7911	16	8	2	1	1	7
155.83	155.84	0.7877	16	0	8	2	0	4
156.71	157.10	0.7864	13	9	4	1	0	1
157.05		0.7860	19	3	5	1	0	2
157.13		0.7859	4	12	2	1		3
158.61	158.78	0.7839	7	11	4	1	1	2
158.83		0.7836	8	8	8	5		5
160.39	160.42	0.7817	15	9	1	2	0	12
161.29	161.30	0.7806	12	0	10	7	1	18
162.72	163.42	0.7791	17	3	7	4	1	10
163.31		0.7785	8	2	11	1		1
163.43		0.7784	6	6	10	13		34
163.90		0.7779	0	0	12			12
164.12		0.7777	6	12	1	2		5
164.24		0.7776	16	2	8	1		3

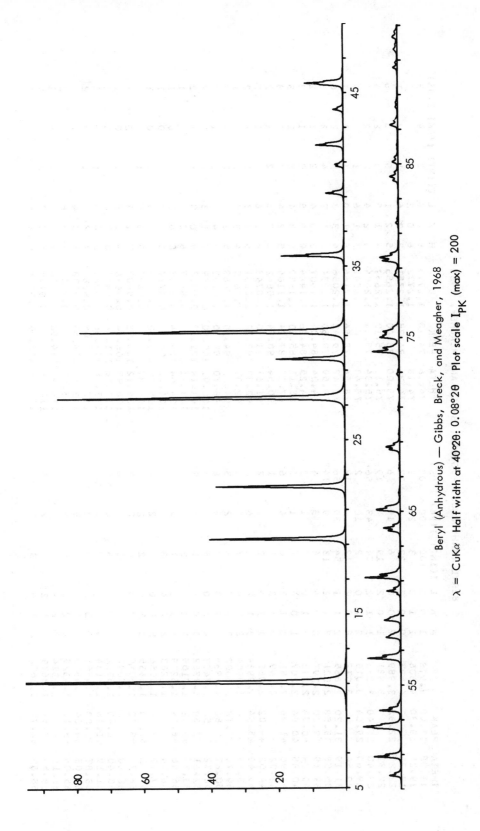

Beryl (Anhydrous) — Gibbs, Breck, and Meagher, 1968

λ = CuKα Half width at 40°2θ: 0.08°2θ Plot scale I$_{PK}$ (max) = 200

823

BERYL (ANHYDROUS) — GIBBS, BRECK AND MEAGHER, 1968

2THETA	PEAK	D	H	K	L	I(INT)	I(PK)	I(DS)
11.08	11.08	7.978	1	0	0	100	100	100
19.25	19.30	4.606	1	1	0	3	20	3
19.31		4.593	0	0	2	22		22
22.27	22.32	3.989	2	0	0	6	19	6
22.31		3.981	1	1	2	19		19
27.40	27.40	3.253	2	1	0	54	43	56
29.60	29.64	3.015	2	0	2	12	10	2
29.64		3.012	1	1	2	54	39	56
31.19	31.20	2.865	2	1	1	13	9	14
35.58	35.58	2.521	2	2	0	3	3	4
39.20	39.20	2.297	1	2	0	2	1	1
40.74	40.74	2.213	3	0	0	1	4	6
40.85	40.84	2.207	2	1	4	6		1
41.96	41.96	2.151	3	1	1	1		2
42.02		2.149	2	1	3	2	2	2
44.02	44.02	2.055	1	1	4	2	6	8
45.46	45.54	1.9934	3	1	2	7		3
45.53		1.9904	2	0	2	3	2	8
49.78	49.78	1.8302	3	2	0	5	4	6
50.87	50.86	1.7935	3	1	3	8	4	9
52.52	52.52	1.7409	4	1	0	5	6	6
52.61	52.60	1.7382	4	0	4	1	3	1
53.53	53.52	1.7105	3	1	1	8	5	9
56.48	56.54	1.6279	4	1	2	3		4
56.54		1.6263	2	2	4	2	2	2
57.73	57.74	1.5956	5	0	0	4	2	4
57.81	57.80	1.5935	3	2	4	2		2
58.72	58.72	1.5710	3	2	5	5	2	4
58.80		1.5690	2	1	6	6		12
60.40	60.40	1.5312	4	0	3	3	2	4
61.19	61.18	1.5134	3	3	0	10	5	4
63.87	63.88	1.4561	3	2	2	3	2	5
64.04	64.04	1.4530	4	2	2	3	3	2
65.05	65.10	1.4325	4	2	4	3	4	5
65.11		1.4314	3	2	4	4		5
68.54	68.54	1.3679	5	1	5	3	2	3
68.69	68.70	1.3652	2	1	6	3	2	4
74.16	74.18	1.2775	5	2	0	2	4	3
74.18		1.2772	6	0	2	6		7
74.33	74.38	1.2751	5	2	1	1	2	1
75.00	75.00	1.2653	5	2	1	5	3	6
75.11	75.10	1.2637	4	1	5	3	2	4
75.34	75.34	1.2604	4	2	4	2	2	3
75.43		1.2591	3	1	6	1		2
78.56	78.56	1.2166	6	1	0	5	1	6
79.42	79.42	1.2056	4	3	3	4	2	6
79.59	79.60	1.2034	2	1	7	2	3	2
84.04	84.04	1.1507	6	0	4	2	1	3
84.25	84.26	1.1484	4	0	8	2	1	3
85.11	85.10	1.1389	5	3	4	2	1	3
87.25	87.24	1.1164	3	2	6	1	1	2
90.54	90.54	1.0842	4	3	6	2	1	2
91.61	91.60	1.0743	5	2	5	1	1	2
92.30	92.32	1.0681	4	3	1	2	1	2
92.37		1.0675	5	3	5	2		2
94.51	94.50	1.0489	5	4	0	1	1	1
98.70	98.70	1.0152	7	1	2	1	1	2
100.91	100.90	0.9989	8	0	3	1	1	2
101.14	101.18	0.9972	6	2	0	2		2
101.21		0.9967	8	0	2	1		1
104.44	104.44	0.9745	8	2	8	2	0	1
104.71	104.72	0.9727	3	4	2	1	0	1
105.29	105.30	0.9690	5	4	8	1	0	1
105.37		0.9685	5	3	5	1		1
106.72	106.72	0.9600	7	1	4	1	0	1
106.93	106.94	0.9586	4	1	8	1	0	1
111.16	111.18	0.9337	8	1	0	1	0	1
111.46	111.46	0.9321	5	0	8	1	0	1
112.06	112.06	0.9287	7	2	3	2	0	2
113.55	113.60	0.9208	6	3	4	2	1	2
113.64		0.9203	4	4	6	1		1
114.47	114.58	0.9160	7	1	5	1	1	1
114.58		0.9154	5	2	7	3		3
114.95	115.00	0.9135	4	2	8	1	1	1

BERYL (ANHYDROUS) - GIBBS, BRECK AND MEAGHER, 1968

2THETA	PEAK	D	H	K	L	I(INT)	I(PK)	I(DS)
117.03	117.04	0.9032	5	5	2	1	0	1
117.50	117.50	0.9010	1	1	10	1	0	2
118.30	118.30	0.8972	7	2	4	1	0	1
119.14	119.14	0.8933	7	3	1	3	1	5
119.26		0.8928	5	4	5	1		2
122.02	122.02	0.8806	4	1	9	2	1	3
122.44	122.48	0.8788	2	1	10	1	0	1
128.48	128.50	0.8552	8	2	2	1	0	3
128.82	128.82	0.8540	8	2	8	2	1	1
130.04	130.04	0.8497	5	4	6	1	0	3
134.37	134.38	0.8356	8	0	6	2	1	3
134.54		0.8351	6	1	8	1		2
138.72	138.76	0.8231	7	1	7	1	0	1
142.19	142.22	0.8142	7	4	2	2	1	4
142.28		0.8140	8	2	4	1		3
142.88	143.04	0.8125	4	1	10	1	0	1
145.33	145.40	0.8069	8	3	1	2	1	4
145.39		0.8068	9	1	3	2		4
145.52		0.8065	7	3	5	2		3
145.70		0.8061	5	4	7	2		4
149.36	149.38	0.7986	7	4	3	2	0	3
149.70	149.92	0.7980	6	3	7	1	0	1
149.84		0.7977	8	3	2	1		2
149.98		0.7975	5	2	9	2		3
155.40	155.40	0.7883	3	3	10	1	0	2
157.01	157.12	0.7860	10	0	2	1	0	1
157.14		0.7858	9	1	4	2		4
158.11	158.14	0.7845	4	2	10	4	0	6
159.12	159.18	0.7832	9	2	3	3	1	5
159.22		0.7831	8	3	3	3		5
160.64	160.72	0.7814	3	1	11	2	0	4
163.49	163.52	0.7783	7	4	4	1	0	2
164.24	164.12	0.7776	7	1	8	1	0	2

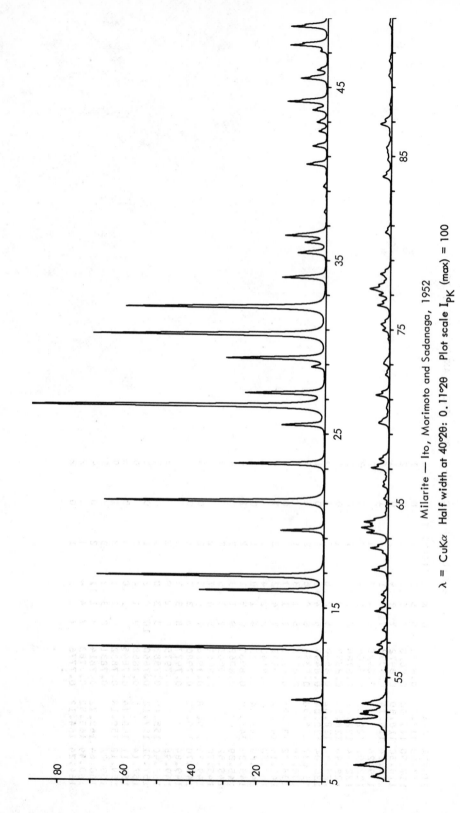

Milarite — Ito, Morimoto and Sadanaga, 1952

λ = CuKα   Half width at 40°2θ: 0.11°2θ   Plot scale I_{PK} (max) = 100

MILARITE -ITO, MORIMOTO AND SADANAGA, 1952

2THETA	PEAK	D	H	K	L	I(INT)	I(PK)	I(DS)
9.68	9.68	9.128	1	0	0	7	9	7
12.67	12.68	6.980	0	0	2	56	71	53
15.97	15.98	5.545	1	1	0	31	38	30
16.81	16.82	5.270	1	0	2	58	68	55
19.43	19.44	4.564	2	0	0	11	13	11
21.11	21.12	4.206	1	1	2	61	66	59
23.27	23.28	3.820	2	0	2	26	27	26
25.50	25.50	3.490	0	0	4	13	13	13
26.59	26.60	3.349	2	1	1	100	100	100
27.33	27.34	3.260	1	1	4	24	24	24
28.84	28.84	3.093	2	1	2	4	4	4
29.33	29.34	3.043	3	0	0	32	30	32
30.70	30.70	2.910	1	1	4	74	70	76
32.26	32.26	2.772	2	2	0	53	60	55
32.27		2.771	3	1	0	14		14
33.99	34.00	2.635	2	2	2	15	13	15
35.43	35.42	2.532	3	1	2	10	9	10
36.02	36.02	2.491	3	1	1	7	6	7
36.41	36.42	2.465	2	2	4	14	12	15
39.25	39.26	2.293	3	0	1	1	1	1
40.53	40.54	2.224	2	1	3	8	6	8
41.57	41.58	2.170	2	1	5	5	4	5
42.43	42.44	2.128	1	1	6	3	3	3
42.97	42.98	2.103	2	2	4	4	3	4
43.67	43.68	2.071	3	1	1	5	4	6
44.16	44.16	2.049	3	2	0	15	12	17
45.50	45.50	1.9919	4	1	1	10	8	12
45.99	45.98	1.9719	4	1	1	4	3	4
47.07	47.08	1.9290	2	1	6	2		3
47.42	47.42	1.9154	4	1	2	15	11	17
48.50	48.50	1.8754	3	0	5	15	11	17
49.26	49.26	1.8482	3	1	6	6	4	7
49.91	49.92	1.8256	5	0	0	13	10	15
52.01	52.02	1.7567	3	0	8	1	1	1
52.39	52.40	1.7450	0	2	6	9	16	11
52.42		1.7441	2	2	2	13		16
52.88	52.88	1.7299	4	1	4	10	8	12

2THETA	PEAK	D	H	K	L	I(INT)	I(PK)	I(DS)
53.04	53.02	1.7250	4	2	0	3	7	4
53.41	53.48	1.7140	1	0	8	2	7	3
53.48		1.7120	4	2	1	9		11
56.40	56.42	1.6299	2	0	8	2		2
56.43		1.6292	4	0	6	3	4	4
56.87	56.88	1.6176	5	1	2	3	3	4
57.71	57.72	1.5960	3	3	4	2	1	3
58.80	58.80	1.5691	5	3	0	5	4	7
59.32	59.32	1.5565	3	2	6	2	2	2
59.75	59.74	1.5464	4	0	4	2	2	2
61.17	61.18	1.5137	3	3	1	7	5	9
62.16	62.16	1.4920	6	0	2	2	2	3
62.42	62.42	1.4864	4	2	5	8	5	10
63.32	63.32	1.4675	5	2	2	9	6	12
63.60	63.60	1.4616	5	2	0	10	8	14
63.93	63.94	1.4549	2	2	8	9		13
63.99		1.4537	5	1	3	1	7	1
65.27	65.28	1.4282	4	3	1	2	1	3
65.98	65.98	1.4147	4	1	7	2	2	3
66.26	66.24	1.4093	3	1	6	2	2	2
66.65	66.66	1.4020	6	0	4	7	1	11
67.05	67.06	1.3946	4	0	8	6	5	2
67.51	67.56	1.3862	6	0	1	4	3	4
67.57		1.3851	4	1	0	3		5
69.61	69.64	1.3495	6	1	10	4	3	4
69.69		1.3482	5	1	4	3	1	1
70.48	70.48	1.3349	2	0	10	1	4	10
71.24	71.24	1.3226	3	1	9	9	1	
72.41	72.40	1.3040	7	0	1	1	2	3
73.02	73.04	1.2946	4	4	2	2		2
73.13		1.2929	4	4	4	4		5
74.96	74.96	1.2658	6	1	0	2	2	2
75.27	75.30	1.2614	6	0	8	8	3	5
75.32		1.2606	6	2	1	1		2
76.38	76.38	1.2458	6	1	5	6	4	9
76.40		1.2455	6	2	2	1		2
76.97	76.98	1.2377	5	2	6	5	3	8

MILARITE —ITO, MORIMOTO AND SADANAGA, 1952

2THETA	PEAK	D	H	K	L	I(INT)	I(PK)	I(DS)
77.28	77.34	1.2336	2	2	10	2	6	3
77.35		1.2326	4	4	4	9		13
77.79	77.78	1.2268	4	2	8	2	2	3
78.19	78.18	1.2214	6	2	3	4	3	5
80.59	80.60	1.1911	2	1	11	2	2	4
83.80	83.80	1.1534	4	1	2	4	3	7
84.79	84.80	1.1424	7	1	9	2	1	3
85.22	85.22	1.1378	4	3	8	1	1	3
85.38	85.48	1.1360	1	1	12	2	2	2
85.50		1.1347	6	3	2	1		2
86.20	86.22	1.1273	2	0	12	1		2
86.85	86.86	1.1205	5	3	8	8	3	13
89.62	89.66	1.0929	3	3	10	1	1	2
89.69		1.0922	6	3	4	3		2
90.51	90.52	1.0844	7	2	3	1	1	2
92.62	92.64	1.0652	8	1	1	2	1	3
93.90	93.90	1.0540	5	5	0	3	1	5
96.06	96.06	1.0360	6	1	9	2	1	4
96.98	97.00	1.0286	6	0	10	2	1	3
97.48	97.42	1.0246	6	2	8	1	1	2
101.06	101.10	0.9978	8	1	5	3	1	5
101.15		0.9971	0	0	14	1		2
101.32	101.34	0.9959	8	2	0	2	1	3
101.62		0.9938	7	1	8	1		2
102.36	102.38	0.9886	3	1	13	1	0	1
102.75	102.76	0.9860	8	2	2	1	1	2
103.52	103.54	0.9807	6	2	9	1	1	2
103.57		0.9804	6	4	5	1		2
107.00	107.04	0.9582	4	4	10	2	1	3
107.08		0.9577	8	2	4	1		1
107.55	107.46	0.9548	6	4	6	1	1	1
108.76	108.74	0.9476	3	0	14	1	0	1
111.36	111.34	0.9326	2	2	14	1	0	1
112.80	112.90	0.9248	8	0	12	2		2
112.92		0.9241	6	0	12	2		3
114.55	114.58	0.9156	8	2	6	3	1	5
114.87	114.96	0.9139	7	1	10	1	1	1

2THETA	PEAK	D	H	K	L	I(INT)	I(PK)	I(DS)
114.95		0.9135	7	4	4	2		4
115.32	115.32	0.9116	4	2	13	1	1	3
115.86	115.84	0.9089	8	3	3	2	1	3
117.24	117.24	0.9022	5	5	8	1	0	1
118.23	118.18	0.8975	9	2	1	1	1	1
120.40	120.42	0.8876	6	3	10	2	0	4
121.44	121.44	0.8831	10	0	4	1	1	1
122.19	122.22	0.8798	8	0	9	3	0	5
122.25		0.8796	8	3	5	1	1	2
123.72	124.04	0.8735	3	1	15	1		2
123.77		0.8733	4	3	13	1		2
123.97		0.8725	0	0	16	3		5
124.08		0.8720	4	2	12	3		1
124.22		0.8715	6	6	2	2		2
125.14	125.06	0.8678	6	4	9	1	0	2
125.87	125.86	0.8650	8	2	8	3	1	6
127.24	127.22	0.8598	10	2	2	1	0	2
128.25	128.26	0.8561	9	2	5	3	1	5
129.46	129.48	0.8518	6	6	4	2	1	5
129.89	129.94	0.8502	6	1	13	1	1	1
132.61	132.68	0.8412	5	5	10	2	0	2
132.70		0.8409	10	1	4	1	0	3
133.51	133.38	0.8383	7	1	12	1		1
134.92	134.86	0.8340	6	0	14	2		1
136.85	136.88	0.8283	8	2	16	2	0	2
138.49	138.50	0.8237	10	2	14	4	0	9
140.31	140.38	0.8189	7	5	13	3	1	6
140.49		0.8184	8	3	12	1	1	3
140.71		0.8178	9	0	12	1		1
142.02	142.02	0.8146	8	0	9	1		1
142.11		0.8144	7	6	9	1		3
144.56	144.54	0.8086	7	6	1	2		4
149.07	149.58	0.7992	2	1	16	2	0	4
149.24		0.7989	2	1	17	2		4
149.48		0.7984	8	0	12	1		2
149.69		0.7980	10	2	4	2	1	6
149.71		0.7980	7	1	3	1		2

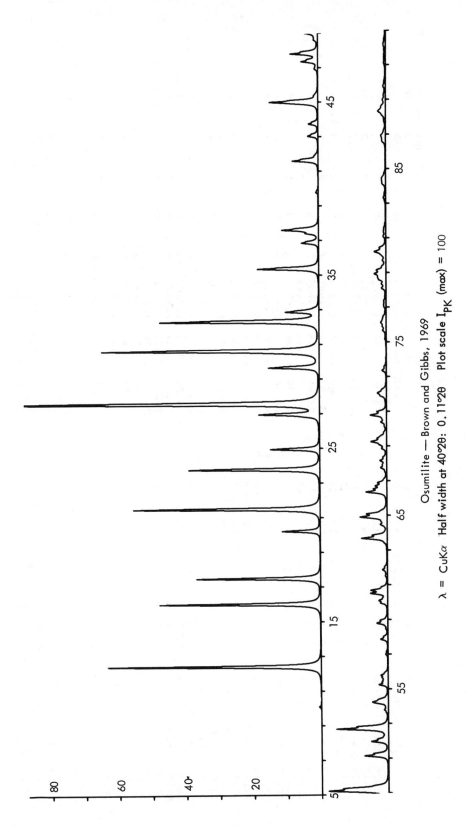

Osumilite — Brown and Gibbs, 1969

λ = CuKα Half width at 40°2θ: 0.11°2θ Plot scale I$_{PK}$ (max) = 100

OSUMILITE – BROWN AND GIBBS, 1969

2THETA	PEAK	D	H	K	L	I(INT)	I(PK)	I(DS)
12.38	12.40	7.142	0	0	2	50	63	44
15.97	15.98	5.544	1	0	0	39	48	35
17.45	17.46	5.077	1	1	0	30	37	28
20.18	20.18	4.397	2	0	0	10	11	9
21.45	21.46	4.138	1	1	1	49	55	46
23.74	23.74	3.744	2	0	2	36	39	35
24.91	24.92	3.571	0	0	4	14	15	13
26.92	26.92	3.309	1	0	4	16	18	16
27.53	27.54	3.237	2	1	1	100	100	100
29.62	29.62	3.014	2	1	2	15	15	15
30.58	30.58	2.921	1	1	4	64	65	67
32.27	32.26	2.772	2	0	4	51	47	54
32.83	32.84	2.726	2	1	3	10	10	11
35.32	35.32	2.539	2	2	0	20	18	22
36.82	36.82	2.439	3	1	1	6	4	6
37.37	37.38	2.404	3	0	4	4	4	4
37.57	37.56	2.392	2	2	2	12	11	14
41.56	41.56	2.171	3	1	3	9	8	11
43.01	43.00	2.101	4	0	2	4	3	5
43.71	43.72	2.069	2	2	4	3	3	3
44.89	44.98	2.018	3	0	0	5	3	4
44.97		2.014	3	1	4	15	15	19
46.90	46.90	1.9355	2	1	0	6	1	1
47.33	47.32	1.9191	4	1	1	6	5	8
47.78	47.78	1.9020	4	1	1	10	8	13
48.59	48.60	1.8722	4	0	4	1	1	1
49.07	49.08	1.8550	3	1	5	18	18	24
49.11		1.8534	4	1	0	7		9
49.26	49.20	1.8480	3	0	6	6	14	8
51.11	51.12	1.7855	3	0	8	9	7	13
51.94	51.94	1.7589	0	0	6	6	5	9
52.66	52.66	1.7366	5	0	0	21	16	28
53.76	53.76	1.7037	3	1	6	1	1	1
54.14	54.20	1.6925	3	3	0	3	5	4
54.21		1.6905	4	1	4	5		7
55.22	55.22	1.6620	4	2	0	3	3	5
55.50	55.50	1.6543	2	0	8	1	2	3

2THETA	PEAK	D	H	K	L	I(INT)	I(PK)	I(DS)
55.62	55.64	1.6509	4	2	1	1	2	2
55.77	55.76	1.6469	3	3	2	1	2	3
57.83	57.84	1.5930	4	1	1	3	2	3
58.76	58.76	1.5700	5	1	1	4	3	4
59.92	59.94	1.5423	5	1	2	2	2	6
60.06	60.06	1.5392	5	1	6	2	3	4
60.48	60.48	1.5294	3	0	8	7	5	3
60.68	60.66	1.5249	5	1	4	4	5	10
61.83	61.84	1.4992	4	2	8	2	1	5
63.66	63.66	1.4605	5	2	3	11	8	18
64.45	64.44	1.4445	4	2	2	3	2	5
64.85	64.88	1.4366	4	1	8	3	8	5
64.88		1.4358	6	0	0	9		14
65.07	65.06	1.4322	5	1	2	2	6	4
66.32	66.32	1.4082	5	2	0	2	6	15
66.68	66.68	1.4015	5	1	7	4	4	7
66.87	66.86	1.3980	4	1	5	2	3	2
67.73	67.72	1.3823	5	1	5	2	2	3
68.14	68.14	1.3750	5	1	10	3	2	5
68.83	68.84	1.3628	4	2	6	2	2	3
69.23	69.24	1.3560	6	0	4	7	5	11
69.55	69.42	1.3504	5	2	3	2	3	3
70.35	70.34	1.3371	3	1	8	2	2	4
70.76	70.76	1.3303	3	1	9	8	5	13
71.88	71.90	1.3124	3	1	10	2	2	3
72.02	72.02	1.3101	5	1	4	10	3	6
73.32	73.32	1.2900	4	3	5	3	1	4
75.63	75.64	1.2564	7	0	2	2	1	2
76.09	76.10	1.2498	4	4	0	1	2	3
78.07	78.08	1.2231	4	4	1	2	1	4
78.57	78.58	1.2165	6	2	9	2	2	4
78.67	78.70	1.2152	6	2	8	1	3	2
78.70		1.2148	5	3	13	3		3
78.91	78.92	1.2121	2	1	6	4	4	7
79.11	79.12	1.2095	1	11	11	3	3	5
79.69	79.70	1.2022	2	2	2	1	1	3
80.04	80.18	1.1978	4	0	10	1	3	3

OSUMILITE – BROWN AND GIBBS, 1969

2THETA	PEAK	D	H	K	L	I(INT)	I(PK)	I(DS)
80.18		1.1961	4	4	4	6		11
81.07	81.08	1.1851	7	0	1	1	1	1
84.10	84.14	1.1501	5	3	3	1	1	1
86.55	86.66	1.1236	4	3	4	1	1	2
86.66		1.1225	5	1	5	3		3
88.14	88.32	1.1074	7	1	4	1	2	3
88.31		1.1057	5	0	8	4		9
88.96	88.96	1.0993	8	0	0	2	1	4
89.76	89.76	1.0916	3	3	10	1	1	3
92.22	92.20	1.0688	4	3	9	1	1	2
93.28	93.28	1.0594	5	1	10	2	1	3
97.69	97.70	1.0230	6	0	0	1	1	4
98.04	98.02	1.0203	0	0	14	1	1	3
99.89	99.90	1.0063	6	4	1	1	0	2
102.30	102.32	0.9890	8	1	4	1	0	2
103.72	103.72	0.9793	7	2	6	1	1	3
105.39	105.40	0.9684	8	1	5	2	1	4
108.14	108.18	0.9512	6	4	5	1		4
108.18		0.9510	8	2	2	2		4
108.54	108.54	0.9488	4	4	10	1	1	5
108.90	108.88	0.9467	2	2	14	1	0	3
109.80	109.80	0.9415	6	3	8	1	1	1
112.29	112.44	0.9275	6	2	10	1		2
112.44		0.9267	8	2	4	1		7
112.94	112.90	0.9240	6	0	12	3	1	4
113.58	113.34	0.9206	7	4	2	2	1	2
114.01	114.00	0.9184	5	4	9	1	1	3
114.58	114.46	0.9154	2	1	15	1	0	1
115.55	115.60	0.9105	3	2	14	1	0	2
116.87	117.24	0.9040	9	1	0	1	1	2
117.13		0.9027	7	1	10	1		2
117.29		0.9020	5	1	13	1		2
119.60	119.86	0.8912	8	3	1	2	0	2
119.87		0.8900	8	1	6	1		3
120.53	120.50	0.8871	3	1	15	1	0	3
121.56	121.54	0.8826	5	0	14	1	0	2
122.56	122.66	0.8783	6	4	8	1		2

2THETA	PEAK	D	H	K	L	I(INT)	I(PK)	I(DS)
122.72		0.8776	8	4	3	1		2
124.90	124.96	0.8688	7	4	5	1	1	3
125.02		0.8683	4	4	12	1		3
125.53	125.50	0.8663	8	3	4	1		1
126.23	126.18	0.8636	8	1	9	1	0	2
127.26	127.26	0.8597	9	0	6	1	0	2
127.94	127.88	0.8572	9	1	8	1	0	3
129.31	129.50	0.8523	8	3	5	2	1	4
129.57		0.8514	6	6	0	1		5
131.06	131.34	0.8462	6	4	9	1		2
131.37		0.8452	8	2	8	1		3
132.86	132.88	0.8404	6	6	2	1		2
133.80	133.74	0.8374	6	0	14	1		2
134.55	134.54	0.8351	8	1	1	1		2
135.13	135.16	0.8333	10	1	2	1		2
136.57	136.60	0.8291	10	1	1	1		2
137.27	137.56	0.8271	5	2	13	4	1	11
137.58		0.8262	6	6	4	1		2
138.59	138.28	0.8234	6	1	8	1		2
140.20	140.18	0.8192	2	1	17	1		4
142.01	142.70	0.8146	7	3	10	1		2
142.52		0.8134	9	1	4	1		2
142.65		0.8131	10	1	4	1		2
142.75		0.8128	9	3	1	3		3
143.21		0.8117	4	1	16	1		2
144.19	144.06	0.8095	8	4	4	1	1	2
144.22		0.8094	9	3	3	1		2
147.92	147.98	0.8014	8	2	10	3		10
150.49	150.52	0.7965	8	5	4	1		4
151.84	151.66	0.7941	7	3	14	2		6
153.07	153.06	0.7920	5	0	6	2		5
153.07		0.7920	7	1	14	1		2
153.88	156.50	0.7907	7	4	9	1		2
156.30		0.7870	7	3	11	3		2
156.73		0.7864	5	4	13	1		2
160.11	160.84	0.7820	9	3	5	5		15
161.02	161.32	0.7809	2	0	18	1	0	3

TABLE 42. MISCELLANEOUS FRAMEWORK SILICATES

Variety	Bertrandite	Bazzite	Zn-Chkalovite	Bavenite	Phenacite
Composition	$Be_4Si_2O_7(OH)_2$	$(Sc,Fe^3,Al)_2Si_3[Si_6 O_{18}](Y,Na,Ca,H_2O)$	$Na_4(Zn,Cd)_2[Si_2O_6]_2$	$Ca_4Be(OH)_{2-x}Al_{2-x} Si_9O_{26-x}$	Be_2SiO_4
Source	not given	Baveno, Italy	Synthetic	Baveno, Italy	not given
Reference	Solov'eva & Belov, 1965	Peyronel, 1956	Simonov & Belov, 1965	Cannillo, Coda and Fagnani, 1966	Bragg & Zacharia-sen, 1930
Cell Dimensions					
a Å	8.73	9.51	21.513	23.19	7.70A (7.68 kx)
b Å	15.31	9.51	7.125	5.005	7.70A (7.68 kx)
c Å	4.56	9.11	7.401	19.39	7.70A (7.68 kx)
α	90	90	90	90	108.0
β deg	90	90	90	90	108.0
γ	90	120	90	90	108.0
Space Group	$Cmc2_1$	P6/mcc	Fdd2	Cmcm	$R\bar{3}$
Z	4	2	4	4	6

TABLE 42. (cont.)

Variety	Bertrandite	Bazzite	Zn-Chkalovite	Bavenite	Phenacite
Site Occupancy		Scattering factors used: (Sc,Fe,Al) = $f Sc^{+3}$ and $(Y,Na,Ca,H_2O) = f Fe^{+2}$			
Method	3-D, film (?)	2-D, film	3-D, counter	3-D, film	2-D, film
R & Rating	0.089 (2)	0.16 (av.) (3)	0.116 (2)	0.066 (1)	--- (2)
Cleavage and habit	{110} perfect Tabular parallel to {001} and sometimes {010}	None Vesicular habit, barrel-shaped prisms	Prismatic parallel to c axis	{001}very good;{100} good. Plates parallel to {100} & elongated to b axis	{11$\bar{2}$0}moderate Rhombohedral
Comment		A Sc-beryl not chemically analyzed		X in formula is 0.1	B estimated
μ	56.5	193.7	124.1	144.0	66.7
ASF	0.6941×10^{-1}	0.1889	0.3437	0.9253×10^{-1}	0.7342×10^{-1}
Abridging factors	1:1:0.5	1:1:0.5	0.5:1:0.3	1:1:0.5	1:1:0.5

833

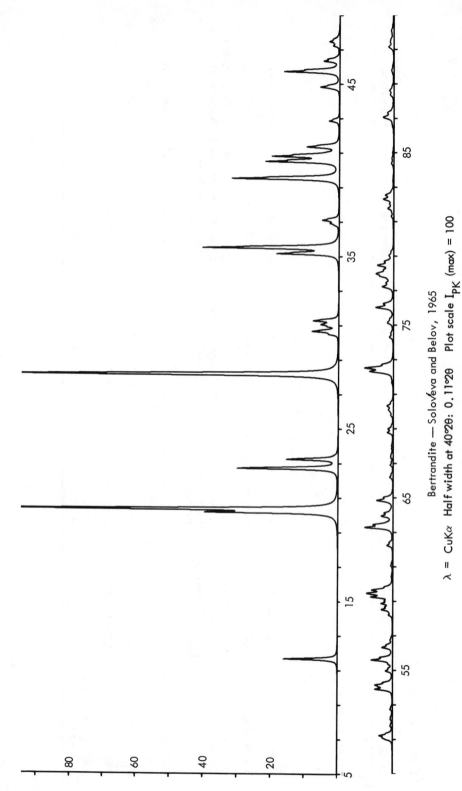

Bertrandite — Solovéva and Belov, 1965

$\lambda = CuK\alpha$ Half width at $40°2\theta$: $0.11°2\theta$ Plot scale I_{PK} (max) = 100

834

BERTRANDITE - SOLOV'EVA AND BELOV, 1965

2THETA	PEAK	D	H	K	L	I(INT)	I(PK)	I(DS)
11.66	11.66	7.584	1	1	0	12	16	12
20.14	20.14	4.406	1	3	0	30	40	30
20.33	20.34	4.365	2	0	0	87	100	86
22.68	22.68	3.918	0	2	1	26	30	26
22.73		3.908	1	1	0	5		4
23.22	23.22	3.828	0	4	0	14	16	14
28.14	28.14	3.168	1	3	1	100	99	100
30.46	30.48	2.932	0	4	1	2	3	2
30.64	30.64	2.916	2	2	1	8	8	8
30.92	30.94	2.889	1	5	0	3	4	3
31.05	31.04	2.878	2	4	0	4	5	4
31.26	31.26	2.859	3	1	0	8	8	8
35.14	35.14	2.552	0	6	0	20	19	20
35.48	35.48	2.528	3	3	0	45	41	45
36.90	36.90	2.434	2	4	1	5	5	5
37.08	37.08	2.422	3	1	1	5	5	5
39.49	39.50	2.280	0	0	2	38	32	39
40.47	40.48	2.227	0	6	1	25	22	26
40.78	40.78	2.211	3	3	1	22	20	22
40.93	40.88	2.203	2	6	0	3	14	3
41.28	41.34	2.185	0	2	2	1	10	1
41.31		2.183	1	1	2	1		1
41.33		2.182	4	0	0	9		10
42.83	42.84	2.109	3	5	0	4	3	4
44.72	44.82	2.025	1	3	2	1	6	2
44.81		2.021	2	0	2	6		6
45.70	45.70	1.9836	2	6	1	21	17	22
46.31	46.32	1.9588	0	4	2	5	5	6
47.21	47.22	1.9236	1	7	1	2	2	2
47.45	47.46	1.9145	3	5	1	2	3	2
47.47		1.9138	0	8	0	2		2
50.98	51.08	1.7899	1	5	2	2	3	2
51.06		1.7871	2	4	2	2		3
51.20	51.20	1.7825	3	1	2	3	4	4
53.88	53.88	1.7002	0	6	2	7	5	7
54.12	54.12	1.6931	3	3	2	6	5	6
54.94	54.94	1.6697	1	9	0	3	2	3
55.34	55.36	1.6586	4	6	0	1	1	1
55.58	55.58	1.6520	5	3	0	9	6	9
56.31	56.30	1.6325	3	7	0	9	3	9
58.18	58.18	1.5842	2	6	2	4	1	5
58.49	58.50	1.5766	4	0	1	2	3	2
58.85	58.84	1.5679	1	9	1	5	3	5
59.23	59.24	1.5587	4	6	1	4	7	9
59.46	59.46	1.5532	5	3	1	9	8	9
59.66	59.64	1.5484	3	5	2	8	6	5
62.21	62.22	1.4909	0	2	3	5	2	2
63.27	63.26	1.4686	3	9	0	13	8	14
63.40	63.42	1.4658	0	8	2	2	6	2
63.93	63.92	1.4550	6	0	0	5	4	5
64.11	64.10	1.4514	0	10	1	1	3	1
64.83	64.84	1.4369	1	3	3	7	5	8
66.18	66.18	1.4109	2	2	3	1	1	1
66.87	66.88	1.3979	3	9	1	3	1	2
68.76	68.76	1.3640	6	2	1	3	2	3
68.99	68.96	1.3600	6	4	0	1	2	1
69.75	69.76	1.3471	1	9	2	1	1	1
70.05	70.06	1.3421	3	1	3	1	2	1
70.31	70.30	1.3377	5	3	2	3	7	2
72.20	72.30	1.3073	5	1	1	10	10	11
72.29		1.3059	0	6	3	9		9
72.50	72.50	1.3026	3	3	3	2	2	2
74.27	74.28	1.2758	6	0	0	6	2	2
75.09	75.12	1.2640	1	5	3	2	2	2
75.17		1.2628	2	6	3	2		2
76.00	76.00	1.2511	3	9	2	9	5	10
77.20	77.20	1.2346	6	2	2	5	3	6
77.80	77.82	1.2265	2	12	0	7	4	5
77.95	77.96	1.2246	6	6	0	4	5	7
78.42	78.44	1.2184	5	6	1	2	4	5
78.45		1.2180	6	6	1	1		1
79.03	79.02	1.2105	3	11	1	2	1	2
81.27	81.28	1.1827	2	12	1	1	1	1
81.74	81.74	1.1771	5	9	0	3	2	4

2THETA	PEAK	D	H	K	L	I(INT)	I(PK)	I(DS)
82.27	82.28	1.1709	7	3	1	6	3	6
82.52	82.52	1.1680	6	4	2	2	3	2
84.36	84.36	1.1471	3	7	3	1	1	
87.04	87.04	1.1186	5	5	3	7	3	8
88.34	88.34	1.1055	6	6	3	2	1	3
88.58	88.60	1.1030	2	0	4	3	2	2
91.12	91.12	1.0788	12	2	2	1	1	4
91.58	91.42	1.0746	5	3	2	1	1	2
92.10	92.08	1.0698	7	1	2	2	1	2
93.34	93.36	1.0589	3	3	4	1	1	1
93.65	93.66	1.0562	1	9	2	1	1	2
95.41	95.44	1.0413	6	2	3	1	1	2
101.91	101.92	0.9918	4	12	2	1	0	2
102.74	102.96	0.9860	0	14	2	1	0	1
102.94		0.9846	3	13	1	1		1
103.63	103.66	0.9799	8	6	1	1	1	1
104.04	104.04	0.9772	0	12	3	1	1	1
104.85	104.86	0.9718	6	6	3	2	1	2
105.44	105.42	0.9680	3	11	3	1	2	1
107.75	107.80	0.9536	2	12	3	1	1	1
107.80		0.9533	8	4	2	1	1	1
107.90		0.9527	5	1	4	1		
108.23	108.20	0.9507	5	9	3	1	1	1
108.79	108.76	0.9474	7	3	3	1	0	1
109.64	109.66	0.9424	3	15	1	4	0	4
110.27	110.08	0.9387	6	12	1	1	1	2
110.67	110.66	0.9365	0	16	1	1	1	1
111.33	111.32	0.9328	9	3	2	1	0	2
113.65	113.66	0.9202	7	0	2	1	0	1
114.79	114.78	0.9144	0	10	4	1	0	1
116.54	116.54	0.9056	0	2	5	1	0	1
117.59	117.60	0.9005	3	9	4	2	0	2
118.26	118.24	0.8974	6	0	4	2	1	2
119.19	119.18	0.8931	1	3	5	1	0	1
119.63	119.62	0.8911	9	1	2	1	0	1
120.49	120.54	0.8872	3	15	2	1		1
120.61		0.8867	2	2	5	1		1
121.02	121.06	0.8849	5	11	3	1	0	1
121.18		0.8842	6	12	2	1		1
122.02	122.34	0.8806	8	2	3	2		2
123.28	123.74	0.8792	9	8	2	1	1	1
123.68		0.8753	8	8	2	1		1
123.84		0.8737	6	4	0	1		1
124.49	124.44	0.8730	10	0	1	1		1
124.88	124.82	0.8704	9	7	1	1	0	1
125.47	125.48	0.8689	3	1	5	1	0	1
125.82	125.82	0.8665	6	10	3	1	0	1
127.51	127.70	0.8651	5	15	1	1	1	2
127.76		0.8588	0	15	5	2		1
128.02		0.8579	3	3	5	1		1
129.94	129.92	0.8569	9	5	2	1	0	1
130.98	130.98	0.8501	0	12	4	1	0	1
131.92	132.12	0.8465	6	15	4	1	0	1
132.15		0.8434	1	15	3	1		2
133.35	133.84	0.8426	2	6	5	1		2
133.81		0.8388	7	9	3	1	1	2
134.21		0.8374	8	6	3	3		4
134.62	134.50	0.8361	0	18	1	1		1
135.42	135.34	0.8349	2	18	0	2	1	2
136.18	136.64	0.8324	5	9	4	1	0	1
136.50		0.8302	7	3	0	1	1	1
136.74		0.8293	8	12	0	2		3
139.16	139.34	0.8286	9	9	1	1		3
139.42		0.8219	5	15	2	2		1
141.74	142.48	0.8212	2	18	1	1		2
142.44		0.8153	10	0	2	1	2	1
142.77		0.8136	3	15	3	9		11
142.89		0.8128	0	6	1	2		2
143.42	143.26	0.8125	6	8	4	1		1
144.05		0.8112	6	12	3	2	1	2
146.24	146.18	0.8098	0	16	2	2		1
149.09	149.48	0.8049	3	17	5	1	0	2
149.47		0.7992	4	6	5	1	1	1
		0.7984	5	3	5	3		4

BERTRANDITE - SOLOVÉVA AND BELOV, 1965

2THETA	PEAK	D	H	K	L	I(INT)	I(PK)	I(DS)
150.28	150.32	0.7969	0	18	2	1	0	1
153.01	154.16	0.7921	4	12	4	4	1	5
154.08		0.7904	9	9	2	7		10
154.41		0.7899	7	15	0	1		1
155.32	155.24	0.7885	3	13	4	2	1	3
155.43		0.7883	8	0	4	2		2
155.99		0.7875	6	16	1	1		2
158.54	158.56	0.7840	9	11	2	2	0	3
158.54		0.7840	2	18	2	4		6
158.88		0.7835	0	10	5	1		1
161.08		0.7809	11	1	1	1		2
161.14		0.7808	4	18	1	1		1

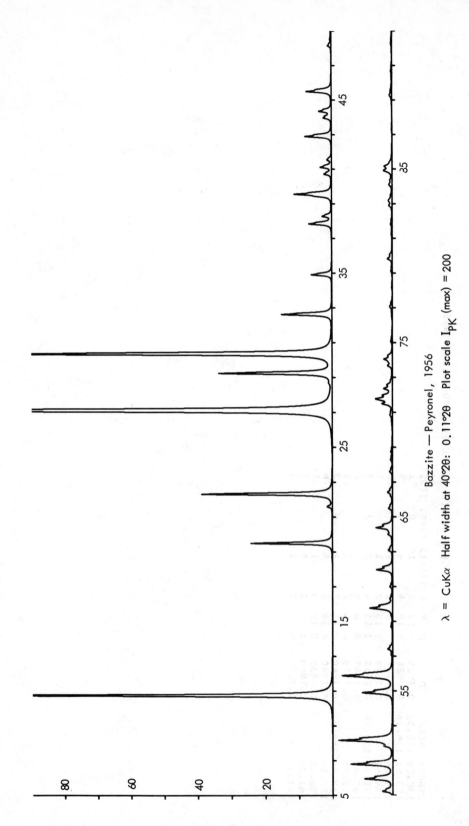

838

Bazzite — Peyronel, 1956

λ = CuKα Half width at 40°2θ: 0.11°2θ Plot scale I$_{PK}$ (max) = 200

BAZZITE – PEYRONEL, 1956

2THETA	PEAK	D	H	K	L	I(INT)	I(PK)	I(DS)
10.73	10.74	8.236	1	0	0	48	64	37
19.47	19.48	4.555	0	0	2	10	12	9
22.28	22.28	3.986	1	1	2	18	19	16
27.09	27.08	3.289	1	1	2	100	100	100
29.21	29.22	3.055	2	0	0	17	17	18
30.32	30.32	2.946	2	0	1	54	53	57
32.59	32.60	2.745	3	0	0	8	8	9
34.88	34.88	2.570	2	1	2	3	3	4
37.81	37.82	2.377	2	2	0	4	3	5
38.24	38.24	2.351	2	2	0	2	1	2
39.53	39.54	2.277	0	0	4	5	6	7
40.69	40.68	2.216	3	1	1	1	1	2
41.08	41.08	2.195	1	0	4	2	2	3
42.87	42.88	2.108	2	2	0	5	4	7
43.94	43.94	2.059	4	0	0	1	1	2
44.32	44.32	2.042	3	1	2	2	2	3
45.47	45.48	1.9930	2	0	4	5	4	7
49.21	49.20	1.8501	3	2	1	2	2	3
49.92	49.92	1.8254	3	1	3	5	4	8
50.75	50.76	1.7972	4	1	0	8	6	13
51.80	51.82	1.7632	4	1	1	1	1	2
52.14	52.14	1.7528	3	0	4	11	8	17
54.87	54.88	1.6718	4	1	2	6	5	11
55.76	55.86	1.6472	5	0	0	3	8	4
55.85		1.6447	2	2	2	9		16
59.74	59.74	1.5466	4	1	3	5	3	9
61.93	61.94	1.4970	3	3	1	4	2	7
64.35	64.36	1.4464	1	1	6	4	3	7
71.47	71.48	1.3188	5	2	0	2	2	5
71.76	71.76	1.3143	6	0	1	4	3	8
72.33	72.34	1.3052	5	2	1	2	1	5
73.65	73.66	1.2850	4	2	2	1	1	3
74.03	74.02	1.2795	4	1	5	1	1	2
79.81	79.80	1.2007	2	1	7	1	1	3
84.08	84.08	1.1502	4	4	2	1	1	3
84.89	84.90	1.1413	5	2	4	3	1	7
103.05	103.06	0.9838	7	1	4	1	0	2
126.67	126.68	0.8619	5	2	8	1	0	3

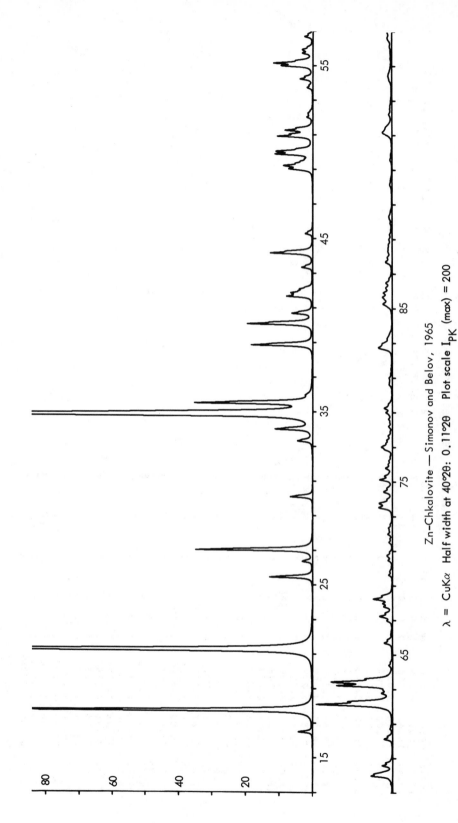

Zn–Chkalovite — Simonov and Belov, 1965

λ = CuKα   Half width at 40°2θ: 0.11°2θ   Plot scale I$_{PK}$ (max) = 200

840

ZN- CHKALOVITE - SIMONOV AND BELOV, 1965

2THETA	PEAK	D	H	K	L	I(INT)	I(PK)	I(DS)
16.47	16.48	5.378	4	0	0	2	2	2
17.75	17.76	4.993	1	1	1	49	52	48
21.27	21.28	4.174	3	1	1	100	100	100
25.43	25.44	3.499	2	0	2	7	6	7
26.33	26.34	3.382	2	2	0	2	2	2
27.02	27.02	3.297	5	1	1	19	17	20
30.06	30.06	2.970	4	2	0	4	3	4
33.29	33.30	2.689	8	0	0	3	2	3
33.97	33.98	2.637	7	1	1	6	6	3
34.81	34.90	2.575	6	0	2	63	56	70
34.93		2.566	0	2	2	35	56	39
35.49	35.50	2.527	6	2	0	21	18	24
35.94	35.94	2.496	2	2	2	1	1	1
38.82	38.84	2.318	1	1	3	7	9	8
38.85		2.316	4	2	1	3		6
40.06	40.06	2.249	1	3	1	5	10	15
40.66	40.66	2.217	3	1	3	13	3	5
41.64	41.64	2.167	9	1	1	4	5	6
41.85	41.84	2.157	3	3	1	5	4	
42.06	42.06	2.146	8	2	0	3	3	4
43.32	43.32	2.087	6	2	2	2	2	2
44.15	44.14	2.050	5	1	3	9	6	3
45.26	45.26	2.002	10	3	1	2	1	11
48.93	49.00	1.8598	7	1	2	9	6	1
49.00		1.8573	8	0	3	2	1	4
49.02		1.8566	0	2	4	3		
49.20	49.20	1.8502	10	0	4	5	4	6
49.44	49.44	1.8416	11	1	1	2	2	2
49.45		1.8416	7	3	0	8	6	10
49.85	49.86	1.8276	12	0	0	8	5	6
49.86		1.8276	0	4	0			
50.03	50.02	1.8214	4	0	4	7	6	6
50.89	50.90	1.7927	2	4	0	8	5	10
50.90		1.7812	4	4	0	6	4	7
51.24	51.24	1.7573	1	1	3	1	1	2
51.99	52.00	1.7496	1	3	3	1	1	1
52.00		1.7056	4	4	0	3	2	4
52.24	52.24	1.6909	4	4	0	3	2	
53.69	53.70							
53.70								
54.20	54.20	1.6909	4	4	0			
54.97	54.98	1.6689	9	1	3	7	5	9
54.98								

2THETA	PEAK	D	H	K	L	I(INT)	I(PK)	I(DS)
55.14	55.12	1.6643	3	3	3	3	6	7
55.70	55.70	1.6487	10	2	2	2	2	3
55.92	55.90	1.6427	2	4	1	2	2	2
56.66	56.66	1.6232	2	2	3	1	1	3
57.96	57.96	1.5899	5	3	3	4	3	6
58.05	58.06	1.5874	2	4	2	2	3	3
58.56	58.56	1.5750	13	1	1	2	2	3
60.11	60.12	1.5380	4	0	4	1	1	3
60.70	60.70	1.5243	8	0	4	3	0	1
62.03	62.12	1.4949	7	0	3	4		4
62.12		1.4929	6	2	4	3		25
62.76	62.76	1.4793	11	3	1	18	2	3
63.21	63.22	1.4697	12	2	2	1	8	19
63.44	63.42	1.4649	6	4	2	13	9	15
64.37	64.38	1.4460	1	1	5	10	1	1
65.67	65.68	1.4205	3	1	5	2	1	3
66.96	66.96	1.3963	1	5	3	3	1	5
67.23	67.22	1.3913	9	3	3	3	2	5
67.78	67.78	1.3813	15	1	1	2	1	3
67.96	67.96	1.3782	8	4	2	1	2	2
68.22	68.22	1.3734	5	1	5	2		3
68.23		1.3734	3	5	1	3	3	4
69.61	69.62	1.3494	13	1	3	1	1	2
70.45	70.44	1.3355	13	3	1	1	1	2
70.74	70.72	1.3307	5	5	1	1	1	2
71.50	71.50	1.3184	14	2	0	1	1	1
71.98	71.98	1.3108	7	1	5	2	2	2
72.33	72.32	1.3052	10	2	4	5	1	1
73.49	73.50	1.2875	12	0	4	4	2	5
73.56		1.2864	10	4	0	3		
73.77	73.76	1.2832	0	4	4	2	2	2
74.39	74.42	1.2742	2	4	4	1	1	2
74.43		1.2735	7	5	1	1		2
75.12	75.12	1.2636	12	4	0	4	2	6
75.79	75.78	1.2541	1	5	5	5	1	3
76.21	76.20	1.2482	4	4	4	4	1	1
76.86	76.88	1.2393	9	1	5	5	1	3

ZN- CHKALOVITE - SIMONOV AND BELOV, 1965

2THETA	PEAK	D	H	K	L	I(INT)	I(PK)	I(DS)
77.00	77.00	1.2374	3	3	5	2	2	4
77.40	77.40	1.2319	1	5	3	1	1	2
78.18	78.18	1.2215	15	5	3	1	1	2
78.60	78.60	1.2161	3	1	3	1	1	2
78.98	78.98	1.2112	15	3	1	2	1	3
79.26	79.24	1.2076	9	5	2	1	1	2
80.59	80.60	1.1910	16	2	2	1	1	1
80.99	80.98	1.1861	5	5	3	1	0	1
82.65	82.72	1.1664	6	5	0	1	2	3
82.72		1.1656	0	2	6	3		6
82.83		1.1644	11	4	6	1		2
83.25	83.24	1.1596	4	6	0	1	1	1
83.38		1.1581	8	4	4	1		1
85.09		1.1392	4	2	6	2	1	1
85.26	85.26	1.1373	18	0	2	2		4
85.65	85.64	1.1331	18	2	0	3	1	5
85.88	85.90	1.1307	0	6	2	1	1	3
86.20	86.18	1.1273	6	6	0	1	1	2
86.46	86.46	1.1245	2	6	2	1	1	3
86.47		1.1222	17	1	3	1	1	2
87.67	87.68	1.1120	9	3	5	1	1	2
88.23	88.28	1.1065	4	6	2	1	1	1
88.31		1.1057	19	1	1	1		1
88.45	88.46	1.1043	17	3	1	1	1	1
88.68	88.66	1.1021	10	4	4	1	1	1
88.98	88.96	1.0991	15	1	3	1	0	1
89.90	89.90	1.0903	13	1	5	1	0	1
90.32	90.30	1.0863	8	6	0	1	0	2
92.15	92.14	1.0695	8	2	6	1	0	1
93.56	93.56	1.0569	11	3	5	1	0	2
95.01	95.16	1.0446	11	3	7	1	1	2
95.14		1.0436	11	5	3	3		2
95.15		1.0435	12	4	4	1		5
95.29		1.0423	8	6	2	1		2
96.20	96.20	1.0349	3	1	7	1	0	2
98.15	98.42	1.0194	15	1	5	1	1	2
98.41		1.0174	17	3	3	1	1	2

2THETA	PEAK	D	H	K	L	I(INT)	I(PK)	I(DS)
100.12	100.12	1.0046	21	3	1	1	1	1
100.54	100.62	1.0016	15	5	1	1	1	2
100.63		1.0009	10	6	2	1		1
100.67		1.0006	13	3	5	1		1
100.95	100.96	0.9986	5	5	1	1	1	1
100.96		0.9985	3	7	1	0		1
101.44	101.34	0.9951	3	2	6	1	0	2
101.87	101.82	0.9920	20	2	2	0	0	1
102.15	102.26	0.9901	7	1	7	0	1	1
102.27		0.9892	13	1	5	1		3
102.89	102.88	0.9850	14	4	4	0	0	1
103.24	103.28	0.9826	4	6	0	0	0	1
103.36		0.9818	5	5	1	1		1
104.04	104.06	0.9772	12	2	6	3		6
104.25	104.26	0.9758	6	4	6	2	1	4
104.57		0.9737	6	5	5	1	1	3
105.71	105.72	0.9663	18	7	1	3		7
106.28	106.24	0.9627	6	0	4	2	1	3
106.94	106.98	0.9586	18	4	2	3	1	6
107.01		0.9581	9	1	7	1	1	2
107.15		0.9573	3	7	7	0		2
107.29	107.28	0.9564	12	3	3	7	1	3
108.54	108.54	0.9489	8	6	2	1	0	1
109.32	109.56	0.9442	19	4	6	1	1	1
109.49		0.9433	9	3	3	1		1
109.54		0.9430	22	2	2	1		2
109.63	109.98	0.9424	5	5	3	2	1	1
110.12		0.9396	14	6	0	1	0	1
110.30	110.40	0.9386	17	5	1	0	1	2
110.43		0.9379	21	3	3	3	0	2
110.86	110.82	0.9354	15	5	3	3	0	5
111.26	111.28	0.9331	21	3	1	1	0	1
111.31		0.9329	8	7	7	1		2
112.14	112.12	0.9283	16	4	4	1	0	2
112.73	112.64	0.9251	0	0	8	1	0	1
113.27	113.38	0.9222	11	1	7	1	0	2

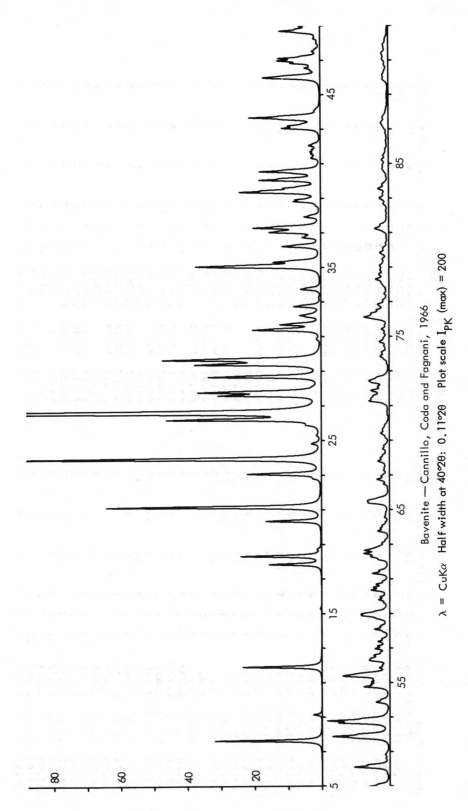

Bavenite — Cannillo, Coda and Fagnani, 1966

λ = CuKα Half width at 40°2θ: 0.11°2θ Plot scale I_{PK} (max) = 200

BAVENITE — CANNILLO, CODA AND FAGNANI, 1966

2THETA	PEAK	D	H	K	L	I(INT)	I(PK)	I(DS)
7.62	7.62	11.595	2	0	0	12	16	10
11.89	11.90	7.438	2	0	2	9	12	8
17.81	17.82	4.976	4	0	2	7	8	6
18.29	18.30	4.847	0	1	4	11	12	10
20.31	20.32	4.368	1	1	2	7	8	7
21.13	21.14	4.201	3	1	0	29	32	28
22.99	23.06	3.865	6	0	0	8	11	8
23.05		3.855	3	1	2	3		3
23.91	23.92	3.719	4	0	4	45	46	43
24.78	24.78	3.590	6	0	2	1	2	1
25.85	25.86	3.443	2	0	4	1	2	1
26.17	26.18	3.402	5	1	0	21	23	21
26.58	26.58	3.351	5	1	1	100	100	100
27.58	27.58	3.232	0	0	6	14	15	14
27.77	27.76	3.210	2	1	6	17	18	18
28.65	28.66	3.113	1	0	6	17	17	18
29.36	29.36	3.039	1	1	6	19	19	19
29.65	29.66	3.010	5	1	5	22	24	23
31.36	31.36	2.850	5	1	3	11	10	12
31.67	31.68	2.823	4	0	5	6		7
32.12	32.20	2.785	5	1	4	2	3	2
32.20		2.777	8	0	2	2		2
32.72	32.72	2.735	7	1	1	4	4	5
33.70	33.70	2.657	7	1	2	3	3	4
35.00	35.06	2.562	3	1	6	10	19	11
35.06		2.557	5	1	5	14		15
35.30	35.30	2.540	7	0	3	6	7	7
36.16	36.20	2.482	0	2	6	1	6	2
36.20		2.479	6	0	0	5		6
36.71	36.72	2.446	2	2	1	3	3	3
37.01	37.02	2.427	2	2	8	7	8	8
37.06		2.424	0	2	7	1		1
37.27	37.28	2.410	1	0	10	11	10	12
37.89	37.90	2.372	2	2	8	2	2	2
37.90		2.372	2	0	2	1		1
38.80	38.80	2.319	10	0	0	4	4	5
38.91	38.90	2.313	3	1	7	1	4	2
39.29	39.36	2.291	9	1	0	2	12	2
39.35		2.288	2	2	1	13		15
39.58	39.58	2.275	9	1	1	4	6	2
40.04	40.04	2.250	7	1	5	10	9	12
40.53	40.54	2.224	0	2	4	2	9	11
41.31	41.30	2.184	1	2	2	2	1	2
41.54	41.54	2.172	1	1	8	1	1	2
42.03	42.02	2.148	5	1	7	1	2	2
43.02	43.04	2.101	6	2	0	2	6	2
43.04		2.100	7	1	8	1		2
43.05		2.099	3	1	4	4		5
43.21	43.16	2.092	10	0	4	3	5	4
43.55	43.66	2.076	4	2	4	5	11	6
43.66		2.071	9	1	8	10		12
45.97	45.98	1.9740	5	1	8	5	9	7
45.99		1.9724	9	1	5	5		2
46.38	46.38	1.9717	1	1	9	2		7
46.52	46.52	1.9560	7	1	7	5	2	1
46.81	46.82	1.9504	2	2	6	1	2	2
46.96	46.96	1.9390	0	1	10	2	5	6
46.98		1.9332	11	0	1	4	6	5
47.11	47.10	1.9325	12	0	0	2		2
47.50	47.50	1.9274	6	2	4	5	6	6
48.58	48.68	1.9124	2	0	10	1	1	2
48.68		1.8726	4	2	6	3	6	3
48.95	48.94	1.8689	9	1	6	6		8
50.07	50.08	1.8591	8	2	2	3	3	4
51.77	51.88	1.8201	8	1	4	7	5	9
51.87		1.7643	6	2	6	4	9	9
52.64	52.66	1.7612	11	1	5	9		5
52.77	52.78	1.7371	6	1	10	9	8	12
53.85	53.84	1.7331	10	2	0	7	9	13
54.73	54.74	1.7010	10	0	10	7	2	10
54.74		1.6756	10	0	8	3	4	3
55.02	55.02	1.6754	10	2	2	3		4
55.34	55.38	1.6675	4	2	8	2	3	3
		1.6586	12	0	6	4	7	5

BAVENITE - CANNILLO, CODA AND FAGNANI, 1966

2THETA	PEAK	D	H	K	L	I(INT)	I(PK)	I(DS)
55.37		1.6579	1	3	0	3		5
55.42		1.6564	14	0	4	2		6
55.84	55.84	1.6449	10	2	3	2	2	2
56.24	56.28	1.6342	8	2	2	2	3	3
56.30		1.6327	0	2	9	1		2
56.37		1.6308	3	0	0	1		2
56.58	56.38	1.6250	3	3	1	2	3	2
56.59	56.58	1.6158	0	0	12	2	2	2
56.94	56.94	1.6050	10	2	4	2	2	3
57.36	57.36		7	1	10	3	2	4
58.07	58.08	1.5871	14	0	4	4	4	5
58.87	58.88	1.5674	5	1	11	3	4	4
58.96	58.96	1.5651	3	3	4	2	3	5
59.78	59.78	1.5457	0	3	10	4	2	2
60.34	60.34	1.5327	1	3	5	1	3	3
60.49	60.50	1.5292	12	2	1	2	2	5
60.68	60.66	1.5248	12	2	4	3	2	2
61.30	61.30	1.5108	5	3	4	2	3	3
62.09	62.12	1.4935	7	1	0	2	3	3
62.25	62.26	1.4901	7	3	1	4	4	6
62.46	62.46	1.4857	4	2	10	3		4
62.64	62.64	1.4818	9	1	10	2	4	4
62.72		1.4800	12	2	4	2		2
63.75	63.76	1.4586	7	3	3	1	2	4
64.08	64.06	1.4520	1	1	7	4	1	2
65.36	65.38	1.4264	7	3	4	2	3	6
65.48	65.48	1.4243	3	1	13	1	1	3
66.46	66.46	1.4056	9	3	0	3	2	4
66.74	66.74	1.4004	14	2	0	2	1	2
67.79	67.78	1.3813	12	0	10	1	1	4
68.49	68.50	1.3688	16	3	6	4	3	3
69.39	69.38	1.3531	9	0	12	2	3	7
71.24	71.26	1.3225	17	1	1	2	3	4
71.37	71.38	1.3204	7	1	13	1		2
71.83	71.86	1.3131	11	1	11	2	3	3
71.87		1.3125	17	1	1	1		
72.32		1.3055	11	1	11	3	3	5
72.42	72.42	1.3038	6	0	14	2		4

2THETA	PEAK	D	H	K	L	I(INT)	I(PK)	I(DS)
73.81	73.82	1.2828	5	1	14	2	1	3
74.66	74.66	1.2701	17	4	4	1	1	2
75.41	75.40	1.2594	14	1	10	1	1	2
76.09	76.10	1.2498	9	3	7	1	4	3
76.10		1.2497	8	0	14	3		2
76.51	76.50	1.2440	2	4	0	1	2	5
76.83	76.82	1.2396	11	3	5	1	2	2
78.29	78.28	1.2201	12	0	12	1	2	2
79.07	79.06	1.2100	5	3	10	2	1	4
79.47	79.46	1.2050	1	3	11	1	1	3
79.79	79.80	1.2009	2	4	4	1	1	2
81.20	81.22	1.1836	12	1	10	1	2	5
83.00	82.98	1.1625	19	1	1	3	1	3
83.51	83.52	1.1566	13	3	5	2	2	5
84.65	84.66	1.1439	17	0	8	1	1	3
85.20	85.20	1.1380	7	3	6	1	1	3
87.19	87.34	1.1170	6	3	11	1	1	3
87.35		1.1153	13	2	6	1		2
89.85	89.86	1.0907	0	2	16	2	1	3
91.03	91.04	1.0796	18	2	6	1	1	3
91.40	91.38	1.0762	11	2	15	1	1	2
94.34	94.34	1.0503	12	4	0	2	1	4
100.57	100.56	1.0013	16	0	14	1	1	3
100.91	100.88	0.9989	12	4	6	1	1	1
102.92	102.96	0.9847	9	3	14	1	1	2
104.00	104.02	0.9775	5	0	19	1	1	2
105.72	105.70	0.9663	24	4	0	1	1	2
106.12	106.08	0.9637	12	4	8	1	1	2
106.85	106.88	0.9591	5	5	4	1	1	1
107.19	107.18	0.9570	7	5	1	1	0	2
108.37	108.38	0.9499	12	2	16	1	1	2
109.89	109.94	0.9409	12	0	18	1	0	1

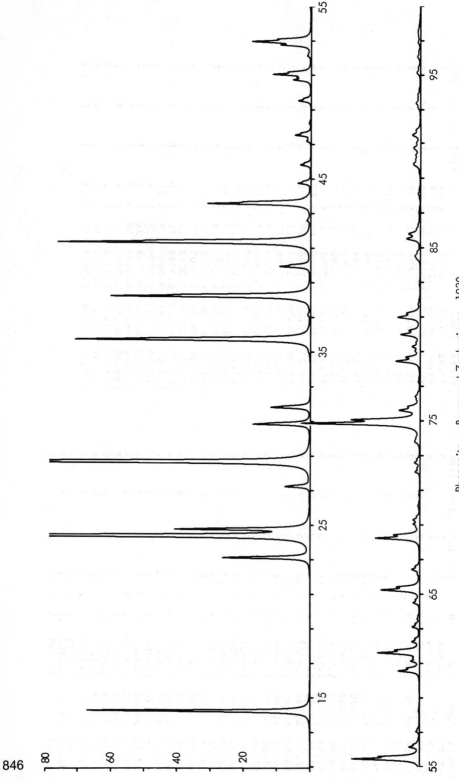

Phenacite — Bragg and Zachariasen, 1930

$\lambda = CuK\alpha$ Half width at $40°2\theta$: $0.11°2\theta$ Plot scale I_{PK} (max) = 150

PHENACITE – BRAGG AND ZACHARIASEN, 1930

2THETA	PEAK	D	H	K	L	I(INT)	I(PK)	I(DS)
14.21	14.22	6.229	-1	1	0	39	45	38
23.08	23.08	3.850	-1	1	1	17	17	17
24.33	24.34	3.655	-2	1	0	52	100	52
24.33		3.655	0	-1	2	49		49
24.73	24.74	3.597	-2	1	1	26	27	25
27.21	27.22	3.275	-2	0	0	5	5	5
28.63	28.64	3.115	-2	2	0	100	92	100
30.82	30.82	2.899	2	1	-1	7	11	7
30.82		2.899	-1	1	2	6		6
31.78	31.78	2.813	-2	2	1	5	8	5
31.78		2.813	1	1	-2	4		4
35.68	35.68	2.514	2	1	0	29	47	29
35.68		2.514	0	1	2	29		29
38.19	38.20	2.355	-3	2	-1	3	40	3
38.19		2.355	1	2	-3	47		48
39.91	39.92	2.257	-2	2	2	8	6	8
41.32	41.32	2.183	3	0	0	32	51	33
41.32		2.183	2	2	-1	33		34
43.55	43.54	2.076	-3	3	0	28	21	29
44.74	44.74	2.024	-3	1	1	3	3	3
45.80	45.80	1.9794	-3	3	1	3	2	3
47.17	47.18	1.9250	2	2	0	1	1	1
47.52	47.52	1.9117	-4	1	1	4	1	5
49.52	49.52	1.8392	0	1	3	4	3	2
49.52		1.8392	3	1	0	2		2
50.72	50.72	1.7983	-4	2	2	5	4	5
51.04	51.04	1.7878	0	-1	4	5	8	6
51.04		1.7878	-1	1	3	5		5
52.74	52.74	1.7341	-4	2	0	6	6	6
52.74		1.7341	0	-3	4	2		2
52.95	52.94	1.7277	-4	3	1	13	12	14
52.95		1.7277	1	3	-4	2		2
55.42	55.42	1.6566	4	-2	1	10	14	11
55.42		1.6566	1	-2	4	10		11
56.12	56.12	1.6375	4	0	0	3	2	3
60.53	60.52	1.5284	3	2	0	4	4	4
60.53		1.5284	0	2	3	3		3

2THETA	PEAK	D	H	K	L	I(INT)	I(PK)	I(DS)
61.58	61.58	1.5047	-5	1	1	7	8	7
61.58		1.5047	-3	3	3	8		8
62.24	62.24	1.4903	3	3	-2	2	2	3
63.09	63.10	1.4722	1	3	-5	2	2	2
65.23	65.22	1.4291	-5	3	2	6	8	7
65.23		1.4291	2	3	-5	7		8
66.41	66.42	1.4064	2	4	-4	2	2	2
68.21	68.20	1.3738	2	2	2	16	9	18
72.04	72.04	1.3098	1	-4	5	1	1	2
72.56	72.56	1.3016	-4	4	5	2	1	2
74.82	74.82	1.2679	-4	4	3	15	24	17
74.82		1.2679	4	2	1	15		18
74.82		1.2679	2	1	-6	7		8
74.82		1.2679	2	2	-4	7		8
75.58	75.58	1.2569	3	4	0	4	4	4
76.37	76.38	1.2459	0	2	4	3		4
78.42	78.42	1.2184	-5	5	0	1	1	1
78.42		1.2184	-5	2	2	5		6
79.01	79.02	1.2107	0	-3	6	5	5	5
79.27	79.26	1.2075	0	-2	-2	1	1	1
79.96	79.96	1.1989	5	2	-6	1	2	2
80.95	80.96	1.1866	-6	3	3	8	4	9
81.05		1.1854	-1	3	4	4		5
84.49	84.48	1.1457	6	-4	0	4	5	5
85.50	85.50	1.1347	5	2	-1	1	1	2
85.50		1.1347	6	1	-4	3		4
85.82	85.78	1.1313	6	-4	-1	3	3	4
85.82		1.1313	1	2	4	1		2
89.85	89.84	1.0907	4	2	-1	1	2	1
90.35	90.34	1.0860	2	2	-7	2	1	2
90.76	90.76	1.0821	2	5	-6	1	1	1
90.76		1.0821	-5	5	-1	1		1
91.50	91.50	1.0753	-7	3	1	1	2	2
91.50		1.0753	5	3	-2	2		2
93.10	93.12	1.0610	-1	1	6	1	1	1

PHENACITE — BRAGG AND ZACHARIASEN, 1930

2THETA	PEAK	D	H	K	L	I(INT)	I(PK)	I(DS)
93.35	93.36	1.0588	-7	4	1	2		1
95.11	95.12	1.0438	-6	4	4	2		1
101.94	102.10	0.9915	4	5	-5	2		3
102.10		0.9904	5	4	-3	2		
102.10		0.9904	0	-1	7	2		
102.10		0.9904	-7	1	0	2		
102.20		0.9904	-3	4	5	2		
103.47	103.46	0.9897	2	6	-6	1		1
104.95	104.96	0.9810	-1	2	2	1	0	1
104.95		0.9712	7	-5	1	1	0	
104.95		0.9712	1	-5	7	1		1
105.55	105.60	0.9673	3	-5	6	1	1	1
107.38	107.38	0.9559	-8	2	2	2	1	2
108.35	108.34	0.9500	-7	6	1	2	1	2
108.35		0.9500	1	6	-7	2		2
109.22	109.22	0.9448	-8	4	1	1	0	1
110.47	110.48	0.9376	-8	4	1	1	1	1
110.47		0.9376	-6	6	3	1		1
110.47		0.9376	1	4	-8	1		1
110.47		0.9376	-6	6	-6	1		1
111.87	111.90	0.9298	3	6	1	1	0	1
111.90		0.9296	6	1	-1	1		1
112.69	112.70	0.9253	7	1	-1	1	0	1
113.15	113.14	0.9229	2	-3	7	1	0	1
113.15		0.9229	7	-3	2	1		1
115.17	115.18	0.9124	0	-3	8	1	1	1
115.17		0.9124	-8	3	0	1		1
116.44	116.48	0.9061	4	3	2	1	1	1
116.44		0.9061	2	3	4	1		1
117.02	117.02	0.9033	5	5	-4	4	2	1
117.02		0.9033	-8	1	1	4		5
117.89	118.16	0.8991	-8	4	4	4	2	5
118.16		0.8978	-8	5	0	3		4
118.16		0.8978	0	-5	8	1		4
118.16		0.8978	2	-6	7	3		4
118.16		0.8978	7	-6	2	1		1
119.89	119.88	0.8899	-8	5	3	1	1	2

2THETA	PEAK	D	H	K	L	I(INT)	I(PK)	I(DS)
119.89	119.89	0.8899	3	5	-8	2	1	3
120.42	120.38	0.8875	5	2	2	1		1
120.42		0.8875	4	4	1	1		1
122.59	122.60	0.8782	-3	4	6	1	0	2
126.16	126.14	0.8639	-8	6	2	2	0	2
128.67	128.70	0.8545	-9	3	3	2	1	2
128.67		0.8545	-7	5	5	1		2
130.95	130.96	0.8466	1	-6	8	2	1	2
130.95		0.8466	8	-6	1	2		2
133.62	133.62	0.8380	6	3	0	2	1	3
133.62		0.8380	0	3	6	2		3
137.69	138.28	0.8259	7	1	-2	1		5
137.98		0.8251	1	1	-8	4	1	2
138.35		0.8241	1	-5	9	2		2
138.35		0.8241	3	7	-7	2		2
141.41	141.44	0.8161	5	5	1	1	1	2
141.41		0.8161	7	1	-1	1		1
142.39	142.28	0.8137	-6	6	5	1	0	1
142.39		0.8137	5	6	-6	1		1
147.61	147.66	0.8021	-4	4	-4	1	0	2
147.61		0.8021	7	4	-7	1		1
148.60	148.62	0.8001	-9	6	1	1	0	1
149.91	149.84	0.7976	-9	5	4	2	0	1
149.91		0.7976	4	5	-9	1		1
154.67	154.92	0.7895	-5	3	7	1	0	1
154.67		0.7895	3	1	-1	1		1
154.90		0.7891	-9	2	0	1		2
154.90		0.7891	0	-2	9	1		1
155.45		0.7883	6	2	2	1		1
156.16		0.7872	-8	7	3	2		2
160.19	160.80	0.7819	-7	7	4	1	1	1
160.66		0.7814	1	-4	9	3		4
160.66		0.7814	-8	5	3	3		4
160.66		0.7814	3	-5	8	4		6
160.66		0.7814	9	-4	1	4		6
163.12		0.7787	-8	8	0	5		8

848

TABLE 43   LIST OF FIVE STRONGEST LINES

D SPACING					I (PEAK)					MINERAL
17.990 - 16.000										
17.547	8.272	6.882	3.078	3.258	81	100	83	54	36	PAULINGITE
15.990 - 14.000										
14.242	7.121	3.560	4.747	2.554	64	100	53	44	29	CORUNDOPHILLITE
14.166	7.083	2.519	3.540	2.145	38	100	57	34	25	RIPIDOLITE
14.164	7.082	3.541	4.282	4.526	88	100	46	32	26	PROCHLORITE
14.333	2.393	4.471	3.583	2.593	100	6	3	3	3	MG-VERMICULITE-2 LAYER
13.990 - 12.000										
12.280	8.683	5.492	7.090	3.702	100	61	24	16	16	NA-TYPE A ZEOLITE
11.990 - 11.000										
11.985	7.433	4.305	3.186	3.350	100	8	8	8	7	SEPIOLITE
10.990 - 10.000										
10.410	3.500	2.733	9.674	2.259	100	16	13	11	9	ASTROPHYLLITE
10.111	3.117	2.911	4.509	2.460	100	25	18	10	8	CELADONITE
10.163	2.671	3.388	2.479	3.450	100	24	17	16	9	FERRI-ANNITE
10.252	2.642	3.442	3.188	3.417	100	89	75	63	57	PHLOGOPITE
10.463	2.636	3.025	3.368	2.116	100	93	90	45	38	KORNERUPINE
10.007	2.557	3.336	4.448	3.476	87	100	83	78	62	MUSCOVITE 2M1 (1)
9.999 - 9.500										
9.674	10.410	3.500	2.733	2.259	11	100	16	13	9	ASTROPHYLLITE
9.656	4.392	2.524	3.219	2.532	87	100	79	77	72	PARAGONITE
9.990	3.343	2.560	2.330	4.450	100	93	90	86	83	MUSCOVITE 3T
9.567	3.189	7.722	3.519	2.943	100	52	39	39	32	NEPTUNITE
9.983	2.611	3.379	3.131	3.328	100	62	55	50	44	FLUORPHLOGOPITE
9.923	2.566	4.471	3.487	2.982	100	87	79	54	53	PHENGITE
9.890	2.549	4.438	4.448	3.463	99	100	93	67	57	MUSCOVITE 2M1 (2)
9.499 - 9.000										
9.024	8.406	2.705	3.119	3.380	65	100	94	67	55	TREMOLITE
9.151	8.334	2.764	3.076	2.514	30	100	38	33	24	GRUNERITE
9.095	8.284	2.749	3.058	3.252	75	100	73	62	48	CUMMINGTONITE
9.007	8.248	3.048	3.231	2.834	84	100	83	67	58	ANTHOPHYLLITE
9.460	3.509	6.838	4.160	3.661	100	40	38	20	16	LAUMONTITE
9.450	2.769	2.611	2.476	1.633	37	100	78	47	46	VESUVIANITE
9.004	2.720	8.276	3.061	2.505	94	100	93	77	63	TIRODITE
8.999 - 8.500										
8.683	12.280	7.090	5.492	3.702	61	100	24	16	16	NA-TYPE A ZEOLITE
8.542	4.087	3.384	3.134		67	100	38	31	30	LOW CORDIERITE

TABLE 43   LIST OF FIVE STRONGEST LINES

D SPACING					I (PEAK)					MINERAL
8.910	3.977	7.945	2.959	5.127	100	79	62	58	47	HEULANDITE
8.530	2.709	3.388	2.547	3.176	66	100	95	82	67	K-RICHERITE
8.499 – 8.000										
8.284	9.095	2.749	3.058	3.252	100	75	73	62	48	CUMMINGTONITE
8.248	9.007	3.048	3.231	2.834	100	84	83	67	58	ANTHOPHYLLITE
8.460	8.542	4.087	3.384	3.134	100	67	38	31	30	LOW CORDIERITE
8.272	6.882	17.547	3.078	3.258	100	83	81	54	36	PAULINGITE
8.008	4.182	2.669	4.004	2.832	100	76	66	36	27	ZUNYITE
8.025	3.396	3.199	4.422	3.537	55	100	67	59	30	GILLESPITE
8.236	3.289	2.946	3.986	3.055	64	100	53	20	17	BAZZITE
8.316	2.901	2.882	3.363	2.922	93	100	48	37	30	BATISITE
8.166	2.886	2.925	3.515	3.455	59	100	83	50	49	BULTFONTEINITE
8.271	2.823	3.642	3.064	4.455	100	79	76	71	50	PROTOAMPHIBOLE
8.334	2.764	3.076	9.151	2.514	100	38	33	30	24	GRUNERITE
8.307	2.732	3.072	3.251	2.513	100	50	41	30	30	MN-CUMMINGTONITE
8.276	2.720	9.004	3.061	2.505	93	100	94	77	63	TIRODITE
8.429	2.719	3.130	4.510	2.541	100	29	22	14	14	RIEBECKITE
8.425	2.706	3.127	3.381	2.552	100	77	61	59	56	HORNBLENDE
8.406	2.705	3.119	9.024	2.380	100	94	67	65	55	TREMOLITE
8.217	2.685	3.044	4.435	2.516	100	60	47	34	31	GLAUCOPHANE
7.999 – 7.500										
7.722	9.567	3.189	3.519	2.943	39	100	52	39	32	NEPTUNITE
7.945	8.910	3.977	2.959	5.127	62	100	79	58	47	HEULANDITE
7.970	4.023	4.385	2.530	2.776	28	100	29	27	15	CLINOHEDRITE
7.790	3.951	2.968	2.647	2.199	39	100	63	33	21	ALPHA BAAL2SI2O8
7.796	3.575	2.496	2.980	2.481	100	49	44	33	31	APOPHYLLITE
7.978	3.252	2.865	4.594	3.981	100	42	39	20	19	BERYL (ANHYDROUS)
7.499 – 7.000										
7.121	14.242	3.560	4.747	2.554	100	64	53	44	29	CORUNDOPHILLITE
7.082	14.164	3.541	4.282	4.526	100	88	46	32	26	PROCHLORITE
7.090	12.280	8.683	5.492	3.702	16	100	61	24	16	NA-TYPE A ZEOLITE
7.433	11.985	4.305	3.186	3.350	8	100	8	8	7	SEPIOLITE
7.085	5.369	3.187	4.186	4.983	100	57	55	39	26	PHILLIPSITE
7.285	4.267	4.916	3.184	3.131	100	95	60	60	54	GISMONDITE
7.232	3.984	2.882	5.813	2.861	100	83	53	43	31	STOKESITE
7.186	3.593	4.167	4.383	4.352	100	63	58	54	45	NACRITE
7.162	3.579	4.123	4.324	3.792	100	58	50	48	36	DICKITE

TABLE 43 LIST OF FIVE STRONGEST LINES

D SPACING					I (PEAK)					MINERAL
7.131	3.566	4.361	4.158	2.333	100	57	47	45	29	KAOLINITE
7.110	3.555	2.370	2.481	4.581	100	54	33	27	24	AL-SERPENTINE
7.142	3.238	2.921	4.138	5.544	63	100	65	56	48	OSUMILITE
7.050	3.237	5.160	6.582	3.107	100	88	72	64	44	ELPIDITE
7.145	3.214	2.771	3.834	2.440	27	37	35	33	21	EUCLASE
7.320	3.207	2.885	3.249	2.665	100	100	77	28	28	HARADAITE
7.053	3.051	3.239	2.800	3.564	63	100	70	59	56	XONOTLITE
7.300	2.719	2.534	2.676	6.535	100	63	63	40	31	ILVAITE
7.305	2.609	4.913	4.077	3.200	100	41	34	27	18	DIOPTASE
7.085	2.559	3.543	2.170	3.948	100	80	53	37	28	CRONSTEDTITE (1 LAYER)
7.083	2.519	14.166	3.540	2.145	100	57	38	34	25	RIPIDOLITE
7.020	2.483	3.510	4.370	2.309	100	55	53	49	34	AMESITE (2 LAYER)
6.999 – 6.500										
6.838	9.460	3.509	4.160	3.661	38	100	40	20	16	LAUMONTITE
6.882	8.272	17.547	3.078	3.258	83	100	81	54	36	PAULINGITE
6.535	7.300	2.719	2.534	2.676	31	100	63	63	40	ILVAITE
6.582	3.237	3.237	5.160	3.107	64	100	88	72	44	ELPIDITE
6.527	5.890	2.865	4.356	2.845	55	100	51	49	48	NATROLITE
6.783	5.006	2.895	3.158	4.809	40	100	72	57	47	LAVENITE
6.975	4.645	5.814	3.047	3.219	31	100	85	66	30	YUGAWARALITE
6.867	3.872	4.923	3.433	2.917	67	100	83	75	52	EPISTILBITE
6.518	3.349	3.472	2.581	3.021	71	100	76	55	54	CELSIAN (BA-FELDSPAR)
6.980	3.177	2.910	5.270	4.206	71	100	70	68	66	MILARITE
6.539	3.102	2.833	2.727	2.816	59	100	92	73	58	AFWILLITE
6.598	3.286	3.067	5.360	2.400	76	100	66	58	49	HEMIMORPHITE
6.578	2.987	2.872	4.517	2.545	100	69	34	30	28	NAIN(SIO3)2
6.592	2.986	2.706	2.577	2.434	33	100	60	28	23	CA-SEIDOZERITE
6.542	2.697	2.752	3.175	2.646	67	100	99	60	56	SPURRITE
6.499 – 6.000										
6.375	4.660	3.240	3.681	2.760	81	100	89	54	32	CANCRINITE
6.468	4.608	2.910	2.593	2.519	92	100	97	96	94	FERROSILITE
6.391	4.013	3.799	3.595	2.993	41	100	75	61	49	PARACELSIAN
6.229	3.655	3.115	2.183	2.514	45	100	92	51	47	PHENACITE
6.284	3.628	2.095	2.565	2.810	100	77	24	22	17	HYDROSODALITE
6.272	3.621	2.091	2.371	2.560	36	100	25	17	15	SODALITE
6.150	3.519	3.567	2.500	2.026	41	100	37	15	14	TUGTUPITE
6.465	3.276	2.752	3.175	3.446	100	52	49	44	41	KAINOSITE

TABLE 43   LIST OF FIVE STRONGEST LINES

D SPACING					I (PEAK)					MINERAL
6.362	2.982	2.899	2.524	2.472	59	100	47	37	34	ACMITE
6.413	2.975	2.581	4.605	4.235	58	100	89	54	47	TOURMALINE
6.313	2.958	2.879	2.512	2.457	58	100	51	42	34	UREYITE
6.299	2.918	4.655	4.522	5.069	100	84	75	71	63	BREWSTERITE
5.999 – 5.750										
5.813	3.984	2.882	2.861	7.232	43	100	83	53	31	STOKESITE
5.890	2.865	4.356	2.845	6.527	100	55	51	49	48	NATROLITE
5.832	5.727	3.542	3.539	3.348	74	100	94	91	75	ALAMOSITE
5.814	4.645	3.047	6.975	3.219	85	100	66	31	30	YUGAWARALITE
5.975	3.622	2.424	3.942	4.130	100	93	85	84	69	BETA NA2SI2O5
5.865	3.604	3.070	4.187	4.323	72	100	81	77	59	DALYITE
5.825	3.364	2.913	4.513	2.975	53	100	86	67	64	BARYLITE
5.849	3.226	3.442	2.082	3.454	100	65	49	48	47	DUMORTIERITE
5.946	2.756	2.968	3.576	2.318	20	100	35	34	28	LEUCOPHANITE
5.749 – 5.500										
5.548	4.526	2.774	3.927	2.171	81	100	50	30	30	ANDALUSITE
5.724	3.707	2.734	2.862	3.305	26	100	62	22	17	BENITOITE
5.727	3.542	3.539	3.348	5.832	100	94	91	75	74	ALAMOSITE
5.605	3.432	2.927	4.854	2.693	78	100	47	15	14	ANALCIME (DISORDERED)
5.527	3.260	2.955	2.904	5.000	70	100	96	96	67	KENTROLITE
5.544	3.238	2.921	7.142	4.138	48	100	65	63	56	OSUMILITE
5.687	2.602	2.607	5.041	4.660	100	39	38	31	24	FERROCARPHOLITE
5.499 – 5.250										
5.369	3.187	4.106	3.261	4.983	57	100	55	39	26	PHILLIPSITE
5.363	3.974	3.392	4.107	3.394	100	76	74	68	58	NARSARSUKITE
5.401	3.431	2.210	2.548	3.400	83	100	60	53	44	MULLITE 1.92-1
5.396	3.426	2.207	2.545	3.400	99	100	57	53	44	MULLITE 2.0-1
5.384	3.433	2.206	2.542	3.388	74	100	66	56	45	MULLITE 1.71-1
5.270	6.980	2.910	4.206	4.108	68	100	71	70	66	MILARITE
5.360	3.102	3.286	2.400	6.598	58	100	76	66	49	HEMIMORPHITE
5.492	12.280	8.683	3.702	7.090	24	100	61	16	16	NA-TYPE A ZEOLITE
5.249 – 5.000										
5.127	8.910	7.945	2.959	3.977	47	100	79	62	58	HEULANDITE
5.169	7.050	6.582	3.107	3.237	72	100	88	64	44	FLPIDITE
5.069	6.299	4.655	4.522	2.918	63	100	84	75	71	BREWSTERITE
5.041	5.687	2.607	4.660	2.602	31	100	39	38	24	FERROCARPHOLITE
5.000	3.260	2.955	2.904	5.527	67	100	96	96	70	KENTROLITE

TABLE 43   LIST OF FIVE STRONGEST LINES

D SPACING					I (PEAK)					MINERAL
5.213	3.211	2.926	4.995	3.344	61	100	90	77	66	VLASOVITE
5.240	3.028	2.410	2.564	3.551	40	100	52	49	3	NA2SIO3
5.006	2.895	3.158	4.809	6.783	100	72	57	47	40	LAVENITE
5.016	2.893	2.801	2.593	3.480	37	100	38	38	34	EPIDOTE
5.016	2.891	2.792	2.586	2.673	32	100	38	36	31	CLINOZOISITE
5.018	2.866	2.698	2.780	2.723	36	100	89	41	36	ZOISITE
4.999 – 4.900										
4.913	7.305	2.609	4.077	3.200	34	100	41	27	18	DIOPTASE
4.916	7.285	4.267	3.184	3.131	60	100	95	60	54	GISMONDITE
4.983	7.085	5.369	3.187	4.106	26	100	57	55	39	PHILLIPSITE
4.993	4.174	2.575	2.569	2.527	52	100	57	56	18	ZN-CHKALOVITE
4.923	3.872	3.433	6.867	2.917	75	100	83	67	52	EPISTILBITE
4.940	3.248	2.999	2.613	2.597	26	100	66	61	32	SPHENE
4.995	3.211	2.926	3.344	5.213	77	100	90	66	61	VLASOVITE
4.937	2.842	2.942	2.976	3.243	54	100	87	67	52	WOHLERITE
4.899 – 4.800										
4.809	5.006	2.895	3.158	6.783	47	100	72	57	40	LAVENITE
4.854	5.432	5.605	2.927	2.693	15	100	78	47	14	ANALCIME (DISORDERED)
4.889	3.190	3.033	2.852	3.151	33	100	51	48	30	LARSENITE
4.799 – 4.700										
4.747	7.121	14.242	3.560	2.554	44	100	64	53	29	CORUNDOPHILLITE
4.718	2.584	2.359	2.889	1.544	15	100	71	33	28	PYROPE
4.779	2.247	2.932	2.978	3.141	41	100	87	77	49	RHODONITE
4.699 – 4.600										
4.655	6.299	2.918	4.522	5.069	75	100	84	71	63	BREWSTERITE
4.645	5.814	3.047	6.975	3.219	100	85	66	31	30	YUGAWARALITE
4.660	5.687	2.602	2.607	5.041	24	100	39	38	31	FERROCARPHOLITE
4.608	3.486	3.913	3.165	1.879	9	100	33	12	9	BETA SPODUMENE
4.682	3.309	2.706	2.339	1.560	100	98	68	67	24	LI2SIO3
4.660	3.240	6.375	3.681	2.760	100	89	81	54	32	CANCRINITE
4.605	3.035	2.909	2.603	2.161	58	100	60	58	51	CLINOFERROSILITE
4.605	2.975	2.581	6.413	4.235	54	97	89	58	47	TOURMALINE
4.608	2.910	2.593	2.519	6.468	100	100	96	94	92	FERROSILITE
4.671	2.900	2.632	3.781	1.595	65	100	59	50	49	PUMPELLYITE
4.599 – 4.500										
4.509	10.111	3.117	2.911	2.460	10	100	25	18	8	CELADONITE
4.510	8.429	2.719	3.130	2.541	14	100	29	22	14	RIEBECKITE

TABLE 43   LIST OF FIVE STRONGEST LINES

D SPACING					I (PEAK)					MINERAL
4.594	7.978	3.252	2.865	3.981	20	100	42	39	19	BERYL (ANHYDROUS)
4.581	7.110	3.555	2.370	2.481	24	100	54	33	27	AL-SERPENTINE
4.526	7.082	14.164	3.541	4.282	26	100	88	46	32	PROCHLORITE
4.517	6.578	3.067	2.987	2.545	30	100	69	34	28	NAIN(SIO3)2
4.522	6.299	2.918	4.655	5.069	71	100	84	75	63	BREWSTERITE
4.526	5.548	2.774	3.927	2.171	100	81	50	30	30	ANDALUSITE
4.513	3.364	2.913	2.975	5.825	67	100	86	64	53	BARYLITE
4.499 -										
4.471	14.333	2.393	3.583	2.593	3	100	6	3	3	MG-VERMICULITE-2 LAYER
4.450	9.990	3.343	2.560	2.330	83	100	93	90	86	MUSCOVITE 3T
4.471	9.923	2.566	3.487	2.982	79	100	87	54	53	PHENGITE
4.455	8.271	2.823	3.642	3.064	50	100	79	76	71	PROTOAMPHIBOLE
4.435	8.217	2.685	3.044	2.516	34	100	60	47	31	GLAUCOPHANE
4.406	8.365	3.168	2.528	2.280	40	100	99	41	32	BERTRANDITE
4.417	3.787	2.827	2.550	3.680	33	100	66	58	43	CAAL2SI2O8
4.495	3.421	3.726	3.333	3.109	11	100	41	22	15	KEATITE SIO2
4.422	3.396	3.199	8.025	3.537	59	100	67	55	30	GILLESPITE
4.451	3.308	2.530	1.718	2.071	73	100	75	64	20	ZIRCON
4.448	2.557	10.007	3.336	3.476	78	100	87	83	62	MUSCOVITE 2M1 (1)
4.438	2.549	9.890	3.297	3.463	93	100	99	67	57	MUSCOVITE 2M1 (2)
4.450	2.311	2.362	3.604	2.637	100	18	17	16	14	CHLORITOID
4.399 -										
4.305	11.985	7.433	3.186	3.350	8	100	8	8	7	SEPIOLITE
4.392	9.656	2.524	3.219	2.532	100	87	79	77	72	PARAGONITE
4.352	7.186	2.593	4.167	4.383	45	100	63	58	54	NACRITE
4.383	7.186	3.593	4.167	4.352	54	100	63	58	45	NACRITE
4.361	7.131	3.566	4.158	3.333	47	100	57	45	29	KAOLINITE
4.370	7.020	2.483	3.510	2.309	49	100	55	53	34	AMESITE (2 LAYER)
4.356	5.890	6.527	2.865	2.845	49	100	55	51	48	NATROLITE
4.385	6.527	7.970	2.530	2.776	29	100	28	27	15	CLINOHEDRITE
4.367	3.906	3.089	2.521	2.424	100	34	28	11	8	HIGH TRIDYMITE (200C)
4.323	3.604	3.070	4.187	5.865	100	100	81	77	72	DALYITE
4.339	3.399	1.847	1.571	1.396	18	100	13	5	5	HIGH QUARTZ
4.365	3.168	2.528	4.406	2.280	100	99	41	40	32	BERTRANDITE
4.345	3.115	3.941	2.580	2.473	18	63	41	53	19	KALSILITE
4.357	2.739	3.040	2.774	3.833	33	100	79	63	36	GAMMA CA2SIO4
4.390	2.523	3.189	3.121	1.479	52	100	52	42	41	MARGARITE

TABLE 43 LIST OF FIVE STRONGEST LINES

D SPACING					I (PEAK) / I					MINERAL
4.299 – 4.200										
4.267	7.285	3.131	3.184		95	100	60	60	54	GISMONDITE
4.282	14.164	4.526	3.541		32	100	88	46	26	PROCHLORITE
4.201	7.082	3.402	3.010		32	100	46	24	23	BAVENITE
4.206	3.719	5.270	2.910		66	100	71	70	68	MILARITE
4.255	3.343	2.457	1.541		19	100	11	8	7	ALPHA QUARTZ
4.217	3.783	3.278	3.225		62	100	79	77	65	HIGH SANIDINE
4.217	3.232	3.283	3.775		61	100	82	75	66	ORTHOCLASE
4.225	3.239	3.287	3.768		62	100	100	72	72	ADULARIA
4.205	3.240	3.468	3.335		59	100	97	53	47	INTERMED. MICROCLINE
4.213	2.753	2.020	2.102		86	100	93	37	30	EULYTITE
4.203	3.286	3.831	3.368		53	100	49	43	31	MAXIMUM MICROCLINE
4.235	3.852	2.353	3.277		54	100	81	65	44	NEPHELINE
4.281	2.581	4.605	6.413		47	100	89	58	54	TOURMALINE
4.264	2.825	2.414	2.487		24	100	72	47	30	JADEITE
4.264	2.697	3.015	2.462	1.612	37	100	84	78	73	ANDRADITE
4.199 – 4.100										
4.160	9.460	3.509	3.661	3.838	20	100	40	38	16	LAUMONTITE
4.182	8.008	2.669	2.832	4.004	76	100	66	36	27	ZUNYITE
4.167	7.186	4.383	4.352	3.593	58	100	63	54	45	NACRITE
4.123	7.162	3.579	3.792	2.324	50	100	58	48	36	DICKITE
4.158	7.131	3.566	2.333	4.361	45	100	57	47	29	KAOLINITE
4.106	7.085	5.369	4.983	3.187	39	100	57	55	26	PHILLIPSITE
4.130	5.975	3.622	3.942	2.424	69	100	93	85	84	BETA NA2SI2O5
4.107	5.363	3.974	3.392	3.261	58	100	76	74	68	NARSARSUKITE
4.187	3.604	3.070	4.323	5.865	77	100	81	72	59	DALYITE
4.150	3.343	3.538	2.767	3.096	100	50	42	39	33	SANBORNITE
4.138	3.238	2.921	5.544	7.142	56	100	65	63	48	OSUMILITE
4.174	2.914	2.789	3.185	2.449	46	100	56	36	27	SPODUMENE
	2.575	2.569	2.527	4.993	100	57	56	52	18	ZN-CHKALOVITE
4.099 – 4.000										
4.087	8.460	3.384	3.134	3.542	38	100	67	31	30	LOW CORDIERITE
4.004	8.008	2.669	2.832	4.182	36	100	76	66	27	ZUNYITE
4.077	7.305	4.913	3.200	2.609	27	100	41	34	18	DIOPTASE
4.023	4.385	2.530	2.776	7.970	100	29	28	27	15	CLINOHEDRITE
4.013	3.799	2.993	6.391	3.595	100	75	61	49	41	PARACELSIAN
4.042	3.206	3.725	3.149	3.184	53	100	62	24	24	OLIGOCLASE AN 29

TABLE 43   LIST OF FIVE STRONGEST LINES

	D SPACING				I (PEAK)					MINERAL
4.032	3.203	3.212	3.178	3.751	80	100	85	82	58	HIGH ALBITE AN 0
4.040	3.197	3.182	3.761	3.913	81	100	84	48	25	BYTOWNITE AN80
4.041	3.196	3.184	3.261	3.125	50	100	97	50	39	TRANSITIONAL ANORTHITE
4.035	3.194	3.180	3.210	3.261	52	100	100	58	53	ANORTHITE AN100
4.027	3.188	3.124	3.658	3.150	67	100	55	37	30	LOW ALBITE AN 0
4.079	3.172	2.551	2.908	3.377	50	100	51	40	39	PROTOENSTATITE
4.040	2.487	2.842	3.136	1.871	100	12	11	9	4	LOW CRISTOBALITE
3.999 - 3.900										
3.977	8.910	7.945	2.959	5.127	79	100	62	58	47	HEULANDITE
3.981	7.978	3.252	2.865	4.594	19	100	42	39	20	BERYL (ANHYDROUS)
3.984	7.232	2.882	5.813	2.861	83	100	53	43	31	STOKESITE
3.948	7.085	2.559	3.543	2.170	28	100	80	53	37	CRONSTEDTITE (1 LAYER)
3.942	5.975	3.622	2.424	4.130	84	100	93	85	69	BETA NA2SI205
3.974	5.363	3.261	3.392	4.107	76	100	74	68	58	NARSARSUKITE
3.927	5.548	3.261	2.774	2.171	30	100	81	50	30	ANDALUSITE
3.906	4.367	3.089	2.521	2.424	34	100	11	8	5	HIGH TRIDYMITE (200C)
3.913	3.486	3.165	4.608	1.879	33	100	12	9	9	BETA SPODUMENE
3.986	3.289	8.236	2.946	3.055	20	100	64	53	17	BAZZITE
3.913	3.197	3.182	4.040	3.761	25	100	84	81	48	BYTOWNITE AN80
3.941	3.115	2.580	2.473	4.345	63	100	53	19	18	KALSILITE
3.951	2.968	7.790	2.647	2.199	100	63	39	33	21	ALPHA BAAL2SI208
3.899 - 3.800										
3.834	3.214	2.771	2.440		33	100	37	35	21	EUCLASE
3.814	3.060	2.687	3.015		49	100	74	47	40	MARIALITE-SCAPOLITE
3.872	4.923	6.867	2.917		100	83	75	67	52	EPISTILBITE
3.831	4.213	3.286	3.368		31	100	53	49	43	MAXIMUM MICROCLINE
3.874	3.038	2.841	3.137		100	79	53	48	41	REEDMERGNERITE
3.852	3.277	4.203	2.353		81	100	65	54	44	NEPHELINE
3.844	3.323	3.089	3.518		20	100	32	29	27	ALPHA-WOLLASTONITE
3.806	2.976	2.904	3.173		33	100	90	58	35	ARDENNITE
3.833	2.739	2.774	4.357		36	100	79	63	33	GAMMA CA2SIO4
3.896	2.462	2.777	1.752	2.274	58	100	71	60	41	FORSTERITE
3.799 - 3.700										
3.702	12.280	8.683	5.492	7.090	16	100	61	24	16	NA-TYPE A ZEOLITE
3.792	7.162	3.579	4.123	2.324	36	100	58	50	48	DICKITE
3.799	4.013	3.595	2.993	6.391	75	100	61	49	41	PARACELSIAN
3.728	3.661	3.679	3.522	1.935	100	54	52	17	8	PETALITE

TABLE 43 LIST OF FIVE STRONGEST LINES

D SPACING					I (PEAK)					MINERAL
3.726	3.421	3.109	4.495		41	100	22	15	11	KEATITE SIO2
3.719	3.351	3.010	4.402		46	100	32	24	23	BAVENITE
3.783	3.314	3.278	4.217		79	100	77	65	62	HIGH SANIDINE
3.775	3.310	3.283	4.217		75	100	82	66	61	ORTHOCLASE
3.768	3.307	3.287	4.215		72	100	100	72	62	ADULARIA
3.725	3.206	4.042	3.149		24	100	62	53	24	OLIGOCLASE AN 29
3.751	3.203	3.178	4.032		58	100	85	82	80	HIGH ALBITE AN 0
3.761	3.197	4.040	3.913		48	100	84	81	25	BYTOWNITE AN80
3.755	3.109	2.242	2.989		61	100	80	49	46	DATOLITE
3.781	2.900	2.632	1.595		50	100	65	59	49	PUMPELLYITE
3.715	2.873	2.482	1.761		61	100	73	39	36	HARDYSTONITE
3.739	2.865	3.231	3.632		42	100	83	49	48	HODGKINSONITE
3.726	2.856	1.763	2.435		43	100	22	21	17	GEHLENITE
3.787	2.827	3.680	4.417		100	66	58	43	33	CAAL2SI2O8
3.707	2.734	2.862	3.305		100	62	26	22	17	BENITOITE
3.699 −	3.600									
3.642	9.401	6.838	4.160		16	100	40	38	20	LAUMONTITE
3.628	8.271	3.064	4.455		76	100	79	71	50	PROTOAMPHIBOLE
3.621	6.284	2.565	2.810		77	100	24	22	17	HYDROSODALITE
3.622	6.272	2.371	2.560		100	36	25	17	15	SODALITE
3.681	5.975	3.942	4.130		93	100	85	84	69	BETA NA2SI2O5
3.604	4.660	6.375	2.760		54	100	89	81	32	CANCRINITE
3.680	4.450	2.362	2.637		16	100	18	17	14	CHLORITOID
3.661	3.787	2.550	4.417		43	100	66	58	33	CAAL2SI2O8
3.679	3.728	3.522	1.935		54	100	52	17	8	PETALITE
3.659	3.728	3.522	1.935		52	100	54	17	8	PETALITE
3.655	3.423	2.919	1.739		36	100	43	25	16	POLLUCITE
3.604	3.188	3.124	3.150		37	100	67	55	30	LOW ALBITE AN 0
3.692	3.115	2.514	6.229		100	92	51	47	45	PHENACITE
3.632	3.070	5.865	4.323		100	81	77	72	59	DALYITE
3.602	2.936	2.105	2.360		80	100	92	60	47	TOPAZ
3.635	2.865	3.231	3.739		48	100	83	49	42	HODGKINSONITE
	2.759	1.704	2.321		39	100	42	24	23	MELIPHANITE
	2.668	1.818	2.937		81	100	65	58	47	MONTICELLITE
3.599 −	3.500									
3.583	14.333	4.471	2.593	2.393	3	100	6	3	3	MG-VERMICULITE-2 LAYER
3.500	10.410	9.674	2.259	2.733	16	100	13	11	9	ASTROPHYLLITE

TABLE 43   LIST OF FIVE STRONGEST LINES

D SPACING					I (PEAK)					MINERAL
3.519	9.567	3.189	7.722	2.943	39	100	52	39	32	NEPTUNITE
3.509	9.460	6.838	4.160	3.661	40	100	38	20	16	LAUMONTITE
3.575	7.796	2.496	2.980	2.481	49	100	44	33	31	APOPHYLLITE
3.593	7.186	4.167	4.383	4.352	63	100	58	54	45	NACRITE
3.579	7.162	4.123	2.324	3.792	58	100	50	48	36	DICKITE
3.566	7.131	4.361	4.158	2.333	57	100	47	45	29	KAOLINITE
3.560	7.121	14.242	4.747	2.554	53	100	64	44	29	CORUNDOPHILLITE
3.555	7.110	2.370	2.481	4.581	54	100	33	27	24	AL-SERPENTINE
3.543	7.085	2.559	2.170	3.948	53	100	80	37	28	CRONSTEDTITE (1 LAYER)
3.540	7.083	2.519	14.166	2.145	34	100	57	38	25	RIPIDOLITE
3.541	7.082	14.164	4.282	4.526	46	100	88	32	26	PROCHLORITE
3.510	7.020	2.483	4.370	2.309	53	100	55	49	34	AMESITE (2 LAYER)
3.519	6.150	3.567	2.500	2.026	100	41	37	15	14	TUGTUPITE
3.539	5.727	3.542	3.348	5.832	91	100	94	75	74	ALAMOSITE
3.542	5.727	3.539	3.348	5.832	94	100	91	75	74	ALAMOSITE
3.538	4.150	3.343	3.096	2.767	42	100	50	39	33	SANBORNITE
3.595	4.013	3.799	2.993	6.391	61	100	75	49	41	PARACELSIAN
3.522	3.728	3.661	3.679	1.935	17	100	54	52	8	PETALITE
3.567	3.519	6.150	2.500	2.026	37	100	41	15	14	TUGTUPITE
3.535	3.475	3.068	2.556	3.304	44	100	95	83	61	PREHNITE
3.568	3.447	2.963	3.240	2.657	100	80	79	63	57	DANBURITE
3.537	3.396	3.199	4.422	8.025	30	100	67	59	55	GILLESPITE
3.537	3.063	3.215	2.617	1.830	100	74	53	49	38	YODERITE
3.564	3.051	3.239	7.053	2.800	56	100	70	63	59	XONOTLITE
3.538	3.037	2.918	2.675	2.673	21	100	39	30	23	RINKITE
3.551	3.028	2.410	2.564	5.240	34	100	52	49	40	NA2SIO3
3.518	2.976	3.323	3.089	3.844	27	100	32	29	20	ALPHA-WOLLASTONITE
3.530	2.935	2.958	2.676	1.938	31	100	89	38	31	PERRIERITE
3.515	2.886	2.925	8.166	3.455	50	100	83	59	49	BULTFONTEINITE
3.576	2.756	2.968	2.318	5.946	34	100	35	28	20	LEUCOPHANITE
3.555	2.502	2.826	2.564	1.778	75	100	99	65	62	FAYALITE
3.525	2.481	2.797	2.539	1.764	55	100	86	66	60	HORTONOLITE
3.525	2.479	2.794	2.536	1.762	54	100	86	70	61	HYALOSIDERITE
3.499 - 3.450										
3.450	10.163	2.671	3.388	2.479	9	100	24	17	16	FERRI-ANNITE
3.487	9.923	2.566	4.471	2.982	54	100	87	79	53	PHENGITE
3.454	5.849	3.226	3.442	2.082	47	100	65	49	48	DUMORTIERITE

TABLE 43　LIST OF FIVE STRONGEST LINES

D SPACING					I (PEAK)					MINERAL
3.486	3.913	3.165	4.608	1.879	100	33	12	9	9	BETA SPODUMENE
3.472	3.349	6.518	2.581	3.021	76	100	71	55	54	CELSIAN (BA-FELDSPAR)
3.468	3.290	3.240	4.225	3.335	47	100	97	59	53	INTERMED. MICROCLINE
3.475	3.068	2.556	3.304	3.535	100	95	83	61	44	PREHNITE
3.460	3.060	3.814	2.687	3.015	100	74	49	47	40	MARIALITE-SCAPOLITE
3.480	2.893	2.801	2.593	5.014	34	100	38	38	37	EPIDOTE
3.455	2.886	2.925	8.166	3.515	49	100	83	59	50	BULTFONTEINITE
3.476	2.557	10.007	3.336	4.448	62	100	87	83	78	MUSCOVITE 2M1 (1)
3.463	2.549	9.890	4.438	3.297	57	100	99	93	67	MUSCOVITE 2M1 (2)
3.449 – 3.400										
3.417	10.252	2.642	3.442	3.188	57	100	89	75	63	PHLOGOPITE
3.442	10.252	2.642	3.188	3.417	75	100	89	63	57	PHLOGOPITE
3.446	6.465	3.276	2.752	3.175	41	100	52	49	44	KAINOSITE
3.442	5.849	3.226	2.082	3.454	49	100	65	48	47	DUMORTIERITE
3.432	5.605	2.927	4.854	2.693	100	78	47	15	14	ANALCIME (DISORDERED)
3.401	5.401	3.431	2.210	2.548	100	83	60	53	44	MULLITE 1.92-1
3.400	5.396	3.426	2.207	2.545	100	99	57	53	44	MULLITE 2.0-1
3.433	3.872	4.923	6.867	2.917	83	100	75	67	52	EPISTILBITE
3.421	3.726	3.333	3.109	4.495	100	41	22	15	11	KEATITE SIO2
3.447	3.568	2.963	3.240	2.657	80	100	79	63	57	DANBURITE
3.431	3.401	5.401	2.210	2.548	60	100	83	53	44	MULLITE 1.92-1
3.426	3.400	5.396	2.207	2.545	57	100	99	53	44	MULLITE 2.0-1
3.433	3.388	5.384	2.206	2.542	66	100	74	56	45	MULLITE 1.71-1
3.414	3.364	2.204	2.540	1.519	80	100	58	43	29	SILLIMANITE
3.402	3.351	3.719	4.201	3.010	23	100	46	32	24	BAVENITE
3.427	3.106	3.106	3.082	3.444	31	100	99	53	35	COESITE
3.444	3.106	3.106	3.082	3.427	35	100	99	53	31	COESITE
3.443	2.921	2.895	3.234	3.032	34	100	89	53	48	BUSTAMITE
3.423	2.919	3.659	2.420	1.739	100	43	36	25	16	POLLUCITE
3.399 – 3.350										
3.350	11.985	7.433	4.305	3.186	7	100	8	8	8	SEPIOLITE
3.368	10.463	2.636	3.025	2.116	45	100	93	90	38	KORNERUPINE
3.388	10.163	2.671	2.479	3.450	17	100	24	16	9	FERRI-ANNITE
3.379	9.983	2.611	3.131	3.328	55	100	62	50	44	FLUORPHLOGOPITE
3.384	8.460	8.542	4.087	3.134	31	100	67	38	30	LOW CORDIERITE
3.381	8.425	2.706	3.127	2.552	59	100	77	61	56	HORNBLENDE
3.380	8.406	2.705	3.119	9.024	55	100	94	67	65	TREMOLITE

TABLE 43   LIST OF FIVE STRONGEST LINES

D SPACING					I (PEAK)					MINERAL
3.388	5.384	3.433	2.206	2.542	100	74	66	56	45	MULLITE 1.71-1
3.392	5.363	3.974	3.261	4.107	68	100	76	74	58	NARSARSUKITE
3.399	4.339	1.847	1.571	1.396	100	18	13	5	5	HIGH QUARTZ
3.351	3.719	4.201	2.010	3.402	100	46	32	24	23	BAVENITE
3.364	3.414	2.204	2.540	1.519	100	80	58	43	29	SILLIMANITE
3.368	3.247	4.213	3.286	3.831	43	100	53	49	31	MAXIMUM MICROCLINE
3.396	3.199	4.422	8.025	3.537	100	67	59	55	30	GILLESPITE
3.352	3.184	1.929	1.375	1.377	56	100	58	52	47	KYANITE
3.377	3.172	2.551	4.908	2.908	39	100	51	50	40	PROTOENSTATITE
3.364	2.913	4.513	2.975	5.825	100	86	67	64	53	BARYLITE
3.363	2.901	8.316	2.882	2.922	37	100	93	48	30	BATISITE
3.388	2.709	2.547	3.176	8.530	95	100	82	67	66	K-RICHERITE
3.349 - 3.300										
3.343	9.990	2.560	2.330	4.450	93	100	90	86	83	MUSCOVITE 3T
3.328	9.983	2.611	3.379	3.131	44	100	62	55	50	FLUORPHLOGOPITE
3.349	6.980	2.910	5.270	4.206	100	71	70	68	66	MILARITE
3.348	5.727	3.542	3.539	5.832	75	100	94	91	74	ALAMOSITE
3.309	4.682	2.706	2.339	1.560	98	100	68	67	24	LI2SIO3
3.343	4.255	1.817	1.541	2.457	100	19	11	8	7	ALPHA QUARTZ
3.343	4.150	3.538	3.096	2.767	50	79	42	39	33	SANBORNITE
3.314	3.783	3.225	3.278	4.217	100	100	77	65	62	HIGH SANIDINE
3.305	3.707	5.724	2.862	2.862	17	100	62	26	22	BENITOITE
3.304	3.475	3.068	3.535	3.021	61	100	95	83	44	PREHNITE
3.349	3.472	6.518	2.581	4.495	100	76	71	55	54	CELSIAN (BA-FELDSPAR)
3.333	3.421	3.109	4.495	3.468	22	100	41	15	11	KEATITE SIO2
3.335	3.290	3.240	4.225	3.468	53	100	97	59	47	INTERMED. MICROCLINE
3.307	3.239	3.768	3.287	4.215	100	100	72	72	62	ADULARIA
3.310	3.211	3.775	3.283	4.217	66	82	75	66	61	ORTHOCLASE
3.344	2.926	4.995	5.213	3.844	32	100	90	77	61	VLASOVITE
3.323	3.089	3.518	3.266	3.059	30	100	29	27	20	ALPHA-WOLLASTONITE
3.307	3.080	3.266	3.059	3.476	83	100	53	25	22	PECTOLITE
3.336	10.007	4.448	3.476	2.071	100	100	87	78	62	MUSCOVITE 2M1 (1)
3.308	4.451	1.718	2.071	2.071	100	75	73	64	20	ZIRCON
3.299 - 3.250										
3.251	8.307	3.072	2.732	2.513	30	100	50	41	30	MN-CUMMINGTONITE
3.252	8.284	2.749	9.095	3.058	48	100	75	73	62	CUMMINGTONITE
3.258	8.272	6.882	17.547	3.078	36	100	83	81	54	PAULINGITE

TABLE 43 LIST OF FIVE STRONGEST LINES

D SPACING					I (PEAK)					MINERAL
3.289	3.236	2.946	3.986	3.055	100	64	53	20	17	BAZZITE
3.252	7.978	2.865	4.594	3.981	42	100	39	20	19	BERYL (ANHYDROUS)
3.276	6.465	2.752	3.175	3.446	52	100	49	44	41	KAINOSITE
3.261	5.363	3.974	3.392	4.107	74	100	76	68	58	NARSARSUKITE
3.278	3.314	3.783	3.225	4.217	65	100	79	77	62	HIGH SANIDINE
3.283	3.310	3.232	3.775	4.217	66	100	82	75	61	ORTHOCLASE
3.287	3.307	3.239	3.768	4.215	72	100	100	72	62	ADULARIA
3.286	3.247	4.213	3.368	3.831	49	100	53	43	31	MAXIMUM MICROCLINE
3.290	3.240	4.225	3.335	3.468	100	97	59	53	47	INTERMED. MICROCLINE
3.261	3.196	3.184	4.041	3.125	50	100	97	50	39	TRANSITIONAL ANORTHITE
3.261	3.194	3.180	3.210	4.035	53	100	58	58	52	ANORTHITE AN100
3.286	3.102	6.598	5.360	2.400	66	100	76	58	49	HEMIMORPHITE
3.298	3.076	2.694	1.873	2.150	44	100	22	20	19	FRESNOITE
3.277	3.017	3.852	4.203	2.353	65	100	81	54	44	NEPHELINE
3.286	2.980	2.877	3.173	2.462	36	100	77	37	37	CLINOENSTATITE
3.260	2.955	2.904	5.527	5.000	100	96	96	70	67	KENTROLITE
3.266	2.908	3.080	3.307	3.059	25	100	53	30	22	PECTOLITE
3.257	2.753	4.205	2.102	2.020	100	93	86	37	30	EULYTITE
3.297	2.549	9.890	4.438	3.463	67	100	99	93	57	MUSCOVITE 2M1 (2)
3.249 - 3.200										
3.231	8.248	9.007	3.048	2.834	67	100	84	83	58	ANTHOPHYLLITE
3.200	7.305	2.609	4.913	4.077	18	100	41	34	27	DIOPTASE
3.214	7.145	2.771	3.834	2.440	37	100	35	33	21	EUCLASE
3.237	7.050	5.160	6.582	3.107	88	100	72	64	44	ELPIDITE
3.226	5.849	3.442	2.082	3.454	65	100	49	48	47	DUMORTIERITE
3.240	4.660	5.814	3.681	2.760	89	100	81	54	32	CANCRINITE
3.219	4.645	6.375	3.047	6.975	30	100	85	66	31	YUGAWARALITE
3.219	4.392	9.656	2.524	2.532	77	100	87	79	72	PARAGONITE
3.247	4.213	3.286	3.368	3.831	100	53	49	43	31	MAXIMUM MICROCLINE
3.240	3.568	3.447	2.963	2.657	63	100	80	79	57	DANBURITE
3.215	3.537	3.063	2.617	1.830	53	100	74	49	38	YODERITE
3.225	3.314	3.783	3.278	4.217	77	100	79	65	62	HIGH SANIDINE
3.232	3.310	3.775	3.283	4.217	82	100	75	66	61	ORTHOCLASE
3.239	3.307	3.768	3.287	4.215	100	100	72	72	62	ADULARIA
3.240	3.290	4.225	3.335	3.468	97	100	59	53	47	INTERMED. MICROCLINE
3.203	3.212	4.032	3.178	3.751	100	85	82	80	58	HIGH ALBITE AN 0
3.249	3.207	2.885	2.665	7.320	28	100	77	28	27	HARADAITE

TABLE 43   LIST OF FIVE STRONGEST LINES

D SPACING					I (PEAK)					MINERAL
3.212	3.203	3.178	4.032	3.751	85	100	82	80	58	HIGH ALBITE AN 0
3.210	3.194	3.180	3.261	4.035	58	100	100	53	52	ANORTHITE AN100
3.206	3.184	4.042	3.725	3.149	100	62	53	24	24	OLIGOCLASE AN 29
3.239	3.051	7.053	2.800	3.564	70	100	63	59	56	XONOTLITE
3.208	3.018	2.901	2.576	2.486	59	66	52	48	47	FERRO-PIGEONITE
3.248	2.999	2.613	2.597	4.940	100	90	61	32	26	SPHENE
3.211	2.926	4.995	3.344	5.213	100	65	77	66	61	VLASOVITE
3.238	2.921	7.142	4.138	5.544	100	100	63	56	48	OSUMILITE
3.234	2.921	2.895	3.032	3.443	53	77	89	48	34	BUSTAMITE
3.207	2.885	2.665	3.249	7.320	100	100	77	28	27	HARADAITE
3.231	2.865	2.955	3.632	3.739	49	100	83	48	42	HODGKINSONITE
3.243	2.842	2.942	2.976	4.937	52	100	87	67	54	WOHLERITE
3.217	2.557	2.110	3.046	1.500	48	100	46	41	40	XANTHOPHYLLITE
3.199 - 3.150										
3.186	11.985	4.305	7.433	3.350	8	100	8	8	7	SEPIOLITE
3.188	10.252	3.442	2.642	3.417	63	100	89	75	57	PHLOGOPITE
3.189	9.567	3.519	7.722	2.943	52	100	39	39	32	NEPTUNITE
3.184	7.285	4.916	4.267	3.131	60	100	95	60	54	GISMONDITE
3.187	7.085	4.106	5.369	4.983	55	100	57	39	26	PHILLIPSITE
3.175	6.465	2.752	3.276	3.446	44	100	52	49	41	KAINOSITE
3.158	5.006	2.895	4.809	6.783	57	100	72	47	40	LAVENITE
3.168	4.365	2.528	4.406	2.280	99	100	41	40	32	BERTRANDITE
3.188	4.027	3.124	3.658	3.150	100	67	55	37	30	LOW ALBITE AN 0
3.165	3.486	3.913	4.608	1.879	12	100	33	9	9	BETA SPODUMENE
3.199	3.396	4.422	3.537	3.149	67	100	59	55	30	GILLESPITE
3.184	3.206	4.042	3.725	3.149	62	100	53	24	24	OLIGOCLASE AN 29
3.178	3.203	3.212	4.032	3.751	82	100	85	80	58	HIGH ALBITE AN 0
3.182	3.197	3.761	3.913	3.913	84	100	81	48	25	BYTOWNITE AN80
3.184	3.196	4.041	3.261	3.125	97	100	50	50	39	TRANSITIONAL ANORTHITE
3.180	3.194	4.035	4.035	4.035	100	100	58	53	52	ANORTHITE AN100
3.151	3.190	3.210	2.852	4.889	30	100	51	48	33	LARSENITE
3.150	3.188	4.027	3.124	3.658	30	100	67	55	37	LOW ALBITE AN 0
3.196	3.184	4.041	3.261	3.125	100	97	50	50	39	TRANSITIONAL ANORTHITE
3.197	3.182	4.040	3.761	3.913	100	84	81	48	25	BYTOWNITE AN80
3.194	3.180	3.210	4.889	4.035	100	100	53	53	52	ANORTHITE AN100
3.190	3.033	2.852	4.889	3.151	100	51	48	33	30	LARSENITE
3.173	2.980	2.877	2.462	3.286	37	100	77	37	36	CLINOENSTATITE

TABLE 43 LIST OF FIVE STRONGEST LINES

D SPACING					I (PEAK)					MINERAL
3.173	2.940	2.598	2.904	3.806	35	100	90	58	33	ARDENNITE
3.195	2.936	2.692	2.105	2.360	92	100	80	60	47	TOPAZ
3.185	2.914	2.789	4.193	2.449	27	100	56	46	36	SPODUMENE
3.177	2.833	2.727	6.539	2.816	100	92	73	59	58	AFWILLITE
3.176	2.709	3.388	2.547	8.530	67	100	95	82	66	K-RICHERITE
3.172	2.551	4.079	2.908	3.377	100	51	50	40	39	PROTOENSTATITE
3.182	2.542	2.888	2.560	2.489	88	100	60	57	55	HYPERSTHENE
3.189	2.523	4.390	3.121	1.479	52	100	52	42	41	MARGARITE
3.184	1.929	3.352	1.375	1.377	100	58	56	52	47	KYANITE
3.149 – 3.100										
3.117	10.111	2.911	4.509	2.460	25	100	18	10	8	CELADONITE
3.131	9.983	2.611	3.379	3.328	50	100	62	55	44	FLUORPHLOGOPITE
3.134	8.460	8.542	4.087	3.384	30	100	67	38	31	LOW CORDIERITE
3.130	8.429	2.719	4.510	2.541	22	100	29	14	14	RIEBECKITE
3.127	8.425	2.706	3.381	2.552	61	100	77	59	56	HORNBLENDE
3.119	8.406	2.705	9.024	3.380	67	100	94	65	55	TREMOLITE
3.131	7.285	4.267	4.916	3.184	54	100	95	60	55	GISMONDITE
3.107	7.050	3.237	5.160	6.582	44	100	88	72	64	FLPIDITE
3.102	6.598	2.286	5.360	6.400	100	76	66	58	49	HEMIMORPHITE
3.136	4.040	2.487	2.842	1.871	9	100	12	11	4	LOW CRISTOBALITE
3.115	3.941	2.580	2.473	4.345	100	63	53	19	18	KALSILITE
3.137	3.874	3.073	3.038	2.841	41	100	79	79	48	REEDMERGNERITE
3.115	3.655	2.183	2.514	6.229	92	100	51	47	45	PHENACITE
3.109	3.421	3.726	3.333	4.495	15	100	41	22	11	KEATITE SIO2
3.149	3.206	3.184	4.042	3.725	24	100	62	53	50	OLIGOCLASE AN 29
3.125	3.196	3.188	4.041	3.261	39	100	97	50	50	TRANSITIONAL ANORTHITE
3.124	3.188	4.031	3.658	3.150	55	100	67	37	30	LOW ALBITE AN O
3.141	3.112	2.931	2.174	1.642	48	100	53	18	17	THORTVEITITE
3.112	3.106	3.438	3.444	3.427	99	100	35	35	31	COESITE
3.106	3.100	3.431	3.444	3.427	100	99	35	35	31	COESITE
3.106	2.931	2.680	2.174	1.642	100	53	35	18	17	THORTVEITITE
3.112	2.906	3.426	2.997	2.628	45	100	99	63	59	CASIO3 (HIGH PRESSURE)
3.147	2.853	3.755	2.242	2.989	100	80	61	49	46	DATOLITE
3.121	2.523	4.390	3.189	1.479	42	100	52	52	41	MARGARITE
3.141	2.247	2.932	2.978	4.779	49	100	87	77	41	RHODONITE
3.099 – 3.050										
3.076	8.334	2.764	9.151	2.514	33	100	38	30	24	GRUNERITE

TABLE 43   LIST OF FIVE STRONGEST LINES

	D SPACING				I (PEAK)					MINERAL
3.072	8.307	2.732	3.251	2.513	41	100	50	30	30	MN-CUMMINGTONITE
3.058	8.284	9.095	2.749	3.252	62	100	75	73	48	CUMMINGTONITE
3.078	8.272	6.882	17.547	3.258	54	100	83	81	36	PAULINGITE
3.064	8.271	2.823	3.642	4.455	71	100	79	76	50	PROTOAMPHIBOLE
3.067	6.578	2.987	4.517	2.545	69	100	34	30	28	NAIN(SIO3)2
3.089	4.367	3.906	2.521	2.424	11	100	34	8	5	HIGH TRIDYMITE (200C)
3.096	4.150	3.343	2.538	2.767	39	100	50	42	33	SANBORNITE
3.073	3.874	3.038	2.841	3.137	79	100	79	48	41	REEDMERGNERITE
3.070	3.604	4.187	5.865	4.323	81	100	77	72	59	DALYITE
3.063	3.537	3.215	2.617	1.830	74	100	53	49	38	YODERITE
3.068	3.475	2.556	3.304	3.535	95	100	83	61	44	PREHNITE
3.060	3.460	3.814	2.687	3.015	74	100	49	47	40	MARIALITE-SCAPOLITE
3.076	3.298	2.694	1.873	2.150	100	44	22	20	19	FRESNOITE
3.055	3.289	2.236	2.946	3.986	17	100	64	53	20	RAZZITE
3.051	3.239	7.053	2.800	3.564	100	70	63	59	56	XONOTLITE
3.082	3.106	3.100	3.444	3.427	53	100	99	35	31	COESITE
3.089	2.976	3.323	3.518	3.844	29	100	32	27	20	ALPHA-WOLLASTONITE
3.080	2.908	3.307	3.266	3.059	53	100	30	25	22	PECTOLITE
3.059	2.908	3.080	3.307	3.266	22	100	53	30	25	PECTOLITE
3.057	2.906	2.997	2.628	3.147	99	100	63	59	45	CASIO3 (HIGH PRESSURE)
3.087	2.873	3.715	2.482	1.761	73	100	61	39	36	HARDYSTONITE
3.075	2.856	3.726	1.763	2.435	22	100	43	21	17	GEHLENITE
3.061	2.720	9.004	8.276	2.505	77	100	94	93	63	TIRODITE
3.060	2.233	2.640	1.726	2.255	100	70	63	57	38	NORBERGITE
3.049 – 3.000										
3.025	10.463	2.636	3.368	2.116	90	100	93	45	38	KORNERUPINE
3.048	8.248	9.007	3.231	2.834	83	100	84	67	58	ANTHOPHYLLITE
3.044	8.217	2.685	4.435	2.516	47	100	60	34	31	GLAUCOPHANE
3.047	4.645	5.814	6.975	3.219	66	100	85	31	30	YUGAWARALITE
3.038	3.874	3.073	2.841	3.137	79	100	79	48	41	REEDMERGNERITE
3.017	3.852	3.277	4.203	2.353	100	81	65	54	44	NEPHELINE
3.015	3.460	3.060	3.814	2.687	40	100	74	49	47	MARIALITE-SCAPOLITE
3.010	3.351	3.719	4.201	3.402	24	100	46	32	23	RAVENITE
3.021	3.349	3.472	6.518	2.581	54	100	76	71	55	CELSIAN (BA-FELDSPAR)
3.018	3.208	2.901	2.486	2.576	100	59	52	48	47	FERRO-PIGEONITE
3.033	3.190	2.852	4.889	3.151	51	100	48	33	30	LARSENITE
3.025	3.048	2.572	2.550	2.620	40	100	41	36	29	JOHANNSENITE

TABLE 43   LIST OF FIVE STRONGEST LINES

D SPACING					I (PEAK)					MINERAL
3.032	2.921	2.895	3.234	3.443	48	100	89	53	34	BUSTAMITE
3.037	2.918	2.675	2.673	3.538	100	39	30	23	21	RINKITE
3.035	2.909	4.605	2.603	2.161	100	60	58	58	51	CLINOFERROSILITE
3.040	2.739	2.774	3.833	4.357	79	100	63	36	33	GAMMA CA2SIO4
3.015	2.697	1.612	2.462	4.264	84	100	78	73	37	ANDRADITE
3.048	2.572	3.025	2.550	2.620	100	41	40	36	29	JOHANNSENITE
3.002	2.558	2.528	2.961	2.568	100	46	42	33	31	FASSITE (AL-AUGITE)
3.046	2.557	3.217	2.110	1.500	41	100	48	46	40	XANTHOPHYLLITE
3.028	2.410	2.564	5.240	3.551	100	52	49	40	34	NA2SIO3
3.007	2.399	2.693	1.979	2.356	80	100	82	53	49	STAUROLITE
2.999 - 2.950										
2.982	9.923	2.566	4.471	3.487	53	100	87	79	54	PHENGITE
2.959	8.910	3.977	7.945	5.127	58	100	79	62	47	HEULANDITE
2.980	7.796	3.575	2.496	2.481	33	100	49	44	31	APOPHYLLITE
2.987	6.578	3.067	4.517	2.545	34	100	69	30	28	NAIN(SIO3)2
2.982	6.362	2.899	2.524	2.472	100	59	47	37	34	ACMITE
2.958	6.313	2.879	2.512	2.457	100	58	51	42	34	UREYITE
2.993	4.013	3.799	3.595	6.391	49	100	75	61	41	PARACELSIAN
2.968	3.951	7.790	2.647	2.199	53	100	39	33	21	ALPHA BAAL2SI208
2.963	3.568	3.447	3.240	2.657	79	100	80	63	57	DANBURITE
2.975	3.364	2.913	4.513	5.825	64	100	86	67	53	BARYLITE
2.976	3.323	3.089	3.518	3.844	100	32	29	27	20	ALPHA-WOLLASTONITE
2.955	3.260	2.904	5.527	5.000	96	100	96	70	67	KENTROLITE
2.999	3.248	2.613	2.597	4.940	66	100	61	32	26	SPHENE
2.989	3.109	2.853	3.755	2.242	46	100	80	61	49	DATOLITE
2.961	3.002	2.558	2.528	2.568	33	100	46	42	31	FASSITE (AL-AUGITE)
2.951	2.990	2.517	2.892	2.892	30	100	55	46	39	DIOPSIDE
2.958	2.935	2.676	3.530	1.938	89	100	38	31	31	PERRIERITE
2.997	2.906	3.057	2.628	3.147	63	100	99	59	45	CASIO3 (HIGH PRESSURE)
2.980	2.877	3.173	2.462	3.286	100	77	37	37	36	CLINOENSTATITE
2.986	2.872	6.592	2.577	2.434	60	100	33	28	23	CA-SEIDOZERITE
2.955	2.865	3.231	3.632	3.739	83	100	49	48	42	HODGKINSONITE
2.976	2.842	2.942	4.937	3.243	67	100	87	54	52	WOHLERITE
2.971	2.759	3.602	1.704	2.321	42	100	39	24	23	MELIPHANITE
2.968	2.756	3.576	2.318	5.946	35	100	34	28	23	LEUCOPHANITE
2.984	2.669	1.595	2.436	1.655	71	100	57	49	20	UVAROVITE
2.964	2.651	1.584	1.644	1.923	36	100	35	22	20	GROSSULAR

TABLE 43   LIST OF FIVE STRONGEST LINES

D SPACING					I (PEAK)					MINERAL
2.975	2.581	6.413	4.605	4.235	100	89	58	54	47	TOURMALINE
2.966	2.518	2.889	2.469	2.870	100	43	38	38	34	OMPHACITE
2.990	2.517	2.526	2.892	2.951	100	55	46	39	30	DIOPSIDE
2.996	2.456	2.017	2.841	2.013	66	100	73	47	47	SAPPHIRINE
2.978	2.247	2.932	3.141	4.779	77	100	87	49	41	RHODONITE
2.955	1.530	1.979	2.247	1.478	100	42	39	25	15	STISHOVITE
2.949 - 2.900										
2.911	10.111	3.117	4.509	2.460	18	100	25	10	8	CELADONITE
2.943	9.567	3.189	7.722	3.519	32	100	52	39	39	NEPTUNITE
2.901	8.316	2.882	3.363	2.922	100	93	48	37	30	BATISITE
2.918	6.299	4.655	4.522	5.069	84	100	75	71	63	BREWSTERITE
2.900	4.671	2.632	3.781	1.595	100	65	59	50	49	PUMPELLYITE
2.910	4.608	2.593	2.519	6.468	97	100	96	94	92	FERROSILITE
2.917	3.872	3.433	4.923	6.867	52	100	83	75	67	EPISTILBITE
2.927	3.432	5.605	4.854	2.693	47	100	78	15	14	ANALCIME (DISORDERED)
2.919	3.423	3.659	2.420	1.739	43	100	36	25	16	POLLUCITE
2.913	3.364	4.513	2.975	5.825	86	100	67	64	53	BARYLITE
2.910	3.349	6.980	5.270	4.206	70	100	71	68	66	MILARITE
2.946	3.289	8.236	3.986	3.055	53	100	64	20	17	BAZZITE
2.904	3.260	2.955	5.527	5.000	96	100	96	70	67	KENTROLITE
2.921	3.238	7.142	4.138	5.544	65	100	63	56	48	OSUMILITE
2.921	3.211	4.995	3.344	5.213	90	100	77	66	61	VLASOVITE
2.926	3.195	3.692	2.105	2.360	100	92	80	60	47	TOPAZ
2.936	3.172	2.551	4.079	3.377	40	100	51	50	39	PROTOENSTATITE
2.908	3.141	3.141	2.174	1.642	53	100	48	18	17	THORTVEITITE
2.931	3.080	3.307	3.266	3.059	100	53	30	25	22	PECTOLITE
2.908	3.057	2.997	2.628	3.147	100	99	63	59	45	CASIO3 (HIGH PRESSURE)
2.906	3.037	2.675	2.673	3.538	39	100	30	23	21	RINKITE
2.918	3.035	3.208	2.603	2.161	60	100	58	58	51	CLINOFERROSILITE
2.909	3.018	2.676	2.576	1.938	52	100	59	48	47	FERRO-PIGEONITE
2.901	2.958	2.598	3.530	1.938	100	89	38	31	31	PERRIERITE
2.935	2.940	2.876	3.173	3.806	58	100	90	35	33	ARDENNITE
2.904	2.901	8.316	2.882	3.363	30	100	93	48	37	BATISITE
2.922	2.895	3.234	3.032	3.443	100	89	53	48	34	RUSTAMITE
2.925	2.886	8.166	3.515	3.455	83	100	59	50	49	BULTFONTEINITE
2.942	2.842	2.976	4.937	3.243	87	100	67	54	52	WOHLERITE
2.917	2.825	2.487	2.414	4.281	100	72	47	30	24	JADEITE

TABLE 43   LIST OF FIVE STRONGEST LINES

D SPACING					I (PEAK)					MINERAL
2.914	2.789	4.193	2.449	3.185	100	56	46	36	27	SPODUMENE
2.937	2.668	3.635	2.588	1.818	47	100	81	65	58	MONTICELLITE
2.940	2.598	2.904	3.173	3.806	100	90	58	35	33	ARDENNITE
2.932	2.247	2.978	3.141	4.779	87	100	77	49	41	RHODONITE
2.899 – 2.850										
2.865	7.978	3.252	4.594	3.981	39	100	42	20	19	BERYL (ANHYDROUS)
2.882	7.232	3.984	5.813	2.861	53	100	83	43	31	STOKESITE
2.861	7.232	3.984	2.882	5.813	31	100	83	53	43	STOKESITE
2.865	5.890	6.527	4.356	2.845	51	100	55	49	48	NATROLITE
2.895	5.006	3.158	4.809	6.783	72	100	57	47	40	LAVENITE
2.856	3.726	3.075	1.763	2.435	100	43	22	21	17	GEHLENITE
2.862	3.707	2.734	5.724	3.305	22	100	62	26	17	BENITOITE
2.885	3.207	3.249	2.665	7.320	77	100	28	28	27	HARADAITE
2.852	3.190	3.033	4.889	3.151	48	100	51	33	30	LARSENITE
2.853	3.109	3.755	2.242	2.989	80	100	61	49	46	DATOLITE
2.873	3.087	3.715	2.482	1.761	100	73	61	39	36	HARDYSTONITE
2.892	2.990	2.517	2.526	2.951	39	100	55	46	30	DIOPSIDE
2.872	2.986	6.592	2.577	2.434	100	60	33	28	23	CA-SEIDOZERITE
2.899	2.982	6.362	2.524	2.472	77	100	59	37	34	ACMITE
2.877	2.980	3.173	2.462	3.286	37	100	37	37	36	CLINOENSTATITE
2.870	2.966	2.518	2.889	2.469	34	100	43	38	38	OMPHACITE
2.889	2.966	2.518	2.469	2.870	38	100	43	38	34	OMPHACITE
2.879	2.958	6.313	2.512	2.457	51	100	58	42	34	UREYITE
2.865	2.955	7.231	3.632	3.739	100	83	49	48	42	HODGKINSONITE
2.886	2.925	8.166	3.515	3.455	100	83	59	50	49	BULTFONTEINITE
2.895	2.921	3.234	3.032	3.443	89	100	53	48	34	BUSTAMITE
2.882	2.901	8.316	3.363	2.922	48	100	93	37	30	BATISITE
2.893	2.801	2.593	5.014	3.480	100	38	38	37	34	EPIDOTE
2.891	2.792	2.586	5.016	2.673	100	38	36	32	31	CLINOZOISITE
2.866	2.698	2.780	5.018	2.723	100	89	41	36	36	ZOISITE
2.889	2.584	2.359	1.544	4.718	33	100	71	28	15	PYROPE
2.865	2.562	1.531	2.443	1.589	63	100	59	41	38	PYROPE (SYNTHETIC)
2.888	2.542	3.182	2.560	2.489	60	100	88	57	55	HYPERSTHENE
2.849 – 2.800										
2.823	8.271	3.642	3.064	4.455	79	100	76	71	50	PROTOAMPHIBOLE
2.834	8.248	9.007	3.048	3.231	58	100	84	83	67	ANTHOPHYLLITE
2.832	8.008	4.182	2.669	4.004	27	100	76	66	36	ZUNYITE

TABLE 43 LIST OF FIVE STRONGEST LINES

d	D SPACING				I (PEAK)					MINERAL
2.810	6.284	3.628	2.095	2.565	17	100	77	24	22	HYDROSODALITE
2.845	5.890	6.527	2.865	4.356	48	100	55	51	49	NATROLITE
2.842	4.040	2.487	3.136	1.871	11	100	12	9	4	LOW CRISTOBALITE
2.841	3.874	3.073	3.038	3.137	48	100	79	79	41	REEDMERGNERITE
2.827	3.787	2.550	3.680	4.417	66	100	58	43	33	CAAL2SI2O8
2.816	3.177	2.833	2.727	6.539	58	100	92	73	59	AFWILLITE
2.833	3.177	2.727	2.816	6.539	92	100	73	59	58	AFWILLITE
2.800	3.051	3.239	7.053	3.564	59	100	70	63	56	XONOTLITE
2.842	2.942	2.976	4.937	3.243	100	87	67	54	52	WOHLERITE
2.825	2.917	2.487	2.414	4.281	72	100	47	30	24	JADEITE
2.801	2.893	2.593	5.014	3.480	38	100	38	37	34	EPIDOTE
2.826	2.502	3.555	2.564	1.778	99	100	75	65	62	FAYALITE
2.841	2.456	2.017	2.996	2.013	47	100	73	66	47	SAPPHIRINE
2.799 - 2.750										
2.764	8.334	3.076	9.151	2.514	38	100	33	30	24	GRUNERITE
2.771	7.145	3.214	3.834	2.440	35	100	37	33	21	EUCLASE
2.752	6.465	3.276	3.175	3.446	49	100	52	44	41	KAINOSITE
2.760	4.660	3.240	6.375	3.681	32	100	89	81	54	CANCRINITE
2.774	4.526	5.548	3.927	2.171	50	100	81	30	30	ANDALUSITE
2.767	4.150	3.343	3.538	3.096	33	100	50	42	39	SANBORNITE
2.776	4.023	4.385	7.970	2.530	15	100	29	28	27	CLINOHEDRITE
2.753	3.257	3.602	1.704	2.020	93	100	86	37	30	EULYTITE
2.759	2.971	3.576	2.318	2.321	100	42	39	24	23	MELIPHANITE
2.756	2.968	4.193	2.449	5.946	100	35	34	28	20	LEUCOPHANITE
2.789	2.914	2.586	5.016	3.185	56	100	46	36	27	SPODUMENE
2.792	2.891	2.698	5.018	2.673	38	100	36	32	31	CLINOZOISITE
2.780	2.866	2.718	2.610	2.723	41	100	89	36	36	ZOISITE
2.793	2.748	3.040	3.833	2.189	91	100	99	78	64	LARNITE (BETA CA2SIO4)
2.774	2.739	2.476	1.633	4.357	63	78	79	36	33	GAMMA CA2SIO4
2.769	2.611			9.450	100	100	47	46	37	VESUVIANITE
2.797	2.481	2.539	1.764	3.525	86	100	66	60	55	HORTONOLITE
2.794	2.479	2.536	1.762	3.525	86	100	70	61	54	HYALOSIDERITE
2.777	2.462	1.752	3.896	2.274	71	100	60	58	41	FORSTERITE
2.749 - 2.700										
2.733	10.410	3.500	9.674	2.259	13	100	16	11	9	ASTROPHYLLITE
2.720	9.004	8.276	3.061	2.505	100	94	93	77	63	TIRODITE
2.719	8.429	3.130	4.510	2.541	29	100	22	14	14	RIEBECKITE

868

TABLE 43   LIST OF FIVE STRONGEST LINES

D SPACING					I (PEAK)					MINERAL
2.706	8.425	3.127	3.381	2.552	77	100	61	59	56	HORNBLENDE
2.705	8.406	3.119	9.024	3.380	94	100	67	65	55	TREMOLITE
2.732	8.307	3.072	3.251	2.513	59	100	41	30	30	MN-CUMMINGTONITE
2.749	8.284	9.095	3.058	3.252	73	100	75	62	48	CUMMINGTONITE
2.719	7.300	2.534	2.676	6.535	63	100	63	40	31	ILVAITE
2.706	4.682	3.309	2.339	1.560	68	100	98	67	24	LI2SIO3
2.734	3.707	5.724	2.862	3.305	62	100	26	22	17	BENITOITE
2.709	3.388	2.547	3.176	8.530	100	95	82	67	66	K-RICHERITE
2.727	3.177	2.833	6.539	2.816	73	100	92	59	58	AFWILLITE
2.739	3.040	2.774	3.833	4.357	100	79	63	36	33	GAMMA CA2SIO4
2.723	2.866	2.698	2.780	5.018	36	100	89	41	36	ZOISITE
2.718	2.748	2.793	2.610	2.189	99	100	91	78	64	LARNITE (BETA CA2SIO4)
2.748	2.718	2.793	2.610	2.189	100	99	91	78	64	LARNITE (BETA CA2SIO4)
2.706	2.697	6.942	2.673	2.646	100	99	67	60	56	SPURRITE
2.721	2.619	2.630	2.433	1.549	84	100	87	80	76	LAWSONITE
2.699 - 2.650										
2.671	10.163	3.388	2.479	3.450	24	100	17	16	9	FERRI-ANNITE
2.685	8.217	3.044	4.435	2.516	60	100	47	34	31	GLAUCOPHANE
2.669	8.008	4.182	4.004	2.832	66	100	76	36	27	ZUNYITE
2.676	7.300	2.719	2.534	6.535	40	100	63	63	31	ILVAITE
2.668	3.635	2.588	1.818	2.937	100	81	65	58	47	MONTICELLITE
2.657	3.568	3.447	2.963	3.240	57	100	80	79	63	DANBURITE
2.687	3.460	3.060	3.814	3.015	47	100	74	49	40	MARIALITE-SCAPOLITE
2.693	3.432	5.605	2.927	4.854	14	100	78	47	15	ANALCIME (DISORDERED)
2.665	3.207	2.885	3.249	7.320	28	100	77	28	27	HARADAITE
2.694	3.076	3.298	1.873	2.150	22	100	44	20	19	FRESNOITE
2.675	3.037	2.918	2.675	2.538	30	100	39	23	21	RINKITE
2.673	3.015	2.918	2.675	3.538	23	100	30	23	21	RINKITE
2.697	3.015	1.612	2.462	4.264	100	84	78	73	37	ANDRADITE
2.669	2.984	1.595	2.436	1.655	100	71	57	49	23	UVAROVITE
2.651	2.964	1.584	1.644	1.923	100	36	35	22	20	GROSSULAR
2.676	2.935	2.958	3.530	1.938	38	100	89	31	31	PERRIERITE
2.673	2.891	2.792	2.586	5.016	31	100	38	36	32	CLINOZOISITE
2.698	2.866	2.780	5.018	2.723	89	100	41	36	36	ZOISITE
2.697	2.706	6.942	2.673	2.646	99	100	67	60	56	SPURRITE
2.673	2.706	2.697	6.942	2.646	60	100	99	67	56	SPURRITE
2.693	2.399	3.007	1.979	2.356	82	100	80	53	49	STAUROLITE

TABLE 43   LIST OF FIVE STRONGEST LINES

D SPACING					I (PEAK)					MINERAL
2.677	2.261	1.746	2.267	2.627	58	100	97	91	55	CHONDRODITE
2.649 –										
2.636	2.600	3.025	3.368	2.116	93	100	90	45	38	KORNERUPINE
2.642	10.463	3.442	3.188	3.417	89	100	75	63	57	PHLOGOPITE
2.611	10.252	3.379	3.131	3.328	62	100	55	50	44	FLUORPHLOGOPITE
2.609	9.983	4.913	4.077	3.200	41	100	34	27	18	DIOPTASE
2.607	7.305	2.607	2.607	4.660	39	100	38	31	24	FERROCARPHOLITE
2.607	5.687	2.602	5.041	4.660	38	100	39	31	24	FERROCARPHOLITE
2.637	5.687	2.311	2.362	3.604	14	100	18	17	16	CHLORITOID
2.647	4.450	2.968	7.790	2.199	33	100	63	39	21	ALPHA BAAL2SI2O8
2.617	3.951	3.063	3.215	1.830	49	100	74	53	38	YODERITE
2.613	3.537	2.999	2.597	2.597	61	100	66	32	26	SPHENE
2.640	3.248	2.233	1.726	2.255	63	100	70	57	38	NORBERGITE
2.620	3.060	2.572	3.025	2.550	29	100	41	40	36	JOHANNSENITE
2.603	3.048	2.909	4.605	2.161	58	100	60	58	51	CLINOFERROSILITE
2.628	3.035	3.057	2.997	2.997	59	100	99	63	45	CASIO3 (HIGH PRESSURE)
2.632	2.906	4.671	3.781	1.595	59	100	65	50	49	PUMPELLYITE
2.611	2.900	2.476	1.633	9.450	78	100	47	46	37	VESUVIANITE
2.610	2.769	2.718	2.793	2.189	78	100	99	91	64	LARNITE (BETA CA2SIO4)
2.646	2.748	2.697	6.942	2.673	56	100	99	67	60	SPURRITE
2.619	2.706	2.721	2.433	1.549	100	87	84	80	76	LAWSONITE
2.630	2.630	2.721	2.433	1.549	87	100	84	80	76	LAWSONITE
2.627	2.619	1.746	2.267	2.677	55	100	97	91	58	CHONDRODITE
2.599 –										
2.593	14.333	2.393	4.471	3.583	3	100	6	3	3	MG-VERMICULITE-2 LAYER
2.557	10.007	3.336	4.448	3.476	100	87	83	78	62	MUSCOVITE 2M1 (1)
2.566	9.923	4.471	3.487	2.982	87	100	79	54	53	PHENGITE
2.552	8.425	2.706	3.127	3.381	56	100	77	61	59	HORNBLENDE
2.554	7.121	14.242	3.560	4.747	29	100	64	53	44	CORUNDOPHILLITE
2.559	7.085	3.543	2.095	3.948	80	100	53	37	28	CRONSTEDTITE (1 LAYER)
2.565	6.284	2.095	2.330	2.810	22	100	77	24	17	HYDROSODALITE
2.560	9.990	3.343	2.519	4.450	90	100	93	86	83	MUSCOVITE 3T
2.593	4.608	2.910	2.910	6.468	96	100	97	94	92	FERROSILITE
2.569	4.174	2.575	4.993	2.527	56	100	57	52	18	ZN-CHKALOVITE
2.575	4.174	2.569	4.993	2.527	57	100	56	52	18	ZN-CHKALOVITE
2.550	3.787	2.827	3.680	4.417	58	100	66	43	33	CAAL2SI2O8
2.560	3.621	6.272	2.091	2.371	15	100	36	25	17	SODALITE

TABLE 43   LIST OF FIVE STRONGEST LINES

D SPACING					I (PEAK)					MINERAL
2.556	3.475	3.068	3.304	3.535	100	95	61	44	83	PREHNITE
2.581	3.349	3.472	6.518	3.021	100	76	71	54	55	CELSIAN (BA-FELDSPAR)
2.597	3.248	2.999	2.613	4.940	66	66	61	26	32	SPHENE
2.557	3.217	2.110	3.046	1.500	100	46	41	40	100	XANTHOPHYLLITE
2.551	3.172	4.079	2.908	3.377	100	50	40	39	51	PROTOENSTATITE
2.580	3.115	3.941	2.473	4.345	100	63	19	18	53	KALSILITE
2.572	3.048	3.025	2.550	2.620	100	40	36	29	41	JOHANNSENITE
2.550	3.048	2.572	3.025	2.620	100	41	40	29	36	JOHANNSENITE
2.564	3.028	2.410	5.240	3.551	100	52	40	34	49	NA2SIO3
2.576	3.018	3.208	2.901	2.486	100	59	52	47	48	FERRO-PIGEONITE
2.568	3.002	2.558	2.528	2.961	100	46	42	33	31	FASSITE (AL-AUGITE)
2.558	3.002	2.528	2.961	2.568	100	42	33	31	46	FASSITE (AL-AUGITE)
2.581	2.975	6.413	4.605	4.235	100	58	54	47	89	TOURMALINE
2.593	2.940	2.801	3.173	3.806	100	58	35	33	90	ARDENNITE
2.586	2.893	2.792	5.014	3.480	100	38	37	34	38	EPIDOTE
2.577	2.891	2.986	5.016	2.673	100	38	32	31	36	CLINOZOISITE
2.562	2.872	1.531	6.592	2.434	63	60	33	23	28	CA-SEIDOZERITE
2.588	2.865	3.635	2.443	1.589	100	59	41	38	65	PYROPE (SYNTHETIC)
2.560	2.668	3.182	1.818	2.937	100	81	58	47	57	MONTICELLITE
2.564	2.542	2.826	2.888	2.489	100	88	60	55	65	HYPERSTHENE
2.584	2.502	2.889	3.555	1.778	100	99	75	62	100	FAYALITE
2.577	2.359	2.267	1.544	4.718	71	33	28	15	80	PYROPE
2.577	1.747	2.267	1.742	2.454	100	95	91	75	80	HUMITE
2.549 - 2.500										
2.549	9.890	4.438	3.297	3.463	99	93	67	57	100	MUSCOVITE 2M1 (2)
2.541	8.429	2.719	3.130	4.510	100	29	22	14	14	RIEBECKITE
2.514	8.334	2.764	3.076	9.151	100	38	33	30	24	GRUNERITE
2.513	8.307	2.732	3.072	3.251	100	50	41	30	30	MN-CUMMINGTONITE
2.516	8.217	2.685	3.044	4.435	100	60	47	34	31	GLAUCOPHANE
2.534	7.300	2.719	2.676	6.535	100	63	40	31	63	ILVAITE
2.519	7.083	14.166	3.540	2.145	100	38	34	25	57	RIPIDOLITE
2.545	6.578	3.067	2.987	4.517	100	69	34	30	28	NAIN(SIO3)2
2.519	4.608	2.910	2.593	6.468	100	97	96	92	94	FERROSILITE
2.519	4.392	9.656	2.532	2.532	100	87	77	77	79	PARAGONITE
2.532	4.392	2.524	2.524	3.219	100	87	79	72	72	PARAGONITE
2.523	4.390	3.189	3.121	1.479	52	52	42	41	100	MARGARITE
2.521	4.367	3.906	3.089	2.424	100	34	11	5	8	HIGH TRIDYMITE (200C)

TABLE 43 LIST OF FIVE STRONGEST LINES

D SPACING					I (PEAK)					MINERAL
2.528	3.168	4.406	2.280		41	100	99	40	32	BERTRANDITE
2.527	2.575	2.569	4.993		18	100	57	56	52	ZN-CHKALOVITE
2.530	4.385	7.970	2.776		27	100	29	28	15	CLINOHEDRITE
2.514	3.115	2.183	6.229		47	100	92	51	45	PHENACITE
2.500	6.150	3.567	2.026		15	100	41	37	14	TUGTUPITE
2.548	5.401	3.431	2.210		44	100	83	60	53	MULLITE 1.92-1
2.545	5.396	3.426	2.207		44	100	99	57	53	MULLITE 2.0-1
2.542	5.384	3.433	2.206		45	100	74	66	56	MULLITE 1.71-1
2.540	3.414	2.204	1.519		43	100	80	58	29	SILLIMANITE
2.530	4.451	1.718	2.071		75	100	73	64	20	ZIRCON
2.542	2.888	2.560	2.489		100	88	60	57	55	HYPERSTHENE
2.528	3.002	2.961	2.568		42	100	46	33	31	FASSITE (AL-AUGITE)
2.517	2.990	2.892	2.951		55	100	55	39	30	DIOPSIDE
2.526	2.990	2.517	2.951		46	100	55	39	30	DIOPSIDE
2.524	2.982	2.899	2.472	6.362	37	100	59	47	34	ACMITE
2.518	2.966	2.889	2.469	2.870	43	100	38	38	34	OMPHACITE
2.512	2.958	6.313	2.879	2.457	42	100	58	51	34	UREYITE
2.502	2.826	3.555	2.564	1.778	100	99	75	65	62	FAYALITE
2.505	2.720	9.004	8.276	3.061	63	100	94	93	77	TIRODITE
2.547	2.709	3.388	3.176	8.530	82	100	95	67	66	K-RICHERITE
2.539	2.481	2.797	1.764	3.525	66	100	86	60	55	HORTONOLITE
2.536	2.479	2.794	1.762	3.525	70	100	86	61	54	HYALOSIDERITE
2.499 – 2.450										
2.479	10.163	3.388	3.450		16	100	24	17	9	FERRI-ANNITE
2.460	10.111	2.911	4.509		8	100	25	18	10	CELADONITE
2.496	7.796	2.980	2.481		44	100	49	33	31	APOPHYLLITE
2.481	7.796	2.496	2.980		31	100	49	44	33	APOPHYLLITE
2.481	7.110	3.555	4.581		27	100	54	33	24	AL-SERPENTINE
2.483	7.020	3.510	4.370		55	100	53	49	34	AMESITE (2 LAYER)
2.487	4.040	2.842	3.136	1.871	12	100	11	9	4	LOW CRISTOBALITE
2.457	3.343	4.255	1.817	1.541	7	100	19	11	8	ALPHA QUARTZ
2.473	3.115	3.941	2.580	4.345	19	100	63	53	18	KALSILITE
2.486	3.018	3.208	2.901	2.576	47	100	59	52	48	FERRO-PIGEONITE
2.472	2.982	6.362	2.899	2.524	34	100	59	47	37	ACMITE
2.462	2.980	2.877	3.173	3.286	37	100	77	37	36	CLINOENSTATITE
2.469	2.966	2.889	2.518	2.870	38	100	43	38	34	OMPHACITE
2.457	2.958	6.313	2.879	2.512	34	100	58	51	42	UREYITE

TABLE 43   LIST OF FIVE STRONGEST LINES

D SPACING					I (PEAK)					MINERAL
2.487	2.917	2.825	2.414	4.281	47	100	72	30	24	JADEITE
2.482	2.873	3.087	3.715	1.761	39	100	73	61	36	HARDYSTONITE
2.481	2.797	2.539	1.764	3.525	100	86	66	60	55	HORTONOLITE
2.479	2.794	2.536	1.762	3.525	100	86	70	61	54	HYALOSIDERITE
2.462	2.777	1.752	3.896	2.274	100	71	60	58	41	FORSTERITE
2.476	2.769	2.611	1.633	9.450	47	100	78	46	37	VESUVIANITE
2.462	2.697	3.015	1.612	4.264	73	100	84	78	37	ANDRADITE
2.489	2.542	3.182	2.888	2.560	55	100	88	60	57	HYPERSTHENE
2.456	2.017	2.996	2.841	2.013	100	73	66	47	47	SAPPHIRINE
2.454	1.747	2.267	1.742	2.577	75	100	95	91	80	HUMITE
2.449 – 2.400										
2.440	7.145	3.214	2.771	3.834	21	100	37	35	33	EUCLASE
2.424	5.975	3.622	3.942	4.130	85	100	93	84	69	BETA NA2SI205
2.424	4.367	3.906	3.089	2.521	5	100	34	11	8	HIGH TRIDYMITE (200C)
2.420	3.423	2.919	3.659	1.739	25	100	43	36	16	POLLUCITE
2.400	3.102	6.598	3.286	5.360	49	100	76	66	58	HEMIMORPHITE
2.410	3.028	2.564	5.240	3.551	52	100	49	40	34	NA2SIO3
2.414	2.917	2.825	2.487	4.281	30	100	72	47	24	JADEITE
2.449	2.914	2.789	4.193	3.185	36	100	56	46	27	SPODUMENE
2.434	2.872	2.986	6.592	2.577	23	100	60	33	28	CA-SEIDOZERITE
2.435	2.856	3.726	3.075	1.763	17	100	43	22	21	GEHLENITE
2.436	2.669	2.984	1.595	1.655	49	100	71	57	23	UVAROVITE
2.433	2.619	2.630	2.721	1.549	80	100	87	84	76	LAWSONITE
2.443	2.562	2.865	1.531	1.589	41	100	63	59	38	PYROPE (SYNTHETIC)
2.399 – 2.350										
2.393	14.333	4.471	3.583	2.593	6	100	3	3	3	MG-VERMICULITE-2 LAYER
2.370	7.110	3.555	2.481	4.581	33	100	54	27	24	AL-SERPENTINE
2.362	4.450	2.311	3.604	2.637	17	100	18	16	14	CHLORITOID
2.371	6.272	3.636	2.091	2.560	17	100	36	25	15	SODALITE
2.353	3.017	3.852	3.277	4.203	44	100	81	65	54	NEPHELINE
2.360	2.936	3.195	3.692	2.105	47	100	92	80	60	TOPAZ
2.399	2.693	3.007	1.979	2.356	100	82	80	53	49	STAUROLITE
2.359	2.584	2.889	1.544	4.718	71	100	33	28	15	PYROPE
2.356	2.399	2.693	3.007	1.979	49	100	82	80	53	STAUROLITE
2.349 – 2.300										
2.330	9.990	3.343	2.560	4.450	86	100	93	90	83	MUSCOVITE 3T
2.324	7.162	3.579	4.123	3.792	48	100	58	50	36	DICKITE

TABLE 43   LIST OF FIVE STRONGEST LINES

D SPACING					I (PEAK)					MINERAL
2.333	7.131	3.566	4.361	4.158	29	100	57	47	45	KAOLINITE
2.309	7.020	2.483	3.510	4.370	34	100	55	53	49	AMESITE (2 LAYER)
2.339	4.682	3.309	2.706	1.560	67	100	98	68	24	LI2SIO3
2.311	4.450	2.362	3.604	2.637	18	100	17	16	14	CHLORITOID
2.321	2.759	2.971	3.602	1.704	23	100	42	39	24	MELIPHANITE
2.318	2.756	2.968	3.576	5.946	28	100	35	34	20	LEUCOPHANITE
2.299 - 2.250										
2.259	10.410	3.500	2.733	9.674	9	100	16	13	11	ASTROPHYLLITE
2.280	4.365	3.168	2.528	4.406	32	100	99	41	40	BERTRANDITE
2.255	3.060	2.233	2.640	1.726	38	100	70	63	57	NORBERGITE
2.274	2.462	2.777	1.752	3.896	41	100	71	60	58	FORSTERITE
2.267	2.261	1.746	2.677	2.627	91	100	97	58	55	CHONDRODITE
2.266	1.750	1.746	2.272	2.266	74	100	93	90	75	CLINOHUMITE
2.272	1.750	1.746	2.272	2.262	75	100	93	75	74	CLINOHUMITE
2.267	1.750	1.746	2.266	2.262	90	100	93	80	74	CLINOHUMITE
2.261	1.747	1.742	2.577	2.454	95	100	91	91	75	HUMITE
2.261	1.746	2.267	2.677	2.627	100	97	91	58	55	CHONDRODITE
2.249 - 2.200										
2.210	3.401	5.401	3.431	2.548	53	100	83	60	44	MULLITE 1.92-1
2.207	3.400	5.396	3.426	2.545	53	100	99	57	44	MULLITE 2.0-1
2.206	3.388	5.384	3.433	2.542	56	100	74	66	45	MULLITE 1.71-1
2.204	3.364	3.414	2.540	1.519	58	100	80	43	29	SILLIMANITE
2.242	3.109	2.853	3.755	2.989	49	100	80	61	46	DATOLITE
2.233	3.060	2.640	1.726	2.255	70	100	63	57	38	NORBERGITE
2.247	2.955	1.530	1.979	1.478	25	100	42	39	15	STISHOVITE
2.247	2.932	2.978	3.141	4.779	100	87	77	49	41	RHODONITE
2.199 - 2.150										
2.170	7.085	2.559	3.543	3.948	37	100	80	53	28	CRONSTEDTITE (1 LAYER)
2.171	4.526	5.548	2.774	3.927	30	100	81	50	30	ANDALUSITE
2.199	3.951	2.968	7.790	2.647	21	100	63	39	33	ALPHA BAAL2SI2O8
2.183	3.655	3.115	2.514	6.229	51	100	92	47	45	PHENACITE
2.174	3.076	2.931	3.141	1.642	18	100	53	48	17	THORTVEITITE
2.150	3.012	2.694	2.694	1.873	19	100	44	22	20	FRESNOITE
2.161	3.035	2.909	4.605	1.603	51	100	60	58	58	CLINOFERROSILITE
2.189	2.748	2.718	2.793	2.610	64	100	99	91	78	LARNITE (BETA CA2SIO4)
2.149 - 2.100										
2.116	10.463	2.636	3.025	3.368	38	100	93	90	45	KORNERUPINE

TABLE 43 LIST OF FIVE STRONGEST LINES

	D SPACING					I (PEAK)					MINERAL
2.145	7.083	2.519	14.166	3.540		25	100	57	38	34	RIPIDOLITE
2.102	3.257	2.753	4.205	2.020		37	100	93	86	30	EULYTITE
2.105	2.936	3.195	3.692	2.360		60	100	92	80	47	TOPAZ
2.110	2.557	3.217	3.046	1.500		46	100	48	41	40	XANTHOPHYLLITE
2.099 -	2.050										
2.095	6.284	3.628	2.565	2.810		24	100	77	22	17	HYDROSODALITE
2.082	5.849	3.226	3.442	3.454		48	100	65	49	47	DUMORTIERITE
2.091	3.621	6.272	2.371	2.560		25	100	36	17	15	SODALITE
2.071	3.308	2.530	4.451	1.718		20	100	75	73	64	ZIRCON
2.049 -	2.000										
2.026	3.519	6.150	3.567	2.500		14	100	41	37	15	TUGTUPITE
2.020	3.257	2.753	4.205	2.102		30	100	93	86	37	EULYTITE
2.017	2.456	2.996	2.841	2.013		73	100	66	47	47	SAPPHIRINE
2.013	2.456	2.017	2.996	2.841		47	100	73	66	47	SAPPHIRINE
1.999 -	1.950										
1.979	2.955	1.530	2.247	1.478		39	100	42	25	15	STISHOVITE
1.979	2.399	2.693	3.007	2.356		53	100	82	80	49	STAUROLITE
1.949 -	1.900										
1.935	3.728	3.661	3.679	3.522		8	100	54	52	17	PETALITE
1.929	3.184	3.352	1.375	1.377		58	100	56	52	47	KYANITE
1.938	2.935	2.958	2.676	3.530		31	100	89	38	31	PERRIERITE
1.923	2.651	2.964	1.584	1.644		20	100	36	35	22	GROSSULAR
1.899 -	1.850										
1.871	4.040	2.487	2.842	3.136		4	100	12	11	9	LOW CRISTOBALITE
1.879	3.486	3.913	3.165	4.608		9	100	33	12	9	BETA SPODUMENE
1.873	3.076	3.298	2.694	2.150		20	100	44	22	19	FRESNOITE
1.849 -	1.800										
1.830	3.537	3.063	3.215	2.617		38	100	74	53	49	YODERITE
1.847	3.399	4.339	1.571	1.396		13	100	18	5	5	HIGH QUARTZ
1.818	3.343	4.255	1.541	2.457		11	100	19	8	7	ALPHA QUARTZ
1.818	2.668	3.635	2.588	2.937		58	100	81	65	47	MONTICELLITE
1.799 -	1.750										
1.761	2.873	3.087	3.715	2.482		36	100	73	61	39	HARDYSTONITE
1.763	2.856	3.726	3.075	2.435		21	100	43	22	17	GEHLENITE
1.778	2.502	2.826	3.555	2.564		62	100	99	75	65	FAYALITE
1.764	2.481	2.797	2.539	3.525		60	100	86	66	55	HORTONOLITE
1.762	2.479	2.794	2.536	3.525		61	100	86	70	54	HYALOSIDERITE

TABLE 43   LIST OF FIVE STRONGEST LINES

D SPACING					I (PEAK)					MINERAL
1.752	2.462	2.777	3.896	2.274	60	100	71	58	41	FORSTERITE
1.750	1.746	2.272	2.266	2.262	100	93	90	75	74	CLINOHUMITE
1.749 - 1.700										
1.739	3.423	2.019	3.659	2.420	16	100	43	36	2.5	POLLUCITE
1.718	3.308	2.530	4.451	2.071	64	100	75	73	20	ZIRCON
1.726	3.060	2.233	2.640	2.255	57	100	70	63	38	NORBERGITE
1.704	2.759	2.971	3.602	2.321	24	100	42	39	23	MELIPHANITE
1.747	2.267	2.267	2.577	2.454	100	95	91	80	75	HUMITE
1.746	2.261	2.267	2.677	2.627	97	100	91	58	55	CHONDRODITE
1.746	1.750	2.272	2.266	2.262	93	100	90	75	74	CLINOHUMITE
1.742	1.747	2.267	2.577	2.454	91	100	95	80	75	HUMITE
1.699 - 1.650										
1.655	2.669	2.984	1.595	2.436	23	100	71	57	49	UVAROVITE
1.640 - 1.600										
1.642	3.112	2.931	3.141	2.174	17	100	53	48	18	THORTVEITITE
1.633	2.769	2.611	2.476	9.450	46	100	78	47	37	VESUVIANITE
1.612	2.697	3.015	4.462	4.264	78	100	84	73	37	ANDRADITE
1.644	2.651	2.964	1.584	1.923	22	100	36	35	20	GROSSULAR
1.599 - 1.550										
1.560	4.682	3.309	2.706	2.339	24	100	98	68	67	LI2SIO3
1.571	3.399	4.339	1.847	1.396	5	100	18	13	5	HIGH QUARTZ
1.505	2.900	4.671	2.632	3.781	49	100	65	59	50	PUMPELLYITE
1.595	2.669	2.984	2.436	1.655	57	100	71	49	23	UVAROVITE
1.584	2.651	2.964	1.644	1.923	35	100	36	22	20	GROSSULAR
1.589	2.557	2.865	1.531	2.443	38	100	63	59	41	PYROPE (SYNTHETIC)
1.549 - 1.500										
1.519	3.364	3.414	2.204	2.540	29	100	80	58	43	SILLIMANITE
1.541	3.343	4.255	1.817	2.457	8	100	19	11	7	ALPHA QUARTZ
1.530	2.955	1.979	2.247	1.478	42	100	39	25	15	STISHOVITE
1.549	2.619	2.630	2.721	2.433	76	100	87	84	80	LAWSONITE
1.544	2.584	2.359	2.889	4.718	28	100	71	33	15	PYROPE
1.531	2.562	2.865	2.443	1.589	59	100	63	41	38	PYROPE (SYNTHETIC)
1.500	2.557	3.217	2.110	3.046	40	100	48	46	41	XANTHOPHYLLITE
1.499 - 1.450										
1.478	2.955	1.530	1.979	2.247	15	100	42	39	25	STISHOVITE
1.479	2.523	4.390	3.189	3.121	41	100	52	52	42	MARGARITE

TABLE 43 LIST OF FIVE STRONGEST LINES

D SPACING					I (PEAK)					MINERAL
1.399	1.350									
1.396	3.399	4.339	1.847	1.571	5	100	18	13	5	HIGH QUARTZ
1.377	3.184	1.929	3.352	1.375	47	100	58	56	52	KYANITE
1.375	3.184	1.929	3.352	1.377	52	100	58	56	47	KYANITE

REFERENCES — PART II

ABRASHEV, K. K., ILYUKHIN, V. V. and BELOV, N. V., 1965, Crystal structure of barylite, $BaBe_2Si_2O_7$ (Use of difference F^2 — syntheses to solve for light atoms in the presence of comparatively heavy ones): Soviet Physics — Crystallography, Trans. from Kristallogr. 9, No. 6 816-827 p. 691-699

APPLEMAN, D. E. and CLARK, J. R., 1965, Crystal structure of reedmergnerite, a boron albite, and its relation to feldspar crystal chemistry: Am. Mineralogist, v. 50, p. 1827-1850

ARAKI, T. and ZOLTAI, T., 1969, Refinement of a coesite structure: Zeit. Krist. (in press)

BAKAKIN, V. V. and BELOV, N. V., 1961, Crystal structure of paracelsian: Soviet Physics — Crystallography, v. 5, no. 6, p. 826 -829 (Trans. from Kristallografiya, 1960, v. 5, no. 6, p. 864-868)

BARTL, H. and FISCHER, K. F., 1967, Untersuchung der Kristallstruktur des Zeolithes Laumontit: Neus. Jahrb. Mineral Monatshefte, p. 33-42

BEATTIE, I. R., 1954, The structure of analcite and ion-exchanged forms of analcite: Acta Cryst., v. 7, p. 357-359

BEGER, R. M. and BUERGER, M. J., 1967, The crystal structure of the mineral pollucite: Proc. Natl. Acad. Sci., v. 58, p. 853-854

BELOV, N. V. and MOKEJEVA, V. I., 1954, The crystal structure ilvaite: Trudy Inst. Krist. Akad. Nauk. SSSR, v. 9, p. 47-102 (in Russian)

BIRLE, J. D., GIBBS, G. V., MOORE, P. B., and SMITH, J. V., 1968, Crystal structures of natural olivines: Am. Mineralogist, v. 53, p. 807-824

BOUCHER, M. L. and PEACOR D. R., 1968, The crystal structure of alamosite, $PbSiO_3$: Zeit. Krist., v. 126, p. 98-111

BRAGG, W. L. and ZACHARIASEN, W. H., 1930, The crystalline structure of phenacite, Be_2SiO_4 and willemite, Zn_2SiO_4: Zeit. Krist., v. 72, p. 518-528

BRAUNER, K. and PREISINGER, A., 1956, Struktur und Entstehung des Sepioliths: Tscherm. Min. Petr. Mitt., v. 6, p. 120-140

BROWN, B. E. and BAILEY, S. W., 1964, The structure of maximum microcline: Acta Crystallogr., v. 17, p. 1391-1400

BROWN, G. E. and GIBBS, G. V., 1969, Refinement of the crystal structure of osumilite : Am. Mineralogist, v. 54, p. 101-116

BUERGER, M. J., BURNHAM, C. W. and PEACOR, D. R., 1962, Assessment of the several structures proposed for tourmaline: Acta Cryst., v. 15, p. 583-590

BUERGER, M. J. and PREWITT, C. T., 1961, The crystal structures of wollastonite and pectolite: Proc. Natl. Acad. Sci., v. 47, No. 12, p. 1884-1888

BUKIN, V. I. and MAKAROV, Y. S., 1967, Crystal structure of hydrosodalite according to neutron diffraction analysis: Geochem. Intern., v. 4, no. 1, p. 19-28 (Transl. from Geokhimiya, 1967, no. 1, p. 31-40)

BURNHAM, C. W., 1963a, Refinement of the crystal structure of sillimanite: Zeit. Krist., v. 118, p. 127-148

——1963b, Refinement of the crystal structure of kyanite: Zeit. Krist., v. 118, p. 337-360

——1964, The crystal structure of mullite: Carnegie Inst. Wash. Year Book, v. 62, p. 158-165

——1965, The crystal structure of mullite: Carnegie Inst. Wash. Year Book, v. 63, p. 223-228

——1967, Ferrosilite: Carnegie Inst. Wash. Year Book, v. 65, p. 285-293

——and BUERGER, M. J., 1961, Refinement of the crystal structure of andalusite: Zeit. Krist., v. 115, p. 269-290

——and RADOSLOVICH, E. W., 1965, Crystal structures of coexisting muscovite and paragonite: Carnegie Inst. Wash. Year Book, v. 63, p. 232-236

——and RADOSLOVICH, E. W., 1968, Personal communication

CANNILLO, E., CODA, A. and FAGNANI, G., 1966, The crystal structure of bavenite: Acta Cryst., v. 20, p. 301-309

——, GIUSEPPETTI, G. and TAZZOLI, V., 1967, The crystal structure of leucophanite: Acta Cryst., v. 23, p. 255-259

——, MAZZI, F. and ROSSI, G., 1966, The crystal structure of neptunite: Acta Cryst., v. 21, p. 200-208

CERVEN, J. F. and FANG, J. H., 1968, Private communication

CHRISTENSEN, A. N. and HAZELL, R. G., 1967, The crystal structure of $NaIn(SiO_3)_2$: Acta Chem. Scand., v. 21, p. 1425-1429

CLARK, J. R., APPLEMAN, D. D. and PAPIKE, J. J., 1968a, Bonding in crystal structures of ordered clinopyroxenes isostructural with diopside, $CaMgSi_2O_6$: Contri. Mineral. and Petrol., v. 20, p. 81-85

——, APPLEMAN, D. D. and PAPIKE, J. J., 1969, Private communication

——and PAPIKE, J. J., 1968, Crystal-chemical characterization of omphacites: Am. Mineralogist, v. 53, p. 840-868

COLE, W. F., H. SORUM and O. KENNARD, 1949, The crystal structures of orthoclase and sanidinized orthoclase, Acta Cryst. 2, 280-287

COLVILLE, A. A., 1968, Private communication

——and ANDERSON, C. P., 1968, The crystal structure of apophyllite: (Abst.), G. S. A. Program Annual Meeting, Mexico City, p. 60

——and GIBBS, G. V., 1965, Refinement of the crystal structure of riebeckite: (Abst.) Geol. Soc. Am. Spec. Pap., v. 82, p. 31

879

——and GIBBS, G. V., 1968, Personal communication

——and RIBBE, P. H., 1966, The crystal structure of oligoclase: (Abst.) Geol. Soc. Am. Spec. Pap., v. 101, p. 41

——and RIBBE, P. H., 1968, The crystal structure of an adularia and a refinement of the structure of orthoclase: Am. Mineral., v. 53, p. 25-37

CRUICKSHANK, D. W. J., LYNTON, H. and (in part) BARCLAY, G. A., 1962, A reinvestigation of the crystal structure of thortveitite, $Sc_2Si_2O_7$: Acta Cryst., v. 15, p. 491-498

DAL NEGRO, A., ROSSI, G. and UNGARETTI, L., 1967, The crystal structure of meliphanite: Acta Cryst., v. 23, p. 260-264

DANØ, M., 1966, The crystal structure of tugtupite—a new mineral, $Na_8Al_2Be_2Si_8O_{24}$ $(Cl, S)_2$: Acta Cryst., v. 20, p. 812-816

DEER, W. A., HOWIE, R. A., and ZUSSMAN, J., 1962, Rock-forming minerals, 1: John Wiley and Sons, Inc., N.Y., p. 333

DOLLASE, W. A., 1965, Reinvestigation of the structure of low cristobalite: Zeit. Krist., v. 121, p. 369-377

——1967, The crystal structure at 220°C of orthorhombic high tridymite from the Steinbach meteorite: Acta Cryst., v. 23, p. 617-623

——1968, Refinement and comparison of the structure of zoisite and clinozoisite: Am. Mineralogist, v. 53, p. 1882-1898

DONNAY, G., MORIMOTO, N., TAKEDA, H. and DONNAY, J. D. H., 1964, Trioctahedral one-layer micas. I. Crystal structure of a synthetic iron mica: Acta Cryst., v. 17, p. 1369-1373

——and ALLMANN, R., 1968, Si_3O_{10} groups in the crystal structure of ardennite: Acta Cryst., v. 24, p. 845-855

DOUGLASS, R. M., 1958, The crystal structure of sanbornite, $BaSi_2O_5$: Am. Mineralogist, v. 43, p. 517-536

DRITS, V. A. and KASHAEV, A. A., 1960, An x-ray study of a single crystal of kaolinite: Soviet Physics-Crystallography, v. 5, no. 2, p. 207-210 (Trans. from Kristallografiya, v. 5, no. 2, p. 224-227)

DUROVIĆ, S., 1962, A statistical model for the crystal structure of mullite: Soviet Physics-Crystallography, v. 7, no. 3, p. 271-278 (Trans. from Kristallografiya, v. 7, no. 3, p. 339-349)

EULER, F. and BRUCE, J. A., 1965, Oxygen coordinates of compounds with garnet structure: Acta Cryst., v. 19, p. 971-978

FERGUSON, R. B., TRAILL, R. V. and TAYLOR, W. H., 1958, The crystal structures of low-temperature and high-temperature albites. Acta Cryst. 11, 331-348

FINGER, L., 1968, Personal Communication

FISCHER, K., 1963, The crystal structure determination of the zeolite gismondite. $CaAl_2Si_2O_8 \cdot 4H_2O$: Am. Mineralogist, v. 48, p. 664 -672

FISCHER, K. F., 1966, A further refinement of the crystal structure of cummingtonite, $(Mg, Fe)_7(Si_4O_{11})_2(OH)_2$: Am. Mineralogist, v. 51, p. 814-818

FLEET, S. G., 1965, The crystal structure of dalyite: Zeit. Krist., v. 121, p. 349-368

FLEET, S. G., CHANDRASEKHAR, S. and MEGAW, H. D., 1966, The structure of bytownite ('body-centered anorthite'): Acta Cryst., v. 21, p. 782-801

FREED, R. L. and PEACOR, D. R., 1967, Refinement of the crystal structure of johannsenite: Am. Mineralogist, v. 52, p. 709-720

FRONDEL, C., 1962, The system of mineralogy. III. Silica minerals, 7th ed., John Wiley and Sons, Inc., N.Y., p. 334

—— and KLEIN, C. Jr., 1965, Ureyite, $NaCrSi_2O_6$: A new meteoritic pyroxene: Science, v. 149, p. 742-744

GABRIELSON, O., 1961, The crystal structures of kentrolite and melanotekite: Arkiv Mineral. Och Geologi, v. 3, no. 7, p. 141-151

GALLI, E., 1965, Affinamento della struttura della perrierite: Mineral. Petrog. Acta, v. 11, p. 39-48

GATINEAU, L., 1963, Localisation des remplacements isomorphiques dans la muscovite: Comptes Rendus, v. 256, p. 4648-4649

GHOSE, S., 1961, The crystal structure of a cummingtonite: Acta Cryst., v. 14, p. 622-627

—— 1965, Mg^{2+}-Fe^{2+} order in an orthopyroxene, $Mg_{0.93}Fe_{1.07}Si_2O_6$: Zeit. Krist., v. 122, p. 81-99

—— and HELLNER, E., 1959, The crystal structure of grunerite and observations on the Mg-Fe distribution: J. Geol., v. 67, no. 6, p. 691-701

GIBBS, G. V., 1964, Crystal structure of protoamphibole (Abst): Geol. Soc. America Spec. Pap. 82, p. 71-72

—— 1966, The Polymorphism of Cordierite I: The Crystal Structure of Low Cordierite. Am. Mineralogist 51, 1068-1087

GIBBS, G. V., 1968, Personal communication

——, BLOSS, F. D. and SHELL, H. R., 1960, Proto-amphibole, a new polytype: Am. Mineralogist, v. 45, p. 974-989

——, BRECK, D. W. and MEAGHER, E. P., 1968, Structural refinement of hydrous and anhydrous synthetic beryl, $Al_2(Be_3Si_6)O_{18}$ and emerald, $Al_{1.9}Cr_{0.1}(Be_3Si_6)O_{18}$: Lithos, v. 1, p. 275-285

—— and RIBBE, 1969, The crystal structures of the humite minerals: I. Norbergite Am. Mineralogist, v. 54, p. 376-390

—— and SMITH, J. V., 1965, Refinement of the crystal structure of synthetic pyrope: Am. Mineralogist, v. 50, p. 2023-2039

GOLOVASTIKOV, N. I., 1965, The crystal structure of dumortierite $(Al, Fe)_7O_3[BO_3][SiO_4]_3$: Soviet Physics Crystallography, v. 10, no. 6, p. 493-495 (Trans. from Dokl. Akad. Nauk. SSSR, v. 162, no. 6, p. 1284-1287)

GORDON, E. K., SAMSON, S. and KAMB, W. B., 1966, Crystal structure of the zeolite paulingite: Science, v. 154, p. 1004-1007

GOTTARDI, G., 1960, The crystal structure of perrierite: Am. Mineralogist, v. 45, p. 1-14

——, 1965, Die Kristallstruktur von Pumpellyit: Tschermaks Mineral Petrog. Mitt., v. 10, p. 115-119

GUNTHER, H. VON, BOLL-DORNBERGER, K., THILO, E. and THILO, E. M., 1955, Die Struktur des Dioptas, $Cu_6(Si_6O_{18}) \cdot 6H_2O$: Acta Cryst., v. 8, p. 425-430

GÜVEN, N., 1968, The crystal structures of $2M_1$ phengite and $2M_1$ muscovite: Carnegie Inst. Wash. Year Book, v. 66, p. 487-494

—— and BURNHAM, C. W., 1967, The crystal structure of 3T muscovite: Zeit. Krist., v. 125, p. 163-183

HAHN, T. and BUERGER, M. J., 1955, The detailed structure of nepheline, $KNa_3Al_4Si_4O_{16}$: Zeit. Krist., v. 106, p. 308-338

HARRISON, F. W. and BRINDLEY, G. W., 1957, The crystal structure of chloritoid: Acta Cryst., v. 10, p. 77-82

HEIDE, H. G. and BOLL-DORNBERGER, K., 1955, Die struktur des dioptas, $Cu_6(Si_6O_{18}) \cdot 6H_2O$: Acta Cryst., v. 8, p. 425-430

ITO, T. and MORI, H., 1953, The crystal structure of datolite: Acta Cryst., v. 6, p. 24-32

——, MORIMOTO, N. and SADANAGA, R., 1952, The crystal structure of milarite: Acta Cryst., v. 5, p. 209-213

JAHANBAGLOO, I. C. and ZOLTAI, T., 1968, The crystal structure of a hexagonal Al-serpentine: Am. Mineralogist, v. 53, p. 14-24

JARCHOW, O., 1965, Atomanordnung und Strukturverfeinerung von Cancrinit: Zeit. Krist., v. 122, p. 407-422

JOHANSSON, G., 1959, A refinement of the crystal structure of danburite: Acta Cryst., v. 12, p. 522-525

KAMB, W. B., 1960, The crystal structure of zunyite: Acta Cryst., v. 13, p. 15-24

KLEIN, C., JR., 1964, Cummingtonite-grunerite series: A chemical, optical and x-ray study: Am. Mineralogist, v. 49, p. 963-982

KNOWLES, C. R., RINALDI, F. F. and SMITH, J. V., 1965, Refinement of the crystal structure of analcime: Ind. Mineral., nos. 1 & 2, p. 127-140

KRSTANOVIĆ, I., 1958, Redetermination of the oxygen parameters in zircon ($ZrSiO_4$): Acta Cryst., v. 11, p. 896-897

LADELL, J., 1965, Redetermination of the crystal structure of topaz: A preliminary account: Norelco Reporter, v. 12, p. 34-47

LEIMER, H. W. and SLAUGHTER, M., 1969, Determination of refinement of the crystal structure of yugawaralite: Zeit. Krist. (in press)

882

LI, CHI-FANG, and PEACOR, D. R., 1968, The crystal structure of LiAlSi$_2$O$_6$-II ("β spodumene"): Zeit. Krist., v. 126, p. 46-65

LIEBAU, F., 1961, Untersuchungen an Schichtsilikaten des Formeltyps A$_m$(Si$_2$O$_5$)$_n$. III. Zur Kristallstruktur von Petalit, LiAlSi$_4$O$_{10}$: Acta Cryst., v. 14, p. 399-406

LÖNS, J. and SCHULZ, H., 1967, Strukturverfeinerung von Sodalith, Na$_8$Si$_6$Al$_6$O$_{24}$Cl$_2$: Acta Cryst., v. 23, p. 434-436

MAKSIMOV, B. A., KHARITONOV, Y. A., ILYUKHIN, V. V. and BELOV, N. V., 1968, Crystal structure of Li$_2$SiO$_3$: Soviet Phys. -Dokl., v. 13, no. 2, p. 85-88 (Trans. from Dokl. Akad. Nauk. SSSR, v. 178, no. 6, p. 1309-1312)

MAMEDOV, K. S. and BELOV, N. V., 1955, Xonotlite: Dokl. Akad. Nauk. SSSR, v. 104, no. 4, p. 615-618 (in Russian)

MEGAW, H. D., 1952, The structure of afwillite, Ca$_3$(SiO$_3$OH)$_2 \cdot$ 2H$_2$O: Acta Cryst., v. 5, p. 477-491

——, KEMPSTER, C. J. E. and RADOSLOVICH, E. W., 1962, The structure of anorthite, CaAl$_2$Si$_2$O$_8$. II. Description and discussion: Acta Cryst., v. 15, p. 1017-1035

MEIER, W. M., 1960, The crystal structure of natrolite: Zeit. Krist., v. 113, p. 430-444

MERKLE, A. B. and SLAUGHTER, M., 1968, Determination and refinement of the structure of heulandite: Am. Mineralogist, v. 53, p. 1120-1138

MIDGLEY, C. M., 1952, Crystal structure of β-dicalcium silicate: Acta Cryst., v. 5, p. 307-312

MOORE, P. B., 1968, Crystal structure of sapphirine: Nature, v. 218, p. 81-82

—— and BENNETT, J. M., 1968, Kornerupine: Its crystal structure: Science, v. 159, p. 524-526

MOORE, P. B. and LOUISNATHAN, J., 1967, Fresnoite: Unusual titanium coordination: Science, v. 156, p. 1361-1362

MORIMOTO, N. and GÜVEN, N., 1968, Refinement of the crystal structure of pigeonite, (Mg$_{0.39}$Fe$_{0.52}$Ca$_{0.09}$)SiO$_3$: Carnegie Inst. Wash. Year Book, v. 66, p. 494-497

——, APPLEMAN, D. E. and EVANS, H. T., JR., 1960, The crystal structures of clinoenstatite and pigeonite: Zeit. Krist., v. 114, p. 120-147

MROSE, M. E. and APPLEMAN, D. E., 1962, The crystal structures and crystal chemistry of väyrynenite, (Mn,Fe)Be(PO$_4$)(OH), and euclase, AlBe(SiO$_4$)(OH): Zeit. Krist., v. 117, p. 16-36

MacGILLAVRY, C. H., KORST, W. L., MOORE, E. J. W. and VAN DER PLAS, H. J., 1956, The crystal structure of ferrocarpholite: Acta Cryst., v. 9, p. 773-776

McCAULEY, J. W., 1968, Crystal structures of the micas $KMg_3AlSi_3O_{10}F_2$ and $BaLiMg_2AlSi_3O_{10}F_2$: Ph.D. thesis, Pennsylvania State University

McDONALD, W. S. and CRUICKSHANK, D. W. J., 1967, A reinvestigation of the structure of sodium metasilicate, Na_2SiO_3: Acta Cryst., v. 22, p. 37-43

—— and CRUICKSHANK, D. W. J., 1967, Refinement of the structure of hemimorphite: Zeit. Krist., v. 124, p. 180-191

McIVER, E. J., 1963, The structure of bultfonteinite, $Ca_4Si_2O_{10}F_2H_6$: Acta Cryst., v. 16, p. 551-558

NERONOVA, N. N. and BELOV, N. V., 1965, Crystal structure of elpidite, $Na_2Zr[Si_6O_{15}] \cdot 3H_2O$: Soviet Physics-Crystallography, v. 9, no. 6, p. 700-705 (Transl. from Kristallografiya, v. 9, no. 6, (1964), p. 828-834)

NEWNHAM, R. E., 1961, A refinement of the dickite structure and some remarks on polymorphism in kaolin minerals: Mineral. Mag., v. 32, p. 683-704

—— and MEGAW, H. D., 1960, The crystal structure of celsian (barium feldspar): Acta Cryst., v. 3, p. 303-312

NIKITIN, A. V. and BELOV, N. V., 1962, Crystal structure of batisite $Na_2BaTi_2Si_4O_{14} = Na_2BaTi_2O_2[Si_4O_{12}]$: Dokl. Akad. Nauk. SSSR -Earth Science Section, v. 146, p. 142-143 (Trans. from Dokl. Akad. Nauk. SSSR, v. 146, no. 6, p. 1401-1403

—— and BELOV, N. V., 1963, Crystal structure of clinohedrite $Ca_2Zn_2(OH)_2Si_2O_7 \cdot H_2O = 2CaZn[SiO_4] \cdot H_2O$: Dokl. Akad. Nauk. SSSR -Earth Science Section, v. 148, p. 118-120 (Transl. from Dokl. Akad. Nauk. SSSR, v. 148, no. 6, p. 1386-1388)

NOVACK, G. A. and GIBBS, G. V., 1968, Personal communication

ONKEN, H., 1964, Verfeinerung der Kristallstruktur von Monticellit: Die Naturwissenschaften, p. 334

PABST, A., 1943, Crystal structure of gillespite, $BaFeSi_4O_{10}$: Am. Mineral., v. 28, p. 372-390

PANT, A. K., 1968, A reconsideration of the crystal structure of β-$Na_2Si_2O_5$: Acta Cryst., v. 24, p. 1077-1083

—— and CRUICKSHANK, D. W. J., 1967, A reconsideration of the structure of datolite, $CaBSiO_4(OH)$: Zeit. Krist., v. 125, p. 286-297

PAPIKE, J. J. and CLARK, J. R., 1967, Crystal-chemical role of potassium and aluminum in a hornblende of proposed mantle origin (Abst.): Geol. Soc. America Spl. Paper 115, p. 171

—— 1968a The crystal structure and cation distribution of glaucophane: Am. Mineralogist, v. 53, p. 1156-1173

—— 1968b Personal communication

884 —— and ZOLTAI, T., 1965, The crystal structure of a marialite scapolite: Am. Mineralogist, v. 50, p. 641-655

—— CLARK, J. R. and HUEBNER, J. S., 1968, Potassic richterite, $KNaCaMg_5Si_8O_{22}(OH,F)_2$: Crystal chemistry and sodium-potassium (Abst.): Geol. Soc. America Annual Meeting, Mexico City, p. 230

PEACOR, D. R., 1967, Refinement of the crystal structure of a pyroxene of formula $M_I M_{II} (Si_{1.5}Al_{0.5})O_6$: Am. Mineralogist, v. 52, p. 31-41

—— and BUERGER, M. J., 1962, The determination and refinement of the structure of narsarsukite, $Na_2TiOSi_4O_{10}$: Am. Mineralogist, v. 47, p. 539-556

—— and BUERGER, J. J., 1962, Determination and refinement of the crystal structure of bustamite, $CaMnSi_2O_6$: Zeit. Krist., v. 117, p. 331-343

—— and NIIZEKI, N., 1963, The redetermination and refinement of the crystal structure of rhodonite, $(Mn, Ca)SiO_3$: Zeit. Krist., v. 119, p. 98-116

PERROTTA, A. J., 1967, The crystal structure of epistilbite: Mineral. Mag., v. 36, p. 480-490

—— and SMITH, J. V., 1964, The crystal structure of brewsterite, $(Sr, Ba, Ca)(Al_2Si_6O_{16}) \cdot 5H_2O$: Acta Cryst., v. 17, p. 857-862

—— 1965, The crystal structure of kalsilite, $KAlSiO_4$: Mineral. Mag., v. 35, p. 588-595

PEYRONEL, G., 1956, The crystal structure of Baveno bazzite: Acta Cryst., v. 9, p. 181-186

PRANDL, W., 1966, Verfeinerung der Kristallstruktur des Grossulars mit Neutronen- und Röntgenstrahlbeugung: Zeit. Krist., v. 123, p. 81-116

PREISINGER, A., 1962, Struktur des Stishovits, höchstdruck-SiO_2: Naturwiss., v. 15, p. 345

—— 1965, Prehnit—ein Schichtsilikattyp: Tschermaks Mineral Petrog. Mitt., v. 10, p. 491-504

PREWITT, C. T., 1967, Refinement of the structure of pectolite, $Ca_2NaHSi_3O_9$: Zeit. Krist., v. 125, p. 298-316

—— and BURNHAM, C. W., 1966, The crystal structure of jadeite, $NaAlSi_2O_6$: Am. Mineralogist, v. 51, p. 956-975

—— KIRCHNER, E. and PREISINGER, A., 1967 Crystal structure of larsenite $PbZnSiO_4$: Zeit. Krist., v. 124, p. 115-130

QUARENI, S. and DE PIERI, R., 1966, La struttura dell'andradite: Atti. e Memorie dell' Accademia Patavina di Science Lettere ed Arti, no. 2, p. 153-170

RAAZ, F., 1930, Über den Feinbau des Gehlenit. ein Beitrag zur Kenntnis der Melilithe: S.-B. Akad. Wiss. Wien, math.-naturwiss. Kl. Abt. I, v. 139, p. 645-672

RABBITT, J. C., 1948, A new study of the anthophyllite series: Am. Mineralogist, v. 33, p. 263-323

RADOSLOVICH, E. W., 1960, The structure of muscovite, $KAl_2(Si_3Al)O_{10}(OH)_2$: Acta Cryst., v. 13, p. 919-932

RENTZEPERIS, P. J., 1963, The crystal structure of hodgkinsonite, $Zn_2Mn[(OH)_2SiO_4]$: Zeit. Krist., v. 119, p. 117-138

RIBBE, P. H., 1963, A refinement of the crystal structure of sanidinized orthoclase: Acta Cryst., v. 16, p. 426-427

—— and MEGAW, H. D., 1963, The structure of transitional anorthite. A comparison with primitive anorthite: Norsk Geol. Tdssk., v. 42, Feldspar Volume, p. 158-167

——, MEGAW, H. D., and TAYLOR, W. H., 1969, The albite structure: Acta Cryst. (in press)

ROSS, M., SMITH, W. L. and ASHTON, W. H., 1968, Triclinic talc and associated amphiboles from Gouverneur Mining District, New York: Am. Mineralogist, v. 53, p. 751-769

RUMANOVA, I. M. and SKIPETROVA, T. I., 1959, The crystal structure of lawsonite: Soviet Phys.-Doklady, v. 4, p. 20-23

——, VOLODINA, G. F. and BELOV, N. V., 1967, The crystal structure of the rare earth ring silicate kainosite, $Ca(Y,Tr)_2[Si_4O_{12}]CO_3 \cdot H_2O$: Soviet Phys.-Cryst., v. 11, no. 4, p. 485-491 (Trans. from Kristallografiya, v. 11, no. 4, p. 549-558, 1966)

SADANAGA, R., TOKONAMI, M. and TAKÉUCHI, Y., 1962, The structure of mullite, $2Al_2O_3 \cdot SiO_2$, and relationship with the structures of sillimanite and andalusite: Acta Cryst., v. 15, p. 65-68

SEGAL, D. J., SANTORO, R. P. and NEWNHAM, R. E., 1966, Neutron-diffraction study of $Bi_4Si_3O_{12}$: Zeit. Krist., v. 123, p. 73-76

SHIBAYEVA, R. I. and BELOV, N. V., 1962, Crystal structure of wöhlerite, $Ca_2Na(Zr,Nb)[Si_2O_7](O,F)_2$: Dokl. Akad. Nauk. SSSR, v. 146, Earth Science Section, p. 128-130 (Trans. from Dokl. Akad. Nauk. SSSR, v. 146, p. 897-900)

SHIROZU, H. and BAILEY, S. W., 1965, Chlorite polytypism: III. Crystal structure of an orthohexagonal iron chlorite: Am. Mineralogist, v. 50, p. 868-885

—— 1966, Crystal structure of a two-layer Mg-vermiculite: Am. Mineralogist, v. 51, p. 1124-1143

SHROPSHIRE, J. KEAT, P. P. and VAUGHAN, P. A., 1959, The crystal structure of keatite, a new form of silica: Zeit. Krist., v. 112, p. 409-413

SIMONOV, V. I. and BELOV, N. V., 1960, Crystal structure of lovenite: Dokl. Akad. Nauk. SSSR, Earth Science Section, v. 130, p. 167-170 (Trans. from Dokl. Akad. Nauk. SSSR, Earth Science Section, v. 130, no. 6, p. 1333-1336)

—— 1965, Crystal structure of the Na, Zn, Cd metasilicate $Na_4ZnCd[Si_2O_6]_2$: Dokl. Akad. Nauk. SSSR, Earth Science Section,

v. 164, p. 123-126 (Transl. from Krist. Dokl. Akad. Nauk. SSSR, 1965, v. 164, no. 2, p. 406-409)

SKSZAT, S. M. and SIMONOV, V. I., 1966, The structure of calcium seidozerite: Soviet Phys.-Cryst., v. 10, no. 5, p. 505-508 (Transl. from Kristallografiya, v. 10, no. 5, p. 591-595)

SMITH, D. K., MAJUMDAR, A. J. and ORDWAY, F., 1965, The crystal structure of γ-dicalcium silicate: Acta Cryst., v. 18, p. 787-795

SMITH, J. V., 1959, The crystal structure of proto-enstatite, $MgSiO_3$: Acta Cryst., v. 12, p. 515-519

—— 1968, The crystal structure of staurolite: Am. Mineralogist, v. 53, p. 1139-1155

—— and DOWELL, L. G., 1968, Revised crystal structure of dehydrated Na-type A zeolite: Zeit. Krist., v. 126, p. 135-142

——, KARLE, I. L., HAUPTMAN, H. and KARLE, J., 1960, The crystal structure of spurrite, $Ca_5(SiO_4)_2CO_3$. II. Description of structure: Acta Cryst., v. 13, p. 454-458

SOLOV'EVA, L. P. and BELOV, N. V., 1965, Precise determination of the crystal structure of bertrandite $Be_4[Si_2O_7](OH)_2$: Soviet Phys.-Cryst., v. 9, no. 4, p. 458-460 (Transl. from Krist., v. 9, no. 4, p. 551-553, 1964)

STEADMAN, R. and NUTTALL, P. M., 1963, Polymorphism in cronstedtite: Acta Cryst., v. 16, p. 1-8

STEINFINK, H., 1958a, The crystal structure of chlorite. I. A monoclinic polymorph: Acta Cryst., v. 11, p. 191-194

—— 1962a, A correction—the crystal structure of chlorite. I. A monoclinic polymorph: Acta Cryst., v. 15, p. 1310

—— 1958b, The crystal structure of chlorite. II. A triclinic polymorph: Acta Cryst., v. 11, p. 195-198

—— 1961, Accuracy in structure analysis of layer silicates: Some further comments on the structure of prochlorite: Acta Cryst., v. 14, p. 198-199

—— 1962b, Crystal structure of a trioctahedral mica: Phlogopite: Am. Mineralogist, v. 47, p. 886-896

—— 1962c, The crystal structure of the zeolite, phillipsite: Acta Cryst., v. 15, p. 644-651

—— and BRUNTON, G., 1956, The crystal structure of amesite: Acta Cryst., v. 9, p. 487-492

STEWART, D. B., 1967, Four-phase curve in the system $CaAl_2Si_2O_8$ - SiO_2 - H_2O between 1 and 10 kilobars: Schweiz. Min. u. Petr. Mitt., v. 47, p. 35-59

SWANSON, H. E., FUYAT, R. K. and UGRINIC, G. M., 1954, Standard x-ray diffraction powder patterns: N. B. S. Circular 539, v. 3, p. 24

TAKÉUCHI, Y., 1958, A detailed investigation of the structure of
hexagonal $BaAl_2Si_2O_8$ with reference to its α-β inversion: Min.
Jour., v. 2, no. 5, p. 311-332

—— 1966, Structures of brittle micas: Proc. 13th Nat'l. Conf. on Clay
and Clay Minerals, v. 25, p. 1-25, Pergamon Press, New York

TAKÉUCHI, Y. and DONNAY, G., 1959, The crystal structure of
hexagonal $CaAl_2Si_2O_8$: Acta Cryst., v. 12, p. 465-470

—— and JOSWIG, W., 1967, The structure of haradaite and a note on the
Si-O bond lengths in silicates: Min. Jour., v. 5, no. 2, p. 98-123

—— and SADANAGA, R., 1966, Structural studies of brittle micas (I) The
structure of xanthophyllite refined: Min. Jour., v. 4, no. 6,
p. 424-437

——, KAWADA, I., and SADANAGA, R., 1968, Personal communication

TAYLOR, W. H., 1930, The structure of analcite $(NaAlSi_2O_6 \cdot H_2O)$:
Zeit. Krist., v. 74, p. 1-19

—— 1962, The structure of the principal felspars: Norsk. Geol. Tidsskr.,
v. 42, p. 1-24

—— and WEST, J., 1928, The crystal structure of the chondrodite series:
Proc. Roy Soc., v. 117, Series A, p. 517-532

TÊ-YÜ, L., SIMONOV, V. I. and BELOV, N. V., 1965, Crystal
structure of rinkite $Na(Na,Ca)_2(Ca,Ce)_4(Ti,Nb)[Si_2O_7]_2(O,F)_2F_2$:
Soviet Physics-Dokl., v. 10, p. 496-498 (Transl. from Dokl. Akad.
Nauk. SSSR, v. 162, no. 6, p. 1288-1291)

THREADGOLD, M., 1963, Crystal structures of hellyerite and nacrite:
Ph.D. thesis, University of Wisconsin

TROJER, F. J., 1968, Crystal structure of a high-pressure polymorph of
$CaSiO_3$: Naturwiss., v. 55, p. 442

VORMA, A., 1963, Crystal structure of stokesite, $CaSnSi_3O_9 \cdot 2H_2O$:
Mineral. Mag., v. 33, p. 615-617

VORONKOV, A. A. and PYATENKO, Y. A., 1962, The crystal structure
of vlasovite: Soviet Phys.-Cryst., v. 6, no. 6, p. 755-760 (Transl.
from Kristallografiya, 1961, v. 6, no. 6, p. 937-943)

WARREN, B. and BRAGG, W. L., 1929, The structure of diopside,
$CaMg(SiO_3)_2$: Zeit. Krist., v. 69, p. 168-193

WARREN, B. E. and MODELL, D. I., 1931, The structure of vesuvianite
$Ca_{10}Al_4(Mg,Fe)_2Si_9O_{34}(OH)_4$: Zeit. Krist., v. 78, p. 422-432

—— and TRAUTZ, O. R., 1930, The structure of hardystonite,
$Ca_2ZnSi_2O_7$: Zeit. Krist., v. 75, p. 525-528

WOODROW, P. J., 1967, The crystal structure of astrophyllite: Acta
Cryst., v. 22, p. 673-678

YOUNG, R. A., 1962, Mechanism of the phase transtion in quartz: Final
Report. Project No. A-447, Solid State Sci. Div. A. F. Office of
Sci. Research

ZACHARIASEN, W. H., 1930, The crystal structure of titanite: Zeit. Krist., v. 73 p. 7-16

—— 1930, The crystal structure of benitoite, $BaTiSi_3O_9$: Zeit. Krist., v. 74, p. 139-146

—— and PLETTINGER, H. A., 1965, Extinction in quartz: Acta Cryst., v. 18, p. 710-714

ZEMANN-HEDLIK, A. and ZEMANN, J., 1955, Die Kristallstrucktur von Petalit, $LiAlSi_4O_{10}$: Acta Cryst., v. 8, p. 781-787

ZVYAGIN, B. B., 1957, Determination of the structure of celadonite by electron diffraction: Soviet Phys.-Cryst., v. 2, no. 3, p. 388-394

889

Page

895

DATE DUE

GAYLORD

PRINTED IN U.S.A.